Ullmann's Encyclopedia of Industrial Chemistry

Volume B 6

Analytical Methods II and Process Control Engineering

VCH

Ullmann's Encyclopedia of Industrial Chemistry

Volumes A 1 – A 28: alphabetically arranged articles
Volumes B 1 – B 8: basic knowledge

Ullmann's Encyclopedia of Industrial Chemistry

Fifth, Completely Revised Edition

Editors: Barbara Elvers, Stephen Hawkins, William Russey

Volume B 6:

Analytical Methods II

Volume Editor: Helmut Günzler

Institut für Analytische Chemie und Radiochemie
Universität Innsbruck

Process Control Engineering

Volume Editor: Martin Polke
with the collaboration of Ulrich Epple and Michael Heim

Lehrstuhl für Prozeßleittechnik, RWTH Aachen

VCH

Numerical data, descriptions of methods or equipment, and other information presented in this book have been carefully checked for accuracy. Nevertheless, authors and publishers do not assume any liability for misprints, faulty statements, or other kinds of errors. Persons intending to handle chemicals or to work according to information derived from this book are advised to consult the original sources as well as relevant regulations in order to avoid possible hazards.

Production Director: Maximilian Montkowski
Editorial Director: Christina Dyllick-Brenzinger
Editorial Assistants: Ilse Bedrich, Louise Elsam, Reinhilde Gutsche, Monika Pikart-Müller, Karin Sora

Library of Congress Card No. 84-25-829

Deutsche Bibliothek, Cataloguing-in-Publication Data:

Die Deutsche Bibliothek – CIP-Einheitsaufnahme

Ullmann's encyclopedia of industrial chemistry / ed.: Barbara Elvers ... [Ed. advisory board Hans-Jürgen Arpe ...]. — Weinheim ; Basel (Switzerland) ; Cambridge ; New York, NY ; Tokyo : VCH.
Teilw. executive ed.: Wolfgang Gerhartz
Bis 4. Aufl. u. d. T.: Ullmanns Encyklopädie der technischen Chemie
ISBN 3-527-20100-9

NE: Gerhartz, Wolfgang [Hrsg.]; Elvers, Barbara [Hrsg.]; Encyclopedia of industrial chemistry

Vol. B. Basic knowledge.
6. Analytical methods. – 2. / Vol. ed.: Helmut Günzler [u. a.]. – 5., completely rev. ed. – 1994
ISBN 3-527-20136-X

NE: Günzler, Helmut [Hrsg.]

A catalogue record for this book is available from the British Library.

Printed on acid-free and low chlorine paper

Distribution
VCH Verlagsgesellschaft, P.O. Box 10 11 61, D-69451 Weinheim (Federal Republic of Germany)
Switzerland: VCH Verlags-AG, P.O. Box, CH-4020 Basel (Switzerland)
Great Britain and Ireland: VCH Publishers (UK) Ltd., 8 Wellington Court, Wellington Street, Cambridge CB1 1HZ (Great Britain)
USA and Canada: VCH Publishers, 220 East 23rd Street, New York NY 10010-4606 (USA)
Japan: VCH, Eikow Building, 10-9 Hongo l-chomc, Bunkyo-ku, Tokyo 113, Japan

Cover design: Wolfgang Schmidt
Composition, printing, and bookbinding: Graphischer Betrieb Konrad Triltsch, D-97070 Würzburg
Printed in the Federal Republic of Germany

Contents

Symbols and Units

Symbols and units agree with SI standards (for conversion factors see pp. VIII–IX). The following list gives the most important symbols used in the encyclopedia. Articles with many specific units and symbols have a similar list as front matter.

Symbol	Unit	Physical Quantity
a_B		activity of substance B
A_r		relative atomic mass (atomic weight)
A	m^2	area
c_B	mol/m^3, mol/L (M)	concentration of substance B
C	C/V	electric capacity
c_p, c_v	$J\,kg^{-1}K^{-1}$	specific heat capacity
d	cm, m	diameter
d		relative density (ϱ/ϱ_{water})
D	m^2/s	diffusion coefficient
D	Gy (= J/kg)	absorbed dose
e		elementary charge
E	J	energy
E	V/m	electric field strength
E	V	electromotive force
E_A	J	activation energy
f		activity coefficient
F	C/mol	Faraday constant
F	N	force
g	m/s^2	acceleration due to gravity
G	J	Gibbs free energy
h	m	height
h	$W \cdot s^2$	Planck constant
H	J	enthalpy
I	A	electric current
I	cd	luminous intensity
k	(variable)	rate constant of a chemical reaction
k	J/K	Boltzmann constant
K	(variable)	equilibrium constant
l	m	length
m	g, kg, t	mass
M_r		relative molecular mass (molecular weight)
n_D^{20}		refractive index (sodium D-light, 20 °C)
n	mol	amount of substance
N_A	mol^{-1}	Avogadro constant (6.023×10^{23} mol^{-1})
p	Pa; bar*	pressure
Q	J	quantity of heat
r	m	radius
R	$J\,K^{-1}mol^{-1}$	gas constant
R	Ω	electric resistance
S	J/K	entropy
t	s, min, h, d, month, a	time
t	°C	temperature
T	K	absolute temperature
u	m/s	velocity

* The official unit of pressure is the pascal (Pa).

Symbols and units (continued from p. VII)

Symbol	Unit	Physical Quantity
U	V	electric potential
U	J	internal energy
V	m^3, L, mL	volume
w		mass fraction
W	J	work
x_B		mole fraction of substance B
α		cubic expansion coefficient
α	$W\,m^{-2}K^{-1}$	heat-transfer coefficient (heat-transfer number)
α		degree of dissociation of electrolyte
$[\alpha]$	$10^{-2}deg\,cm^2g^{-1}$	specific rotation
η	$Pa \cdot s$	dynamic viscosity
θ	°C	temperature
$\varkappa$		c_p/c_v
λ	$W\,m^{-1}K^{-1}$	thermal conductivity
λ	nm, m	wavelength
μ		chemical potential
ν	Hz; s^{-1}	frequency
ν	m^2/s	kinematic viscosity (η/ϱ)
π	Pa	osmotic pressure
ϱ	g/cm^3	density
σ	N/m	surface tension
τ	Pa (N/m^2)	shear stress
φ		volume fraction
χ	Pa^{-1} (m^2/N)	compressibility

Conversion Factors

SI unit	Non-SI unit	From SI to non-SI multiply by
Mass		
kg	pound (avoirdupois)	2.205
kg	ton (long)	9.842×10^{-4}
kg	ton (short)	1.102×10^{-3}
Volume		
m^3	cubic inch	6.102×10^4
m^3	cubic foot	35.315
m^3	gallon (U.S., liquid)	2.642×10^2
m^3	gallon (Imperial)	2.200×10^2
Temperature		
°C	°F	°C × 1.8 + 32
Force		
N	dyne	1.0×10^5

Conversion factors (continued from p. VIII)

SI unit	Non-SI unit	From SI to non-SI multiply by
Energy, Work		
J	Btu (int.)	9.480×10^{-4}
J	cal (int.)	2.389×10^{-1}
J	eV	6.242×10^{18}
J	erg	1.0×10^{7}
J	kW · h	2.778×10^{-7}
J	kp · m	1.020×10^{-1}
Pressure		
MPa	at	10.20
MPa	atm	9.869
MPa	bar	10
kPa	mbar	10
kPa	mm Hg	7.502
kPa	psi	0.145
kPa	torr	7.502

Prefixes for Powers of Ten

E (exa)	10^{18}	M (mega)	10^{6}	d (deci)	10^{-1}	n (nano)	10^{-9}
P (peta)	10^{15}	k (kilo)	10^{3}	c (centi)	10^{-2}	p (pico)	10^{-12}
T (tera)	10^{12}	h (hecto)	10^{2}	m (milli)	10^{-3}	f (femto)	10^{-15}
G (giga)	10^{9}			μ (micro)	10^{-6}	a (atto)	10^{-18}

Abbreviations

The following is a list of the abbreviations used in the text. Common terms, the names of publications and institutions, and legal agreements are included along with their full identities. Other abbreviations will be defined wherever they first occur in an article. For further abbreviations, see p. VII (Symbols and Units), p. XIV (Companies and Country Codes in Patent References). The names of periodical publications are abbreviated exactly as done by Chemical Abstracts Service.

abs. absolute
a.c. alternating current
ACGIH American Conference of Governmental Industrial Hygienists
ACS American Chemical Society
ADI acceptable daily intake
ADN accord européen relatif au transport international des marchandises dangereuses par voie de navigation interieure (European agreement concerning the international transportation of dangerous goods by inland waterways)
ADNR ADN par le Rhin (regulation concerning the transportation of dangerous goods on the Rhine and all national waterways of the countries concerned)
ADP adenosine 5′-diphosphate
ADR accord européen relatif au transport international des marchandises dangereuses par route (European agreement concerning the international transportation of dangerous goods by road)
AEC Atomic Energy Commission (United States)
AIChE American Institute of Chemical Engineers

AIME	American Institute of Mining, Metallurgical, and Petroleum Engineers
AMP	adenosine 5′-monophosphate
APhA	American Pharmaceutical Association
API	American Petroleum Institute
ASTM	American Society for Testing and Materials
ATP	adenosine 5′-triphosphate
BAM	Bundesanstalt für Materialprüfung (Federal Republic of Germany)
BAT	Biologischer Arbeitsstoff-Toleranzwert (biological tolerance value for a working material, established by MAK commission, see MAK)
Beilstein	Beilstein's Handbook of Organic Chemistry, Springer, Berlin – Heidelberg – New York
BET	Brunauer – Emmett – Teller
BGA	Bundesgesundheitsamt (Federal Republic of Germany)
BGBl.	Bundesgesetzblatt (Federal Republic of Germany)
BIOS	British Intelligence Objectives Subcommitee Report (see also FIAT)
BOD	biological oxygen demand
bp	boiling point
B.P.	British Pharmacopeia
BS	British Standard
ca.	circa
calcd.	calculated
CAS	Chemical Abstracts Service
cat.	catalyst; catalyzed
cf.	compare
CFR	Code of Federal Regulations (United States)
Chap.	chapter
ChemG	Chemikaliengesetz (Federal Republic of Germany)
C.I.	Colour Index
CIOS	Combined Intelligence Objectives-Subcommitee Report (see also FIAT)
CNS	central nervous system
Co.	Company
COD	chemical oxygen demand
conc.	concentrated
const.	constant
Corp.	Corporation
crit.	critical
CTFA	The Cosmetic, Toiletry and Fragrance Association (United States)
DAB 9	Deutsches Arzneibuch, 9th ed., Deutscher Apotheker-Verlag, Stuttgart 1986
d.c.	direct current
decomp.	decompose, decomposition
DFG	Deutsche Forschungsgemeinschaft (German Science Foundation)
dil.	dilute, diluted
DIN	Deutsche Industrie Norm (Federal Republic of Germany)
DMF	dimethylformamide
DNA	deoxyribonucleic acid
DOE	Department of Energy (United States)
DOT	Department of Transportation – Materials Transportation Bureau (United States)
DTA	differential thermal analysis
EC	European Community
ed.	editor, edition, edited
e.g.	for example
emf	electromotive force
EmS	Emergency Schedule
EN	European Standard (European Community)
EPA	Environmental Protection Agency (United States)
EPR	electron paramagnetic resonance
Eq.	equation
ESCA	electron spectroscopy for chemical analysis
esp.	especially
ESR	electron spin resonance
Et	ethyl substituent ($-C_2H_5$)
et al.	and others
etc.	et cetera
EVO	Eisenbahnverkehrsordnung (Federal Republic of Germany)
exp (...)	$e^{(...)}$, mathematical exponent
FAO	Food and Agriculture Organization (United Nations)
FDA	Food and Drug Administration (United States)
FD & C	Food, Drug and Cosmetic Act (United States)
FHSA	Federal Hazardous Substances Act (United States)
FIAT	Field Information Agency, Technical (United States reports on the chemical industry in Germany, 1945)
Fig.	figure

fp	freezing point
Friedländer	P. Friedländer, Fortschritte der Teerfarbenfabrikation und verwandter Industriezweige, Vol. 1–25, Springer, Berlin 1888–1942
FT	Fourier transform
(g)	gas, gaseous
GC	gas chromatography
GefStoffV	Gefahrstoffverordnung (regulations in the Federal Republic of Germany concerning hazardous substances)
GGVE	Verordnung in der Bundesrepublik Deutschland über die Beförderung gefährlicher Güter mit der Eisenbahn (regulation in the Federal Republic of Germany concerning the transportation of dangerous goods by rail)
GGVS	Verordnung in der Bundesrepublik Deutschland über die Beförderung gefährlicher Güter auf der Straße (regulation in the Federal Republic of Germany concerning the transportation of dangerous goods by road)
GGVSee	Verordnung in der Bundesrepublik Deutschland über die Beförderung gefährlicher Güter mit Seeschiffen (regulation in the Federal Republic of Germany concerning the transportation of dangerous goods by sea-going vessels)
GLC	gas-liquid chromatography
Gmelin	Gmelin's Handbook of Inorganic Chemistry, 8th ed., Springer, Berlin–Heidelberg–New York
GRAS	generally recognized as safe
Hal	halogen substituent (–F, –Cl, –Br, –I)
Houben-Weyl	Methoden der organischen Chemie, 4th ed., Georg Thieme Verlag, Stuttgart
HPLC	high performance liquid chromatography
IAEA	International Atomic Energy Agency
IARC	International Agency for Research on Cancer, Lyon, France
IATA-DGR	International Air Transport Association, Dangerous Goods Regulations
ICAO	International Civil Aviation Organization
i.e.	that is
i.m.	intramuscular
IMDG	International Maritime Dangerous Goods Code
IMO	Inter-Governmental Maritime Consultive Organization (in the past: IMCO)
Inst.	Institute
i.p.	intraperitoneal
IR	infrared
ISO	International Organization for Standardization
IUPAC	International Union of Pure and Applied Chemistry
i.v.	intravenous
Kirk-Othmer	Encyclopedia of Chemical Technology, 3rd ed., J. Wiley & Sons, New York–Chichester–Brisbane–Toronto 1978–1984
(l)	liquid
Landolt-Börnstein	Zahlenwerte u. Funktionen aus Physik, Chemie, Astronomie, Geophysik u. Technik, Springer, Heidelberg 1950–1980; Zahlenwerte und Funktionen aus Naturwissenschaften und Technik, Neue Serie, Springer, Heidelberg, since 1961
LC_{50}	lethal concentration
LCLo	lowest published lethal concentration
LD_{50}	lethal dose
LDLo	lowest published lethal dose
ln	logarithm (base e)
LNG	liquefied natural gas
log	logarithm (base 10)
LPG	liquefied petroleum gas
M	mol/L
M	metal (in chemical formulas)
MAK	Maximale Arbeitsplatz-Konzentration (maximum concentration at the workplace in the Federal Republic of Germany); cf. Deutsche Forschungsgemeinschaft (ed.): Maximale Arbeitsplatzkonzentrationen (MAK) und Biologische Arbeitsstoff-

	Toleranz-Werte (BAT), VCH Verlagsgesellschaft, Weinheim (published annually)
max.	maximum
MCA	Manufacturing Chemists Association (United States)
Me	methyl substituent ($-CH_3$)
Methodicum Chimicum	Methodicum Chimicum, Georg Thieme Verlag, Stuttgart
MFAG	Medical First Aid Guide for Use in Accidents Involving Dangerous Goods
MIK	maximale Immissionskonzentration (maximum immission concentration)
min.	minimum
mp	melting point
MS	mass spectrum, mass spectrometry
NAS	National Academy of Sciences (United States)
NASA	National Aeronautics and Space Administration (United States)
NBS	National Bureau of Standards (United States)
NCTC	National Collection of Type Cultures (United States)
NIH	National Institutes of Health (United States)
NIOSH	National Institute for Occupational Safety and Health (United States)
NMR	nuclear magnetic resonance
no.	number
NRC	Nuclear Regulatory Commission (United States)
NRDC	National Research Development Corporation (United States)
NSC	National Service Center (United States)
NSF	National Science Foundation (United States)
NTSB	National Transportation Safety Board (United States)
OECD	Organization for Economic Cooperation and Development
OSHA	Occupational Safety and Health Administration (United States)
p., pp.	page, pages
Patty	G. D. Clayton, F. E. Clayton (eds.): Patty's Industrial Hygiene and Toxicology, 3rd ed., Wiley Interscience, New York
PB report	Publication Board Report (U.S. Department of Commerce, Scientific and Industrial Reports)
PEL	permitted exposure limit
Ph	phenyl substituent ($-C_6H_5$)
Ph. Eur.	European Pharmacopoeia, 2nd. ed., Council of Europe, Strasbourg 1981 –
phr	part per hundred rubber (resin)
PNS	peripheral nervous system
ppm	parts per million
q. v.	which see (quod vide)
ref.	refer, reference
resp.	respectively
R_f	retention factor (TLC)
R. H.	relative humidity
RID	règlement international concernant le transport des marchandises dangereuses par chemin de fer (international convention concerning the transportation of dangerous goods by rail)
RNA	ribonucleic acid
R phrase (R-Satz)	risk phrase according to ChemG and GefStoffV (Federal Republic of Germany)
rpm	revolutions per minute
RTECS	Registry of Toxic Effects of Chemical Substances, edited by the National Institute of Occupational Safety and Health (United States)
(s)	solid
SAE	Society of Automotive Engineers (United States)
s.c.	subcutaneous
SI	International System of Units
SIMS	secondary ion mass spectrometry
S phrase (S-Satz)	safety phrase according to ChemG and GefStoffV (Federal Republic of Germany)
STEL	Short Term Exposure Limit (see TLV)
STP	standard temperature and pressure (0° C, 101.325 kPa)
T_g	glass transition temperature
TA Luft	Technische Anleitung zur Reinhaltung der Luft (clean air regulation in Federal Republic of Germany)
TA Lärm	Technische Anleitung zum Schutz gegen Lärm (low noise regulation in Federal Republic of Germany)
TDLo	lowest published toxic dose
THF	tetrahydrofuran

TLC	thin layer chromatography
TLV	Threshold Limit Value (TWA and STEL); published annually by the American Conference of Governmental Industrial Hygienists (ACGIH), Cincinnati, Ohio
TOD	total oxygen demand
TRK	Technische Richtkonzentration (lowest technically feasible level)
TSCA	Toxic Substances Control Act (United States)
TÜV	Technischer Überwachungsverein (Technical Control Board of the Federal Republic of Germany)
TWA	Time Weighted Average
Ullmann	Ullmanns Encyklopädie der Technischen Chemie, 4th ed., Verlag Chemie, Weinheim 1972–1984; 3rd ed., Urban und Schwarzenberg, München 1951–1970
USAEC	United States Atomic Energy Commission
USAN	United States Adopted Names
USD	United States Dispensatory
USDA	United States Department of Agriculture
U.S.P.	United States Pharmacopeia
UV	ultraviolet
UVV	Unfallverhütungsvorschriften der Berufsgenossenschaft (workplace safety regulations in the Federal Republic of Germany)
VbF	Verordnung in der Bundesrepublik Deutschland über die Errichtung und den Betrieb von Anlagen zur Lagerung, Abfüllung und Beförderung brennbarer Flüssigkeiten (regulation in the Federal Republic of Germany concerning the construction and operation of plants for storage, filling, and transportation of flammable liquids; classification according to the flash point of liquids, in accordance with the classification in the United States)
VDE	Verband Deutscher Elektroingenieure (Federal Republic of Germany)
VDI	Verein Deutscher Ingenieure (Federal Republic of Germany)
vol	volume
vol.	volume (of a series of books)
vs.	versus
WHO	World Health Organization (United Nations)
Winnacker-Küchler	Chemische Technologie, Carl Hanser Verlag, München
wt	weight
$	U.S. dollar, unless otherwise stated

Abbreviations for the Names of Frequently Cited Companies

Air Products	Air Products and Chemicals
Akzo	Algemene Koninklijke Zout Organon
Alcoa	Aluminum Company of America
Allied	Allied Corporation
Amer. Cyanamid	American Cyanamid Company
BASF	BASF Aktiengesellschaft
Bayer	Bayer AG
BP	British Petroleum Company
Celanese	Celanese Corporation
Daicel	Daicel Chemical Industries
Dainippon	Dainippon Ink and Chemicals Inc.
Dow Chemical	The Dow Chemical Company
DSM	Dutch Staats Mijnen
Du Pont	E.I. du Pont de Nemours & Company
Exxon	Exxon Corporation
FMC	Food Machinery & Chemical Corporation
GAF	General Aniline & Film Corporation
W.R. Grace	W.R. Grace & Company
Hoechst	Hoechst Aktiengesellschaft
IBM	International Business Machines Corporation
ICI	Imperial Chemical Industries
IFP	Institut Français du Pétrole
INCO	International Nickel Company
3M	Minnesota Mining and Manufacturing Company
Mitsubishi Chemical	Mitsubishi Chemical Industries
Monsanto	Monsanto Company
Nippon Shokubai	Nippon Shokubai Kagaku Kogyo
PCUK	Pechiney Ugine Kuhlmann
PPG	Pittsburg Plate Glass Industries
Searle	G.D. Searle & Company
SKF	Smith Kline & French Laboratories
SNAM	Societá Nazionale Metandotti
Sohio	Standard Oil of Ohio
Stauffer	Stauffer Chemical Company
Sumitomo	Sumitomo Chemical Company
Toray	Toray Industries Inc.
UCB	Union Chimique Belge
Union Carbide	Union Carbide Corporation
UOP	Universal Oil Products Company
VEBA	Vereinigte Elektrizitäts- und Bergwerks-AG
Wacker	Wacker Chemie GmbH

Country Codes

The following list contains a selection of standard country codes used in the patent references.

AT	Austria
AU	Australia
BE	Belgium
BG	Bulgaria
BR	Brazil
CA	Canada
CH	Switzerland
CS	Czechoslovakia
DD	German Democratic Republic
DE	Federal Republic of Germany (and Germany before 1949)*
DK	Denmark
ES	Spain
FI	Finland
FR	France
GB	United Kingdom
GR	Greece
HU	Hungary
ID	Indonesia
IL	Israel
IT	Italy
JP	Japan*
LU	Luxembourg
MA	Morocco
NL	Netherlands*
NO	Norway
NZ	New Zealand
PL	Poland
PT	Portugal
SE	Sweden
SU	Soviet Union
US	United States of America
YU	Yugoslavia
ZA	South Africa
EP	European Patent Office*
WO	World Intellectual Property Organization

* For Europe, Federal Republic of Germany, Japan, and the Netherlands, the type of patent is specified: EP (patent), EP-A (application), DE (patent), DE-OS (Offenlegungsschrift), DE-AS (Auslegeschrift), JP (patent), JP-Kokai (Kokai tokkyo koho), NL (patent), and NL-A (application).

Periodic Table of Elements

1A ("European" group designation according to old IUPAC recommendation)
1 (group designation according to 1985 IUPAC proposal)
IA ("American" group designation, also used by the Chemical Abstracts Service until the end of 1986)

	2A 2 IIA	3A 3 IIIB	4A 4 IVB	5A 5 VB	6A 6 VIB	7A 7 VIIB	8 8 VIII	8 9 VIII	8 10 VIII	1B 11 IB	2B 12 IIB	3B 13 IIIA	4B 14 IVA	5B 15 VA	6B 16 VIA	7B 17 VIIA	0 18 VIIIA
1.0079 $_{1}$**H**																	4.0026 $_{2}$**He**
6.941 $_{3}$**Li**	9.0122 $_{4}$**Be**											10.811 $_{5}$**B**	12.011 $_{6}$**C**	14.007 $_{7}$**N**	15.9994 $_{8}$**O**	18.998 $_{9}$**F**	20.180 $_{10}$**Ne**
22.990 $_{11}$**Na**	24.305 $_{12}$**Mg**											26.982 $_{13}$**Al**	28.086 $_{14}$**Si**	30.974 $_{15}$**P**	32.066 $_{16}$**S**	35.453 $_{17}$**Cl**	39.948 $_{18}$**Ar**
39.098 $_{19}$**K**	40.078 $_{20}$**Ca**	44.956 $_{21}$**Sc**	47.88 $_{22}$**Ti**	50.942 $_{23}$**V**	51.996 $_{24}$**Cr**	54.938 $_{25}$**Mn**	55.847 $_{26}$**Fe**	58.933 $_{27}$**Co**	58.69 $_{28}$**Ni**	63.546 $_{29}$**Cu**	65.39 $_{30}$**Zn**	69.723 $_{31}$**Ga**	72.61 $_{32}$**Ge**	74.922 $_{33}$**As**	78.96 $_{34}$**Se**	79.904 $_{35}$**Br**	83.80 $_{36}$**Kr**
85.468 $_{37}$**Rb**	87.62 $_{38}$**Sr**	88.906 $_{39}$**Y**	91.224 $_{40}$**Zr**	92.906 $_{41}$**Nb**	95.94 $_{42}$**Mo**	98.906 $_{43}$**Tc** *	101.07 $_{44}$**Ru**	102.91 $_{45}$**Rh**	106.42 $_{46}$**Pd**	107.87 $_{47}$**Ag**	112.41 $_{48}$**Cd**	114.82 $_{49}$**In**	118.71 $_{50}$**Sn**	121.75 $_{51}$**Sb**	127.60 $_{52}$**Te**	126.90 $_{53}$**I**	131.29 $_{54}$**Xe**
132.91 $_{55}$**Cs**	137.33 $_{56}$**Ba**		178.49 $_{72}$**Hf**	180.95 $_{73}$**Ta**	183.85 $_{74}$**W**	186.21 $_{75}$**Re**	190.2 $_{76}$**Os**	192.22 $_{77}$**Ir**	195.08 $_{78}$**Pt**	196.97 $_{79}$**Au**	200.59 $_{80}$**Hg**	204.38 $_{81}$**Tl**	207.2 $_{82}$**Pb**	208.98 $_{83}$**Bi**	208.98 $_{84}$**Po** *	209.99 $_{85}$**At** *	222.02 $_{86}$**Rn** *
223.02 $_{87}$**Fr** *	226.03 $_{88}$**Ra** *																

138.91 $_{57}$**La**	140.12 $_{58}$**Ce**	140.91 $_{59}$**Pr**	144.24 $_{60}$**Nd**	146.92 $_{61}$**Pm** *	150.36 $_{62}$**Sm**	151.97 $_{63}$**Eu**	157.25 $_{64}$**Gd**	158.93 $_{65}$**Tb**	162.50 $_{66}$**Dy**	164.93 $_{67}$**Ho**	167.26 $_{68}$**Er**	168.93 $_{69}$**Tm**	173.04 $_{70}$**Yb**	174.97 $_{71}$**Lu**
227.03 $_{89}$**Ac** *	232.04 $_{90}$**Th** *	231.04 $_{91}$**Pa** *	238.03 $_{92}$**U** *	237.05 $_{93}$**Np** *	244.06 $_{94}$**Pu** *	243.06 $_{95}$**Am** *	247.07 $_{96}$**Cm** *	247.07 $_{97}$**Bk** *	251.08 $_{98}$**Cf** *	252.08 $_{99}$**Es** *	257.10 $_{100}$**Fm** *	258.10 $_{101}$**Md** *	259.10 $_{102}$**No** *	260.11 $_{103}$**Lr** *

* Elements with unstable isotopes; mass of most important isotope given

Analytical Methods II

Thermal Analysis and Calorimetry

STEPHEN B. WARRINGTON, Thermal Analysis Consultancy Service, Leeds Metropolitan University, Leeds, United Kingdom (Chap. 1)

GÜNTHER W. H. HÖHNE, Sektion für Kalorimetrie, Universität Ulm, Ulm, Federal Republic of Germany (Chap. 2)

1. Thermal Analysis

1.1. General Introduction

1.1.1. Definitions

Thermal analysis (TA) has been defined as "a group of techniques in which a physical property of a substance and/or its reaction products is measured as a function of temperature while the substance is subjected to a controlled temperature programme" [1]. The formal definition is usually extended to include isothermal studies, in which the property of interest is measured as a function of time.

The definition is a broad one, and covers many methods that are not considered to fall within the field of thermal analysis as it is usually understood. The present chapter will be restricted to the major techniques as currently practiced. Since all materials respond to heat in some way, TA has been applied to almost every field of science, with a strong emphasis on solving problems in materials science and engineering, as well as fundamental chemical investigations. TA is applicable whenever the primary interest is in determining the effect of heat upon a material, but the techniques can also be used as a means of probing a system to obtain other types of information, such as composition.

The aim of this chapter is to give an overview of the main TA methods and their applications, with sufficient references to provide reader access to more detailed information. The following list summarizes properties subject to investiga-

0-89573-541-5/94/$5.00 + .50

tion and the corresponding thermal analysis techniques:

Mass	Thermogravimetry (TG)
Temperature	Differential thermal analysis (DTA)
Enthalpy	Differential scanning calorimetry (DSC)
Evolved gas	Evolved gas analysis (EGA)
Physical dimensions	Thermodilatometry (TD)
Mechanical properties	Thermomechanical analysis (TMA)
	Dynamic mechanical analysis (DMA)
Optical properties	Thermoptometry
Electrical properties	Thermoelectrometry
Magnetic properties	Thermomagnetometry

1.1.2. Sources of Information

Two journals (the *Journal of Thermal Analysis* and *Thermochimica Acta*) devote their contents entirely to TA; the *Proceedings* of the (now) four-yearly Conferences of the International Confederation for Thermal Analysis and Calorimetry (ICTAC) constitute an excellent additional source of research papers. Specific information regarding *Proceedings* volumes for the nine ICTAC Conferences between 1965 and 1991 is available in [1]. The most complete listing of worldwide TA literature is also found in the ICTAC handbook [1], which in addition gives addresses for national TA societies and important equipment suppliers. The most useful textbooks include [2]–[5].

1.2. Thermogravimetry

1.2.1. Introduction

Thermogravimetry (TG) is used to measure variations in mass as a function of temperature (or time). Processes amenable to study in this way are listed in Table 1. TG is one of the most powerful TA techniques from the standpoint of quantitative data, and for this reason it is often employed in combination with other measurements.

1.2.2. Instrumentation

The instrument used is a *thermobalance*. The schematic diagram in Figure 1 presents the main components of a typical modern unit. Component details vary according to the design, and choice of a particular instrument is usually dictated by requirements of the problem under investigation (temperature range, sensitivity, etc.). The balance mechanism itself is usually of the null-deflection type to ensure that the sample's position in the furnace will not change. The balance transmits a continuous measure of the mass of the sample to an appropriate recording system, which is very often a computer. The resulting plot of mass vs. temperature or time is called a *TG curve*. Balance sensitivity is usually of the order of one microgram, with a total capacity of as much as a few hundred milligrams. Furnaces are available that operate from subambient (e.g., −125 °C) or room temperature up to as high as 2400 °C. A furnace programmer normally supports a wide range of heating and cooling rates, often in combination, as well as precise isothermal control. The programming functions themselves are increasingly being assumed by com-

Table 1. Processes that can be studied by thermogravimetry

Process	Weight gain	Weight loss
Ad- or absorption	*	
Desorption		*
Dehydration/desolvation		*
Sublimation		*
Vaporization		*
Decomposition		*
Solid–solid reactions		*
Solid–gas reactions	*	*

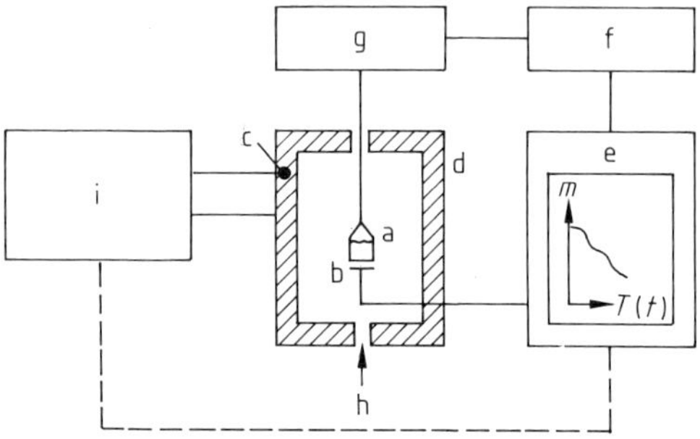

Figure 1. Schematic diagram of a thermobalance
a) Sample; b) Sample temperature sensor; c) Furnace temperature sensor; d) Furnace; e) Recorder or computer, logging sample mass, temperature, and time; f) Balance controller; g) Recording microbalance; h) Gas; i) Furnace temperature programmer
The computer may also control the furnace programmer

puters. Most applications involve heating rates of 5–20 °C min^{-1}, but the ability to heat a sample as rapidly as 1000 °C min^{-1} can be useful in the simulation of certain industrial processes, for example, or in flammability studies. Special control methods, grouped under the term *controlled-rate thermal analysis* (CRTA), are receiving increasing attention [6], and offer advantages in resolving overlapping processes and in kinetic studies.

The quality of the furnace atmosphere deserves careful attention. Most commercial thermobalances operate at atmospheric pressure. Vacuum and high-pressure studies normally require specialized equipment, either commercial or home-made, as do experiments with corrosive gases [7]. The ability to establish an inert (oxygen-free) atmosphere is useful, as is the potential for rapidly changing the nature of the atmosphere.

The mode of assembly of the components varies; for example, the furnace might be above, below, or in line with the balance. Sample containers also vary widely in design; cylindrical pans are common, typically 5–8 mm in diameter and 2–10 mm high, though flat plates and semisealed containers may be used to investigate the effects of atmospheric access to a sample. Compatibility between the construction materials and the system under investigation must be carefully considered. Materials commonly available include aluminum, platinum, alumina, and silica. Temperature indication is normally provided by a thermocouple located near the sample container. Because of inevitable thermal gradients within the apparatus, an indicated temperature can never be taken as an accurate reflection of the temperature of the sample. Reproducible location of the thermocouple is vital; recommended calibration procedures have been described in [8]. Only in simultaneous TG–DTA instruments is direct measurement of the temperature possible.

1.2.3. Factors Affecting a TG Curve

Wendlandt [2] has listed 13 factors that directly affect a TG curve, both sample- and instrument-related, some of which are interactive. The primary factors are heating rate and sample size, an increase in either of which tends to increase the temperature at which sample decomposition occurs, and to decrease the resolution between successive mass losses. The particle size of the sample, the way in which it is packed, the crucible shape, and the gas flow rate also affect the progress of a thermal reaction. Careful attention to consistency in experimental details normally results in good repeatability. On the other hand, studying the effect of deliberate alterations in such factors as heating rate can provide valuable insights into the nature of observed reactions. All these considerations are equally applicable to other techniques as well, including DTA (Section 1.3).

1.2.4. Applications

TG has been applied extensively to the study of analytical precipitates for gravimetric analysis [9]. One example is calcium oxalate, as illustrated in Figure 2. Information such as extent of hydration, appropriate drying conditions, stability ranges for intermediate products, and reaction mechanisms can all be deduced from appropriate TG curves. Figure 2 also includes the first derivative of the TG curve, termed the *DTG curve*, which is capable of revealing fine details more clearly.

A TG method has also been devised for the proximate analysis of coal, permitting four samples per hour to be analyzed with results of the same precision and accuracy as BS and ASTM methods [10]. A typical curve is shown in Figure 3. The approach taken here can be generalized to the compositional analysis of many materials [11]. TG also facilitates comparisons of the relative stabilities of polymers (Fig. 4). An analysis of curves prepared at different heating rates

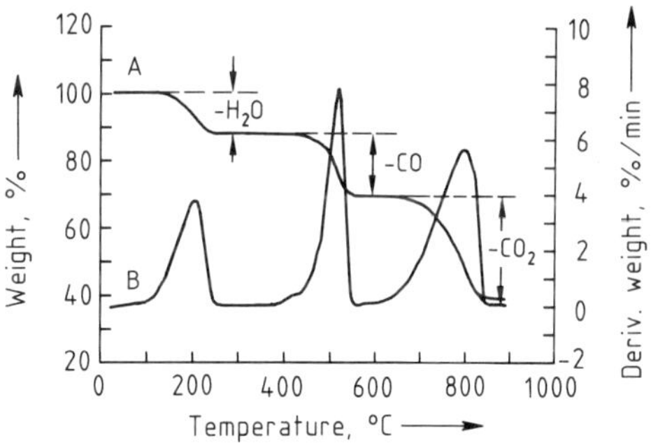

Figure 2. TG (A) and DTG (B) curves for calcium oxalate monohydrate
Sample mass 30 mg, heating rate 20 °C min^{-1}, argon atmosphere

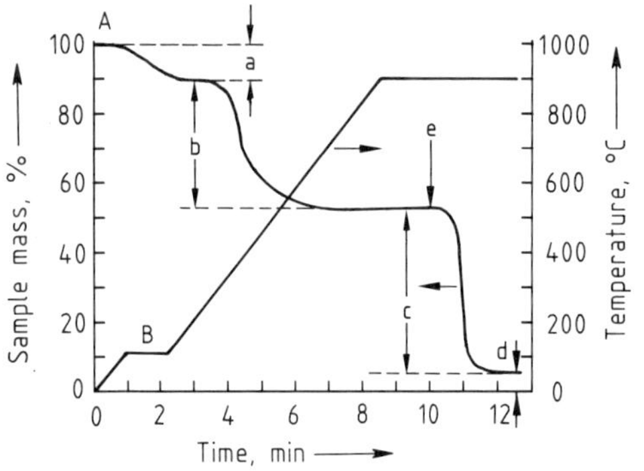

Figure 3. TG curve (A) illustrating a programmed-temperature method for the proximate analysis of coal, where the line B defines the temperature program
a) Moisture; b) Volatiles; c) Fixed carbon; d) Ash; e) Point at which the atmosphere was changed to air

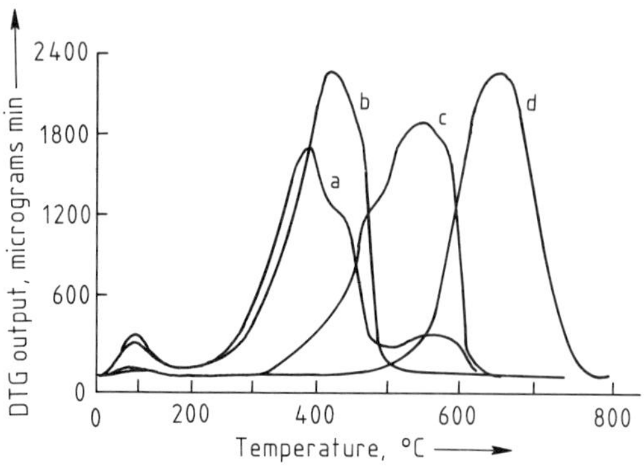

Figure 5. DTG curves for four ASTM standard coal samples heated in air [12]
a) Lignite "A"; b) Sub-bituminous "A"; c) High-volatility bituminous "A"; d) Buckwheat anthracite

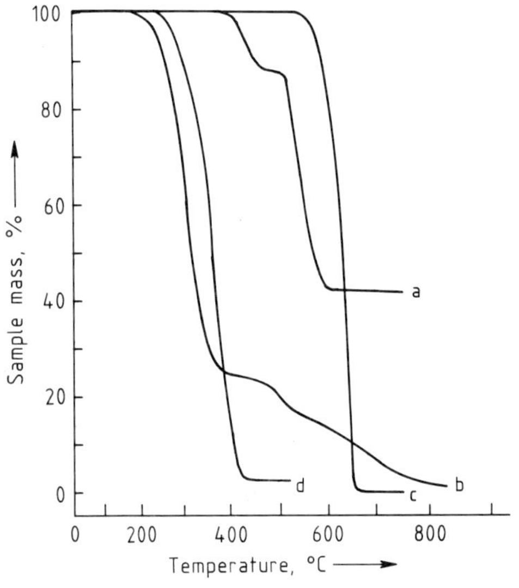

Figure 4. TG curves for four polymers in air, showing their relative stabilities
a) Silicone rubber; b) PVC; c) PTFE; d) Perspex

makes it possible to extract kinetic data for estimating the service lifetimes of such materials. In another coal-related application, an empirical correlation has been established between characteristic temperatures obtained from DTG curves of coals examined under air and the performance of the same coals in steam-raising plants [12], permitting the rapid assessment of new fuels (Fig. 5). The simulation of industrial processes, particularly heterogeneous catalysis, is another important area of application.

1.3. Differential Thermal Analysis and Differential Scanning Calorimetry

1.3.1. Introduction

Both differential thermal analysis (DTA) and differential scanning calorimetry (DSC) are concerned with the measurement of energy changes, and as such are applicable in principle to a wider range of processes than TG. From a practical standpoint DSC may be regarded as the method from which quantitative data are most easily obtained. The use of DSC to determine absolute thermodynamic quantities is discussed in Sections 2.3.2 and 2.4.1. Types of processes amenable to study by these methods are summarized in Table 2.

Table 2. Processes that can be studied by DTA/DSC

Process	Exothermic	Endothermic
Solid–solid transitions	*	*
Solid–liquid transitions	*	*
Vaporization		*
Sublimation		*
Adsorption	*	
Desorption		*
Desolvation		*
Decomposition	*	*
Solid–solid reactions	*	*
Solid–liquid reactions	*	*
Solid–gas reactions	*	*
Curing	*	
Polymerizations	*	
Catalytic reactions	*	*

1.3.2. Instrumentation

Figure 6 shows the components of a typical DTA apparatus. The sample and an inert reference material (commonly alumina) are subjected to a common temperature program while the temperature difference between the two is monitored. In an ideally designed instrument, the temperature difference ΔT remains approximately constant if no reaction is taking place in the sample. When an endothermic process such as melting occurs, the sample temperature T_s lags behind that of the reference T_r; $\Delta T = T_s - T_r$ is thus negative, so a (negative) peak is produced on the corresponding *DTA curve*, which is a record of ΔT against either temperature or time (Fig. 7). Exothermic processes lead to peaks in the opposite direction. The peak *area* is related to both the energy change and the sample size according to the relationship

$$\Delta H \cdot m = K \int \Delta T \cdot \mathrm{d}t$$

where m is the sample mass, and ΔH is the enthalpy of reaction. The proportionality constant K is temperature-dependent, and with classical DTA it is also influenced by thermal properties of the sample and container. Certain apparatus designs minimize these influences, so that once the characteristics of $K = f(T)$ have been determined the ΔT signal can be conditioned (electronically or digitally) to give an output signal calibrated directly in power units; i.e., the sensitivity is nominally independent of temperature. This is the principle underlying heat-flux DSC. For a more complete description of this technique, as well as power-compensated DSC, see Section 2.3.2. A wide variety of commercial devices covers the temperature range from ca. −150–2500 °C, but high-temperature instruments are less sensitive than low-temperature designs due to constraints imposed by thermocouple technology. DSC instruments are usually limited to a maximum temperature of 700 °C, though heat-flux DSC has been extended to 1400 °C. DTA head designs vary enormously, but most accommodate samples in the range 5–200 mg in metal or ceramic containers. The thermocouples usually take the form of beads that fit into recesses in the containers, or plates upon which the containers rest. Nowadays the thermocouple is never actually placed in the sample, in contrast to the "classical" arrangement shown in Figure 6.

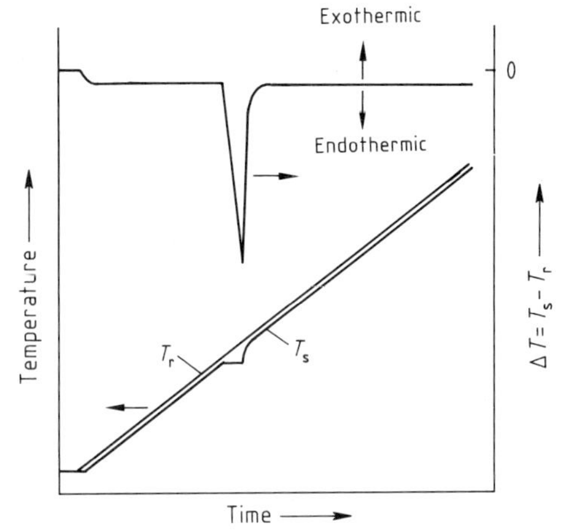

Figure 7. Typical DTA curve for the melting of a pure material

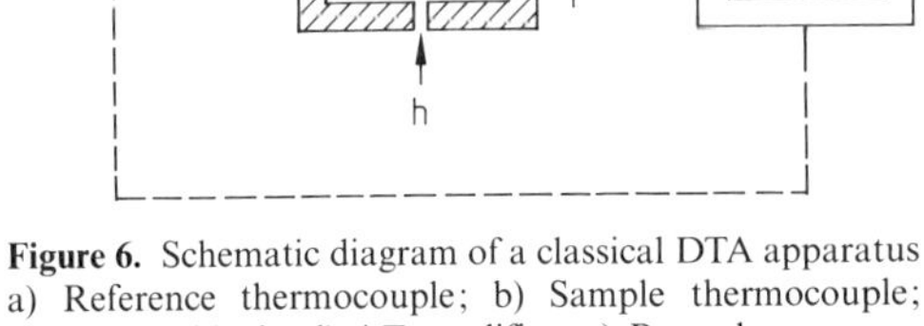

Figure 6. Schematic diagram of a classical DTA apparatus
a) Reference thermocouple; b) Sample thermocouple; c) Heating block; d) ΔT amplifier; e) Recorder or computer, logging T_s, ΔT, and time; f) Furnace; g) Temperature programmer, which may be linked to the computer; h) Gas inlet
T_s, sample temperature; T_r, reference temperature; $\Delta T = T_s - T_r$

1.3.3. Applications

DSC is used extensively in polymer science [13] (→Plastics, Properties and Testing, **A 21**, pp. 29 f). A generalized DSC curve for a polymer is shown in Figure 8. Most polymers display a *glass transition*, in the course of which the material passes from a glassy to a rubbery state with a simultaneous increase in specific heat capacity (→ Plastics, Properties and Testing, **A 21**, pp. 5 ff). Glass-transition temperature measurements are

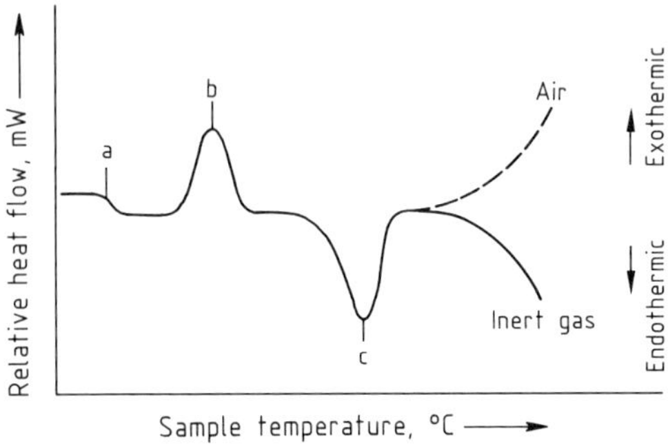

Figure 8. Generalized DSC curve for a polymer
a) Glass transition; b) Crystallization; c) Melting
The subsequent reactions may be endothermic or exothermic

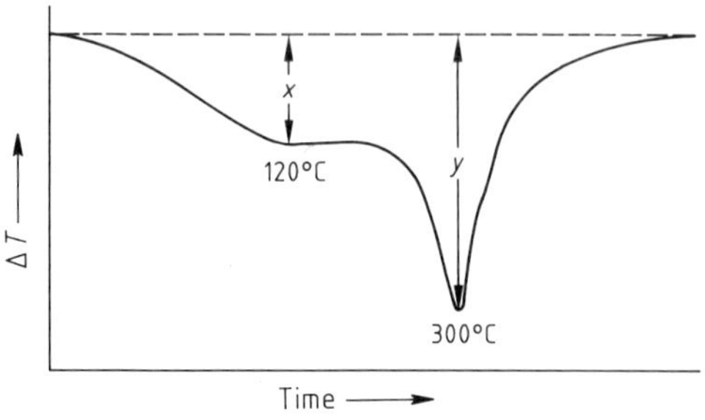

Figure 9. DTA curve for a partially converted high-alumina cement concrete from a test drilling
Degree of conversion $= 100\ y/(x + y)$

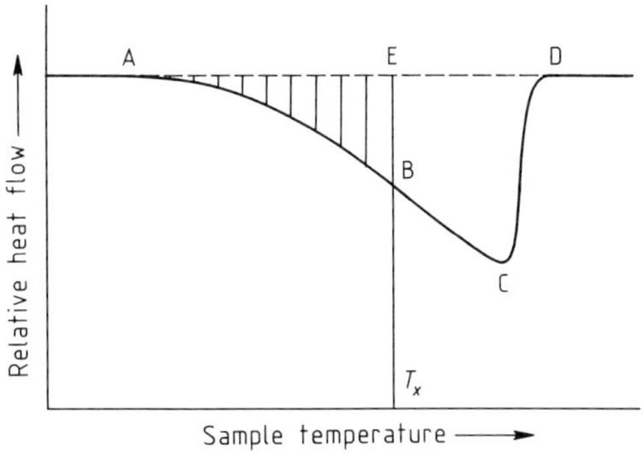

Figure 10. Generalized DSC curve for the melting of an impure material
Fraction melted at temperature T_x = (area *ABE*)/(area *ADC*)

used for material characterization and comparison, often providing information about such factors as thermal history, crystallinity, extent of cure, and plasticizer content of a polymer. Amorphous polymers usually exhibit a crystallization exotherm, and thermosetting polymers a curing exotherm, both of which can be analyzed kinetically to establish the thermal treatment required to obtain desired properties in a particular product. The melting endotherm is a measure of the extent of crystallinity, an important parameter related to mechanical properties. The temperature at which exothermic degradation begins under an oxidizing atmosphere is used to compare oxidative stabilities of different polymers or different stabilizers. Discrimination can be improved by isothermal measurements, in which the material under investigation is heated initially in an inert gas, after which the atmosphere is changed to air or oxygen and the time lapse prior to exothermic deflection is measured. The same approach is applicable to studies of oils, greases, fats, etc., though high-pressure cells may be necessary to prevent volatilization.

The particular temperatures at which peaks are observed can be used for the identification of components in a mixture, and the size of a particular peak can be used for quantitative evaluation. Examples include the determination of quartz in clays and analysis of the constituents of polymer blends, both of which would be difficult to achieve by other means. DTA has been used to determine the presence and degree of conversion of high-alumina cement concrete (HAC), a material subject to severe loss of strength under certain atmospheric conditions due to solid-state reactions in the cement matrix. Figure 9 shows a DTA curve for partially converted HAC together with the way the data are used for a conversion measurement.

An analysis of the shape of the fusion peak from a relatively pure material (e.g., a pharmaceutical agent) can, with certain restrictions, lead to an estimate of purity [14]. Figure 10 shows an idealized melting curve for an impure material. By calculating the fraction (F) of material melted at a series of temperatures, and applying several simplifying assumptions, the mole fraction of an impurity can be calculated from the expression:

$$T_s = T_0 - \frac{R T_0^2 X_2}{\Delta H_f} \cdot \frac{1}{F}$$

where T_s = sample temperature, T_0 = melting point of the pure material, R = gas constant, X_2 = mole fraction of impurity, ΔH_f = enthalpy

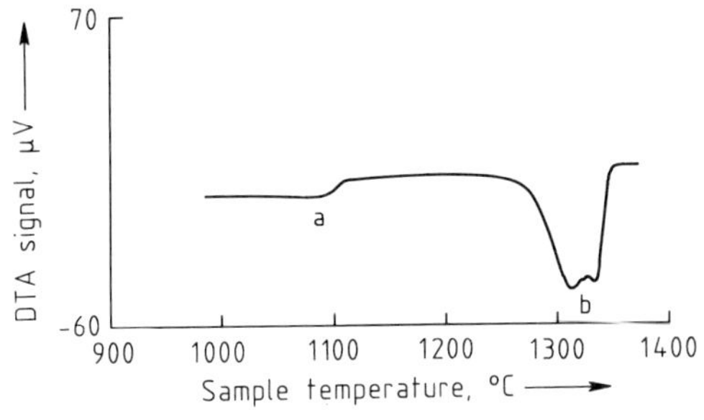

Figure 11. DTA curve for a nickel-based superalloy
a) Solution of the γ' phase; b) Melting
250 mg sample heated at 10 °C min^{-1} in argon

of fusion, and F = the fraction melted at temperature T_s. The calculations are readily performed by commercial software.

DTA curves can be used to construct phase diagrams for mixtures. Measurement of a single point on the phase diagram may be sufficient for quality-control purposes. Figure 11 shows the DTA curve for a nickel-based superalloy, where the solvus temperature (the temperature corresponding to complete solution of one solid phase in another) of the γ' phase characterizes the heat treatment to which the alloy has been subjected.

Solid–solid reactions have also been successfully studied by these techniques, as have decompositions of high explosives. The applicability of DTA/DSC to very small samples is of obvious value here. Similarly, preliminary screening of potentially hazardous reaction mixtures is conveniently carried out in this way as a guide to the likelihood of exothermic reaction, the corresponding temperature range and magnitude, and possible effects of pressure and atmospheric conditions (see also Section 2.4.5).

1.4. Simultaneous Techniques

1.4.1. Introduction

TG or DTA/DSC alone rarely gives sufficient information to permit a complete interpretation of the reactions in a particular system; results must usually be supplemented by other thermal methods and/or general analytical data. Alternative thermal methods are best applied simultaneously, leading, for example, to TG and DTA information from the same sample under identical experimental conditions. This avoids ambiguity caused by material inhomogeneity, but also problems attributable to different experimental conditions with different instruments, which sometimes markedly affects the correspondence between TG and DTA curves. Other advantages include: (1) indication of the thermal stability of materials examined by DTA, which in turn makes it possible to correct measured heats of reaction for partially decomposed samples; (2) accurate temperature measurement in TG work, which is vital in kinetic studies; and (3) the detection of unsuspected transitions in a condensed phase, which may help explain puzzling features of the corresponding TG curve.

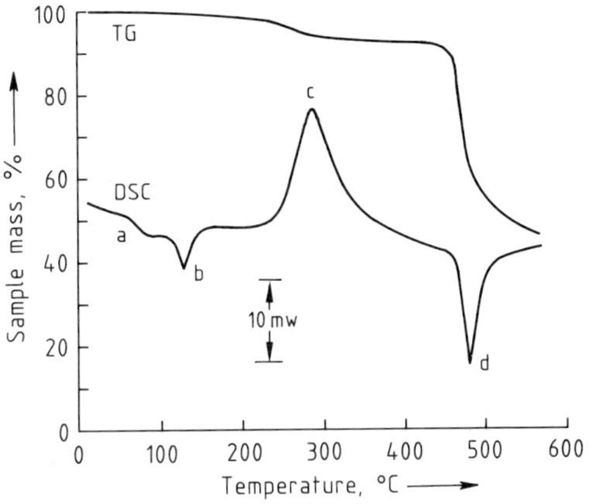

Figure 12. Simultaneous TG–DSC curves for an uncured polyimide resin
a) Glass transition; b) Melting; c) Curing; d) Degradation
10 mg sample heated at 10 °C min^{-1} in argon

The range of simultaneous techniques has been reviewed by Paulik [15]. TG–DTA and TG–DSC are the commonest simultaneous methods, followed by evolved gas analysis (Section 1.5).

1.4.2. Applications

TG–DSC curves showing the curing and decomposition of a polyimide resin are presented in Figure 12. Following a glass transition (a) and subsequent fusion of the material at ca. 120 °C (b), which are detected by DSC, the TG curve reveals the initial stages of an exothermic curing reaction (c) that occurs with a 7% mass loss. Endothermic decomposition of this resin does not take place until ca. 430 °C, as shown again by TG (d), a finding that permitted assignment of the appropriate baseline for measuring the heat of curing.

1.5. Evolved Gas Analysis

Evolved gas analysis (EGA) measures the nature and/or quantity of volatile products released on heating. In practice it is often performed simultaneously with TG or TG–DTA, and is particularly useful in furnishing direct chemical information to augment the physical measurements of TG or DTA. Additional advantages that have emerged include specificity—a single decomposition can be followed against a background of concurrent processes—and sensitivity, which can be far greater than with TG alone [16], [17].

Instrumentation. Almost every imaginable type of gas detector or analyzer has been utilized in EGA, including hygrometers, nondispersive infrared analyzers, and gas chromatographs. Absorption of the products into solution permits analysis by coulometry, colorimetry, ion selective electrode measurements, or titrimetry. The most important analyzers are Fourier transform infrared (FTIR) spectrometers and, preeminently, mass spectrometers [18]. The two latter methods can be used to record spectra repetitively, thereby producing a time-dependent record of the composition of the gas phase, from which EGA curves can be constructed for selected species.

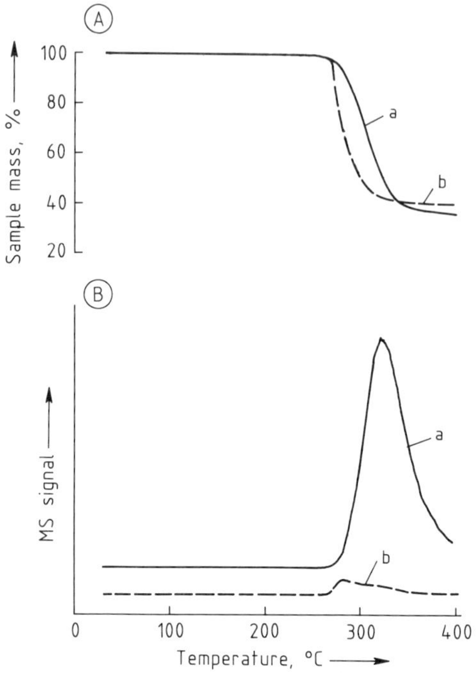

Figure 14. Simultaneous TG–MS curves showing the evolution of benzene (78 amu)
A) TG curve; B) MS curve
a) PVC without a smoke retardant; b) PVC with a smoke retardant (4% MoO_3)
15 mg samples heated at 10 °C min^{-1} in argon

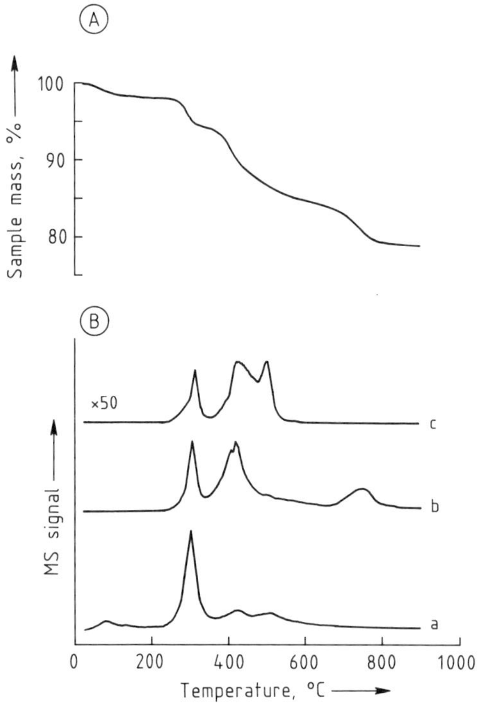

Figure 13. Simultaneous TG–MS curves for a brick clay
A) TG curve; B) MS data
Evolution of a) Water (18 amu); b) Carbon dioxide (44 amu); c) Sulfur dioxide (64 amu)
40 mg sample heated at 15 °C min^{-1} in air

Applications. Simultaneous TG–MS curves for a brick clay are shown in Figure 13, which simulates a set of firing conditions [19]. After an initial loss of moisture, combustion of organic matter at ca. 300 °C gives characteristic peaks for CO_2, H_2O, and SO_2. More SO_2 results from the oxidation of iron sulfides. Clay dehydroxylation occurs in the range 350–600 °C, and calcite dissociates to give the CO_2 peak observed at higher temperatures. EGA is of great value in the interpretation of complex TG and DTA curves.

In another example (Fig. 14), the evolution of benzene from two samples of PVC, one containing a smoke retardant, was compared quantitatively by injecting known amounts of benzene into the purge gas stream before and after the experiments, thereby providing a standard for estimating the amount of benzene produced by the polymer [19]. The smoke retardant was found in this case to reduce benzene evolution by a factor of ca. 20.

Emanation Thermal Analysis (ETA) [20] is a special case of EGA in which one establishes the rate of release of an inert gas from a heated material. The gas (argon, radon, or krypton), which is usually radioactive to allow for very sensitive detection, must first be incorporated into the solid being investigated. One of the main applications of ETA involves the characterization of powders. The rate of gas release from a powder is related to surface area, thereby permitting phenomena such as sintering to be detected.

1.6. Mechanical Methods

This section covers a family of techniques leading to dimensional or mechanical data for a material as a function of temperature or time. For a succinct review of the application areas see [21].

Thermodilatometry (TD) measures dimensional changes as a function of temperature in materials subject to negligible loads. A probe, which is held in light contact with the heated sample, is connected to a sensitive position sensor, usually a linear variable differential transformer (LVDT). In addition to providing expansion coefficients the technique can also indicate phase changes, sintering, and chemical reactions. Major application areas include metallurgy and ceramics.

Thermomechanical Analysis. The equipment used in thermomechanical analysis (TMA) is similar in principle to that for TD, but provision is made for applying various types of load to the specimen, so that penetration, extension, and flexure can be measured. This approach to analyzing such modes of deformation is illustrated schematically in Figure 15. The technique finds most use in polymer studies, as in the determination of glass-transition and softening temperatures for thin films and shrinkage characteristics of fibers.

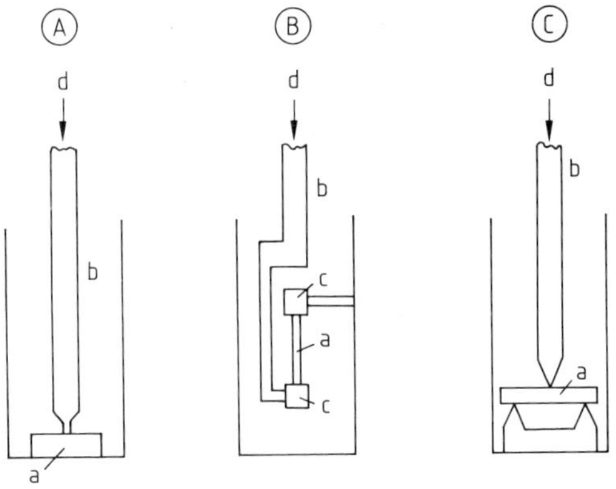

Figure 15. Modes of deformation subject to examination by TMA
A) Penetration; B) Extension; C) Flexure
a) Sample; b) Probe; c) Clamps; d) Load

Dynamic Mechanical Analysis. Polymers exhibit both elastic and viscous behavior under stress. In dynamic mechanical analysis (DMA) a sinusoidally modulated stress, often flexural (though measurement is also possible under tensile, compressive, shear, or torsional conditions) is applied to a specimen of material maintained under a specified temperature regime. Displacement transducers measure strain induced in-phase with the stress, as well as strain that lags behind. The former gives a measure of the sample's modulus, or stiffness, and the latter reflects damping characteristics. Like most other thermoanalytical techniques, DMA can be used to generate quantitative physical data, but it is also invoked in a comparative sense: to monitor the effects of additives, for example, or in quality control. It is particularly well suited to the examination of engineering composites, since gelation and curing behavior is sometimes more easily followed by DMA than by alternative techniques such as DSC.

1.7. Less Common Techniques

Thermoptometry. This group of techniques includes:

1) *Thermomicroscopy*, in which a material is observed under reflected, transmitted, or polarized transmitted light. The corresponding apparatus is referred to as a *hot stage*, which may also incorporate a DSC sensor for additional information.

2) *Thermophotometry*, in which the intensity of reflected or transmitted light is measured.
3) *Thermoluminescence*, which provides a measure of the intensity of light emitted by the material itself.

Electrical and Magnetic Techniques. *Thermomagnetometry* is used to study the magnetic properties of materials by revealing apparent weight changes that accompany phase transitions in the presence of an applied magnetic field, or variations in the amount of a magnetic substance.

Thermoelectrometry comprises methods for the measurement of resistance or capacitance during heating. Some equipment in this category provides remote probes for in situ monitoring of the curing of thermosets.

Acoustic Techniques. *Thermosonimetry* and *thermoacoustimetry* are used to measure, respectively, sound that is emitted or transmitted by a heated sample. The former technique has been used to study internal crystalline rearrangements and annealing, while the latter is useful for determining compressive and shear moduli.

2. Calorimetry

2.1. Introduction

Calorimetry in the broadest sense means the quantitative measurement of energy exchanged in the form of heat during a reaction of any type. By contrast, *thermal analysis* (Chap. 1) is concerned only with the measurement and recording of temperature-induced changes or temperature differences. Since all chemical reactions and many physical changes (e.g., deformation, phase transformations) are associated with the uptake or release of heat, the quantitative investigation of heat exchange is a relatively simple and universal method for characterizing particular processes both in an overall sense and with respect to time.

Nevertheless, only in recent decades has calorimetry emerged from the laboratories of a few thermodynamicists and specialists to become a widespread, convenient analytical method. The development of commercial calorimeters since the 1950s has led to rapid dissemination and application of the method even beyond the bounds of universities.

The explanation of this phenomenon, as with many other apparatus-based developments, lies in the refinement in measurement techniques made possible by modern electronics. As a rule, heat cannot be measured directly, but must be determined instead on the basis of a temperature change in the system under investigation. Very accurate classical thermometers tend to be very slow measuring devices. Resistance thermometers and thermocouples respond much more rapidly, but they require electronic amplification, and amplifiers with the required precision have become available only in recent decades.

A further explanation for the increasing importance of calorimetry is the development of personal computers, which relieve the user of the often laborious task of evaluating the experimental results, thereby opening the way for the method to become a routine laboratory technique.

In the sections that follow, calorimetry and the present state of instrument development will be described in such a way as to reveal both the possibilities and the limits of the method. The scope of the chapter rules out exhaustive treatment of the subject, so the goal is rather to delineate the method and guide the interested reader to other relevant technical literature.

2.2. Methods of Calorimetry

As noted above, heat that is released or consumed by a process cannot be measured directly. Unlike material quantities such as amount of substance, which can be determined with a balance, or the Volume of a liquid, which can be established with a liter measure, the quantity heat must be measured indirectly through its effect on a substance whose temperature it either raises or lowers. The fundamental equation for all of calorimetry describes the relationship between heat exchanged with a calorimeter substance and a corresponding temperature change:

$$\Delta Q = C(T) \cdot \Delta T \quad (1)$$

where ΔQ = exchanged heat (J), ΔT = observed temperature change (K), and $C(T)$ = heat capacity (J/K).

For heats (and therefore temperature changes) that are not too large, the temperature change is directly proportional to the exchanged heat. The proportionality constant is the heat capacity of the calorimeter substance (previously referred to as the "water value"). However, if the temperature change exceeds a few Kelvin, the temperature dependence of heat capacity stands in the way of a linear relationship, and a knowledge of the temperature function of the particular heat capacity in question is required in order to determine heat on the basis of a measured temperature difference.

The relationship above leads directly to the standard calorimetric methods for determining the heat of a process: either temperature is held constant by appropriate compensation for the heat effect, and the required compensation power is measured, or a temperature change is determined and used to calculate a corresponding value for exchanged heat. A precondition for the latter approach is accurate knowledge of the heat capacity as well as the heat transport properties of the measurement system.

2.2.1. Compensation for Thermal Effects

In the first method of heat measurement, temperature changes in the calorimeter substance are avoided by supplying or dissipating heat exactly equal in magnitude (but opposite in sign) to that associated with the process under investigation. As a rule, electrical energy is used to provide this compensation, either by introduction in the form of Joulean heat or dissipation through the Peltier effect. A combination approach is also possible, with an appropriate constant level of cooling and simultaneous controlled electrical heating. The requisite amount of compensation power can today be determined with a high degree of precision. It was once customary to invoke for compensation purposes "latent" heat derived from a phase change; in other words, heat released in a process was measured by using it to melt a substance such as ice, with subsequent weighing of the resulting water and conversion of the data into energy units based on the known heat of fusion for ice.

The method of compensation is advantageous, since it permits measurements to be carried out under quasi-isothermal conditions, thereby avoiding heat loss from the calorimeter to the surroundings by heat transport processes. Furthermore, there is no need for a calibrated thermometer, only a sensor sufficiently sensitive to temperature changes to provide adequate control over the compensation power.

2.2.2. Measurement of a Temperature Difference

In this alternative method of heat measurement, which is also indirect, a measured temperature difference is used to calculate the amount of heat exchanged. A distinction can be made between temporal and spatial temperature-difference measurements. In *temporal* temperature-difference measurement the temperature of a calorimeter substance is measured before and after a process, and a corresponding heat is calculated on the basis of Equation (1) (which presupposes an accurate knowledge of the heat capacity). In the *spatial* method a temperature difference between two points within the calorimeter (or between the calorimeter substance and the surroundings) is the quantity of interest. The basis for interpretation in this case is the equation for stationary heat conduction (Newton's law of cooling):

$$\Phi = \lambda(T) \cdot A \cdot \Delta T \cdot l^{-1} \qquad (2)$$

where Φ = heat flow rate (W), $\lambda(T)$ = thermal conductivity ($Wm^{-1}K^{-1}$), A = cross-sectional area (m^2), ΔT = temperature difference (K), and l = length (m).

From this expression it follows that the heat flow rate through a heat-conducting material is proportional to the corresponding temperature difference. If this temperature difference is recorded as a function of time, then for a given thermal conductivity one acquires a measure of the corresponding heat flow rate, which can be integrated to give a total heat. The technique itself is very simple, but the result will be correct only if the measured temperature difference accurately reflects the total heat flow rate and no heat is lost through undetected "heat leaks". Careful calibration and critical tests of the calorimeter are absolutely necessary if one hopes to obtain reliable results.

A fundamental disadvantage of the temperature-difference method is the fact that nonisothermal conditions prevail, so the precondition of a stationary state is fulfilled only approximately. The potential therefore exists for nonlinear phenomena in the heat-transport process, which might result in conditions outside the range of validity for Equation (2). Furthermore, with nonisothermal studies it is impossible to rule out completely the presence of undetected leaks. It must also be pointed out that in the case of nonisothermal studies the thermodynamic quantities of interest cannot be established according to the methods of reversible thermodynamics, at least not without further consideration. Strictly speaking, *irreversible* thermodynamics are applicable here, although the distinction is generally unimportant in calorimetric practice.

2.3. Calorimeters

There are many diverse types of calorimeters, differing with respect to measuring principle, mode of operation, and general construction. Measuring principles have been described in the preceding sections. In terms of mode of operation, two approaches can be distinguished: *static calorimeters*, in which the temperature remains constant or changes only in consequence of the heat of reaction, and *dynamic calorimeters*, in which the temperature of the calorimeter substance is varied deliberately according to a prescribed program (usually linear with time).

With regard to construction, a distinction can be made between *single* and *differential* calorimeters. Twin design is characteristic of the latter: that is, all components are present in duplicate, and the two sets are arranged as identical as possible. The reaction under investigation takes place in one side of the device, while the other side contains an inert reference substance. The output signal represents the difference between signals originating in the sample and reference side. With such a difference signal all symmetrical effects cancel (e.g., heat leaks to the surroundings).

2.3.1. Static Calorimeters

Considering first the static calorimeters, three types can be distinguished depending upon whether they are operated in an isothermal, isoperibolic (isoperibolic = uniform surroundings), or adiabatic manner.

2.3.1.1. Isothermal Calorimeters

In the isothermal mode of operation it is imperative that all thermal effects be somehow compensated. This is achieved either electrically or with the aid of a phase transition for some substance. Only phase-transition calorimeters can be regarded as strictly isothermal. In this case thermodynamics ensures that the temperature will remain precisely constant since it is controlled by a two-phase equilibrium of a pure substance. The most familiar example is the ice calorimeter, already in use by the end of the 18th century and developed further into a precision instrument about 100 years later by BUNSEN (Fig. 16). The liquid–gas phase transition has also been used for thermal compensation purposes; in this case a heat of reaction can be determined accurately by measuring the volume of a vaporized gas.

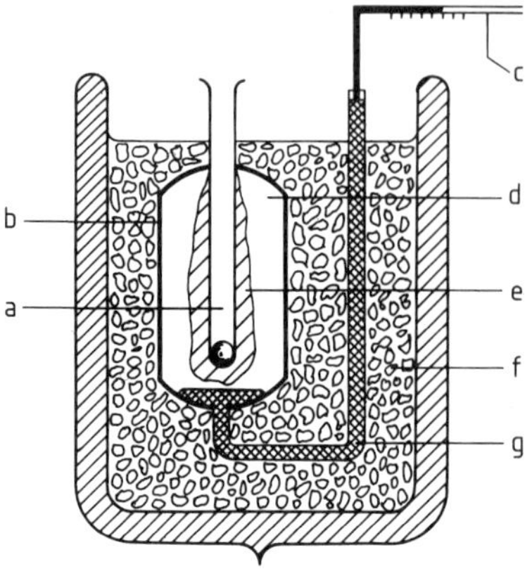

Figure 16. Ice calorimeter (based on [29])
a) Sample holder; b) Calorimeter vessel; c) Capillary with mercury thread; d) Water; e) Ice; f) Ice–water mixture; g) Mercury

Phase-transformation calorimeters are easy to construct, and for this reason they are not available commercially. Such a device permits very precise determination of the heat released during a process. An important drawback is the fact that the isothermal method is limited to the few fixed temperatures corresponding to phase transitons of suitable pure substances.

This disadvantage does not apply to an isothermal device based on electrical compensation. Nevertheless, calorimeters of the latter type operate only in a quasi-isothermal mode, since electronic control systems depend for their response upon small deviations from an established set point, and a certain amount of time is required for changing the prevailing temperature. The use of modern circuits and components ensures that errors from this source will be negligible, however. Electrical compensation makes it possible to follow both endothermic and exothermic processes. In both cases the compensation power is readily measured and recorded or processed further with a computer. Isothermal calorimeters are used quite generally for determining heats of mixing and solution. Commercial devices are available that also support the precise work required for multiphase thermodynamics. As a rule such calorimeters are of the single rather than the differential type.

2.3.1.2. Isoperibolic Calorimeters

An isoperibolic calorimeter is distinguished by the fact that it is situated within a thermostated environment. As a result, the temperature of the surroundings is kept constant, as is the effect of the surroundings on the calorimeter vessel. A well-defined thermal conduction path is maintained between the calorimeter vessel and its surroundings so that the temperature difference between the calorimeter vessel and the environment will be proportional to the heat flow rate. Temperature equilibration usually adheres to an exponential time function, and an integral taken over the entire course of this function is proportional to the heat of reaction. Such instruments often have exceptionally large time constants; that is, temperature equilibration occurs very slowly. Indeed, the greater the equilibration time, the more sensitive is the calorimeter. It is of course difficult with an exponential function to identify precisely the end point, so the integral over the measurement function and therefore the heat of interest are subject to corresponding degrees of uncertainty. The advantages lie in simplicity of construction and ease of producing a measurement signal (a straightforward temperature difference), which might for example be derived from a differential thermocouple and a simple set of amplifiers.

The best-known calorimeter of this type was developed by TIAN and CALVET (Fig. 17). Here the defined heat-conduction path to the thermostated surroundings (a large aluminum block) consists of a large number of differential thermocouples. This arrangement permits optimum determination of the heat flow rate to the surroundings, and such an instrument can be very sensitive (microcalorimeter).

If heat transport to the surroundings is restricted by suitable insulation, temperature equilibration becomes so slow that—with small corrections—temperature changes in the calorimeter substance alone can serve as the basis for determining the heat of a reaction. Calorimeters in this category include the classical mixing calorimeter (with heat exchange) in which the sample substance is brought into contact with a (usually) colder calorimeter substance, whereupon an intermediate "mixing temperature" is established (Fig. 18). Other examples of mixing calorimeters include the simpler types of combustion calorimeters, in which the calorimeter substance is water, and the *drop calorimeter* (see Fig. 24). The latter consists of a metal block whose temperature change subsequent to introduction of a sample of accurately known temperature provides information about the sample's heat capacity. The category of isoperibolic calorimeters also includes flow calorimeters, in which two fluid media (usually liquids containing the reactants) are first brought to the temperature of the thermostated surroundings and then combined. The ensuing reaction causes a measurable temperature change, the extent of which is proportional to the heat of the reaction (Fig. 19).

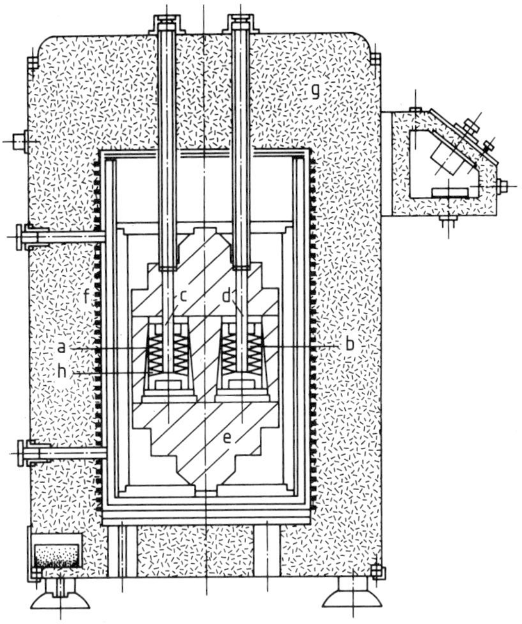

Figure 17. Tian–Calvet calorimeter
a, b) Heat conduction paths between measuring cells and block; c, d) Sample and reference containers; e) Isoperibolic block; f) Thermostatic jacket; g) Thermal insulation; h) Thermopiles

2.3.1.3. Adiabatic Calorimeters

An adiabatic calorimeter is equipped with an electronic control system designed to ensure to the greatest extent possible that the surroundings of the calorimeter cell remain at all times at precisely the same temperature as the substance under examination. The object is to prevent virtually all heat transport to the surroundings, thereby making it possible to establish the total heat effect from the change in temperature of the calorimeter substance. Good adiabatic calorimeters are difficult to construct, and they are not available commercially. Nevertheless, precision instruments of this type permit quantities such as specific heat capacities, as well as associated anomalies, to be measured very accurately (Fig. 20). The method entails an extraordinary level of effort, and it is practiced only by specialists working in a few laboratories in the world.

Simpler adiabatic calorimeters are distributed commercially in the form of adiabatic mixing and combustion calorimeters. These constitute very satisfactory routine instruments, especially since the adiabatic mode of operation (which entails measurement of only a single temperature difference) facilitates automated data acquisition and interpretation.

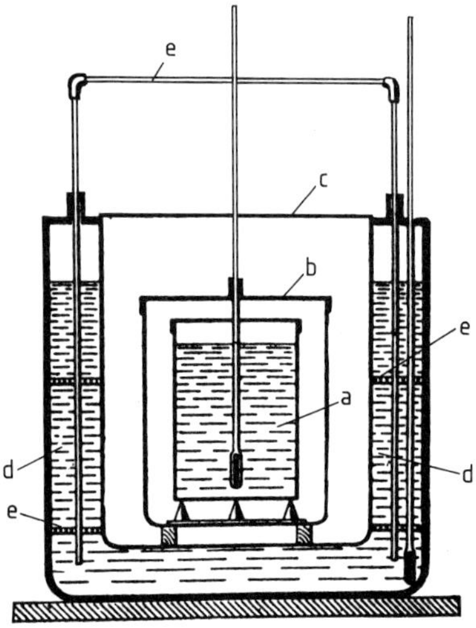

Figure 18. A classical mixing calorimeter (from Meyers Conversationslexicon, 1893)
a) Calorimeter vessel; b) Convection and radiation shield; c) Cover; d) Jacketing vessel (thermostated); e) Lifting mixer

2.3.2. Scanning Calorimeters

Scanning calorimeters are devices in which the sample temperature is deliberately changed during the course of an experiment according to a prescribed program. As a rule, heating or cooling of the sample is caused to occur linearly with time, and the heat flow rate required to accom-

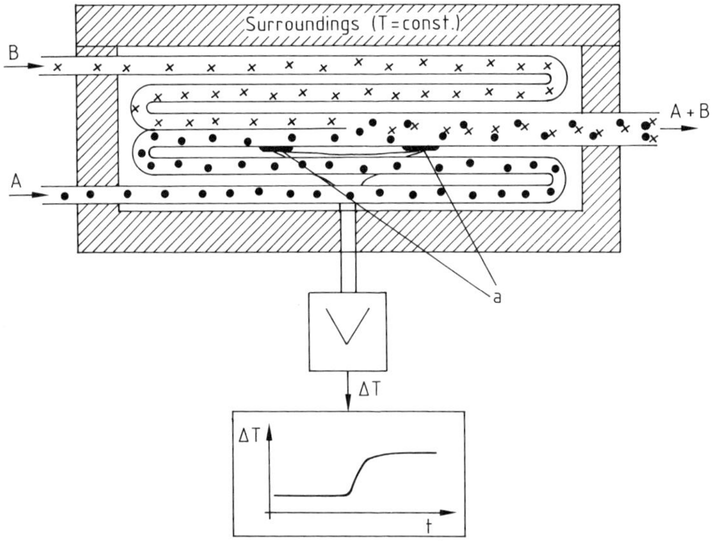

Figure 19. Flow calorimeter, with a chemical reaction occurring in the liquid
A, B reactants, A + B reaction product; a) Differential thermocouple

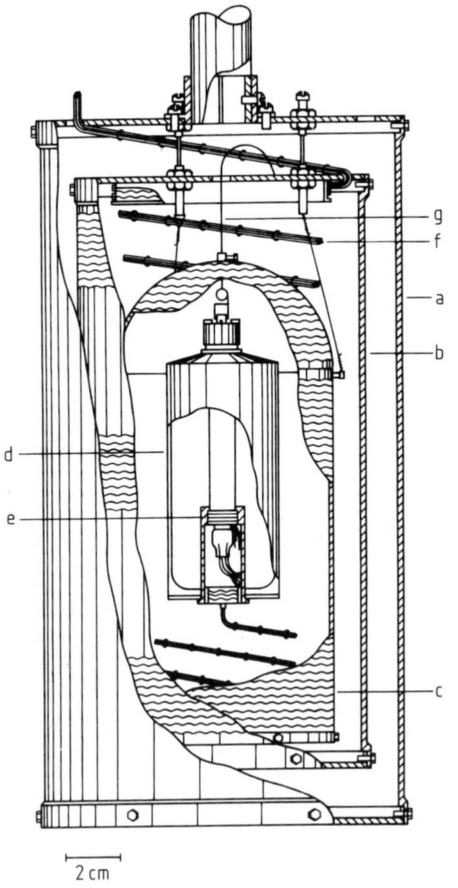

Figure 20. Construction of an adiabatic calorimeter (based on [30])
a) External radiation shield; b) Internal radiation shield; c) Adiabatic jacket; d) Calorimeter vessel; e) Sample suspension platform with resistance thermometer and heater; f) Electrical leads; g) Calorimeter suspension system

plish the desired change is determined and output as a function of time or temperature. Any chemical or physical change in the sample at a particular temperature will manifest itself as a change in the heat flow rate, leading to a "peak" in the measurement curve whose area is proportional to the heat of the process (see also Section 1.3.2).

Such instruments have been on the market since the 1950s. Their great advantage is that they permit heats of reaction and transformations of every type to be measured both rapidly and with reasonable precision. Heating rates usually fall in the range 5–10 K/min, so a measurement extending over 300 K can be complet-

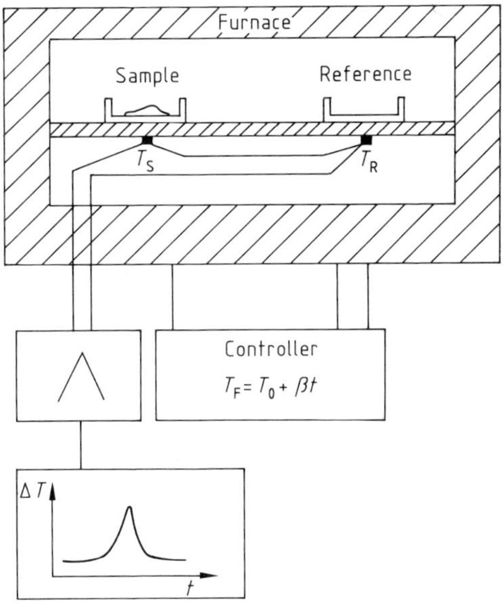

Figure 21. Operating principle of a temperature-difference scanning calorimeter (heat-flux calorimeter)
T_F furnace temperature, T_S sample temperature, T_R reference temperature, T_0 initial temperature, ΔT differential temperature $T_S - T_R$, β heating rate, t time

ed within an hour, providing a reliable indication of all changes in the sample over this temperature range, including heats of reaction. The rapid development of calorimetry in recent years is due largely to the perfection of this type of calorimeter.

Scanning calorimeters are designed almost without exception as differential devices. With regard to the principle of measurement, two varieties can be distinguished: *differential-temperature* and *power-compensated* calorimeters. Although the two types rely upon very different approaches to determining heat flow rates with respect to the sample, and are therefore subject to different sources of error, they are marketed under the single heading *differential scanning calorimeter* (DSC).

2.3.2.1. Differential-Temperature Scanning Calorimeters

In a differential-temperature scanning calorimeter the heat necessary for raising the sample temperature reaches the sample position via one well-defined heat conduction path (cf. Fig. 21). The measured signal is derived from a temperature difference generated along this path, which is proportional to the heat flow rate. This measuring and operating principle leads to the alternative designation *heat-flux calorimeter* for such a device. In order to minimize errors from heat leaks and other influences of the surroundings, calorimeters of this type are always based on a differential design, with full duplex construction. Many commercial models are available.

Any temperature difference between the sample and the reference is directly proportional to the corresponding differential heat flow rate, and so this is the quantity to be measured. The appropriate functional relationship follows directly from Equation (2):

$$\Phi = K(T) \cdot \Delta T \quad (3)$$

The heat flow rate of interest, Φ (in watts), is proportional to the measured temperature difference ΔT. Unfortunately, the proportionality constant K (the calibration factor for the calorimeter) is itself a function of the thermal conductivity of the sample (see Eq. 2), and therefore temperature dependent. Moreover, temperature differences between the sample and the reference result in variable amounts of heat loss to the surroundings via inevitable heat leaks. Heat exchange with the environment also occurs via radiation and convection. Since the two transport processes are linked to temperature differences in a nonlinear way, the calibration factor in the presumably linear calorimeter equation necessarily depends on ΔT, which is the quantity to be measured. Consequently, the calibration factor depends upon such sample parameters as mass, thermal conductivity, and specific enthalpy difference [31], so heats measured with these instruments must realistically be assigned an uncertainty of about five percent unless considerable effort has been expended in calibration [32], [33]. Repeatability of the results is generally many times better, but the user should not make the mistake of assuming an equally high level of certainty. Another source of systematic error may develop during the measurement if some reaction causes a change in the heat capacity and/or the extent of heat transfer to the sample holder. Both phenomena contribute to changes in the baseline, with the result that the course of the latter cannot be plotted accurately (Fig. 22). This in turn means that the area corresponding to the heat of reaction also cannot be determined exactly, leading to uncertainty in the associated heat [34]. Despite these restrictive comments, differential-temperature scanning calorimeters offer the advantages of a simple, transparent mode of operation and relatively low cost, and for this reason they account for a large share of the market. With appropriately critical supervision they are capable of fulfilling their assignment admirably.

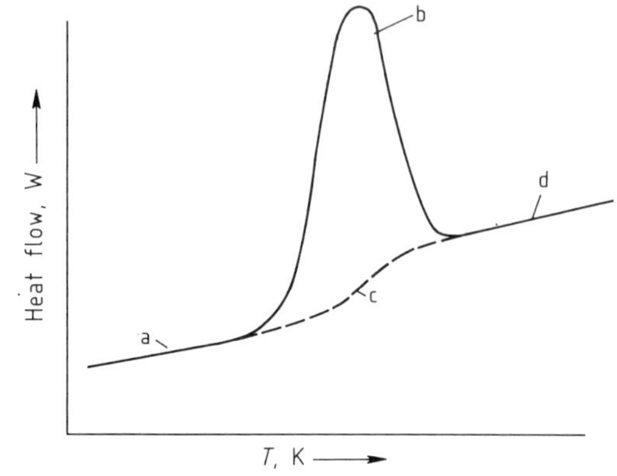

Figure 22. Output curve from a scanning calorimeter showing a shift in the baseline caused by variability in either c_p or the coefficient of heat transfer
a) Original baseline; b) Peak; c) Unknown baseline path; d) Final baseline

2.3.2.2. Power-Compensated Scanning Calorimeters

Like differential-temperature devices, power-compensated scanning calorimeters always feature duplex construction, with sample and reference holders heated in such a way that at every instant the temperature of each corresponds almost exactly to a programmed set temperature. Any temperature differences that arise between sample and reference by reason of thermal events in the sample are compensated immediately by appropriate changes in the electrical heating (Fig. 23). The output signal is proportional to the instantaneous differential heat flow rate. Since temperature differences between sample and reference are essentially compensated at once, measurement errors due to undetected heat leaks can be largely avoided. Instruments of this type, the most widely distributed of which are manufactured by Perkin–Elmer, require a very complicated electronic control system, and are therefore rather expensive. The principal disadvantage lies in the fact that the operation of the

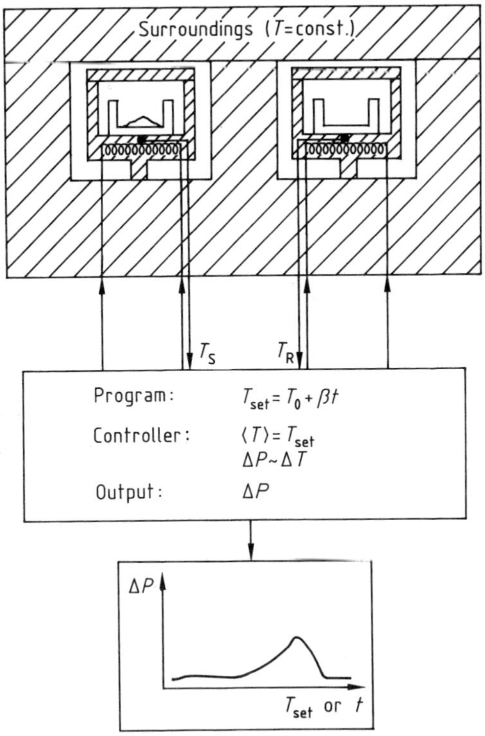

Figure 23. Operating principle of a power-compensated differential scanning calorimeter (Perkin–Elmer)
T_0 initial temperature; T_S sample temperature; T_R reference temperature; $\langle T \rangle$ mean temperature between sample and reference; ΔP differential power $P_S - P_R$; β heating rate; t time

electronic system is a complete mystery to the average layman, thereby obscuring possible sources of error. These instruments also require that there be some deviation, albeit minimal, from the set temperature to guide the control circuit, so just as in the preceding case some systematic error must be anticipated (although certainly to a lesser degree), associated once again with temperature differences between sample and reference and nonlinear heat exchange. Accordingly, the level of systematic uncertainty for heats measured by power-compensated DSC can be estimated at 1–3 %.

2.4. Applications of Calorimetry

Modern calorimeters permit relatively rapid and truly precise measurement of heat exchanges in a wide variety of reactions. Since heat evolution is proportional to the conversion (extent of reaction) in a chemical, physical, or biological reaction, calorimetric measurement constitutes one method for quantitative evaluation of the reaction itself. Measurement is possible not only of the total heat (and therefore the total conversion) of a reaction but also of the course of the reaction with respect to time. This opens the way to studies of kinetics, problems related to stability and safety, and other areas in which temperature and time-dependencies in a reaction play a part.

The sections that follow describe several possible applications of calorimetry, although these should be regarded only as illustrative, since the scope of the present article rules out any treatment that is even approximately exhaustive. No such limits apply to the reader's imagination, however, and any process that generates (or consumes) heat can in principle be followed with a calorimeter. Most recently, enzymatic and bacterial manufacturing processes have assumed increasing importance in industry, and interest in biocalorimetry is growing accordingly [35].

2.4.1. Determination of Thermodynamic Functions

Various thermodynamic potential functions (enthalpy, entropy, free energy, etc.) can be determined by integration on the basis of specific heat capacities. The accurate establishment of specific heat capacities at constant pressure, c_p, as a function of temperature is thus of fundamental importance. Different types of calorimeters are used for determinations of c_p depending upon accuracy requirements and the temperature range in question.

The most accurate results are obtained with adiabatic calorimeters, which are useful over a temperature range from fractions of a Kelvin to about 1000 K. These instruments permit specific heat capacities to be determined with very small uncertainties (< 1 %). Nevertheless, such measurements are so complex that they are rarely applicable to normal laboratory practice in industry, especially given the lack of commercial instruments.

The standard approach to c_p measurement therefore relies on the less precise but more easily constructed and operated mixing calorimeters, in particular drop calorimeters (Fig. 24). A (mean) specific heat capacity is calculated in this

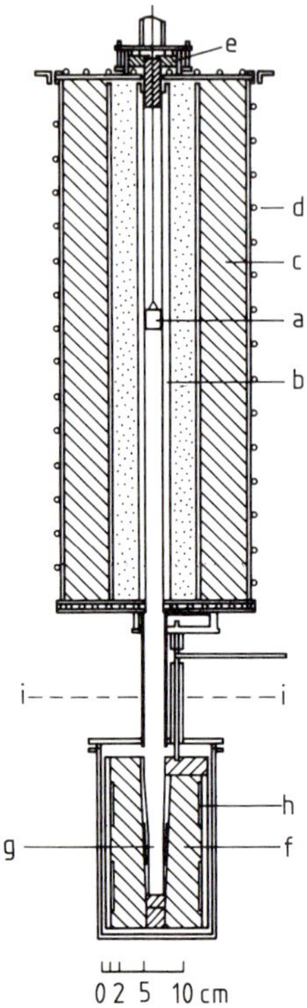

Figure 24. High-temperature drop calorimeter (based on [36])
a) Sample; b) Heated corundum tube; c) Magnesia insulation; d) Cooling coils; e) Sample mounting device; f) Copper block; g) Hole for receiving sample; h) Resistance thermometer; i) Level of the thermostated oil bath

case from the heat released when a sample of known temperature is introduced (dropped) into an isoperibolic calorimeter. Charging temperatures may exceed 2500 K. Determining the temperature dependence of c_p requires many individual measurements at various charging temperatures. The uncertainty in such measurements is usually 3–5%.

A truly elegant method for determining c_p at < 1000 K takes advantage of a scanning calorimeter, preferably a power-compensated DSC. The procedure involves raising the temperature of a sample in the calorimeter at a defined rate (typically 10 K/min). The same measurement is subsequently repeated, this time with empty crucibles. The difference between the two measurement curves provides a record of the heat flow rate into the sample, from which the specific heat capacity can be calculated. This method makes it possible conveniently to measure the $c_p(T)$ of a material over a temperature range of several hundred Kelvins within the course of a single day. However, it does demand a very high level of repeatability (baseline stability) for the DSC in question. In the case of differential-temperature (heat-flux) DSC—even with careful calibration—uncertainties of 5–10% in the results must be anticipated. With power-compensated DSC the systematic uncertainties tend to be in the range of 3–5%. Uncertainties in the thermodynamic potential functions of interest are comparable. Moreover, a calorimeter operating dynamically always indicates a temperature that deviates to a greater or lesser extent from the true sample temperature, resulting in further uncertainty. It must in any case be borne in mind that potential functions determined on the basis of thermodynamic equilibrium, and thus defined statically, may differ from functions determined dynamically, since equilibrium conditions are not maintained in DSC, and in the case of a reaction even stationary conditions are lacking.

2.4.2. Determination of Heats of Mixing

For liquid systems consisting of several components, multiphase thermodynamics provides the theoretical background for computing phase diagrams from heats of mixing for the relevant components. Phase diagrams in turn are interesting not only to thermodynamicists; they are also of great importance in chemical process engineering practice.

The determination of heats of mixing is a classical field of investigation for calorimetry, one that at the required level of precision is today incomparably less laborious than previously. Enthalpies of mixing were formerly determined in some cases indirectly via vapor pressure measurements, but today they are measured directly in (quasi)isothermal or isoperibolic calorimeters.

The art of conducting such an experiment successfully lies in rigorously excluding vaporization during the measurement, since specific heats of vaporization are orders of magnitude larger than the effects of mixing that are to be measured. As excess quantities (which characterize deviations from ideal behavior), heats of mixing are invariably small, and they must be determined with high precision in order to permit the construction of sufficiently precise phase diagrams. Isothermal calorimeters are therefore definitely to be preferred, since here the problem of incomplete determination of a heat of mixing—owing perhaps to undetected heat leaks—is less severe than with an isoperibolic instrument. Moreover, the isothermal heat of mixing is defined more precisely, and it lends itself more readily to mathematical treatment than in the case of parallel temperature changes.

Phase transitions in solid mixtures can be measured directly in a scanning calorimeter, since vapor pressures associated with the components are here so low that vaporization plays virtually no role and therefore does not interfere. Corresponding phase diagrams (i.e., liquidus–solidus curves) can be inferred from heat flow rate curves obtained by controlled heating and cooling of appropriate mixtures (cf. Section 1.3.3). Thermodynamically relevant quantities can in turn be determined indirectly by comparing measured phase diagrams with suitable computed diagrams.

2.4.3. Combustion Calorimetry

The determination of heats of combustion for fuels of all types is prescribed in detail by government regulation, since the "calorific value" or "heat content" of a particular coal, oil, or natural or manufactured gas determines its economic value. Combustion calorimeters or "bomb calorimeters" (usually operated adiabatically) are used for such measurements, whereby the sample is combusted in a "calorimetric bomb" (likewise standardized; cf. Fig. 25) in pure oxygen at $3 \cdot 10^6$ Pa to a defined set of end products. Methods of measurement and interpretation are also standardized, and they have been simplified to such an extent that the entire procedure can be automated.

The same instruments are useful for determining heats of combustion for a wide variety of (usually organic) substances, from which standard enthalpies of formation can be determined. For these calorimetric methods errors in measurement are very small (< 1 ‰), with a level of precision that is also appropriate for analytical purposes (e.g., purity or composition checks on batches of chemicals). Combustion calorimetry is useful in biology as well. For example, in determinations of the nutrient content of leaves and fruits as a function of season and place based on appropriate heats of combustion.

Many combustion calorimeters are available commercially. Instruments utilizing water as the calorimeter substance generally operate adiabatically, thereby avoiding the somewhat complicated Regnault–Pfaundler procedure (cf. [27]) for determining and interpreting prior and subsequent temperature patterns. "Dry" combustion calorimeters that avoid the use of water or other liquids have also recently come onto the market.

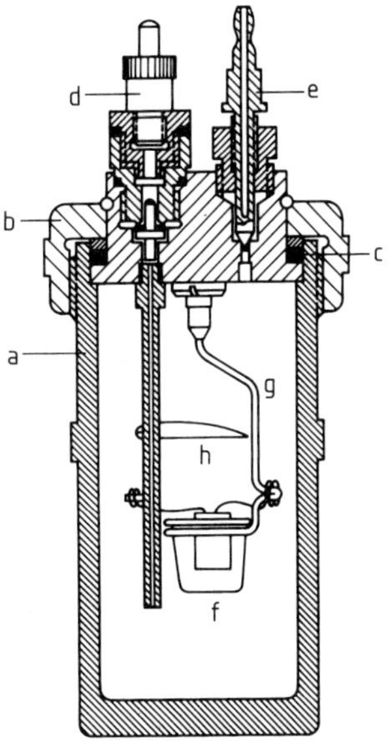

Figure 25. Example of a calorimetric bomb for use in a combustion calorimeter
a) Pressure vessel; b) Screw cap; c) Sealing ring; d) Inlet valve; e) Outlet valve; f) Sample crucible with sample; g) Crucible holder; h) Heat shield

2.4.4. Reaction Calorimetry

Every chemical process is associated with some type of heat effect. Accurate knowledge of heat released or consumed is of great importance in process engineering. Moreover, heat evolution is strictly coupled with the particular course of a reaction. If one is able to determine unambiguously the time dependence of the heat flow rate associated with a particular sample, one also gains access to the corresponding reaction rate law, which in turn opens the way to the reaction kinetics. Knowledge of the appropriate kinetic parameters permits one to predict the course of the reaction under other sets of conditions, which is of great practical value. Heat of reaction determinations as well as kinetic studies are best carried out under isothermal (and isobaric) conditions, and for this reason it is advisable to select a (quasi)isothermal calorimeter with compensation for effects of the measurement process. One practical difficulty lies in the fact that reaction begins immediately after introduction of the sample at the appropriate temperature, even though the calorimeter itself is not yet at thermal equilibrium (and therefore not ready for use). One way out of this dilemma is to devise a system for bringing the components of the reaction together only after isothermicity has been achieved. With liquids the challenge is not particularly great, since pumps and stirrers can ensure rapid (and thorough) mixing of the reactants at any desired time. The piercing of a membrane (or an ampoule) is also a common expedient. Another possibility is the use of a specially designed reaction calorimeter. With one device of this type the first reactant is present from the outset in a suitable glass or metallic vessel, to which a second reactant is added with mixing after stationary conditions have been established (Fig. 26). With the aid of electronic temperature control the reaction can then be carried out in any way desired: isothermally, adiabatically, or under the influence of some temperature program. Measurement of the relevant quantities, process control, and thermochemical data interpretation are all accomplished with the aid of a personal computer. A second type of special reaction calorimeter, suitable only for liquid components, is so constructed that each reactant is pumped in through its own system of tubing, with the streams uniting after temperature equilibration (Fig. 27). The ensuing reaction causes a temperature change, from which the corresponding heat of reaction can be calculated.

Transport and mixing of liquid or gaseous components is accomplished rather easily, but the situation is more complex for solid substances or mixtures. Here there is little hope of satisfactorily mixing the components inside a normal calorimeter, so one is forced to resort to

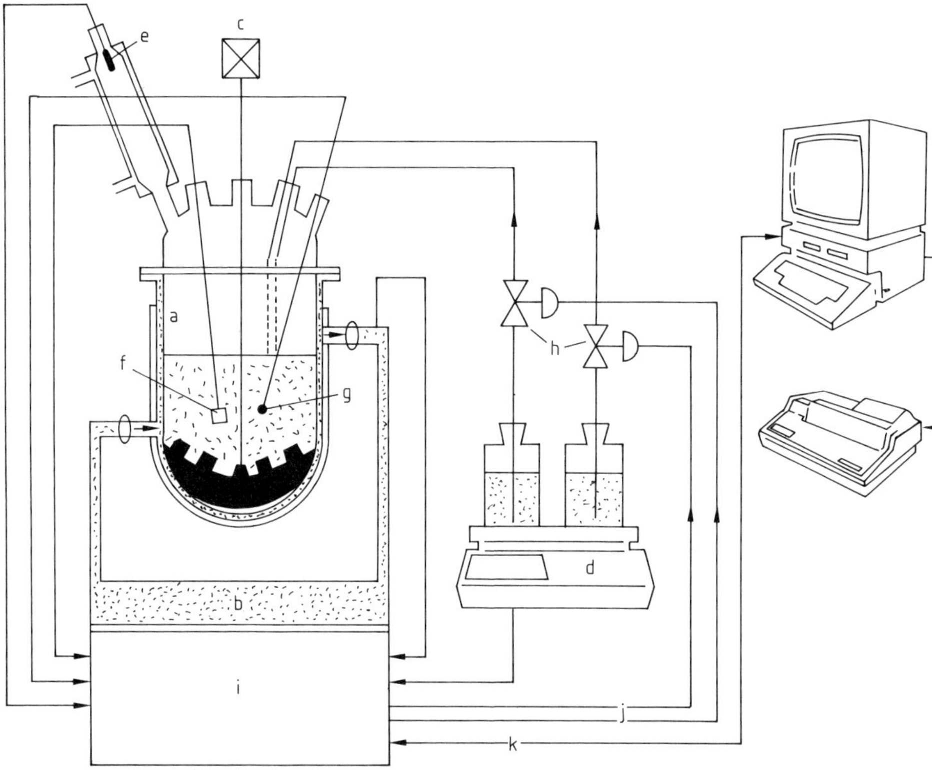

Figure 26. Operating principle of a reaction calorimeter (Contraves)
a) Reaction vessel; b) Heating and cooling bath; c) Stirrer; d) Balance; e) Pressure sensor for internal pressure; f) pH probe; g) Temperature sensor; h) Dosing valve; i) Control unit; j) Control and monitoring signals; k) Measured signal

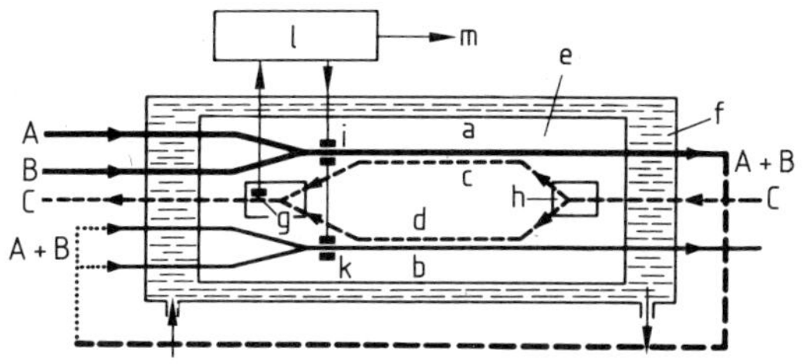

Figure 27. Schematic diagram of a flow calorimeter after P. PICKER (Setaram)
A, B) Reactants; A + B) Reaction mixture; C) Heat-exchange liquid
a) Reaction tube (sample tube); b) Reference tube; c, d) Heat-exchange tubes; e) Vacuum jacket; f) Thermostated liquid; g) Thermistor; h) Flow divider; i, k) Heating elements; l) Measurement and control electronics; m) Output signal

a scanning calorimeter. The calorimeter is charged at a temperature below that at which the substances react, after which the temperature is increased linearly with time until reaction is complete. From a heat flow rate–time function acquired at a well-defined heating rate it is possible to establish both the heat of reaction and, if need be, kinetic parameters for the process. One prerequisite, however, is that the recorded function reproduces in an unbiased way the true heat production of the sample. This presents serious problems, since heat must inevitably flow from the point at which it is generated to the site of the temperature sensors before any measurement signal can appear. In the case of a fairly rapid chemical process, where the rate constant is comparable to the time constant of the calorimeter, the measurement signal from the recording device must be treated mathematically ("desmeared") [37] before it can be utilized for a kinetic analysis. Owing to the smaller time constant and more precise linearity of the corresponding temperature–time program, a scanning calorimeter with power compensation is preferred over a temperature-difference DSC.

2.4.5. Safety Studies

Calorimetric methods applicable to safety studies are also concerned with clarifying the kinetic behavior of substances, but in this case from a safety engineering viewpoint. The important role assigned recently to such studies justifies their being treated as a separate topic.

The vast majority of chemical substances are *endothermic* materials; that is, they release heat when they decompose, and their decomposition is accelerated by heat. Chemical reaction kinetics teaches that any reaction is capable of proceeding at temperatures considerably lower than the temperature at which the maximum reaction rate is achieved. Usually, however, the reaction rate under these conditions is so low that the process is not apparent. The more unstable a substance, the lower is its decomposition temperature, and therefore the greater its decomposition rate at room temperature. In the event that heat evolved during storage—attributable to traces of reaction occurring under storage conditions—exceeds the level at which heat is conducted away, then the decomposition rate may increase very rapidly, possibly to the point of an explosion.

Thorough knowledge of decomposition kinetics is therefore essential for ensuring safe storage and handling. The requisite information can, in principle, be gained through the methods described in Section 2.4.4, but it is often impossible to avoid an uncontrolled outcome and simultaneous destruction of the measuring system. This obviously rules out the use of expensive apparatus, which has led to the development of special safety calorimeters that are either inexpensive and disposable or designed to withstand the effects of violent reactions.

Simple arrangements of this type clearly preclude precision measurements, but they are entirely adequate from the specific viewpoint of assessing storage safety at particular temperatures or with particular types of packaging. The simplest example is an exceedingly primitive isoperibolic calorimeter consisting of a glass ampoule, either open or fitted with a pressure-tight closure, mounted together with a thermocouple inside a small furnace—a device that anyone can easily assemble and calibrate. More ambitious designs differ from standard calorimeters in that the sample holder (or measuring head) is incorporated into an autoclave of sufficient stability to permit studying the thermal behavior even of explosives. Included in this category are the so-called *accelerating rate calorimeters* (ARC), in which a sample is heated very slowly under quasiadiabatic conditions until the reaction intensifies by itself as a result of exothermic decomposition. The resulting temperature–time function can then be used to establish the corresponding heat effect and stability behavior.

3. References

References for Chapter 1

[1] J. O. Hill (ed.): "For Better Thermal Analysis and Calorimetry," 3rd ed., International Confederation for Thermal Analysis and Calorimetry, 1991.
[2] W. Wendlandt: *Thermal Analysis,* 3rd ed., Wiley-Interscience, New York 1986.
[3] M. E. Brown: *Introduction to Thermal Analysis,* Chapman and Hall, London 1988.
[4] R. C. Mackenzie (ed.): *Differential Thermal Analysis,* Academic Press, London, vol. 1, 1970, vol. 2, 1972.
[5] F. Paulik, J. Paulik in C. L. Wilson, D. W. Wilson (eds.): *Comprehensive Analytical Chemistry,* vol. 12, part A, Elsevier, Amsterdam 1981.
[6] M. Reading in E. L. Charsley, S. B. Warrington (eds.): *Thermal Analysis – Techniques and Applications,* Royal Society of Chemistry, London 1992.
[7] S. P. Wolsky, A. W. Czanderna (eds.): *Microweighing in Controlled Environments,* Elsevier, Amsterdam 1980.
[8] ASTM Standard E914-83.
[9] C. L. Duval: *Inorganic Thermogravimetric Analysis,* Elsevier, Amsterdam 1953.
[10] M. R. Ottaway, *Fuel* **61** (1982) 713.
[11] C. M. Earnest (ed.): *Compositional Analysis by Thermogravimetry,* ASTM STP 997, Philadelphia 1988.
[12] J. W. Cumming, J. McLaughlin, *Thermochim. Acta* **57** (1982) 253.
[13] E. Turi (ed.): *Thermal Characterization of Polymeric Materials,* Academic Press, New York 1981.
[14] R. L. Blaine, C. K. Schoff (eds.): *Purity Determination by Thermal Methods,* ASTM STP 838, Philadelphia 1984.
[15] F. Paulik, J. Paulik, *Analyst (London)* **103** (1978) 417.
[16] W. Lodding (ed.): *Gas Effluent Analysis,* Marcel Dekker, New York 1967.
[17] S. B. Warrington in E. L. Charsley, S. B. Warrington (eds.): *Thermal Analysis – Techniques and Applications,* Royal Society of Chemistry, London 1992.
[18] M. R. Holdiness, *Thermochim. Acta* **75** (1984) 361.
[19] E. L. Charsley, C. Walker, S. B. Warrington, *J. Therm. Anal.* **40** (1993) 983 (Proc. 10th ICTAC Conf., Hatfield, 1992).
[20] V. Balek, J. Tolgyessy in C. L. Wilson, D. W. Wilson (eds.): *Comprehensive Analytical Chemistry,* vol. 12, part C, Elsevier, Amsterdam 1984.
[21] M. Reading in E. L. Charsley, S. B. Warrington (eds.): *Thermal Analysis – Techniques and Applications,* Royal Society of Chemistry, London 1992.

General References for Chapter 2

[22] W. Hemminger, G. Höhne: *Calorimetry, Fundamentals and Practice,* Verlag Chemie, Weinheim 1984.
[23] E. Koch: *Non-Isothermal Reaction Analysis,* Academic Press, London 1977.
[24] J. P. McCullough, D. W. Scott (eds.): *Experimental Thermodynamics,* vol. 1: "Calorimetry of Non-reacting Systems," Butterworths, London 1968.
[25] B. Le Neindre, B. Vodar (eds.): *Experimental Thermodynamics,* vol. 2, Butterworths, London 1975.
[26] H. Spink, I. Wadsö: "Calorimetry as an Analytical Tool in Biochemistry and Biology," in D. Glinck (ed.): *Methods of Biochemical Analysis,* vol. 23, J. Wiley & Sons, New York 1976.
[27] S. Sunner, M. Mansson (eds.): "Combustion Calorimetry," *Experimental Chemical Thermodynamics,* vol. 1, Pergamon Press, Oxford 1979.
[28] H. Weber: *Isothermal Calorimetry,* Herbert Lang, Bern, Peter Lang, Frankfurt 1973.

Specific References for Chapter 2

[29] R. W. Bunsen, *Ann. Phys.* **141** (1870) 1.
[30] E. D. West, E. F. Westrum Jr. in J. P. McCullough, D. W. Scott (eds.): *Experimental Thermodynamics,* vol. 1, Butterworths, London 1968, p. 354.
[31] G. W. H. Höhne: "Remarks on the Calibration of Differential Scanning Calorimeters," *J. Therm. Anal.* **37** (1991) 1987–2000.
[32] G. W. H. Höhne et al.: "The Temperature Calibration of Scanning Calorimeters," Part I: *Thermochim. Acta* **160** (1990) 1–12; Part II: H. K. Cammenga et al., *Thermochim. Acta* **219** (1993) 333.
[33] S. M. Sarge et al.: "The Heat Calibration of Scanning Calorimeters," *Thermochim. Acta,* in preparation (1994).
[34] W. F. Hemminger, S. M. Sarge: "The Baseline Construction and its Influence on the Measurement of Heat with Differential Scanning Calorimeters," *J. Therm. Anal.* **37** (1991) 1455–1477.
[35] J. Lamprecht, W. Hemminger, G. W. H. Höhne (eds.): "Calorimetry in the Biological Sciences," *Thermochim. Acta* **193** (Special Issue) (1991) 1–452.
[36] T. B. Douglas, E. G. King in J. P. McCullough, D. W. Scott (eds.): *Experimental Thermodynamics,* vol. 1, Butterworths, London 1968, p. 297.
[37] G. W. H. Höhne, J. Schawe: "Dynamic Behaviour of Power Compensated Differential Scanning Calorimeters," *Thermochim. Acta* **229** (1993) 27.

Surface Analysis

John C. Rivière, Harwell Laboratory, AEA Technology (retired), Didcot, United Kingdom

0-89573-541-5/94/$ 5.00 + .50

1. Introduction

Wherever the properties of a solid surface are important, it is also important to have the means to measure those properties. The surfaces of solids play an overriding part in a remarkably large number of processes, phenomena, and materials of technological importance. These include catalysis; corrosion, passivation, and rust; adhesion; tribology, friction, and wear; brittle fracture of metals and ceramics; microelectronics; composites; surface treatments of polymers and plastics; protective coatings; superconductors; and solid surface reactions of all types with gases, liquids, or other solids. The surfaces in question are not always external; processes occurring at inner surfaces such as interfaces and grain boundaries are often just as critical to the behavior of the material. In all the above cases, the nature of a process or of material behavior can be understood completely only if information about both surface composition (i.e., types of atoms present and their concentrations) and surface chemistry (i.e., chemical states of the atoms) is available. Occasionally, knowledge of the arrangement of surface atoms (i.e., the surface structure) is also necessary.

First of all, what is meant by a *solid surface*? Ideally the surface should be defined as the point at which the solid terminates, that is, the last atom layer before the adjacent phase (vacuum, vapor, liquid, or another solid) begins. Unfortunately such a definition is impractical because the effect of termination extends into the solid beyond the outermost atom layer. Indeed, the current definition is based on that knowledge, and the surface is thus regarded as consisting of that number of atom layers over which the effect of termination of the solid decays until bulk properties are reached. In practice, this decay distance is of the order of 5–20 nm.

By a fortunate coincidence, the depth into the solid from which information is provided by the techniques described here matches the above definition of a surface almost exactly. These techniques are therefore *surface specific*; in other words, their information comes only from that very shallow depth of a few atom layers. There are other techniques that can be *surface sensitive*, in that they would normally be regarded as techniques for bulk analysis, but have sufficient sensitivity for certain elements that those elements can be analyzed with these techniques even if they are present only on the surface.

Why should surfaces be so important? The answer is twofold. Firstly, because the properties of surface atoms are usually different from those of the same atoms in the bulk, and secondly, because in any interaction of a solid with another phase the surface atoms are the first to be encountered. Even at the surface of a *perfect single crystal* the surface atoms behave differently from those in the bulk simply because they do not have the same number of nearest neighbors; their electronic distributions are altered and hence their reactivity. Their structural arrangement is often also different. When the surface of a *polycrystalline or glassy multielemental solid* is considered, such as that of an alloy or a chemical compound, the situation can be very complex. The processes of preparation or fabrication may produce a material whose surface composition is quite different from that of the bulk, in terms of both constituent and impurity elements. Subsequent treatments (e.g., thermal and chemical) will almost certainly change the surface composition to something different again. The surface is highly unlikely to be smooth, and roughness at both micro- and macrolevels may be present, leading to the likelihood that many surface atoms will be situated at corners and edges and on protuberances (i.e., in positions of increased reactivity). Surfaces exposed to the atmosphere, which include many of those of technological interest, will acquire a contaminant layer 1–2 atom layers thick, containing principally carbon and oxygen but also other impurities present in the local environment. Atmospheric exposure may also cause oxidation. Because of all these possibilities the surface region must be considered as a separate entity, effectively a separate quasi-two-dimensional phase overlaying the normal bulk phase. Analysis of the properties of such a quasi phase necessitates the use of techniques in which the information provided originates only or largely within the phase (i.e., the surface-specific techniques described in this article).

Nearly all these techniques involve interrogation of the surface with a particle probe. The function of the probe is to excite surface atoms into states giving rise to emission of one or more of a variety of secondary particles such as electrons, photons, positive and secondary ions, and neutrals. Since the primary particles used in the probing beam may also be either electrons or photons, or ions or neutrals, many separate techniques are possible, each based on a different primary–secondary particle combination. Most of these possibilities have now been established, but in fact not all the resultant techniques are of general application, some because of the restricted or specialized nature of the information obtained and others because of difficult experimental requirements. In this article, therefore, most space is devoted to those surface analytical techniques that are widely applied and readily available commercially, whereas much briefer descriptions are given of the many others whose use is less common but that—in appropriate circumstances, particularly in basic research—can provide vital information.

Because the various types of particle can appear in both primary excitation and secondary emission, most authors and reviewers have found it convenient to group the techniques in a matrix, in which the columns refer to the nature of the exciting particle and the rows to the nature of the emitted particle [1]–[9]. Such a matrix of techniques is given in Table 1, in terms of the acronyms now accepted. The meanings of the acronyms, together with some of the alternatives that have appeared in the literature, are in Listing 1.

Listing 1. Meanings of surface analysis acronyms, and their alternatives, that appear in Tables 1 and 2.

1) Photon Excitation

RAIRS	Reflection–absorption infrared spectroscopy (or IRAS IR reflection–absorption spectroscopy)
SERS	Surface-enhanced Raman scattering
SHG	Second harmonic generation
XPS	X-ray photoelectron spectroscopy (or ESCA electron spectroscopy for chemical analysis)
XAES	X-ray (excited) Auger electron spectroscopy
UPS	Ultraviolet photoelectron spectroscopy (or PES photoemission spectroscopy)
SRPS	Synchrotron radiation photoelectron spectroscopy
PSD	Photon-stimulated desorption

Table 1. Surface-specific analytical techniques using primary particle excitation*

Emission	Excitation**			
	Photons $h\nu$	Electrons e^-	Ions n^+	Neutrals n^0
$h\nu$	RAIRS SERS ellipsometry SHG	SXAPS DAPS CLS IPES BIS	IBSCA GDOS	
e^-	XPS XAES UPS SRPS	AES SAM ELS HREELS CEELS AEAPS LEED RHEED	IAES PAES INS MQS	
n^+, n^-	PSD	ESD ESDIAD	SSIMS ISS RBS GDMS	FABMS
n^0	PSD	ESD	SNMS	

* Some techniques in Table 1 have angle-resolved variants, with the prefix AR, thus giving ARUPS, ARAES, ARELS, etc.
** For meanings of acronyms, see Listing 1.

2) Electron Excitation

SXAPS	Soft X-ray appearance potential spectroscopy (or APS appearance potential spectroscopy)
DAPS	Disappearance potential spectroscopy
CLS	Cathodoluminescence spectroscopy
IPES	Inverse photoemission spectroscopy
BIS	Bremsstrahlung isochromat spectroscopy
AES	Auger electron spectroscopy
SAM	Scanning Auger microscopy
ELS	(electron) Energy loss spectroscopy
HREELS	High-resolution electron energy loss spectroscopy (or LEELS low-energy electron loss spectroscopy)
CEELS	Core-electron energy loss spectroscopy (or ILS ionization loss spectroscopy)
AEAPS	Auger electron appearance potential spectroscopy
LEED	Low-energy electron diffraction
RHEED	Reflection high-energy electron diffraction
ESD	Electron-stimulated desorption (or EID electron-induced desorption)
ESDIAD	Electron-stimulated desorption ion angular distribution

3) Ion Excitation

IBSCA	Ion beam spectrochemical analysis (or SCANIIR surface composition by analysis of neutral and ion impact radiation) (or BLE bombardment-induced light emission)
GDOS	Glow discharge optical spectroscopy
IAES	Ion (excited) Auger electron spectroscopy
PAES	Proton (or particle) (excited) Auger electron spectroscopy
INS	Ion neutralization spectroscopy
MQS	Metastable quenching spectroscopy
SSIMS	Static secondary-ion mass spectrometry
ISS	Ion scattering spectroscopy (or LEIS low-energy ion scattering)
RBS	Rutherford backscattering spectroscopy (or HEIS high-energy ion scattering)
GDMS	Glow discharge mass spectrometry
SNMS	Secondary neutral mass spectrometry

4) Neutral Excitation

FABMS	Fast-atom bombardment mass spectrometry

5) Thermal Excitation

TDS	Thermal desorption spectroscopy

6) High Field Excitation

IETS	Inelastic electron tunneling spectroscopy
APFIM	Atom probe field-ion microscopy
POSAP	Position-sensitive atom probe
STM	Scanning tunneling microscopy
STS	Scanning tunneling spectroscopy

7) Mechanical Force

AFM	Atomic force microscopy

A few techniques, including one or two important ones, cannot be classified according to the nature of the exciting particle, since they do not employ primary particles but depend instead on the application either of heat or a high electric field. The latter techniques are listed in Table 2.

2. X-Ray Photoelectron Spectroscopy (XPS)

X-ray photoelectron spectroscopy is currently the most widely used surface analytical technique, and is therefore described here in more detail than any of the other techniques. At its inception by SIEGBAHN and coworkers [10] it was called ESCA (electron spectroscopy for chemical analysis), but the name ESCA is now considered too general, since many surface electron spectroscopies exist, and the name given to each one

Table 2. Surface-specific analytical techniques using non-particle excitation

Emission	Excitation*		
	Heat kT	High electrical field F	Mechanical force
n^+	TDS	APFIM POSAP	
n^-	TDS		
e^-		IETS STM, STS	
(Displacement)			AFM

* For meanings of acronyms, see Listing 1.

must be precise. Nevertheless, the name ESCA is still used in many places, particularly in industrial laboratories and their publications. Briefly, the reasons for the popularity of XPS are the exceptional combination of compositional and chemical information that it provides, its ease of operation, and the ready availability of commercial equipment.

2.1. Principles

The surface to be analyzed is irradiated with soft X-ray photons. When a photon of energy $h\nu$ interacts with an electron in a level with binding energy E_B, the entire photon energy is transferred to the electron, with the result that a photoelectron is ejected with kinetic energy

$$E_{Kin} = h\nu - E_B - e\varphi \qquad (1)$$

where $e\varphi$ is a small, almost constant, work function term.

Obviously $h\nu$ must be greater than E_B. The ejected electron may come from a core level or from the occupied portion of the valence band, but in XPS most attention is focused on electrons in core levels. Since no two elements share the same set of electronic binding energies, measurement of the photoelectron kinetic energies provides an *elemental analysis*. In addition, Equation (1) indicates that any changes in E_B are reflected in E_{Kin}, which means that changes in the chemical environment of an atom can be followed by monitoring changes in the photoelectron energies, leading to the provision of *chemical information*. XPS can analyze all elements in the periodic table except hydrogen and helium.

Although XPS is concerned principally with photoelectrons and their kinetic energies, ejection of electrons by other processes also occurs. An ejected photoelectron leaves behind a core hole in the atom. The sequence of events following the creation of the core hole is shown schematically in Figure 1. In the example, the hole has been created in the K shell, giving rise to a photoelectron whose kinetic energy would be $(h\nu - E_K)$, and is filled by an electronic transition from the L_2 shell. The energy $E_K - E_{L_2}$ associated with the transition can then either be dissipated as a characteristic X-ray photon or given up to an electron in a third shell, shown in this example as the L_3. The second of these possibilities is called the Auger process after its discoverer

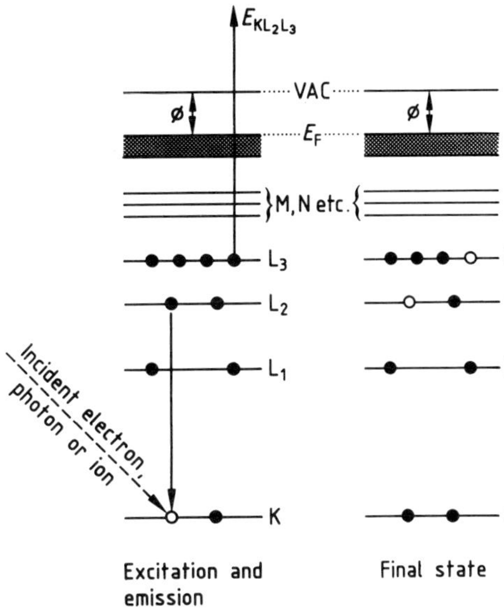

Figure 1. Auger process in an atom in a solid

[11], and the resultant ejected electron is called an Auger electron and has an energy given by

$$E_{KL_2L_3} = E_K - E_{L_2} - E^*_{L_3} \qquad (2)$$

$E^*_{L_3}$ is starred because it is the binding energy of an electron in the L_3 shell in the presence of a hole in the L_2 shell, that is, not quite the same as E_{L_3}.

X-ray photon emission (i.e., X-ray fluorescence) and Auger electron emission are obviously competing processes, but for the shallow core levels involved in XPS and AES the Auger process is far more likely.

Thus in all X-ray photoelectron spectra, features appear due to both photoemission and Auger emission. In XPS, the Auger features can be useful but are not central to the technique, whereas in AES (see Chap. 3), Equation (2) forms the basis of the technique.

At this point the nomenclature used in XPS and AES should be explained. In XPS the *spectroscopic notation* is used, and in AES the *X ray notation*. The two are equivalent, the different usages having arisen for historical reasons, but the differentiation is a convenient one. They are both based on the so-called *j–j* coupling scheme describing the orbital motion of an electron around an atomic nucleus, in which the total angular momentum of an electron is found by summing vectorially the individual electron spin and angular momenta. Thus if *l* is the electronic angular momentum quantum number and s the electronic spin momentum quantum number, the total angular momentum for each electron is given by $j = l + s$. Since l can take the values 0, 1, 2, 3, 4,... and $s = \pm\frac{1}{2}$, clearly $j = \frac{1}{2}, \frac{3}{2}, \frac{5}{2}$, etc. The principal quantum number n can take values 1, 2, 3, 4,.... In *spectroscopic notation*, states with $l = 0, 1, 2, 3,\ldots$ are designated $s, p, d, f,\ldots$, respectively, and the letter is preceded by the number n; the j values are then appended as suffixes. Therefore one obtains $1s$, $2s$, $2p_{1/2}$, $2p_{3/2}$, $3s$, $3p_{1/2}$, $3p_{3/2}$, etc.

In *X-ray notation*, states with $n = 1, 2, 3, 4,\ldots$ are designated K, L, M, N,..., respectively, and states with various combinations of $l = 0, 1, 2, 3,\ldots$ and $j = \frac{1}{2}, \frac{3}{2}, \frac{5}{2}$, are appended as the suffixes 1, 2, 3, 4,.... In this way one arrives at K, L_1, L_2, L_3, M_1, M_2, M_3, etc. The equivalence of the two notations is set out in Table 3.

In X-ray notation the Auger transition shown in Figure 1 would therefore be labeled KL_2L_3. In this coupling scheme, six Auger transitions would be possible in the KLL series. Obviously, many other series are possible (e.g., KLM, LMM, MNN). These are discussed more fully in Chapter 3, dealing with AES.

The reasons why techniques such as XPS and AES, which involve measurement of the energies of ejected electrons, are so surface specific, should be examined. An electron with kinetic energy E moving through a solid matrix M has a probability of traveling a certain distance before losing all or part of its energy as a result of an inelastic collision. Based on that probability, the average distance traveled before such a collision is known as the inelastic mean free path (imfp) $\lambda_M(E)$. The imfp is a function only of M and of E. Figure 2 shows a compilation of measurements of λ made by SEAH and DENCH [12], in terms of atomic monolayers as a function of kinetic energy. Note that both λ and energy scales are logarithmic. The important consequence of the dependence of λ on kinetic energy is that in the ranges of secondary electron kinetic energies used in XPS and AES, the values of λ are very small. In XPS, for example, typical ranges are 250–1500 eV, corresponding to a range of λ from about four to eight monolayers, while in AES, the energy range is typically 20 to 1000 eV, in which case λ would range from about two to six monolayers. What this means in practice is that if the photoelectron or the Auger electron is to escape into a vacuum and be detected, it must originate at or very near the surface of the solid. This is the reason why the electron spectroscopic techniques are surface specific.

Table 3. Spectroscopic and X-ray notation

Quantum numbers			Spectroscopic state	X-ray suffix	X-ray state
n	l	j			
1	0	1/2	$1s$	1	K
2	0	1/2	$2s$	1	L_1
2	1	1/2	$2p_{1/2}$	2	L_2
2	1	3/2	$2p_{3/2}$	3	L_3
3	0	1/2	$3s$	1	M_1
3	1	1/2	$3p_{1/2}$	2	M_2
3	1	3/2	$3p_{3/2}$	3	M_3
3	2	3/2	$3d_{3/2}$	4	M_4
3	2	5/2	$3d_{5/2}$	5	M_5
	etc.		etc.	etc.	etc.

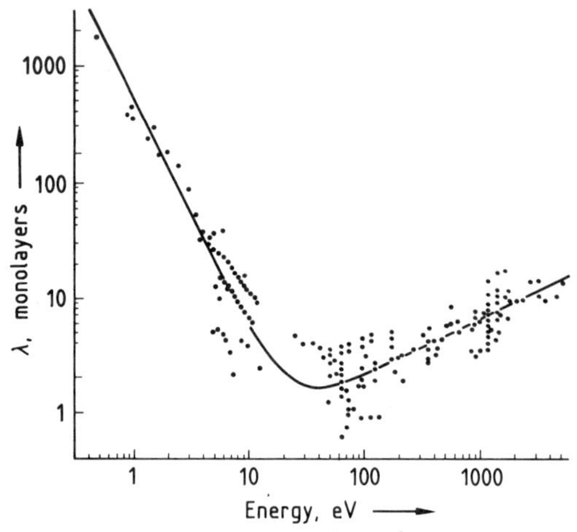

Figure 2. Compilation by SEAH and DENCH [12] of measurements of inelastic mean free path as a function of electron kinetic energy
The solid line is a least-squares fit

2.2. Instrumentation

2.2.1 Vacuum Requirements

Electron spectroscopic techniques require vacua of the order of 10^{-8} Pa for their operation. This requirement arises from the extreme surface specificity of these techniques, mentioned above. With sampling depths of only a few atom layers, and elemental sensitivities down to 10^{-5} atom layers (i.e., one atom of a particular element in 10^5 other atoms in an atomic layer), the techniques are clearly very sensitive to surface contamination, most of which comes from the residual gases in the vacuum system. According to gas kinetic theory, to have enough time to make a surface analytical measurement on a surface that has just been prepared or exposed, before contamination from the gas phase interferes, the base pressure should be 10^{-8} Pa or lower, that is, in the region of ultrahigh vacuum (UHV).

The requirement for the achievement of UHV conditions imposes restrictions on the types of material that can be used for the construction of surface analytical systems, or inside the systems, because UHV can be achieved only by accelerating the rate of removal of gas molecules from internal surfaces by raising the temperature of the entire system (i.e., by baking). Typical baking conditions are 150–200 °C for several hours. Construction materials should not be distorted, lose their strength, produce excessive gas, or oxidize readily under such treatment, which should be repeated after each exposure of the inside to atmospheric pressure. Thus, the principal construction material is stainless steel, with mu-metal (76% Ni, 5% Cu, 2% Cr) used occasionally where magnetic screening is needed (e.g., around electron energy analyzers). For the same reasons, metal seals, not elastomers, are used for the demountable joints between individual components—the sealing material usually being pure copper although gold is sometimes used. Other materials that may be used between ambient atmosphere and UHV are borosilicate glass or quartz for windows, and alumina for electrical insulation for current or voltage connections. Inside the system, any material is permissible that does not produce volatile components either during normal operation or during baking. Thus, for example, brass that contains the volatile metal zinc could not be used.

The production of vacua in the UHV region has been routine for many years now, but success still depends on attention to details such as the maintenance of strict cleanliness. For descriptions of the modes of operation of the four types of pump commonly used (i.e., diffusion, sputter ion, turbomolecular, and sublimation), see →21. Vacuum Technology, **B3**.

2.2.2. X-Ray Sources

The most important consideration in choosing an X-ray source for XPS is energy resolution. Equation (1) gives the relationship between the kinetic energy of the photoelectron, the energy of the X-ray photon, and the binding energy of the core electron. Since the energy spread, or linewidth, of an electron in a core level is very small, the linewidth of the photoelectron energy depends on the linewidth of the source, if no undue broadening is introduced instrumentally. Now in XPS the analyst devotes much effort to extracting chemical information by means of detailed study of individual elemental photoelectron spectra. Such a study needs an energy resolution better than 1.0 eV if subtle chemical effects are to be identified. Thus the linewidth of the X-ray source should be significantly smaller than 1.0 eV if the resolution required is not to be limited by the source itself.

Other considerations are that the source material, which forms a target for high-energy electron bombardment leading to the production of X rays, should be a good conductor—to allow rapid removal of heat—and should also be compatible with UHV.

Table 4 lists the energies and linewidths of the characteristic X-ray lines from a few possible candidate materials.

Not many sources with sufficiently narrow lines are available. Y Mζ and Zr Mζ would be suitable, but their line energies are too low to be useful for general application, although they have been used for special purposes. Silicon is a poor thermal conductor and is not easy to apply as a coating on an anode, so only the Mg K_α and Al K_α lines remain, which are in fact the two used universally in XPS.

For efficient production of X rays by electron bombardment, exciting electron energies that are at least an order of magnitude higher than the line energies must be used, so that in Mg and Al sources accelerating potentials of 15 kV are employed. Modern sources are designed with dual anodes, one anode face being coated with mag-

Table 4. Energies and linewidths of some characteristic low-energy X-ray lines

Line	Energy, eV	Width, eV
Y M_ζ	132.3	0.47
Zr M_ζ	151.4	0.77
Nb M_ζ	171.4	1.21
Mg K_α	1253.6	0.70
Al K_α	1486.6	0.85
Si K_α	1739.5	1.00
Y L_α	1922.6	1.50
Zr L_α	2042.4	1.70

nesium and the other with aluminum, and with two filaments, one for each face. Thus, a switch from one type of X irradiation to the other can be made very quickly.

To protect the sample from stray electrons from the anode, from heating effects, and from possible contamination by the source enclosure, a thin (≈2-μm) window of aluminum foil is interposed between the anode and the sample. For optimum X-ray photon flux on the surface (i.e., optimum sensitivity), the anode must be brought as close to the sample as possible, which means in practice a distance of ≈2 cm. The entire X-ray source is therefore retractable via a bellows and a screw mechanism.

The X radiation from magnesium and aluminum sources is quite complex. The principal K_α lines are in fact unresolved doublets and should correctly be labeled $K_{\alpha_{1,2}}$. Many satellite lines also exist of which the more important in practice are the $K_{\alpha_{3,4}}$, separated from the principal line by ≈10 eV and roughly 8 % of its intensity, and the K_β, lying about 70 eV higher in energy. In addition, the spectrum is superimposed on a continuous background called bremsstrahlung radiation, arising from inelastic processes.

Removal of satellites, elimination of the bremsstrahlung background, and separation of the $K_{\alpha_{1,2}}$ doublet can be achieved by monochromatization, shown schematically in Figure 3. The X-ray source is positioned at one point on a spherical surface, called a Rowland sphere after its originator, and a very accurately ground and polished quartz crystal is placed at another point. X rays from the source are diffracted from the quartz, and by placing the sample at the correct point on the Rowland sphere, the K_{α_1} component can be selected to be focused on it, according to the well-known Bragg relation for diffraction. Quartz is a very convenient diffracting medium for Al K_α since the spacing between the $10\bar{1}0$ planes is exactly half the wavelength of the X radiation, and for first-order diffraction the Bragg angle is 78.5°. Since the width of the Al K_{α_1} line is <0.4 eV, the energy dispersion needed around the surface of the sphere implies that the Rowland sphere should have a diameter of at least 0.5 m. Accurate positioning of source, crystal, and sample is essential. Although an XPS spectrum will be much "cleaner" when a monochromator is used, because satellites and background have been removed, the photon flux at the sample is much lower than that from an

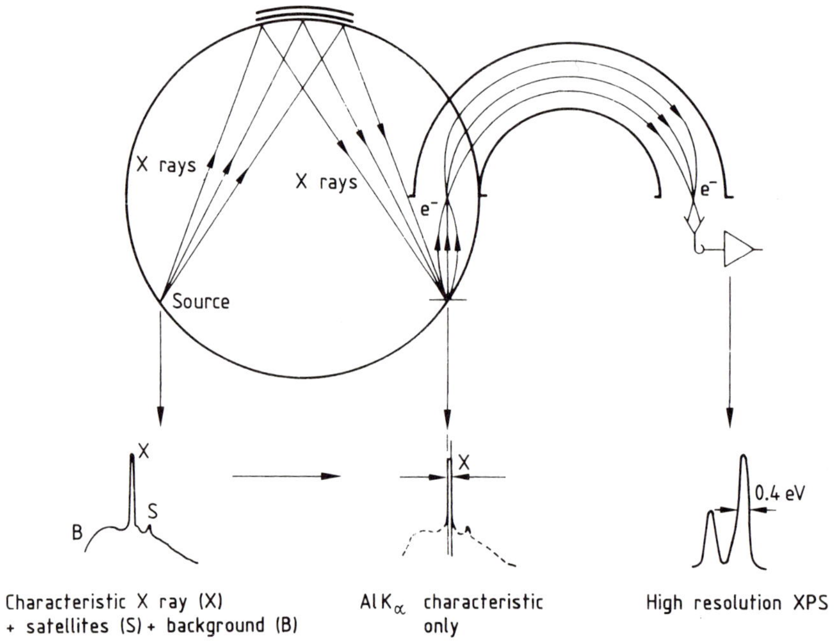

Figure 3. Schematic of X-ray monochromatization to remove satellites, eliminate bremsstrahlung background, and separate the Al $K_{\alpha_{1,2}}$ doublet
Courtesy of Kratos Analytical.

unmonochromatized source operating at the same power. Against this must be set the greatly improved signal-to-background level in a monochromatized spectrum.

2.2.3. Synchrotron Radiation

The discrete line sources described above for XPS are perfectly adequate for most applications, but some types of analysis require that the source be tunable (i.e., that the exciting energy be variable). The reason is to allow the photoionization cross section of the core levels of a particular element or group of elements to be varied, which is particularly useful when dealing with multielement semiconductors. Tunable radiation can be obtained from a synchrotron.

In a synchrotron, electrons are accelerated to near-relativistic velocities and constrained magnetically into circular paths. When a charged body is accelerated, it emits radiation, and when the near-relativistic electrons are forced into curved paths, they emit photons over a continuous spectrum. The general shape of the spectrum is shown in Figure 4. For a synchrotron with an energy of several gigaelectronvolts and a radius of some tens of meters, the energy of the emitted photons near the maximum is of the order of 1 keV (i.e., ideal for XPS). As can be seen from the universal curve, plenty of usable intensity exists down into the UV region. With suitable monochromators on the output to select a particular wavelength, the photon energy can be tuned continuously from about 20 to 500 eV. The available intensities are comparable to those from conventional line sources.

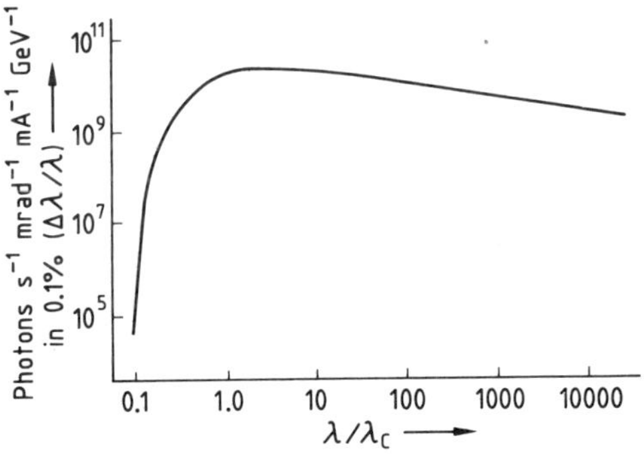

Figure 4. Normalized spectrum of photon energies emitted from a synchrotron
λ_c = wavelength characteristic of the individual synchrotron

2.2.4. Electron Energy Analyzers

In electron spectroscopic techniques—among which XPS is the most important—analysis of the energies of electrons ejected from a surface is central. Because of the low kinetic energies involved in the techniques, analyzers using magnetic fields are undesirable. Therefore the energy analyzers used are exclusively of the electrostatic deflection type. The two that are now universally employed are the concentric hemispherical analyzer (CHA) and the cylindrical mirror analyzer (CMA). Since both have been used in XPS, both are described here, although in practice the CHA is more suitable for XPS, and the CMA for AES.

Energy analyzers cannot be discussed without a discussion of energy resolution, which is defined in two ways. *Absolute resolution* is defined as ΔE, the full width at half-maximum (FWHM) of a chosen peak. *Relative resolution* is defined as the ratio R of ΔE to the kinetic energy E of the peak energy position (usually its centroid), that is, $R = \Delta E/E$. Thus absolute resolution is independent of peak position, but relative resolution can be specified only by reference to a particular kinetic energy.

In XPS, closely spaced peaks at any point in the energy range must be resolved, which requires the same absolute resolution at all energies. In AES, on the other hand, the inherent widths of peaks in the spectrum are greater than in XPS and the source area on the surface is far smaller, both of which lead to operation at constant relative resolution.

Concentric Hemispherical Analyzer (CHA). The CHA is shown in schematic cross section in Figure 5 [13]. Two hemispheres of radii R_1 (in-

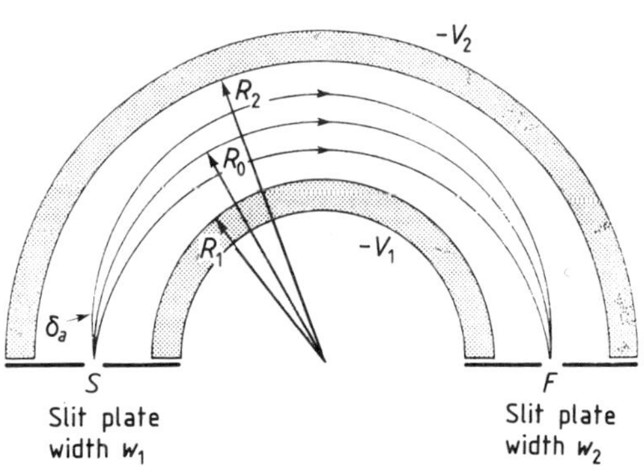

Figure 5. Schematic of a concentric hemispherical analyzer (CHA) [13]

ner) and R_2 (outer) are positioned concentrically. Potentials $-V_1$ and $-V_2$ are applied to the inner and outer hemispheres, respectively, with V_2 greater than V_1. The source S and the focus F are in the same plane as the center of curvature, and R_0 is the radius of the equipotential surface between the hemispheres. If electrons of energy $E = eV_0$ are injected at S along the equipotential surface, they will be focused at F if

$$V_2 - V_1 = V_0\left(\frac{R_2}{R_1} - \frac{R_1}{R_2}\right) \quad (3)$$

If electrons are injected not exactly along the equipotential surface, but with an angular spread $\delta\alpha$ about the correct direction, then the energy resolution is given by

$$\Delta E/E = (W_1 + W_2)/4\,R_0 + (\delta\alpha)^2 \quad (4)$$

where W_1 and W_2 are the respective widths of the entrance and exit slits. In most instruments, for convenience in construction purposes, $W_1 = W_2 = W$, whereupon the resolution becomes

$$\Delta E/E = W/2\,R_0 + (\delta\alpha)^2 \quad (5)$$

In XPS the photoelectrons are retarded to a constant energy, called the pass energy, as they approach the entrance slit. If this were not done, Equation (5) shows that to achieve an absolute resolution of 1 eV at the maximum kinetic energy of about 1500 eV (using Al K_α radiation), and with a slit width of 2 mm, would require an analyzer with an average radius of about 300 cm, which is impracticable. Pass energies are selected in the range 20–100 eV for XPS, which allows the analyzer to be built with a radius of 10–15 cm.

Modern XPS spectrometers employ a lens system on the input to the CHA, which has the effect of transferring an image of the analyzed area on the sample surface to the entrance slit of the analyzer. Such a spectrometer is illustrated schematically in Figure 6.

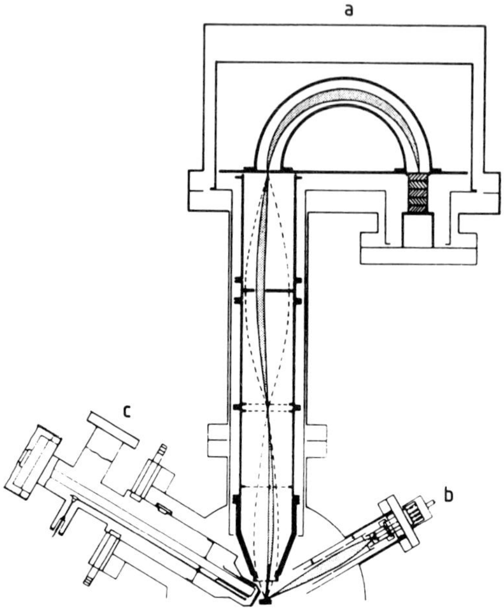

Figure 6. Typical configuration in a modern XPS spectrometer
Courtesy of Leybold
a) XPS–Auger analyzer; b) Auger electron gun; c) X-ray source

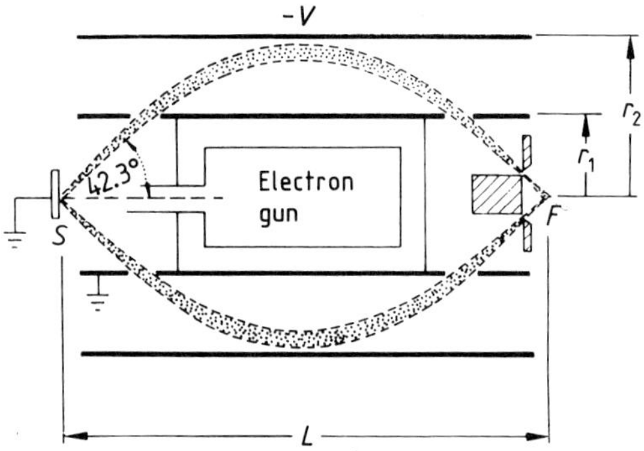

Figure 7. Schematic of a cylindrical mirror analyzer (CMA)

Cylindrical Mirror Analyzer (CMA). In the CMA shown schematically in Figure 7, two cylinders of radii r_1 (inner) and r_2 (outer) are accurately positioned coaxially. Annular entrance and exit apertures are cut in the inner cylinder, and a deflection potential $-V$ is applied between the cylinders. Electrons leaving the sample surface at the source S with a particular energy E on passing into the CMA via the entrance aperture are then deflected back through the exit aperture to the focus F. For the special case in which the acceptance angle $\alpha = 42.3°$, the first-order aberrations vanish, and the CMA becomes a second-order focusing device. The relationship between the electron energy E and the deflection potential is then

$$\frac{E}{eV} = \frac{1.31}{\ln(r_2/r_1)} \quad (6)$$

As in the CHA, therefore, scanning the deflection potential $-V$ and recording the signal as a function of electron energy provide the distribution in energy of electrons leaving the sample surface.

If the angular spread of the acceptance angle is $\delta\alpha$, then the relative energy resolution of a CMA is

$$\Delta E/E = 0.18\ W/r_1 + 1.39\ (\delta\alpha)^3 \qquad (7)$$

where W is the effective slit width. For $\delta\alpha \approx 6°$, a typical value, W can be replaced by the source size (i.e., the area on the sample from which electrons are accepted).

In *AES* the inner cylinder is at earth potential, and primary electrons originate in an integral electron gun built into the middle of the analyzer. The source size in AES is very small, because of the focused primary beam, and the important property of the CMA for AES is therefore its very high transmission. In *XPS* the entire sample surface is normally flooded with X rays, in which case the source size is the entire acceptance area of photoelectrons, and Equation (7) shows that the relative energy resolution would be too poor. The important parameter in XPS is not the transmission alone, but the luminosity, which is the product of transmission, acceptance area, and acceptance solid angle.

The single-stage CMA depicted in Figure 7 is therefore not suitable for XPS, but a variation of the CMA that is suitable is the double-pass CMA shown in Figure 8 [14]. Two CMAs are placed in series, so that the exit aperture for the first is the entrance aperture for the second. Electrons from the sample surface are retarded to a constant pass energy, as in the CHA, by spherical retardation grids in front of the entrance aperture, centered on the source area on the sample. The first mesh is grounded, and the second mesh is at the potential of the inner cylinder, which in the XPS mode of operation is not grounded, as in AES, but floating at the scanning potential. Adequate energy resolution is achieved in the second stage because the effective source area, being defined not by the irradiated sample surface but by the size of the first stage exit aperture, is sufficiently small. Entrance and exit apertures can be changed in the second stage by an externally operated mechanism.

2.2.5. Spatial Resolution

The principal disadvantage of conventional XPS is the lack of spatial resolution; the spectral information comes from an analyzed area of several square millimeters and is, therefore, an average of the compositional and chemical analyses from that area. On the other hand, the majority of samples of technological origin are inhomogeneous on a scale much smaller than that of normal XPS analysis, and obtaining chemical information on the same scale as the inhomogeneities would be very desirable.

Major steps have been taken recently in the direction of improved XPS spatial resolution by making use of the focusing properties of the CHA described above. One of the more successful configurations that have evolved for this purpose is shown in Figure 9, from the work of Coxon et al. [15]. The instrument incorporating it is called the ESCASCOPE. Compared to a conventional XPS system, it has additional lenses (3 and 5) on the input and output sides of the

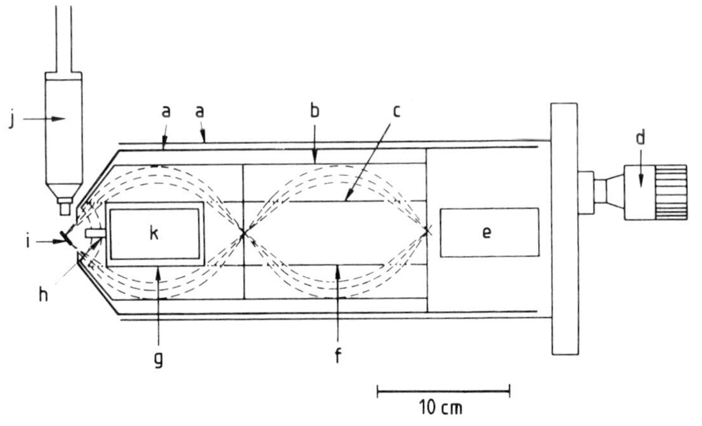

Figure 8. Schematic of double-pass CMA designed by Palmberg [14] for either XPS or AES
a) Inner and outer magnetic shields; b) Outer cylinder; c) Inner cylinder; d) Rotary motion for aperture change; e) Electron multiplier; f) Second stage; g) First stage; h) Spherical retarding grids; i) Sample; j) X-ray source; k) Electron gun

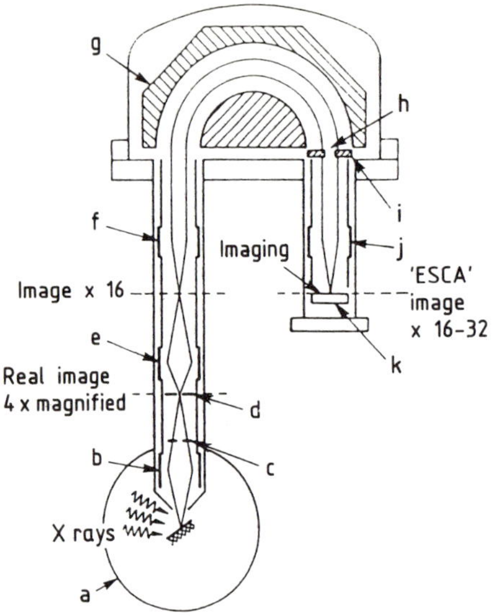

Figure 9. Schematic of the electron-optical arrangement designed by Coxon et al. [15] for the production of energy-resolved two-dimensional images in XPS
a) Spherical mu-metal chamber; b) Lens 1; c) Objective aperture; d) Field aperture; e) Lens 2; f) Lens 3; g) 180° Hemispherical analyzer; h) Hole; i) Spectrum detector 1; j) Lens 5; k) Image detector 2

CHA. Lenses 1 and 2 collect electrons from the analyzed area and present a real image of the sample, magnified about 16 times, at the object plane of lens 3.

Lens 3 is so arranged that the distance to its object plane is equal to its focal length, so that electrons reach the object plane (also the entrance slit of the analyzer) traveling in straight lines. This means that electrons leaving a point on the sample surface within a narrow angular range, pass into the analyzer in a parallel beam inclined at a particular angle to the object plane, with no intensity variation in the plane; another way of putting this is to say that a Fourier transformation of the surface image exists at the entrance slit.

The CHA (lens 4) then forms an image of the Fourier-transformed surface image in its own image plane (the exit slit) and introduces energy dispersion at the same time, as discussed above. The angular information is conserved. Photoelectrons pass through a hole in the center of the image plane and into lens 5, which is arranged similarly to lens 3. The Fourier-transformed image is then inverted so that a real image is formed in the image plane of lens 5, and this image will be an energy-filtered two-dimensional image of the sample surface. The image is detected via a position-sensitive detector consisting of two 40-mm-diameter channel plates, a phosphor screen, and a TV camera. Since the photoelectron kinetic energies can be selected at will in the CHA, images in electron energies characteristic of a particular element, or of a particular chemical state of an element, can be formed.

An example of chemical state imaging in the ESCASCOPE is given in Figure 10, also from [15]. The material is a contaminated fluoropolymer, the contamination being in the form of dark spots of 10–80-μm size. Both images in Figure 10 are in the C 1*s* photoelectron peak, but of different chemical states. The upper is in that contribution arising from carbon bound to fluorine, and the lower from carbon bound to carbon (i.e., graphitic in nature). The complementarity of the images indicates that the contamination is graphitic in nature.

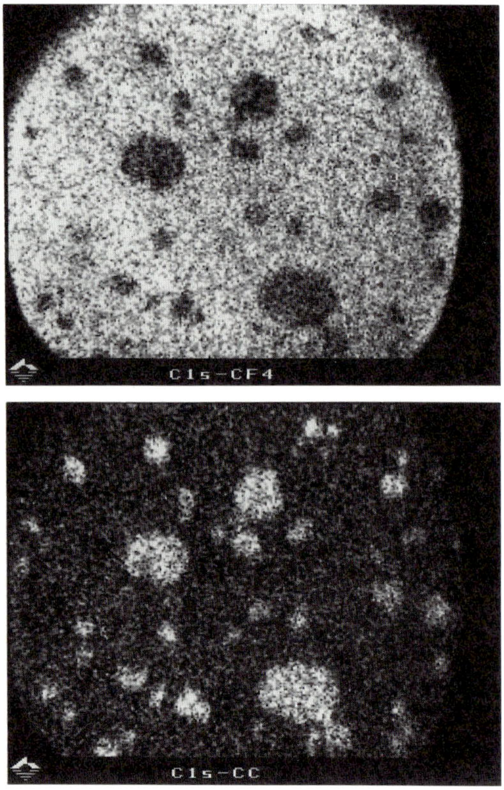

Figure 10. Chemical state images obtained with the ESCASCOPE in Figure 9, from a contaminated fluoropolymer [15]
A) Image in contribution to C 1*s* from C–F bonding; B) Image in contribution to C 1*s* from C–C bonding

2.3. Spectral Information and Chemical Shifts

Figure 11 shows a *wide* or *survey XPS spectrum*, that is, one recorded over a wide range of energies, in this case 1000 eV. The radiation used was unmonochromatized Al K_α, at 1486.6 eV, and the surface is that of almost clean copper. Such a spectrum reveals the major features to be found, but to investigate minor or more detailed features, spectra are acquired over much more restricted energy ranges, at better energy resolution; the latter spectra are called *narrow* or *detail scans*.

The most prominent features in Figure 11, and in most photoelectron spectra, are peaks arising from excitation of core-level electrons according to Equation (1). Thus, at the left-hand end, two intense peaks are found at about 553 and 533 eV, corresponding to photoelectrons from the $2p_{3/2}$ and $2p_{1/2}$ levels of copper, respectively, the separation of 20 eV being the spin–orbit splitting. At the right-hand, high kinetic energy end, are three other copper photoelectron peaks at about 1363, 1409, and 1485 eV, which arise from the $3s$, $3p$, and $3d$ levels, respectively. The other prominent features associated with copper are the three Auger peaks $L_3M_{4,5}M_{4,5}$, $L_3M_{2,3}M_{4,5}$, and $L_3M_{2,3}M_{2,3}$ appearing at 919, 838, and 768 eV, respectively. As pointed out in Section 2.1, the creation of a core hole by any means of excitation can lead to ejection of an Auger electron, so that in XPS Auger features form a significant contribution to the spectrum. If Auger and photoelectron peaks happen to overlap in any spectrum, they can always be separated by changing the excitation (e.g., from Al K_α to Mg K_α, or vice versa) since the Auger peaks are invariant in energy whereas the photoelectron peaks must shift with the energy of the exciting photons according to Equation (1).

In addition to features due to copper in Figure 11 are small photoelectron peaks at 955- and 1204-eV kinetic energies, arising from the oxygen and carbon $1s$ levels, respectively, due to the presence of some contamination on the surface. Other minor features can be seen associated with the major peaks. Those situated about 10 eV toward higher kinetic energy occur because unmonochromatized X radiation was used; they are photoelectron peaks excited by the $K_{\alpha3,4}$ satellite already mentioned (see Section 2.2.2.). They would disappear if a monochromator were used. Toward the lower kinetic energy of the major peaks (e.g., visible some 10–30 eV below the Cu $2p$ peaks) are the so-called *plasmon loss peaks*. Peaks of this type are shown more clearly in Figure 12, from BARRIE [16], in narrow scans from an aluminum surface. They arise when outgoing electrons of sufficient energy, originating in any way, lose energy by exciting collective oscillations in the conduction electrons in a solid. The oscillation frequency, and hence the energy involved, is characteristic of the nature of the

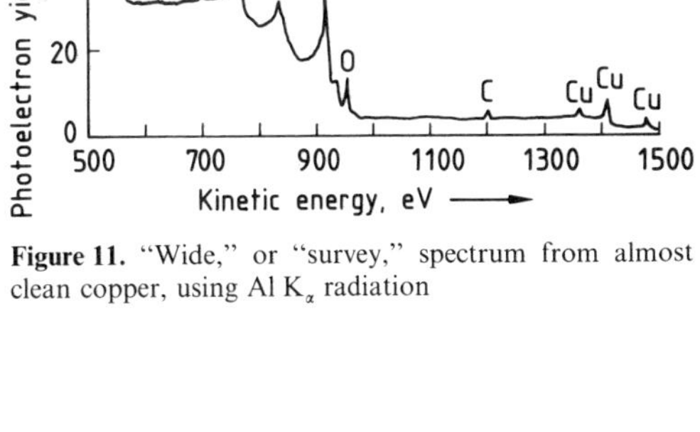

Figure 11. "Wide," or "survey," spectrum from almost clean copper, using Al K_α radiation

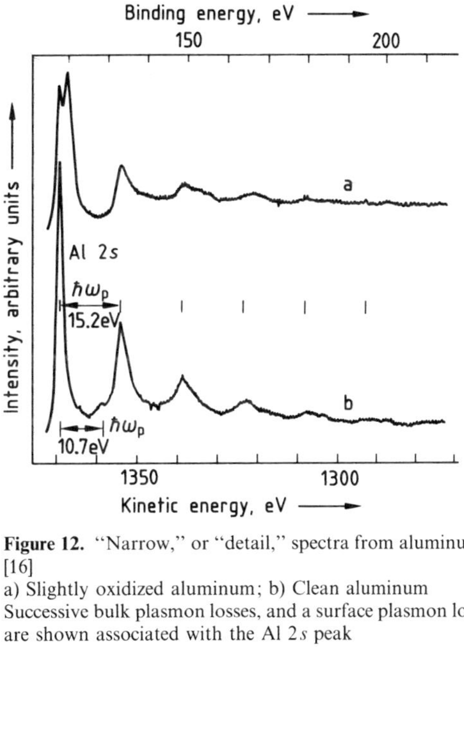

Figure 12. "Narrow," or "detail," spectra from aluminum [16]
a) Slightly oxidized aluminum; b) Clean aluminum
Successive bulk plasmon losses, and a surface plasmon loss are shown associated with the Al $2s$ peak

solid and is called the plasmon frequency (or energy). The plasmon loss determined by the three-dimensional nature of the solid is called a "bulk" or "volume" loss, but when it is associated with the two-dimensional nature of the surface, it is called a "surface" loss. In Figure 12 (b), which is from clean aluminum, examples of both types of plasmon loss from the Al 2*s* peak can be seen—a bulk loss of 15.2 eV and a surface loss of 10.7 eV. Theoretically, the surface plasmon loss energy should be a factor of $2^{1/2}$ less than the bulk plasmon loss energy. Where the bulk loss is very pronounced, as in a free-electron-like metal such as aluminum, a succession of peaks can be found at multiples of the loss energy, as indicated in Figure 12, with progressively decreasing intensity. Spectrum (a) in Figure 12, from slightly oxidized aluminum, reveals the rapid quenching of the surface plasmon on oxidation and also shows a chemical shift in the Al *2s* photoelectron peak.

Chemical shift is the name given to the observed shift in energy of a photoelectron peak from a particular element when the chemical state of that element changes. When an atom enters into combination with another atom or group of atoms, an alteration occurs in the valence electron density, which might be positive or negative according to whether charge is accepted or donated, causing a consequent alteration in the electrostatic potential affecting the core electrons. Therefore the binding energies of the core electrons change, giving rise according to Equation (1) to shifts in the corresponding photoelectron peaks. Tabulation of the chemical shifts experienced by any one element in a series of pure compounds of that element thus enables its chemical state to be identified during analysis of unknown samples. Many such tabulations have appeared, and a major collation of them is found in [2]. The identification of chemical state in this way is the principal advantage of XPS over other surface analytical techniques.

An example of a spectrum exhibiting a chemical shift is that of the tin 3*d* peaks in Figure 13. A thin layer of oxide on the metallic tin surface allows photoelectrons from both the underlying metal and the oxide to appear together. Resolution of the doublet $3d_{5/2}$, $3d_{3/2}$ into the components from the metal (Sn^0) and from the oxide Sn^{n+} is shown in Figure 13 B. The shift in this case is 1.6–1.7 eV. Curve resolution is an operation that can be performed routinely in data processing systems associated with photoelectron spectrometers.

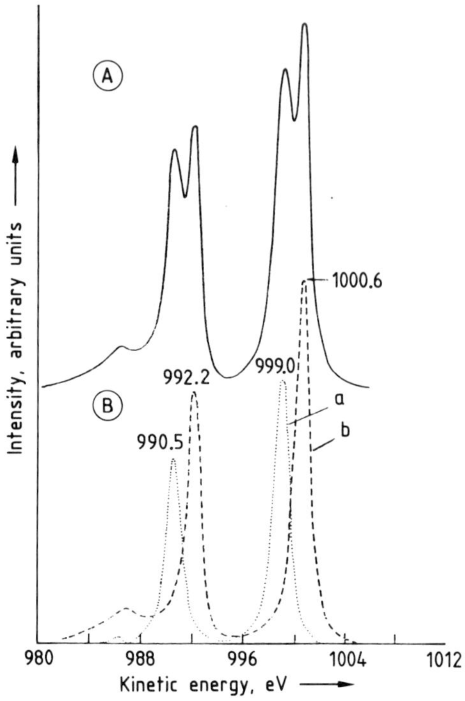

Figure 13. Example of a chemical shift in the Sn 3*d* peak, exhibited by a very thin layer of Sn oxide on Sn metal
A) Spectrum after linear background subtraction; B) Spectrum resolved into its respective components
a) Sn^{n+}; b) Sn^0

Another type of satellite often seen in XPS spectra is the so-called *shake-up satellite*. Such satellites occur when formation of a compound leads to the presence of unpaired electrons in 3*d* or 4*f* levels; they can also occur in organic compounds according to the degree of conjugation in C=C double bonds, particularly in aromatic compounds. Figure 14, from Scrocco [17], compares the copper 2*p* photoelectron spectra from CuO and Cu_2O. Since Cu^{2+} as in CuO has an open shell configuration, the 2*p* spectrum shows strong shake-up satellites, whereas that from Cu_2O does not because Cu^+ has a closed shell. In many cases, such satellite structures can be used to diagnose the oxidation state(s).

Other spectral features occasionally seen in XPS spectra arise from *multiplet splitting*. The latter occurs when an element has unpaired electrons either in the valence band or in localized core levels very close to the valence band (as in, for example, some of the rare-earth elements). In the 4*f* electron levels of the rare earths, correlation between spin and angular electron momenta

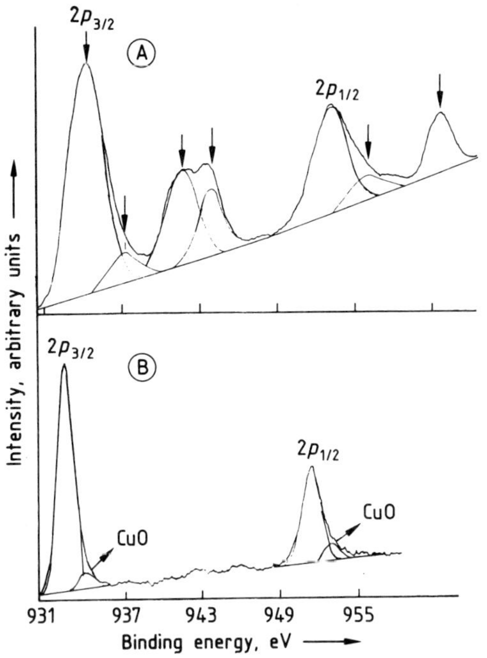

Figure 14. Example of Cu2p "shake-up" peaks [17]
A) Present in a Cu^{2+} compound (CuO); B) Absent in a Cu^{+} compound (Cu_2O)

can occur in many different ways, leading to multiplets at different energies, most of which are resolvable [18] in XPS as a multipeak spectrum.

Photoelectrons of highest kinetic energy in a spectrum (i.e., those of lowest binding energy) originate in the shallowest levels of the solid, that is, in the valence band. Thus the photoelectron peak labeled 3d in the copper spectrum in Figure 11 corresponds to the valence band spectrum, although no detailed structure can be seen in such a "survey" spectrum because the energy resolution used there is not adequate. If much better resolution is used together with a monochromatized source, the valence bands of many metals reveal interesting detail, as shown [19] in Figure 15 for gold. However, for a systematic study of valence band structure, XPS is not ideal, because of both photoionization cross-section effects and the high kinetic energies involved, and the techniques of ultraviolet photoelectron spectroscopy (UPS) and synchrotron radiation photoelectron spectroscopy (SRPS) are preferred.

2.4. Quantification and Depth Profiling

2.4.1. Quantification

If an X-ray photon of energy $h\nu$ ionizes a core level X in an atom of element A in a solid, the photoelectron current I_A from X in A is

$$I_A(X) = K \cdot \sigma(h\nu : E_B) \cdot \beta(h\nu : X) \cdot \bar{N}_A \cdot \lambda_M(E_x) \cdot \cos\theta \quad (8)$$

where $\sigma(h\nu : E_B)$ is the photoelectric cross section for ionization of X (binding energy E_B) by photons of energy $h\nu$; $\beta(h\nu : X)$ the asymmetry parameter for emission from X by excitation with photons of energy $h\nu$; $\bar{N}_A$ the atomic density of A averaged over depth of analysis; $\lambda_M(E_x)$ the inelastic mean free path in matrix M containing A, at kinetic energy E_x, where $E_x = h\nu - E_B$; and θ the angle of emission to surface normal. K is a

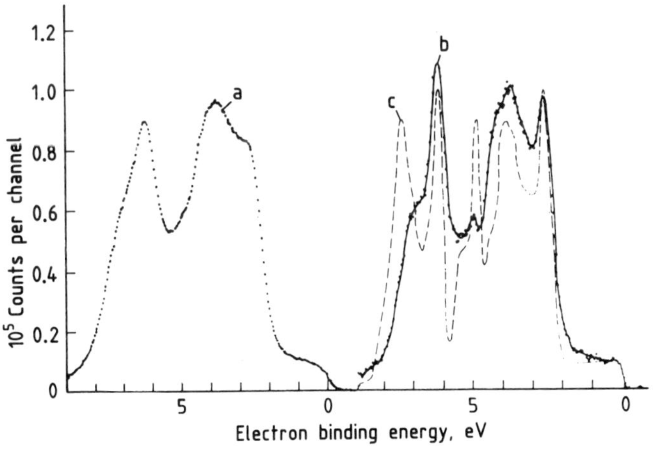

Figure 15. Valence band spectra of gold [19]
a) Using monochromatized Al K_α radiation, and high-energy resolution; b) Result of deconvoluting instrumental broadening from the raw data of a); c) Result of a theoretical calculation of the valence band structure

constant of proportionality that contains fixed operational parameters such as incident X-ray flux, transmission of analyzer at kinetic energy E_x, and efficiency of detector at kinetic energy E_x, kept fixed during any one analysis.

Values of the total cross section σ for Al K_α radiation, relative to the carbon 1*s* level, have been calculated by SCOFIELD [20] (Fig. 16), and of the asymmetry parameter β by REILMAN et al. [21] (Fig. 17, plotted by SEAH [22]). SEAH and DENCH [12] have compiled many measurements of the inelastic mean free path, and for elements the best-fit relationship that they found was

$$\lambda_M(E) = 538\ E^{-2} + 0.41\ (aE)^{1/2} \text{ monolayers} \quad (9)$$

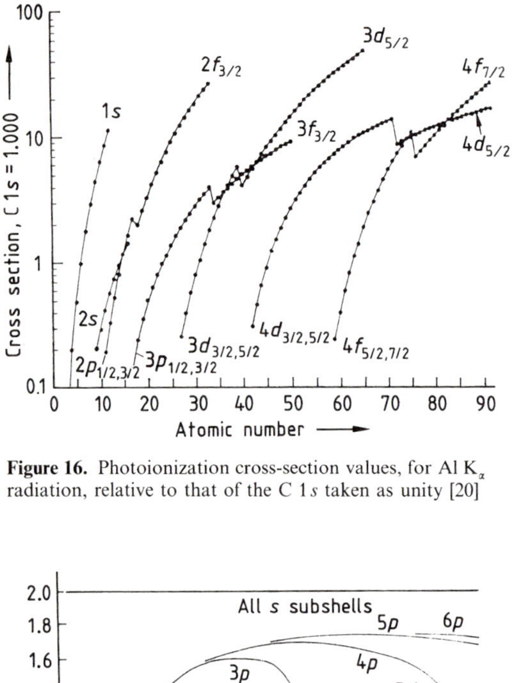

Figure 16. Photoionization cross-section values, for Al K_α radiation, relative to that of the C 1*s* taken as unity [20]

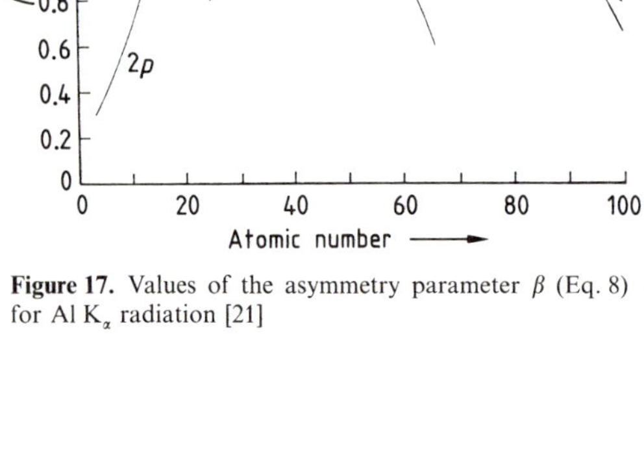

Figure 17. Values of the asymmetry parameter β (Eq. 8) for Al K_α radiation [21]

where a is the monolayer thickness in nanometers.

In principle, therefore, the surface concentration of an element can be calculated from the intensity of a particular photoelectron emission, according to Equation (8). In practice, the *method of relative sensitivity factors* is in common use. The intensity of a particular elemental photoelectron peak, normally the F 1*s*, is taken as standard and put equal to unity, and all intensity measurements are referred to it. Thus, from Equation (8), the ratio $I_A(X)/I_F(1s)$, with $I_F(1s)$ equal to unity, is

$$I_A(X) = \left[\frac{\sigma(h\nu:E_x)\cdot\beta(h\nu:X)\cdot\lambda_A(E_x)}{\sigma(h\nu:E_{1s})\cdot\beta(h\nu:1s)\cdot\bar{N}_F\cdot\lambda_F(E_{1s})}\right]\bar{N}_A \quad (10)$$

When $I_A(X)$ is measured from pure A and a well-characterized fluorine standard is used, the expression in square brackets is the relative sensitivity factor of A. Tables of such empirically derived sensitivity factors have been published by several authors, the most comprehensive being that in [23].

2.4.2. Depth Profiling

Very often, in addition to analytical information about the original surface, information is required about the distribution of chemical composition to depths considerably greater than the inelastic mean free path. Nondestructive methods such as variation of emission angle or variation of exciting photon energy are possible, but they are limited in practice to evaluation of chemical profile within depths of only ca. 5 nm. To obtain information from greater depths requires destructive methods, of which the one used universally is removal of the surface by ion bombardment, also called sputtering. The combination of removal of the surface by sputtering and of analysis by either XPS, AES, or SSIMS is termed depth profiling. In AES and SSIMS the sputtering and analytical operations can occur simultaneously, so that a continuous composition profile is obtained if a single element is being monitored, or a quasi-continuous one if several elements are monitored. In XPS, which is inherently slower and where chemical information is required along with the depth, profiling is recorded in a stepwise fashion, with alternate cycles of sputtering and analysis. Examples of profiles through oxide films on pure iron and on Fe–12Cr–1Mo alloy are shown in Figure 18, in which the respective contributions from the metallic and oxide components of the iron and chromium spectra have been quantified [24]. In these examples the oxide films were only ≈5 nm thick on iron and ≈3 nm thick on the alloy.

Current practice in depth profiling is to use positively charged argon ions at energies between 0.5 and 10 keV, focused into a beam of 2–5-μm diameter, which is then rastered over the area to be profiled. In XPS the bombarded area should be significantly greater than the analyzed area so that only the flat bottom of the sputtered crater is analyzed. Ion current densities are variable between 5 and 50 μA/cm². Uniformity of sputtering, and therefore the depth resolution of the profile, can be improved by rotating the sample during bombardment.

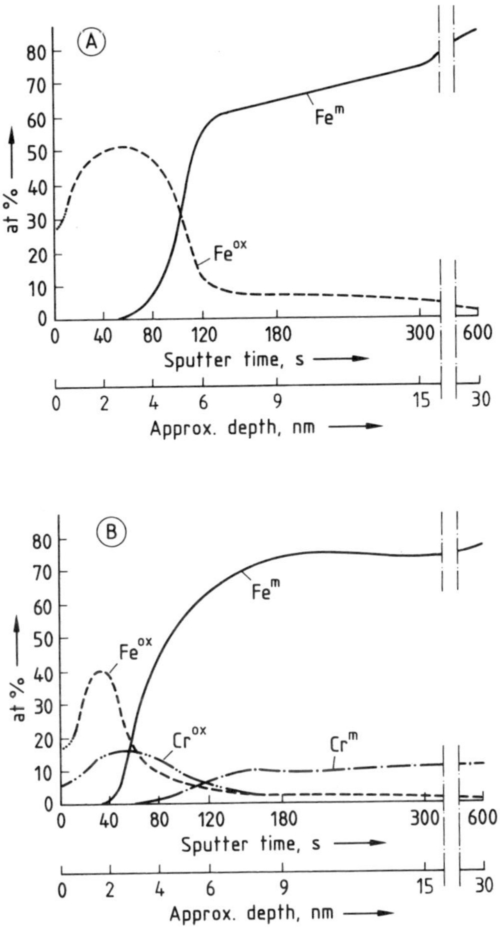

Figure 18. Depth profiles of thin (3–5-nm) oxide films [24] on
A) Pure iron; B) Fe–12Cr–1Mo alloy

During the sputtering process, many artifacts are introduced that can affect quantification; some of these derive from the nature of the sample itself (e.g., roughness, crystalline structure, phase distribution, electrical conductivity), and others are radiation induced (e.g., atomic mixing, preferential sputtering, compound reduction, diffusion and segregation, sputter-induced topography). This is an extensive subject in its own right; a discussion of all aspects is given in [25].

2.5. The Auger Parameter

Section 2.3 shows that in an XPS spectrum, X-ray-excited Auger peaks are often as prominent as the photoelectron peaks themselves. For many elements, the chemical shifts in Auger peaks are actually greater than the shifts in photoelectron peaks. The two shifts can be combined in a very useful quantity called the Auger Parameter α^*, first used by WAGNER [26] and defined in its modified form [27] as

$$\alpha^* = E_K(jkl) + E_B(i) \tag{11}$$

where $E_K(jkl)$ is the kinetic energy of the Auger transition jkl, and $E_B(i)$ is the binding energy of an electron in level i. The Auger and photoelectron peak energies must be measured in the same spectrum.

The Parameter α^* is independent of photon energy and has the great advantage of being independent of any surface charging. It is therefore measurable for all types of material from metals to insulators. Extensive tabulations of α^* have appeared, based on a large number of standard materials, and the most reliable have recently been collected by WAGNER [28].

As defined in Equation (11), α^* is purely empirical; any prominent and conveniently situated Auger and photoelectron peaks can be used. If α^* is defined slightly more rigorously, by replacing $E_B(i)$ by $E_B(j)$ in Equation (11), it is related closely to a quantity called the extraatomic relaxation energy R^{ea}, since $\Delta\alpha^*$ can be shown to equal $2\Delta R^{ea}(j)$. In other words, the change in Auger Parameter between two chemical states is equal to twice the change in the extraatomic relaxation energies in the two states, associated with the hole in the core level *j*. Thus α^* has an important physical basis as well as being useful analytically.

2.6. Applications

XPS has been applied in virtually every area in which the properties of surfaces are important. Given below are a few selected examples from some typical areas; for more comprehensive discussions of the applications of XPS, see [1], [3]–[9], [23], [29]–[33].

2.6.1. Catalysis

An understanding of the basic mechanisms of catalysis has been one of the major aims of XPS from its inception. To this end the technique has been used in two ways: (1) in the study of "real" catalysts closely resembling those used industrially; (2) in the study of "model" catalysts, where the number of parameters has been reduced by the use of well-characterized crystal surfaces.

For industrial catalysts the chemical information obtainable from XPS is all important, since the course and efficiency of the catalyzed reaction will be governed by the states of oxidation of the elements at the surface. Thus XPS is used to analyze both elemental and chemical composition as a function of bulk composition, of surface loading where the active material is deposited on a support, of oxidation and reduction treatments, and of time in reactor under operating conditions. Figure 19 demonstrates the effect on the chemical state of molybdenum in the surface of Mo–Sm–O catalysts of various compositions [34]. For the highest concentration of molybdenum (spectrum d) the Mo 3*d* doublet is very similar to that of MoO_3, but at the samarium-rich composition (spectrum a), it is more like that of a molybdate or a mixed oxide. The effect of successive reduction treatments on a cobalt–molybdenum–alumina catalyst is shown in the changes in the Mo 3*d* spectra of Figure 20 [35]. The air-fired catalyst in spectrum a contains entirely Mo(VI) at the surface, but with repeated hydrogen reduction the spectrum changes radically and can be curve resolved (Fig. 21) into Mo(VI), (V), and (IV) contributions. The chemical state of the surface in spectrum f contains approximately equal proportions of the three oxidation states.

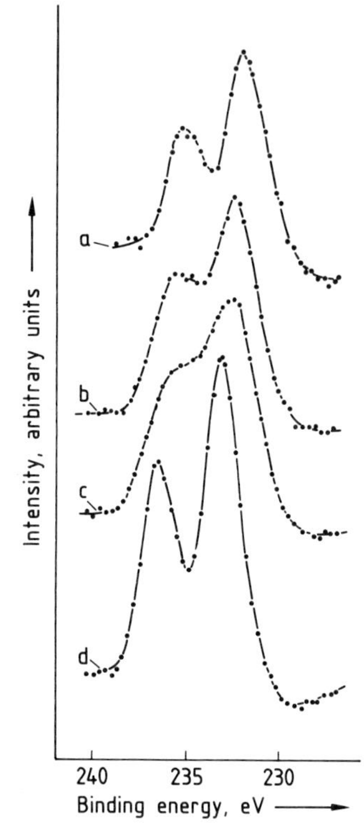

Figure 19. The Mo 3*d* XPS spectra of model catalysts in the Mo–Sm–O system [34]
Mo:Sm ratios are a) 0.25; b) 1.33; c) 4.00; d) 8.00

Other recent applications of XPS to catalytic problems include catalyst–support interactions [36]–[38], variation of surface catalyst loading [39]–[42], effect of reduction [43]–[45], and characterization of catalyst active surface species [46]–[51].

2.6.2. Corrosion and Passivation

Corrosion products that are formed as thin layers on metal surfaces in either aqueous or gaseous environments, along with the nature and stability of passive and protective films on metals and alloys, have also been major areas of application of XPS. Just as in catalysis, XPS has been used in two ways, one in which materials corroded or passivated in the natural environment are analyzed, and another in which well-characterized, usually pure metal surfaces are studied after exposure to controlled conditions. Unlike catalysis, however, in which the two approaches tend to run parallel and not to overlap, in corrosion

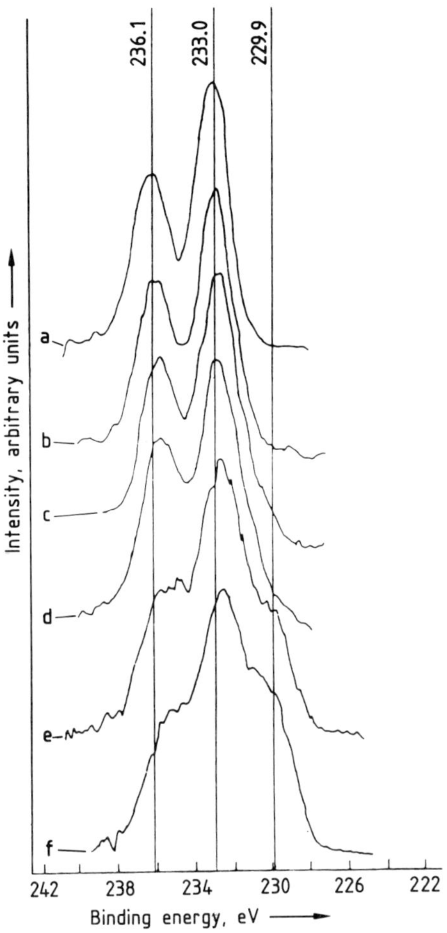

Figure 20. Changes in the Mo $3d$ XPS spectrum from a cobalt–molybdenum–alumina catalyst during successive reduction treatments in hydrogen at 500 °C [35]
a) Air-fired catalyst; b) Reduction time 15 min; c) 50 min; d) 60 min; e) 120 min; f) 200 min

significant overlap and feedback from one to the other occur.

More effort has probably been devoted to study of the corrosion and passivation properties of Fe–Cr–Ni alloys such as stainless steels and other transition-metal alloys than to any other metallic system. The type of spectral information obtainable from an Fe–Cr alloy of technological origin, carrying an oxide and contaminant film after corrosion, is shown schematically in Figure 22 [52]. Actual XPS spectra recorded from a sample of the alloy Stellite-6 (a Co–Cr alloy containing some tungsten) treated in lithiated and borated water at 300 °C are shown as a function of treatment time in Figure 23. Stellite-6 is a hard alloy used in nuclear reactors. The high-temperature aqueous solution simulates light-water reactor conditions. The spectra reveal the removal of tungsten from the surface with increasing treatment time, a decrease in cobalt, and the appearance of nickel and iron deposited from stainless steel components elsewhere in the water circuit.

Spectra from an Fe–Cr alloy first polished, then passivated, and finally ion etched to remove the passive film are shown in Figure 24. Each spectrum has been resolved into the inelastic background and the various oxidation states of the element. The authors, MARCUS and OLEFJORD [53] were even able to separate the oxide and hydroxide contributions to the chromium spectrum from the passive film.

An example [54] of typical spectra recorded during comparison of the gaseous oxidation of pure chromium and of an Ni–Cr alloy is shown in Figure 25. There the Cr $2p$ spectra from the two materials have been recorded as a function of exposure to oxygen. Note that pure chromium does not oxidize as rapidly at room temperature as chromium in the alloy.

XPS has also been applied recently in this field to copper films for the study of microbial effects [55], the behavior of UO_2 in mineral water [56], corrosion-resistant amorphous alloys [57]–[59], and the hydrolysis of fluorozirconate glasses [60].

2.6.3. Adhesion

The adhesion of metal and ink to polymer, as well as the adhesion of paint and coating to metal, are of vital importance in several technologies. In the aircraft industry, aluminum-to-aluminum adhesion is being employed increasingly. The strength and durability of an adhesive bond depend absolutely on the way in which the adhesive compound interacts with the surfaces to which it is supposed to adhere, which in turn often involves pretreatment of the surfaces to render them more reactive. The nature and degree of the reactivity are functions of the chemical states of the adhering surfaces, states that can be monitored by XPS.

In electronic packaging technology, metal–polymer adhesion has been important for some years. Many studies have been devoted to the interactions between metals and polymers, and that of CHIN-AN CHANG et al. [61] is a good example. They deposited 200-nm-thick films of copper, chromium, titanium, gold, and aluminum on various fluorocarbon polymer films, and found that adhesion in terms of peel strength was highest for titanium. Figure 26 shows the reason why on the perfluoroalkoxy polymer tested, the C 1*s* spectra after deposition reveal that for titanium, and to a lesser extent chromium, a strong carbide bond is formed as evidenced by the peak at 282–283 eV. With the other metals, very low adhesion was found, and no carbide-like peaks occurred.

The path of failure of an adhesive joint can give information about the mechanism of failure if analysis of the elemental and chemical composition can be carried out along the path. Several authors have performed such analyses by loading the adhesive joint until it fractures and then using XPS to analyze each side of the fracture. Long-term weathering of the joint can be simulated by immersion in water at elevated temperature. The work of BOERIO and ONDRUS [62] demonstrates the nature of the results obtainable, an example of which is given in Figure 27. Silicon, carbon, and oxygen spectra were recorded from each side of the fracture face of an iron–epoxy resin joint before and after weathering in

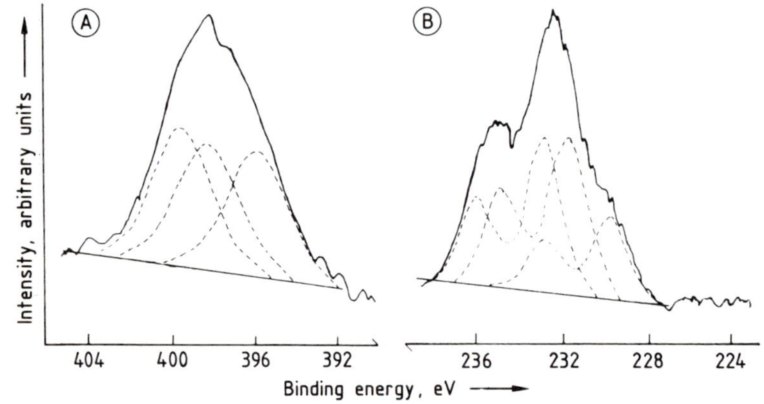

Figure 21. Examples of curve resolution of the Mo 3*p* (A) and 3*d* (B) spectral envelopes carried out by PATTERSON et al. [35] during the catalyst reduction treatment described in Figure 20

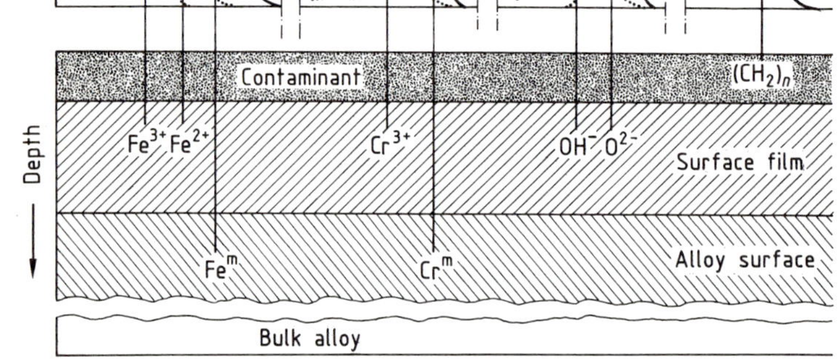

Figure 22. Schematic of type of information obtainable from XPS spectra from an Fe–Cr alloy with an oxide film underneath a contaminant film [52]

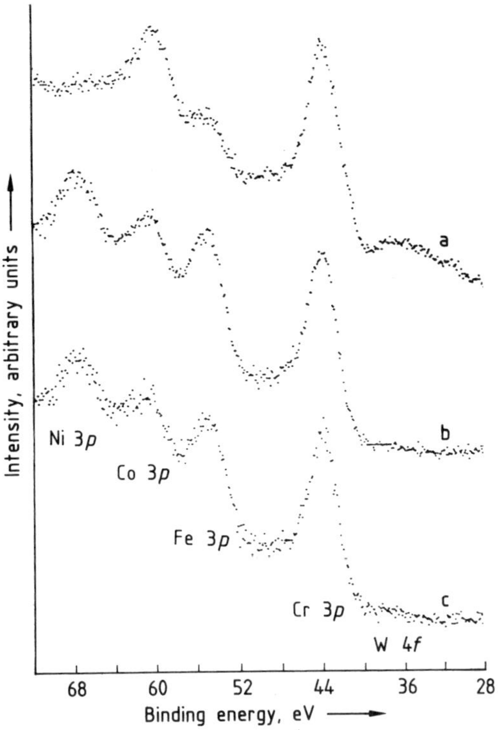

Figure 23. Transition-metal XPS spectra recorded from the surface of the alloy Stellite-6 after treatment for increasing lengths of time in lithiated and borated water at 300 °C [52]
a) 4 h; b) 168 h; c) 360 h

water for 7 d at 60 °C. Before weathering the failure path was as much through the adhesive itself as through the iron oxide–epoxy resin interface, but after weathering, failure occurred almost entirely near the interface.

Also studied by XPS recently have been the measurement of energies of adhesion of aluminum and silicon on molybdenum [63], the ion-enhanced adhesion of nickel to glassy carbon [64], the effects of pretreatments on paint–metal adhesion [65], and the adhesion of titanium thin films to various oxide surfaces [66].

2.6.4. Superconductors

In the years since the discovery of superconducting oxides in 1986, more papers have probably been published describing the application of XPS to superconductors than in any other field. One reason for this frenzied activity has been the search for the precise mechanism of superconductivity in new materials, and an essential piece of information concerns the chemical states of the constituent elements, particularly copper, which is common to all of them. Some theories predict that small but significant amounts of the Cu^{3+} state should occur in superconductors, but other models disagree. Comparison by Steiner et al. [67] of the Cu $2p_{3/2}$ spectra from copper compounds containing copper as Cu^0 (metal), Cu^+, Cu^{2+}, and Cu^{3+}, with the spectra from the superconducting oxide $YBa_2Cu_3O_7$ and from the base material La_2CuO_4, showed conclusively that the Cu^{3+} contribution to the superconducting mechanism must be negligible. Some of their comparative spectra are shown in Figure 28. The positions of the Cu $2p_{3/2}$ peak in $YBa_2Cu_3O_7$ and La_2CuO_4, along with the satellite structure, are very similar to that in CuO (i.e., Cu^{2+}, rather than in $NaCuO_2$, which contains Cu^{3+}).

Recent effort in superconductor research has been devoted mostly to attempts to fabricate thin films with the same superconducting properties as the bulk material, and to prepare suitable electrical contact to the surfaces of superconductors. XPS has been of great use in both areas through its ability to monitor both composition and chemical state during the processes either of thin superconducting film formation, or of the deposition of thin conducting and semiconducting films onto a superconducting surface. Typical of the latter studies is that of Ziegler et al. [68], who deposited increasing amounts of silicon onto epitaxial films of $YBa_2Cu_3O_{7-x}$. Figure 29 from their paper shows the Si $2p$ and Ba $4d$ spectra during the deposition of silicon and after heating the films in oxygen at 250 °C once 6 nm of silicon had accumulated. The Ba $4d$ spectrum at the bottom can be resolved into three $4d_{5/2,3/2}$ doublets, corresponding—with increasing energy—to barium in the bulk superconductor, in the surface of the superconductor, and as carbonate. With increasing silicon thickness the barium spectrum attenuates but does not change in character, but silicon does not appear in the spectrum until nearly 5 nm has been deposited, suggesting migration into the superconductor at room temperature. At a thickness of 6 nm, the Si $2p$ spectrum can be separated into suboxide contributions, indicating reaction of silicon with the $YBa_2Cu_3O_{7-x}$ surface. After heating in oxygen the Si $2p$ spectrum changes to that characteristic of SiO_2 and silicates, while the Ba $4d$ spectrum

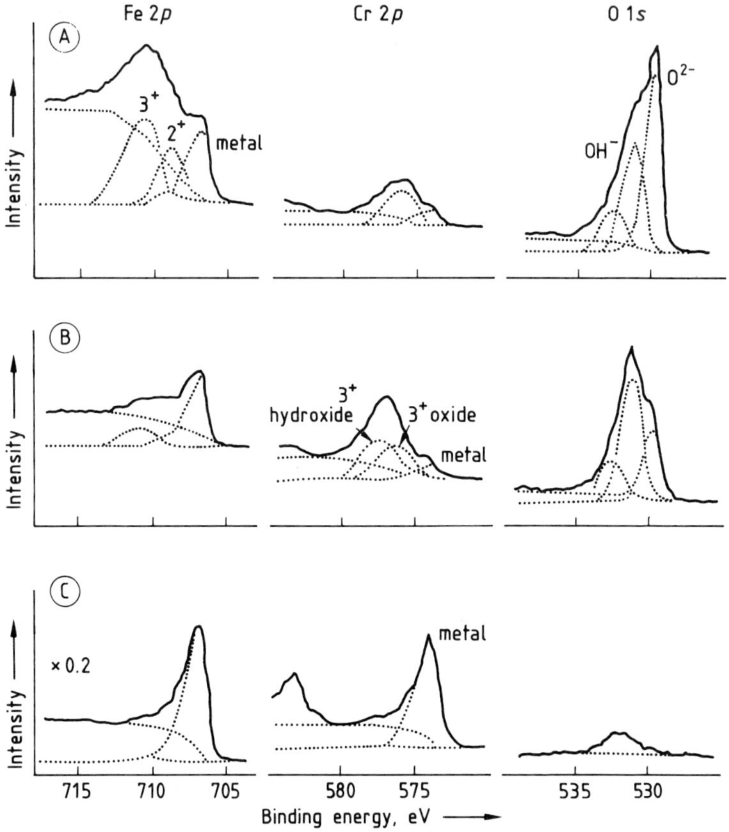

Figure 24. XPS spectra from an Fe–Cr alloy [53]
A) Mechanically polished, exposed in H_2O and air; B) Passivated in 0.5 M H_2SO_4 at 740-mV standard hydrogen electrode for 1 h; C) Ion etched ≈10 nm

also changes by increasing in intensity and conforming mostly to that expected of a barium silicate. As a result of the latter changes the superconducting properties of the film were destroyed. The Y $3d$ and Cu $2p$ spectra establish that yttrium and copper oxides are formed as well.

Other surface reactions studied have been those of gold [69] and silver [70] with $Bi_2Sr_2CaCu_2O_{8+y}$, of niobium with $YBa_2Cu_3O_{7-x}$ [71], and of lead with $YBa_2Cu_3O_{7-x}$ [72]. Characteristics of superconductors in the form of thin films prepared in several ways have been recorded by XPS for $YBa_2Cu_3O_{7-x}$ [73]–[77], $GdBa_2Cu_3O_{7-x}$ [78], $Nd_{2-x}Ce_xCuO_{4-y}$ [79], $Rb_xBa_{1-x}BiO_3$ [80], and $Tl_2Ca_2Ba_2Cu_3O_{10}$ [81]. Detailed electronic studies have been made of the bulk superconductors $Bi_2Sr_2Ca_{1-x}Nd_xCu_2O_y$ [82]–[83], $Nd_{2-x}Ce_xCuO_{4-y}$ [84]–[87], $ErBa_2Cu_4O_8$ [88], and $Tl_2Ba_2CaCu_2O_8$ [89].

2.6.5. Interfaces

The chemical and electronic properties of elements at the interfaces between very thin films and bulk substrates are important in several technological areas, particularly microelectronics, sensors, catalysis, metal protection, and solar cells. To study conditions at an interface, depth profiling by ion bombardment is inadvisable since both composition and chemical state can be altered by interaction with energetic positive ions. Normal procedure is therefore to start with a clean or other well-characterized substrate and deposit the thin film onto it slowly at a chosen temperature while XPS monitors the composition and chemical state by recording selected characteristic spectra. The procedure continues until no further spectral changes occur, as a function either of film thickness, of time elapsed since deposition, or of changes in substrate temperature.

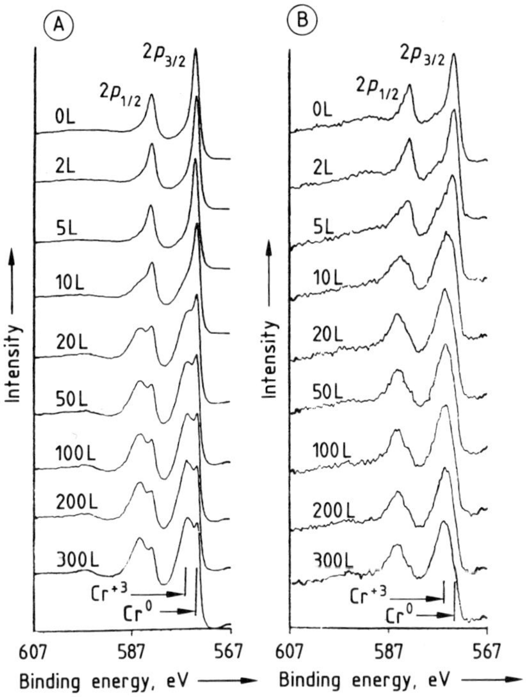

Figure 25. The Cr $2p$ XPS spectra during exposure to oxygen at room temperature [54] of
A) Pure Cr; B) Ni–Cr alloy
L = Langmuir (10^{-6} Torr · s)

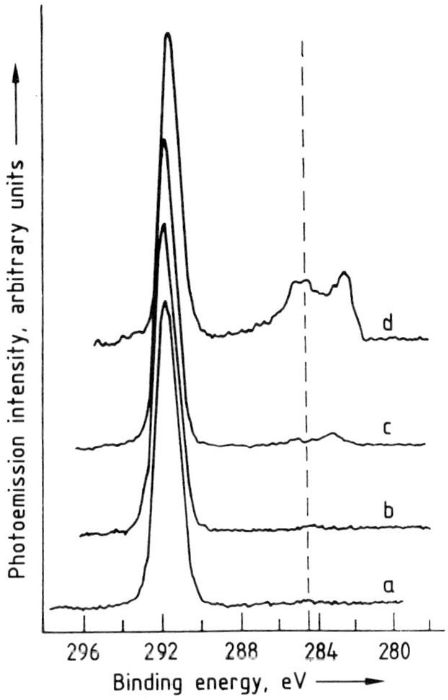

Figure 26. The C 1*s* XPS spectra recorded by CHIN-AN CHANG et al. [61] from perfluoroalkoxy polymer (PFA)
a) Before deposition; b) Deposition of copper; c) Deposition of chromium; d) Deposition of titanium

A good example is the study of the reaction at the interface of the rare-earth element thulium with the silicon (111) surface, carried out by GOKHALE et al. [90]. Figure 30 shows the Si $2p$ spectrum from the substrate as thulium is deposited on it at room temperature. With the Mg K_α radiation used for the analysis, the kinetic energy of the Si $2p$ electrons is about 1155 eV, corresponding to an inelastic mean free path of about 2 nm. The Si $2p$ peak is still plainly detectable after 12.5 nm of thulium have been deposited, and in addition a new feature appears on the low binding energy side of the peak, increasing in magnitude and shifting to lower energy with thickness. Raising the temperature in stages of the thickest film and use of X-ray diffraction, established that the energetic final position labeled R_1 corresponded to Tm_5Si_3 and the initial energetic position found at a thickness of 0.5–1.0 nm, to $TmSi_2$. However, the magnitude of the residual Si $2p$ peak was too great at all stages to be accounted for by compound formation alone, indicating the formation of thulium clusters followed by their coalescence into islands.

Measurement of the Tm $4d_{5/2}$ binding energy during the interaction also showed that the valence of thulium was 3+ at all temperatures and coverages.

Recent studies of interfaces by XPS have also included those of SiO_2 [91], [92], samarium [93], and amorphous hydrogenated carbon [94] on silicon; of silver, indium, and aluminum on GaP [95]; of tin [96] and RbF–Ge interlayers [97] on GaAs; of platinum on TiO_2 [98]; of titanium, niobium, and nickel on Al_2O_3 [99]; of chromium and iron on W (110) [100]; and of ZrO_2 on nickel [101].

Other important areas of application of XPS have included polymers [102]–[106], lubricating films [107], [108], glasses [109]–[113], ion implantation [114]–[118], and cleaning [119]–[121] and passivation [122]–[123] of semiconductor surfaces.

3. Auger Electron Spectroscopy (AES)

After XPS, AES is the next most widely used surface analytical technique. As an accepted surface technique, AES actually predates XPS by two to three years, since the potential of XPS as a surface-specific technique was not recognized immediately by the surface science community. Pioneering work was carried out by HARRIS [124] and by WEBER and PERIA [125], but the technique as it is known today is basically the same as that established by PALMBERG et al. [126].

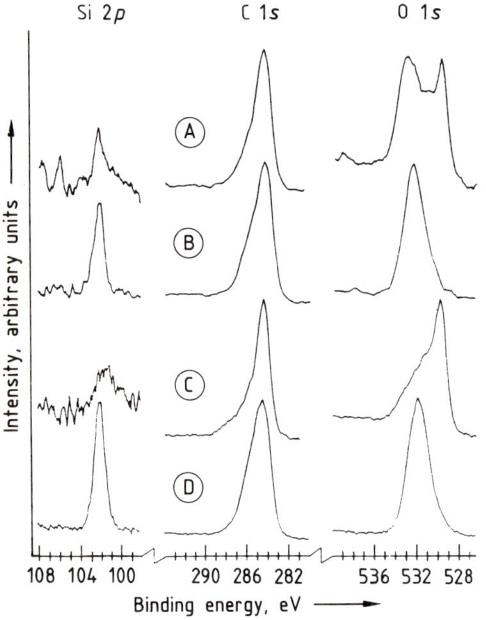

Figure 27. The Si $2p$, C $1s$, and O $1s$ XPS spectra from the adherend and adhesive sides of an iron–epoxy resin joint before and after weathering treatment [62]
A) and B) Adherend and adhesive before weathering; C) and D) Adherend and adhesive after immersion in water at 60 °C for 7 d

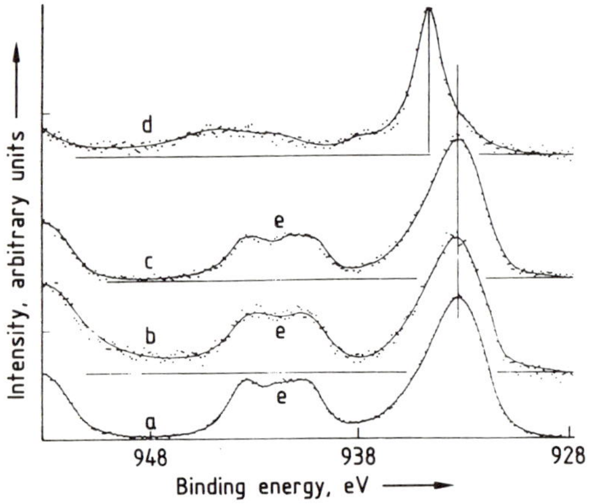

Figure 28. Comparison of the Cu $2p_{3/2}$ and satellite XPS spectra from several copper compounds with the spectrum from the superconducting oxide $YBa_2Cu_3O_7$ [67]
a) CuO, $\Gamma = 3.25$ eV; b) La_2CuO_4, $\Gamma = 3.30$ eV; c) $YBa_2Cu_3O_7$, $\Gamma = 3.20$ eV; d) $NaCuO_2$, $\Gamma = 1.60$ eV; e) Cu^{2+} satellites

3.1. Principles

The surface to be analyzed is irradiated with a beam of electrons of sufficient energy, typically in the range of 5–30 keV, to ionize one or more core levels in surface atoms. After ionization the atom can then relax by either of the two processes described in Section 2.1 for XPS, that is, ejection of a characteristic X-ray photon (fluorescence) or ejection of an Auger electron. Although these are competing processes, for shallow core levels ($E_B < 2$ keV) the probability of the Auger process is far higher. The Auger process is described schematically in Figure 1, which points out that the final state of the atom is doubly ionized and that the kinetic energy of the

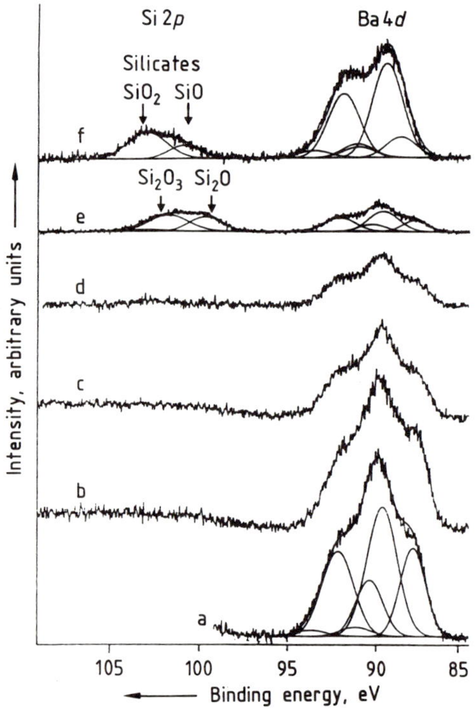

Figure 29. The Si $2p$ and Ba $4d$ XPS spectra [68] from the a) Clean surface of the superconducting oxide $YBa_2Cu_3O_{7-x}$; b)–e) Same surface after increasing deposition of b) 0.8 nm Si; c) 2.8 nm Si; d) 4.4 nm Si; e) 6.0 nm Si; f) Deposited surface after heating to 250 °C in oxygen (8×10^{-3} mbar)

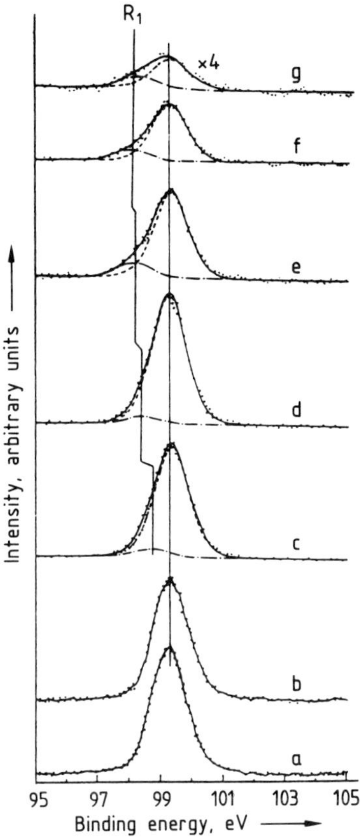

Figure 30. The Si $2p$ XPS spectra with Mg K_α excitation during deposition of thulium to a thickness of 12.5 nm at room temperature [90]
a) Clean silicon; b) 0.2 nm Tm; c) 0.5 nm Tm; d) 2 nm Tm; e) 4 nm Tm; f) 6.5 nm Tm; g) 12.5 nm Tm

electron arising from the Auger transition ABC is given by

$$E_{ABC} = E_A - E_B - E_C^* \quad (12)$$

where E_A, E_B, and E_C are binding energies of the A, B, and C levels in the atom, E_C^* being starred because it is the binding energy of C modified by an existing hole in B.

Equation (12) shows that the Auger energy is a function only of atomic energy levels. Since no two elements have the same set of atomic binding energies, analysis of Auger energies provides elemental identification. Even if levels B and C are in the valence band of the solid, analysis is still possible because the dominant term in Equation (12) is always the binding energy of A, the initially ionized level.

Equation (12) also shows that the heavier the element (i.e., the greater the number of atomic energy levels), the more numerous are the possible Auger transitions. Fortunately, large differences exist in the probabilities between different Auger transitions, so that even for the heaviest elements, only a few intense transitions occur, and analysis is still possible.

In principle, chemical as well as elemental information should be available in AES, since the binding energies appearing in Equation (12) are subject to the same chemical shifts as measured in XPS. In practice, since the binding energies of three levels are involved, extracting chemical information from Auger spectra is, in most cases, still too difficult. However, significant progress is being made in this area, and improved theory and modern data processing methods are likely to enable valuable chemical information to be derived soon.

The reasons AES is one of the surface-specific techniques have been given in Section 2.1, with reference to Figure 2. The normal range of kinetic energies recorded in an AES spectrum would typically be from 20 to 1000 eV, corresponding to inelastic mean free path values of 2 to 6 monolayers.

The nomenclature used in AES has also been mentioned in Section 2.1. The Auger transition in which initial ionization occurs in level A, followed by the filling of A by an electron from B and ejection of an electron from C, would therefore be labeled ABC. In this rather restricted scheme, one would thus find in the KLL series the six possible transitions KL_1L_1, KL_1L_2, KL_1L_3, KL_2L_2, KL_2L_3, and KL_3L_3. Other combinations could be written for other series such as the LMM, MNN, etc.

3.2. Instrumentation

3.2.1. Vacuum Requirements

The same considerations discussed in Section 2.2.1 for XPS apply to AES.

3.2.2. Electron Sources

The energy of the exciting electrons does not enter into Equation (12) for the Auger energy, unlike that of the exciting X-ray photon in XPS; thus the energy spread in the electron beam is irrelevant. What is relevant is the actual energy of the primary electrons because of the dependence on that energy of the cross section for ionization by electron impact. For an electron in a core level of binding energy E_B, this dependence increases steeply with primary energy above E_B, passes through a maximum about 3–5 times E_B, and then decreases to a fairly constant plateau. It is illustrated in Figure 31, in which various theoretical and experimental cross sections are shown for ionization of the Ni K shell, from CASNATI et al. [127]. The implication of this dependence is that for efficient ionization of a particular core level, the primary energy should be about five times the binding energy of an electron in that level. Since in most samples for analysis, several elements are found in the surface region with core-level binding energies extending over a large range, choice of a primary energy that is sufficiently high to ionize all the core levels efficiently is advisable. Hence for *conventional AES*, primary energies are typically in the range 5–10 keV. In *scanning Auger microscopy* (*SAM*), much higher primary energies are used, in the range of 25–50 keV, because in that version of AES as small an electron spot size must be achieved on the sample as possible, and the required focusing can be effected only at such energies.

For the primary energy range 5–10 keV used in *conventional AES*, the electron emitter is thermionic, usually a hot tungsten filament, and focusing of the electron beam is carried out electrostatically. Typically, such an electron source would be able to provide a spot size on the specimen of about 0.5 µm at 10 keV and a beam current of about 10^{-8} A. The beam can normally be

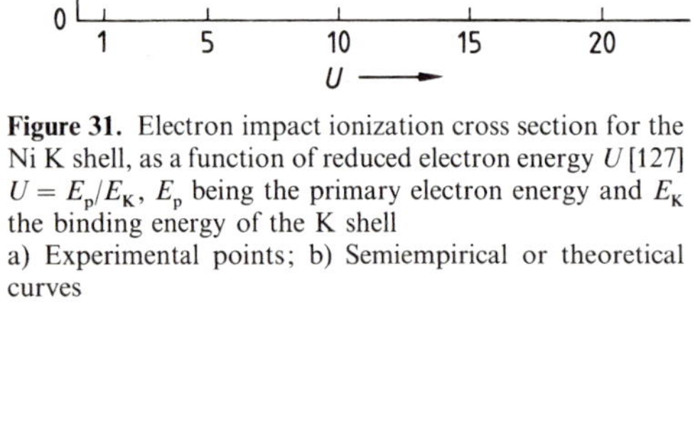

Figure 31. Electron impact ionization cross section for the Ni K shell, as a function of reduced electron energy U [127] $U = E_p/E_K$, E_p being the primary electron energy and E_K the binding energy of the K shell
a) Experimental points; b) Semiempirical or theoretical curves

rastered over the specimen surface, but such a source would not be regarded as adequate for SAM. Sources for *SAM* may be of either the thermionic or the field emission type, the latter being particularly advantageous in achieving minimum spot sizes since it is a high-brightness source. Focusing is performed electromagnetically, and at optimum performance a modern electron gun for SAM would be capable of a spot size of 15 nm at 30 keV and a beam current of about 10^{-10} A. The beam is rastered over the surfaces at scan rates variable up to TV rates, with the dimensions of the scanned area variable as well.

3.2.3. Electron Energy Analyzers

The electron energy analyzer found to be most suitable for AES, the cylindrical mirror analyzer (CMA), has already been described in Section 2.2.4. The important property of this analyzer for AES is its very high transmission, arising from the large solid angle of acceptance; for a typical angular spread $\delta\alpha$ of 6°, the transmission is about 14%. Since the source area (i.e., spot size) in AES is so small, as many Auger electrons as possible must be accepted into the analyzer, compatible with adequate resolution.

An example of the arrangement of a single-stage CMA in a modern system for AES analysis is shown in Figure 32, taken from [128]. The double-pass CMA described in Section 2.2.4 (see Fig. 8) can also be used for AES by grounding the spherical retardation meshes and the inner cylinders and by setting the variable internal apertures to minimum. Although the double-pass CMA is not ideal for either AES or XPS, use of the same analyzer, in its different modes of operation, for both techniques is very convenient. Similarly, the concentric hemispherical analyzer (CHA), also described in Section 2.2.4, is often used for both techniques for the sake of convenience, even though it is much more suitable for XPS than AES.

Until recently, all Auger spectra were presented in energy distributions differentiated with respect to energy, rather than in the direct energy distributions used in XPS [i.e., $dN(E)/dE$ rather than $N(E)$]. The effect of differentiating is to enhance the visibility of Auger features, as demonstrated in Figure 33, in which the differential distribution of secondary electron energies from boron is shown above the corresponding

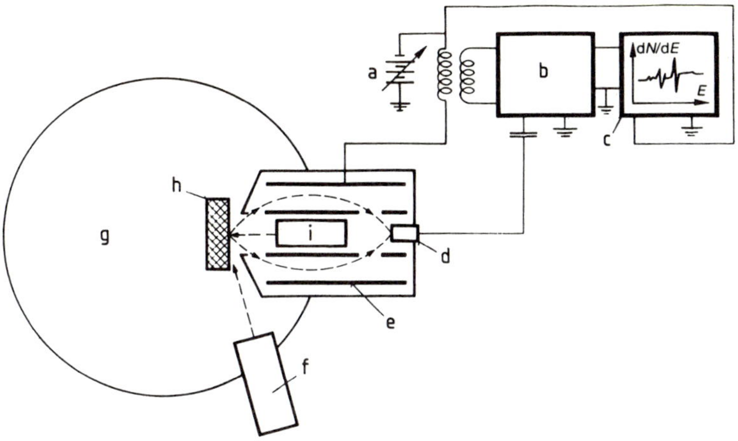

Figure 32. Typical experimental arrangement for AES using a single-pass CMA [128]
a) Voltage sweep; b) Lock-in amplifier; c) Display; d) Electron multiplier; e) Cylindrical mirror analyzer; f) Ion gun; g) UHV chamber; h) Sample; i) Electron gun

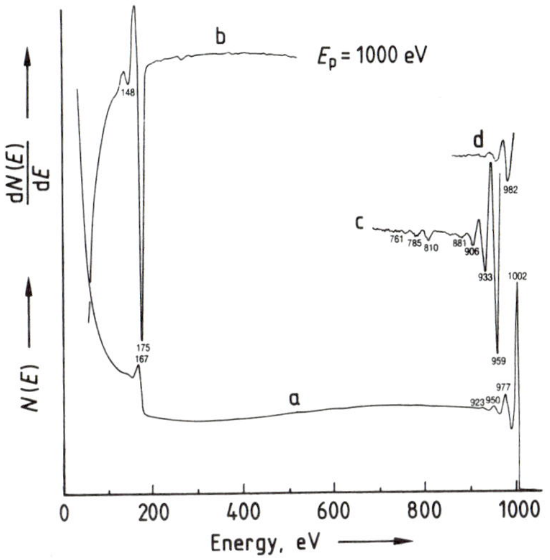

Figure 33. Secondary electron distribution from boron
a) In the $N(E)$ mode; b)–d) In the $dN(E)/dE$ mode, where the differentiation reveals the plasmon satellites

N(E) distribution. The differentiation was performed by a synchronous detection technique, in which a small voltage modulation at high frequency was superimposed on the deflecting potential as the latter was scanned, and then the second harmonic of the detected signal was amplified preferentially. An indication of this technique is given in Figure 32. Nowadays the $N(E)$ distribution is usually recorded digitally, and the differentiation is carried out by computer. However, with the rapid improvement in the signal-to-noise characteristics of detection equipment, the trend in presenting Auger spectra is increasingly toward the undifferentiated $N(E)$ distribution.

Note from Figure 33 that although the true position of the boron KLL Auger peak in the $N(E)$ distribution is at 167 eV, the position in the $dN(E)/dE$ distribution is taken for purely conventional reasons to be that of the negative minimum (i.e., at 175 eV). The other spectral features in Figure 33 are described in Section 3.3.

3.3. Spectral Information

As stated above, the range of primary energies used typically in conventional AES is 5–10 keV, and the ionization cross section passes through a maximum at 3–5 times the binding energy of an electron in the ionized core level. Therefore the core levels that can be ionized efficiently are limited. Thus K-shell ionization efficiency is adequate up to $Z \approx 14$ (silicon), L_3 up to $Z \approx 38$ (strontium), M_5 up to $Z \approx 76$ (osmium), and so on (for the much higher primary energies used in SAM, the ranges are correspondingly extended). This means that in various regions of the periodic table, characteristic Auger transitions occur that are most prominent under typical operating conditions.

Figures 34–36 show some of these prominent Auger features. They have all been taken from [129], and recorded using a CMA, in the differential distribution. Examples of the KLL Auger

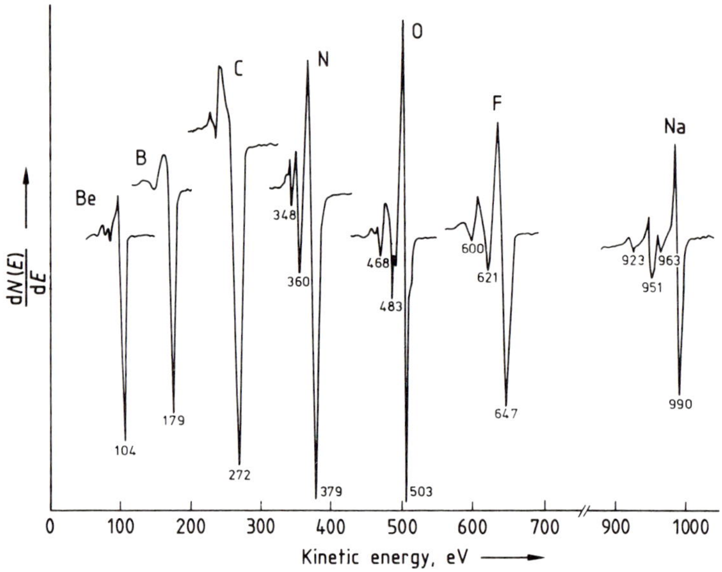

Figure 34. KLL Auger series characteristic of the light elements [129]

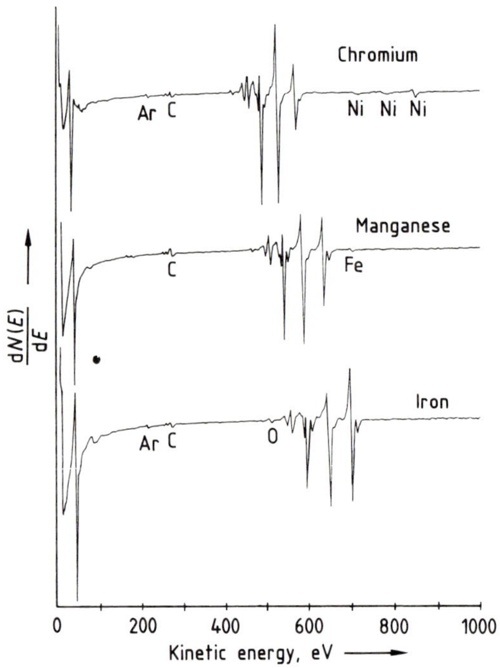

Figure 35. LMM and MMM Auger series in the middle of the first series of transition elements [129]

series are shown in Figure 34, of the LMM series in Figure 35, and of the MNN series in Figure 36. In the KLL series the most prominent feature is a single peak arising from the $KL_{2,3}L_{2,3}$ transition, with minor features from other Auger transitions to lower energy. The LMM series is characterized by a triplet arising from $L_3M_{2,3}M_{2,3}$, $L_3M_{2,3}M_{4,5}$, and $L_3M_{4,5}M_{4,5}$ transitions in ascending order of kinetic energy, with another intense peak, the $M_{2,3}M_{4,5}M_{4,5}$, appearing at very low energy. The characteristic doublet seen in the MNN series arises from the $M_{4,5}N_{4,5}N_{4,5}$ transitions, in which the doublet separation is that of the core levels M_4 and M_5.

Note that AES transitions involving electrons in valence band levels are often written with a V rather than the full symbol of the level. Thus $L_3M_{2,3}M_{4,5}$ and $L_3M_{4,5}M_{4,5}$ would often appear as $L_3M_{2,3}V$ and L_3VV, respectively, and similarly $M_{2,3}M_{4,5}M_{4,5}$ as $M_{2,3}VV$. In Figure 35 the increase in intensity of the L_3VV peak relative to the other two, upon going from chromium to iron, is due to the progressive increase in the electron density in the valence band.

Chemical effects in Auger spectra are quite commonly observed, but difficult to interpret compared to those in XPS since additional core levels are involved in the Auger process. Some examples of the changes to be seen in the KLL spectrum of carbon in different chemical environments are given in Figure 37 [130].

Associated with prominent features in an Auger spectrum are the same types of energy loss

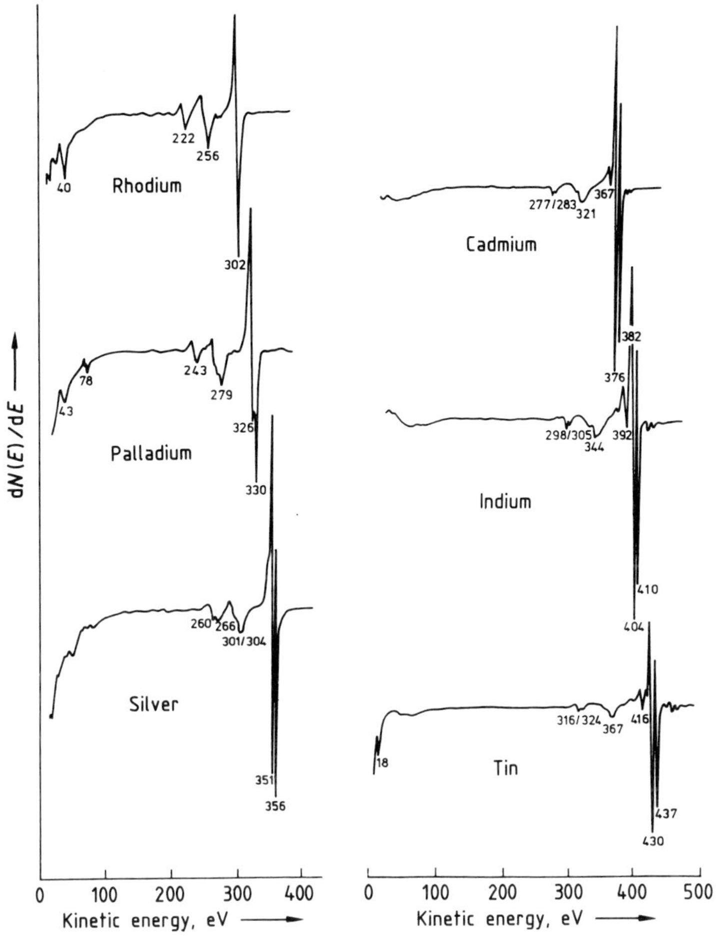

Figure 36. MNN Auger series in the second series of transition elements [129]

feature, the plasmon losses, that are found associated with photoelectron peaks in an XPS spectrum. As described in Section 2.3, plasmon energy losses arise from excitation of modes of collective oscillation of the conduction electrons by outgoing secondary electrons of sufficient energy. Successive plasmon losses suffered by the backscattered primary electrons can be seen in both the $N(E)$ and the $dN(E)/dE$ spectra of Figure 33, from boron. The same magnitude of plasmon loss, ≈27 eV, is also associated with the KLL Auger peak of boron. Plasmon losses are also present in Figures 34–37, but they are difficult to disentangle from minor Auger features.

Another type of loss feature not often recorded in Auger spectra can also be seen in Figure 33, in the $dN(E)/dE$ spectrum at 810 eV. This type arises from core level ionization and forms the basis for the technique of core electron energy loss spectroscopy (CEELS). A primary electron interacting with an electron in a core level can cause excitation either to an unoccupied continuum state, leading to complete ionization, or to localized or partly localized final states, if available. Since the primary electron need not give up all its energy in the interaction, the loss feature appears as a step in the secondary electron spectrum, with a tail decreasing progressively to lower energies. Differentiation, as in Figure 33, accentuates the visibility. In that figure, the loss at 810 eV corresponds to the K-shell ionization edge of boron at about 190 eV.

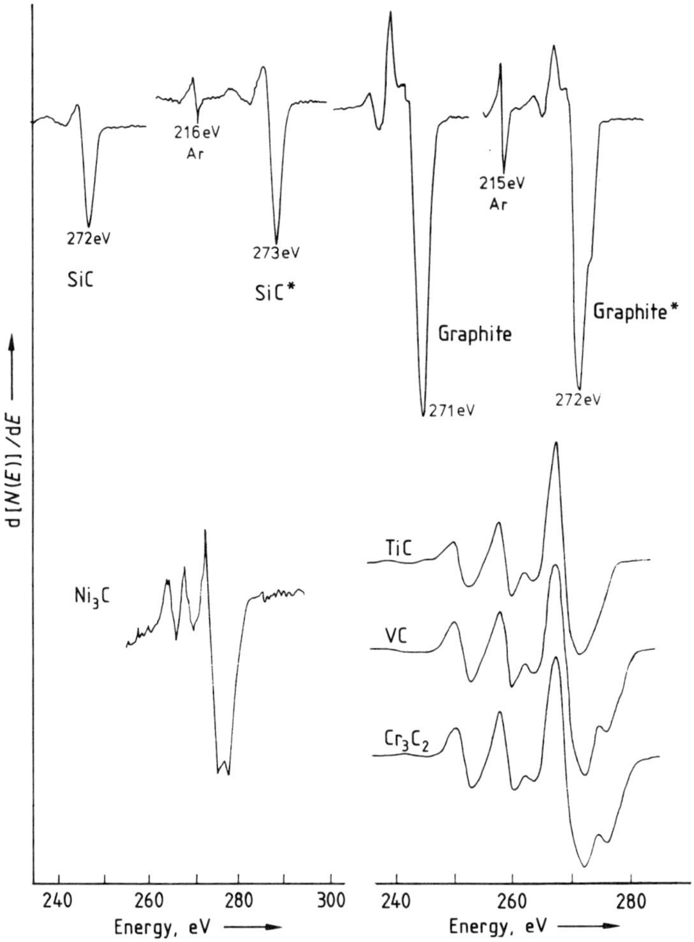

Figure 37. Examples of the effect of different chemical states on the KLL Auger spectrum of carbon [130] (SiC* and graphite* denote Ar^+-bombarded surfaces of SiC and graphite, respectively)

3.4. Quantification and Depth Profiling

3.4.1. Quantification

If ionization of a core level X in an atom A in a solid matrix M by a primary electron of energy E_p gives rise to the current I_{XYZ} of electrons produced by the Auger transition XYZ, then the Auger current from A is

$$I_{XYZ} = K \cdot \sigma(E_p{:}E_B) \cdot \{1 + r_M(E_B{:}\alpha)\}\bar{N}_A \cdot \lambda_M(E_{XYZ}) \cdot \cos\theta \quad (13)$$

where $\sigma(E_p{:}E_B)$ is the cross section for ionization of level X with binding energy E_B by electrons of energy E_p; r_M the backscattering factor that takes account of the additional ionization of X with binding energy E_B by inelastic electrons with energies between E_p and E_B; α the angle of incident electron beam to surface normal; $\bar{N}_A$ the atomic density of A averaged over depth of analysis; $\lambda_M(E_{XYZ})$ the inelastic mean free path in matrix M containing A, at the Auger kinetic energy E_{XYZ}; and θ the angle of Auger electron emission to surface normal.

Similarly to Equation (8), K is a constant of proportionality containing fixed operational parameters such as incident electron current density, transmission of the analyzer at the kinetic energy E_{XYZ}, efficiency of the detector at the kinetic energy E_{XYZ}, and the probability of the Auger transition XYZ.

The cross section for electron impact ionization has already been mentioned in Section 3.2.2 in connection with electron sources, and various experimental and theoretical cross sections have been shown in Figure 31 for the particular case of the K shell of nickel. The expression for the cross section derived by CASNATI et al. [127] gives reasonably good agreement with experiment, with the earlier expression of GRYZINSKI [131] also being useful.

For the inelastic mean free path, Equation (9) given for XPS also applies.

The new factor in Equation (8) compared to that for XPS is the backscattering factor r_M. ICHIMURA and SHIMIZU [132] have carried out extensive Monte Carlo calculations of the backscattering factor as functions of primary beam energy and angle of incidence, of atomic number and of binding energy, and a selection of their results for $E_p = 10$ keV at normal incidence is shown in Figure 38. The best fit of their results to experiment gives the following relationship:

$$r = (2.34 - 2.1\,Z^{0.14})U^{-0.35} + (2.58\,Z^{0.14} - 2.98) \quad \text{for } \alpha = 0°$$

$$r = (0.462 - 0.777\,Z^{0.20})U^{-0.32} + (1.5\,Z^{0.20} - 1.05) \quad \text{for } \alpha = 30°$$

$$r = (1.21 - 1.39\,Z^{0.13})U^{-0.33} + (1.94\,Z^{0.13} - 1.88) \quad \text{for } \alpha = 45° \quad (14)$$

where Z is the atomic number and $U = E_p/E_B$.

Thus, as for XPS, in principle the average surface concentration $\bar{N}_A$ can be calculated from the measurement of the Auger current, according to Equation (13). Again, as in XPS, relative sensitivity factors are generally used. The Auger current $\bar{N}_A{}^s$ for the same transition XYZ in a standard of pure A is measured under the same experimental conditions as in the analysis of A in M, whereupon the ratio of Auger currents is

$$\bar{N}_A/\bar{N}_A{}^s = \frac{I_{XYZ}(1 + r_S)\lambda_s}{I^s_{XYZ}(1 + r_M)\lambda_M} \quad (15)$$

since the ionization cross section σ is the same for both standard and sample. Then, as before, Equation (15) can be reduced to

$$\bar{N}_A = S_A I_{XYZ} \quad (16)$$

where

$$S_A = \frac{\bar{N}_A{}^s(1 + r_s)\lambda_s}{I^s_{XYZ}(1 + r_M)\lambda_M} \quad (17)$$

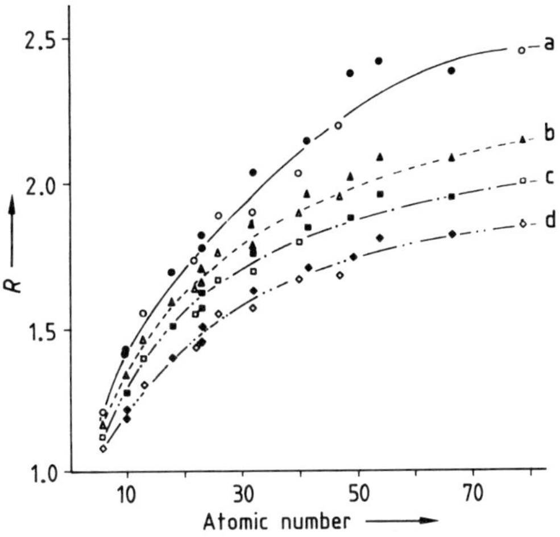

Figure 38. Values of the backscattering factor R (i.e., $1 + r$) calculated as a function of atomic number for an electron beam of 10 keV at normal incidence by ICHIMURA et al. [132]
a) $E_B = 0.1$ keV; b) $E_B = 0.5$ keV; c) $E_B = 1.0$ keV; d) $E_B = 2.0$ keV

S_A is then the relative sensitivity factor. Normally, values of S_A are derived empirically or semiempirically. Tables of such sensitivity factors have been published by PAYLING [133] and by MROCZKOWSKI and LICHTMAN [134] for differential Auger spectra, and for the direct, undifferentiated, spectra by SATO et al. [135]. A comprehensive discussion of the assumptions and simplifications involved and the corrections that should be applied is given in [136], [22]. Because many workers still measure Auger current as the peak-to-peak height in the differential distribution, which takes no account of peak width or area, quantification in AES is still not as accurate as that in XPS.

3.4.2. Depth Profiling

As in XPS, elemental distributions near the surface, but to depths greater than the analytical depth resolutions, are frequently required when using AES. Again, the universally employed technique is that of progressive removal of the surface by ion sputtering, with simultaneous or quasi-simultaneous analysis of the exposed surface. Since the peak-to-peak height of an Auger peak in the differential distribution can be recorded very quickly by a CMA, it is normal in AES depth profiling, when only 10–12 individual peaks need to be recorded, to perform simultaneous sputtering and analysis. Such an operation is called multiplexing; the CMA is programmed to switch rapidly from one energetic peak position to the next, spectral recording being within a chosen energy window at each position. Should many more peaks be needed, or an entire energy spectrum, the ion erosion must be stopped during acquisition, and the profiling is quasi simultaneous as in XPS.

In modern ion guns using differential pumping, the pressure of argon (the gas normally used) in the analysis chamber can be maintained below 5×10^{-5} Pa during depth profiling, thus reducing the possibility of damage to the electron detector and the electron gun filament in the CMA. Such a gun is shown schematically in Figure 39. The source region around the filament (g) and anode (f) is differentially pumped, and positive ions are extracted through a cone-shaped extractor (b), then accelerated through a small aperture (e) to an energy selected between 0.5 and 5 keV. Focusing occurs in two tubular lenses (c) down to a spot size of 10 µm at 5 keV, and the beam can be rastered by two pairs of deflector plates (d) at the end of the gun. The area of raster on the specimen can be chosen between 1×1 mm^2 and 10×10 mm^2. According to the emission from the coated nonburnout iridium filament, the beam current can be selected from 1 to 10 µA.

Since the ion intensity is not uniform across the beam from an ion gun, but approximately Gaussian in distribution, in AES depth profiling either the area of the focused electron beam on the sample must be coincident with the region of maximum ion intensity in the focused ion beam or, preferably, the ion beam must be rastered. The analyzed area should then be centered within the rastered area to ensure that the flat bottom of the sputtered crater is analyzed and not the sloping sides.

The depth resolution achievable during profiling depends on many variables, and the reader is referred, as before, to the comprehensive discussion in [25].

A variation on depth profiling that can be performed by modern scanning Auger instruments (see Section 3.5) is to program the incident electron beam to jump from one preselected position on a surface to each of many others in turn, with multiplexing at each position. This is called *multiple point analysis*, which provides an elemental map after each sputtering step or each period of continuous sputtering. The set of maps can then be related to each other in a computer frame-store system to derive a three-dimensional analysis of a selected micro volume.

3.5. Scanning Auger Microscopy (SAM)

If an incident electron beam of sufficient energy for AES is rastered over a surface in a manner similar to that in a scanning electron microscope (SEM), and if the analyzer, usually a CMA, is set to accept electrons of Auger energies characteristic of a particular element, then a visual display whose intensity is modulated by the Auger peak intensity will correspond to the distribution of that element over the surface. The result is called an *Auger map or image*. Care must be taken in interpreting the intensity distribution, since the Auger intensity depends not only

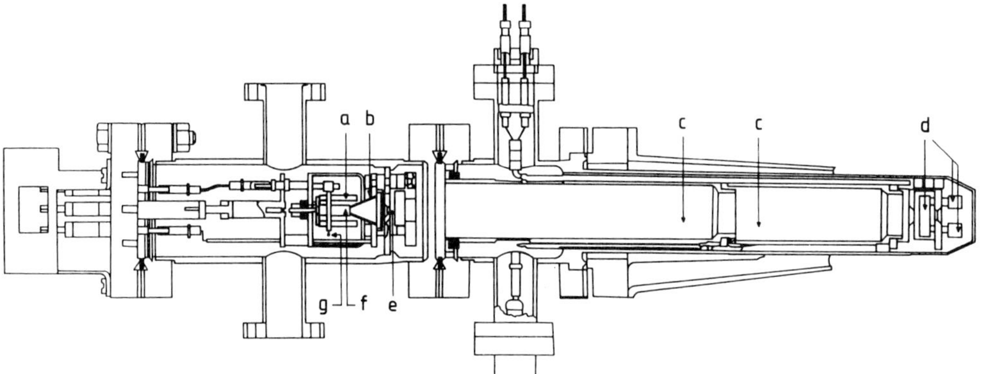

Figure 39. Ion source using a thermionic emitter in a collision chamber for ionization
Courtesy of Leybold
a) Electron repeller; b) Extractor; c) Tubular lenses; d) Deflector plates; e) Aperture; f) Anode; g) Filament

on the local concentration of the element but on the topography as well, because surface roughness can affect the inelastic background underneath the Auger peak. Therefore Auger maps are customarily presented as variations of the ratio of peak intensity divided by the magnitude of the background on the high-energy side, which can be carried out easily by computer.

In SAM the thrust of development is toward ever-better spatial resolution, which means the smallest possible focused spot size on the sample compatible with adequate Auger signal-to-noise ratio in the image, for acquisition within a reasonable length of time. Electrostatic focusing of the incident electron beam is inadequate on its own, and the electron guns used for SAM all employ electromagnetic focusing. Similarly, focusing electrons of low energy into a sufficiently small spot is difficult, so the energies that must be used in SAM are in the range of 25–50 keV, with beam currents of the order of 1 nA or less. An electron gun operating at such energies and currents with electromagnetic focusing would be capable of producing a minimum spot size of about 15 nm. An illustration of the application of SAM at moderate resolution is shown in Figure 40 [137]. The surface was that of a fractured compact of SiC to which boron and carbon had been added to aid the sintering process. The aim of the analysis was to establish the uniformity of distribution of the additives and the presence or absence of impurities. The *Auger maps* show not only very nonuniform distribution of boron (Fig. 40 A) but also a strong correlation of boron with sodium (Fig. 40 C), and a weaker correlation of boron with potassium (Fig. 40 B). Point analyses shown to the right at points A and B marked on the images reveal the presence of sulfur and calcium in some areas as well. In this case the sintering process had not been optimized.

3.6. Applications

Like XPS, the application of AES has been very widespread, particularly in the earlier years of its existence; more recently, the technique has

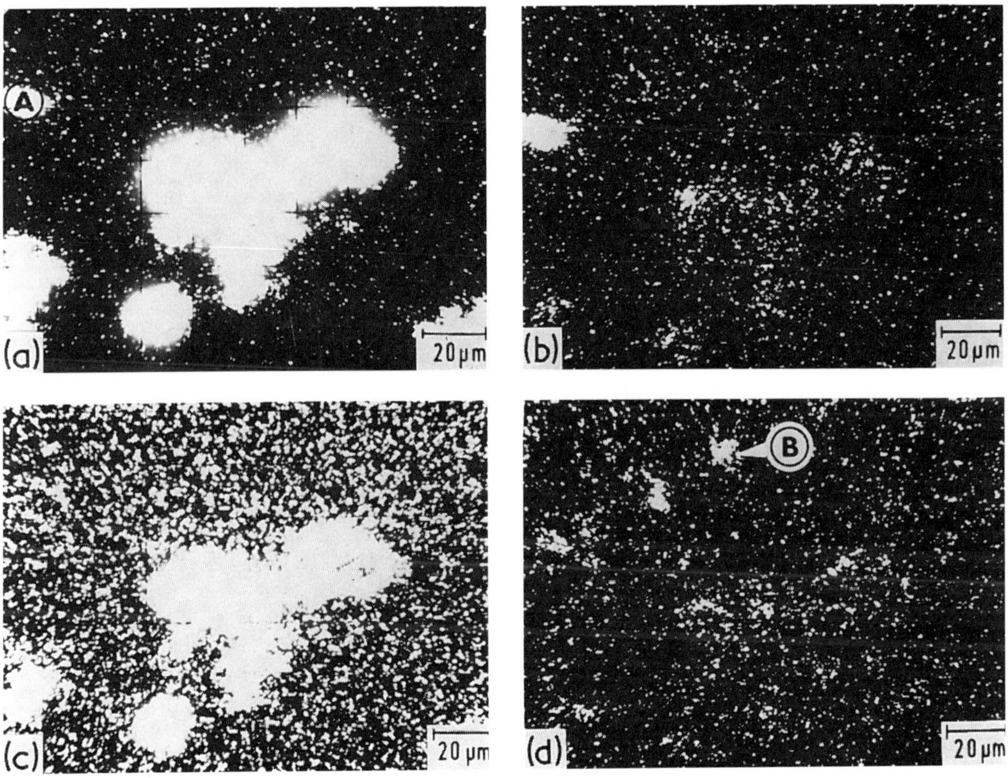

Figure 40. SAM map of fractured SiC after sintering with B addition [137]
a)–d) Elemental maps in boron, potassium, sodium, and oxygen, respectively; E), F) Point analyses at points Ⓐ and Ⓑ, respectively

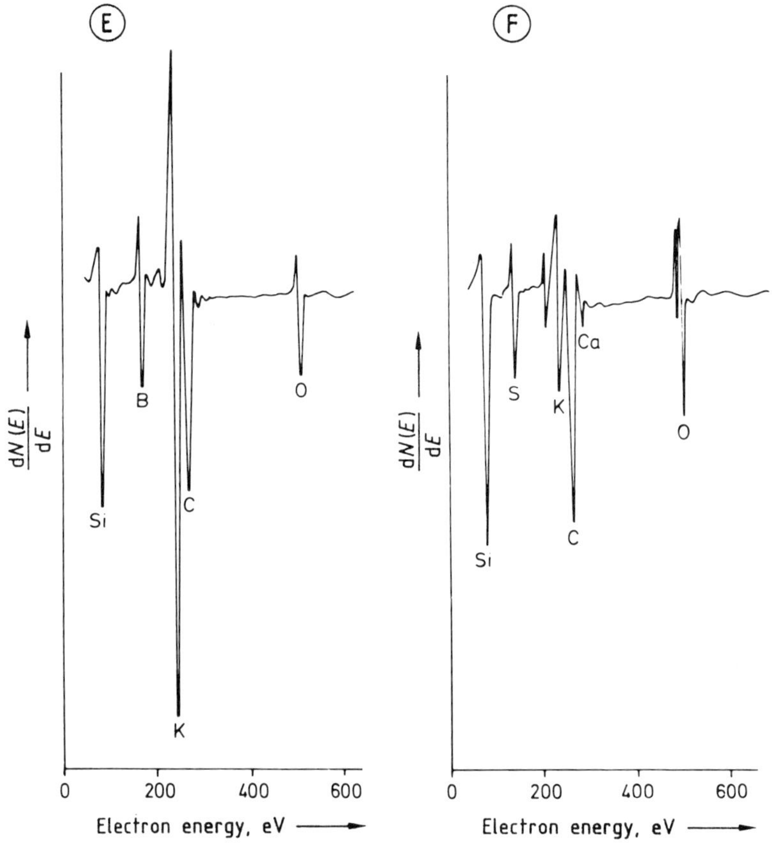

Figure 40. (continued)

been applied increasingly to those problem areas that need the high spatial resolution that AES can provide and XPS as yet cannot. Since data acquisition in AES is faster than in XPS, it is also employed widely in routine quality control by surface analysis of random samples from production lines of, for example, integrated circuits.

3.6.1. Grain Boundary Segregation

One of the original applications of AES, and still one of the most important, is the analysis of grain boundaries in metals and ceramics. Very small amounts of impurity or dopant elements in the bulk material can migrate under appropriate temperature conditions to the boundaries of the grain structure and accumulate there. In that way the concentration of minor elements at the grain boundaries can become much higher than in the bulk, and the cohesive energy of the boundaries can be so altered that the material becomes brittle. Knowledge of the nature of the segregating elements and their grain boundary concentrations as a function of temperature can be used to modify fabrication conditions and thereby improve the strength performance of the material in service.

GORETZKI [138] has discussed the importance and the analysis of internal surfaces such as grain boundaries and phase boundaries, and given several examples. He emphasizes that to avoid ambiguity in interpretation of the analysis, the internal surfaces must be exposed by fracturing the material under UHV conditions inside the electron spectrometer. If this is not done, atmospheric contamination would immediately change the surface condition irrevocably. One of the examples from GORETZKI's paper is given in Figure 41, in which Auger spectra in the differential mode from fracture surfaces in the embrittled (Fig. 41 A) and unembrittled (Fig. 41 B) states of a 12% Cr steel containing phosphorus as an impurity, are compared. Embrittlement occurs when the steel is held at 400–600 °C for a long

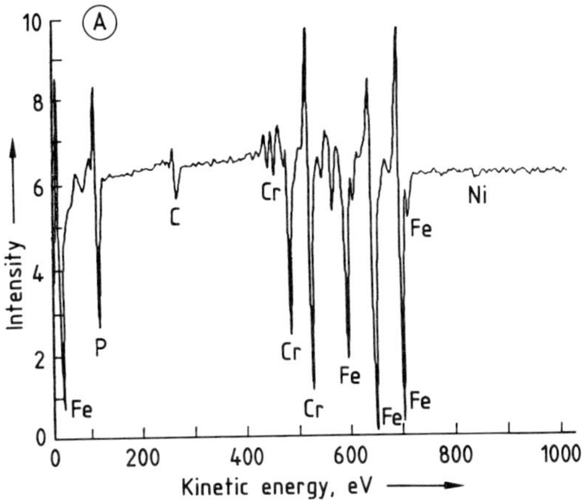

Figure 41. Auger spectra from fracture surfaces of a 12% Cr steel [138]
A) Embrittled state; B) Unembrittled state

time. Note in the spectrum from the embrittled material not only the greatly enhanced phosphorus concentration at the grain boundary, but also increased amounts of chromium and nickel with respect to iron. By setting the analyzer energy first to that of the phosphorus peak, and then to that of the principal chromium peak, Auger maps could be recorded showing the distribution of phosphorus and chromium over the fracture surface. The two maps showed high correlation of the two elements, indicating association of phosphorus and chromium at the grain boundary, possibly as a compound.

An example of the application of SAM to fracture in ceramic materials has already been given in Section 3.5 (Fig. 40). Other recent applications in the study of grain boundary segregation in metals have included the effects of boron, zirconium, and aluminum on Ni_3Al intermetallics [139], the segregation of phosphorus and molybdenum to grain boundaries in the superalloy Nimonic PE 16 as a function of aging treatment [140], and the effect of addition of aluminum on the concentration of oxygen at grain boundaries in doped and sintered molybdenum rods [141].

3.6.2. Semiconductor Technology

With ever-increasing miniaturization of integrated circuits goes a need for the ability to analyze ever smaller areas so that the integrity of fabrication at any point on a circuit can be checked. This checking process includes not only microdetails such as continuity of components in the form of thin films and establishment of the correct elemental proportions in contacts, diffusion barriers, and Schottky barriers, but also the effectiveness of more macrotreatments such as surface cleaning. In all these types of analyses, depth profiling (see Section 3.4.2) is used extensively because, in general, elemental compositional information is required, not information about chemical state. As mentioned previously, chemical state information tends to become blurred or destroyed by ion bombardment.

Various materials have been prescribed as *diffusion barriers* to prevent one circuit material from diffusing into another (usually Si) at the temperatures needed for preparation of certain components such as oxide films in integrated circuits. Thus a Ta–N film has been found to act as an efficient barrier between the metallizing alloy Al_3Ta and a silicon substrate, although SASAKI et al. [142] have shown that the barrier properties depend on the stoichiometry of the Ta–N film. These authors prepared structures of 30 nm of Ta–N on silicon, followed by 200 nm of Al_3Ta on top of the Ta–N, with the Ta–N intermediate having various stoichiometries. The composite structures were then heated to 500 and 600 °C, and analyzed by AES with depth profiling. The results for a Ta–N intermediate barrier formed by sputtering in an Ar + 5% N_2 mixture are shown in Figure 42. The profiles before heat treatment (Fig. 42 A) and after 550 °C treatment (Fig. 42 B) indicate sharp interfaces between the various layers, with little interdiffusion. After 600 °C heat treatment (Fig. 42 C), however, very substantial intermixing has taken place, and the

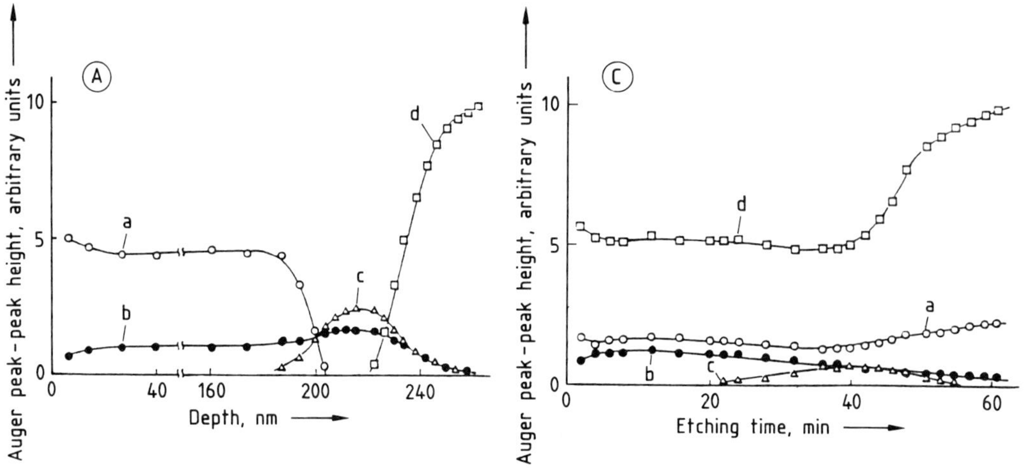

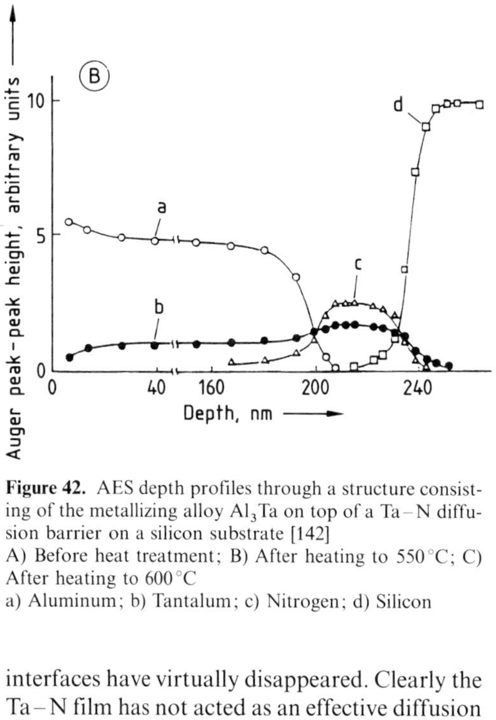

Figure 42. AES depth profiles through a structure consisting of the metallizing alloy Al_3Ta on top of a Ta–N diffusion barrier on a silicon substrate [142]
A) Before heat treatment; B) After heating to 550 °C; C) After heating to 600 °C
a) Aluminum; b) Tantalum; c) Nitrogen; d) Silicon

interfaces have virtually disappeared. Clearly the Ta–N film has not acted as an effective diffusion barrier above 550 °C. On increasing the N_2 content in the sputtering gas to Ar + 10 % N_2, and performing the same profiles, no intermixing was evident even at 600 °C, demonstrating that Ta–N can certainly be used as a diffusion barrier, but that care must be taken to achieve a minimum N:Ta stoichiometry. Ti–N is also used as a diffusion barrier between aluminum and silicon, and the analysis of Ti–N stoichiometry in contact holes in advanced integrated circuits has been studied by PAMLER and KOHLHASE [143]. Since the holes are themselves of 2–3-μm diameter, analysis of their interior surface requires special techniques.

The nature of *metallic contacts to compound semiconductors* has also been an interesting problem. Ohmic contacts must be made to semiconducting surfaces, otherwise circuits could not be fabricated. However, during fabrication, excessive interdiffusion must not occur, since otherwise the electrical characteristics of the semiconductor cannot be maintained. Among the many studies of the reactions of metallizing alloys with compound semiconductors are those of PROCOP et al. [144] on Au–Ge contacts to GaAs, of REIF et al. [145] on Ti–Pt–Au contacts to GaAs, and of STEIN et al. [146] on Au–Pt–Ti contacts to GaAs and InP.

AES has also been applied to a study of the oxidation of copper in printed circuit boards [147], of sulfur passivation of GaAs [148], of cleaning of germanium and GaAs with hydrogen dissociated by a noble-gas discharge [149], and of segregation and corrosion of silver layers electroplated onto light emitting diode (LED) lead frames as part of the final packaging [150].

3.6.3. Thin Films and Interfaces

The nature of the interface formed between very thin metallic films and substrates of various types has been studied extensively by AES, just as it has by XPS (see Section 2.6.5). The degree and extent of interaction and interdiffusion can be established by AES with depth profiling, although chemical information is normally absent, but the development of the interface is usually studied by continuous recording of Auger spectra during film deposition. Changes in the $M_{4,5}VV$ Auger spectrum from a palladium substrate as europium was deposited on it are shown in Figure 43 [151]. Up to a coverage of about two monolayers, no change occurred in the position or shape of the palladium spectrum, but for greater thicknesses the spectrum became narrower, indicating a decrease in the width of the palladium valence band arising from filling of the band. Such filling is due to an interdiffusion process between europium and palladium occurring even at room temperature.

In some cases, interaction of film with substrate does not occur at room temperature but occurs only after annealing at high temperature. A good example is that of a 10-nm-thick titanium film on an SiO_2 substrate. In Figure 44 [152], the Ti LMM Auger spectra are shown in the $N(E)$ (i.e., undifferentiated) mode, rather than the more usual differentiated $dN(E)/d(E)$ mode seen in Figures 41 and 43. Before annealing, the titanium film shows the typically sharp structure in the LMM region, but after annealing to 900 °C the spectra change character completely. The small L_3VV peak near 450 eV has disappeared; the L_3MV peak has reduced in intensity, with the

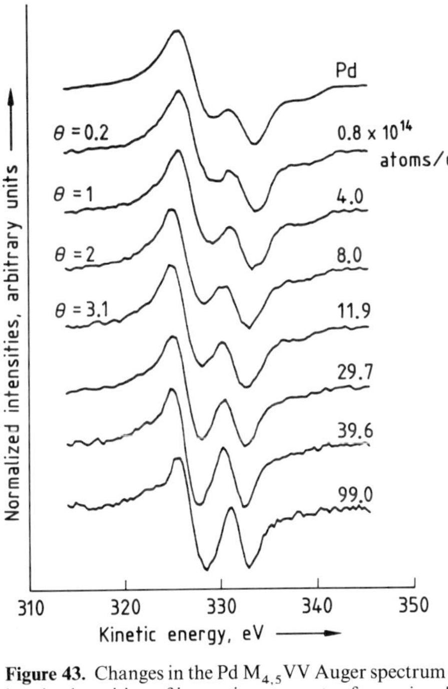

Figure 43. Changes in the Pd $M_{4,5}VV$ Auger spectrum during the deposition of increasing amounts of europium [151] θ = coverage (monolayers)

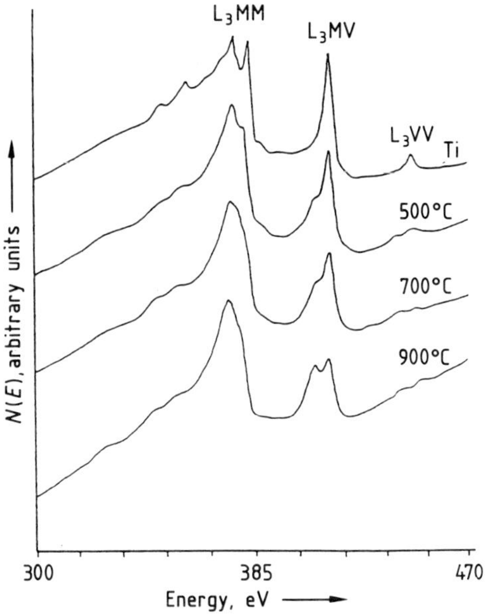

Figure 44. Changes in the Ti LMM Auger spectra, in the $N(E)$ mode, from a 10-nm titanium film on SiO_2 after annealing to increasingly higher temperatures [152]

appearance of an additional peak at lower kinetic energy; and the LMM peak has lost its fine structure and broadened. All these features, like the measured O:Ti ratio, are consistent with the appearance of TiO_2, indicating complete oxidation of titanium as a result of reaction with SiO_2.

Other interesting thin-film studies using AES have included the growth and interaction of silicon on nickel [153], and of palladium on silicon [154], and the preparation and characterization of radio-frequency (rf) sputtered TiN_x [155], WC [156], and Al_3Ta [157] films.

3.6.4. Superconductors

Although AES has not been applied quite as extensively as XPS to the new high-T_c superconducting oxides, the number of applications since their discovery has also been large. The majority of applications have consisted of routine depth profiling measurements aimed at checking the extent of reaction of various materials with the superconductor. Recently such measurements, routine though they may be, have become even more important in analyzing uniformity and concentration during attempts to prepare superconductors in the form of thin films. This is because the thin films must of course be deposited on a substrate and must also be annealed to achieve their correct superconducting properties. At the annealing temperature, reaction with the substrate may occur, which must be avoided if possible. A favored substrate, because of its alleged inertness, is $SrTiO_3$. Typical of the depth profiles recorded to check possible interaction are those shown in Figure 45 [158]. The $YBa_2Cu_3O_{7-x}$ film was deposited by ion beam codeposition from sources of Cu_2O, Y_2O_3, and $BaCO_3$, together with oxygen ion bombardment, onto an $SrTiO_3$ substrate held at 350 °C, to a thickness of 100 nm. Figure 45 A shows the peak-to-peak Auger intensities as a function of sputter time, and Figure 45 B the same after conversion of yttrium, barium, copper, and oxygen intensities to atomic concentrations. Notice in the upper profile the sharpness of the interface, with relatively little diffusion of strontium and titanium into the superconducting film and none of yttrium, barium, and copper into the $SrTiO_3$.

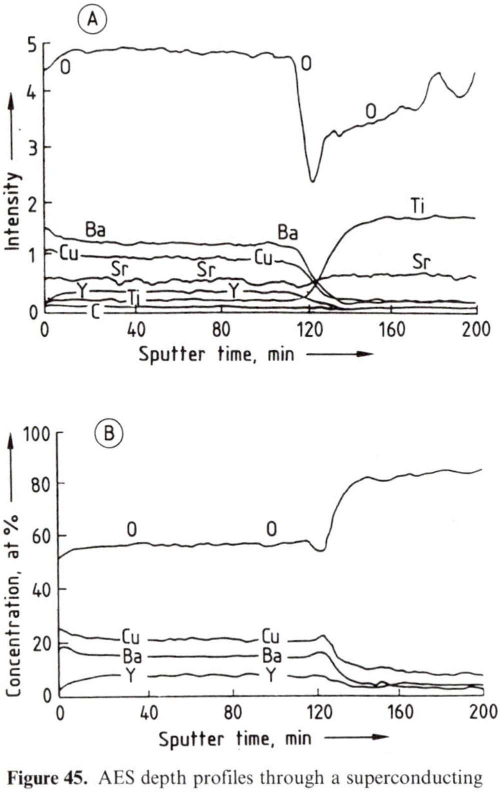

Figure 45. AES depth profiles through a superconducting $YBa_2Cu_3O_{7-x}$ film 100 nm thick on an $SrTiO_3$ substrate held at 350 °C [158]
A) Raw peak-to-peak (p-p) intensity data; B) Conversion of raw data to atomic concentrations

In the lower profile, note the almost correct stoichiometry of the film and the uniformity through the film.

The effectiveness of so-called *buffer layers* has also been examined by AES; these are inert layers interposed between the superconducting film and a reactive substrate such as silicon. ZrO_2 is a much-used buffer material, and its role in preventing interdiffusion between $YBa_2Cu_3O_{7-x}$ films and silicon has been studied by AARNINK et al. [159] using not only AES but RBS as well. Another possibility is RuO_2, whose effectiveness was established by JIA and ANDERSON [160]. Other types of study of superconductors using AES have included the surfaces produced by fracture of sintered $YBa_2Cu_3O_{7-x}$ [161], a comparison of the surfaces of pure $YBa_2Cu_3O_{7-x}$ and its silver-added composite [162], and a quantitative comparison of superconducting bulk crystals and thin films [163].

3.6.5. Surface Segregation

The surface composition of alloys, compounds, and intermetallics may change profoundly during heat treatment as a result of segregation of either or both constituent and impurity elements. Surface enhancement may also occur through chemical forces during oxidation or other reaction. The mechanism of segregation is similar to that of grain boundary segregation already discussed, but of course at the free surface material is much more likely to be lost through reaction, evaporation, etc., so that excessive continued segregation can change the bulk properties as well as those of the surface. Nearly all such studies have been carried out on metals and alloys, and AES has been applied to the problem almost from its inception.

Sometimes the segregation of metallic and nonmetallic elements together to surfaces can result in the formation of surface compounds, as demonstrated by UEBING [164]. He heated a series of Fe–15% Cr alloys containing either 30 ppm nitrogen, 20 ppm carbon, or 20 ppm sulfur to 630°–800 °C and obtained the typical Auger spectra shown in Figure 46. In each case an excess of both chromium and the nonmetal occurs in the spectrum, over that expected if the bulk concentration were maintained. Furthermore, some chemical information available from AES can be used here, because the fine structure associated with the nitrogen and carbon KLL

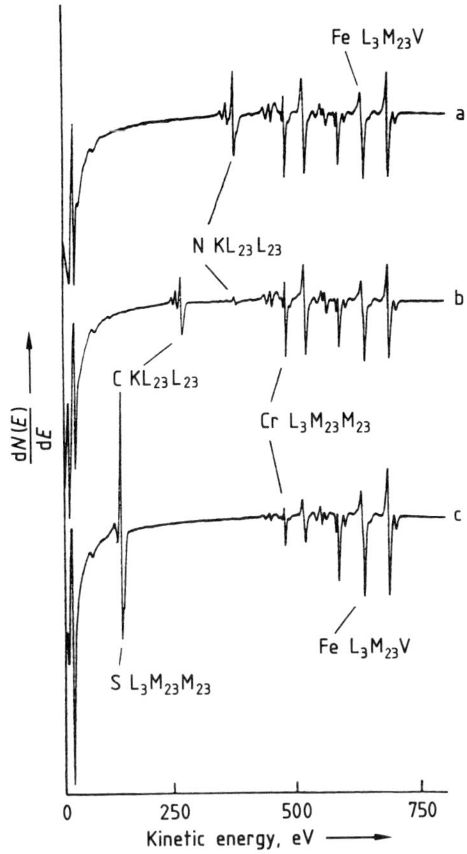

Figure 46. Formation of surface compounds on Fe–15% Cr alloys by cosegregation of chromium and a nonmetallic element [164]
a) Nitride; b) Carbide; c) Sulfide

spectra, if not that of sulfur, is characteristic of those elements as the nitride and carbide, respectively. Thus the cosegregation of chromium with nitrogen and with carbon has led to the formation of surface nitrides and carbides, with the same conclusion likely but not so certain, for sulfur. The surface nitride was identified as CrN and the carbide as CrC, the latter being hypothetical but probably stable as a surface compound.

Other recent surface segregation studies involving AES, often with depth profiling, have been carried out on the alloys Ag–Pd [165]; $Au_{0.7}Cu_{0.3}$ [166]; Fe–40Cr–3Ru, Fe–40Cr–3Pt, and Fe–40Cr, with a platinum overlayer [167]; Cr–5Pt and Cr–15Pt [168]; and phosphorus segregation in pure iron using XPS and LEED as well [169].

AES has also been used to study the implantation of nickel in silicon [170], the biological corrosion of condenser tubes [171], the surfaces of amorphous metals [172], [173], and superhard a-C:H coatings [174].

4. Static Secondary Ion Mass Spectrometry (SSIMS)

SSIMS originated with Benninghoven [175] in 1969 (i.e., at roughly the same time as AES). The prefix "static" was added by him to distinguish the technique from "dynamic" SIMS—the difference between the two lying in the incident current densities used, of the order of 1–10 nA/cm² for SSIMS but much higher for dynamic SIMS. In addition, dynamic SIMS usually operates in an imaging or scanning mode, whereas scanning SSIMS has only recently appeared.

SSIMS ranks with XPS and AES as one of the principal surface analytical techniques; its use, although expanding, is probably still rather less than that of the other two. Problems in interpretation continue, but its sensitivity for some elements is much greater than that of the other two techniques; in addition, a large amount of chemical information is available from it in principle.

4.1. Principles

A beam of positive ions irradiates a surface, leading to interactions that cause the emission of a variety of types of secondary particle, including secondary electrons, Auger electrons, photons, neutrals, excited neutrals, positive secondary ions, and negative secondary ions. SSIMS is concerned with the last two of these, *positive* and *negative secondary ions*. The emitted ions are analyzed in a mass spectrometer, giving rise to a positive or negative mass spectrum consisting of parent and fragment peaks characteristic of the surface. The peaks may be identifiable as arising from the substrate material itself, from contamination and impurities on the surface, or from deliberately introduced species adsorbed on the surface.

When a heavy energetic particle such as an argon ion at several kiloelectronvolts encounters a surface, it cannot be stopped immediately by the first layer of atoms but continues into the surface until it comes to a halt as a result of losing energy by atomic and electronic scattering. Along the way to its stopping place, the ion displaces some atoms from their normal positions in the solid structure, and as they recoil, these atoms in turn displace others, which also displace additional atoms, and so on, resulting in a complex sequence of collisions. Depending on the energy absorbed in an individual collision, some atoms are displaced permanently from their normal positions, whereas others return elastically after temporary displacement. The whole sequence is called a collision cascade and is illustrated schematically, along with some other processes, in Figure 47 [176]. As the cascade spreads out from the path of the primary ion, its effect may eventually reach atoms in the surface layer, and if by that stage enough kinetic energy is left in the collisional interaction, bonds may be broken and material leave the surface as atoms or clusters. The material is then said to be sputtered; the above model of the sputtering process is from SIGMUND [177].

Although SIGMUND's collision cascade model is able to explain much about the parameters involved in the removal of material from a surface by sputtering, it cannot predict the degree of positive or negative ionization of the material. In truth, a unified theory of secondary ion formation in SSIMS does not yet exist, although many models have been proposed for the process. Since the secondary-ion yields (i.e., the probabilities of ion formation) can vary by several orders of magnitude for the same element in different matrices, or for different elements across the periodic table, lack of a means of predicting ion yields in given situations is a hindrance to both interpretation and quantification. Some theoretical and semitheoretical models have had success in predictions over very restricted ranges of experimental data, but they fail when extended further. This is not an appropriate place to describe the theoretical models; for further information see, e.g., [1].

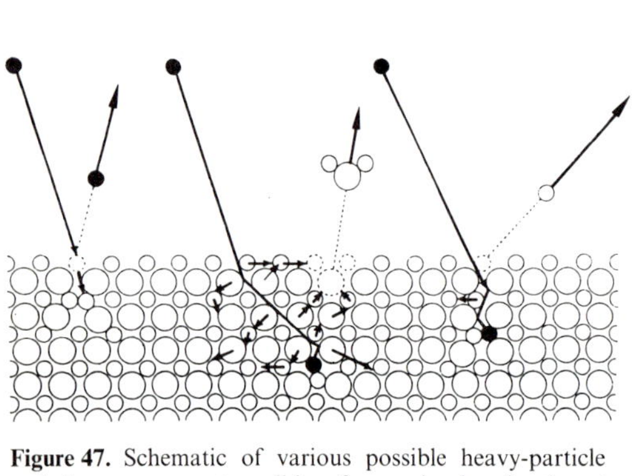

Figure 47. Schematic of various possible heavy-particle emission processes at a solid surface on ion bombardment [176]

4.2. Instrumentation

4.2.1. Ion Sources

Two types of source are in use according to the type of instrument employed for SSIMS. In the first, *positive ions from a noble gas* (usually argon) are produced either by electron impact or in a plasma created by a discharge. The ions are then extracted from the source region, accelerated to the chosen energy, and focused in an electrostatic ion optical column. An ion gun of this type using electron impact ionization is shown in Figure 39.

In the second type, *positive ions from a liquid metal* (almost always gallium) are produced in the manner shown schematically in Figure 48 [178]. A fine needle (f) (tip radius ≈5 μm) of refractory metal passes through a capillary tube (d) into a reservoir of liquid metal (e). Provided the liquid metal wets the needle, which can be ensured by fabrication procedures, the liquid is drawn through the tube and over the tip by capillary action. Application of 5–10 kV between the needle and an extractor electrode (b) a short distance away draws the liquid up into a cusp because of the combined forces of surface tension and electrostatic stress. In the region just above the cusp, ions are formed and are acceler-

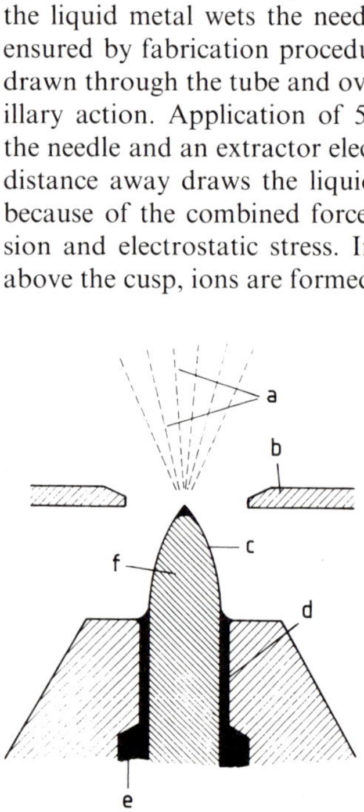

Figure 48. Schematic of the design and operation of a liquid-metal ion source (LMIS) [178]
a) Metal ions; b) Extractor; c) Liquid-metal film; d) Capillary tube; e) Liquid metal; f) Needle

ated through an aperture in the extractor, from where they may be focused in an ion optical column onto the sample surface. Gallium is invariably used as the liquid metal because its melting point is 35 °C and it remains liquid at room temperature; other metals have been tried but require heating of the reservoir to melt them.

Argon-ion guns of the electron impact type operate in the energy range 0.5–5 keV at 1–10 μA, and provide optimum spot sizes on a surface of about 10 μm. The discharge-type argon-ion guns, often using the duoplasmatron arrangement, operate up to 10 keV, with currents variable up to 20 μA and, according to the particular focusing lens system used, can provide spot sizes down to about 2 μm. The *gallium liquid metal ion source* (*LMIS*) is what is known as a *high-brightness source*, which means that ion production occurs in a very small volume. As a result the ion beam can be focused to a fine spot, and at 8–10 keV a spot size of 0.2 μm can be achieved, while at 30 keV it reduces to about 50 nm. All the ion gun optical columns provide deflection plates for rastering the ion beam over areas adjustable from many square millimeters to a few tens of square micrometers.

4.2.2. Mass Analyzers

As with ion sources, two types of ion mass analyzer are in use, and indeed each type of source goes with each type of analyzer. The argon-ion sources are used with quadrupole mass spectrometers and the gallium LMIS with the time-of-flight (TOF) mass spectrometer.

Quadrupole Mass Spectrometer. The quadrupole mass spectrometer (see also, →Mass Spectrometry, **B5**) has been in use for many years as a residual gas analyzer (with an ionizing hot filament) and in desorption studies and SSIMS. It consists of four circular rods, or poles, arranged so that they are equally spaced in a rectangular array and exactly coaxial. Figure 49 indicates the arrangement, as depicted by KRAUSS and GRUEN [179]. Two voltages are applied to the rods, a d.c. voltage and an rf voltage. When an ion enters the space between the rods, it is accelerated by the electrostatic field, and for a particular combination of d.c. and rf voltages, the ion has a stable trajectory and passes to a detector. For other combinations of voltages, the trajectory diverges rapidly and the ion is lost either by hitting one of the poles or by passing between them to another part of the system. The mass resolution is governed by the dimensions of the mass spectrometer, the accuracy of construction, and the stability and reproducibility of the ramped voltage. The quadrupole mass spectrometer is compact, does not require magnets, and is entirely ultrahigh vacuum compatible—hence its popularity. It does have disadvantages, however, in that its transmission is low and decreases with increasing mass number.

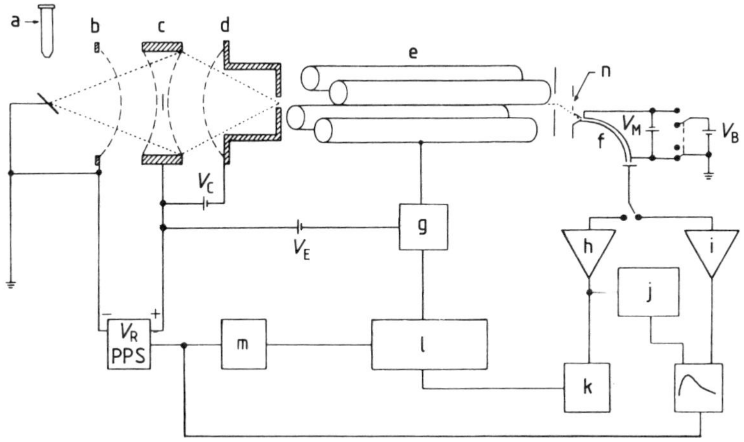

Figure 49. Experimental arrangement for SSIMS used by KRAUSS and GRUEN [179] employing a quadrupole mass spectrometer for mass analysis and a retarding-field analyzer for prior energy selection
a) Ion gun; b)–d) Lenses 1–3; e) Quadrupole mass spectrometer; f) Charge detecting electron multiplier; g) Quadrupole power supply; h) Pulse amplifier; i) Iso amplifier; j) Rate meter; k) Multichannel analyzer; l) Mass programmer; m) Sawtooth generator; n) Exit hole for neutrals

Time-of-Flight Mass Spectrometer. Because the quadrupole mass spectrometer is unsuitable for the analysis of large molecules on surfaces or of heavy metals and alloys, the TOF mass spectrometer has been developed for use in SSIMS by BENNINGHOVEN and coworkers [180]. The TOF mass spectrometer works on the principle that the time taken for a charged particle such as an ion to travel in a constant electrostatic and magnetic field from its point of origin to a detector is a function of the mass-to-charge ratio of the ion. For the same charge, light ions travel the fixed distance more rapidly than heavy ions. For good mass resolution, the flight path must be sufficiently long (1–1.5 m), and very sophisticated high-frequency pulsing and counting systems must be employed to time the flight of the ion to within a few nanoseconds.

The original design [180] used pulsed beams of argon ions, but with the commercial development of the LMIS (see Section 4.2.1) the capability of the TOF system has been extended significantly. The principle of LMIS operation allows the beam of $^{69}Ga^+$ and $^{71}Ga^+$ ions to be focused to a probe of 50-nm minimum diameter, while being pulsed at frequencies up to 20 kHz and rastered at the same time. Mass filtering selects either of the isotopes before the beam strikes the sample. When an LMIS is operated at beam currents of 10–100 pA, the possibility is opered up of performing scanning SSIMS and of producing secondary-ion images in selected masses. A system designed for such *TOF imaging SSIMS* is shown schematically in Figure 50, from ECCLES and VICKERMAN [181]. Pulsing of the ion source is achieved by rapid deflection of the ion beam across a small aperture, the pulse length being variable between 4 and 50 ns. The flight path length is 1.57 m, part of which includes an energy compensator to ensure that all ions of the same mass, but of different energies, arrive at the same time. Secondary electron detectors (SED) allow topographical images produced by ion-induced secondary electrons to be generated.

With such a TOF imaging SSIMS instrument, the useful mass range is extended beyond 10 000 amu; parallel detection of all masses occurs simultaneously; up to four secondary-ion images can be acquired simultaneously; and within each image, up to 20 mass windows (very narrow mass ranges) can be selected. The amount of data generated in a short time is enormous, and very sophisticated and capacious systems are required to handle and process the data.

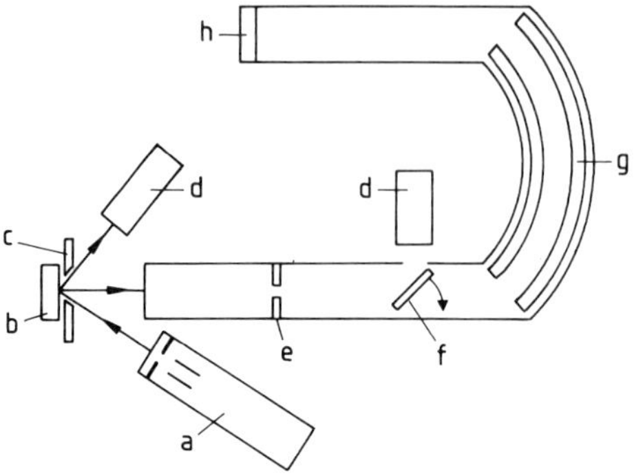

Figure 50. Schematic of the imaging time-of-flight SSIMS system designed by ECCLES and VICKERMAN [181]
a) Pulsed ion gun; b) Target; c) Extraction electrode; d) SED; e) Variable iris; f) Movable plate; g) Energy-compensated flight tube; h) Detector

4.3. Spectral Information

A SSIMS spectrum, like any other mass spectrum, consists of a series of peaks of varying intensity (i.e., ion current) appearing at certain mass numbers. The masses can be allocated on the basis either of atomic mass and charge, in the simple case of ions of the type $M_i^{n\pm}$, where both n and i can each equal 1, 2, 3, ..., or of molecular mass and charge, in the more complex case of cluster ions of type $M_iX_j^{n\pm}$, where again n, i, and j can each equal 1, 2, 3, ... In SSIMS, many of the more prominent secondary ions happen to have $n = i = 1$, which makes allocation of mass numbers to fragments slightly easier. The masses can be identified as arising either from the substrate material itself, from deliberately introduced molecular or other species on the surface, or from contamination and impurities on the surface. Complications in allocation often arise from isotopic effects. Although some elements have only one principal isotope, in many others the natural isotopic abundance can cause difficulties in identification.

An example showing both the allocation of peaks and the potential isotopic problems appears in Figure 51, in which the positive and negative SSIMS spectra from a nickel specimen have been recorded by BENNINGHOVEN et al. [182]. The specimen surface had been cleaned by standard procedures, but SSIMS reveals many impurities on the surface, particularly alkali and alkaline-earth metals, for which it is especially sensitive. For reasons discussed in Section 4.4, the peak heights cannot be taken to be directly

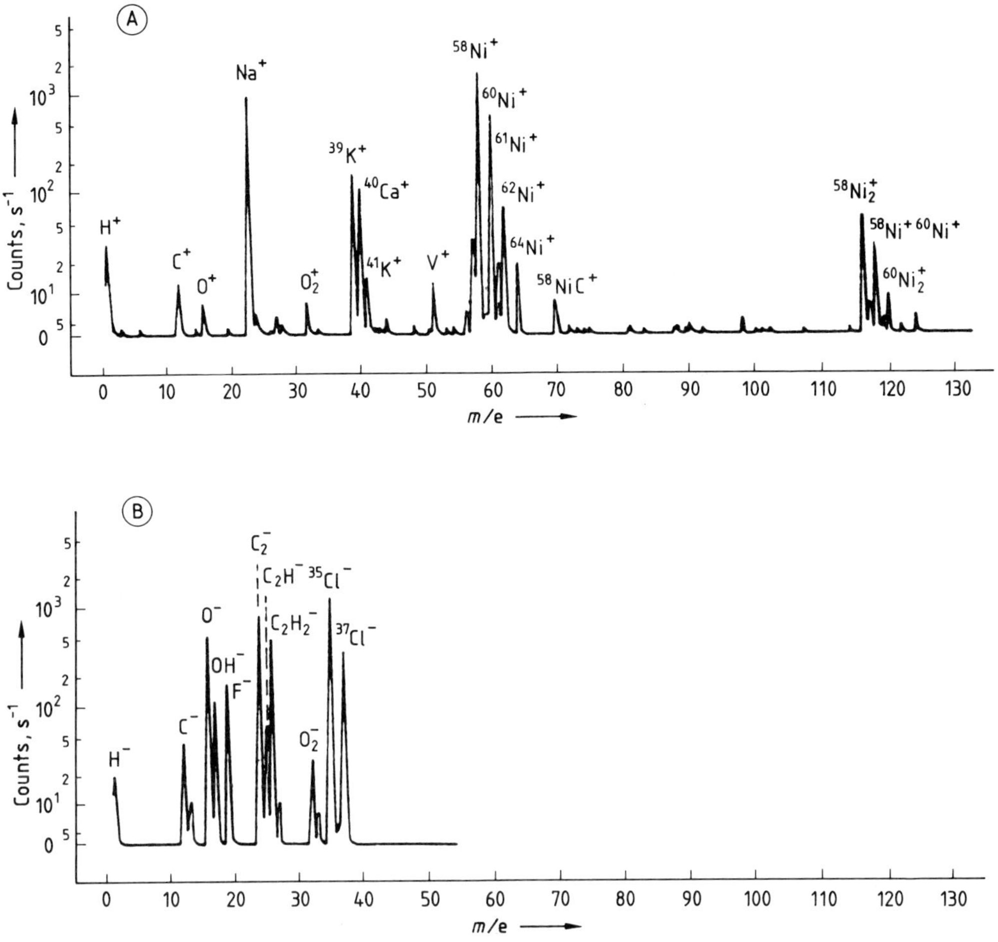

Figure 51. SSIMS spectra from a "cleaned" nickel target, under conditions of 3 keV primary argon ion energy and 5×10^{-9} A primary current [182]
A) Positive SSIMS spectrum; B) Negative SSIMS spectrum

proportional to the concentrations on the surface; thus, for example, the similar heights of the Na^+ and $^{58}Ni^+$ peaks should not lead to the conclusion that there is as much sodium as nickel on the surface. Note in Figure 51 the appearance of isotopes of nickel, potassium, and chlorine in their correct natural abundances and also, in the positive ion spectrum, peaks due to M_2^+ ions, also with isotopic effects. Since the intensity scale is logarithmic, the minor peaks are exaggerated, but identification problems do exist.

The relationship between what is recorded in a SSIMS spectrum and the chemical state of the surface is not as straightforward as in XPS and AES. Because of the large number of cluster ions that appear in any SSIMS spectrum from a multicomponent surface (e.g., during the study of a surface reaction), a great deal of chemical information is obviously available in SSIMS, potentially more than in XPS. The problem in using the information from a cluster ion lies in the uncertainty of knowing whether or not the cluster is representative of a group of the same atoms situated next to each other in the surface. For some materials, such as polymers, the clusters observed are definitely characteristic of the material, as seen in Figure 52 from Bletsos et al. [183], where the SSIMS spectrum from polystyrene measured in a TOF mass spectrometer reveals peaks spaced at regular 104 mass unit intervals, corresponding to the polymeric repeat unit. On the other hand, for example during

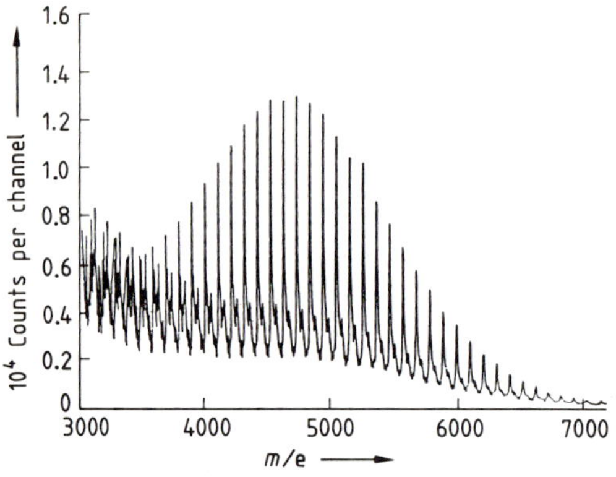

Figure 52. Positive SSIMS spectrum of polystyrene (M_n = 5100, $M_{n,\,calc.}$ = 4500, $M_{n,\,corr.}$ = 5600) from time-of-flight mass spectrometer [183]

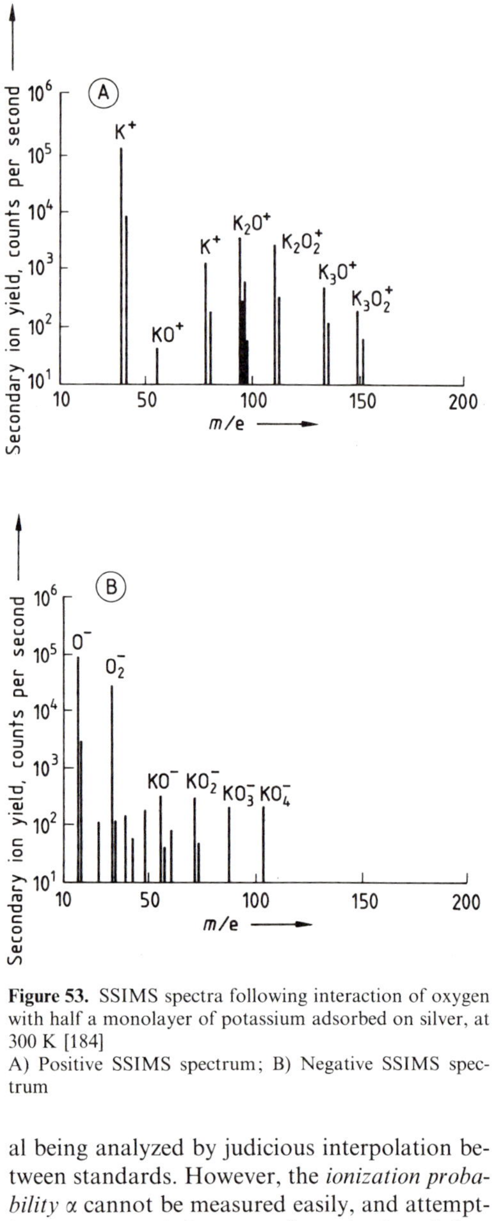

Figure 53. SSIMS spectra following interaction of oxygen with half a monolayer of potassium adsorbed on silver, at 300 K [184]
A) Positive SSIMS spectrum; B) Negative SSIMS spectrum

oxidation studies, many metal–oxygen clusters are normally found in both positive and negative secondary-ion spectra, whose relationship to the actual chemical composition is problematic. Figure 53 from McBreen et al. [184] shows the SSIMS spectra recorded after interaction of oxygen with half a monolayer of potassium adsorbed on a silver substrate. In both positive and negative spectra K_iO_j clusters are found that the authors were unable to relate to the surface condition, although quite clearly much chemical information is available in the spectra. The use of pure and well-characterized standards can sometimes help in the interpretation of complex spectra.

4.4. Quantification

The secondary-ion current at a particular mass number from element A in a matrix B can be written as

$$I^{\pm}_{\mathrm{A(B)}} = K \cdot S \cdot C_{\mathrm{A}} \cdot \alpha^{\pm}_{\mathrm{A(B)}} \qquad (18)$$

where S is the total sputtering yield of B; C_A the concentration of A in B; and $\alpha^{\pm}_{A(B)}$ the probability of positive or negative ionization of element A in matrix B.

As for XPS and AES (Eq. 8 and Eq. 13, respectively) K is a constant containing the various instrumental and other factors that are maintained approximately fixed during an analysis.

The *sputtering yield S* can be measured from elemental and multielemental standards under various operational conditions and can therefore be known to reasonable accuracy for the material being analyzed by judicious interpolation between standards. However, the *ionization probability* α cannot be measured easily, and attempting to measure it for a set of standards and then trying to interpolate would be of no use, because it is very matrix and concentration dependent and can vary very rapidly. In fact, it represents the principal obstacle in achieving proper quantification in SSIMS.

Because of problems in the measurement of α, quantification is normally carried out either by relative sensitivity factor methods or by semiempirical approximations to α based on one of the ionization theories.

Relative Sensitivity Factors. In SSIMS, unlike XPS and AES, elemental sensitivities are taken as relative to the intensity of the signal from an internal reference element, rather than to that of a single chosen element such as fluorine or carbon. The internal reference would normally be the major constituent of the material. If the signal from the reference is $I^{\pm}_{R(B)}$, it can be written

$$I^{\pm}_{A(B)}/I^{\pm}_{R(B)} = \frac{K \cdot S \cdot C_A \cdot \alpha^{\pm}_{A(B)}}{K \cdot S \cdot C_R \cdot \alpha^{\pm}_{R(B)}} \tag{19}$$

or

$$I^{\pm}_{A(B)}/I^{\pm}_{R(B)} = \varrho_{AR} C_A / C_R \tag{20}$$

where ϱ_{AR} is the relative sensitivity factor of A with respect to R. Of course ϱ_{AR} contains matrix-dependent quantities, but their variations are damped to some extent by virtue of taking ratios, and in practice ϱ_{AR} is assumed constant for low concentrations of A (e.g., <1 at%). It can be evaluated from measurements on a well-characterized set of standards containing A in known dilute concentrations. The accuracy of the method is not as good as XPS and AES when relative sensitivity factors are used.

Semiempirical Approximation. The method that adopts a semiempirical approximation to α uses the *Andersen–Hinthorne local thermodynamic equilibrium* (*LTE*) model [185] for estimating the degree of ionization. The model does not take account of any details of individual ionization processes but assumes that the region at and near the surface involved in sputtering can be approximated by a dense plasma in local thermodynamic equilibrium. The plasma has an associated temperature T. The ratio of the concentrations of two elements X and Y in a matrix B, if sufficiently dilute, can then be written as

$$C_X/C_Y \sim (I^{\pm}_{X(B)}/I^{\pm}_{Y(B)}) \exp\{(I_{P(X)} - I_{P(Y)})/kT\} \tag{21}$$

where $I^{\pm}_{X(B)}$, $I^{\pm}_{Y(B)}$ are the positive secondary-ion currents of X and Y, respectively, in B; and $I_{P(X)}$, $I_{P(Y)}$ the first ionization potentials of neutral atoms X and Y.

T is the only adjustable parameter in Equation (21). Measurement of the secondary-ion currents of all components then provides a set of coupled equations for $C_X/C_Y, \ldots, C_N/C_Y$, which, together with the identity $C_X + C_Y + \ldots C_N = 100\,\%$, enables the concentrations to be found. The model does not have a sound physical basis, but the method has been found to be surprisingly accurate for some materials (e.g., steels). However, the values of T, the effective plasma temperature, that are necessary for agreement are sometimes ridiculously high.

4.5. Applications

In general, SSIMS has not been used for quantification of surface composition because of all the uncertainties described above. Its application has been more qualitative in nature, with emphasis on its advantages of high surface specificity, very high sensitivity for certain elements, and multiplicity of chemical information. Increasingly, it is being used in the high spatial resolution TOF imaging mode and also in parallel with a complementary technique, such as XPS, on the same system.

4.5.1. Oxide Films

The higher surface specificity of SSIMS compared to XPS and AES (i.e., 1–2 monolayers versus 2–8 monolayers) can be useful in establishing more precisely the chemistry of the outer surface than would be possible with the latter techniques. Although from details of the O 1*s* spectrum XPS could give the information that OH was present on a surface as well as oxide, and from the C 1*s* spectrum that hydrocarbons as well as carbides were present, only SSIMS could say which hydroxide and which hydrocarbons. In the growth of oxide films for various purposes (e.g., passivation or anodization), such information is valuable since it provides a guide to the quality of the film and the nature of the growth process.

One important area in which the properties of very thin films play a significant role is adhesion of aluminum to aluminum in the aircraft industry. For bonds of the required strength to be formed, the aluminum surfaces must be pretreated by anodizing them, which should create not only the right chemical conditions for adhesion but the right structure in the resultant oxide film as well. Films formed by anodizing aluminum in phosphoric acid under a.c. conditions have been studied by TREVERTON et al. [186] using SSIMS and XPS together. Knowledge of both the chemistry of the outer monolayers in terms of aluminum oxides and hydroxides, and of the nature and degree of any contamination introduced by the anodizing process, was important. Figure 54 from their paper compares the positive and negative SSIMS spectra after anodizing for 3 and 5 s at 20 V and 50 Hz. Both sets of spectra show that the longer anodizing treatment caused significant reduction in hydrocarbon contamination (peaks at 39, 41, 43, 45 amu in the positive spectrum, and 25, 41, 45 amu in the negative), and marked increases in oxide- and hydroxide-associated cluster ions at 44, 61, 70, and 87 amu in the positive spectrum and 43 and 59 amu in the negative. At higher masses, peaks allocated to $Al_2O_4H^-$, $Al_2O_3^-$, and $Al_3O_5^-$ were also identified. Residual PO_2^- and PO_3^-

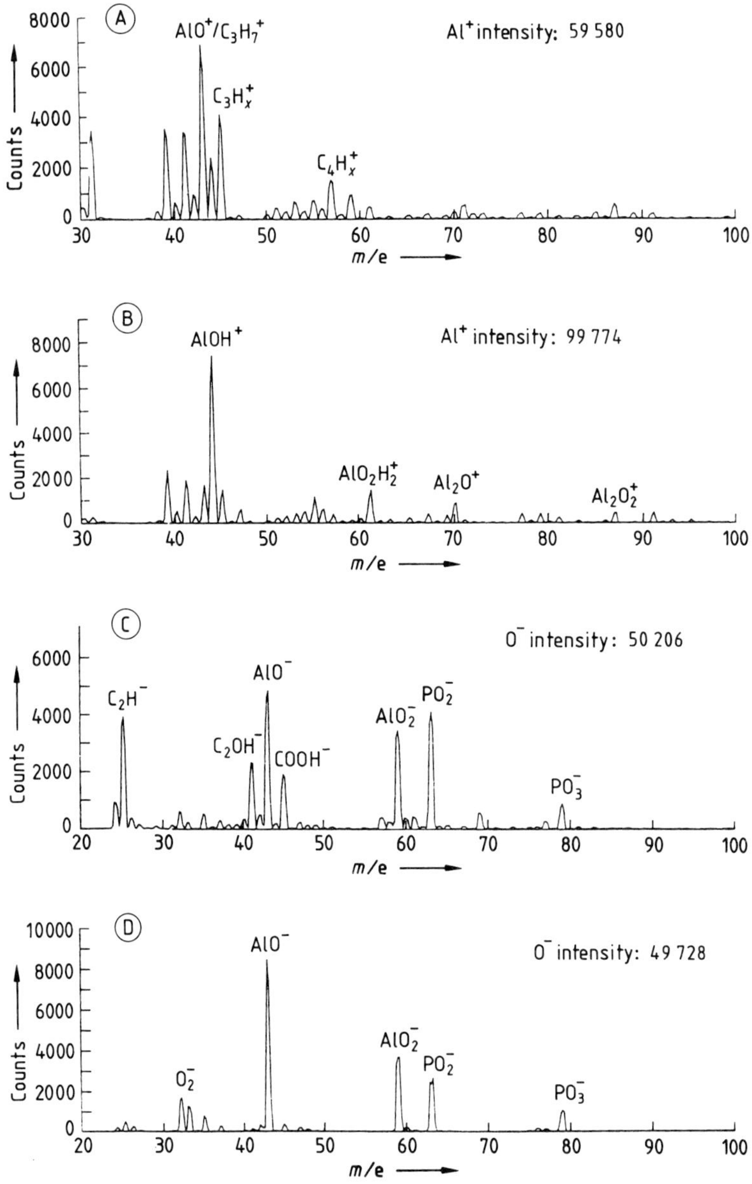

Figure 54. SSIMS spectra from oxide films formed on aluminum after anodization treatment in phosphoric acid for 3 or 5 s [186]
A) Positive SSIMS spectrum, 3 *s*; B) Positive SSIMS spectrum, 5 *s*; C) Negative SSIMS spectrum, 3 *s*; D) Negative SSIMS spectrum, 5 *s*

from the anodizing bath were reduced at the longer anodizing time. The low level of contaminant is important since the adhesive coating can then wet the surface more effectively, leading to improved adhesion. In addition, the change in aluminum chemistry from AlO^+ (mass 43) to $AlOH^+$ (mass 44) on going from 3- to 5-s anodizing was significant, and not an effect that could have been established by XPS.

4.5.2. Interfaces

The protection of steel surfaces by paint depends significantly on the chemistry of the paint–metal interface. The system has many variables since the metal surface is normally pretreated in a variety of ways, including galvanizing and phosphating. Indeed, several interfaces of importance probably exist, and corrosion protection may be a function of conditions at all of them. To study the effects of the variables without having to wait for natural weathering processes to cause corrosion, normal procedure is to apply an accelerated cyclic corrosion test (CCT) that involves scribing through the paint to the bare metal followed by salt spray, drying, and exposure to high humidity.

SSIMS has been applied by van Ooij et al. [187] to study of the interfaces in a paint–metal system; both quadrupole and TOF SSIMS were employed, as well as XPS. Figure 55 from their paper shows the positive SSIMS spectra from the metal (Fig. 55 A) and paint (Fig. 55 B) sides of an interface exposed by soaking samples in *N*-methylpyrrolidone, which removes the paint by swelling but does not attack it chemically. The metal surface had been phosphate treated before paint application. Both spectra are identified in terms of organic materials; in other words the failure was in an organic layer close to, but not at, the metal surface. This identification is supported by the observations that the mass peak at 52 amu, characteristic of chromate, was negligible and that in the negative spectra the intensity of PO_x^- ions was also low. Note that epoxy-related peaks at 59 and 135 amu were observed on both sides, as well as aromatic fragments at 77, 91, 115, 128, 152, and 178 amu, the latter being more intense on the metal side. When the spectra were compared with those from an unphosphated and painted metal surface that had suffered delamination, the characteristic epoxy peaks were not seen on the metal side, indicating that

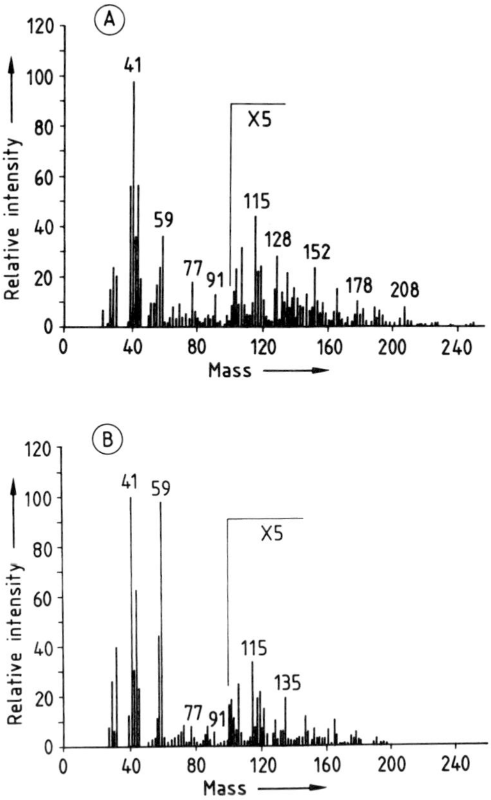

Figure 55. Positive SSIMS spectra from each side of the metal–paint interface on a hot-dip galvanized steel surface after accelerated cyclic corrosion testing [187]
A) Metal; B) Paint

in the absence of the phosphate layer the adhesion of the epoxyurethane to the substrate was lost.

4.5.3. Polymers

Treatment of polymer surfaces to improve their properties with respect to wetting or water repulsion and to adhesion, is by now a standard procedure. The treatment is designed to change the chemistry of the outermost groups in the polymer chain, without affecting bulk polymer properties. Any study of the effects of treatment therefore requires a technique that is specific mostly to the outer atom layers, which is why SSIMS has been used extensively in this area.

One of the favored forms of surface treatment is that of plasma etching. Depending on the conditions of the plasma discharge (i.e., nature of discharge gas, gas pressure, rate of gas flow,

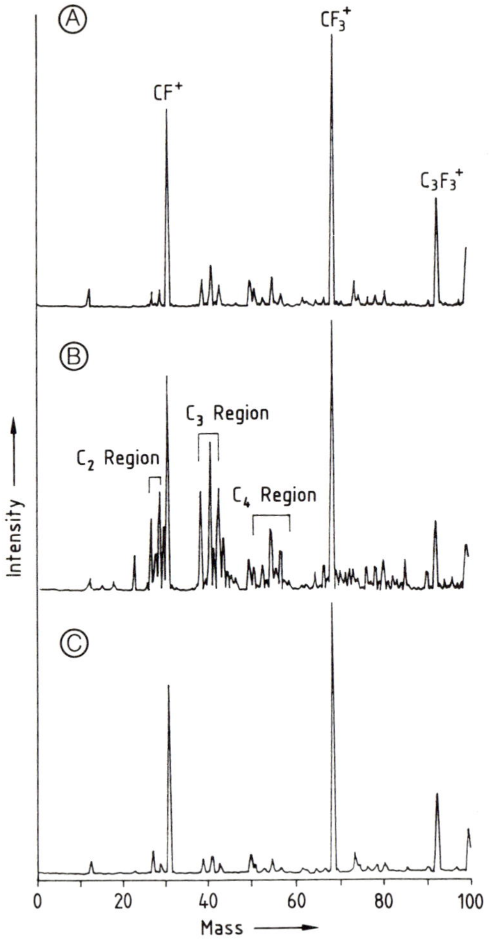

Figure 56. Positive SSIMS spectra from PTFE [188]
A) Untreated; B) 0.5-min oxygen plasma treatment; C) 15-min oxygen plasma treatment

discharge power and frequency, and substrate polarity), plasma etching is able to alter the chemical characteristics of the surface over a wide range. By its nature a plasma discharge is also both highly controllable and reproducible. The effects of plasma treatment on the surface of polytetrafluoroethylene (PTFE) have been studied by MORRA et al. [188] using combined SSIMS and XPS. They found, as evidenced in Figure 56, that not only did the plasma change the surface chemistry, but the surface emission fragments also changed the plasma discharge conditions temporarily. Thus the untreated (Fig. 56 A) and 15-min oxygen plasma treated (Fig. 56 C) positive-ion spectra are virtually identical, whereas the 0.5-min oxygen plasma treated spectrum (Fig. 56 B) is quite different. The initial breakup of the PTFE surface produced hydrocarbon fragments which reacted with the oxygen to produce water vapor, thus changing the local nature of the discharge gas. The important deduction to be made is that one of the vital parameters in plasma treatment is the length of time in contact with the plasma; in this case, too long a contact time resulted in no surface changes.

Other SSIMS studies of polymer surfaces have included perfluorinated polyether [189], low-density polyethylene [190], poly(ethylene terephthalate) [191], and the oxidation of polyetheretherketone [192].

4.5.4. Surface Reactions

Because of the inherently destructive nature of ion bombardment, the use of SSIMS alone in study of the reactions of surfaces with gases and vapors must be viewed with caution, but in combination with other surface techniques it can provide valuable additional information. The parallel techniques are most often XPS, TDS, and LEED, and the complementary information required from SSIMS normally refers to the nature of molecules on surfaces and with which, if any, other atoms they are combined.

A typical SSIMS spectrum of an organic molecule adsorbed on a surface is that of thiophene on ruthenium at 95 K, shown in Figure 57, from the study of COCCO and TATARCHUK [193]. The exposure was only 0.5 Langmuir (i.e., 5×10^{-7} torr · sec $\triangleq$ 37 Pa · s) and the principal positive ion peaks are those from ruthenium, consisting of a series of seven isotopic peaks around 102 amu. However, ruthenium–thiophene complex fragments are found at ca. 186 and 160 amu, each showing the same complicated isotopic pattern, indicating that even at 95 K interaction between the metal and the thiophene had occurred. In addition, thiophene and protonated thiophene peaks are observed at 84 and 85 amu, respectively, with the implication that no dissociation of the thiophene had taken place. The smaller masses are those of hydrocarbon fragments of different chain lengths.

SSIMS has also been used in the study of the adsorption of propene on ruthenium [194], the decomposition of ammonia on silicon [195], and the decomposition of methanethiol on nickel [196].

Recently SSIMS has been applied in the TOF–SSIMS imaging mode to the study of very

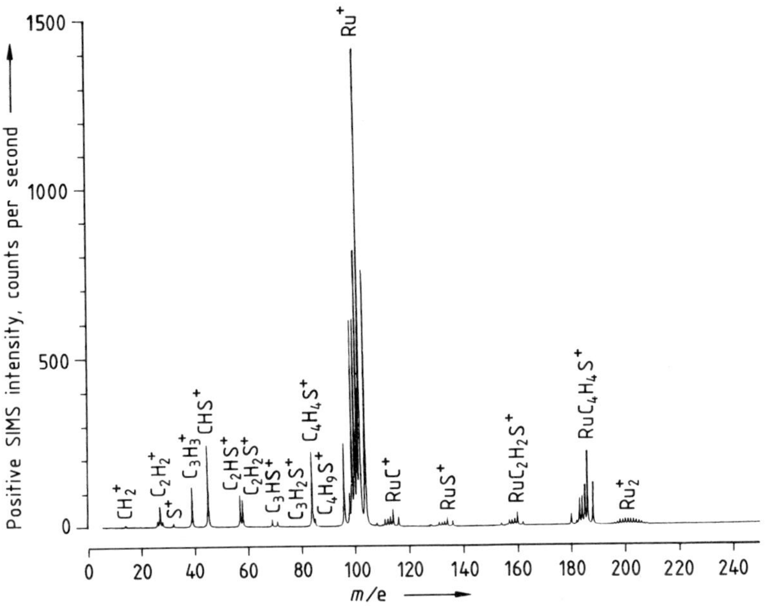

Figure 57. Positive SSIMS spectrum from Ru (0001) after exposure to 0.5 L thiophene at 95 K [193]

thin layers of organic materials [197], polymeric insulating materials [198], and carbon fiber and composite fracture surfaces [199]. In these studies a spatial resolution of ca. 200 nm in mass-resolved images was achieved.

5. Ion Scattering Spectroscopies (ISS and RBS)

Analysis of surfaces by ion scattering has a long history since several techniques are included according to the energy of the primary ion. They have been classified by Armour [200] as *low-energy ion scattering* (LEIS) for the range 100 eV to 10 keV, *medium-energy ion scattering* (MEIS) for the range 100 to 200 keV, and *high-energy ion scattering* (HEIS) for the range 1 to 2 MeV. However, LEIS is more often called *ion scattering spectroscopy* (ISS), whereas HEIS is always called *Rutherford backscattering spectroscopy* (RBS). ISS is invariably surface specific, while RBS is normally surface sensitive.

5.1. Ion Scattering Spectroscopy (ISS)

Although treated here as one of the major surface analytical techniques, ISS, like SSIMS, is used much less extensively than XPS or AES. Again, like SSIMS, it is often used in conjunction with other techniques and is valuable as a complement to the others for its two advantages of high surface specificity and ease of quantification.

5.1.1. Principles

A beam of inert gas ions, usually He^+ or Ne^+, of typical energy 0.5–4 keV and current 10^{-8}–10^{-9} A, is directed at a surface, and the energies of the ions of the same mass and charge that are scattered from the surface are analyzed. Because inert gas ions have a very strong electron affinity, the probability is high that on collision with a surface the majority will be neutralized. After two collisions the number of original ions left will be negligible. Thus analysis of scattered ion energies will refer mostly to those ions that have suffered only one collision (i.e., from the outermost layer of atoms). Hence ISS is very surface specific. In addition, with primary ions of low mass and energy, the ion–atom interaction time is much shorter than the characteristic time of vibration of an atom in a solid, so that the collision can be approximated as that between the ion and a free atom. This means that binary collision (or "billiard-ball") theory can be applied, in which case the relationship between the energy E_1 after collision and the energy E_0 before collision, of the ion of mass M_1 with a surface atom of mass M_2, is

$$E_1/E_0 = \{1 + (M_2/M_1)\}^{-2}[\cos\theta \pm \{(M_2/M_1)^2 - \sin^2\theta\}^{1/2}]^2 \quad (22)$$

where θ is the scattering angle (i.e., angle between incident and scattered beams).

If θ is chosen to be 90°, then Equation (22) reduces to the simple relationship

$$E_1/E_0 = (M_2 - M_1)/(M_2 + M_1) \quad (23)$$

Equation (22) shows that if the energy and mass of the incident ion, and the scattering angle, are all fixed, E_1 depends only on M_2, the mass of the surface atom. Thus the energy spectrum measured in ISS can be converted directly into a mass spectrum, providing an elemental compositional analysis of the surface. Because only the mass of the surface atom is involved, no chemical state information is available.

Note from Equations (22) and (23) that as M_2 increases, the ratio E_1/E_0 tends to unity, which means that for a light ion such as He^+, heavy elements are difficult to resolve in the energy spectrum. ISS is therefore most suitable for the analysis of light elements on surfaces. The mass resolution can be improved by increasing the mass of the ion (e.g., by going to Ar^+ or Xe^+), but then problems of sputtering intrude. Another way of improving the resolution is by using scattering angles >90°, and in most ISS geometries the angle is ≈135°.

5.1.2. Instrumentation

As with other surface-specific techniques, the vacuum requirements for ISS are stringent, vacua better than 10^{-8} Pa being needed. With the single monolayer specificity mentioned above, good UHV conditions are vital.

The basic instrumental requirements are an ion source, a sample target, and an energy analyzer. The source is normally a commercial ion gun, for example that shown in Figure 39, operated at a few kiloelectronvolts and with beam currents of He^+ or Ne^+ of 1–100 nA. For scattered ion energy analysis, a standard CHA or CMA with reversed polarity can be used, or alternatively a special analyzer may be constructed. The basic arrangement is shown in Figure 58 from Martin and Netterfield [201]; also shown is an auxiliary electron source used for charge neutralization by flooding the surface with low-energy electrons. In particularly sophisticated experiments the incident ion beam is mass-filtered before striking the target, and both target and analyzer are rotatable around the point of impact to allow angular resolution.

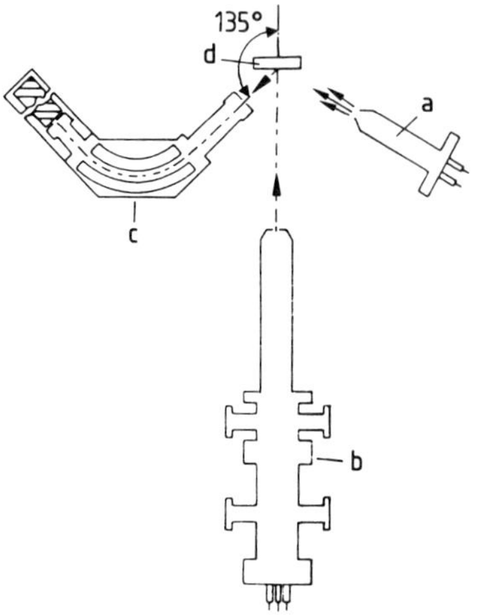

Figure 58. Experimental arrangement of ISS system used by Martin and Netterfield [201]
a) Electron flood gun; b) Ion gun; c) ISS analyzer; d) Sample

Similar to developments in SSIMS, significant improvements in ISS have been obtained by using *time-of-flight techniques*. The additional sensitivity available with a TOF system allows much lower incident ion current densities to be employed, thereby reducing the effects of beam damage to the surface. Such damage is of course negligible when using He^+, but if heavier ions such as Ar^+ are required for better mass resolution, the current density must be reduced. Another major advantage of a TOF energy analyzer is that it enables energy analysis not only of ions but of neutrals ejected from the surface, since the flight time is independent of charge state. This has led to the introduction of a structural technique called *noble-gas impact collision ion scattering spectroscopy* with neutral detection (NICISS), by Niehus [202].

5.1.3. Spectral and Structural Information

Spectra in ISS are recorded in the form of scattered intensity versus the ratio E_1/E_0, so that the range of the latter will be from 0 to 1. Some

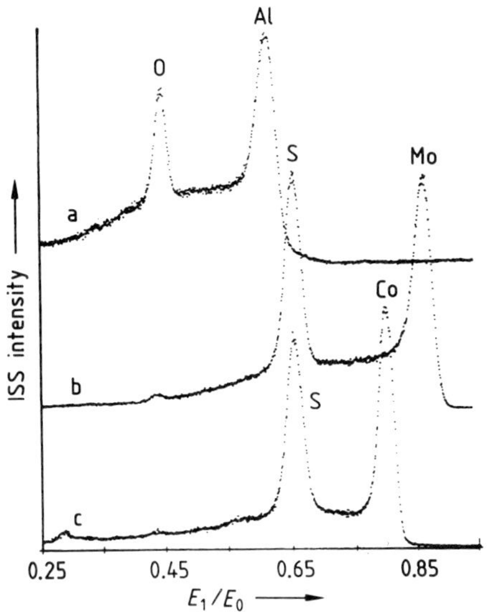

Figure 59. Examples of ISS spectra for a set of standard materials used in the study of cobalt–molybdenum sulfide catalysts [203]
a) Al_2O_3; b) MoS_2; c) Co_9S_8

typical examples are shown in Figure 59, from BEUKEN and BERTRAND [203], for a set of Al_2O_3, MoS_2, and Co_9S_8 standards used by them in their study of cobalt–molybdenum sulfide catalysts. The conditions—He^+ at 2 keV and 15 nA, and a scattering angle of 138°—are also typical. Acquisition time per spectrum was about 5 min, and an electron flood gun was used for charge neutralization.

As seen in Figure 59, the E_1/E_0 ratio can be converted directly to mass and therefore to elemental identification. Notice also the disadvantage mentioned above, that as $E_1/E_0 \rightarrow 1$, the mass separation becomes poor; i.e., if cobalt ($M = 59$) and molybdenum ($M = 96$) had appeared in the same spectrum they would have only just been resolved under the conditions. However, the resolution for light elements is good.

In addition to elemental compositional information, ISS can also be used to provide information about surface structure, and in this application it is complementary to LEED. The reason for the structural possibility is indicated schematically in Figure 60, which shows the ion trajectories calculated by VAN DER VEEN [204] for a parallel beam of ions hitting an atom at the surface. Most ions are scattered in the forward direction, with only a few making an interaction

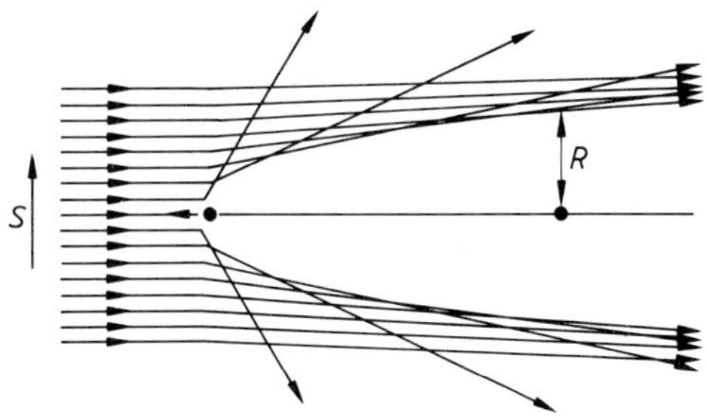

Figure 60. Ion trajectories for a parallel beam of light ions hitting an atom [204]

strong enough to be scattered back toward the detector. Thus a "shadow cone" arises behind the atom, R being the radius of the cone at the position of the next atom in the row. Since R is of the same order as interatomic spacing, atoms in a row lying immediately behind the scattering atom are "invisible" to the primary beam. S is the so-called impact parameter, which depends on ion energy and distance of nearest approach between ion and atom.

If the experimental geometry is varied, by varying either the angle of primary ion incidence or the target azimuthal angle, while the scattering angle is kept constant, the shadow cone is effectively swept through the rows of atoms behind a surface atom. The scattered intensity will then vary periodically according to whether the scattering is mostly from surface atoms or whether at certain angles contributions to the scattering come from atoms further along a row. If the near-surface crystallinity is good, then the variations in angular intensity can be used to derive or confirm a surface structure. The technique is particularly valuable when using scattering angles close to 180°, since the problem of multiple scattering (i.e., successive scattering off two or more atoms) is minimized, and the dependence of the scattered intensity on the angle of incidence is a direct representation of the shape of the shadow cone. This variation of ISS is called *impact collision ion scattering spectroscopy* (ICISS). Figure 61 from NIEHUS and COMSA [205] compares the ICISS measurements from a Cu (110) crystal surface before (Fig. 61 A) and after adsorption of oxygen (Fig. 61 B). Various models had been proposed for the structure of the oxygen-reacted surface, but only the "missing row" model shown on the lower right can explain the observed angular dependence.

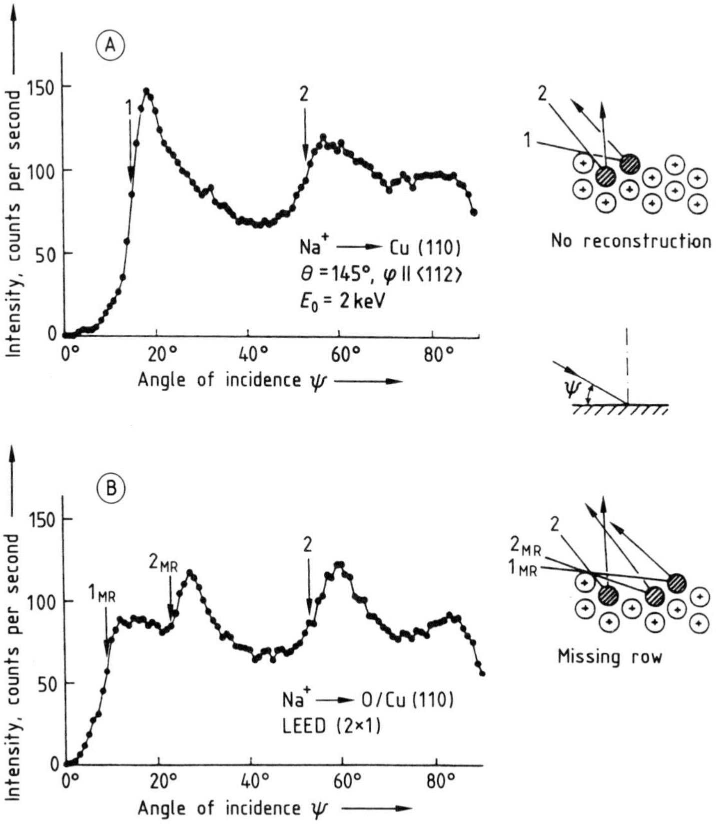

Figure 61. Impact collision ISS of Na^+ scattering from a Cu (110) surface before (A) and after (B) surface reconstruction due to oxygen adsorption, as a function of angle of incidence [205]

5.1.4. Quantification and Depth Information

If the scattering process is considered as a simple binary collision, the intensity I_A^+ of primary ions scattered from atoms of type A in the surface is given by

$$I_A^+ = I_O^+ N_A \sigma_A (1 - P) S \cdot \Delta\Omega \cdot T \qquad (24)$$

where I_O^+ is the intensity of the primary beam; N_A the density of scattering atoms A; σ_A the scattering cross section of atoms A; P the neutralization probability of a primary ion; S the shadowing factor; $\Delta\Omega$ the acceptance angle of the analyzer; and T the transmission function of the analyzer.

In an analysis, the primary ion current, the experimental geometry, and the transmission function are kept constant, so that to relate I_A^+ to N_A requires knowledge of σ_A and P.

Scattering Cross Section. To evaluate σ_A, information is needed about the physical nature of the interaction between the ion and the atom. In the ion energy range used in ISS the interaction can be considered in terms of purely repulsive potential (i.e., a Coulomb repulsion with inclusion of some screening, since some electronic screening occurs during the interaction). The general form of the interaction potential is thus

$$V(r) = \frac{Z_1 Z_2 e^2}{r} \varphi(r/a) \qquad (25)$$

where Z_1, Z_2 are the atomic numbers of ion and scattering atom; r is the ion–atom distance; and a the characteristic screening length.

The most widely used screening function is that due to Molière [206] using Thomas–Fermi statistics, which is approximated by

$$\varphi(r/a) = \varphi(x) = 0.35\exp(-0.3x) + 0.55\exp(-1.2x) + 0.1\exp(-6x) \qquad (26)$$

with

$$a = 0.4685[Z_1^{2/3} + Z_2^{2/3}]^{-1/2} \quad (27)$$

Thus $V(r)$ can be computed and hence the so-called scattering integral

$$\psi = \pi - 2p \int_{r_{min}}^{\infty} dr/r^2 \{1 - (p^2/r^2) - V(r)/E\}^{1/2} \quad (28)$$

where $\psi = \theta + \sin^{-1}\{(\sin\theta)/A\}$, θ being the scattering angle, and $A = M_2/M_1$; p is the impact parameter; and $E = \{M_2/(M_1 + M_2)\}E_0$.

From Equation (28) can be calculated the differential scattering cross section for scattering into unit solid angle, which is

$$d\sigma(\theta)/d\Omega = (p/\sin\theta)\,(dp/d\theta) \quad (29)$$

and hence the cross section.

Neutralization Probability. The neutralization probability P in Equation (24) is not quite so straightforward to compute. As stated earlier, neutralization is a highly probable event, because the approaching ion represents a deep potential well compared to the work function of a solid. Based on the theory of neutralization by Auger electron ejection developed by HAGSTRUM [207], the probability of an ion escaping neutralization has the form

$$(1 - P) = \exp\left\{\left(\frac{1}{V_{IT}} + \frac{1}{V_{ST}}\right)V_c\right\} \quad (30)$$

where V_{IT}, V_{ST} are the components of velocities of initial and scattered ions, respectively, normal to the surface; and V_c is the characteristic velocity.

V_c is dependent on the electron transition rate, which in turn depends, among other things, on the chemical state of the surface. It is approximated by

$$V_c = \int_{r_0}^{\infty} R(r)\,dr \quad (31)$$

where $R(r)$ is the transition rate and r_0 the distance of closest approach of ion to surface.

V_c is thus a complex quantity, and difficult to estimate, which makes $(1 - P)$ difficult to calculate. However, most estimates indicate that $(1 - P) \approx 10^{-3}$.

Quantification. As with the other surface analytical techniques, in principle one can therefore quantify by calculation, and indeed in a few cases, such as the quantitative study of potassium segregation to the V_6O_{13} (001) surface carried out by DE GRYSE et al. [208], such an approach has been successful. In general, however, the uncertainties in some of the quantities in Equation (24) make such a calculation too difficult, and quantification is based on standards. Thus, BRONGERSMA et al. [209], in their study of surface enrichment of copper in a Cu–Ni alloy, calibrated the ISS intensities against pure copper and nickel targets, whereas NELSON [210] used gold and silver elemental standards in ISS analyses of a range of Au–Ag alloys. He found a linear relationship between the surface concentrations of the alloys after prolonged sputtering to equilibrium composition, and the bulk concentrations measured by an electron microprobe, as shown in Figure 62. BEUKEN and BERTRAND [203] reduced each of the spectra from the standard materials in Figure 59 to components consisting of Gaussian peaks and inelastic tails, and then added the components in different proportions to synthesize the observed spectra from their sulfided Co–Mo catalysts. The result of such synthesis when applied to the ISS spectrum from the catalyst after heating to 800 °C is given in Figure 63. From such data processing the relative concentrations of the metallic elements on the catalyst surface after various heat treatments could be derived very accurately.

Depth Profiling. Unlike SSIMS, ISS is not inherently destructive under normal conditions of operation (i.e., He^+ at a few kiloelectronvolts and 1–10 nA) since the sputtering rate is negli-

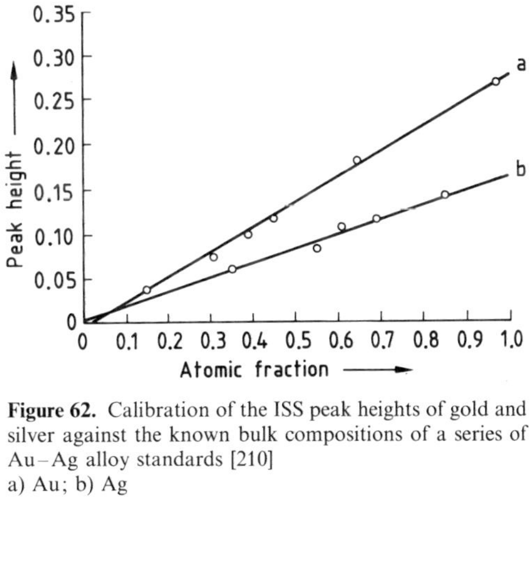

Figure 62. Calibration of the ISS peak heights of gold and silver against the known bulk compositions of a series of Au–Ag alloy standards [210]
a) Au; b) Ag

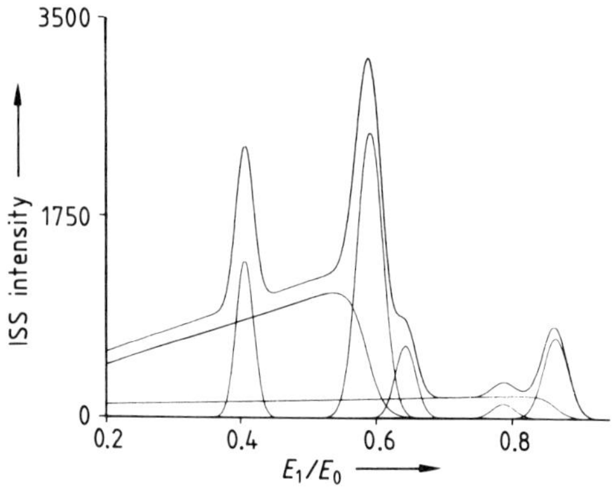

Figure 63. Curve synthesis method applied to the ISS spectra from the catalyst CoMoS–γ-Al_2O_3 after heat treatment at 800 °C [203]

gible. If information in depth is required, then surface erosion must be introduced, as for XPS and AES. One obvious way of performing depth profiling is to use the same primary ion for both ISS and sputtering, but at quite different power densities for the two operations. Thus Rossi et al. [211] used Ne^+ at 45 nA and 3 keV (i.e., 1.35×10^{-4} W) for ISS, and at 300 nA and 5 keV (i.e., 1.5×10^{-3} W) for sputtering, in their study of passive film formation on Fe53–Ni. Figure 64 shows the Fe–Ni regions of ISS spectra obtained by them after successively longer sputtering times. Inelastic backgrounds have been subtracted from the spectra, and the peaks have been resolved into iron and nickel contributions. After 5 min sputtering, the iron contribution is greatly enhanced, indicating an Fe-rich layer within the passive film, about 1 nm below the surface. On prolonged sputtering the bulk composition is reached.

Another method of depth profiling with ISS, used by Kang et al. [212], is to bombard the surface simultaneously with He^+ and Ar^+ ions. ISS is performed using He^+, while the surface is eroded with Ar^+. Because the gases were mixed together in the same ion source, a duoplasmatron, the energies of measurement and sputtering were the same, in the range 1.5–2.0 keV. According to the mixing ratio used, the relative currents of He^+ and Ar^+ could be varied.

The much smaller sampling depth of ISS, of the order of one atomic layer, compared to those of XPS and AES, implies that the depth resolution in ISS during profiling will be better. This improved depth resolution has been demonstrated by Haeussler [213] in comparisons of ISS, XPS, and AES profiles of the same materials.

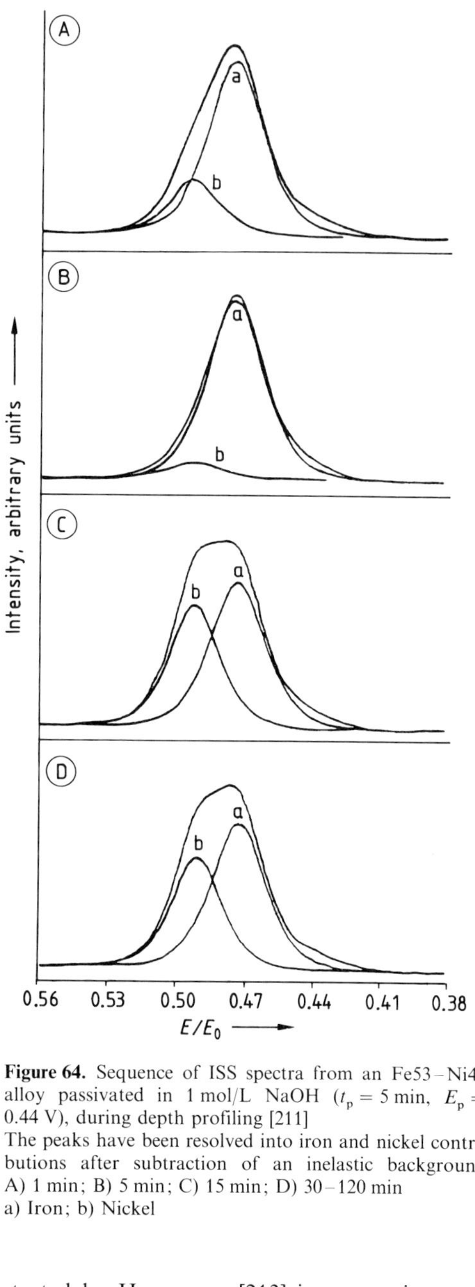

Figure 64. Sequence of ISS spectra from an Fe53–Ni47 alloy passivated in 1 mol/L NaOH (t_p = 5 min, E_p = 0.44 V), during depth profiling [211]
The peaks have been resolved into iron and nickel contributions after subtraction of an inelastic background
A) 1 min; B) 5 min; C) 15 min; D) 30–120 min
a) Iron; b) Nickel

5.1.5. Applications

Because of its high surface specificity relative to XPS and AES, most published applications of ISS have been concerned with basic experiments on clean surfaces, usually of single crystals. The negligible sputtering rate when using low-energy He^+ also means that gas–solid interactions can be studied without worry about alteration of the surface during measurement.

Typical of recent basic studies has been that of SHEN et al. [214] of oxygen adsorption on Ni_3Al (001). The intermetallic compound Ni_3Al is known to have excellent oxidation resistance in air at high temperature, forming an adherent aluminum oxide film, so information about the initial stages of oxidation is of interest. The Ni_3Al (001) surface was cleaned by Ne^+ ion bombardment, and its stability as a function of temperature checked by ISS, which showed that up to 900 °C the Ni–Al ratio remained unchanged. Figure 65 shows a selection of ISS spectra from the alloy surface as a function of oxygen exposure at 700 °C, using 2-keV He^+ at current densities between 2×10^{-8} and 8×10^{-8} A/cm^{-2}. The oxygen peak increases, as expected, and both the aluminum and the nickel peaks decrease. However, the nickel peak intensity decreases at a much faster rate than that of aluminum, and the peak eventually disappears. Plots of normalized nickel and oxygen peak intensities indicated that the nickel peak disappeared when one monolayer of Al_2O_3 was formed, confirming that the protective mechanism operates in the earliest stages of interaction. The authors also used ISS structurally to establish that the growing oxide had no order and was not in coherence with the underlying crystal.

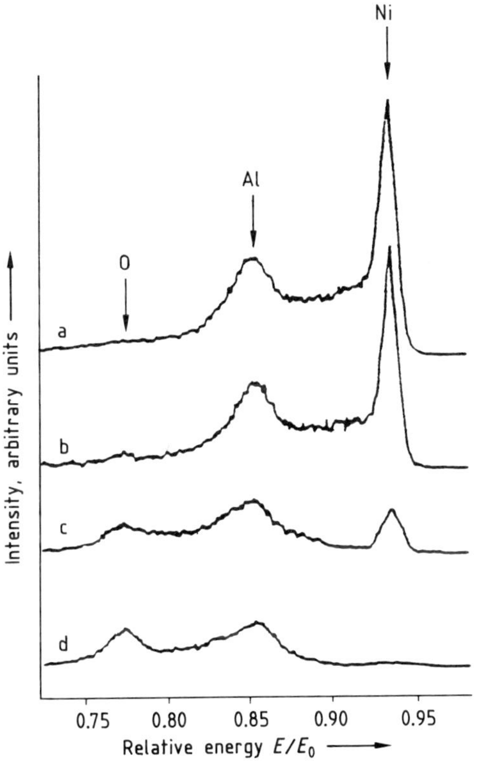

Figure 65. ISS spectra from Ni_3Al (001) after exposure to oxygen at 2×10^{-5} Pa and 700 °C [214]
a) Clean; b) 1-s Exposure; c) 15-s Exposure; d) 100-s Exposure

A more technological application to which ISS is particularly suited is that of the surface composition of dispenser cathodes. These cathodes are believed to achieve optimum operation when an approximately monolayer coverage of barium and oxygen exists at the surface. When using XPS and AES, establishing that any barium signal comes only from the outer atomic layer of the surface is not easy, but the single-layer sensitivity of ISS should be able to provide that information. A problem in applying ISS to the cathode system is that the active barium–oxygen layer is supported on a tungsten substrate, and that both barium and tungsten are heavy elements whose resolution using He^+ would be difficult. On the other hand, oxygen must also be analyzed. The problem was solved by MARRIAN et al. [215] by performing ISS with a mixture of He^+ and Ne^+, so that oxygen would be analyzable by the He^+, while the use of Ne^+ would help separate barium and tungsten. A spectrum from their paper is given in Figure 66, showing that the analyses using the mixture did in fact provide the analytical information needed. The authors were able to establish the existence of fractional monolayer coverage of barium and oxygen on the cathode surface.

Other recent applications of ISS have included the study of thin ZrO_2 films [201], a comparison of various techniques for cleaning InP substrates before epitaxial growth [216], analysis of conversion coatings on aluminum alloys [217], and study of alkali-induced sintering of an iron oxide monolayer on platinum [218].

5.2. Rutherford Backscattering Spectroscopy (RBS)

RBS has been in use since the 1960s for bulk materials analysis, and increasingly since the 1970s for thin-film analysis, particularly in the semiconductor field. The overall number and

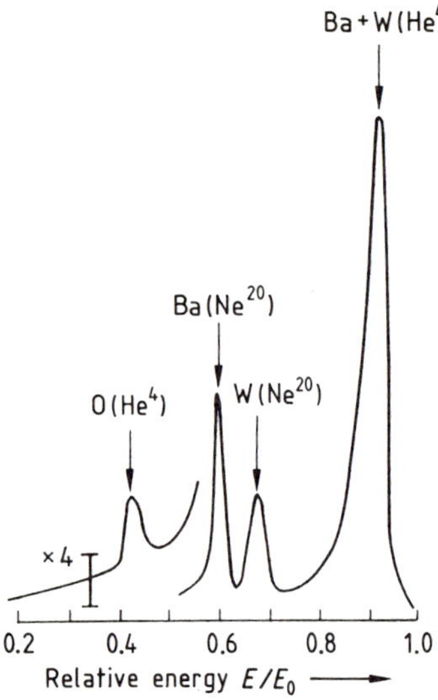

Figure 66. ISS spectra from a tungsten surface containing oxygen and barium, using a mixture of He^+ and Ne^+ to analyze for oxygen as well as separate barium and tungsten [215]
×4 = amplification factor

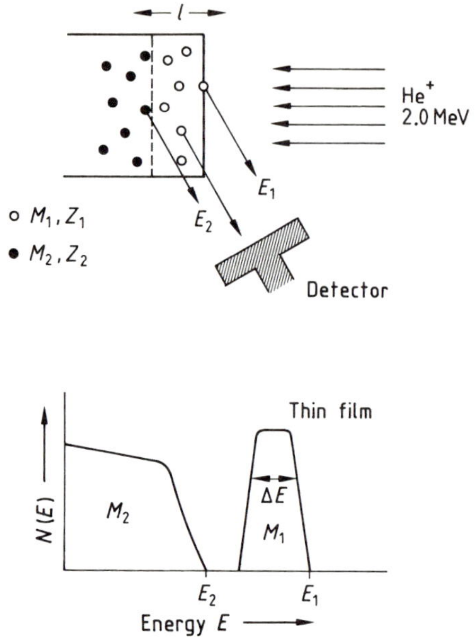

Figure 67. Illustration [219] of the principle of Rutherford backscattering spectroscopy (RBS) from a solid consisting of a thin film of thickness l composed of atoms of mass M_1 and atomic number Z_1 situated on a substrate of atoms of mass M_2 and atomic number Z_2

range of applications are enormous, but those dealing specifically with surface problems are less numerous. Nevertheless RBS ranks along with SSIMS and ISS as one of the major surface analytical techniques.

5.2.1. Principles

As in ISS, elemental compositional analysis is achieved by measuring the energy losses suffered by light primary ions in single binary collisions. Although in ISS the primary species is usually He^+, in RBS either H^+ or He^+ is used exclusively, and at very much higher energies, typically 1–2 MeV. Consequently, due to the penetration of the primary ion, both outer atomic layers and subsurface layers of a material are accessible; the surface sensitivity comes when the material in the outer few monolayers consists of heavy elements.

The energy of the backscattered ion, and hence elemental analysis, is governed by the same relationship (Eq. 22) already described for ISS. However, in RBS the ion penetrates the target and may be backscattered by the target atoms at any point along its path. Because of its very high energy, the ion can lose energy by collisional interactions with electrons and plasmons along either or both incoming and outgoing paths, which are straight lines. Thus, instead of the single peak observed by scattering from a particular element in the first atom layer in an ISS spectrum, the additional losses in energy cause a much broadened peak to be observed in an RBS spectrum. In RBS the energy spectrum is in fact a superposition of mass (i.e., elemental) analysis and depth analysis.

The principles have been illustrated elegantly by VAN OOSTROM [219], as shown in Figure 67. There a thin film consisting of atoms of mass M_1, atomic number Z_1, and thickness l is situated on a substrate of atoms of mass M_2 and atomic number Z_2. The scattered ion energies from M_1 and M_2 are E_1 and E_2, respectively. If $M_1 > M_2$, then the energy $E_1 > E_2$, in which case the peak corresponding to M_1 is well separated from M_2, and the peak width ΔE is related directly to l. Because the atoms M_1 are in the surface, the threshold E_1 will be given by Equation (22) as in ISS, but since the atoms M_2 are below the thin film the threshold E_2 may be lower than that given by Equation (22) if energy losses occur in the film. The energy spectrum produced by scattering from atoms M_2 will start at the threshold E_2 and then increase into a very broad band since M_2 forms the substrate of effectively infinite depth. If $M_1 < M_2$, the peak due to M_1 merges with the broad band from M_2, and may or may not appear as a much smaller feature on top of the band.

The above makes clear that for surface analysis by RBS, conditions must be such that $M_1 \gg M_2$ for the peak due to M_1 to be completely resolved. Under these circumstances, the sensitivity of RBS is comparable to AES and XPS (i.e., of the order of 10^{-5} atomic layers). As in ISS, no chemical information is available.

5.2.2. Instrumentation

He^+ or H^+ ions with energies in the megaelectronvolt region are produced in particle accelerators. These may take the form of a tandem electrostatic Van de Graaff accelerator; a cyclotron, which uses electromagnetic confinement to a circular orbit; or a linear accelerator, in which the ion is accelerated linearly across gaps between drift tubes by electromagnetic fields. The reader is referred to [220] for greater detail about accelerator technology and about accelerator-based methods of analysis. Accelerators normally produce ions in several charge states, so that the required ion is chosen by magnetic selection and often by velocity selection as well. The purified ion beam passes immediately into a UHV environment via differentially pumped apertures, since the vacuum environment in most accelerators is not that of UHV. For general analysis, in which channeling of the beam along particular crystal directions is not present, ions strike the specimen at normal incidence. When structural information is sought, however, in the same manner as in ISS, the specimen may be rotated about the point of ion impact to vary the angle of incidence about channeling directions.

Detection of backscattered ions is normally performed by a silicon surface barrier detector at high scattering angles for increased surface sensitivity, as indicated in Figure 67. The detector position is often variable about the specimen position to allow various scattering angles to be chosen.

The ion beams used in RBS can be focused to small spots if required, although conventional analysis would normally be performed over an area of $\approx 1\ mm^2$. Spot sizes of 5–10 μm can be obtained by suitable lens systems, but scanning at such high energies is too difficult. Selected area analyses using small spot beams is carried out by specimen manipulation.

5.2.3. Spectral and Structural Information

An RBS spectrum, like that of an ISS spectrum, consists of the energy distribution of backscattered primary ions. Unlike ISS, however, the energies of the backscattered ions are governed not only by the mass of the scattering atom but also by the depth of that atom.

An example of the nature of the compositional information obtainable from an RBS spectrum is based on Figure 68, from the work of Kuiper et al. [221]. They grew a 90-nm-thick film of SiO_2 on a silicon substrate, then deposited a 110-nm-thick film of titanium on top of the SiO_2, and heated the sandwich to various temperatures. In Figure 68 the dotted line is the spectrum recorded before heating; the solid line, after heating to 575 °C for 1 h; and the dashed line, after heating to 800 °C for 1 h. The vertical arrows are positioned at the backscattered energies expected for atoms of the indicated elements if they were in the surface layer.

For the titanium film, some titanium atoms always exist in the outer layer, but on heating to successively higher temperatures, titanium is found at progressively greater depths (i.e., it is redistributing). The width of the peak before heating corresponds exactly to the thickness of the deposited film. In the case of silicon, at no

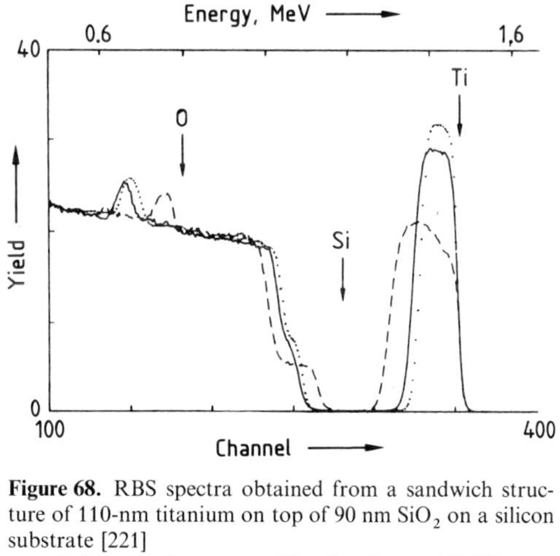

Figure 68. RBS spectra obtained from a sandwich structure of 110-nm titanium on top of 90 nm SiO_2 on a silicon substrate [221]
... = Before heating; — = After heating to 575 °C; ---- = After heating to 800 °C

stage do any silicon atoms appear in the outer layer, although after heating to 800 °C the silicon edge is near the surface. The double edge of the silicon signal is due to the fact that, after heating to 800 °C, silicon is in two chemical states, one as a silicide in combination with titanium, and the other in the underlying SiO_2, which has reduced in thickness to 14 nm. After 800 °C heating, the oxygen atoms have mostly moved nearer to the surface, with little change in their distribution after 575 °C treatment.

For relatively simple situations such as that of Figure 68, a spectrum simulation program can be used, in which the adjustable parameters are the stoichiometries of the likely compounds and their distribution with depth. Thus after 575 °C heating, the simulation gave the result that the surface consisted of titanium on top of a layer of $TiO_{0.23}$ overlaying 72 nm of SiO_2, with negligible silicide. After 800 °C, however, the fit was to a layer of Ti_2O_3 on top of the silicide Ti_5Si_3, covering the remaining 14 nm of SiO_2. The results agreed with depth profiling by AES performed at the same time.

RBS can be used to detect and measure the thickness of an amorphous layer on an otherwise crystalline substrate. The method is illustrated schematically in Figure 69, from WERNER [222]. If the incident ion beam is directed onto the crystal in an "open" or channeling direction (Fig. 69 A), it will penetrate more deeply into the crystal than if it is aligned along a nonchanneling or "random" direction. Thus the overall backscattered intensity in the channeling direction will be much lower than in the random direction, and in addition the small surface peak arising from collisions at the surface will be visible, as indicated in schematic spectra in Figure 69 B. Should the surface be covered with an amorphous layer (Fig. 69 C), no channeling is possible in the layer, so that more collisions will take place with atoms near the surface, and the intensity of the backscattered signal from the surface will increase (for the same incident beam direction), as shown schematically in Figure 69 D.

An application of this method of obtaining surface structural information has been demonstrated by BHATTACHARYA and PRONKO [223]. Implantation in compound semiconductors for production of the required electrical characteristics is usually accompanied by crystalline damage, and in particular a surface amorphous layer may be formed. Figure 70 from their paper shows the results of the RBS channeling measurements on GaAs implanted with silicon before and after annealing at 950 °C. The upper spectrum (a) is taken with the ion beam in a nonchanneling or "random" direction, and has the expected structureless shape. When the beam was aligned with a channeling direction, the spectrum before annealing (b) was also of high intensity and revealed a broad peak near the edge, while after annealing the intensity dropped substantially and a much narrower near-edge peak remained (c). The broad peak in the unannealed state showed that the surface had become amorphous to a thickness of ≈140 nm, while the spectrum after annealing resembled closely that

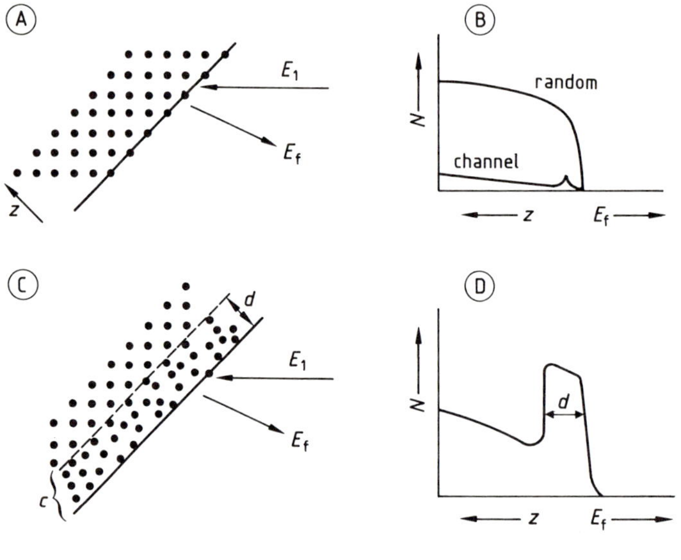

Figure 69. Illustration of RBS expected according to direction of incident beam [222]
A) Beam along a channeling (open) direction; B) Spectra expected from either nonchanneling (random) or channeling (open) directions; C) Beam incident on an amorphous layer of thickness d on surface; D) Spectrum expected from amorphous layer

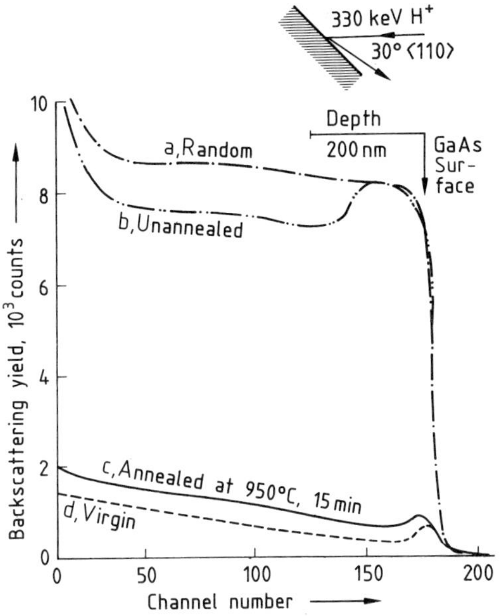

Figure 70. RBS spectra from GaAs implanted with silicon before and after annealing at 950 °C (120 keV Si^+–GaAs, dose: 5×10^{15} cm^{-2}) [223]
Uppermost spectrum is taken in a nonchanneling direction; the others are in channeling directions

of the crystalline unimplanted crystal (d). Further structural information was obtained by recording the backscattered intensity as a function of angle around the GaAs [100] and [110] crystal directions, from which it was possible to calculate that about 70 % of the implanted silicon occupied substitutional lattice sites.

5.2.4. Quantification

The intensity of the current of high-energy ions backscattered from surface atoms of type A is very similar to Equation (24) for ISS, except that of course the neutralization probability is not relevant. Equation (24) then becomes for RBS

$$I_A^+ = I_o^+ N_A \sigma_A S \cdot \Delta\Omega \cdot T \qquad (32)$$

where the symbols have the same meanings as before. Here σ can be derived from the Rutherford differential scattering cross section $d\sigma$ given by

$$d\sigma = \frac{Z_1 Z_2 e^2}{4E^2} \cdot \frac{\cos\theta - \{1 - [(M_1/M_2)\sin\theta]^2\}^{1/2}}{\sin^4\theta\{1 - [(M_1/M_2)\sin\theta]^2\}^{1/2}} \qquad (33)$$

θ being the scattering angle. For a particular primary ion and fixed experimental conditions, the differential cross section is thus a function only of the mass and atomic number of the scattering atom A, and can be calculated. Thus in principle, all the terms in Equation (32), except the required N_A, are known. In practice, the effective solid angle $\Delta\Omega$ of the analyzer is not well known, and using standards against which to calibrate the peak in the RBS spectrum arising from surface atoms (the so-called surface peak) is more common than attempting to derive calculated compositions. Such a peak can be seen clearly in the lower spectra of Figure 70.

Calibration is achieved by either implanting into or depositing on a light element such as silicon, a known amount of a much heavier element. Then if N_S is the known density of the standard element, one can write

$$N_A = (\sigma_S/\sigma_A) \cdot (I_A^+/I_S^+) N_S \qquad (34)$$

given the same experimental conditions of measurement on the standard and the unknown. Then, in terms of the usual relative sensitivity factors, Equation (34) can be written as

$$N_A = R(A:S) I_A^+ \qquad (35)$$

The accuracy depends on being able to measure the areas of the surface peaks in both standard and unknown with sufficient accuracy, which is often easy. Accuracies in RBS under favorable conditions can be better than 1 %.

5.2.5. Applications

Since RBS has operated as a major compositional and structural technique for bulk analysis for about 30 years, the number of published applications in that area is now enormous. Its applications in the field of specifically surface studies have not been so numerous, but they are increasing in number with the advent of improved vacuum technology. Many have been of the type already discussed in Section 5.2.3, in which the reaction or interdiffusion of one or more heavy metals on a semiconductor surface has been studied.

Typical of such studies is that of Nava et al. [224] on the interaction of an Ni–Pt alloy with crystalline silicon. The system is a particularly favorable one since both nickel and platinum are much heavier than silicon, and platinum is in turn much heavier than nickel. Alloy films about 145 nm thick were deposited on silicon by co-deposition of nickel and platinum in approximately equal amounts. Annealing of the film and substrate was then carried out at progressively increasing temperature. Figure 71 shows the RBS spectra before annealing (solid line), after 30 min at 400 °C (dashed line), and after 30 min at 700 °C (dotted line). As before, the vertical arrows indicate the energy positions of the backscattered edges to be expected if the indicated elements are in the surface layer.

As shown, the as-deposited film contained both platinum and nickel atoms in the surface layer, and covered the silicon completely, since no silicon atoms appeared at the surface layer energy position. The sharpness of the trailing edges of the platinum and nickel peaks, as well as that of the leading silicon edge, showed that the interface between film and substrate was itself sharp. The film composition, based on relative peak area under the platinum and nickel peaks, was $Ni_{55}Pt_{45}$. Annealing at 400 °C causes little movement of the platinum, but segregation of the nickel into the interface, coupled with movement of silicon in the opposite direction (i.e., also into the interface). The compound formed there is identified as NiSi from the silicon step height. At 700 °C platinum has also shifted considerably, along with further movement of nickel, until they are in fact completely intermixed. The silicon edge has now moved up to the surface position, indicating that it is mixed with both platinum and nickel. Careful quantification establishes that both PtSi and NiSi have been formed in approximately equal quantities. Confirmation of the formation of these compounds came from X-ray diffraction. Thus the alloy–silicon reaction occurred first by the formation of NiSi, with a Pt-rich layer left above it, followed at slightly higher temperature by the reaction of platinum with diffusing silicon, and finally at the highest temperatures by complete mixing of PtSi and NiSi.

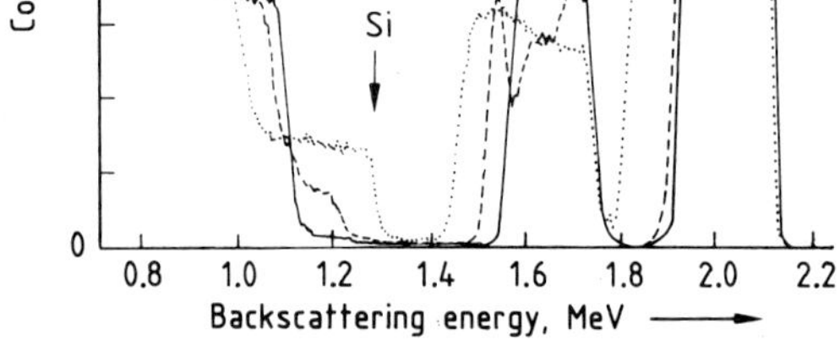

Figure 71. RBS spectra of a 145-nm-thick Ni–Pt alloy film on silicon before and after annealing [224]
— = Before annealing; --- = Annealed at 400 °C; ... = Annealed at 700 °C

Similar metal–Si reaction studies have been performed by DE NIJS and VAN SILFHOUT [225] on Ti–Si, by FASTOW et al. [226] on pulsed ion beam irradiation of Ni–Cr on silicon, by ELLWANGER et al. [227] of $Pt_{0.4}Ni_{0.6}$ on silicon, and by OBATA et al. [228] of TiC on carbon and molybdenum.

RBS has also been used to characterize palladium and tin catalysts on polyetherimide surfaces [229], titanium nitride thin films [230], silicon oxynitride films [231], and silicon nitride films [232], and to study the laser mixing of Cu–Au–Cu and Cu–W–Cu thin alloy films on Si_3N_4 substrates [233], and the annealing behavior of GaAs after implantation with selenium [234].

6. Scanning Tunneling Methods (STM, STS, AFM)

Scanning tunneling microscopy (STM) was invented by BINNIG and ROHRER in 1983 [235], for which they received the Nobel Prize. The ability to produce images related directly to topography on an atomic scale was a major step forward for surface science, which until then had had to deduce surface atomic structure from techniques such as LEED and ISS. Since the appearance of STM, many sister techniques using similar principles have been developed for the study either of special materials, of surfaces in nonvacuum ambients, or of the surface distribution of effects other than purely topographical.

6.1. Principles

In *scanning tunneling microscopy* a metal tip is brought so close to a conducting surface (of the order of 0.5–2.0 nm) that an electron tunneling current becomes measurable at applied voltages in the range 2 mV to 2 V. The tip is then scanned over the surface. The tunneling current is monitored while scanning and used in either of two ways: in one, a feedback circuit changes the height of the tip above the surface in order to maintain the current constant (*constant current mode*); in the other, the height of the tip is kept constant while variations in current are recorded (*constant height mode*). In the simplest, one-dimensional view, the dependence of tunneling current I_T on applied voltage V_T and tip–surface separation s is given by the approximation

$$I_T \approx V_T \exp(-A\varphi^{1/2}s) \qquad (36)$$

where A is a constant and φ is the effective local work function. Equation (36) shows that for a typical work function value of about 4.5 eV, the current changes by about an order of magnitude for each change in the separation of 0.1 nm, when close to the surface. Therefore in the constant current mode of operation the tip follows the surface contours by rising or falling, whereas in the constant height mode the current rises or falls. In either case an image related to surface topography is obtained by constructing a map from line scans displaced progressively in one direction by equal amounts. The principle is illustrated schematically in Figure 72, taken from Hansma and Tersoff [236] and Hansma et al. [237].

For constant local work function φ, Equation (36) predicts a linear dependence of I_T on V_T, but this is obeyed only at very low values of V_T, since the equation is approximated from a more complex and rigorous expression. At higher V_T, the one-dimensional tunneling model on which Equation (36) is based breaks down, and the three-dimensional electronic structure of the surface region (i.e., the local band structure) must be considered. The tunneling current must then be calculated by first-order perturbation theory; if the assumptions are made that the tip can be replaced by a point and that the tip wave functions are localized (i.e., a constant density of surface states), then the current is given by

$$I_T \sim \sum_s |\psi_s(r)|^2 \delta(E_s - E_F) \qquad (37)$$

where ψ_s is the wave function of surface state s; E_s the energy associated with ψ_s; and E_F the Fermi energy. The quantity on the right of Equation (37) is simply the surface local density of states (LDOS). Thus the tunneling current is proportional to the surface LDOS at the position of the point. For a fixed applied voltage V_T, the image in STM is therefore not just a function of topographical variations, but is essentially an image of constant surface LDOS. If the tip is now positioned over a particular point on the surface, and

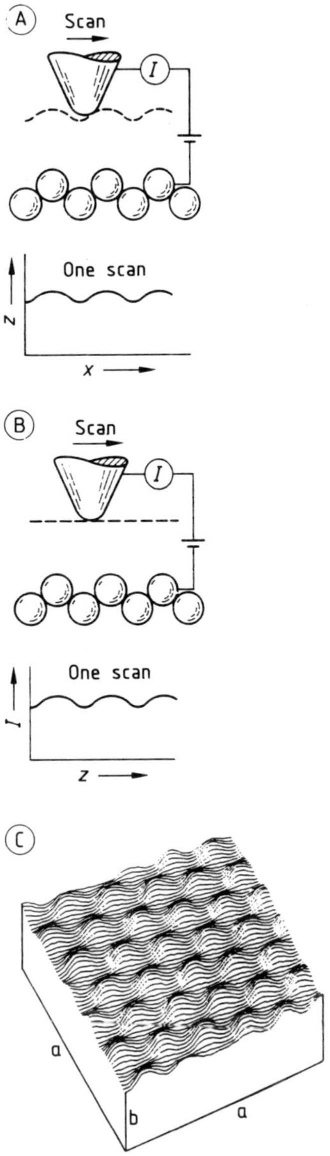

Figure 72. Illustration of modes used in STM [236], [237]
A) Constant current mode; B) Constant height mode; C) Surface topographical-spectroscopic image produced by scanning along progressive incremental distances in the y direction
a) 1.3 nm; b) 0.5 nm

the current recorded as a function of V_T, the $I-V$ dependence will mirror the LDOS, that is, peaks will occur near saddle points in the surface band structure, along with features due to localized states (e.g., so-called dangling bonds). The $I-V$ dependence is therefore providing a tunneling spectroscopy, and according to whether $V_T < 0$ or $V_T > 0$, occupied or unoccupied states, respectively, are explored. Once a spectral feature has been found at a particular value of V_T, if the applied voltage is set to that value and the tip is scanned as before, a two-dimensional distribution of that feature is obtained. This is known as *scanning tunneling spectroscopy* (STS).

Because STM and STS are basically the same, in that topography and electronic distribution at the surface both contribute in varying proportions to the image that is actually seen, other methods of recording pure topography have been sought. In addition, STM (STS) requires that the surface be conducting, whereas the topography of many nonconducting surfaces would be of interest. In 1986, BINNIG et al. [238] calculated that a cantilever could be made with a spring constant lower than that of the equivalent spring between atoms. Thus the vibrational frequencies of atoms in a molecule or solid combined with the atomic mass lead to interatomic spring constants of about 10 N/m, compared with spring constants of 0.5–1 N/m in narrow strips of soft metal such as aluminum. A cantilever with a very fine tip virtually in contact with the surface, with actual applied forces in the range of $10^{-13}-10^{-6}$ N, and a means of sensing and monitoring the tip deflection, will therefore measure tip–surface interactions arising from van der Waals, electrostatic, magnetic, and other weak forces. This is known as *atomic force microscopy* (AFM). It does not depend on the conducting nature of the surface, and the information is purely topographical; a vacuum environment is not necessary, and indeed its application in a variety of media has led to a number of variants of the technique.

6.2. Instrumentation

All *scanning tunneling microscopes* have been constructed as variations on the schematic shown in Figure 72. Although several commercial instruments are available, the basic construction is relatively simple, so that many individual designs have appeared in the literature. An example is shown in Figure 73A, from HANSMA et. al. [237]. The sample (d) is fixed to a table mounted on three advance screws (e, f). The tip (a) is attached to the end of a piezoelectric translator (b), the tripod arrangement of Figure 72 being replaced by a segmented tube since the latter is more rigid and therefore less sensitive to vibration. The sample is brought near the tip first by manual operation of the coarse advance screws while the tip is viewed through a microscope, until the separation is about 10 μm. Then a stepping motor drives the fine advance screw to bring the sample more slowly toward the tip, the final approach being monitored via the tunneling current itself. Since the ends of the coarse advance screws form the pivotal axis for the fine screw, the motion of the latter is reduced by a factor of 10, allowing very fine approach adjustment.

The tip is scanned linearly across the surface by ramped voltage applied to the x piezosegment, with progressive shifts in the y direction to create a rastered pattern of movement. Distance to the sample surface is controlled by a voltage applied to the z segment, the voltage being determined in a feedback circuit by the tunneling current. Images are accumulated by multiple scans, each scan consisting of the voltage applied to the z segment as a function of that applied to the x, and displaced successively by y shifts.

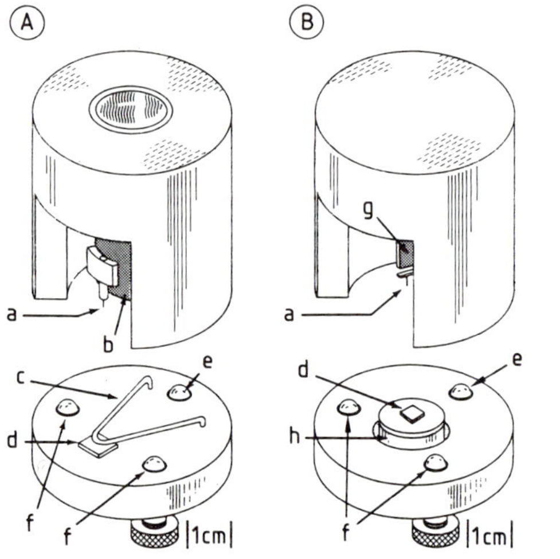

Figure 73. Schematics of the designs for an STM and an AFM used by HANSMA et al. [237]
A) STM; B) AFM
a) Tip; b) Single-tube xyz piezo; c) Sample clip; d) Sample; e) Fine advance screw; f) Coarse advance screws; g) Spring deflector sensor; h) xyz Piezo

Fabrication of tips does not require any difficult technology. The reason for this is that any fine point fabricated by mechanical or electrochemical means will not be smooth on an atomic scale, but will have many "microtips" of near-atomic dimensions. When the tip is brought close enough to the surface, the tunneling current automatically selects the microtip that is nearest. Occasionally tips compete, or undesirable whiskers may be present, in which case an increase in voltage beyond the tunneling range will usually rectify the situation by removal of one or the other.

The other constructional requirement is rigidity. Low-frequency vibrations such as those found in buildings must not reach the tip–sample region; so, as indicated in Figure 73, components must be robust and rigid in order that such vibrations are damped.

In an *atomic force microscope* (an example is given in Fig. 73 B), the deflection of the cantilever spring is normally sensed by reflection of a laser beam from a mirror on the back of the cantilever into a two-segment photodiode, as shown in Figure 74 from WEISENHORN et al. [239]. The sample is positioned on a piezoelectric translator, and the signal from the photodiode is sent to a feedback circuit controlling the voltage applied to the *z* piezoelement, which moves the sample up or down in response to small changes in cantilever deflection. In this way an almost constant force can be maintained on the tip, which is important when imaging vulnerable biological material. For the latter the surface is scanned at the smallest force compatible with tip contact. Rastering to produce images is performed by the *x* and *y* piezoelements of the translator.

Since the AFM does not use tunneling currents to sense the deflection of the cantilever and does not require a vacuum environment for operation, it has been used very successfully under water and other liquids. Figure 74 B shows the type of liquid cell employed by HANSMA and coworkers [239]–[243]. The tip with its mirror is attached to the underside of a strip of poly(methyl methacrylate) that also carries a circular groove for a small O-ring. Once the sample has made contact with the O-ring, the cell is sealed, but the O-ring material is sufficiently flexible to allow movement of the sample under the tip in the usual way. A recent modification [244] of the design allows the fluid in the cell to be circulated in order to simulate flow conditions, while inserted electrodes allow cyclic voltammetry to be performed.

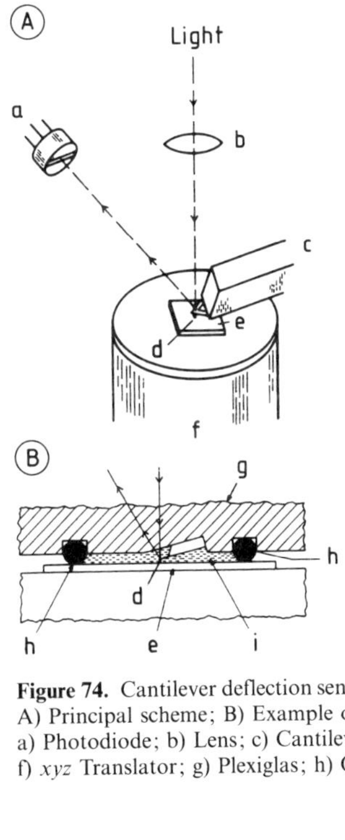

Figure 74. Cantilever deflection sensing used in AFM [239]
A) Principal scheme; B) Example of use in a liquid cell
a) Photodiode; b) Lens; c) Cantilever; d) Tip; e) Sample; f) *xyz* Translator; g) Plexiglas; h) O-ring; i) Liquid

6.3. Spatial and Spectroscopic Information

Since the inception of STM, and latterly of AFM, a large number of elegant and often beautiful images have appeared in the literature, increasingly in color to emphasize the contrast. Although atomic resolution seems to have been achieved frequently with *STM*, care must be taken in the interpretation of images because of the overlap of topographical and electronic distribution variations. Thus in Figure 75, from EBERT et al. [245], imaging of the InP (110) surface reveals the positions either of the phosphorus atoms (A) or the indium atoms (B), depending on the applied voltage. In (A), at $V_T = -2.6$ V, the occupied state density is mapped, and in (B), at $V_T = +2.2$ V, the unoccupied state density. In both cases the resolution is atomic, but the spatial localization of the states is quite different.

With *AFM*, the images are unambiguous, provided the tip is very sharp so that multiple microtip effects are absent. Several materials have been used as tests of resolution, a favorite

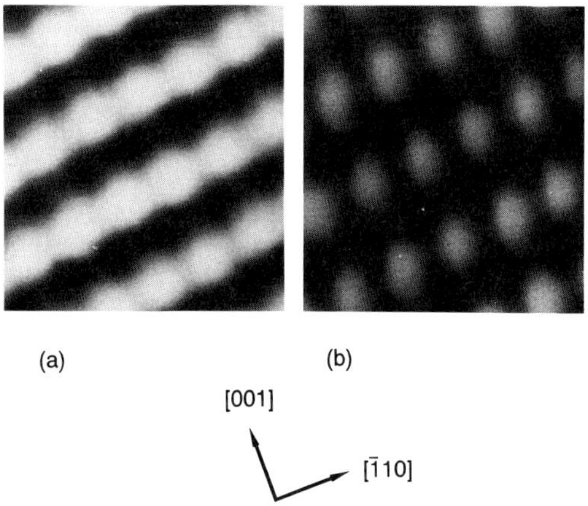

Figure 75. STM images of the InP (110) surface over an area of 2×2 nm^2 [245]
A) Tunneling voltage −2.6 V; B) Tunneling voltage +2.2 V. Images at the two voltages reveal the occupied and unoccupied state densities, respectively

Figure 76. AFM image of graphite in air [247]
Image size 1.17 nm. Carbon atoms can be seen as darker regions around central holes

being freshly cleaved, highly oriented pyrolytic graphite, in which the average distance between carbon atoms is 0.14 nm. Calculations by GOULD et al. [246] show that every surface carbon atom should be imaged with equal intensity, and Figure 76 from GOULD et al. [247] confirms this prediction. The field of view in the figure is 1.17 nm across, and individual carbon atoms can be seen as alternate light and dark regions around each hole.

According to the applied voltage range, as indicated above for InP, STM, or rather, STS, can give information on both occupied and unoccupied surface states. The ST spectra over an extended range of V_T are therefore likely to be comparable with those from other techniques, although the comparison may not be exact because of the very localized nature of the tunneling that takes place in STS. Only if the bulk band structure is similar to the surface density of states (DOS) is agreement likely, and this has been found by KAISER and JAKLEVIC [248] for the clean Pd (111) surface. In Figure 77 from their paper, the $I-V$ spectrum from −2.4 to +2.4 V (Fig. 77 A) is compared with the experimentally determined band structure (Fig. 77 B) and the UPS spectrum (Fig. 77 C). The STS peaks at −1.4 and +1.0 V correspond to the positions of the *d*-band edge at about −1.2 eV and of the saddle point at +0.94 eV in B, while in the UPS spectrum (Fig. 77 C) surface-state peaks are found at about −0.5 and −2.1 eV, which match the peaks at −0.6 and −2.1 V. The minor feature in A near +0.5 V may also be a surface state. STS has also been compared with IPES [249] over the unoccupied range of surface states of graphite.

6.4. Applications

The ability of *AFM* to provide topographical images of all types of surface at spatial resolutions varying from the atomic to the micrometer scale, in air, vacuum, or under liquids, has led to a great variety of applications. They can be divided roughly into inorganic and organic (biological) applications.

An example of the way in which AFM has been applied to *inorganic material* surfaces is provided by the study of GRATZ et al. [250] on the dissolution of quartz in potassium hydroxide solution. Such a study forms part of a general attempt to understand the phenomena of the dissolution and growth of crystal surfaces, either of natural origin or grown synthetically. The authors etched optical-quality natural quartz for 4 h at 148 °C in KOH solution and then examined the surface with AFM, in this case in ethanol, which was found to keep the surface clean. Quartz is an insulator and could not therefore be studied by STM.

The etching created prismatic etch pits with smooth floors. Figure 78 shows some of the results. Image A, of area 1490×1490 nm^2, reveals

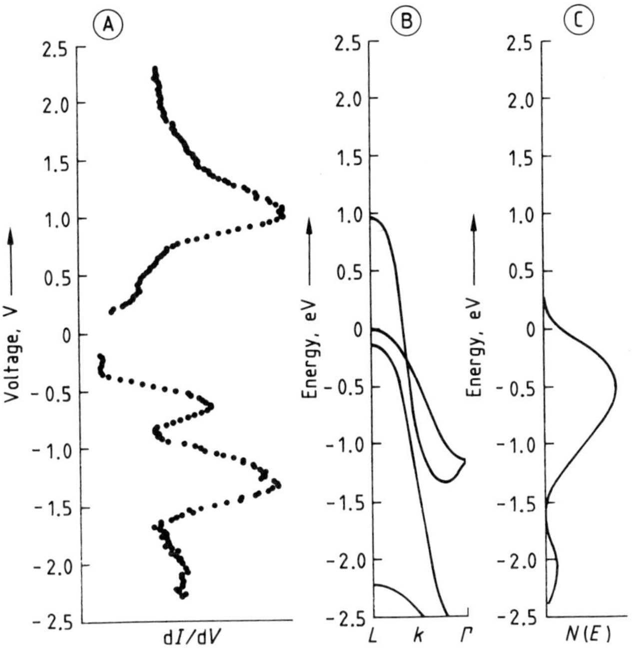

Figure 77. Comparison of
A) STS spectrum; B) Experimental band structure; and C) UPS spectrum, of palladium (111) [248]
L, k, Γ = directions of symmetry in three-dimensional space

the contours in the pit; the height separation from top to bottom of the pit is about 20 nm. The *c* and *a* axes of the quartz are indicated by the long and short arms of the cross, respectively. In image B, one quadrant of the pit is shown at higher magnification and in three dimensions, the field of view being 650×650 nm^2. On the vertical scale the range is only 14 nm. The ledges visible in B are shown in C to be of only 0.7-nm height, to be atomically flat, and to be about 30 nm wide. Thus the etch pit can be described as an elongated amphitheatre, with a series of flat terraces of individual height only a few tenths of a nanometer. The results are said to be in agreement with the classical model of crystal dissolution.

Other recent inorganic material applications have included synthetic ultrafiltration membranes in air and in water [251], clay mineral surfaces [252]–[253], binding of molecules to zeolite surfaces [239], and electrochemical deposition of copper on gold in sulfate and perchlorate solutions [244].

Both AFM and STM have been able to image *biological materials* without apparent damage, which might be expected for AFM but is surprising for STM. In fact the imaging conditions used in STM must be kept as weak as possible to avoid damage. Thus, to obtain the image shown in Figure 79, from ARSCOTT and BLOOMFIELD [254], of a series of parallel fibers of Z-DNA, tunneling current and voltage levels were 1–2 nA and −100 to +100 mV, respectively. The characteristic left-handed twist of the Z-DNA helix is obvious in the image, and the measured helical periodicity of 4.19 nm is only 8 % less than the crystallographic value, indicating minimal contraction due to dehydration. The measurements were performed in air.

STM has also successfully imaged phosphorylase from muscle [255] and biomembranes [256]–[257], while AFM has produced images of proteins in a buffer solution [258], immunoglobulin [259]–[260], DNA [261]–[252], and bacteria [263].

Although many surface analytical techniques, including some of those already described, can provide valuable information about the nature of surface reactions, the information is invariably averaged over a larger number of reaction sites. With STS, reactions and reactivity can be studied at selected individual, even atomic, surface sites, which is an exciting capability. The observations of WOLKOW and AVOURIS [264] on the reaction of NH_3 with the Si(111) surface illustrate the possibilities elegantly. Figure 80 from their paper shows a series of images of the Si (111)–(7 × 7) surface. In (A), the clean surface

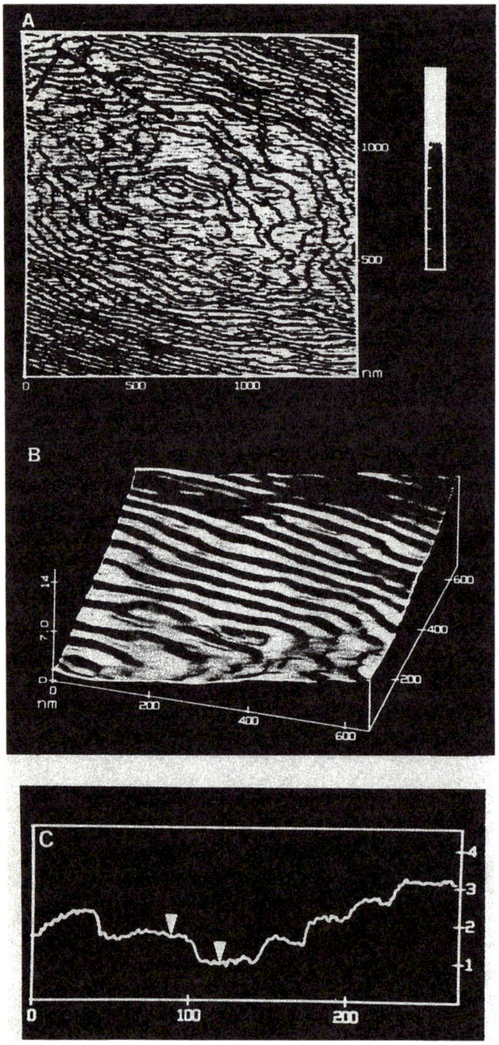

Figure 78. AFM images of an etch pit formed on quartz by etching in KOH at 148 °C [250]
The images were recorded under ethanol
A) Vertical view showing contours; B) Three-dimensional view of one pit quadrant; C) Profile across pit floor; height between cursors is 0.7 nm

is imaged at +0.8 V, revealing the characteristic hexagonal surface structure, while in (B), the same tunneling voltage of +0.8 V is used, but after exposure to 1 Langmuir (i.e., 10^{-6} Torr · s or 1.33×10^{-4} Pa · s) of NH_3. In (B), about half the surface atoms seem to be missing. If the tunneling voltage is increased to +3.0 V, as in (C), the missing atoms reappear, but with different contrast from (A). The apparent disappearance is not real, but is an electronic effect due to the

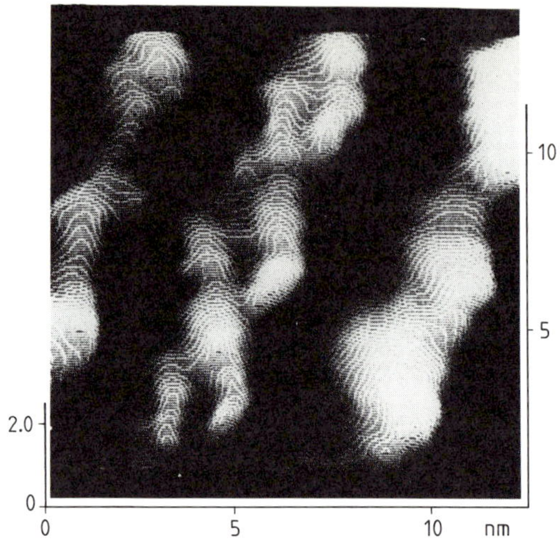

Figure 79. STM image in air of parallel fibers of Z-DNA [254]
Tunneling current and voltage were kept very low to minimize damage

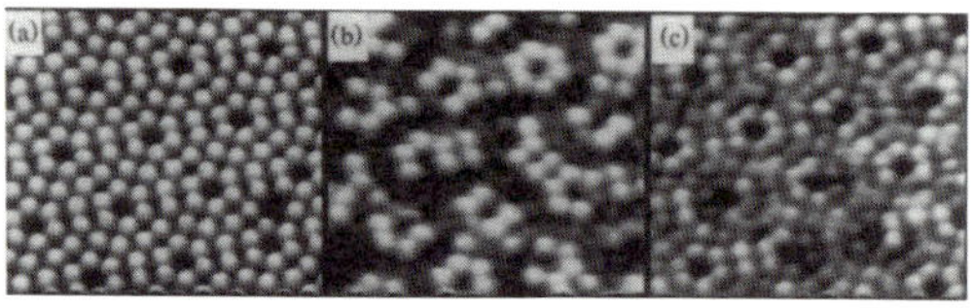

Figure 80. Series of STM images of the Si(111)–(7 × 7) surface [264]
A) Clean surface at 0.8 V; B) After exposure to 1 L of NH_3, also at 0.8 V; C) Same as B), but at 3.0 V

reaction of some of the surface atoms with NH_3; the basic structure is preserved, changes being those in the local DOS at certain atoms.

The accepted model for the (7 × 7) reconstruction of the Si (111) surface, that of TAKAYANAGI et al. [265], has silicon atoms in three special positions in the unit cell, those (in position A) underneath holes in a regular array in the outer layer, and others (B and C) present as adatoms (i.e., situated on top of the outer layer). Atoms B are adjacent to the holes at A; atoms C are in the center of the side of the unit cell. With STS the tip can be stationed over each of these atomic sites in turn, and the differences in local DOS recorded, both before and after reaction with NH_3. Figure 81 A shows the spectra from the clean surface, Figure 81 B, after the reaction. The quantity $(dI/dV)/(I/V)$ is proportional to the local DOS.

Spectra from the clean surface show a very strong band at ≈ -0.8 eV from atoms A, sug-

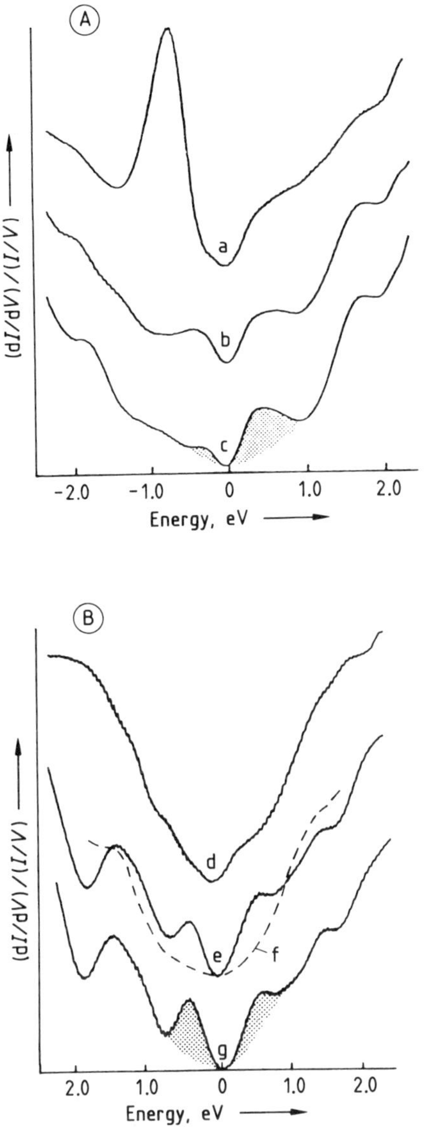

Figure 81. STS spectra recorded from the Si(111)–(7 × 7) surface [264]
A) In the clean condition; B) After reaction with 1 L of NH_3
a) Atom A; b) Atom B; c) Atom C; d) Atom A; e) Atom B not itself reacted but situated next to reacted atoms; f) Reacted atom B; g) Atom C not itself reacted but situated next to reacted atoms

gesting a fully occupied dangling bond surface state. Since atoms B and C have much weaker occupied bands, the implication is that charge transfer occurs from B and C to A, with more being contributed by C than B. Both B and C have stronger unoccupied bands, confirming the transfer. After reaction with NH_3, the strong occupied dangling bond state at A has vanished, and at B (the dashed line) the small occupied band has also been lost. Elimination of these states accounts for the apparent disappearance of reacted atoms from the image in Figure 80 B. The other spectra in Figure 81 B, the solid lines for B and C, are for atoms of those types not themselves reacted but adjacent to reacted A atoms. In both cases, but particularly for C, a transfer of intensity has occurred from the unoccupied to the occupied states, suggesting that during reaction a charge transfer takes place from A to C, and to a lesser extent B, allowing A to react with NH_3.

The above experiment demonstrates clearly how STS can follow changes in electronic distribution, and differences in reactivity, on an atomic scale—an unique capability. Although the Si (111) surface is probably one of the best candidates for such a demonstration, being atomically flat and possessing strong dangling bond states, the possibilities that are opened up by this technique in general are very clear.

STS has also been applied to a study of the unoccupied surface states of graphite [249], hydrogen-like image states on clean and oxygen-covered nickel and on gold epitaxed on silicon (111) [266], [267], the superconducting energy gap in Nb_3Sn [268], the electronic structure of the InP (110) surface [245], and copper phthalocyanine adsorbed on Cu (100) [269].

7. Other Surface Analytical Methods

7.1. Ultraviolet Photoelectron Spectroscopy (UPS)

In its principle of operation, UPS is similar to XPS, in that a surface is irradiated with photons and the energies of the ejected photoelectrons are analyzed. The physical relationship involved is the same as that of Equation (1) for XPS, but in UPS the energy $h\nu$ of the exciting photons is much lower since the photons are derived from a gaseous discharge that produces hard UV radiation. The gas normally used is helium, which, depending on pressure conditions in the discharge, will provide line sources of energy 21.21 eV (He I) or 40.82 eV (He II) with very narrow linewidths (≈20 meV). Because of these low exciting energies, the binding energies E_B that appear in Equation (1) refer not to core levels, as in XPS, but to shallow valence band levels and to other shallow levels, such as those of adsorbates, near the valence band. For energy analysis, the CHA is again used, but the pass energies employed for UPS are much lower than in XPS since better energy resolution is required.

UPS is not an analytical technique in the sense that it can provide quantified elemental concentrations on a surface as XPS and AES can, but it is a powerful tool in basic research and is another of the group of most frequently used techniques. It is used in two ways. In its *angle-integrated* form it has been employed very extensively in studying the interactions of gaseous molecules with surfaces, usually in combination with other techniques such as XPS, AES, LEED, and TDS. For complex molecules, interpretation of the UPS spectra during interaction is often possible in terms of the molecular orbital structure of the free molecule, as demonstrated in Figure 82 from NETZER and MACK [270]. The other way in which it has been used is in *angle-resolved UPS (ARUPS)*. In this form the energy analyzer, a CHA of small radius, is mounted in such a way mechanically that it can traverse the space around the sample in both polar and azimuthal directions, while the angle of photon incidence is kept constant. The angle of incidence is then changed by rotation of the sample. If the sample is a single crystal with a clean surface, the dispersion of spectral features (i.e., their movement along the energy axis and their increase or decrease in magnitude) can be used theoretically to plot the band structure in the surface region. The changes in surface-associated band structure during reaction with gaseous molecules can then be followed, leading to additional information about the nature of the reaction.

7.2. Light-Spectroscopic Methods (RAIRS, SERS, Ellipsometry, SHG, IBSCA, GDOS)

The light-spectroscopic techniques have one thing in common: information is provided by them in the form of light, visible or IR, that has either had its spectroscopic properties changed by reflection from a surface or is emitted from atoms in excited states ejected from the surface. Four of the techniques (RAIRS, SERS, ellipsometry, SHG) also use light as the incident probe, the other two (IBSCA, GDOS) use incident ions.

Reflection–Absorption Infrared Spectroscopy. In RAIRS, IR radiation is reflected from a plane surface through an adsorbed layer, the

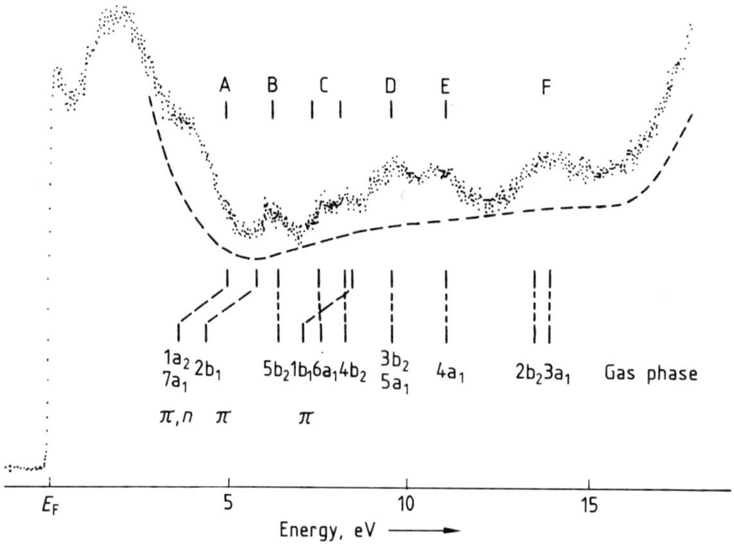

Figure 82. UPS He II spectrum from 5 Langmuirs pyridine adsorbed on Pd (111), with probable molecular orbital assignments ($\alpha = 60°$, $\theta = 50°$) [270]
The dashed line is part of the spectrum from clean Pd (111)
A–F = designation of spectral position of peaks

reflected light losing intensity at those frequencies at which the light frequency coincides with a vibrational mode at the surface. The vibrations may be those within the adsorbed species itself, or they may arise from interaction with surface atoms. The substrate surface must be a good reflector. The incident IR beam is plane polarized and is arranged to strike the surface near grazing incidence, typically 80–82°; both plane polarization and grazing incidence are required theoretically. For adequate signal-to-noise ratio in the reflected signal, the absorption cross section of the adsorbed layer at the characteristic frequencies must be high enough to be able to use good energy resolution.

Most earlier research using RAIRS was carried out with dispersive spectrometers to analyze the reflected light, but Fourier transform methods have largely taken over now, leading to the acronym FT-RAIRS. The advantages of using FT methods are, firstly, that either much improved signal-to-noise ratios can be obtained for the same acquisition time or data can be acquired faster for the same signal-to-noise ratio; and, secondly, that all spectral features are recorded simultaneously.

FT-RAIRS is closely related to high-resolution electron energy loss spectroscopy (HREELS) (see Section 7.3), and they are often used together in the same system. The two techniques are complementary, since FT-RAIRS has high energy resolution (≈ 2 cm^{-1}) and speed of acquisition, but its effective spectral lower limit is ≈ 800 cm^{-1} due to the transmission characteristics of the IR transparent optical components, while HREELS can cover the entire spectral range down to 0 cm^{-1} but has inferior energy resolution (≈ 20–60 cm^{-1}). A comparison of spectra obtained by the two techniques is shown in Figure 83, from ERLEY [271], for the adsorption of CH_3CN on Cu (111) at 100 K. The bands labeled ν_3, ν_6, ν_1, and ν_5 can be resolved by FT-RAIRS but not by HREELS, but HREELS can record vibrational bands at wavenumbers below the FT-RAIRS cutoff.

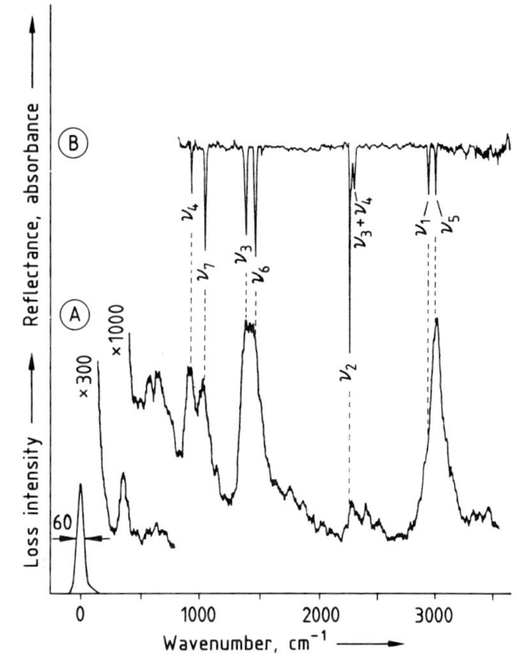

Figure 83. Comparison of HREELS and RAIRS vibrational spectra of CH_3CN adsorbed on Cu (111) at 100 K [271]
A) HREELS; B) RAIRS

Surface-Enhanced Raman Scattering. Conventional Raman spectroscopy is beginning to be used more often in the study of very thin surface films, with the availability of improved detectors, but it still suffers from low intensity in the scattered signal. The discovery by FLEISCHMANN et al. [272] of the *surface enhancement effect* has enabled the Raman scattering effect to be applied more widely, even though restrictions exist. When the surfaces of certain metals, particularly silver, copper, and gold, are roughened on an atomic or near-atomic scale, the intensity of the scattered light is enhanced by factors of 10^5–10^6. Roughening is usually performed electrochemically and is subject to some variation. Various theories have been put forward for the effect, without general agreement as yet. Despite the restrictions on the nature of the substrate, SERS has found very many applications, some of which have been carried out in UHV (e.g., the reactions of N_2 and CO_2 on silver, copper, and potassium surfaces at 40 K) [273]. For a thorough review of the physics of the SERS phenomenon and a discussion of applications, see [274].

Ellipsometry is a method of measuring the thicknesses, and changes in thickness, of very thin films on surfaces, to very high sensitivity, of the order of 5×10^{-3} nm. Polarized light, of wavelength within the UV to IR range of the spectrum, is specularly reflected from a surface with one or more layers of refractive index different from that of the substrate. The latter must be flat and highly reflecting. After the reflection the light is no longer plane polarized but elliptically polarized. The parameters measured are $\tan\psi$, the amplitude ratio of the resolved components of the electric vector of the reflected light parallel to and perpendicular to the plane of incidence, and $\cos\Delta$, the phase difference of the two components. *Fixed-photon-energy ellipsometry* can provide only a measure of the thickness, but if the elliptical polarization parameters are measured as a function of the light energy and the angle of incidence, the technique is known as *spectroscopic ellipsometry*. The observed dispersion with energy can be related to the dielectric loss function; since the energy dependence of loss functions varies from one material to another, the contributions of different surface layers can in principle be distinguished. Ellipsometry in either form is therefore not a compositional analytical technique, but it can provide valuable additional physical information when used with one of the compositional techniques such as AES or XPS.

Optical second harmonic generation (SHG) is another light reflection technique of fairly recent origin that is being applied increasingly in UHV, although it can be used in almost any medium. Like ellipsometry, to which it is related, SHG is very sensitive to the interfacial regions between thin films and substrates, and can be used to study adsorption, molecular orientation, and surface structure. If a linearly polarized wave at a frequency ω is incident on a surface whose dielectric constant is different from that of the ambient medium, then a nonlinear source polarization is induced at the surface, which radiates at frequency 2ω in both reflected and transmitted directions. Inversion symmetry dictates that the second-order nonlinear susceptibility vanishes in the bulk of the solid medium, but at the surface the inversion symmetry is broken, leading to electric dipole contributions to the second-order nonlinearity. By using a laser pulsed at high frequency, time-resolved reactions at surfaces with time scales in the range 10^{-14}–10^{-6} s can be studied. Enhancement of the SHG response can be achieved by tuning the laser to an electronic transition in the substrate or an optical transition of an adsorbate. The technique has recently been reviewed at length [275].

Ion Beam Spectrochemical Analysis. As remarked earlier, ion irradiation of a surface produces visible light photons as well as many other types of particle. The light emission arises from the deexcitation of ions and excited neutrals, and is characterized by a spectrum consisting of a series of sharp lines at wavelengths associated with the type of ionized or excited atom. Spectral analysis thus provides elemental identification by reference to standard emission wavelength tables. The technique is called ion beam spectrochemical analysis (IBSCA) but has several other names as well. Typical operating conditions include bombardment by Ar^+ ions at energies in the range 2–20 keV, with current densities of 10–100 μA/cm^2; the light emitted just above the target surface is then focused onto the entrance slit of a scanning optical monochromator, which analyzes the light spectrum. Spatial resolution depends on the type of ion source being used, and can be of the order of 2–5 μm with a suitable type such as a duoplasmatron. Because emission takes place from particles ejected from the surface layer, the surface specificity is similar to SSIMS (i.e., one atom layer). Quantification can in principle be carried out theoretically (e.g., based on the Andersen–Hinthorne LTE model [185] mentioned in Chapter 4) but is normally performed by relative intensity measurements, that is, by measuring the intensity of emission from a pure elemental standard under the same instrumental conditions of analysis as used for an unknown. Applications have included studies of the oxidation of chromium [276], the effect of segregation of carbon and sulfur to the surfaces of Fe–Cr–Mo crystals [277], the reaction between a thin PbO film and fused silica [278], and the protonation of lithium silicate glasses [279].

Glow Discharge Optical Spectroscopy. Unlike many of the other techniques described in this section, GDOS is widely used industrially and instruments are available commercially. In the great majority of applications, however, GDOS is used not as a surface-specific analytical technique, but as a rapid and quantifiable technique for determination of elemental concentration profiles through films often tens of micrometers thick. GDOS has certain similarities to IBSCA. The surface to be analyzed is made the cathode in a diode-type discharge cell, whose design originated with GRIMM [280], and argon is flowed through the cell at pressures of 10^3–10^4 Pa. Application of a d.c. voltage of ≈ 1 kV between an anode and the sample cathode establishes a glow discharge. Positive ions produced in the discharge sputter the sample surface and erode it continuously, most of the sputtered material leaving the surface as neutral atoms. The emitted atoms pass into the discharge, in which they are excited by multiple collisions, and the excited states then decay by the emission of visible light, whose spectrum is analyzed in a multichannel spectrophotometer. According to the chosen operating conditions, the rate of erosion of the sample can be varied between 30 nm/min and 1 μm/min. The inherent sensitivity of the technique is high so that, if required, thin surface layers could be analyzed.

In arriving at a quantification of GDOS intensities, one must realize that as a result of the rapid uniform sputtering of the sample surface, both the surface composition (due to preferential sputtering) and the surface geometry (due to crater formation) are changing during the analysis. The latter change alters the glow discharge operating conditions since during sputtering the cathode surface is receding from the discharge region. BENGTSON [281] has shown how to obtain reasonable quantification on the basis of empirical expressions for the excitation function of the discharge and the sputtering rate in the discharge. If the intensity I_λ of an emission at wavelength λ from an atom of type A is measured, then the intensity is related to the concentration c_A of A by

$$I_\lambda = k_{\lambda A} c_A C i^2 (V - V_0)^{x(\lambda,A)} \qquad (38)$$

where $k_{\lambda A}$ is a constant containing instrumental and atomic factors; C a constant dependent on cell geometry; i the bombardment current; V the applied cell voltage; V_0 the discharge threshold voltage, also dependent on geometry; and $x(\lambda,A)$ a constant characteristic of the emission line.

Equation (38) can be used to measure the concentration of an element in an unknown material if the emission parameters from a standard are known, since by taking the ratio of currents from unknown and standard, some of the parameters in the expression disappear. For elements in low concentration in alloys or compounds the accuracy is good, typically ±5%.

Although GDOS is normally applied as a very fast profiling technique, leading to the publication of many profile diagrams of films up to ≈20 μm thick, its operating principles indicate that it is inherently surface-specific, and it can indeed be operated at a much lower erosion rate to study the near-surface region. GDOS also has an advantage over most of the other principal surface analytical techniques in that it can analyze for hydrogen. This ability and also its operation in a surface-specific manner are demonstrated in Figure 84, from BERNERON and CHARBONNIER [282], where the profiles by GDOS (A) and by AES (B) of an anodic passive film on iron are compared. The film was only ≈5 nm thick. GDOS shows a pile-up of hydrogen at the film–metal interface and is able to detect boron (from the boric acid buffer) beyond the point at which AES cannot. The profiles are qualitatively similar. Note that GDOS gives purely elemental information, with no direct chemical information available.

Other near-surface applications have included the implantation profiles of phosphorus in iron and FeO [283]; TiN and copper coatings on steel [284]; chromate layers on aluminum [285]; and implantation profiles of phosphorus and boron in iron and nickel [286].

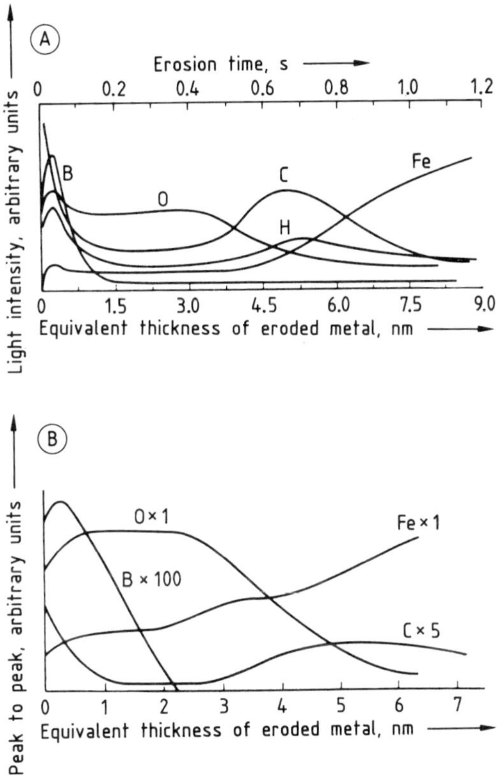

Figure 84. Comparison of depth profiles through a passive anodic film on iron [282]
A) GDOS; B) AES

7.3. Electron Energy Loss Methods (ELS, CEELS, HREELS, IETS)

Energy Loss Spectroscopy and Core-Electron Energy Loss Spectroscopy. The techniques of ELS and CEELS can be discussed together because they use the same experimental arrangement, basically that of conventional AES. A beam of primary electrons interacts with electrons in surface atoms, and the energy losses suffered by the primary electrons are measured by recording the appropriate part of the secondary electron spectrum in an energy analyzer. In *ELS* the losses are relatively small being due to the energies required to stimulate electronic transitions within the valence band of the solid, and collective oscillations of the conduction electrons (plasmon oscillations). For each solid the plasmons have energies characteristic of the bulk structure and of the surface structure, the surface plasmon having theoretically an energy lower than that of the bulk by a factor of $2^{1/2}$. In *CEELS* the primary electron interacts with a core level, usually but not always leading to ionization, and the losses are in general much greater than in ELS. Thus in ELS the observed losses are normally within 30–50 eV of the primary energy, whereas in CEELS they may be several hundred electronvolts away.

Since all loss features are by definition associated with the primary energy, features characteristic of the surface can be compared to those of the bulk by variation of the primary energy (i.e., by effectively varying the electron escape depth). Such variation is standard practice in ELS, in which the range of primary energies used is typically 100–1000 eV. In CEELS, on the other hand, although the relation between primary energy and escape depth still applies, the primary energy must be high enough to ionize the core level with adequate efficiency, as discussed in Section 3.2.2 for AES. Thus for CEELS the primary energy range would be the same as for conventional AES (i.e., 2–5 keV).

Applications. ELS cannot be used to extract compositional information, but it has been used extensively in conjunction with other techniques, particularly AES and UPS, in basic studies of reactions occurring at clean or otherwise well-characterized surfaces. Changes in both plasmon and valence band losses occurring during reaction, particularly as a function of primary energy, provide useful information complementary to that offered by the other techniques. A typical example is shown in Figure 85, from the work of Rosina et al. [287], for the interaction of H_2 with the clean Ce (001) surface. A primary energy of 102.9 eV was used for most of the measurements, to ensure maximum surface sensitivity, with a switch to 1000 eV occasionally to emphasize more bulk-related features. In the clean surface spectrum (curve e) the surface plasmon at 8.2 eV is dominant, with a shoulder at ≈12.5 eV due to

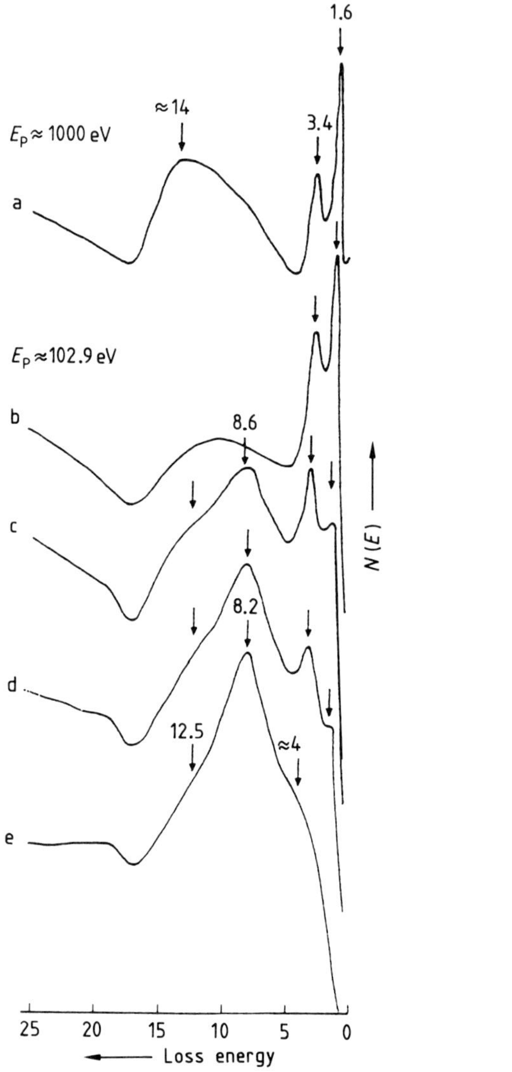

Figure 85. ELS spectra at two primary energies (E_p) recorded during interaction of hydrogen with the Ce (001) surface [287]
a) +900 L H_2; b) +900 L H_2; c) +600 L H_2; d) +300 L H_2; e) Clean

the bulk plasmon and another shoulder at ≈4 eV probably arising from an intrinsic surface state. On hydrogenation, two additional features at 1.6 and 3.4 eV appear simultaneously with suppression of the 4-eV shoulder. The new features increase with H_2 exposure, and become dominant at 900 L and a primary energy of 102.9 eV. When the primary energy is changed to 1000 eV, the bulk plasmon is seen to have shifted to ≈14 eV from ≈12.5 eV. Both the features at 1.6 and 3.4 eV and the new position of the bulk plasmon are characteristic of the formation of CeH_2. The increase in the 1.6-eV loss relative to that at 3.4 eV in going to the higher primary energy is indicative of a hydride phase increasing in hydrogen concentration with depth. This particular example is also important because it illustrates that ELS can give information about surface interactions with H_2 that neither XPS nor AES can give.

The many other recent applications of ELS include the interaction of oxygen with titanium [288], [289]; of oxygen [290] and of hydrogen [291] with Co–Ti; of oxygen with tin [292]; of pyrazine with silver [293]; and of ytterbium films with aluminum [294].

CEELS has been applied more to the study of core-electron excitation to localized final atomic states than to excitation to the continuum (i.e., complete ionization). Core-level ionization thresholds can in principle be used to obtain an elemental composition, but the spectral features are weak and the analysis is performed far better by XPS. On the other hand, where shallow, highly localized final states exist, CEELS has been able to provide useful information. Such localized states are plentiful throughout the lanthanide rare-earth series, in some cases being very close to the valence band. Figure 86, from STRASSER et al. [295], shows some examples of the fine structure to be found in CEELS above the main $4d$ excitation threshold, in the lighter lanthanides. The structure arises from strong coupling between the $4d$ holes left after excitation and the $4f$ electrons or holes, the $4f$ states being highly localized. Where the $4f$ states are close to the valence band, their occupancy is dependent on the atomic valence, and valence changes can be monitored exactly by recording the $4d$ CEELS spectra during treatments such as oxidation that usually alter the valence. This analytical variation has been explored exhaustively by NETZER and coworkers [296]–[298].

High-Resolution Electron Energy Loss Spectroscopy. HREELS has already been mentioned in the description of RAIRS as being a vibrational spectroscopy. In this technique, electrons of very low primary energy, typically 3–5 eV, are directed at a surface carrying adsorbed molecules. As the electrons approach closely enough, they interact with the molecular field and are reflected (i.e., scattered) having lost discrete amounts of energy corresponding to characteristic vibrational frequencies, either within the molecules or between the molecules and the surface. Analysis of the scattered electron energy distribution then reveals the energy losses, and hence the vibrational frequencies. If the molecules are attached to the surface only loosely (*physisorption*), many of the characteristic vibrations within a molecule are similar to those observable in the gas phase. These consist of various stretching and deformation modes of vibration. In addition, other modes appear as a result of attachment to the surface. In the gas phase a molecule can translate and rotate freely, but if one end is fixed to a surface it can no longer do so, and therefore certain translational and rotational modes of vibration become observable that cannot be observed in the free molecule. Should the molecule be attached strongly to the surface (*chemisorption*), perhaps with the loss of reactive groups, and certainly with a transfer of electronic charge, then the characteristic free-molecule stretching vibrations can become so distorted that they no longer resemble those seen in the gas phase. Thus the HREELS spectrum from a complex polyatomic molecule adsorbed on a surface can itself be very complex and contain a great deal of information concerning the way in which the molecule has reacted with the surface and also with neighboring molecules. The orientation of the molecule can often be deduced as well, particularly when HREELS is used with other techniques such as LEED, UPS, and TDS.

Because the energy widths of the characteristic vibrational energies are small, <0.1 meV, for optimum energy resolution one must achieve as narrow an energy spread in the incident electron beam as possible. In all HREELS equipment, therefore, extreme measures are taken to reduce the spread in a variety of ways. Most importantly, the incident electrons are energy selected in a monochromator that is basically an electron energy analyzer used in reverse (see, for example, [299]). Monochromators properly constructed can reduce the energy spread in the incident beam to 2–3 meV, but to maintain that width in the beam at the sample surface requires the complete elimination of stray magnetic and electrostatic fields around the sample. Thus very careful screening of the entire HREELS analytical arrangement is necessary. Many experiments are also carried out at low temperature to reduce the thermal broadening of characteristic vibrations.

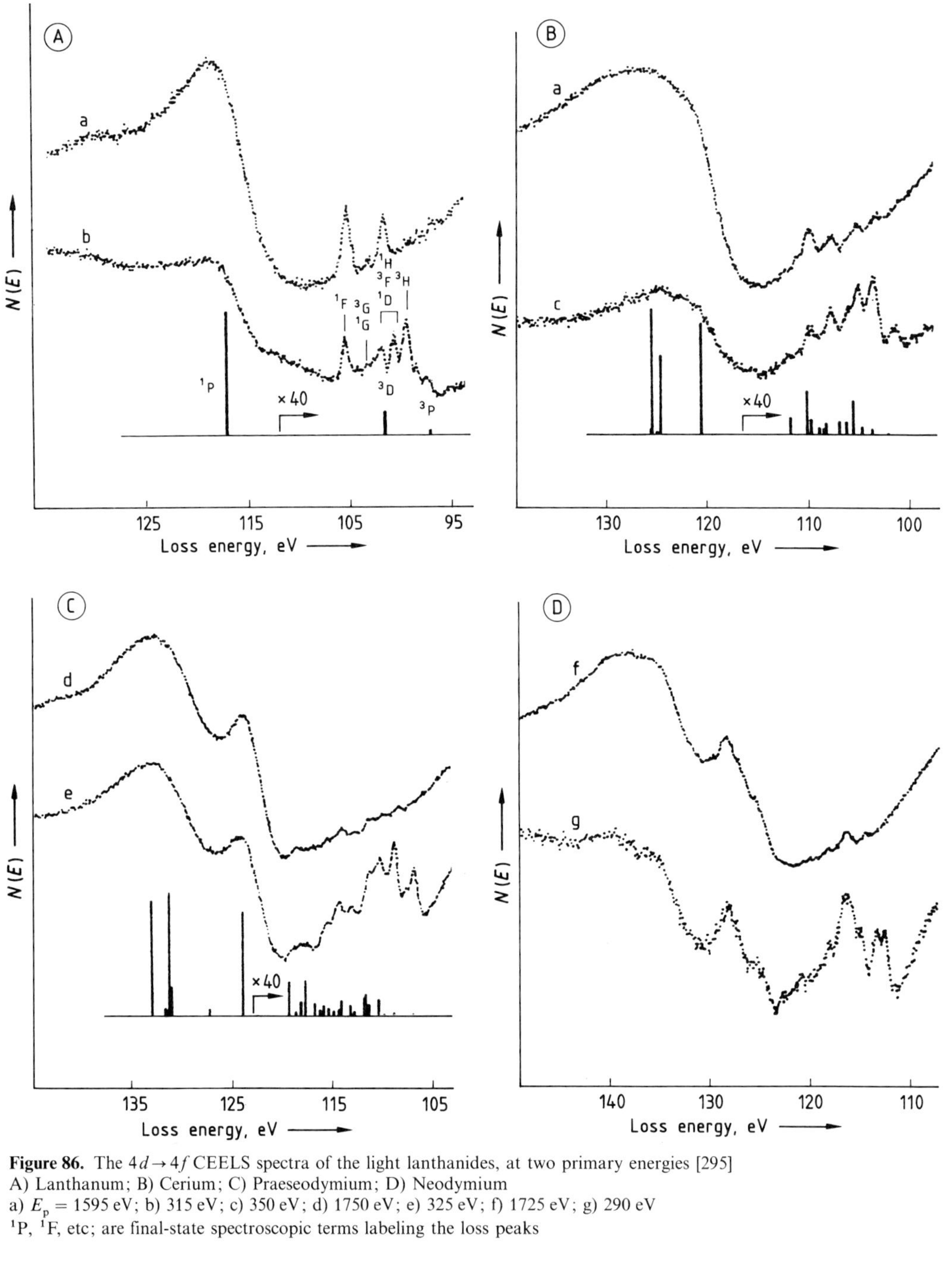

Figure 86. The $4d \rightarrow 4f$ CEELS spectra of the light lanthanides, at two primary energies [295]
A) Lanthanum; B) Cerium; C) Praeseodymium; D) Neodymium
a) E_p = 1595 eV; b) 315 eV; c) 350 eV; d) 1750 eV; e) 325 eV; f) 1725 eV; g) 290 eV
^{1}P, ^{1}F, etc; are final-state spectroscopic terms labeling the loss peaks

The above description indicates that the applications of HREELS have been entirely basic and concerned with the interactions of molecules with clean or well-characterized surfaces. A very typical example of the HREELS spectra observed during increasing coverage of an adsorbed molecule is shown in Figure 87, from BARTKE et al. [300], for the adsorption of NO on Nb (100) at 20 K. The energy loss scales are expressed both in millielectronvolts and reciprocal centimeters for convenience, and the energy resolution, as governed by the width of the primary beam, is ≤4 meV.

The loss spectrum from the "clean" surface (curve a) is featureless apart from a tiny peak at 63 meV due to atomic oxygen, undetectable by AES. Initial adsorption of NO causes the peak to grow and others to appear; with increasing coverage the oxygen peak (arising from lattice Nb–O vibrations) shifts to 72 meV. The new peaks at 57 and 116 meV are ascribed to atomic nitrogen bonded in a "bridged" position (i.e., across two Nb atoms) and to atomic oxygen bonded on top of a niobium atom, respectively. Further new peaks at 200, 211, and 228 meV, after 2 Langmuirs adsorption, are associated with stretching vibrations within the NO molecule adsorbed on top of a niobium atom. The slight differences in loss energy are explained by different angles of adsorption of the NO (i.e., tilted or bent). With further NO adsorption (curves f–h) the NO stretching vibration shifts to 220 meV and becomes very strong, with a simultaneous decrease in the atomic vibration, indicating a preponderance of molecular adsorption. Additional peaks at 231 meV and at 22 and 33 meV are attributed to vibrations within an ON–NO dimer on the surface. Thus, with correct interpretation, a few spectra such as those in Figure 87 can provide a large amount of information on the course of a reaction and the nature of the entities actually on the surface.

HREELS has also been used recently to study the adsorption systems 2,3-dimethyl-2-butene and 2-butyne on Ni (111) [301]; straight-chain alcohols on Ag (100) [302]; CO and potassium on Cu (111) [303]; O_2 on cleaved and sputtered graphite [304]; CO on copper-covered

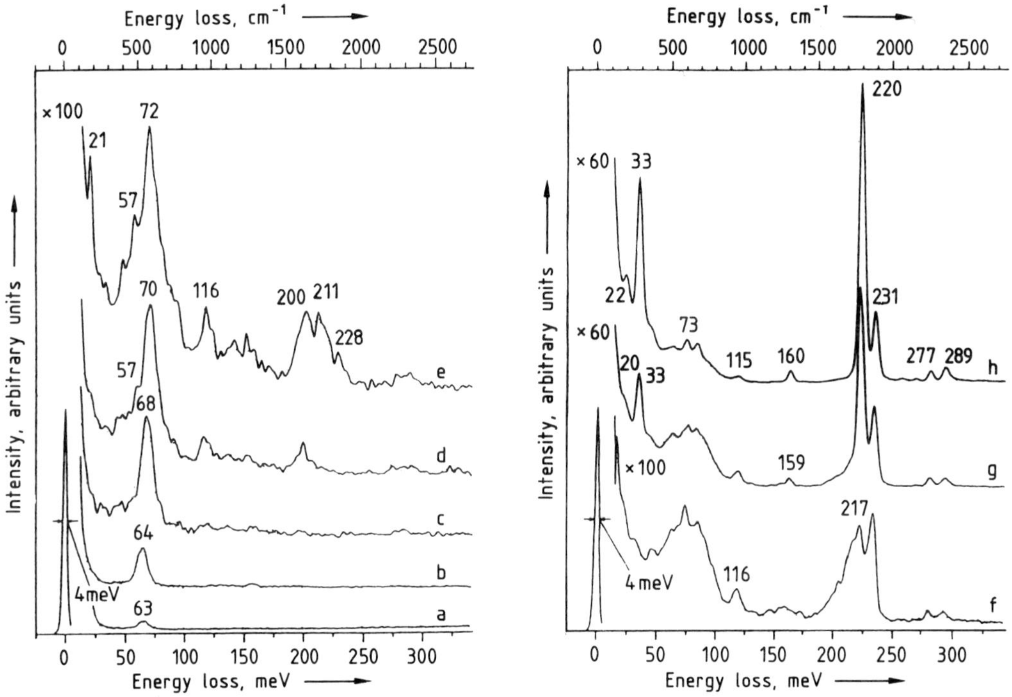

Figure 87. HREELS spectra recorded during the adsorption of NO on Nb (110) at 20 K (primary energy 2.5 eV, $I_0 = I_s = 52°$) [300]
Intensities are normalized to the elastic peak, and spectra were taken in the specular reflection mode
a) 0 L; b) 0.25 L; c) 0.5 L; d) 1 L; e) 2 L; f) 4 L; g) 8 L; h) 16 L; L = Langmuir (10^{-6} Torr · s)

Ni (111) [305]; O_2 on Fe (100) [306]; and CO on Ru (10$\bar{1}$0) [307]. Many of these systems are models for catalytic reactions of technological interest.

Inelastic Electron Tunneling Spectroscopy. IETS is unique in that it is entirely surface specific, but does not require a vacuum environment and has almost never been performed in UHV. Its principle is relatively simple. If two metals are separated by a thin (≈ 3 nm) insulating layer, and a voltage is applied across them, electrons can tunnel from one to the other through the insulator, and if no energy is lost by the electrons, the process is known as elastic tunneling. However, should discrete impurity states occur in the interface between either metal and the insulator, or molecules exist in such an interface that possess characteristic vibrational energies, then the tunneling electron can give up some of this energy either to the state or to the vibrational mode, before reaching the other metal. This is called inelastic tunneling. Obviously, the applied voltage must be greater than the state or vibrational energies. If the current across the metal–insulator–metal sandwich is recorded as a function of applied voltage, the current increases as the threshold for each state or vibrational mode is crossed. The increases in current are in fact very small, and for improved detectability, the current is double differentiated with respect to voltage, thereby providing in effect a vibrational spectrum that can be compared directly with free-molecule IR and Raman spectra.

The metal–insulator–metal sandwich is known as a tunnel junction, and its preparation is all-important. The standard junction consists of an aluminum strip ca. 60–80 nm thick and 0.5–1.0 mm wide deposited in a very good vacuum onto scrupulously clean glass or ceramic, the surface of the strip then being oxidized either thermally or by glow discharge. The resultant oxide layer is extremely uniform and about 3 nm thick. Introduction of adsorbed molecules onto the oxide layer, or "doping" as it is called, is then effected either by immersion in a solution followed by spinning to remove excess fluid, or by coating from the gas phase. The final stage in preparation is deposition of the second metal (invariably lead) of the sandwich; this deposition is carried out in a second vacuum system (i.e., not the one used for aluminum deposition), the final thickness of lead being about 300 nm, and the width of the lead strip being the same as that of the aluminum strip. The reason for using lead is that all IETS measurements are carried out at liquid-helium temperature, 4.2 K, to optimize the energy resolution by reducing the contribution of thermal broadening to the linewidth, and of course lead is superconducting at that temperature.

Normally not one, but several, junctions are fabricated simultaneously, since the likelihood of junction failure always exists. After fabrication the electrical connections are made to the metal electrodes, the junctions are dropped into liquid helium, and measurements are commenced at once.

Although insulators other than aluminum oxide have been tried, aluminum is still used almost universally because it is easy to evaporate and forms a limiting oxide layer of high uniformity. To be restricted therefore to adsorption of molecules on aluminum oxide may seem like a disadvantage of the technique, but aluminum oxide is very important in many technical fields. Many catalysts are supported on alumina in various forms, as are sensors, and in addition the properties of the oxide film on aluminum metal are of the greatest interest in adhesion and protection.

Penetration of moisture into an adhesion bond to aluminum is one of the reasons for ultimate failure of such a bond, and the problem is minimized by treatment of the aluminum oxide surface with a "hydration inhibitor" before adhesion. Some of the most successful inhibitors have been aminophosphonates, the suggestion being that P–O–Al bonds are formed that prevent the ingress of moisture. The interaction of various phosphorus acids with alumina has been studied by Ramsier et al. [308]. Figure 88 compares the IETS spectra (spectra b) with the IR absorption spectra (spectra a) for phosphonic acid (A) and phosphinic acid (B). In both cases the strong IR peaks at 979 and 1185 cm^{-1} for phosphinic acid and at 946 and 1179 cm^{-1} for phosphonic acid, characteristic of P–OH and P=O stretching vibrations, respectively, are completely absent in the IETS spectra. These absences indicate that the acids are adsorbed on the alumina in resonance-stabilized forms reached via a condensation mechanism in which a water molecule is formed, which then disperses into solution.

Other applications of IETS have included complexation reactions of $CoBr_2$ and $CoCl_2$ on silane-modified alumina [309], silylation of plasma-grown aluminum oxide [310], the effect of hydrogen on the growth of MgO [311], and adsorption of silane adhesion promoters on aluminum oxide [312].

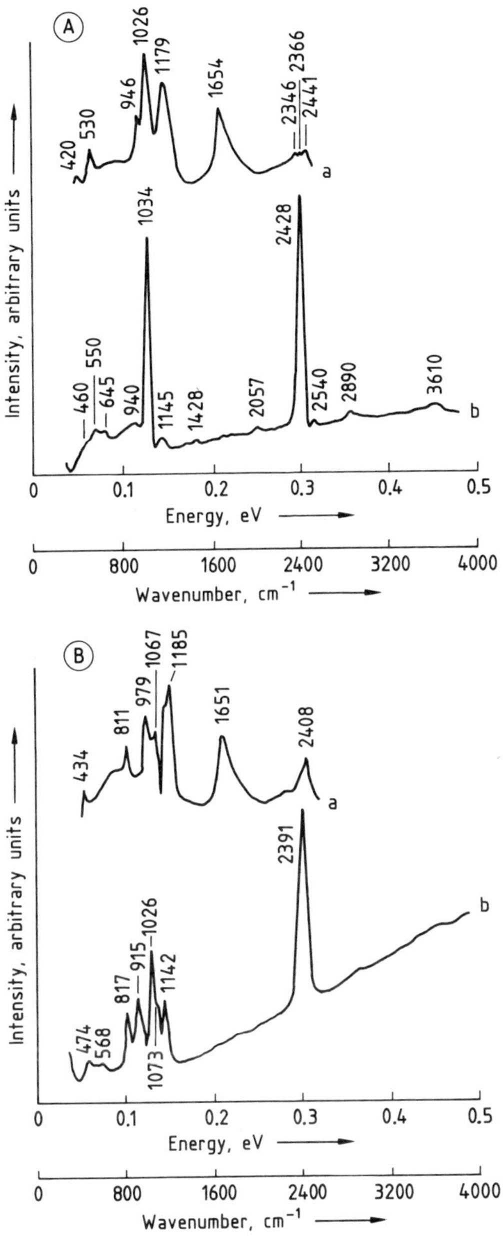

Figure 88. Comparison of IR absorption spectra in water with IETS spectra when adsorbed on aluminum oxide [308], of
A) Phosphonic acid; B) Phosphinic acid
a) IR spectrum; b) IETS spectrum

7.4. Appearance Potential Methods (SXAPS, AEAPS, DAPS)

Appearance potential methods all depend on detecting the threshold for ionization of a shallow core level and the fine structure near the threshold, differing only in the way in which detection is performed. In all of them the primary electron energy is ramped upward from near zero to whatever is appropriate to the sample material, while the primary current to the sample is kept constant. As the incident energy is increased, it passes through successive thresholds for the ionization of core levels of atoms in the surface. An ionized core level, as discussed earlier, can recombine by the emission either of a characteristic X-ray photon or of an Auger electron.

Soft X-Ray Appearance Potential Spectroscopy. In SXAPS the X-ray photons emitted by the sample are detected, normally by letting them strike a photosensitive surface from which photoelectrons are collected, but also—with the advent of X-ray detectors of increased sensitivity—by direct detection. Above the X-ray emission threshold from a particular core level the excitation probability is a function of the densities of unoccupied electronic states. Since two electrons are involved, the incident and the excited, the shape of the spectral structure is proportional to the self-convolution of the unoccupied state densities.

Auger Electron Appearance Potential Spectroscopy. Due to the emission of an Auger electron as an alternative to soft X-ray photons, the total secondary electron yield will show an increase as an ionization threshold is crossed. It is the total secondary yield that is monitored in AEAPS, in a rather simple experimental arrangement. The secondary current arises not just from the Auger electrons themselves, but also from inelastic scattering of Auger electrons produced at greater depths below the surface. Since the yield change is basically a measure of the probability of excitation of a core-level electron to an empty state above the Fermi level, the fine structure above the threshold will be similar to that seen in SXAPS. One of the problems in AEAPS is that of a poorly behaved background, which means that SXAPS is preferred.

Disappearance Potential Spectroscopy. Crossing an ionization threshold means that electrons are lost from the primary beam as a result of ionization of a core hole. Thus if the reflected current of electrons at the primary energy, more usually termed the *elastically reflected current*, is monitored as a function of energy, it should show a sharp decrease as a threshold is crossed. This is the principle of operation of DAPS. It is in a sense the inverse of AEAPS, and indeed if spectra from the two techniques from the same surface are compared, they can be seen to be mirror images of each other. Background problems occur also in DAPS.

The principal advantages of AEAPS and DAPS over SXAPS is that they can be operated at much lower primary electron currents, thus causing less disturbance to any adsorbed species.

7.5. Inverse Photoemission (IPES, BIS)

Irradiation of a surface with electrons leads to the emission not only of X-ray photons of energies characteristic of the material, but also of a continuous background of photons called *bremsstrahlung radiation*. If a detector is set to detect only those background photons of a particular energy, and the primary electron energy is ramped upward from zero, then the variations in photon flux should mirror the variations in the densities of unoccupied electron states. The process is called inverse photoemission because it is clearly the opposite of ordinary photoemission as in XPS and UPS, and is illustrated in Figure 89 from WOODRUFF et al. [313]. Although the name inverse photoemission spectroscopy (IPES) should apply to all forms of the inverse photoemission technique, in fact it is confined by usage to the form in which the energy at which the photons are detected is in the UV region, typically with $h\nu = 9.7$ eV. If on the other hand a crystal monochromator normally used to provide monochromatized X-rays for XPS is used in reverse, with a detector placed where the electron source is in Figure 3, then detection is at the X-ray energy $h\nu = 1486.6$ eV, and the technique is called bremsstrahlung isochromat spectroscopy (BIS).

Since IPES maps the densities of unoccupied states, it is related to other techniques that do the same (i.e., STS and SXAPS). When used in conjunction with a technique that maps the densities

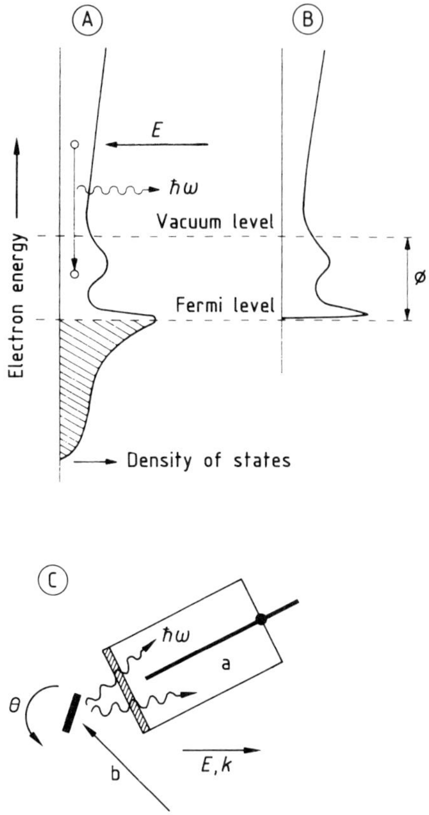

Figure 89. Schematic of the principles of IPES (A) and BIS (B) [313]; C) Experimental arrangement
a) Geiger–Müller counter; b) Electron beam

of occupied surface states, such as UPS or ELS, a continuous spectrum of state density from occupied to unoccupied can be obtained. Just as in UPS, in which angular resolution allows elucidation of the three-dimensional occupied band structure, so in IPES angular resolution allows mapping of the three-dimensional unoccupied band structure. This version is called *KRIPES* (i.e., *K-resolved IPES*).

The various versions of inverse photoemission have been applied to purely basic problems, just as with UPS and ELS. Recent typical examples include the bonding of H_2 on Ni (110) [314]; the oxidation of a titanium thin film [315]; the effects on H_2 chemisorption of a monolayer of nickel on Cu (111) [316]; unoccupied electronic structure in La_2CuO_4 [317] and in $Bi_2Sr_2CaCu_2O_8$ [318]; and KRIPES studies of the Ge (113) (2 × 1) [319] and $TiN_{0.83}$ (100) [320] surfaces.

7.6. Ion Excitation Methods (IAES, INS, MQS, SNMS, GDMS)

Ion (Excited) Auger Electron Spectroscopy. Auger emission following creation of a core hole by electron or photon irradiation has been described under AES and XPS. Incident ions can also create core holes, so that Auger emission as a result of ion irradiation can also occur, giving rise to the technique of IAES. The ions used are normally noble-gas ions, but protons and α-particles have occasionally been used as well. In addition to the normal Auger features found in AES and XPS spectra, peaks are found in IAES spectra arising from Auger transitions apparently taking place in atoms or clusters sputtered from the surface. These peaks do not always coincide with those found in gas-phase excitation, and others are found that are not present in gas-phase measurements. Because of the complexity of the spectra, IAES cannot be used directly as an analytical technique, but it is very useful in basic physics experiments that study Auger processes occurring in excited atoms. Good examples of ion-excited Auger spectra and their interpretation can be found in [321]–[327].

Ion Neutralization Spectroscopy. The technique of INS is probably the least used of all those described here, because of experimental difficulties, but it is also one of the physically most interesting. Ions of He^+ of a chosen low energy in the range 5–10 eV approach a metal surface and within an interaction distance of a fraction of a nanometer form ion–atom pairs with the nearest surface atoms. The excited quasi molecule so formed can deexcite by Auger neutralization. If unfilled levels in the ion fall outside the range of filled levels of the solid, as is the case for He^+, then an Auger process can take place in which an electron from the valence band of the solid fills the core hole in the ion and the excess energy is given up to another valence electron, which is then ejected. Since either of the valence electrons can come from anywhere within the valence band, the observed energy spectrum reflects the local density of states at the solid surface, but is a self-convolution of the LDOS and of transition probabilities across the valence band. The technique of INS was originated by HAGSTRUM [328], who has given a very detailed exposition of the complex mathematical treatment necessary for deconvolution of the spectra [329], [330]. Because the technique of STS is also capable of extracting the LDOS, the results from INS and STS should be comparable, but such a comparison does not seem to have been attempted yet. HAGSTRUM has used INS extensively to study nickel, both the differences in the various low-index faces and the effects of adsorption of the chalcogen elements oxygen, sulfur, and selenium [331]–[333].

Metastable Quenching Spectroscopy. MQS is in a sense an extension of INS. Instead of He^+ ions, helium atoms in metastable states are used in the incident beam, at the same low energies. The excited singlet state $He^*(2^1S_1)$ has an energy of 20.62 eV and a lifetime of 2×10^{-2} s, and so is suitable. It is produced by expanding helium gas at high pressure through a nozzle into a cold cathode discharge sustained by combined electrostatic and magnetic fields. The high fields prevent ions and fast neutrals from leaving the discharge, and the beam is then nearly all $He^*(2^1S_1)$. As the metastable ion approaches the surface, either of two mechanisms can lead to deexcitation, both resulting in Auger emission similar to INS. If the excited level in the atom *can resonate with empty states at the Fermi level* of the surface, electron transfer from the atom to the surface can occur, leading to resonance ionization of the helium atom. Auger neutralization then takes place as in INS, and the resultant spectrum is again a self-convolution of the LDOS. If, however, the *excited level cannot resonate with empty surface states*, then direct Auger deexcitation can occur, in which the hole in the inner shell of the metastable helium atom is filled from a surface state of the sample, followed by ejection of the excited electron from the helium atom. The process is also called Penning ionization. In the latter process, only one electron is ejected, and the resultant spectrum is thus an unconvoluted reflection of the LDOS.

MQS was first demonstrated by CONRAD et al. [334], and the mechanisms involved in deexcitation have been discussed by CONRAD et al. [335] and by BOZSO et al. [336]. Even the brief description given above makes clear that during reaction at a surface, leading to changes in the surface electronic structure, the situation can easily change from resonance to nonresonance of the excited atomic level with surface states, in which case the deexcitation then switches from resonance ionization + Auger neutralization to Penning ionization. The resultant spectrum then changes in character from one similar to INS to one similar to UPS. Such a switch has been observed by BOZSO et al. [337]. Like INS, MQS is highly surface specific. Its applications have been entirely basic, for example, the adsorption of NH_3 on Ni (111) [338] and Ni (110) [339]; of CO, PF_3, NH_3, C_2H_2, and C_6H_6 on palladium and copper surfaces [340]; of CO on copper and on copper-covered ruthenium [341]; and of potassium on hydrogen-covered ruthenium [342].

Secondary Neutral Mass Spectrometry and Glow Discharge Mass Spectrometry. SNMS and GDMS are grouped together because of the close similarities in principle between them. In both techniques the surface to be analyzed is sputtered and the contribution to the sputtered flux of secondary neutrals is measured. They differ in the methods used to measure the secondary neutral flux. In *GDMS*, as in GDOS, atoms are sputtered via a glow discharge in which the surface is the cathode, and on entering the discharge region, many of the atoms are ionized. The ions are then extracted from the discharge and passed via differential pumping stages into a region at much lower pressure where they can be mass analyzed. Either magnetic or quadrupole mass spectrometers are used for mass analysis. In *SNMS*, three different approaches to postionization of neutrals have been adopted. The first uses *electron impact ionization* (i.e., direct ionization of neutrals by an electron beam from a hot filament as they leave the surface). Secondary ions produced as a result of sputtering are deflected out of the secondary particle flux before they reach the postionization region. If the ion deflection and hot filament are both switched off, the instrument can be used as a normal SSIMS analyzer. The second postionization approach is that of *plasma ionization* in a high-frequency, low-pressure gas discharge (i.e., under conditions very different from the higher pressure discharge used for GDMS and GDOS). Sputtering can be performed either by accelerating positive ions from the discharge to the sample surface or by an external ion gun. In either case, neutrals produced by sputtering pass into the plasma where some are ionized and are thus extracted directly into a quadrupole mass spectrometer. Again, with an external ion gun and without the plasma, SSIMS can be performed. This variation of SNMS has been pioneered by OECHSNER and coworkers [343], and Figure 90 shows schematically some of the experimental arrangements developed by OECHSNER [344]. Thirdly, postionization can be

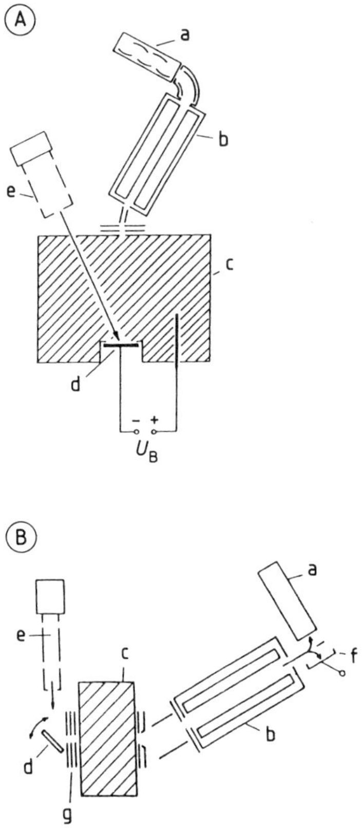

Figure 90. Schematics of designs for SNMS [344]
A) Direct bombardment mode via noble-gas ions in the plasma, or separate bombardment mode via rastered ion beam traversing the plasma; B) External bombardment mode via external ion gun
a) Multiplier; b) Quadrupole mass spectrometer; c) High-frequency plasma; d) Sample; e) Ion gun; f) Faraday cup; g) Electrical diaphragm

achieved by *high-intensity laser beams*, involving either resonant or nonresonant multiphoton ionization. In the resonant form, a tunable dye laser is tuned to exactly the correct wavelength to excite a chosen atom to an excited state; while in that state, interaction with a second photon leads to ejection of an electron (i.e., ionization). In the nonresonant form, excimer lasers of such high power are used that even if the laser frequency is a long way off resonance for many atoms, adequate ionization efficiency is achieved. The latter variations of SNMS have been given their own acronyms, surface analysis by resonance ionization of sputtered atoms (*SARISA*) and surface analysis by laser ionization (*SALI*), respectively.

Like SSIMS and GDOS, both SNMS and GDMS have been developed commercially, and

GDMS has also had wide industrial application. Most of the applications have, like GDOS, consisted of rapid profiling of surface films many micrometers thick, with special emphasis on the depth distribution of trace elements, and would not be regarded as surface specific. Nevertheless, again like GDOS, GDMS is in principle a surface-specific technique and has occasionally been used [345] in an ultrahigh-vacuum system and at very low erosion rates.

One of the attractions of SNMS and GDMS is the relative ease and accuracy of quantification using standard materials and relative sensitivity factors. For most elements, particularly the metallic and semiconducting elements, the spread of sensitivity factors is less than an order of magnitude, comparable to the situation in XPS. In addition, the actual sensitivities are significantly higher than in XPS or AES, with analysis at the part-per-million level being routine and at the part-per-billion level being quite possible. A typical profile using GDMS to detect low concentrations is shown in Figure 91 from Sanderson et al. [346], through two layers of silicon-doped GaAs grown epitaxially on bulk GaAs, the silicon levels being different in each layer and in the bulk. Each epilayer can be distinguished easily in terms of the silicon concentration at the part-per-million level, and the semiquantitatively derived silicon concentrations agree well with the amounts known to be incorporated during growth.

A demonstration of both the high depth resolution obtainable in SNMS and the quantification of the technique is given by Figure 92 from Wucher et al. [347]. It shows a profile through a sandwich structure consisting of 39 alternate layers of chromium and nickel, each 15 nm thick, recorded using a commercial instrument with both bombardment and neutral postionization provided by a high-frequency plasma with an applied voltage of 515 V. Figure 92 A shows the raw intensity data; Figure 92 B, the measured total sputter yield, in which the chromium yield is markedly lower than that of nickel; and Figure 92 C, the intensities corrected for differing sputter yields and then quantified. The sharpnesses of the interlayer interfaces allowed the thickness-

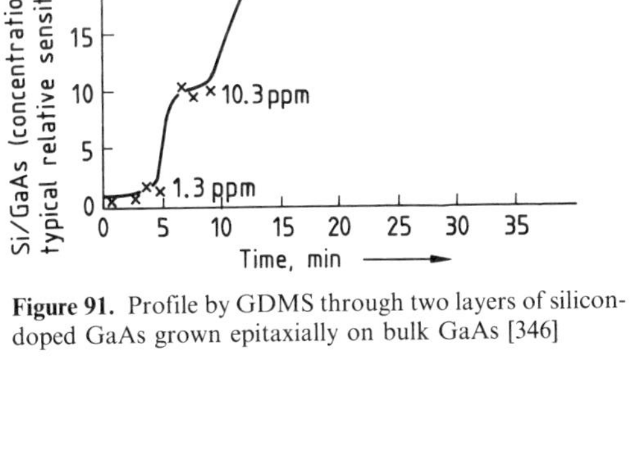

Figure 91. Profile by GDMS through two layers of silicon-doped GaAs grown epitaxially on bulk GaAs [346]

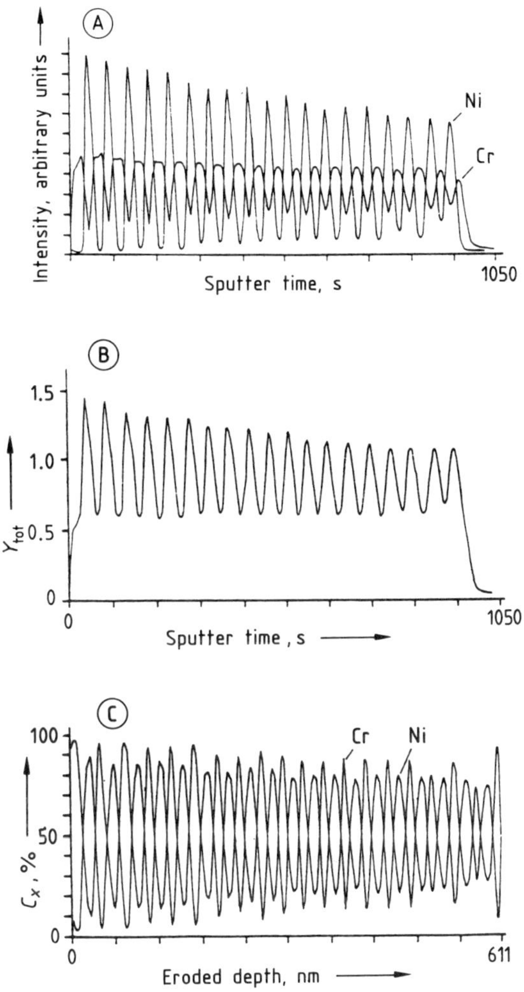

Figure 92. Profile by SNMS of multilayer Ni–Cr sandwich system, with layer thickness of 15 nm [347]
A) Raw intensity data; B) Time-dependent total sputter yield; C) Plot of concentration versus depth evaluated from A) and B)

es of the individual layers to be determined with a standard deviation of only 10 %.

GDMS has been applied extensively in the steel industry for trace-element analysis and for profiling surface films and coatings. Some examples are given in [348], e.g., detection of uranium and thorium in concentrations of a few parts per billion in aluminum; of some 16 trace elements in concentrations of fractions of a part per billion in indium; of more than 25 impurity elements in a high-temperature alloy in concentrations ranging from 0.2 to 2000 ppm; and of 5 nonmetallic impurity contents in low-alloy steels. Good accuracy was achieved throughout. Applications of SNMS have included determination of hydrogen profiles in a-Si:H [349]; analysis of insulators such as SiO_2, Si_3N_4, PTFE, and glasses [350]; the study of aminoacid layers on metals [351]; depth profiles of silicides [352]; analysis of bulk polymers [353]; thin-film systems [354]; and Cu–Al alloy [355] by SALI, and various metallic and nonmetallic materials by SARISA [356].

7.7. Neutral Excitation (FABMS)

Fast-atom bombardment mass spectrometry is very similar to SSIMS in practice, the only difference being that instead of using positively charged ions as the primary probe, a beam of energetic neutral atoms is used. Secondary ions are emitted as in SSIMS and analyzed in a mass spectrometer, usually of the quadrupole type. The beam of fast atoms is produced by passing a beam of ions through a charge-transfer cell, which consists of a small volume filled with argon to a pressure of about 10^2 Pa. Charge transfer occurs through resonance between fast argon ions and argon atoms with thermal energy, with efficiencies of 15–20 % if geometric and pressure conditions are optimized. Residual ions are removed by electrostatic deflection.

FABMS has two advantages over SSIMS, both arising from the use of neutral rather than charged particles. Firstly, little or no surface charging of insulating materials occurs, so that organics such as polymers can be analyzed without the need to employ auxiliary electron irradiation to neutralize surface charge. Secondly, the extent of beam damage to a surface, for the same particle flux, is much lower using FABMS than SSIMS, thus allowing materials such as inorganic compounds, glasses, and polymers to be analyzed with less worry about damage introducing ambiguity into the analysis. An example illustrating these advantages is given in Figure 93 from Michael and Stulik [357]. There the SSIMS (Fig. 93 A) and the FABMS (Fig. 93 B) secondary positive ion spectra are compared for PTFE; the surface of the PTFE has to be coated with a gold film for SSIMS measurement but not for FABMS. More fragment ions occur in the SSIMS spectrum than in FABMS, and in the FABMS spectrum the size of the peak at $m/e = 12$ (C^+) relative to the principal peak at $m/e = 69$ (CF_3^+) is much smaller than in SSIMS, indicating much less damage of the polymer surface during FABMS analysis.

FABMS was developed as an alternative to SSIMS by Vickerman and coworkers; for a detailed description, see [358]–[362]. Applications have included analyses of tin and lead oxides

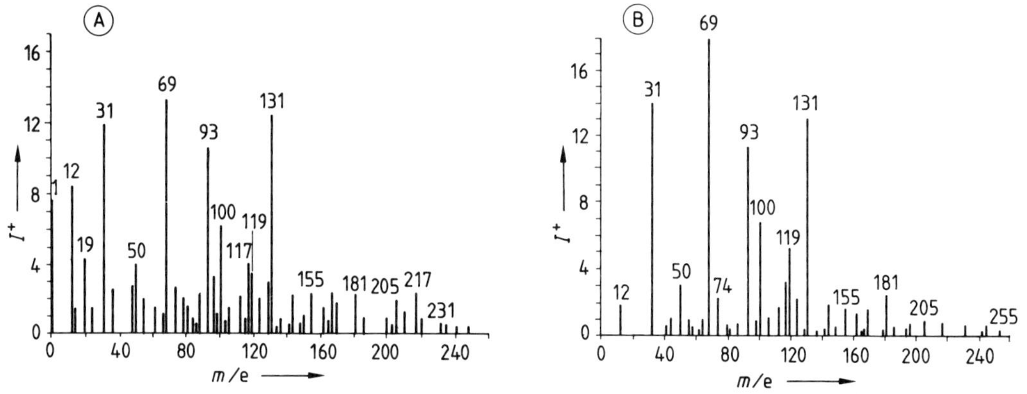

Figure 93. Comparison of SSIMS (A) and FABMS (B) positive secondary-ion spectra from PTFE [357]

[363], of poly(methyl methacrylate)–PTFE sandwich structures [364], of multicomponent glasses [358], of silica [358], and of phosphate- and chromate-treated aluminum [360].

7.8. Atom Probe Field-Ion Microscopy (APFIM, POSAP, PLAP)

The atom probe field-ion microscope (APFIM) and its subsequent developments, the *position-sensitive atom probe* (*POSAP*) and the *pulsed laser atom probe* (*PLAP*), have the ultimate sensitivity in compositional analysis (i.e., single atoms). The APFIM was developed by Müller et al. [365], having grown out of *field-ion microscopy* (*FIM*) which also originated earlier with Müller [366].

FIM is purely an imaging technique in which the specimen in the form of a needle with a very fine point (radius 10–100 nm) is at low temperature (liquid nitrogen or helium) and surrounded by a noble gas (He, Ne, or Ar) at 10^{-2}–10^{-3} Pa. A fluorescent screen or a microchannel plate is situated a few centimeters from the needle. A high positive voltage is applied to the needle, which causes noble-gas atoms approaching the needle to be ionized over points of local field enhancement (i.e., over prominent atoms at the surface); the ions are then repelled from these points and travel in straight lines to the screen or plate, where an atomic image of the needle tip is formed. This is called a field-ion image, and the voltage at which the image is optimized is called the *best imaging voltage*. The latter is in the range 5–20 kV.

Atom Probe Field-Ion Microscopy. If the voltage is raised further above the imaging voltage, the cohesive energy that binds atoms to the surface can be exceeded, and atoms can be removed by what is termed *field evaporation*. The evaporation field required is a function of the sample material and the crystallographic orientation of the needle. The removal of atoms is the basis of APFIM and its daughter techniques. If the evaporation field is pulsed with very short (≤ 10 ns) voltage pulses of extremely sharp rise times (≈ 1 ns), and the time taken for a field-evaporated atom to travel from the tip to a detector is measured, then the mass-to-charge ratio of the ionized atom can be established; i.e., elemental identification of the individual atom is possible. The region on the tip from which atoms are removed is selected by rotating or tilting the tip until the desired region, as viewed in the FIM image, falls over an aperture of about 2-mm diameter in the fluorescent screen. Most field-evaporated atoms will strike the screen, but those from the selected area will pass through the aperture into the time-of-flight mass spectrometer. For an aperture of 2 mm, the area analyzed on the tip is ca. 2-nm in diameter.

A modern instrument using an energy-compensated TOF mass spectrometer is shown in Figure 94 from Miller [367]. Atoms may also be removed from a tip by focusing a pulsed laser onto it, as an alternative to field evaporation, leading to the variation of APFIM called the *pulsed laser atom probe* (*PLAP*).

Position-Sensitive Atom Probe. Pulsed operation of the APFIM leads to analysis of all atoms within a volume consisting of a cylinder of ca. 2-nm diameter along the axis of the tip and aperture, with single atomic layer depth resolution but no indication of just where within that volume any particular atom originated. The advent of position-sensitive detectors has allowed APFIM to be extended so that three-dimensional compositional variations within the analyzed volume can be determined. The development has been pioneered by Smith and coworkers [368], [369], and is called position-sensitive atom probe (POSAP). The aperture and single-ion detector of APFIM are replaced by a wide-angle double channel plate, with a position-sensitive anode just behind the plate. Field- or laser-evaporated ions strike the channel plate, releasing an electron cascade, which is accelerated toward the anode. The impact position is located by the division of electric charge between three wedge-and-strip electrodes. From this position the point of origin of the ion on the tip surface can be determined, since the ion trajectories are radial. Thus after many evaporation pulses, leading to removal of a volume of material from the tip, both the identities and the positions of all atoms within that volume can be mapped in three dimensions. Because an evaporated volume may contain many thousands of atoms, the data collection and handling capabilities must be particularly sophisticated.

POSAP is the only technique available for identifying and locating precipitates, second phases, particles, and interfaces on an atomic scale, and has therefore had considerable application in metallurgical and semiconductor prob-

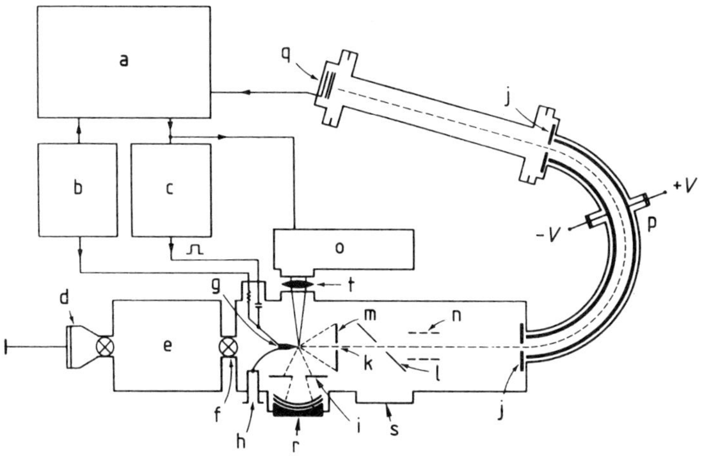

Figure 94. Energy-compensated time-of-flight mass spectrometer for APFIM measurements [367]
a) Computer-controlled fast digital CAMAC system; b) d.c. High voltage; c) High-voltage pulse; d) Airlock; e) Preparation chamber; f) Isolation value; g) Specimen; h) Cryostat; i) Variable iris; j) Herzog electrode; k) Probe aperture; l) Mirror; m) Channel plate; n) Einzel lens; o) Pulsed laser; p) Energy-compensating lens; q) Single-atom detector; r) Atom probe imaging viewpoint; s) FIM viewpoint; t) Focusing lens

lems. A typical example of a POSAP three-dimensional analysis is given by CEREZO [370] for the ferrite phase of an aged duplex stainless steel CF 3 containing 21 % chromium and 9 % nickel as well as manganese, silicon, molybdenum, and carbon. The Fe–Cr system has a spinodal region in which it is thermodynamically unstable with respect to small compositional fluctuations, leading to chromium-rich and iron-rich phases forming a complex interconnected morphology. Figure 95 from CEREZO shows a series of sections by POSAP, each image being about 15 nm in diameter, with a vertical separation between sections of about 3 nm. The images are in chromium, on a gray scale, with white representing ≤ 5 at% Cr, and black ≥ 50 at% Cr. The sequence shows clearly that the chromium-rich phase is interconnected in depth. From such a sequence, with the help of powerful data handling, a three-dimensional reconstruction can be derived, as shown in Figure 96, where the chromium isosurface for concentrations about halfway between the maximum and minimum values is illustrated. The view is the same as in Figure 95 (i.e., along the direction of analysis). The isosurface represents the approximate interface between the chromium-rich and chromium-depleted regions, and shows clearly the nature of the interconnected structure with an average width of only 2–3 nm.

POSAP has also been applied to the phase chemistry of AlNiCo 2 magnet alloys; the precipitation of copper in iron-based alloys; surface segregation in ceramic oxide superconductors; heterostructures in compound semiconductors [369]; and Pt–Rh catalysts and multiquantum well structures [371]. Applications of *APFIM* have included finely dispersed carbides in a 14Co–10Ni–2Cr–1Mo high-strength steel [372]; the distribution of vanadium atoms in the ordered alloy Fe–Co–2V [373]; the location of hafnium atoms in the ordered alloy Ni_3Al [367]; the formation of precipitates and segregation of elements in a nickel-based superalloy [374]; and metal–semiconductor and $Si–SiO_2$ interfaces [375].

7.9. Desorption Methods (ESD, ESDIAD, TDS)

Electron-Stimulated Desorption and Electron-Stimulated Desorption Ion Angular Distribution. Electron irradiation of a surface, particularly one covered with one or more adsorbed species, can give rise to many types of secondary particles, including positive and negative ions. In ESD and ESDIAD, the surface is irradiated with electrons of energies in the range of 100–1000 eV, and the ejected positive-ion currents of selected species are measured in a mass spectrometer. If the angular distribution of the secondary ions is also measured, either by display on a screen or by using position-sensitive detec-

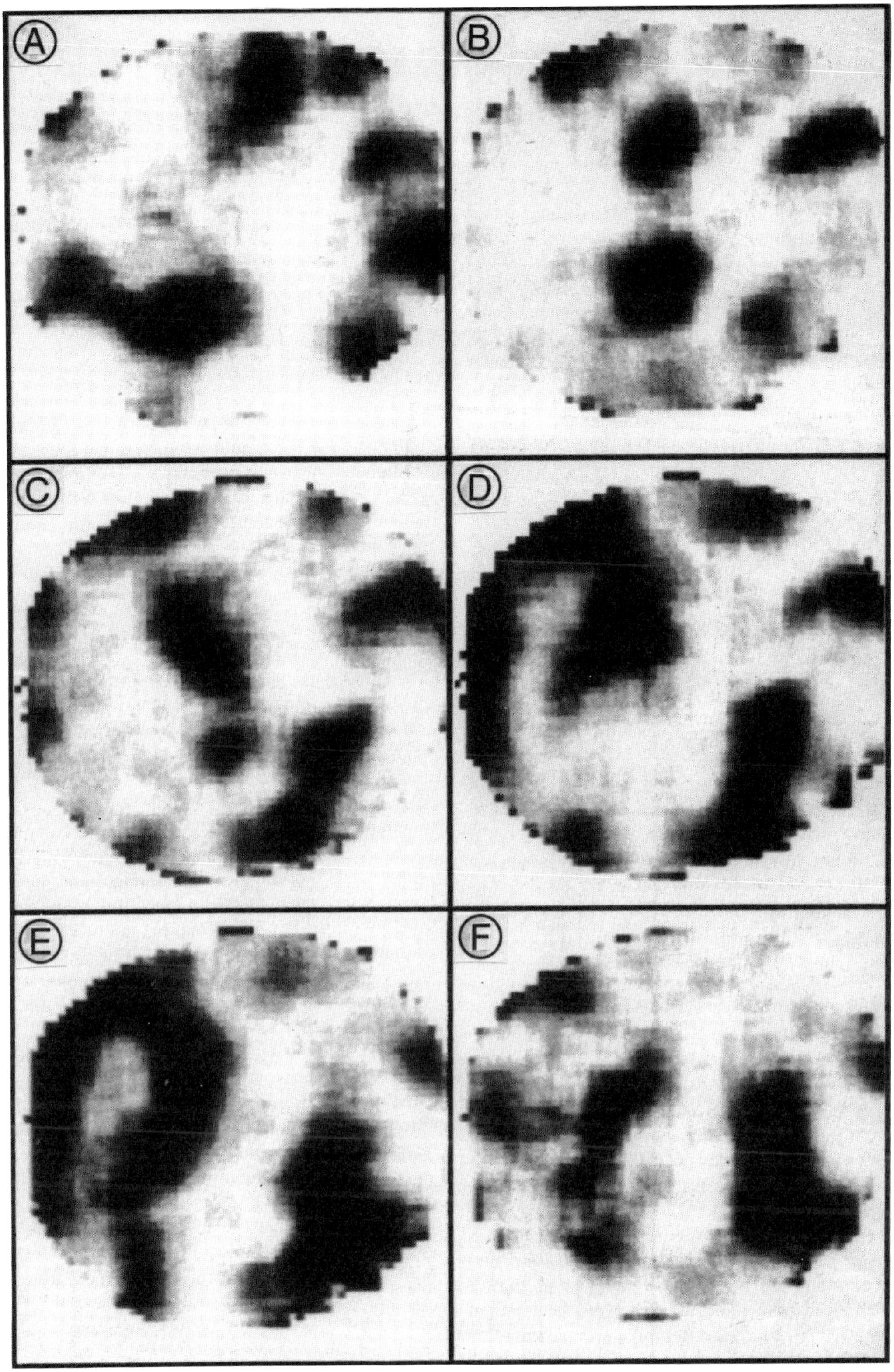

Figure 95. Sequence of images A–F obtained in POSAP by Cerezo [370] from the ferrite phase of an aged duplex stainless steel CF3
Each image is 15 nm in diameter, and the amount removed between each successive image corresponds to a vertical separation of ca. 4 nm
White areas represent ≤ 5 at% Cr, dark areas ≥ 50 at% Cr

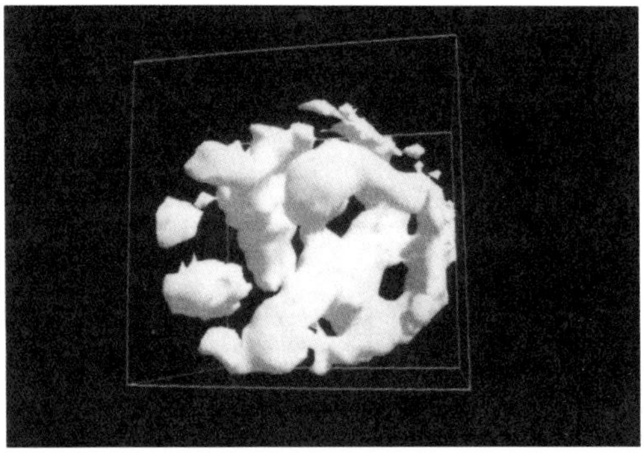

Figure 96. Three-dimensional reconstruction of data of Figure 95 [370]
The isosurface has been drawn through all points halfway between maximum and minimum chromium concentrations. The interconnected structure is Cr-rich

tion, then electron-stimulated desorption (ESD) becomes electron-stimulated desorption ion angular distribution (ESDIAD).

The electron desorption techniques are not used, and probably cannot be used, for compositional analysis, but they provide valuable information on the nature of electronic interactions leading to the breaking of bonds and, in the angle-resolved form, on the geometry of surface molecules and the orientation of broken bonds. The primary electrons do not, at the energies employed, succeed in breaking molecular or surface-to-molecule bonds or in knocking ions out of the surface directly, but the process is one of initial electronic excitation. An electron is absorbed by a surface–adsorbate complex or an adsorbed molecule itself, leading to excitation to an excited state by a Franck–Condon process. If the excited state is antibonding and the molecule or radical is already far enough from the surface, desorption can occur. Since return to the ground state after excitation is a much more probable process, the cross sections for ion desorption are low, 10^{-20}–10^{-23} cm^2. If core-level ionization is involved in the initial interaction with the incident electron, rather than valence levels as in the Franck–Condon-type excitation, then a desorption mechanism based on Auger decay has been proposed by KNOTEK and FEIBELMAN [376]. The core level left behind after interatomic Auger decay creates a positive ion, which is then expelled by the repulsive Madelung potential.

The most common ions observed as a result of electron-stimulated desorption are atomic (e.g., H^+, O^+, F^+), but molecular ions such as OH^+, CO^+, H_2O^+, and CO_2^+ can also be found in significant quantities following adsorption of H_2O, CO, CO_2, etc. Substrate metallic ions have never been observed, which means that ESD is not applicable for surface compositional analysis. The most important application of ESD in the angularly resolved form ESDIAD is in determining the *structure and mode of adsorption* of adsorbed species. This derives from the fact that the ejection of positive ions in ESD is not isotropic. Instead the ions desorb along certain directions only, characterized by the orientation of the molecular bonds that are broken by electron excitation. This orientational dependence is exemplified in Figure 97 from MADEY [377], illustrating the H^+ ESDIAD patterns and suggested models for the adsorption of H_2O on Ag (110). Series A is for adsorption of H_2O on clean silver, B for adsorption on silver with a preadsorbed bromine layer, and C for adsorption on silver with a preadsorbed oxygen layer. The top row consists of three-dimensional perspective plots of the ion intensities; the second row shows two-dimensional intensity contour maps; and the bottom row shows schematic models. On clean silver, the random emission centered about the surface normal indicates that the adsorbed H_2O is disordered locally, with no preferred orientation of the broken bonds. When bromine is present, the H_2O molecules are believed to form hydration shells around each bromine atom, leading to pronounced ordering and hence preferred bond orientation, which is reflected in the very distinct four-beam ESDIAD pattern seen in B. With preadsorbed oxygen, another type of ordering is found, caused by the dissociation of H_2O and reaction with the adsorbed oxygen to form adsorbed OH species. These OH radicals are thought to be tilted along (001) azimuths, giving rise to the intense two-beam pattern, with much weaker beams along the $(1\bar{1}0)$ azimuths, seen in C. The weaker beams are due to OH-stabilized H_2O in low concentration.

ESDIAD is obviously not a diffraction technique such as LEED, but it gives direct information about surface structure in real space. The sensitivity is to local bonding geometry, and long-range order is not necessary as in LEED. ESDIAD is especially sensitive to the orientation of hydrogen atoms in surface complexes, which is difficult to observe by any other technique.

Thermal desorption spectroscopy (TDS), sometimes called temperature-programmed desorption (TPD), is simple in principle. A gas or

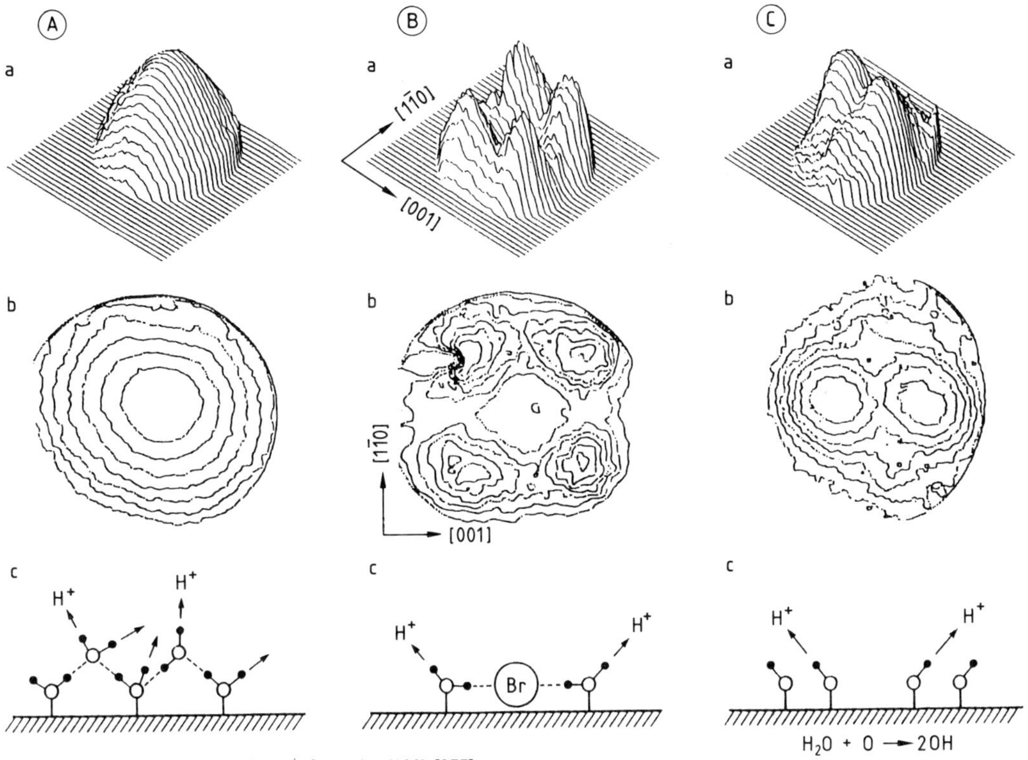

Figure 97. ESDIAD pattern in H^+ from Ag (100) [377]
A) Water-adsorbed; B) Water on preadsorbed Br; C) Water on preadsorbed O
a) Three-dimensional perspective plots; b) Two-dimensional contour maps; c) Schematic models

mixture of gases is allowed to adsorb on a clean metal foil for a chosen length of time; then, after the gas is pumped away, the foil is heated at a strictly linear rate to a high temperature, during which the current of a particular ion or group of ions is monitored as a function of temperature. The ion masses are selected in a quadrupole mass spectrometer. As the binding energy thresholds of the adsorbed species on the surface are crossed, peaks in the desorbed ion current appear at characteristic temperatures. From the characteristic temperatures and the shape of the desorption peak above the threshold, the activation energies for desorption can be obtained, along with information about the nature of the desorption process. The mass spectrum from the mass spectrometer, of course, provides information about the species that actually exist on the surface after adsorption.

Although simple in principle, experimental artefacts that are possible in TDS must be avoided. Thus, ions accepted by the mass spectrometer must originate only from the surface of the foil, and in addition, the temperature distribution across the foil should be uniform to avoid the overlapping of desorption processes occurring at different temperatures. To ensure that these experimental requirements are met, the angle of acceptance into the spectrometer is restricted by placing a drift tube with an aperture between the foil and the spectrometer, and the foil itself is usually in the form of a long thin ribbon, from whose center section only are ions accepted. In addition, the heating rate must be sufficiently fast that the desorbed species accepted into the mass spectrometer is characteristic of the desorption process, but not so fast that the pressure rise around the foil is too great.

With correct experimental procedure, TDS is straightforward to use and has been applied extensively in basic experiments concerned with the nature of reactions between pure gases and clean solid surfaces. The majority of these applications have been catalysis related (i.e., performed on surfaces acting as models for catalysts), and in every case TDS has been used with other techniques such as UPS, ELS, AES, and LEED. To a certain extent it is quantifiable, in that the area under a desorption peak is proportional to the number of ions of that species desorbed in that temperature range, but measurement of the area is not always easy if several processes overlap.

Figure 98 from KRÜGER and BENNDORF [378] shows some typical desorption spectra, following the adsorption of ethylene oxide on Ag (110) at 110 K at exposures increasing from 0.22 to 13.3 Langmuirs. The oxidation of ethylene to ethylene oxide is an industrial process for which silver has important catalytic properties. At very low coverage, only a single mass 29 (the major fragment from the cracking of C_2H_4O in the spectrometer ionization source) desorption peak is found at 170 K, which increases to saturation at about 3 Langmuir, and shifts down to 160 K. Beyond 3 Langmuir exposure, peaks b and c appear at lower temperatures, and c in particular continues to increase with exposure without reaching saturation. For the latter reason, peak c is obviously due to multilayer condensed ethylene oxide. Peaks a and b originate from the adsorption of ethylene oxide in two different sites on the Ag (110) surface, site a being filled first, followed by b when all a sites are occupied. Figure 98 C shows the total integrated area of all peaks as a function of exposure, a strictly linear dependence indicating that the occupancy of sites a and b and condensation into a multilayer all have equal probabilities. Data such as these, when taken with measurements from complementary techniques, are invaluable in establishing reaction mechanisms.

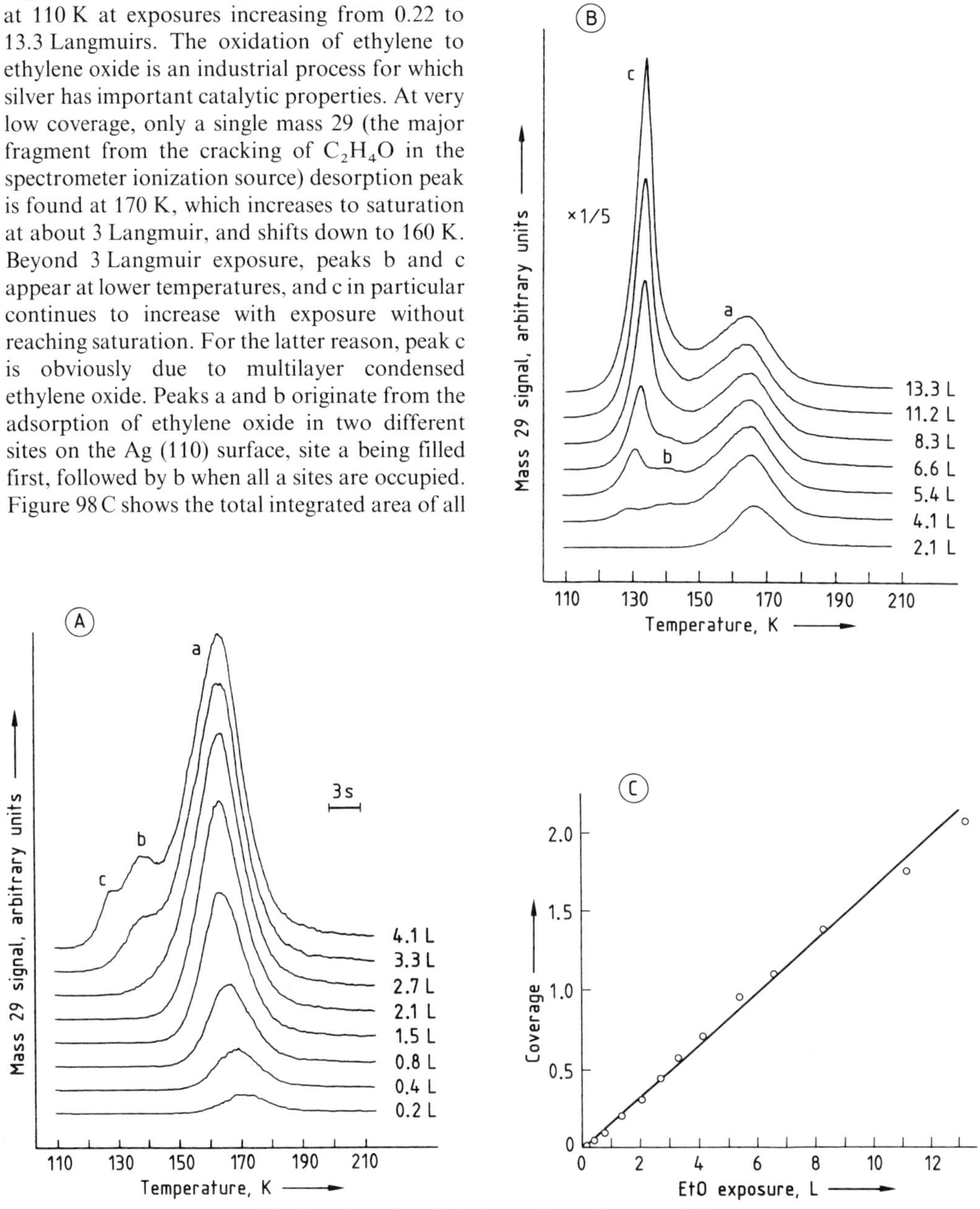

Figure 98. Desorption spectra following adsorption of ethylene oxide on silver at 110 K up to an exposure of 13.3 L [378]
A) 0.2–4.1 L; B) 2.1–13.3 L; C) Relation between surface coverage and exposure of ethylene oxide

7.10. Diffraction Methods (LEED, RHEED) (see also →Structural Analysis by Diffraction, **B5**)

Both LEED and RHEED are probes of surface structure that depend on the existence of long-range order of periodic structure in the surface. They do not provide compositional information and are always used along with other techniques such as AES that can analyze for composition. LEED is probably the oldest surface-specific technique, having been discovered by DAVISSON and GERMER in 1927 [379].

Low-Energy Electron Diffraction. In LEED a collimated and monoenergetic beam of electrons of variable energies between about 20 and 1000 eV strikes the surface of a single crystal at normal incidence. An energy selector, almost universally consisting of a set of spherical grids centered on the specimen, repels all secondary electrons except those elastically scattered (i.e., those at the primary energy). The elastically scattered electrons are then accelerated to a spherical fluorescent screen, on which the diffraction pattern can be seen. Measurement of diffraction spot intensity can be made either with a photometer or with a movable Faraday cup.

Diffraction occurs due to the wave nature of electrons. The waves represented by the incident electron beam interfere as a result of reflection by a periodic crystal structure in such a way that the reflected beams are anisotropic in direction and are in fact concentrated in certain directions related directly to the crystal structure. Where the reflected beams are intercepted by the fluorescent screen, a spot pattern results. The purely geometric relationship between the observed pattern and the surface structure is governed by the well-known Bragg equation, which connects the directions of the interference maxima with the surface lattice spacing and the primary energy. The observed diffraction pattern is a direct representation of the reciprocal lattice of the surface. Since the electron wavelength is of the same order of magnitude as the surface lattice spacing, only a few diffraction spots appear at low primary energies (<100 eV), but as the energy is increased, further spots appear and move with energy toward the normally deflected (i.e., undiffracted) beam. At the same time, the existing spots vary in intensity, appearing and disappearing on the fluorescent screen as they move with energy inward along radial directions.

The simple geometric, or kinematic, representation of the LEED pattern is valid only for scattering from those atoms actually in the surface (i.e., exposed to the incident electron beam). However, the incident beam has a finite depth of penetration, which of course increases with increasing energy; therefore the periodicity of the crystal normal to the surface must also be taken into account. This can be explored by recording the intensity of a particular diffraction spot as a function of primary electron energy (i.e., wavelength). The normally reflected beam will have intensity maxima due to the vertical periodicity, which are called primary Bragg peaks if the maxima are predicted by the Bragg equation, and secondary Bragg peaks if they are not. The latter are caused by multiple scattering effects, which can be accounted for only in a dynamic theory of scattering. The latter is complicated and can be applied only by using computer simulation, in which a feasible structure for the surface is guessed, from which diffraction intensities are calculated and compared with experiment. In this way a surface structure can be solved completely, but ambiguities cannot be eliminated in most cases.

Reflection High-Energy Electron Diffraction. The same diffraction mechanism is used in RHEED, but the experimental arrangement is quite different, and the observed pattern differs as well. Much higher primary electron energies are used than in LEED, in the range of 15–30 keV, and the primary beam is directed onto the crystal surface at near grazing incidence, at angles $<5°$. As a result, the diffraction pattern is a mixture of spots and streaks. For such low angles of incidence the surface must be very smooth and flat, which is a limitation of the technique, but the requirement also means that any growth feature appearing above the original surface is easily identified since it produces its own characteristic diffraction pattern in transmission. In a variation of RHEED, the angle of the primary beam is varied backward and forward through a few degrees, which is called *rocking*, by analogy with the rocking patterns recorded in X-ray diffraction. Analysis of surface structure, with the assistance of sophisticated data processing, is less ambiguous when the rocking variation is used.

Applications. In virtually all basic work on interactions with, and reactions on, single-crystal surfaces over the last 50 years *LEED* has been employed to provide essential structural information. The structures proposed would in general be ambiguous, in the sense of the existence of several possibilities, if LEED were used on its own, but when the LEED data are taken in conjunction with those from other techniques applied at the same time, they have proved invaluable. The number of publications that have included LEED is probably greater than for any of the other techniques described here.

RHEED has also been used in basic studies, although to a far lesser extent because of its inherent restrictions, but in recent years it has found technological application as well. Semiconductor surfaces are usually flat, and epitaxial films grown on them are often flat as well. In those circumstances RHEED has proved to be a useful tool in the quality control of the structural condition of a semiconductor substrate and then of the perfection or otherwise of any films or layers grown or deposited on it. Thus each of the large UHV systems built for molecular beam epitaxy (MBE) contains a RHEED diagnostic system as one of the standard inspection techniques of the epilayer during and after growth.

8. Summary and Comparison of Techniques

Important surface analytical techniques are summarized and compared in Table 5.

9. Surface Analytical Equipment Suppliers

By now a large number of suppliers of surface analytical equipment can be found; a few are able to provide any combination of several techniques; the majority specialize in just one or two.

Supplier	Techniques		
Kratos Analytical Ltd., Barton Dock Road, Urmston, Manchester M31 2LD, United Kingdom	XPS SSIMS	AES UPS	SAM ISS
Ulvac Corporation, 10-3, 1-Chome Kyobashi, Chuo-Ku, Tokyo 104, Japan	AES	SSIMS	LEED
Riber, 133–137, Boulevard National, B.P. 231, F-92503 Rueil Malmaison, France	XPS SAM	AES	SSIMS
V. G. Microtech, Bellbrook Business Park, Bell Lane, Uckfield, East Sussex TN22 1BR, United Kingdom	XPS HREELS SNMS UPS AFM	AES SSIMS ISS BIS APFIM	SAM GDMS STM
Perkin Elmer, Physical Electronics Division, 6509 Flying Cloud Drive, Eden Prairie MN 55344, United States	XPS SSIMS	AES UPS	SAM
SPECS GmbH, Voltastrasse 5, 13355 Berlin, Germany	SNMS	SSIMS	
Scienta Instruments AB, Seminariegatan 33H, S-75228 Uppsala, Sweden	XPS		
Omicron Vakuumphysik GmbH, Idsteinerstrasse 78, 65232 Taunusstein, Germany	STM	AFM	LEED
Oxford Applied Resarch Ltd., Crawley Mill, Witney, Oxon. OX8 5TJ, United Kingdom	RHEED		
Staib Instrumente, Obere Hauptstrasse 45, 85354 Freising, Germany	RHEED		
Princeton Research Instruments, Inc., P.O. Box 1174, Princeton NJ 08542, United States	LEED		
National Electrostatics Corp., Graber Road, Box 310, Middleton WI 53562, United States	RBS		
J. A. Woollam Co., 650 J Street, Suite 39, Lincoln NE 68508, United States	ELL		
Rudolph Research, One Rudolph Road, Box 1000, Flanders NJ 07836, United States	ELL		

Table 5. Summary and comparison of the more important techniques of surface analysis

Technique	Primary probe	Secondary particle(s)	Quantity measured	Elemental range	Type of information	Depth of information	Lateral resolution	Sensitivity, at %	Ease of quantification	Insulator analysis	Destructive	UHV environment
XPS	X-ray photons (Al or Mg K_α)	e^- (photo.)	E	all except H, He	elemental, chemical	4–8 monolayers	1 mm (standard), 5 μm (imaging)	10^{-4}	good	yes	hardly at all	yes
XAES	X-ray photons (Al or Mg K_α)	e^- (Auger)	E	all except H, He	Auger parameter	4–8 monolayers	1 mm	10^{-4}	N/A*	yes	hardly at all	yes
AES	e^- (5–10 keV)	e^- (Auger)	E	all except H, He	elemental	2–6 monolayers	0.5 μm	10^{-4}	moderate	difficult	often	yes
SAM	e^- (20–50 keV)	e^- (Auger)	E	all except H, He	elemental	2–6 monolayers	30 nm	10^{-4}	moderate	no	often	yes
SSIMS	n^+ (Ar^+, Ga^+)	n^+, n^-, clusters	M	all	chemical	1–2 monolayers	1 mm (standard), 50 nm (LMIS)	10^{-2}–10^{-6}	poor	difficult	yes	yes
ISS	n^+ (He^+, Ne^+)	n^+ (He^+, Ne^+)	E	all except H, He	elemental, structural	1–2 monolayers	≈1 mm	10^{-4}	good	yes	not with He^+, possibly with Ne^+	yes
RBS	n^+ (H^+, He^+)	n^+ (H^+, He^+)	E	high Z on low Z	elemental, depth distribution	1 monolayer–2 μm	5 μm–1 mm	10^{-6}	good	yes	no	no
UPS	UV photons (He I, He II)	e^- (photo.)	E	(adsorbed molecules)	chemical (valence band)	2–10 monolayers	≈2 mm	10^{-3}	N/A*	yes	no	yes
RAIRS	IR photons	IR photons (reflected)	wave-number	(adsorbed molecules)	chemical (vibrational)	1–2 monolayers	≈1 mm	10^{-4}	N/A*	no	no	no
SERS	laser photons	photons (scattered)	wave-number	restricted	chemical (vibrational)	10–100 nm	≈0.5 mm	10^{-6}	N/A*	no	no	yes
GDOS	n^+ (Ar^+)	$n^* \rightarrow h\nu$	wave-length	all	elemental (profile)	10 nm–5 μm (profile)	none	10^{-6}	good	yes	yes	no
ELS	e^- (50–2000 eV)	e^- (loss)	E	(adsorbed molecules)	chemical	1–6 monolayers	≈0.5 mm	10^{-3}	N/A*	yes	no	yes
HREELS	e^- (2–5 eV)	e^- (loss)	E	(adsorbed molecules)	chemical (vibrational)	1 monolayer	≈1 mm	10^{-3}	N/A*	yes	no	yes
IETS	(high field)	e^- (loss)	I (tunneling)	(adsorbed molecules)	chemical (vibrational)	1 monolayer	N/A*	10^{-3}	moderate	no	no	no
IPES	e^- (ramped)	photons (UV or X-ray)	photon flux	all, except H, He	chemical (unoccupied states)	1–6 monolayers	0.5–5 μm	10^{-3}	N/A*	yes	possibly	yes

Table 5. (Continued)

Technique	Primary probe	Secondary particle(s)	Quantity measured	Elemental range	Type of information	Depth of information	Lateral resolution	Sensitivity, at%	Ease of quantification	Insulator analysis	Destructive	UHV environment
SNMS	n^+ (Ar^+)	n^0	*M*	all	elemental	1–2 monolayers	≈1 mm	10^{-2}–10^{-6}	moderate	yes	yes	yes
GDMS	n^+ (Ar^+)	$n^* \rightarrow n^+$	*M*	all	elemental	10 nm–5 µm (profile)	none	10^{-6}	good	yes	yes	no
APFIM	(high field) (laser)	n^+	*M*	all	elemental	1 monolayer	2 nm	(1 atom)	very good	no	yes	yes
POSAP	(high field)	n^+	*M*	all	elemental	1–100 monolayers (profile)	2–20 nm	(1 atom)	very good	no	yes	yes
STM, STS	(high field)	e^-	*I* (tunneling)	all	topographic, chemical	1 monolayer	1 atom	(1 atom)	N/A*	no	possibly	yes
AFM	mechanical force		mechanical deflection	all	topographic	1 monolayer	1 atom	(1 atom)	N/A*	yes	no	no
LEED	e^- (200–1000 eV)	e^- (elastic)	*I* (diffracted)	all	structural	1–4 monolayers	≈0.5 mm	10^{-3}	N/A*	no	no	yes
RHEED	e^- (15–30 keV)	e^- (elastic)	*I* (diffracted)	all	structural	1–30 monolayers	30 nm	10^{-3}	N/A*	no	possibly	yes
TDS	T (ramped)	n^+, n^-	*I* (desorbed)	(adsorbed molecules)	chemical	1–2 monolayers	≈2–3 mm	10^{-4}	moderate	no	yes	yes

* N/A = not applicable.

Sub Monolayer Science, 486 Ellis Street, Mountain View CA 94043, United States	SSIMS		
McAllister Technical Services, West 280 Prairie Avenue, Coeur d'Alene ID 83814, United States	STM		
Park Scientific Instruments, 1171 Borregas Avenue, Sunnyvale CA 94089, United States	STM	AFM	
Topometric, 1505 Wyatt Drive, Santa Clara CA 95054, United States	STM	AFM	
Digital Instruments, 6780 Cortona Drive, Santa Barbara CA 93117, United States	STM	AFM	
Burleigh Instruments Inc., Burleigh Park, Fishers NY 14453, United States	STM		
LK Technologies Inc., 3910 Roll Avenue, Bloomington IN 47403, United States	STM		
W. A. Technology Ltd., Chesterton Mills, French's Road, Cambridge CB4 3NP, United Kingdom	STM	AFM	LEED

10. References

[1] J. C. Rivière: *Surface Analytical Techniques,* Oxford University Press, Oxford 1990.

[2] M. P. Seah, D. Briggs in D. Briggs, M. P. Seah (eds.): *Practical Surface Analysis,* 2nd ed., vol. 1, J. Wiley & Sons, Chichester 1990, p. 13.

[3] D. M. Hercules, S. H. Hercules, *J. Chem. Educ.* **61** (1984) 402–409.

[4] M. J. Higatsberger, *Adv. Electron. Electron. Phys.* **56** (1981) 291–358.

[5] H. W. Werner, *Mikrochim. Acta Suppl.* **8** (1979) 25–50.

[6] A. E. Morgan, H. W. Werner, *Phys. Scr.* **18** (1978) 451–463.

[7] J. Tousset, *Le Vide* **33** (1978) 201–211.

[8] G. B. Larrabee, *Scanning Electron Micros.* part I **1977**, 639–650.

[9] A. Benninghoven, *Appl. Phys.* **1** (1973) 3–16.

[10] K. Siegbahn et al.: *ESCA: Atomic, Molecular, and Solid State Structure Studied by Means of Electron Spectroscopy,* Almqvist and Wiksells, Uppsala 1967.

[11] P. Auger, *J. Phys. Radium* **6** (1925) 205–209.

[12] M. P. Seah, W. A. Dench, *SIA Surf. Interface Anal.* **1** (1979) 2–11.

[13] M. P. Seah in J. M. Walls (ed.): *Methods of Surface Analysis,* Cambridge University Press, Cambridge 1989.

[14] P. W. Palmberg, *J. Electron Spectrosc. Relat. Phenom.* **5** (1974) 691–703; *J. Vac. Sci. Technol.* **12** (1975) 379–384.

[15] P. Coxon, J. Krizek, M. Humpherson, I. R. M. Wardell, *J. Electron Spectrosc. Relat. Phenom.* **51/52** (1980) 821–836.

[16] A. Barrie, *Chem. Phys. Lett.* **19** (1973) 109–113.

[17] M. Scrocco, *Chem. Phys. Lett.* **63** (1979) 52–56.

[18] M. Campagna, G. K. Wertheim, Y. Baer in L. Ley, M. Cardona (eds.): *Photoemission in Solids II,* Springer Verlag, Berlin 1979, chap. 4.

[19] G. K. Wertheim, D. N. E. Buchanan, N. V. Smith, M. M. Traum, *Phys. Lett. A* **49 A** (1974) 191–192.

[20] J. H. Scofield, *J. Electron. Spectrosc. Relat. Phenom.* **8** (1976) 129–137.

[21] R. F. Reilman, A. Msezane, S. T. Manson, *J. Electron Spectrosc. Relat. Phenom.* **8** (1976) 389–394.

[22] M. P. Seah in D. Briggs, M. P. Seah (eds.): *Practical Surface Analysis,* 2nd ed., vol. I, J. Wiley & Sons, Chichester 1990, p. 222.

[23] D. Briggs, M. P. Seah (eds.): *Practical Surface Analysis,* 2nd ed., vol. I, J. Wiley & Sons, Chichester 1990, p. 635.

[24] P. Brüesch, K. Müller, A. Atrens, H. Neff, *Appl. Phys. A* **A 38** (1985) 1–18.

[25] S. Hoffmann in D. Briggs, M. P. Seah (eds.): *Practical Surface Analysis,* 2nd ed., vol. I, J. Wiley & Sons, Chichester 1990, p. 143.

[26] C. D. Wagner, *Faraday Discuss. Chem. Soc.* **60** (1975) 291–300.

[27] C. D. Wagner, L. H. Gale, R. H. Raymond, *Anal. Chem.* **51** (1979) 466–482.

[28] C. D. Wagner in D. Briggs, M. P. Seah (eds.): *Practical Surface Analysis,* 2nd ed., vol. I, J. Wiley & Sons, Chichester 1990, p. 595.

[29] J. M. Walls (ed.): *Methods of Surface Analysis,* Cambridge University Press, Cambridge 1989.

[30] D. P. Woodruff, T. A. Delchar: *Modern Techniques of Surface Analysis,* Cambridge University Press, Cambridge 1988.

[31] J. F. Watts: *An Introduction to Surface Analysis by Electron Spectroscopy,* Oxford University Press, Oxford 1990.

[32] G. Ertl, J. Küppers: *Low Energy Electrons and Surface Chemistry,* 2nd ed., VCH, Weinheim 1985.

[33] D. Briggs (ed.): *Handbook of X-Ray and Ultraviolet Photoelectron Spectroscopy,* Heyden, London 1977.

[34] J. M. López Nieto, V. Cortés Corberán, J. L. G. Fierro, *SIA Surf. Interface Anal.* **17** (1991) 940–946.

[35] T. A. Patterson, J. C. Carver, D. E. Leyden, D. M. Hercules, *J. Phys. Chem.* **80** (1976) 1700–1708.

[36] G. Morea et al., *J. Chem. Soc. Faraday Trans.* **85** (1989) 3861–3870.

[37] A. Cimino, D. Gazzoli, M. Inversi, M. Valigi, *SIA Surf. Interface Anal.* **10** (1987) 194–201.

[38] A. Katrib et al., *Surf. Sci.* **189/190** (1987) 886–893.

[39] K. Balakrishnan, J. Schwank, *J. Catal.* **127** (1991) 287–306.

[40] B. L. Gustafson, P. S. Wehner, *Appl. Surf. Sci.* **52** (1991) 261–270.

[41] Yong-Xi Li, J. M. Stencel, B. H. Davis, *Appl. Catal.* **64** (1990) 71–81.

[42] G. Moretti, *SIA Surf. Interface Anal.* **17** (1991) 745–750.
[43] M. A. Baltanas, J. H. Onuferko, S. T. McMillan, J. R. Katzer, *J. Phys. Chem.* **91** (1987) 3772–3774.
[44] V. Di Castro, C. Furlani, M. Gargano, M. Rossi, *Appl. Surf. Sci.* **28** (1987) 270–278.
[45] V. Lindner, H. Papp, *Appl. Surf. Sci.* **32** (1988) 75–92.
[46] M. C. Burrell, G. A. Smith, J. J. Chera, *SIA Surf. Interface Anal.* **11** (1988) 160–164.
[47] D. L. Cocke, M. S. Owens, R. B. Wright, *Langmuir* **4** (1988) 1311–1318.
[48] J. L. G. Fierro, J. M. Palacios, F. Tomas, *SIA Surf. Interface Anal.* **13** (1988) 25–32.
[49] C. S. Kuivila, J. B. Butt, D. C. Stair, *Appl. Surf. Sci.* **32** (1988) 99–121.
[50] F. Lange, H. Schmelz, H. Knözinger, *J. Electron Spectrosc. Relat. Phenom.* **57** (1991) 307–315.
[51] J. Z. Shyu, K. Otto, *Appl. Surf. Sci.* **32** (1988) 246–252.
[52] K. Asami, K. Hashimoto, *Langmuir* **3** (1987) 897–904.
[53] P. Marcus, I. Olefjord, *Corros. Sci.* **28** (1988) 589–602.
[54] Shin-Puu Jeng, P. H. Holloway, C. D. Batich, *Surf. Sci.* **227** (1990) 278–290.
[55] P. J. Bremer et al., *SIA Surf. Interface Anal.* **17** (1991) 762–772.
[56] M.-P. Lahalle, R. Guillaumont, G. C. Allen, *J. Chem. Soc. Faraday Trans.* **86** (1990) 2641–2644.
[57] B.-P. Zhang et al., *Corros. Sci.* **33** (1992) 103–112.
[58] H. Yoshioka et al., *Corros. Sci.* **32** (1991) 313–325.
[59] R. B. Diegle et al., *J. Electrochem. Soc.* **135** (1988) 1085–1092.
[60] C. G. Pantano, R. K. Brow, *J. Am. Ceram. Soc.* **71** (1988) 577–581.
[61] Chin-An Chang, Yong-Kil Kim, A. G. Schrott, *J. Vac. Sci. Technol. A* **8** (1990) 3304–3309.
[62] F. J. Boerio, D. J. Ondrus, *J. Adhes.* **22** (1987) 1–12.
[63] J. N. Andersen et al., *J. Phys. Condens. Matter* **1** (1989) 7309–7313.
[64] A. A. Galuska, *Appl. Surf. Sci.* **40** (1989) 19–31, 33–40, 41–51.
[65] W. J. van Ooij, A. Sabata, A. D. Appelhans, *SIA Surf. Interface Anal.* **17** (1991) 403–420.
[66] Y.-H. Kim, Y. S. Chaug, N. J. Chou, J. Kim, *J. Vac. Sci. Technol. A* **5** (1987) 2890–2893.
[67] P. Steiner et al., *Z. Phys. B* **67** (1987) 497–502.
[68] C. Ziegler et al., *Fresenius J. Anal. Chem.* **341** (1991) 308–313.
[69] Y. Hwu et al., *Appl. Phys. Lett.* **59** (1991) 979–981.
[70] Y. Hwu et al., *Appl. Phys. Lett.* **57** (1990) 2139–2141.
[71] Q. Y. Ma et al., *J. Vac. Sci. Technol. A* **9** (1991) 390–393.
[72] R. P. Vasquez, M. C. Foote, B. D. Hunt, L. Bajuk, *J. Vac. Sci. Technol. A* **9** (1991) 570–573.
[73] H. Behner, G. Gieres, B. Sipos, *Fresenius J. Anal. Chem.* **341** (1991) 301–307.
[74] W. A. M. Aarnink, J. Gao, H. Rogalla, A. van Silfhout, *J. Less Common Met.* **164/165** (1990) 321–328.
[75] C. C. Chang et al., *J. Appl. Phys.* **67** (1990) 7483–7487.
[76] G. Frank, C. Ziegler, W. Göpel, *Phys. Rev. B. Condens. Matter* **43** (1991) 2828–2834.
[77] S. Tanaka et al., *Jpn. J. Appl. Phys.* **30** (1991) L 1458–L 1461.
[78] S. Kohiki et al., *Appl. Phys. A* **A 50** (1990) 509–514.
[79] S. Kohiki et al., *J. Appl. Phys.* **68** (1990) 1229–1232.
[80] M. A. Sobolewski, S. Semancik, E. S. Hellman, E. H. Hartford, *J. Vac. Sci. Technol. A* **9** (1991) 2716–2720.
[81] G. Subramayam, F. Radpour, V. J. Kapoor, G. H. Lemon, *J. Appl. Phys.* **68** (1990) 1157–1163.
[82] K. Tanaka, H. Takaki, K. Koyama, S. Noguchi, *Jpn. J. Appl. Phys.* **29** (1990) 1658–1663.
[83] H. Yamanaka et al., *Jpn. J. Appl. Phys.* **30** (1991) 645–649.
[84] T. Suzuki et al., *Phys. Rev. B Condens. Matter* **42** (1990) 4263–4271.
[85] A. Fujimori et al., *Phys. Rev. B Condens. Matter* **42** (1990) 325–328.
[86] M. Klauda et al., *Physica C* **173** (1991) 109–116.
[87] K. Okada, Y. Seino, A. Kotani, *J. Phys. Soc. Jpn.* **60** (1991) 1040–1050.
[88] P. Adler, H. Buchkremer-Hermanns, A. Simon, *Z. Phys. B Condens. Matter* **81** (1990) 355–363.
[89] K. H. Young, E. J. Smith, M. M. Eddy, T. W. James, *Appl. Surf. Sci.* **52** (1991) 85–89.
[90] S. Gokhale et al., *Surf. Sci.* **257** (1991) 157–166.
[91] N. Aoto, E. Ikawa, N. Endo, Y. Kurogi, *Surf. Sci.* **234** (1990) 121–126.
[92] M. Niwano et al., *J. Vac. Sci. Technol. A* **9** (1991) 195–200.
[93] J. Onsgaard, M. Christiansen, F. Ørskov, P. J. Godowski, *Surf. Sci.* **247** (1991) 208–214.
[94] M. Kawasaki, G. J. Vandentop, M. Salmeron, G. A. Somorjai, *Surf. Sci.* **227** (1990) 261–267.
[95] M. Alonso et al., *J. Vac. Sci. Technol. B* **8** (1990) 955–963.
[96] M. Tang et al., *J. Vac. Sci. Technol. B* **8** (1990) 705–709.
[97] R. Klauser, M. Oshima, H. Sugahara, Y. Murata, *Surf. Sci.* **242** (1991) 319–323.
[98] U. K. Kirner, K. D. Schierbaum, W. Göpel, *Fresenius J. Anal. Chem.* **341** (1991) 416–420.
[99] F. S. Ohuchi, M. Kohyama, *J. Am. Ceram. Soc.* **74** (1991) 1163–1187.
[100] N. D. Shinn, C. H. F. Peden, K. L. Tsang, P. J. Berlowitz, *Phys. Scr.* **41** (1990) 607–611.
[101] S. Harel, J.-M. Mariot, E. Beauprez, C. F. Hague, *Surf. Coat. Technol.* **45** (1991) 309–315.
[102] G. Beamson, A. Bunn, D. Briggs, *SIA Surf. Interface Anal.* **17** (1991) 105–115.
[103] W. F. Stickle, J. F. Moulder, *J. Vac. Sci. Technol A* **9** (1991) 1441–1446.
[104] A. Naves de Brito et al., *SIA Surf. Interface Anal.* **17** (1991) 94–104.
[105] A. Chikoti, B. D. Ratner, D. Briggs, *Chem. Mater.* **3** (1991) 51–61.
[106] W. Pamler, F. Bell, L. Mühlhoff, H. J. Barth, *Mater. Sci. Eng.* **A 139** (1991) 364–371.
[107] L. L. Cao, Y. M. Sun, L. Q. Zheng, *Wear* **140** (1990) 345–357.
[108] S. Noël, L. Boyer, C. Bodin, *J. Vac. Sci. Technol. A* **9** (1991) 32–38.
[109] A. Osaka, Y.-H. Wang, Y. Miura, T. Tsugaru, *J. Mater. Sci.* **26** (1991) 2778–2782.
[110] S. M. Mukhopadhyay, S. H. Garofalini, *J. Non Cryst. Solids* **126** (1990) 202–208.
[111] E.-T. Kang, D. E. Day, *J. Non Cryst. Solids* **126** (1990) 141–150.
[112] Z. Hussain, E. E. Khawaja, *Phys. Scr.* **41** (1990) 939–943.
[113] R. K. Brow, Y. B. Peng, D. E. Day, *J. Non Cryst. Solids* **126** (1990) 231–238.

[114] B. A. van Hassel, A. J. Burggraaf, *Appl. Phys. A* **A 52** (1991) 410–417.
[115] P. Prieto, L. Galán, J. M. Sanz, *Surf. Sci.* **251/252** (1991) 701–705.
[116] K. Oyoshi, T. Tagami, S. Tanaka, *J. Appl. Phys.* **68** (1990) 3653–3660.
[117] Z.-H. Lu, A. Yelon, *Phys. Rev. B Condens. Matter* **41** (1990) 3284–3286.
[118] A. Carnera et al., *J. Non Cryst. Solids* **125** (1990) 293–301.
[119] R. W. Bernstein, J. K. Grepstad, *J. Appl. Phys.* **68** (1990) 4811–4815.
[120] A. Ermolieff et al., *Semicond. Sci. Tech.* **6** (1991) 98–102.
[121] S. V. Hattangady et al., *J. Appl. Phys.* **68** (1990) 1233–1236.
[122] Maria Faur et al., *SIA Surf. Interface Anal.* **15** (1990) 641–650.
[123] H. Hasegawa et al., *J. Vac. Sci. Technol. B* **8** (1990) 867–873.
[124] L. A. Harris, *J. Appl. Phys.* **39** (1968) 1419–1427, 1428–1431.
[125] R. E. Weber, W. T. Peria, *J. Appl. Phys.* **38** (1967) 4355–4358.
[126] P. W. Palmberg, G. K. Bohm, J. C. Tracy, *Appl. Phys. Lett.* **15** (1969) 254–255.
[127] E. Casnati, A. Tartari, C. Baraldi, *J. Phys. B* **15** (1982) 155–167.
[128] D. Landolt, H. J. Mathieu, *Oberfläche Surf.* **21** (1980) 8–15.
[129] L. E. Davis et al.: *Handbook of Auger Electron Spectroscopy*, Physical Electronic Ind., Minnesota 1976.
[130] (a) E. Kny, *J. Vac. Sci. Technol.* **17** (1980) 658–660.
(b) J. Kleefeld, L. L. Levinson, *Thin Solid Films* **64** (1979) 389–393.
(c) M. A. Smith, L. L. Levinson, *Phys. Rev. B Solid State* **16** (1977) 1365–1369.
[131] M. Gryzinski, *Phys. Rev.* **138** (1965) A 305–321.
[132] S. Ichimura, R. Shimizu, *Surf. Sci.* **112** (1981) 386–408.
[133] R. Payling, *J. Electron Spectros. Relat. Phenom.* **36** (1985) 99–104.
[134] S. Mroczkowski, D. Lichtman, *J. Vac. Sci. Technol. A* **3** (1985) 1860–1865.
[135] T. Sato et al., *SIA Surf. Interface Anal.* **14** (1989) 787–793.
[136] M. P. Seah, *SIA Surf. Interface Anal.* **9** (1986) 85–98.
[137] R. Hamminger, G. Grathwohl, F. Thümmler, *J. Mater. Sci.* **18** (1983) 353–364.
[138] H. Goretzki, *Fresenius Z. Anal. Chem.* **329** (1987) 180–189.
[139] T. H. Chuang, *Mat. Sci. Eng.* **A 141** (1991) 169–178.
[140] D. J. Nettleship, R. K. Wild, *SIA Surf. Interface Anal.* **16** (1990) 552–558.
[141] C. Setti, *Mikrochim. Acta* **1** (1987) 437–444.
[142] K. Sasaki, A. Noya, T. Umezawa, *Jpn. J. Appl. Phys.* **29** (1990) 1043–1047.
[143] W. Pamler, A. Kohlhase, *SIA Surf. Interface Anal.* **14** (1989) 289–294.
[144] M. Procop, H. Raidt, B. Sandow, *Phys. Status Solidi A* **99** (1987) 573–580.
[145] A. Reif, P. Streubel, A. Meisel, D. Zeissig, *Phys. Status Solidi A* **122** (1990) 331–340.
[146] S. Stein et al., *Fresenius J. Anal. Chem.* **341** (1991) 66–69.
[147] A. Manara, V. Sistori, *SIA Surf. Interface Anal.* **15** (1990) 457–462.
[148] J. Shin, K. N. Geib, C. W. Wilmsen, Z. Lilliental-Weber, *J. Vac. Sci. Technol. A* **8** (1990) 1894–1898.
[149] S. V. Hattangady et al., *Mater. Res. Soc. Symp. Proc.* **165** (1990) 221-226.
[150] A. Hupfer, J. Albrecht, D. Dietrich, *Fresenius J. Anal. Chem.* **341** (1991) 439–444.
[151] F. Bertran et al., *Surf. Sci. Lett.* **245** (1991) L 163–L 169.
[152] X. Wallart et al., *J. Appl. Phys.* **69** (1991) 8168–8176.
[153] V. M. Bermudez, *Surf. Sci. Lett.* **230** (1990) L 155–L 161.
[154] H. Roux, N. Boutaoui, M. Tholomier, *Thin Solid Films* **172** (1989) 141–148.
[155] B. J. Burrow, A. E. Morgan, R. C. Ellwanger, *J. Vac. Sci. Technol. A* **4** (1986) 2463–2469.
[156] K. Machida, M. Enyo, I. Toyoshima, *Thin Solid Films* **161** (1988) L 91–L 95.
[157] K. Sasaki, A. Noya, T. Umezawa, *Jpn. J. Appl. Phys.* **27** (1988) 1190–1192.
[158] A. Gauzzi, H. J. Mathieu, J. H. James, B. Kellett, *Vacuum* **41** (1990) 870–874.
[159] W. A. M. Aarnink et al., *Appl. Surf. Sci.* **47** (1991) 195–203.
[160] Q. X. Jia, W. A. Anderson, *Appl. Phys. Lett.* **57** (1990) 304–306.
[161] J. Colino, J. L. Sacedon, L. Del Olmo, J. L. Vincent, *J. Vac. Sci. Technol. A* **8** (1990) 4021–4025.
[162] Udayan De, S. Natarajan, E. W. Seibt, *Physica C* **183** (1991) 83–89.
[163] E. W. Seibt, A. Zalar, *Mater. Lett.* **11** (1991) 1–5.
[164] C. Uebing, *Surf. Sci.* **225** (1990) 97–106.
[165] F. Reniers, M. Jardinier-Offergeld, F. Bouillon, *SIA Surf. Interface Anal.* **17** (1991) 343–351.
[166] S. Nakanishi et al., *Surf. Sci. Lett.* **247** (1991) L 215–L 220.
[167] M. J. van Staden, J. P. Roux, *Appl. Surf. Sci.* **44** (1990) 271–277.
[168] R. F. Visser, J. P. Roux, *Appl. Surf. Sci.* **51** (1991) 115–124.
[169] W. Arabczyk, H.-J. Müssig, F. Storbeck, *Surf. Sci.* **251/252** (1991) 804–808.
[170] A. Schönborn, H. Bubert, E. H. de Kaat, *Fresenius J. Anal. Chem.* **341** (1991) 241–244.
[171] J.-R. Chen et al., *Appl. Surf. Sci.* **33/34** (1988) 212–219.
[172] S. K. Sharma, S. Hofmann, *Appl. Surf. Sci.* **51** (1991) 139–155.
[173] S. Badrinarayanan, A. B. Mandale, S. Sinha, *Surf. Coat. Technol.* **34** (1988) 133–139.
[174] R. Hauert, J. Patscheider, R. Zehringer, M. Tobler, *Thin Solid Films* **206** (1991) 330–334.
[175] A. Benninghoven, *Z. Phys.* **220** (1969) 159–180.
[176] J.-C. Pivın, *J. Mater. Sci.* **18** (1983) 1267–1290.
[177] P. Sigmund in N. H. Tolk, J. C. Tully, W. Heiland, C. W. White (eds.): *Inelastic Ion-Surface Collisions*, Academic Press, New York 1977.
[178] P. D. Prewett, D. K. Jefferies, *J. Phys. D* **13** (1980) 1747–1755.
[179] A. R. Krauss, D. M. Gruen, *Appl. Phys.* **14** (1977) 89–97.
[180] P. Steffens et al., *J. Vac. Sci. Technol. A* **3** (1985) 1322–1325.
[181] A. J. Eccles, J. C. Vickerman, *J. Vac. Sci. Technol. A* **7** (1989) 234–244.

[182] A. Benninghoven, K.-H. Müller, M. Schemmer, P. Beckman, *Proc. Int. Conf. Solid Surf.* **3rd** (1977) 1063–1066.
[183] I. V. Bletsos, D. M. Hercules, A. Benninghoven, D. Greifendorf in A. Benninghoven, R. J. Colton, D. S. Simons, H. W. Werner (eds.): *SIMS V*, Springer Verlag, Berlin 1986.
[184] P. H. McBreen, S. Moore, A. Adnot, D. Roy, *Surf. Sci.* **194** (1988) L 112–L 118.
[185] C. A. Andersen, J. R. Hinthorne, *Anal. Chem.* **45** (1973) 1421–1438.
[186] J. A. Treverton et al., *SIA Surf. Interface Anal.* **15** (1990) 369–376.
[187] W. J. van Ooij, A. Sabata, A. D. Appelhans, *SIA Surf. Interface Anal.* **17** (1991) 403–420.
[188] M. Morra, E. Occhiello, F. Garbassi, *SIA Surf. Interface Anal.* **16** (1990) 412–417.
[189] D. E. Fowler, R. D. Johnson, D. van Leyen, A. Benninghoven, *SIA Surf. Interface Anal.* **17** (1991) 125–136.
[190] G. J. Leggett, D. Briggs, J. C. Vickerman, *SIA Surf. Interface Anal.* **17** (1991) 737–744.
[191] W. D. Ramsden, *SIA Surf. Interface Anal.* **17** (1991) 793–802.
[192] D. J. Pawson et al., *SIA Surf. Interface Anal.* **18** (1992) 13–22.
[193] R. A. Cocco, B. J. Tatarchuk, *Surf. Sci.* **218** (1989) 127–146.
[194] B. H. Sakakini et al., *Surf. Sci.* **271** (1992) 227–236.
[195] X.-L. Zhou, C. R. Flores, J. M. White, *Surf. Sci. Lett.* **268** (1992) L 267–L 273.
[196] M. E. Castro, J. M. White, *Surf. Sci.* **257** (1991) 22–32.
[197] D. Briggs, M. J. Hearn, *SIA Surf. Interface Anal.* **13** (1988) 181–185.
[198] D. Briggs et al., *SIA Surf. Interface Anal.* **15** (1990) 62–65.
[199] M. J. Hearn, D. Briggs, *SIA Surf. Interface Anal.* **17** (1991) 421–429.
[200] D. G. Armour, *Vacuum* **31** (1981) 417–428.
[201] P. J. Martin, R. P. Netterfield, *SIA Surf. Interface Anal.* **10** (1987) 13–16.
[202] H. Niehus, *Appl. Phys. A* **53** (1991) 388–402.
[203] J.-M. Beuken, P. Bertrand, *Surf. Sci.* **162** (1985) 329–336.
[204] J. F. van der Veen, *Surf. Sci. Rep.* **5** (1985) 199–287.
[205] H. Niehus, G. Comsa, *Surf. Sci.* **140** (1984) 18–30.
[206] G. Molière, *Z. Naturforsch.* **2 A** (1947) 133–143.
[207] H. D. Hagstrum, *Phys. Rev.* **123** (1961) 758–765.
[208] R. De Gryse, J. Landuyt, L. Vandenbroucke, J. Vennik, *SIA Surf. Interface Anal.* **4** (1982) 168–173.
[209] H. H. Brongersma, M. J. Sparnaay, T. M. Buck, *Surf. Sci.* **71** (1978) 657–666.
[210] G. C. Nelson, *J. Vac. Sci. Technol.* **13** (1976) 974–975.
[211] A. Rossi, C. Calinski, H.-W. Hoppe, H.-H. Strehblow, *SIA Surf. Interface Anal.* **18** (1992) 269–276.
[212] H. J. Kang, R. Shimizu, T. Okutani, *Surf. Sci.* **116** (1982) L 173–L 178.
[213] E. N. Haeussler, *SIA Surf. Interface Anal.* **2** (1980) 134–139.
[214] Y. G. Shen, D. J. O'Connor, R. J. MacDonald, *SIA Surf. Interface Anal.* **17** (1991) 903–910.
[215] C. R. K. Marrian, A. Shih, G. A. Haas, *Appl. Surf. Sci.* **24** (1985) 372–390.
[216] S. J. Hoekje, G. B. Hoflund, *Thin Solid Films* **197** (1991) 367–380.
[217] H. Puderbach, *Fresenius Z. Anal. Chem.* **314** (1983) 260–264.
[218] G. H. Vurens, D. R. Strongin, M. Salmeron, G. A. Somorjai, *Surf. Sci.* **199** (1988) L 387–L 393.
[219] A. van Oostrom, *Vacuum* **34** (1984) 881–892.
[220] T. W. Conlon, *Contemp. Phys.* **26** (1985) 521–558.
[221] A. E. T. Kuiper, M. F. C. Willemsen, J. C. Barbour, *Appl. Surf. Sci.* **35** (1988–1989) 186–198.
[222] H. W. Werner in H. Oechsner (ed.): *Thin Film and Depth Profile Analysis*, Springer Verlag, Berlin 1984.
[223] R. S. Bhattacharya, P. P. Pronko, *Appl. Phys. Lett.* **40** (1982) 890–892.
[224] F. Nava et al., *Thin Solid Films* **89** (1982) 381–385.
[225] J. M. M. de Nijs, A. van Silfhout, *Appl. Surf. Sci.* **40** (1990) 349–358.
[226] R. Fastow et al., *J. Vac. Sci. Technol. A* **5** (1987) 164–168.
[227] R. C. Ellwanger, A. E. Morgan, W. T. Stacy, Y. Tamminga, *J. Vac. Sci. Technol. B* **1** (1983) 533–539.
[228] T. Obata et al., *Thin Solid Films* **87** (1982) 207–214.
[229] M. C. Burrell, G. A. Smith, J. J. Chera, *SIA Surf. Interface Anal.* **11** (1988) 160–164.
[230] H. Z. Wu et al., *Thin Solid Films* **191** (1990) 55–67.
[231] A. E. T. Kuiper, S. W. Koo, F. H. P. M. Habraken, Y. Tamminga, *J. Vac. Sci. Technol. B* **1** (1983) 62–66.
[232] F. H. P. M. Habraken et al., *J. Appl. Phys.* **53** (1982) 404–415.
[233] Z. L. Wang, J. F. M. Westendorp, F. W. Saris: "Ion Beam Modification of Materials," in B. Biasse, G. Destefanis, J. P. Gailliard (eds.): *Proc. 3rd Int. Conf.*, Grenoble 1982.
[234] R. S. Bhattacharya et al., *J. Phys. Chem. Solids* **44** (1983) 61–69.
[235] G. Binnig, H. Rohrer, *Surf. Sci.* **126** (1983) 236–244.
[236] P. K. Hansma, J. Tersoff, *J. Appl. Phys.* **61** (1987) R 1–R 23.
[237] P. K. Hansma, V. B. Elings, O. Marti, C. E. Bracker, *Science (Washington D.C.)* **242** (1988) 209–216.
[238] G. Binnig, C. F. Quate, C. Gerber, *Phys. Rev. Lett.* **56** (1986) 930–933.
[239] A. L. Weisenhorn et al., *Science (Washington D.C.)* **247** (1990) 1330–1333.
[240] B. Drake et al., *Science (Washington D.C.)* **243** (1989) 1586–1589.
[241] H. G. Hansma et al., *Clin. Chem. (Winston Salem N.C.)* **37** (1991) 1497–1501.
[242] S. Manne, H.-J. Butt, S. A. C. Gould, P. K. Hansma, *Appl. Phys. Lett.* **56** (1990) 1758–1759.
[243] H.-J. Butt et al., *J. Struct. Biol.* **105** (1990) 54–61.
[244] S. Manne et al., *Science (Washington D.C.)* **251** (1991) 183–186.
[245] Ph. Ebert, G. Cox, U. Poppe, K. Urban, *Surf. Sci.* **271** (1992) 587–595.
[246] S. A. C. Gould, K. Burke, P. K. Hansma, *Phys. Rev. B Condens. Matter* **40** (1989) 5363–5366.
[247] S. A. Gould et al., *Ultramicroscopy* **33** (1990) 93–98.
[248] W. J. Kaiser, R. C. Jaklevic, *IBM J. Res. Dev.* **30** (1986) 411–416.
[249] B. Reihl, J. K. Gimzewski, J. M. Nicholls, E. Tosatti, *Phys. Rev. B Condens. Matter* **33** (1986) 5770-5773.
[250] A. J. Gratz, S. Manne, P. K. Hansma, *Science (Washington D.C.)* **251** (1991) 1343–1346.
[251] P. Dietz et al., *Ultramicroscopy* **35** (1991) 155–159.
[252] H. Hartman et al., *Clays Clay Miner.* **38** (1990) 337–342.

[253] H. Lindgreen et al., *Am. Mineral.* **76** (1991) 1218–1222.
[254] P. G. Arscott, V. A. Bloomfield, *Ultramicroscopy* **33** (1990) 127–131.
[255] R. D. Edstrom et al., *Ultramicroscopy* **33** (1990) 99–106.
[256] K. A. Fisher et al., *Ultramicroscopy* **33** (1990) 117–126.
[257] J. A. N. Zasadzinski et al., *Science (Washington D.C.)* **239** (1988) 1013–1015.
[258] A. L. Weisenhorn et al., *Biophys. J.* **58** (1990) 1251–1258.
[259] H. G. Hansma et al., *Clin. Chem. (Winston Salem N.C.)* **37** (1991) 1497–1501.
[260] J. N. Lin et al., *Langmuir* **6** (1990) 509–511.
[261] H. G. Hansma et al., *J. Vac. Sci. Technol. B* **9** (1991) 1282–1284.
[262] A. L. Weisenhorn et al., *Langmuir* **7** (1991) 8–12.
[263] H.-J. Butt et al., *J. Struct. Biol.* **105** (1990) 54–61.
[264] R. Wolkow, Ph. Avouris, *Phys. Rev. Lett.* **60** (1988) 1049–1052.
[265] K. Takayanagi, Y. Tanishiro, M. Takahashi, S. Takahashi, *J. Vac. Sci. Technol. A* **3** (1985) 1502–1506.
[266] G. Binnig et al., *Phys. Rev. Lett.* **55** (1985) 991–994.
[267] A. Baratoff et al., *Surf. Sci.* **168** (1986) 734–743.
[268] S. A. Elrod, A. L. de Lozanne, C. R. Quate, *Appl. Phys. Lett.* **45** (1986) 1240–1242.
[269] P. Sautet, C. Joachim, *Surf. Sci.* **271** (1992) 387–394.
[270] F. P. Netzer, J.-U. Mack, *Chem. Phys. Lett.* **95** (1983) 492–496.
[271] W. Erley, *J. Electron Spectros. Relat. Phenom.* **44** (1987) 65–78.
[272] M. Fleischmann, P. J. Hendra, A. J. McQuillan, *Chem. Phys. Lett.* **26** (1976) 163–174.
[273] W. Akemann, A. Otto, *Surf. Sci.* **272** (1992) 211–219.
[274] A. Otto, I. Mrozek, H. Grabhorn, W. Akemann, *J. Phys. Condens. Matter* **4** (1992) 1143–1212.
[275] G. L. Richmond, J. M. Robinson, V. L. Shannon, *Prog. Surf. Sci.* **28** (1988) 1–70.
[276] R. J. MacDonald, P. J. Martin, *Surf. Sci.* **67** (1977) 237–250.
[277] J. C. Hamilton, R. J. Anderson, *Surf. Sci.* **149** (1985) 81–92.
[278] H. Bach, F. G. K. Baucke, *J. Am. Ceram. Soc.* **65** (1982) 527–533.
[279] F. G. K. Baucke, H. Bach, *J. Am. Ceram. Soc.* **65** (1982) 534–539.
[280] W. Grimm, *Naturwissenschaften* **54** (1967) 586; *Spectrochim. Acta Part B* **23 B** (1968) 443–448.
[281] A. Bengtson, *Spectrochim. Acta Part B* **40 B** (1985) 631–639.
[282] R. Berneron, J. C. Charbonnier, *SIA Surf. Interface Anal.* **3** (1981) 134–141.
[283] J. Pons-Corbeau, *SIA Surf. Interface Anal.* **7** (1985) 169–176.
[284] A. Bengtson, L. Danielson, *Thin Solid Films* **124** (1985) 231–236.
[285] A. Quentmeier et al., *Mikrochimica Acta Suppl.* **11** (1985) 89–102.
[286] J. Takadoum et al., *SIA Surf. Interface Anal.* **6** (1984) 174–183.
[287] G. Rosina, E. Bertel, F. P. Netzer, *Phys. Rev. B Condens. Matter* **34** (1986) 5746–5753.
[288] B. M. Biwer, S. L. Bernasek, *Surf. Sci.* **167** (1986) 207–230.
[289] E. Bertel, R. Stockbauer, T. E. Madey, *Surf. Sci.* **141** (1984) 355–387.
[290] J. C. Rivière, F. P. Netzer, G. Rosina, *SIA Surf. Interface Anal.* **18** (1992) 333–344.
[291] F. P. Netzer, J. C. Rivière, G. Rosina, *Phys. Rev. B* **38** (1988) 7453–7460.
[292] C. M. Stander, *Appl. Surf. Sci.* **16** (1983) 463–468.
[293] A. Otto, B. Reihl, *Surf. Sci.* **178** (1986) 635–645.
[294] J. Onsgaard, I. Chorkendorff, O. Ellegaard, O. Sørensen, *Surf. Sci.* **138** (1984) 148–158.
[295] G. Strasser, G. Rosina, J. A. D. Matthew, F. P. Netzer, *J. Phys. F Met. Phys.* **15** (1985) 739–751.
[296] F. P. Netzer, G. Strasser, J. A. D. Matthew, *Solid State Commun.* **45** (1983) 171–174.
[297] G. Strasser, F. P. Netzer, *J. Vac. Sci. Technol. A* **2** (1984) 826–830.
[298] G. Strasser, E. Bertel, J. A. D. Matthew, F. P. Netzer in P. Wachter, H. Boppart (eds.): *Valence Instabilities,* North Holland Publ., Amsterdam 1982.
[299] M. Nishijima, S. Masuda, H. Kobayashi, M. Onchi, *Rev. Sci. Instrum.* **53** (1982) 790–796.
[300] T. U. Bartke, R. Franchy, H. Ibach, *Surf. Sci.* **272** (1992) 299–305.
[301] A. Fricke, H. Graupner, L. Hammer, K. Müller, *Surf. Sci.* **272** (1992) 182–188.
[302] Q. Dai, A. J. Gellman, *Surf. Sci.* **257** (1991) 103–112.
[303] S. Bao et al., *Surf. Sci.* **271** (1992) 513–518.
[304] M. J. Nowakowski, J. M. Vohs, D. A. Bonnell, *Surf. Sci. Lett.* **271** (1992) L 351–L 356.
[305] X. H. Feng et al., *J. Chem. Phys.* **90** (1989) 7516–7522.
[306] J.-P. Lu, M. R. Albert, S. L. Bernasek, D. J. Dwyer, *Surf. Sci.* **215** (1989) 348–362.
[307] G. Lauth, T. Solomun, W. Hirschwald, K. Christmann, *Surf. Sci.* **210** (1989) 201–224.
[308] R. D. Ramsier, P. N. Henriksen, A. N. Gent, *Surf. Sci.* **203** (1988) 72–88.
[309] K. W. Hipps, U. Mazur, *Surf. Sci.* **207** (1989) 385–400.
[310] P. N. T. van Velzen, M. C. Raas, *Surf. Sci.* **161** (1985) L 605–L 613.
[311] M. C. Gallagher, Y. B. Ning, J. G. Adler, *Phys. Rev. B Condens. Matter* **36** (1987) 6651–6656.
[312] D. M. Brewis et al., *SIA Surf. Interface Anal.* **6** (1984) 40–45.
[313] D. P. Woodruff, P. D. Johnson, N. V. Smith, *J. Vac. Sci. Technol. A* **1** (1983) 1104–1110.
[314] G. Rangelov, N. Memmel, E. Bertel, V. Dose, *Surf. Sci.* **236** (1990) 250–258.
[315] R. Konishi, S. Ikeda, T. Osaki, H. Sasakura, *Jpn. J. Appl. Phys.* **29** (1990) 1805–1806.
[316] K. H. Frank, R. Dudde, H. J. Sagner, W. Eberhardt, *Phys. Rev. B Condens. Matter* **39** (1989) 940–948.
[317] N. B. Brookes et al., *Phys. Rev. B Condens. Matter* **39** (1989) 2736–2739.
[318] T. Watanabe et al., *Physica C* **176** (1991) 274–278.
[319] J. M. Nicholls, B. Reihl, *Surf. Sci.* **218** (1989) 237–245.
[320] P. L. Wincott, K. L. Håkansson, L. I. Johansson, *Surf. Sci.* **211/212** (1989) 404–413.
[321] E. W. Thomas, *Vacuum* **34** (1984) 1031–1044.
[322] R. Whaley, E. W. Thomas, *J. Appl. Phys.* **56** (1984) 1505–1513.
[323] R. A. Baragiola, L. Nair, T. E. Madey, *Nucl. Inst. Methods Phys. Res.* **B58** (1991) 322–327.
[324] R. H. Milne, E. A. Maydell, D. J. Fabian, *Appl. Phys. A* **A52** (1991) 197–202.

[325] H. Brenten, H. Müller, D. Kruse, V. Kempter, *Nucl. Inst. Methods Phys. Res.* **B58** (1991) 328–332.
[326] C. Benazeth, N. Benazeth, M. Hou, *Surf. Sci.* **151** (1985) L137–L143.
[327] M. Nègre, J. Mischler, N. Benazeth, *Surf. Sci.* **157** (1985) 436–450.
[328] H. D. Hagstrum, *Phys. Rev.* **150** (1966) 495–515.
[329] H. D. Hagstrum, G. E. Becker, *Phys. Rev. B Solid State* **4** (1971) 4187–4202.
[330] H. D. Hagstrum, G. E. Becker, *Phys. Rev. B Solid State* **8** (1973) 1592–1603.
[331] H. D. Hagstrum, G. E. Becker, *J. Chem. Phys.* **54** (1971) 1015–1032.
[332] G. E. Becker, H. D. Hagstrum, *Surf. Sci.* **30** (1972) 505–524.
[333] H. D. Hagstrum, G. E. Becker, *J. Vac. Sci. Technol.* **14** (1977) 369–371.
[334] H. Conrad et al., *Phys. Rev. Lett.* **42** (1979) 1082–1086.
[335] H. Conrad et al., *Surf. Sci.* **100** (1980) L461–L466.
[336] F. Bozso et al., *J. Chem. Phys.* **78** (1983) 4256–4269.
[337] F. Bozso et al., *Surf. Sci.* **136** (1984) 257–266.
[338] F. Bozso et al., *Surf. Sci.* **138** (1984) 488–504.
[339] L. Lee et al., *Surf. Sci.* **165** (1986) L95–L105.
[340] W. Sesselmann et al., *Surf. Sci.* **146** (1984) 17–42.
[341] G. Rocker, H. Tochihara, R. M. Martin, H. Metiu, *Surf. Sci.* **181** (1987) 509–529.
[342] G. H. Rocker et al., *Surf. Sci.* **208** (1989) 205–220.
[343] H. Oechsner, E. Stumpe, *Appl. Phys.* **14** (1977) 43–47.
[344] H. Oechsner, *Nucl. Inst. Methods Phys. Res.* **B33** (1988) 918–925.
[345] J. W. Coburn, E. Taglauer, E. Kay, *J. Appl. Phys.* **45** (1974) 1779–1786.
[346] N. E. Sanderson et al., *Microchim. Acta* **1** (1987) 275–290.
[347] A. Wucher, H. Oechsner, F. Novak, *Thin Solid Films* **174** (1989) 133–137.
[348] P. M. Charalambous, *Steel Res.* **5** (1987) 197–203.
[349] J. Sopka, H. Oechsner, *J. Non Cryst. Solids* **114** (1989) 208–210.
[350] H. Oechsner, *Scanning Microscopy* **3** (1989) 411–418.
[351] A. Benninghoven, M. Kempken, P. Klüsener, *Surf. Sci.* **206** (1988) L927–L933.
[352] P. Beckmann, M. Kopnarski, H. Oechsner, *Mikrochim. Acta Suppl.* **11** (1985) 79–88.
[353] U. Schühle, J. B. Pallix, C. H. Becker, *J. Vac. Sci. Technol. A* **6** (1988) 936–940.
[354] J. B. Pallix, C. H. Becker, N. Newman, *J. Vac. Sci. Technol. A* **6** (1988) 1049–1052.
[355] E. Kawatoh, R. Shimizu, *Jpn. J. Appl. Phys.* **30** (1991) 832–836.
[356] D. M. Gruen et al., *Nucl. Inst. Methods Phys. Res.* **B58** (1991) 505–511.
[357] R. Michael, D. Stulik, *App. Surf. Sci.* **28** (1987) 367–381.
[358] D. J. Surman, J. A. van den Berg, J. C. Vickerman, *SIA Surf. Interface Anal.* **4** (1982) 160–167.
[359] A. Brown, J. C. Vickerman, *Analyst (London)* **109** (1984) 851–857.
[360] A. Brown, J. C. Vickerman, *SIA Surf. Interface Anal.* **6** (1984) 1–14.
[361] A. Brown, J. A. van den Berg, J. C. Vickerman, *Spectrochim. Acta Part B* **40B** (1985) 871–877.
[362] A. Brown, *Eur. Spectros. News* **81** (1988) 13–21.
[363] W. Unger, W. Pritzkow, *Int. J. Mass. Spec. Ion Process.* **75** (1987) 15–26.
[364] R. Michael, D. Stulik, *Appl. Surf. Sci.* **28** (1987) 53–62.
[365] E. W. Müller, J. A. Panitz, S. B. McLane, *Rev. Sci. Instrum.* **39** (1968) 83–86.
[366] E. W. Müller, *Z. Phys.* **131** (1951) 136–142.
[367] M. K. Miller, *Int. Mat. Rev.* **32** (1987) 221–240.
[368] A. Cerezo, T. J. Godfrey, G. D. W. Smith, *Rev. Sci. Instrum.* **59** (1988) 862–866.
[369] A. Cerezo et al., *J. Micros. (Oxford)* **154** (1989) 215–225.
[370] A. Cerezo, *Vacuum* **42** (1991) 605–611.
[371] A. Cerezo et al., *EMSA Bull.* **20** (1990) 77–83.
[372] Li Chang, G. D. W. Smith, G. B. Olson, *J. Phys. Colloq.* **C2**, Suppl. 3, **47** (1986) 265–275.
[373] S. S. Brenner, M. K. Miller, *J. Met.* **35** (1983) 54–63.
[374] G. D. W. Smith, *Metals Handbook*, **10** (1986) 9th ed., 583–602.
[375] A. Cerezo, G. D. W. Smith, *Materials Forum* **10** (1987) 104–116.
[376] M. L. Knotek, P. J. Feibelman, *Phys. Rev. Lett.* **40** (1978) 964–967.
[377] T. E. Madey, *Science (Washington D.C.)* **234** (1986) 316–322.
[378] B. Krüger, C. Bendorff, *Surf. Sci.* **178** (1986) 704–715.
[379] C. J. Davisson, L. H. Germer, *Phys. Rev.* **30** (1927) 705–740.

Chemical and Biochemical Sensors

KARL CAMMANN (Chap. 1, Sections 2.1, 2.3, 2.4, 4, 5), BERND ROSS (Sections 2.2, 2.2.1, 2.2.1.2), WOLFGANG HASSE (Section 2.2.1.3), CHRISTA DUMSCHAT (Section 2.2.1.4), ANDREAS KATERKAMP (Section 2.2.2), JÖRG REINBOLD (Sections 2.1.2, 2.2.3), GUDRUN STEINHAGE (Section 2.2.4), BERND GRÜNDIG (Sections 3.1–3.2.3, 3.5), REINHARD RENNEBERG (Sections 3.1, 3.2.4–3.4), Institut für Chemo- und Biosensorik, Münster, Federal Republic of Germany

NORBERT BUSCHMANN (Section 2.2.1.1), Westfälische Wilhelms-Universität Münster, Federal Republic of Germany

089573-541-5/94/$5.00 + .50

1. Introduction to the Field of Sensors and Actuators

When a person suffers the loss of one of the five senses it is recognized as a serious obstacle to carrying out everyday activities. In the case of blindness and deafness the expression "disabled" is even used as an indication of the severity of the loss, just as with those whose limbs (legs or arms) or other crucial body components are incapable of functioning properly. In the animate world generally, sensing and then acting or reacting is closely linked with the very essence of life, where input from some type of sensory system always precedes intelligent action. From the earliest days of mankind it was the availability of a powerful "coping mechanism" that constituted the driving force leading first to intelligent action and later to the unbiased accumulation of knowledge.

Any intellectual information structure capable of producing theories is dependent upon a reliable input of information. Imperfect observations, or perceptual processes subject to frequent errors, will never provide a satisfactory basis for working theories subject to verification by others. In this respect analytical chemistry can be regarded as one of the cognitive means available to the chemist interested in ascertaining the composition of the material world and the distribution and arrangement of atoms and molecules within a particular sample. Since both the nature and the three-dimensional arrangement of atoms determine the characteristics of matter, reliable information with respect to both is crucial for understanding the whole of the material world. This observation is equally valid for the inorganic, the organic, and also the living world. A high level of structured information is essential to the work of most natural scientists. The ability to describe in detail a complex chemical process (e.g., the combustion of fuel in an engine or waste in a municipal incinerator, the synthesis of minerals deep within the earth, or the course of biochemical reactions taking place in living systems) is crucially dependent on the quality of the information available. High technology is impossible without detailed information on the state of the system in question, as has been amply demonstrated in semiconductor technology, where knowledge of the purity of the materials and the state of cleanliness in the production plant is absolutely essential.

Complex and especially dynamic systems involving chemical processes are often described in terms of simplified models. The more chemical information is available, the better will be the model—and consequently the more complete the understanding and control over the system. Chemical analysis constitutes the primary source of information for characterizing the various states of such a system.

As a result of a tremendous increase in the demand for analytical chemical information, the number of analyses conducted for acquiring information about particular samples is increasing dramatically. It is thought that worldwide more than 10^9 analyses are now performed every day—with a double-digit growth rate. Many complex systems are governed by time-dependent parameters, which makes it essential that as much information as possible be acquired within a particular period of time so that the various reactions taking place within the system can be followed, and to provide insight into the dynamics of the mechanisms involved. Devices designed to permit continuous measurement of a quantity are called *sensors*. Various types of sensors have been developed for monitoring many important *physical* parameters, including temperature, pressure, speed, flow, etc. The number of potential interferences is rather limited in these cases, so it is not surprising that the corresponding sensors offer a level of reliability considerably higher that that for sensors that measure *chemical* quantities (e.g., the concentration of a certain species, the activity of an ion, or the partial pressure of a gaseous material).

There are two major fields of application for chemical sensors: environmental studies and biomedical analysis. In the field of environmental analysis there is a great demand for the type of continuous monitoring that only a sensor can provide, since the relevant parameter for toxicological risk assessment is always the *dose* (i.e., a concentration multiplied by an exposure time). The study of environmental chemistry also depends upon a continuous data output that provides not only baseline levels but also a reliable record of concentration outbursts. *Data networks* can be used to accumulate multiple determinations of a particular analyte regardless of whether these are acquired simultaneously or at different times within a three-dimensional sampling space (which may have dimensions on the order of miles), thereby providing information about concentration gradients and other factors essential for pinpointing the location of an emission source. The inevitable exponential increase in the number of chemical analyses associated with this work threatens to create serious financial and also ecological problems, since most classical analytical techniques themselves produce a certain amount of chemical waste.

Our society is characterized by a population with an average age that is continuously increasing. This has stimulated the widespread biomedical application of chemical sensors for monitoring the health of the elderly without confining them to hospitals. Sensors for determining the concentrations of such key body substances as glucose, creatinine, or cholesterol could prove vitally important in controlling and maintaining personal health, at the same time enhancing the quality of the patient's life. The same principle underlies sensor applications in intensive care wards. Sensors also offer the potential for significant cost reductions relative to conventional laboratory tests, and they can even improve the quality of the surveillance because continuous monitoring can lower the risk of complications. For example, the monitoring of blood potassium levels can give early warning of the steady increase that often precedes an embolism, providing sufficient time for clinical countermeasures.

It should be noted at the outset, however, that the development costs associated with a chemical sensor are typically many orders of magnitude higher than those for a sensor measuring a physical quantity. A satisfactory return on investment therefore presupposes either mass production in the case of low-cost sensors or special applications that would justify an expensive sensor.

Another important field of application for chemical sensors is *process control*. Here the sensor is expected to deliver some crucial signal to the actuators (valves, pumps, etc.) that control the actual process. Fully automated process control is feasible with certain types of feedback circuits. Since key chemical parameters in many chemical or biochemical processes are not subject to sufficiently accurate direct determination by the human senses, or even via the detour of a physical parameter, sensor technology becomes the key to automatic process control. Often the performance of an entire industrial process depends on the quality and reliability of the sensors employed. Successful adaptation of a batch process to *flow-through reactor production technology* is nearly impossible without chemical sensors for monitoring the input, the product quality, and also—an increasingly important consideration—the waste. *Municipal incinerators* would certainly be more acceptable if they offered sensor-based control of SO_2, NO_x, and, even more important, the most hazardous organic compounds, such as dioxines. Compared to the high construction costs for a plant complete with effective exhaust management, an extra investment in sensor research can certainly be justified.

These few examples illustrate the increasing importance of chemical sensor development as a key technology in various closed-looped sensor/actuator-based process control applications. Rapid, comprehensive, and reliable information regarding the chemical state of a system is indispensable in many high technology fields; for this reason there is also an increasing need for intelligent or "smart" sensors capable of controlling their own performance automatically.

Finally, there is an urgent and growing demand for analytical chemists and engineers skilled in producing reliable chemical sensors delivering results that are not only reproducible but also accurate (free of systematic errors attributable to interferences from the various sample matrices).

2. Chemical Sensors [1]–[5]

Until the introduction in the 1980s of the lambda probe [6]–[8] for oxygen measurement in the exhaust systems of cars employing catalytic convertors (see Section 2.1.3), the prototype for the most successfully applied chemical sensor was certainly the electrochemical pH electrode, introduced over 50 years ago and discussed in detail in Section 2.2.1.1. This device can be used to illustrate the identification and/or further optimization of all the critical and essential features characterizing sensor performance, including reversibility, selectivity, sensitivity, linearity, dynamic working range, response time, and fouling conditions.

Electrochemical electrodes for chemical analysis have actually been known for over 100 years (the Nernst equation was published as early as 1889), and they can therefore be considered as the forerunners of chemical sensors. In general, a chemical sensor is a small device placed directly on or in a sample and designed to produce an electrical signal that can be correlated in a specific way with the concentration of the *analyte*, the substance to be determined. If the analyte concentration varies in the sample under investigation, the sensor should faithfully follow the variation in both directions; i.e., it should function in a reversible way, and it should deliver analytical data at a rate greater than the rate of change of the system (quasi-reversible case). From the standpoint of practical applications, complete thermodynamic reversibility is relatively unimportant so long as measurements are acquired with sufficient rapidity compared to changes in the analyte concentration. Another restriction arises from the demand for a reagent-free measurement: an extremely selective or specific cooperation between the analyte atom, ion, or molecule and the surface of a receptor-like sensor element should be the sole basis for gaining information about the analyte concentration.

The general construction principle of a chemical sensor is illustrated in Figure 1. The most important part is the *sensing element* (receptor) at which the molecular or ionic recognition process takes place, since this defines the overall selectivity of the entire sensor. The sensing-element surface is sometimes covered by an additional surface layer acting as a protective device to improve the lifetime and/or dynamic range of the sensor, or to prevent interfering substances from reaching the sensing-element surface. The analyte recognition process takes place either at the surface of the sensing element or in the bulk of the material, leading to a concentration-dependent change in some physical property that can be transformed into an electrical signal by the appropriate transducer. Since there should be no direct contact between the sample and the transducer, the latter cannot itself improve the selectivity, but it is responsible for the sensitivity of the sensor, and it functions together with the sensing element to establish a dynamic concentration range. The electrical signal of the transducer is usually amplified by a device positioned close to the transducer or even integrated into it. If extensive electrical wiring is required between the sensor and the readout–control unit, the electrical signal might also be digitized as one way of minimizing noise pick-up during signal transmission. Several chemical signals produced by multiple sensing elements, each connected to its own specific transducer, can be transmitted via modern multiplexing techniques, which require only a two-wire electrical connection to the control unit.

If a biochemical mechanism is used in the molecular recognition step, the sensor is called a *biosensor*. Systems of this type are dealt with in Chapter 3.

Key features of chemical sensors subject to specification are, in order of decreasing importance:

1) Selectivity (accuracy; better: truthfulness of the result)
2) Limit of detection (LOD) for the analyte
3) Sensitivity
4) Dynamic response range
5) Stability
6) Response time
7) Reliability (maintenance-free working time)
8) Lifetime

Selectivity. The *selectivity* of a chemical sensor toward the analyte is often expressed in terms of a dimension that compares the concentration of the corresponding interfering substance with an analyte concentration that produces the same sensor signal. This factor is obtained by dividing the sensitivity of the sensing device toward the analyte (= slope of the calibration curve; see below) by the sensitivity for the corresponding in-

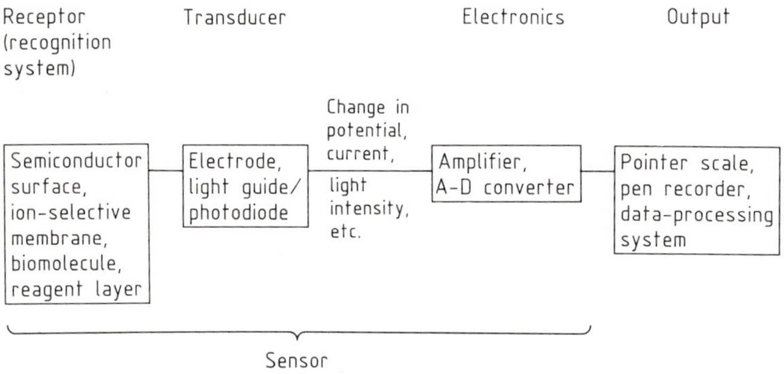

Figure 1. Schematic arrangement of a typical chemical or biochemical sensor [9]

terfering substance. Typical selectivities range from 10^{-3} to $<10^{-6}$. In order to estimate the error attributable to limited selectivity the concentration of the interferent must be multiplied by the selectivity coefficient. If, for example, a sodium-selective glass electrode shows a typical selectivity coefficient toward potassium ions on the order of 10^{-3}, a thousand-fold excess of potassium- over sodium-ion concentration would produce an equivalent sensor signal, therefore doubling the reading and leading to an error of 100 %. It should be noted, however, that such selectivities are themselves concentration- and matrix-dependent. This means that near the limit of detection of a particular sensor an interfering species will disrupt the measurement of an analyte to a greater extent than it will at higher concentrations. Therefore, true response factors can rarely be used for correction purposes.

The selectivity is the most important parameter associated with a chemical sensor since it largely determines the truthfulness of the analytical method. Truthfulness (or *accuracy*) is more than simply precision, repeatability, or reproducibility: it represents agreement with the true content of the sample. All the other parameters mentioned are subject to systematic error, which can only be recognized if an analyte is determined by more than one method. Since selectivity is always limited, all chemical sensors are prone to report higher concentrations than a sample actually contains. In the case of environmental analysis this positive error can be regarded as a safety margin if relevant interferents are also present. In such a situation the sensor acts as a probe for establishing when a sample should be taken to be analyzed in the laboratory. The reproducibility of chemical sensor measurements is often well below 1 % relative.

However, it should also be noted that sensing elements are subject to poisoning, resulting in a diminished response for the analyte. Sensor poisons are substances that either cover the sensing surfaces, thereby influencing the analyte recognition process or, in the case of catalytic sensors, decrease the activity of the catalyst. Since these fouling effects are not linear, predictable, or mathematically additive, great care is necessary whenever data from a sensor array are treated by the methods of mathematical data management (e.g., pattern recognition, neural networks, etc.). The influence of sensor poisoning can be completely different and unpredictable for every single sensing element. There is no simple theory that permits the transfer of poisoning effects with respect to one sensing element to other parts of a sensing array as a way of avoiding recalibration. Thus, any attempt to overcome insufficient selectivity by the use of a sensor array with multiple sensor elements displaying slightly differing selectivities for the analyte and the interferents, followed by the application of chemometrics to correct for selectivity errors, is very dangerous in the context of unknown and variable sample matrices. Unfortunately, this is usually the situation in environmental analysis. In the learning and/or calibration phase of sensor-array implementation it is essential that the sample matrix be understood. In the case of a known matrix and a closed system, and if the number of possible interferents is > 5, the number of calibration steps (involving different mixtures of the interfering compounds, selected to cover all possible concentration and mixture variations) required to provide reliable information on the 5-dimensional response space of the sensor array for a $5 + N$ array is extremely large (usually several hundred). With real samples the time necessary for calibrating the sensors might exceed the period of stability. Since with open environmental systems the type and number of interferents is not known, any attempt to correct for inadequate sensor selectivity by use of a sensor array increases the risk of systematic errors. This assertion is valid for every type of mathematical data treatment (pattern recognition, neural networks, etc.) for samples of unknown and/or variable matrices. However, for trace analysis in the sub-ppm range all analytical methods (with the exception of isotope dilution) are subject to systematic errors, as has been demonstrated with interlaboratory comparison tests using even the most sophisticated techniques [10].

Limit of Detection (LOD) for the Analyte. In analytical chemistry the *limit of detection* (LOD) is exceeded when the signal of the analyte reaches at least three times the general noise level for the reading. No quantitative measurement is possible at this low concentration; only the presence of the analyte can be assumed with a high probability (> 99 %) in this concentration range. The range of quantification ends at ten times the LOD. Therefore, the LOD of an analytical instrument must always be ten times smaller than the lowest concentration to be quantified. It is worth mentioning in this context that LOD values reported by most instrument manufacturers disregard the influence of interfering compounds. Furthermore, the LOD is often worse when an analyte must be determined within its characteristic matrix.

In some cases with extremely sensitive sensors a zero-point calibration might be difficult to perform, because traces of the analyte would always be present or might easily be carried into the calibration process by solvents or reagents. The lowest *measurable* level is called the *blank value*. The LOD is then defined as three times the standard deviation of the blank value, expressed in concentration units (not in the units of the signal, as is sometimes improperly suggested).

Normally the LOD becomes worse as a sensor ages.

Sensitivity. As mentioned above, the *sensitivity* of a sensor is defined by the signal it generates, expressed in the concentration units of the substance measured. This corresponds to the slope of the corresponding calibration curve when the substance is the analyte, or to the so-called response curve for interferents. With some sensors the sensitivity rises to a maximum during the device's lifetime. A check of the sensitivity is therefore a valuable quality-assessment step. Intelligent sensors are expected to carry out such checks automatically in the course of routine performance tests. Since in most cases the sensitivity depends on such other parameters as the sample matrix, temperature, pressure, and humidity, certain precautionary measurements are necessary to ensure that all these parameters remain constant both during calibration and in the analysis of real samples.

Dynamic Response Range. The *dynamic response range* is the concentration range over which a calibration curve can be described by a single mathematical equation. A potentiometric sensing device follows a logarithmic relationship, while amperometric and most other electrochemical sensors display linear relationships. Both types of signal-to-concentration relationship are possible with optical sensors; in absorption measurements it is the absorbance with its logarithmic base that is the determining factor, whereas a fluorescence measurement can be described by a linearized function. The broader the concentration range subject to measurement with a given sensing system, the less important are dilution or enrichment steps during sample preparation. The measurement range is limited by the LOD at the low concentration end, and by saturation effects at the highest levels. Modern computer facilities make it possible also to use the nonlinear portion of a calibration curve as a way of saving sample preparation time, but analytical chemists are very reluctant to follow this course because experience has shown that the initial and final parts of a calibration curve are generally most subject to influence by disturbances (electronic and chemical). A good chemical sensor should function over at least one or two concentration decades. Sensors with excellent performance characteristics include the lambda oxygen probe and the glass pH electrode, both of which cover analyte concentration ranges exceeding twelve decades!

Stability. Several types of signal variation are associated with sensors. If the signal is found to vary slowly in two directions it is unlikely to be regarded as having acceptable stability and reproducibility, especially since it would probably not be subject to electronic correction.

Output variation in a single direction is called *drift*. A steadily drifting signal can be caused by a drifting zero point (if no analyte is present) and/or by changes in sensor sensitivity (i.e., changes in the slope of the calibration curve). Drift in the sensor zero-concentration signal (*zero-point drift*) can be corrected by comparison with a signal produced by a sensor from the same production batch that has been immersed in an analyte-free solution.

When drift observed in a sensor signal is attributable exclusively to zero-point drift, and there has been no change in the sensitivity, a one-point calibration is permissible: that is, either the instrumental zero point must itself be adjusted prior to the measurement, or else the sensor signal for a particular calibration mixture must be recorded. Assuming there is no change in the slope of the calibration curve, this procedure is equivalent to effecting a parallel shift in the calibration curve. If the sensor is routinely stored in a calibrating environment between measurements then such a one-point calibration becomes particularly easy to perform.

Special attention must be directed toward the physical and chemical mechanisms that cause a sensor signal to drift. Differential techniques are satisfactory only if the effect to be compensated has the same absolute magnitude and the same sign with respect to both devices. A consideration often neglected in potentiometry is proper functioning of the reference electrode. Several authors [11]–[13] have suggested that in ion-selective potentiometry a "blank membrane" electrode (one with the same ion-selective membrane composition as the measuring electrode but without the selectivity-inducing electroactive compound) would create the same phase-boundary potential for all interfering compounds as the measuring electrode itself. This is definitely not the case, however. At the measuring electrode, analyte ions establish a relatively high exchange-current density, which fixes the boundary potential. The effect of a particular interfering ion will depend on its unique current–voltage characteristics at such an interface. Therefore, interfering ions may establish different phase-boundary potentials depending on the particular mixed potential situations present at the membranes in question, which means that a differential measuring technique will almost certainly fail to provide proper compensation.

If both the zero point of a sensor and its sensitivity change with time, a two-point calibration is necessary. The combination of a changing slope for the calibration curve and a change in the intercept is generally unfavorable, but this is often the situation with catalytic gas sensors. If the slope of the calibration curve also depends on the sample matrix, only standard addition techniques will help.

The stability of a chemical sensor is usually subject to a significant aging process. In the course of aging most sensors lose some of their selectivity, sensitivity, and stability. Some sensors, like the glass pH electrode, can be rejuvenated, while others must be replaced if certain specifications are no longer fulfilled.

Response Time. The *response time* is not defined in an exact way. Some manufacturers prefer to specify the time interval over which a signal reaches 90% of its final value after a ten-fold concentration increase, while authors sometimes prefer the 95% or even 99% level. The particular percentage value chosen represents a pragmatic decision, since most signal–time curves follow an exponential increase of the form:

$$\text{signal} \times (1 - e^{-kt})$$

where the true final value is unknown and/or will never be reached in a mathematical sense. Therefore, specification in terms of the time constant k is clearer. This corresponds to the time required for a signal to reach about 67% of its final value, which can be ascertained without waiting for a final reading: as the slope of the curve ln(signal) vs. time. If the response is constant and independent of the sample matrix, this equation can be used to calculate a final reading immediately after the sensor has been introduced into the sample.

Typical response times for chemical sensors are in the range of seconds, but some biosensors require several minutes to reach a final reading. At the other extreme, certain thin-film strontium titanate oxygen sensors show response times in the millisecond range.

The response time for a sensor is generally greater for a decreasing analyte concentration than for an increasing concentration. This effect is more pronounced in liquids than in gaseous samples. Both the surface roughness of the sensor and/or the dead volume of the measuring cell have some influence on the response time. Small cracks in the walls of a measuring cell can function as analyte reservoirs and diminish the rate of analyte dilution.

With certain sensors the response time can also depend on the sample matrix. In the presence of strongly interfering substances the response time for a chemical sensor might increase as a result of an increase in the time required to reach final equilibrium.

Reliability. There are various kinds of reliabilities. One involves the degree of trust that can be placed in an analytical result delivered by a particular chemical sensor. Another is a function of the real time during which the sensor actually performs satisfactorily without a breakdown and/or need for repair.

There is no way to judge fairly the analytical reliability of a chemical sensor, since this depends strongly upon the expert ability of the analyst to choose a suitable sensor and a suitable sample preparation routine in order to circumvent predictable problems.

With respect to the first generation of sensors, the ion-selective electrodes, the experience of many users has led to the following "reliability hit list", with the most reliable devices cited first: glass pH electrodes, followed by fluoride, sodium glass, valinomycin-based potassium, sulfide, and iodide electrodes, and culminating in the electrodes for divalent ions. In a general way, sometimes very personal judgements regarding reliability have always been linked to a particular sensor's specificity toward the analyte in question.

Reliability can be improved considerably if analytical conclusions are based on measurements obtained by different methods. For example, it is possible to determine sodium either with an appropriate glass-membrane electrode or with a neutral carrier-based membrane electrode. If both give the same analytical results, the probability of a systematic error is very low. Likewise, potassium might be measured using selective membrane electrodes with different carrier molecules. This is of course different from analysis based on a sensor array, in which the selectivities of the individual sensors differ only slightly, and different also in the sense that no correction procedure is invoked in case two results are found to disagree. Use of the Nernst–Nikolsky equation (see Section 2.2.1.1), together with the introduction of additional electrodes for determining the main interferents, has been shown to be an effective way of correcting errors in the laboratory with synthetic samples. Since selectivity numbers (i.e., parameters like the selectivity coefficient) are influenced by the sample matrix through its ionic strength, content of surface-active or lipophilic compounds, or interfering ions, it is dubious whether data from real environmental samples should be corrected by this method.

The length of time over which a chemical sensor can be expected to function reliably can be remarkably great (in the range of years), as in the case of glass-membrane or solid-state membrane electrodes, the lambda probe (Section 2.1.3), and the Taguchi gas sensor (see Section 2.1.2). On the other hand, a biosensor that depends upon a cascade of enzymes to produce an analyte signal will usually have a short span of proper functioning (a few days only). Potentiometric ion-selective membrane electrodes and optical chemosensors have lifetimes of several months. In the case of biosensors it should be noted that anything that changes the quarternary space-orientation of the recognition biomolecules will destroy the proper functioning of the sensor. Enzymes can be influenced by such factors as pH, certain heavy-metal ions, certain inhibitors, and high temperature (resulting in denaturation).

Lifetime. As already mentioned above, some rugged solid-state sensors have *lifetimes* of several years. Certain sensors can also be regenerated when their function begins to deteriorate. The shortest lifetimes are exhibited by biosensors. Because of the huge amount of effort invested worldwide in building an implantable glucose sensor it has been possible to develop a biosensor for this purpose that can operate satisfactorily for nearly six months provided certain precautions are taken. Ion-selective electrodes and optical sensors based on membrane-bound recognition molecules often lose their ability to function by a leaching-out effect. In optical sensors the photobleaching effect may also reduce the lifetime to less than a year. On the other hand, amperometric cells work well for many years, albeit with restricted selectivities.

Problems associated with inadequate lifetimes are best overcome with mass-produced miniaturized replacement sensors based on inexpensive materials. Minimizing replacement costs may well represent the future of biosensors. Installation of a new sensor to replace one that is worn-out can also circumvent surface fouling, interfering layers of proteins, and certain drift and poisoning problems.

Comparison of Sensor Data with that Obtained by Traditional Analytical Methods. It is extremely difficult to compare sensor data with traditional data; indeed, generalizations of this type are rarely possible anywhere in the field of analytical chemistry. Sensor developers are often confronted with the customer's tendency to consider use of a sensor only if all else has failed. This means that the most adverse conditions imaginable are sometimes proposed for the application of a chemical sensor. It is also not fair to compare a device costing a few dollars with the most expensive and sophisticated instrumentation available, nor is it appropriate to compare the performance of a chemical sensor with techniques involving time-consuming separations. In this case only the corresponding detector should be compared with the sensor. Sometimes the very simple combination of a selective chemical sensor with an appropriate separation technique is the most effective way to obtain the redundant data that offer the highest reliability.

Chemical sensors are generally superior to simple photometric devices because they are more selective or faster (as in the case of optical sensors based on the photometric method), more flexible, more economical, and better adapted to continuous sensing. The latter advantage can also be achieved by traditional means via flow-through measuring cells, but this leads to a waste of material and sample, and also to the production of chemical waste. Reagent-free chemical sensors show their greatest advantages in continuous-monitoring applications, in some of which they are called upon to fulfill a control function without necessarily reporting an analytical result, as in the case of the lambda probe described below (Section 2.1.3).

2.1. Molecular Recognition Processes and Corresponding Selectivities

In any sensing element the functions to be fulfilled include sampling, sample preparation, separation, identification, and detection. Therefore, successful performance in these various capacities largely determines the quality of the chemical sensor as a whole. Selective recognition of an analyte ion or molecule is not an easy task considering that there are more than five million known compounds. Recognition can be accomplished only on the basis of unique characteristics of the analyte in question. Several different recognition processes are relevant to the field of chemical sensors, ranging from energy differences (in spectroscopy) to thermodynamically determined variables (in electrochemistry), including kinetic parameters (in catalytic processes). The most specific interactions are those in which the form and the spatial arrangement of the various atoms in a molecule play an important role. This is especially true with biosensors based on the complementary (lock-and-key) principle. Here the analyte molecule and its counterpart have exactly complementary geometrical shapes and come so close together that they interact on the basis of the weak van der Waals interaction forces.

2.1.1. Catalytic Processes in Calorimetric Devices

Pellistors are chemical sensors for detecting compounds that can be oxidized by oxygen. A catalyst is required in this case because the activation energy for splitting a doubly-bonded oxygen molecule into more reactive atoms is too high for the instantaneous "burning" of oxidizable molecules. In most cases platinum is the catalyst of first choice because of its inertness. The principle underlying oxidizable-gas sensors involves catalytic burning of the gaseous analyte, which leads to the production of heat that can in turn be sensed by various temperature-sensitive transducers. Often what is actually measured is the increase in electrical resistance of a metal wire heated to an elevated temperature (ca. 300–400 °C) by the current flowing through it. However, it is also possible to use more sensitive semiconductor devices or even thermopiles in order to register temperature changes of $< 10^{-4}$ °C.

In order to understand the selectivity displayed by a pellistor toward various flammable compounds it is necessary to consider the elementary steps in the corresponding catalytic oxidation of analyte at the catalyst surface. Since this is in fact a surface reaction, various adsorption processes play a dominant role. First, oxygen must be adsorbed, permitting the oxygen double bond to be weakened by the catalyst. Then the species to be oxidized must be adsorbed onto the same catalyst surface where it can subsequently react with the activated oxygen atoms. The process of adsorption may follow one of two known types of adsorption isotherms, as reflected in the calibration curves for these devices. An equilibrium consisting of adsorption, catalyzed reaction, and desorption of the oxidized product leads to a constant signal at a constant analyte concentration. The sensor response function is influenced by any change in the type or number of active surface sites, since this in turn affects both the adsorption processes and the catalytic efficiency. Likewise, compounds that have an influence on any of the relevant equilibria will also alter the calibration parameters. Especially problematic are strongly adsorbed oxidation products, which lower the turnover rate of analyte molecules and thus the sensitivity of the gas sensor.

Given the sequence of events that must occur when an organic molecule is oxidized by atmospheric oxygen, thereby delivering the heat that is actually to be measured, one can readily understand the importance of changes in the adsorption and desorption equilibria. The selectivity observed with such a calorimetric device arises not because some gas reactions are associated with larger enthalpy changes than others, but rather because those gas molecules that exhibit the most rapid adsorption and desorption kinetics are associated with the highest turnover rates. The latter of course depend on molecular-specific heats of ad- and desorption, which have a major influence on the overall reaction kinetics. A change in these specific heats always results in a corresponding change in sensor selectivity. Consequently, any change in the catalyst material, its physical form, or its distribution within the mostly ceramic pellet will in turn alter the gas selectivity and sensitivity. The same is true for variations in the working temperature. On the other hand, nonreacting compounds can also influence a sensor's response and thereby the calibration function if they in some way affect one of the relevant equilibria and/or the catalytic power of the catalyst. Such catalyst poisons as hydrogen sulfide show a strongly detrimental effect on sensor response.

Summarizing the selectivity characteristics of these calorimetric gas sensors, any gas will be subject to detection if it can be catalytically oxidized with a high turnover rate by atmospheric oxygen at elevated temperature. From a kinetic point of view, smaller molecules like carbon monoxide or methane are favored. In contrast to biosensors there is here no precise molecular "tight-fit" recognition of the geometrical form of the analyte. The selectivity of this type of gas sensor is therefore rather limited. Differences in the often reaction-rate controlled adsorption and desorption processes for different oxidizable gas molecules are not sufficiently large to allow selective detection of only a single compound. However, this is not necessarily a disadvantage in certain sensor applications, such as detecting the absence of explosive gases (especially important in the mining industry) or carbon monoxide in an automobile garage. In the latter case any positive error resulting from gasoline interference could in fact be regarded as providing a safety margin, since unburned gasoline should be absent from such locations as well.

2.1.2. Reactions at Semiconductor Surfaces and Interfaces Influencing Surface or Bulk Conductivities

Introduction. Since the early 1960s it has been known that the electrical conductance of certain semiconductor materials such as binary and ternary metal oxides (e.g., SnO_2 [14], ZnO [15], Fe_2O_3 [16]—all of which are n-semiconductor materials—and CuO or NiO [17]—*p*-semiconductors) depends on the adsorption of gases on their solid surfaces. The underlying principle here involves a transfer of electrons between the semiconductor surface and adsorbed gas molecules, together with charge transduction in the interior of the material. Typical gases detected by semiconducting devices include oxidizable substances such as hydrogen, hydrogen sulfide, carbon monoxide, and alkanes (SnO_2, ZnO, etc.), as well as reducible gases like chlorine, oxygen, and ozone (NiO, CuO).

In 1967 both Shaver [18] and Loh [19] described effects achievable with oxide semiconductors modified by the addition of noble metals (e.g., Pt, Pd, Ir, Rh), and since that time the sensitivity and selectivity of semiconductor sensing devices has been significantly enhanced. Intense efforts in this direction, coupled with the further addition of metal oxides [20]–[22], resulted in widespread application of semiconductor gas sensors beginning in the 1970s.

One of the earliest SnO_2 sensors, designed by N. Taguchi, is referred to as the "Figaro sensor" (see below). Sensors of this type make it possible to detect as little as 0.2 ppm of an oxidizable compound such as carbon monoxide or methane [23]. Nevertheless, certain details of the associated sensing mechanism are still not understood theoretically. An important aim of current research is to overcome limitations of the present generation of sensors, especially instability, irreproducibility, and nonselectivity [24].

Construction and Characterization of Semiconductor Sensors. Semiconductor gas sensors are characterized by their simple construction. A schematic overview of the construction principle of a homogeneous semiconducting gas sensor is provided in Figure 2A [25]. Sensor operation is based on a change in the resistance (or conductance) of an oxidic semiconductor in the presence of interacting gases. A time-dependent record reflecting transient exposure of a sensor to a gas leading to an increase in conductance is illustrated in Figure 2B. The sensitivity of a semiconductor sensor is strongly affected by its operating temperature, which is normally in the range 200–400 °C. Chemical regeneration of the oxide surface is possible by a reheating process.

A commercial sensor of the "Figaro" type (TGS 813) is shown in Figure 3A; 400000 such sensors were sold in 1988 [26].

Miniaturization leads to a more modern version of the SnO_2 sensor, normally prepared by thick-film techniques (Fig. 3B) in which thin SnO_2 films, insulator layers (SiO_2), and integrated heating films are sputtered onto silicon substrates. This approach is compatible with high rates of heating and low-cost production [9].

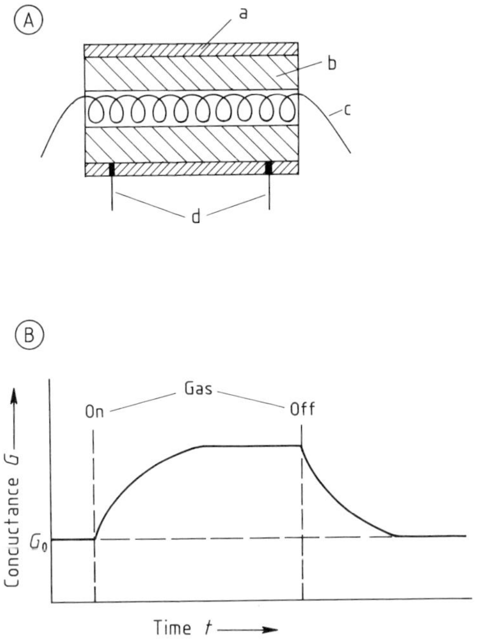

Figure 2. A) Schematic drawing of a homogeneous semiconducting gas sensor; B) Result of transient exposure of the sensor to a gas that increases the conductance [25]
a) Semiconductor; b) Ceramic; c) Heater; d) Contacts
(with permission from Elsevier, Amsterdam)

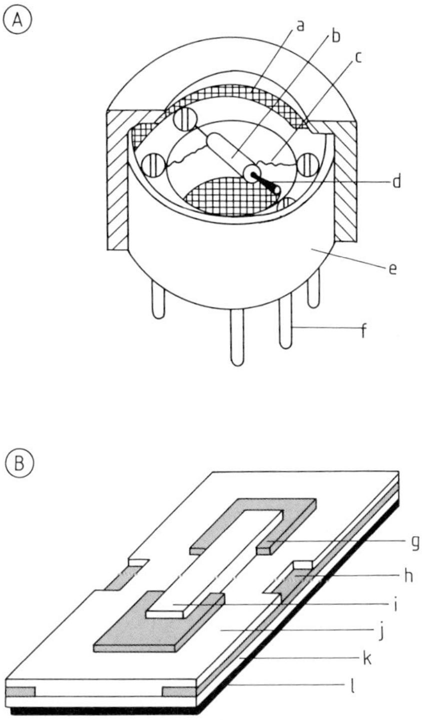

Figure 3. Design characteristics of gas sensors (with permission from VDE-Verlag, Berlin)
A) Figaro Type TGS 813 [26]; B) SnO_2 thick-film type [9]
a) Stainless steel screen; b) Sensor element; c) Leads; d) Heater; e) Epoxy resin; f) Nickel contact; g) Gold; h) Heat contact; i) SnO_2; j) Insulating layer; k) Heater; l) Substrate

Working Principles and Theory. The mechanisms responsible for semiconductor gas-sensor operation can be divided into two classes. The first class involves changes in bulk conductance (transducer function), while the second relies on changes in surface conductance (receptor function). The physical phenomena associated with these two mechanisms are shown schematically in Figure 4 A [27]; Figure 4 B addresses the same problem at the microstructural level [28].

The description of functional principles that follows relates directly to *n*-semiconductors, but its application to *p*-semiconductors is straightforward. In a first step, oxygen molecules from the air form a layer of more or less strongly adsorbed (ionisorbed) oxygen molecules at the surface, resulting in a local excess of electrons. In other words, oxygen acts as a *surface acceptor*, binding electrons from the surface space–charge layer. With respect to the principal energy states (levels) of the electrons in the surface space–charge layer, ionisorption results in a decrease in the electron concentration and an increase in the electronic energy (Fig. 5) [26].

Subsequent reaction with reducing gases (e.g., CH_4 or CO) leads to an increase in charge density (and therefore an increase in conductivity), associated with three possible mechanisms [17], [25], [29]:

1) Adsorption of the reducing molecules as donors, causing electrons to be shifted into the conductance band of the oxide
2) Reaction of the reducing molecules with ionisorbed oxygen under conditions leading to the production of bound electrons
3) Reduction of oxidic oxygen by the reducing molecules, resulting in oxygen vacancies which act as donors, thereby increasing the conductance

Cases 1) and 2) alter the amount of charge stored in the *surface states*, and therefore the amount of charge of opposite sign in deeper parts of the region. For a theoretical derivation of the relationship between the conductance of a semi-

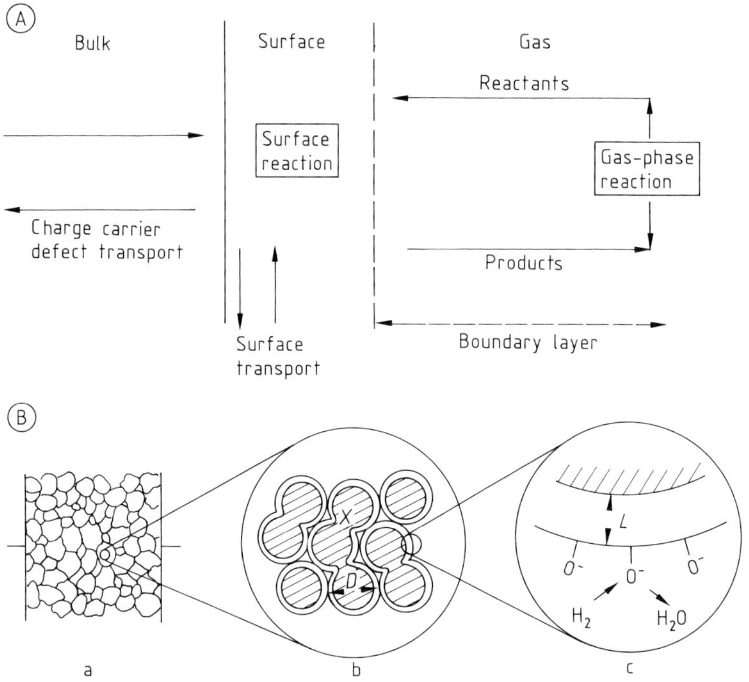

Figure 4. A) Illustration of the physical and chemical phenomena involved in the transduction of a change in the gaseous atmosphere into a change in conductance of a ceramic oxide (with permission from Adam Hilger, Bristol); B) Surface reactions (receptor function) may occur between adsorbed species and defects in the solid (transducer function), between different adsorbed species, or between gas molecules and adsorbed species (or any or all of these together), and the surface reaction may involve a catalyst [27], [28] (with permission from Elsevier, Amsterdam)
a) Element (output resistance change); b) Microstructure (transducer function); c) Surface (receptor function)
D = Particle size; X = Neck size; L = Thickness of space–charge layer

conducting oxide layer and the composition of the gaseous surroundings, the following facts must be considered [24], [30]:

Oxygen must be present; that is, these sensors respond only to nonequilibrium gas mixtures containing both combustible gases (CO, hydrocarbons, H_2, etc.) and oxygen.

There exists a temperature of maximum response; that is, the relative change in conductivity upon introduction of a combustible gas increases with increasing temperature, but falls to zero at sufficiently high temperature

With respect to the relationship between conductivity (σ) and gas partial pressure, Equation (1) has been found to apply:

$$\sigma \sim p^{\beta} \qquad (1)$$

where p is the partial pressure of the combustible gas and β is generally in the range 0.5–1.0 depending on the mechanism.

This microscopic model is limited to idealized thin films (10–100 nm of, for example, SnO_2). Here the gas–solid interactions can be described in terms of the electronic surface states, which are related to energy levels in the band gap model (Fig. 5). Other surface phenomena involved are lattice defect points, trace impurities, and material segregations near the surface. Crystal dislocations and adsorbed atoms or molecules also play an important role in states at the gas–solid interface. As far as n-semiconductors are concerned, both adsorbed hydrogen and oxygen vacancies can function as surface donors, whereas ionisorbed oxygen is a surface acceptor. A surface potential difference develops as a consequence of negatively charged adsorbed oxygen (O_2^- and O^{2-}) and positively charged oxygen vacancies within the space–charge region below the surface layer. Increasing the amount of negatively charged oxygen at the surface causes the surface potential to increase up to the Fermi level (the highest level occupied by electrons). This defines the surface potential and represents the surface state, but the precise magnitude of the potential is a function of the oxygen partial pressure. The

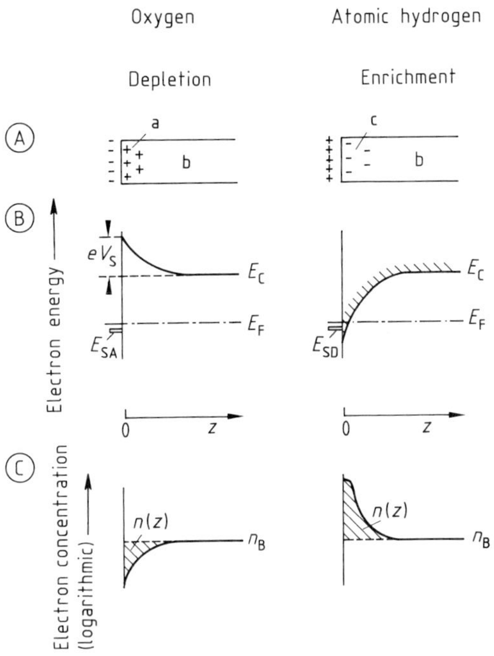

Figure 5. Space–charge layer at the surface of ZnO after exposure to oxygen and hydrogen (with permission from VDE-Verlag, Berlin)
A) Vacuum; B) Band structure in the vicinity of the conducting edge; C) Concentration $n(z)$ of electrons in the conductance band [26]
a) Ionized donors; b) Crystal; c) Electrons
E_{SA} = Acceptor level; E_{SD} = Donor level; n_B = Bulk electron density; z = Distance from surface; eV_S = Surface energy barrier height; E_C = Conductance band edge; E_F = Fermi level

distribution of the various adsorbed oxygen species depends on the temperature, and is influenced by the presence of hydrogen and other gaseous compounds. With respect to the two possible charged oxygen species, O_2^- and O^-, it can be assumed that only O^- is reactive, and that the rate of interconversion of the species is low compared to the rate of the surface combustion reaction, consistent with the following kinetic scheme [27], [31]:

$$\begin{array}{ccccccccc} & & & & & & 2\,OH^- & & \\ & & & & & & \updownarrow H_2O & & \\ O_2(g) & \rightleftarrows & O_2(s) & \rightleftarrows & O_2^- & \rightleftarrows & O^- & \rightleftarrows & O^{2-} \\ & & & & & & \downarrow R & & \\ & & & & & & RO & & \end{array}$$

At low temperature the adsorbed species is mainly O_2^-, which is converted into O^- when the temperature is increased above 450 K.

If a combustible gas R reacts with the adsorbed oxygen species, a steady state occupancy Θ of the surface state is established, which is less than the equilibrium occupancy in air. The following mechanism can be assumed:

$$n' + 1/2\,O_2(g) \underset{k_{-1}}{\overset{k_1}{\rightleftarrows}} O^-(s)$$

$$R(g) + O^-(s) \xrightarrow[k_2]{} RO(g) + n'$$

where n' denotes a conductance electron.

Necessary conditions for sensitivity with respect to the partial pressure p of the combustible gas are then:

$$k_{-1} \ll k_2\,p(R) \quad \text{and} \quad k_1\,p(O_2)^{1/2} \ll k_2\,p(R) \tag{2}$$

in which case

$$\Theta = \frac{k_1\,p(O_2)^{1/2}}{k_{-1} + k_1 (O_2)^{1/2} + k_2\,p(R)} \tag{3}$$

The third case described above is somewhat different: here the observed effect is related to a change in bulk conductance as a function of stoichiometric changes in oxygen activity in the interior of the crystal lattice. The change in conductivity can be described by the relationship [27]:

$$\sigma = \sigma_0 \exp\left(E_A \big/ kT\right) p(O_2)^{\frac{1}{n}} \tag{4}$$

where σ_0 denotes the electrical conductivity in air, k denotes the Boltzmann constant, T is the temperature in Kelvin, E_A is an activation energy, and the sign and value of n depend on the nature of the point defects arising when oxygen is removed from the lattice. At low temperature (< 500 °C) the rate-determining step is interfacial combustion, whereas oxygen vacancies within the lattice dominate at higher temperatures below ca. 1000 °C.

Thus, the observed overall conductance represents a combination of surface effects (both electronic and ionic) together with grain-boundary and volume-lattice effects. The principles elaborated above have been applied widely in the control of oxygen in combustion processes at high temperatures, as in automobile engines.

Parameters. Under a constant set of conditions (temperature, air, pressure, humidity, etc.) the ratio of the conductance in sample air to that in pure air (G/G_0) is proportional to the analyte concentration (Fig. 6). Apart from the concentration of adsorbed gases, other parameters, both physical and chemical, may also have an influ-

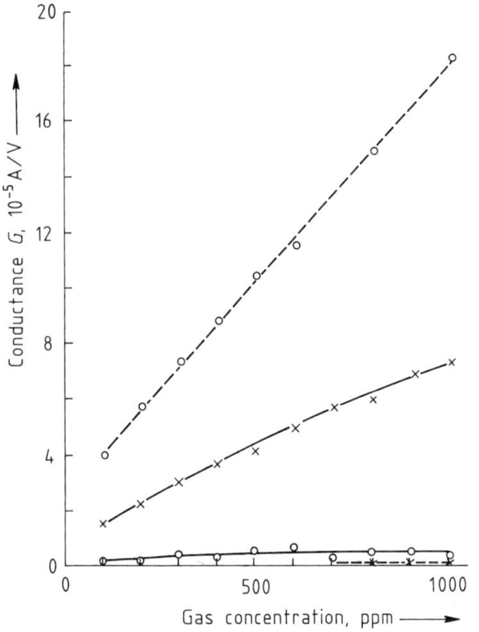

Figure 6. Sensitivity data from commercial SnO_2-based sensors, with sintered layers containing stabilizing additives like SiO_2 (conductance as a function of CO and CH_4 concentration in air); the Type 711 sensor probably operates at a lower temperature and incorporates different catalytically active additives than the Type 813 sensor [25], [32] (with permission from Elsevier, Amsterdam)
----- Sensor Type 711; ——— Sensor Type 813; × = CH_4; o = CO

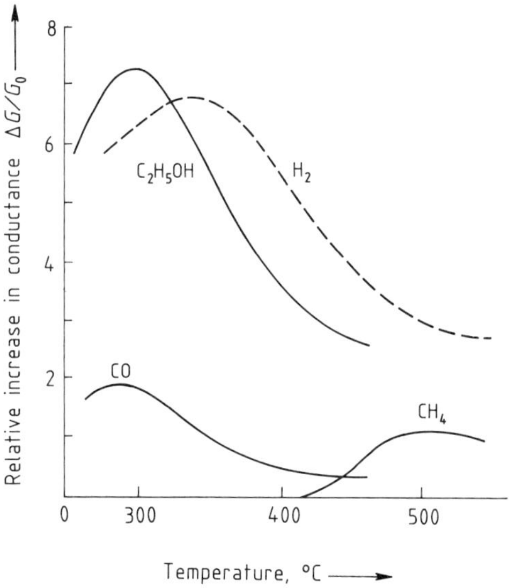

Figure 7. Sensor response to reducing gases (1000 ppm in air) as a function of temperature (sintered layer of SnO_2, thickness 0.05 mm, 0.05 wt% Sb; based on [33], [25]) (with permission from Elsevier, Amsterdam)

ence on the conductance and the optimum working conditions. Considering both components of the conductance change described above, the most important factors influencing conductance are:

1) The microstructure of the semiconducting particles and the surface composition, both of which are characterized by the crystallite size D, the grain-size distribution, and the coagulation structure, all subject to some control through the inclusion of additives
2) The working temperature of the material, which has a unique optimum (maximum) value for a particular material composition, and must therefore be determined separately for every analyte gas and every substrate composition (Fig. 7)

Tin dioxide has become the favorite material for sensor applications of this type because of its simple preparation and high sensitivity at low operating temperature. Commercially available sensors are usually sintered $\geq 700\,°C$ to ensure sufficient mechanical strength; the cross-sectional diameter of the SnO_2 particles is typically ≥ 20 nm. Unfortunately there is a sharp decrease in gas sensitivity (corresponding to a decrease in the resistance of the SnO_2) when the particle size D increases beyond about 6 nm (Fig. 8 A) [28]. Additives such as barium, phosphorus, or lanthanum oxides can be used to restrict thermal growth of the SnO_2 particles to a limit of ca. 10 nm (Fig. 8 B) [28].

Various approaches have been applied to interpreting the parameters influencing gas sensitivity. Models that relate gas sensitivity to grain size (D) and the geometry of contacts between the SnO_2 particles include the intergrain model and the neck model. An extended approach taking into account both models has been developed by YAMAZOE [28], in which gas sensitivity is related to electron charge transfer within the microstructure of the polycrystalline elements (cf. Fig. 4 B).

Selectivity. The selectivity of all semiconductor gas sensors is very limited, but it can be optimized within narrow limits by choosing the best operating temperature and the most appropriate dopant. As in the case of pellistors, overall selectivity is controlled by the ultimate rate-determining step; i.e., the compound that shows the highest oxidation rate determines the selectivity.

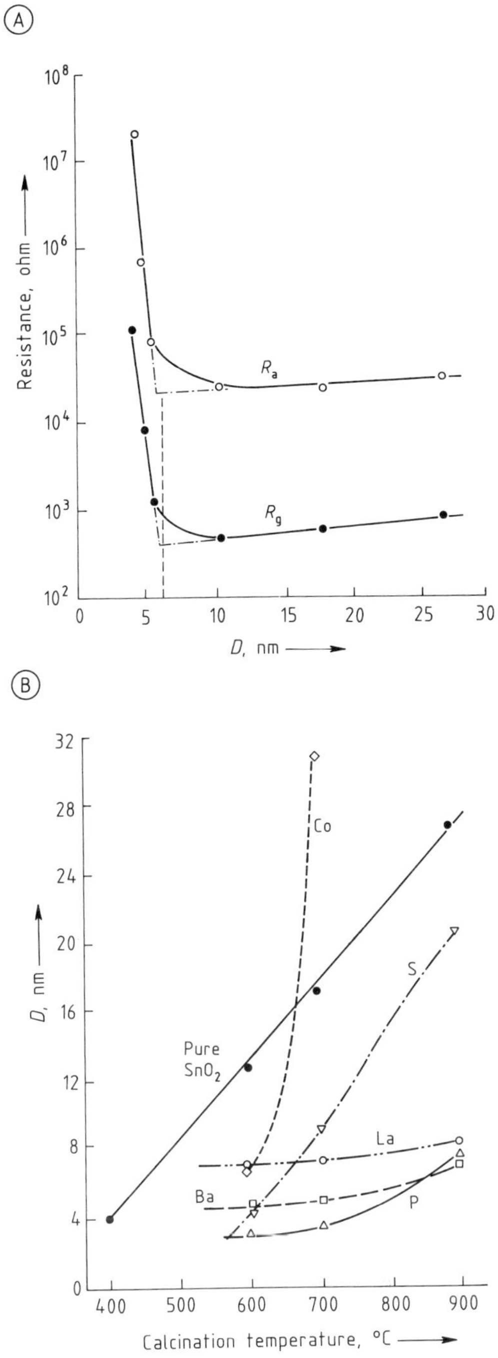

Figure 8. A) Influence of crystallite size (*D*) on electrical resistance for an SnO_2-based sensor in dry air (R_a) and in air containing 800 ppm H_2 (R_g) at 300 °C (elements sintered at 400 °C); B) Changes in SnO_2 crystallite size as a function of calcination temperature for pure SnO_2 samples and SnO_2 impregnated (5%) with oxides of various foreign metals [28] (with permission from Elsevier, Amsterdam)

As is generally the case, the slowest step controls the overall reaction rate. In order to construct a sensor that selectively transforms only the analyte into another product, side reactions must be prevented between the compound that is actually detected (here oxygen) and compounds other than the analyte. This is nearly impossible to achieve in the case of a considerable excess of interfering compounds. Since the physicochemical phenomenon underlying heterogeneous catalysis is based on adsorption–desorption steps and subject to the kinetics of the heterogeneous chemical reaction itself, no known mechanism displays exceptional selectivity. It is not generally possible for a solid-state surface to recognize a single class of molecules because the forces involved in the ad- and desorption steps are rather strong, and are influenced only by overall molecular design and not by details of molecular structure, in contrast to the situation with biocatalysts. It must also be kept in mind that every compound can be considered as a potential interferant, since it might influence the heterogeneous rate constant for catalysis even if it does not itself react with oxygen. This also means that every catalyst poison, or even dust blocking the active surface, is likely to interfere in a measurement, which in turn limits the possibility of using chemometric methods to correct for errors caused by lack of selectivity.

2.1.3. Selective Ion Conductivities in Solid-State Materials

The lambda probe, based on selective oxygen ion diffusion through solid yttrium-doped ZrO_2 above 400 °C, is an extremely selective sensor for gaseous oxygen. Frequently used for measuring the residual oxygen content in exhaust gases from internal combustion engines, the lambda probe is a potentiometric device that functions as a concentration cell in which gaseous oxygen is in equilibrium with lattice oxygen in the solid electrolyte (Fig. 9). A higher concentration of oxygen on one side of the membrane leads to a potential difference which, according to the Nernst equation, is proportional to the ratio of the two oxygen concentrations. Likewise the fluoride ion-selective electrode based on selective fluoride-ion diffusion through solid-state (crystalline) LaF_3 is an extremely selective sensor for fluoride ions in solution.

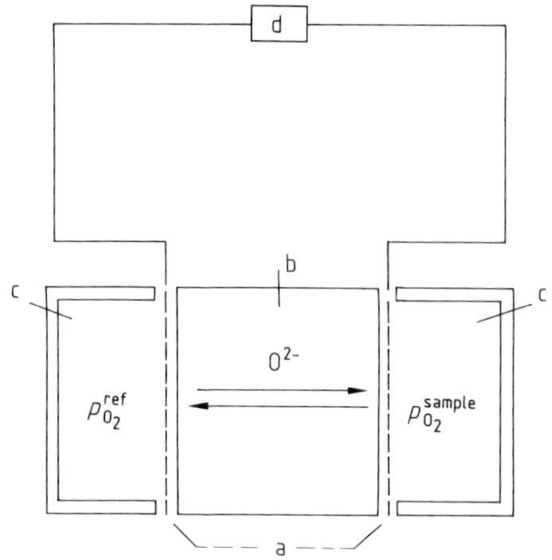

Figure 9. Schematic representation of a lambda probe operated as a potentiometric sensor
a) Porous electrode; b) ZrO_2, doped; c) Oxygen; d) Voltage measurement system

Selectivity in both of these cases is controlled by the size and the charge of the diffusing ion. The diffusion process takes place via discrete ion jumps to nonoccupied lattice positions and/or from and to interstitial positions. Ion selection here can be regarded as a filtering process, with the additional requirement for an appropriate ionic charge to prevent overall electroneutrality. Because the analyte molecules and ions involved are rather small, they can be separated quite readily from larger molecules or ions. Thus, the solid-state material functions like an ideal permselective membrane. Any solid state material characterized by exceptionally high ionic conductivity, such as Ag_2S (which has extremely high Ag^+-ion conductivity), will be a very good sensor for that ion. In the case of the ion-selective membrane electrodes used as Nernstian potentiometric sensors in solution, however, more detailed consideration is necessary (see Section 2.2.1.1).

Since the analyte recognition process is here rather straightforward, and because the filtering effect is accompanied by additional restrictions (appropriate charge and multiple straining), the selectivities obtained through this approach to measuring are extremely high. Many of the compounds that interfere do so not through direct participation in the solid-state diffusion process but rather by affecting the analyte concentration on the sensor surface by other means (e.g., reactions with the analyte, etc.).

2.1.4. Selective Adsorption–Distribution and Supramolecular Chemistry at Interfaces

Supramolecular chemistry has been a source of fascinating new molecular recognition processes. At the active center of a biocatalyst (enzyme) and/or in the binding region of a large antibody macromolecule, the recognition process is attributable mainly to a perfect bimolecular approach of complementary structures. Association or binding occurs only if the relatively weak intermolecular forces involved (dipole–dipole and van der Waals forces) are further supported by a uniquely favorable bimolecular approach that encourages interatomic interactions. A selective reaction of this type requires that the host molecule be carefully tailored so that it matches the guest molecule perfectly. Molecular modeling with attention to the activation energies for closest approach is an especially valuable tool in the design of synthetic host molecules for particular analytes. The analyte can be either a charged or an uncharged species. In the case of an uncharged molecule, however, selective binding of the analyte to the host molecule must be detected with the aid of a mass-sensitive transducer.

Further research in the field of supramolecular chemistry for sensors seems warranted given the prospect for improved selectivity based on custom-tailored synthetic guest molecules for the selective binding of particular host molecules (analytes). Even differentiation between enantiomers may be achievable.

2.1.5. Selective Charge-Transfer Processes at Ion-Selective Electrodes (Potentiometry)

An important aim of quantitative analytical chemistry is the selective detection of specific types of ions in aqueous solution. An electrode capable of selectively detecting one type of ion is called an *ion-selective electrode* (ISE). The operating principle of a particular ISE depends upon the nature of the interaction between ions in solution and the surface of that electrode. Consistent with thermodynamic distribution, the ion in question crosses the phase boundary and interacts with the electrode phase after first discarding its hydrate shell, which results in a net flow of ions across the boundary. Ideally, the counterion would not take part in this flow, so a charge separation would develop, causing the counterion to remain in the neighboring border area. The resulting electrode potential (difference) can be described thermodynamically and kinetically—again, in the ideal case—by the Nernst equation.

The selectivity of such a potentiometric device is that of an ideal Nernstian sensor, the potential of which remains stable irrespective of whatever current may be flowing through the sensor interface. An ideal Nernstian sensor is characterized by a current–voltage curve that is very steep and nearly parallel to the current axis.

The selectivity of a potentiometric ion-selective interface is controlled largely by interfacial charge-transfer kinetics and not by the stability constant for reaction of the analyte ion with the electroactive compound (e.g., valinomycin in the case of potassium ion). The exchange-current density for the analyte ion relative to interfering ions determines the extent of potentiometric selectivity. Any factor that increases the former (conditioning of the electrode with the measuring ion, ion-pair formation by lipophilic counterions, etc.) leads to a better and more selective Nernstian sensor. The ideal electroactive compound apparently behaves like a selective charge-transfer catalyst for the analyte ion. Advantages that might perhaps be gained from tailored host molecules for accelerating the transfer of analyte ions into the surface of the electrode have not yet been fully investigated. Recognition of an analyte ion depends upon size and charge as well as the speed with which the ion loses its solvent sphere prior to entering the electrode phase. Therefore, the electroactive compounds most suitable for embedding in the membrane of an ion-selective electrode (usually consisting of PVC containing a plasticizer) are those that play an active role in successive replacement of the often firmly held hydrate shells of the analyte ions.

2.1.6. Selective Electrochemical Reactions at Working Electrodes (Voltammetry and Amperometry)

Depending on the potential range at the working electrode, as many as about ten different electrode processes can be separated voltammetrically based on the typical resolution of a polarographic curve, which requires a difference voltage of 200 mV for the half-wave potential and a concentration-dependent limiting current. Selectivity in the process is therefore rather limited, and depends on such thermodynamic parameters as the reaction enthalpy for reduction or oxidation at the electrode surface. The potential of the working electrode determines whether a particular compound (analyte) can be oxidized or reduced, and compounds with similar redox characteristics might interfere.

The problem of interference is of course greatest when extremely positive or negative working electrode potentials are applied at the working electrode, since at high potential the small differences between electrochemical reaction rates become negligable. The "resolution" of a typical current–voltage curve is diminished if the electrochemical reaction is kinetically hindered, resulting in so-called irreversible current steps that are less steep. Irreversible electroactive species require a greater overpotential before the diffusion-controlled limiting current plateau is reached, therefore increasing the chance that an interfering compound will become electrochemically active.

In certain biosensors an irreversible heterogeneous reaction is transformed into a homogeneous redox reaction via a very reversible redox system. The latter can be transformed back at the working electrode by a much lower overpotential, thereby reducing the chance of co-oxidation or co-reduction of interfering compounds. Since the limiting current is strictly proportional to the analyte concentration and also to the concentrations of interfering compounds, chemometric data treatment and/or differential measurements may be used to correct for errors.

2.1.7. Molecular Recognition Processes Based on Molecular Biological Principles

Metabolic biosensors are based on special enzyme-catalyzed reactions of the analyte. The recognition process, also known as the lock-and-key principle, is extremely selective, permitting differentiation even between chiral isomers of a molecule. Recognition in this case demands a perfect geometrical fit as well as the appropriate dipole and/or charge distribution to permit binding of the analyte molecule inside the generally much larger biomolecule (host). Since fitting is associated with all three dimensions, a very large biomolecule with its stabilized structure is more effective than the smaller host molecules commonly encountered in supramolecular chemistry. The selectivity of a specific biocatalytic reaction can be further improved by ensuring detection of only the transformed analyte or a stoichiometric partner molecule through a chemical transducer located behind the recognition layer. In a certain sense, this type of biosensor represents the highest possible level of selectivity, and comparable results would be difficult to achieve by other means.

Immunosensors based on mono- or polyclonal antibodies constitute the second best choice with respect to high selectivity. Depending on the amount of effort expended in screening and choosing the desired antibodies from among the millions of different antibody molecules available, the selectivity can be remarkably high. On the other hand, it is also possible to choose antibody molecules that bind only to a certain region (epitope) of the analyte molecule, leading to biomolecules capable of recognizing all members of some class of compounds sharing similar molecular structures. This type of sensor, though it cannot be calibrated, is very valuable for screening purposes. The application of antibody molecules has been considerably simplified by the large-scale preparation of monoclonal versions with precisely identical features. It has even become possible to construct immunosensor arrays based on monoclonal antibody molecules with exactly matched selectivities. Results obtained from such arrays must be evaluated by modern methods of pattern recognition.

In order to overcome the limited stability of large protein molecules, research is currently in progress to isolate only the binding region of the F_{ab} part (antibody-binding fragment) of an antibody molecule. Smaller molecules are of course likely to display somewhat more limited recognition ability, however, with selectivities approaching those of supramolecular host molecules synthesized in the traditional way.

2.2. Transducers for Molecular Recognition: Processes and Sensitivities

2.2.1. Electrochemical Sensors

Electrochemical sensors constitute the largest and oldest group of chemical sensors. Although many such devices have reached commercial maturity, others remain in various stages of development. Electrochemical sensors will be discussed here within the broadest possible framework, with electrochemistry interpreted simply as any interaction involving both electricity and chemistry. Sensors as diverse as enzyme electrodes, high-temperature oxide sensors, fuel cells, and surface-conductivity sensors will be included. Such sensors can be subdivided based on their mode of operation into three categories: *potentiometric* (measurement of voltage), *amperometric* (measurement of current), and *conductometric* (measurement of conductivity) [9], [34].

Electrochemistry implies the transfer of charge from an electrode to some other phase, which may be a solid or a liquid. During this process chemical changes take place at the electrodes, and charge is conducted through the bulk of the sample phase. Both the electrode reactions themselves and/or the charge transport phenomenon can be modulated chemically to serve as the basis for a sensing process. Certain basic rules apply to all electrochemical sensors, the cardinal one being the requirement of a closed electrical circuit. This means that an electrochemical cell must consist of at least two electrodes, which can be regarded from a purely electrical point of view as a sensor electrode and a signal return.

Another important general characteristic of electrochemical sensors is that charge transport within the transducer portion of the sensor and/or inside the supporting instrumentation (which constitutes part of the overall circuit) is always electronic. On the other hand, charge transport in the sample under investigation may be electronic, ionic, or mixed. In the latter two cases electron transfer takes place at the electrode – sample interface, perhaps accompanied by electrolysis, and the corresponding mechanism becomes one of the most critical aspects of sensor performance.

The overall current – voltage relationship is complex, and it can vary as the conditions change. The relationship is affected primarily by the nature and concentration of the electroactive species, the electrode material, and the mode of mass transport. A total observed current can be analyzed in terms of its cathodic and anodic components. If the two currents are equal in magnitude but opposite in sign, there must be an *exchange current* passing through the electrode-surface/sample interface.

Both the charge-transfer resistance and the exchange-current density are critically important parameters in the operation of most electrochemical sensors. These parameters are directly proportional to the electrode area, so the smaller the electrode the higher will be the resistance, all other parameters remaining unchanged. Therefore, it is the electrochemical process taking place at the smaller electrode that determines the absolute value of the current flowing through the entire circuit. The auxiliary electrode will begin to interfere only if its charge-transfer resistance becomes comparable to that of the working electrode. Microelectrodes represent one approach to avoiding such interferences (see Section 2.2.1.2, p. 151).

On the other hand, in zero-current potentiometry the relative sizes of the two electrodes is immaterial. Acquiring useful information in this case requires only that the potential of the working electrode be measured against a well-defined and stable potential from a reference electrode. Any foreign potential inadvertently present within the measuring circuit can contaminate the information, so it is mandatory that the reference electrode be placed as close to the working electrode as is practically possible. Thus, in an amperometric (or conductometric) measurement the source of information can be localized by choosing a small working electrode, whereas it cannot be localized with zero-current potentiometric measurements.

2.2.1.1. Self-Indicating Potentiometric Electrodes

Fundamental Considerations. Within the context of this chapter "potentiometry" is understood to mean the measurement of potential differences across an indicator electrode and a reference electrode under conditions of zero net electrical current. Such a measurement can be used either for determining an analyte ion directly (direct potentiometry) or for monitoring a titration (see below).

In recent years potentiometry has proven to be well suited to the routine analysis of a great number and variety of analytes [4], [35]–[48]. Ion-selective electrodes (ISEs) are commercially available for many anions and cations, as indicated in Tables 3a–d, p. 143. Other analytes can be determined using ISEs in an indirect way. Chapter 3 deals with various types of potentiometric biosensors. A special advantage of ion-selective potentiometry is the possibility of carrying out measurements even in microliter volumes without any loss of analyte.

At present, most ISEs unfortunately do not provide absolute selectivity for a single ion. This fact requires that one have access to very detailed information regarding the nature of samples to be analyzed so that interferences can be either eliminated or otherwise taken into account.

The theory of (ion-selective) potentiometry has already been introduced in Section 2.1.5. Here it is necessary only to remind the reader of the Nernst equation and the Nernst–Nikolsky equation, both of which are used in the evaluation of potentiometric measurements.

The *Nernst equation* is valid only under ideal conditions (with no interfering ions, etc.):

$$E = E^0 + (RT/zF)\ \ln a_M \qquad (5a)$$

$$= E^0 + (0.059/z)\ \log a_M \qquad (5b)$$

The *Nernst–Nikolsky equation* on the other hand takes into consideration the influence of interfering ions on the potential of an ISE:

$$E = E^0 + (RT/z_M F) \ln [a_M + \sum K_{MI} (a_I)^{z_M/z_I}] \qquad (6)$$

where

E = Potential of the ISE, measured with zero net current
E^0 = Standard potential of the ISE
R = Gas constant (8.314 J mol^{-1} K^{-1})
T = Absolute temperature (K)
F = Faraday constant (96485 C mol^{-1})
a_M = Activity of the ion to be measured
a_I = Activity of an interfering ion
z_M, z_I = Electrical charges of the measured and interfering ions
K_{MI} = Selectivity coefficient

The value of K_{MI} depends both on the activity a_M and on the particular combination of analyte ion and interfering ion.

Types of Ion-Selective Electrodes. Simple metal-ion electrodes such as those based on silver, gold, platinum, etc., will not be dealt with in this discussion since they are used mainly for indicating purposes in potentiometric titrations. Apart from these, the following types of ISEs can be distinguished:

1) Glass-membrane electrodes
2) Solid-membrane electrodes
3) Liquid-membrane electrodes (including PVC-membrane electrodes)
4) ISEs based on semiconductors (ion-selective field-effect transistors, etc.)

These four types will be the subject of more detailed consideration.

Glass-Membrane Electrodes [41]. The best-known glass-membrane electrode is the pH electrode. Its most important component is a thin glass membrane made from a sodium-rich type of glass. Depending on the intended application the membrane may take one of several geometric forms: spherical, conical, or flat. For applications in process streams, special pH electrodes have been devised that are resistant to high pressures, and sterilizable pH electrodes are available for biotechnological applications.

Examples of the different types of pH electrodes are illustrated in Figure 10. The *spherical* type is most frequently used for the direct measurement of pH or for acid – base monitoring. It is robust and appropriate for most routine applications. *Conical* electrodes can be used as "stick-in electrodes" for pH measurements in meat, bread, cheese, etc. A conical membrane can easily be cleaned, which is important if measurements are to be made in highly viscous or turbid media. A *flat* membrane facilitates pH measurements on surfaces, such as on human skin.

The pH electrode must be combined with a reference electrode [42] that provides a constant potential for completing the electrical circuit. A suitable reference system is frequently integrated into the body of the pH electrode itself. Combined electrodes of this type are particularly easy to handle. pH Measurements by means of glass electrodes are subject to interference by alkali-

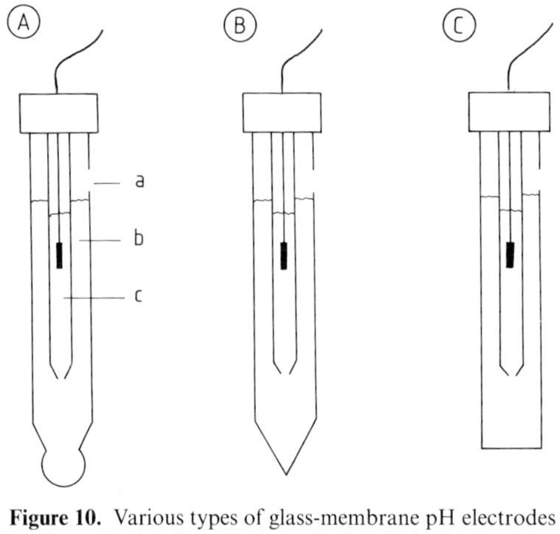

Figure 10. Various types of glass-membrane pH electrodes (with permission from Springer-Verlag, Heidelberg)
A) Spherical membrane; B) Conical membrane; C) Flat membrane
a) Fill hole; b) Inner filling solution (buffer); c) Reference system

metal ions, especially Na^+. The selectivity coefficient K_{MI} for interference by the sodium ion is 10^{-13}. This means that a 10^{13}-fold concentration of sodium is required to produce the same potential as a given concentration of protons.

It is important that the surface of a glass membrane be swollen before measurements are taken. Prior to first use the electrode should therefore be allowed to swell overnight in a 3 mol/L KCl solution. For the same reason the electrodes should also be stored in a solution of this type between measurements and after measurements have been conducted in nonaqueous solvents.

Apart from the pH-selective glass-membrane electrodes, other commercially available glass-membrane electrodes are responsive to sodium ions, although the selectivity of a pNa electrode is much lower than that of a pH electrode. Approximate selectivity coefficients K_{MI} in this case are 10^3 for interference by H^+ and 10^{-3} for interference by Li^+, K^+, and Cs^+.

Solid-Membrane Electrodes. The sensing element of a solid-membrane electrode consists of a material showing ionic conductivity for the particular ions that are to be determined. Such a device may be manufactured from a single crystal, as in the well-known fluoride-selective electrode, where the sensing element is a single crystal of lanthanum fluoride doped with europium(II) fluoride to lower the electrical resistance. Alternatively, a membrane can be produced by grinding the crystalline sensor material together with an appropriate additive (PTFE powder, silicone rubber), followed by solidification either by application of high pressure or by addition of a cross-linking agent. The membrane is then fixed to the electrode body. Electrical contact is established either by a conducting adhesive or by an internal reference system. Figure 11 illustrates schematically the construction of two types of solid-membrane electrodes. A detailed compilation of commercially available solid-state electrodes is provided in Tables 3 a – d at the end of this section.

Generally speaking, interference must be anticipated from any ion that tends to react with the membrane material to form compounds less soluble than the membrane itself. Thus, the chloride electrode, which consists of a $AgCl/Ag_2S$ membrane, is poisoned by mercury(II) ions. Due to the solid state of the membrane the reaction rate is rather low, so a poisoned membrane can be repaired by carefully polishing it with an ap-

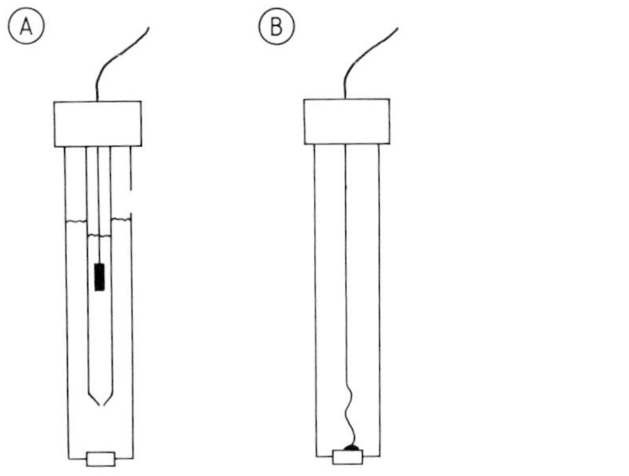

Figure 11. Two types of solid-membrane electrodes (with permission from Springer-Verlag, Heidelberg)
A) With internal filling solution and reference system; B) With solid contact

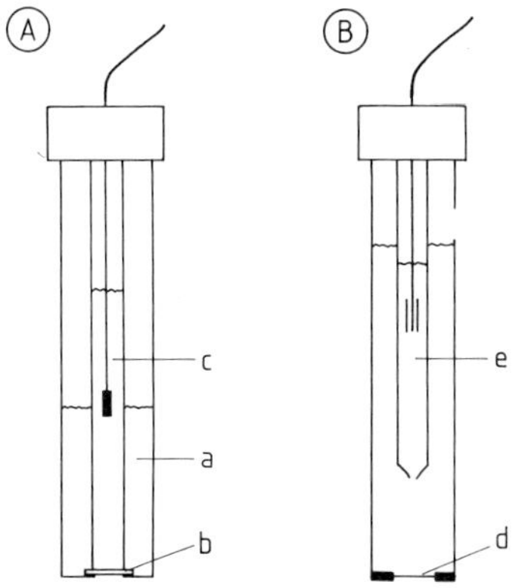

Figure 12. Two types of liquid-membrane electrodes (with permission from Springer-Verlag, Heidelberg)
A) Porous-membrane electrode; B) PVC-membrane electrode
a) Reservoir for liquid ion carrier; b) Porous membrane; c) Reference system; d) PVC membrane; e) Reference system

propriate polishing powder (as described in instruction manuals furnished with the electrodes).

Liquid-Membrane Electrodes. The electroactive phase in this type of electrode consists of a liquid that has been stabilized mechanically. For this purpose the electroactive components may be dissolved in a highly viscous and apolar organic solvent such as one of the higher alcohols or paraffin. A porous material (e.g., a membrane filter) is then soaked with this solution and introduced in place of the solid membrane in an electrode of the type described above. More frequently, however, the liquid is not fixed in a porous material, but rather embedded into a high molecular mass polymer like PVC. In this case the plasticizer acts as a solvent for the electroactive component. Simplified construction schemes for these two types of electrodes are illustrated in Figure 12.

For applications like those discussed in Sections 2.1.4 and 2.1.5 the electroactive component of a liquid-membrane electrode would consist of a (charged) ion exchanger or an (uncharged) ion carrier. In both cases the electroactive component must have a rather high affinity for the ion to be determined, but at the same time display the greatest possible selectivity with respect to interfering ions. Table 1 lists ion exchangers and ion carriers frequently used in liquid-membrane ISEs.

ISEs Based On Semiconductors (ISFETs). Devices based on ion-selective field-effect transistors (ISFETs) are sufficiently distinctive to warrant separate treatment in Section 2.2.1.4.

Table 1. Ion exchangers and ion carriers frequently encountered in liquid-membrane electrodes

Ion	Electroactive compound
Cations	
K^+	valinomycine
NH_4^+	nonactine–monactine
Li^+	lithium carrier, e.g., ETH 1644 (Fluka)
Ca^{2+}	calcium salt of didecylphosphonic acid
Anions	
NO_3^-	Ni(*o*-phen)$_3^{2+}$
ClO_4^-	Fe(*o*-phen)$_3^{2+}$
BF_4^-	Fe(*o*-phen)$_3^{2+}$

Instrumentation. The discussion here of instrumentation required for potentiometric measurements will be limited to a brief overview of suitable reference electrodes [42]. The conventional reference electrode consists of an electrode of the second kind, like the Ag/AgCl/KCl system. Much effort has been devoted to the development of miniaturized reference systems suited to direct integration into ISFETs, but most systems of this type fail to fulfill all the requirements for a "true" reference electrode with a well-defined and stable potential difference relative to the standard hydrogen electrode. Reference electrodes with gel-stabilized internal electrolytes are produced for applications in process analysis.

Table 2 provides an overview of the most frequently used reference electrodes.

Table 2. Common reference electrodes

Electrode	Potential vs. NHE	Useful temperature range, °C	Interferences
$Hg/Hg_2Cl_2/KCl$ (0.1 mol/L)	+ 0.334 V		
$Hg/Hg_2Cl_2/KCl$ (1.0 mol/L)	+ 0.280 V	15–70	
$Hg/Hg_2Cl_2/KCl$ (satd.)	+ 0.241 V		
$Ag/AgCl/KCl$ (0.1 mol/L)	+ 0.290 V		complexing agents (CN^-), S^{2-}, strong oxidizing or reducing agents
$Ag/AgCl/KCl$ (1.0 mol/L)	+ 0.222 V	15–110	
$Ag/AgCl/KCl$ (satd.)	+ 0.197 V		
$Hg/Hg_2SO_4/K_2SO_4$ (0.5 mol/L)	+ 0.682 V	15–70	
$Hg/Hg_2SO_4/K_2SO_4$ (satd.)	+ 0.650 V		
$Tl(Hg)/TlCl/KCl$ (3.5 mol/L)	− 0.575 V	15–120	S^{2-}, strong oxidizing agents; no interference from complexing agents

Evaluation of Potentiometric Results. The aim of all methods for the evaluation of potentiometric data is the computation of substrate activities from potential differences observed between an ISE and a reference electrode. Generally speaking, two different approaches are possible: (1) the direct potentiometric measurement of activities, and (2) monitoring during the course of a titration.

A determination by *direct potentiometric measurement* is accomplished either by calibrating the electrode with solutions of known concentration, or by using the techniques of standard addition or standard subtraction. Since calibration with standard solutions usually involves solutions of pure salts, thorough knowledge of the samples to be analyzed is a prerequisite to avoiding erroneous results due to interfering ions present in the sample matrix. Influences attributable to other substances present in the sample that do not interfere directly with the determination but produce changes in ionic strength can be minimized by adding a total ionic strength adjusting buffer ("TISAB solution") to both standard solutions and samples. Standard-addition or -subtraction methods should be invoked whenever it is not possible to make appropriate standard solutions (owing to a very complex matrix, for example). Figure 13 illustrates the evaluation of a standard-addition analysis.

Monitoring a titration by means of an ISE has the advantage over direct potentiometric measurement that the accuracy is determined largely by the titration reaction and not simply by the calibration function of the electrode. This is especially important for ISEs with changing or unknown slopes. For this reason the determination of a surfactant should always be carried out as a titration and not as a direct measurement.

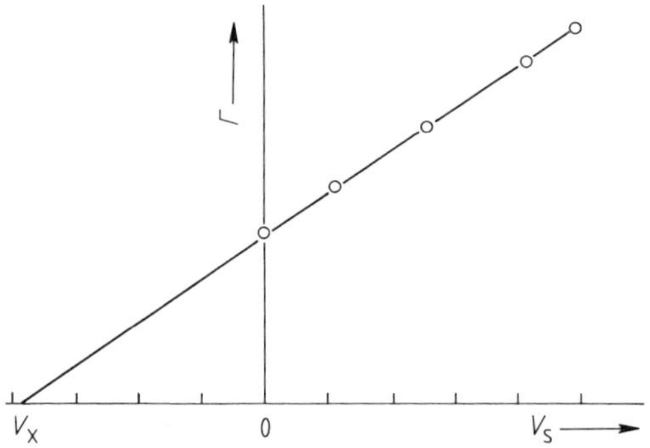

Figure 13. Plot for the evaluation of a standard-addition analysis
V_S = Volume of standard solution; V_x = Volume of analyte
$\Gamma = pE \cdot V_{total}$, with $pE = 10^{\Delta E \cdot (z/S)}$ (ΔE = Potential difference; z = Charge of the ion, including sign; S = Slope of the electrode, including sign)
The unknown concentration of the analyte c_x is $c_x = (-c_s \cdot V_x)/V_{initial}$, where c_s = Concentration of the standard solution and $V_{initial}$ = Initial volume of the analyte

Commercially Available ISEs. Tables 3a–d provide an overview of ISEs commercially available in early 1993. ISEs described only in the literature, and which must therefore be self-constructed, have not been included in the tables. Instructions for the preparation of PVC-membrane elecrodes can be found in the literature [3], [4], [11].

2.2.1.2. Voltammetric and Amperometric Cells

Information is obtained with this type of electrochemical sensor from either the combined current/potential–concentration relationship (voltammetry) or from the current–concentration relationship alone (amperometry).

Table 3a. Commercially available ISEs with selectivity for cations

ISE type	MT[a]	Concentration range, mol/L[b]	Useful pH range[b]	Temp. range, °C[b]	Membrane resistance, MΩ[b,c]	Suppliers[d]	Interfering ions[b,e] in order of decreasing interference; miscellaneous remarks
pNa^+	G	$2 \cdot 10^{-6}$–1	pH = pNa + 4	0–60	100–200	CIMORWΩ	H^+, K^+, NH_4^+, Li^+; Ag^+ must be absent
pK^+	P	10^{-6}–1	2–12	0–40	10–20	CIMORWΩ	Rb^+, Cs^+, NH_4^+, Tl^+, H^+, Ag^+, $Tris^+$, Li^+, Na^+; cationic surfactants must be absent
pCa^{2+}	P	$5 \cdot 10^{-7}$–1	2.5–11	0–40	1–4	CIMORWΩ	Pb^{2+}, Hg^{2+}, H^+, Sr^{2+}, Fe^{2+}, Cu^{2+}, Ni^{2+}, NH_4^+, Na^+, $Tris^+$, Li^+, K^+, Ba^{2+}, Zn^{2+}, Mg^{2+}; cationic surfactants must be absent
pCd^{2+}	S	10^{-7}–10^{-1}	2–12	0–100	≤ 1	MOWΩ	Cu^{2+}, Hg^{2+}, Ag^+ must be absent; Fe^{2+}/Fe^{3+} interfere if present in concentrations > $c[Cd^{2+}]$
pPb^{2+}	S	10^{-6}–10^{-1}	4–7	0–100	≤ 1	MORWΩ	Cu^{2+}, Hg^{2+}, Ag^+ must be absent; Fe^{2+}/Fe^{3+} and Cd^{2+} interfere if present in concentrations > $c[Pb^{2+}]$
pCu^{2+}	S	10^{-8}–10^{-1}	2–12	0–80	≤ 1	CMORW	Hg^{2+}, Ag^+, S^{2-} must be absent; Cl^-, Br^-, Fe^{3+}, Cd^{2+} interfere if $c > c[Cu^{2+}]$; Anions forming complexes with Cu^{2+} (Hal^-, OAc^-) interfere; this ISE is sensitive to light, so all measurements should be carried out in the dark
pAg^+	S	10^{-7}–10^{-1}	2–12	0–80	≤ 1	CIMORWΩ	Hg_2^{2+}, Hg^{2+}, proteins
pNH_4^+	P	< 10^{-6}–1	4–7	0–50	≤ 10	IR	cationic surfactants must be absent; interferences from K^+, Rb^+, H^+, Cs^+, Na^+/Li^+, Sr^{2+}, Ba^{2+}, Mg^{2+}

Table 3b. Commercially available ISEs with selectivity for anions

ISE type	MT[a]	Concentration range, mol/L[b]	Useful pH range[b]	Temp. range, °C[b]	Membrane resistance MΩ[b,c]	Suppliers[d]	Interfering ions[b,e] in order of decreasing interference; miscellaneous remarks
pF^-	S	10^{-6}–sat. soln.	5–11	0–100	$\approx 0.2/\leq 1$	CIMORWΩ	pH < 5, pH > 11
pCl^-	S	$5\cdot10^{-5}$–1	2–12	0–50	≤ 0.1	CIMORWΩ	CN^-, I^-, Br^-, $S_2O_3^{2-}$, NH_3, OH^-, metal ions forming complexes with Cl^-; S^{2-} must be absent
pBr^-	S	$5\cdot10^{-6}$–1	2–12	0–50	≤ 0.1	CIMORWΩ	CN^-, I^-, $S_2O_3^{2-}$, NH_3, Cl^-, metal ions forming complexes with Br^-; S^{2-} must be absent
pI^-	S	10^{-7}– > 1	2–12	0–80	≤ 0.1	CIMORWΩ	S^{2-}, CN^-, Br^-, Cl^-, $S_2O_3^{2-}$, metal ions forming complexes with I^-
pCN^-	S	$< 10^{-6}$–10^{-2}	11–13	0–80	$\leq 30/\leq 0.5$	CIMORWΩ	S^{2-} must be absent; interferences from I^-, CrO_4^{2-}, $S_2O_3^{2-}$, Br^-, Cl^-, metal ions forming complexes with CN^-
$pSCN^-$	S	$5\cdot10^{-6}$–1	2–12	0–50	≤ 0.1	MOW	$S^{2-} \approx I^-$, Br^-, CN^-, $S_2O_3^{2-}$, Cl^-, OH^-
$pClO_4^-$	P	$7\cdot10^{-6}$–1		0–40		O	not sorted: I^-, NO_3^-, Br^-, ClO_3^-, CN^-, NO_2^-, HCO_3^-, CO_3^{2-}, Cl^-, $H_xPO_4^{x+3-}$, OAc^-, F^-, SO_4^{2-}
pNO_3^-	P	10^{-5}–1	3–12	0–50	1	CIMORWΩ	anionic surfactants must be absent; interference from $SCN^- \approx MnO_4^-$, $NO_2^- \approx ClO_4^-$, I^-, Br^-, HCO_3^-, $F^- \approx SO_4^{2-}$
pBF_4^-	P	$7\cdot10^{-6}$–1	2.5–11	0–40		MOW	ClO_4^-, I^-, ClO_3^-, CN^-, $Br^- \approx NO_2^-$, NO_3^-, Cl^-, PO_4^{3-}, OAc^-, F^-, SO_4^{2-}
pS^{2-}	S	10^{-6}–10^{-1}	12–14	0–80		CIMORWΩ	Hg_2^{2+}, Hg^{2+}, proteins

Table 3c. Commercially available ISEs with selectivity for gases

ISE type	MT[a]	Concentration range, mol/L[b]	Useful pH range[b]	Temp. range, °C[b]	Membrane resistance, MΩ[b,c]	Suppliers[d]	Interfering ions[b,e] in order of decreasing interference; miscellaneous remarks
pNH_3	M	$< 10^{-6}$–$5 \cdot 10^{-2}$	alkaline soln.	0–50		IMOWΩ	all types of detergents and wetting agents must be absent; interferences from volatile amines
pCO_2	M	$7 \cdot 10^{-6}$–$2 \cdot 10^{-2}$	acidic soln.	0–50		IOΩ	all types of detergents and wetting agents must be absent; interferences from SO_2, NO_x, H_2S
pNO_x	M	10^{-6}–$5 \cdot 10^{-3}$	acidic soln.	0–50		IO	all types of detergents and wetting agents must be absent; interferences from SO_2, CO_2, volatile carbonic acids
pCl_2	M	10^{-7}–$3 \cdot 10^{-4}$		0–50		O	oxidizing agents

Table 3d. Commercially available ISEs with selectivity for miscellaneous species

ISE type	MT[a]	Concentration range, mol/L[b]	Useful pH range[b]	Temp. range, °C[b]	Membrane resistance, MΩ[b,c]	Suppliers[d]	Interfering ions[b,e] in order of decreasing interference; miscellaneous remarks
Hard[f]	P	$7 \cdot 10^{-6}$–1		0–50		O	not sorted: Na^+, Cu^{2+}, Zn^{2+}, Fe^{2+}, Ni^{2+}, Sr^{2+}, Ba^{2+}, K^+
Surf[g]	P	$\approx 10^{-6}$[h]	1–13	0–40		MO	oppositely charged bulky ions interfere with the titration reaction; high concentrations of nonionic surfactants may interfere

[a] MT = Membrane type; G = Glass-membrane electrode; P = PVC-membrane electrode; S = Solid-state membrane electrode; M = gas-permeable membrane electrode. Glass and solid-state membranes are more resistant to certain organic solvents (acetone, methanol, benzene, dioxane, etc.). PVC-membrane electrodes must not be allowed to come into contact with any organic solvent, because this dramatically reduces the lifetime of the electrode. [b] Mean values as provided by the suppliers. [c] Only if different suppliers provide very different values for membrane electrical resistance are two values listed in the table. [d] This list was compiled in Spring 1993 and should not be regarded as complete; C = Ciba Corning; I = Ingold Messtechnik; M = Metrohm; O = Orion; R = Radiometer; W = WTW; Ω = Omega. [e] A very detailed compilation of selectivity coefficients is available in [17]. [f] Hard: water hardness; i.e., divalent cations. [g] Surf = ionic surfactants. [h] This type of ISE should be used only for the monitoring of titrations; the value cited refers to the mimimum titratable concentration of an anionic surfactant (pure solution).

A voltammetric measurement is accomplished by scanning the potential difference across an electrochemical cell from one preset value to another and recording the cell current as a function of the applied potential. A curve so generated is known as a *voltammogram*. Amperometric measurements are made by recording the current flow in a cell at a single applied potential.

The essential operational function of a transducer for this purpose is the transfer of one or more electrons to or from the transducer surface. The flow of these electrons is what constitutes the output signal. Voltammetric and amperometric devices are also capable of conferring a degree of selectivity on the overall sensing process. Furthermore they are relatively simple: in the most elementary case, the transducer (electrochemical cell) consists of nothing more than two electrodes immersed in a suitable electrolyte. A more complex arrangement might involve the use of a three-electrode cell, with one of the electrodes providing a reference potential.

In spite of the inherent diagnostic advantages of voltammetry, a transducer based on this technique represents a rather cumbersome approach to sensing, mainly because of the electronic circuitry required to scan the applied potential. Accordingly, most sensing applications involve cells operating in an amperometric mode.

Voltammetry. When a slowly changing potential is applied to an electrode immersed in an electrolyte solution containing a redox species, a current will be observed to flow as soon as the applied potential reaches a certain value. This current arises from a heterogeneous electron transfer between the electrode and the redox couple, resulting in either oxidation or reduction of the electroactive species. At a sufficiently oxidizing or reducing potential the magnitude of the current may become a function of mass transfer of the redox species to the electrode. Redox couples that give rise to such currents are frequently referred to as *reversible couples*. If the rate of electron transfer between the redox couple and the electrode is high compared with the rate of mass transfer, then the electrode reaction is reversible. Under these circumstances the concentration ratio for oxidized and reduced forms of the couple at the electrode surface is described by the Nernst equation (Eqs. 5a and 5b).

If electron transfer between the redox species and electrode is very slow relative to the mass transport of solution species to the electrode, then the observed current will not be a function of mass transport. In this case, the low rate of electron transfer results in a concentration ratio of the two forms of the redox couple at the electrode surface that no longer conforms to the Nernst equation. Current – voltage curves of these so-called irreversible processes are of limited analytical utility.

An intermediate situation arises when the electron-transfer and mass-transport rates are comparable. Such quasi-reversible electrode reactions are quite common, and their analytical utility depends to a large extent on careful control of the mass-transfer rate in the electroanalytical method used for their study.

Instruments suitable for voltammetry and amperometry consist of three basic components: a wave-form generator, some form of potential control, and an electrochemical cell. Modern electroanalytical systems employ a three-electrode arrangement for the electrochemical cell. A device called a *potentiostat* is used to maintain a programmed or fixed potential difference between the two current-carrying electrodes (the working electrode and the auxiliary electrode) relative to a third electrode (reference electrode), the function of which is to provide a fixed potential reference in the cell [49], [50].

Both potentiostats and waveform generators have benefited substantially in recent years from the introduction of operational amplifiers, as well as from the availability of desktop computers. Electrochemical cell design is now well-established both for static systems and for detectors situated in flowing reagent streams.

Despite the fact that solid electrodes have been in use for electroanalytical purposes for some time and are well documented, selection of an appropriate electrode material and its preparation before use remains an area of considerable interest and great controversy among electroanalytical chemists [51]. Solid electrodes for analytical use should always be prepared in a rigorous and reproducible manner. This generally involves polishing the electrode physically with successively finer grades of carborundum or diamond paste. The polished electrode should then be rinsed thoroughly to remove as completely as possible all traces of the polishing materials.

Many different applied potential waveforms have been employed for analytical purposes, but only a few are of mainstream significance [52], [53]. These are presented in Table 4.

Table 4. Voltammetric techniques (adapted from [20])

Technique	Applied potential*	Measured signal	Comments
Linear-sweep voltammetry	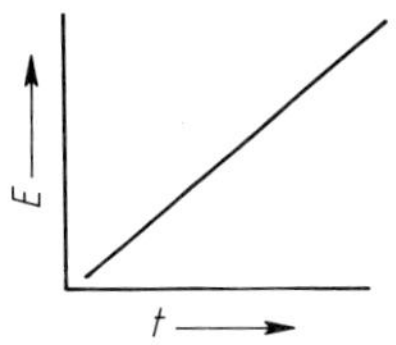	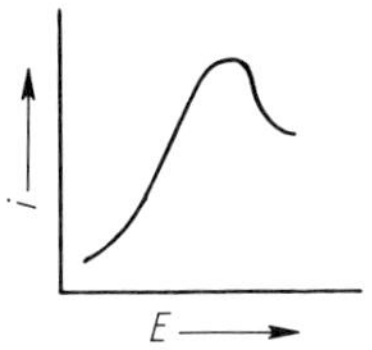	used mainly for diagnostic purposes; current measured throughout the scan
Cyclic voltammetry	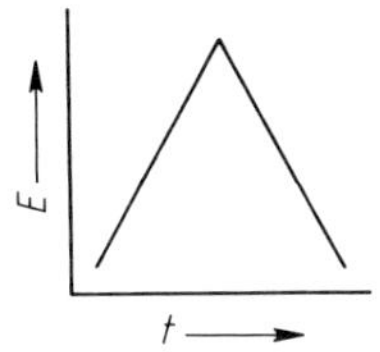	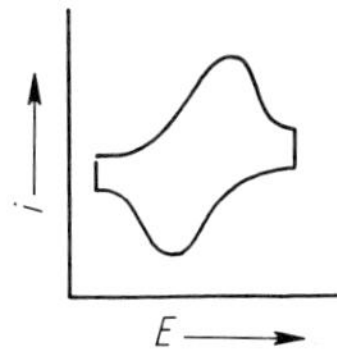	reversal of the potential at the end of the forward scan makes this a powerful diagnostic tool; current is measured throughout the scan
Differential-pulse voltammetry	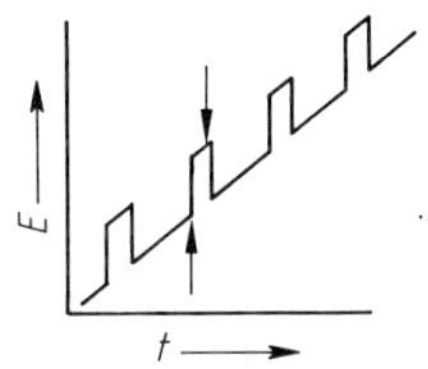	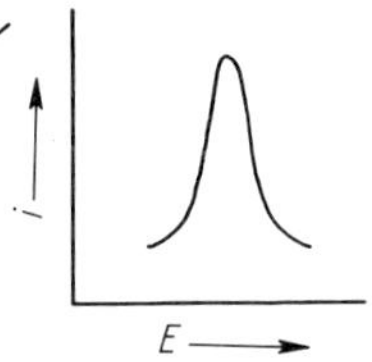	the displayed signal is the difference between the current just prior to and at the end of the applied pulse; low scan rates, but good limits of detection
Square-wave voltammetry	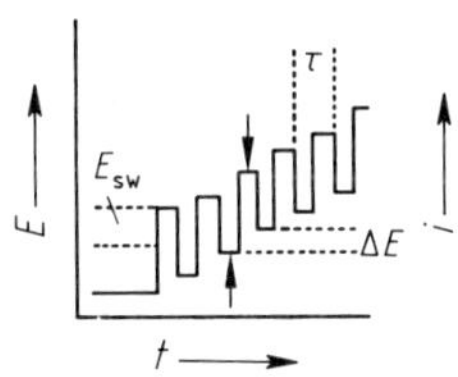	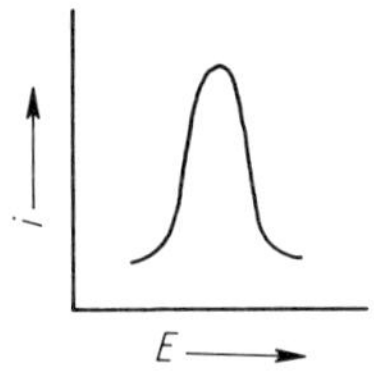	the displayed signal is the difference between currents measured on the forward and reverse pulses; rapid scans and good detection limits

* → = Point at which the current is measured; τ = Portion of the staircase during which pulses of width $\tau/2$ are superimposed; E_{sw} = Square-wave amplitude.

In *linear-sweep* voltammetry the current is recorded while the potential of the working electrode is swept from one selected limit to another at a rate between 1 mV/s and 1000 mV/s. In *cyclic* voltammetry, two linear sweeps are recorded, as shown in Table 4. The distinguishing feature of cyclic voltammetry is that electrogenerated species formed in the forward sweep are subject to the reverse electrochemical reaction in the return sweep [54]. Figure 14 shows a typical cyclic voltammetric curve together with the important parameters. In both methods, as the potential is swept to the electroactive region of the redox couple, the current response rises to a peak value before decaying. This decay is caused by depletion of the electroactive species in the zone close to the electrode surface, which means that the diffusion zone spreads out further into the bulk solution.

Pulse and square-wave voltammetry [55] are much better candidates for incorporation into a chemical sensor than linear-sweep or cyclic voltammetry, primarily because of the possibility of discriminating between faradaic and capacitive currents. When a potential pulse is applied to an electrode, the capacitive current that flows is proportional to the magnitude of the pulse, and it decays exponentially with time. The faradaic current, on the other hand, decays according to the square root of time. Figure 15 illustrates the expected decreases in faradaic and capacitive currents.

Proper selection of the measuring time permits a signal-to-noise ratio to be improved dramatically. This characteristic is exploited in normal-pulse, differential-pulse, and square-wave voltammetry. In the first of these techniques it represents the only mechanism for decreasing the effect of capacitive current. Further elimination of the capacitive current can be achieved in differential-pulse voltammetry by limiting the duration of the applied pulse and by subtracting the current observed immediately prior to the imposition of the pulse. Increased rejection of the charging current leads to improved detection

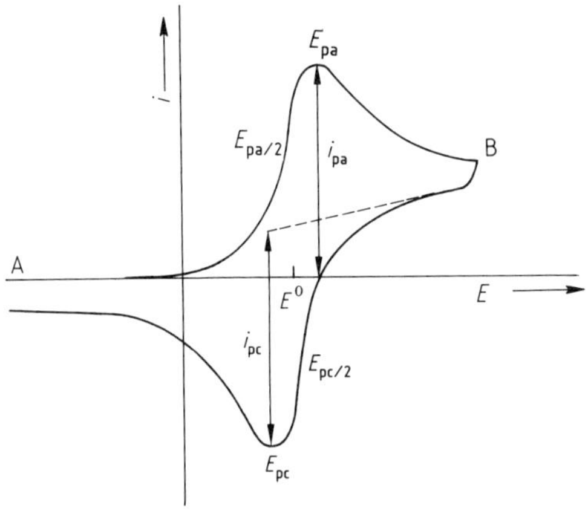

Figure 14. Typical potential vs. current (*E* vs. *i*) curve from a cyclic voltammetric experiment with a reversible redox couple
A) Starting potential; B) Reverse potential
E^0 = Standard potential; E_{pa} = Anodic peak potential; E_{pc} = Cathodic peak potential; $E_{pa/2}$ = Anodic half-wave potential; $E_{pc/2}$ = Cathodic half-wave potential; i_{pc} = Cathodic peak current; i_{pa} = Anodic peak current

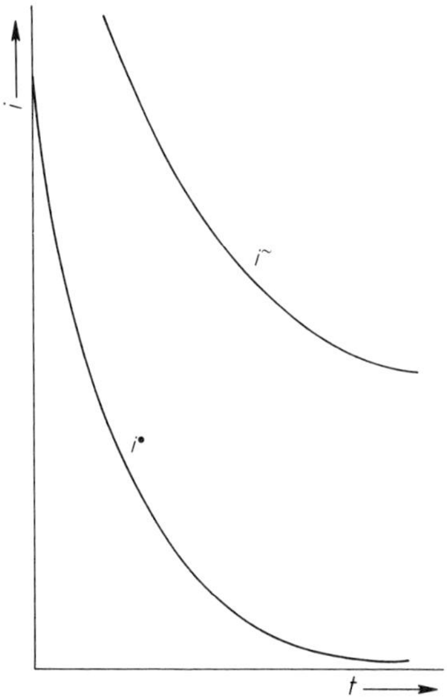

Figure 15. Decrease in faradaic and capacitive current with elapsed time from the imposition of a potential pulse
$i^{\bullet}$ = Faradaic current; $i^{\sim}$ = Capacitive current

limits; indeed, this variant of voltammetry permits concentrations as low as $10^{-8}-10^{-7}$ mol/L to be measured quite readily. In order to avoid problems associated with the electroactive species, the delay between application of successive pulses must be approximately one-half second. This in turn imposes a limit on the scan rate in differential-pulse voltammetry, thereby decreasing its usefulness in sensor applications.

The technique of square-wave voltammetry offers greater promise as a voltammetric method for probing selective chemistry because of the high rate at which the corresponding scan can be executed. The analytical signal in this technique constitutes a difference between the current for the forward pulse and the current for the reverse pulse. Because of the large amplitude of the square wave in a reversible reduction, a reduced electroactive species formed at the electrode during the forward pulse is reoxidized by the reverse pulse. Consequently, the sensitivity of the method is enhanced relative to differential-pulse voltammetry.

This element of speed is crucial to square-wave voltammetry, because, like all voltammetric techniques based on pulse waveforms, the measured current is proportional to $t^{-1/2}$. However, in contrast to the other pulsed voltammetric techniques, square-wave voltammetry causes very little of the depletion that gives rise to distortion of the current–voltage waveform. Accordingly, square-wave voltammetry is uniquely capable of benefiting from high scan rates. A typical compromise frequency for the waveform is 200 Hz. Coupled with a dE value of 10 mV, this gives rise to a scan rate of 2 V/s. The scan rate is limited by a concomitant increase in the capacitive current. At a solid electrode, the square-wave method is no better than pulsed-voltammetric techniques with respect to rejection of the capacitive current.

Another significant advantage of square-wave technology is the possibility of rejecting a wide range of background currents. With respect to capacitive currents this is achieved by the subtraction of two currents in a manner analogous to differential-pulse voltammetry. Slowly varying capacitive currents that arise when surface groups reorganize on certain types of solid electrodes are also eliminated provided the rate of variance is sufficiently low.

Amperometry. Amperometry has traditionally been concerned with maintaining a fixed potential between two electrodes, but pulsed techniques have recently attracted considerable attention as well. The applied potential at which current measurements are made is usually selected to correspond to the mass-transport-limited portion of the corresponding voltammetric scan. Table 5 summarizes the various amperometric methods.

Theoretically, the current obtained in a quiescent solution at a conventional set of electrodes gradually decays to zero according to the Cottrell equation [56]:

$$i_j = nFAC\,(D/\pi t)^{1/2} \qquad (7)$$

where t is the elapsed time from application of the potential pulse.

The observed decrease in current is due to a slow spread of the diffusion layer out into the bulk solution, with a concomitant decrease in the concentration gradient. In practice, this process continues for ca. 100 s, after which random convection processes in the solution take over, putting an end to further movement of the diffusion layer. Waiting nearly two minutes to obtain a steady-state current is not a particularly attractive alternative. Accordingly, in amperometric measurements for sensor applications the spread of the diffusion layer is controlled by invoking one or more of the following mechanisms:

1) *Convective Diffusion.* From a practical point of view, convective diffusion can be achieved in two ways: by moving the solution relative to the electrode, or moving the electrode relative to the solution. Of the systems developed to move the electrode, only one merits consideration here, the *rotating disc electrode* (RDE), but there are few possibilities for realizing this electrode in a sensor.

2) *Imposition of a Physical Barrier in the Form of a Membrane.* There are four primary advantages to covering the electrodes of an amperometric device with a membrane permeable only to the analyte: (a) poisoning of the electrodes by electroactive or surface active species is limited; (b) the resolution of the system is enhanced, because extraneous electroactive species that might otherwise undergo electron transfer at the electrode are excluded; (c) the composition of the electrolyte occupying the space between the membrane and the electrodes remains constant; and (d) the membrane forms a physical barrier to prevent the diffusion layer from spreading into the bulk solution.

The limiting current obtained at a membrane-covered amperometric device is a function

Table 5. Amperometric methods (adapted from [20])

Method	Applied potential	Measured signal	Comments
Chronoamperometry	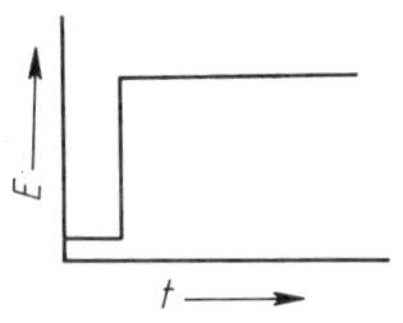	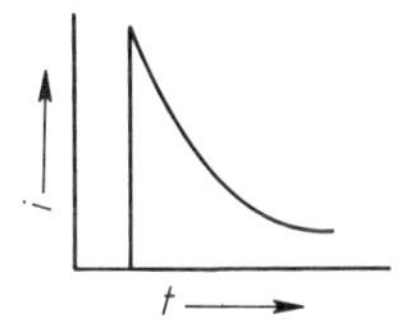	the electrode potential is pulsed to a region in which the analyte is electroactive; a decaying current reflects the growth of the diffusion layer
Pulsed amperometry	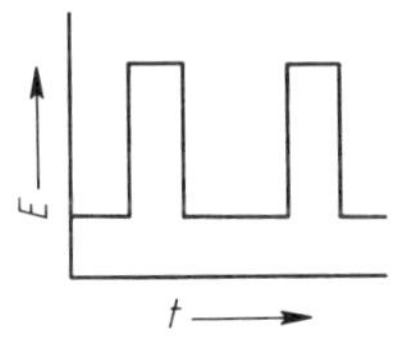	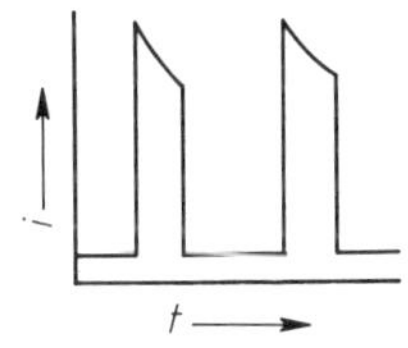	the electrode potential is pulsed briefly to a region in which the analyte is electroactive; between pulses the diffusion layer may be eliminated by forced or natural convection
Pulsed amperometric detection	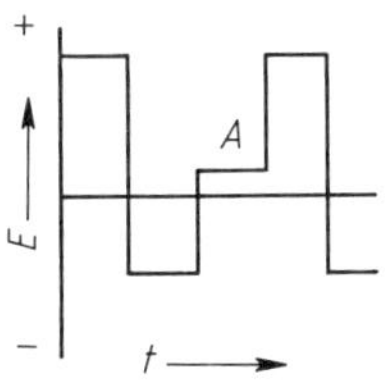	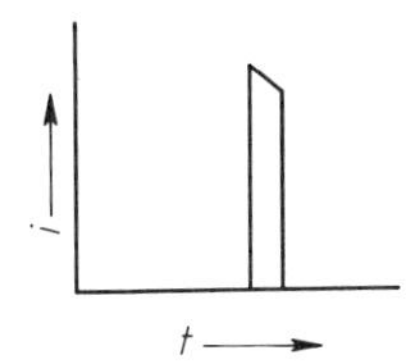	electrode conditioning, analyte sorption, and catalytic electrooxidation are all promoted by the use of this waveform; the current is measured only during the last part of the cycle (region *A*)

of time, reaching in due course a steady-state value. The current–time transient cannot be described by a single equation, because different transport mechanisms operate during this time interval. For a typical membrane-covered amperometric oxygen detector with an electrolyte thickness of 10 µm, a membrane thickness of 20 µm, an electrode radius > 2 mm, and a diffusion coefficient for the analyte two orders of magnitude lower in the membrane than in the electrolyte solution, it can be shown that the current–time transient can be described by three equations. For times > 20 s after the imposition of the potential pulse to the electroactive region, the limiting current is given by

$$i_j = nFAC(P_m/d) \quad (8)$$

where P_m is the permeability coefficient of the membrane and d is its thickness.

An important consequence of this equation is that the limiting current is a function of the permeability of the membrane. Unfortunately, membrane properties such as permeability are themselves a function of time and ambient conditions, particularly temperature. Accordingly, there are advantages to operating membrane-covered detectors away from this steady-state condition. For a typical detector, the limiting current < 100 ms after imposition of the measurement potential pulse is given by

$$i_j = nFAC(K_b/K_0)(D_e/\pi t)^{1/2} \quad (9)$$

where t is elapsed time since imposition of the measuring potential, D_e is the diffusion coefficient for the analyte in the electrolyte, and K_b is the distribution coefficient for the analyte at the internal electrolyte–membrane interface. The ratio K_b/K_0 expresses the salting-out effect of the electrolyte solution.

In effect this equation describes the diffusion-limited current arising within the electrolyte layer. In other words, the diffusion layer has not had sufficient time to spread to the membrane, and so the limiting current is completely independent of the membrane. This has the advantage that alterations in membrane characteristics, as well as alterations in ambient conditions that might affect membrane performance, have no effect whatsoever on the limiting current. One of the factors to which this applies is stirring. Under steady-state conditions, stirring of some type is necessary to prevent changes that would otherwise be caused by the diffusion layer moving out into the bulk solution, although the rate of such stirring is not critical. When transient measurements are made, no stirring whatsoever is required. The true situation regarding transient measurements with amperometric devices is actually not quite so simple as depicted here; for example, there is also a need to take into account contributions from the capacitive current.

The classic amperometric membrane-covered device is the Clark oxygen sensor [57], which consists of a platinum working electrode, an Ag/AgCl reference electrode, a KCl electrolyte solution, and an oxygen-permeable membrane made of teflon or silicone rubber. Figure 16 is a schematic diagram of such a sensor. Dissolved oxygen passes through the permeable membrane, after which it is reduced at the working electrode. The current flow is proportional to the amount of oxygen present in the solution. The following reactions take place at the surface of the working electrode:

$$O_2 + 2H_2O + 2e \longrightarrow [H_2O_2] + 2OH^-$$
$$[H_2O_2] + 2e \longrightarrow 2OH^-$$

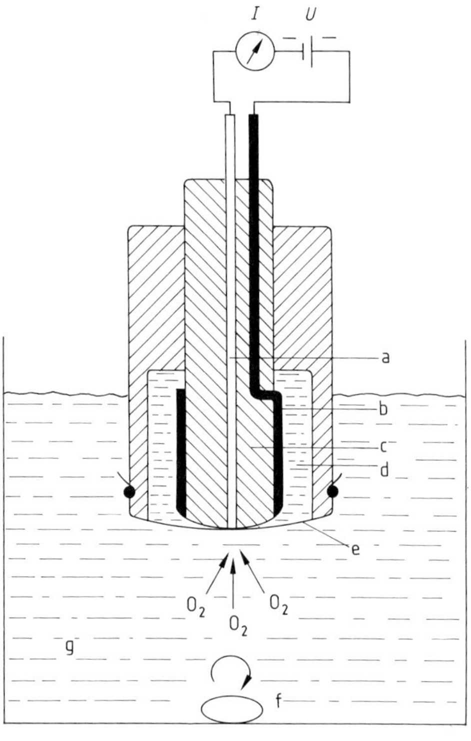

Figure 16. Schematic diagram of a Clark-type oxygen sensor
a) Cathode; b) Anode; c) Insulator; d) Electrolyte; e) Membrane; f) Stirrer; g) Sample for analysis

A separate polarographic wave corresponding to the reduction of hydrogen peroxide is observed only with a mercury cathode, which produces two waves for the reduction of oxygen. At other electrodes (Pt, Au, C) only one four-electron wave is obtained, corresponding to complete reduction to four hydroxide ions [58]. The potential E of the working electrode (at the diffusion-limiting current plateau) usually lies between − 0.6 and − 0.9 V vs. Ag/AgCl.

Progress in semiconductor technology during recent years has made it possible to miniaturize the classical oxygen sensor [59]–[62] for use as a transducer in several chemical and biochemical sensors.

3) *Pulsing the Electrode from a Region of Electroinactivity to a Region of Transport-Controlled Electroactivity, after which the Potential is Restored to the Electroinactive Region.* For electrochemical sensors without membranes, applying a potential pulse from a region of electroinactivity to a region of electroactivity has several advantages. For example, the duty cycle of the pulse can be arranged in such a way that the electrode spends the majority of its time at the electroinactive potential. Under these circumstances electrode reactions give rise to products that poison the electrode, but these products are of minimal significance because they are produced at a very low rate. Moreover, natural processes of convection are able to redistribute the electroactive species during the time that the electrode is at the rest potential, thus limiting the spread of the diffusion layer into the bulk solution. The drawback to this technique is the appearance of a capacitive current. Limiting the size of the pulse decreases the contribution of this current, but a small pulse from a rest potential to the electroactive potential is achievable only for reversible electrode reactions. The alternative is to allow sufficient time for the capacitive current to decay away between imposition of the potential pulse and measurement of the current.

4) *Pulsed Amperometric Detection* [63]. This is a relatively new technique that has made possible the direct amperometric determination of many compounds formerly considered unsuitable for analysis in this way. The corresponding waveform is illustrated in Table 5. This rather complex waveform serves several purposes. On platinum, for example, and at the extreme anodic potential, surface-confined oxidation products that tend to poison the electrode are desorbed. This process occurs concurrently with the formation of PtO_2 on the surface. At the cathodic potential the surface oxide is subsequently reduced, and a clean platinum surface is regenerated. In some cases, sorption of the analyte occurs at this cathodic potential, so the analytical signal is recorded at an intermediate potential. Several different mechanisms can give rise to this signal. Electroactive species that are not adsorbed at the electrode may undergo conventional oxidative electrode reactions. Alternatively, electroactive species that are adsorbed at the electrode during the cathodic swing of the cleaning cycle may undergo either conventional oxidative desorption or oxidative desorption catalyzed by an oxide species on the surface of the electrode. For many compounds the latter reaction is much more favorable from a thermodynamic point of view than oxidation in the absence of a catalyst. This type of reaction occurs most readily when both the analyte and the oxide are confined to the surface. Finally, adsorbed electroinactive analytes may also be detected by virtue of their ability to block the electrode sites at which the PtOH formation occurs. Consequently, any faradaic current arising from the oxidation of platinum is diminished in the presence of adsorbed electroinactive analyte.

5) *Microelectrodes.* Electroanalytical applications using microelectrodes have become more common in recent years [64]–[68] as silicon technology has opened new possibilities for constructing electrodes with micrometer and submicrometer dimensions [69], [70]. The rate of electrolysis at such electrodes is approximately equivalent to the rate of diffusion. Linear-sweep voltammetry in this case gives rise to the same S-shaped current–voltage curves observed with rotating disc or dropping mercury electrodes. The time-dependent current arising at a microelectrode is given by the equation for the chronoamperometric current at a conventional electrode under spherical diffusion conditions:

$$i_t = nFACD\,[\{1/(tD\pi)^{1/2}\} + \{1/r\}] \qquad (10)$$

where r is the electrode radius. Indeed, a sphere or hemisphere represents a good model for the diffusion zone that surrounds such an electrode, and because of enhanced mass transport to the electrode, a steady-state current is achieved very rapidly once a potential pulse has been applied to the electrode.

Although the diffusion zone is large in this case relative to the electrode surface, it is small compared to the zone surrounding a conventional electrode. This small size means that the faradaic current associated with a microelectrode is relatively immune to the effects of convection in the bulk solution, and the current in a flowing stream is independent of the rate of flow. Decreased effects due to capacitance and resistance at microelectrodes, coupled with high mass-transport rates, make it possible for electrochemical measurements to be obtained in this way in cells containing highly resistive solutions. In fact, provided sufficient free carrier is available to charge the double layer, microelectrodes can be used for direct electrochemical measurements in the gas phase [71].

2.2.1.3. Conductance Devices

Conductance is used as a sensor signal in two different areas of sensor technology. On one hand it is possible to determine the electrolytic conductance of electrolytes, which permits conclusions to be reached on a wide range of electrolyte properties. Moreover, partial pressures of gaseous compounds can be determined in the gas phase by examining their influence on the electronic conductance of such materials as semiconducting oxides [72]–[75].

Conductometric Sensors for Monitoring Electrolytic Conductance. *Introduction.* Electrolytic conductance serves as a useful signal of the electrical conductivity of an electrolyte, and *conductometry* has been developed into an electrochemical analytical method based on measuring the conductance in electrolyte solutions. All ions in such a solution act as current carriers as a function of their mobility, charge, and concentration. For this reason conductivity measurements are inherently nonselective with respect to any system to which a voltage difference has been applied. Despite this disadvantage, conductivity measurements are indispensable sources of information, especially for monitoring the chemical purity of water samples. In pollution control, conductometric data provide a reliable measure of the total content of ionic pollutants, which is sometimes all that is required. Advantages of conductivity detectors include their universal applicability, simplicity, low price, and wide effective concentration range. Conductivity measurement is one of the most commonly used electrochemical techniques in the control of industrial chemical processes. Conductivity detectors are also employed in ion chromatography and HPLC as a way of detecting separated ions in the eluate from a column [76]. Furthermore, conductometric analyzers are used in the monitoring of atmospheric pollutants such as acidic or alkaline gases (e.g., HCl, CO_2, SO_2, SO_3, NH_3) after these have been transferred into aqueous solution [77], [78]. Conductivity measurements are nondestructive with respect to the sample, and they are always advantageous when electrolytes must be determined in a medium of low self-conductivity.

Definitions. The electrolytic conductivity σ of a solution is defined as the conductance G of the electrolyte between two electrodes, each with an area A, separated by a distance d. The conductivity σ is then given by

$$\sigma = G \cdot d/A \tag{11}$$

The traditional unit of electrolytic conductivity, Siemens per centimeter (S/cm), has recently been replaced by S/m [79], but the new unit has not yet been accepted widely by the instrument-producing industry, nor by the users of conductivity measurement devices. The English-language literature has historically made use of the "mho" (reciprocal ohm), which is another way of expressing the conductance G. The quotient d/A (cm^{-1}) is defined as the cell constant k, and depends on the geometry of the cell. The electric field is never homogeneous over the cross section of a real cell, so it is not in fact possible to calculate a cell constant k directly from geometric dimensions; it must instead be determined with calibration solutions of known conductivity [80].

The temperature dependence of σ can be deduced from ionic mobilities. This dependence is expressed in terms of the temperature coefficient $\alpha = (d\sigma/dT)(1/\sigma)$, which generally has a positive value between 1 and 3% per degree Celsius. Knowledge of α in turn makes it possible to calculate conductivities at temperatures other than the temperature of measurement. Whenever the need arises to determine the conductivity of a solution with a very low electrolyte concentration it is essential also to take into account the self-dissociation of pure water, which has a conductivity $\sigma_{25} = 0.05483$ µS/cm.

Measurement Principles. The equipment used for measuring conductance usually consists of a conductivity sensor, a measuring attachment, and a temperature-compensation unit. Conductance values are obtained by establishing either the current flow between a set of electrodes subjected to a constant AC voltage, or the current induced in the secondary coil of a transformer connected to the primary coil by the electrolyte of interest. Capacitive coupled cells have been developed as well, but because of their expensive construction they are no longer of interest [81]–[84].

Two-Electrode Cells. Two-electrode cells are equipped with a pair of electrodes across which both the voltage and the current can be measured. Use of an AC voltage source is a convenient way of avoiding polarization effects and electrolyte decomposition. Figure 17 illustrates a typical construction together with the corresponding equivalent circuit. The advantage of this design is the simplicity of both the cell and the measuring attachment.

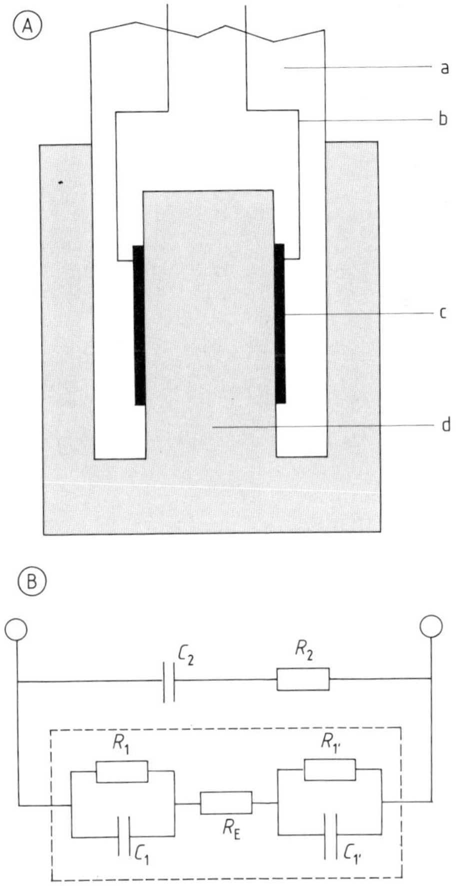

Figure 17. Depiction of a two-electrode cell (A) together with the corresponding equivalent circuit (B)
The dashed rectangle includes all network components involving the electrolyte in contact with the electrodes.
a) Cell body; b) Leads; c) Electrode; d) Sample
R_E = Electrolyte resistance; R_1, $R_{1'}$ = Charge-transfer resistances at the interface electrode–electrolyte; C_1, $C_{1'}$ = Capacitance of the electrochemical double layer; R_2, C_2 = Shunt circuit in the cell

Disadvantages are associated with the need for measuring both the current and the voltage at a single electrode, which may lead to electrode polarization and a measured conductivity lower than the true value. The resistance contributed by a cell in contact with a sample must also be taken into consideration in conjunction with other resistance and capacitance values. At high conductivities ($\sigma > 1$ mS/cm) the resistance of the connecting wires is another significant factor, and electrode reactions can cause polarization of the electrode–electrolyte interface. These problems can be overcome most effectively with an AC voltage oscillating at 1–5 kHz and electrode materials with minimum polarizability, such as platinum covered with a layer of platinum black. At low conductivities ($\sigma < 10$ μS/cm) measured conductivity values may be too high because of capacitive–ohmic shunts caused by the cell body or the connecting wires ("Parker effect"). Contamination by adsorbed ions is also a problem with porous electrodes. The measurement frequency should therefore be kept as low as possible, and the electrodes should be smooth. Resistive precipitates at the electrode surface are sometimes a cause of artificially low conductivity.

For the reasons mentioned above, the "cell constant" is in fact constant only over a limited conductivity range. Suitable values for the cell constant depend on the anticipated level of conductivity:

Expected conductivity	Cell constant
< 0.1 μS/cm–1 mS/cm	0.1 cm^{-1}
1 μS/cm–100 mS/cm	1.0 cm^{-1}
10 μS/cm–1 S/cm	10 cm^{-1}

Four-Electrode Cells. The four-electrode cell was developed as a way of overcoming the problems inherent in two-electrode cells. Two current electrodes in this case generate an electric field, and two separate voltage electrodes are provided for measuring the voltage drop with the aid of a high-impedance amplifier. This design is insensitive to polarization effects, and it also avoids interference from insulating films or precipitates because these are formed outside the zone of measurement. The result is a single system that can be used to measure any conductivity > 1 μS/cm.

Inductively Coupled Cells. The principle underlying inductively coupled conductivity cells is very simple. Two coils of a transformer are shielded from each other magnetically such that the only coupling loop is formed by an insulating tube filled with the sample. An AC voltage in the primary coil induces an AC current in the sample, the magnitude of which depends on the dimensions of the loop and the conductivity of the sample. The voltage induced in the second coil depends only on the input voltage of the primary coil and the conductance of the sample. Typical working frequencies are 80–200 Hz. An advantage of these cells is that they permit contactless measurement. Very high conductivities (> 1 S/cm) can be measured without polarization errors because of the contactless decoupling of the signal. Even thin resistive films within the tube fail to falsify the measured conductivity. Another advantage associated with cells of this type is the possibility of encapsulating the entire device in a polymer offering high chemical resistance. Inductively coupled conductivity cells are thus of special interest for measurements in highly conducting and aggressive media [85].

Conductometric Sensors for Monitoring Gaseous Compounds. *Introduction.* Gaseous compounds in the gas phase are detected with chemically sensitive devices that respond electrically to changes in conductance. Brattain and Bardeen (using germanium, in 1952) [86] and Heiland (zinc oxide, 1954) [87] were the first to observe modulation of the electrical conductance of a solid-state device by gases from the surrounding atmosphere. The most commonly used material for conductometric solid-state gas sensors is now tin dioxide [88]. As noted previously (Section 2.1.2) commercial sensors based on polycrystalline SnO_2 ("Taguchi sensors") [89] are produced by companies such as Figaro Inc., for a wide range of industrial and domestic applications, including gas monitors, leak detectors, and alarm systems for toxic or inflammable gases. Apart from SnO_2, many other inorganic semiconducting oxides can serve as sensitive materials in gas sensors, including TiO_2, ZnO, and Fe_2O_3 with various additives and dopants [9], [90]. Certain organic semiconductors such as phthalocyanines with appropriate central metal atoms, or polymers like polypyrrole or polysiloxane, also show sensitivity to gaseous compounds. These sensors are often referred to as homogeneous or heterogeneous semiconductors, or as dielectric sensors. Detailed lists of materials sensitive to detectable gaseous compounds together with the corresponding operating temperatures are provided in [72]–[75], [90]. The design principle upon which such sensors are based is quite simple: a sensitive semiconducting sample is placed between two metallic contacts, and the combination is equipped with a resistance heater capable of maintaining the appropriate working temperature, usually between 100 °C and 700 °C. Semiconductor sensors have dimensions in the range of millimeters, and are easily fabricated by either thick- or thin-film technology [91].

Heterogeneous Semiconductor Sensors. The response of a heterogeneous semiconductor sensor is based on changes in surface conductance, mainly due to changes in the free electron concentration within a thin surface layer of the semiconductor as a consequence of charge exchange with adsorbed species accompanying chemisorption or heterogeneous reactions [27] (see Section 2.1.2). The most commonly used semiconducting materials are the two very stable oxides SnO_2 [92], [93] and ZnO [94]. Because the conductance change is a surface phenomenon, the ratio of surface to bulk material should be made as large as possible (using thin layers) to maximize the effect. For sintered polycrystalline materials the grain size should also be as small as possible.

At a given temperature the number of gas molecules adsorbed at a semiconductor surface—hence the conductance—correlates exponentially with the partial pressure of the gas. Oxidizing gaseous compounds such as oxygen, nitrogen dioxide, and chlorine cause a resistance increase by extracting electrons from the conduction band of an *n*-type semiconductor, while reducing gases like hydrogen, carbon monoxide, methane, hydrocarbons, or alcohols lead to a resistance decrease. Humidity decreases the resistance of an *n*-type semiconductor, causing considerable interference.

Response times for sensors of this type are usually in the range of seconds or minutes. For some gases it has been found that catalytic effects at the semiconductor surface can have a significant impact on the response. Thus, sensor properties such as response and selectivity can be improved by adding catalytically active metals like palladium, rhodium, silver, or platinum. This doping of the semiconductor surface is one important way of modifying the selectivity of a sensor [95]–[97], but another possibility for adjusting sensor properties for a particular application is careful selection of the working temperature [98].

The influence of temperature on the selectivity of a sensor depends on the fact that each gas has its own *characteristic temperature*: a temperature at which the reversible relationship between gas concentration and the concentration of electrons at the semiconductor surface (and therefore the conductance) reaches a maximum [92]. Working temperatures for SnO_2-based sensors usually lie between 200 °C and 450 °C depending on the application.

Various techniques are used for the fabrication of semiconductor sensors. Conductance sensors from structurized sintered polycrystalline ceramics can be produced by thick- or thin-film technology. Chemically sensitive materials in the form of single crystals or whiskers can be attached to electrodes by thin- or thick-film techniques as well. Mass production of sensors requires that the resulting devices be characterized by a defined level of conductance. For example, the conductance of polycrystalline SnO_2 can be adjusted by subsequent thermal treatment > 800 °C under a controlled partial pressure of oxygen. Another approach to defined conductance involves doping the semiconductor with antimony or fluorine. The reproducibility and stability of a sensor signal is generally diminished by irreversible changes in the defect structure of the semiconductor surface.

Homogeneous Semiconductor Sensors. The effectiveness of this type of semiconductor sensor depends on changes in bulk conductance. The stoichiometry and concentration of bulk point defects, and thus the conductance of such materials, is a function of the partial pressure of a gaseous analyte. Conductance changes arise through reversible changes in small deviations from ideal stoichiometry, which in turn influence the concentrations of bulk point defects and free electrons [27]. The deviation from ideal stoichiometric composition is usually very small; for example, in $Cu_{2-x}O$ the deficit x in the Cu ratio is on the order of 0.001. Since conductance and deviation from ideal stoichiometry are proportional, a measured conductance can be used to determine a corresponding partial pressure. Regarding the kinetics of these sensors, an important factor is diffusion of a compound like oxygen or the corresponding lattice defects that compensate for nonideal stoichiometry within the bulk of the semiconductor. Any gaseous reaction involving point defects at the surface of the semiconductor must also be appropriately fast. Both requirements can be met through the combination of high temperature and a catalytic coating (e.g., Pt).

Most bulk-conductance sensors consist of semiconducting oxides the conductance of which changes with the partial pressure of oxygen. They are used primarily for monitoring oxygen partial pressures in the air or in combustion gases. Titanium dioxide (TiO_2) is one of the most commonly used homogeneous sensor materials for the determination of oxygen partial pressures [99]. At 900 °C and an oxygen partial pressure of 0.1 MPa, TiO_2 has precisely the stoichiometric composition. At lower oxygen partial pressures TiO_2 has an oxygen deficit x expressed by the formula TiO_{2-x}. This oxygen deficit specifies the number of unoccupied oxygen sites and the corresponding excess of trivalent titanium in the lattice. Both defects act as electron donors, thereby contributing to the concentration of free electrons in the conductance band of the semiconductor. The oxygen deficit, and thus the conductance, varies inversely with the partial pressure of oxygen in a characteristic way. Thus, the conductance σ_e of an n-type semiconducting oxide like TiO_{2-x}, CeO_{2-x}, or Nb_2O_{5-x} is given by the equation:

$$\sigma_e = \mathrm{const}(T) \cdot p(O_2)^{-1/m}, \qquad (12)$$

where const(T) is a temperature-dependant constant and m usually varies between 4 and 8 depending on the nature of the disorder in the oxide structure.

On the other hand, p-type semiconducting oxides like $Cu_{2-x}O$, $Ni_{1-x}O$, and $Co_{1-x}O$ have a deficit x of metal relative to ideal stoichiometry. This deficit increases with increasing oxygen partial pressure. Ionic lattice defects occurring in oxides of this type act as electron acceptors. In this case the concentration of the defects, and therefore the conductance, increases with increasing oxygen partial pressure according to the expression:

$$\sigma_e = \mathrm{const}(T) \cdot p(O_2)^{1/m} \qquad (13)$$

Here again, m usually varies between 4 and 8.

The strong exponential dependence of const(T) on temperature is a great disadvantage with homogeneous semiconductor sensors, one that can lead to large measurement errors. Nevertheless, appropriate sensor design makes it possible to compensate for the temperature dependence. In the case of a TiO_2 combustion-gas sensor, for example, the conductance of an oxygen-sensitive TiO_2 sample can be compared with that of a second sample displaying the same response to temperature but coated in such a way as to eliminate its sensitivity to gas composition. The conductance difference between the two samples is then a function only of the gas composition, not of fluctuations in the temperature.

An important requirement for any metal oxide under consideration for use as a homogeneous semiconducting sensor is that it display a sufficiently high bulk diffusion coefficient $\tilde{D}$ for the conductance-determining species. Response times t for oxide layers with a thickness y can be estimated from the Einstein equation:

$$y^2 = 2\tilde{D} \cdot t \tag{14}$$

For example, a Cu_2O layer with a thickness of 1 μm reaches a response time of ca. 1 ms, corresponding to a diffusion coefficient of ca. 10^{-5} $cm^2 s^{-1}$, only upon heating to 700–800 °C. The diffusion coefficient can be altered by doping, but it also depends on the extent of deviation from ideal stoichiometry.

A number of oxides (e.g., TiO_2, CeO_2, Nb_2O_5, CoO) or titanates with a perovskite structure ($Ba/Sr/CaTiO_3$) have been successfully tested for application as oxygen sensors [100], [101]. Most of these investigations involved determination of the oxygen partial pressure in exhaust gases from combustion processes at partial pressures between 10 and 10^{-18} kPa. At low oxygen partial pressures and high temperature, certain oxides (e.g., Cu_2O, CoO, NiO) are unsuitable because of their reduction to pure metal. Sensors based on TiO_2 ceramics are by far the most highly developed devices. These consist of porous sintered ceramic materials with a platinum surface doping, and are used to determine air–fuel ratios in the exhaust gases of combustion engines.

Apart from TiO_2, a number of other semiconducting oxides might be useful as sensor materials for special applications over limited ranges of oxygen partial pressure, as in monitoring the level of oxygen in inert gases. There is also a good possibility of optimizing various ternary and higher oxide combinations for sensor application by suitable choice of composition with respect to stability, diffusion, and surface properties [102]–[104]. Thus, barium and strontium ferrates show conductances with considerably less temperature dependence relative to TiO_2 or CoO.

Organic Conductometric Sensors. For several years, certain organic semiconducting polymers have attracted attention in the search for new chemical sensors [105]. The possibility of utilizing conductivity changes induced by gases adsorbed on organic materials is interesting for a number of reasons. For example, it might permit the detection of very low levels of atmospheric pollutants. Moreover, organic materials are much more easily modified than inorganic materials with respect to such characteristics as sensitivity, working temperature, and selectivity. The greatest disadvantage of organic materials in gas detection is that they are usually very poor conductors, and conductivity measurements would be correspondingly difficult. Also, organic materials are thermally unstable, so it is often impossible to use them at temperatures at which gas–solid interactions proceed rapidly and reversibly.

Phthalocyanines constitute one group of organic materials with thermal stability > 400 °C, and they are usually considered to have semiconducting rather than insulating properties. Thin layers of phthalocyanine complexes based on metals like lead [106] and copper [107] have been particularly carefully studied as potential gas sensors [108]. The conductance of these films is a very sensitive function of gases with electron-acceptor properties, including nitrogen dioxide, chlorine, and iodine. Films of phthalocyanines with various central metal atoms can be deposited on substrates by sublimation or screen printing. Changing the nature of the central metal atom makes it possible to adjust the sensitivity and selectivity of a phthalocyanine film in accordance with the proposed application [109]–[114].

Another organic semiconductor with gas-sensitive properties is polypyrrole. Chemiresistors with ambient-temperature gas sensitivity to both electron-donating and electron-accepting gases have been achieved using thin polypyrrole films prepared by electrochemical polymerization on electrodes. Sensitivities have been investigated toward such gases as nitrogen dioxide [115], ammonia [116], and methanol [117], [118]. A wide variety of polymers of this type is available, including substituted polypyrroles, polythiophenes, polyindoles, and polyanilines.

Humidity Sensors. Humidity sensors are devices that respond to changes in water-vapor pressure by a change in electrical resistance, capacitance, or a combination of the two as reflected in the impedance. The sensitive material might be an electrolyte, an organic polymer, or a metal oxide subject to a change in ionic–electronic conductivity or capacitance with respect to humidity as a consequence of physical adsorption or absorption of water molecules. In general these devices offer a very cost-effective continuous or spot-check approach to monitoring humidity in air and other gases [119], [120].

Electrolyte humidity sensors. Electrolyte humidity sensors are based on an electrolyte solution (e.g., LiCl) the ion conductivity of which changes as a result of evaporation or condensation of water in response to changes in the relative humidity of the surrounding atmosphere. The fabrication of these sensors is quite straightforward. A porous substrate or an organic binder is impregnated with the electrolyte in such a way as to prevent electrolyte from flowing out of the sensor even at very high humidity. Contact is maintained with platinum electrodes on both sides of the substrate. Even thin films of potassium metaphosphate or barium fluorite are suitable for humidity measurements. Despite their slow response and limited linear range these sensors have an outstanding reputation due to good reproducibility, long-term stability, and low cost.

Ceramic Humidity Sensors. The first widely-used ceramic humidity sensor was the aluminum oxide sensor, in which aluminum is oxidized anodically in an acidic solution to form a porous oxide film, sometimes less than 1 µm thick, that displays the necessary hygroscopicity. Adsorption of water molecules causes a decrease in resistance and an increase in capacitance, which is reversed by subsequent desorption. Other materials are also used in humidity sensors, including ZnO with added Li_2O and V_2O_5, colloidal Fe_3O_4, α-Fe_2O_3, mixed oxides of TiO_2 and SnO_2, $MgCr_2O_4$–TiO_2, TiO_2–V_2O_5, $ZnCr_2O_4$–$LiZnVO_4$, $MgAl_2O_3$, $MgFe_2O_4$, or H_3PO_4–$ZrSiO_2$, as well as $PbCrO_4$ with added alkali oxide [121]. Sometimes the signal of such a sensor is subject to drift as a result of changes in the surface structure or contamination from other adsorbed gases. Nevertheless, reproducibility can be restored by heat cleaning.

The most important considerations with respect to sensor characteristics are surface properties (hydrophilic–hydrophobic), pore-size distribution, and electrical resistance. To ensure adequate sensitivity and response a sensor of this type should consist of a very thin film of porous ceramic with a porosity $> 30\%$. The contacting electrodes may be interdigitated or porous sandwich-type structures made from noble metals (e.g., Pd, Pt, Au) and so constructed that they do not obstruct the pores of the oxide film. Humidity affects not only the resistance of a porous ceramic but also its capacitance by extending the surface area in contact with the electrodes. The high dielectric constant of adsorbed water molecules also plays an important role.

Polymeric Humidity Sensors. Certain polymers are capable of taking up water as a function of the relative humidity of the surrounding atmosphere, leading to changes in such mechanical and electrical properties as volume, resistance, and dielectric characteristics. A number of polymers have been used as sensitive materials in this regard, including ion-exchange resins, polymer electrolytes, hydrophilic polymers, and hygroscopic resins [122]. Hydrophilic polymers [e.g., poly(ethylene oxide)–sorbitol] swell by absorption of water. If such a polymer is filled (mixed) with carbon or metal powder of very small particle size, the resistance observed depends on the distance and the number of particles that are in contact with each other. Uptake of water leads to swelling of the polymer and disturbance of the ohmic contacts between conducting particles, which in turn increases the resistance. For this reason the resistance correlates with humidity. Useful materials include cellulose esters and polyalcohols.

The resistance of polymers with dissociable groups (e.g., sulfonates, amines, amides, or even hydroxides) also depends on humidity. The resistance of such polymers in fact decreases with absorption of water because of ionic dissociation and an increase in ionic conduction.

The absorption of water by polymers is especially suited to the development of a capacitive humidity sensor based on the high dielectric constant of water, since capacitance depends on the area, the thickness, and the dielectric constant of a dielectric. Uptake of water causes an increase in the dielectric constant followed by an increase in capacitance. Polyimide is a suitable sensitive material, but so are certain inorganic ceramics such as Al_2O_3 [123] and low-density Ta_2O_5, which change their capacitance with the formation of water dipoles. Low-density Ta_2O_5 layers can be formed by anodic oxidation of sputtered tantalum films, whereas polyimide films are deposited by spin coating and subsequent polymerization.

2.2.1.4. Ion-Sensitive Field-Effect Transistors (ISFETs)

As described in Section 2.2.1.1, with conventional ion-sensitive electrodes (ISEs) the ion-sensitive membrane is placed between the analyte and the internal reference solution. An internal reference electrode makes electrical connection to the internal reference solution, and the measurement system is completed by a second reference electrode in contact with the analyte solution. In any modern pH meter the internal reference electrode is connected to the gate metal of a metal-oxide semiconductor field-effect transistor (MOSFET), which acts as an impedance-converting device (Fig. 18). Normally, a conventional ISE functions very well. However, in certain applications problems can arise due to large size, high cost, and sensitive construction.

These problems might be resolved by the ion-sensitive field-effect transistor (ISFET). In an ISFET the ion-sensitive membrane is placed directly on the gate insulator of the field-effect transistor; alternatively, the gate insulator itself, acting as a pH-sensitive membrane, may be exposed to the analyte solution (Fig. 19). If one compares an ISFET with the conventional ISE measurement system, the gate metal, connecting leads, and internal reference system have all been eliminated. An ISFET is a small, physically robust, fast potentiometric sensor, and it can be produced by microelectronic methods, with the future prospect of low-cost bulk production. More than one sensor can be placed within an area of a few square millimeters. The first ISFETs were described independently by Bergveld [124] and Matsuo, Esashi, and Inuma [125].

ISFET Operation. The ISFET can be regarded as a special type of MOSFET. To understand the operation of an ISFET it is therefore necessary first to understand the operation of a MOSFET.

An n-channel MOSFET (Fig. 20) consists of a *p*-type silicon substrate with two *n*-type diffusions, the source (c) and the drain (d). The struc-

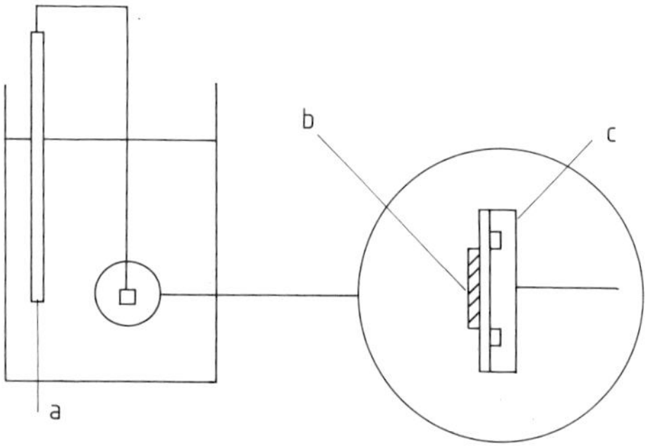

Figure 19. Schematic diagram of an ISFET measuring circuit and cell assembly
a) Reference electrode; b) Ion-selective membrane; c) ISFET

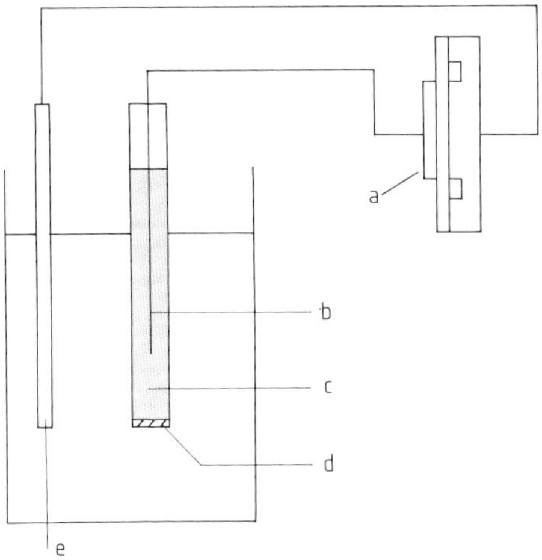

Figure 18. Schematic diagram of a membrane-electrode measuring circuit and cell assembly
a) MOSFET; b) Internal reference half cell; c) Internal filling solution; d) Ion-selective membrane; e) Reference electrode

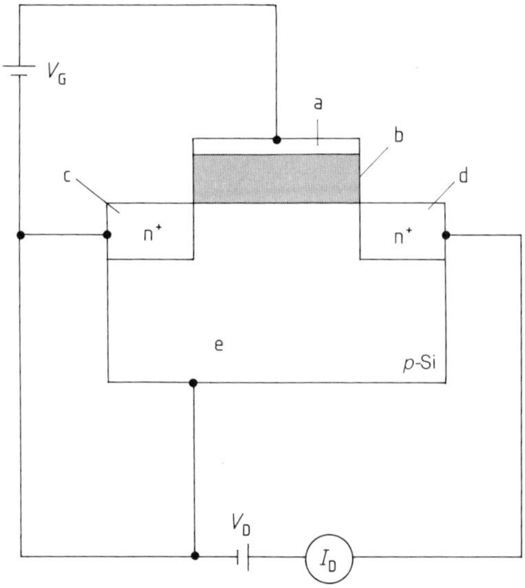

Figure 20. Schematic diagram of an n-channel MOSFET
a) Gate metal; b) Insulator; c) Source; d) Drain; e) Bulk

ture is covered with an insulating layer (b), on top of which a gate electrode (a) is deposited over the area between the source and the drain. A voltage is applied to the gate that is positive with respect to both the source and the bulk material. This voltage produces a positive charge at the gate electrode and a negative charge at the surface of the silicon under the gate. The negative charge is formed by electrons, which are the minority carriers in the substrate, and this in turn creates a conducting channel between the source and the drain. The density of electrons in the channel, and thus the conductivity of the channel, is modulated by the applied gate voltage (V_G). If a constant voltage (V_D) is applied between the source and the drain a change in the conductivity of the channel leads to a change in current between source and drain. The drain current I_D in the unsaturated region ($V_D < V_G - V_T$) of a MOSFET is equal to:

$$I_D = \mu C_{ox} \frac{W}{L}\left[(V_G - V_T)V_D - \frac{1}{2}V_D^2\right] \tag{15}$$

μ = Electron mobility in the channel
C_{ox} = Oxide capacitance per unit area
W/L = Ratio of channel width to length
V_T = Threshold voltage

A detailed description of the MOSFET is provided in [126].

In an ISFET (Fig. 21) the gate metal is replaced by a sensitive membrane and an electrolyte solution, or simply by the electrolyte solution, which is in contact with a reference electrode. The potential of the gate with respect to the substrate is the sum of the potentials of the various electronic and electrochemical processes that occur in the external path between the silicon substrate and the gate; i.e., the reference electrode half-cell potential E_{ref}, the solution–membrane or solution–insulator interfacial potential φ_M, and the metal–silicon contact potential. These voltages are added to the externally applied voltages V_g, and the resulting voltage has the same significance and function as that defined in the theory of MOSFET operation. V_{T*}, the threshold voltage of the ISFET, can be described as:

$$V_{T*} = V_T - E_{ref} - \varphi_M. \tag{16}$$

From Equation (15) we obtain the following relationship:

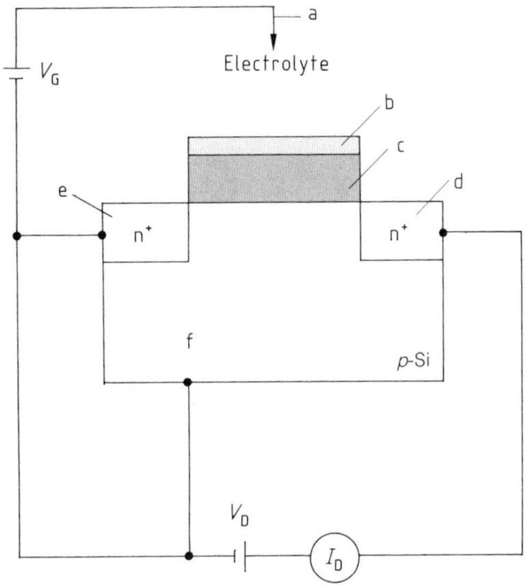

Figure 21. Schematic diagram of an ISFET
a) Reference electrode; b) Ion-selective membrane; c) Insulator; d) Drain; e) Source; f) Bulk

$$I_D = \mu C_{ox} \frac{W}{L}\left[(V_G - V_T + E_{ref} - \varphi_M)V_D - \frac{1}{2}V_D^2\right]. \tag{17}$$

φ_M, the solution–membrane or solution–insulator interfacial potential, depends on the activity of the ions in the analyte solution according to the Nernst–Nikolsky equation [Eq. (6), Section 2.2.1.1] for ISFETs with an additional membrane, or according to a similar equation (see below) for ISFETs with a solution–insulator interface.

If all parameters in equation (17) apart from φ_M are kept constant, changes in I_D are caused exclusively by changes in the activity of ions in the analyte solution. Thus, in an ISFET, the drain current is modulated by the ionic activity in the analyte solution. A detailed description of ISFET operation is presented in [127].

pH-Sensitive ISFETs. Gate insulators such as Si_3N_4, Al_2O_3, and Ta_2O_3 have been found to be pH sensitive, so FETs in which bare gate insulators of this type are exposed to a solution respond to changes in pH. The pH response can be explained by the "site-binding theory." This model assumes that there are ionizable binding sites present at the surface of an insulator, and that these can exhibit amphoteric behavior with respect to the potential-determining ion (i.e., H^+):

$$A - OH \rightleftharpoons A - O^- + H_S^+$$
$$A - OH + H_S^+ \rightleftharpoons A - OH_2^+$$

According to this model the pH dependence of φ_M is given by the following equation:

$$\varphi_M = 2.3 \frac{kT}{q} \left[\frac{\beta}{\beta + 1} \right] (pH_{pzc} - pH) \tag{18}$$

under the condition that $\beta \gg q\varphi_M/kT$. pH_{pzc} is the pH value for which $\varphi_M = 0$; β is a parameter that reflects the chemical sensitivity of the outer gate insulator, with a value dependent on the density of hydroxyl groups and the surface reactivity. Surfaces with a high β value show Nernstian behavior. For a detailed description of the site-binding model, see [127].

ISFETs Sensitive to Other Ions. Deposition of an ion-selective membrane on top of the gate insulator opens the way to measurement of ions other than H^+. Most of the sensitive materials for ISE applications described in Section 2.2.1.1 have been used in conjunction with ISFETs, providing sensors covering a wide variety of species. Thus, ISFETs with solid membranes have been described with AgCl–AgBr membranes sensitive to Ag^+, Cl^-, and Br^- [128], for example, or with LaF_3 membranes sensitive to fluoride [129]. Ion-sensitive polymer-matrix membranes (liquid membranes with a polymer matrix) have also been used as sensitive membranes for ISFETs. The first experiments involved mainly PVC membranes [130], [131], but these membranes show poor adhesion and poor mechanical strength. To improve membrane adhesion, modified PVC was utilized as a matrix material [132]. The use of silicones as matrix materials has made it possible to prepare very durable ISFETs with polymer-matrix membranes [133]. Other reports describe the use of photopolymerized polymers as matrix materials [134], [135].

Problems. The commercial development of ISFETs has been slow despite intensive research. From the standpoint of broad commercialization the ISFET is in competition with conventional ISEs with respect to price, performance, and reliability, and it is essential that some means be devised for producing ISFETs by commercially acceptable methods. Reaching this stage requires that several problems be solved, however.

ISFETs have so far been used primarily in conjunction with conventional macro reference electrodes. This fact greatly limits the potential benefits to be gained from miniaturization of the sensor. An optimal reference electrode for ISFETs should be miniaturizable and display long-term stability, and it should be subject to fabrication with microelectronic techniques. Attempts have been made to miniaturize macro reference electrodes [136], or to use modified surfaces with extremely low surface-site densities as reference FETs [137], but to date the above-mentioned requirements have not been met.

A further problem relates to encapsulation. The ion-sensitive gate area is exposed to the analyte solution, but all other parts of the sensor must be insulated from the solution. In most cases encapsulation is carried out by hand using epoxy resins, but it is impossible to produce ISFETs at low cost by this technique. Both photolithographic methods [138], [139] and an electrochemical method [140] have been proposed for the solution of this problem. One very promising development is the backside-contacted ISFET [141]. The electrical contacts in this case are protected from the analyte solution by means of O-rings, thereby circumventing the need for resins.

As noted previously, ISFETs sensitive to ions other than H^+ are prepared by deposition of an additional membrane on top of the gate insulator. If such ISFETs are to be commercialized this membrane must be deposited by a mass-production technique, which will be difficult to achieve with polymer membranes. One possible solution is the development of photolithographically patternable ion-sensitive membranes [142], but such membranes have so far been developed only for a few ions, and problems in the photolithographic deposition of more than one type of membrane on a single chip have not been resolved. The interface between an insulator and a sensitive membrane is also ill-defined in a thermodynamic sense, but this problem was solved by introducing a hydrogel between the sensing membrane and the insulator. Futhermore, ISFETs suffer from drift. The influence of this effect would be eliminated if the sensor were to be used in flow-injection analysis (FIA), so successful applications can be anticipated for ISFETs in the field of dynamic measurement [143].

Only if all the problems cited are successfully resolved will the ISFET live up to the promises that have been held out for it for more than two decades: low cost, multifunctionality, and a robust and reliable construction.

2.2.2. Optical Sensors

2.2.2.1. Fiber-Optical Sensors

In recent years the use of optical fibers has become increasingly important in the field of chemical and biochemical sensors. This is clearly due to the fact that fiber-optical sensors combine the strengths of well-known bulk optical analytical methods with the unique advantages of optical fibers. More specifically:

1) An enormous technology base already exists for familiar optical methods of chemical analysis. There is probably no class of chemical analyte that has not at some time been the subject of optical determination through absorption or fluorescence spectroscopy. Methods also exist for labeling a target analyte with an appropriate chromophore or a fluorophore, a common practice in immunoassays.
2) Fiber-optic techniques offer increased sensitivity relative to conventional bulk optic approaches. For example, evanescent wave (EW) spectroscopy is significantly more sensitive than bulk attenuated total reflection spectroscopy. Fiber-optical EW spectroscopy is the only technique suitable for use with highly absorbing or scattering media.
3) The interrogating light in a fiber-optics system remains guided, so no coupling optics are required in the sensor region, and an all-fiber approach is feasible. This represents a sharp contrast to bulk optics, where proper optical adjustment is extremely critical. Fiber optics also provides a good solution to the problem of wide separation between the sensing area and the detector, as in the in vivo monitoring of medical parameters.
4) Recent developments in optical communications and integrated optics have led to optical materials that permit significant miniaturization of the sensing and detecting devices.
5) Since the measurement signal is optical, electrical interference is avoided, and the absence of electrical connectors ensures safe operation in hazardous environments, such as those containing explosive vapors.

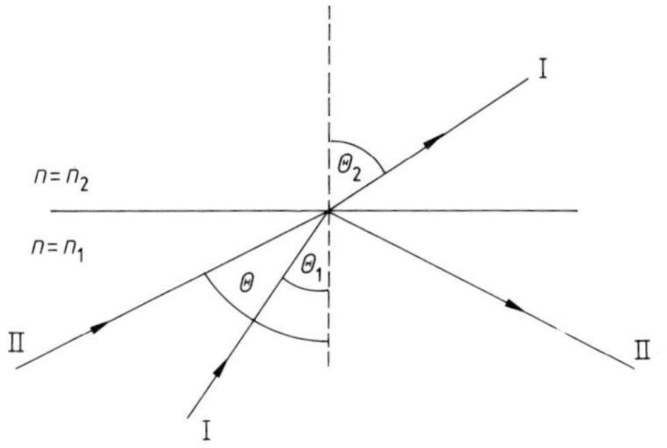

Figure 22. Refraction and total internal reflection
n = Refractive index

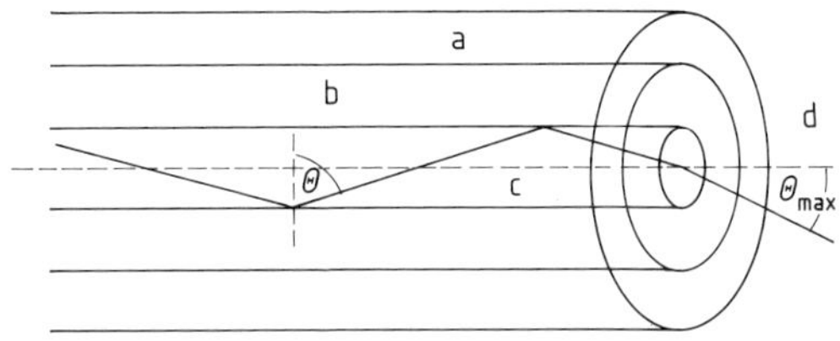

Figure 23. Structure of an optical fiber
a) Jacket; b) Cladding, $n = n_{clad}$; c) Core, $n = n_{core}$; d) Environment, $n = n_{env}$

The light inside an optical fiber is guided by the principle of total internal reflection, as described by Snell's law and depicted in Figure 22. A ray of light striking the interface between two media with differing refractive indices n_1 and n_2 is refracted (ray I) according to the relationship $\sin(\Theta_1) \times n_1 = \sin(\Theta_2) \times n_2$. If the condition $n_1 > n_2$ is fulfilled, total internal reflection will be observed for several angles of incidence. Thus, at an angle close to the horizontal the incident beam will no longer be refracted into the second medium, but will instead be totally reflected internally at the interface (ray II in Fig. 22). The *critical angle of incidence* $\Theta_c = \arcsin(n_2/n_1)$, which defines the onset of total internal reflection, is determined by the refractive indices of the two media, and whenever $\Theta > \Theta_c$ light is guided inside the medium characterized by n_1.

The general structure of an optical fiber is illustrated in Figure 23. Light entering from the left is confined within the core, which is surrounded by a cladding of lower refractive index (n_{clad}). The whole arrangement is surrounded by a nonoptical jacket, usually made from plastic and designed to provide both stabilization and protection. The fibers themselves can be constructed from a wide variety of transparent materials with differing refractive indices, including various glasses, fused silica, plastics, and sapphire. Most commercially available fibers have glass cores with n_{core} in the vicinity of 1.48 and diameters ranging from 5–200 µm; typical claddings have $n_{clad} < 1.48$ and thicknesses of 50–1000 µm. An optical fiber is usually characterized by its numerical aperture $NA = (n_{core}^2 - n_{clad}^2)^{1/2}$. The value of NA determines the cone angle Θ_c within which incident light will become entrapped by the fiber, where $\Theta_c = \arccos(NA)$; $\Theta_{max} = \arcsin(NA/n_{env})$ for the cone angle Θ_{max} with which light leaves the fiber at its end (see Fig. 23).

The simple ray-optical description of light-guiding presented above fails to account for several phenomena important with respect to fiber-optical sensors. The most important of these is the *evanescent wave*. The intensity of light that is totally reflected internally does not in fact fall abruptly to zero at the core–cladding interface. Instead, the electromagnetic field intensity decays exponentially as a function of distance from the interface, thereby extending into the medium of lower refractive index. This field is called the *evanescent field*, and its penetration depth d is defined by the distance from the interface within which it decays to a fraction 1/e of its value at the interface. The angle of incidence Θ strongly influences the penetration depth. Values for d vary from 50–200 nm, increasing as Θ decreases; d reaches its maximum when Θ is very close to the critical angle.

Fiber-optical sensors can be constructed in either of two fundamentally different optical configurations with respect to the sensing area. The first is a wave-guide-binding configuration, in which recognition occurs at the surface of the fiber core via the evanescent field. This is achieved by removing the jacket and cladding from the fiber and causing the analyte molecules to bind to the core surface, as in the example described below. Such an arrangement is described as an *evanescent wave sensor* or *intrinsic fiber-optical sensor*. The resulting sensitivity depends greatly on the value of NA [144].

Alternatively, the fiber-end configuration might be considered, in which analyte is released for recognition purposes into the illuminated space at the end of the fiber. In this case the fiber acts only as a light pipe, resulting in what is known as an *extrinsic fiber-optical sensor*.

Both configurations entail similar optical systems: a spectral light source for excitation (e.g., a xenon or tungsten lamp equipped with a monochromator or optical filter, an LED, or a laser) and a spectral light-detection device consisting of a monochromator or filter together with a photomultiplier or photodiode. A schematic diagram applicable to the optical systems for these two sensor configurations is presented in Figure 24.

More detailed information regarding the physical principles of fiber-optical chemical and biochemical sensors and the corresponding detection systems is available in [145] and [146].

Biochemical Sensors. Antibodies are widely used as recognition elements in fiber-optical biosensor applications. This is due to the high specificity of the binding reaction between an antibody and an analyte (antigen). To determine this reaction opticaliy, labeling agents such as fluoresceine, rhodamine, or phycoerythrine are used, all of which show absorption and emission bands in the visible region of the spectrum. Linear amplification can be achieved by binding more than one label. Depending on the circumstances, either the antigen or the antibody might be tagged (see below).

The evanescent wave (intrinsic) configuration is most important for biosensors because it permits the antibody or antigen to be immobilized directly on the core surface, and it extends the path for possible interaction between the light and the recognition elements. Accordingly, only tagged molecules inside the evanescent field are subject to excitation. The amount of light

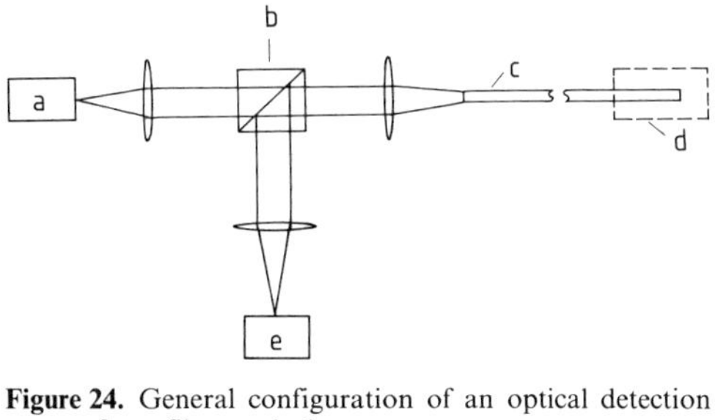

Figure 24. General configuration of an optical detection system for a fiber-optical sensor
a) Source; b) Beam splitter; c) Fiber; d) Sensing area; e) Detector
The excitation system consists of a xenon or tungsten lamp together with a monochromator or optical filter, an LED, or a laser, whereas the light detection system might be a monochromator or filter with a photomultiplier or photodiode.

energy absorbed, or the light emitted from these molecules and collected by the fiber, gives rise to the measurement signal. In contrast to the extrinsic sensor approach, the intrinsic arrangement therefore reduces noise that would otherwise result from free labeled molecules or intrinsic fluorescence within the environment [146]. Despite these advantages, however, extrinsic biosensors have also been constructed, with binding devices placed in front of the fiber as a way of fixing the recognition elements. These devices can be either glass plates or transparent membranes. This offers the advantage of easy removal and replacement of the recognition elements [147].

Figure 25 provides an example of a biosensor of the intrinsic type. Here the antigen has been immobilized at the fiber core surface where it has been allowed to bind with a labeled antibody. Energy from the evanescent wave is absorbed by the labels, generating a detectable fluorescence signal. Addition of unlabeled analyte to the surrounding solution leads to competition for the binding sites. This permits some of the labeled antibody molecules to diffuse away from the surface, causing the fluorescence signal to decrease. The same general approach is also applicable to immobilized antibodies and tagged antigens [148].

Competition between a labeled antigen and an unlabeled analyte can also be effected in an external column, where the labeled antigen has previously been bound to an antibody. Once the competition reaches equilibrium any labeled antigen that has been released can be detected with antibody immobilized on a fiber. In this case the fluorescence signal would be expected to increase with an increased concentration of analyte.

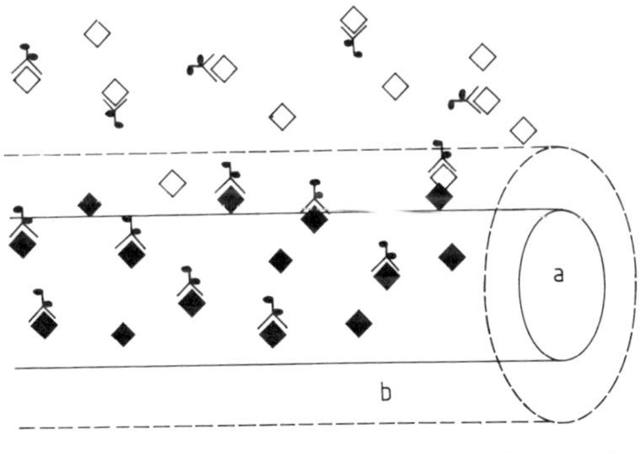

Figure 25. A fiber-optical immunosensor, in which labeled antibody molecules are gradually replaced by free analytes, causing the fluorescence signal to decrease
a) Core; b) Region of evanescent wave
◆ Fixed analyte; ◇ Free analyte; ⅄ Tagged antibody

Enzymes are in some cases also useful for recognition and labeling purposes. For example, glucose oxidase has been used to determine glucose on the basis of free oxygen that is consumed, where the disappearance of oxygen is measured with a fiber-optical oxygen subsensor [149]. Labeling of antibodies or antigens with urease or alkaline phosphatase and subsequent addition of a fluoro- or chromogenic substrate leads to detectable production of fluorophores or chromophores. The signal in this case is subject to multiple enhancement since it is proportional to the concentrations of both labeled antigen and antibody as well as to incubation time.

Chemical Sensors. Recognition with a *chemical sensor* is achieved via an analyte-specific reaction involving chemical compounds located inside the sensing area. The reaction must be one that is accompanied by changes in absorbance or fluorescence characteristics. Several examples are described below. A common physical arrangement utilizes a fiber-end approach and two single fibers or two fiber bundles, one for guiding light into the sensing area and the other to carry it away. The chemical reagents are confined within a transparent membrane or a small segment of the fiber itself, near its end and shielded by a membrane.

A particularly wide variety of fiber-optical sensors has been developed for measuring pH values, and many of these sensors are also used as subsensors for other analytes. Sensors for pH determination can be divided into two general types. The first relies on colorimetric acid–base indicators to produce changes in absorbance. For example, the dye phenol red can be copolymerized with acrylamide and bisacrylamide to yield dyed polyacrylamide microspheres. These are in turn packed into a length of cellulose dialysis tubing placed at the end of a pair of fibers that can serve as a vehicle for monitoring changes in absorbance [150]. Bromothymol blue immobilized on styrene–divinylbenzene copolymer has been similarly employed [151]. Alternatively, fluorometric acid–base indicators can be used to detect changes in fluorescence intensity that accompany protonation. Thus, one pH sensor was based on fluorescence quenching by H^+ of the excited state of immobilized fluoresceinamine [152]. 8-Hydroxypyrene-1,3,6-trisulfonate (HPTS) fixed on ethylene vinyl acetate has also been invoked as a fluorescence pH sensor [153].

Chemical sensors for other ions have been reported as well, including the halides and pseudohalides [154]. A sensor of this type is based on the dynamic quenching of fluorescence emission from glass-immobilized, heterocyclic, acridinium and quinolinium indicators. The fluorometric process involved is known to obey the Stern–Volmer equation [155]. Al(III) ions have been determined with morin (3,5,7,2′,4′-pentahydroxyflavone) [156]. Thus, aluminum displaces the hydrogen ion when it binds to morin, leading to a change in fluorescence properties. Be(II) has also been determined using morin [157]. Furthermore, reports have appeared on sensors for Mg(II), Zn(II), and Cd(II) based on quinoline-8-ol sulfonate immobilized electrostatically on an anion-exchange resin [158].

Gas sensors for carbon dioxide [159] and ammonia [160] in aqueous solution have been constructed on the basis of a gas-permeable membrane and an optical pH subsensor. Gases diffusing through the membrane produce a change in the pH in the sensing region of a fiber-optical pH sensor. Sulfur dioxide [161] and oxygen [162] sensors take advantage of the quenching of the energy transfer (Förster transfer) between excited pyrene and perylene, both of which can be immobilized on a silicone matrix. Other O_2 sensors rely on the effect of fluorescence quenching as described by the Stern–Volmer equation. These generally involve metal complexes like tris(2,2′-bipyridine)ruthenium(II) [163]. In most cases fluorescence quenching also leads to a shortening of the fluorescence lifetime, which can be determined, for example, by phase-sensitive detection [164]. The advantage of this method is that fluorescence lifetime is independent of the fluorophore concentration, and there is no signal drift due to bleaching.

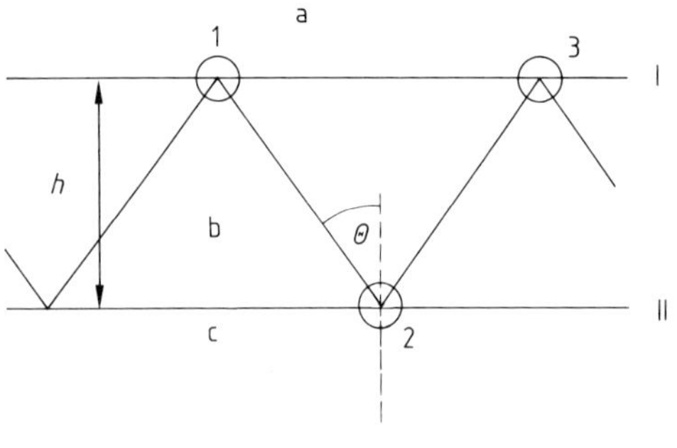

Figure 26. Propagation of a confined light ray within a planar waveguide mounted on a substrate
Most substrate materials are glasses or crystals with a refractive index of 1.4–2.4, and the cover material is air; the thickness of the film ranges from 0.1 to several micrometers.
a) Cover, $n = n_c$; b) Film, $n = n_f$; c) Substrate, $n = n_s$

2.2.2.2. Integrated Optical Chemical and Biochemical Sensors

Generally speaking, integrated optical devices can be constructed using techniques similar to those for manufacturing semiconductor devices. This offers great potential for cost reduction and facilitates the miniaturization often required in sensor applications. Furthermore, such sensors retain all the advantages of the well-known fiber-optical sensors.

The propagation of a light ray confined within a planar dielectric film (or *waveguide*) is depicted in Figure 26. Guidance of the light is possible only if total internal reflection occurs at the interfaces film–cover (I) and film–substrate (II). This condition is met if the refractive index of the film n_f is greater than that of both the cover n_c and the substrate n_s. Moreover, the angle of incidence Θ must be greater than the critical angles for total internal reflection at the interfaces I (Θ_{cI}) and II (Θ_{cII}), given by the expressions $\sin(\Theta_{cI}) = n_c/n_f$ and $\sin(\Theta_{cII}) = n_s/n_f$. To fulfill the self-consistency condition within the waveguide, the phase shift of the light wave as it travels from point 1 to point 3 must be an integral multiple m of 2π, where m is called the *mode number*.

$$\left(\frac{2\pi}{\lambda}\right) \cdot L(n_f, h, \Theta) + \Phi_I(n_f, n_c, \Theta) + \Phi_{II}(n_f, n_s, \Theta) = 2\pi m \qquad (19)$$

Φ_I and Φ_{II} are the phase shifts caused by reflection at the interfaces I and II, and $(2\pi/\lambda)L$ is the shift accompanying travel along the optical path $L = 2h\cos(\Theta)$. The values of n_f, n_c, n_s, and m are usually fixed by the fabrication process for the waveguide, and λ is established by the light source. The value of Θ can therefore be calculated from Equation (19). It is useful to define an *effective refractive index* n_{eff} for the guided mode, defined as $n_{eff} = n_f \sin(\Theta) = n_{eff}(n_f, n_c, n_s, h, m, \lambda)$, which, apart from λ, depends only on the material and geometric parameters of the waveguide. The parameter n_{eff} can be used to treat all types of light-propagation effects within the waveguide.

Even though light is totally reflected internally at the interfaces, its intensity does not fall to zero outside the waveguide. Just as with optical fibers, the small part of the guided light located outside the waveguide is referred to as the evanescent field, with a penetration depth ranging from 100–300 nm depending on the waveguide performance. A more detailed mathematical treatment of planar and strip waveguides is presented in [165].

Waveguides intended as chemical or biochemical transducers can be operated in either of two modes. A comprehensive physical description of these operating modes and the corresponding efficiencies is provided in [166].

First, it is possible to use a waveguide for acquiring refractive index data. This can be accomplished by coating the waveguide with a reactive but transparent chemical layer of a thickness greater than the penetration depth of the evanescent field. In this case the layer can be treated as if it were a cover material. Any chemical reaction with the layer will result in a change in the refractive index n_c, which in turn leads to a change in the effective refractive index n_{eff}. This approach has been used successfully in the production of sensors for CO_2 and SO_2 [167].

The adsorption and desorption of molecules is depicted in Figure 27 as an example of the second operational mode. The waveguide in this case is coated with a thin chemical layer consisting, for example, of an antibody. When the antibody–antigen binding reaction occurs, the effective thickness [166] of the entire waveguide increases, which again causes a change in the effective refractive index of the guide. Direct immunosensors based on this method have often been reported in the literature [168], [169].

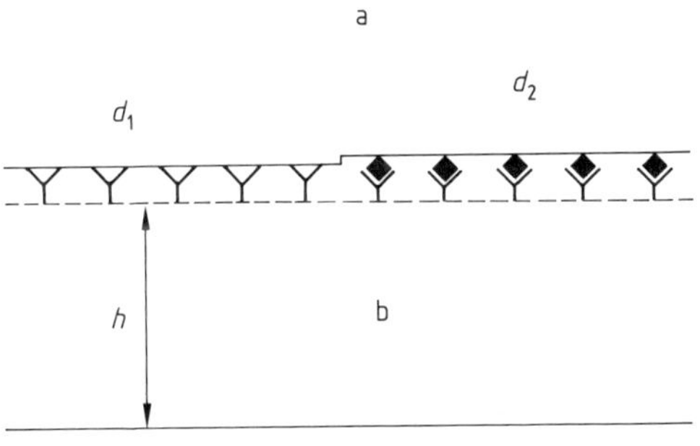

Figure 27. Specific antibody–antigen binding at the surface of a planar waveguide, causing the thickness of the adlayer to change
a) Cover, $n = n_c$; b) Film, $n = n_f$; c) Substrate, $n = n_s$
d_1, d_2 = Layer thicknesses before and after binding

Detecting a change in n_{eff} is a problem that can be solved in various ways. An interferometer approach analogous to the Mach–Zehnder interferometer [170], [171], the Fabry–Perot interferometer [167], or the difference interferometer [172] has often been used. The general arrangement of a Mach–Zehnder interferometer is illustrated in Figure 28 A. The diameter of the strip waveguide is usually on the order of 5 µm, with a distance between the two arms of < 1 mm, and a total length for the device of < 3 cm. Only the signal arm with its reactive chemical layer is exposed; the reference arm is covered by a dielectric film with a refractive index lower than that of the waveguide. The output power P_{out} of the device depends on the extent of the light phase shift $\Delta\Phi = (2\pi/\lambda) L n_{eff}$ between the signal arm and the reference arm; e.g., $P_{out} = P_{in}\cos^2(\Delta\Phi/2)$.

Grating couplers can also be used to determine the value of n_{eff}, as depicted schematically in Figure 28 B. For efficient light *incoupling* the angle α must be consistent with the expression $n_{eff} = n_0 \sin(\alpha) + l(\lambda/\Lambda)$, where l is the diffraction order and Λ the grating period. According to this relationship, a change in n_{eff} leads to a change in the angle α that will reinforce light guided within the waveguide. A grating coupler can also be used in an *outcoupling* arrangement. In this case it is the angle for the outcoupled light that would conform to the grating equation. A detailed description of grating-coupler methods is presented elsewhere [173].

2.2.2.3. Surface Plasmon Resonance

Transducers for optical sensing systems in chemical and biochemical sensors have also been developed based on the sensitivity of surface plasmon resonance (SPR) devices. Since first de-

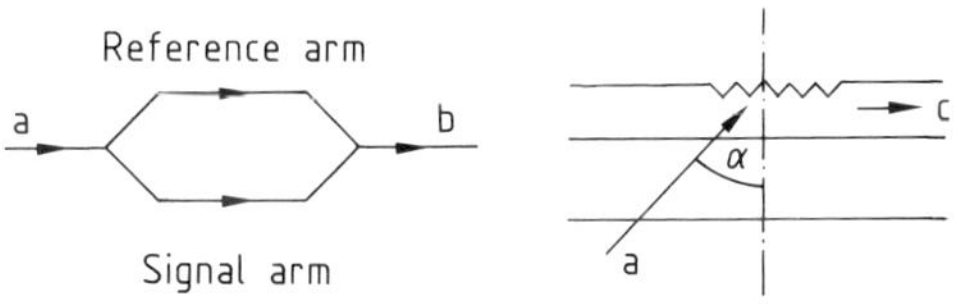

Figure 28. General arrangement of a Mach–Zehnder interferometer (A) and an input grating coupler (B)
In A only the signal arm is exposed to an active chemical layer while the reference arm is covered, whereas in B the grating is provided with an active layer and exposed to the sample.
a) Input beam; b) Output beam; c) Incoupled beam

scribed in 1983 [174], SPR transducers have been used in many different applications, including immunoassays [174], [175] and sensors for gases [174], [176] and liquids [177].

Electron charges located at a metal–dielectric interface can be caused to undergo coherent fluctuations known as *surface plasma oscillations* or *surface plasmons* (SP). The frequency ω of such longitudinal oscillations is related to the wave vector $\boldsymbol{k}_{sp}$ by the dispersion relationship.

$$\boldsymbol{k}_{sp} = \left(\frac{\omega}{c}\right) \cdot \sqrt{\frac{\varepsilon_M \cdot \varepsilon_D}{\varepsilon_M + \varepsilon_D}} \tag{20}$$

where $\varepsilon_M(\omega)$ is the permittivity of the metal and $\varepsilon_D(\omega)$ that of the surrounding dielectric (cf. Fig. 29); c is the velocity of light in a vacuum. The permittivity of the metal ($\varepsilon_M = \varepsilon_{M1} + i \cdot \varepsilon_{M2}$) is complex and the value of $\varepsilon_D = n_D^2$ is real, where n_D is the refractive index of the dielectric. A surface plasmon can be excited only if the conditions $\varepsilon_{M1} < 0$ and $|\varepsilon_{M1}| > \varepsilon_{M2}$ are fulfilled. For example, gold, silver, and palladium are potentially applicable metals with negative values of ε_{M1} for frequencies corresponding to visible light [178].

SP excitation can be accomplished in several ways. A very simple and often used approach is illustrated in Figure 29. RATHER [178] has described this as the *attenuated total reflection* (ATR) coupler. Here the base of a prism is coated with a thin metal film approximately 50 nm thick. Incident monochromatic light is reflected at this interface in such a way that a projection of the light vector $\boldsymbol{k}_0$ on the interface becomes $\boldsymbol{k}_x = \boldsymbol{k}_0 n_p \sin(\Theta)$. If the resonance (Re) condition $\boldsymbol{k}_x = \mathrm{Re}\{\boldsymbol{k}_{sp}\}$ is fulfilled, the evanescent field of the incident light excites an SP at the metal–dielectric interface, although this is only possible with light polarized parallel to the plane of incidence (p-polarized). The excitation is observed as a sharp minimum in the intensity of the reflected light, and it can be detected by scanning the incident angle.

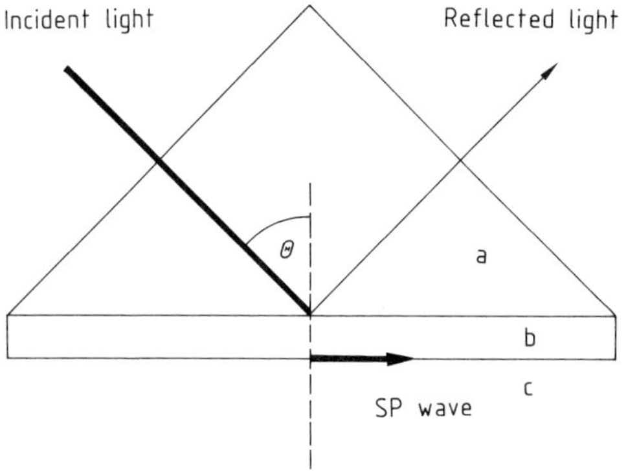

Figure 29. Attenuated total reflection coupler used to excite surface plasmons at the metal–dielectric interface; excitation is observed as a minimum in the intensity of the reflected light by scanning the angle of incidence
a) Prism; b) Metal; c) Dielectric

Using such an ATR device, small changes in $\varepsilon_D = n_D^2$ and the build-up of a thin layer at the metal–dielectric interface can easily be detected by measuring the shift in the *resonance angle* Θ.

For an immunosensing application the metal layer would be coated with appropriate antibodies, either immobilized covalently or embedded into a matrix layer. Upon exposure to a specific antigen the resonance angle should shift relative to that observed in an analyte-free environment due to the increase in layer thickness that accompanies binding of the analyte. For example, a thickness change of 0.1 nm results in a 0.01° change in the resonance angle, a shift that can be measured quite accurately (see also Section 3.3.1).

If an ATR device is to be used for chemical sensing, the metal surface must be coated with a thick dielectric layer of a reactive chemical. Chemical reaction within the layer then produces changes in n_D that can be detected by the corresponding shift in Θ.

2.2.3. Mass-Sensitive Devices

2.2.3.1. Introduction

Mass changes are usually detected with the aid of a gravimetric device, such as a balance. Ideally, balance sensitivity can be extended down into the microgram range, but the desire for further improvements in sensitivity and miniaturization of the experimental setup has led to the development of other mass-sensitive devices: the so-called quartz microbalance (QMB), which is the best-known system, and other, more recent devices such as the surface acoustic wave oscillator (SAW), the Lamb wave oscillator (LW), and the acoustic plate-mode oscillator (APM), which represent the present state of the science (Section 2.2.3.2). These devices are all based on the *piezoelectricity* of solids, (see Section 2.2.3.3), a phenomenon that permits the generation of vibrations or the propagation of waves. A piezoelectric substrate, usually a quartz plate, is stimulated to vibration by an oscillating electric circuit. This corresponds to an interconversion of electrical and mechanical vibrations (deformations of the crystal), resulting in a device that can be regarded as either an oscillator or a resonator.

Bulk waves, surface waves, and waves of other types (see Section 2.3.3.3) are associated with characteristic resonance frequencies corresponding to the most stable frequencies of vibration. Such a resonance frequency is itself highly dependent on the mass of the oscillating plate. A slight change in the mass of the plate will result in a considerable change in the resonance frequency, which can be measured very accurately. An oscillator of this type can therefore be used as a mass-sensitive transducer. Other factors that may influence the observed frequency shift include changes in the viscosity and density of surface layers as well as such physical parameters as forces, temperature, and pressure.

To construct a mass-sensitive sensor the surface of a piezoelectric material must be coated with an appropriate chemically or biochemically reactive layer capable of interacting with the proposed analyte and—ideally—recognizing it as well (see Section 2.3.3.4). Such an interaction causes the mass of the crystal to increase, resulting in a measurable shift in the resonance frequency. A relationship thus exists between the amount of analyte adsorbed or absorbed and the frequency shift of the oscillating system. Resonance frequency changes can be established very accurately, so very small changes in mass can be detected. A theoretical approach applicable to simple cases is described in [179].

To date, mass-sensitive devices have been used mainly for sensing analytes in the gas phase; they are also applicable in principle to liquids, but this presupposes a more sophisticated physical arrangement and complicated signal control. Many published papers have dealt with the potential use of QMB, SAW, and, more recently, APM and LW devices as mass-sensitive detectors in chemical and biochemical sensors, but no such device has yet established itself commercially—with the exception of QMBs employed for thin-layer thickness measuring, which cannot be regarded as chemosensors in the strictest sense of the term [180].

Historical Development. The first report concerning the use of piezoelectric resonance from a quartz crystal in a gravimetric microbalance is by J. Strutt (Lord Rayleigh), who in 1885 described a shift in resonance frequency accompanying an infinitesimal mass change in a mechanical oscillator [181]. Fundamental studies into the mass sensitivity of quartz resonators and their potential application for measuring the thickness of thin layers were reported in 1959 by Sauerbrey [179], [182]. King was the first to use a quartz microbalance (in 1964) as a gas-phase detector in an analytical application [183], [184]. Numerous more recent papers discuss the fundamentals [185]–[192] and applicability [193]–[196] of bulk acoustic wave (BAW) piezoelectric resonators as chemical and biochemical sensors in the gas phase as well as in liquids, but physical devices for the measurement of pressure, temperature, or distance have attracted greater commercial interest [197], [198]. Nevertheless, quartz microbalances based on bulk acoustic waves do find frequent use in analytical research and development.

Rayleigh in an early paper [181] also noted the existence of waves that are propagated only in relatively thin surface layers, but it was not until 1979 that Wohltjen and Dessy adapted these so-called surface acoustic waves (SAWs) for incorporation into a chemical gas sensor [199]–[201]. The chief advantages of SAWs are increased sensitivity relative to BAWs and a greater potential for miniaturization. On the other hand, SAWs are associated with a variety of circuitry problems resulting from the relatively high operating frequencies required for adequate sensor sensitivity. A modification of this type of wave was introduced by Chuang and White [202], [203], who adapted a SAW sensor by creating an exceptionally thin membrane-like piezoelectric region in the substrate directly beneath the acoustic path. This in turn led to the production of *Lamb waves*, which proved to be particularly suitable for chemical sensor applications, opening the way to the so-called plate-mode oscillator, a device representing the present state of the art.

2.2.3.2. Fundamental Principles and Basic Types of Transducers

Solids and other vibrationally active mechanical structures are capable of being stimulated by electrical, thermal, or optical means either to vibration or to the further propagation of waves. Piezoelectric solids, preferably quartz (SiO_2), are especially susceptible to emission and reception of mechanical vibrations and waves. The anisotropic nature of quartz supports various types of waves, each with its characteristic frequency, amplitude, and propagation properties.

Survey of the Different Types of Piezoelectric Substrates. A summary of the various types of transducers involving piezoelectric solids, including their chemical and physical applications, is presented in Figure 30. Acoustic gravimetric sensors can be categorized generally on the basis of their design, their wave mode (BAW, SAW, APM, LW), and their frequency range. Figure 31 offers a schematic overview of the principles and design characteristics of various piezoelectric transducers. Criteria for establishing the usefulness of a particular vibrational system for the determination of mass include [185]:

1) The system should be readily excitable, preferably by electrical means
2) The system must lend itself to coupling with a frequency- (or period-) measuring device without significant disturbance
3) The system must be associated with sharply defined resonant frequencies so that the frequency (or period) of vibration can be determined precisely within a reasonably short period of time
4) In order to provide the required mass sensitivity, the mass-induced change in resonant frequency must be greater than the instability level of the resonant frequency and such uncertainties in the frequency (or period) as may accompany the measurement technique employed
5) Resonant frequency changes due to environmental disturbances (e.g., fluctuations in temperature or pressure, the presence of electrical or magnetic fields, or external mechanical stress) must be small relative to those caused by mass changes
6) An equation must be available in analytical form relating mass changes to the corresponding shifts in resonant frequency

Bulk Acoustic Wave Transducers. Waves of the bulk acoustic type (BAWs) can be invoked for microgravimetric applications through the use of either AT- or BT-cut plates derived from a single crystal of α-quartz, where "AT" and "BT" refer to cuts with specific orientations relative to the main axis of the crystal (see below and [205]). Figure 32 A represents a schematic diagram of such a quartz crystal with its associated electrodes and sensing layers. This diagram also indicates how a species to be detected might be absorbed into the sensing layer. Both sides of the quartz are in contact with electrodes. The corre-

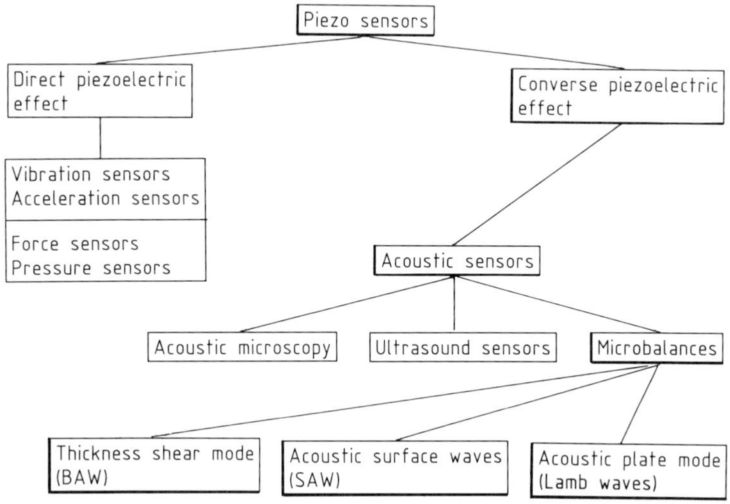

Figure 30. Selected applications of piezoelectric transducers with respect to both the direct piezoelectric effect (active transducer) and the converse piezoelectric effect (passive transducer) [198] (with permission from Springer-Verlag, Heidelberg)

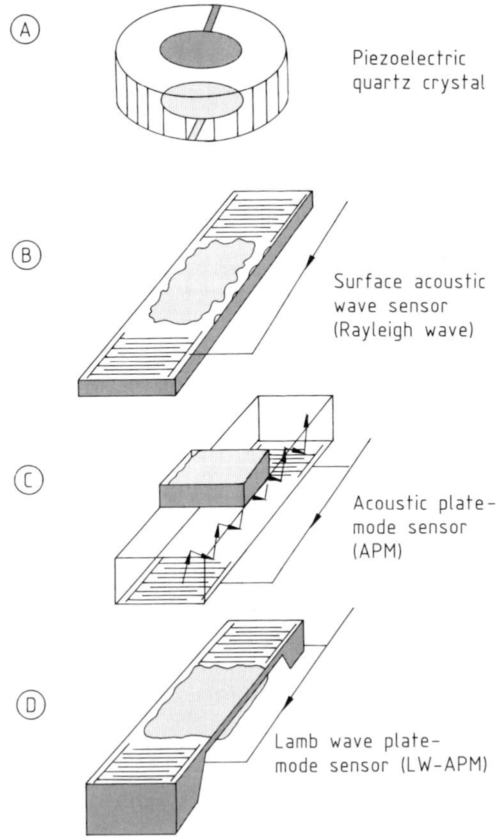

Figure 31. Schematic overview of acoustic gravimetric devices, illustrating the principles of wave propagation and design [204] (with permission from VDE-Verlag, Berlin)
A) BAW; B) SAW; C) APM; D) LW substrates

sponding electrodes are usually 300–1000 nm thick and 3–8 mm in diameter, and are made of gold, nickel, silver, or aluminum [193]. Introduction of oscillating electrical energy through an appropriate electrical circuit causes the mechanical structure to begin vibrating. In the case of AT-cut quartz the crystal vibrates in the thickness shear mode, as shown in Figure 32 B (see also Fig. 47).

Theory. Microgravimetric application of acoustic devices presupposes a quantitative relationship between an observed relative shift in the resonance frequency and an added mass. In the case of a quartz plate vibrating in a liquid an extended approach is required with respect to changes in the viscosity and density. The first useful relationship of this type was developed by SAUERBREY, who treated the added mass as an "added thickness" of the oscillator. The relation-

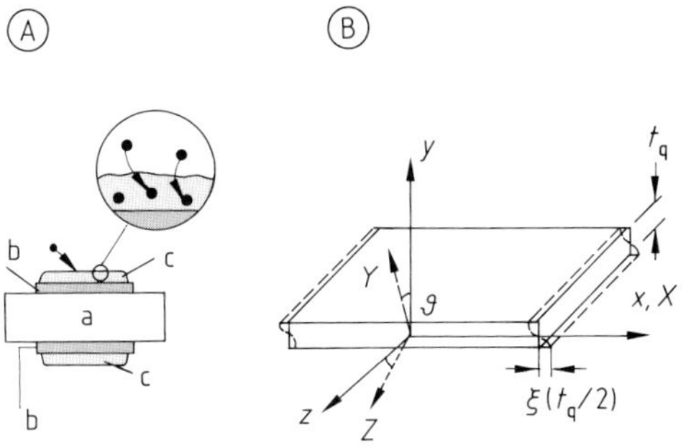

Figure 32. Schematic diagram of a quartz crystal augmented with electrodes and sensing layers, illustrating the way in which the species to be detected is absorbed into the sensing layer (A) and the ideal thickness shear mode of a quartz plate (B) [206], [179]
a) Quartz crystal; b) Electrodes; c) Sensing layers
x, y, z = Coordinates of the quartz plate; X, Y, Z = (IEEE-standard) coordinates for a single crystal of α-quartz, where X is the polar axis and Z is the optical axis; t_q = Thickness of the quartz plate, where the wavelength $\lambda = 2\,t_q$; ξ = Shear amplitude; ϑ = Angle of rotation

ship itself was based on a quartz crystal vibrating in its thickness shear mode in the gas phase (Fig. 33). The observed oscillation frequency (f_q) is inversely proportional to the thickness t_q of the crystal; $f_q = N/t_q$, where N is a frequency constant (N = 0.168 MHz cm for AT-cut quartz at room temperature). The mass of the crystal is $m = \varrho A t_q$, where ϱ is the crystal density and A its cross-sectional area. In this derivation the mass increment due to a foreign mass Δm is treated as an equivalent change in the mass of the crystal. Invoking the approximation $\varrho_q = \varrho_f$ leads to the equation:

$$\Delta f_q = -f_q \frac{\Delta m}{m} \quad (21)$$

which finally provides the following expression for an AT-cut crystal:

$$\Delta f_q = -2.3 \times 10^6 f_q^2 \frac{\Delta m}{A} \quad (22)$$

where f_q is expressed in megahertz, Δm in grams, and A in square centimeters. Since f_q for a quartz crystal is typically 10 MHz, and a frequency change of 0.1 Hz can be readily detected by modern electronics, it is possible in this way to detect mass changes of as little as about 10^{-10} g/cm² [206].

LU has developed a model involving propagation velocities for sound waves in the substrate

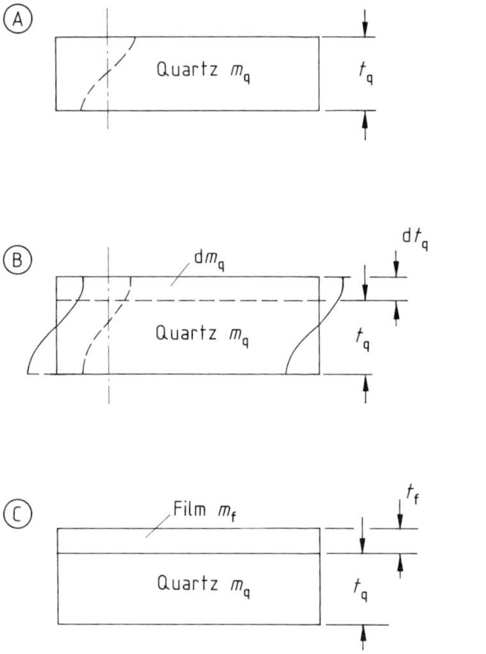

Figure 33. Shear wave (with permission from Plenum Publishing, New York)
A) In a quartz crystal; B) After an incremental increase in the crystal thickness; C) With the addition of a film
m_q, t_q = Mass and thickness of the quartz; m_f, t_f = Mass and thickness of the film

(for quartz, v_q) and in the sensing layer (v_f) [192]. Reflection and refraction occur at the interface between the crystal (q) and the film (f) (Fig. 33) analogous to the optical reflection and refraction observed at a boundary between two materials with different optical densities. The shear velocities in the quartz crystal and the film coating, respectively, are:

$$\text{quartz} \qquad v_q = \sqrt{\frac{\mu_q}{\varrho_q}} \tag{23}$$

$$\text{film} \qquad v_f = \sqrt{\frac{\mu_f}{\varrho_f}}$$

Given a knowledge of the shear moduli μ_f and μ_q, Equation (23) can be used to calculate the shear velocity of the wave. Ideal wave-propagation behavior is observed with a crystal upon which such a film has been deposited provided there is a negligible amount of frictional loss. In this case the resonance condition is valid, which leads to Equation (24):

$$\tan\left(\frac{\pi \cdot f_c}{f_q}\right) = \left(\frac{\varrho_f \cdot v_f}{\varrho_q \cdot v_q}\right) \tan\left(\frac{\pi \cdot f_c}{f_f}\right) \tag{24}$$

where $f_c = \omega/2\pi$ is the resonance frequency of the crystal upon which material has been deposited. The terms $\varrho_f \cdot v_f$ and $\varrho_q \cdot v_q$ are the acoustic impedances Z_f and Z_q of the film and crystal, respectively. Their ratio, $Z = Z_f/Z_q$, is an important parameter with respect to acoustic matching of the materials. For optimum resonance conditions Z should be as close as possible to unity, because its value affects the resilience of the whole assembly, and maximum stability corresponds to $Z \sim 1$ [207].

For example, Z has a value of 1.08 for a piezoelectric crystal covered with aluminum electrodes (shear-mode impedance of SiO_2: $Z_{SiO_2} = 8.27 \times 10^6$ kg s m^{-2}; that of Al: $Z_{Al} = 8.22 \times 10^6$ kg s m^{-2}), indicative of the good match associated with this material combination. Deposition of gold electrodes on an SiO_2 surface leads to a Z-value of 0.381, which is barely acceptable (shear-mode impedance of gold: $Z_{Au} = 23.2 \times 10^6$ kg s m^{-2}) [207].

From a practical point of view it is important to develop a relationship involving more experimentally common parameters influencing the frequency response. The frequency response of a piezoelectric sensor in the presence of a gas at a particular pressure can be related generally to three effects: the hydrostatic effect (p), the impedance effect (x), and the sorption effect (m):

$$-\frac{\Delta f}{f_0} = \left(\frac{\Delta f}{f}\right)_p + \left(\frac{\Delta f}{f}\right)_x + \left(\frac{\Delta f}{f}\right)_m \tag{25}$$

For gravimetric measurements and sensor applications, the latter effect (that is, to added mass) is the most important, and the other two effects can be viewed as nonspecific interferences. Because both of these terms are three orders of magnitude smaller than the mass term, the mass relationship alone is sufficient for characterizing measurements in the gas phase. However, situations do exist in which the other two effects cannot be ignored. For instance, if the pressure is changed from that of the surrounding air to a vacuum the first term is no longer negligible, and in the case of liquids any change in viscosity or density alters the impedance, so this term must be taken into consideration.

Other theoretical approaches relevant to applications in the gas phase and dealing with thickness-shear mode acoustic wave quartz sensors are presented in [189], [208]–[213].

Bulk Acoustic Waves in Liquids. When a quartz crystal is placed in a liquid, there is a significant change in both the density and the viscosity of the surrounding medium. Both become much larger than in the gas phase, and the consequences must be taken into account with respect to the second term in Equation (25). Thus, the Sauerbrey equation is no longer applicable if the viscoelastic properties of the contacting medium (liquid and/or film) change during the course of an experiment. The liquid now represents an additional mass load, which produces coupling between the substrate elastic shear wave and the liquid. KANAZAWA derived the first theoretical relationship linking the properties of a liquid to the frequency response of a piezoelectric crystal. In a liquid medium, the shear motion of the crystal causes motion to occur in the adjacent layer of solvent molecules (usually water). Up to a layer thickness of approximately 1 μm the solvent molecules move with a certain degree of "slip," which corresponds to a phase shift in the third and subsequent layers. Therefore, vibrational energy from the oscillator is continuously dissipated into the liquid. If only one side of the crystal is coupled to the liquid, the result is a standing wave perpendicular to the substrate surface. For liquids with Newtonian behavior this wave can be described theoretically as a damped wave. As is usual for this type of wave, a decay constant k is used to describe the magnitude of the damping. In this case the constant k can be related to $(\omega/2\nu)^{1/2}$, where ν is the kinematic viscosity. The damping distance (limit of wave propagation) is on the order of micrometers, while the frequency shift can be expressed [214] as:

$$\Delta f = f_0^{3/2}\left(\frac{\nu}{\pi\mu_q\varrho_q}\right)^{1/2} \qquad (26)$$

Figure 34 shows the effect of microscopic kinematic viscosity at the crystal–liquid interface, where the magnitude of the observed frequency shift has been plotted against the corresponding increase in density for a glucose solution.

The frequency shift for quartz is highly dependent on changes in the viscosity and density of a surrounding liquid. This situation can be discussed in a straightforward way on the basis of the *quartz equivalent circuit* [207], [215], shown in Figure 35, which consists of the following elements:

1) The L'–C' combination, which determines the resonance frequency f_m of the motional arm
2) A parallel capacitance C_p, caused by the electrodes on the quartz and any stray capacitance
3) A resistive component R'

The resistance R' is a measure of mechanical losses from the vibrating quartz, and it therefore reflects the viscoelastic properties of the contacting liquid [216], [217] or a deposited film [218]. Surface roughness also influences the value of R' [219]. Changes in the vibrating mass correspond to changes in the self-inductance L' [220], [221]. Thus, the resonance frequency of the motional arm (f_m; see Fig. 35) changes in proportion to the mass. In deriving Equation (26) KANAZAWA considered only those L' values for the motional arm that contribute to a frequency shift, but in a real experiment any change in $\Delta R'$ is normally accompanied by a change in L' (typical values for R' in air and in water are 7 Ω and 350 Ω, respectively, for a 6 MHz quartz crystal). Thus, when a quartz crystal is immersed in liquids of different viscosities, both L' and R' can be expected to vary. Use of an impedance analyzer makes it possible to determine the increase in R' associated with operation of a quartz crystal in air versus a liquid. The frequency shift of the motional arm can be calculated from the values of C' and L' using the formula $f_m = 1/[2\pi(L'\ C')^{1/2}]$ [215].

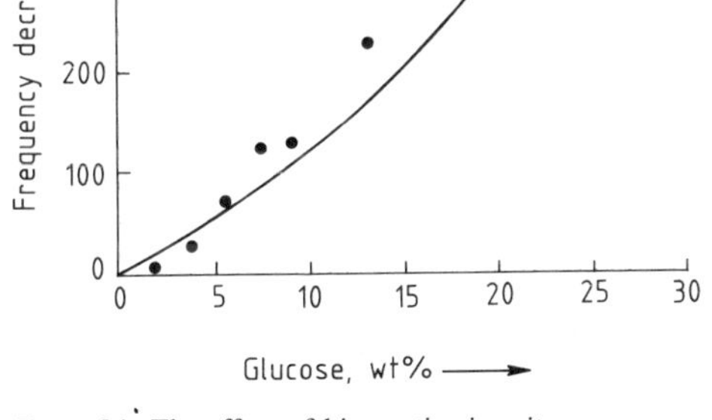

Figure 34. The effect of kinematic viscosity on resonance frequency as a function of glucose concentration (in water) for a 5-MHz quartz crystal
—— = Calculated; ● = Measured (with permission from Am. Chem. Soc., Washington)

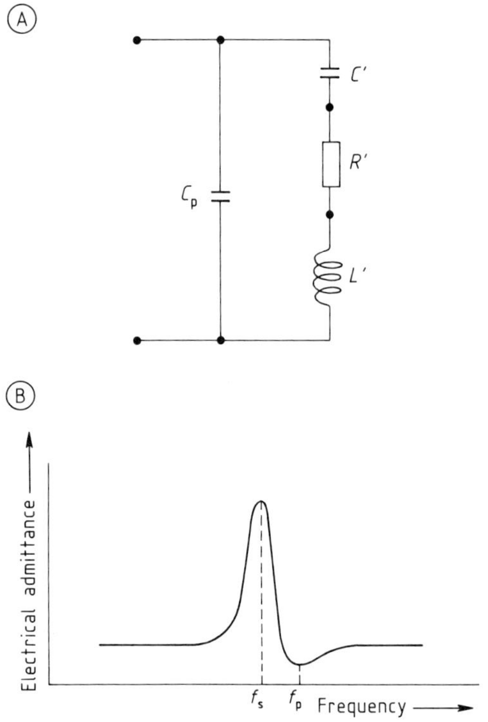

Figure 35. Equivalent electrical circuit (A) and resonance frequency of a mounted piezoelectric crystal oscillator (B) [207], [215]
C_p = Capacitance of the mounted crystal; C', L' = Rigidity and mass of the added material; R' = Equivalent to the mechanical loss, which also represents the acoustic (mechanical) load; f_s, f_p = Series and parallel resonance frequencies
(with permission from Plenum Publishing, New York)

SAW Transducers. Surface acoustic wave (SAW) sensors have been designed with sensitivity to many different physical quantities, including force, acceleration, hydrostatic pressure, electric field strength, dew point, and gas concentration [222]. The operating principle of an SAW for use with gases is conceptually quite simple. A surface acoustic wave—a periodic deformation perpendicular to the material surface—is transmitted across the surface of some appropriate solid that has been exposed to the atmosphere subject to analysis. With a homogeneous substrate, usually a piezoelectric solid like quartz or $LiNbO_3$, the required SAW (sometimes also called a Rayleigh wave) is generated by means of an interdigital transducer (IDT).

In contrast to a quartz microbalance, the requisite IDT components are deposited only on one side of the crystal. These take the form of planar, interleaved, metal electrode structures,

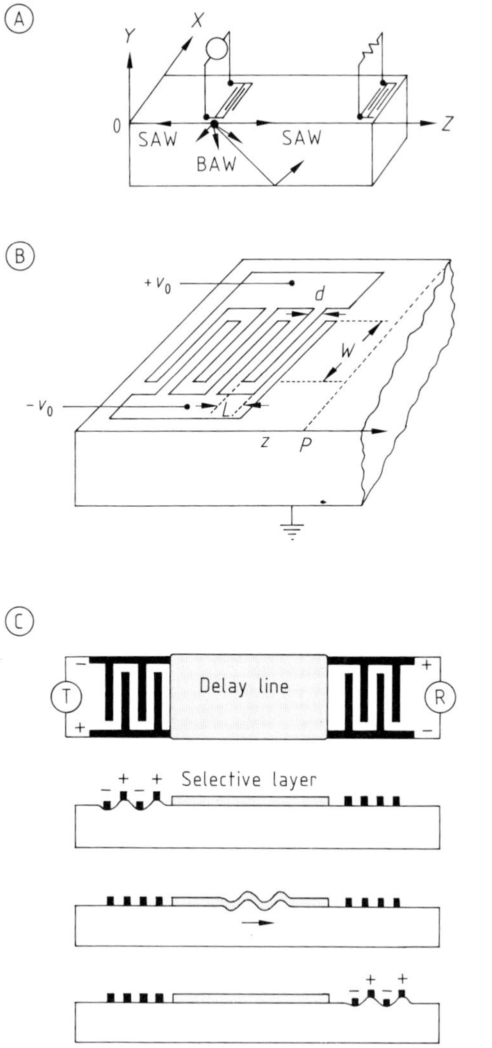

Figure 36. Interdigital metal pattern of a uniform transducer; the IDT behaves like a sequence of ultrasonic sources or receivers [223]
A) An SAW device with input and output transducers; B) A uniform transducer defined by half-spatial period L, electrode width d, and aperture width W; the observation line P used for derivation of the impulse response is at a distance $z + L$ from the first source; C) Schematic diagram of an SAW sensor with a transmitter T, a receiver R, and a chemically selective layer deposited in the form of a delay line [207]
(with permission from Plenum Publishing, New York)

where adjacent electrodes are supplied with equal but opposite potentials (Fig. 36 A). Application of a time-varying r. f. potential causes the crystal to undergo physical deformations, and if these are confined to the surface region of the crystal, the result is a surface acoustic wave (Fig. 36 B) [224]. Such vibrations will interfere constructively only if the distance $L/2$ between two adjacent "fingers" (see Fig. 28 A, p. 165) is equal to one-half the elastic wavelength. The frequency $f = v_R/L$ that corresponds to such a cumulative effect is called the *synchronous frequency* or *resonance frequency* (f) [206].

The waves that are generated have their mechanical and electrical components in a single plane, the *sagittal* plane, which is perpendicular to the substrate surface. The phase velocity (v_R) and the amplitude (A) of an SAW are determined by elastic, piezoelectric, dielectric, and conductive properties of the substrate as well as by its mass. If one of these parameters can be modulated in an appropriate way by the quantity of interest, a sensing device can be constructed. Such a modulation could be induced in the transducer itself and/or in the transmission region (waveguide or *delay line*). With layered substrates the physical properties of each layer as well as the thicknesses of the layers determine the phase velocity and the amplitude of the resulting SAW [206]. Various energy-loss mechanisms also operate on the propagating SAW: scattering loss due to finite grain size, thermoelastic loss due to nonadiabatic behavior of the acoustic conductor, viscous loss caused by the dissipation of energy in a direction perpendicular to the transducer–environment interface, and hysteresis absorption due to irreversible coupling between SAW energy and an adsorbate [207]. Any addition of mass during the sensing step may change the magnitude of one or more of these phenomena, and may also lead to changes in v_R and A, especially a frequency shift (Δf) or a relative frequency shift ($\Delta f/f$). An exact description of the physical processes involved in the generation, propagation, and detection of SAWs is complex, and is available from such sources as [224], [225].

The considerations presented above provide the basis for using SAWs in gas-monitoring devices, because observed frequency shifts are proportional to added mass. However, the characteristic frequency of an SAW is also sensitive to changes in temperature or pressure. The ideal substrate would be one with a zero temperature delay coefficient but a high piezoelectric coupling effect. Despite extensive materials research efforts, it is currently necessary to choose between these parameters; e.g., ST-cut quartz is the preferred substrate if a zero temperature coefficient is most important, and cuts of lithium niobate are selected for high piezoelectric coupling.

Sensor Configurations. A single SAW sensor may sometimes suffice for a gas sensing application (Fig. 37 A), but often the sensors display undesirable sensitivity to such effects as temperature, pressure, and ambient humidity. Overcoming these problems requires the establishment of a relative signal output. For example, VETELINO et al. have described a dual-line SAW gas sensor [206], [226], [227]. One delay line in this case is used for measuring, while the other acts as a reference (see Fig. 37 B). The advantage of a dual delay-line configuration is that it transforms the output signal into a relative change in oscillator frequency, which can be attributed only to the effects of the analyte. D'AMICO et al. have described a system based on a three-transducer de-

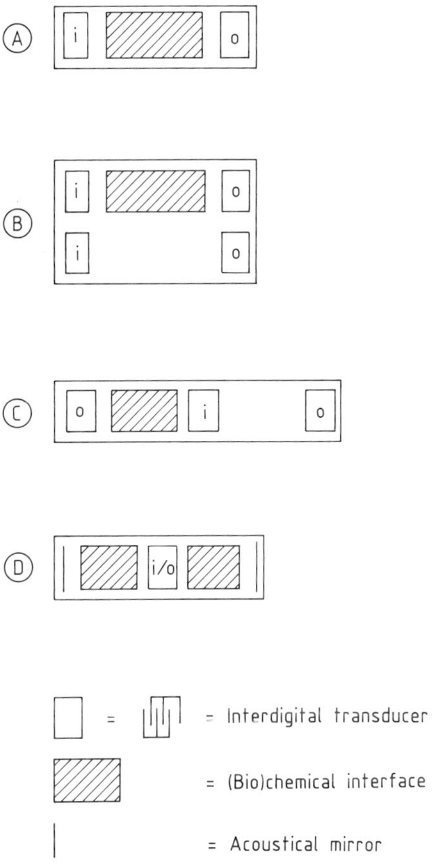

Figure 37. Various possible sensor configurations for SAW chemical sensors [206]
A) Single delay line; B) Dual delay line; C) Three-transducer device; D) Resonator
i = Input transducer; o = Output transducer

lay line (see Fig. 37 C). This device also has two dual-delay lines, but the paths are established in this case by one input IDT and two output IDTs. Again, one path of the delay line is coated and the other remains uncoated as a way of minimizing the effects of both temperature and pressure [228].

The first use of an *SAW resonator* sensor was reported by MARTIN et al. [229]. The corresponding resonator incorporated a single IDT located between two acoustically reflecting mirrors. The distance between the mirrors was so adjusted as to ensure constructive interference between successive reflections, resulting in a maximum for one particular frequency (Fig. 37 D).

Theory. The SAW velocity v_R can be affected by many factors, each of which is associated with a potential sensor response:

$$\frac{\Delta v}{v_R} = \frac{1}{v_R}\left(\frac{\partial v}{\partial m}\Delta m + \frac{\partial v}{\partial c}\Delta c + \frac{\partial v}{\partial \sigma}\Delta \sigma + \frac{\partial v}{\partial \varepsilon}\Delta \varepsilon + \frac{\partial v}{\partial T}\Delta T + \frac{\partial v}{\partial P}\Delta P + \ldots\right) \quad (27)$$

where v_R is the phase velocity unperturbed by such external factors as m, c, σ, and ε (the mass, elastic constant, electrical conductivity, and dielectric constant of the solid medium, respectively), as well as T and p, the environmental temperature and pressure [222]. Three types of films have been considered for use as reactive layers:

1) Nonconductive isotropic overlay film
2) Electrically conducting overlay film
3) Metal-oxide semiconducting film

A theoretical treatment can be developed for each of these film types with respect to its properties, involving, in the nonconductive case, the thickness t_f, the mass density ϱ, λ (the Lamé constant), and μ (the shear modulus).

Regarding the relationship $\Delta v/v = -\Delta f/f$ between SAW velocity and the variation in the SAW dual-delay line oscillator frequency, perturbation analysis leads to the following result for the relative frequency shift [225]:

$$\Delta f = (k_1 + k_2) f_0^2 t_f \varrho - k_2 f_0^2 t_f \left[\frac{4\mu(\lambda + \mu)}{v_r^2(\lambda + 2\mu)}\right] \quad (28)$$

where k_1 and k_2 are material constants for the SAW substrate. Additional information with respect to these equations is available from [200], [222], [225].

Thus, the film mass per unit area is $t_f \varrho$. If chemical interaction does not alter the mechanical properties of the film, the second term in Equation (28) can be neglected, in which case the frequency shift Δf can be attributed exclusively to the added mass:

$$\Delta f = (k_1 + k_2) f_0^2 t_f \varrho \quad (29)$$

Coating a polymer film 1 μm thick onto a quartz SAW sensor operating at 31 MHz should therefore cause a frequency shift $\Delta f = -130$ kHz [207].

Plate-Mode Oscillators (PMOs). *Lamb Wave Oscillators.* In all the SAW devices discussed above, the acoustic wave propagates in a slab of material whose thickness is infinitely larger than the wavelength λ of the propagating wave. When the thickness of the plate is reduced to such an extent that it becomes comparable to λ, the entire plate becomes involved in the periodic motion, producing a symmetric and antisymmetric *Lamb wave* (LW) (Fig. 38). This behavior is observed in plate-mode oscillators with thicknesses of a few micrometers. A cross-sectional view through a typical PMO is shown in Figure 39.

Certain performance factors make the LW-PMO potentially attractive as a chemical sensor. Both surfaces contribute to the signal, so the observed sensitivity is greater than for a corresponding SAW device. However, the most important advantage follows from the fact that the velocity of the lowest-order wave in the antisymmetric mode is much lower than in the case of the corresponding SAW oscillator. This may be important for applications involving mass-sensing in liquids, which are problematic with high-frequency SAW devices. As the frequency of the

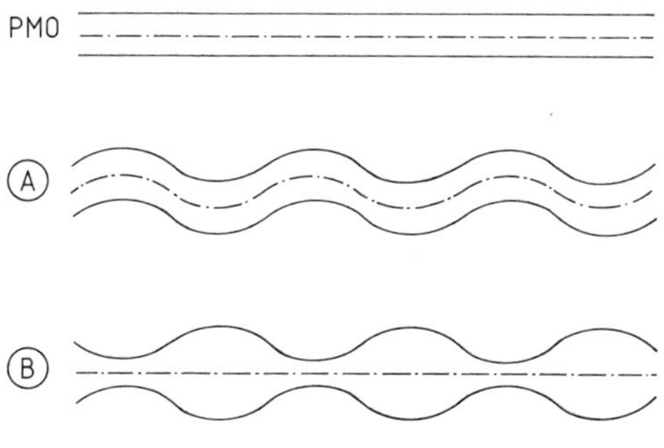

Figure 38. Lamb waves propagating in a thin plate-mode oscillator (PMO) [202], [207] (with permission from Plenum Publishing, New York)
A) Symmetric waves; B) Antisymmetric waves

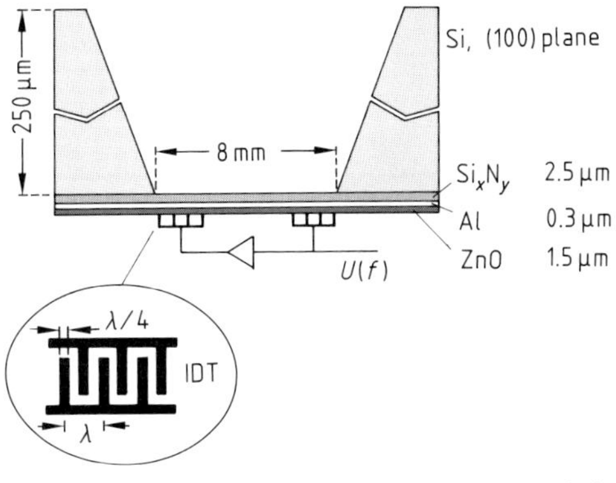

Figure 39. Cross section through a Lamb wave PMO, including the corresponding interdigital transducers [204]

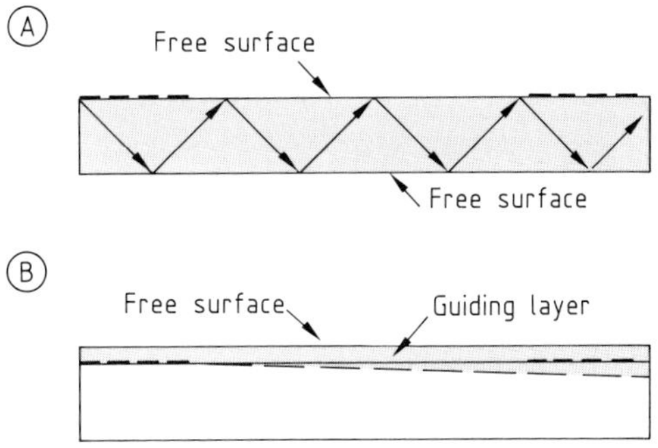

Figure 40. Cross-sectional view of (A) APM and (B) Love plate devices [230]

Lamb wave decreases below the velocity of the compressional wave in the liquid, energy loss in the perpendicular direction decreases as well. A Lamb wave also has a greater frequency dispersion than the corresponding surface acoustic wave, resulting in increased resolution and sensitivity.

However, there are certain practical problems. The fragility of the thin plate and its consequent sensitivity to external pressure is particularly serious. Performance analyses as well as details regarding fabrication of the device can be found in the original paper [203].

Acoustic Plate-Mode Oscillators and Love Plate Devices [230]. These two types of acoustic PMOs, like BAW sensors, are characterized by surface-guided, shear-horizontally (SH) polarized waves, which have a horizontally polarized component of displacement (see also Fig. 47). In principle the performance of such APMs is similar to that of the Lamb wave sensors described above, but the thickness of the substrate is greater, so waves propagate from the surface into the bulk of the material. Energy is confined mainly in the bulk of the plate as the wave propagates through multiple reflections, generating displacements in both the upper and lower surfaces. The Love plate sensor includes a waveguide structure, whereby the SH wave is confined within an elastic layer deposited on a SAW substrate capable of supporting SH waves. Figure 40 provides cross-sectional views of APM and Love plate devices.

Acoustic sensors based on SH waves have become rather widespread in recent years due to their ability to function in liquid media.

Relative Sensitivities of BAW, SAW, LW, APM, and Love Plate Devices. Because of excitation and propagation differences among the various transducers described, the mass-deposition sensitivity of the several devices would be expected to vary considerably, with some geometries proving more attractive than others for particular sensing applications. Table 6 offers a performance comparison for these transducers. Table 6 also summarizes the calculated frequency-to-mass sensitivities for the various devices. Sensitivity variations can be explained on the basis of energy confinement on the sensing surface and a particular device's operating frequency. A decrease in the thickness of the substrate leads to a sensitivity increase, which can be realized with either an acoustic plate-mode or a Lamb wave device. These are therefore associated with the highest sensitivities, followed by the Love plate, which also has a very thin layer. Less sensitivity is available with BAW devices, for which the sensitivity depends on the thickness of the crystal.

2.2.3.3. Theoretical Background

Piezoelectricity. *The Piezoelectric Effect.* Because all acoustic gravimetric sensors are based on the phenomenon of piezoelectricity, it seems appropriate to discuss briefly the effect itself. Piezoelectricity was first observed by the Curie brothers (Jaques and Pierre) in 1880 [233]. It is a reversible phenomenon, consisting of linear electromechanical interactions between mechanical and electrical properties in certain crystals (Fig. 41).

Table 6. Comparison of sensitivities and other characteristics of acoustic gravimetric sensors [204], [231], [232]; international symbols (short form) have been used to denote specific cuts of the various materials

Acoustic wave type	Sensitivity formula, $cm^2/\mu g$	Calculated sensitivity, $cm^2/\mu g$	Frequency, MHz	Thickness, µm	Material
Bulk wave resonator (BAW)	$S_m = -\frac{2}{\varrho\lambda} = -\frac{1}{\varrho t_f}$	− 23	10	165	AT-cut quartz
Surface (Rayleigh, SAW)	$S_m = -\frac{K(\sigma)}{\varrho\lambda}$	− 129	100	NA*	ST-cut quartz, $LiNbO_3$
	$0.8 < K < 2.2$ $0 < \sigma < 0.5$	− 516	400	NA*	$K = 1.16$; $\sigma = 0.35$
Lamb plate (LW), A_0 mode	$S_m = \frac{1}{\varrho t_f}$	− 450 − 951	4.7 2.6	3 (not given)	ZnO on SiN
Acoustic plate mode (APM, SH)		− 3000 − 300 − 30	NA* NA* NA*	1 10 100	ST-cut quartz
Love plate (transverse surface)		− 37 − 182	100 500	NA* NA*	$LiNbO_3$

* not available.

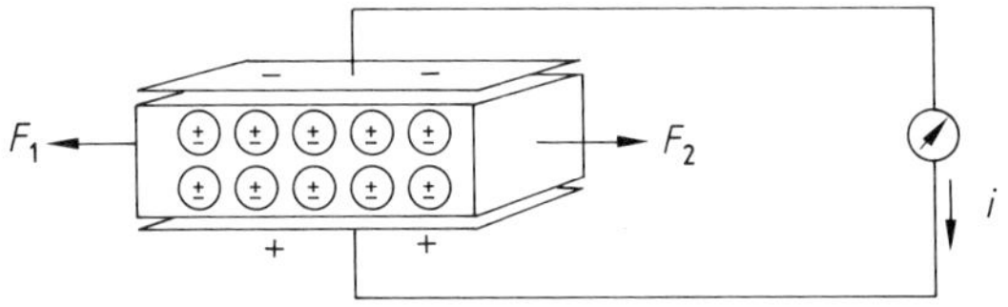

Figure 41. Schematic illustration of the way mechanical and electrical energy can be interconverted with the aid of an appropriate crystal [234]
F_1, F_2 = Mechanical forces; i = Electrical current

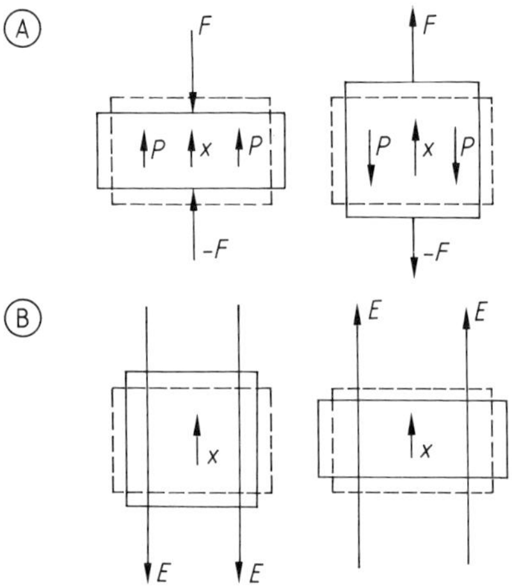

Figure 42. Schematic representation of the direct piezoelectric effect (A) and the converse (reciprocal) piezoelectric effect (B) for an X-quartz plate (right-handed quartz) [198]
F = Applied force; P = Polarization; E = Applied electrical field
(with permission from Springer-Verlag, Heidelberg)

A piezoelectric effect occurs only with those ionic crystalline solids whose crystals contain a polar axis along which the physical properties are not constant. The *direct* piezoelectric effect is observed when an applied force (F) produces an electrical polarization (P), and a *converse* (reciprocal) piezoelectric effect results when an applied electrical field (E) induces a strain (S, deformation of the crystal) (Fig. 42).

A simplified view of piezoelectricity at the microscopic level is provided by the example of α-quartz (Fig. 43). Here a net dipole moment will arise if the hexagonal structure is stretched or compressed along a direction parallel (Fig. 43 B) or perpendicular (Fig. 43 C) to one of the three in-plane symmetry axes. When symmetry is achieved, the net dipole moment of the molecule vanishes.

Structural Aspects of Piezoelectric Solids. In order for a crystal to be piezoelectric it must be noncentrosymmetric; i.e., it cannot contain a center of symmetry. Solids are usually grouped for structural characterization purposes into seven crystal systems (from lowest to highest symmetry: triclinic, monoclinic, orthorhombic, tetragonal, trigonal, hexagonal, and cubic), which can in turn be divided into 32 point groups, depending upon point symmetry. Of these, 11 classes are centrosymmetric and 20 are piezoelectric. An exceptional case is class 432 from the cubic crystal system, because it is neither centrosym-

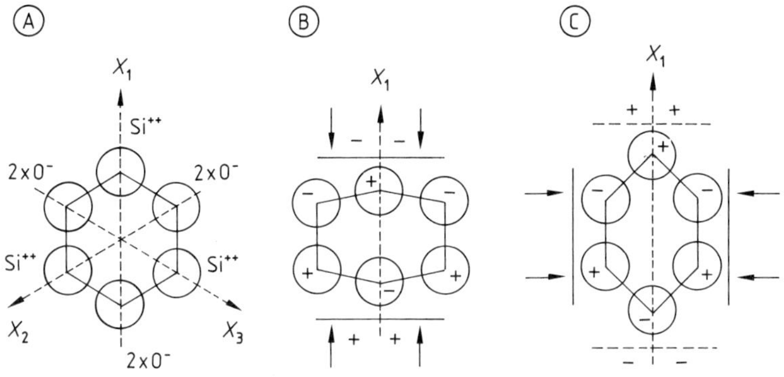

Figure 43. Physical interpretation of the piezoelectric effect [235]
A) Simplified microscopic model of a quartz crystal lattice; B) Longitudinal effect; C) Transverse effect
"Si^{++}" and "$2O^{-}$" refer to centers of gravity (circles) for charges associated with the two types of atoms, where the tetrahedral "SiO_4" structure has been projected onto a plane (as a hexagon)

metric nor piezoelectric. In addition to the 21 noncentrosymmetric classes, 11 classes can be distinguished on the basis that they possess no plane of symmetry. This means that they are associated with both right- and left-handed forms, which cannot be interconverted by simple rotation, a phenomenon known as enantiomorphism [236].

Table 7 summarizes on the basis of international point-group symbols the structural distribution with respect to piezoelectric classes, centrosymmetric classes, and enantiomorphism. For example, $LiNbO_3$, which belongs to the class $3m$, is piezoelectric, but not enantiomorphic, whereas α-quartz, in class 32, is both piezoelectric and enantiomorphic. The latter characteristic is illustrated in Figure 44.

Not every solid material associated with a piezoelectric class is suitable for practical applications, because in some cases the piezoeffect is too weak. Large piezoelectric coefficients are a prerequisite for an electromechanical transducer, like those found, for example, with α-quartz or $LiNbO_3$. Therefore, these are the solids most often used as mass-sensitive plates or substrates.

Table 7. Distribution of crystal structures (point groups) with respect to centrosymmetric, piezoelectric, and enantiomorphic classes [236]

Crystal system	Centrosymmetric classes	Piezoelectric classes	Classes with enantiomorphism
Triclinic	$\bar{1}$	1	1
Monoclinic	$2/m$	2, m	2
Orthorhombic	mmm	222, $mm2$	222
Tetragonal	$4/m$, $4/mmm$	4, $\bar{4}$, 422, $4mm$, $\bar{4}2m$	4, 422
Trigonal	$\bar{3}$, $\bar{3}m$	3, 32, $3m$	3, 32
Hexagonal	$6/m$, $6/mmm$	6, $\bar{6}$, 622, $6mm$, $6m2$	6, 622
Cubic	$m3$, $m3m$	23, $\bar{4}3m$	23, 432

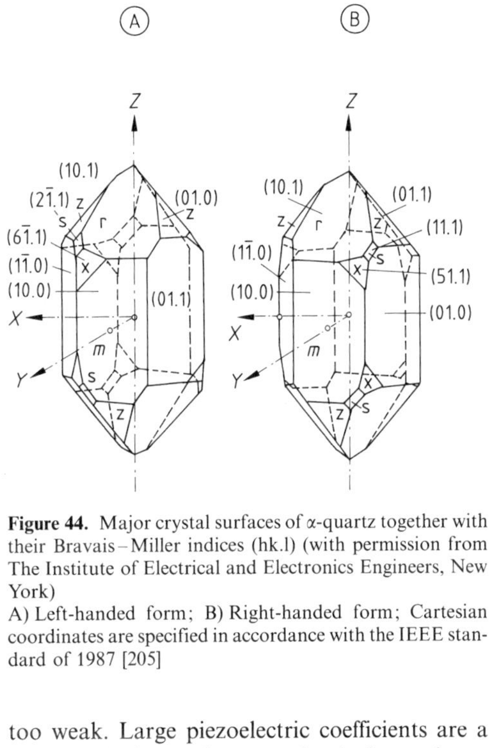

Figure 44. Major crystal surfaces of α-quartz together with their Bravais–Miller indices (hk.l) (with permission from The Institute of Electrical and Electronics Engineers, New York)
A) Left-handed form; B) Right-handed form; Cartesian coordinates are specified in accordance with the IEEE standard of 1987 [205]

Theory and Physics of Piezoelectricity. The discussion that follows constitutes a very brief introduction to the theoretical formulation of the physical properties of crystals. If a solid is piezoelectric (and therefore also anisotropic), acoustic displacement and strain will result in electrical polarization of the solid material along certain of its dimensions. The nature and extent of the changes are related to the relationships between the electric field (**E**) and electric polarization (**P**), which are treated as vectors, and such elastic factors as stress (**T**) and strain (**S**), which are treated as tensors. In piezoelectric crystals an applied stress produces an electric polarization. Assuming the dependence is linear, the direct piezoelectric effect can be described by the equation:

$$\boldsymbol{P}_{\rm i} = d_{\rm ijk}\,\boldsymbol{T}_{\rm jk} \quad \text{or} \quad \boldsymbol{P}_{\rm i} = e_{\rm ijk}\,\boldsymbol{S}_{\rm jk} \tag{30}$$

The quantities $d_{\rm ijk}$ and $e_{\rm ijk}$ are known as the *piezoelectric strain* and *piezoelectric stress* coefficients, respectively.

Conversely, if such a crystal is placed in an electrical field it will become deformed, a phenomenon known as the converse or reciprocal piezoelectric effect (Fig. 42 B). The contributions of stress and strain in this case can be expressed:

$$\boldsymbol{S}_{\rm jk} = d_{\rm ijk}\,\boldsymbol{E}_{\rm i} \quad \text{or} \quad \boldsymbol{T}_{\rm ij} = e_{\rm ijk}\,\boldsymbol{E}_{\rm i} \tag{27}$$

Since the stress and strain tensors are symmetric, the piezoelectric coefficients can be converted from tensor to matrix notation. Table 8 provides the piezoelectric matrices for α-quartz together with several values $d_{\rm ijk}$ and $e_{\rm ijk}$, including the corresponding temperature coefficients [234], [236].

For sensor applications, the magnitudes of $d_{\rm ijk}$ and $e_{\rm ijk}$ should be as great as possible, and the temperature dependence should be as small as possible. Although lithium niobate displays larger stress and strain constants than α-quartz, the temperature influence in this case is also great. Therefore, α-quartz is the material most often selected for piezoelectric sensor applications. Because of the anisotropic behavior of a piezoelectric material, properties like resonance frequency and temperature dependence can be optimized by cutting a plate from a single crystal in a particular way. Figure 45 shows two possible cuts with respect to a natural α-quartz crystal. For use as acoustic gravimetric sensors, only AT- and BT-cut quartz plates are useful. These *y*-ro-

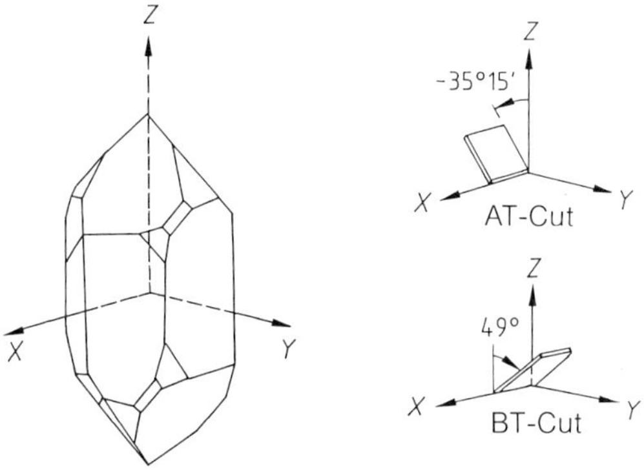

Figure 45. Oriented quartz crystal cuts in relation to the natural crystal [185] (with permission from Elsevier, Amsterdam)

Table 8. Piezoelectric matrices for α-quartz and $LiNbO_3$, together with the corresponding piezoelectric stress and strain coefficients [234], [236]

Classification	α-Quartz: Trigonal system, class 32		$LiNbO_3$: Trigonal system, class 3 m			
Matrix	$\begin{pmatrix} d_{11} & -d_{11} & 0 & d_{14} & 0 & 0 \\ 0 & 0 & 0 & 0 & -d_{14} & -2d_{11} \\ 0 & 0 & 0 & 0 & 0 & 0 \end{pmatrix}$		$\begin{pmatrix} 0 & 0 & 0 & 0 & d_{15} & -2d_{22} \\ -d_{22} & d_{22} & 0 & d_{15} & 0 & 0 \\ d_{31} & d_{31} & d_{33} & 0 & 0 & 0 \end{pmatrix}$			
Piezoelectric strain constants	d_{11}	d_{14}	d_{22}	d_{31}	d_{33}	d_{15}
	2.3	−0.67	21	−1	6	68
	2.31	−0.670	22.4	−1.2	18.8	78.0
	2.31	−0.727	20.8	−0.85	6.0	69.2
Temperature coefficients of piezoelectric strain constants (×10⁻⁴/°C)	$T_{d_{11}}$	$T_{d_{14}}$	$T_{d_{22}}$	$T_{d_{31}}$	$T_{d_{33}}$	$T_{d_{15}}$
	−2.0	17.7	2.34	19.1	11.3	3.45
	−2.15	12.9				
Piezoelectric stress constants	e_{11}	e_{14}	e_{22}	e_{31}	e_{33}	e_{15}
	0.173	0.04	2.5	0.2	1.3	3.7
	0.171	0.0403	2.43	0.23	1.33	3.76
Temperature coefficients of piezoelectric stress constants (×10⁻⁴/°C)	$T_{e_{11}}$	$T_{e_{14}}$	$T_{e_{22}}$	$T_{e_{31}}$	$T_{e_{33}}$	$T_{e_{15}}$
	−1.6	−14.4	0.79	2.21	8.87	1.47

tated cuts provide two different high-frequency plates that vibrate in a shear mode along an axis parallel to the major surface, as indicated previously (see also Fig. 32 B, p. 169). Minimizing temperature effects requires that the plate be cut at a very precise orientation. The temperature dependence of the relative frequency for various cuts of quartz crystal is shown as a function of the angle of cut in Figure 46 [237].

More detailed theoretical treatments of piezoelectric solids and their properties are provided by [198], [238]–[240]. Another more sophisticated approach based on dynamic quantum mechanics is presented in [241]. Interest has also developed in recent years in piezoelectric polymers, especially as a result of their strongly piezoelectric properties [198].

Vibrations and Waves. The theory of acoustic waves in solids is well understood, and many comprehensive descriptions exist [225], [242]. The description presented here should suffice to explain sensor function despite the fact that it is limited and qualitative.

Acoustic waves in solids consist of oscillatory motions of the material comprising the solid. The stress–strain relationship treated in the preceding section leads ultimately to two types of acoustic displacements: *longitudinal* or *tensile* motions that cause the dimensions of the solid body to change along a single direction without accompanying changes along the orthogonal directions, and *shear* motion, which produces changes in the shape of the solid body. Figure 47 illustrates these two types of motion [232].

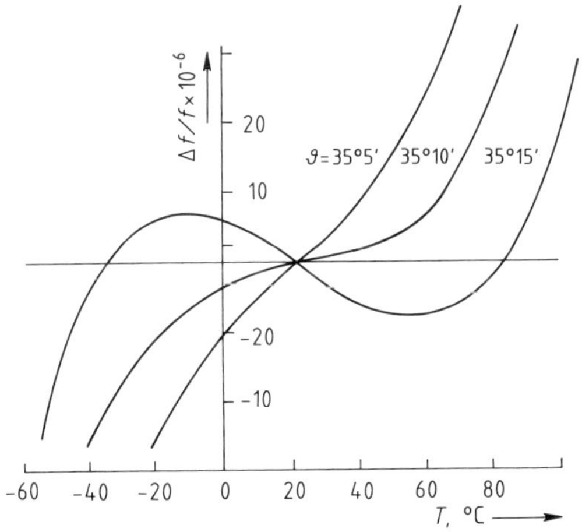

Figure 46. Temperature dependence of the relative vibrational frequency for different AT-cuts of a quartz crystal [237] (with permission from Elsevier, Amsterdam)

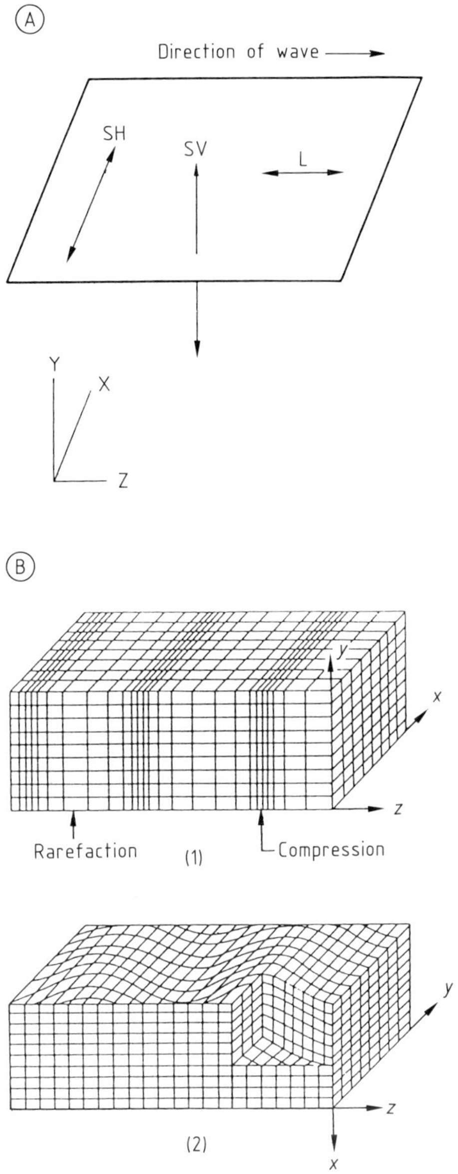

Figure 47. A) Schematic representation of longitudinal and shear displacement [232]; (with permission from Elsevier, Amsterdam) B) Grid diagrams for plane uniform waves propagating along the z axis in a material of infinite extent [224] (with permission from J. Wiley & Sons, New York):
(1) Tensile motion (one dimension); (2) Shear motion (two-dimensional; y-polarized in this case)
L = Longitudinal (tensile) motion; SH = x-Polarized shear (horizontal); SV = y-Polarized shear (vertical)

In piezoelectric solids, an applied force generally creates both quasi-longitudinal and quasi-shear waves, where the acoustic displacement in a wave will be largely either longitudinal or shear. Pure longitudinal (L) and shear waves (SH = horizontally polarized shear waves; SV = vertically polarized shear waves) are generated if wave propagation occurs along certain crystallographic axes or with specific orientations. The generation of mixed wave types is also possible, but normally not desirable for sensor applications.

As far as microgravimetric sensors are concerned, two types of wave-propagation devices can be distinguished: those characterized by wave propagation perpendicular to the plate surface (BAW) and those in which the waves propagate along the surface (SAW) or in the bulk of the substrate (LW, APM). Transducers of the former type are normally resonators. A given device might be designed to utilize either longitudinal or shear waves propagated between the faces, as in microbalances based on AT- or BT-cut quartz plates operating in their SH high-frequency fundamental mode. Only acoustic waves satisfying boundary conditions for the resonator surfaces will propagate efficiently across such a device. The waves can be visualized as standing sinusoidal waves with displacement nodes at the resonator surfaces (see p. 168).

Transducers in the second category include wave delay-line oscillators (see Figs. 31 and 36), which are based primarily on two types of waves. One is a coupled linear combination of two wave-equation solutions, an L-wave and an SV-wave. In this wave, also called a Rayleigh wave, particle displacement takes the form of an elliptical motion in the plane perpendicular to the surface (sagittal plane) and containing the direction of propagation. Some displacement also occurs outside the sagittal plane, but the deviations are minor (Fig. 48). Two of the most important cuts providing these wave types, and which are used in chemical sensors, have been illustrated in Figure 45. Waveguides that rely on Rayleigh wave devices for their operation are often referred to as surface acoustic wave (SAW) devices, and they have been extensively employed as sensors.

The second type of wave in this context is a horizontal shear (SH) wave, whose acoustic displacement is in the plane of the waveguide. An SH wave generally travels at a different velocity than a Rayleigh wave, and for piezoelectric waveguides the SH wave becomes a true surface wave provided the acoustic impedance at the surface differs from that in the substrate bulk. Such a difference can be caused by piezoelectric stiffening of the SH mode, deposition of a layer of dissimilar material on the surface, the presence of a surface grating, or liquid in contact with the surface. Depending on their origin, SH surface waves are referred to in the literature as Bleustein – Gulayev (BG) waves, Stoneley waves, SH-SAW, or surface transverse waves. Pure surface transverse waves (shear waves) appear to show promise for applications in liquids, especially biosensor applications, because the dissipation of energy across the phase boundary into the liquid is approximately zero, which greatly reduces the generation of longitudinal sound waves in the liquid itself.

Plate-mode waves are analogous in form to Rayleigh and SH waves in a semi-infinite waveguide. Two groups of plate modes can be distinguished: the first consists of Lamb waves or flexural waves, where the acoustic displacement is a combination of longitudinal and vertical shear motion; a second family involves SH acoustic displacement [232]. Because of their frequency range (100 kHz to the GHz region) these sound waves fall in the ultrasound category, with a propagation velocity of 1 – 10 km/s (cf. electrical signals, which travel at 100000 – 300000 km/s).

2.2.3.4. Technical Considerations

The (Bio)Chemical Interface. A piezoelectric transducer is transformed into a mass-sensitive sensor by coating it with an appropriate (bio)chemical film or layer, which may be inorganic, organic, or even metallic in nature. Ideally the film should be strongly adherant and deposited in such a way as to obtain a plane surface. The thickness of such a layer is usually on the order of 1 – 100 μm depending upon the nature of the transducer. The chemically sensitive layer serves as an interface between the environment and a data-processing system, and it is responsible for both the sensitivity and the selectivity of the sensor. This layer should display the following properties:

1) Reversibility, sensitivity, and selectivity
2) Inertness with respect to chemical cross-influences, such as humidity or carbon dioxide
3) Ease of coating, strong adhesion, and long life

Cut	Designation	Type of wave	Operating frequency
AT-Quartz	(YXl)-35°	Thickness shear mode	0.5-100 MHz

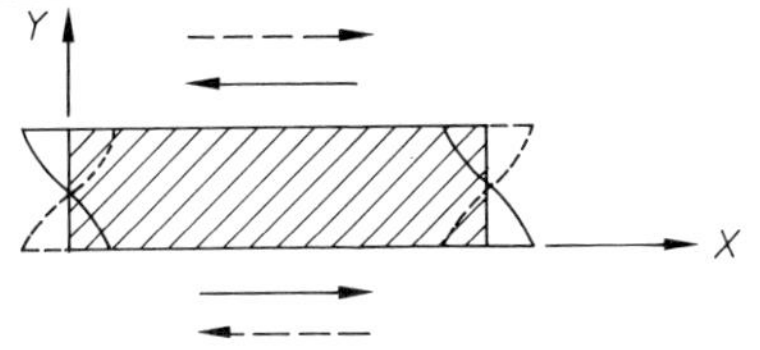

Cut	Designation	Type of wave	Operating frequency
GT-Quartz	(YXlt)-51°/-45°	Longitudinal extension	50-550 kHz

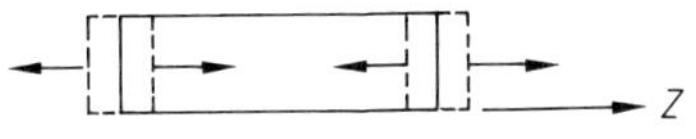

Cut	Designation	Type of wave	Operating frequency
ST-Quartz	(YXl)-42.5°	Surface wave, vertically polarized	50-1500 MHz

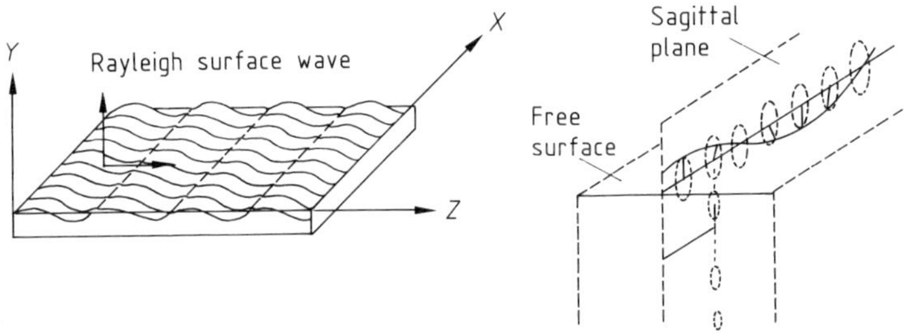

Cut	Designation	Type of wave	Operating frequency
Various Y-rotated quartz cuts		Surface wave, horizontally polarized	e.g., 330 Mz

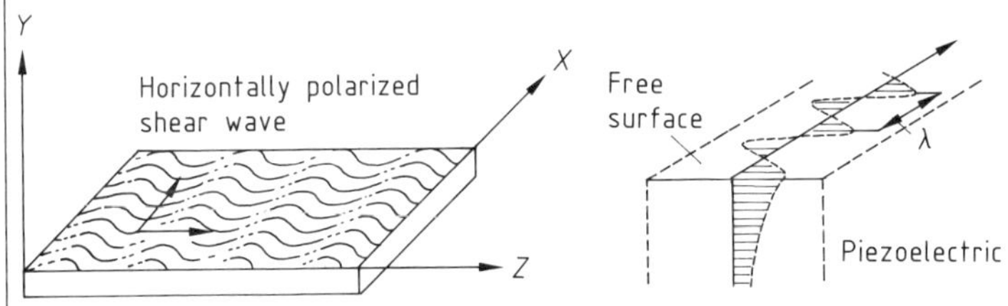

Figure 48. Wave properties associated with several different types of quartz cuts [235]; see also Figure 37

Most chemically sensitive layers are insulators, such as organic substances with special functionalities, but conducting materials such as metal films are also useful. Examples include:

1) Donor–acceptor functionalities; e.g., carbocations [243], M(II)porphyrins [244], or betaines [245]
2) Host–guest binding systems (supramolecular compounds); e.g., molecular cavities [246], clathrate systems [247]
3) Pure metallic layers; e.g., palladium for the detection of hydrogen [248]

Interaction of the Chemical Layer with an Analyte. Interaction with an analyte can take many forms ranging from adsorption–absorption to chemisorption, including the compromise of coordination chemistry. The following types of chemical interactions can be distinguished [249]:

1) Absorption
   Here the analyte is distributed between the chemical interface (a liquid film, amorphous solid, or polymer, analogous to the stationary phase in gas chromatography) and the surrounding medium. In the case of a gas interacting with the chemical interfacial layer, once equilibrium has been achieved the amount of analyte present in the layer is a function of the partition coefficient between the gas and the interface (as specified by Raoult's and Henry's laws) as well as the thickness and area of the interface. Because distribution tends to be determined mainly by the polarity of the chemical interface and the analyte, little selectivity is expected.
2) Adsorption
   The analyte in this case interacts only at the surface of the interfacial layer, and no chemical bonds are formed or broken. Attraction energies are usually in the range 0–10 kJ/mol (i.e., attributable to van der Waals forces), extending in the case of hydrogen bonds to as much as 40 kJ/mol. Adsorption may result from physical attraction between a nonreactive gaseous analyte and a metal surface, for example, or formation of a donor–acceptor association (H^+ or e^-). Adsorption is a universal phenomenon, and it is quite nonselective. The basic principles of gas adsorption have been described by LANGMUIR [250] and DE BOER [251].
3) Chemisorption [252]
   Chemisorption involves very strong interaction—up to ca. 300 kJ/mol—developed at the surface of the chemical interface, including the formation and breaking of chemical bonds. High selectivities can therefore be anticipated, but there is a simultaneous loss of reversibility. Because of the partially irreversible nature of bond formation, such a sensor would actually behave more like a dosimeter, although this might be advantageous in some applications.
4) Coordination chemistry [249]
   A compromise offering both selectivity and reversibility of analyte binding is presented by the area of coordination chemistry. A typical coordination compound consists of a central metal ion M surrounded by a neutral or charged (often organic) ligand. The extent of selectivity can often be influenced by the choice of the metal ion as well as by the choice of the ligand—taking into account both electronic and steric factors. Complexation with an analyte produces changes in the properties of the coordination compound, and these changes can be subject to detection. The selectivity is a function of structural, topological, and polarity parameters.

Other types of chemical interfaces have also been used for microgravimetric gas detection, including polymeric phthalocyanines [253], porphyrins [254], ferrocenes [255], metal clusters [256], crown ethers [257], and cyclodextrins [258].

In many cases it has proven to be very difficult to find a selective chemical interface for a particular analyte, so the use of sensor arrays has been proposed. Numerous papers have dealt with this approach, invoking both BAW and SAW sensors. Pattern-recognition techniques employing increasingly powerful microprocessors are required to generate the appropriate signals [259].

Coating Methods. The stability of a (bio)chemical interface depends heavily on the attachment of the material to the surface of the sensor device. One or more of the following coating methods is employed according to the nature of the sensitive layer:

Smearing [200]
Spin coating [260]
Solvent evaporation [261]
Spraying [262]
Langmuir–Blodgett film formation [263]
Physical vapor deposition [264]
Sputtering [265]
Chemical immobilization [266]

The first four methods listed involve simple coating procedures in which the interfacial material (usually nonpolar) is dissolved in an organic solvent and then deposited directly onto the substrate surface. Evaporation of the solvent leaves a film that adheres to the surface of the substrate by physical bonds. The Langmuir – Blodgett technique makes it possible to achieve very thin films. This is a dipping procedure in which molecules are transferred with great accuracy from the surface of a liquid to the surface of the substrate. Even monolayers can be prepared in this way. More complicated methods include physical vapor deposition (PVD) and chemical vapor deposition (CVD). Very planar layers can be achieved in this way, but the reaction conditions are usually quite drastic, so the technique is essentially restricted to materials that are relatively inert, such as inorganic compounds or metals.

2.2.3.5. Specific Applications

Having reviewed the fundamental aspects of microgravimetric transducers and piezoelectric solids with respect to the generation and propagation of vibrations and waves, it is now appropriate to turn to the practical considerations of instrumentation and application, addressing first the BAW devices.

Bulk Acoustic Wave Sensors. *Experimental Arrangement.* A typical experimental setup involving a piezoelectric mass-sensitive quartz crystal detector is depicted schematically in Figure 49 [193]. The piezoelectric quartz crystal (a) is

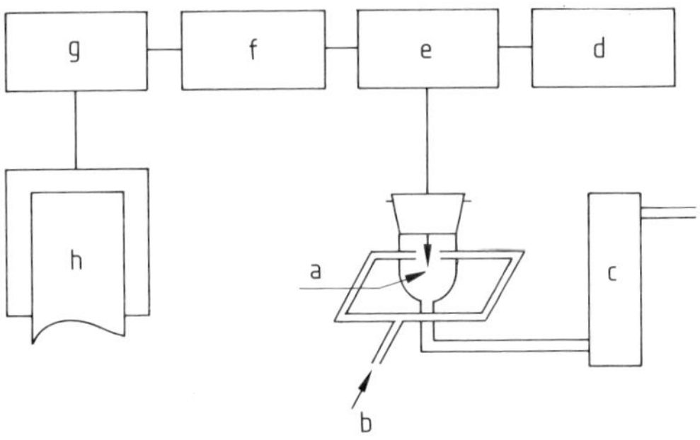

Figure 49. Typical experimental apparatus based on a piezoelectric quartz-crystal detector [193] (with permission from Elsevier, Amsterdam)
a) Piezoelectric quartz crystal; b) Sample; c) Flow meter; d) Power supply; e) Oscillator; f) Frequency counter; g) Digital-to-analog converter; h) Recorder

shown here inserted directly in a gas stream (b) with a flow velocity of ca. 10 – 100 mL/min, but a measuring cell containing a stationary atmosphere of the analyte gas is also useful. The oscillator (e) is usually powered by a regulated power supply (d; e.g., 5 – 15 V) that drives the quartz crystal. The frequency output from the oscillator is monitored with a frequency counter (f), which should be modified by a digital-to-analog converter (g) to permit the frequency data to be recorded.

Parameters that adversely affect the performance of a piezoelectric sensor include the mass of the chemical layer, built-in stress produced by the chemical layer, stress from the electrodes (clamping), and changes in temperature. A more sophisticated device such as a microprocessor-controlled dual-crystal instrument would be required to eliminate these undesirable effects [267].

The device described is designed for measurements in the gas phase; applications in a liquid medium demand an extended version, including a liquid-tight box, an oscilloscope or network analyzer, and provisions for more detailed signal analysis (impedance analysis). Improvements in the oscillating circuit are particularly important for obtaining satisfactory results with liquids [268] – [270].

Applications. This section will be limited to a brief consideration of the use of BAW sensors in gas-sensing devices and as sensors for liquid phases. Piezoelectric microbalances can also be used for monitoring such heterogeneous samples as aerosols and suspensions, as discussed in [237].

Gas-Phase Sensing. In principle, both inorganic and organic analytes are subject to detection. Figure 50 provides an example of results obtained from a piezoelectric quartz microbalance (QMB) resonator used to detect gaseous tetrachloroethylene (C_2Cl_4). Observed response curves are illustrated as a function of time and for various analyte concentrations. A wide variety of solvent vapors can be detected with high sensitivity (in the ppm range), indicating the potential of this approach to gas-phase monitors and sensors for applications in environmental analysis or process control.

Table 9 provides an overview of the detection of selected gaseous analytes. Information regarding interfering gases can be found in the original literature. The effect of changes in the relative humidity of the environment tends to be a general problem.

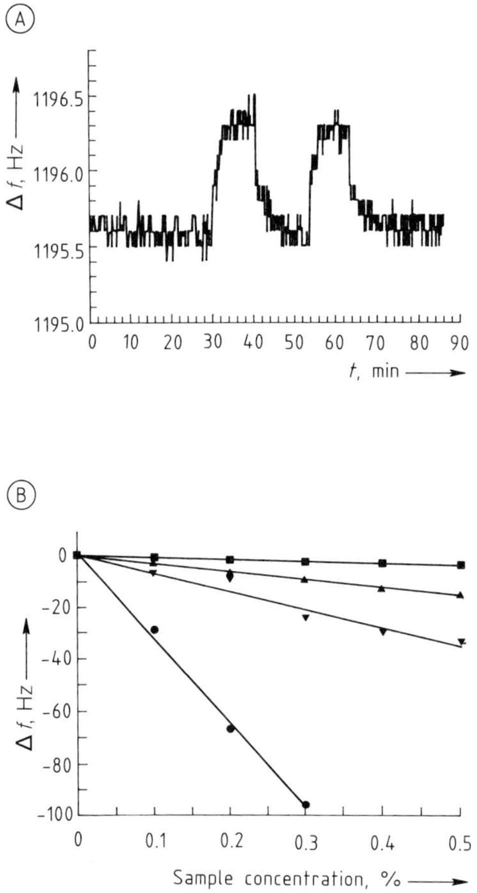

Figure 50. A) QMB sensor response (frequency shift) from 30-ppm pulses of tetrachloroethylene; the quartz crystal was operated in its fundamental thickness shear mode at an initial resonance frequency of 10 MHz, and was coated on both sides with an organic host compound (tetratosylazacyclophane) to a layer thickness of ca. 20 nm (resolution frequency 0.1 Hz); B) QMB-sensor characteristics of tetratosylazacyclophane with respect to different analytes (● tetrachloroethylene, ▾ benzene, ▴ acetone, ■ methanol); 70-nm coating on each side of the quartz [271]
(with permission from Elsevier, Amsterdam)

Applications in Liquids. Apart from the classical sensors designed for detecting ions, piezoelectric crystal resonators are the preferred devices for conducting biochemical measurements in liquids. Figure 51 illustrates results for the detection of human immunodeficiency virus (HIV) antibodies by means of a synthetic HIV peptide (p 24) immobilized on a piezoelectric quartz sensor. The data refer to the frequency shift observed upon addition of authentic HIV antibody. The quartz device was operated in this case in its fundamental thickness shear mode at 20 MHz, with an oscillator circuit stability of ca. 0.5 Hz. The quartz plate was coated with HIV peptide (p 24) dissolved in phosphate buffer at a dilution of 1:1000. Figure 52 shows the corresponding response as a function of different concentrations of monoclonal HIV antibodies.

Surface Acoustic Wave Sensors. Many reports have appeared regarding the use of SAWs as gas detectors. These reports also discuss problems associated with the choice of a coating, achieving adequate specificity, selecting a coating thickness, and depositing reproducible levels of polymer coating on the surface of the sensing element. The variety of coatings used is as diverse as the nature of the analytes, covering the range from simple polymers through metallocyanines to metal films [283]. The application of SAW sensors in liquid systems is still an active area of research, so this topic will not be treated. Information regarding liquid applications for mass-sensitive devices, especially SAW, APM, and LW devices, is provided in [284], [285].

Experimental Arrangement. A schematic representation of a SAW sensing device is provided in Figure 53. This particular resonator was used by D'AMICO et al. [286] for the detection of hydrogen. It consists of a dual delay-line structure in which one of the propagation paths (l_S) is coated with a thin palladium film to act as the sensing channel, while the other (l_R) is uncoated and serves as a reference.

Applications. The type of results to be expected with a variety of analytes in the gas phase is illustrated by the example of an NO_2-SAW sensor. The device in question was developed by NIEUWENHUIZEN and VENEMA [287], and had an operating frequency of 79 MHz with a metal-free phthalocyanine coating. A frequency change of ca. 670 Hz was observed upon exposure of the sensor to 88 ppm of nitrogen dioxide in air at 150 °C. Figure 54 A demonstrates the sensitivity of the device as a function of nitrogen dioxide concentration, while Figure 54 B shows the response of a typical sensor as a function of time with alternating exposure to nitrogen dioxide and air. Possible interferences were also investigated, as described in a similar report published earlier by the same authors [288].

A summary of selected papers describing the use of SAWs for gas sensing is provided in Table 10, subdivided according to the type of analyte.

Table 9. Survey of selected reports on BAW sensors for inorganic and organic analytes [206]

Analyte	Range, ppm	Chemical interface	References
Inorganics			
CO	1–50	HgO/Hg→Au	[272]
CO_2	≤100000	7,10-dioxa-3,4-diaza-1,5,12,16-hexadecatrol	[273]
$COCl_2$	8–200	methyltrioctylphosphonium dimethylphosphonate	[274]
HCN	13–93	bis(pentane-2,4-dionato)nickel(II)	[275]
Organics			
Nitroaromatics	0.001–100	carbowax 100	[276], [277]
Dimethylhydrazine	NR*	polybutadienes	[278]
Acetoin	0.008–0.120	tetrabutylphosphonium chloride	[279]
Formaldehyde	0.010–100	formaldehyde dehydrogenase	[280]
Vinylchloride	≤80	amine 220	[281]

* not reported.

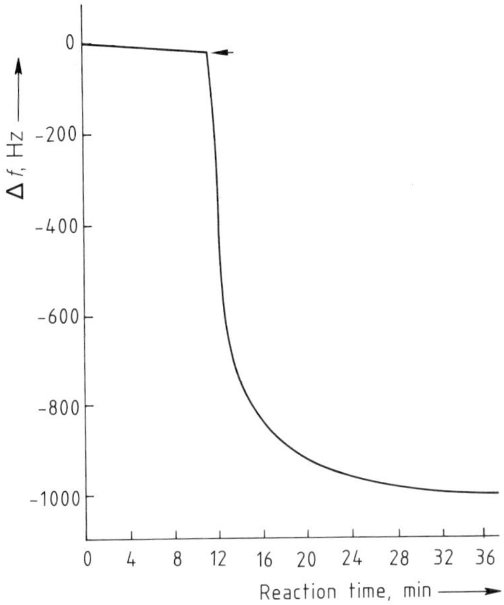

Figure 51. Response for the reaction of anti-HIV antibodies (dilution 1:1000) with adsorbed HIV peptide; the saturation value for the frequency shift is reached ca. 20 min after sample introduction (indicated by the arrow) [282] (with permission from Elsevier, Amsterdam)

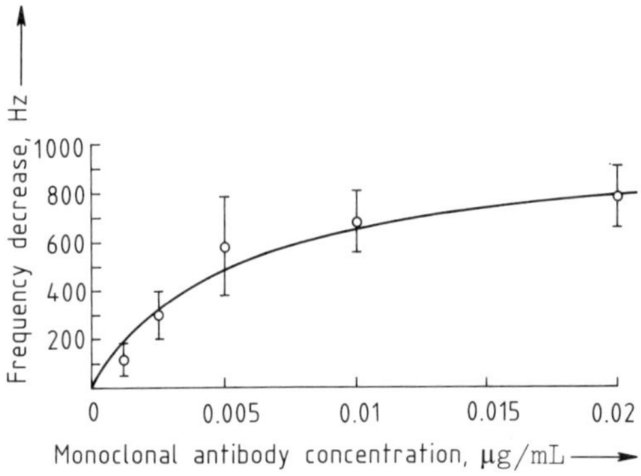

Figure 52. Frequency response for the reaction of HIV antibodies at several concentrations with adsorbed HIV peptide (p 24) [282] (with permission from Elsevier, Amsterdam)

2.2.3.6. Conclusions and Outlook

Mass-sensitive devices offer great promise as chemical sensors, particularly since they rely on the most basic physical effect that accompanies the interaction of one chemical substance with another: the mass effect.

Both BAWs and SAWs have already been successfully applied to the monitoring of many gases, vapors, and more recently, liquids. One of the most critical aspects of chemical sensor research is the search for a suitable chemical interface. The selectivity and sensitivity of mass-sensitive devices can be improved by the use of more carefully designed polymers and other special coating materials. Correction algorithms might also be developed to compensate for variations due to changes in humidity, temperature, pressure, etc. Humidity poses one of the greatest problems with respect to most sensing devices. Better design of the electronic oscillator circuits might also increase the stability of the signal in response to a gas or vapor, thereby improving the signal-to-noise ratio and raising the level of sensitivity.

Research on chemical sensors tends to be unusually multidisciplinary. Such new techniques as the use of Love plates and LW-APM make it possible to combine the advantages of microcompatibility, high sensitivity, and suitability for work in both the gas and liquid phases. Nevertheless, anyone proposing to conduct investigations in this area should be aware of the complexity of interpreting data accumulated from multicomponent gas mixtures over a prolonged period of time, and of parameter identification and interpretation in the case of liquids [206], [283].

2.2.4. Calorimetric Devices

Introduction. Most chemical as well as enzyme-catalyzed reactions are accompanied by changes in enthalpy. For this reason, calorimetric transducers represent a universally applicable approach to chemical or biochemical sensors.

Thermal devices for detecting an enthalpimetric effect (either exo- or endothermic) are insensitive to changes in optical properties of the sample, so colored solutions and turbidity do not constitute interferences. There is also no need for

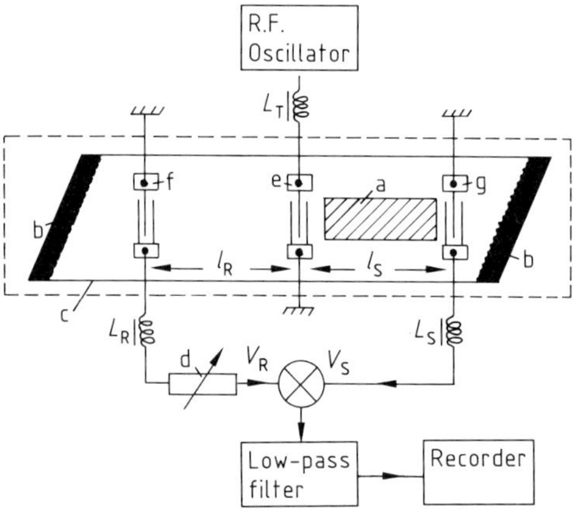

Figure 53. Schematic representation of a SAW hydrogen sensor [286] (with permission from Elsevier, Amsterdam)
a) Palladium film; b) Acoustic absorbers; c) YZ-$LiNbO_3$; d) "Stretched line," an adjustable resistance for matching the frequency of V_R to that of V_S; e) Input transducer T; f) Output transducer T_R; g) Output transducer T_S;
l_R, l_S = Propagation path (l_R is the distance between T_R and T; l_S between T_S and T); L_T, L_R, L_S = Series inductors (establishing a fixed e.g., 50 Ω electric input–output impedance); V_R, V_S = Output voltage of the dual delay line

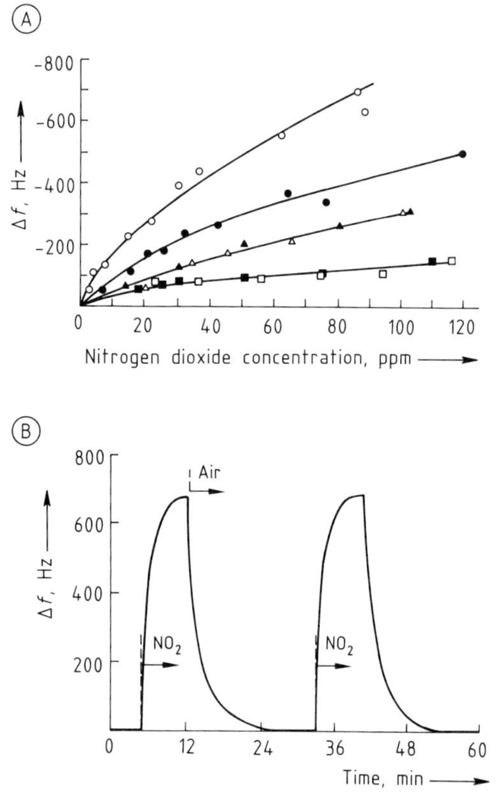

Figure 54. A) Sensitivity of an NO_2-SAW resonator at 120 °C (●, ■, ▲) and 150 °C (○, □, △) with three different metal-free pthalocyanine layers: PC 42 (●, ○), PC 46 (■, □) and PC 47 (▲, △) [287]; B) Sensor output as a function of time at 150 °C (PC 42, 88 ppm of NO_2)
(with permission from Elsevier, Amsterdam)

Table 10. Survey of selected reports on SAW gas sensors [283] (with permission from Adam Hilger, Bristol)

Analyte	Sensor material	Frequency, MHz	Temperature, °C	Chemical interface	References
Inorganics					
H_2	$LiNbO_3$	75		Pd	[286], [289]
H_2	quartz	23		Pd	[290]
Halogens	quartz	50–80	30–150	Cu–PC[a]	[291]
H_2O	$LiNbO_3$	75	0–100	polymer	[292], [293]
NO_2	$LiNbO_3$	150	70	Cu–PC[a]	[294]
Organics					
DMMP[c]	quartz	31–300	20–80	polymers	[295]
Styrene	quartz	30		Pt complex	[261]
Cyclopentadiene	quartz	31	25	PEM[b]	[296]
Various	ZnO	109	24	ZnO	[297]–[299]
Various	quartz	158	35	polymers	[300]

[a] PC = Phthalocyanine. [b] PEM = Poly(ethylene maleate). [c] DMMP = Dimethyl methylphosphonate.

a detectable reaction product, whereas amperometric sensors, for example, require the formation of some product subject to transformation at an electrode. Amplification can be achieved by immobilization of a catalyzing reagent that enhances the rate of enthalpy change. Thermal transducers are especially suitable for continuous measurements in flow systems. An important disadvantage, however, is the requirement for thermostatic control of the device. For this reason calorimetric systems tend to be rather large. In addition, calorimetric sensors are relatively complicated to use [301], [302]. Table 11 provides molar enthalpies for several important types of chemical reactions.

The measurement signal in this case, ΔT, is related to the change in enthalpy ΔH, but also to the heat capacity of the system c_s, according to the following equation:

$$\Delta T = n\Delta H / c_s$$

n = Moles of product

This means that the heat capacity of the sensor itself should be minimized to ensure sensitive measurements. A temperature change can easily be transduced into an electrical signal. The usual measuring device is based on a reference thermal transducer incorporated into a Wheatstone bridge.

Thermal Transducers Commonly Utilized in Chemical and Biochemical Sensors. *Metal-Resistance Thermometers.* The electrical resistance of many metals rises sharply with increasing temperature. Resistors made of platinum (called "Pt 100" because they provide a resistance of 100 Ω at 0 °C) are often used for temperature measurement in the range from − 220 °C to + 750 °C. Other metals like nickel or copper are also applicable for temperatures ≤ 150 °C. Self-heating, caused by a current flow, interferes with the measurement, and the response time of a metal resistor is relatively long (ca. 5 s). The advantages of such resistors are high sensitivity and long-term stability. Platinum resistance devices are especially common in catalytic chemical sensors. A reducing gas may be oxidized in the presence of a heated catalyst (e.g., Rh, Pd, or Pt) in order to increase the reaction rate. In 1962 BAKER [305] described a device called a *pellistor* that effectively separates the catalyst from the platinum wire (see Fig. 55). The detection limit for methane with this device was found to be 20 ppm.

Table 11. Molar reaction enthalpies for selected processes [303], [304]

Reaction	Examples	Enthalpy $-\Delta H_R$, kJ/mol
Oxidation	methane; O_2	800
Neutralization	NaOH, HCl	55
Protonation	tris(hydroxymethyl) aminomethane; H^+	47
Enzyme catalysis	enzyme: catalase; substrate: H_2O_2	100
	enzyme: glucose oxidase; substrate: glucose	80

Thermistors. These are semiconductor resistances with temperature coefficients sufficiently high to make them suitable for use in temperature measurement. Many semiconductors have negative temperature coefficients (NTC), which means that their resistance decreases with increasing temperature. Thermistors with reproducible temperature coefficients are difficult to produce, and self-heating within the sensor is always a problem. On the other hand, thermistors are usually both inexpensive and sensitive, and they can be used over a wide temperature range (0–1000 °C).

The thermistor most commonly used in chemical or biochemical sensors, known as an *enzyme thermistor* (ET), was designed in 1974 by MOSBACH et al. [306]; it consists of an enzyme reactor specially adapted for use in a flow system. A split-flow arrangement was suggested in 1976 [307], whereby one portion of the analyte solution serves as a reference and flows through a blank column, while another passes through the enzyme reactor itself. The reference signal is subtracted from the signal produced by the enzyme reactor (Fig. 56) [307 a]. Samples are analyzed at a rate of ca. 20/h. Since the enzyme reaction and the detecting system are separated

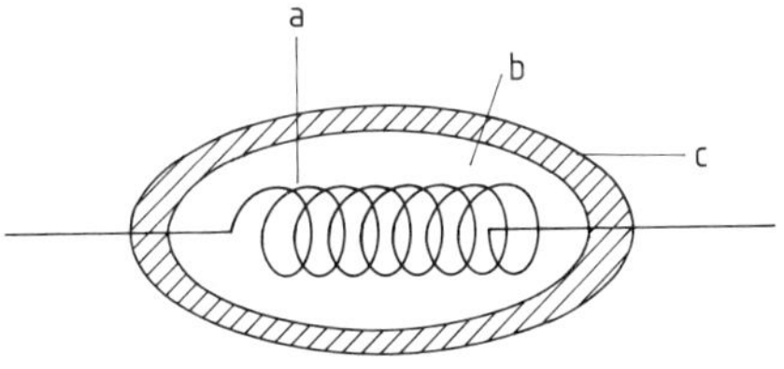

Figure 55. Pellistor device
a) Metal resistor; b) Inactive material; c) Catalyst

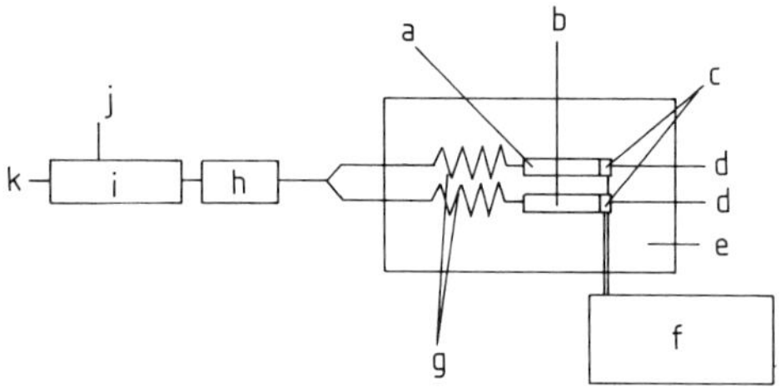

Figure 56. Enzyme thermistor system (with permission from Academic Press, Orlando)
a) Reference reactor; b) Enzyme reactor; c) Thermistors; d) Waste; e) Insulation; f) Wheatstone bridge; g) Heat exchangers; h) Pump; i) Injector; j) Sample; k) Buffer

in space, an enzyme sensor of this type cannot be strictly interpreted as a "biosensor" according to the definition provided in Chapter 3.

Many other devices featuring lower detection limits have been developed in recent years, including a four-channel enzyme-based thermistor that permits the detection of four substrates simultaneously [308]. Another approach involves fabrication of a thermal microbiosensor on a silicon chip. In this case the sensor consists of a micromachined enzyme reactor, inlet and outlet flow channels, and a microthermistor [309].

Thermocouples/Thermopiles. Thermocouples operate on the basis of the Seebeck effect. If wires fabricated from two different metals or semiconductors are soldered together to form a circuit, any temperature difference that exists between the joined points leads to a measurable potential difference, the magnitude of which depends on the extent of the temperature difference and the materials involved. The use of several thermocouples connected in series (*thermopiles*) increases the sensitivity of the sensor, but if a single thermocouple is damaged the entire sensor is affected. It is more difficult to miniaturize thermoelectric sensors than resistance devices, and the long-term stability of thermocouples is often not good. A typical response time is less than one second, and sensors of this type can be used over a temperature range from − 200 °C to + 1600 °C.

Thermal devices based on thermocouples have been extensively investigated in recent years. Guilbeau et al. described in 1987 a thermoelectric sensor for the measurement of glucose [310]. This device incorporated an antimony–bismuth thin-film thermopile with enzyme immobilized on the side containing the active junctions but not on that of the reference junctions

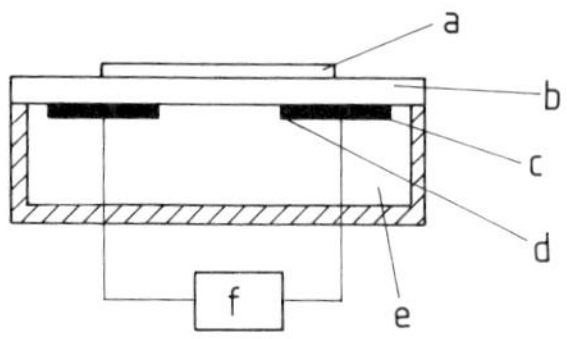

Figure 57. Thermopile-based biosensor (with permission from J. B. Lippincott Company, Philadelphia)
a) Enzyme layer; b) Mylar; c) Reference junction; d) Active junction; e) Insulation; f) Voltmeter

(see Fig. 57). Exposing the sensor to a solution containing the substrate causes the enzyme-catalyzed reaction to take place, resulting in heat exchange that is detected by the thermoelectric transducer.

Summary. Most chemical calorimetric sensors in use today are based on the pellistor device, and are designed for monitoring gases; sensor systems of this type have been subjected to intensive investigation [311]. In general, however, sensor determination of the heat of a reaction is rare, primarily because of the availability of other, more sensitive transducers. On the other hand, thermometric biosensors are applicable to a wide range of analytes, including enzyme substrates, enzymes, vitamins, and antigens. Many applications have been reported in clinical analysis, process and fermentation control, and environmental analysis [312]. Thermal biosensors are seldom used in industry because they are difficult to control and relatively expensive. Universal applicability and new techniques for fabricating microthermosensors suggest that these may find wider use in the future.

2.3. Problems Associated with Chemical Sensors

All nonbiochemical sensors suffer from a lack of that selectivity which would permit their wide application without detailed knowledge regarding the matrix composition of the samples under investigation. Application is therefore restricted to those areas in which the sample matrix is thoroughly understood and not subject to change. Compared to the selectivity problem, all other problems (e.g., stability, sensitivity, lifetime) are relatively unimportant, since they can be resolved by incorporating the sensor into some type of intelligent instrument capable of carrying out all necessary corrections, recalibrations, and performance tests. In some cases insufficient selectivity of a chemical sensor can be circumvented by integrating a selective chemical reaction into the instrumental design. In this way any interfering compounds are eliminated and/or the analyte of interest is selectively transformed into a species capable of being detected unambiguously. A simple approach to this end takes advantage of the flow-injection analysis (FIA) technique.

Any chemometric method of error correction with respect to lack of selectivity must be checked very carefully, especially if it is to be applied to an open system with varying and/or unknown interfering compounds. In some cases a prior chemical separation and/or selection step is required.

Despite the superb selectivity or even specificity of biochemical sensors for recognizing particular analyte molecules, interfering compounds must be dealt with even here in real sample applications. The compounds in this case do not disturb the actual measurement of the analyte; instead they influence the recognition process. For this reason it is important to ensure the use of a stable and well-defined host molecule with an analyte-fitting geometry that is not subject to change. Anything that might alter the host would influence analyte recognition. Thus, any compound or measuring condition capable of changing the tertiary structure of the sensitive protein skeleton of some biomolecule used for recognition purposes would disturb the measurement: first by modifying the selectivity, but also by altering the sensitivity. It is well established that most selective biomolecules are very stable only in a physiological environment similar to the one in which they were produced; even changes in the ionic strength can influence the tertiary structure of a delicate protein macromolecule. Furthermore, certain metal ions (e.g., Hg^{2+}, Pb^{2+}, Ag^{+}, Cu^{2+}) are known to degrade recognition and thus strongly interfere. A similar effect is produced by surface-active molecules (detergents). Bearing these complications in mind, the superb selectivity of biochemical sensors must be evaluated in a broader perspective.

2.4. Multisensor Arrays and Pattern-Recognition Analysis

Theoretically, and under ideal conditions, a sensor array containing several similar sensors with slightly different selectivities and sensitivities towards the analyte and its main interferences should make it possible to compensate for errors caused by the limited selectivity of each individual sensor. One additional but slightly different sensor is required for each interfering compound. The task then becomes one of solving n equations for n unknowns, which can be accomplished with an algorithm for matrix calculations and an electronic calculator. However, there are several restrictions that limit the use of this compensating technique.

First, the interfering components must be known in advance so that one can prepare a set of calibration mixtures for obtaining the individual analytical functions for each sensor under variable interference (matrix) concentrations. The approach is therefore limited to a closed sample location, and it presupposes complete information regarding potential interferents within the volume in question. This may prove applicable in certain production control situations, but certainly not in cases where the sample matrix is not sufficiently known a priori, as in the case of most environmental and some clinical analyses. Such precautions cannot be circumvented even with the use of the most sophisticated modern approaches to pattern-recognition analysis and/or neural-network treatment. All potentially interfering compounds must be known in advance and introduced into the array during the so-called learning phase through a permutation of all possible types of mixtures. Much time may be consumed in the preparation of the necessary mixtures and in performing the measurements, and the drift associated with most sensors and the limited lifetimes of biochemical sensors could render the whole procedure meaningless!

Second, the linearity and true independence of all effects cannot be guaranteed. In some cases of sensor poisoning all analytical functions (selectivities and sensitivities) obtained during prior calibration and/or learning are rendered invalid for every individual sensor. How is a sensor array to distinguish between irreversible poisoning effects and the normally reversible disturbances attributable to ordinary interferents? How should it differentiate between the effects of a positive error (where a portion of the signal is due to interferents) and general matrix effects resulting in diminished sensitivity? Especially in environmental trace analysis (<10 ppm) each sample is very likely to contain more than ten interfering compounds. Intimate knowledge of each, as well as the availability of appropriate standards to permit extensive permutation at various concentrations presupposes the preparation of countless mixtures to be utilized in the training process for the sensor array. It must also be recognized that even trace amounts of certain compounds may totally disrupt the analyte recognition process.

Taking all this into consideration, as well as the fact that most chemists lack fundamental experience with sensor arrays in real open-environmental trace analysis, considerable caution is warranted with respect to this approach to increasing the selectivity of sensors and correcting for interferants. In general, application of a simple and rapid preliminary separation step will be more productive with respect to the goal of achieving a reliable analytical result. Literature examples of successful sensor-array applications related to the identification of beverages or the characterization of odors [313]–[315] should not be extrapolated to the quantitative analysis of a single analyte. Many of the sensor-array applications described in the literature [316]–[324] involve problems that could have been solved by much simpler means, such as exact density measurement, possible now with extreme accuracy. Differentiation among various samples with similar compositions in the context of a closed system (i.e., without the variability associated with an open matrix) can usually be achieved more economically on the basis of simple physical measurements.

3. Biochemical Sensors (Biosensors)

3.1. Definitions, General Construction, and Classification

Biosensors are devices based on a direct spatial coupling between a biologically active compound and a signal transducer equipped with an electronic amplifier (Fig. 58). Biological systems at various levels of integration are used to recognize a particular substance that is to be determined in a specific way. Different types of biologically sensitive materials can be used for selective recognition, including enzymes, multienzyme systems, organelles, photosensitve membranes from plants and bacteria, membranes generally, protoplasts, whole intact cells, tissue slices, antibodies, lectins, DNA, and transport protein or

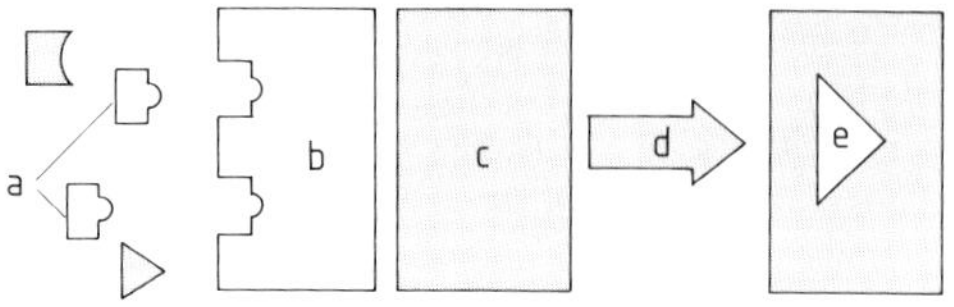

Figure 58. The basic principles of a biosensor
a) Analyte; b) Biological receptor; c) Transducer; d) Electrical signal; e) Signal processing

receptor systems isolated from cell membranes. The macromolecular biological compound is ordinarily immobilized in close proximity to the transducer surface, thereby facilitating direct or mediated signal transfer to the transducer. Besides fixing the receptor on the transducer surface this immobilization also serves the important function of stabilizing the biological material.

The first indicating step is specific complex formation between the immobilized biologically active compound (in a broader sense, the receptor R) and the analyte (A). This selective and usually reversible interaction produces secondary chemical or physical signals that are recognizable by a suitable electrochemical, optical, enthalpimetric, conductometric, or piezoelectric transducer. The transducer of choice depends on the nature of the molecular-level binding affinity with respect to the analyte, which may cause changes in such parameters as the redox state of the receptor compounds, the concentrations of electrochemically or optically active substances that are consumed or produced, or heat, mass, or electromagnetic radiation (see Table 12) [325]–[328].

Depending upon the mechanism of biochemical interaction between the receptor and the analyte, two basic types of biosensors can be distinguished: *biocatalytic* (metabolic) sensors, and *bioaffinity* sensors (Table 13).

Among the analytes detectable with biosensors are carbohydrates, organic acids, alcohols, phenols, amines, heterocyclic compounds, proteins, enzymes, cofactors, inorganic ions, hormones, vitamins, peptides, drugs, xenobiotics, microorganisms, gases, and pesticides. Analyte concentrations measurable with various types of biosensors range from 10^{-7} mol L^{-1} (biocatalytic sensors) to as little as 10^{-15} mol L^{-1} (affinity sensors). Based on their level of integration, biosensors can be subdivided into three generations:

1) *First Generation.* The receptor is entrapped between or bound to membranes, and the combination is fixed on the surface of an appropriate transducer.
2) *Second Generation.* The receptor is bound covalently to the transducer's surface, thereby eliminating the need for a semipermeable membrane.
3) *Third Generation.* The receptor is bound to an electronic device that transduces and amplifies the signal (e.g., the gate of a field-effect transistor, an array of amperometric ultramicroelectrodes, or interdigitated microelectrodes), utilizing the techniques of microelectronics and micromechanics.

3.2. Biocatalytic (Metabolic) Sensors

Metabolic biosensors are based on the specific recognition of some enzyme substrate and its subsequent chemical conversion to the corresponding product(s). The biocatalytic reaction itself accomplishes specific binding between the active site of the enzyme and the substrate, conversion of the substrate to product, and release of the product from the active center, which in turn leads to regeneration of the binding affinity and catalytic state of the enzyme.

The linear measuring range for metabolic sensors depends on the "apparent" Michaelis–Menten constant (K_{Mapp}) of the immobilized en-

Table 12. Transducer systems used in the design of biosensors

Transducer	Indicating a change in:	Caused by:
Amperometric electrodes	electron transport	enzyme-catalyzed oxidation or reduction of an analyte
Potentiometric electrodes, ISFETs, MOSFETs, conductometric devices	ion concentration	enzyme-catalyzed oxidation, hydrolysis, or cleavage of C–C, C–O, C–N, or other bonds
Calorimetric devices (thermistors, thermopiles)	enthalpy	enzyme-catalyzed reactions
Piezoelectric crystals	mass	bioaffinity reaction (between Ab–Ag, Ab–hapten, an analyte and receptor protein, or analyte and lectin
Fluorometric fiber optics	fluorescing concentration of enzyme substrates/products or immunolabels	enzyme and bioaffinity reactions
Waveguides	optical path (adlayer)	bioaffinity reaction
Devices based on surface plasmon resonance	resonance (adlayer)	bioaffinity reaction

Table 13. Biosensor specification according to the nature of the primary sensing reaction [327]

Indicating reaction:*

Biocatalytic (metabolic) sensors:

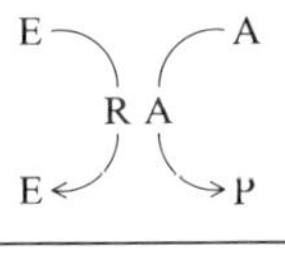

Bioaffinity sensors:

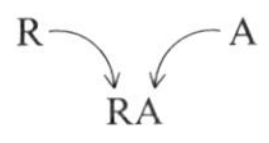

Biocatalytic (metabolic) sensors: Analytes	Receptors	Bioaffinity sensors: Analytes	Receptors
Substrate	enzyme	protein	dye
Cosubstrate	enzyme system	glycoprotein	lectin
Cofactor	organelle	cells	antigen
Prosthetic group	microorganism	antigen/hapten	antibody
	tissue slice	antibody	RNA/DNA
		neurotransmitter	receptor protein
		pesticide	
		enzyme (system)	

* E = Enzyme; A = Analyte; P = Product; R = Receptor; RA = Receptor–analyte complex.

zyme, which describes a particular substrate concentration associated with half the maximum rate of the enzyme reaction. The reaction rate of an immobilized receptor enzyme is controlled by the rate of substrate and product diffusion through both the semipermeable membrane covering the enzyme and the layer of immobilized enzyme itself, resulting in a higher K_M value than that anticipated under conditions of kinetic control. Enzymes, coupled enzyme sequences, organelles, microorganisms, or tissue slices can all be used as receptor materials for metabolic sensors. Under certain conditions, cosubstrates, effectors (inhibitors, activators), and enzyme activities can also be determined by metabolic sensors.

3.2.1. Monoenzyme Sensors

Enzyme sensors rely on an immobilized enzyme as the biologically active component. Many biocatalytic reactions, especially those of oxidoreductases, hydrolases, and lyases, are associated with the consumption or formation of electroactive substrates or products in concentrations proportional to the analyte concentration. For this reason most commercially available biosensors are enzyme sensors based on electrochemical transducers.

Typical electrochemically detectable (co-) substrates and products include oxygen, hydrogen peroxide, hydrogen ion, ammonia, carbon dioxide, reduced or oxidized cofactors, and redox-active (oxidized or reduced) prosthetic groups in oxidoreductases, all of which can easily be converted into electrical signals by suitable transducers (e.g., amperometric or potentiometric electrodes or conductometric sensors, Table 14).

Immobilization of an enzyme in or at an artificial matrix or a membrane usually stabilizes the biocompound, thereby increasing the half-life for activity. Techniques of enzyme immobilization include physical or chemical adsorption, ionic and covalent bonding (perhaps to functionalized transducer surfaces), crosslinking by bifunctional agents, and entrapment in such polymer matrices as natural or artificial gels and conducting polymers. The first biosensor was described by Clark and Lyons in 1962, who introduced the term "enzyme electrode" [329]. In this first enzyme sensor, an oxidase enzyme in a sandwich membrane was placed next to a platinum redox electrode. A platinum anode polarized at +0.6 V (vs. SCE) responded to peroxide produced by the resulting enzyme reaction with the substrate. Application of a polarization voltage of − 0.6 V permitted the amperometric detection

Table 14. Common configurations for monoenzyme sensors

Enzyme class	Detectable compound involved in the enzyme reaction	Typical transducer
Oxidoreductases		
Oxidases	O_2, H_2O_2	pO_2 electrode, noble-metal–carbon redox electrode
Oxygenases	O_2, quinones, catechol	pO_2 electrode, (amperometric) carbon electrode
NAD(P)-dependent dehydrogenases	NAD(P)H/NAD(P)	(amperometric) mediator-modified redox electrode, fluorometer, photometer
PQQ-dependent dehydrogenases	PQQ/$PQQH_2$	
Reductases	NAD(P)H/NAD(P) Heme-, Cu-, Mo-, FeS-containing centers	
Hydrogenases	H_2, NAD(P)H, NAD(P), $FADH_2$, $FMNH_2$, FeS-, Ni-containing centers	
Hydrolases	$NH_3/(NH_4)$	(potentiometric) ammonia-sensitive electrode
	CO_2	(potentiometric) pCO_2 electrode
	H^+	pH electrode, ISFETs, coulometer, conductometric and enthalpimetric devices
	redox-active products (phenolic, thio-group-containing compounds)	(amperometric) redox electrodes

of oxygen consumption. The primary target substrate was β-D-glucose (Fig. 59). Stoichiometric turnover of the glucose oxidase-catalyzed reaction yields a sensor response proportional to the glucose concentration up to the apparent K_M value of the immobilized enzyme. This in turn led to the first glucose analyzer for measuring glucose in dilute blood on the basis of an amperometric enzyme sensor. A commercial analog, the Yellow Springs Instrument Model 23 YSI, appeared on the market in 1974. The same indicating technique has more recently been applied to many other oxygen-mediated oxidase systems, which constitute the most widely exploited enzyme classes for the construction of biosensors.

Amperometric Biosensors Using Artificial Electron Mediators. Amperometric biosensors for monitoring oxidase-catalyzed oxygen consumption or the production of hydrogen peroxide are, under certain circumstances, restricted in their use because of a limited availability of dissolved oxygen. This may necessitate careful pretreatment of the sample with rigorous exclusion of oxygen-consuming compounds, as well as application of a high polarization voltage, which may cause interfering currents in complex matrices.

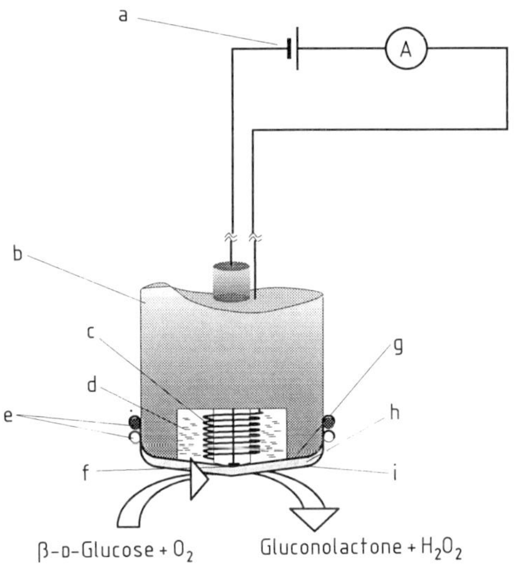

Figure 59. Configuration of an amperometric enzyme electrode for glucose determination
a) Polarization voltage (−600 mV or +600 mV vs. Ag/AgCl); b) Electrode body; c) Ag/AgCl pseudo-reference electrode; d) Electrolyte solution; e) O-Rings; f) Platinum working electrode; g) O_2/H_2O_2-permeable membrane; h) Glucose oxidase (immobilized layer); i) Semipermeable membrane

It should be possible to overcome such problems by direct electron transfer between the redox-active prosthetic group of an enzyme and the electrode surface. In a few cases, especially ones involving small redox proteins such as cytochrome c, efficient electron transfer has been achieved in this way with promotors that direct the active redox centers to the electrode surface. Macromolecular redox proteins exhibit rather low rates of heterogeneous electron transfer, and the transfer is often accompanied by a large overpotential, but the use of redox mediators may increase the electron-transfer rate dramatically. These low-molecular-mass (in)organic redox species undergo facile reversible redox reactions with both the prosthetic groups of the enzyme and the electrode surface according to the general scheme depicted in Figure 60. The resulting electron shuttle reduces the overvoltage of the biological redox system to the formal potential of the mediator. Suitable redox mediators include quinoid redox dyes, quinones, organometallic compounds, and fulvalenes (see Table 15) [330]–[334].

In constructing such a biosensor the redox mediators can be immobilized at the electrode surfaces in various ways, including chemisorption or covalent bonding, chemical or physical deposition of a polymer film containing the mediator molecules, or physical incorporation of the redox mediators (which are only slightly water soluble) directly into the electrode material. One ingenious approach involves covalent binding of the mediator to the electrode via an appropriate *spacer*, or electrical "wiring" of the enzyme through an electron-relaying redox polymer net-

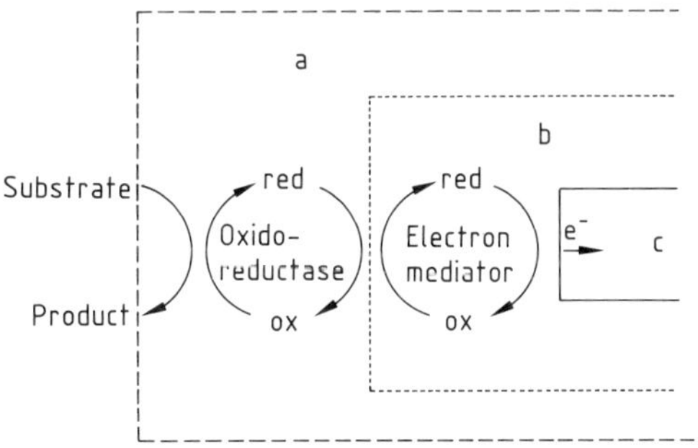

Figure 60. Operating principle of a mediator-modified amperometric biosensor, involving an oxidase or a dehydrogenase (in the case of a reductase, electron transfer would proceed from the electrode to the analyte substrate)
a) Amperometric enzyme sensor; b) Mediator-modified electrode; c) Redox electrode

Table 15. Redox mediators leading to efficient electron transfer with redox proteins

Mediator type	Examples	Structures
Quinones	benzoquinone	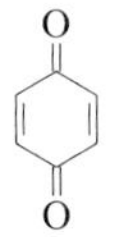
Quinodimethanes	tetracyanoquinodimethane (TCNQ)	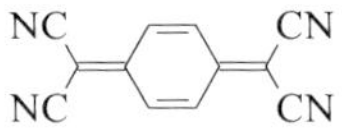
Tetrathiafulvalene (TTF) and its derivatives		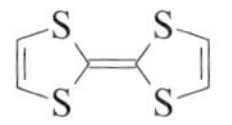
Metallocenes and their derivatives	ferrocene	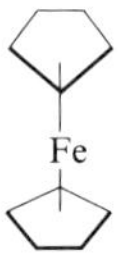
Phthalocyanines (Co, Fe, Mn, Cu)	cobalt phthalocyanine (Co-PC)	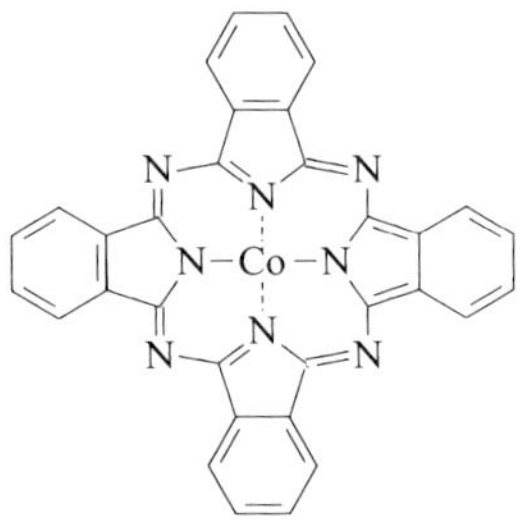
Quinoid redox dyes		
Phenazines	*N*-methylphenazonium (NMP^+) salts	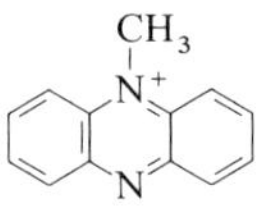
Phenoxazines	meldola blue	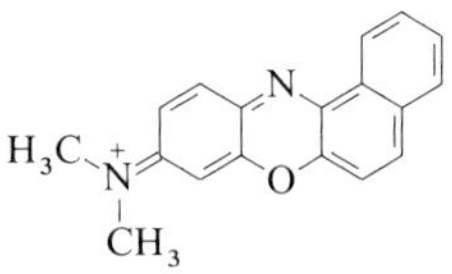
Phenothiazines	methylene blue	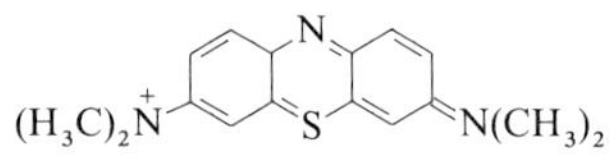
Indamines	toluylene blue	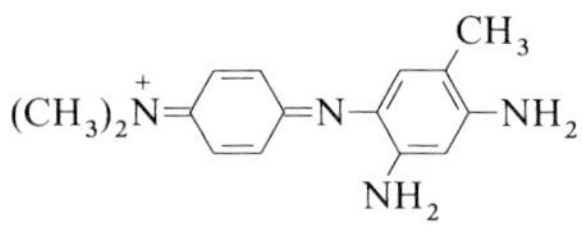
Viologens	methyl viologen	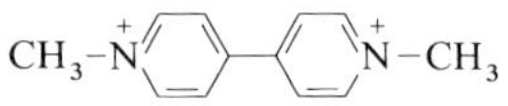
Complex compounds		
Ruthenium complexes		$[Ru(NH_3)_6]^{3+/2+}$, $[Ru(en)_3]^{2+/3+}$
Osmium complexes		$[Os(bpy)_2Cl]^{3+/2+}$
Cobalt complexes		$[Co(bpy)_3]^{3+/2+}$, $[Co(en)_3]^{2+/3+}$
Iron complexes		$[Fe(CN)_6]^{3+/4+}$, $[Fe(phen)_3]^{2+/3+}$

work. In both cases electron transfer from the redox centers to the electrode surface is thought to occur by electron tunneling [335]–[337]. The use of a redox mediator increases the dynamic range of a sensor, minimizes interferences, and provides an indicator reaction that is usually independent of the oxygen content of the matrix. Furthermore, such mediators undergo reversible redox reactions with redox-active coenzymes, dehydrogenases, reductases, and hydrogenases, as well as the redox-active products of enzyme-catalyzed reactions, which opens many new possibilities for incorporating these classes of oxidoreductases into new types of biosensors.

3.2.2. Multienzyme sensors

Coupled enzyme reactions mimic certain metabolic situations in organelles and cells that have themselves already been exploited in biosensors [338], [339]. The use of enzyme sequences facilitates:

1) Broadening the scope of detectable analytes
2) Improving selectivity
3) Increasing both sensitivity and dynamic range
4) Eliminating interferences

Depending on the target analyte, sample matrix, required measuring range, etc., enzymes can be coupled to produce reaction sequences operating either in series or in parallel. An alternative approach to realizing multienzyme sensors is the use of cell complexes, whole cells, or tissue slices comprising physiologically integrated enzyme systems (see below).

Enzyme Sequences Operating in Series. The number of substances that can be determined directly by monoenzyme sensors is limited, because many enzyme-catalyzed reactions involve no physicochemically detectable substances. To form readily detectable compounds it is often necessary, therefore, to couple together several different enzyme reactions in a sequential way. The analyte of interest is first converted by some specifically catalyzing enzyme into a product that is itself convertable in the course of one or more subsequent enzyme reactions involving detectable secondary (co)substrates or products. Biosensors based on such enzyme sequences have been developed for the determination of many substrates, including disaccharides (sucrose, maltose, cellulose, lactose), cholesterol, creatinine, and nucleic acids [327].

With certain enzyme sequences it is possible to achieve chemical signal amplification through the accumulation of a particular product by cycling the substrates or cofactors. The sensitivity of some biosensors can be enhanced in this way by as much as three orders of magnitude. Signal-amplifying enzyme systems have been based on oxidase–dehydrogenase, dehydrogenase–transferase, and phosphorylase–phosphatase–oxidase sequences, permitting the determination, for example, of lactate, ATP, and phosphate at the picomole per liter level (Fig. 61) [328]–[340].

Enzyme Sequences Operating in Parallel. Multienzyme sequences operating in parallel are used for the elimination of interfering agents, expansion of the dynamic measuring range, and (in principle) the indirect detection of an electro-

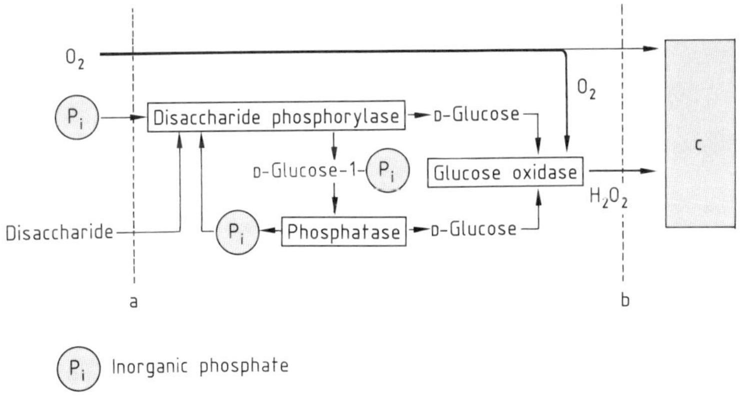

Figure 61. Enzyme sequence for the detection of inorganic phosphate, based on the principle of chemical signal amplification [340]
a) Dialysis membrane; b) H_2O_2/O_2 permeable membrane; d) Working electrode

chemically inactive reaction catalyzed by one of the enzymes in a substrate competition. *Anti-interference* enzyme systems are the most important. Substances interfering with an indicating reaction can be eliminated as one way of improving the analytical quality of the measurement by covering the immobilized enzyme with an anti-interference enzyme layer, in which a second enzyme converts the disturbing substance into an inert product. This strategy has made it possible to remove such interfering agents as ascorbic acid and endogenous glucose [338].

Biosensors Based on Integrated Biological Enzyme Systems. Cellular organelles and membranes, as well as whole cells or tissue slices from animal or plant sources comprising physiologically coupled multienzyme systems are often employed as receptor materials in biosensors based on electrochemical detection. Native physiologically integrated enzyme systems display enhanced stability due to their in vivo environment. Biological material of this type offers the potential for low cost and simple preparation. Furthermore, this opens the way to analytes that are converted to electrochemically detectable products not by simple enzyme reactions, but rather by complex multienzyme reactions within an intact biological structure. Analytes subject to such receptor systems include chloroaromatics, caprolactam, naphthalene, tryptophan, several hormones, and acetic acid [325], [341].

3.2.3. Enzyme Sensors for Inhibitors

Sensors for inhibitors are based on competitive inhibition between a substrate and an inhibitor for an enzyme binding site, inhibition of an enzyme reaction by product accumulation, irreversible binding of an inhibitor to an active binding site, or redox-active heme or SH groups of enzymes that lead to the blocking of enzyme activity.

When thylakoid membranes, protoplasts, or whole cells are used as receptor materials, various inhibitors cause metabolic "short circuits," or inhibition of the (photosynthetic) electron-transport system, which results in a decrease in such metabolic parameters as oxygen consumption or production, photocurrent development, dehydrogenase activity, or fluorescence. In contrast to metabolic sensors, the first sensing step here is a result of binding of the inhibitor to an active binding site or redox center on the enzyme. Sensors of this type usually operate at low enzyme-loading factors and with substrate saturation in order to ensure a highly sensitive signal of limitation in the kinetically controlled reaction due to the presence of an inhibitor. Examples of such inhibitors (analytes) include toxic gases, pesticides, chloroorganic compounds, phenols, and heavy-metal ions [342], [343].

3.2.4. Biosensors Utilizing Intact Biological Receptors

Molecular receptors are cellular proteins (often membrane-bound) that bind specific chemicals in such a way as to produce a conformational change in the protein structure. This conformational change in turn triggers a cellular response: opening of an ion channel, for example, or secretion of an enzyme. In general, molecular receptors can be distinguished from other types of receptors (including larger multicomponent systems) on the basis of their composition, which is a single protein, though this protein may contain more than one subunit. Among the important receptors known to date are receptors for hormones (control agents for a wide range of cellular and body functions), amino acids, insulin, and neurochemical transmitters, in addition to receptors capable of binding synthetic bioactive chemicals, such as drugs [344], [345].

Molecular receptors have two important properties relevant to their possible incorporation into biosensor devices: intrinsic signal amplification and high specificity. The specificity is a consequence of a highly evolved binding region within the receptor that is fully optimized for binding one particular ligand by a variety of electrostatic, hydrogen bonding, van der Waals, and hydrophobic forces.

Limitations affecting the study and use of molecular receptors include:

1) Relatively difficult experimental procedures for the isolation of the receptor proteins (even prior to purification), which are both time-consuming and expensive, and often require animal sources
2) An absence of techniques for obtaining more than very small amounts of pure receptors
3) The rapid loss of biological function that usually follows isolation

Molecular biology has brought considerable progress in the isolation and purification of receptors, while advances in artificial membrane technology and reconstitution have recently reduced the problem of instability.

Comparatively few biosensors of this type have been reported [344], [345], primarily because of problems associated with the successful isolation of molecular receptors (in sufficient quantity and purity) and their subsequent stabilization in artificial environments in such a way that appreciable biological activity is retained.

A clear exception, however, is a fiber-optic sensor based on evanescent fluorescence developed by ELDEFRAWI et al. [345], [346] that contains nicotinic acetylcholine receptor (nAChR). This was the first neurotransmitter receptor to be purified and characterized in vitro, and it is involved in chemical signalling. The sensor in question consisted of nAChR immobilized on the surface of a quartz fiber. Determination of the concentration of acetylcholine and other natural neurotransmitters (e.g., nicotine) was accomplished by measuring the extent to which they inhibited the binding of a standard solution of a fluorescence-labeled toxin (fluorescein isothiocyanate-α-bungarotoxin, FITC-α-BGT). Binding determinations were based on measurements of fluorescence intensity in the region of the evanescent wave (that is, at the fiber – solution interface, see Section 2.2.2.1), where the nAChR was immobilized. The sensor had an acceptable signal-to-noise ratio of 10^2 and an analysis time < 10 min (based on the initial rate of signal increase rather than a steady-state value), and it was capable of operating under conditions of high ionic strength. However, it suffered from the problem of nonregeneration of the sensing element because of the irreversibility of FITC-α-BGT binding to the nAChR.

A similar receptor-based biosensor using electrochemical detection was reported for the detection and measurement of riboflavin (vitamin B_2) [344], [345]. This device is based on the competition for aporiboflavin-binding protein (apoRBP) between riboflavin present in the sample subject to analysis and a riboflavin analogue bound in a membrane. In the absence of riboflavin all the apoRBP is bound to the analogue on the membrane. When riboflavin solution is added, the five-fold greater affinity of apoRBP for its natural cofactor (riboflavin) results in displacement of apoRBP from the membrane into the solution, causing a change in the electrical potential of the membrane. A riboflavin measurement range of $0.1 - 2 \times 10^{-6}$ mol/L was reported.

Further progress in molecular receptor-based sensors will be a function of advances in molecular biology with respect to enhancing the quantity, purity, and stability of receptors that can be used in vitro.

3.3. Affinity Sensors

Affinity sensors take advantage of physico-chemical changes that accompany complex formation (e.g., changes in layer thickness, refractive index, light absorption, or electrical charge). Once a measurement has been made the initial state must be regenerated by splitting off the complex.

Examples include immunosensors utilizing antibodies or antigens for recognition purposes, lectins, and "true" biological receptors in matrix-bound form (see Section 3.2.4).

Immunosensors Based on Antibody Molecules. Biosensors based on the antigen – antibody interaction are called *immunosensors*. The high specificity and high sensitivity that typifies an antigen – antibody (Ag – Ab) reaction has been used in a vast range of laboratory-based tests incorporating antibody components (immunoassays). The analytes detected and measured have included many medical diagnostic molecules, such as hormones (e.g., pregnancy-related steroids), clinical disease markers, drugs (therapeutic and abused), bacteria, and such environmental pollutants as pesticides. A crucial distinction must be made between an immunoassay and an immunosensor, however, with respect to the technology relevant to the analysis. The ultimate aim is a low-cost, disposable immunosensor, with the entire device being used only once. For example, it has now proven possible in practice to develop an instrument containing a disposable "sensor chip" that can be used for Ab or Ag tests.

Two different types of immunosensors have been described:

1) *Direct immunosensors* register immunochemical complex formation at the transducer surface via electrochemical or optical changes. The advantage of a direct immunosensor is that measurement of the antigen – antibody interaction can be accomplished immediately without any need for additional antibodies or markers (enzymes, fluorescent labels, etc.).

2) *Indirect immunosensors* are based (like homogeneous and heterogeneous immunoassays) on labeling of one of the partners in an immune interaction. One class of indirect immunosensors derived mainly from enzyme immunoassays (EIA) encompasses the enzyme immunosensors. These combine the high selectivity of an immunoreaction with the high sensitivity characteristic of the enzymatic amplification effect.

3.3.1. Direct-Sensing Immunosensors without Marker Molecules

Among direct immunosensors the most favored are optical sensors, because they incorporate the advantages of fiber optics or optoelectronic transducers. The nonelectrical properties of optical systems offer important safety advantages for medical, industrial, and military applications. Remote sensing is clearly advantageous, for example, for explosives and toxins. Optical-fiber probes are also mechanically flexible, small, inexpensive, and disposable.

The detection of an evanescent field in an optical waveguide (see Section 2.2.2.1) is a highly sensitive technique especially suitable for monitoring the antibody–antigen interaction [344], [345]. One partner of the immuno pair (e.g., the antibody) is immobilized at the surface of the waveguide, and its reaction with the other partner in the test sample is monitored. An evanescent wave generally penetrates a distance of several hundred nanometers extending the full length of an antibody–antigen complex, so in contrast to conventional techniques there is no need to carry out a prior separation of the nonspecific biological components.

One example of a commercial immunosensor is produced by Serono Diagnostics. This fluorescence-based evanescent-wave immunosensor [344], [345], [347] incorporates a novel capillary-fill design. The system consists of two glass plates separated by a narrow capillary gap of 100 μm. The lower plate acts as an optical waveguide and contains on its surface an immobilized layer of antibodies. The system benefits from its capillary-fill system, by means of which a fixed volume of sample is drawn into the space between the plates regardless of the volume of the bulk sample (a blood droplet for example). Delivery of a highly reproducible volume of sample eliminates a significant source of error in analyte measurement. The Serono immunosensing format constitutes a competitive assay, though a sandwich-type assay could also be used. The upper plate contains on its surface a layer of fluorescence-labeled analyte trapped in a water-soluble matrix. On addition of sample the labeled analyte is released into solution, where it competes for antibody sites on the lower plate with any unlabeled analyte present in the sample. After a fixed incubation period the fluorescence signal of the evanescent waves is measured and related to analyte concentration.

This technology does have shortcomings, however, some of which are related to the practical problems associated with carrying out measurements on whole blood. These include poor capillary flow because of the "stickiness" of blood compared with water, increased incubation time resulting from slow dissolution of the labeled analyte matrix, and the presence of fluorescent and highly colored molecules in blood, such as hemoglobin. Drawbacks include the necessity for an incubation period of several minutes (due to Ag–Ab interaction) and increased manufacturing costs (each analyte to be tested requires a dedicated sensor, which involves labeling and immobilization prior to physical construction).

The technique of surface plasmon resonance (SPR) represents a very sensitive alternative, and its feasibility has been adequately demonstrated [348]. As noted previously (Section 2.2.2.3), SPR is a phenomenon produced by a beam of light directed onto a glass–metal interface, usually a glass prism in contact with a layer of gold or silver (Fig. 62). At some specific angle (the resonance angle) a component of the electromagnetic light wave propagates in the metal along the plane of the interface, taking the form of surface plasmons. The resonance angle is quite sensitive to changes in refractive index and dielectric constant at the interface up to a distance of several hundred nanometers from the actual metal surface, with the sensitivity decreasing exponentially as a function of distance from the surface. Immobilization of an antibody on the surface causes a measurable shift in the resonance angle, and binding of a high-molecular antigen to the immobilized antibody leads to a further change. For typical biological systems this binding-induced shift in resonance angle (expressed in resonance units) is approximately linearly proportional to the concentration of bound antigen (or antibody if antigen has been pre-immobilized).

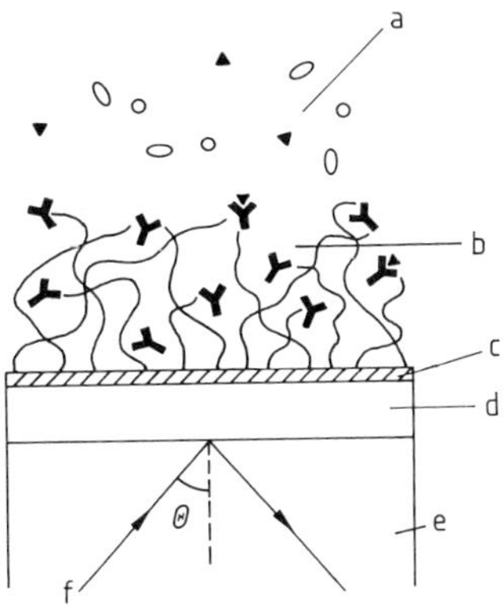

Figure 62. Operating principle of an SPR immunosensor
a) Bulk solution; b) Carboxymethylated dextran matrix; c) Linker layer, $HO-(CH_2)_n-SH$; d) Gold film; e) Glass substrate; f) Light
(with permission from GEC Journal of Research, Essex)

The SPR-based biosensor systems BIAcore and BIAlite, developed by Pharmacia Biosensor AB and now available commercially, represent a breakthrough in immunosensor technology, particularly in the use of SPR techniques. Apart from high cost, the only limitation associated with this technique is that the sensitivity depends on the molecular mass of the adsorbed layer, which in turn controls the optical thickness, so low concentrations of small molecules (molecular mass < 250, including most haptens) are unlikely to be measurable directly without additional labeling. In addition to SPR, Mach–Zehnder interferometers and so-called grating couplers are also used for optical immunosensing.

Alternative transducer technologies that have been applied in immunosensor research include piezoelectric and electrochemical systems. *Piezoelectric* immunosensors [349], [350] (see Section 2.2.3) tend to suffer from significant levels of nonspecific binding to the piezoelectric substrate, which makes accurate analyte quantification difficult. However, this type of immunosensor may have an important role to play in the search for traces of volatile drugs in security monitoring, and it offers the advantages of relatively small size and low manufacturing cost. The main disadvantage from the standpoint of immunosensing is that reproducible measurements have so far been achieved only in the gas phase, whereas the Ab–Ag interaction is restricted to aqueous solution.

Potentiometric immunosensors [327] (see Section 2.2.1.1) are unfortunately associated with a low signal-to-noise ratio as a result of the low charge density on most biomolecules relative to such background interferences as ions. These sensors also show a marked dependence of signal response on sample conditions, including pH and ionic strength. Similarly, *immunoFET* devices (see Section 2.2.1.4), which measure very small changes in an electric field when an Ab–Ag binding reaction occurs, tend to suffer from practical problems associated with membrane performance and various artifacts.

Amperometric immunotechnology (immunoelectrodes, for example) has shown more promising results at the research prototype stage than other types of electrochemical systems. The preferred technique continues to be the indirect amperometric immunosensor approach described below, which uses enzymes as markers.

3.3.2. Indirect-Sensing Immunosensors using Marker Molecules

The most satisfactory marker molecules for use in this context have been fluorophores and enzymes. As regards the sensors themselves, several fundamental types exist.

Reactor immunosensors are based on a cartridge or microcolumn packed with appropriate immobilized antibodies (which serve as so-called catching antibodies). As in the case of enzyme immunoassays, the competition, sandwich, and replacement principles are all subject to exploitation. The amount of bound (or replaced) antigen is determined via the extent of enzyme activity remaining after substrate addition, which can be established with an electrochemical or optical sensor. For many applications (on-line measurements, coupling with flow-injection analysis), repeated and automated measurements are required. Because of the extremely high loading of catching antibodies, reactor immunosensors have the advantages of high sensitivity, regenerability, and reproducibility. On average they can be reused 50–100 times without loss of activity.

Membrane immunosensors employ antibodies immobilized directly at the surface of a sensor, or at a membrane covering the sensor, as shown in Figure 63. An example of the use of this type of immunosensor is illustrated schematically in Figure 64. The analyte to be detected is an early myocardial infarction marker, the fatty acid binding protein (FABP). A modified Clark electrode is covered by a membrane bearing immobilized monoclonal catching antibodies directed against FABP. After binding of the analyte and washing, a secondary polyclonal antibody is added to the solution. This secondary antibody is conjugated with glucose oxidase (GOD). The next step is incubation followed by further washing, after which glucose is added to the measuring solution. Glucose is oxidized by GOD in an amount directly proportional to the amount of FABP, yielding H_2O_2, the concentration of which is monitored with an amperometric sensor, as shown in Figure 65.

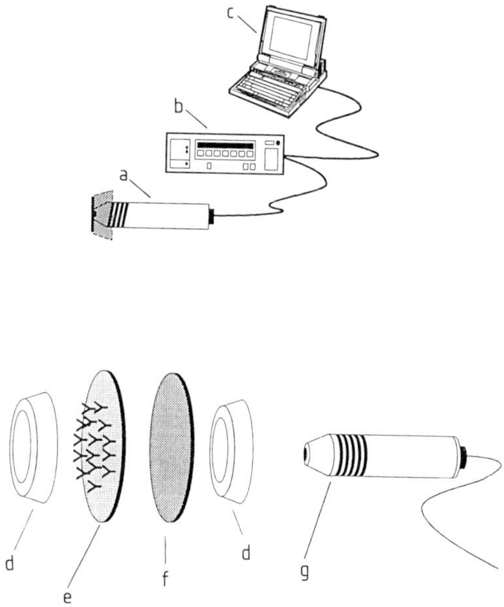

Figure 63. Design principle of an immunosensor system
a) Immunosensor; b) Electronic unit; c) Computing unit; d) O-Rings; e) Catching antibody membrane; f) Dialysis membrane; g) Pt vs. Ag/AgCl electrode

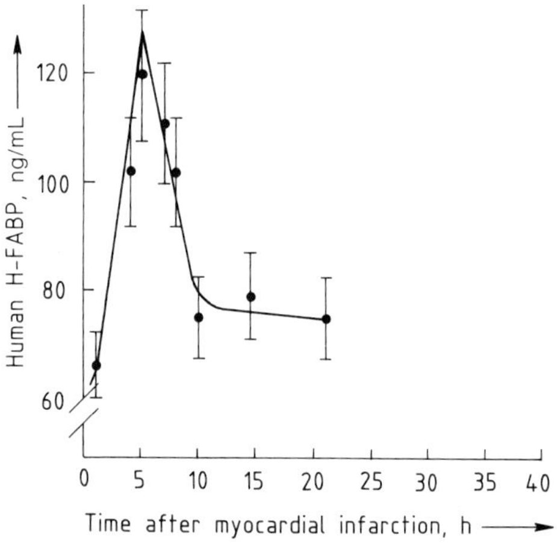

Figure 65. Analytical results obtained with the human heart FABP sensor described in Figure 64 as a function of time after onset of a myocardial infarction

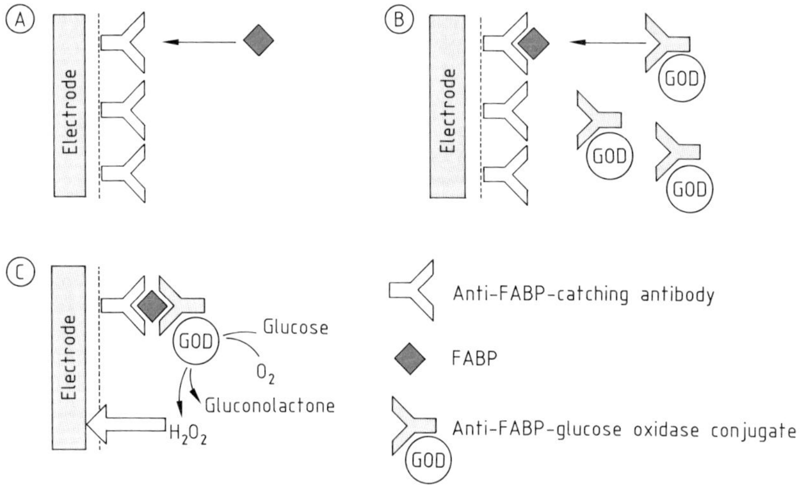

Figure 64. Schematic representation of the operation of an immunosensor for detecting the fatty acid binding protein (FABP)
A) Analyte binding step; B) Introduction of a secondary polyclonal antibody conjugated with glucose oxidase (GOD); C) Determination of GOD activity after addition of glucose

3.4. Whole-Cell Biosensors

Biological cells can be regarded as small membrane-bound structures containing a high concentration of chemicals, including enzymes, nucleic acids, ions, many types of protein, small organic molecules, and a host of others. A major challenge in exploiting whole (intact) cells in biosensor research is the problem of achieving selective transduction of a specific biochemical event or process within the context of this complex cellular chemistry [328], [341], [344], [345], [351] (Fig. 66).

The sensitivity and specificity of a particular cell with respect to a particular chemical varies enormously depending on the type, source, and environment of the cell and the precise nature of the chemical itself. It is not possible to provide any universal guidelines beyond such generalizations as the observation that most mammalian cells respond to a range of highly toxic chemicals, including cyanides. Even in this case it is known that hydrogen cyanide can interact with a number of different cellular components (such as electron transport proteins and enzymes) because of its small molecular size (which facilitates diffusion through biological membranes) and the ability of the cyanide anion to bind strongly to a wide variety of biomolecules.

The stability of biological cells is of critical importance in biosensor applications. In general, cells derived from higher organisms are rather difficult to isolate and maintain in a viable form for the extended periods of time required in most sensor formulations. The most biologically robust cells are derived from less complex organisms, particularly yeast, bacteria, and algae. Cells from these organisms can generally be stabilized in an in vitro environment more easily than complex cells; indeed, many of the organisms themselves are unicellular (yeast, bacteria, algae), leading to biosensors incorporating truly "living" biological components.

Natural cell differentiation and the growth characteristics of such whole-cell biosensors sometimes cause problems associated with signal drift over time. Nevertheless, there are potential advantages to be gained by exploiting whole cells in a sensor format, including:

1) Versatility (cells respond to a wide range of chemicals)
2) The presence in situ of biological structures supporting multicomponent or multistep biological reaction sequences useful for detection purposes and ensuring optimum enzyme activity (many enzymes lose activity when isolated in vitro)
3) Low preparation cost; i.e., growth in culture

For the reasons outlined above, nearly all whole-cell biosensors reported to date have employed relatively simple bacterial or algal cells

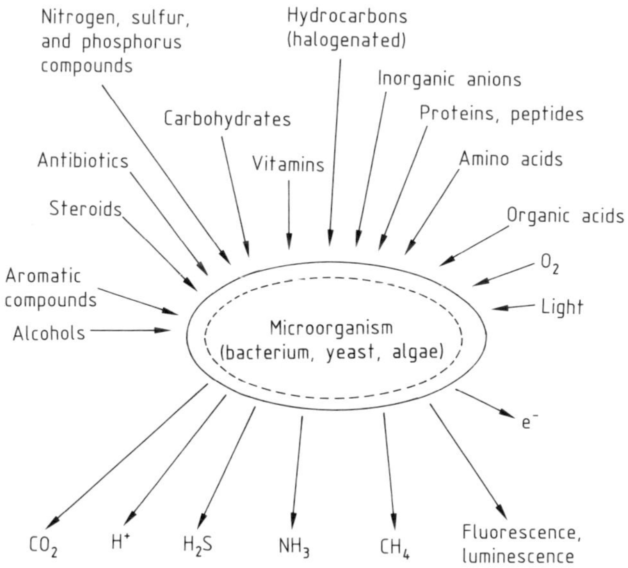

Figure 66. Microorganisms as receptor systems in biosensors useful for detecting specific groups of compounds

[343], [352], [353]. Existing whole-cell biosensor technology has also been based exclusively on electrochemical transduction of cellular responses.

The increasing interest in whole-cell biosensors since the first reports in 1975 [352] has led to the development of a large number of analyte-specific amperometric and potentiometric sensors. In these devices an immoblized layer of whole cells (entrapped in a gel, for example, or membrane-supported) is placed at the surface of an electrode. The cells then function as an enzyme source, where the enzyme of interest is present at a high relative concentration and is effectively able to function in its natural optimized environment. Bacterial cells in particular may contain very large amounts of particular enzymes relative to all others, especially if they are grown on a single source of carbon or nitrogen.

Rechnitz and Karube [352] have pioneered in the development of many bacterial electrodes based on this principle using various bacteria depending on the target analyte. Ion-selective electrodes for NH_3, O_2, CO_2, H_2S and H^+ have all been used in conjunction with immobilized whole cells. Riedel et al. [353] have recently demonstrated that preincubation of certain bacterial electrodes with the desired analyte (substrate) can enhance the sensitivity of a sensor toward that chemical by a factor of as much as 25. This induction approach may prove to be widely applicable. The shelf-life of a whole-cell electrochemical sensor can extend up to several weeks with fully optimized storage conditions (low temperature, for example). Microbe thermistors (sensors that respond to the heat evolved during bacterial metabolism of a substrate) have also been developed, but these present problems once again with respect to analyte specificity.

A significant advance in whole-cell biosensor technology accompanied the development by Rawson et al. [343] of a mediated electrochemical sensor employing a layer of blue-green algae (cyanobacteria). These authors demonstrated that a low molecular mass redox-active mediator, such as $Fe(CN)_6^{3-}$ or benzoquinone, could accept electrons from the photosynthetic energy chain of a cyanobacterium and relay them, via diffusion, to an electrode. The sensor proved capable of detecting aqueous herbicides on the basis of their inhibition of electron transport in the cells. This resulted in a dramatic fall in current, since fewer electrons reached the mediator and hence the electrode. The advantage of this approach to the use of whole cells is that it permits reasonably specific detection of a group of chemicals (herbicides, in this case) rather than limited-range detection of a single substance.

The principal application at the present time for cell sensors is in the measurement of biochemical oxygen demand (BOD) in wastewater. For this purpose yeast and/or bacterial cells (mainly the yeasts *Trichosporon* sp. or *Issatchenkia* and the bacterium *Rhodococcus erythropolis* [341], [351]) are immobilized on top of a Clark oxygen electrode and the oxygen consumption is measured. The great advantage of BOD sensors is a short measurement time (20 s to several minutes) compared with the classical BOD test in wastewater treatment, which requires five days. The main disadvantage is that a BOD sensor reflects only low molecular mass and easily assimilable substrates. High molecular mass substrates such as proteins and starch cannot penetrate the protective dialysis membrane of the sensor. Pretreatment with hydrolytic enzymes could solve this problem. In general, however, BOD sensors, represent a promising approach to the control of wastewater facilities.

3.5. Problems and Future Prospects

Compared with traditional sensors and classical methods of analysis, biosensors offer an attractive alternative for the rapid, selective, and highly sensitive assay of a wide variety of analytes. Further advantages of this cheap and economical tool emerge with the possibility of direct measurements within complex matrices, demanding only a rudimentary sample pretreatment, and the prospect of an uncomplicated measuring procedure that avoids the use of toxic reagents and organic solvents. Potential fields of application for biosensors are suggested in Table 16.

Current biosensor research is concerned with a variety of unsolved problems posing barriers to further expansion within the analytical market. Until recently, biosensors were available for only a few analytes (roughly ten). This is mainly a consequence of insufficient reliability associated with poor stability of the biomaterial, including a dependence on physical and chemical parameters in the environment and interferences within the transducers. In the case of implantable biosensors for in vivo monitoring of metabolites or feedback control of artificial organs (e.g., the pancreas), a lack of biocompatibility with the body environment, organic interferences, and inadequate stability of the biological receptor have so far prevented long-term implantation. Nevertheless, with respect to rapid, sensitive, and selective low-cost determination of a great variety of relevant compounds no suitable alternative to the principle of biosensors is apparent at this time.

Constructing biosensors with greater reliability requires a better understanding of the basic principles of enzyme catalysis and immunochemical reactions, protein structures, pathways for receptor-based signal amplification, and the interfacial behavior of biocompounds at the artificial transducer surface. Biosensor research must therefore be directed toward the following points:

1) Protein engineering, with the aim of enhancing the stability of enzymes, catalytic antibodies, and bifunctional enzymes
2) Screening stable enzymes from microorganisms (extremophiles)
3) Utilization of antibody fragments (Fab, Fab′), bispecific antibodies, and single-chain and single-domain antibodies (tailored through genetic engineering and its expression in bacteria)
4) Modification of the efficiency, specificity, and modulation of antibodies by site-directed mutagenesis
5) Use of DNA libraries for the production of monoclonal antibodies to meet specific demands of the desired transducer

Table 16. Potential fields of application for biosensors

Field of application	Biosensor opportunities
Medicine	patient self-control, home health care (glucose, lactate, creatinine, phenylalanine, histamine) in vivo analysis long-term in vivo control of metabolites and drugs control element (biotic sensor) for artifical prosthesis and organs rapid analysis at intensive-care units surface imaging of organs during implantation bedside monitoring
Clinical chemistry	diagnostics for metabolites, drugs, enzymes, vitamins, hormones, allergies, infectious diseases, cancer markers, pregnancy, etc. laboratory safety
Environmental protection	pollution control monitoring/screening of toxic compounds in water supplies, solid and liquid wastes, soil and air (e.g., pesticides, inorganic ions, explosives, oils, PAHs, PCBs, microorganisms, volatile vapors and gases) self control of industrial companies and farms alarm systems for signaling hazardous conditions determination of organic load (BOD)
Chemical, pharmaceutical, food industry	monitoring and control of fermentation processes and cell cultivation (substrates, metabolites, products) food quality control (screening/detection of microbial contaminations; estimation of freshness, shelf life; olfactory qualities and flavor; rancidity; analysis of fats, proteins, carbohydrates in food) study of efficiency of drugs detection of leakage and hazardous concentrations of liquids and gases in buildings and mine shafts indoor air quality checks location of oil deposits
Agriculture	quality control of soils estimation of degradation/rottage (e.g., of biodegradable waste, or in wood or plant storage) rapid determination of quality parameters of milk
Military	detection of chemical and biological warfare agents (e.g., nerve gases, pathogenic bacteria, viruses)

6) Selection and specification of biological receptor systems applicable to the quantification of olfaction and taste
7) Manipulating enzymes and cofactors by electrically wiring redox-active centers to transducer surfaces, utilizing the principles of both molecular electronics and biological electron transfer
8) Investigation of interfacial effects and interactions between biocompounds and transducer surfaces at the molecular and atomic levels by exploiting new methods of surface analysis (e.g., scanning microscopy techniques, fast-atomic-bombardment mass spectrometry, laser-assisted mass spectrometry, time-of-flight secondary ion mass spectrometry, Fourier-transform infrared spectrometry, ellipsometry, X-ray photoelectron spectroscopy, and electron microscopy)
9) Materials research, including biocompatible membranes (based on compounds found within the body) and redox-active electrically conducting (metal)organics and conjugated polymers
10) Development of (multifunctional) silicon-based ultramicroelectrodes covalently linked to biomaterials

Concerning methodological applications of biosensors, most of the recent developments relate to enzyme electrodes integrated into laboratory analyzers for the determination of low molecular mass compounds in the medical and biotechnological fields. A further important line of development is directed toward disposable biosensors for screening and semiquantitative analysis in environmental control, patient home diagnostics, and food-quality control. Direct testing would avoid the time-consuming procedures of sample storage and transfer as well as expensive laboratory analyses. A medium-term application is expected to be the use of biosensors in (quasi)continuously operating random-process analyzers based on micromachined fluidics elements, together with low-maintenance systems for the control and monitoring of water supplies and clinical bedside monitoring. Finally, a long-term target is the development of highly reliable implantable biosensors for long-term implantation. Biosensor applications in the field of instrumentation should also contribute to the solution of many important problems in conventional measuring techniques, opening major new areas to modern analysis.

4. Actuators and Instrumentation

The instrumentation associated with chemical sensors is rather simple and inexpensive, especially in the case of electrochemical transducers. All that is normally required is a high-ohmic voltmeter (pH meter), a conductometer, and/or polarographic equipment. Fiber-optical devices, on the other hand, can be quite expensive depending on quality requirements with respect to the monochromatic light source. Spectrophotometers with fiber-optical inputs and outputs are already commercially available. A laser source is required in certain cases of sensitive fluorescence measurements with small-volume samples. Exploiting the modern techniques of acoustic wave modulation also requires more than a simple frequency counter. For example, a network analyzer can be of great help in following the shift in a resonance curve caused by interaction of an analyte with the transducer surface.

The highest costs are usually related to development of a selective analyte recognition layer, perhaps the most important part of a complete chemical sensor. Transducer technology is much more highly developed; indeed, there seems to be little need here for further improvement without equivalent improvements in the selectivity of sensor elements.

One of the most important areas of current sensor research is microsystem technology (mst). A microsystem is defined as one too small to be fabricated by traditional miniaturized mechanical techniques, especially one that is dependent on modern mass-production techniques involving silicon or LIGA technology. A *system* must, by definition, consist of at least two interactive processes. A typical example is the antiblocking brake system (ABS) in an automobile, in which acceleration sensors are capable of detecting a blocked wheel and selectively controlling the brake power to that wheel via actuators.

Examples of analogous complete systems in chemistry are still rare. The best example is perhaps the "artificial pancreas," in which a glucose sensor measures current blood-glucose concentrations in order to control an appropriate actuator, the insulin pump. Bedside devices for this purpose have operated successfully for many years, but all attempts at implantation have so far failed, mainly because of the limited lifetime of the glucose biosensor.

A complete system providing both a sensor and an actuator would be ideal in the field of process control, but because of a lack of truly reliable chemical sensors on the market the concept has not been widely implemented. One exception relates to the analytical method of coulometry, a technique that offers great potential for delivering chemical compounds to a controlled reaction. Especially attractive in this context is the method of constant-current coulometry, which can be carried out with an end-point sensor and a coulometric actuator for maintaining a generator current until the end-point has been reached. In this case both of the required devices can be miniaturized and constructed with the same technology.

5. Future Trends and Outlook

Surveys that have been conducted on the subject of chemical sensors all predict a growing market during the late 1990s and the early years of the 21st century [354]–[356]. Anticipated developments in automation, environmental protection, health care, and high technology ensure a strong demand for sensors. However, devices of this type must become increasingly reliable and even simpler to use. Japanese researchers see the first area for mass application of sensors to be the health care industry, especially with respect to the elderly. The demographic prospect of an increasing number of retirees and a lack of sufficient care centers may trigger the use of many sensors for monitoring the health of people living for as long as possible in their own homes.

With respect to environmental chemistry, controls over the emission of organic compounds will undoubtedly be strengthened in the future. For example, municipal incinerators without provisions for the effective measurement of all hazardous organic compounds will soon be strictly forbidden everywhere. It seems likely that a large fraction of the moneys previously spent for defense purposes will be redirected toward improvements in continuous measurement technology. Sensors will be required to compete in this respect with such fast separation methods as capillary electrophoresis, which permits the separation of most compounds within ca. 10 min.

One critical remark is in order regarding papers published in the field of chemical sensors. Many such reports reflect a lack of input from qualified analytical chemists during the development and testing stages. Insufficient understanding is often apparent with respect to the analytical process itself and the difficulties involved in obtaining a truly reliable result; indeed, many sensor researchers seem to be unaware of such basic problems as sample–matrix effects. Anyone who claims to have developed a new sensor should be required to demonstrate its effectiveness with a real sample under real conditions! Publications that fail to reveal the most important features of a sensor are also essentially worthless. Chemical sensors are analytical devices, so they should be subject to the same standards of characterization as other instrumental tools. Quality-control considerations mandate critical evaluation of any new molecular sensor presented to the scientific community. Every report concerning a new sensor development should clearly delineate the crucial features, with none omitted.

The most important parameters for characterizing a chemical sensor or a biosensor include:

1) **Selectivity.** The selectivity with respect to important known interferents for the particular application in question should always be expressed numerically; expressions like "relative selectivity" toward the analyte, or "not affected by typical concentrations of interfering compounds" are inappropriate.
2) **Sensitivity and Limit of Detection (LOD).** The slope of a calibration curve expressed in the appropriate units (e.g. microamps, millivolts, or hertz) per millimole or micromole should always be provided both for a matrix-free solution and for a sample in its matrix. The corresponding LOD should also be reported in both cases.
3) **Stability of the Analytical Function (Drift).** A zero-point drift value should be provided for a sensor in a matrix-free solution and also with respect to real samples involving a matrix. The same applies to the stability of the slope (varying sensor sensitivity).
4) **Response Time.** Since response time is also a function of the sample matrix it should be measured both in pure calibration solutions and with real samples. Generally two different response times are observed, one for a concentration rise and one for decreasing analyte concentration, with the latter usually slightly greater. Since IUPAC has not yet recommended a specific procedure for measuring this parameter, one should express the response time in the form of a time constant

(= 1/e of the final value), since this is the approach taken with electronic circuits. Only in the case of a sensor that never reaches its final value should other approaches be considered for characterizing the rate of response to a changing analyte concentration.

5) **Sensor Lifetime.** Especially in the field of biosensors the lifetime of a sensor in contact with a calibration solution or a real sample, or during storage in a refrigerator with an optimal conditioning solution, is of considerable interest, and it should always be mentioned. Conditions leading to rapid sensor fouling should also be identified in order to permit appropriate precautions to be taken. Examples include the fouling of certain enzymes by inhibitors or heavy metals, or microbiological attack. Any use of conditioning solutions containing metal complexation reagents, inhibitor scavengers, sodium azide, etc., should be clearly described.

Other important information might include total power consumption, auxiliary equipment costs, total elapsed time between failures, temperature compensation measures, or maintenance requirements.

Finally, sensor developers must remain aware of progress within instrumental analytical chemistry generally, especially in the area of fast high-resolution chromatography. Indeed, progress has already been so great in the latter field that more than 20 anions (including phosphate) can now be separated completely in less time than would be required for a response from a multienzyme phosphate biosensor! The serious challenge this poses to sensor advocates will be obvious.

6. References

[1] W. Göpel, J. Hesse, J. N. Zemel (eds.): *Sensors – A Comprehensive Survey*, vols. 1–7, VCH Verlagsgesellschaft, Weinheim 1989–1991.
[2] R. W. Murray et al.: "Chemical Sensors and Microinstrumentation," *ACS Symp.* Ser. **403** (1989).
[3] J. Janata: *Principles of Chemical Sensors*, Plenum Press, New York 1989.
[4] K. Cammann: *Das Arbeiten mit ionenselektiven Elektroden*, 2nd ed., Springer Verlag, Berlin 1977.
[5] J. Janata, R. Huber: *Solid State Chemical Sensors*, Academic Press, New York 1985.
[6] T. Takeuchi, *Sens. Actuators* **14** (1988) 109.
[7] G. Hötzel, H. M. Wiedemann, *Sens. Rep.* **4** (1989) 32.
[8] J. P. Pohl, *GIT Fachz. Lab.* **5** (1987) 379.
[9] K. Cammann et al., *Angew. Chem.* **103** (1991) 519; *Angew. Chem. Int. Ed. Engl.* **30** (1991) 516–539.
[10] W. Masing: *Handbuch der Qualitätssicherung*, Hanser Verlag, München 1988.
[11] K. Cammann: *Working with Ion-Selective Electrodes*, Springer Verlag, Berlin 1979.
[12] A. K. Corington: *Ion Selective Electrode Methodology*, vols. I and II, CRC Press, Boca Raton, Fla., 1979.
[13] H. Freiser (ed.): *Ion Selective Electrodes in Analytical Chemistry*, Plenum Press, New York 1979.
[14] N. Taguchi, JP 4538200, 1962.
[15] T. Seiyama, A. Kato, K. Fujiishi, M. Nagatani, *Anal. Chem.* **34** (1962) 1502.
[16] Y. Nakatani, M. Sakai, M. Matsuoka in T. Seyama et al. (eds.): *Proc. Int. Meet. Chem. Sensors (Fukuoka, Japan)*, Elsevier, Amsterdam, Sept. 1983, pp. 147–152.
[17] S. R. Morrison: "Semiconductor Gas Sensors," *Sens. Actuators* **2** (1982) 329.
[18] P. J. Shaver, *Appl. Phys. Lett.* **11** (1967) 255.
[19] J. C. Loh, JP 4328560; FR 1545292, 1967.
[20] N. Yamazoe, Y. Krakowa, T. Seiyama: "Effects of Additives on Semiconductor Gas Sensors," *Sens. Actuators* **4** (1983) 283.
[21] S. Matsushima, Y. Teraoka, N. Miura, N. Yamazoe, *Jpn. J. Appl. Phys.* **27** (1988) 1798.
[22] M. Nitta, S. Kanefusa, S. Ohtani, M. Haradome, *J. Electr. Mater.* **13** (1984) 15.
[23] A. Chiba: "Development of the TGS Gas Sensor," in S. Yamauchi (ed.): *Chemical Sensor Technology*, vol. 4, Elsevier, Amsterdam 1992, pp. 1–18.
[24] T. A. Jones: "Characterisation of Semiconductor Gas Sensors," in P. T. Moseley, B. C. Tofield (eds.): *Solid State Gas Sensors*, Adam Hilger, Bristol 1987.
[25] G. Heiland, D. Kohl, *Sens. Actuators* **8** (1985) 227.
[26] W. Weppner: "Halbleiter-Sensoren" in R. Grabowski (ed.): *Sensoren und Aktoren*, VDE-Verlag, Berlin 1991, pp. 167–183.
[27] D. E. Williams: "Characterization of Semiconductor Gas Sensors," in P. T. Moseley, B. C. Tofield (eds.): *Solid State Gas Sensors*, Adam Hilger, Bristol 1987.
[28] N. Yamazoe, N. Miura: "Some Basic Aspects of Semiconductor Gas Sensors," in S. Yamauchi (ed.): *Chemical Sensor Technology*, vol. 4, Elsevier, Amsterdam 1992, pp. 19–41.
[29] G. Heiland: "Homogeneous Semiconducting Gas Sensors," *Sens. Actuators* **2** (1982) 343.
[30] H. Windischmann, P. Mark, *J. Electrochem. Soc.* **126** (1979) 627.
[31] S. Chang, *J. Vac. Sci. Technol.* **17** (1980) 366.
[32] J. Watson: "The Tin Dioxide Gas Sensor and Its Applications," *Sens. Actuators* **5** (1984) 29.
[33] N. Komori, S. Sakai, K. Konatsu in T. Seyama et al. (eds.): *Proc. Int. Meet. Chem. Sensors (Fukuoka, Japan)*, Elsevier, Amsterdam, Sept. 1983, p. 57.
[34] M. L. Hitchmann, H. A. O. Hill, *Chem. Br.* **22** (1986) 1177.
[35] D. Ammann: *Ion Selective Microelectrodes – Principles, Design and Application*, Springer Verlag, Berlin 1986.
[36] P. L. Bailey: *Analysis with Ion-Selective Electrodes*, Heyden & Sons, London 1976.
[37] R. Bock: "Nachweis- und Bestimmungsmethoden," in *Methoden der Analytischen Chemie*, vol. 2, part 2, Verlag Chemie, Weinheim 1984, pp. 33–83.

[38] K. Cammann: "Fehlerquellen bei Messungen mit ionenselektiven Elektroden," in *Analytiker-Taschenbuch*, vol. 1, Springer Verlag, Berlin 1980.
[39] R. A. Durst: "Ion-Selective Electrodes," *NBS Special Publication No. 314*, 1969.
[40] S. Ebel, W. Parzefall: *Experimentelle Einführung in die Potentiometrie*, Verlag Chemie, Weinheim 1975.
[41] G. Eisenman: *Glass Electrodes for Hydrogen and Other Cations, Principles and Practice*, Marcel Dekker, New York 1967.
[42] D. J. G. Ives, J. Janz: *Reference Electrodes*, Academic Press, New York 1961.
[43] R. Koryta: *Ion Selective Electrodes*, Cambridge University Press, Cambridge 1975.
[44] T. S. Ma, S. S. M. Hassan: *Organic Analysis Using Ion-Selective Electrodes*, 2 vols., Academic Press, New York 1982.
[45] G. J. Moody, J. D. R. Thomas: *Selective Ion Electrodes*, Merrow Publishing, Watford 1977.
[46] F. Oehme: *Ionenselektive Elektroden: CHEMFETs – ISFETs – pH-FETs, Grundlagen, Bauformen und Anwendungen*, Hüthig Verlag, Heidelberg 1991.
[47] E. Pungor: *Ion-Selective Electrodes*, Akadémiai Kiadó, Budapest 1973.
[48] K. Umezawa, Y. Umezawa: *Selectivity Coefficients for Ion Selective Electrodes*, University of Tokyo Press, Tokyo 1983.
[49] A. J. Bard, L. R. Faulkner: *Electrochemical Methods*, J. Wiley, New York 1980.
[50] A. M. Bond, H. B. Greenhill, I. D. Heritage, J. B. Reust, *Anal. Chim. Acta* **165** (1984) 209.
[51] A. J. Bard: *Electroanalytical Chemistry*, vol. 9, Marcel Dekker, New York 1976.
[52] G. Henze, R. Neeb: *Elektrochemische Analytik*, Springer Verlag, Berlin 1985.
[53] K. Cammann: "Elektrochemische Verfahren," in H. Naumer, W. Heller (eds.): *Untersuchungsmethoden in der Chemie*, Thieme Verlag, Stuttgart 1986.
[54] P. T. Kissinger, W. R. Heinemann, *J. Chem. Educ.* **60** (1983) 702.
[55] J. Osteryoung, J. J. O'Dea in A. J. Bard (ed.): *Electroanalytical Chemistry*, vol. 14, Marcel Dekker, New York 1986, pp. 209–308.
[56] P. Vadgama, G. Davis, *Med. Lab. Sci.* **42** (1985) 333.
[57] L. C. Clark, US 2913386, 1958.
[58] H. Suzuki et al., *Sens. Actuators B* **2** (1990) 297.
[59] S. C. Cha, M. J. Shao, C. C. Liu, *Sens. Actuators B* **2** (1990) 239.
[60] W. Sansen et al., *Sens. Actuators B* **1** (1990) 298.
[61] H. Suzuki et al., *Sens. Actuators B* **1** (1990) 528.
[62] H. Suzuki et al., *Sens. Actuators B* **1** (1990) 275.
[63] D. S. Austin et al. *J. Electroanal. Chem. Interfacial Electrochem.* **168** (1984) 227.
[64] R. M. Wightman, *Anal. Chem.* **53** (1981) 1125.
[65] T. Hepel, J. Osteryoung, *J. Electrochem. Soc.* **133** (1986) 752.
[66] L. Zhaohui et al., *J. Electroanal. Chem. Interfacial Electrochem.* **259** (1989) 39.
[67] C. L. Colyer, J. C. Myland, K. B. Oldham, *J. Electroanal. Chem. Interfacial Electrochem.* **263** (1989) 1.
[68] J. Osteryoung, S. T. Singleton, *Anal. Chem.* **61** (1989) 1211.
[69] R. E. Howard, E. Hu: *Science and Technology of Microfabrication*, Materials Research Society, Pittsburgh 1987.
[70] C. D. Baer, N. J. Stone, D. A. Sweigart, *Anal. Chem.* **58** (1988) 78.
[71] J. Ghoroghchiana et al., *Anal. Chem.* **58** (1988) 2278.
[72] W. Göpel, J. Hesse, J. N. Zemel: *Sensors*, vol. 2, VCH Verlagsgesellschaft, Weinheim 1991.
[73] W. Göpel, J. Hesse, J. N. Zemel: *Sensors*, vol. 3, VCH Verlagsgesellschaft, Weinheim 1991.
[74] P. T. Moseley, B. C. Tofield: *Solid State Gas Sensors*, Adam Hilger, Bristol 1987.
[75] P. T. Moseley, J. O. W. Norris, D. E. Williams: *Techniques and Mechanisms in Gas Sensing*, Adam Hilger, Bristol 1991.
[76] H. Emons, G. Jokuszies, *Z. Chem.* **28** (1988) 197.
[77] R. Kalvoda, *Electroanalysis* **2** (1990) 341.
[78] F. Opekar, A. Trojanek, *Anal. Chim. Acta* **203** (1987) 1.
[79] ISO/TC 147 (draft): *Electrolytic Conductivity*, Beuth Verlag, Berlin.
[80] DIN 38404: Deutsche Verfahren zur Wasser-, Abwasser- und Schlammuntersuchung, VCH Verlagsgesellschaft, Weinheim 1993.
[81] F. Oehme in W. Göpel, J. Hesse, J. N. Zemel (eds.): *Sensors*, vol. 2, VCH Verlagsgesellschaft, Weinheim 1991.
[82] F. Oehme, R. Bänninger: *ABC der Konduktometrie*, Polymetron AG, Mönchaltorf (Switzerland).
[83] K. Rommel, *VGB Kraftwerkstech.* **65** (1985) 417.
[84] K. Rommel: *Kleine Leitfähigkeitsfibel*, Wissenschaftlich-Technische Werke, Weilheim 1988.
[85] F. Oehme, M. Jola, *Betriebsmeßtechnik*, Hüthig Verlag, Heidelberg 1982.
[86] W. H. Brattain, J. Bardeen, *Bell Syst. Tech. J.* **32** (1952) 1.
[87] G. Heiland, *Z. Phys.* **138** (1954) 459.
[88] J. Watson, *Sens. Actuators B***8** (1992) 173.
[89] N. Taguchi, US 3644795, 1972.
[90] D.-D. Lee, D.-H. Choi, *Sens. Actuators B***1** (1990) 231.
[91] B. C. Tofield in P. T. Moseley, B. C. Tofield (eds.): *Solid State Gas Sensors*, Adam Hilger, Bristol 1987.
[92] R. Lalauze, C. Pijolat, S. Vincent, L. Bruno, *Sens. Actuators B***8** (1992) 237.
[93] J. Watson, *Sens. Actuators* **5** (1984) 29.
[94] R. M. Geatches, A. V. Chadwick, J. D. Wright, *Sens. Actuators B***4** (1991) 467.
[95] S. R. Morrison, *Sens. Actuators* **12** (1987) 425.
[96] S. Matsushima et al., *Sens. Actuators B***9** (1992) 71.
[97] T. Maekawa et al., *Sens. Actuators B***9** (1992) 63.
[98] S. Nakata, H. Nakamura, K. Yoshikawa, *Sens. Actuators B***8** (1992) 187.
[99] U. Kirner et al., *Sens. Actuators B***1** (1990) 103.
[100] N. Li, T.-C. Tan, *Sens. Actuators B***9** (1992) 91.
[101] N. Koshizaki, K. Yasumoto, K. Suga, *Sens. Actuators B***9** (1992) 17.
[102] M. Egashira, Y. Shimizu, Y. Takao, *Sens. Actuators B***1** (1990) 108.
[103] P. T. Moseley, D. E. Williams, *Sens. Actuators B***1** (1990) 113.
[104] E. Ye. Gutman, I. A. Myasnikov, *Sens. Actuators B***1** (1990) 210.
[105] R. S. Tieman, W. R. Heineman, J. Johnson, R. Seguin, *Sens. Actuators B***8** (1992) 199.
[106] Y. Sadaoka, T. A. Jones, W. Göpel, *Sens. Actuators B***1** (1990) 148.
[107] S. Dogo, J. P. Germain, C. Maleysson, A. Pauly, *Sens. Actuators B***8** (1992) 257.
[108] J. W. Gardner, M. Z. Iskandarani, B. Bott, *Sens. Actuators B***9** (1992) 133.

[109] M. Trometer et al., *Sens. Actuators B*8 (1992) 129.
[110] J. D. Wright, *Prog. Surf. Sci.* **31** (1989) 1.
[111] T. A. Temofonte, K. F. Schoch, *J. Appl. Phys.* **65** (1989) 1350.
[112] S. Kanefusa, M. Nitta, *Sens. Actuators B*9 (1992) 85.
[113] W. Göpel, K. D. Schierbaum, D. Schmeisser, D. Wiemhöfer, *Sens. Actuators* **17** (1989) 377.
[114] B. Bott, S. C. Thorpe in P. T. Moseley, J. O. W. Norris, D. E. Williams (eds.): *Techniques and Mechanisms in Gas Sensing*, Adam Hilger, Bristol 1991.
[115] T. Hanawa, S. Kuwabata, H. Yoneyama, *J. Chem. Soc. Faraday Trans. 1* **84** (1988) 1587.
[116] C. Nylander, M. Armgarth, I. Lundström, *Anal. Chem. Symp. Ser.* **17** (1983) 203.
[117] P. N. Bartlett, P. B. M. Archer, S. K. Ling-Chung, *Sens. Actuators* **19** (1989) 125.
[118] P. N. Bartlett, P. B. M. Archer, S. K. Ling-Chung, *Sens. Actuators* **19** (1989) 141.
[119] H. Arai, T. Seiyama: "Humidity Control," in W. Göpel, J. Hesse, J. N. Zemel (eds.): *Sensors*, vol. 2, VCH Verlagsgesellschaft, Weinheim 1991.
[120] A. K. Michell in P. T. Moseley, J. O. W. Norris, D. E. Williams (eds.): *Techniques and Mechanisms in Gas Sensing*, Adam Hilger, Bristol 1991.
[121] T. Nenov, S. Yordanov, *Sens. Actuators B*8 (1992) 117.
[122] Y. Sakai, Y. Sadaoka, H. Fukumoto, *Sens. Actuators* **13** (1988) 243.
[123] P. Bergveld, *IEEE Trans. Biomed. Eng.* **BME-17** (1970) 70.
[124] J. E. Moneyron, A. De Roy, J. P. Besse, *Solid State Ionics* **46** (1991) 175.
[125] T. Matsuo, M. Esashi, K. Inuma, *Proc. Dig. Joint Meeting*, Tohoku Sect. IEEEJ, Oct. 1971.
[126] S. M. Sze: *Physics of Semiconductor Devices*, Wiley, New York 1981.
[127] P. Bergveld, A. Sibbald: *Analytical and Biomedical Applications of Ion Sensitive Field Effect Transistors*, Elsevier, Amsterdam 1988.
[128] Y. G. Vlasov, D. E. Hackleman, R. P. Buck, *Anal. Chem.* **51** (1979) 1579.
[129] W. Moritz, I. Meierhöffer, L. Müller, *Sens. Actuators* **15** (1988) 211.
[130] S. D. Moss, J. Janata, C. C. Johnson, *Anal. Chem.* **47** (1975) 2238.
[131] A. Sibbald, P. D. Whalley, A. K. Covington, *Anal. Chim. Acta* **159** (1984) 47.
[132] D. J. Harrison et al., *J. Electrochem. Soc.* **135** (1988) 2473.
[133] P. D. van der Wal et al., *Anal. Chim. Acta* **231** (1990) 41.
[134] A. van den Berg, A. Griesel, E. Verney-Norberg, *Sens. Actuators B*4 (1991) 235.
[135] R. W. Catrall, P. I. Iles, J. C. Hamilton, *Anal. Chim. Acta* **169** (1985) 403.
[136] R. Smith, D. C. Scott, *IEEE Trans. Biomed. Eng.* **BME 33** (1986) 83.
[137] A. van den Berg, P. Bergveld, D. N. Reinhoudt, E. J. R. Sudhölter, *Sens. Actuators* **8** (1985) 129.
[138] N. J. Ho, J. Kratochvil, G. F. Blackburn, J. Janata, *Sens. Actuators* **4** (1983) 413.
[139] C. Dumschat et al., *Sens. Actuators B*2 (1990) 271.
[140] K. Domansky, J. Janata, M. Josowicz, D. Petelenz, *Analyst (London)* **118** (1993) 335.
[141] H. H. van den Vlekkert et al., *Sens. Actuators* **14** (1988) 165.
[142] C. Dumschat et al., *Anal. Chim. Acta* **243** (1991) 179.
[143] P. Bergveld, *Sens. Actuators B*4 (1991) 125.
[144] W. F. Love, L. J. Button, R. E. Slovacek: "Optical Characteristics of Fiberoptic Evanescent Wave Sensors," in D. L. Wise, L. B. Wingard (eds.): *Biosensors with Fiberoptics*, Humana Press, Clifton, N.J., 1991.
[145] S. J. Lackie, T. R. Glass, M. J. Block: "Instrumentation for Cylindrical Waveguide Evanescent Fluorosensors," in D. L. Wise, L. B. Wingard (eds.): *Biosensors with Fiberoptics*, Humana Press, Clifton, N.J., 1991.
[146] I. M. Walczak, W. F. Love, T. A. Cook, R. E. Slovacek: "The Application of Evanescent Wave Sensing to a High-sensitivity Fluoroimmunoassay," *Biosensors Bioelectron.* **7** (1992) 39–47.
[147] B. J. Tromberg, M. J. Sepaniak, T. Vo-Dinh, G. D. Griffin: "Fiberoptic Probe for Competitive Binding Fluoroimmunoassay," *Anal. Chem.* **59** (1987) 1226–1230.
[148] R. B. Thompson, F. S. Ligler: "Chemistry and Technology of Evanescent Wave Biosensors," in D. L. Wise, L. B. Wingard (eds.): *Biosensors with Fiberoptics*, Humana Press, Clifton, N.J., 1991.
[149] W. Trettnak, P. J. M. Leiner, O. S. Wolfbeis: "Fiberoptic Glucose Biosensor with an Oxygen Optrode as the Transducer," *Analyst (London)* **113** (1988) 1519–1523.
[150] I. J. Peterson, S. R. Goldstein, R. V. Fitzgerald, D. K. Buckhold: "Fiberoptic pH Probe for Physiological Use," *Anal. Chem.* **52** (1980) 864–869.
[151] G. F. Kirkbright, R. Narayanaswamy, N. A. Welti: "Studies with Immobilized Chemical Reagents Using a Flow-cell for the Development of Chemically Sensitive Fiberoptic Devices," *Analyst (London)* **109** (1984) 12–19.
[152] L. A. Saari, W. R. Seitz: "pH Sensor Based on Immobilized Fluoresceinamine," *Anal. Chem.* **54** (1982) 821–823.
[153] Z. Zhujun, W. R. Seitz: "A Fluorescence Sensor for Quantifying pH in the Range from 6.5 to 8.5," *Anal. Chim. Acta* **160** (1984) 47–55.
[154] E. Urbano, M. Offenbacher, O. S. Wolfbeis: "Optical Sensor for Continuous Determination of Halides," *Anal. Chem.* **56** (1984) 427–430.
[155] O. S. Wolfbeis, E. Urbano: "Fluorescence Quenching Method for Determination of Two or Three Components in Solution," *Anal. Chem.* **55** (1983) 1904–1907.
[156] L. A. Saari, W. R. Seitz: "Immobilized Morin as Fluorescense Sensor for Determination of Aluminum(III)," *Anal. Chem.* **55** (1983) 667–670.
[157] L. A. Saari, W. R. Seitz: "Optical Sensor for Beryllium Based on Immobilized Morin Fluorescence," *Analyst (London)* **109** (1984) 655–657.
[158] Z. Zhujun, W. R. Seitz: "A Fluorescent Sensor for Aluminum(III), Magnesium(II), Zinc(II) and Cadmium(II) Based on Electrostatically Immobilized Quinolin-8-ol-sulfonate," *Anal. Chim. Acta* **171** (1985) 251–258.
[159] Z. Zhujun, W. R. Seitz: "A Carbon Dioxide Sensor Based on Fluorescence," *Anal. Chim. Acta* **170** (1984) 209–216.
[160] M. A. Arnold, T. J. Ostler: "Fiberoptic Ammonia Gas Sensing Probe," *Anal. Chem.* **58** (1986) 1137–1140.

[161] A. Sharma, O. S. Wolfbeis: "Fiberoptic Fluorosensor for Sulfur Dioxide Based on Energy Transfer and Exciplex Quenching," *Proc. SPIE Int. Soc. Opt. Eng.* **990** (1989) 116–120.
[162] A. Sharma, O. S. Wolfbeis: "Fiberoptic Oxygen Sensor Based on Fluorescence Quenching and Energy Transfer," *Appl. Spectrosc.* **42** (1988) 1009–1011.
[163] O. S. Wolfbeis, M. J. P. Leiner, H. E. Posch: "A New Sensing Material for Optical Oxygen Measurement, with Indicator Embedded in an Aqueous Phase," *Mikrochim. Acta* **1986**, no. 3, 359–366.
[164] K. W. Berndt, J. R. Lakowiz: "Electroluminescent Lamp-Based Phase Fluorometer and Oxygen Sensor," *Anal. Biochem.* **201** (1992) 319–325.
[165] "Integrated Optics" in T. Tamir (ed.): *Topics in Applied Physics*, Springer Verlag, Berlin 1975.
[166] K. Tiefenthaler, W. Lukosz: "Sensitivity of Grating Couplers as Integrated-optical Chemical Sensors," *J. Opt. Soc. Am. B Opt. Phys.* **B6** (1975) 209–220.
[167] A. Brandenburg, R. Edelhauser, F. Hutter: "Integrated Optical Gas Sensors Using Organically Modified Silicates as Sensitive Films," *Sens. Actuators B* **11** (1993) 361–374.
[168] D. Schlatter et al.: "The Difference Interferometer: Application as a Direct Immunosensor," in: *The Second World Congress on Biosensors*, Oxford, Elsevier Advanced Technology, 1992, pp. 347–355.
[169] Ph. Nellen, K. Tiefenthaler, W. Lukosz: "Integrated Optical Input Grating Couplers As Biochemical Sensors," *Sens. Actuators B* **15** (1988) 285–295.
[170] R. G. Heidemann, R. P. H. Kooyman, J. Greve: "Performance of a Highly Sensitive Optical Waveguide Mach–Zehnder Interferometer Immunosensor," *Sens. Actuators B* **10** (1993) 209–217.
[171] N. Fabricius, G. Gaulitz, J. Ingenhoff: "A Gas Sensor Based on an Integrated Optical Mach–Zehnder Interferometer," *Sens. Actuators B* **7** (1992) 672–676.
[172] Ch. Fattinger, H. Koller, P. Wehrli, W. Lukosz: "The Difference Interferometer: A Highly Sensitive Optical Probe for Molecular Surface-Coverage Detection," in: *The Second World Congress on Biosensors*, Oxford, Elsevier Advanced Technology, 1992, pp. 339–346.
[173] K. Tiefenthaler: "Integrated Optical Couplers As Chemical Waveguide Sensors," *Adv. Biosensors* **2** (1992) 261–289.
[174] B. Liedberg, C. Nylander, I. Lundström: "Surface Plasmon Resonance for Gas Detection and Biosensing," *Sens. Actuators* **4** (1983) 299–304.
[175] P. B. Daniels, J. K. Deacon, M. J. Eddowes, D. Pedley: "Surface Plasmon Resonance Applied to Immunosensing," *Sens. Actuators* **14** (1988) 11–17.
[176] J. Gent et al.: "Optimization of a Chemooptical Surface Plasmon Resonance Based Sensor," *Appl. Opt.* **29** (1990) 2343–2849.
[177] K. Matsubaru, S. Kawata, S. Minami: "Optical Chemical Sensor Based on Surface Plasmon Measurement," *Appl. Opt.* **27** (1988) 1160–1163.
[178] H. Rather: *Surface Plasmons*, Springer Verlag, Berlin 1988.
[179] G. Sauerbrey, *Z. Phys.* **155** (1959) 206–222.
[180] H. K. Pulker, J. P. Decostered in C. Lu, A. W. Czanderna (eds.): "Applications of Piezoelectric Quartz Crystal Microbalances," *Methods and Phenomena*, vol. 7, Elsevier, Amsterdam 1984, Chap. 3.
[181] Lord Rayleigh, *Proc. London Math. Soc.* **17** (1885) 4–11.
[182] G. Sauerbrey, *Z. Phys.* **178** (1964) 457.
[183] W. H. King, Jr., *Anal. Chem.* **36** (1964) 1735–1739.
[184] W. H. King, Jr., US 3164004, 1965.
[185] A. W. Czanderna, C. Lu: "Applications of Piezoelectric Quartz Crystal Microbalances," in C. Lu, A. W. Czanderna (eds.): *Methods and Phenomena*, vol. 7, Elsevier, Amsterdam 1984. chap. I.
[186] M. Thompson et al. *Analyst* (*London*) **116** (1991) 881–890.
[187] L. Wimmer, S. Hertl, J. Hemetsberger, E. Benes, *Rev. Sci. Instrum.* **55** (1984) 605–609.
[188] C.-S. Lu, O. Lewis, *J. Appl. Phys.* **43** (1972) 4385–4390.
[189] J. G. Miller, D. I. Bolef, *J. Appl. Phys.* **39** (1968) 5815–5816.
[190] E. Benes, *J. Appl. Phys.* **56** (1984) 608–626.
[191] D. R. Denison, *J. Vac. Sci. Technol.* **10** (1973) 126–129.
[192] C.-S. Lu, *J. Vac. Sci. Technol.* **12** (1975) 578–583.
[193] G. G. Guilbault: "Applications of Piezoelectric Quartz Crystal Microbalances," in C. Lu, A. W. Czanderna (eds.): *Methods and Phenomena*, vol. 7, Elsevier, Amsterdam 1984, chap. 8.
[194] H. Beitnes, K. Schrøder, *Anal. Chim. Acta* **158** (1984) 57–65.
[195] A. Kindlund, H. Sundgren, I. Lundström, *Sens. Actuators* **6** (1984) 1–17.
[196] G. G. Guilbault: *Ion-Selective Electrode Review*, vol. 2, Pergamon Press, Oxford 1980, pp. 3–16.
[197] R. M. Langdon: "Resonator Sensors—a Review," *J. Phys. E* **18** (1985) 103–115.
[198] J. Tichy, G. Gautschi: *Piezoelektrische Meßtechnik*, Springer Verlag, Berlin 1980.
[199] H. Wohltjen, R. Dessy, *Anal. Chem.* **51** (1979) 1458–1464.
[200] H. Wohltjen, R. Dessy, *Anal. Chem.* **51** (1979) 1465–1470.
[201] H. Wohltjen, R. Dessy, *Anal. Chem.* **51** (1979) 1470–1478.
[202] C. T. Chang, R. M. White: "Excitation and Propagation of Plate Mode in an Acoustically Thin Membrane," *Proc. IEEE Ultrasonics Symp.* **1982**, 295–298.
[203] R. M. White, P. W. Wicher, S. W. Wenzel, E. T. Zellers: "Plate-Mode Ultrasonic Oscillator Sensors," *IEEE Trans. Ultrason. Dev. Ferroelectr. Freq. Contr.* **UFFC-34** (1987) 162–171.
[204] T. G. Giesler, J.-U. Meyer: *Fachbeilage Mikroperipherik*, vol. 6 (1992) XLVI.
[205] IEEE Standard on Piezoelectricity (ANSI/IEEE Std. 176-1987), The Institute of Electrical and Electronics Engineers, New York 1987.
[206] M. S. Nieuwenhuizen, A. Venema: *Sensors–A Comprehensive Survey*, vol. II, VCH Verlagsgesellschaft, Weinheim 1991, p. 652.
[207] J. Janata: *Principles of Chemical Sensors*, Plenum Press, New York 1989, pp. 55–80.
[208] J. G. Miller, D. I. Bolef, *J. Appl. Phys.* **39** (1968) 45.
[209] K. H. Behrndt, *J. Vac. Sci. Technol.* **8** (1971) 622.
[210] A. P. M. Glassford in A. M. Smith (ed.): *Progress in Astronautics and Aeronautics*, vol. 56, American Institute of Aeronautics and Astronautics, New York 1977, p. 175.
[211] A. P. M. Glassford, *J. Vac. Sci. Technol.* **15** (1978) 1836.
[212] V. Mecea, R. V. Bucur, *Thin Solid Films* **60** (1979) 73.
[213] R. A. Crane, G. Fischer, *J. Phys. D* **12** (1979) 2019.

[214] K. K. Kanazawa, J. G. Gordon, *Anal. Chem.* **57** (1985) 1770.
[215] C. Fruböse, K. Doblhofer, C. Soares, *Ber. Bunsen-Ges. Phys. Chem.* **97** (1993) 475–478.
[216] R. Beck, U. Pittermann, K. G. Weil, *Ber. Bunsen-Ges. Phys. Chem.* **92** (1988) 1363–1368.
[217] F. Eggers, Th. Funck, *J. Phys. E* **20** (1987) 523–530.
[218] A. Glidle, R. Hillman, S. Bruckenstein, *J. Electroanal. Chem. Interfacial Electrochem.* **318** (1991) 411–420.
[219] R. Beck, U. Pittermann, K. G. Weil, *J. Electrochem. Soc.* **139** (1992) no. 2, 453–461.
[220] R. Beck, U. Pittermann, K. G. Weil, *Ber. Bunsen-Ges. Phys. Chem.* **92** (1988) 1363–1368.
[221] D. Salt: *Handbook of Quartz Crystal Devices*, van Nostrand Reinhold, New York 1987.
[222] Z. P. Khlebarov, A. I. Stoyanova, D. I. Topalova, *Sens. Actuators B***8** (1992) 33–40.
[223] V. M. Ristic: *Principles of Acoustic Devices*, J. Wiley and Sons, New York 1983.
[224] P. Hauptmann, *Sens. Actuators A* **25** (1991) 371–377.
[225] B. A. Auld: *Acoustic Fields and Waves in Solids*, vol. II, J. Wiley and Sons, New York 1973.
[226] J. F. Vetelino, D. L. Lee, WO 831511, 1983.
[227] A. Bryant, D. L. Lee, J. F. Vetelino, *Ultrason. Symp. Proc.* **1981**, 171–174.
[228] A. D'Amico, *Sens. Actuators* **3** (1982/83) 31–39.
[229] S. J. Martin, S. S. Schwartz, R. L. Gunshor, R. F. Pierret, *J. Appl. Phys.* **54** (1983) 561–569.
[230] E. Gizeli, A. C. Stevenson, N. J. Goddard, C. R. Lowe: *Chemical Sensors*, Waseda University Int. Conference Center, Tokyo, Sept. 13–17, 1992.
[231] P. Hauptmann: *Sensoren, Prinzipien und Anwendungen*, Hanser Verlag, München 1990.
[232] G. J. Bastiaans in S. Yamauchi (ed.): *Chemical Sensor Technology*, Elsevier, Amsterdam 1992, pp. 181–204.
[233] J. Curie, P. Curie, *Bull. Soc. Min. Paris* **3** (1880) 90.
[234] R. A. Heising: *Quartz Crystal for Electrical Circuits – Their Design and Manufacture*, van Nostrand, New York 1947, p. 16.
[235] S. Büttgenbach: *Mikromechanik*, Teubner, Stuttgart 1991, p. 39.
[236] L. E. Halliburton, J. J. Martin in E. A. Gerber, A. Ballato (eds.): *Precision Frequency Control*, vol. 1, Acoustic Resonators and Filters, Academic Press, New York 1985.
[237] M. H. Ho: "Applications of Quartz Crystal Microbalances in Aerosol Measurements," in C. Lu, A. W. Czanderna (eds.): "Applications of Piezoelectric Quartz Crystal Microbalances," *Methods and Phenomena*, vol. 7, Elsevier, Amsterdam 1984.
[238] W. G. Cady: *Piezoelectricity*, McGraw Hill, New York 1946.
[239] J. F. Nye: *Physical Properties of Crystals*, Oxford University Press, London 1957.
[240] J. C. Brice: "Crystals for Quartz Resonators," *Rev. Mod. Phys.* **57** (1985) 105.
[241] M. Born, K. Huang: *Dynamical Theory of Crystal Lattices*, Clarendon Press, Oxford 1954.
[242] E. K. Sittig: "Acoustic Wave Devices," in *Encyclopedia of Physical Science and Technology*, vol. 1, Academic Press, New York 1987, pp. 83–109.
[243] F. L. Dickert, M. Vonend, H. Kimmel, G. Mages, *Fresenius Z. Anal. Chem.* **333** (1989) 615–618.
[244] A. W. Snow, J. R. Griffith, N. P. Marullo, *Macromolecules* **17** (1984) 1614.
[245] F. L. Dickert, *Chem. Unserer Zeit* **26** (1992) 138–143.
[246] B. A. Cavic-Vlasak, L. J. Rajakovic, *Fresenius J. Anal. Chem.* **343** (1992) 339–347.
[247] E. Weber, A. Ehlen, C. Wimmer, J. Bargon, *Angew. Chem.* **105** (1993) 116–117.
[248] I. Lundström, M. S. Shivaraman, C. M. Stevenson, *J. Appl. Phys.* **46** (1975) 3876–3881.
[249] M. S. Nieuwenhuizen, W. Barendsz, *Sens. Actuators* **11** (1987) 45–62.
[250] I. Langmuir: "The Adsorption of Gases on Plane Surfaces of Glass, Mica and Pt," *J. Am. Chem. Soc.* **40** (1918) 1361.
[251] J. H. de Boer: *The Dynamical Aspects of Adsorption*, Academic Press, Oxford 1953.
[252] A. Clark: *The Chemisorptive Bond*, Academic Press, New York 1974.
[253] F. M. Moser, A. L. Thomas: *Phthalocyanine Compounds*, Reinhold Publ. Co., New York 1963, chap. 8.
[254] L. F. Barringer, D. P. Rillema, J. H. Ham IV, *J. Inorg. Biochem.* **21** (1984) 195.
[255] J. R. Lenhard, R. W. Murvay: "Chemically Modified Electrodes," *J. Am. Chem. Soc.* **100** (1978) 7880.
[256] J. S. Miller (ed.), *ACS Symp. Ser.* **182** (1981) part 15.
[257] M. T. Reed et al.: "Thermodynamic and Kinetic Data for Cation Macrocyclic Interactions," *Chem. Rev.* **85** (1985) 271.
[258] D. W. Armstrong, *J. Liq. Chromatogr.* **7**, Suppl. 2 (1984) 353–376.
[259] M. S. Nieuwenhuizen, A. Venema: "Mass-Sensitive Devices," in W. Göpel et al. (eds.): *Sensors*, vol. II, chap. 13, VCH Verlagsgesellschaft, Weinheim 1991, pp. 657–658.
[260] J. G. Brace, T. S. Sanfelippo, S. Joshi, *Sens. Actuators* **14** (1988) 47–68.
[261] E. T. Zellers, R. M. White, S. M. Rappaport, S. W. Wenzel, *Proc. Int. Conf. Sens. Actuators 4th* **1987**, Tokyo: *IEE Japan*, pp. 459–461.
[262] M. S. Nieuwenhuizen, A. J. Nederlof, *Sens. Actuators B* **2** (1990) 97–101.
[263] A. W. Snow et al., *Langmuir* **2** (1986) 513–519.
[264] A. J. Ricco, S. J. Martin, T. E. Zipperian, *Sens. Actuators* **8** (1985) 105–111.
[265] A. Byrant et al., *Sens. Actuators* **4** (1983) 105–111.
[266] M. Janghorbani, H. Freund, *Anal. Chem.* **45** (1973) 325–332.
[267] A. Kindlund, H. Sundgren, I. Lundström, *Sens. Actuators* **6** (1984) 1.
[268] D. Soares, *J. Phys. E*, in press.
[269] C. Barnes, *Sens. Actuators A* **29** (1991) 59–69.
[270] S. Bruckenstein, M. Shay, *Electrochim. Acta* **30** (1985) 1295–1300.
[271] F. L. Dickert, A. Haunschild, V. Maune, *Sens. Actuators B* **12** (1993) 169–173.
[272] M. Ho, G. G. Guilbault, E. P. Scheide, *Anal. Chem.* **52** (1982) 1998–2002.
[273] J. M. Jordan, Dissertation, University of New Orleans 1985.
[274] A. Suleiman, G. G. Guilbault, *Anal. Chim. Acta* **162** (1984) 97–102.
[275] J. F. Alder, A. E. Bentley, P. K. P. Drew, *Anal. Chim. Acta* **182** (1986) 123–131.
[276] Y. Tomita, M. H. Ho, G. G. Guilbault, *Anal. Chem.* **51** (1979) 1475–1478.
[277] J. A. O. Sanchez-Pedreno, P. K. P. Drew, J. F. Alder, *Anal. Chim. Acta* **182** (1986) 285–291.
[278] G. M. Varga, Jr., US NTIS AD Report 780171/5GA, 1974.

[279] E. C. Hahn, A. Suleiman, G. G. Guilbault, J. R. Cananaugh, *Anal. Chim. Acta* **197** (1987) 195–202.
[280] G. G. Guilbault, *Anal. Chem.* **55** (1983) 1682–1684.
[281] M. J. van Sant, Dissertation, University of New Orleans 1985.
[282] C. Kößlinger et al., *Biosens. Bioelectron.* **7** (1992) 397–404.
[283] C. G. Fox, J. F. Alder: "Surface Acoustic Wave Sensors for Atmospheric Gas Monitoring," in P. T. Moseley, J. Norris, D. E. Williams (eds.): *Techniques and Mechanism in Gas Sensing*, Adam Hilger, Bristol 1991, chap. 13.
[284] J. Auge, P. Hauptmann, F. Eichelbaum, S. Rösler, *Sens. Actuators*, in press.
[285] J. W. Grate, S. J. Martin, R. M. White, *Anal. Chem.* **65** (1993) 987–996.
[286] A. D'Amico, A. Palma, E. Verona, *Sens. Actuators* **3** (1982/83) 31–39.
[287] A. Venema et al., *Sens. Actuators* **10** (1986) 47.
[288] A. Venema et al., *Electron. Lett.* **22** (1986) 184.
[289] A. D'Amico, A. Palma, E. Verona, *Appl. Phys. Lett.* **41** (1982) 300–301.
[290] A. D'Amico, M. Gentilli, P. Veradi, E. Verona, *Proc. Int. Meet. Chem. Sens. 2nd* **1986**, 743–746.
[291] A. W. Snow et al., *Langmuir* **2** (1986) 513–519.
[292] J. G. Brace, T. S. Sanfelippo, S. Joshi, *Sens. Actuators* **14** (1988) 47–68.
[293] J. G. Brace, T. S. Sanfelippo, *Proc. 4th Int. Conf. Sens. Actuators* **1987**, 467–470.
[294] M. S. Nieuwenhuizen, A. J. Nederlof, M. J. Vellekoop, A. Venema, *Sens. Actuators* **19** (1989) 385–392.
[295] H. Wohltjen, *Sens. Actuators* **5** (1984) 307–325.
[296] A. W. Snow, H. Wohltjen, *Anal. Chem.* **56** (1984) 1411–1416.
[297] S. J. Martin, S. S. Schwartz, R. L. Gunshor, R. F. Pierret, *J. Appl. Phys.* **54** (1983) 561–569.
[298] S. J. Martin, K. S. Schweizer, S. S. Schwartz, R. L. Gunshor, *Ultrason. Symp. Proc.* **1984**, 207–212.
[299] A. J. Ricco, S. J. Martin, T. E. Zipperian, *Sens. Actuators* **8** (1985) 319–333.
[300] J. W. Grate et al., *Anal. Chem.* **60** (1988) 869–875.
[301] K. Mosbach, *Biosens. Bioelectron.* **6** (1991) 179.
[302] P. T. Walsh, T. A. Jones in W. Göpel et al. (eds.): *Sensors: A Comprehensive Survey*, vol. 2, VCH Verlagsgesellschaft, Weinheim 1991, p. 529.
[303] A. F. Hollemann, E. Wiberg: *Lehrbuch der anorganischen Chemie*, De Gruyter, Berlin 1985.
[304] B. Danielsson, F. Winquist in A. E. Cass (ed.): *Biosensors: A Practical Approach*, Oxford University Press, Oxford 1989, p. 191.
[305] A. R. Baker, GB 892530, 1962.
[306] K. Mosbach et al., *Biochim. Biophys. Acta* **364** (1974) 140.
[307] B. Mattiasson et al., *Anal. Lett.* **9** (1976) 217.
[307a] B. Danielsson et al., *Appl. Biochem. Bioeng.* **3** (1981) 103.
[308] H. G. Hundeck et al., *GBF Monogr. Ser.* **17** (1992) 321.
[309] B. Xie et al., *Sens. Actuators B* **6** (1992) 127.
[310] E. J. Guilbeau et al., *Trans. Am. Soc. Artif. Intern. Organs* **33** (1987) 329.
[311] M. G. Jones, T. G. Nevell, *Sens. Actuators* **16** (1989) 215.
[312] B. Danielsson et al., *Methods Enzymol.* **137** (1988) 181.
[313] S. Wold et al.: *Food Research and Data Analysis*, Applied Sciences, Barking 1983, pp. 147–188.
[314] T. Nakamoto, K. Fukunishi, T. Moriizumi, *Sens. Actuators B* **1** (1989) 473.
[315] K. Ema et al., *Sens. Actuators B* **1** (1989) 291.
[316] J. R. Stetter, P. C. Jurs, S. L. Rose, *Anal. Chem.* **58** (1986) 860.
[317] U. Weimar et al., *Sens. Actuators B* **1** (1990) 93.
[318] W. P. Carey et al., *Sens. Actuators* **9** (1986) 223.
[319] D. Wienkc, K. Danzer, *Anal. Chim. Acta* **184** (1986) 107.
[320] B. R. Kowalski, *Anal. Chem.* **47** (1975) 1152.
[321] B. R. Kowalski, *Anal. Chem.* **52** (1980) 112 R.
[322] E. Frank, B. R. Kowalski, *Anal. Chem.* **54** (1982) 232 R.
[323] K. R. Beebe, B. R. Kowalski, *Anal. Chem.* **59** (1987) 1007 A.
[324] D. W. Osten, B. R. Kowalski, *Anal. Chem.* **57** (1985) 908.
[325] A. F. P. Turner, I. Karube, G. S. Wilson (eds.): *Biosensors – Fundamentals and Applications*, Oxford University Press, Oxford 1987.
[326] H.-L. Schmidt, R. Kittsteiner-Eberle, *Naturwissenschaften* **73** (1986) 314–321.
[327] F. Scheller, F. Schubert: *Biosensors, Techniques and Instrumentation in Analytical Chemistry*, vol. 11, Elsevier, Amsterdam 1992.
[328] E. A. H. Hall: *Biosensors*, Open University Press, Buckingham 1990.
[329] L. C. Clark, Jr., C. Lyons: *Ann. N. Y. Acad. Sci.* **102** (1962) 29–45.
[330] J. J. Kulys, A. S. Samalius, G. J. S. Sviermikas, *FEBS Lett.* **114** (1980) 7–10.
[331] M. J. Eddows, H. A. O. Hill, *Biosci. Rep.* **1** (1981) 521–532.
[332] M. L. Fultz, R. A. Durst, *Anal. Chim. Acta* **140** (1982) 1–18.
[333] L. Gorton, *J. Chem. Soc. Faraday Trans. 1* **82** (1986) 1245–1258.
[334] B. Gründig et al., in F. Scheller, R. D. Schmid (eds.): "Biosensors: Fundamentals, Technologies and Applications," *GBF Monogr.* **17** (1992) 275–285.
[335] Y. Degani, A. Heller, *J. Phys. Chem.* **91** (1987) no. 6, 1285–1289.
[336] Y. Degani, A. Heller, *J. Am. Chem. Soc.* **111** (1989) 2357–2358.
[337] W. Schuhmann, T. J. Ohara, H.-L. Schmidt, A. Heller, *J. Am. Chem. Soc.* **113** (1991) no. 4, 1394–1397.
[338] F. W. Scheller, R. Renneberg, F. Schubert, in S. Colowick, N. O. Kaplan, K. Mosbach (eds.): *Methods in Enzymology*, vol. 137, Academic Press, San Diego 1988, pp. 29–44.
[339] F. Schubert, D. Kirstein, K. L. Schröder, F. W. Scheller, *Anal. Chim. Acta* **169** (1985) 391–406.
[340] A. Warsinke, B. Gründig, DE 4227569, 1993.
[341] K. Riedel, B. Neumann, F. Scheller, *Chem. Ing. Tech.* **64** (1992) no. 6, 518–528.
[342] M. H. Smit, A. E. G. Cass, *Anal. Chem.* **62** (1990) 2429–2436.
[343] D. M. Rawson, A. J. Willmer, *Biosensors* **4** (1989) 299–311.
[344] R. S. Sethi: *Biosensors & Bioelectronics*, in press.
[345] M. P. Byfield, R. A. Abuknesha: "Biochemical Aspects of Biosensors," *GEC J. Res.* **9** (1991) 97–117.
[346] M. E. Eldefrawi et al., *Anal. Lett.* **21** (1988) 1665–1680.
[347] R. A. Badley et al., *Phil. Trans. R. Soc. London B* **316** (1987) 143–160.

[348] S. Lofas, B. Johnsson, *J. Chem. Soc. Chem. Commun.* **1990**, 1526–1528.
[349] J. E. Roederer, G. J. Bastiaans, *Anal. Chem.* **55** (1983) 2333–2336.
[350] G. G. Guilbault, J. H. Luong, *J. Biotechnol.* **9** (1988) 1–10.
[350a] M. P. Byfield, R. A. Abuknesha, *GEC J. of Res.* **9** (1991) 108.
[351] M. Hikuma et al. *Biotechnol. Bioeng.* **21** (1979) 1845–1850.
[352] I. Karube, S. Suzuki, *Ion Sel. Electrode Rev.* **6** (1984) 15–58.
[353] K. Riedel, R. Renneberg, F. Scheller, *Anal. Lett.* **53** (1990) 757–770.
[354] Frost & Sullivan: *The European Market for Industrial Gas Sensors* and *The European Market for Portable and Transportable Analytical Instruments*, Frost & Sullivan Eigenverlag, Frankfurt 1992.
[355] INFRATEST: *Chemische und biochemische Sensoren – Marktübersicht*, Infratest Industria, 1992.
[356] MIRC: *The European Market for Process Control Equipment and Instrumentation*, Frost & Sullivan Eigenverlag, Frankfurt 1992.

Microscopy

ANDRES KRIETE, Klinikum der Justus-Liebig-Universität Gießen, Institut für Anatomie und Zytobiologie, Gießen, Federal Republic of Germany and HEINZ GUNDLACH, Carl Zeiss, Oberkochen, Federal Republic of Germany (Chap. 1)

SEVERIN AMELINCKX, Universiteit Antwerpen (RUCA), Antwerpen, Belgium (Section 2.1)

LUDWIG REIMER, Physikalisches Insitut Universität Münster, Münster, Federal Republic of Germany (Section 2.2)

0-89573-541-5/94/$ 5.00 + .50

1. Modern Optical Microscopy

1.1. Introduction

Optical microscopy goes back as far as the 16th century, when magnifiying glasses and optical lenses became available. ANTONI VAN LEEUWENHOEK (1623–1723) observed structures lying beyond the resolution limit of the eye, such as bacteria, with a 270-times magnification single lens. The exact design of the microscope was established from a theory for image formation developed by ABBE in 1873 [1.1]. With the introduction of apochromatic lenses (→ Optical Materials, **A 18**, p. 201) and oil immersion, a theoretical resolution limit of 0.2 μm became available at this early stage. Light microscopy advanced further by new lens designs, optical contrast enhancement methods, and application of fluorescent and immunofluorescent dyes, also in combination with inverted microscope designs. Optoelectronic methods such as video techniques have further broadened the spectrum of light microscopic applications. Finally, laser scanning techniques and confocal imaging allowed the investigation of microscopic structures three-dimensionally. Compared to a standard microscope, the confocal microscope has enhanced lateral and axial resolution, as well as improved contrast, and in particular it can remove out-of-focus blur. Thus, an optical, noninvasive sectioning capability is achieved of thick specimens. The final breakthrough of confocal microscopy in biomedicine and material science boosted the development of a variety of commercial confocal laser scanning microscopes. Highly developed optoelectronic components such as laser light sources and sensitive detectors, as well as the application of specific fluorescent dyes, enabled wide acceptance of this new imaging technique together with effective use of powerful computers to digitally store, visualize, and analyze three-dimensional data.

1.2. Basic Principles of Light Microscopy

1.2.1. Optical Ray Path

The rigid *stand* of the microscope is the "carrying" element, with mechanical and optical parts attached. The *objective* is the central optical unit, generating an intermediate image that is magnified by the *eyepiece* for visual observation [1.2]–[1.4]. *Condenser* and *collector* optimize specimen illumination. Lens designs corrected for a *finite length* form a mirrored, magnified, aerial (intermediate) image, which is projected infinity by the eyepiece (Fig. 1.1). The distance between the back focal plane of the objective and the primary image plane (which is the front focal plane of the eyepiece) is defined as the *optical tube length*. The *mechanical tube length*, measured from the mounting plane of the objective to the mounting plane of the eyepiece, is 160 mm for currently used microscopes. Any aberations present in the intermediate image, such as the lateral chromatic aberation (LCA), are compensated by the eyepiece. Optical–mechanical components such as the Bertrand lens, or epi-illuminating reflectors mounted within the optical path require a so-called telan system. Objectives of *infinity-corrected designs* generate an infinite image projected into a real image by the tube lens (Fig. 1.2.). The infinity color-corrected system (ICS) optic is an example of how the tube lens together with the objective can correct aberrations, particularly chromatic ones [1.5]. The intermediate image is monochromatic; conse-

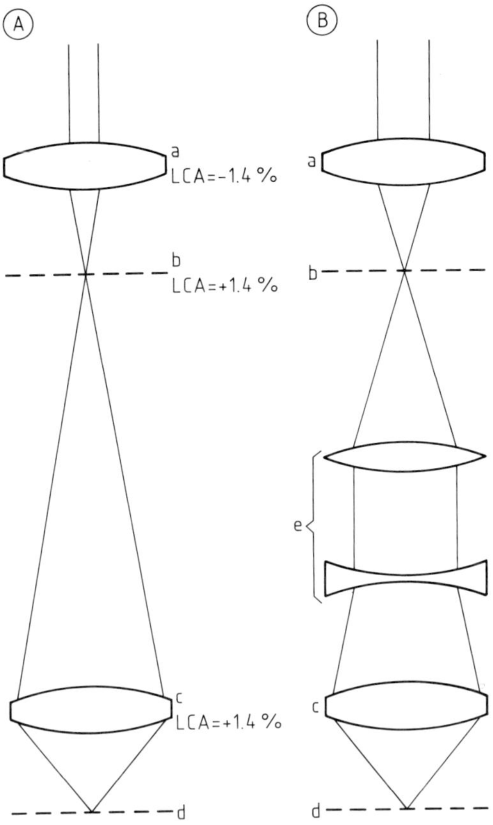

Figure 1.1. Schematic optical path for objectives of finite length; compensating eyepieces correct for chromatic aberations
A) Conventional microscope with finite optics; B) Microscope with finite optics and telan optics for insertion of optical components
a) Eyepiece; b) Primary intermediate image; c) Objective; d) Object plane; e) Telan system (incident light fluorescence, Optovar, Bertrand lens)

quently, photographic and video recording do not require compensating lenses. Moreover, the path between lens and tube lens may be used by all optical–mechanical components for various illuminating and contrast generating methods without telan optics (Fig. 1.2).

1.2.2. Imaging Performance and Resolution

The image generation in microscopy cannot be explained sufficiently by means of geometric optics alone; wave optics must also be considered [1.1], [1.6]. Imaging a point source generates a diffraction pattern in the focal plane. The central parts of this Airy disk contain most of the intensity; therefore, the diameter of this kernel

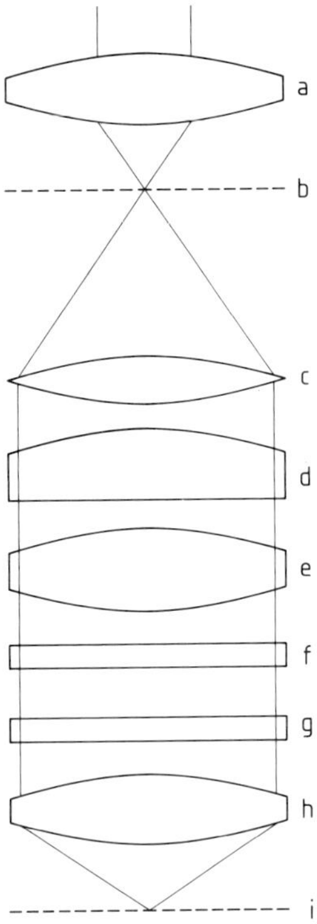

Figure 1.2. Schematic optical path for objectives corrected for infinity with color-corrected intermediate image
a) Eyepiece; b) Plane of intermediate image; c) Tube lens; d) Bertrand lens slider; e) Optovar 1 X, 1.25 X, 1.6 X, reflector for incident light; f) $\lambda/4$ plate, plate, compensators; g) Differential interference contrast (DIC) slider; h) Objective; i) Specimen plane

can be defined as the smallest resolvable image element (Fig. 1.3 A–D). The diameter of the kernel d_k is given by

$$d_k = 1.22\ \lambda/n \cdot \sin\alpha \qquad (1.1)$$

with λ being the wavelength, n the refractive index of the material, α the angle between the optical axis and the marginal beam of the light cone that enters the objective, and $n \cdot \sin\alpha$ the numerical aperture NA.

The resolution limit for two points is reached when the maximum of the first diffraction spot coincides with the first minimum of the second one. ABBE developed this theory and proved it experimentally in his historical papers [1.1]. The resolution is calculated from the available numerical aperture and the wavelength λ by

$$d = 1.22\ \lambda/NA + NA \qquad (1.2)$$

As an example, a lens and condenser of $NA = 1.4$ and $\lambda = 530$ nm give a smallest separation of $d = 200$ nm. Resolution can be improved either by a higher NA, which is difficult to obtain in lens design, or by using shorter wavelength, as in ultraviolet microscopy [1.7].

Both the size of the diffraction spot and the resolution of the eye define the utilizable magnification. Since the resolution of the eye is ca. 2–4 arcminutes, total magnification should not exceed 500–1000 times the numerical aperture.

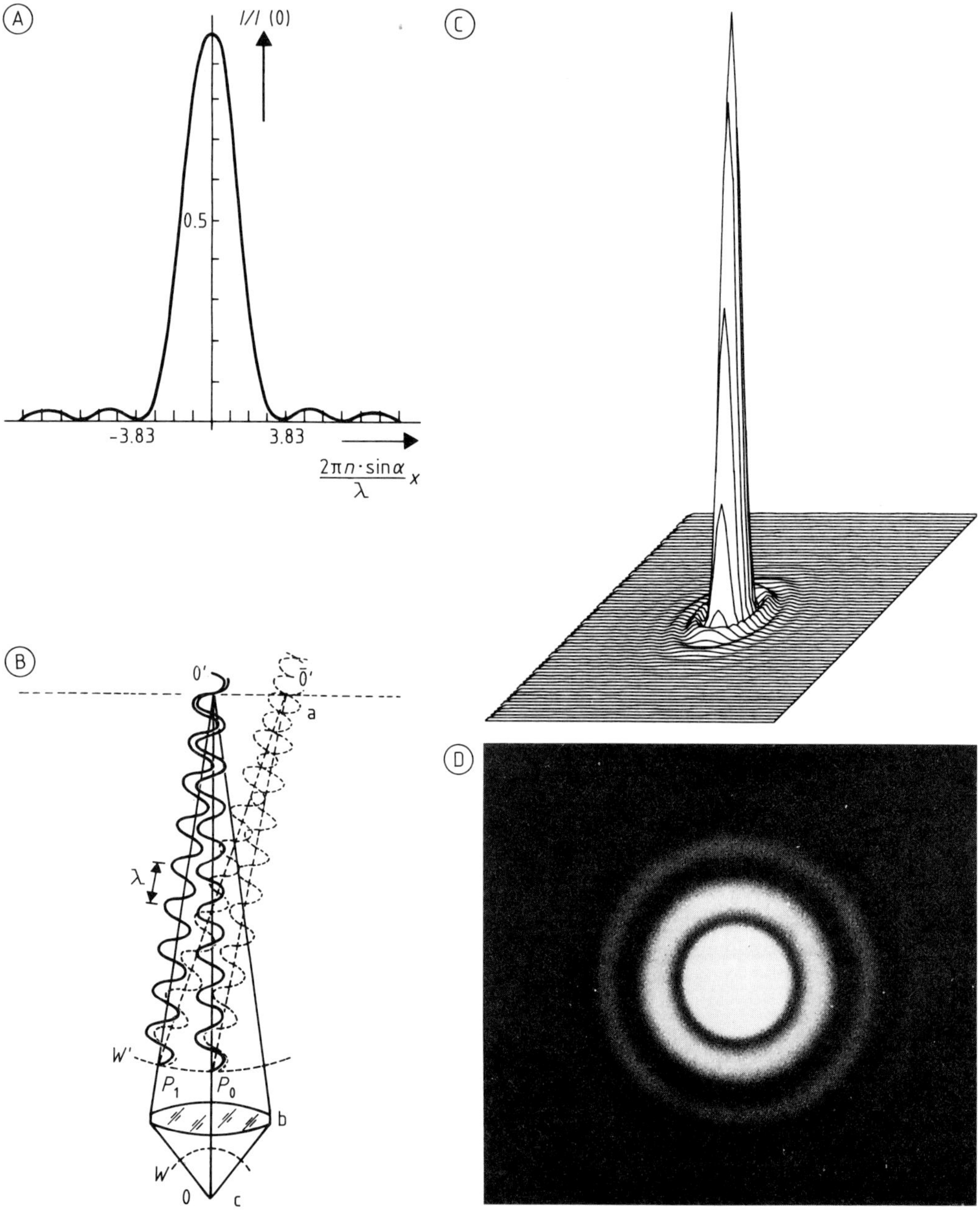

Figure 1.3. A) and B) Intensity distributions in the image of a luminous point; C) Computed intensity distribution; D) Airy disk
a) Image plane; b) Lens; c) Objects plane

Magnification above this limit does not deliver any additional information, but may be used for automatic measuring procedures or counting.

1.2.3. Characteristics and Classification of Lenses

The imperfections of optical lenses give rise to various distortions, either monochromatic or chromatic, (see also → Optical Materials, **A 18**, pp. 197–201) that are visible in the diffraction pattern. Monochromatic distortions include aberration, curvature, astigmatism, and coma. Chromatic aberrations are caused by differences in the refractive index of glasses. These types of errors are minimized by combining several pieces of glass with different form and refractive index, or by using diaphragms. The three main categories of lenses are

1) Achromatic objectives
2) Fluorite or semiapochromatic objectives
3) Apochromatic objectives

Fluorite and apochromatic lenses are mostly designed as plane (i.e., corrected for a flat field of view). Computers allow calculation of the diffraction pattern under various imaging conditions to optimize imaging performance for a wide range of applications. The present advantage in modern lens design and production is indicated by the image quality of the ICS Plan Neofluar class. At a field-of-view number of 25, color photography, as well as all optical contrast enhancement methods, are supported with achromatic correction comparable to apochromatic lenses with a good transmission down to UV at 340 nm. These objectives are generally applicable for all fluorescent methods (see Sections 1.3.2, 1.6.4). Plan apochromates feature a still higher aperture: chromatic rendition is improved, as well as correction at the field-of-view boundary; therefore these objectives are very suitable for color microphotography. In addition, a complete correction not only at the focus level but also for a limited range above and below the focal plane is achieved. For specific applications, the following lenses are also available:

1) Multi-immersion objectives of type Plan Neofluar, corrected for use with oil, glycerol, or water, with or without a cover glass
2) Objectives for specimen without cover slide
3) Objectives with a long working distance
4) Objectives with iris diaphragm
5) Objectives for phase contrast investigation
6) Objectives, strain free for polarized light microscopy
7) Ultrafluars, corrected between 230 and 700 nm

1.2.4. Eyepieces and Condensers

The main characteristics of *eyepieces* are magnification and type of correction, such as color compensation and correction of viewing angle. The correction of the *condensers* must match the properties of the lenses. Special or combined condensers are required for certain methods of contrast generation or for longer working distance with inverted microscopes.

1.3. Illumination and Contrast Generation

Illumination and contrast are central points in the discussion of imaging performance. Without sufficient contrast, neither the required magnification nor a maximal resolution can be obtained. A. KÖHLER developed optical illumination for microscopy in 1893 [1.8]. The main advantage of Köhler illumination is that in spite of the small filament of the light bulb, the entire area of the lamp field stop, as well as the illuminated part of the specimen, have a uniform luminance (Fig. 1.4). The twofold ray path of diaphragms and lenses suggests that adjustment of the condenser iris, being the illumination aperture, controls resolution, contrast, and resolution depth. This principle is also important for correct setup of the confocal microscope as well as electron microscopy.

1.3.1. Optical Contrast Generation

For the following considerations the principles of wave optics are used, which describe image formation as a result of diffraction and interference.

Dark-Field Illumination. In dark-field illumination the illuminating aperture is always set higher than the lens aperture. Only light waves being diffracted by the microstructures are captured by the lens. Objects appear as bright diffraction patterns on a dark background.

Phase Contrast. Based on ABBE's theory the Dutch physicist FRITS ZERNIKE (Nobel prize in 1953) developed phase contrast imaging in 1935, to convert the invisible phase shift within the

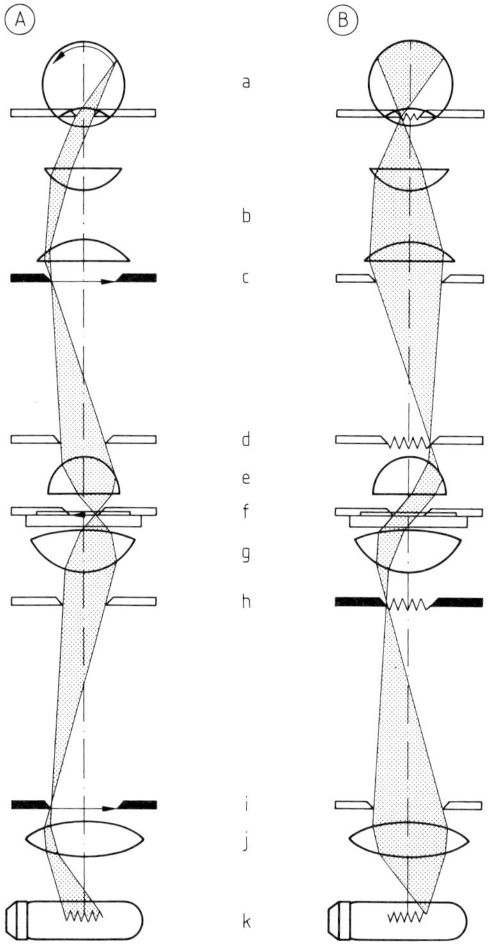

Figure 1.4. Schematic of Köhler illumination
A) Image-forming ray path; B) Illuminating ray path
a) Eye; b) Eyepiece; c) Eyepiece field stop; d) Exit pupil of objective; e) Objective; f) Specimen; g) Condenser; h) Aperture diaphragm; i) Lamp field stop; j) Light collector; k) Light source

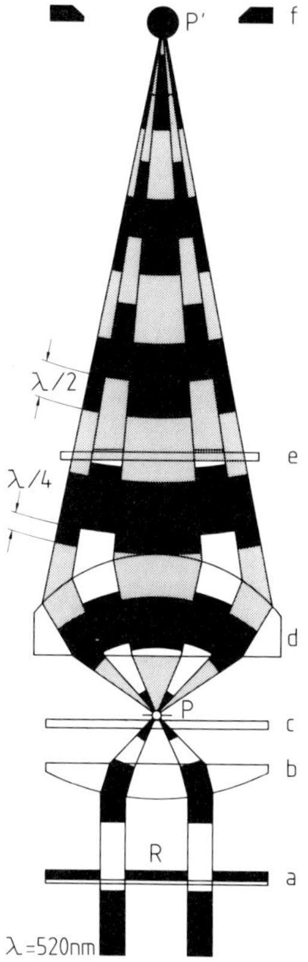

Figure 1.5. Image path of the phase contrast microscope, converting phase shifts into a noticeable contrast
a) Illuminating diaphragm ring; b) Condenser; c) Specimen generating a phase shift; d) Objective plate; e) Phase plate; f) Intermediate image

specimen into recognizable contrast [1.9]. A phase plate located in the lens modifies the phase and amplitude of the direct (iluminating) wave in such a way that a phase shift of $\lambda/2$ is obtained (Fig. 1.5). The amplitudes of the direct diffracted waves are equalized, and by destructive interference, dark-phase objects appear on a bright background. The application of this technique ranges from routine diagnosis to cell biology research [1.6], [1.10]. Today this method is also used advantageously in combination with epifluorescence. The specimen is adjusted under phase contrast before switching to fluorescence light in order to avoid bleaching, or both imaging modes are combined to match the fluorescent signal with the morphological image.

Nomarski Differential Interference Contrast (DIC). In 1955 the physicist George Nomarski simplified the two-beam interference microscope in a way that it became available for routine microscopy [1.11] (Fig. 1.6). DIC uses modified wollaston prisms (Fig. 1.6 B), lying outside the focal areas of condenser and objective [1.11]–[1.13]. Normal lenses are used; the contrast can be adjusted by shifting the lens-sided prisms. As in phase contrast, differences in the optical path length are visualized. Objects appear as a relief, and thick specimens lack the halo-typical pattern of phase contrast imaging. Full aperture of both the objective and the condenser allow optical sectioning (Fig. 1.7). Combined with inverted microscopes (see Section 1.4) and contrast-enhanced video microscopy (see Section 1.5), ultra-

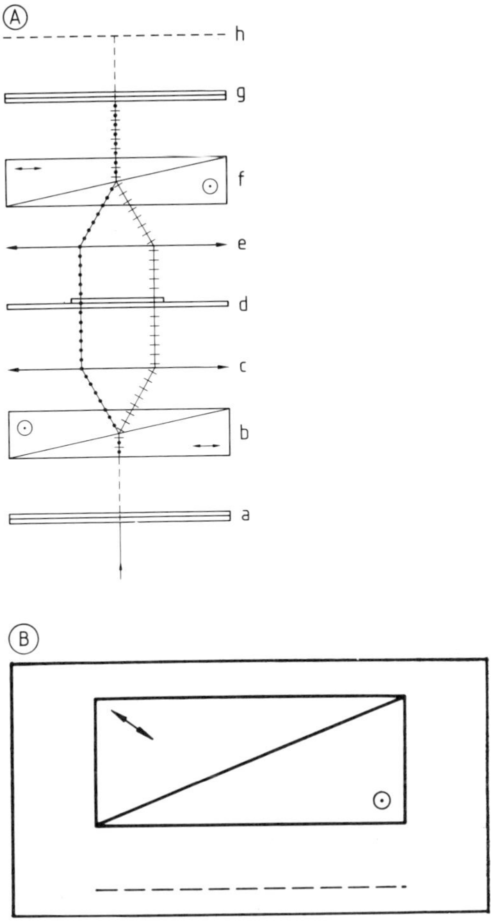

Figure 1.6. A) Image path of a two-beam interference microscope; B) Nomarski prism with direction of optical axes and position of interference plane indicated
a) Polarizer (45°); b) First Wollaston prism; c) Condenser; d) Specimen; e) Objective; f) Second Wollaston prism; g) Analyzer (135°); h) Intermediate image
Contrast is generated by a phase shift of the polarized and prism-split illuminating beams, which are combined by the second prism

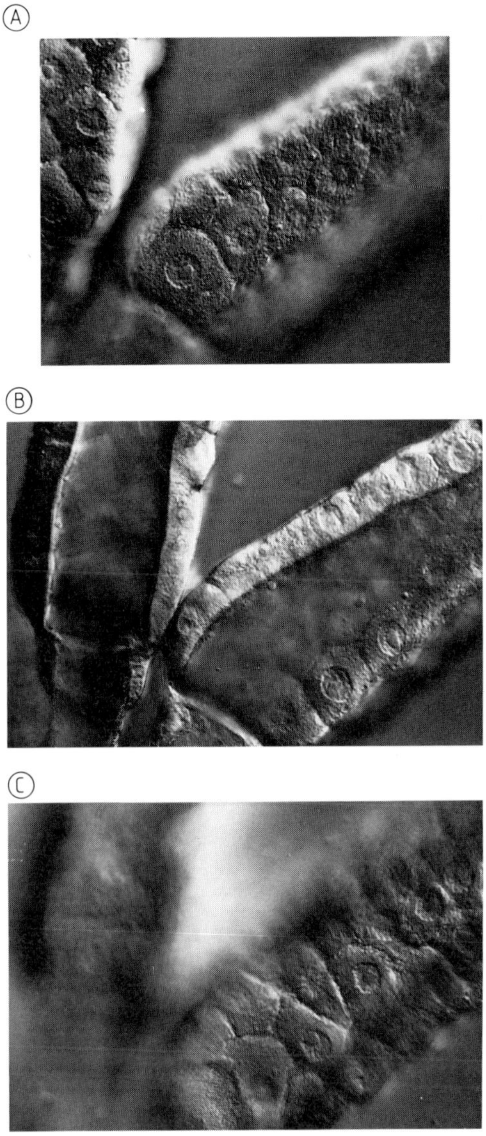

Figure 1.7. A)–C) Optical sectioning of salivary gland of drosophila by DIC Nomarski (Specimen courtesy of L. Rensing, Universität Bremen, Germany)

structures in cells and tissues can be imaged. Certain applications combine DIC with phase contrast. For routine applications, phase contrast lenses may give better results at lower magnification. The DIC image is well suited to be combined with a fluorescent image as well.

Polarization and Reflection Contrast Microscopy. Polarization microscopy is used preferentially for certain applications in cell research, to enhance double diffraction (anisotropic) structures such as muscle fibers or the mitotic apparatus. Quantitative studies require strain-free polarization lenses for the polarized light used. With the help of reflection contrast microscopy [1.14], [1.15], visible interference is generated at epi-illumination, for example, to study adhesive properties of living cells on glass surfaces [1.16]. Moreover, this method may be combined with phase contrast or Nomarski DIC contrast. For the evaluation of immunogold-marked specimens, immunogold staining (IGS), reflection as well as transmission contrast is feasible.

1.3.2. Fluorescence Microscopy

The development of immunofluorescence greatly influenced medical practice and basic biomedical research [1.17]–[1.19 a]. Many biomedical specimens emit *autofluorescence* if illuminated properly. Today a wide range of fluorescent microstructures can be excited by application of fluorochromes (*secondary fluorescence*). Immunofluorescence is based on antigen–antibody connectivity [1.19 b]. This couples the high specificity of the immune reaction with the extreme sensitivity of fluorescence microscopy. Various markers for antibodies exist and derivates of fluorescein, mainly fluorescein–isothiocyanate (FITC), are the preferred fluorochromes. For this kind of microscopy, special lamps and filters are required. Light sources used include high-pressure bulbs as well as lasers. Fluorescence microscopy is used either in transmission or in epi-illumination; the latter combines phase contrast and Nomarski DIC [1.20], [1.21]. Fluorescence microscopic objectives with high aperture and a good transmission between 340 and 700 nm are preferred. Multi-immersion lenses of plan-type Neofluar have been designed specifically for such applications, as well as low-magnification lenses featuring high apertures (primary magnification 40, *NA* 1.3 oil). Plan apochromates are more suitable for illumination in visible light. The brightness b of fluorescent images increases proportionally to the square of the numerical aperture of lens L and condenser C, and decreases inversely with the square of the total magnification M

$$b \sim \frac{NA_{\mathrm{L}}^2 + NA_{\mathrm{C}}^2}{M^2} \qquad (1.3)$$

This makes ocular lenses with low primary magnification an ideal choice. At epi-illumination the brightness depends on the power of four of the aperture, since the lens acts as the condenser as well:

$$b \sim \frac{NA_{\mathrm{L}}^4}{M^2} \qquad (1.4)$$

In the 1980s, fluorescence methods became a major tool for sensitive and highly specific detection in biomedical research and medical diagnosis. The development of reagents based on fluorescence includes the use of biological molecules: antibodies, RNA, DNA, proteins, and other macromolecules and physiologic indicators [1.17]–[1.19]. These methods have enabled new findings in cell biology, molecular biology, developmental biology, and cancer research. The cytoskeleton (Fig. 1.8) is responsible for the cell shape, cell division, capacity of the cell to move, and the intracellular transport of cell organelles [1.22], [1.23]. Multiparameter fluorescence microscopy uses a broad selection of fluorescent dye molecules to examine several properties or components of a single cell simultaneously (Fig. 1.9).

A series of new probes suitable for this purpose has been developed by extending the usable spectrum to the near infrared (Fig. 1.10). *Fluorescence-based in situ hybridization* (FISH) is rapidly being recognized as a powerful tool for detecting and quantifying genetic sequences in cells and isolated chromosomes [1.24]. FISH is based on the highly specific binding of pieces of DNA, labeled with a fluorescent probe, to

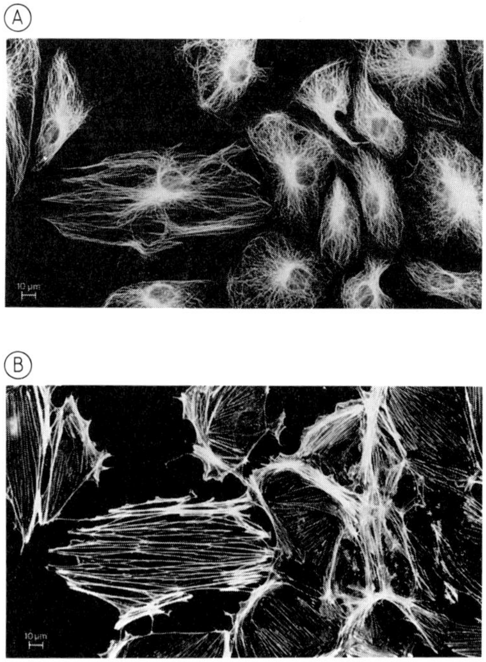

Figure 1.8. Double fluorescence micrograph of epithelial kidney cells
A) Microtubules, fluorescent dye FITC; B) Actin filament, fluorescent dye tetramethylrhodamine-B-isothiocyanate (TRITC)

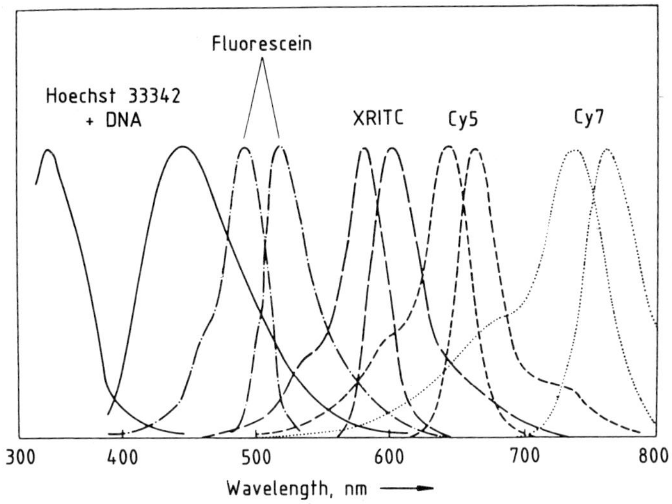

Figure 1.9. Excitation and emission curves of various fluorochromes
Filter curves depict dyes Hoechst 33342 (suitable for DNA), FITC (for actin), rhodamine (endosomes), and cyanine 5 (mitochondria)
(Courtesy of L. Taylor, University of Pittsburgh, United States)

unique sequences of DNA or RNA in interphase nucleus or metaphase chromosomes. Recent developments enable the simultaneous visualization of three or more DNA regions (chromosome painting) [1.24a].

New applications also rely on specifically developed fluorescent markers, such as biosensors or environmental markers. A fluorescent dye is chemically linked to a macromolecule, creating an analogue of the natural molecule. The fluorescent analogue is specifically designed to measure chemical properties of the macromolecule, instead of tracking its exact location within the cell. Calcium regulatory proteins are such a class of biosensors [1.19]. The mobility of molecules can be studied with a method called *fluorescence redistribution after photobleaching* (FRAP), in which a small volume of the specimen is monitored after bleaching with a high-intensity light. The recovery of fluorescence within the volume is a measure of the diffusion coefficient. In particular, the mobility of lipids and proteins at membranes has been studied. Fluorescence is also the preferred imaging mode in confocal microscopy (see Section 1.6.4).

1.4. Inverted Microscopy

Inverted microscopes were originally designed for the evaluation and counting of plankton and sediments [1.25]. The first inverted research microscope which included photomicrography, microphotometry, cinemicrography, and video imaging was the Axiomat [1.4], [1.26]–[1.28]. Inverted microscopes in U-design (Fig. 1.11) with a rigid stage are of great advantage for investigations in cell and developmental biology, as well as in cell physiology experiments that are performed in optical microscopes [1.28]–[1.31]. Identification and selective micropreparation on live nuclear components in oocytes of *Xenopus* in the DIC Nomarski technique facilitate high-resolution optical microscopy in different planes of the specimen (optical sectioning), which can be isolated by selective micropreparation for subsequent analysis in the electron microscope [1.32].

In the life cycle of the green alga *Acetabularia*, high-resolution phase and DIC microscopy have made part of the endoplasmic reticulum and lampbrush chromosomes visible [1.33], [1.34]. The distinct advantage of inverted compared to upright microscopes is the unhindered application of cell and tissue culture chambers in conjunction with micromanipulators, micropipettes, microelectrodes (patch clamp method), and injection systems for experimental cell research. Computer-controlled microinjection can be performed automatically [1.35]. Manipulation of cells, organelles, and genomes is also possible by means of laser microbeam and optical trapping [1.23].

1.5. Optoelectronic Imaging

Electronic imaging methods such as video microscopy and laser scanning microscopy can extend resolution slightly, but the ability to

A)

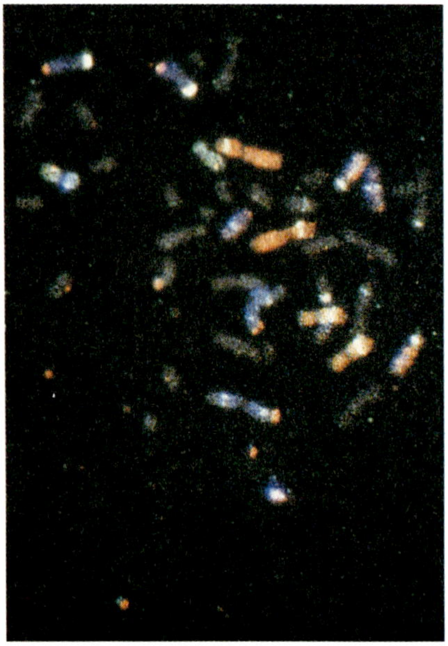

B)

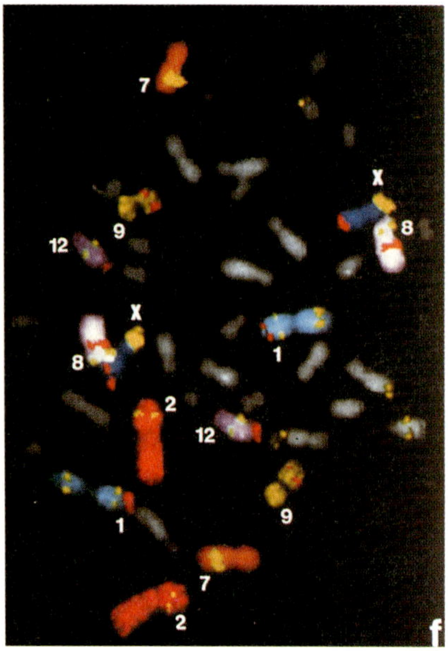

Figure 1.10. A) Micrograph by triple exposure of color slide film using conventional fluorescence microscopy; B) Pseudocolored digital image recorded with the CCD camera for chromosome 1, 2, 7, 8, 9, 12 and the human X chromosome
Green = FITC; Red = TRITC; Blue = DAPI;
(Courtesy of T. Cremer, Heidelberg, Germany and Oxford University Press)

extend the detectibility of subresolved structures and subresolution movements is more important. A video camera can generally be mounted to the same port employed for normal photomicrography. In *video-enhanced contrast microscopy* (VEC) (Fig. 1.12), high-resolution video cameras based on a vidicon tube are used. This principle of adjusting gain, and offset and digital image processing is also used in laser scanning microscopes (see Section 1.6). Digital imaging and image processing make it possible to see structures that are smaller than the resolution limit [1.12], [1.36]–[1.39]. By means of this method, single microtubules can be observed [1.40], [1.41]. The size of each microtubule is only one-tenth of the visible-light resolution limit. In the 1980s, solid-state cameras were introduced into microscopy [1.42], [1.43]. Most of them are based on a *charge coupled device* (CCD), which produces digital data directly for computer processing and image analysis. For low light level (LLL) or video intensification microscopy (VIM), i.e., weak fluorescence or small FISH signals (see Fig. 1.10 B), cooled, sensitive CCD cameras for black and white or color are available now and have replaced tube cameras [1.44]. Digital imaging techniques have gained increasing attention as supplements to conventional epifluorescence and photomicrography [1.45]. *Microphotometry* permits spectrophotometric analysis of substances, as well as statistical and kinetic measurement and compilation of two-dimensional intensity profiles. [1.31].

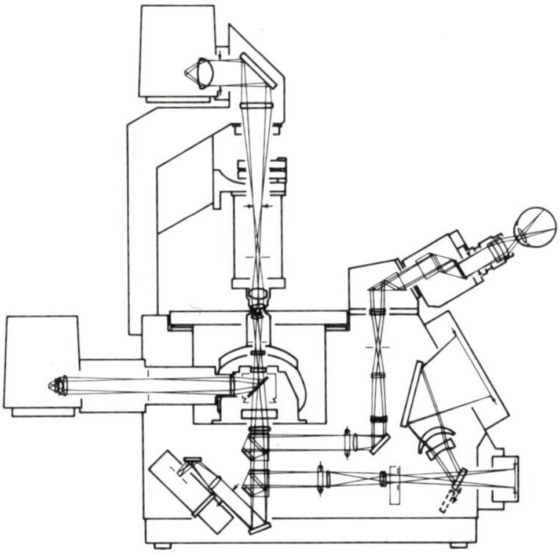

Figure 1.11. Light path of an inverted microscope

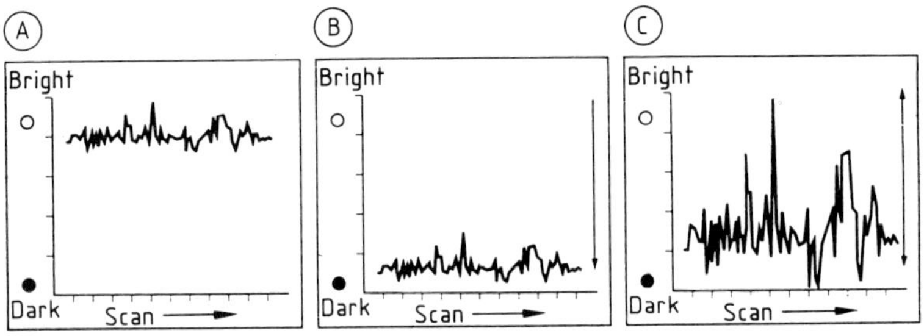

Figure 1.12. Principle of video-enhanced contrast microscopy
A) Original image; B) Offset (brightness); C) Contrast gain

1.6. Confocal Laser Scanning Microscopy

1.6.1. Basic Principles

The fundamentals of confocal microscopy go back to 1959, when MINSKY described a new kind of microscopic apparatus [1.46]. The historical development of confocal microscopy is reviewed elsewhere [1.47], [1.48]. Two different kinds of confocal instruments are available: the tandem scanning microscope and the laser scanning microscope. The *tandem scanning microscope* uses normal light chopped by a rotating Nipkow disk [1.49]. The light beams reflected by the specimen cross the rotating disk on its opposite side (thus tandem), and due to the symmetric pattern of holes, out-of-focus blur is removed. This system allows real-time and real-color imaging, but is limited in sensitivity because of the fixed size of the pinholes.

The more comonly applied *confocal laser scanning microscope* (CLSM) features a variable pinhole and allows adjustment of magnification by modifying scan angles. A schematic example of the principle of confocal laser scanning is given in Figure 1.13. The laser light is focused via the scanner (b) through the tube lens (c) and the objective (d), and illuminates a small spot in the specimen (e). Emitted light emanating from the focal plane and the planes above and below (dotted and dashed lines) is directed via the scanner to the dichroic beam splitter (a) where it is decoupled and directed onto a photomultiplier (i). A pinhole (h) in front of the photomultiplier is positioned at the crossover of the light beams emerging from the focal point. This plane corresponds to the intermediate image of the Köhler illumination described in Section 1.3. Light emanating from above and below the focal point has its crossover behind and before the pinhole plane

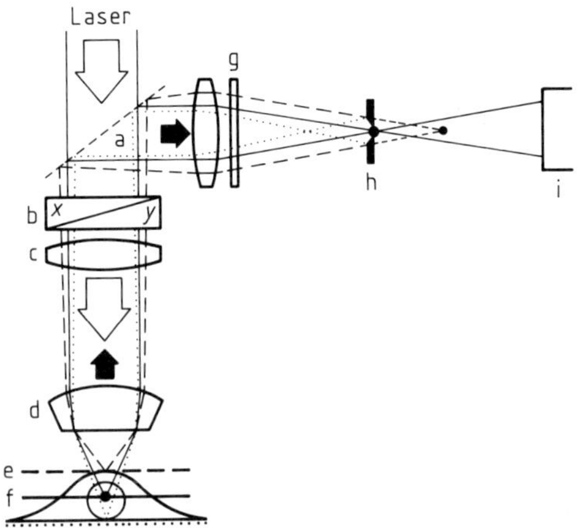

Figure 1.13. Ray path in the confocal microscope
a) Dichroic beam splitter; b) Scanner; c) Tube lens; d) Objective; e) Specimen; f) Focal plane; g) Filter; h) Pinhole; i) Detection system (photomultiplier)

so that the pinhole acts as a spatial filter. Numerous papers elucidate the basic aspects of confocal image formation [1.50]–[1.55].

Since only a small specimen point is recorded, the entire image of a specimen is generated by either moving the specimen (*stage scanning*) or moving the beam of light across the stationary specimen (*beam scanning*). The latter is utilized in all commercial instruments (Fig. 1.14). Theoretical considerations have shown that both scanning forms give identical resolution. The optical sectioning capability at lateral scanning opens a way to image a three-dimensional structure with a slight axial stepping of the stage (see Fig. 1.15).

1.6.2. Imaging Performance

The imaging characteristic in confocal microscopy is given by the properties of the specimen weighted by the applicable spatial confocal response function. An explanation based on wave optics suggests that the point spread functions of the illuminating path and the detection path (resembling the photon detection possibility) are convolved, consequently resulting in a narrow point resolution [1.56].

Describing image formation of an reflecting object and the corresponding coherent imaging properties must be distinguished from a fluorescent object viewed in incoherent imaging because in the latter the illuminating wavelength and the detected wavelength are different. In terms of resolution, the Rayleigh criterion determines the minimum resolvable distance between two points of equal brightness (see Eq. 2). For *incoherent imaging* using a lens with a certain numerical aperture NA at a wavelength λ, the *lateral resolution* d_l is [1.53]:

$$d_l = 0.46\ \lambda/NA \qquad (1.5)$$

With a wavelength of 514 nm (green argon laser line) and a numerical aperture of 1.4, this equation gives a resolution of $d_l = 157$ nm. Compared to conventional imaging (see Eq. 1.2), the resolution is ca. 32% higher [1.57].

An approximation of the *axial resolution* d_a is given by

$$d_a = 1.4\ \lambda/NA^2 \qquad (1.6)$$

The values (514 nm and $NA = 1.4$) give a d_a of 367 nm.

One way to measure the axial response of a confocal microscope is to record light intensity from a thin layer with infinite lateral extension. In a conventional microscope, no difference—and therefore no depth discrimination—are obtained. However, in confocal microscopy, a strong falloff occurs if the object is out of focus. The free width at half maximum (FWHM) of such a function is proportional to the optical section thickness [1.55], [1.58]. The optical section thickness also depends on the diameter of the pinhole. The optimum pinhole size D with a maximum deterioration of 10% in reflection is given by [1.54]

$$D = 0.95\ \lambda\ M/NA \qquad (1.7)$$

with magnification M, λ, and D given in micrometers. In fluorescence microscopy the size has to be reduced 10–30%, depending on the ratio of excitation to emission wavelengths. Another way to describe the imaging performance is by the modulation transfer function (MTF) (see also, → Photography; **A 20**, pp. 114–116); such measured and theoretically derived three-dimensional functions have been published [1.47], [1.59].

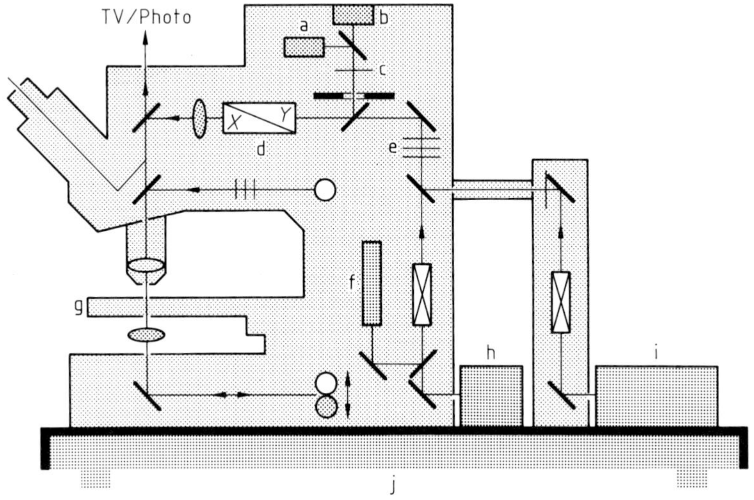

Figure 1.14. Setup of the confocal microscope, based on a research light microscope (Zeiss)
a), b) Photomultiplier tubes; c) Confocal variable pinhole; d) Scanners; e) Barrier filter; f) He–Ne laser (543 nm, 633 nm); g) Stage; h) Ar laser (488 nm, 514 nm); i) UV laser (350–360 nm); j) Antivibration table
Light of internal (He–Ne) and external (Ar ion or UV) lasers passes the scanner unit; switchable mirrors and barrier filters permit multiwavelength operation. Additional barrier filters are located between the confocal variable pinhole and the beam splitter adjacent to the photomultipliers.

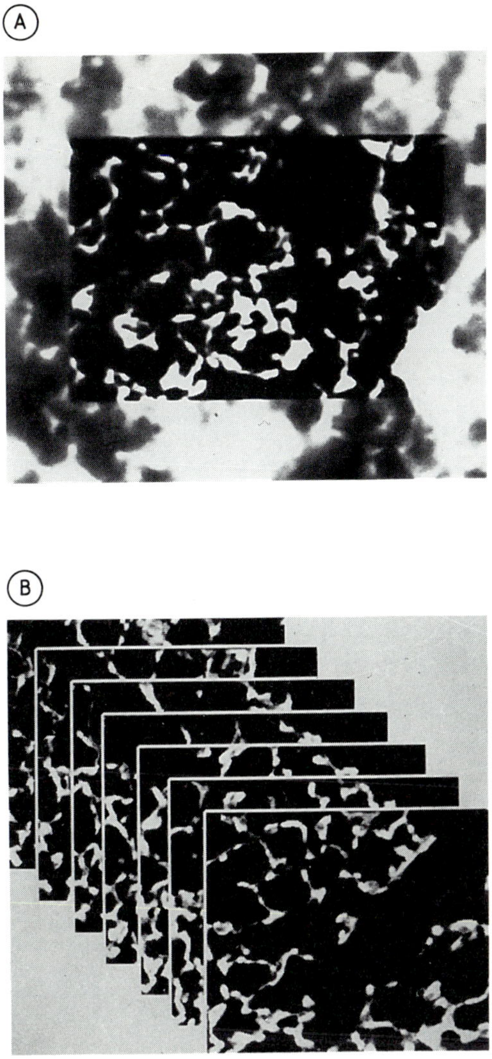

Figure 1.15. A) Conventional versus confocal imaging (inserted region of interest) of a lung structure in fluorescence; B) Multislice image sequence obtained by optical sectioning

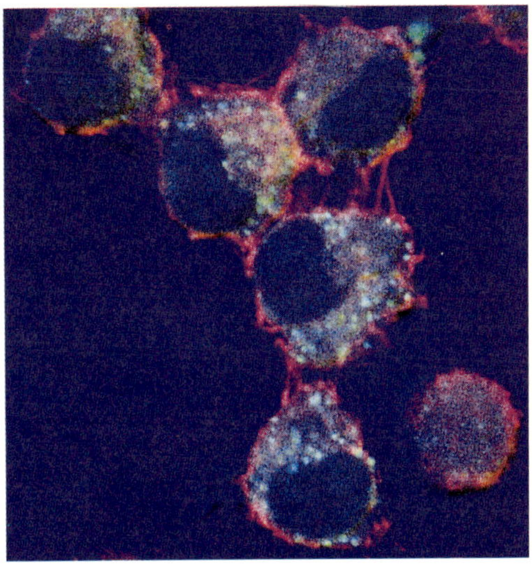

Figure 1.16. Multichannel confocal image of macrophages (J774, mouse cell line)
Extrathyroidal release of thyroid hormones from thyroglobulin is marked by fluorescence dye (first channel, green); actin cytoskeleton is labeled with rhodamine-phalloidin (second channel, red). Morphological information (third channel), providing a dark background in the cells, is detected in transmission.
(Permission of K. Brix, V. Herzog, Institut für Zellbiologie, Bonn, Germany)

1.6.3. Instrumentation and New Developments

Confocal microscopic equipment is available as an add-on to a standard microscope or in the form of more fully integrated and stable devices, including computer control equipment. Both upright and inverted types of microscopes are available. The inverted confocal microscope allows bigger specimens to be studied and offers easy handling of micromanipulating or injecting devices (see also Section 1.4). Beam scanning is obtained by rotating galvanometric mirrors. Axial control is realized by a stepping motor drive or a piezoelectric translator. Most instruments are equipped with several detectors to record multifluorescence events. Some also offer transmission detectors. Signals from different channels can then be mixed electronically (see Fig. 1.16). Recently, confocal transmission microscopy has also been studied. Improvement of contrast and resolution depending on the pinhole size of a thin specimen has been documented in a confocal transmission arrangement using differential interference contrast [1.60].

Variations of the confocal design have been developed, such as bilateral scanning in real time with a scanning slit [1.61]. Because of the fast scanning, TV detectors such as a sensitive CCD camera are best suited. Slit-scan confocal microscopes offer a better signal-to-noise ratio and real-time imaging capabilities but a slightly reduced resolution compared to a conventional CLSM.

Various types of lasers can be linked to the microscope [1.62], but up to now only continous-wave gas lasers have been used routinely. This includes argon-ion lasers having excitation lines at 488 and 514 nm and helium – neon lasers with a major line at 633 nm and a weaker line at 543 nm. The use of UV lasers requires achromatic lenses and specifically designed scanning mirrors. Lasers used in the UV range are argon-ion UV or argon-ion tunable lasers, as well as krypton-ion and helium – cadmium lasers. Depending on the emission lines of the fluorescent dyes, various filters are used in front of the photomultiplier, characterized by quantum efficiency and spectral response. The total sensitivity of a system is given by the transfer efficiency of the various optical components [1.63] and, in addition, the quantum efficiency of the detector and its associated electronics [1.64], [1.65].

Simplified types of laser scanning microscopes, such as the *fiber scanning optical microscope* have also been proposed [1.61]. Here the fiber can act both as a light delivering source and as a pinhole. A totally different physical approach to reduce out-of-focus blur is based on a physical phenomenon called *double-photon excitation*, whereby fluorescence is generated by simultaneous absorption of two photons of long wavelength at the size of a single molecule. The necessary energy emitted by pulse lasers is present only at the focal plane; thus a "confocal" effect is produced without any additional special optical equipment [1.66].

The diffraction pattern of the illuminating beam is elongated axially compared to the lateral direction (see Eq. 1.5 versus 1.6). This anisotropy is caused basically by the limited numerical aperture of the microscopic lens. An optimal diffraction pattern would be achieved by an unlimited, 360° aperture. This concept is known as the *4-pi microscope*. A prototype version uses two lenses centered around the specimen. The object in focus is illuminated from two sides, and fluorescence emanating from the object is detected through both lenses [1.56].

1.6.4. Imaging Modalities and Biomedical Applications

Imaging Modalities. Confocal contrast generation include reflectance, rescattering, and fluorescence. The first results in confocal imaging were obtained on unstained tissue in *reflectance* [1.67]. Reflectance is also the preferred imaging modality used in industrial inspection of semiconductors, often combined with methods such as optical beam induced conductivity (OBIC).

In biomedical science, the study of unstained tissue is limited due to reflectance. The reflectance must be intensified by metal impregnation (e.g., Golgi stain); one such example is given in Figure 1.17. Other examples include peroxidase – labeling with nickel intensification in neurobiology [1.68] or silver-stained nucleolar organizer region-associated proteins (AgNORs). Gold immunolabeling has also been used [1.69]. In *rescattering*, variations in the refractive index can give rise to reasonable contrast. One particular example is in ophthalmology [1.70] where structures in the transparent cornea such as cell nuclei, keratocytes, neurons, or the fibers of the lens can be visualized (see Fig. 1.18).

Fluorescence is the most commonly used imaging mode in confocal microscopy, such as, applications in cell biology [1.71], [1.72], which include the use of autofluorescence, specific dyes in combination with antibodies, and in situ hybridization. Comprehensive reviews of available fluorochromes are published elsewhere [1.69], [1.73] – [1.75]. UV lasers broaden the spectrum of available fluorescence dyes; even the autofluourescence present in reduced pyridine nucleotides [NAD(P)H] can be monitored [1.76].

Biomedical Applications. Optical sectioning devices, being noninvasive and having low radia-

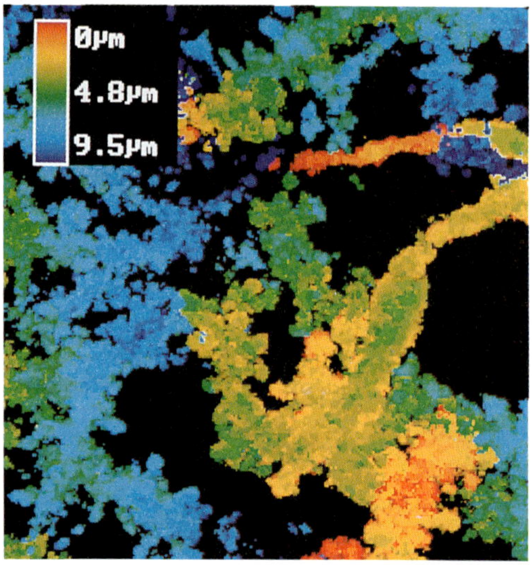

Figure 1.17. Golgi-stained neuron: three-dimensional display of a confocal data set acquired in reflection
Color depth coding visualizes the three-dimensional neuron; various spines of different size give a brushlike impression.
(Specimen courtesy of B. FORTMANN, Institut für Ernährungswissenschaften, Gießen, Germany)

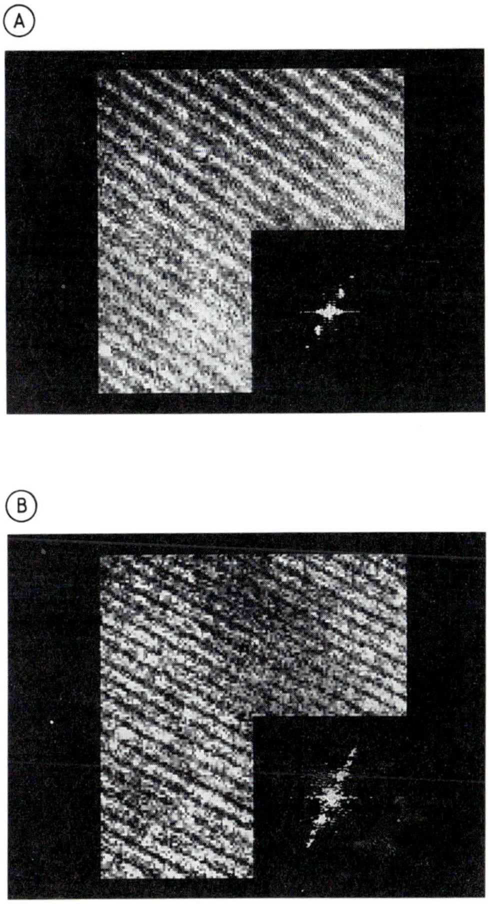

Figure 1.18. Confocal images of the eye's lens fibers, resulting from rescattering at microstructures with different indices of refraction
A) Imaging through the 400 μm transparent cornea; B) Imaging with cornea removed
The inserted Fourier power spectrum documents the loss in spatial frequency content.
(Specimen courtesy of B. R. MASTERS, USUHS, Bethesda, United States)

tion damage, are ideal for studying living specimen, organs, and tissues. With confocal microscopes, structures can be observed in vivo at full three-dimensional microscopic resolution. Small organisms or organisms in an early stage of development can be observed, as in embryogenesis [1.77], [1.78]. Concerning individual structures, morphological changes, such as the movement of cell compartments (see Fig. 1.19), growth of neurons [1.79], or synaptic plasticity in the retina, have also been studied [1.80]. Moreover, confocal arrangements allow optical trapping (i.e., user-controlled movement of small spheres filled

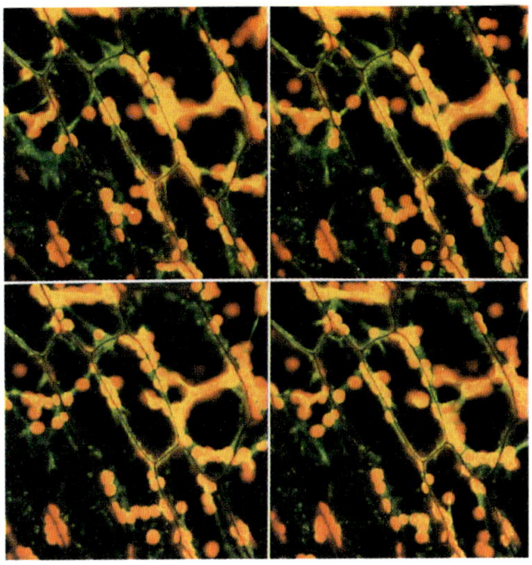

Figure 1.19. Application of confocal microscopy to plant cells (*Egeria densa Planchon*), in vivo
Cell boundaries and the movement of chloroplasts are visualized by autofluorescence (green) and FITC labeling (red).
(Permission of I. DAHSE, Institut für Biophysik, Jena, Germany)

with biochemical solution into and within cells). A new application of confocal microscopy is in clinical imaging, which includes the observation of wound healing, flow processes in veins, and opthalmology [1.70], [1.75] (see Fig. 1.18).

1.6.5. Computer Applications in Confocal Three-Dimensional Imaging

The imaging process of the confocal microscope samples small volumes sequentially, which are stored in computer memory. Subsequently, these volume elements or voxels are processed by computers to enhance intensities, to visualize structures three-dimensionally, and to segment and measure specific volumetric subtleties.

1.6.5.1. Visualization

Visualization of the optical serial sections is the most frequently used computer technique in confocal microscopy, independent of specific applications [1.72], [1.81], [1.82]. Most commercial instruments have three-dimensional software implementations ready for the user. Three-dimensional objects are displayed on the computer screen by staking up the individual sections. Such techniques have been reviewed in several articles [1.83]–[1.87]. Besides surface rendering based on contours, volumetric representations that rely on the intensities of the voxels are the preferred techniques in confocal microscopy. A comparative assessment of both techniques is given in [1.88]. Reconstructions are presented three-dimensionally on a computer graphics screen by using either visual cues, animation, of stereoscopic displays.

One method used in *volume rendering* is to define imaginary rays through the volume (*ray-casting techniques*) [1.89], [1.90]. The way of casting is accomplished by the definition of certain projection geometry. Given a central projection, the angles of the rays are defined by the matrix of the screeen and the center of projection. A common implementation traverses the data in a front-to-back order, which has the advantage that the algorithm can stop when a predefined property of the accumulation process has been reached. This class of methods is also referred to as *image order rendering*, since the matrix of the screen is used as the starting point for the accumulation process. Because of the projection geometry, such a ray-firing technique does not necessarily pierce the center of a voxel, and an interpolation has to be taken into account. To improve the output of ray-casting renderings, preprocessing of the voxel scenes may be performed as, for example, in the renditions of complex chromatin arrangements [1.91]. Modalities available in image order rendering include maximum projection (Fig. 1.20), integration, and surface mode (Fig. 1.18).

A much simpler geometry is realized in a parallel projection, such as in *object order rendering*. Here the volume is rotated according to the current viewing angle and then traversed in a back-to-front order [1.92]. During traversal, each voxel is projected to the appropiate screen pixel where the contributions are summed or blended, a situation that is referred to as compositing. During the accumulation process other properties of voxels besides the intensities may be considered, such as opacity, color, gradient values [1.93], [1.94], or the simulation of fluorescence processes (SFP) [1.95]. However, since volume rendering relies entirely on voxel attributes, it has some disadvantages if analytical representations are required. Therefore, elements of computer graphics are combined with volumetric renditions. Examples can be found in vector descriptions of volumetric data sets [1.96] or in geometric attributes expressing local properties

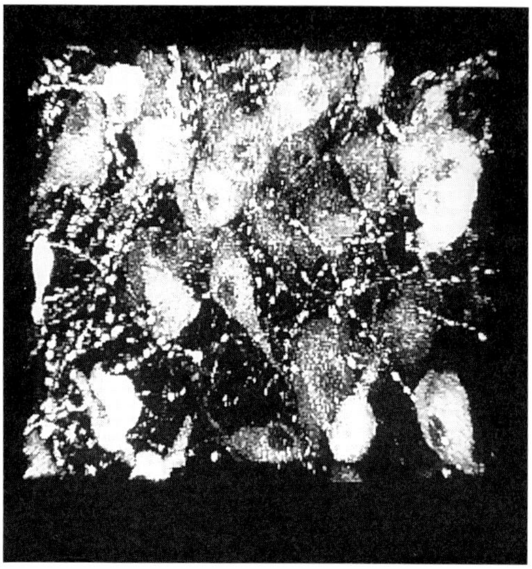

Figure 1.20. Volume rendering of a confocal sequence The neural network of immunohistochemically stained cells, together with connecting fibers in the neural optical tract of hamsters is visualized by this projection (Specimen courtesy of H. KORF, Zentrum für Morphologie, Frankfurt, Germany)

and topologies that can be embedded in the three-dimensional reconstruction [1.86], [1.94], [1.97]. The vast amount of data available in four-dimensional studies initiated new concepts in computing and visualization that utilize special data compression techniques [1.80].

1.6.5.2. Image Analysis and Measurement

Analysis applied to confocal microscopical data sets has as its aim to directly measure specific structures of any form and orientation in three dimensions. In many cases, undesirable phenomena must be corrected first. These include attenuation of the laser beam in thick specimens, photobleaching or saturation of the fluorescence, autofluorescence, and background noise, and the distortions caused by the embedding media [1.69], [1.75], [1.97 a]. The loss of resolution in thick specimens is shown in Figure 1.18. The correction of such phenomena is often limited by the underlying theoretical models selected, since the effects are highly object dependent.

Undesired effects in fluorescence confocal microscopy can be corrected by ratio imaging. Ratioing (i.e., the compensation of two channels by digital division) can substantially reduce cross-talk effects in double detectors [1.54]. Environmental markers indicating the pH value or the level of free calcium ions (Ca^{2+}) also require ratio imaging to compensate for structural densities or section thickness [1.19].

Due to the anisotropic resolution of the data sets, various ways of improving axial resolution are available; *deconvolving methods* use the point spread function of the optics, which has to be measured first [1.98]. The algorithms can be differentiated further by their ability to incorporate noise. The classical filter is the Wiener filter; new developments include the iterative *maximum likelihood estimation* [1.99]. This method was extended with a blind deconvolution algorithm, using certain constraints to model the point spread function, which is a promising approach for biomedical users (see Fig. 1.21). However, calculation times are still in the range of 1 h for some hundred interations. A different approach is the *nearest-neighbor deconvolution* developed for bright-field imaging [1.100], by taking into account the blur present in the planes above and below the focal plane. More recently, this method has been contrasted with a no-neighbor deconvolution method [1.101]. Besides directly improving data, isotropy can also be obtained by interpolating the lateral dimension [1.69]. Transformation of data from a cubic format into other lattices has also been proposed, based on mathematical morphology [1.102].

Sets in digital image processing to enhance, filter, segment, and identify structures are frequently applied to three-dimensional microscopic data sets. This can be done either section by section or by using the corresponding three-dimensional extension of digital filters, such as the Gauss, Laplace, Sobel, and Median filter or noise removal algorithms [1.91], [1.103]–[1.105]. Segmentation often requires interaction of the user, since automatic segmentation algorithms generally rely on the intensities of the voxels only. Once the binary images have been obtained, binary morphological operations using structuring elements for erosion and dilatation can be performed [1.106].

Digital measurements address the size and volume of structures, quantities, or forms and relation between volumetric subtleties. To measure volumes, all voxels that identify a certain structure are summed up. It is particularly difficult to measure the surface of structures, and most methods offer an estimation based only on stereological methods. Object counting in volumes usually requires the definition of guard boxes, which are moved throughout the volume. Objects falling into such boxes are evaluated by following certain stereological rules to avoid any bias. Stereological considerations are also used to estimate spatial statistics, such as numerical densities or nearest-neighbor distances; such methods have been used extensively in three-dimensional cytometry [1.69].

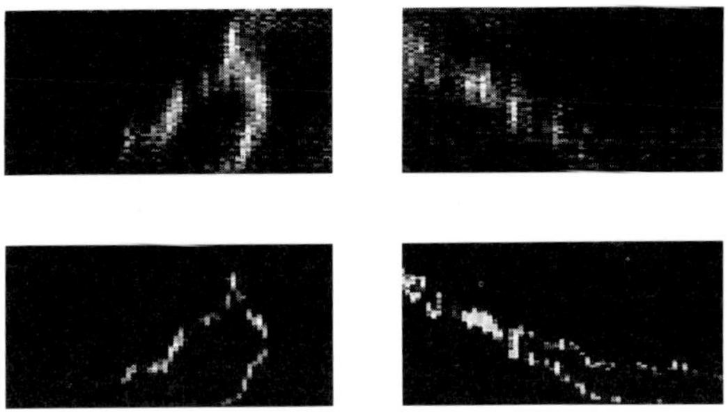

Figure 1.21. Projectional side views of rat hippocampal dendrites stained with Lucifer yellow
Top: Original confocal data set; Bottom: Restored data following 100 iterations of the maximum likelihood blind deconvolution algorithm
(Permission of T. HOLMES, Rensselaer Insitute, Troy, United States)

2. Electron Microscopy

2.1. Transmission Electron Microscopy

2.1.1. Introduction

The images produced in transmission electron microscopy are essentially due to local diffraction phenomena; absorption contrast plays only a minor role. Not only is electron diffraction responsible for the image formation; it also allows establishment of the orientation relationship between direct space, as observed in the image, and reciprocal space as imaged in the diffraction pattern (see → Structural Analysis by Diffraction, **B5**). Images and the corresponding diffraction patterns are equally important for a detailed interpretation; both should always be produced from the same selected area, with the specimen orientation unchanged. For this reason, most commercial microscopes have the ability to switch easily from the diffraction mode to the imaging mode and vice versa (see Section 2.1.5).

2.1.2. Scattering by Atoms: Atomic Scattering Factor

Electrons are scattered by atoms as a result of Coulomb interaction with the nucleus and the electron cloud. The atomic scattering factor $f_e(\theta)$ thus contains two terms of opposite sign

$$f_e(\theta) = \frac{me^2\lambda}{2h^2}[z - f_x(\theta)]/\sin^2\theta \tag{2.1}$$

where z is the atomic number; $f_x(\theta)$ the scattering factor for X rays; m the electron mass; e the electron charge; λ the wavelength of the electrons; and h Planck's constant [2.14].

The first term clearly relates to the nucleus, whereas the second term is due to the electron cloud. The interaction with matter is stronger ($\times 10^4$) for electrons than for X rays or neutrons, which interact only with the electron cloud or with the nucleus, respectively. As a result multiple scattering will not be negligible in electron diffraction experiments. Moreover, electron scattering is oriented mainly in the forward direction. Tables of $f_e(\theta)$ for different atoms are given in [2.15].

2.1.3. Kinematic Diffraction by Crystals

2.1.3.1. Lattice, Reciprocal Lattice

The amplitude diffracted by an assembly of atoms results from interference of the waves scattered by the atoms (i.e., the scattered amplitude in any given direction is obtained by summing the amplitudes scattered in that direction by the individual atoms and taking into account the phase differences due to path differences resulting from the geometry of the assembly). For a physical and mathematical description of the important terms lattice and reciprocal lattice, see [2.16] and →Structural Analysis by Diffraction, **B5**.

2.1.3.2. Geometry of Diffraction

Bragg's Law [2.17]. Like other diffraction phenomena in periodic structures, electron diffraction can be described in either direct or reciprocal space. Bragg's diffraction condition describes diffraction in direct space. A diffracted beam will be formed whenever the incident beam encloses with the set of lattice planes an angle $\theta_{\mathbf{H}}$, such that the path difference $2d_{\mathbf{H}} \sin\theta_{\mathbf{H}}$ between waves diffracted by successive lattice planes is an integer n of wavelengths λ of the radiation used:

$$2d_{\mathbf{H}} \sin\theta_{\mathbf{H}} = n\lambda \tag{2.2}$$

where **H** denotes the Miller indices of the plane. In Figure 2.1 this path difference is indicated by a thicker line. For 100-kV electrons, $\lambda = 4$ pm, and as a result, the Bragg angles $\theta_{\mathbf{H}}$ are very small—a few degrees at most.

Ewald Construction [2.18]. Diffraction conditions can also be formulated in terms of reciprocal space as

$$\bar{\boldsymbol{k}} = \bar{\boldsymbol{k}}_0 + \bar{\boldsymbol{g}} \tag{2.3}$$

where $\bar{\boldsymbol{k}}_0$ is the wave vector of the incident electron beam and $\bar{\boldsymbol{k}}$ the wave vector of the diffracted beam, where $|\bar{\boldsymbol{k}}| = |\bar{\boldsymbol{k}}_0| = 1/\lambda$; $\bar{\boldsymbol{g}}$ is a reciprocal lattice vector. Equations (2.2) and (2.3) have the same physical content. They can be obtained by expressing the conservation of energy and of momentum of the incident electrons on scattering. More details on Ewald's construction are given in [2.19] and in → Structural Analysis by Diffraction, **B5**.

Diffraction by a Thin Foil. Since electrons are strongly "absorbed" in solids the specimen must be a thin foil (< 300 nm); otherwise no diffracted beams will be transmitted. The number of unit cells along the normal to the foil is thus finite, whereas along directions parallel to the foil plane, the number of unit cells can be considered infinite. In such specimens the diffraction conditions are relaxed, and diffraction also occurs for

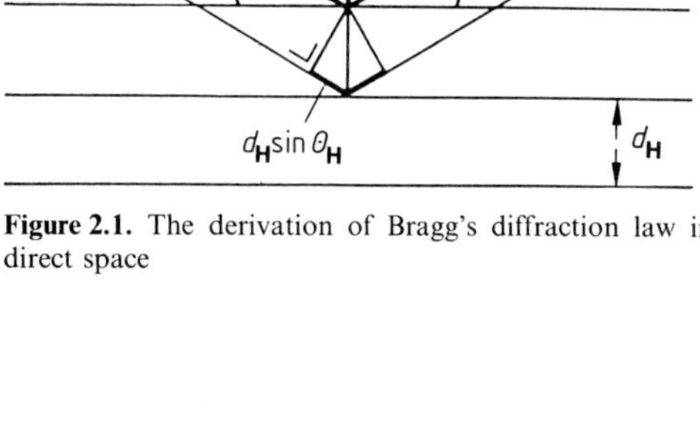

Figure 2.1. The derivation of Bragg's diffraction law in direct space

angles of incidence deviating somewhat from the Bragg angle. In reciprocal space this relaxation results in a transformation of the sharp reciprocal lattice nodes into thin rods (so-called relrods) perpendicular to the foil plane. The Ewald sphere intersects such a rod to produce a diffracted beam. The direction of this beam is obtained by joining the center of Ewald's sphere with this intersection point. The resulting diffraction pattern can to a good approximation be considered to be a planar section of the reciprocal lattice. Indexing of the diffraction spots is simple and unambiguous.

The distance $\bar{s}_{\bar{g}}$ by which the lattice node G is missed by Ewald's sphere is called the *excitation error*. It is a vector along the direction of the foil normal joining the reciprocal lattice node G to the intersection point with Ewald's sphere. It is positive when pointing in the sense of the incident beam and negative in the opposite case.

Column Approximation [2.20]. The small magnitude of the Bragg angles causes the electrons to propagate along narrow columns parallel to the beam direction. The intensity observed in a point at the exit face of the specimen is thus determined by the amplitude scattered by the material present in the column located at that point. In a perfect foil the intensity does not depend on the considered column, but when defects are present, it does.

Kinematic Rocking Curve. The amplitude of the beam scattered by a foil of thickness z_0 is obtained by summing the amplitudes scattered by the volume elements along a column perpendicular to the foil surface and taking into account the phase differences due to their depth position in the column. For a perfect foil the intensity of the scattered beam, I_s for reflection **H** is given by

$$I_s(s, z_0) = F_{\mathbf{H}}^2 \sin^2 \pi s z_0/(\pi s)^2 \qquad (2.4)$$

where $F_{\mathbf{H}}$ is the structure factor [2.3] with indices **H**.

The dependence of I_s on s (i.e., on the direction of the incident beam and on the foil thickness z_0) is called the *rocking curve* (see Fig. 2.2). The separation of the zeros is $1/z_0$, and the full width of the central peak is $2/z_0$.

Bent Contours [2.21]. For a curved foil, s varies along the specimen; the loci of constant s, imaged as bright and dark fringes, are called bent extinction contours. They form as shown schematically in Figure 2.3 for a cylindrically bent foil.

Thickness Contours. The thickness dependence of I_s is periodic with period $1/s_{\bar{g}}$: the period thus diverges for $s_{\bar{g}} = 0$ (i.e., for the exact Bragg condition). The loci of equal thickness are also imaged as lines of equal intensity; they are called thickness extinction contours or wedge fringes.

Kikuchi Lines [2.22]. In rather thick specimens a second type of diffraction pattern becomes prominent, which consists mainly of lines rather than spots. These lines occur in parallel pairs; the line closest to the center of the diffraction pattern is darker, whereas the one further from the center is brighter than the background (Fig. 2.4). These lines are the intersections of

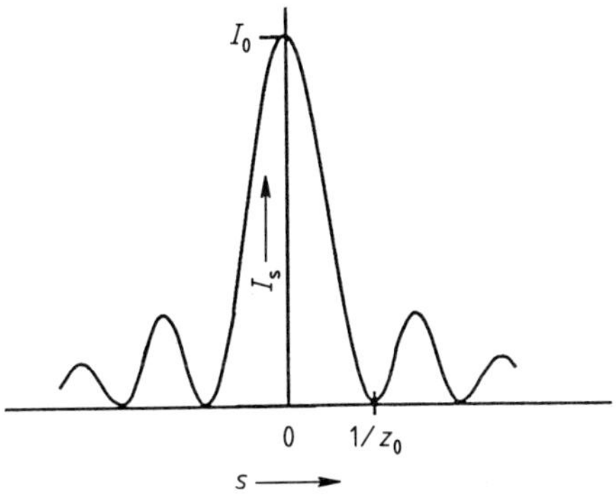

Figure 2.2. Plot of scattered intensity I_s versus excitation error, i.e., the "rocking curve" according to the kinematic approximation

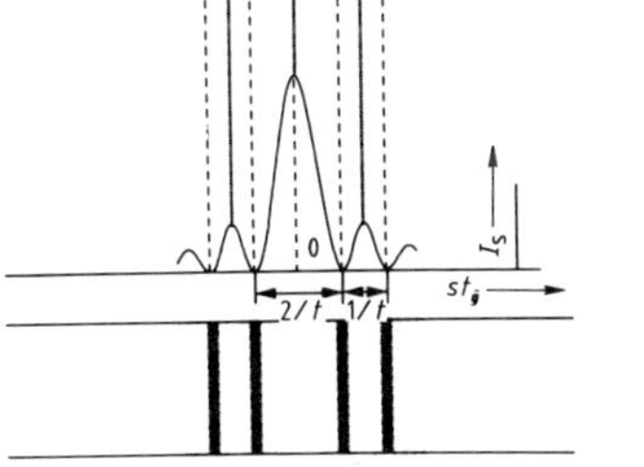

Figure 2.3. Formation of bent extinction contours by a cylindrically bent foil

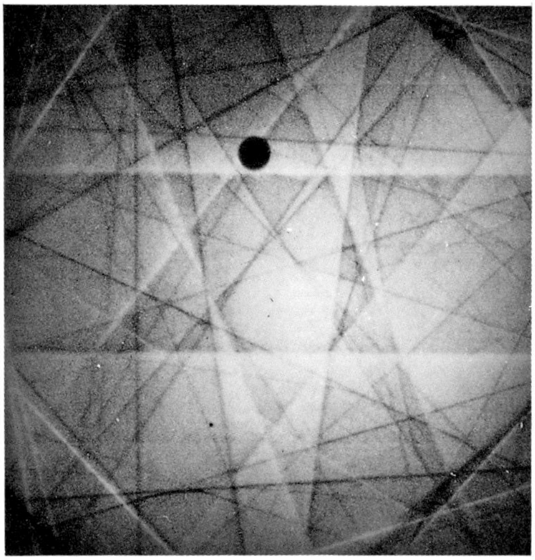

Figure 2.4. Kikuchi line pattern produced by a silicon foil Note the parallel pairs of bright (excess) and dark (deficiency) lines. The deficiency lines are close to the center. The excess lines are separated in angle by $2\theta_H$ from the corresponding deficiency lines

Ewald's sphere with rather flat double cones having semiapex angles equal to $90° - \theta_H$, where θ_H is the Bragg angle associated with the pair of lines ($\theta_H \approx 1-2°$). The angular separation of the bright–dark line pair is $2\theta_H$.

If the Bragg condition for a set of lattice planes is satisfied exactly, the bright Kikuchi line associated with that set of planes will pass through the corresponding Bragg spot, whereas the dark one will pass through the origin. The separation of a Bragg spot from its corresponding Kikuchi line is a measure of the excitation error s of that reflection. As a result, the Kikuchi line pattern provides a means to measure s [2.23].

2.1.4. Dynamic Diffraction by Crystals

2.1.4.1. General Considerations

According to kinematic theory the intensity of diffraction spots is proportional to the square of the structure factor (Eq. 4). This simple proportionality is lost in reality because of multiple diffraction effects, which are taken into account adequately in dynamic theory [2.24].

Although kinematic theory does not predict the intensities of diffraction spots correctly, it is useful because the geometric features of diffraction are well described. For large s or very small foil thicknesses the relative spot intensities are qualitatively reproduced. For $s = 0$, kinematic theory breaks down since a diffracted beam with diffraction vector $\bar{g}_1$ may then acquire an amplitude comparable to that of the incident beam, and thus act as an incident beam and give rise to diffraction in the reverse sense (i.e., with diffraction vector $-\bar{g}_1$).

For most orientations of the incident beam, many diffracted beams are excited simultaneously and a section of reciprocal space is imaged. However by carefully orienting a sufficiently thick specimen the intensity of a single diffracted beam can be maximized: this is called a *two-beam case*. Such specimen orientations should be used, whenever possible, in the quantitative study of defects.

2.1.4.2. Basic Equations

The two-beam dynamic theory describes the interplay between transmitted and diffracted beams by means of a set of coupled differential equations of the form [2.4], [2.25].

$$\begin{cases} \mathrm{d}T/\mathrm{d}z = (\pi i/t_{-g}) S \exp 2\pi i s z & (2.5\,\mathrm{a}) \\ \mathrm{d}S/\mathrm{d}z = (\pi i/t_g) T \exp -2\pi i s z & (2.5\,\mathrm{b}) \end{cases}$$

where T and S are the complex amplitudes of the transmitted and scattered beams, respectively; z measures the depth in the foil. The exponential factors take into account the growing phase shift due to the excitation error, with increasing distance z behind the entrance face. The factor i takes into account the phase jump on scattering; t_g and t_{-g} are called the *extinction distances*. The extinction distance, which has the dimension of length, is a measure of the strength of the reflection $\bar{g}$. For low-order reflections of elemental metals it is of the order of a few tens of nanometers, but for weak superstructure reflections it may be several hundred nanometers. Whether a foil is "thick" or "thin" depends on the number of extinction distances in the foil thickness. Kinematic theory is valid only for foils with a thickness that is a fraction of an extinction distance.

Absorption can phenomenologically be accounted for by assuming the extinction distances to be complex [2.26]. Formally $1/t_g$ is replaced by $1/t_{\bar{g}} + i/\tau_g$, where τ_g is called the absorption length. Empirically, it is found that τ_g ranges from $5\,t_{\bar{g}}$ to $15\,t_{\bar{g}}$. The substitution can be made, either in the Equation set (2.5) or directly in the solution of this set.

2.1.4.3. Dynamic Rocking Curve

By ignoring absorption and taking into account the fact that $T = 1$ and $S = 0$ at $z = 0$ (entrance face), integration of Equation set (2.5) gives

$$T(s, z) = [\cos \pi \sigma z - i(s/\sigma) \sin \pi \sigma z] \exp \pi i s z \quad (2.6\,a)$$

$$S(s, z) = (i/\sigma t_g) \sin \pi \sigma z \exp \pi i s z \quad (2.6\,b)$$

and thus

$$I_s = S S^* = (1/\sigma t_g)^2 \sin^2 \pi \sigma z \quad (2.7)$$

$$I_T = 1 - I_s \quad (2.8)$$

where

$$1/\sigma = t_g/(1 + s^2 t_g^2)^{1/2} \quad (2.9)$$

For $s = 0$ this reduces to $\sigma = 1/t_g$ (i.e., the depth period is now $t_{\bar{g}}$ and for large s: $\sigma = s$). The divergence of kinematic theory for $s = 0$ is now removed. The depth variation of the transmitted and scattered beams, leading to thickness extinction contours, is represented schematically in Figure 2.5 for $s = 0$ as well as for $s \neq 0$.

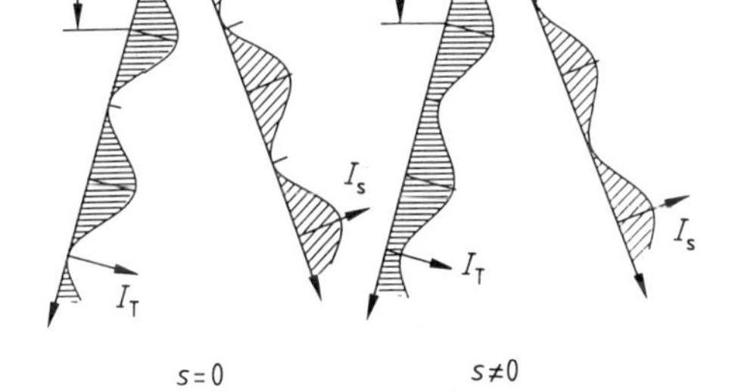

Figure 2.5. Interplay between incident and diffracted beam according to the dynamic theory
A) Bragg's condition is satisfied exactly, the depth period is t_g; I_S and I_T pass periodically through zero; B) $s \neq 0$ the depth period is now $1/\sigma_g$; I_T varies periodically but does not pass through zero; I_S and I_T are complementary since absorption was neglected

2.1.4.4. Anomalous Absorption, Bormann Effect [2.26], [2.27]

Taking absorption into account gives more complex rocking curves of the type shown in Figure 2.6 for a foil of thickness $t = 3\,t_{\bar{g}}$ and $\tau_g = 10\,t_{\bar{g}}$. The curve for the diffracted beam (b) is symmetric, whereas the curve representing the transmitted beam (a) is asymmetric in s. For the same absolute value of s, the transmitted intensity is larger for $s > 0$ than for $s < 0$. This asymmetry is known as the Bormann effect; it was demonstrated initially for X-ray diffraction under dynamic conditions [2.27]. The steep change of I_T with s in the vicinity of $s = 0$ causes a large sensitivity of I_T to small changes in orientation.

2.1.4.5. Lattice Fringes [2.28]

The Electron Wave Function at the Exit Face. The amplitude of the transmitted wave for a unit incident wave can be represented by $T \exp 2\pi i \bar{\boldsymbol{K}} \cdot \bar{\boldsymbol{r}}$, where $\bar{\boldsymbol{K}}$ is the wave vector of the incident wave inside the crystal (i.e., corrected for refraction at the vacuum–crystal interface). Similarly the diffracted wave can be represented by $S \exp 2\pi i (\bar{\boldsymbol{K}} + \bar{\boldsymbol{g}}) \cdot \bar{\boldsymbol{r}}$, where $\bar{\boldsymbol{g}}$ is the diffraction vector (i.e., the reciprocal lattice vector corresponding to the excited reflection). The wave function of the electrons emerging at the exit face of the foil is then

$$\psi = (T + S \exp 2\pi i \bar{\boldsymbol{g}} \cdot \bar{\boldsymbol{r}}) \exp 2\pi i \bar{\boldsymbol{K}} \cdot \bar{\boldsymbol{r}} \quad (2.10)$$

The observed intensity

$$I_\psi \equiv \psi \psi^* = I_T + I_S + 2\sqrt{I_T I_S} \sin(2\pi g x + \varphi) \quad (2.11)$$

with

$$\tan \varphi = (s/\sigma) \tan \pi \sigma t\,; \quad (2.12)$$

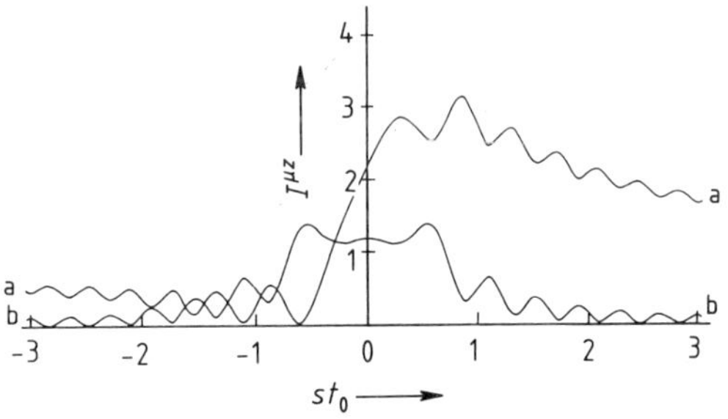

Figure 2.6. Rocking curves for transmitted (a) and scattered (b) beams according to the two-beam dynamic theory, with anomalous absorption taken into account

x is measured along the exit surface in the direction of $\bar{g}$: further, the following abbreviations are used: $I_T \equiv TT^*$; $I_S \equiv SS^*$. This expression represents sinusoidal fringes with a period $1/|\bar{g}|$; they can be considered as forming an image of the lattice planes $\bar{g}$.

2.1.4.6. Faulted Crystals

Planar Interfaces. Planar interfaces are imaged as fringes parallel to the foil surface; their characteristics depend on their geometric features. *Translation interfaces* (stacking faults, out-of-phase boundaries, etc.) separate two crystal parts related by a parallel translation $\bar{\boldsymbol{R}}$ (Fig. 2.7 A), and *domain boundaries* separate domains that differ slightly in orientation (i.e., for which the excitation errors are different $\Delta s = s_1 - s_2$; (Fig. 2.7 B) [2.29].

Diffraction Equations for Faulted Crystals. The displacement over R_0 (R_0 = constant) of the exit part of a column with respect to the entrance part (Fig. 2.7 A), the two parts being separated by a fault plane, can be accounted for by noting that a supplementary phase change $\alpha = 2\pi\bar{g} \cdot \bar{\boldsymbol{R}}$ occurs on diffraction but not on transmission (see Fig. 2.8). The path difference Δ between diffracted waves 1 and 2 is $\Delta = 2R_0 \sin\theta_H$, where θ_H is the Bragg angle and hence $2d_H \sin\theta_H = \lambda$ with $d_H = 1/|\bar{g}|$. This gives for the phase change $\alpha = (2\pi/\lambda)\Delta = 2\pi \boldsymbol{g} R_0$, where R_0 is in general the component of $\bar{\boldsymbol{R}}$ along $\bar{g}$ (i.e., $gR_0 = \bar{g} \cdot \bar{\boldsymbol{R}}$).

The equations describing diffraction by the displaced exit part are obtained by the substitution $S \to S \exp -i\alpha$ in Equations (2.5). The amplitudes T and S emerging from a column intersecting such an interface are then obtained by integrating the Equation (2.5) set along a column down to the level of the interface and subsequently further down to the exit face, using the adapted form of Equation (2.5) (Fig. 2.8).

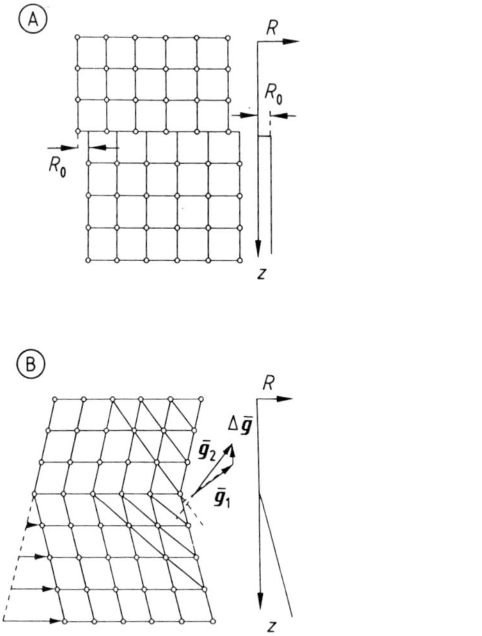

Figure 2.7. Schematic of two different types of interfaces A) Translation interfaces with constant displacement vector $\bar{\boldsymbol{R}}$; B) Domain boundary between two regions for which the excitation error is slightly different; the displacement R increases with increasing distance from the interface (the homologous diffraction vectors $\bar{g}_1$ and $\bar{g}_2$ are slightly different: $\Delta\bar{g} = \bar{g}_2 - \bar{g}_1$)

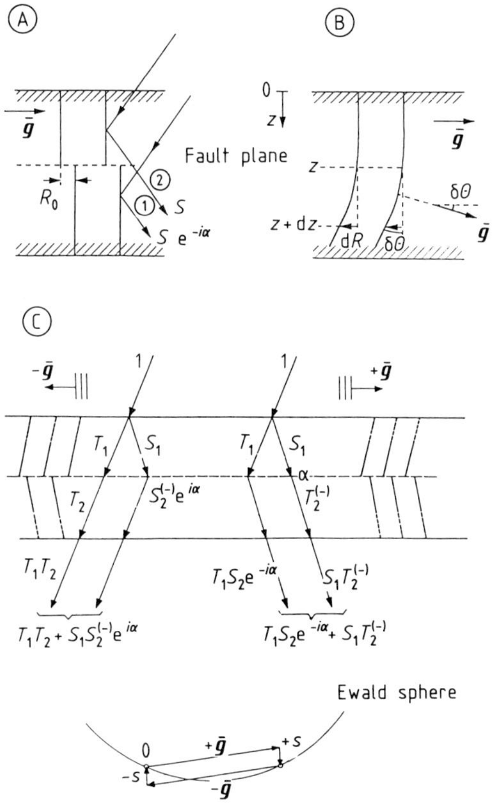

Figure 2.8. Diffraction by a faulted crystal
A) Effect of a translation interface with vector $\bar{\boldsymbol{R}}_0$; B) Effect of a displacement field $\bar{\boldsymbol{R}}(\bar{r})$; C) Schematic of the expressions for transmitted and scattered beam amplitude for a foil containing a general interface parallel to the foil surfaces

The effect on T and S of the presence of a defect described by a displacement field $\bar{R}(\bar{r})$ (for instance due to a dislocation) (Fig. 2.8) can be taken into account in Equations (2.5) by replacing the constant s by an effective local value s_{eff} which now depends on z, i.e., on the depth along the columns, and which is given by $s_{eff} = s + \bar{g}(d\bar{R}/dz)$. This can be shown rigorously but it can also be understood intuitively by the following geometrical considerations (Fig. 2.8 B).

Let the diffraction vector of the considered set of lattice planes be $\bar{g}$ at the level z below the entrance face. At level $z + dz$ the displacement $\bar{R}$ has increased by $d\bar{R}$ and as a result the local vector $\bar{g}$ is rotated over a small angle $\delta\theta \simeq \tan\theta = dR/dz$. The change in s corresponding to this rotation of $\bar{g}$ over $\delta\theta$ is given by $\delta s = g\,\delta\theta$, i.e., by $g(dR/dz)$ or somewhat more general by $d(\bar{g}\cdot\bar{R})/dz$, expressing that only displacements which change the orientation of the vector $\bar{g}$ are operative.

In the particularly simple case of a domain boundary as modeled in Figure 2.7 B, $R = kz$ (k = constant) in the exit part. Hence, at the level of the domain boundary, s changes abruptly into $s + k$ along the integration column.

Diffraction by a Crystal Containing a Planar Interface [2.30]. The amplitudes of the transmitted and scattered beams for a foil containing a planar interface, with displacement vector $\bar{R}$ (Fig. 8 B), can be formulated as

$$T(s_1, s_2, z_1, z_2, \alpha) = T_1(s_1, z_1)\,T_2(s_2, z_2) + S_1(s_1, z_1)\,S_2(-s_2, z_2)\exp i\alpha \qquad (2.13)$$

$$S(s_1, s_2, z_1, z_2, \alpha) = T_1(s_1, z_1)\,S_2(s_2, z_2)\cdot\exp(-i\alpha) + S_1(s_1, z_1)\,T_2(-s_2, z_2) \qquad (2.14)$$

with $\alpha = 2\pi\bar{g}\cdot\bar{R}$. These relations can also be derived from the adapted set of Equation (2.5).

The first equation states that the amplitude of the transmitted beam T results from the interference between the doubly transmitted beam (i.e., the beam transmitted by the first part T_1 and subsequently transmitted by the second part: $T_1 \times T_2$) and the doubly scattered beam $S_1 \times S_2^{(-)} \exp i\alpha$. The factor $\exp i\alpha$ takes into account the phase shift on the amplitude S_2 scattered by the second part, resulting from the translation of the second part of the foil.

A similar interpretation can be given to the second expression. The amplitudes $T_1(s_1, z_1)$, $T_2^{(-)}$, $S_1^{(-)}$, $S_2^{(-)}$ are given by Equations (2.6 a) and (2.6 b). The indices 1 and 2 refer to the entrance and exit parts, respectively: the minus superscript means that in the corresponding Equations (2.6 a) and (2.6 b), s must be replaced by $-s$, because diffraction occurs from the negative side of the set of lattice planes (i.e., the diffraction vector is $-\bar{g}$).

If $s_1 = s_2$ but $\alpha \neq 0$, Equations (2.13) and (2.14) refer to a pure translation interface. If $s_1 \neq s_2$ and $\alpha = 0$, the expressions are applicable to a domain boundary. For mixed interfaces, $s_1 \neq s_2$ as well as $\alpha \neq 0$. If the interfaces are parallel to the foil surfaces, transmitted and scattered intensities are constant over the specimen area, but they vary in a pseudoperiodic fashion with the thicknesses z_1 and z_2 of entrance and exit parts, with $z_1 + z_2 = z_0$ (total thickness). The expressions describing this variation explicitly are obtained by substituting Equations (2.6 a) and (2.6 b) into (2.13) and (2.14).

If the interfaces are inclined with respect to the foil surfaces this depth variation is displayed as a pattern of fringes parallel to the foil surfaces in the area where the two wedge-shaped crystal parts overlap.

The fringe profiles of α-fringes ($s_1 = s_2$; $\alpha \neq 0$) and δ-fringes ($\Delta s = s_1 - s_2 \neq 0$; $\alpha = 0$) in bright field (BF) and dark field (DF), exhibit different properties. The symmetry properties allow the fringe patterns due to the two kinds of interfaces to be distinguished. The properties of α-fringes with $\alpha = \pi$ are singular [2.31].

2.1.4.7. Moiré Patterns [2.32]–[2.34].

The superposition of two identical crystal foils having a small orientation difference θ (rotation moiré) or of two parallel crystal foils with a small difference in lattice parameters (parallel moiré) gives rise to a fringe pattern, which can to a good approximation be considered as the coincidence pattern of the lattice fringes corresponding to the diffraction vectors $\bar{g}_1$ and $\bar{g}_2$ active in the two crystal foils.

The fringes are parallel to the average direction $\bar{g}$ of the two diffraction vectors $\bar{g}_1$ and $\bar{g}_2$ in the case of rotation moiré patterns and perpendicular to the common direction of the two diffraction vectors in the case of a parallel moiré. From the moiré spacing Δ, small differences can be deduced in the lattice parameters d_1 and d_2 of the two components of the sandwich since $\Delta = d_1 d_2/(d_1 - d_2)$, or the small orientation difference $\theta = |\Delta\bar{g}|/|\bar{g}|$ in the case of a rotation moiré. Figure 2.9 shows an optical analogue for the two types of moiré fringes produced by the superposition of two line patterns.

2.1.5. Operating Modes of the Electron Microscope [2.4], [2.6], [2.7]

2.1.5.1. Microscope Optics

The ray paths in an electron microscope are shown in Figure 2.10, according to the geometrical optics approximation, for the two main operating modes: high-magnification, high-resolution imaging, and selected area diffraction.

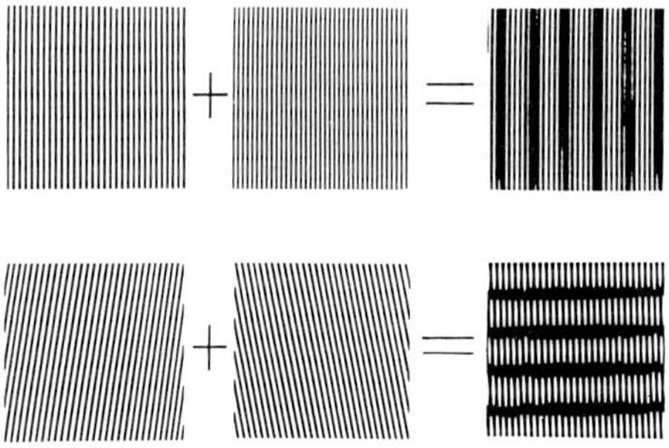

Figure 2.9. Optical analogue for the formation of moiré fringes in a sandwich crystal
(Upper) The two crystals are parallel, but the lattice parameters are slightly different; (Lower) The two crystal parts exhibit a slight orientation difference

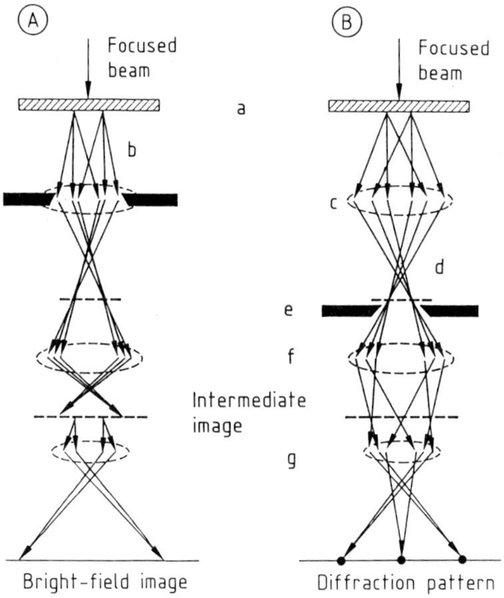

Figure 2.10. Schematic of the beam path in an electron microscope according to the geometrical optics approximation
A) High-magnification, high-resolution mode; B) Selected area diffraction mode
a) Specimen; b) Objective aperture; c) Objective lens; d) Field-limiting aperture; e) Gaussian image plane; f) Intermediate lens; g) Projector lens

The microscope is essentially a three-lens system: an objective lens (c), an intermediate lens (f), and a projector lens (g). Each lens may be a composite lens. A movable selector aperture (b) is present in the image plane of the objective lens; a second aperture (d) is placed at the objective lens, close to the back focal plane. With the first aperture a small area (≤ 1 μm) of the image (i.e., of the specimen) is selected, with the second aperture a single beam is selected or a number of the image-forming diffracted beams.

The characteristics of the objective lens are crucial because they determine to a large extent the image resolution and the contrast. The other two lenses mainly provide the desired magnification.

Although magnetic lenses normally rotate the image about the optical axis, in microscopes designed at the end of the 1980s, these rotations are compensated by a suitable device, and the image and the diffraction pattern have the same orientation.

2.1.5.2. High-Resolution, High-Magnification Mode

In the first (Fig. 2.10) imaging ray path the beam produced by a source (tungsten filament, LaB_6, field emission gun; see also Section 2.2.2.1) and collimated by a condenser lens system, is scattered by the object, and an image is formed in the image plane of the objective lens. By means of the selector aperture, an area of the specimen is selected and magnified by the intermediate lens, which is focused on the image plane of the objective lens and provides a magnified image in the image plane of the intermediate lens. This image serves as the object for the projector lens, which forms the final image either on a fluorescent screen, on a photographic plate, or on an image intensifier followed by a TV camera. In the latter case—which is virtually a requirement for successful high-resolution work—the image can be viewed on a TV monitor and the final adjustments performed before the image is recorded photographically. Electronic recording on tape is of course possible directly via the TV signal; this mode is used mainly for so-called in situ dynamic studies.

2.1.5.3. Diffraction Mode (Figure 2.10 B)

When operating in the diffraction mode the intermediate lens (f) is weakened (i.e., its focal length is enlarged) so as to cause the back focal plane of the objective lens (c) to coincide with the object plane of the projector lens (g). This produces a magnified image of the diffraction pattern. In doing so, the selected area is not changed since only the intermediate lens current is changed; the diffraction pattern is thus representative of the selected area. The area selcted in the diffraction mode is much larger than the field of view under high-resolution conditions.

As the electron beam passes through the specimen, various interactions occur next to diffraction: characteristic X rays are produced, the electrons suffer energy losses, etc. These effects can be used for analytical applications (see Section 2.1.9).

2.1.6. Diffraction Contrast Images [2.4], [2.23]

2.1.6.1. Imaging Modes

Diffraction contrast images are usually obtained under two-beam conditions. By means of the aperture placed close to the back focal plane of the objective lens, either the transmitted or the diffracted beam is selected and the corresponding diffraction spot is highly magnified (Fig. 2.11). The image obtained in this way is a map of the intensity distribution in the selected diffraction spot. The image made in the transmitted beam is a *bright-field image*, whereas the image made in the diffracted beam is a *dark-field image*. Since the electrons are confined to narrow columns, the latter can be considered the picture elements of this image. The amplitude of the transmitted (or scattered beam) associated with a point at the back surface of the specimen is obtained by summing the contributions of the volume elements along a column parallel to the incident beam centered on that point. The amplitude depends on how the excitation error s changes along the column, which in turn varies with changes in the orientation of the diffracting planes along the column. Strain fields cause such orientation changes and can thus be imaged in diffraction contrast, that it allows greater understanding of how defect images come about. Diffraction contrast images do not reveal the crystal structure but are very sensitive to the presence of strain fields caused by defects such as dislocations.

2.1.6.2. Dislocation Contrast [2.1], [2.4]

Consider a foil (Fig. 2.12) containing an edge dislocation in E, with an orientation such that in the perfect parts of the foil the amplitude of the incident beam is approximately equally divided between the transmitted and the diffracted beams. To the left of the dislocation the considered lattice planes are locally inclined in such a way as to better satisfy Bragg's condition (i.e., s is smaller) than in the perfect parts; more electrons are diffracted than in the perfect part. A lack of intensity will then be noted left of the dislocation in the transmitted beam, leading to a dark line in the bright-field image. At the right of the dislocation, Bragg's law is less well satisfied

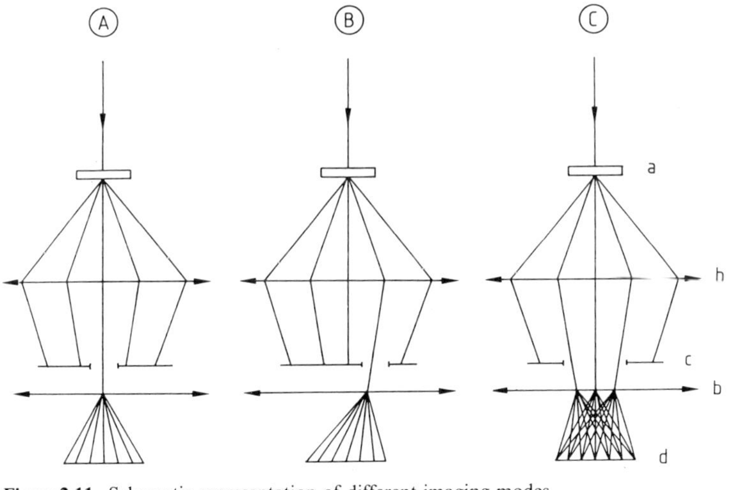

Figure 2.11. Schematic representation of different imaging modes
A) Bright-field diffraction contrast mode; B) Dark-field diffraction contrast mode; C) High-resolution imaging mode
a) Specimen; b) Lens; c) Aperture; d) Screen

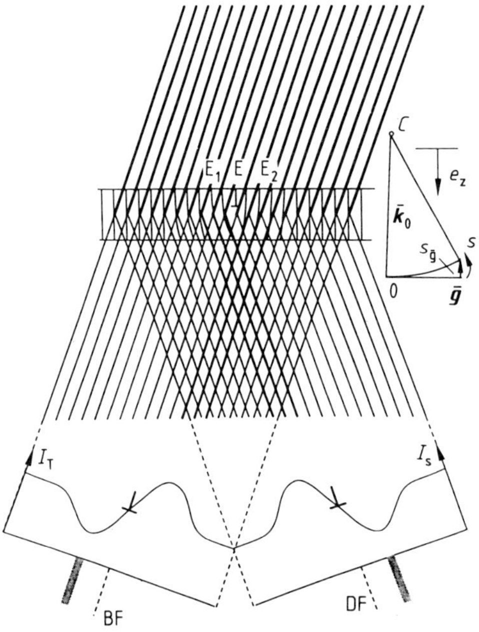

Figure 2.12. Schematic representation of the image formation at an edge dislocation in a thin foil
The line thickness is a measure of the beam intensity. The image profiles for bright-field (BF) and dark-field (DF) images are represented schematically along with diffraction conditions in the perfect part of the foil

than in the perfect part. This leads to a lack of intensity in the scattered beam and thus to a dark line in the dark-field image. The image is on one side of the dislocation: the side on which the dark line occurs in the bright-field image is called the *image side*. According to this model the image side is where the lattice planes are tilted into the Bragg orientation.

The conditions for kinematic diffraction [2.35] are best approximated in the *weak-beam* method, which consists of making a dark-field image in a weakly excited diffraction spot. The dislocation image then consists of a narrow bright line on a darker background.

2.1.6.3. Extinction Conditions for Defects

Diffraction by a family of lattice planes that remain undeformed by the presence of the dislocation will not produce an image of the dislocation [2.4], [2.36].

Quantitative images based on dynamic theory can be computer simulated by using the methods developed in [2.37]. A comparison of simu-

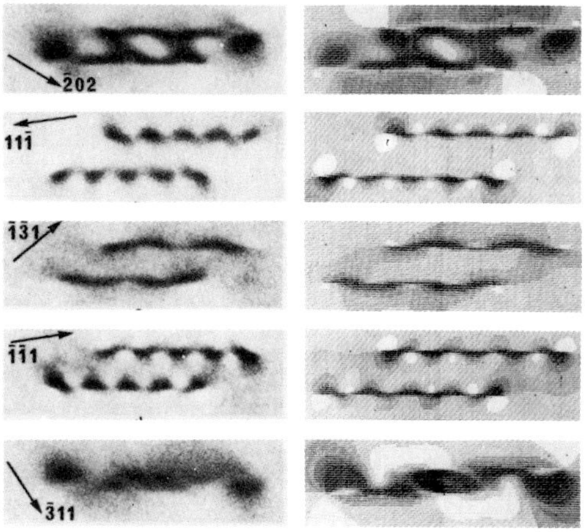

Figure 2.13. Examples of computed images (right) and the corresponding observed images (left) of the same dislocation by using different diffraction vectors [2.37]

lated and observed images allows determination of all the relevant parameters of various dislocation types (Fig. 2.13). More details are given in [2.38].

2.1.6.4. Domain Textures

Phase transitions in crystals are usually accompanied by changes in symmetry, the space group of the low-temperature phase being a subgroup of the space group of the high-temperature phase. On cooling, the crystal breaks up in to *orientation variants* (or twins) related by the lost symmetry elements of the high-temperature phase and *translation variants* (out-of-phase boundaries) related by the lattice translations lost on forming the superlattice [2.39]. The composite diffraction pattern of a domain texture is the superposition of the diffraction patterns due to individual domains. The diffraction pattern taken across a *reflection twin interface* contains one row of unsplit spots, which is perpendicular to the mirror plane. All other diffraction spots are split along a direction parallel to the unsplit row of spots [2.3].

In diffraction contrast images made in a cluster of *split diffraction spots*, orientation domains exhibit differences in brightness (i.e., domain contrast), especially for conditions close to $s = 0$ where the intensity variation with s is steep (Fig. 2.14). The presence of translation variants is not reflected in the diffraction pattern, except by some diffuse scattering.

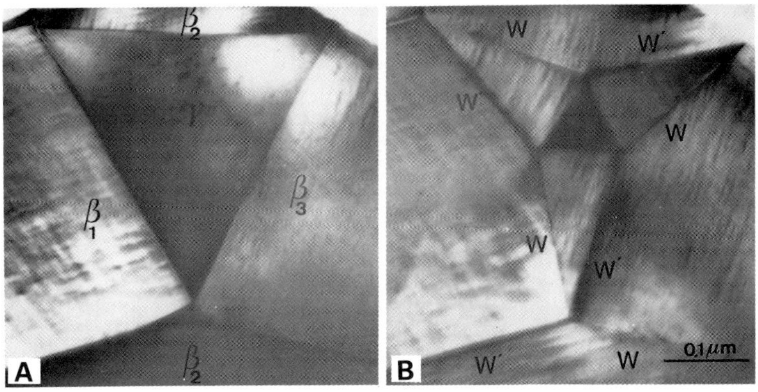

Figure 2.14. Domain texture in β-lead orthovanadate resulting from the $\gamma \rightarrow \beta$ transformation
A) The central triangular area still consists of the high-temperature γ-phase, whereas the other domains are in the low temperature β-phase; B) The same area as A) after additional cooling. The unstable triangular γ-area has been reduced further (Courtesy of C. Manolikas)

Images made in *unsplit reflections* do not reveal brightness differences due to orientation differences of the lattice, but they may exhibit domain contrast due to differences in the structure factor. This is the case for Dauphiné twins in quartz, which are related by a 180° rotation about the threefold axis (Fig. 2.15). The two domains produce the same diffraction pattern since their lattices are common, but certain reflections have structural factors that differ in magnitude [2.40]. A dark-field image made in such a reflection will exhibit a different brightness in the two domains, which is called the structure factor contrast [2.4].

2.1.6.5. Interface Contrast and Domain Contrast

Interface contrast (i.e., an α-fringe pattern) is produced in the projected area of the interface even if the structure factors in the two domains have the same magnitude, but differ in phase by α; no domain contrast is produced under these conditions. Interface contrast is also produced at *orientation domain boundaries*; δ-fringes are produced in the projected area, since the s-values on both sides of the boundary are different in general [2.41]. Extinction of the δ-fringe pattern occurs if the s-values are the same for the two domains.

The presence or absence of domain contrast across a fringe pattern helps to distinguish the two types of interface. If the two domains on either side of the interface exhibit the same brightness for all reflections, the fringes must be due to a translation interface; they must exhibit α-character. Domains exhibiting a difference in brightness are separated by a domain boundary and the fringes are of the δ-type.

2.1.6.6. Strain Field Contrast

The strain field associated with precipitate particles in a matrix whose lattice parameters differ slightly from those of the precipitate can also be revealed as regions of different brightness [2.42]. For example, a spherical particle in an elastically isotropic matrix produces a spherically symmetric strain field. In certain parts of this strain field the vector $\bar{\boldsymbol{R}}$ is locally perpendicular to $\bar{\boldsymbol{g}}$; such areas will show up with the background brightness. In other areas, $\bar{\boldsymbol{g}}$ is parallel to $\bar{\boldsymbol{R}}$; such areas will show up with a brightness that is different from the background, either lighter or darker.

2.1.7. Convergent Beam Diffraction

[2.8], [2.43]

2.1.7.1. Geometry of Convergent Beam Patterns

In the early 1980s, a class of applications based on the use of a convergent beam of electrons (convergent beam diffraction, CBD) was developed. Whereas parallel beams allow high spatial resolution, *high angular resolution* is more easily achievable with convergent beams. For these applications the electrons are incident along directions within a cone of revolution or along the surface of a cone (hollow cone method)

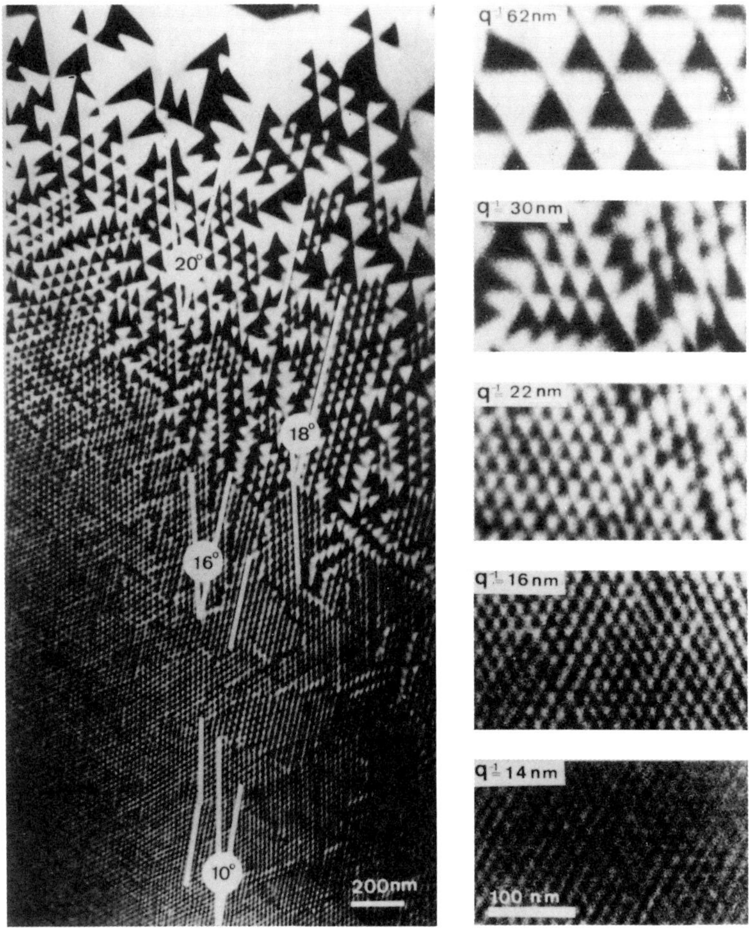

Figure 2.15. Dauphiné twin lattice in α-quartz close to the phase transition into the high-temperature β-phase viewed along the threefold axis
A temperature gradient is established across the specimen area. The α-phase is broken up in regular arrays of columnar domains whose size decreases with increasing temperature [2.40]

having its apex in (or close to) the sample plane. Under these conditions, electrons are incident on the specimen under all possible directions within a certain angular range, which should be of the order of the Bragg angles of the lowest-order reflections of the material under study. In practice, the semiapex angle α varies from 2 to 10 mrad. Also, the corresponding diffracted beams for each reciprocal lattice node form cones. The intersections of these cones with the photographic plate are circular disks whose radii are proportional to the semiapex angle of the cone of incident directions (Fig. 2.16). If the specimen is perfectly flat and defect free, these disks are images of the angular intensity distribution in each particular reflection.

2.1.7.2. Point Group Determination [2.44], [2.45]

When the incident beam is parallel to a zone axis, the symmetry of the intensity distribution in the complete disk pattern, including the fine lines in the 000 disk, reflects the point symmetry of the crystal along this zone axis (Fig. 2.17).

Several techniques can be used to determine the point group [2.44]. The method developed by Buxton et al. [2.45] is based on the use of zone axis patterns and of dark-field diffraction patterns obtained under exact two-beam conditions. By choosing the appropriate optical conditions, these disks do not overlap and the exact Bragg spots remain in the centers of the corresponding disks. The optics required for the application of

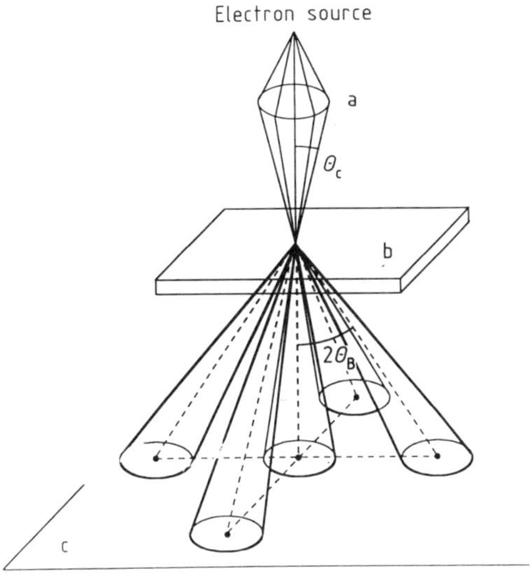

Figure 2.16. Production of a convergent beam electron diffraction pattern
a) Probe forming optics; b) Specimen; c) Film
The incident beam is convergent. Diffraction disks are centered on each diffraction spot.

Figure 2.17. Zone axis convergent beam electron diffraction pattern revealing the presence of a sixfold rotation axis [2.43]

these methods are described in [2.46]. With increasing cone angle the disks start to overlap and the nature of the pattern changes, but the symmetry elements are conserved.

A disk in the CBED pattern usually consists of broad fringes due to the dynamic interaction between beams belonging to the zero-order Laue zone (ZOLZ). Finer lines are formed by the interaction of high-order Laue zone reflections (HOLZ) with the zero-order beams (Fig. 2.18). The geometry of the HOLZ lines is determined by the accelerating potential and the lattice parameters; comparison of computer-simulated and observed patterns allows determination of one parameter if the other is known [2.46], [2.47].

Figure 2.18. Deficiency HOLZ (high-order Laue zone) lines in the central disk of a convergent beam diffraction pattern of silicon [2.43]

2.1.7.3. Space Group Determination

Space group determination is based on the observation of the Gjønnes–Moodie lines [2.48] in certain "forbidden" diffraction disks. Reflections that are kinematically forbidden may appear as a result of double diffraction under multibeam dynamic diffraction conditions. If such forbidden spots are produced along pairs of different symmetry-related diffraction paths that are equally excited, the interfering beams may be exactly in antiphase for certain angles of incidence if the structure factors have opposite signs. Since the convergent beam disks are formed by beams with continuously varying directions of incidence, this condition will always be satisfied somewhere in the disk. The locus of points for which the phase difference is exactly π is a dark fringe along a diameter of the disk for which the Bragg condition is satisfied exactly (Gjønnes–Moodie lines, Fig. 2.19).

2.1.7.4. Foil Thickness Determination

From the dark-field two-beam rocking curve (Section 2.1.7.1) observed in the disk corresponding to the diffracted beam, the specimen thickness can be deduced with high accuracy provided the accelerating voltage and the interplanar spacing of the excited reflection are known [2.49].

2.1.8. High-Resolution Electron Microscopy

In recent years, high-resolution electron microscopy has become an important tool in the study of complicated crystal structures and their defects. Whereas detailed considerations of electron optics are of only marginal importance in the case of diffraction contrast images, they become essential for a discussion of the atomic resolution structure images produced by crystals.

2.1.8.1. Image Formation in an Ideal Microscope (Fig. 2.20)

Let the crystalline object be a thin foil characterized by a two-dimensional transmission function $q(x, y)$ that describes at each point of the exit surface of the specimen the amplitude and phase of the electron beams emerging from the column situated at (x, y) after dynamic diffraction in the foil. The diffraction pattern can be described to a good approximation as the Fourier transform $Q(u, v)$ of the object function $q(x, y)$. This diffraction pattern acts in turn as a source of Huyghens wavelets, which interfere to form the image, after linear magnification by the optical lens systems; the image is, in turn, the Fourier transform $\psi(x, y)$ of the diffraction pattern.

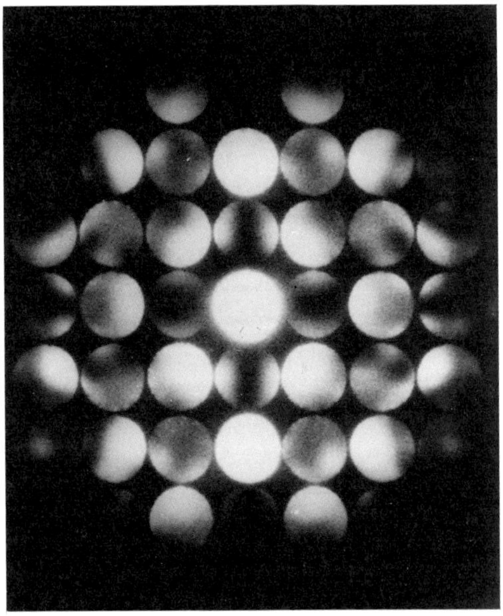

Figure 2.19. Convergent beam electron diffraction pattern exhibiting Gjønnes–Moodie lines due to the presence of glide mirror planes and twofold screw axis [2.43]

2.1.8.2. Image Formation in a Real Microscope

In real microscopes the situation is complicated by the presence of finite apertures and lens aberrations. The apertures truncate the Fourier transform by admitting only a finite number of beams to the image formation process. Lens aberrations depend on the angle β that a diffracted beam encloses with the optical axis of the microscope, hence they introduce angle-dependent phase shifts between diffracted beams.

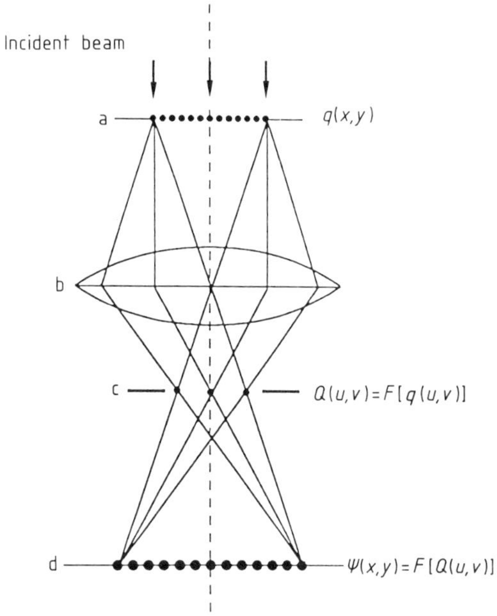

Figure 2.20. Image formation in an ideal microscope
a) Specimen; b) Objective lens; c) Back-focal plane, objective aperture; d) Image plane
The diffraction pattern $Q(u, v)$ is the Fourier transform of the object $q(x, y)$; the image is the inverse Fourier transform $\psi(x, y)$ of the diffraction pattern

Moreover, they cause "blurring" of the image, a point of the object plane being represented as a disk in the image plane and vice versa.

Spherical Aberration. Spherical aberration causes electron beams enclosing an angle β with the optical axis to suffer a phase shift χ_S with respect to the central beam:

$$\chi_S = \tfrac{1}{2}\pi C_S \beta^4/\lambda \tag{2.15}$$

where C_S is the spherical aberration constant.

Defocus. Under exact Gaussian focusing conditions the contrast produced by an object that changes only the phase of the incident electron wave (i.e., a so-called phase grating) is minimal. This is the case for thin crystalline foils. Visual contrast improves when working in underfocused conditions. The phenomenon is in a sense similar to phase contrast in optical microscopy of phase objects. Defocusing causes a relative phase shift of the diffracted beams enclosing different angles with the central beam. The phase shift caused by a defocus distance ε with respect to the Gaussian focus is given by

$$\chi_D = \pi\varepsilon\beta^2/\lambda \tag{2.16}$$

The total phase shift caused by spherical aberration and defocusing is thus:

$$\chi(\beta) = (\tfrac{1}{2}\pi C_S \beta^4 + \pi\varepsilon\beta^2)/\lambda \tag{2.17}$$

$\chi(\beta)$ depends on β in the way shown in Figure 2.21, i.e., there is a flat minimum at $\beta_{min} = (-\varepsilon/C_S)^{1/2}$ and a zero at $\beta = (-2\varepsilon/C_S)^{1/2}$.

Phase Grating Approximation [2.50]. The effect of this phase shift on the image parameter can be understood most easily by discussing the case of a pure phase grating, which is a reasonable model for a very thin crystalline foil. Such a specimen causes a phase shift χ of the electron wave on passing through the foil because the wavelength of the electron is different in vacuum and in the crystal. This phase shift in the point (x,y) can be written as $\sigma\varphi(x,y)$ where φ is the projected potential integrated over the sample thickness along the propagation direction of the electrons, and $\sigma = \pi/\lambda E$ (λ = wavelength of electron in vacuum; E = accelerating voltage). Ignoring absorption, the object function for a thin foil can be written as

$$q(x,y) = \exp i\sigma\varphi(x,y) \simeq 1 + i\sigma\varphi(x,y) \tag{2.18}$$

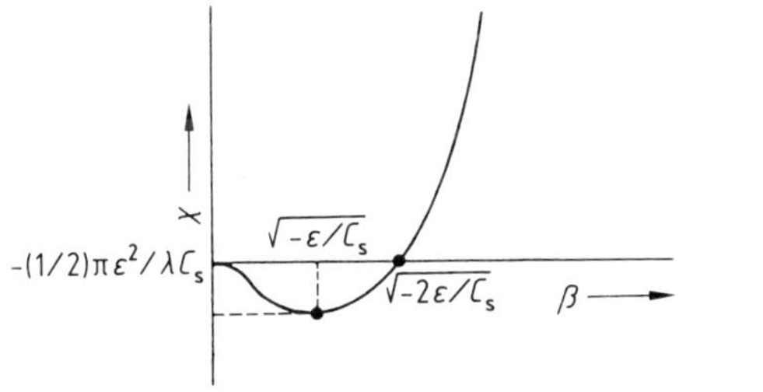

Figure 2.21. Dependence on β of the phase shift $\chi(\beta)$ of a beam enclosing an angle β with the optical axis of the microscope

Taking also into account the phase shift (Eq. 2.17) caused by the lens system on the diffracted beams, and neglecting absorption it can be shown by Fourier transformation [2.51] that the image intensity is finally given by

$$I(x,y) \simeq 1 - 2\sigma\varphi(x,y) \tag{2.19}$$

whereby it was assumed that $\sin\chi = -1$ ($\cos\chi = 0$). The image contrast, defined as $(I - I_0)/I_0 = -2\sigma\varphi(x,y)$, where I_0 is the incident intensity, is thus directly proportional to the projected potential $\varphi(x,y)$ provided $\sin\chi = -1$.

Provided $\sin\chi = -1$, columns of large projected potential are imaged as dark dots. If $\sin\chi = +1$, the same columns would be imaged as bright dots.

Optimum Imaging Conditions. The most stringent requirement is $\sin\chi = -1$; this condition can only be met approximately and only in a limited β interval. The condition $\sin\chi = +1$ can be satisfied only for a limited number of discrete β-values.

The situation is somewhat comparable to positive and negative phase contrast in optical microscopy. The lens imperfections have been used to introduce a phase shift of $\pi/2$ in the same way as the quarter wavelength ring in the optical microscope. Only beams passing through the "window" in which the condition $\sin\chi = -1$ is met, interfere with the required phase relationship that causes the image to represent maxima in projected potential as dark areas.

From Equation (2.17) it is clear that $\chi(\beta)$ depends on the parameters C_S, ε, and λ. The condition $\sin\chi = -1$ can thus be met by adjusting one or several of these parameters. For a given instrument C_S and λ are fixed and the observer can meet this condition by optimizing the defocus ε. From Equation (2.17) it follows that χ adopts the essentially negative value $\chi = -\pi\varepsilon^2/(2\lambda C_S)$ for $\beta = (-2\varepsilon/C_S)^{1/2}$. Sin χ will be -1 if $\pi\varepsilon^2/(2\lambda C_S) = \pi/2$, i.e., for

$$\varepsilon_S = -(\lambda C_S)^{1/2} \tag{2.20}$$

For the defocus $\varepsilon = \varepsilon_S$ the curve $\sin\chi$ versus β will exhibit a rather flat part in the region around

$$\beta = (-\varepsilon/C_S)^{1/2} \tag{2.21}$$

where $\sin\chi = -1$. The underfocus value ε_S corresponding to this optimum is called Scherzer defocus [2.52]. Most high resolution images are made under this defocus condition and in the thinnest part of the sample.

2.1.8.3. Resolution Limiting Factors

Aperture. The presence of an aperture admitting only beams that enclose an angle with the optical axis not exceeding β_A imposes a resolution limit, called the Abbe limit [2.53]. A point in object space is imaged as a circle with a radius

$$\varrho_A = 0.61\,\lambda/\beta_A$$

Chromatic Aberration. Instabilities ΔE in the high voltage E of the microscope cause a spread in the wavelength of the incident electron, which blurs the image. Variation ΔI in the lens current I similarly causes a spread in the focal length of the lenses, which also contributes to such blurring. Furthermore, the inelastic scattering of electrons in the specimen causes slight changes in the wavelength of electrons emerging from the specimen. The net result of these different phenomena can be described as an effective spread in the focal length of the objective lens:

$$\Delta f = C_C[(\Delta E/E)^2 + 4(\Delta I/I)^2]^{1/2} \quad (2.22)$$

where C_C is the chromatic aberration constant; the corresponding disk of least confusion in object space is $\varrho_C = \beta \Delta f$. In concrete cases, $\Delta E/E \leq 10^{-6}$; $\Delta F \approx 10$ mm, $C_C \approx 10$ mm.

Astigmatism. Ideal lenses should have cylindrical symmetry. However, in reality, slight deviations may occur, leading to a dependence of the focal length on the azimuth of the ray path considered. Astigmatism can occur for various reasons, including inhomogeneities in the polepiece material, asymmetry in the windings, or a dirty aperture. As a result of astigmatism a point source is imaged as two different line foci at right angles to each other. At exact focus the image deformation disappears but the image is still blurred. Astigmatism can be corrected by a stigmator—a device (usually an octopole) that makes the lens appear perfectly cylindrical—by applying weak additional magnetic fields. Careful correction of astigmatism is essential for the production of high-quality, high-resolution images.

Beam Divergence. The intense illumination required for high-resolution imaging is obtained by a condenser lens system that produces a slightly conical beam with a typical apex angle of the order of 10^{-3} rad. This also leads to some image blurring since the final image is the superposition of images corresponding to the different directions of incidence.

Mechanical Instability. The obtainable resolution depends not only on the lens characteristics but also to a large extent on the mechanical stability of the microscope, which should be installed on a vibration-free heavy concrete block. In many cases the mechanical stability is actually the resolution-determining factor. The specimen holder should moreover be creep free, which is a problem for hot and cold stages. High-resolution work is therefore performed almost exclusively at room temperature.

Ultimate Resolution. The final disk of confusion due to the different nonmechanical blurring effects is given approximately by

$$\varrho = (\varrho_A^2 + \varrho_S^2 + \varrho_C^2)^{1/2} \quad (2.23)$$

which depends on β. At low angles the aperture effect is usually dominant, whereas at high angles the spherical aberration is the limiting factor (for 100-kV microscopes, $C_S \approx 8.2$ mm and $C_C \approx 3.9$ mm). The optimum radius of least confusion is achieved for a certain β-value and the radius of confusion corresponding to that optimum is

$$\varrho_0 = 0.9\, \lambda^{3/4}\, C_S^{1/2} \quad (2.24)$$

Therefore, a gain in resolution can be achieved by reducing C_S and by reducing λ (i.e., using a higher voltage). At present, many microscopes used in material science and in solid-state chemistry operate at 200–400 kV.

At higher voltages, radiation damage to the specimen severely limits observation time. Advanced designs aim at reducing C_S to allow a decrease of the high voltage for a given resolution. Practical resolution limits are at present ca. 0.16 nm, but new developments tend to peak this value to below 0.10 nm.

Directly interpretable high-resolution images (i.e., having a direct relationship to either the projected lattice potential or the electron density) are formed by the interference between beams passing through the "window" in the image transfer function (ITF). As a result, only low-order reflections (i.e., low-order Fourier components of the lattice potential) generally contribute to the image, which limits the detail that can be resolved (irrespective of the point resolution) since fine details are carried by the high-order Fourier components. However, information transmitted beyond the window (i.e., by the rapidly oscillating part of the image transfer function) can also be used. This information can be extracted by computational techniques, for instance, by using as input a set of images made at a series of closely spaced defocus values [2.54].

2.1.8.4. Image Formation Models

Fourier Model. High-resolution image formation is considered most conveniently as occurring in two steps. In the first step, electrons propagate through the crystal and produce a two-dimensional periodic electron distribution at the exit face, which images the projected lattice potential or the projected electron density. In a second step, this exit face acts as a planar distribution of point sources of spherical electron waves, which interfere behind the foil and form the diffracted beams that move in the lens system of the microscope. Interference of these beams produces the final image, which will be directly interpretable only if the beams interfere with the correct phases.

The resolution of the image depends on the spatial frequencies of the highest-order beams admitted by the selector aperture that still contribute to the image with the correct phase. Only beams passing through the window in the ITF contribute to a directly interpretable image.

The image will exhibit more detail as the order of included reflections increases, but it will give a faithful representation of the structure only if the Fourier components have correct relative phases; this puts a practical limit on the number of useful reflections.

Complete software packages for the simulation of high-resolution images are available commercially (see, e.g., [2.55]).

Channeling Model. An alternative model that emphasizes the particle nature of electrons provides a simple intuitive picture [2.56]. An atomic column parallel to the incident beam is a cylindrically symmetrical potential well, which acts on the moving incident electrons as a succession of alternating convergent and divergent lenses. At the entrance face of the foil the incident electron distribution is uniform, but in the first part of the foil the electrons are attracted toward the core of the atom columns and focusing occurs, transforming part of the potential energy of the electrons into kinetic energy. Subsequently the electrons repel one another and defocusing occurs. The focusing–defocusing motion occurs periodically as a function of the depth in the foil, the depth period being a function of the atomic number of the atoms in the column. The heavier the atoms are, the smaller is the depth period. As a result, for a foil of given thickness, one type of atom column will give rise to a relative maximum in the electron distribution, whereas another type may give rise to a relative minimum. These extremes are not necessarily imaged as extremes of the same nature. A maximum may be imaged either as a bright dot or as a dark dot, depending on the defocus and on the foil thickness (see Fig. 2.22). This model explains why different atom columns produce dots of different brightness. The focusing and defocusing behavior was confirmed by computer simulations, which also showed that the depth periods are in the 4–10-nm range depending on the atomic number.

2.1.9. Analytical Methods

2.1.9.1. X-Ray Microanalysis [2.10]

Principle. An important feature of electron microscopy is the possibility of performing in the same instrument both a qualitative and a quantitative elemental analysis of the same small crystal fragment that produced the electron diffraction pattern and the image. The most commonly used method consists in principle of detecting the characteristic X rays excited by the high-energy electrons passing through the specimen.

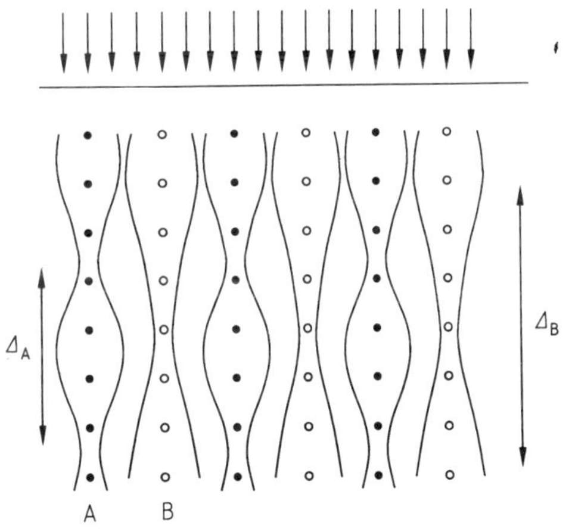

Figure 2.22. Channeling of electrons along columns of atoms parallel to the incident beam
Alternating focusing and defocusing occurs. The focusing distances Δ_A and Δ_B are different for A and B columns

For this purpose, an X-ray spectrometer, mounted sideways on the microscope column at the level of the specimen holder, captures the X rays emitted in all directions by the specimen within a certain solid angle centered around the chosen takeoff direction and transmitted through an X-ray transparent window (Fig. 2.23). The spectrum consists of the characteristic X-ray lines superposed on a continuous background of bremsstrahlung (see also → Surface Analysis, pp. 28–35).

The physical basis of X-ray microanalysis is Moseley's law [2.57], which establishes a direct and systematic relation between the atomic number Z and the characteristic X-ray wavelength λ in a family of X-ray lines (e.g., associated with similar transitions in different elements):

$$\lambda = A/(Z - C)^2 \qquad (2.25)$$

A and C depend on the considered family; K_α, $K_\beta, \ldots, L_\alpha, \ldots$, X-ray lines are produced when an inner shell (e.g., K-shell) electron has been ejected from the atom by absorbing an amount of energy from an incident electron, exceeding the critical ionization energy associated with that shell. On deexcitation, this vacant level is filled by an electron originating from a higher-lying shell or subshell of the same atom. The transition

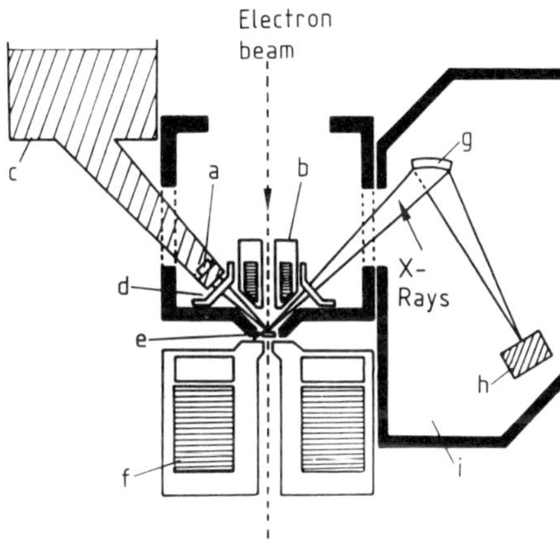

Figure 2.23. Specimen–detector configuration used in X-ray microanalysis
Left: Energy-dispersive X-ray system (EDX) with Si(Li) detector; Right: Wavelength-dispersive spectrometer (WDS) with bent crystal
a) Si(Li) detector; b) Mini lens; c) Liquid-nitrogen reservoir; d) X-ray window; e) Specimen; f) Objective lens; g) Crystal; h) Detector; i) Crystal spectrometer

is accompanied either by the emission of X rays, which is the phenomenon of interest here, or by the ejection of an Auger electron (→ Surface Analysis, pp. 27–28). Equation (2.25) shows that the atomic number can in principle be determined by measuring the characteristic X-ray wavelength.

Wavelength-Dispersive Spectrometry (WDS). Two types of X-ray spectrometers are in use: wavelength-dispersive spectrometers and energy-dispersive spectrometers.

In wavelength-dispersive spectrometers, dispersion is achieved by an analyzing crystal. Since the Bragg angle for a given set of lattice planes depends on the wavelength of the incident radiation, X rays of different wavelengths can be separated spatially by the crystal, and a spectrum can thus be analyzed by changing the orientation of the analyzing crystal by θ and detecting the amount of radiation passing through a slit whose angular position is rotated simultaneously over 2θ. Often the analyzing crystal, which acts as a grating, is cylindrically curved so that focusing along a line occurs. The detector, usually a gas proportional counter, is placed in the focus and connected to a single channel analyzer, followed by standard counting electronics.

Depending on the wavelength range to be detected, different analyzing crystals are necessary; provision is usually made on the equipment to allow switching in several crystals from a built-in carousel. Extension of this method to long-wavelength X rays is hampered by the absence of suitable crystals with large interplanar spacings. Synthetic layered structures, prepared by molecular beam epitaxy, and Langmuir–Blodgett layers have been used for this purpose since the end of the 1980s.

Energy-Dispersive Spectrometer (EDS). In energy-dispersive spectrometers, spectral analysis is based on measuring the X-ray energy rather than the wavelength. As a detector, a solid-state ionization chamber is used (e.g., a *p–i–n* junction in silicon, an intrinsic germanium crystal, or a mercuric iodide crystal). The energy resolution is of the order of 150 eV or less for X-ray energies in the range 1–12 keV. The silicon *p–i–n* junction is produced by diffusing lithium (at elevated temperature) into a *p*-type silicon slab under a reverse bias, creating an intrinsic region via the exact compensation of fixed acceptors by mobile lithium donors, as a result of electromigration of the lithium ions. Under reverse bias an electric field is created in the broad intrinsic region, which acts as the sensitive volume. Such detectors must be kept permanently at liquid-nitrogen temperature since lithium ions migrate at room temperature. When an X-ray photon passes

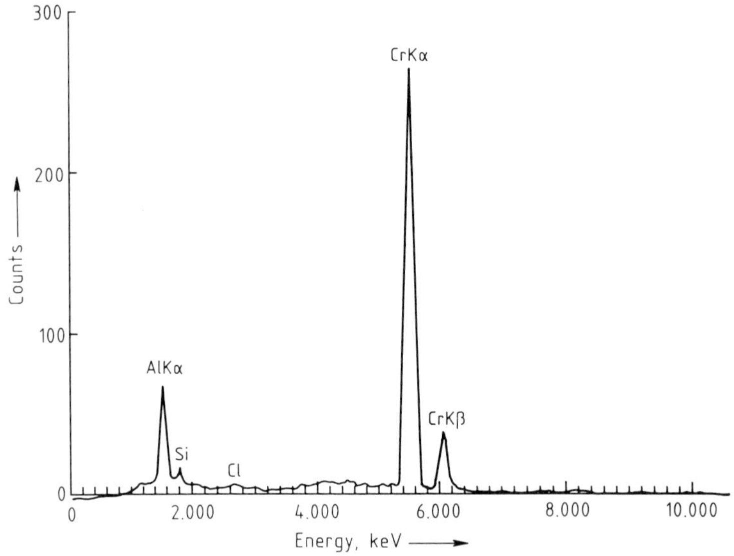

Figure 2.24. Example of X-ray spectrum obtained by means of an energy dispersive spectrometer The material is a chromium aluminum alloy.

through the intrinsic region, its energy is dissipated almost entirely by the formation of electron–hole pairs, which are swept away by the reverse bias field and produce a voltage pulse across the detector crystal that is proportional to the energy of the incoming X-ray photon. The X-ray spectrum is then established by analysis of the pulse heights with a multichannel analyzer. For qualitative analysis, only the line positions are important, but for quantitative analysis the line shape and the peak heights are also used. Fundamental for the interpretation of EDS spectra is the assumption that all the energy of the X-ray photon is deposited in the detector (silicon escape peaks are ignored). The number of charges n created by a photon with energy E is then given by $n = E/E_i$, where $E = 3.6$ eV for silicon. Since the charge pulses to be detected are small, reducing the noise of the detector is essential; this is a second reason for cooling the detector. EDS has many advantages over WDS, the main ones being the much shorter acquisition time and the smaller minimum useful probe size. On the other hand, complications also occur that do not arise in WDS instruments; the major ones are escape peaks, pulse pileups, electron beam scattering, peak overlap, and window absorption effects. In practice, the latter prevents detection of light elements with $Z \leq 10$ (beryllium window) unless a windowless detector is used.

Quantitative Analysis. Quantitative analysis is based on the assumption that the intensity of the Lorentzian X-ray peak is proportional to the number of atoms of the species responsible for the peak. Since the theoretical relations are complicated and involve a large number of parameters, some of which are difficult to evaluate, well-characterized homogeneous standards are generally used for calibration, the composition of which has been determined by classical chemical macroscopic methods. These standards, which should also have similar shapes to the sample, are then measured in EDS equipment under the same well-defined and reproducible conditions as the unknown sample, and the spectra are compared. This calibration leads to the so-called k_i factors (coefficients of proportionality) for the different elements [2.58]:

$$C_{i\,\text{unknown}}/C_{i\,\text{standard}} = I_{i\,\text{unknown}}/I_{i\,\text{standard}} = k_i \quad (2.26)$$

where C_i are weight fractions.

Since k_i may depend on the particular matrices, Equation (2.26) must be corrected as follows [2.10], [2.59]:

$$C_{i\,\text{unknown}}/C_{i\,\text{standard}} = Z_i \times A_i \times F_i \quad (2.27)$$

where the matrix correction factors Z_i, A_i, and F_i still depend on the element under consideration; Z_i is the factor due to atomic number; A_i is due to X-ray absorption; and F_i to X-ray fluorescence. Other correction factors are due to the specimen geometry because X-ray production is depth dependent since it is proportional to the local electron energy. In the thin foils used in TEM the ZAF correction factors can often be neglected. X-ray microanalysis allows quantities of material of the order of 10^{-11} g to be analyzed; the concentration can be as low as 0.01 %, and the precision is of the order of 1 %.

For a more detailed discussion of the method and its technicalities, see [2.58]–[2.60].

Special Effects [2.61]. For foil orientations close to an exact Bragg orientation, the transmitted and scattered beams both propagate along the lattice planes. In a model crystal with a primitive structure, one wavefield has its maximum amplitude along planes coinciding with the atomic planes; the other has its maximum along the planes midway between atomic planes. For $s > 0$, the latter wavefield is excited, and therefore easy transmission occurs since the electrons in this wavefield avoid the atom cores (Bormann effect). For $s < 0$, the former wavefield is excited, but strongly absorbed, since the electrons pass close to the atom cores and can excite X-ray emission. X-ray emission is thus expected to be enhanced on the $s < 0$ side of a bent contour ($s = 0$).

In a crystal whose superstructure consists of alternating A and B atomic layers parallel to the considered lattice planes, which are close to the exact Bragg position, one type of wavefield is peaked along the planes of A atoms, whereas the other is peaked along the B-atom planes. Depending on the sign of s, one of the wavefields is enhanced relative to the other; hence the A and B atomic layers will be excited unequally. Under such conditions, X-ray microanalysis may produce erroneous results for the A/B ratio. Therefore, foil orientations should be avoided that cause an extinction contour to pass through the selected area due to anomalous transmission. On the other hand, this effect has been exploited to locate the site occupied by a third atomic species, in an AB superstructure, using a technique called ALCHEMI (atom location by channeling-enhanced microanalysis) [2.61].

2.1.9.2. Electron Energy Loss Spectrometry (EELS) (→ Surface Analysis, pp. 94–99)

Principles [2.10], [2.62]. Apart from elastic scattering, which is responsible for electron diffraction, inelastic scattering events also occur as electrons pass through the foil. Inelastic processes can be caused by (1) single electron excitations, such as X-ray and Auger electron production; and (2) collective excitations, such as volume and surface plasma oscillations (plasmons) and phonons. Collective excitation can be revealed indirectly as characteristic energy losses of the incident electrons. The energy of elementary excitations (plasmons and phonons) is low compared to the incident electron energy (100–140 kV).

Plasmons have energies in the range of 10–20 eV, whereas phonon energies are of the order of 10 meV. Individual quantized plasmons are thus much easier to detect than phonons. Since the positions of the plasmon loss peaks are characteristic of the materials, they can be used as analytical tools for aluminum and magnesium, for example.

However, for chemical analysis, the absorption edges in EELS curves are more important (Fig. 2.25); they reflect the absorption phenomena leading to X-ray production and exhibit a fine structure on the high-energy side. This fine structure is referred to as EXELFS (extended energy loss fine structure) [2.63], which is the analogue of EXAFS (extended X-ray absorption fine structure). The steep rise on the low-energy side of the absorption edge is due to the excitation of inner-shell electrons and characterizes the element.

The fine structure is produced by the electron wave originating from an inner shell that is partially back-reflected by the surrounding atoms, which leads to a modulation of the excitation probability of inner-shell electrons. This fine structure can therefore provide information not only on the chemical nature of the absorbing atom, but also on the number of nearest neighbors and their distance.

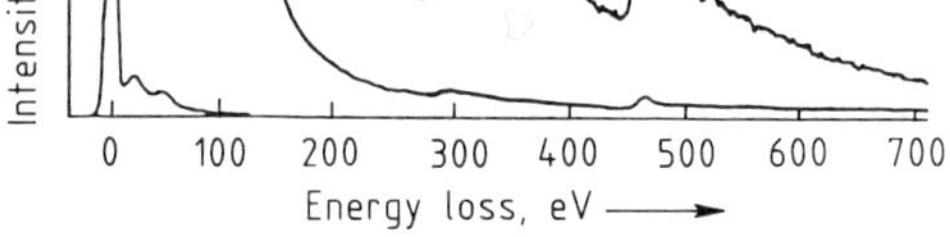

Figure 2.25. Example of EELS (electron energy loss spectrometer) spectrum of TiC
On the low-energy side the plasmon peak is visible. The fine structure on the high-energy end is caused by interference of waves reflected by the neighboring atoms

As an analytical tool, EELS is in a sense complementary to X-ray microanalysis and makes use of the same type of phenomena. EELS works well for light elements, whereas X-ray microanalysis is much more difficult to apply to elements with $Z \leq 11$.

Spectrometers. The simplest system uses a magnetic prism to produce spatial separation of electrons with different energies. The spectrum is scanned by means of a slit, and a phosphor coupled to a photomultiplier is used as the detection system.

One of the most advanced types of equipment is the ω electron spectrometer [2.64]. It consists of four magnetic prisms that bend the incident beam back to its original direction; it can be mounted on an electron microscope column behind the specimen where inelastic scattering is produced. Slits allow a certain energy range to be selected from the spectrum. The spectrometer can thus also act as a filter and can produce, e.g., zero-loss images in which the blurring due to inelastic scattering is eliminated.

Since the 1980s *X-ray microanalysis* based on energy-dispersive spectrometers has become a routine technique. Equipment as well as data processing hardware and software are available commercially as optional accessories for most microscopes. In contrast, *EELS* equipment is still evolving and is not yet standard equipment; its interpretation has not yet reached the same level of confidence as for X-ray microanalysis.

2.1.10. Specimen Preparation

The objective of all specimen preparation methods is to obtain a crystal fragment that is sufficiently thin to transmit electrons with energies of 100 keV or more, and has a sufficiently large lateral size. The numerous methods available depend strongly on the type of material and observation method.

2.1.10.1. Diffraction Contrast Specimens

Specimens should have a thickness of a few hundred nanometers. If too thick, the foil is not transparent; if too thin, achieving a two-beam orientation is difficult; moreover, the sample is not representative of bulk material for a number of applications.

Electrically conducting bulk samples (metals and alloys) can generally be thinned by electropolishing; *bulk ceramic samples* are often chemically thinned and polished. The resulting sample is usually perforated in the center; the edges of the hole are then wedge shaped and have transparent edges.

Insulating thin foils may become charged and deflect the electron beam; this can be avoided by coating the specimen with a thin layer of amorphous carbon, which is electrically conducting.

Thin films of uniform thickness can be obtained by *vapor deposition* under vacuum or by *chemical deposition* techniques; such samples usually have microstructures that are different from those of samples prepared from bulk material.

Layered crystals exhibiting a pronounced cleavage plane, such as graphite or MoS_2, can be thinned by repeated cleavage to produce rather uniformly thin specimens.

A truly universal thinning method does not yet exist. However, *ion beam bombardment under grazing incidence* is applicable to many materials, in particular to brittle ones such as most semiconductors. In the first step, these materials are mechanically polished to produce slabs with dimples; subsequent ion beam thinning under grazing incidence then produces a hole whose edges are wedge shaped. Thinning can be achieved either from one side only or from both sides of the slabs; this is important, for instance, for quantum wells grown by molecular beam epitaxy. Surface damage left after ion bombardment can often be removed by a final chemical polish. Figure 2.26 schematically represents thinning methods used for semiconductor specimens.

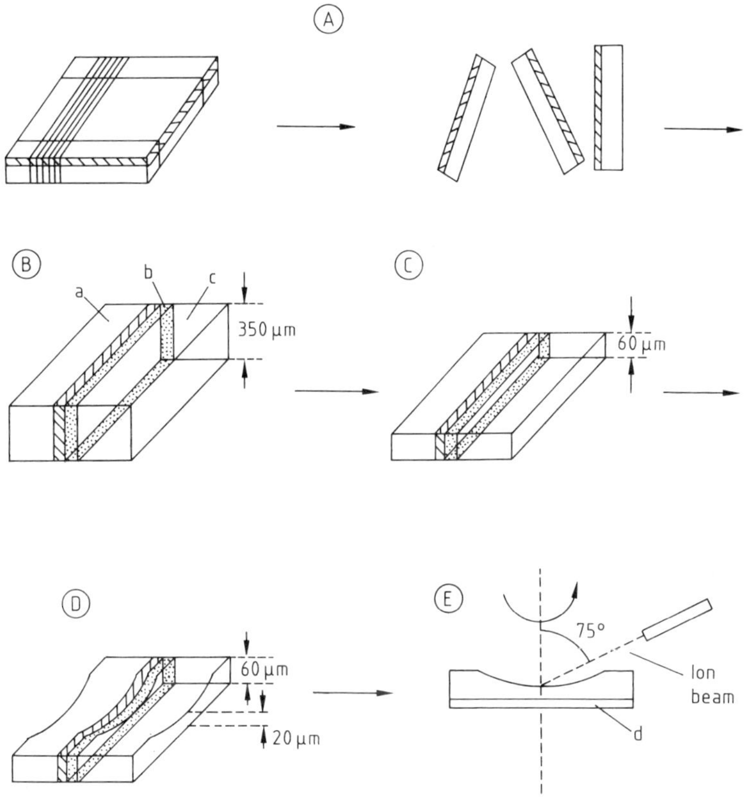

Figure 2.26. Preparation of a cross section of integrated circuits
A) Specimen is cut into thin slabs; B) Slabs are glued on a glass slide; C) Composite specimen is thinned mechanically; D) Thin specimen is dimpled; E) Final thinning is done by ion bombardment from one side while the specimen is rotated
a) Integrated circuit; b) Glue line; c) and d) Glass slide
(Courtesy of IMEC)

2.1.10.2. High-Resolution Specimens

The various methods described above can also of course be used to prepare specimens for use in high-resolution imaging. However much thinner specimens (1 – 10 nm) are required than for diffraction contrast. Nevertheless, more rudimentary methods can often be used successfully, since much smaller lateral dimensions of the specimen can be tolerated. Specimens of brittle materials (e.g., oxide superconductors) are in many cases obtained simply by crushing the material in a quartz mortar under a protecting organic liquid. In such a way the crystal fragments are dispersed in a suitable liquid, and a drop of the suspension is deposited on a grid covered with a holey carbon film and allowed to dry. The edges of fragments protruding over the holes are potentially suitable specimens, but actually only a small fraction of the crystal fragments are sufficiently thin to produce high-quality images. The thickness of the specimen is often the resolution-limiting parameter. Materials that deform plastically when crushed at room temperature often behave as brittle when crushed while immersed in liquid nitrogen.

2.1.11. Applications to Specific Materials and Problems

The number and diversity of applications of different forms of electron microscopy have become so large that an exhaustive survey, even when restricted to chemical aspects, is almost impossible. In particular, its application to structural problems in solid-state chemistry has evolved explosively in recent years. The examples given in what follows attempt to show as large a diversity as possible.

2.1.11.1. Crystal Structures

Only in simple cases can an "ab initio" structural determination be made by electron microscopy and diffraction, essentially because in electron diffraction the intensity of the reflections is not simply proportional to the square of the structure amplitude, but also depends in a complicated manner on the orientation and the thickness of the sample.

However, whereas X-ray and neutron diffraction provide only average structures (→ Structural Analysis by Diffraction, **B5**), electron diffraction and microscopy can provide detailed information on local structures, especially when the average structure is known from other techniques. Also planar defects (i.e., local deviations from the average structure) can be studied to suggest possible new structures in which such defects occur periodically.

Since electron microscopy requires minute samples, the crystallography of materials that are available only as small particles (e.g., catalysts) can be studied in direct and reciprocal space. The early crystallographic studies on C_{60} and C_{70} crystals were for instance, performed by electron diffraction [2.65]. Even fine powders or fine-grained ceramics contain single crystal particles large enough to produce a single crystal electron diffraction pattern; therefore, the lattice parameters and the unit cell can be determined easily, albeit with moderate precision. Subsequently, X-ray powder diffraction patterns can be indexed unambiguously and the lattice parameters determined with high precision. Electron diffraction and especially high-resolution electron microscopy are complementary techniques to X-ray and neutron diffraction.

Imaging Modes for Various Materials. *Polytypes, Mixed-Layer Polytypes.* The structures of polytypes are usually based on the various stacking modes of closely packed layers of spheres. The stacking sequence can best be revealed by imaging the structure along the zone parallel to the direction of the close-packed rows of atoms. Figure 2.27 shows a foil of SiC – 15 R imaged along the close-packed rows by using three different modes; it illustrates clearly the effect on the image detail of increasing the number of beams participating in the image-forming process. In Figure 2.23 A, only two successive spots out of the densely populated row of 000l spots are included; only the superspacing consisting of five-layered SiC lamellae is revealed as a line image. The isolated stacking faults are imaged as anomalously wide line spacings.

Including in the aperture, a cluster of at least five successive spots of the central row between the 0000 and 00015 reflections, but excluding parallel rows, reveals the individual SiC layers and the superperiod due to the five-layered lamellae as lines; it shows in particular that the superperiod contains five layers (Fig. 2.23).

Finally, when spots from neighboring dense rows are also collected, a two-dimensional dot pattern is obtained and the close-packed rows

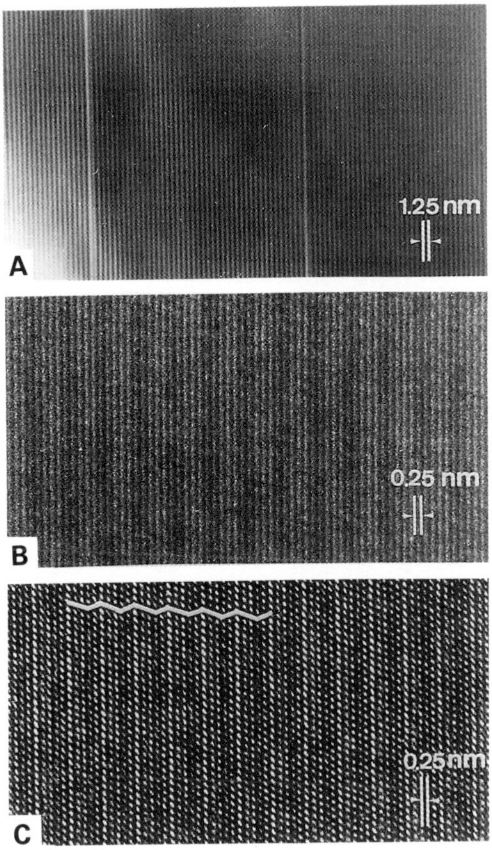

Figure 2.27. Imaging of the 15 R polytype of SiC along the closely packed rows of atoms
A) Two successive spots of the 0001 row contribute to the image; only the five-layer lamellae are imaged as lines; singular line spacings represent stacking faults; B) Cluster of five successive spots of the 0001 row contributes to the image; single SiC layers are revealed; each lamellae of A) is seen to contain five layers; C) Spots from parallel rows are allowed to contribute to the image; the stacking sequence ABCAC can be deduced from the two-dimensional image since close-packed atom columns are imaged as dots

are imaged as bright dots; the stacking sequence ABCAC... is directly revealed (Fig. 2.23).

The same imaging mode is used for the study of mixed-layer compounds based on the epitaxy of close-packed layers [2.66].

Ordered Alloys. The diffraction patterns of ordered binary alloys, derived from the face-centered cubic (fcc) structure, consist of strong basic spots that would also be produced by the disordered fcc lattice, and of weaker superlattice spots characteristic of the superstructure. The first-order superlattice spots correspond to larger interplanar spacings and are thus closer to the origin than the basic spots. Therefore images can be produced by selecting superstructure spots next to the direct beam only and excluding basic spots. The image so obtained (*bright-field superlattice image*) reveals only the configuration of the minority atoms as bright dots, since these atoms determine the superlattice. This information is sufficient to determine completely the projection of the superstructure, provided the basic lattice is known.

An image can also be formed exclusively by means of the superstructure spots situated within a reciprocal unit cell of the basic lattice (*dark-field superlattice image*) [2.67]. This last mode produces the best-contrasted images of the minority atom sublattice (Fig. 2.28).

If basic and superlattice spots are included in the image-forming process the basic lattice is also revealed; minority and majority atoms are imaged as dots with a different brightness. Such images contain information that is redundant if only the superstructure is sought. Examples of superlattice images of ordered alloys obtained by using the bright-field superlattice mode are reproduced in Figure 2.29.

Long-Period Interface Modulated Structures. Many long-period structures are derived from simpler structures by the periodic introduction of planar translation interfaces. Typical examples are the long-period antiphase boundary structures in many ordering alloy systems such as Au–Mn, Cu–Au, and the shear structures derived from a simple transition-metal oxide structure such as WO_{3-x}, TiO_{2-x}. Such compounds produce characteristic diffraction patterns. In combination with their high-resolution images, their structures can usually be elucidated completely, including local deviations from the average structure, which often occur in such phases.

In a model [2.68], the superstructure is assumed to consist of identical slabs of basic structure with thickness D limited by planes normal to $\bar{e}_{\mathrm{n}}$. Usually the slab thickness is equal to an integer number n of unit cell parameters a of the basic structure i.e., $D = na$, but this need not be the case. Successive slabs are separated by planar interfaces (stacking faults, anti-phase boundaries, discommensuration walls, etc.) with a displacement vector $\bar{\boldsymbol{R}}$ and a unit normal $\bar{e}_{\mathrm{n}}$.

The diffraction pattern of such a structure consists of clusters of equally spaced superstructure spots (spacing: $1/D$), called satellites, located around the positions of the basic spots. The

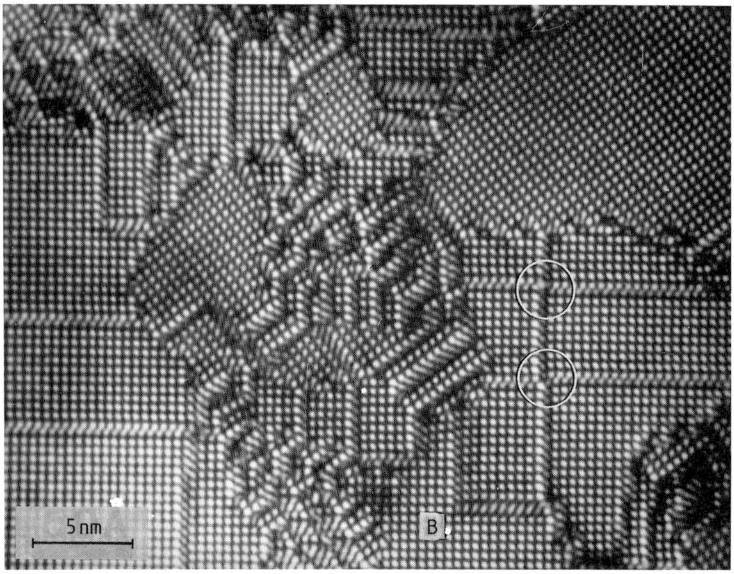

Figure 2.28. Dark-field superlattice image of Au_4Mn
The manganese atoms are represented as bright dots. Note the presence of two orientation variants and several translation variants [2.67], [2.69]

intensity of the satellites decreases rapidly with separation of the satellite from the basic spot to which it belongs, which also determines its position. The positions $\bar{\boldsymbol{H}}$ of satellites are given by the relation [2.68]

$$\bar{\boldsymbol{H}} = \bar{\boldsymbol{h}} + (m + \bar{\boldsymbol{h}} \cdot \bar{\boldsymbol{R}})\,\bar{\boldsymbol{q}} \qquad (2.28)$$

(m = integer; order of the satellite; $\bar{\boldsymbol{q}} = (1/D)\bar{\boldsymbol{e}}_n$. The diffraction pattern thus exhibits main (or basic) spots at $\bar{\boldsymbol{h}}$. With each basic spot $\bar{\boldsymbol{h}}$ a linear sequence of equidistant satellites $m\bar{\boldsymbol{q}}$ is associated. These sequences are perpendicular to the interfaces and they are shifted with respect to the positions of the basic spot over a fraction $\bar{\boldsymbol{h}} \cdot \bar{\boldsymbol{R}}$ of the intersatellite spacing; $\bar{\boldsymbol{q}}$ is the wave vector of the modulation. Provided the basic structure is known, the long-period structure can be determined from the geometry of the diffraction pattern.

If the displacement vector $\bar{\boldsymbol{R}}$ is parallel to the interfaces, the latter are termed conservative since the long-period structure has the same chemical composition as the basic structure.

If $\bar{\boldsymbol{R}}$ is not parallel to the interfaces the superstructure may have a composition different from that of the basic structure; this is the case for shear structures, in which the interfaces perpendicular to $\bar{\boldsymbol{e}}_n$ are generated by removing slabs of material with thickness $\bar{\boldsymbol{R}} \cdot \bar{\boldsymbol{e}}_n$, followed by closing the gaps so created by the displacement $\bar{\boldsymbol{R}}$. If the removed slab has a composition different from that of the basic structure, the overall composition changes in the process.

The alloy, based on a face-centered cubic lattice, with ideal composition $Au_{22}Mn_6$, produces the diffraction pattern shown in Figure 2.30 A and represented schematically in Figure 2.30 B. It consists of sequences of superstructure spots, marked as full dots in Figure 2.30 B, associated with the basic spots marked as small open disks in Figure 2.30 B, which are the spots that would be produced by the Au_4Mn structure. Both types of spots are present simultaneously in Figure 2.30 A because the selected sample contained a small area of Au_4Mn as well as an area exhibiting the $Au_{22}Mn_6$ structure. From the direction of the rows of superstructure spots, their spacing, and the fractional shifts, the model of Figure 2.30 D can be derived [2.69]. In this model, only the manganese columns are represented. The high-resolution image (Fig. 2.30 C) made in the bright-field superlattice mode can be compared directly with the model of Figure 2.30 D; only the minority atom columns are revealed as bright dots.

Molecular Crystals; Fullerites. Organic crystals are usually prone to ionization damage and decompose very rapidly under electron irradiation; they can thus be studied for only a short

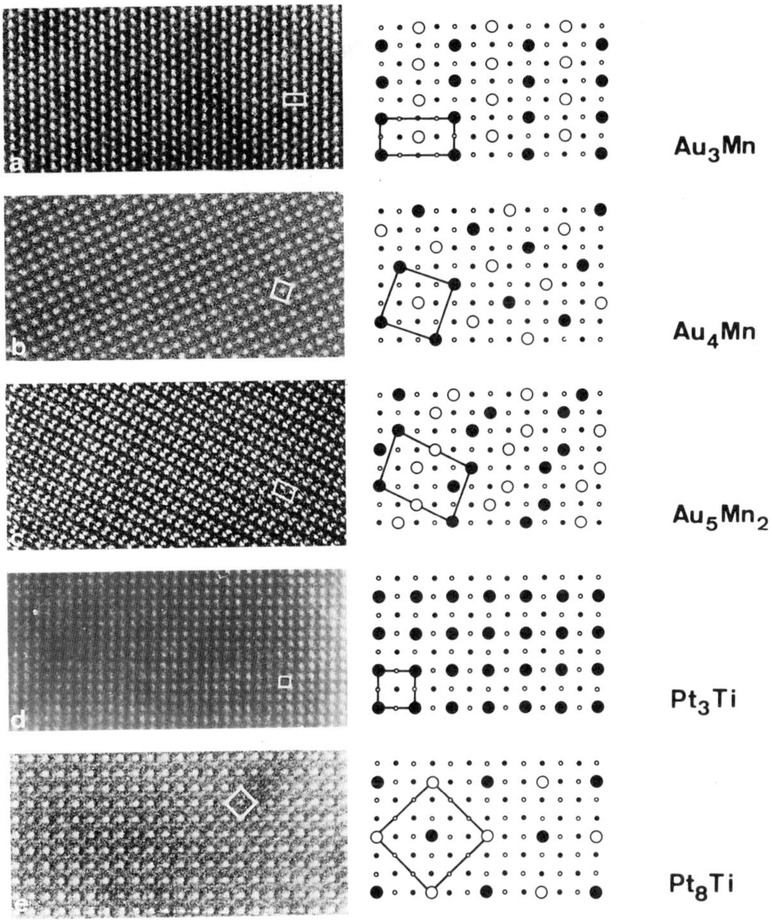

Figure 2.29. Various superstructures in ordered binary alloys imaged using the bright-field superlattice mode and compared to structural models

time (a few seconds) and only with a very low electron beam intensity. Transmission electron microscopy has, therefore, seldom been applied to organic crystals. However, the all-carbon molecules C_{60}, C_{70}, etc., (fullerenes) discovered at the end of the 1980s resist electron radiation fairly well. Early structural studies on the crystalline phases of fullerencs (fullerites) were performed mainly by electron microscopy because only small quantities of sufficiently pure material were available. At room temperature, C_{60} crystals were found to have the fcc structure, often containing intrinsic stacking faults, twins, and other defects characteristic of low stacking fault energy fcc alloys [2.70]. Figure 2.31 shows, for instance, various faults in a crystal of C_{60}: an intrinsic fault containing a dipole of stair rod dislocations in S, and a Frank partial dislocation and its associated stacking fault in A. The diffraction effects and the microstructure caused by the orientational phase transition in C_{60} at 255 K, from the room temperature fcc phase to the simple cubic low-temperature phase, were studied by means of electron microscopy [2.71] (Fig. 2.32).

Vapor-grown C_{70} crystals were found to belong to two different structural types. About 1 % of the vapor-grown crystals are hexagonally close packed; the rest are face-centered cubic. Many crystal fragments exhibit both crystal structures. The hexagonally close-packed crystals undergo several orientational phase transitions on cooling, leading to a monoclinic orientationally ordered superstructure at low temperature [2.72]. The fcc crystals undergo a single sluggish phase transition into a rhombohedral phase.

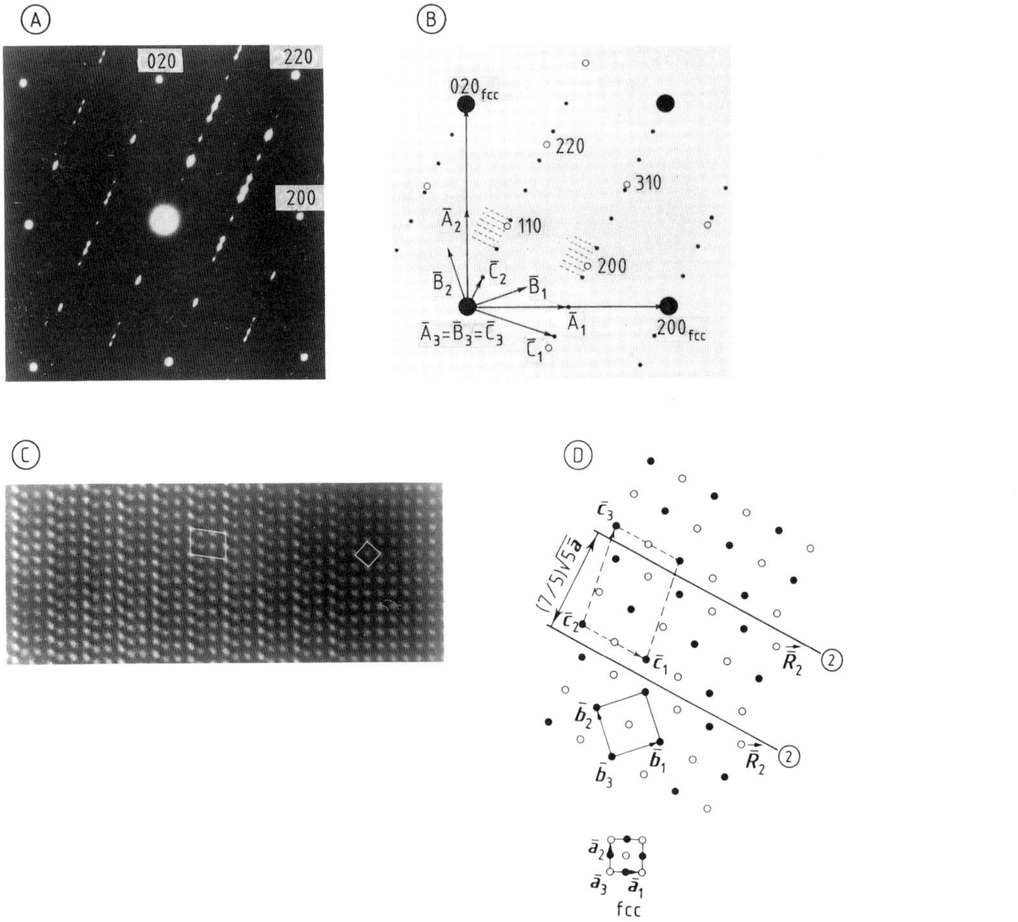

Figure 2.30. One-dimensional nonconservative long-period superstructure of the Au_4Mn structure, with theoretical composition $Au_{22}Mn_6$ [2.69]
A) Composite diffraction pattern along the [001] zone of Au_4Mn; B) Schematic representation of composite diffraction pattern (open dots indicated locations of basic spots, full dots are superstructure spots); C) High-resolution image of the superstructure (on the left, a small area of basic Au_4Mn structure is also visible); D) Model of the one-dimensional superstructure; only manganese columns are represented

Quasi Crystals. Until 1984, one of the most firmly established fundaments of crystallography was the postulate that crystals have three-dimensional translation symmetry, and that only symmetry rotation axes with multiplicities 1, 2, 3, 4, and 6 can occur because they are the only ones that are consistent with homogeneous space filling. However electron microscopy and electron diffraction have shown that 5-, 8-, 10-, and 12-fold symmetry axes occur in the structure of certain, usually rapidly cooled, alloy phases with complicated compositions [2.73], [2.74]. Careful tilting experiments have demonstrated the presence of icosahedral and dodecahedral symmetry groups in many of these phases. Their electron diffraction patterns consist of sharp spots, whose geometry reflects the above-mentioned symmetry elements, but they do not exhibit translation symmetry. The same conclusions follow from high-resolution images [2.75].

At present, numerous alloy systems are known to exhibit noncrystallographic symmetry elements and to lack three-dimensional periodicity; they are called quasi crystals. Figure 2.33 shows the diffraction pattern of a quasi crystal with composition Al–Mn exhibiting a fivefold axis, whereas Figure 2.34 shows the corresponding high-resolution image; pentagonal ar-

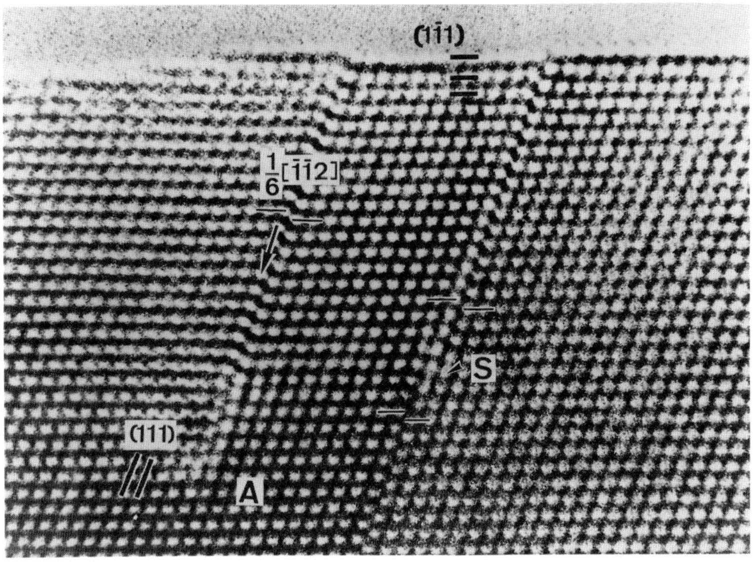

Figure 2.31. High-resolution image of a fragment of a C_{60} molecular crystal: bright dots represent rows of C_{60} molecules Note the presence of a stepped intrinsic stacking fault (in S). Also a Frank-type partial dislocation and its associated stacking fault are present (in A) [2.70]

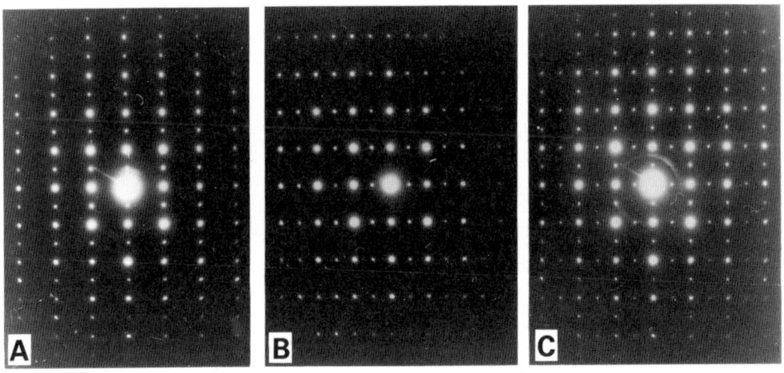

Figure 2.32. Diffraction patterns of the simple cubic phase of C_{60} along a cube zone [2.71]
A) One variant; B) Perpendicular variant to A); C) Both variants present

rangements of bright dots are visible in many places.

Artificial Layer Structures and Semiconductor Devices. In recent years the growth of artificial layer structures, which can be used as quantum wells, has become an important research activity. The properties of such layered structures are determined to a large extent by the perfection of the interfaces, as well as the regularity of their spacing. Both parameters can be studied by means of high-resolution, cross-sectional samples. The steps in the preparation of cross-sectional samples are discussed in Section 2.1.10. The method is clearly destructive and cannot be used for fabrication control. However, it is the most direct way to calibrate other nondestructive methods.

Figure 2.35 shows a high-resolution image of a quantum well consisting of the alternation of GaAs and (GaAl)As. The positions of the interfaces are indicated by arrows. The interfaces are flat to within two or three atomic layers; no interfacial dislocations are observed. The composition profiles across such interfaces can in principle be studied by measuring the brightness of the dots representing the atom columns [2.76].

For the study of semiconductor devices, diffraction contrast images made in a high-voltage

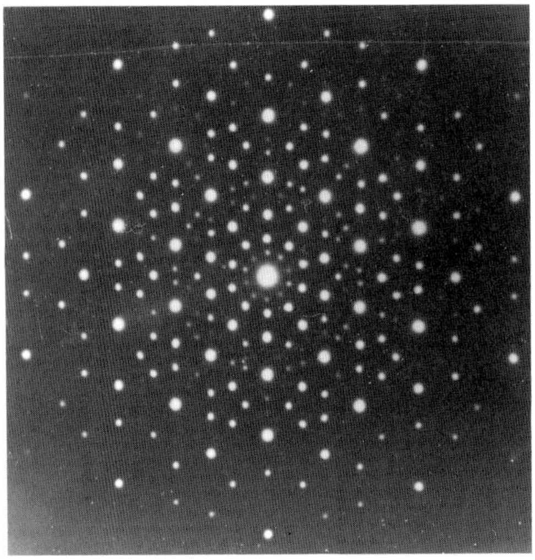

Figure 2.33. Diffraction pattern of the icosahedral phase of quenched Al–Mn alloy exhibiting fivefold symmetry (Courtesy of G. Van Tendeloo)

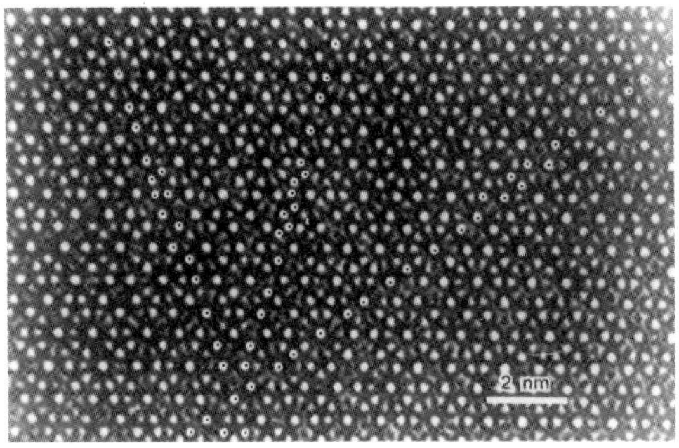

Figure 2.34. High-resolution electron micrograph of rapidly cooled Al–Mn alloy showing the absence of a periodic structure; pentagonal arrangements of dots can be observed (Courtesy of G. Van Tendeloo)

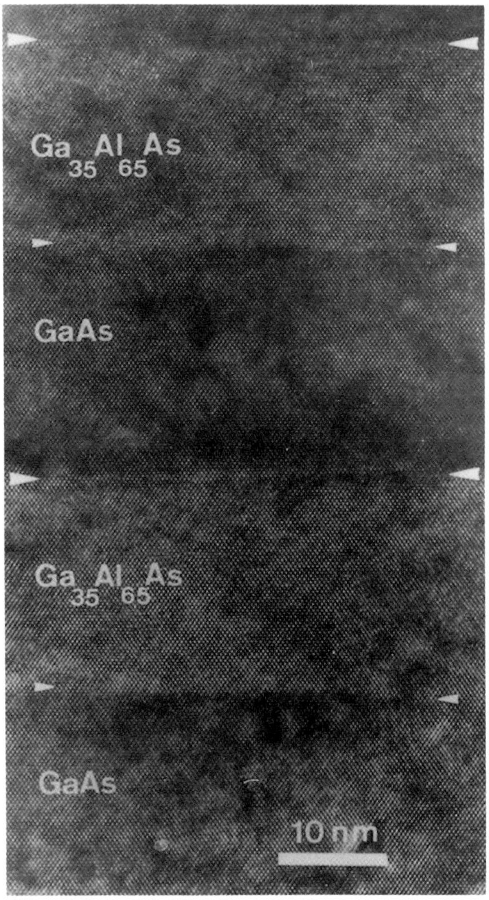

Figure 2.35. High-resolution image of quantum well consisting of alternation of pure GaAs and (GaAl)As

electron microscope (≈1000 kV) are often useful since thicker specimens can be tolerated than in a conventional microscope (≈200 kV). An example is shown in Figure 2.36, which is the cross section of a field effect device. The different layers have been indicated along with the orientation of the silicon substrate. The black dots numbered 1–5 are images of dislocations seen end on, which were nucleated by the stresses set up as a result of the oxidation process. From the spatial distribution of the dislocations the magnitude of the stress can be estimated.

During crystal growth, small oxide particles are often formed in the interior of silicon single crystals. Such particles cause a compressive stress on the surrounding matrix. This may lead to prismatic punching, whereby disks of self-interstitials limited by loops of perfect dislocations are emitted, thereby relieving the stresses. These loops can glide along cylindrical surfaces whose cross section is determined by the size and shape of the particle, and whose generators are parallel to the different glide vectors of silicon, which are of type $^1/_2$[110]. The geometry of the process can be established with diffraction contrast images (see Fig. 2.37).

Carbon Nanotubes, Onions. The preparation of fullerene-containing soot by the vaporization of graphite electrodes in an electric arc leads simultaneously to the formation of very fine *hollow carbon needles* or "tubules" with diameters as low as 1 nm. Electron microscopy and electron diffraction have shown that these needles consist of concentric seamless graphene tubes (2–15 tubes) [2.77]. The most remarkable feature

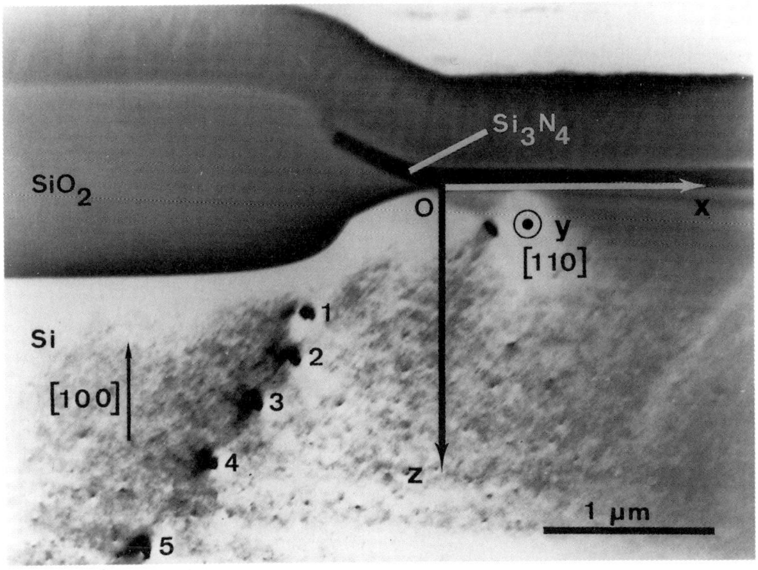

Figure 2.36. Cross section of field effect device
The procession of dislocation 1 ... 5 results from stresses set up by the oxidation process (Courtesy of J. Van Hellemont)

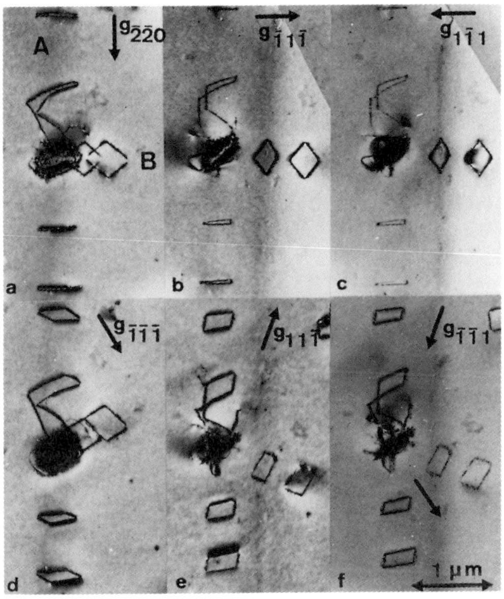

Figure 2.37. Oxide particle in the interior of a silicon single crystal
Stresses have been relieved by prismatic punching. The dislocation loops are imaged using different reflections (Courtesy of H. Bender)

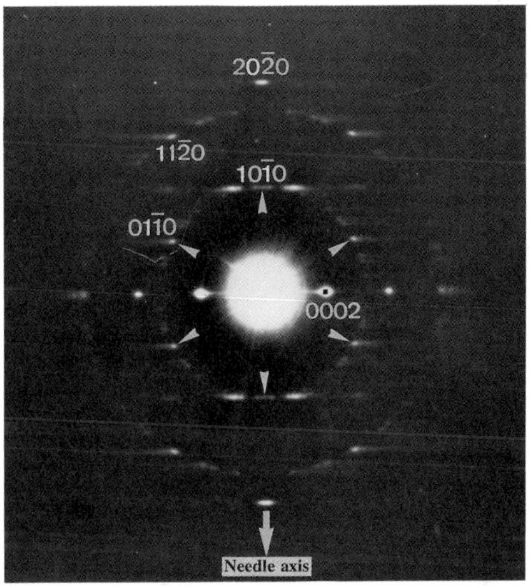

Figure 2.38. Electron diffraction (fiber) pattern produced by a graphite nanotube showing that certain graphene tubes are helically wound, whereas others remain normal
Note the outward streaking of the diffraction spots (Courtesy of X. F. Zhang)

is the helical character of some of the tubes within a tubule, as deduced from fiber diffraction patterns (Fig. 2.38). The helical structure is a consequence of the stepwise increase in circumference of successive concentric tubes by πc (c = lattice parameter of 2 H graphite). Since πc is not commensurate with the a-parameter of graphite the $c/2$ spacing between successive tubes and their seamless character can be reconciled only if some of the tubes at least become helical.

However, a detailed analysis of the fiber pattern has shown that in most tubules the majority of tubes are nonhelical [2.78].

Intense "in situ" electron irradiation of particles of carbon soot for tenths of minutes in the electron microscope produces spherical particles consisting of *onion-like* concentric spherical shells of graphene-like layers [2.79]. The microscope allows the concentric shell structure to be imaged during its formation.

High-T_c Superconductors. High-resolution electron microscopy and electron diffraction have recently been applied extensively in structural studies of high-T_c oxide superconductors. In particular, $YBa_2Cu_3O_{7-\delta}$ and the related 1–2–3 family of compounds have been studied in great detail. Numerous studies of superconducting compounds can be found in *Physica C* from 1989 onward. The generic 1–2–3 compound ($YBa_2Cu_3O_{7-\delta}$) has a structure derived from the triperovskite structure in which the succession of layers along the *c*-direction is CuO–BaO–CuO_2–Y–CuO_2–BaO, etc. The CuO layer consists of Cu–O–Cu–O– chains parallel to the b_0-direction; this feature breaks the tetragonal symmetry and reduces the symmetry to orthorhombic. The average structure was determined by X-ray and neutron diffraction, but several important structural features were discovered by electron microscopy. The oxygen deficiency was found to be accommodated by the absence of oxygen in a fraction of the chains in the CuO layer [2.80]. The structure in which one of two chains is alternately free of oxygen, the so-called $2a_0$ (or ortho II) superstructure, has an ideal composition with $\delta = 0.5$. It occurs in two orientation variants, corresponding to the (110) reflection twin texture of the 1–2–3 compound, which was also discovered by electron microscopy. The presence of the $2a_0$ structure was shown to be responsible for the 60 K plateau in the T_c versus composition curve [2.80]. Locally, structures with a period of $3a_0$ occur [2.81].

Much attention has been devoted to various substitutions in CuO layers, replacing copper ions by other metallic ions such as Fe, Co, etc., but also substituting by complex anions such as CO_3^{2-}, SO_4^{2-}, PO_4^{3-}, or NO_3^-. High-resolution electron microscopy has demonstrated that the substitution occurs in the CuO layers, and has allowed the resulting superstructures to be visualized. In the case of SO_4^{2-} substitution, for instance, the superstructure has been shown not to be commensurate with the basic 1–2–3 lattice [2.82]. Figure 2.39 shows a view along the b_0-direction of the 1–2–3 matrix. The prominent, bright dot squares reveal the SO_4^{2-}-containing rows that replace CuO chains. Their arrangement is not periodic and can best be described as resulting from a concentration wave with a wave vector that is not commensurate with the 1–2–3 lattice along the *a*-direction.

Replacing copper by cobalt or gallium in the CuO layer results in the formation of corner-sharing chains of CoO_4 (GaO_4), tetrahedra along the [110] and [1$\bar{1}$0] directions of the 1–2–3 matrix lattice [2.82]. The resulting structure is still orthorhombic, but with a diagonal unit cell ($a_0 \approx b_0 \approx a_p\sqrt{2}$), where a_p is the lattice parameter of the basic perovskite. Electron microscopy [2.83] showed that a superstructure is formed, which is localized in the Co–O (Ga–O) layers,

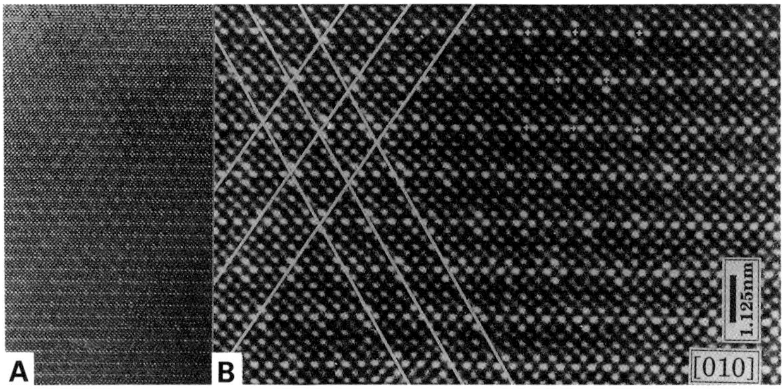

Figure 2.39. High-resolution image along the $[010]_0$ direction of the SO_4^{2-}-substituted 1–2–3 ($YBa_2[Cu_{3-x}(SO_4)_x]O_{7-\delta}$) compound
The SO_4^{2-}-containing rows parallel to [010] are marked as small crosses of prominent bright dots. (Courtesy of T. Krekels) [2.82]

and in which alternating parallel CoO_4 (GaO_4) chains have mirror-related configurations, leading to doubling of the lattice parameter perpendicular to the chain direction. Figure 2.40 is a view of the cobalt-containing material along a $[120]_p$ direction, which allows the period doubling along a_0, as well as the possibility of polytypism along c_0, to be observed.

Superionic Conductors. Superionic conductors often consist of a stable rigid framework forming large channels along which ions can move easily. The channels in such structures can be made visible in high-resolution images made along a zone parallel to the channel direction. Figure 2.41 shows such a high-resolution image of the natural mineral hollandite with idealized composition $Ba_xMn_8O_{16}$ [2.84]. Octagonal channels, formed by interconnected strings of edge and corner-sharing MnO_6 octahedra, are imaged as octagons of black dots. These channels are occupied by barium ions, which are also visible as black dots in the centers of the octagons.

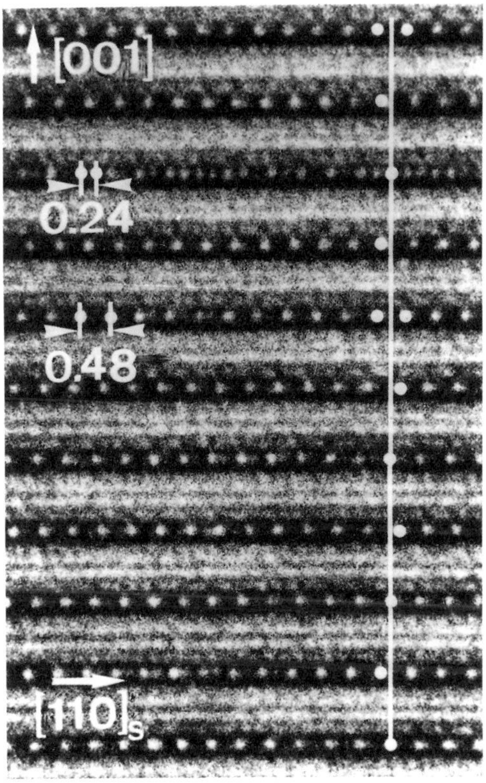

Figure 2.40. High-resolution image along the $[120]_p$ zone of the cobalt-substituted 1–2–3 compound $YBa_2Cu_2CoO_7$ Dark dots reveal indirectly the arrangement of chains of corner-sharing CoO_4 tetrahedra that replace the CuO layers. (Courtesy of T. Krekels) [2.83]

Polymers. Although most organic materials usually deteriorate rapidly from ionization damage in the electron beam, meaningful observations can be made on certain polymers. In Figure 2.42, fiber patterns of a poly(-*p*-phenylene)

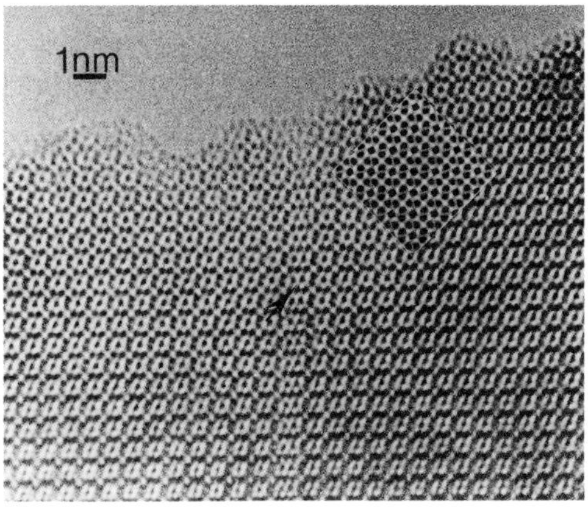

Figure 2.41. Structure of natural monoclinic hollandite $Ba_xMn_8O_{16}$ as viewed along the [010] zone
The image is compared with a simulated image in the inset [2.84].

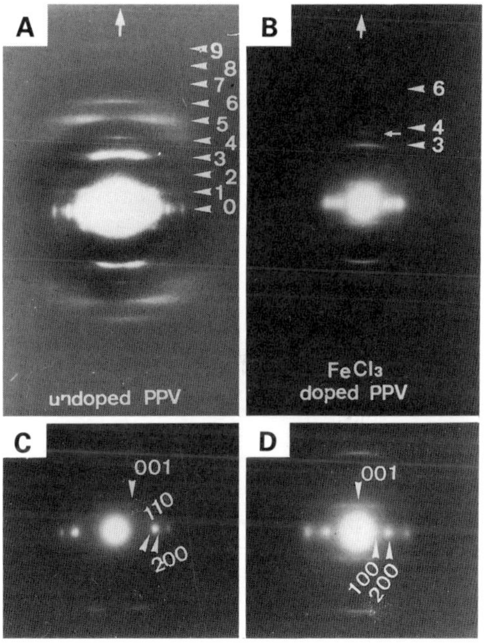

Figure 2.42. Fiber texture pattern of a stretched PPV polymer
A) and C) Pristine polymer; B) and D) After $FeCl_3$ doping
(Courtesy of X. F. Zhang)

stretched six to seven times are reproduced (PPV, an alternating copolymer of *p*-phenylene and acetylene) at two different photographic exposures to reveal the details of the pattern close to the origin [2.85]. The direction of stretching is indicated by arrows. The patterns of the left column (A) and (C) refer to the pristine material. On doping with $FeCl_3$ this material becomes a good electrical conductor. The effect of $FeCl_3$ doping on the diffraction pattern is shown in the right column (B and D). Whereas in the undoped material the 001 diffraction vector encloses an obtuse angle with the row of hoo reflections, this angle becomes 90° in the doped material, which suggests that the symmetry changes from monoclinic to orthorhombic on doping.

2.1.11.2. Defects

Dislocations. Dislocation configurations have been studied mainly by diffraction contrast bright-field imaging since this method is very sensitive to lattice strain fields. Under these conditions, dislocations appear as dark lines when two-beam diffraction conditions close to $s \approx 0$ are used. A network of dislocations in the basal plane (0001) of zinc is visible in Figure 2.43.

Images of dissociated dislocations (i.e., dislocation ribbons) have been used extensively as a means to deduce the stacking fault energy [2.1].

The separation between partial dislocations is usually measured on a weak beam image (see Section 2.1.6.2). The image, which is now a very narrow bright line on a darker background, is essentially kinematic. In the high-resolution image of Figure 2.44, dissociated dislocations in silicon are viewed along the [110] zone. Between the two partial dislocations an intrinsic stacking

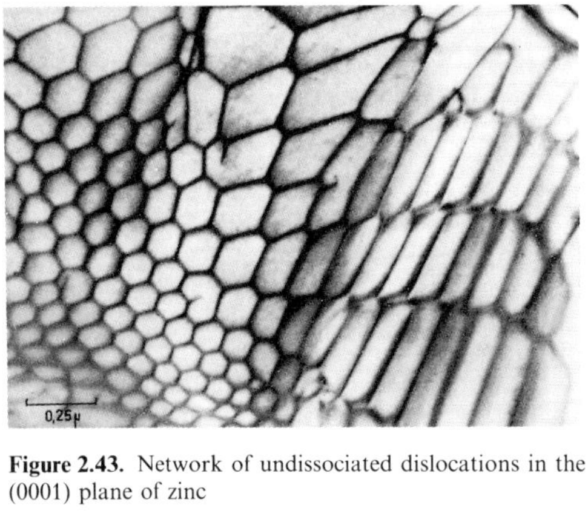

Figure 2.43. Network of undissociated dislocations in the (0001) plane of zinc

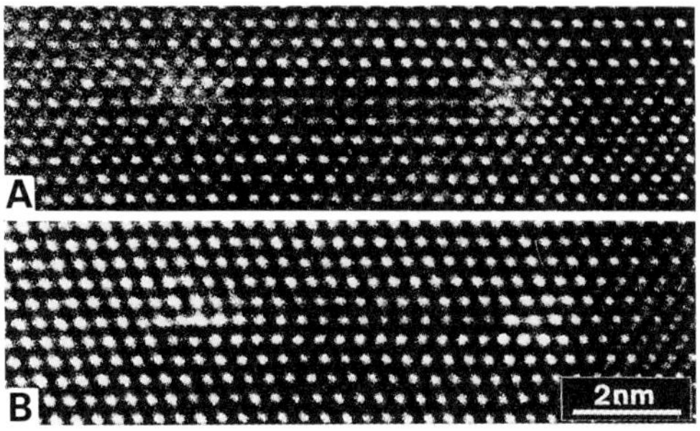

Figure 2.44. Dissociated dislocations in silicon viewed along the [110] zone in the high-resolution mode (Courtesy of H. Bender)

Figure 2.45. High-resolution image of a stacking fault in 2H-wurtzite (Courtesy of H. Bender)

fault is present. Such configurations have been used to deduce stacking fault energies [2.1], [2.51].

Planar Defects. The diffraction contrast images of planar interfaces that intersect the foil surfaces consist of a set of fringes parallel to the closest surface with a depth period given by $1/\sigma_g$ (which is equal to t_g for $s = 0$; see Section 2.1.4.6). If the planar interface is parallel to the foil surfaces, which is often the case in cleaved foils, the faulted area exhibits only a brightness difference.

High-resolution images can reveal stacking faults directly, especially in close-packed structures. In such structures the image made along a zone parallel to the close-packed rows reveals the stacking directly. A stacking fault in a wurtzite crystal is reproduced in Figure 2.45. (Single antiphase boundaries are imaged in Figure 26, whereas periodic antiphase boundaries in Au_4Mn are visible in Figure 28 C).

Domain boundaries and twins in planes that intersect the foil surfaces are imaged as δ-fringes (Section 2.1.4.6) [2.30]. The extinction condition is now $\Delta s = 0$ (or $\Delta \bar{g} = 0$!) which is satisfied for the family of planes common to the two domains. When the interfaces are nearly perpendicular to the foil plane the main effect is domain contrast (i.e., a brightness difference in the domains on either side of the interface) as shown in Figure 2.46, where the same twin boundaries in the high-temperature superconductor $YBa_2Cu_3O_{7-\delta}$ have been imaged in three different modes.

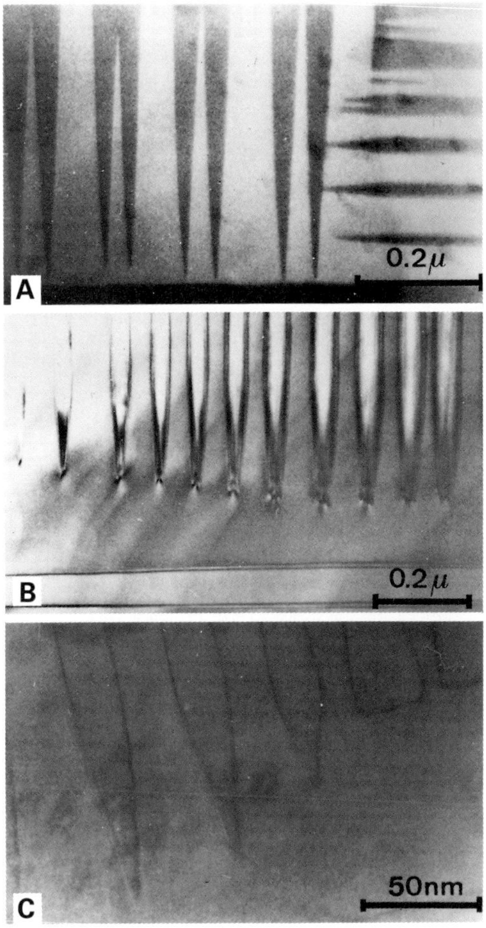

Figure 2.46. Twin boundaries in $YBa_2Cu_3O_{7-y}$
(A) Domain contrast; (B) Interface contrast; (C) High-resolution imaging (Courtesy of H. W. Zandbergen)

2.1.11.3. Small Particles

Like most other electron microscope techniques, transmission electron microscopy allows the particle-size distributions and particle shapes of very fine powders to be studied. Such studies are important in several areas of research: including catalysis, magnetic recording, and photography. Transmission electron microscopy provides, in addition, the possibility of obtaining electron diffraction patterns of single particles as well as high-resolution images. In Figure 2.47 single particles of silver vapor deposited on an amorphous carbon substrate is shown. They are clearly not single crystals but aggregates of multiply twinned fcc crystallites having an overall icosahedral shape.

The topotactic dehydration reaction [2.86] transforming goethite into hematite was studied "in situ" at room temperature and shown to produce small twinned hematite crystals (Fig. 2.48). In the initial stages of the dehydration process, satellite spots appear in the diffraction pattern around the hematite spots; they were found to be due to a texture consisting of a periodic alternation of voids and hematite crystals, rather than to a long-period superstructure.

The crystal habit of the silver halide particles used in photographic emulsions is important since it is one of the parameters determining the photosensitivity of the film. The hexagonal and triangular (111) tabular crystals of AgCl were shown to be in fact multiply twinned on planes parallel to the habit plane. Depending on the parity of the number of twin lamellae the particle grows into either hexagonal or a triangular habit.

When silver halide crystals with a cube habit are exposed to electrons, photolytic metallic silver specks are formed "in situ" in an epitaxial relation to the silver halide substrate. This gives rise to moiré fringes parallel to the cube edges due to the difference in lattice spacing between AgCl and Ag (Fig. 2.49).

2.1.11.4. Surface Studies [2.87]

Crystal surfaces can be studied in a transmission electron microscope by different modes. It is possible to use an electron beam, in reflection under *grazing incidence*, magnifying the Bragg reflected beam due to planes roughly parallel to the surface. The surface of the sample must be almost parallel to the beam. The resulting image is highly distorted in the sense that the magnifi-

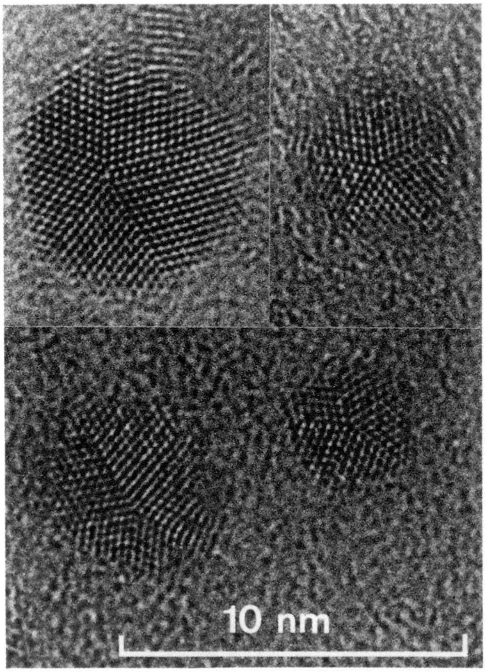

Figure 2.47. Small silver particles consisting of an icosahedral aggregate of multiply twinned face-centered cubic crystals (Courtesy of C. Goessens)

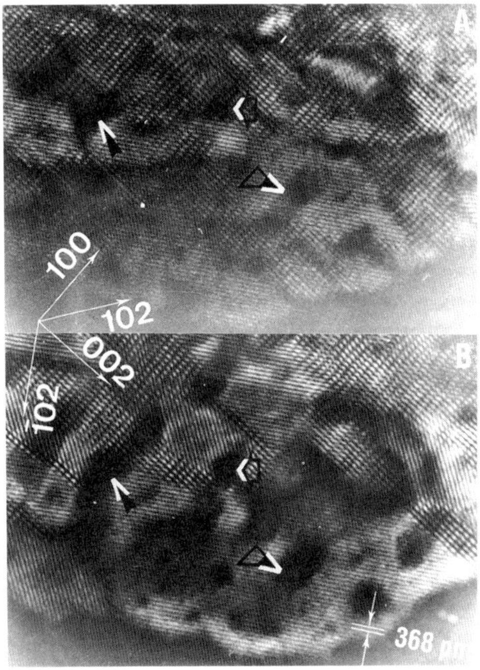

Figure 2.48. Small particles of twinned hematite formed in the electron microscope as a decomposition product of goethite [2.86]

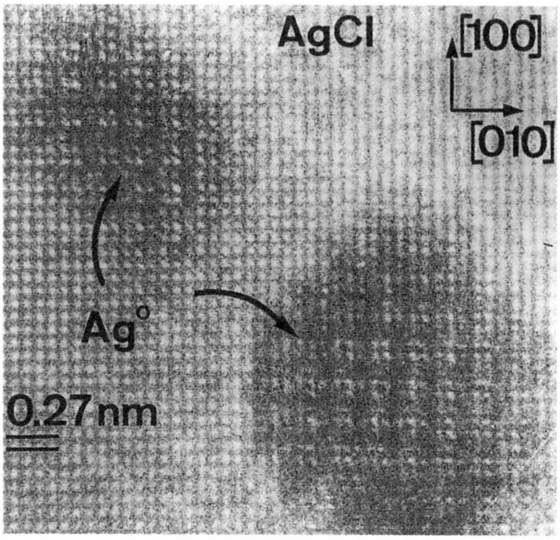

Figure 2.49. Silver chloride crystal viewed along the cube zone

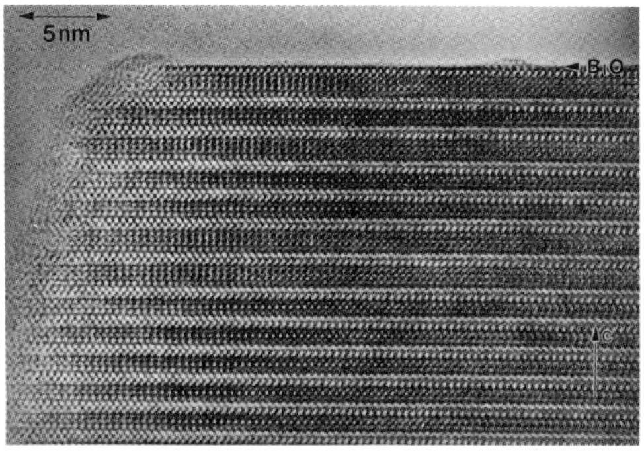

Figure 2.50. High-resolution image of the free cleavage surface of a superconductor ($Bi_2Sr_2Ca_1Cu_2O_x$) Cleavage occurs between the two BiO layers [2.88]

cation in a direction perpendicular to the beam is much lower than the magnification along the projection of the electron beam. This Bragg reflection mode can be used to image surface steps and to detect surface superstructures due to reconstruction or to adsorbed layers by diffraction. A very clean vacuum is required.

A second mode consists of observing the surface profile in transmission with the *electron beam parallel to the surface*. In this way, surface relaxations and surface reconstruction can be observed together with atomic resolution of the substrate. An example of the cleavage surface of a high-T_c superconductor is reproduced in Figure 2.50. It allows the layer within the structure to be located along which cleavage takes place [2.88].

2.1.11.5. "In situ" Studies [2.89]

By using different types of stages the specimen chamber of the microscope can be transformed into a small laboratory. Heating and cooling stages make it possible to study "in situ" the changes in crystal structure and in microtexture accompanying phase transitions, by observing (1) the appearance or disappearance of superstructure spots in the diffraction pattern, and (2) the fragmentation with translation and/or orientation domains in the direct space image.

Environmental cells can be used to study in situ solid–gas reactions, such as oxidation processes or decomposition reactions.

Plastic deformation can be studied and the dislocation propagation observed in real time in a straining stage.

The creation of radiation damage by ionization or by electron–atom collisions can be studied in medium- and high-voltage microscopes. The formation of agglomerates of point defects, such as dislocation loops and stacking fault tetrahedra, can be observed during irradiation with the image-forming electrons. (Light atoms, such as oxygen, are readily displaced by electrons of 400 kV.)

Radiation ordering and disordering in alloy systems have successfully been studied in high-voltage microscopes at different temperatures.

2.2. Scanning Electron Microscopy

2.2.1. Introduction

In a scanning electron microscope (SEM) [2.90]–[2.92], a small electron probe 1–10 nm in diameter scans in a raster across the surface of the specimen (Fig. 2.51). The incident electrons are elastically and inelastically scattered by the specimen. *Elastic scattering* results in large scattering angles and zigzag electron trajectories. Therefore, a fraction of electrons can leave the specimen as backscattered electrons (BSE) (Fig. 2.52). The slowing down of electrons by *inelastic scattering* results in an electron range R. Electrons from the specimen atoms that are excited by inelastic scattering can leave the specimen as *secondary electrons* (SE) from a thin surface layer Λ_{SE} of ca. 1–10 nm. By convention, electrons in the energy spectrum (Fig. 2.53) with $E \leq 50$ eV are called SE. The secondary electrons consist of (1) SE1 excited by the primary

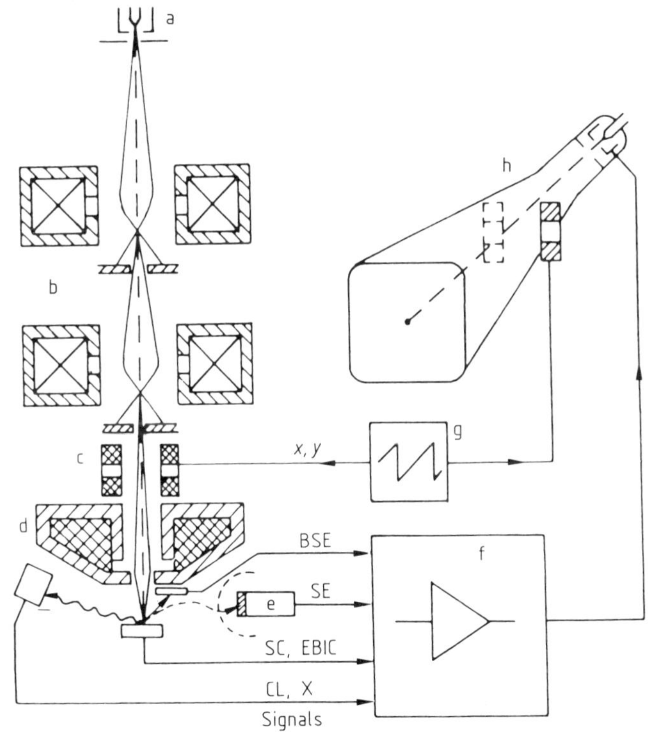

Figure 2.51. Schematic cross section of a scanning electron microscope (SEM)
a) Electron gun; b) Condenser lenses; c) Scan coils; d) Objective; e) Photomultiplier; f) Amplifier; g) Scan generator; h) Cathode-ray tube
BSE = Backscattered electrons; SE = Secondary electrons; SC = Specimen current; EBIC = Electron-beam-induced current; CL = Cathodoluminescence; X = X rays

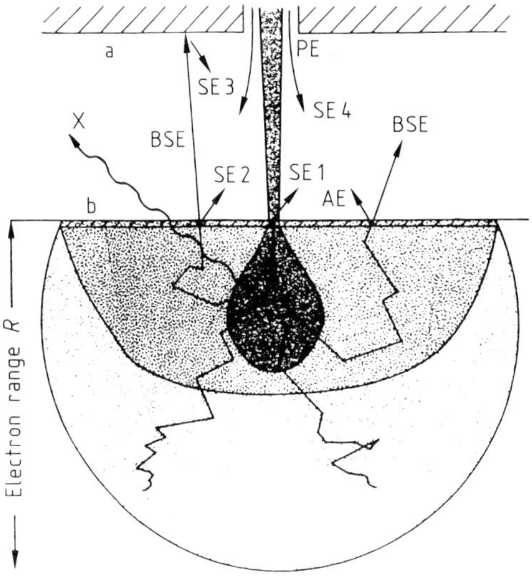

Figure 2.52. Excitation of backscattered electrons (BSE), Auger electrons (AE), and different groups of secondary electrons (SE) by primary electrons (PE)
Volumes of electron diffusion, BSE trajectories, and emitted X rays are shown with increasing density of gray levels
a) Polepiece; b) Specimen

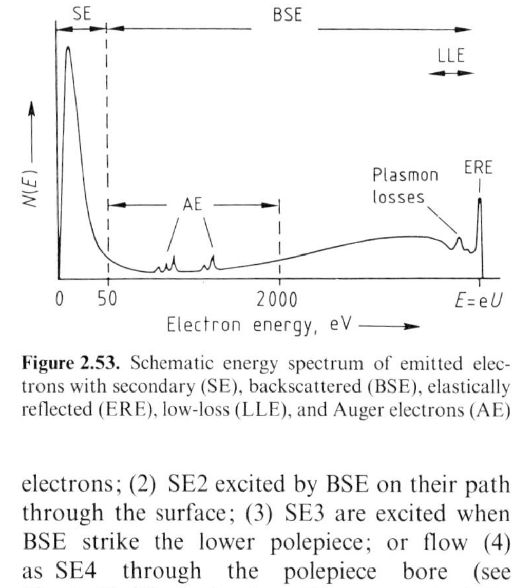

Figure 2.53. Schematic energy spectrum of emitted electrons with secondary (SE), backscattered (BSE), elastically reflected (ERE), low-loss (LLE), and Auger electrons (AE)

electrons; (2) SE2 excited by BSE on their path through the surface; (3) SE3 are excited when BSE strike the lower polepiece; or flow (4) as SE4 through the polepiece bore (see Fig. 2.52). The ionization of inner atomic shells results in the emission either of characteristic X-ray quanta (X) or of Auger electrons (AE).

The *image* is formed by the signal of emitted secondary electrons, backscattered electrons, Auger electrons, absorbed speciment current (SC), or X-ray quanta, which modulate the intensity of a cathode-ray tube rastered in synchronism (Fig. 2.51). Conventional SEMs work with

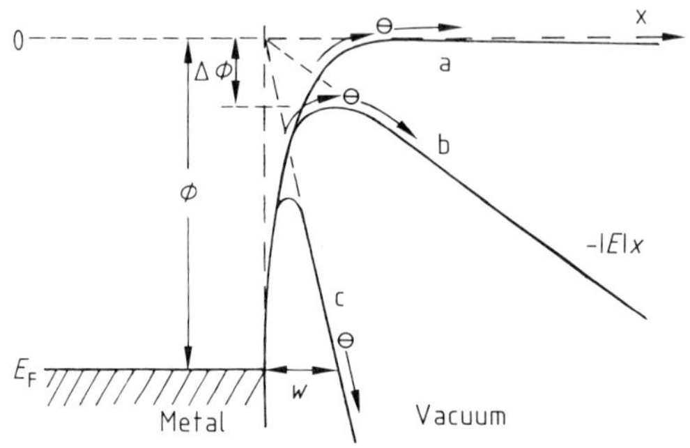

Figure 2.54. Potential at the cathode – vacuum interface for W and LaB_6 thermionic emission, ZrO – W Schottky emission, and field emission guns
a) Thermionic emission; b) Schottky emission; c) Field emission
ϕ = Work function; E_F = Fermi energy

electron acceleration voltages of 5 – 30 kV, whereas a *low-voltage scanning electron microscope* (LVSEM) [2.93], [2.94] uses 0.5 – 5 kV.

2.2.2. Instrumentation

2.2.2.1. Electron Guns

The following types of electron emitters are used in SEM electron guns.

Thermionic cathodes consist of a directly heated tungsten hairpin cathode at $T_c = 2500 - 3000$ K, or an indirectly heated pointed rod of lanthanum or cerium hexaboride (LaB_6, CeB_6) at 1400 – 2000 K. The electrons must overcome the work function Φ of 4.5 eV (W) or 2.7 eV (LaB_6) by thermal activation (Fig. 2.54, curve a). Between the cathode at the potential $-U$ and the grounded anode, a negatively biased Wehnelt electrode forms a crossover of diameter 20 – 50 μm (W) or 10 – 20 μm (LaB_6) as an effective electron source. The emitted electrons show an energy spread $\Delta E = 1 - 2$ eV (W) or 0.5 – 1 eV (LaB_6). A measure of the quality of an electron gun is the axial gun brightness β:

$$\beta = j/\pi\alpha^2 \approx j_c \mathrm{E}/\pi k T_c \quad (2.29)$$

where j is the current density, α the aperture angle, and E the electron energy. The axial gun brightness is constant for all points on the axis regardless of lenses and aperture diaphragms (j_c and T_c are current density and temperature, respectively, at the cathode). It increases proportionally to the electron energy E, with

$\beta \approx 10^5$ A cm^{-2} sr^{-1} (where sr denotes steradian) for tungsten at $E = 20$ keV and about ten times higher values for LaB_6.

Schottky emission cathodes consist of zirconium-doped tungsten tips, with a radius of about 0.5–1 µm, coated with a ZrO layer. This layer decreases the work function from 4.5 to 2.7 eV. A Schottky emission cathode works with a higher electric field strength at the tip that decreases the work function by $\Delta\Phi$ (Schottky effect, Fig. 2.54) and concentrates the emission at the tip with a virtual electron source having a diameter of 15–20 nm. However, the electrons still have to overcome the work function by thermal activation. The energy spread is 0.3–1 eV.

Field-emission guns (FEG) consist of tungsten tips with a radius of ca. 0.01–0.1 µm. An intermediate anode at ca. 1–2 kV extracts electrons by tunneling through the barrier of the work function, which is $w = 1$–10 nm in width (Fig. 2.54). Two types of FEG exist: (1) *cold FEGs* that work with the tip at room temperature, and the tip is flashed once a day, (2) *heated FEGs* work at 1800 K. The energy spread is only 0.2–0.4 eV for cold and 0.5–0.7 eV for heated FEGs. The gun brightness of FEGs $\beta \approx 10^7$–10^8 A cm^{-2} sr^{-1}.

2.2.2.2. Electron Probe Formation

The crossover of a thermionic gun or the virtual point sources of Schottky or field-emission guns are demagnified by two to three electromagnetic lenses (Fig. 2.51), so that a geometric probe size d_g is formed at the specimen with a diameter of 1–10 nm. A diaphragm of ca. 50–200 µm diameter in front of the last lens limits the electron-probe aperture α_p. The electron–probe current I_p, d_g, and α_p cannot be changed independently because of the conservation of gun brightness β in Equation (2.29):

$$I_p \approx \frac{\pi}{4} d_g^2 j_p = \frac{\pi^2}{4} \beta d_g^2 \alpha_p^2 \qquad (2.30)$$

The geometric probe size d_g is superposed by aberration disks of spherical (d_s) and chromatic (d_c) aberration and diffraction (d_d):

$$d_s = 0.5 C_s \alpha_p^4; \quad d_c = C_c(\Delta E/E)\alpha_p; \quad d_d = 0.6\lambda/\alpha_p \qquad (2.31)$$

with the aberration constants C_s and C_c, respectively; ΔE is the energy spread of the electron gun, and λ is the deBroglie wavelength

$$\lambda = h/mv = 1.23/\sqrt{E} \qquad (2.32)$$

with λ in nanometers and E in kiloelectronvolts. An astigmatism of the last probe-forming lens can be compensated using a stigmator, which consists of magnetic quadrupoles.

The (quadratic) superposition of d_g, d_d, d_s, and d_c results in an effective probe diameter d_p (Fig. 2.55), where d_g and d_s dominate for high electron energies and thermionic guns, and d_d and d_c dominate for low-voltage SEM and field-emission guns. This superposition results in a minimum probe size between 1 and 10 nm at optimum apertures of tens of milliradians. An anticipated increase in resolution (decrease of electron probe diameter) requires a compensation of the chromatic aberration by using a combination of electrostatic and magnetic multipoles.

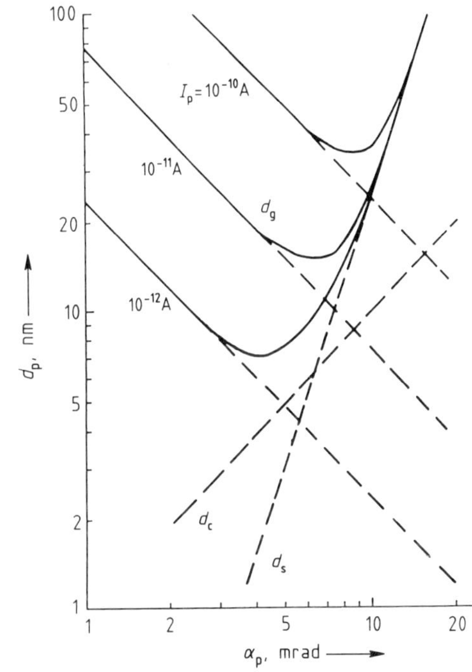

Figure 2.55. Double-logarithmic superposition of geometric probe size d_g for different probe currents I_p and aberration disks d_c and d_s versus the electron probe aperture α_p for a thermionic tungsten cathode
$E = 20$ keV; $\beta = 7 \times 10^4$ A cm^{-2} sr^{-1}; $C_s = 50$ mm; $C_c = 20$ mm; $\Delta E = 1$ eV

The low optimum probe aperture α_p of the SEM results in a large depth of field D_f:

$$D_f = \delta/\alpha_p = \Delta/\alpha_p M \quad (2.33)$$

where $\Delta = \delta M \approx 0.1$ mm is the resolution and M the magnification on the cathode-ray tube (CRT) screen. As a consequence, an SEM has a two-order-of-magnitude greater depth of field than a light microscope even at low magnification.

To decrease the aberration constants C_s and C_c, the focal length of the last probe-forming lens can be reduced by a stronger excitation of the magnetic lens, which means a short working distance between specimen and polepiece for high resolution. The aberration constants can be decreased further if the specimen is positioned in the lens field; this makes through-lens detection of secondary electrons necessary.

2.2.2.3. Detectors

Secondary electrons are most frequently used for image formation and can be detected by a scintillator–photomultiplier combination (*Everhart–Thornley detector*, Fig. 2.56). A large fraction of slow SE is attracted by a positively biased grid (b) and accelerated to a yttrium–aluminum–garnet (YAG) scintillator or a P47 powder scintillator (c) biased at +10 kV. The photons generated are guided in a light pipe (d) to a photomultiplier (f). This SE detector shows a high signal-to-noise ratio and a large bandwidth up to 10 MHz. The collection field of an Everhart–Thornley detector mounted on one side of the specimen can disturb the electron probe in LVSEM. In this case, beam deflection and distortion can be avoided either by a combination of electrostatic and magnetic quadrupoles forming a Wien filter, by a retarding electrostatic lens, or by through-lens detection of SE.

Backscattered electrons have enough energy for direct production of a greater number of photons in a scintillator coupled via light pipe to a photomultiplier. However, the Everhart–Thornley detector cannot be used effectively for faster BSE because only a small fraction hits the scintillator area. Therefore, *annular or semiannular top detectors* collecting BSE with a high takeoff angle or *ring detectors* with low takeoff, are applied to make the best use of the angular characteristics of BSE. Alternatively, *semiconductor detectors* can be used. The BSE produce a large number $n = E/E_i$ of electron–hole pairs, where $E_i = 3.6$ eV is the mean energy per excitation in silicon. These charge carriers can be separated in a p–n junction and form an electron beam-induced current (EBIC). BSE can also be detected by the *conversion of BSE to SE3* at the polepiece and other parts of the specimen chamber, when a negatively biased electrode around the specimen retards SE1 and SE2 from the specimen.

Detectors for LVSEM. Both scintillator and semiconductor detectors show a signal decreasing linearly with decreasing electron energy, and a threshold energy of ca. 1–5 keV. Therefore, their application in LVSEM requires postacceleration. As a recent alternative, *microchannel plates* (MCPS) can be used for the detection of BSE. These consist of a slice from a boule of tightly packed, fused tubes of lead-doped glass with a 10–20 µm inner diameter and a resistance of 10^8–10^9 Ω over their length. Annular disks

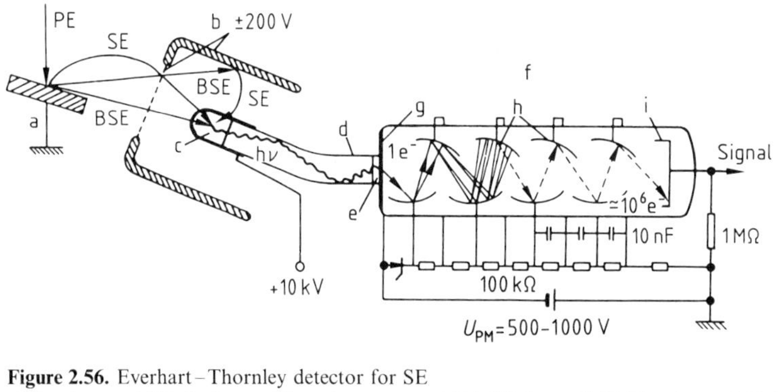

Figure 2.56. Everhart–Thornley detector for SE
a) Specimen; b) Collector grid and screen; c) Scintillator; d) Light pipe; e) Optical contact; f) Photomultiplier; g) Photocathode; h) Dynodes; i) Anode

3–4 mm thick can be mounted below the polepiece. The incident electrons produce SE at the inner tube wall, which are accelerated by a continuous voltage drop along the tube with a bias of 1 keV and are multiplied inside the MCP tubes by a multiplier-like action. The anode plate and a preamplifier at a potential of 1 kV are electrically insulated by an optical decoupler or by transmitting a modulated 30-MHz signal.

2.2.3. Electron–Specimen Interactions

2.2.3.1. Elastic and Inelastic Scattering

Elastic scattering of incident electrons results from the attractive Coulomb force of the nucleus screened by the atomic electron cloud with a differential Mott cross section of

$$\frac{d\sigma_M}{d\Omega} = r(\theta)\,\frac{eZ}{4(4\pi\varepsilon_0)^2 m^2 v^4}\,\frac{1}{\sin^4(\theta/2)} \qquad (2.34)$$

where $r(\theta)$ is the ratio between the Mott and Rutherford cross sections, Ω is the solid angle, Z the atomic number of the nucleus, ε_0 the vacuum permittivity, and m the mass of the electron [2.90]. The strong differences between Mott and Rutherford cross sections result from taking account of the spin–orbit coupling of electrons and solving the relativistic Dirac equation, whereas the Rutherford cross section is only a solution of the Schrödinger equation not containing the spin.

Inelastic scattering results in an excitation of electrons of the solid and a corresponding energy loss ΔE of the incident electron. Information about the differential inelastic cross section $d^2\sigma/d\Omega\, d(\Delta E)$ can be obtained from dielectric theory or experimental electron energy loss spectra (EELS; see Section 2.1.9.2) of high-energy electrons [2.95], [2.96].

A series of inelastic scattering processes with statistical energy losses results in a slowing down of electrons, which can be described by a mean energy loss per unit path length (Bethe stopping power S)

$$S = \left|\frac{dE_m}{ds}\right| = \frac{2\pi e^4 N_A \rho Z}{(4\pi\varepsilon_0)^2 A E}\ln(1.166\,E/J) \qquad (2.35)$$

with the mean ionization potential $J \approx 12.5\,Z$. As energy decreases, fewer subshells are ionized, which can be described by a parabolic decrease of $1/S$ below $E/J = 6.3$.

2.2.3.2. Electron Diffusion

The decrease of mean electron energy E_m with increasing path length s of electron trajectories due to Equation (2.35) results in a mean total path (Bethe range R_B) that increases with increasing Z of the material. In low-Z material with less frequent large-angle scattering, R_B and the practical range R are equal, whereas in high-Z material with more frequent scattering, the trajectories are strongly curled and $R < R_B$, as demonstrated by Monte Carlo simulations in Figure 2.57. The range

$$R \approx 6.6\,E^{5/3} \qquad (2.36)$$

is to a large extent independent of material, when R is measured in units of mass thickness (μg/cm²) and E in keV.

2.2.3.3. Emission of Secondary and Backscattered Electrons

The *backscattering coefficient* η describes the fraction of *primary electrons* leaving the specimen with an energy reduced by energy losses. Emitted electrons with energies of ≤ 50 eV are called secondary electrons (see Section 2.2.1) and are described by the *secondary electron yield* δ. Both quantities depend on electron energy E, atomic number Z of the specimen, and surface tilt angle ϕ ($\phi = 0$: normal incidence), and show characteristic energy and angular distributions that are important to the discussion of image formation in using BSE and SE signals.

Figure 2.58 shows the dependence of the *backscattering coefficient* η on atomic number Z for different electron energies E. For $E > 5$ keV, η increases monotonically with increasing atomic number. For multicomponent targets

$$\eta = \Sigma c_i \eta_i \qquad (2.37)$$

shows a best fit to experiments, where c_i represents the mass fractions. This dependence of η on Z is responsible for the atomic number (compositional) contrast of the BSE signal for $E = 5$–30 keV (see Section 2.2.4.1). The backscattering coefficient η is approximately independent of E in the range 10–100 keV. Below 5 keV, η decreases for $Z > 30$ and increases for $Z < 30$ with decreasing E. The reason for this is the Mott elastic cross section (Eq. 2.34). An increasing tilt angle ϕ of the specimen ($\phi = 0$: normal incidence) results in an increase of η.

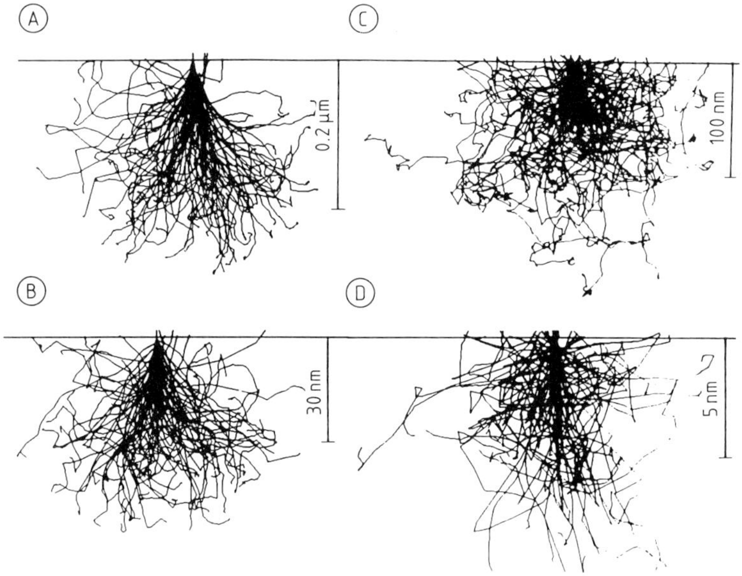

Figure 2.57. Monte Carlo simulations of 10- and 1-keV electron trajectories in carbon and gold
A) C, 10 keV, $\varrho = 1\ g/cm^3$; B) C, 1 keV, $\varrho = 1\ g/cm^3$; C) Au, 10 keV; D) Au, 1 keV

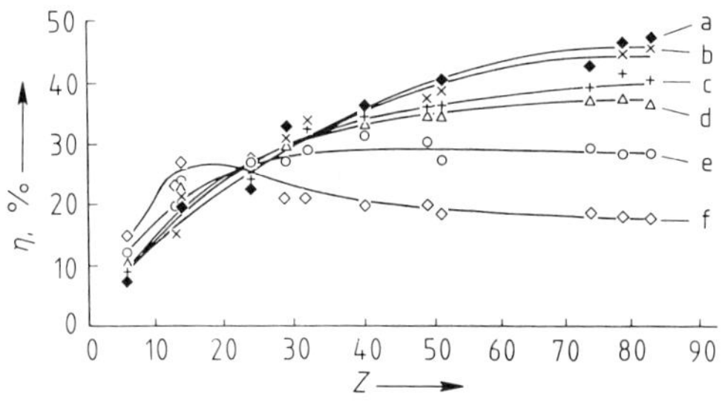

Figure 2.58. Backscattering coefficient η as a function of atomic number Z for different electron energies E
a) $E = 5$ keV; b) $E = 4$ keV; c) $E = 3$ keV; d) $E = 2$ keV; e) $E = 1$ keV; f) $E = 0.5$ keV

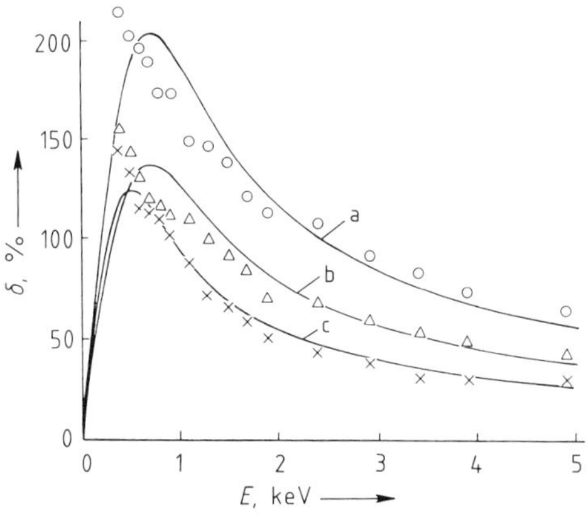

Figure 2.59. Decrease of SE yield δ with increasing electron energy E at normal incidence for
a) Au; b) Cu; c) C

The *secondary electron yield* δ shows a maximum for primary energies E of a few hundred electron volts and decreases $\sim E^{-0.8}$ for higher energies (Fig. 2.59). For $E > 5$ keV, the SE yield contains the contributions of SE1 excited by the primary electrons (PE) (Fig. 2.51) and of SE2 excited by the BSE. For $E < 5$ keV SE1 and SE2 can hardly be distinguished because the exit depths of SE and BSE have the same order of magnitude. The dependence of δ on the surface tilt angle ϕ can be approximated by $\delta \sim \sec^n \phi$, where the exponent n decreases from 1.3 (Be) to 1.1 (Al) and 0.65 (Au).

2.2.3.4. Specimen Charging and Damage

Insulating specimens show a *negative charging* by adsorbed electrons, when the total electron yield $\sigma = \eta + \delta$ is less than unity beyond a critical energy E_2 where σ changes from values larger than to values smaller than unity. Therefore, a conductive coating is necessary for working at high electron energies. With decreasing energy, σ can become greater than unity below

E_2. In this case, more electrons leave the specimen as SE or BSE than are absorbed. The specimens becomes *positively charged* by a few volts only because SE of low energies will be retarded. Therefore, insulating specimens can be observed at low energies without coating, although rough specimens can still show negative charging when, for example, SE cannot escape from holes.

Temperature-sensitive specimens can be damaged by *heat generation*. The increase of surface temperature at the electron probe decreases with decreasing energy, and such specimens can be observed better by LVSEM. Biological specimens are damaged by ionization, which results in a loss of mass and finally a polymerized carbon-enriched conglomerate within a layer of the order of the electron range R.

2.2.4. Image Formation and Analysis

2.2.4.1. Topographic and Material Contrast

Surface topography can be imaged with the SE signal, where the contrast is generated by the dependence of δ on the tilt angle ϕ of a surface element, and by shadowing effects caused by reduced SE collection from surfaces with normals opposite to the direction of SE collection or inside holes and trenches. This results in a contrast nearly equivalent to a light illumination from the detector, which, however, is disturbed by the diffusion contrast caused by SE2 generated by BSE at a greater distance from the electron impact. This typically results in bright zones near edges with a width of about the range R. Therefore, a low-voltage SEM at 1–5 keV decreases the diffusion effect and shows a better topography, as demonstrated in Figure 2.60. With in-lens operation and through-lens collection of SE by an Everhart–Thornley detector (ETD) inside the last lens, the SE signal becomes independent of azimuth, which is an advantage for the metrology of integrated circuits, but a disadvantage for recognizing the topography and distinguishing elevations and indentations, for example.

The dependence of δ and especially of η on atomic number Z (Fig. 2.58) leads to an atomic number or compositional contrast that can be used for the discrimination of phases with different mean atomic numbers, as demonstrated by a eutectic alloy shown in Figure 2.61.

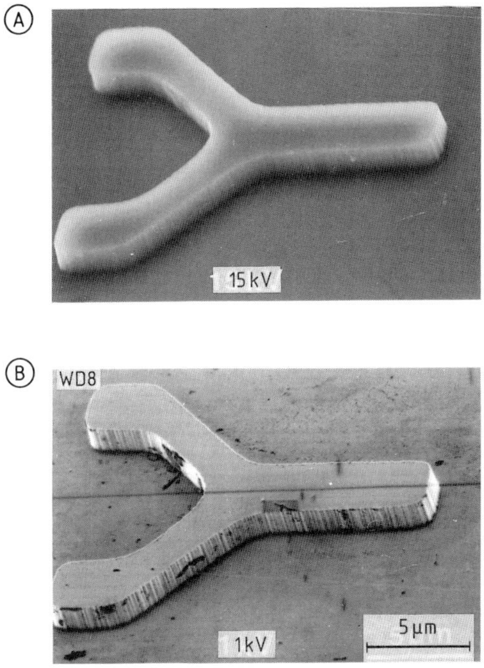

Figure 2.60. SE image of an etched structure (Y) on silicon with
A) 10-keV; B) 1-keV electrons

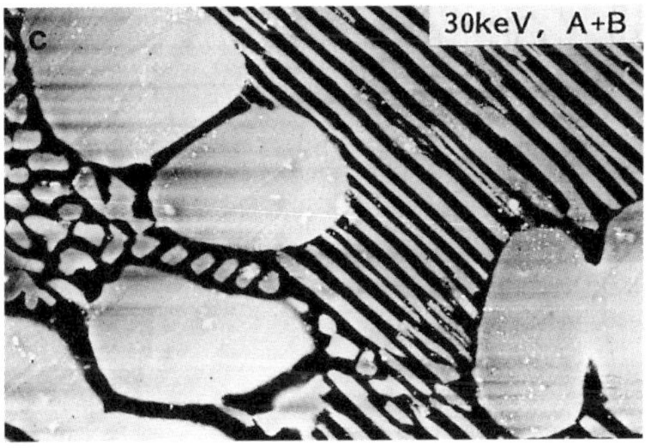

Figure 2.61. Example of material contrast of an Al–Ag eutectic alloy recorded with 30-keV electrons

2.2.4.2. Electron Channeling Effects

Electron waves entering a single crystal propagate as Bloch waves, which results in channeling effects and an orientation dependence of all electron–specimen interactions that are concentrated at the nuclei, such as large-angle scattering for backscattering or inner-shell ionization for X-ray and Auger electron emission.

When rocking an incident electron beam, the backscattering coefficient is modulated by a few percent. This results in an electron channeling pattern (ECP) (Fig. 2.62) containg excess (bright) or defect (dark) Kikuchi lines (see Section 2.1.3.2) that are the intersections of the Kossel cones of apex $90° - \theta_B$ and an axis normal to the lattice planes of distance d. Therefore, such a pattern contains the crystal symmetry, and the distances of opposite Kikuchi lines include an angle $2\theta_B$, which fulfills the Bragg condition

$$2\,d \sin \theta_B = \lambda \qquad (2.38)$$

Scanning a polycrystalline specimen results in changes of brightness forming the crystal orientation or channeling contrast (Fig. 2.63).

Another type of channeling pattern is the electron backscattering pattern (EBSP) in which the intensity modulation can be observed on a fluorescent screen as the dependence of backscattering on takeoff direction. An EBSP has the advantage of covering a large angular range of about $\pm 30°$, whereas the rocking for ECP can be realized only with maximum $\pm 2-3°$.

Figure 2.62. Electron channeling pattern of a (111) silicon surface recorded with 20-keV electrons

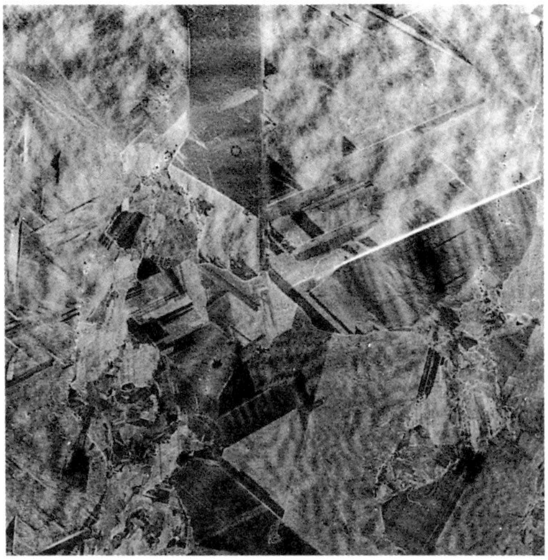

Figure 2.63. Superposed crystal orientation and magnetic contrast type I on polycrystalline cobalt

2.2.4.3. Imaging and Measurement of Surface Potentials

A positively biased or charged part of the specimen retards low-energy SE and appears darker than the surrounding parts at ground potential. Negatively biased or charged parts appear brighter because more SE are repelled and can reach the collection field of the ETD. This results in the voltage contrast. However, no unique relation exists between brightness and bias because the SE signal depends on the surrounding potentials (near-field effect) and the collection efficiency of an ETD (far-field effect).

For quantitative measurement of surface potentials U_s on integrated circuits, a high extraction field strength (400–1000 V/cm) is needed at the surface, which decreases the influence of near fields by neighboring potentials. The potential is measured by the shift in the SE spectrum after passage through a spectrometer of the retarding or deflection type [2.97].

2.2.4.4. Imaging of Magnetic Fields

Magnetic fields of the specimen can act on primary, secondary, and backscattered electrons by the Lorentz force $F = e\,v \times B$. The *magnetic contrast type 1* is caused by the deflection of SE by external magnetic fields, which can be observed for magnetic recording media and uniaxial ferromagnetic materials. The best contrast is observed when only about half of the SE are collected. Then the signal intensity can change by 1–10 % for opposite directions of the stray field. Figure 2.63 shows the superposition of magnetic contrast type 1 and the channeling contrast on a polycrystalline cobalt specimen.

Magnetic contrast type 2 is caused by the deflection of BSE in internal magnetic fields. However, sufficient contrast can be observed only for tilt angles ϕ of 40–50°, and the contrast increases with increasing energy, but is only a few per thousand for opposite directions of B. Another possibility for measuring external stray fields is the deflection of primary electrons, which pass at a short distance parallel to the surface.

2.2.4.5. Electron-Beam Induced Current

The mean number E/E_i of electron–hole pairs generated in semiconductors—with a mean formation energy of $E_i = 3.6$ eV in silicon, for example—normally recombine. The electric field inside depletion layers separates the charge carriers, and minority carriers can diffuse to the depletion layer and contribute to the charge collection I_{cc} or electron-beam induced current. Depletion layers can be formed by p–n junctions parallel or perpendicular to the surface or by Schottky barriers formed by a nonohmic evaporated metal contact. Therefore, a scanning electron probe becomes a useful tool for qualitative and quantitative analysis of junctions and semiconductor parameters [2.98], which is demonstrated by the following examples:

1) Imaging the position and depth of depletion layers below conductive pads and passivation layers, by utilizing the increasing penetration depth (range) with increasing electron energy
2) Measuring the width of depletion layers and their increase with increasing reverse bias
3) Imaging sites of avalanche breakdown in depletion layers
4) Imaging lattice defects (dislocations, stacking faults) and dopant striations that actively influence the diffusion length, for example, by a Cottrell atmosphere of dopant atoms
5) Measuring the diffusion length L from a semilogarithmic plot of the EBIC signal [$\sim \exp(-x/L)$] versus the distance x of the electron probe from a perpendicular p–n junction, and measuring the lifetime τ with a signal [$\sim \exp(-t/\tau)$] when chopping the electron beam
6) Measuring the surface recombination rate S from the variation of EBIC with increasing electron energy and depth of carrier formation

2.2.4.6. Cathodoluminescence

The excitation of light from materials stroked by electrons, known from fluorescent screens of TV tubes for example, is called cathodoluminescence (CL). In semiconductors with a direct band gap, electrons that are excited from the valence to the conduction band can recombine with the holes by emission of radiation, whereas semiconductors with an indirect band gap have a reduced probability of radiative recombination. In semiconductors and most inorganic materials, CL depends strongly on the concentration of dopants, which can either enhance CL by forming luminescence centers for radiative transitions or quench CL by forming centers of nonradiative transitions.

Organic specimens such as anthracene and plastic scintillator material, for example, can also show CL. Fluorescent dyes are used in biology for selective staining. However, all organic material is damaged by electron irradiation, and CL is quenched at incident charge densities a few orders of magnitude lower than those necessary for radiation damage of the crystal structure and loss of mass by ionization processes.

The CL signal can be detected by a photomultiplier or dispersively by using a spectrometer between specimen and detector. As in color TV, the signals from three detectors with color filters can be used for a real-color image [2.99].

2.2.4.7. Special Imaging Methods

The *specimen current mode* uses the current

$$I_s = I_p[1 - (\eta + \delta)] \tag{2.39}$$

to earth, which is complementary to the SE and BSE emission and can be used for imaging the compositional contrast of BSE or the magnetic contrast type 2.

In *environmental* SEM, the partial pressure of water or other gases can be increased near the specimen by differentially pumped diaphragms. This offers the possibility of observing wet specimens without drying, for example [2.100].

In the *thermal-wave acoustical mode*, the electron beam is chopped at frequencies in the 100-kHz to 5-MHz range and produces periodic specimen heating. The periodic changes in thermal expansion excite an acoustic wave that can be picked up by a piezoelectric crystal transducer of lead zirconate titanate.

2.2.5. Elemental Analysis

2.2.5.1. X-Ray and Auger Electron Emission

Either ionization of an inner shell and the subsequent filling of the vacancy by an electron from higher energy states result in the emission of an X-ray quantum, or the energy is transferred to another electron in the higher shell, which leaves the atom as an Auger electron (see also → Surface Analysis, pp. 27–28). The sum of the X-ray fluorescence yield ω and the Auger electron yield is unity, and ω decreases strongly from near unity to low values with decreasing Z and increasing X-ray series $n =$ K, L, M. The X-ray quantum energies of a series increases approximately $\sim \sqrt{Z}$ (Moseley's law), which is basic for elemental analysis by X rays.

2.2.5.2. X-Ray Spectrometers

Two different types of X-ray spectrometer exist: wavelength- and energy-dispersive spectrometers (WDS and EDS, see Section 2.1.9.1), which can be used in either an X-ray microanalyzer (XRMA) or an SEM.

A *wavelength-dispersive spectrometer* can analyze the characteristic lines of all elements down to beryllium when special analyzing crystals with a large lattice-plane spacing are used for Bragg reflection. XRMA with electron energies between 10 and 60 keV can work with WDS for recording three different wavelengths simultaneously, whereas only one WDS is used in an SEM.

The electron–hole pairs produced in the *energy-dispersive spectrometer* are separated by a voltage drop of ca. 1 kV at the crystal and result in a charge pulse proportional to the X-ray quantum energy. This pulse is amplified by a charge-sensitive preamplifier and runs through a unit for pulse shaping so that it can be recorded in a multichannel analyzer (MCA). The number of counts in a channel that is proportional to the quantum energy is increased by one unit. This allows all quantum energies to be recorded simultaneously, and the growth of a spectrum can be followed on the screen of the MCA (Fig. 2.64). Therefore, EDS are often also used in XRMA to survey the emitted X-ray spectrum. The vacuum inside an EDS can be separated from the vacuum of the microscope by a 8–10-μm-thick beryllium foil, which absorbs the characteristic radiation of elements $Z \leq 11$ below $E_x = 1$ keV. An organic foil or a windowless detector can also be used to record the lines of carbon ($E_x = 250$ eV), oxygen, and nitrogen, for example.

WDS and EDS show the following characteristic differences. *WDS* requires very accurate orientation of the electron impact on the Rowland circle within an area a few micrometers in diameter. The specimen must be planar and is shifted mechanically to record a line scan or elemental map. The electron probe can, by contrast, be scanned over an area of a few millimeters

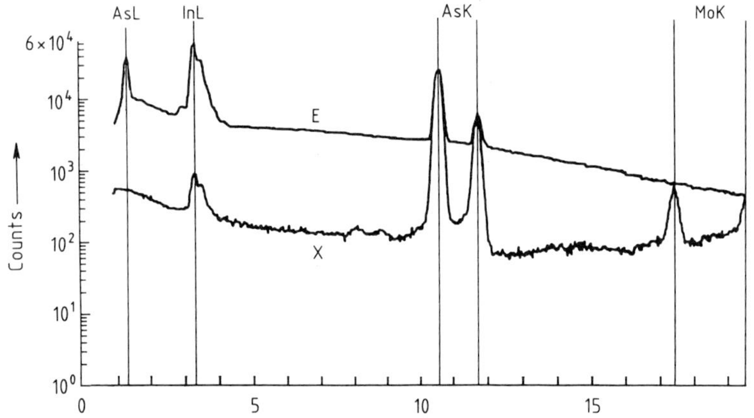

Figure 2.64. Energy-dispersive X-ray spectrum of InAs (logarithmic scale of counts per channel) recorded by excitation with 30-keV electrons (E) and X-ray fluorescence (X) using the molybdenum K radiation excited in a molybdenum foil target in front of the specimen

without any loss of efficiency when an EDS is used. WDS analyze only one line with a resolution of about 10 eV. The proportional counter allows a high counting rate of about 10000 counts per second (cps). Therefore, high probe currents and a large number of counts per unit time can be recorded. The energy resolution of an *EDS* is only about 100–200 eV, and a ten times greater fraction of the continuum below the characteristic X-ray peaks is counted compared to WDS. This decreases the signal-to-background ratio in EDS. The poorer resolution of an EDS can also result in an overlap of neighboring X-ray lines, whereas all lines can normally be separated by a WDS. An EDS must work with lower probe currents because all quanta are counted and only a count rate of ca. 10000 cps guarantees no overlap of sequential pulses. Pulse-rejection electronics are used when the time between two pulses is shorter than the pulse-shaping time necessary for the MCA.

2.2.5.3. X-Ray Microanalysis

For a quantitative XRMA [2.101], the number n_a of counted characteristic X-ray quanta of an element a of concentration c_a in an alloy or compound is compared with the number n_s from a pure elemental standard ($c_s = 1$) or a compound of known composition. Only an approximate value for c_a can be obtained from the k ratio

$$k = n_a/n_s \approx c_a/c_s \qquad (2.40)$$

The exact measurement of concentration needs a *ZAF correction method* (see p. 247). This considers the differences in the stopping power and the backscattering of electrons between specimen and standard (*Z correction*), in the depth distribution of X-ray generation and *absorption* (*A*) of X rays recorded under a takeoff angle of 40–60°, and in the excitation of X-ray quanta of the element of interest by characteristic X rays and continuum of the matrix (*fluorescence F*).

Special correction programs must be used, for example, for tilted specimens, thin film coatings, small particles, and biological specimens.

2.2.5.4. Special X-Ray Techniques

When the specimen is scanned in a raster or along a line, the pulses of selected characteristic X-ray lines can produce an elemental distribution profile or map, respectively.

In a *single crystal*, the excited X rays are Bragg diffracted at the lattice plane, and their isotropic angular characteristics show defect and excess Kossel lines of apex angle $90° - \theta_B$, which can be used for accurate measurement of lattice parameters and strains when the Kossel pattern is recorded on a photographic emulsion.

X-ray fluorescence analysis can be applied in an SEM when generating X rays in a thin foil, which stops the electrons, but the transmitted X rays can excite X-ray fluorescence in the specimen, which is recorded by an energy-dispersive spectrometer. The background of the spectrum is much lower than that from direct electron excitation (Fig. 2.64, curve X versus curve E), which results in a better signal-to-background ratio for trace elements. However, only larger areas 0.1–1 mm in diameter can be analyzed.

The small source of X rays generated by an electron probe on a bulk target can also be used for X-ray projection microscopy.

3. References

[1.1] E. Abbe: "Beiträge zur Theorie des Mikroskops und der mikroskopischen Wahrnehmung," *Arch. Mikrosk. Anat.* **9** (1873) 413.

[1.2] D. Gerlach: *Das Lichtmikroskop*, 2nd ed., Thieme Verlag, Stuttgart 1985.

[1.3] H. Gundlach: *Mikroskope. Handbuch Biomedizinische Technik*, vol. 4, Springer Verlag, TÜV Rheinland, (1991) pp. 1–20.

[1.4] K. Michel: "Axiomat von Zeiss, ein Mikroskop mit neurem Konzept," *Zeiss. Inf.* **20** (1973).

[1.5] F. Muchel: ""ICS"-A New Principle in Optics," *Zeiss Inf.* **30** (1988) 20–27.

[1.6] K. Michel: *Die Grundzüge der Theorie des Mikroskops*, 3rd ed., Wissenschaftliche Verlagsgesellschaft, Stuttgart 1981.

[1.7] A. Köhler: "Mikrophotographische Untersuchungen mit ultraviolettem Licht," *Z. Wiss. Mikrosk. Mikrosk. Tech.* **21** (1904) 129–165, 273–404.

[1.8] A. Köhler: "Ein neues Beleuchtungsverfahren für mikrophotographische Zwecke," *Z. Wiss. Mikrosk. Mikrosk. Tech.* **10** (1893) 433–440.

[1.9] F. Zernike: "Das Phasenkontrastverfahren bei der mikroskopischen Beobachtung," *Phys. Z.* **36** (1935) 848.

[1.10] H. Gundlach: "Neuere Entwicklungen und Anwendungen in der Lichtmikroskopie," *Gynäkologie* **23** (1990) 328–335.

[1.11] G. Nomarski: "Microinterferometre Differentielle a Ondes Polarisees," *J. Phys. Radium* **16** (1955) 9.

[1.12] R. D. Allen et al.: "The Zeiss Nomarski Differential Interference Equipment for Transmitted-Light Microscopy," *Z. Wiss. Mikrosk. Mikrosk. Tech.* **69** (1969) 193–221.

[1.13] W. Lang: "Nomarski Differential Interference Contrast Microscopy," *Zeiss Inf.* **16** (1968) no. 70, 114–120.

[1.14] M. Francon: *Einführung in die neuen Methoden der Lichtmikroskopie*, Verlag Braun, Karlsruhe 1967.

[1.15] W. I. Patzelt: "Reflexionskontrast. Eine neue lichtmikroskopische Technik," *Mikrokosmos* **3** (1977) 78–81.

[1.16] S. Pentz et al.: "Darstellung des Adhäsionsverhaltens kultivierter Leberzellen an Glasoberflächen während der Mitose durch Reflexionskontrast-Mikroskopie," *Zeiss Inf.* **25** (1980) 91, 41–43.

[1.17] H. Gundlach et al.: "Immunfluorescance Microscopy with the New Zeiss Photomicroscope Axiophot," *Zeiss Inf.* **29** (1997) 36–39.

[1.18] A. J. Lacey: *Light Microscopy in Biology*, IRL Press, Oxford University, 1989.

[1.19] D. L. Taylor, M. Nederlof, F. Lanni, A. S. Waggoner: "The New Vision of Light Microscopy," *Am. Sci.* **80** (1992) 322–335.

[1.19 a] D. L. Taylor et al.: *Fluorescence Microscopy of Living Cells in Culture*, Academic Press, San Diego 1989.

[1.19 b] A. H. Coons et al.: "Immunological Properties of an Antibody Containing a Fluorescent Group," *Proc. Soc. Exp. Biol. Med.* **47** (1941) 200–202.

[1.20] J. S. Ploem: "The Use of a Vertical Illuminator with Interchangeable Dichroic Mirrors for Fluorescence Microscopy with Incident Light," *Z. Wiss. Mikrosk. Mikrosk. Tech.* **68** (1967) 129–143.

[1.21] L. Trapp: "Über Lichtquellen und Filter für die Fluoreszenzmikroskopie und über die Auflichtfluoreszenzmethode bei Durchlichtpräparaten," *Acta Histochemica* **7** (1965), pp. 327–338.

[1.22] W. W. Franke: "Different Intermediate-Sized Filaments Distinguished by Immunofluorescence Microscopy," *Proc. Nat. Acad. Sci. USA* **75** (1978) no. 10, 5034–5038.

[1.23] O. Greulich et al.: "The Light Microscope on its Way from an Analytical to a Preparative Tool," *J. Microsc. (Oxford)* **167** (1992) 127–151.

[1.24] P. Lichter et al.: "Analysis of Genes and Chromosomes by Nonisotopic in Situ Hybridization," *Gata* **8** (1991) no. 1, 24–35.

[1.24 a] Chr. Lengauer et al., "Chromosomal Barcodes produced by Multifluorescence in situ Hybridization with Multiple YAG Clones and Whole Chromosomes painting probes," in *Human Molecular Genetics*, 5th ed., vol. 2, Oxford University Press, pp. 505–512.

[1.25] H. Utermöhl: "Neue Wege in der quantitativen Erfassung des Planktons," *Verh. intern. Verein Limnol.* **5** (1931) 567–596.

[1.26] H. Gundlach et al.: "Mikrokinematographie mit dem Mikroskop "Axiomat"," *Res. Film* **9** (1976) no. 1, 30–36.

[1.27] H. Gundlach: *Die Anwendung von inversen Mikroskopen in der Zell- und Entwicklungsbiologie*, Suppl. GIT, Verlag E. Giebler, Darmstadt 1981, pp. 9–15.

[1.28] H. A. Tritthart et al.: "The Zeiss Axiomat Applied to the Examination of the Morphology and Physiology of Myocardial Cells in Culture," *Zeiss Inf.* **25** 1980 no. 90, 10–13.

[1.29] M. Horster et al.: "Application of Differential Interference Contrast with Inverted Microscopes to the Invitro Perfused Nephron," *J. Microsc. (Oxford)* **117** (1979) 375–379.

[1.30] F. K. Möllring: "Inverse Mikroskopie IM 35 und ICN 405 für Biologie, Medizin und Metallographie," *Zeiss Inf.* **23** (1977) 18–19.

[1.31] H. Piller: *Microscope Photometry*, Springer, Berlin 1977.

[1.32] H. Gundlach et al.: "Identification and Selective Micropreparation of Live Nuclear Components with the Zeiss IM 35 Inverted Microscope," *Zeiss Inf.* **25** (1980), no. 91, 36–40.

[1.33] H. Spring et al.: "DNA Contents and Numbers of Nucleoli and Pre-rRNA-Genes in Nuclei of Gemetes and Vegetative Cells of Acetabularia Mediterranea," *Exp. Cell. Res.* **114** (1978) 203–215.

[1.34] H. Spring et al.: "Transcriptionally Active Chromatinin in Loops of Lumpbrush Chromosomes at Physiological Salt Concentration as Revealed Electron Microscopy of Sections," *Eur. J. Cell Biol.* **24** (1981) 298–308.

[1.35] W. Ansorge et al.: "Performance of an Automated System for Capillary Microinjection into Living Cells," *J. Biochem. Biophys. Methods* **16** (1988) 283–292.

[1.36] R. D. Allen et al.: "Video-Enhanced-Contrast Polarization (AVEC-Pol) Microscopy: A New Method Applied to the Detection of Birefringence in the Motile Reticulopodial Network of Allogromia Laticollaris," *Cell Motil.* **1** (1981) 275–289.

[1.37] R. D. Allen et al.: "Video-Enhanced-Contrast, Differential Interference Contrast (AVEC-DIC) Microscopy: A New Method Capable of Analyzing Microtubule-Related Motility in the Reticulopodial Network of Allogromia Laticollaris," *Cell Motil.* **1** (1981) 291–302.

[1.38] S. Inoue: "Foundations of Confocal Scanning Imaging in Confocal Microscopy" in [1.51, pp. 1–14].

[1.39] S. Inoue et al.: "The Acrosonal Reaction of Thyone Sperm Head Visualized by High Resolution Video Microscopy," *J. Cell Biol.* **93** (1982) 812–819.

[1.40] J. H. Hayden et al.: "Detection of Single Microtubules in Living Cells: Particle Transport Can Occur in Both Directions Along the Same Microtubule," *J. Cell Biol.* **99** (1984) 1785–1793.

[1.41] T. Salmon et al.: "Video-Enhanced Differential Interference Contrast Light Microscopy," *Bio Techniques* **7** (1989) 624–633.

[1.42] D. J. Arndt-Jovin et al.: "Fluorescence Digital Imaging Microscopy in Cell Biology," *Science (Washington D.C.)* **230** (1985) 247–256.

[1.43] Y. Hiraoka et al.: "The Use of a Charge-Coupled Device for Quantitative Optical Microscopy of Biological Structures," *Science (Washington D.C.)* **238** (1987) 36–41.

[1.44] H. H. Sedlacek et al.: "The Use of Television Cameras Equipped with an Image Intensifier in the Immunofluorescence Microscopy," *Behring Inst. Mitt.* **59** (1976) 64–70.

[1.45] H. Gundlach et al.: *Electronic Photography-Technology, Systems and Applications in Microbiology. Proc. of Int. Symosium on Electronic Photography*, Cologne The Society for Imaging Sciene and Technology, Springfield, Virginia USA, 1992, pp. 60–65.

[1.46] M. Minsky: US 3013467, 1961.

[1.47] G. Cox: "Photons Under the Microscope. Marvin Minsky – the Forgotten Pioneer?," *Australian EM Newsletter* **38** no. 4, (1993) 4–10.

[1.48] C. J. R. Sheppard: "15 Years of Scanning Optical Microscopy at Oxford," *Proc. Roy. Mic. Soc.* **25** (1990) 319–321.

[1.49] M. Petran, M. Hadravsky, M. D. Egger, R. Galambos: "Tandem-Scanning Reflected Light Microscope," *J. Opt. Soc. Am.* **58** (1968) 661–664.

[1.50] G. J. Brakenhoff, P. Blom, P. Barends: "Confocal Scanning Light Microscopy with High Aperture Immersion Lens," *J. Micros. (Oxford)* **117** (1979) 219–232.

[1.51] J. Pawley: *Handbook of Biological Confocal Microscopy*, Plenum Press, New York 1990.

[1.52] C. J. R. Sheppard, A. Choundhury: "Image Formation in the Scanning Microscope," *Opt. Acta* **24** (1977) 1051–1073.

[1.53] C. J. R. Sheppard, T. Wilson: "Depth of Field in the Scanning Microscope," *Opt. Lett.* **3** (1978) 115–117.

[1.54] P. Wallen, K. Carlsson, K. Mossberg: "CLSM as a Tool for Studying the 3-D Morphology of Nerve Cells," in [1.81, pp. 110–143].

[1.55] T. Wilson: "Optical Sectioning in Confocal Fluorescence Microscopy," *J. Microsc. (Oxford)* **154** (1989) 143–156.

[1.56] S. Hell, E. Stelzer: "Properties of a 4Pi Confocal Fluorescence Microscope," *J. Opt. Soc. Am.*, in press.

[1.57] I. J. Cox, C. J. R. Sheppard, T. Wilson: "Superresolution in Confocal Fluorescent Microscopy," *Optik (Stuttgart)* **60** (1982) 391–396.

[1.58] T. Wilson (ed.): *Confocal Microscopy*, Academic Press, London 1990.

[1.59] C. J. R. Sheppard, M. Gu: "3-D Transfer Functions in Confocal Scanning Microscopy," in [1.81, pp. 251–280].

[1.60] C. J. Cogswell, J. W. O'Byna: "A High Resolution Confocal Transmission Microscope," *SPIE Proceedings* **1660** (1992) 503–511.

[1.61] G. J. Brakenhoff, K. Visscher: "Bilateral Scanning and Array Detectors," *J. Micros. (Oxford)* **165** (1990) 139–146.

[1.62] E. Gratten van der Veen: "Lasers in Confocal Microscopy," in [1.51].

[1.63] K. S. Wells, D. R. Sandison, J. Strickler, W. W. Webb: "Quantitative Fluorescence Imaging with Laser Scanning Confocal Microscopy," in [1.51, pp. 27–39].

[1.64] J. Art: "Photon Detectors for Confocal Microscopy," in [1.51, pp. 127–139].

[1.65] A. Kriete, H.-J. Wagner: "Computerized Spatio-Temporal (4-D) Representation of Confocal Microscopy: Application to Neuroanatomical Plasticity," *J. Microsc. (Oxford)* **169** (1993) 27–31.

[1.66] W. W. Webb: "Two Photon Excitation in Laser Scanning Fluorescence Microscopy," *Trans. Roy. Micr. Soc.*, vol. 1, IOP Publ., Bristol, pp. 445–450.

[1.67] M. D. Egger, M. Petran: "New Reflected Light Microscope for Viewing Unstained Brain and Ganglian Cells," *Science (Washington D.C.)* **157** (1987) 305–307.

[1.68] J. S. Deitch, K. L. Smith, J. W. Swann, J. N. Turner: "Parameters Affecting Imaging of HRP Reaction Product in the Confocal Scanning Laser Microscope," *J. Microsc. (Oxford)* **160** (1990) 265–278.

[1.69] J. P. Rigaut, S. Caravajal-Gonzalez, J. Vassy: "3-D Image Cytometry," in [1.81, pp. 205–237].

[1.70] B. R. Masters, A. Kriete, J. Kukulies: "Ultraviolet Confocal Fluorescence Microscopy of the In-Vitro Cornea: Redox Metabolic Imaging," *Applied Optics* **32** (1993) no. 4, 592–596.

[1.71] D. M. Shotton: "Confocal Scanning Optical Microscopy and its Applications for Biological Specimens," *J. Cell. Sci.* **94** (1989) 175–206.

[1.72] J. N. Turner: "Confocal Light Microscopy. Biological Applications (Special Issue)," *Electron Microsc. Tech.* **18** (1991) 1.

[1.73] R. P. Haugland: *Molecular Probes: Handbook of Fluorescent Probes and Research Chemicals*, Eugene Molecular Probes Inc., 1989.

[1.74] R. Y. Tsien, A. Waggoner: "Fluorophores for Confocal Microscopy: Photophysics and Photochemistry," in [1.51, pp. 169–178].

[1.75] A. Villringer, U. Dirnagl, K. Einhäupl: "Microscopical Visualization of the Brain in Vivo," in [1.81, pp. 161–181].

[1.76] B. R. Masters: (1992) "Confocal Ocular Microscopy – a New Paradigm for Ocular Visualization," in [1.50, pp. 183–203].

[1.77] P. C. Cheng, R. G. Summers: Image Contrast in Confocal Microscopy in [1.51, pp. 179–195).

[1.78] R. G. Summer, P. C. Cheng: "Analysis of Embryonic Cell Division Patterns Using laser Scanning Confocal Microscopy," in G. W. Bailey (ed.): *Proc. Annual meeting Electron Microsc. Soc. Am.* **47** (1989) 140–141.

[1.79] N. O'Rourke, S. E. Fraser: "Dynamic Changes in Optic Fiber Terminal Arbors Lead to Retinotopic Map Formation: An In-Vivo Confocal Microscopic Study," *Neuron* **5** (1990) 159–171.

[1.80] A. Kriete, T. Pepping: "Volumetric Data Representation in Microscopy: Applications to Confocal and NMR-Microimaging," in [1.81, pp. 329–358].

[1.81] A. Kriete: *Visualization in Biomedical Microscopies 3-D Imaging and Computer Applications*, VCH-Verlagsgesellschaft, Weinheim 1992.

[1.82] A. Kriete: "Image Quality Considerations in Computerized 2-D and 3-D Microscopies" in P. C. Cheng, T. H. Lin, W. H. Wu, J. L. Wi (eds.): *Multidimensional Microscopy*, Springer Verlag, New York 1993, pp. 209–230.

[1.83] H. Chen, J. W. Sedat, J. A. Adard: "Manipulation, Display and Analysis of Three-Dimensional Biological Images," in J. Pawley (ed.): *Handbook of Biological Confocal Microscopy*, Plenum Press, New York 1990, pp. 141–150.

[1.84] J. S. Hersh: "A Survey of Modeling Representations and their Application to Biomedical Visualization and Simulation," Conf. VBC, IEEE Comp. Soc. Press, Los Alamitos 1990, pp. 432–441.

[1.85] D. P. Huijsmanns, W. H. Lamers, J. A. Los, J. Strackee: "Toward Computerized Morphometric Facilities: a Review of 58 Software Packages for Computer-Aided 3-D Reconstruction, Quantification, and Picture Generation from Parallel Serial Sections," *Anat. Rec.* **216** (1986) 449–470.

[1.86] A. Kriete, P. C. Cheng (eds.): "3-D Microscopy (Special Issue)," *Computerized Medical Imaging and Graphics*, vol. 17, Plenum Press, N.Y. 1993, p. 8.

[1.87] R. Ware, V. LoPresti: "Three-Dimensional Reconstruction from Serial Sections," *Comp. Graph.* **2** (1975) 325–440.

[1.88] J. K. Udupa, H. M. Hung: "Surface Versus Volume Rendering: a Comperative Assessment," *VBC '90*, Proceedings IEEE, Atlanta 1990, pp. 83–91.

[1.89] S. D. Roth: "Ray-Casting for Solid Modeling," *Comput. Graphics Image Processing* **18** (1982) 109–144.

[1.90] H. K. Tuy, L. T. Tuy: "Direct 2-D Display of 3-D Objects," *IEEE CG & A* **4** (1984) 29–33.

[1.91] M. Montag et al.: "Methodical Aspects of 3-D Reconstruction of Chromatin Architecture in Mouse Trophoblast Giant Nuclei," *J. Microsc. (Oxford)* **158** (1990) 225–233.

[1.92] G. Frieder, D. Gordon, R. A. Reynolds: "Back-to-Front Display of Voxel-Based Objects," *IEEE CG & A* **5** (1985) no. 1, 52–60.

[1.93] R. A. Debrin, L. Carpenter, P. Hanrahan: "Volume Rendering," *Computer Graphics* **22** (1988) no. 4, 65–74.

[1.94] M. Levoy: "A Hybrid Ray-Trace for Rendering Polygon and Volume Data," *IEEE CG & A* **3** (1990) 33–40.

[1.95] H. T. M. van der Voort, G. J. Brakenhoff, M. W. Baarslag: "Three-Dimensional Visualization Methods," *J. Microsc. (Oxford)* **153** (1989) no. 2, 123–132.

[1.96] P.-O. Forsgren: "Visualization and Coding in Three-Dimensional Image Processing," *J. Microsc. (Oxford)* **159** (1990) no. 2, 195–202.

[1.97] M. Levoy et al.: "Volume Rendering in Radiation Treatment Planning," *VBC '90*, Proceedings IEEE, Atlanta 1990, pp. 4–10.

[1.97a] T. Visser, J. L. Oud, G. T. Brakenhoff: "Refractive Index and Axial Distance Measurements in 3-D Microscopy," *Optik (Stuttgart)* **90** (1991) 17–19.

[1.98] J.-A. Conchello, E. W. Hanssen: "Enhanced 3-D Reconstruction from Confocal Scanning Microscope Images. Deterministic and Maximum Likelihood Reconstrauctions," *App. Opt.* **29** (1990) no. 26, 3795–3804.

[1.99] T. J. Holmes, Y.-H. Liu: "Image Restoration for 2-D and 3-D Fluorescence Microscopy," in [1.81, pp. 283–323].

[1.100] D. A. Agard, J. W. Sedat: "Three-Dimensional Architecture of Polytene Nucleus," *Nature (London)* **302** (1983) 676–681.

[1.101] G. Wang, W. S. Liou, T. H. Lin, P. C. Cheng: "Image Restoration in Light Microscopy," in T. H. Lin, W. L. Wu, J. L. Wu (eds.): *Multidimensional Microscopy*, Springer Verlag, New York 1993, 191–208.

[1.102] F. Meyer: "Mathematical Morphology: 2-D to 3-D," *J. Microsc.* **165** (1992) 5–28.

[1.103] P. C. Cheng et al.: "3-D Image Analysis and Visualization in Light Microscopy and X-Ray Micro-Tomography," in [1.81, pp. 361–398].

[1.104] K. Mossberg, U. Arvidsson, B. Ulfhake: "Computerized Quantification of Immunoflourescence Labelled Axon Terminals and Analysis of Co-Localization of Neurochemicals in Axon Terminals with A Confocal Scanning Laser Microscope," *J. Histochem Cyctochem.***38** (1990) 179–190.

[1.105] B. Roysam et al.: "Unsupervised Noise Removal Algorithms for 3-D Confocal Fluorescence Microscopy," *Micron and Micros. Acta* **23** (1992) in press.

[1.106] V. Conan et al.: "Geostatistical and Morphological Methods Applied to 3-D Microscopy," *J. Microsc. (Oxford)* **166** (1992) 169–184.

General References

[2.1] S. Amelinckx: "The Direct Observation of Dislocations," Suppl. 6 in F. Seitz, D. Turnbull (eds.): *Solid State Physics*, Academic Press, London 1964.

[2.2] S. Amelinckx, R. Gevers, J. Van Landuyt (eds): *Diffraction and Imaging Techniques in Material Science*, North-Holland Publishing Company, Amsterdam 1970, 1978.

[2.3] F. R. N. Nabarro (ed.): *Dislocation in Solids*, North-Holland, Amsterdam 1979.

[2.4] P. B. Hirsch, R. B. Nicholson, A. Howie, D. W. Pashley, M. J. Whelan: *Electron Microscopy of Thin Crystals*, Butterworths, London 1965.

[2.5] H. Bethge, J. Heydenreich (eds.): *Electronenmikroskopie in der Festkörperphysik*, Springer Verlag, Berlin 1982.

[2.6] J. C. H. Spence: "Experimental High Resolution Electron Microscopy," *Monographs on the Physics and Chemistry of Materials*, Oxford Science Publications, Clarendon Press, Oxford 1981.

[2.7] G. Thomas: *Transmission Electron Microscopy of Metals*, John Wiley and Sons, New York 1962.

[2.8] J. C. H. Spence, J. M. Zuo: *Electron Microdiffraction*, Plenum Press, New York 1992.

[2.9] R. W. Cahn, P. Haasen, E. J. Kramer (eds.): *Materials Science and Technology*, vol. 2A, VCH Verlagsgesellschaft, Weinheim 1992.

[2.10] D. C. Joy, A. D. Romig, Jr., J. I. Goldstein (eds.): *Principles of Analytical Electron Microscopy*, Plenum Press, New York 1986.

[2.11] L. Reimer: "Transmission Electron Microscopy," *Springer Series in Optical Sciences*, Springer Verlag, Berlin 1984.

[2.12] J. M. Cowley (ed.): "Electron Diffraction Techniques," vols. 1 and 2, *Monographs on Crystallography 3*, International Union of Crystallography, Oxford University Press, Oxford 1992.

[2.13] P. G. Merli, M. Vittori Antisari (eds.): *Electron Microscopy in Materials Science*, World Scientific, Singapore 1992.

Specific References

[2.14] N. F. Mott, H. S. W. Massey: *The Theory of Atomic Collisions*, Clarendon Press, Oxford 1949.

[2.15] J. A. Ibers, B. K. Vainshtein: "Scattering Amplitudes for Electrons," in K. Londsdale (ed.): *International Tables for X-Ray Crystallography*, vol. 3, Kynoch, Birmingham 1962.

[2.16] in [2.9], p. 250.

[2.17] W. L. Bragg, *Nature* (*London*) **124** (1929) 125.

[2.18] P. P. Ewald, *Ann. Phys.* (*Leipzig*) **54** (1917) 519.

[2.19] in [2.9], p. 251.

[2.20] S. Takagi, *Acta Crystallogr.* **15** (1962) 1311.

[2.21] R. D. Heidenreich, *J. Appl. Phys.* **20** (1949) 993.

[2.22] S. Kikuchi, *Jpn. J. Phys.* **5** (1928) 23.

[2.23] in [2.1], p. 125.

[2.24] H. A. Bethe, *Ann. Phys.* (*Leipzig*) **87** (1928) 55.
C. H. MacGillavry, *Physica* (*Amsterdam*) **7** (1940) 329.

[2.25] A. Howie, M. J. Whelan, *Proc. Roy. Soc. London A*, **263** (1961) 217.
A. Howie, M. J. Whelan, *Proc. Roy. Soc. London A*, **267** (1962) 206.

[2.26] H. Yoshioka, *J. Phys. Soc. Jpn.* **12** (1957) 628.
H. Hashimoto, A. Howie, M. J. Whelan, *Proc. Roy. Soc. London A* **269** (1962) 80.

[2.27] G. Bormann, *Z. Phys.* **42** (1941) 157.
G. Bormann, *Z. Phys.* **127** (1950) 297.

[2.28] H. Hashimoto, M. Mannami, T. Naiki, *Philos. Trans. R. Soc. London A* **253** (1961) 459.

J. W. Menter, *Proc. Roy. Soc. London A* **236** (1956) 119.
[2.29] G. Van Tendeloo, S. Amelinckx, *Acta Crystallogr. Sect. A: Cryst. Phys. Diffr. Theor. Gen. Crystallogr.* **A 30** (1974) 431.
[2.30] S. Amelinckx, J. Van Landuyt, in [2.2], p. 107.
H. Hashimoto, M. J. Whelan, *J. Phys. Soc. Jpn.* **18** (1963) 1706.
P. B. Hirsch, A. Howie, M. J. Whelan, *Philos. Trans. R. Soc. London A* **252** (1960) 499.
P. B. Hirsch, A. Howie, M. J. Whelan, *Philos. Mag.* **7** (1962) 2095.
[2.31] C. M. Drum, M. J. Whelan, *Philos. Mag.* **11** (1965) 205.
J. Van Landuyt, R. Gevers, S. Amelinckx, *Phys. Status Solidi* **7** (1964) 519.
[2.32] G. A. Bassett, J. W. Menter, D. W. Pashley, *Proc. Roy. Soc. London A* **246** (1958) 345.
D. W. Pashley, J. W. Menter, G. A. Bassett, *Nature (London)* **179** (1957) 752.
[2.33] H. Hashimoto, R. Uyeda, *Acta Crystallogr.* **10** (1957) 143.
[2.34] R. Gevers, *Philos. Mag.* **7** (1963) 769.
[2.35] D. J. H. Cockayne, I. L. E. Ray, M. J. Whelan, *Philos. Mag.* **20** (1969) 1265.
D. J. H. Cockayne, M. J. Jenkins, I. L. E. Ray, *Philos. Mag.* **24** (1971) 1383.
R. de Ridder, S. Amelinckx, *Phys. Status Solidi B* **43** (1971) 541.
[2.36] S. Amelinckx, P. Delavignette, *J. Appl. Phys.* **33** (1962) 1458.
[2.37] P. Humble, in [2.2], p. 315.
P. Humble, *Aust. J. Phys.* **21** (1968) 325.
A. K. Head, *Aust. J. Phys.* **20** (1967) 557.
A. K. Head, P. Humble, L. M. Clarebrough, A. T. Morton, G. T. Forwood: "Computed Electron Micrographs and Defect Identification," in: S. Amelinckx, P. Gevers, J. Nihoul (eds.): *Defects in Crystalline Solids*, vol. 7, North-Holland, Amsterdam 1973.
[2.38] in [2.9], p. 58.
[2.39] C. Boulesteix, J. Van Landuyt, S. Amelinckx, *Phys. Status Solidi A* **33** (1976) 595.
[2.40] G. Van Tendeloo, J. Van Landuyt, S. Amelinckx, *Phys. Status Solidi A* **33** (1976) 723.
M. Snijkers, R. Serneels, P. Delavignette, R. Gevers, S. Amelinckx, *Chryst. Lattice Defects* **3** (1972) 99.
[2.41] R. Gevers, J. Van Landuyt, S. Amelinckx, *Phys. Status Solidi* **11** (1965) 689.
M. J. Goringe, U. Valdré, *Proc. R. Soc. London A* **295** (1966) 192.
[2.42] M. F. Ashby, L. M. Brown, *Philos. Mag.* **8** (1963) 1083, 1649.
[2.43] J. W. Steeds, E. Carlino, in [2.13], p. 279.
[2.44] P. Goodman, *Acta Crystallogr. Sect. A: Cryst. Phys. Diffr. Theor. Gen. Crystallogr.* **31** (1975) 793; **31** (1975) 804.
M. Tanaka, *J. Electron Microsc. Tech.* **13** (1989) 27.
J. W. Steeds, R. Vincent, *J. Appl. Crystallogr.* **16** (1983) 317.
[2.45] B. F. Buxton, J. A. Eades, J. W. Steeds, G. M. Rackham, *Philos. Trans. R. Soc. London A* **281**, 171.
J. Eades, M. Shannon, B. Buxton in O. Johari (ed.): *Scanning Electron Microscopy,* I.I.I.R. Institute, Chicago 1983, p. 83.
[2.46] M. Tanaka, M. Terauchi, T. Kaneyama: *Convergent Beam Electron Diffraction I and II,* Jeol Ltd., Tokyo.
[2.47] N. S. Blom, E. W. Schapink, *J. Appl. Crystallogr.* **18** (1985) 126.
[2.48] J. Gjønnes, A. F. Moodie, *Acta Crystallogr.* **19** (1965) 6567.
P. Goodman, G. Lempfuhl, *Z. Naturforsch. A* **19** (1964) 818.
[2.49] W. Kossel, G. Möllestedt, *Ann. Phys. Leipzig* **36** (1939) 113.
C. H. MacGillavray, *Physica (Amsterdam)* **7** (1940) 329.
[2.50] G. R. Grinton, J. M. Cowley, *Optik (Stuttgart)* **34** (1971) 221.
J. M. Cowley, S. Iijima: "The Direct Observation of Crystal Structures," in H. R. Wenk (ed.): *Electron Microscopy in Mineralogy,* Springer Verlag, Berlin 1976, p. 123.
[2.51] S. Amelinckx in [2.9], p. 5.
[2.52] O. Scherzer, *J. Appl. Phys.* **20** (1949) 20.
[2.53] E. Abbe, *Arch. Mikrosk. Anat.* **9** (1873) 413.
[2.54] D. van Dyck, M. Op de Beeck, *Proc. Int. Conf. Electron Microsc.* **12th** (1990) vol. 1, 64.
D. van Dyck, W. Coene, *Ultramicroscopy* **15** (1989) 29.
D. van Dyck, M. Op de Beeck, *Proc. Int. Conf. Electron Microsc.* **12th** (1990) 26.
[2.55] J. M. Cowley, A. F. Moodie, *Acta Crystallogr.* **10** (1957) 609.
[2.56] D. van Dyck et al. in W. Krakow, M. O'Keefe (eds.): *Computer Simulation of Electron Microscope Diffraction and Images,* The Minerals, Metals and Materials Society, 1989, p. 107.
[2.57] H. G. J. Moseley, *Philos. Mag.* **26** (1913) 1024; **27** (1914) 703.
[2.58] E. L. Hall, in [2.9], p. 147.
[2.59] M. H. Jacobs, J. Baborovska: *Electron Microscopy,* The Institute of Physics, London 1972, p. 136.
K. F. J. Heinrich: *Electron Beam X-Ray Microanalysis,* Van Rostrand, New York 1981.
[2.60] S. J. B. Reed: *Electron Microprobe Analysis,* Cambridge University Press, London 1975.
[2.61] J. C. Spence, H. Tafto, *J. Microsc. (Oxford)* **130** (1983) 147.
[2.62] B. Jouffrey, in [2.13], p. 363.
[2.63] R. D. Leapman, L. A. Grunes, P. L. Fejes, J. Silcox, in B. K. Teo, D. C. Joy (eds.): *EXALFS Spectroscopy,* Plenum Press, New York 1981, p. 217.
[2.64] G. Zanchi, J. P. Perez, J. Sevely, *Optik* **43** (1945) 495.
W. Pejas, H. Rose, *Electron Microscopy,* Microscopic Society of Canada, Toronto 1978, p. 44.
H. T. Pearce-Percy, D. Krahl, J. Jeager in D. G. Brandon (ed.): *Electron Microscopy,* vol. 1, Tal International, Jerusalem 1976, p. 348.
[2.65] G. Van Tendeloo, M. Op de Beeck, S. Amelinckx, J. Bohr, W. Krätschmer, *Europhys. Lett.* **15** (1991) no. 3, 295.
[2.66] S. Kuypers et al., *J. Solid State Chem.* **73** (1988) 192.
[2.67] S. Amelinckx, *Chim. Scr.* **14** (1978/1979) 197.
[2.68] J. van Landuyt, R. de Ridder, R. Gevers, S. Amelinckx, *Mater. Res. Bull.* **5** (1970) 353.
[2.69] G. van Tendeloo, S. Amelinckx, *Phys. Status Solidi A* **43** (1977) 553; **49** (1978) 337; **65** (1981) 431.
[2.70] S. Muto, G. Van Tendeloo, S. Amelinckx, *Philos. Mag.*, in press.

[2.71] G. van Tendeloo et al., *J. Phys. Chem.* **96** (1992) 7424.
[2.72] G. Van Tendeloo et al., *Europhys. Lett.* **21** (1993) 329. M. A. Verheijen et al., *Chem. Phys.* **166** (1992) 287.
[2.73] D. Schechtman, I. Blech, D. Gratias, J. W. Cahn, *Phys. Rev. Lett.* **53** (1984) 1951.
[2.74] C. Janot: "Quasicrystals, A Primer," *Monographs on the Physics and Chemistry of Materials*, Oxford Science Publishers, Oxford 1992.
[2.75] R. Penrose, *Bull. Inst. Math. Appl.* **10** (1974) 266.
[2.76] A. Ourmazd, D. W. Taylor, J. Cunningham, C. W. Tu, *Phys. Rev. Lett.* **62** (1989) no. 8, 933. A. Ourmazd, R. H. Baumann, M. Bode, Y. Kim, *Ultramicroscopy* **34** (1990) 237.
[2.77] S. Iijima, *Nature (London)* **354** (1991) 56. S. Iijima, T. Ichihashi, Y. Ando, *Nature (London)* **356** (1992) 776.
[2.78] X. F. Zhang et al., *J. Cryst. Growth* **130**, (1993) p. 36ff. (1993).
[2.79] D. Ugarte, *Nature (London)* **359** (1992) 707.
[2.80] G. Van Tendeloo, H. W. Zandbergen, S. Amelinckx, *Solid State Commun.* **63** (1987) no. 5, 389–393; **63** (1987) no. 7, 603–606. H. W. Zandbergen, G. Van Tendeloo, T. Okabe, S. Amelinckx, *Phys. Status Solidi A* **103** (1987) 45–72.
[2.81] T. Krekels et al., *Appl. Phys. Lett.* **59** (1991) 3048. T. Krekels et al., *Solid State Commun.* **79** (1991) 607–614.
[2.82] T. Krekels et al., *Physica C* **210** (1993), 439–446.
[2.83] T. Krekels et al., *J. Solid State Chem.* (1993), in press.
[2.84] L. C. Nistor, G. Van Tendeloo, S. Amelinckx, *J. Solid State Chem.* **105** (1993), 313–335.
[2.85] X. F. Zhang, personal communication.
[2.86] F. Watari, J. van Landuyt, P. Delavignette, S. Amelinckx, *J. Solid State Chem.* **29** (1979) 137–150. F. Watari, P. Delavignette, S. Amelinckx, *J. Solid State Chem.* **29** (1979) 417–427.
[2.87] K. Yagi, in [2.12, vol. 2], p. 261. K. Yagi, *J. Appl. Crystallogr.* **20** (1987) 147. T. Hasegawa et al., *The Structure of Surfaces II*, Springer Verlag, Berlin 1987, p. 43.
[2.88] H. W. Zandbergen et al., *Physica C* **158** (1989) 155. H. W. Zandbergen, W. A. Groen, F. C. Mijlhoff, G. Van Tendeloo, S. Amelinckx, *Physica C* **151** (1988) 325.
[2.89] H. Fujita (ed.): *In-situ Experiments with High Voltage Microscope,* Research Center for High Voltage Electron Microscopy, Osaka 1985.
[2.90] L. Reimer: "Scanning Electron Microscopy. Physics of Image Formation and Microanalysis," *Springer Ser. in Opt. Sciences*, vol. 45, Springer Verlag, Berlin 1985.
[2.91] O. C. Wells: *Scanning Electron Microscopy*, McGraw-Hill, New York 1974.
[2.92] D. B. Holt, M. D. Muir, P. R. Grant, I. M. Boswarva: *Quantitative Scanning Electron Microscopy*, Academic Press, London 1974.
[2.93] J. B. Pawley: "LVSEM for High Resolution Topographic and Density Contrast Imaging," *Adv. Electron. Electron. Phys.* **83** (1992) 203–274.
[2.94] L. Reimer: *Image Formation in Low-Voltage Scanning Electron Microscopy*, SPIE Press, Bellingham 1993.
[2.95] H. Raether: "Excitation of Plasmons and Interband Transitions by Electrons," *Springer Tracts Mod. Phys.* **88**, Springer, Berlin 1980.
[2.96] R. F. Egerton: *Electron Energy-Loss Spectroscopy in the Electron Microscope*, Plenum Publishing, New York 1986.
[2.97] E. Menzel, E. Kubalek, *Scanning* **5** (1983) 151–171.
[2.98] D. B. Holt, D. C. Joy: *SEM Microcharacterization of Semiconductors*, Academic Press, London 1989.
[2.99] G. V. Saparin: "Cathodoluminescence" in P. W. Hawkes, U. Valdrè (eds.): *Biophysical Electron Microscopy*, Academic Press, London 1990, pp. 451–478.
[2.100] G. D. Danilatos: "Foundations of Environmental Scanning Electron Microscopy," *Adv. Electron. Electron. Phys.* **78** (1988) 1–102.
[2.101] K. F. J. Heinrich: *Electron-Beam X-Ray Microanalysis*, Van Nostrand-Reinhold, New York 1981.

Rheometry

TREVOR A. STRIVENS, ICI Paints Plc. (retired), Slough, Berkshire, United Kingdom

1. The Science of Rheology

For basics of rheology, see also →5. Fluid Mechanics, **B 1**; →Plastics, General Survey, **A 20**; →Plastics, Properties and Testing, **A 21**.

1.1. Introduction

Rheology is defined as the science of deformation and flow of matter in response to the application of mechanical force. The etymology of the word emphasizes flow (Greek, ϱεω and λογοσ mean "to flow" and "word," respectively). Although rheology overlaps classical (solid) mechanics at one end of its range and fluid mechanics at the other, it generally includes materials that show both deformation and flow under the influence of mechanical force. This excludes chemically cross-linked materials such as rubber, which may exhibit flow because of chemical degradation resulting from heat generated by the application of excessive mechanical force (thermomechanical degradation). However, it includes materials such as solid emulsion paints or thermoplastic polymers, which become liquid under the influence of stress and increased temperature, respectively. Many of these materials exhibit viscoelasticity; i.e., both irreversible (energy dissipative processes, viscous flow) and reversible (energy storage processes, elasticity) changes. Energy is dissipated mainly in the form of heat or sound, and stored as potential energy by conversion from kinetic energy. The exact balance between irreversible and reversible processes when force is applied to the substance and deformation results, depends on factors such as temperature and the magnitude of the force or deformation applied. It also depends on the time scale of the laboratory experiment relative to the

0-89573-541-5/94/$ 5.00 + .50

characteristic times of internal material processes (e.g., molecular or particle diffusion). Some materials exhibit plastic flow, which is different in character from viscous flow but still irreversible (energy dissipative).

The science of rheology can be said to have been founded with the meeting of an international scientific committee on April 29, 1929 to draw up a constitution for the first society of rheology. Although the systematic study of rheology dates from then, SCOTT-BLAIR [15] has pointed out a number of rheological observations dating back to very ancient times (ca. 1650 B.C.) with the use of water clocks by ancient Egyptians (a forerunner of modern flow cups for measuring viscosity).

In 1660, ROBERT HOOKE discovered the law governing (linear) elastic deformation. In 1687, NEWTON proposed his law relating stress to velocity gradients in fluids. With these two laws, the modern era of rheology began, with substantial contributions over the next 200 years from (among others) EULER, CAUCHY, COULOMB, POISEUILLE, HAGEN, COUETTE, REYNOLDS, and BINGHAM, before the foundation of the first society of rheology in 1929 in the United States.

Because of the wide range of materials and physical concepts involved, rheology is a truly interdisciplinary science. Thus it unites contributions from mathematics, physics, and chemistry, and from specialists in colloid science, materials science, polymers, chemical engineering, and computer modeling.

1.2. Terminology

1.2.1. Rheological Measurement Terms

The two main modes of deformation, or ways in which materials can be changed in shape by the application of force, are stretching (or *extensional deformation*) and *shearing deformation*. These modes can be divided further. For example, extensional deformation can be *uniaxial*, (i.e., stretching in one direction only, as in stretching a thread or filament of material from end to end) or *biaxial* (i.e., stretching simultaneously in two directions, as in inflating a rubber balloon). Shear deformation can be *simple* or *torsional*.

In *torsional deformation*, the shear is rotational around a central axis. *Simple shear deformation* is shown in Figure 1. The upper plate can move parallel to the lower fixed plate under the influence of an applied force F. A thin slice of material of constant thickness h is trapped between the two plates and adheres to both (i.e., no slip occurs as the upper plate moves). The area of the upper plate is A. The *shear stress* σ (sometimes written as τ, which usually denotes relaxation time) is defined as force per unit area (i.e., F/A, measured in pascals). If the material is a *perfect* (or ideal) *elastic solid*, it obeys Hooke's law and is referred to as a *Hookean solid* or body. When force is applied, the upper plate moves a certain distance b, then stops in the new position for as long as the force is applied. The amount of deformation is defined as b/h (providing the angle Θ is small) and is known as the *shear strain* γ. Since the material thickness h remains constant, then according to Hooke's law, the shear strain is proportional to the shear stress applied. The constant of proportionality is the *shear modulus* of *elasticity* G. The unit of G is again pascal because the shear strain is dimensionless. Thus $G = \sigma/\gamma$.

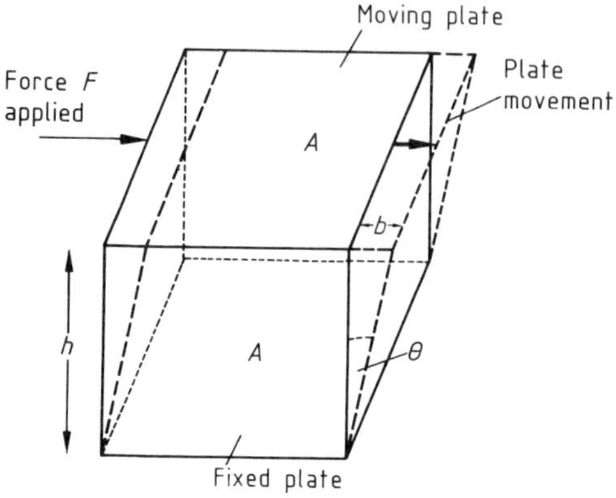

Figure 1. Diagram of simple shear to show basic rheological parameters

If the stress is removed, the upper plate returns immediately to its original position and the mechanical energy stored as potential energy in the stretched material during application of the force is recovered in the kinetic energy of the return movement.

If the slice of the material is a simple *ideal liquid*, it obeys Newton's law and is referred to as a *Newtonian liquid*. When force is applied, the upper plate moves at a constant velocity v for as long as the force is applied. In an analogous fashion, a *shear strain rate*, usually known as the *shear rate*, is defined as v/h (denoted by $\dot{\gamma}$, i.e., $d\gamma/dt$). [Shear rate is sometimes symbolized as D, but $\dot{\gamma}$ is used to avoid confusion with the Deborah number (see Section 1.2.2) or the differential operator d/dt, which are sometimes written as D.] In Figure 1, the liquid velocity is assumed to decrease linearly across the gap between the moving upper and the fixed lower plates. The shear rate is proportional to the stress, and the constant of proportionality is the *shear viscosity* η in units of pascal seconds (Pa · s). If the force is removed, the plate stops moving (unless the inertia due to its mass continues its motion for a short time) but does not return to its initial position. There are no stretched elements in the material that, by relaxing, can pull the plate back toward its original position, and the potential energy of the plate is zero. The mechanical energy applied has been dissipated as heat [16].

Analogously, an *elongation strain* ε and *strain rate* $\dot{\varepsilon}$, as well as an elongation modulus (*Youngs's modulus*, E) and an *elongation viscosity* $\bar{\eta}$, may be defined. Under certain well-defined material conditions, the elongation modulus is three times the shear modulus, and the elongation viscosity three times the shear viscosity (Trouton's law or ratio [17] →5. Fluid Mechanics, **B 1**, pp. **5**-39–**5**-40).

1.2.2. Rheological Classification of Materials

Many materials show characteristics of both ideal solids and viscous liquids: they are *viscoelastic solids* or *liquids* depending on which material characteristic predominates. If the modulus or viscosity is independent of strain or strain rate, respectively (i.e., the materials obey Hooke's or Newton's laws), they are referred to as *linear* materials because their response to deformation is linear. In practice, most materials obey these laws over limited ranges of stress, strain, or strain rate. When the modulus becomes a function of strain and the viscosity a function of strain rate (usually, in both cases a decreasing function) the material response is *nonlinear*. At high shear rates, nonlinear responses can lead to so-called normal forces. These arise because the force exerted by the moving liquids becomes more and more out of alignment with the flow as the shear flow increases (see Section 2.5.5).

Most materials, even if linear, do not respond instantaneously to the application of a force; even if these materials are linear, they show a *time-dependent* response. This time dependence becomes much more obvious if the applied stress or strain itself varies with time, for example, in *oscillatory or dynamic* experiments where the variation is sinusoidal, or in transient experiments where the time scale of the material response is long compared to those of experimental artifacts. Whether a predominantly elastic or viscous response is obtained is often dictated by the relationship between the material response time (material characteristic time) and the time scale of the experimental observation. This time ratio is the Deborah number D, named by M. REINER after the Hebrew prophetess Deborah who sang "the mountains flowed before the Lord" (Judg. 5) [18]. Low Deborah numbers indicate viscous, and high numbers (>1) elastic response [19]. Thus, even simple liquids can exhibit elasticity at very high frequencies [20].

A number of qualitative, classificatory terms are used in rheology to describe various types of material response. If the shear rate is increased linearly with time between zero and a maximum value and then decreased over the same range the resulting viscosity–shear rate plots for the increasing and decreasing shear rate halves of the experiment may either be superimposed on each other (*pseudoplastic behavior*, Fig. 2), or the plot of increasing shear rate may be above the plot of decreasing shear rate, to form a hysteresis loop (*thixotropic behavior*, Fig 3) (see also →5. Fluid Mechanics, **B 1**, pp. **5**-26–**5**-28). Instead of continuous shear rate variation, with modern rheometers a series of measurements can often be programmed at discrete shear rate values (ascending and descending as before), with a predetermined time at each value of shear rate before the viscosity is measured. Whether pseudoplastic or thixotropic behavior is observed depends on the ratio of the time the material requires to reach a steady state to the experimental time at each shear rate.

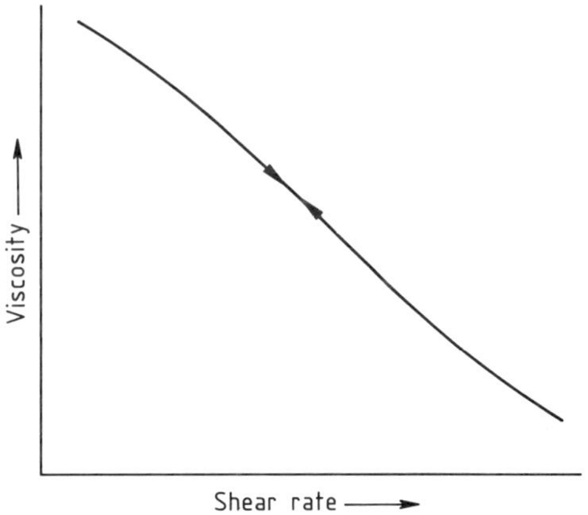

Figure 2. Viscosity–shear rate profile (pseudoplastic)

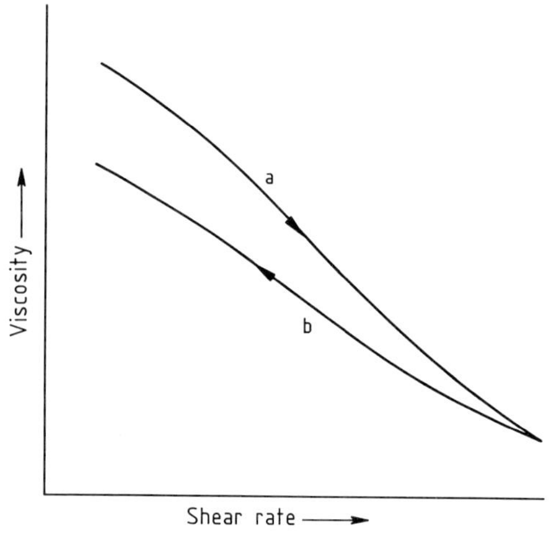

Figure 3. Viscosity–shear rate profile (thixotropy)
a) Increasing shear rate; b) Decreasing shear rate

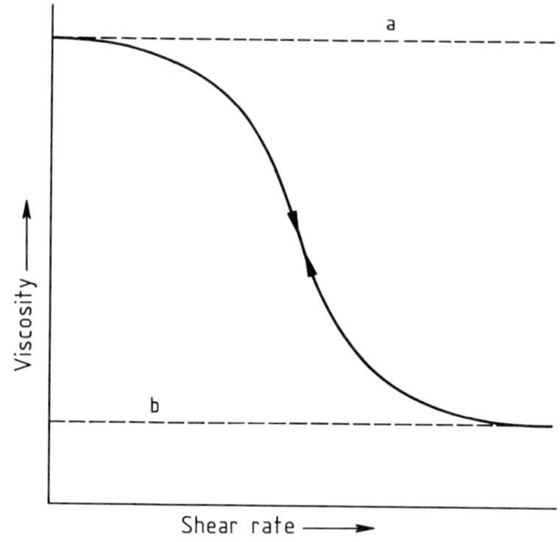

Figure 4. Viscosity–shear rate profile with Newtonian plateaus
a) Lower Newtonian plateau; b) Upper Newtonian plateau

At both low and high shear rates, materials may show regions of Newtonian behavior, with pseudoplastic behavior at intermediate shear rate values (Fig. 4). These regions are the *lower Newtonian* and *upper Newtonian plateau regions*, respectively. The terms lower and upper refer to the shear rate values and not to the Newtonian viscosity values in each region.

Many familiar materials, such as emulsion paints, toothpaste, and cream, exhibit structure at rest (i.e., they have a more or less solid appearance). However, when sheared by stirring they become liquid-like. Although they may not show thixotropy at moderate to high shear rates, they nearly always do at low shear rates. In addition, the viscosity appears to increase more and more rapidly as lower and lower shear rates are used, with no sign of the approach of a lower Newtonian plateau region (Fig. 5). A linear plot of shear stress versus shear rate appears to extrapolate not to the origin but to a positive value of shear stress, the so-called *yield stress* [21]. Experimentally, determination of the yield stress is difficult because of the long times most of these materials require to reach a steady shear stress value at very low shear rates, problems of data extrapolation, and instrumental difficulties, etc. These materials nearly always show an upper Newtonian region because shearing breaks down the structure formed at rest and the material flows like a "normal" liquid. Other materials have different behavior at high shear rates: in particular, the viscosity increases with increasing shear rate.

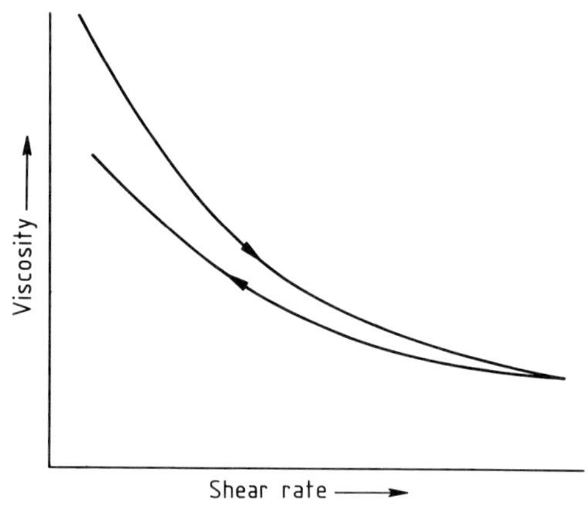

Figure 5. Viscosity–shear rate profile of a structured liquid

This behavior is referred to as (→ 5. Fluid Mechanics, **B 1**, pp. **5**-27–**5**-28):

1) *Shear thickening*, when the increase is due to enhanced flocculation of particles in dispersion (orthokinetic flocculation) [22]
2) *Rheopexy*, when the shear flow orients anisotropic particles in dispersion into favorable positions for flocculation [23]; it is usually time dependent

These phenomena have also been observed in polymer solutions and other materials, but they are associated predominantly with the rheological behavior of dispersions. Flocculation refers to the formation of clusters, chains, or networks of particles due to net attractive forces existing between the particles of dispersions. Such flocculation may also lead to the shear thinning phenomena already discussed [24], [25].

Other terms refer to the approximate or exact conformation of the rheological data to a particular mathematical equation, empirical equation, or theoretical model. Thus, if rheological parameters exhibit a linear dependence on frequency or shear rate on double logarithmic plots, the rheological behavior is referred to as *power law behavior*. If the material shows a yield stress and the square root of the measured viscosity (the so-called apparent viscosity) is proportional to the square root of the shear rate, the material is described as having Casson flow behavior, after the theoretical model of structure breakdown developed by CASSON (see also →5. Fluid Mechanics, **B 1**, pp. **5**-25–**5**-27) [26].

Finally, *reduced viscosity values* are often used. Two kinds of reduced viscosities exist. Firstly, Newtonian plateau viscosity values for *dispersions or polymer solutions* are often divided by the viscosity of the dispersion medium or the solvent, respectively, before the viscosity is related to disperse phase volume or polymer concentration. Secondly, in describing *shear rate dependence*, certain equations such as the Cross equation use a reduced viscosity of the kind

$$\frac{\eta_a - \eta_x}{\eta_o - \eta_x}$$

where η_a is the measured or apparent viscosity, and η_o, and η_x are, respectively, the lower and upper Newtonian plateau values of the viscosity.

A full glossary of rheological terms based on the 1975 British Standard [27] and the U.S. Society of Rheology recommendations [28] is found in [29].

1.2.3. Classification of Rheological Experiments and Measuring Systems

In addition to the simple type of shear deformation described (Fig. 1), two other types are also of considerable importance. These are (1) *pipe, capillary,* or *Poiseuille flow* and (2) *torsional flow*. Capillary flow in so-called *U-tube viscometers* forms the ultimate measurement standard for Newtonian liquids. The flow is characterized by a nonlinear velocity distribution across the tube radius. Further, this distribution depends on the rheology of the test liquid. Since the velocity distribution can be predicted accurately for Newtonian and for a few (mathematically) simple non-Newtonian liquids (e.g., power law fluids), U-tube viscometers are used primarily for Newtonian liquids and for very high shear rate measurements (where liquids tend to be Newtonian, anyway). Since fresh unsheared liquid is passing through the tube continuously, fewer problems occur because of errors due to mechanical energy dissipation as heat at high shear rates than in the case of the small static sample volumes used in other types of shearing deformation.

Simple deformation is really not very practicable for steady shearing measurements, because of the limitation in total strain (or the product of shear rate and time) that may be applied (one plate sliding over another with the material confined between) and the difficulties of confining all but the most viscous liquids in the gap between the plates. A simple way of reducing these difficulties is to use circular plates with a rotational movement around a central axis, instead of a sliding motion between two parallel plates; this is referred to as *torsional shear deformation* or flow. Again, one plate is fixed and the other moves with the gap between them kept constant. This measuring geometry is known as *parallel plate* geometry. However, the angular velocity of the moving plate now varies across the plate radius (zero at the center and maximum at the edge of the plate); consequently, the shear rate does also. By substituting, for one plate, a cone whose tip touches the other plate, the shear rate becomes independent of the radial distance from the center of the moving plate. This occurs because the angular velocity increases linearly with distance from the center of the moving plate, but so does the gap between the plates (Fig. 6). This type of measuring geometry is known as *cone and plate geometry*. In practice, a small part of the cone tip is cut off or truncated (see Section

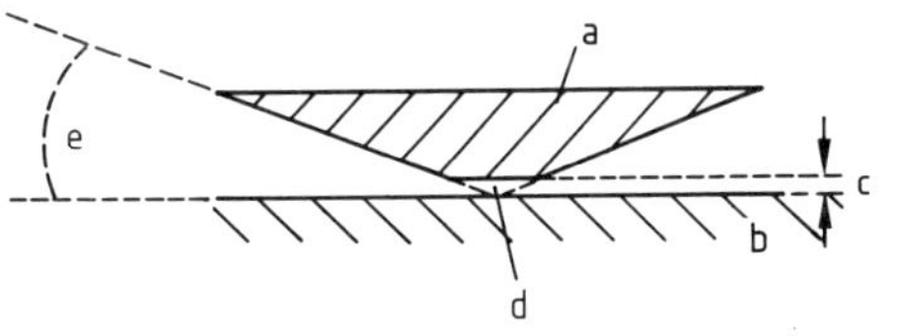

Figure 6. Cone and plate geometry
a) Cone; b) Plate; c) Gap; d) Truncated cone tip; e) Cone angle

2.2.2.2). Another torsional measuring geometry is the so-called *Couette*, *concentric cylinder,* or *cup and bob*, where the material to be measured is confined in the space between the cup and a cylindrical bob inside it. By keeping the radii of the cups large and the gap between the cylinder inner wall and the bob wall small, the difference in radial velocities at the two walls is small. Consequently, the shear rate across the gap is fairly uniform (see Fig. 7), although not as uniform as in the cone and plate geometry. However, large errors may arise from the contribution of the flow between the bob end and the cup wall. Cone angles, as well as the radius of cone or plate and the radii of Couette components, can be used to control the shear range within given limits of rotation speeds.

By rotating one member of the measurement geometry at a given speed, viscosity–shear rate measurements can be obtained. If the time scales are correct (see Section 1.2.2), *steady-state* measurements are obtained. By oscillating one member between limits at a controlled amplitude and frequency, particularly with a sinusoidal relationship to time, *oscillatory* or *dynamic* measurements are performed. Other time dependences, such as square wave or triangular, can also be applied. They are generally more difficult to use, from the point of view of both instrumental capability and mathematical analysis. An *impulsive* or *transient measurement* is made by applying a single square wave input or a linearly or logarithmically increasing input of known level for a certain period of time and measuring the material response as a function of time (see Section 2.5.3).

In *extensional flow*, the measuring geometry and test method depend on the shear viscosity and cohesive strength of the liquid. Thus, for liquids with high shear viscosity and high cohesive strength such as polymer melts, threads can be formed that can be stretched or wound on reels under well-defined conditions. More solid-like melts can be made into sheets for biaxial deformation and into bars or rods for uniaxial deformation. As the shear viscosity is reduced, a thread of liquid can be drawn into a tube from the surface of a liquid sample, and a stable liquid profile can be maintained in this manner. The thread or column can then be stretched by moving the tube vertically away from the liquid surface. However, the presence of a disperse phase, such as solid particles of filler or pigment, or gas bubbles (e.g., air), reduces the cohesive strength of the liquid and leads to instability or breakdown of the liquid column. Further reduction of the shear viscosity will render the formation of liquid columms or threads impossible, because of the very low cohesive strength of liquid threads. However, many low-concentration, high molar mass polymer solutions have extensional viscosities many orders of magnitude greater than their very low shear viscosity of a few millipascal-seconds [30]. Such liquids can be measured by the opposing jet apparatus (commercially available), by the four-roller technique (if the liquid is transparent), or in a semiquantitative manner by pressurized flow cups or thread formation as drops break away from a syringe needle tip (see Section 2.4).

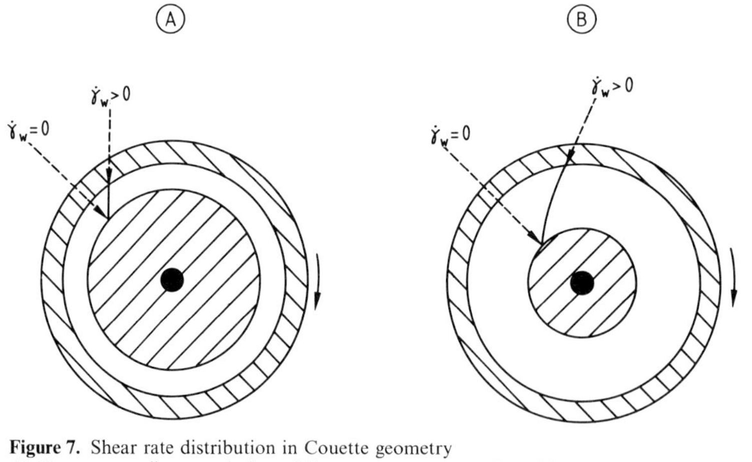

Figure 7. Shear rate distribution in Couette geometry
A) Narrow gap, linear variation of shear rate across gap; B) Wide gap, nonlinear variation of shear rate across gap

1.3. Examples of Rheological Behavior

(→5. Fluid Mechanics, **B 1**)

1.3.1. Newtonian Liquids

The results of oscillatory and creep measurements for a Newtonian oil are given in Figures 8 and 9. In Figure 8, the real and imaginary components (G' and G'') of the complex modulus G^*, as well as the dynamic viscosity η' are plotted as a function of frequency. At low frequencies the dynamic viscosity is equal in value to the (Newtonian) steady rate shear viscosity value. Correspondingly, the imaginary component of the modulus or *loss modulus* G'' is proportional to frequency f and is much greater in value than the real component or *dynamic elasticity modulus* G'. This is typical of the dynamic response of a liquid. The loss modulus represents the portion of the total mechanical energy, supplied during the applied strain cycle, which is dissipated as heat during the cycle. The portion stored as potential energy by elastic deformation of the material is represented by the modulus of elasticity G'. The ratio of the imaginary to the real component of the (complex) modulus of elasticity yields the value of tan δ, where δ is the phase angle difference between the applied stress or strain cycle and the corresponding response (measured) cycle. The modulus of elasticity is very small at low frequencies but increases sharply (to the second power) with increasing frequency. This means that at higher frequencies, the two plots cross and the apparently Newtonian liquid becomes elastic at ca. 40 Hz. Correspondingly, the dynamic viscosity η' decreases as the high-frequency end is approached. Since simple liquids exhibit elasticity only at very high frequencies [20], many orders of magnitude above the ones shown in Figure 8, clearly another source of energy storage exists. This additional source is the measuring system itself, particularly the contribution to the overall response signal from its inertia and elasticity. This contribution becomes a maximum at the so-called natural frequency, which is close to 40 Hz for the particular measuring systems used here. The inertia term increases with the second power of the frequency. If this inertia contribution is subtracted, the elasticity modulus G' is low and virtually independent of frequency. Thus, measurement conditions should be chosen so as to eliminate or minimize the contribution of the measurement system either by appropriate corrections or by replicate measurements using different measurement systems or conditions. It is most important to have access to raw measured data values so that such corrections can be made by the operator and to understand exactly how printed rheological parameters are derived from such data in the computer software of commercial rheometers.

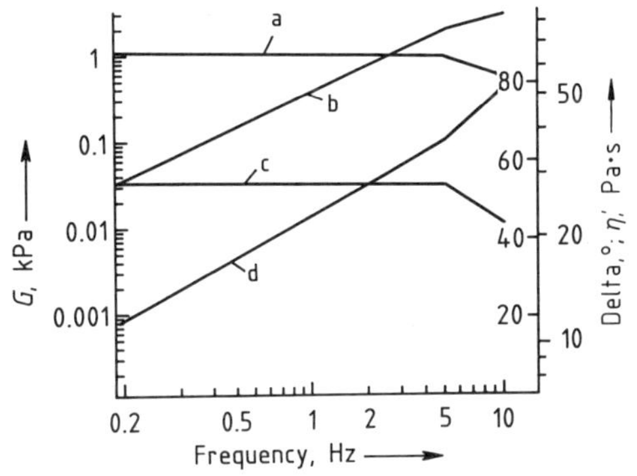

Figure 8. Oscillatory data (Newtonian liquids)
a) Delta; b) Loss modulus G''; c) Dynamic viscosity η'; d) Dynamic modulus of elasticity G'

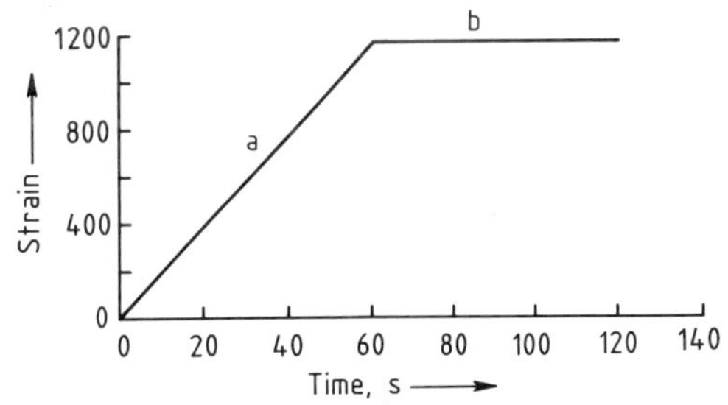

Figure 9. Creep data (Newtonian liquids)
a) Retardation period; b) Relaxation period

Results for creep measurements on the same oil are shown in Figure 9. The creep experiment is a typical transient experiment in which a constant stress is applied virtually instantaneously to the samples and the corresponding strain induced in the sample is measured as a function of time (*retardation period*). The stress is then removed instantaneously, and the sample strain is again measured as a function of time during its recovery from stress application (*relaxation period*). In the case of a Newtonian oil, the strain increases linearly with time during the retardation period. However, during the relaxation period, the strain remains constant.

When the stress is removed, all the mechanical potential energy supplied during the retardation period has dissipated. Consequently, the sample remains at the maximum strain value attained during the retardation period, and no change occurs during the relaxation period measurements.

1.3.2. Elastic Solids

As an example of an elastic solid, Figure 10 and Figure 11 show the results of oscillatory (dynamic) and transient (creep) experiments on a 6% (food) gelatin gel in water at 25 °C. The dynamic data show two striking features: first, the dynamic modulus of elasticity G' (curve a) is independent of frequency, and second, it is much greater than the loss modulus G'' (curve b). The loss modulus G'' exhibits a minimum but very little frequency dependence. The creep experiment (Fig. 11) shows an almost constant strain value during the retardation period, with almost complete and instantaneous recovery to zero strain during the relaxation period. In the figure compliance (measured strain divided by applied stress) is plotted against time. In creep experiments, it is more convenient to use compliance instead of modulus of elasticity. Compliance is denoted by J and related reciprocally to G, i.e., the unit of compliance is Pa^{-1}. Deviations from ideal Hookean behavior can be attributed to either slight irreversible deformation or, more likely, some failure in instrumental or measurement conditions.

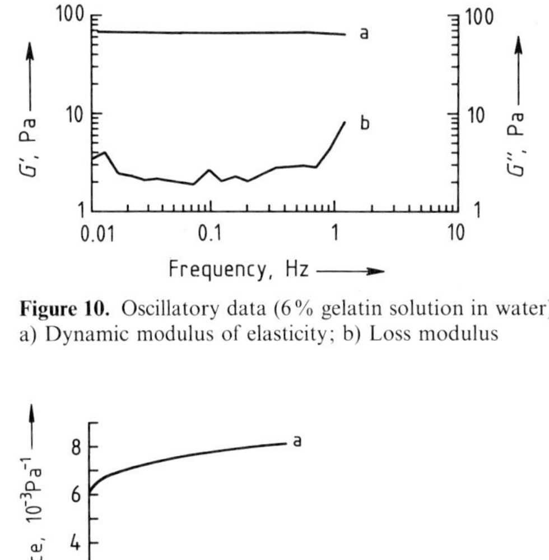

Figure 10. Oscillatory data (6% gelatin solution in water)
a) Dynamic modulus of elasticity; b) Loss modulus

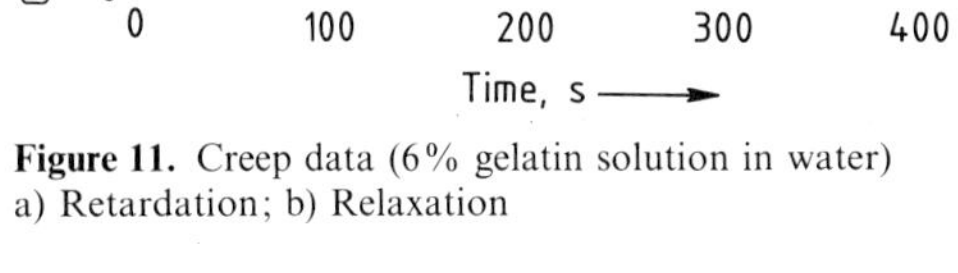

Figure 11. Creep data (6% gelatin solution in water)
a) Retardation; b) Relaxation

1.3.3. Simple Viscoelastic Liquids

The 0.1 mol/L cetyltrimethylammonium bromide (1:1 molar ratio)–salicylic acid solution (CTAS) system in water forms long rod-like micelles that like many other such materials, exhibit strong elasticity while flowing easily like a liquid [31]. The 0.1 mol/L CTAS system is fairly unique because it exhibits virtually ideal behavior, in that it has a single relaxation time and can be modeled by the simple Maxwell model (→Plastics, Properties and Testing, **A21**, pp. 12–13) [32].

As can be shown from viscosity–shear rate measurements, the viscosity is Newtonian at low shear rate values (i.e., it has little or no dependence on shear rate). At higher shear rates, the viscosity decreases rapidly with increasing shear rate. At even higher shear rates the viscosity once again approaches Newtonian behavior (see Fig. 4). To display the full rheological profile of a given sample, data from more than one measurement system must often be combined, as has been done for the concentrated poly(methyl methacrylate) dispersion in the hydrocarbon system shown in Figure 12. Misalignment of data, as shown in this example, can arise from small cali-

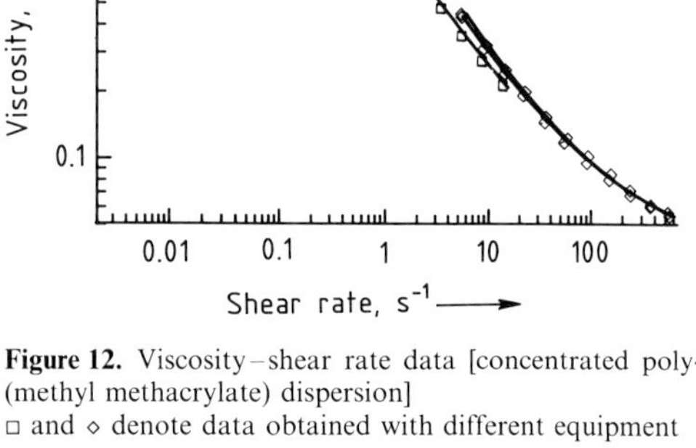

Figure 12. Viscosity–shear rate data [concentrated poly(methyl methacrylate) dispersion]
□ and ◇ denote data obtained with different equipment

bration errors, difficulties in obtaining repeatable samples, or sample handling.

Figure 13 shows oscillatory data for the 0.1 mol/L CTAS system. Now the features of both the Newtonian liquid and the Hookean solid are combined, with a narrow transition region between them in the middle of the frequency range. At low frequency, the dynamic viscosity η' is constant and the loss modulus G'' is greater than the dynamic modulus of elasticity G' (predominantly viscous response). The loss modulus and the modulus of elasticity increase approximately linearly, and to the second power, respectively, with increasing frequency. At the high-frequency end, the response is predominantly elastic. The dynamic modulus of elasticity is approximately constant, and the loss modulus lies below it and decreases rapidly (as does dynamic viscosity). In the narrow frequency region in between, the loss modulus passes through a sharp maximum, at which frequency (the *crossover frequency*) the modulus of elasticity equals the loss modulus and has a value approximately half that of its maximum at higher frequencies. This crossover frequency corresponds to the *relaxation time* (the reciprocal of the frequency divided by 2π). In Figure 13, the relaxation time is several seconds. This finite relaxation time can be demonstrated even more clearly by the (transient) stress relaxation experiment. Here, a sudden strain is applied to the sample, and the resultant torque (stress) exerted by the sample is measured as a function of time. After electronic smoothing, the stress–time curve shows an initial time-independent value (corresponding to the high-frequency modulus G' value of Fig. 13)

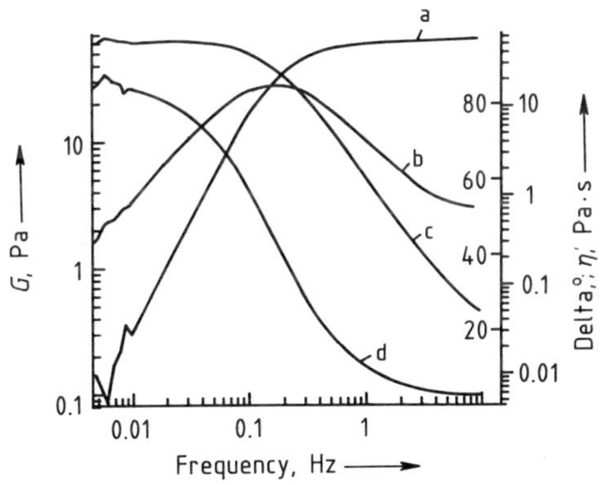

Figure 13. Oscillatory data (0.1 mol/L CTAS)
a) Dynamic modulus of elasticity G'; b) Loss modulus G''; c) Dynamic viscosity η'; d) Delta

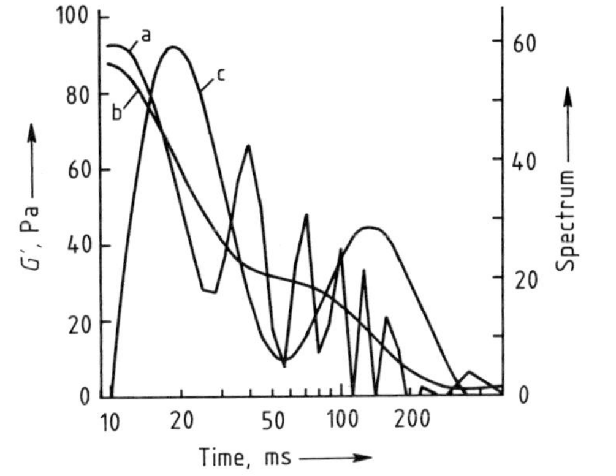

Figure 14. Stress relaxation data (0.1 mol/L CTAS)
a) Dynamic modulus of elasticity G'; b) G' after electronic smoothing; c) First differential of b

and then decreases fairly rapidly toward zero (Fig. 14). The inflection point of the plot G' versus time (curve b) corresponds to the relaxation time. This can be seen more clearly if the first differential of the curve is plotted against time (curve c, relaxation spectrum in Fig. 14). The time corresponding to the peak of this curve is the relaxation time. With identical samples and test conditions (not the case in Figs. 13 and 14), the two methods should agree fairly closely for the relaxation times.

The reason for the necessary electronic smoothing of the relaxation data in Figure 14 is that the elasticity of the sample causes a sudden twisting force (torque) to be applied to the calibrated bar of the measuring system. The system then oscillates at its natural frequency in the same way a pendulum does when moved from its position and released. Like the pendulum, this initial oscillation decays at a rate dependent on the velocity of the dissipation of the kinetic energy of movement by the surrounding medium. In the case of the measuring system it decays rapidly because of the sample viscosity. The initial peak in the spectrum (curve c) at ca. 20 ms represents this initial decay due to the measuring system.

1.3.4. Complex Viscoelastic Liquids

In Figure 15, the results of dynamic (oscillating) measurements on a more complex system are depicted. The system measured here is a concentrated dispersion of poly(methyl methacrylate) (PMMA) in a paraffin. The dispersion is

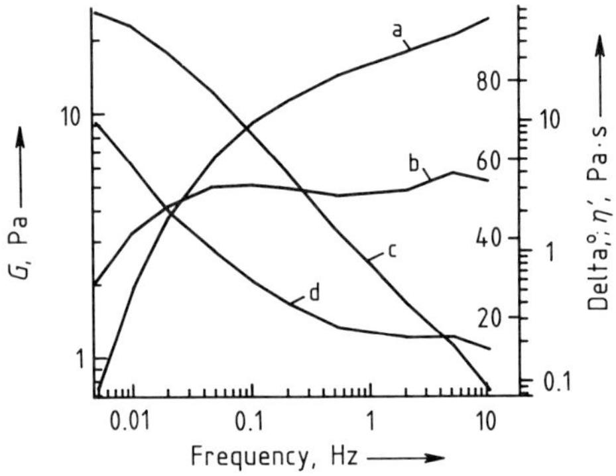

Figure 15. Oscillatory data (concentrated dispersion)
a) Dynamic modulus of elasticity G'; b) Loss modulus G''; c) Dynamic viscosity η'; d) Delta

sterically stabilized with poly(hydroxystearic acid) (PHS)–PMMA graft copolymers (for preparation details, see [33]). Now a much broader transition between liquid-like viscous response and solid-like elastic response is obtained. In particular, the peak in the loss modulus G'' plot becomes very shallow, and the crossover of the loss and elasticity modulus plots occurs lower than the peak frequency. The crossover frequency now corresponds to the longest relaxation time of the system, and the broader transition is due to a range of relaxation times in the system. This contrasts with the simple system in Figure 13, where only one relaxation time occurs. By superimposing many simple systems like that of Figure 13—each with its own relaxation time and appropriately weighted low-frequency plateau viscosity and high-frequency plateau elasticity values—the broader, less definite transition behavior shown in Figure 15 is obtained. However, as Figure 12 shows, the system is still strongly pseudoplastic under continuous shear, with a well-developed lower Newtonian viscosity and the approach of an upper Newtonian region clearly visible at higher shear rate. Although the particle-size distribution is not particularly narrow ($\pm 10\,\%$ of the mean particle diameter), the range of transition times is apparently more dependent on concentration than on particle-size variation. If the dispersion is diluted gradually, the range of relaxation times decreases and the transition region becomes narrower. At any given moment, because of their concentration and lack of free volume, the particles are arranged in random groupings of touching particles, with each group having a different size and, consequently, its own characteristic relaxation time. Thus a range of relaxation times is created. As the system is diluted, Brownian motion and more free volume lead to the breakdown of the largest groups into smaller groups. Thus the range of group sizes becomes smaller, and the range of relaxation times is reduced. In addition, the plateau value of the modulus of elasticity high-frequency decreases sharply. Since this value reflects the total number of particle–particle contacts per unit volume, it supports the particle contact group explanation of the relaxation time range observed [34].

1.3.5. Thixotropic Liquids

In Figure 16, steady shear viscosity measurements are presented for a typical industrial thixotropic material. Here the shear rate is varied continuously through an increasing and a decreasing cycle. Immediately after completion of the first cycle, a second identical cycle of shearing is applied to the sample. Two factors are striking: (1) the first cycle shows a thixotropic loop (i.e., the viscosity–shear rate plot for the descending half of the cycle lies at viscosity values below those for the ascending—increasing shear rate—half of the shear cycle); and (2) in the second cycle, the plots are superimposed on each other. In other words, the sample has now become pseudoplastic in its continuous shear rheology.

These data show two of the characteristics of thixotropic materials. First, an interaction exists between experimental time and material response time. Although not shown here for rea-

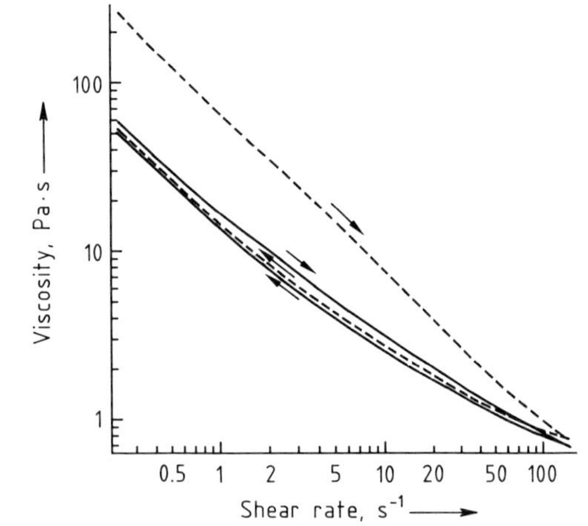

Figure 16. Two shear rate cycles (thixotropic material)
---- = first cycle; —— = second cycle

son of clarity, increasing the cycle time from the 5 min used here decreases the loop width or area enclosed by the ascending and descending plots, whereas decreasing the cycle time increases the loop width. Furthermore, the existence of a previous shear history, as exemplified by the first cycle of shearing, leads to a change in the sample rheology observed.

Second, although an upper Newtonian limit is approached at the high shear rate of the cycle, no corresponding limiting value of the viscosity appears to exist at low shear rates. Instead, if shear stress is plotted versus shear rate and the plot is extrapolated to zero shear rate, it does not go through the origin but through a positive value of the shear stress, the so-called yield stress (see Section 1.2.2). Although experimental problems occur in obtaining reliable yield stress values, and the physical reality of the yield stress has been debated [21], no doubt exists about the practical reality of such an effect. This is evident in the difficulty of producing flow for these materials, particularly in start-up of operations such as pumping or mixing efficiency in large vessels with paddle stirrers. Such practical difficulties are magnified if elasticity is present in the material at rest.

The elasticity effects in thixotropic materials can be demonstrated by dynamic (oscillatory) measurements. Data for the same industrial material are shown in Figure 17. Plots of the modulus of elasticity G' and the loss modulus G'' on double logarithmic axes are parallel straight lines (i.e., both moduli exhibit a power law dependence on frequency). This sort of dependence is characteristic of a wide range of relaxation times, with the maximum possible time being infinite.

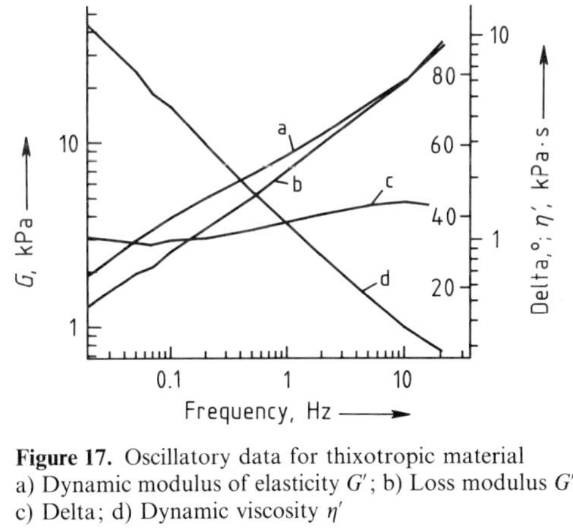

Figure 17. Oscillatory data for thixotropic material
a) Dynamic modulus of elasticity G'; b) Loss modulus G''; c) Delta; d) Dynamic viscosity η'

Such a range of times arises from the tenuous structure in terms of both size and degree of branching, and extent of distribution across the volume of the sample. Such structures are characteristic of associating (flocculating) particles in dispersion, associating polymer chains in solution, or mixtures of the two. Since such associations frequently result from physical rather than chemical bonding, they are susceptible to breakage by temperature increase or small mechanical deformation, thus giving rise to other characteristic features of thixotropic materials. These physically bonded structures may extend throughout the sample, giving rise to modulus of elasticity values greater than the loss modulus. However, because of their tenuous network and readily labile nature, they rarely exhibit the predominantly Hookean-type elasticity behavior characteristic of chemically cross-linked polymer gels or very regularly and highly (physically) cross-linked gels, such as the gelatin system discussed [35].

2. Rheological Measurements

2.1. Basic Deformation Modes for Measurement

Two basic modes of deformation are described in detail, namely, shear and extensional deformation. Other modes exist (e.g., bulk or compressional deformation) that are not discussed here.

2.1.1. Shear Deformation

Three submodes of shear deformation exist—simple, torsional, and capillary, or Poiseuille deformation. *Simple shear flow* has conceptual, but little practical, value. Simple shear flow is limited to highly viscous or stiff materials, such as printing ink or asphalt, and can be used in very limited shear rate or total shear strain ranges. Most measurements are made with torsional shear or capillary flow equipment.

Capillary flow is applied to low-viscosity fluids, and gravitational force is used to drive the flow. However, if air or gas pressure is used to drive the flow, fluids with a much higher viscosity can be measured at much higher shear rates. In this way, polymer solutions or melts can be measured. The combination of high viscosity and high shear rate leads to high rates of energy dissipation as heat. Because of the continuous flow of fresh sample into the capillary and the low residence time of a particular volume element of the sample in the capillary, the high rate of energy dissipation usually does not lead to problems of sample temperature control and consequent errors in viscosity measurements.

Torsional shear measurements, in contrast, use the same aliquot of sample confined between the elements of the measuring geometry. Because of the small gap between the elements of the measuring geometry necessary to achieve high shear rates and the reduced sample size, the energy dissipation rate becomes a problem in terms of temperature control. The narrow gap also presents some severe mechanical design problems. However, the narrow gap combined with the relatively large radius of the cup and cylinder (*Couette geometry*) gives a shear rate that is virtually constant across the gap. Although the *cone and plate geometry* is more suitable for achieving uniform shear rate across the sample, the high rotational speed and consequently strong centrifugal forces tend to force the sample from the gap, which limits the maximum shear rate obtainable.

Higher shear rate values can be obtained with the *Couette geometry,* due to containment of the sample within the cup. By contrast, capillary flow leads to a gradient of shear rate across the radius of the capillary. Whereas the gradient can be predicted for Newtonian liquids, this is more difficult if the viscosity–shear rate relationship is unknown. However, because many liquids show near-Newtonian behavior at very low and very high shear rates, capillary flow is more widely applicable than merely as a standard reference measurement for viscosity, despite the difficulty in calculating the shear rate gradient.

2.1.2. Extensional Deformation

Although a large number of commercial rheometers are available to perform torsional flow measurements, relatively few exist for measurement of extensional flow. This reflects the much greater practical difficulty of making such measurements, particularly at high rates of extension or with very mobile liquids. In addition, the problem of generating flows approximating steady extension is by no means trivial. For simple liquids, Trouton's rule states that the extensional viscosity is three times the shear viscosity value. However, the extensional viscosity of several high molar mass dilute polymer solutions is many orders of magnitude higher than the shear viscosity. For example, METZNER and METZNER [36] measured very dilute aqueous solutions of polyacrylamide (100 mg/L) and found shear viscosities of 1.4 mPa · s and extensional viscosities of 9 Pa · s (Trouton ratio 6500 instead of 3). Other systems showed Trouton ratios as high as 30 000 [37]. Measurements using a different technique substantiate these findings [38]. However, as demonstrated recently, a disparity exists between results obtained with different techniques on the same sample [39]. Practically, only steady-state measurements can be performed because of experimental difficulties. However, stress transients are observable before steady-state flow is attained. Some of the disparities between measurement methods may relate to the failure to obtain true steady-state values or the onset of molecular degradation in high stretching rate flows (dilute polymer solutions).

Two main types of flow in extension have been distinguished by WALTERS [40]. These are (1) uniaxial, and (2) biaxial and two-dimensional flow. Most methodical studies use *uniaxial methods*. Some studies of *biaxial and two-dimensional flow* have been made by DENSON and coworkers [41], in which the formation of bubbles by inflation of small areas of plastic sheet has been studied. Even if the flow may be steady (Eulerian sense), each liquid element must be subjected to a rate of deformation that is independent of time: i.e., transients have died away, and the flow is steady in this sense also (Lagrangian sense). In this latter sense, particularly with low-viscosity liquids at high extension rate, achieving steady extensional flow under practical laboratory conditions is difficult.

Extensional flow has many obvious practical applications in industries based on polymers, for example, fiber spinning, extrusion, vacuum forming, and blow molding. Extensional flow can also be applied in studies, e.g., on surface coatings, atomization during paint application, or the properties of coatings applied by roller coating. Other important areas are lubrication, drag reduction, and flow through porous media (of interest to the oil industry).

2.2. Torsional Shear Deformation

2.2.1. Rheometer Types

Two basic types of rheometer are available for rheological measurement in torsional shear deformation: controlled shear strain–shear (strain) rate and controlled stress instruments. Historically, the former type is older, the first one having been described by COUETTE in 1890.

In controlled *shear strain–shear rate instruments*, the lower member of the measuring geometry is rotated through a small angle (shear strain) or rotated at a controlled speed (shear rate). The sample, confined by the system geometry, exerts a force on the fixed upper member as a result of movement of the lower. This force or torque is measured by means of a sensitive transducer (e.g., a piezoelectric device) or by the (relatively) small deflection of a calibrated torque bar, which couples the upper member to the body of the instrument. The angular deflection or rotation speed is then converted into the applied shear strain or shear rate, and the measured force or torque into the required stress. The applied shear strain or shear rate can be varied in time to increase or decrease linearly, or it can be applied discontinuously with varying rest times between applications.

Controlled stress instruments operate according to a different principle. The lower member of the measuring geometry is rigidly attached to the body of the instrument. By means of a controlled electric current applied to an induction motor, a rotary torque is applied to the measuring head to which the upper member of the geometry is rigidly attached. Movement of the head can be measured by, for example, an optical diffraction grating system, in terms of both angular deflection (shear strain) and angular rotation speed (shear rate). By varying the current, the stress can be controlled as required.

The following applies to both types of instrument:

1) Because of increasing requirements of sensitivity, accuracy, etc., the measuring heads are nearly always supported by air bearings that minimize errors due to friction in bearings.
2) Most modern rheometers incorporate computers that, besides converting measured (raw) data into the required rheological parameters, also control the experimental procedure. This allows considerable flexibility of control and permits a sequence of inputs to be designed to simulate complex practical situations.
3) When measuring transient properties, a potentially important factor is the input time of the two types of instrument. Thus the imposition of a steady torque by electrical means is extremely rapid. Instantaneous (Hookean) elasticity can be measured since measurement can start 1 ms after commencement of stress application with a measurement rate of 120 s^{-1} in a typical commercial instrument. By contrast, a purely mechanical drive system requires a longer time (rise time) to reach a preset strain level (typically 20 ms to reach the preset strain zero). This results in modification of the short-time stress relaxation behavior observed, particularly if fast relaxation processes occur in the sample. With torque bar measurements, rapid application of the strain leads to an initial oscillating stress (the torque bar behaves like a suddenly struck pendulum). The extent of oscillation and rapidity of decay depend not only on the properties of the measuring head and geometry, but also on the energy dissipative capability of the sample. Although electronic filtration can be used to remove the initial oscillation, it may not be entirely satisfactory. With electrically driven instruments, such instrumental transients are small and of very short duration and do not disturb the measurements.

Some constant-stress instruments are equipped with software that, by using feedback control of stress level, can provide controlled shear rate measurements. Despite the apparent duplication of steady-state, transient, and oscillation measurements on both stress and shear rate instruments, the two types of instrument provide very different and complementary kinds of information. The selection of instrument must depend on both the type of flow application to be studied and the sample to be measured. For example, when studying the flow-out of a wet paint film after application, a stress-controlled rheometer is more appropriate because the flow of the wet paint is caused and driven by stresses or forces originating from gravitational forces and from surface tension gradients on the film surface [42]. In contrast, when studying stirring or mixing problems, shear rate controlled rheometers are more appropriate because the rotation speed of a stirrer blade or a mixing head will determine the shear rates applied to the sample.

2.2.2. Measurement Geometries

The most common geometries of commercial rheometers are parallel plate, cone and plate, and concentric cylinder. Others are available, or can be made, but they have rather more specialized uses. A detailed description of all geometries, together with theoretical considerations, can be found in [9], [43], [44].

2.2.2.1. Parallel Plates

The parallel plate rheometer is a logical development of the parallel plate simple shear geometry (see Section 1.2.3). Instead of a sliding motion, however, the plates are replaced by disks a fixed distance apart, which can rotate about a central axis. This means that both the strain and the shear rate vary across the radius of the disk, from zero at the center to a maximum at the disk edge. This renders parallel plates unsuitable for viscosity–shear rate measurement (unless the liquid is Newtonian). However, the geometry is highly suitable for dynamic (oscillatory) measurements. Because the linear response region is generally used, the strain amplitude variation across the radius is immaterial. The advantages of parallel plates are:

1) Correction for fluid inertia effects is relatively easy (this is important for combinations of low-viscosity liquids and high-frequency measurement)
2) The gap can be varied freely

Fluid inertia effects can be significant, since they depend on the magnitude of the product αh, where h is the plate separation and α is given by the equation

$$\alpha^2 = -\frac{\mathrm{i}\,\omega\varrho}{\eta^*} \tag{1}$$

where ω is the angular frequency (2π times the frequency in Hertz), ϱ the liquid density, η^* the complex dynamic shear viscosity, and $\mathrm{i}^2 = -1$. Clearly αh can be reduced by decreasing the plate distance h, but this procedure has limitations. First, when working with dispersion systems, a value of h more than 50 times the average disperse particle diameter should be used to avoid errors due to wall effects [45]. Second, the narrower the gap, the greater is the likelihood of error due to edge effects, which result from the sample drying at the edge. Examples showing the magnitude of fluid inertia effects are found in [46].

Errors due to *wall effects* arise from the presence of a concentration gradient in particles in dispersion systems (sometimes referred to as the depletion layer). This occurs because the wall exerts a repulsive force (hydrodynamic, not colloidal in origin) on the particle as it approaches. *Loss of solvent* by evaporation during the test is always a problem, because it leads to the buildup with time of concentrated viscous or solid (skin) layers at the sample–air interface. This buildup leads to enhanced torque readings and hence measurement errors. Evaporation can be slowed by coating the sample surface with a layer of immiscible and impermeable liquid; however, the best solution is to saturate the atmosphere around the sample surface with solvent vapor.

2.2.2.2. Cone and Plate (Fig. 6, p. 283)

To eliminate variation of shear strain or strain rate across the radius of the parallel plate, one of the disks is replaced with a cone (see Section 1.2.3), thus ensuring that the shear rate remains constant. In addition, a small cone angle is required to achieve this [47]. A small part of the cone center is truncated (see Fig. 6), so that it has no direct contact with the plate. Direct contact would result in the risk of friction and, hence, errors in torque measurement and mechanical wear. The smaller the cone angle (measurement at the rim), the greater is the shear rate for a given rotational speed. Cone angles normally vary between 0.5 and 5°.

In addition to the source of error already mentioned under parallel plates, another is due to *errors in gap setting*. To achieve a uniform shear rate, the gap between the cone and the plate must be adjusted carefully so that the cone tip (afterwards removed by truncation) just touches the center of the plate. Since the gap settings are typically between 15 and 60 µm and many commercial rheometers are equipped with a micrometer, dial gauge, etc., that cannot be set to better than 1 µm, errors can be of the order of 2–6%. If programmed temperature tests are carried out on the same samples, errors due to gap setting variation from thermal expansion can be considerable. *Wall effects* can cause serious errors, particularly with concentrated dispersions [48]. *Fluid inertia effects* are particularly difficult to correct, because the α term now appears in the solution of a partial differential equation instead of the ordinary differential equation required for describing responses in parallel plate or concentric cyclinder geometry. A series solution for this equation have been given with Bessel functions and associated Legendre functions in the product αr, where α is as defined above and r is the cone radius [49]. These equations are so complicated that apparently no one has utilized them, even with the aid of computers.

The cone and plate geometry is the obvious choice for viscosity–shear rate measurements. It is not suitable for oscillatory measurements unless fluid inertia corrections can be neglected.

2.2.2.3. Concentric Cylinder (Cup and Bob, Couette)

The concentric cylinder geometry possesses two main advantages over the cone and plate and the parallel plate geometries: (1) a larger shearing surface, and therefore, greater sensitivity when measuring low-viscosity materials; (2) minimization of errors due to drying or skinning at the sample surface–air interface. This last point is not immediately obvious. In normal sample filling, the sample level should be above the top of the bob but below that of the cup. This ensures that most of the torque produced by the sample originates in its bulk as the sample is sheared in the narrow gap between the outer wall of the bob and the inner wall of the cup. Since the shaft of the bob normally has a much smaller diameter than the bob itself, the torque contribution from any thickening with time on the sample surface is minimal. Evaporation from the sample surface tends to be into a stagnant air layer if the surface level is below the top of the cup, and under such conditions a layer of immiscible, impermeable liquid can easily be placed on the sample surface to prevent evaporation loss. By arranging the system so that the cup and bob radii are large, and the gap between the bob and cup inner wall is small, the shear situation for any small volume element of the sample becomes approximately that of simple shear with uniform shear rate across the gap. In addition, the generally larger sample volumes reduce errors due to heating at high shear rates, and confinement of the sample at high shear rates helps prevent sample loss by centrifugal force effects.

However, the concentric geometry has a major drawback: the corrections required because of end effects (i.e., the fact that some of the torque contribution from the sample is due to the end of the bob, where the shear regime is ill-defined). For example, errors as high as 25% were found with Newtonian standard oils, using a conventional narrow gap cone ended bob [50]. Other options commercially available are bobs with hollow ends — the idea being to trap an air bubble under the bob. Thus the air pocket reduces the torque contribution from the bob end. However, due to problems from the wetting effects, of shear on the bubble and the rim on the end of the bob, the end effect may still be finite but also variable. Other means of minimizing or eliminating end effects are discussed in [51].

Alternatively, end effects may be *calibrated* out by measuring the torque–angular rotation speed ratio for various heights of sample in the cup. By extrapolating the torque to zero, the extra bob height equivalent to the end effect can be determined and allowed for in viscosity calculations. Within experimental error, these end effect corrections are independent of rotation speed. Corrections must be done for each sample; the calibration for a Newtonian liquid cannot be used [51].

Although end effects appear not to be significant for linear viscoelastic materials in dynamic measurements [52], the effects of *fluid inertia* may be significant. Again, the mathematical treatment of these effects, although easier than in cone and plate geometry, is still more difficult than with parallel plate geometry. A method of approximation using a series to obtain the modulus value is given in [53] and the whole topic is discussed in [54].

2.2.2.4. Other Geometries

Other geometries include double cone systems (one cone fitting into another wider-angle cone containing the sample), various squeezing or sliding flow instruments, instruments with noncoaxial or spherical geometries, and various penetrometers. These are all described in [2], [3], [9].

Whereas most of these have limited general applicability, some are commercially available. For example, the *double cone system* is claimed to be useful for elastic samples, which may fracture when sheared in conventional geometries. *Squeezing and sliding flow instruments* are of interest for very viscous materials (e.g., printing inks), while many types of *penetrometer* (instruments, in which depth–time profiles are measured for a suddenly released loaded needle or other probe) are used in industry for measuring solid and semisolid materials, such as lubricating greases, soaps, gels, foods, or paints.

The *noncoaxial geometries* were potentially of great interest in that they allowed dynamic data to be derived from continuous rather than oscillatory motion. However, they do not provide any significant advantage over conventional rheometers in terms of either the accessible frequency range or the amelioration of sources of error [55].

2.3. Capillary (Poiseuille) Shear Flow

Two main type of materials are tested by capillary rheometers: low-viscosity liquids, where the driving pressure is provided by gravitational force (hydrostatic head type), and high-viscosity liquids such as polymer melts, where the driving pressure is provided by gas or hydraulic pressure. All capillary rheometers are easily temperature controlled. Often when using metal capillaries, good control at elevated temperature is easier than in torsional rheometers. Glass U-tube-type viscometers used for capillary measurements on low-viscosity liquids can be immersed in a temperature-controlled water bath for good control of temperature. Besides being more easily thermostated, capillary viscometers are generally simpler, cheaper, and more accurate than torsional viscometers.

During the flow of a liquid through a capillary, three regions of flow can be distinguished: entry flow, viscometric flow, and exit flow. The *entry flow region* represents the "funneling" effect of the liquid as it flows from a relatively wide-diameter reservoir into the capillary, and this effect takes some time to relax to its limiting value. Estimates of the length of the entry region vary, it is generally accepted as about 40 times the capillary diameter. To achieve consistent results, the reservoir diameter should be at least 15 times the capillary diameter [56]. In the *viscometric region*, the stresses produced by flow along the tube axis are constant. Clearly, pressure measurements should be made in this region. The *exit region* is induced by the exit of the free liquid stream from the capillary end. It is usually shorter than the entry region. The free stream may have a larger diameter than that of the capillary (die swell) and is often associated with elasticity in the liquid. Experimentally, the length of the viscometric region is maximized by using a sufficiently long capillary tube for the experiment.

The *experimental procedure* consists of measuring the volume flow rate of the liquid and the total pressure drop between the ends of the capillary tube. If the sample is an unknown material, a number of capillaries of different length and interior diameter may be required to establish the flow curve of the liquid and to make suitable correction for end effects (both entry and exit) and other sources of error.

The greatest problem with *interpreting the results* of capillary flow measurements is calculation of the velocity profile across the diameter of the tube. For a Newtonian fluid, the profile can be predicted to be parabolic; for non-Newtonian fluids the profiles can be calculated only if the viscosity—shear rate relationship is known. This can be done for simple rheological models, such as power law or Bingham fluids (→5. Fluid Mechanics, **B1**, pp. **5**-25–**5**-27). However, other models give intractable or complicated mathematical formulas. Although capillary measurement methods are commonly used for non-Newtonian liquids, the results are often presented as flow curves (i.e., pressure drop–flow volume rate plots). Methods do exist for deriving viscosity–shear rate values from flow curve data, but these are valid, in general, only for fluids in which the departure from Newtonian behavior is small. Nevertheless, despite these problems, capillary flow measurements are attractive not only because of their simplicity and cheapness in combination with accuracy, but also because of their ease of application to very high shear rate measurements. With respect to errors and repeatability, the results of capillary measurements on Newtonian liquids are an order of magnitude better than corresponding results on the best commercial torsional instruments [50]. Furthermore, most torsional rheometers are not capable of measuring at shear rates much above 1000 s^{-1} because of the various errors involved. Although the precise limit depends on the geometry and the sample involved, the essential problem is the low sample volume with small gaps and high rate of energy dissipation at high shear rates, with consequent temperature control problems.

In the case of *hydrostatic-type viscometers* such as the Ostwald U-tube viscometer, the driving pressure is set by filling a reservoir with sample and timing, between two graduation lines, the emptying of the reservoir through the capillary tube. Such capillary instruments, usually made of glass, are low-stress instruments (1–15 Pa) and have maximum shear rate values of ca. 1–20 000 s^{-1} based on efflux times of 200–800 s [57]. Their main use is as secondary standards for viscosity measurements [58], [59], but they are important in establishing the molar mass of polymers or polymer fractions (from gel permeation chromatography separations, for example) by measuring the low shear rate (Newtonian) viscosity of dilute solutions (→Plastics, Analysis, **A 20**, pp. 527–528). Another important industrial application is the calculation of liquid flow in pipes.

Finally, *efflux* or *flow cups* are widely used in industry. Basically, these are hydrostatic head instruments, consisting of cylinders with a conical bottom, pierced at the center by a short capillary whose length does not usually exceed the cup wall thickness. They may be operated at elevated (e.g., the Saybolt and Redwood cups used in the petroleum industry) or ambient (e.g., the Ford or BS cups used in the paint industry) temperature. The cup is filled to a mark with the capillary plugged. The plug is then removed, and the time required to fill a graduated flask placed underneath to a specified volume, or until the fluid stream first breaks into drops is measured. Although these methods give reasonable results for quality control or diluting to a given viscosity (e.g., for spray application of paint), they are not very sound in concept. First, they are totally unsuitable for non-Newtonian liquids because the gravitational driving stress varies considerably due to the large drop in liquid level (ca. 0.06 m) and also the capillary length is so short that viscometric flow is never established.

2.4. Extensional Flow

Extensional flow instruments can be classified according to NAZEM and HILL [60]. Controllable experiments refer to those in which the stress or extensional strain rate is kept constant and the steady value of the other parameters is measured. In practice, controllable experiments are applicable only to high-viscosity materials, such as polymer melts. Noncontrollable techniques must be applied to low-viscosity liquids.

Polymers. Polymer specimens are commonly *stretched between two blocks*, one fixed and one movable. This places a constraint on the maximum strain, and hence strain rate, that can be attained because of the finite length of the equipment and the requirement allowing transient stresses to die away to steady levels. By suitable electronic or mechanical means, a programmed force or length increase can be applied so as to maintain constant stress [61] or constant extensional strain rate [62]–[64]. An improved strain rate controlled instrument has been described that has two sets of rollers at either end of the specimen [65]. This allows a maximum rate of ca. 1 s^{-1}, which is nearly an order of magnitude higher than the other instruments available. Examples of these instruments are commercialy available.

In *biaxial extension*, sheets of polymeric material can be tested either by bubble inflation or by the apparatus developed by MEISSNER [66] in which an octagonal clamping system is driven by a complex servomotor system.

Mobile Liquids. For mobile liquids, steady-state measurements in both the Eulerian and the Lagrangian senses (Section 2.1.2) are impossible with the present state of knowledge [67]. Therefore, noncontrollable methods are applied. Of these, the laboratory equivalent of fiber spinning equipment is commercially available and widely used [68]–[73]. The liquid being tested is extruded through a die and collected on a turning spool. Both the diameter of the threadline as a function of position and the flow rate are measured, and the extensional stress is calculated from the force on either the die or the turning spool. The limitation of this method is probably the cohesive energy of the liquid and whether, in practice, a coherent stable thread can be drawn from the liquid.

Alternative methods applicable to more mobile liquids are the open siphon method [74]–[77], the converging flow method [36], [78]–[84], the triple jet method [38], and the more recent (commercially available) opposing jets method. In the *open siphon* or *Fano flow method*, the sample is sucked into a siphon until a stable column is formed [75]. This has an advantage over methods involving dies, because the liquid has not suffered any (perhaps unknown) deformation before being stretched [73]. *Converging flow techniques* are simple to use in that liquid flows from a reservoir through an orifice into a capillary. However, analysis of the measurements is difficult, and measurement requires that the liquid be transparent. In the *triple jet technique*, two high-velocity jets converge on a central lower-velocity jet of the same liquid, thus causing the central jet to be stretched. Finally, in the *opposing jets technique*, two jets are mounted horizontally: one on a fixed vertical arm and one on a pivot equipped with a transducer. The jets are immersed in the liquid under test, and liquid is sucked in or blown out through the jets at a controlled flow rate by two syringes also filled with test liquid. The flow rate is controlled by a constant-speed stepping motor, which drives a moving block that pulls or pushes the plungers of the two syringes equally. By measuring the deflection of the pivoted jet and applying a measured but variable force to maintain its position, the extensional stress is obtained. Despite the limitation imposed by the maximum travel distance of the plungers in the syringes, extensional viscosities can be measured at strain rates up to 50 000 s^{-1} or more.

2.5. Measurement Procedures

Three basic measurement procedures are described — steady state, dynamic (oscillatory), and transient — together with a statement of the equation for calculating stress, strain, or strain rate for each of the main measurement geometrics (see Table 1 and Table 2). Because both extensional deformation and capillary shear deformation are carried out only under steady-state conditions, the bulk of this section is devoted to torsional measurements. Extensional flow equations tend to depend on the apparatus and technique used. Detailed discussion of the equations along with their theoretical derivation, is given in [2] and [7].

Table 1. Equations for flow shear stress, and shear rate for different measuring geometries

Measuring geometry	Shear stress		Shear rate	
Parallel plate				
(plate radius R; separation h; distance from center of plate r; angular velocity Ω)	total torque G $G = \int_0^R 2\pi r^2 \sigma \, \mathrm{d}r$ for Newtonian liquid	(2)	$\dot\gamma = \frac{\Omega r}{h}$	(3)
	$\eta = \frac{2hG}{\pi \Omega R^4}$	(4)		
Cone and plate				
[symbols as for parallel plate: θ is angle between cone and plate surface (radian)]	$\sigma = \frac{G}{2\pi R^3}$	(5)	$\dot\gamma = \Omega/\theta$	(6)
Concentric cylinder				
(cup radius R; bob radius r; bob height h)	$\sigma = \frac{G}{2\pi R^2 h}$ where $R = R$ or r*	(7)	$\dot\gamma = \frac{r\Omega}{R - r}$	(8)
Capillary flow				
(tube radius a; length l; pressure difference between ends P; volume flow per second V)	$\sigma_w = \frac{aP}{2l}$ Newtonian liquids only	(9)	$\dot\gamma_w = \frac{4V}{\pi a^3}$	(10)
subscript w indicates stress at tube wall	in general, $V = -\frac{\pi a^3}{\sigma^3} \int_0^{\sigma_w} \sigma^2 f(\sigma) \, \mathrm{d}\sigma$ and $\eta = \frac{\sigma_w}{\dot\gamma_w}$			(11)

* If $r/R = 0.98$, the two stress values agree within 4% [85].

2.5.1. Steady-State Measurements

Steady-state conditions are achieved by choosing a value of stress or shear rate to apply to the sample and waiting sufficiently long for the measured shear rate or stress value to become constant before proceeding to the next value. Generally, one starts with low applied values and proceeds stepwise to the maximum required value. The sequence is then reversed to check that shearing has not brought about any irreverisble change in the sample. This is necessary for torsional shear when the same sample volume is used for all measurements. When fresh sample is supplied continuously to the measuring region (e.g., in capillary flow), it is not necessary unless the bulk sample is suspected of changing with time, due to sedimentation of the dense dispersion phase, slow phase separation, etc.

In practice, the more solid-like the material, or the more viscous the liquid, or the closer the material is to phase separation or changes, then the longer are the times required to attain steady state. For example, some structured emulsion paints may require 5 min or more to reach a measured value constant to within 10% over 30 s at a moderate applied stress (100 Pa) [85]. In these cases, thixotropic behavior may have to be accepted for reasons of expediency, i.e., the time required to complete the entire measurement procedure or irreversible changes in the sample over time (e.g., evaporation losses, sedimentation), and the accuracy requirements made less stringent.

The more fluid the material, the more likely is the occurrence of pseudoplastic or Newtonian behavior. In this case, the time to complete the entire measurement may be considerably reduced by using a continuous sweep up and down instead of the stepwise procedure.

2.5.2. Dynamic (Oscillatory) Measurements

Whereas the objective with steady shear measurement is to measure the viscosity over a range of imposed stresses or shear rates, the objective of dynamic measurements is to study the sample as near rest as possible. Shear moduli values are therefore usually measured as a function of applied shear stress or strain at a fixed frequency to define the linear response region (i.e., the range of amplitude values in which the moduli are constant). In practice, the linearity-limiting amplitude does not vary very much with frequency. However, some structured materials have a lin-

Table 2. Equations for oscillatory flow*

Measuring geometry	Without fluid inertia corrections	With fluid inertia corrections
Parallel plate	$\frac{e^{i\delta}}{\theta} = 1 - \frac{i}{\eta^*} S$ (12) with $S = \frac{2h(K - I\omega^2)}{\pi a^4 \omega}$	$\frac{e^{i\delta}}{\theta} = \cos \alpha h + S' \alpha h \sin \alpha h$ (13) with $S' = \frac{2(K - I\omega^2)}{\pi \varrho a^4 h \omega^2}$
Concentric cylinder	as parallel plate, with $S = \frac{(R^2 - r^2)(K - I\omega^2)}{4\pi L R^2 r^2 \omega}$ (14)	$\frac{e^{i\delta}}{\theta} = \frac{\pi r^2}{2R} X$ $\{\alpha[J_2(\alpha r) Y_1(\alpha R) - Y_2(\alpha r) J_1(\alpha R)] + \alpha^2 S'' X$ $[J_1(\alpha r) Y_1(\alpha R) - Y_1(\alpha r) J_1(\alpha R)]\}$ (15) with $S'' = \frac{K - I\omega^2}{2\pi\varrho r^3 L\omega}$
Cone and plate	as parallel plate, with $S = \frac{3\alpha(K - I\omega^2)}{2\pi r^3 \omega}$ (16)	see [49]

* For explanation of symbols, see Section 2.5.2.

ear response region at very low applied stress or strain amplitude values, so compromises must sometimes be made between the demands of linearity and the ability of the rheometer to make accurate measurements at low amplitudes.

Few workers have studied the *nonlinear dynamic viscoelastic* properties of materials. Not only are the measurements more complicated and the data more difficult to analyze, theoretical development is still rudimentary, at least in its application to rheological properties. In this respect the work of ONOGI and his group should be mentioned [86].

The formulas in Table 2 are applicable to controlled-strain amplitude rheometers only. Those for controlled-stress instruments are similar, but differ in detail.

The equations show the relationship between the measured parameters and the complex dynamic shear viscosity η^*. The measured parameters are the phase angle δ and the stress amplitude θ_0. The ratio of the stress amplitude θ_0 to the strain amplitude input θ_1 is denoted by θ [87].

In Table 2, K and I are the torque constant and the moment of inertia about the axis of the torque-measuring element of the rheometer, respectively; $\omega = 2\pi f$ is the angular frequency (radians per second); and f denotes the measuring frequency (hertz). In the fluid inertia correction term $\alpha^2 = i\omega\varrho/\eta^*$, where ϱ denotes the density of the sample and $i^2 = -1$. For parallel plates, h denotes plate separation and a the plate radius; for concentric cylinders, R_0 and R_1 denote the radii of the outside (internal cup radius) and the inside cylinders (bob), respectively; L is the bob length; and J and Y are Bessel functions of the first and second kind. Finally, for the cone and plate geometry, α_0 and r denote, respectively, the angle between the cone and the plate and cone radius.

Resolving Equation (12)

$$\frac{e^{i\delta}}{\theta} = 1 - \frac{i}{\eta^*} S \qquad (12)$$

for the elasticity modulus components yields

$$G' = \frac{\omega S \theta (\cos\delta - \theta)}{\theta^2 - 2\theta\cos\delta + 1} \qquad (17\,a)$$

$$G'' = \frac{-\omega S \theta \sin\delta}{\theta^2 - 2\theta\cos\delta + 1} \qquad (17\,b)$$

The corresponding equations for the controlled stress rheometer are

$$G' = C(\theta\cos\delta - 1) + \frac{C I \omega^2}{K} \qquad (18\,a)$$

$$G'' = C\theta\sin\delta \qquad (18\,b)$$

$$C = \frac{\pi a^4}{2hK}$$

The equation taking account of fluid inertia effects is also simpler for controlled-stress instruments [85]

$$\frac{e^{i\delta}}{\theta} = \frac{K - I\omega^2}{K} + iC\alpha h\eta^*\omega\cot(\alpha h) \qquad (19)$$

Besides fluid inertia effects, other sources of error can occur in dynamic measurements due to natural frequency effects and sample stiffness. The sample stiffness is, however, rarely important. Correction methods have been described [88], but the situation should be avoided by ensuring that θ is always small compared to unity.

The *natural frequency* ω_0 of the measuring system is given by $\omega_0^2 = K/I$. At this frequency, $\theta = 1$ and $\delta = 0°$ for controlled-strain experiments and the values of G' and G'' become indeterminate (Eqs. 17, 18). However, for controlled-stress instruments, $\delta = 90°$ and, again, the G' and G'' values are indeterminate. Although accurate values can be measured on either side of the natural frequency, at higher frequencies the accuracy of measurement becomes more limited as frequency increases beyond the natural frequency value. This is due to the increasing dominance of the inertia term ($I\omega^2$) on the right-hand side of Equations (17) and (18) [89]. A better procedure is to shift the value of the natural frequency by changing either K (by using, for example, a stiffer torque bar) or I (by altering the mass of the measuring geometry—e.g., substituting a glass or plastic cone for a steel cone or using a smaller-diameter cone). By this means, the natural frequency can be shifted out of the required frequency range.

In addition to the forced oscillation (i.e., where an oscillation frequency is imposed on the sample–measuring system combination), the system may be in so-called *free oscillation*. However, this technique should properly be classified as a transient method (see Section 2.5.3).

2.5.3. Transient Methods

In the free oscillation method, the sample–measuring system is deflected (usually in torsion), then released and allowed to oscillate until damping by the sample stops the oscillation (→Plastics, Properties and Testing, **A 21**, pp. 16–17). The oscillation frequency and the amplitude decrease (decrement) per cycle are measured. Because the oscillation frequency can often be arranged by judicious choice of mass, etc., to be in the 0.1–2-Hz range, measurements can be made rapidly. Its speed of measurement and inherent simplicity make this an attractive technique for studying rapid changes in rheological properties during, for example, resin curing or polymer cross-linking. Using a woven braid to support the initially low-viscosity fluid, GILLAM and coworkers have made extensive use of this technique [90]. The other primary transient methods are creep and stress relaxation (→Plastics, Properties and Testing, **A 21**, pp. 14–16). In *creep measurement*, the sample undergoes a stepwise increase in stress, and the resulting strain is measured as a function of time. The stress is then removed, and the decay of strain with time is followed. In the *stress relaxation experiment*, a stepwise strain is applied to the sample, and the resulting stress is measured as a function of time until the stress either decays to zero or becomes constant.

When using, for example, parallel plates in a torsional shear stress relaxation test, the modulus of elasticity $G(t)$, defined as the stress $\sigma(t)$ at time t divided by the applied strain γ, is given by

$$G(t) = \frac{2hT}{\pi R^4 \varphi} \tag{20}$$

where T is the measured torque at time t and φ is the angular deflection (radians). In general, the equations in Table 1 may be used to calculate stress, and by substituting γ for $\dot{\gamma}$ and φ for Ω the strains can be found. In creep measurements the compliance $J(t)$ is often calculated.

The same considerations still apply in choosing measuring geometries, and cone and plate geometry is probably best wherever possible. In addition, the rise time (i.e., the time to reach the chosen stress or strain value from an initial zero value) must be taken into account. Ideally, the sample's relaxation times should be longer than the rise time. The application of sudden deflections to a torque bar can lead to initial free oscillations that can swamp sample response in the initial few seconds of a stress relaxation experiment. The magnitude of these oscillations depends on the brevity of the rise time, the strain value applied, and the stiffness of the sample. This last factor can be reduced by increasing the gap between parallel plates, reducing the cone diameter, etc.

2.5.4. Combined Methods

With the aid of computer software, many commercial torsional shear rheometers can provide any required combination or sequence of basic measurement procedures just described. This is particularly useful in two ways. First, many structured materials are very sensitive to stress or strain and have very long *recovery times*. Consequently, the procedure of sampling from the bulk and loading the sample in the rheometer may impose sufficient stress or strain to change its properties significantly (i.e., the sample may suffer unknown deformations before measurement). The long recovery times may mean that even leaving the sample for a long time before measurement is not sufficient for it to return to its original state. This can be circumvented by imposing an initial deformation history, for example, by applying a shear rate until a constant stress value is obtained. The measurement can then be carried out immediately afterwards, and the effect of varying the initial history on the measurements can be studied. The second possibility is that by varying the time interval between the completion of the initial history and the start of measurement, the recovery process can be determined. Particularly useful in measuring recovery processes are repeated oscillatory measurements over a period of time at single frequencies or (better still) superimposed multifrequency waveforms, which are becoming available in many software packages. Such techniques are also useful if the sample is changing with time because of some other effects than relaxation, e.g., chemical cross-linking.

The second way in which such combined methods are useful is to *simulate complex flow problems*. For example, paint applied to a vertical wall is spread by the high stress exerted by a roller or brush; the wet liquid film should then remain unchanged under the influence of low stresses due to gravity. This unchanged state should remain for up to 5 min, after which viscosity increases or the appearance of elasticity—due to the loss of solvent by evaporation (drying) or the onset of curing—prevents further change. If flow does occur under the influence of gravity, it sometimes does not show immediately, but appears after a few tens of seconds as drips, runs from edges, and wavy irregular patterns of uneven thickness. These, of course, spoil the appearance and protective power of the final paint coating. This process may be simulated by applying a high stress for a short time (brush application), followed by a creep measurement made at small applied stress (0.2–5 Pa), which can be varied to simulate variations in wet-film thickness and specific gravity. The creep measurement will then show whether significant flow occurs at a specific stress (film thickness) and within a certain specified time (drying time).

Other examples stem from the tendency of many disperse systems to form *structures or become solid-like* when at rest for long periods. Whereas such structure usually breaks down when a stress or strain is applied to it and the material liquefies, in the design and operation of pumping and mixing equipment, the pressure (stress) or mixing speed (shear rate) required to liquefy the sample and the length of time for which it should be applied, must be known. Rheological measurement procedures can be designed to simulate these conditions, whereas dynamic measurements can be used to probe the nature and strength of the structure at rest.

2.5.5. Normal Stress Measurements

When an elastic liquid is sheared (e.g., a polymer solution or melt) a significant component of the total force is often developed in the liquid, which is in a direction normal to the direction of shear. Thus, in addition to the shear force usually measured, a significant force may be measured by mounting a suitable transducer in the face of the cone or plate of the measurement geometry. These so-called normal forces lead to rod climbing or Weissenberg effects (→5. Fluid Mechanics, **B 1**, pp. **5**-33–**5**-34) in which such liquids form a convex meniscus around a rotating stirrer rod or shaft [91], [92].

Normal force effects occur in all the basic measurement techniques, provided the material possesses sufficient elasticity and is fluid enough to be sheared without sample fracture. Dispersion systems, particularly well-stabilized ones, do not usually possess sufficient elasticity, even when very concentrated, to exhibit measurable normal force values. Methods for measuring normal force in continuous shear have been described in [93]. In dynamic measurements the measured normal force oscillates at twice the applied frequency [94].

3. Results of Rheological Measurements

Recent reviews of rheological results and theory include [10], [12], [14].

3.1. Steady-State Measurements

At low concentrations, both polymer solutions and dispersions follow Einstein's formula and are Newtonian liquids (viscosity independent of shear rate)

$$\eta_r = \frac{\eta_o}{\eta_s} = 1 + 2.5\varphi \qquad (21)$$

where η_o is the viscosity at low shear rate, η_s is the solvent or dispersion medium viscosity, and φ is the total phase volume occupied by the polymer molecules or particles.

The number of interactions increases with increasing particle or molecular concentration, either by interaction through collision or by modification of the flow fields around the particles and the mutual interaction of these fields. These interactions increase the amount of energy dissipation per unit time during shearing, and consequently, the overall viscosity of the dispersion or solution also increases. Finally, at ca. 20% volume concentration, the number of interactions becomes sufficiently large that the effect of strong shearing is to reduce their number and intensity by changing a random, three-dimensional fluctuating network of interactions into a less random, two-dimensional fluctuating network. Such reduction in interactions results in less energy dissipation and, consequently, a lower viscosity for the whole system. This leads to the appearance of two Newtonian viscosities: a high value corresponding to the three-dimensional random situation at low shear rate (lower Newtonian plateau) and a lower value corresponding to the more ordered situation at high shear rate (upper Newtonian plateau). At intermediate shear rates, the viscosity is shear rate dependent. At these intermediate values, a balance is struck between the shear rate trying to order the system and the thermal motion of the solvent molecules trying to disorder the particles or solute molecules. This balance may be achieved rapidly. If the interactions are relatively strong, the rate at which the balance is reached may be of the same order or greater than the rate of change of shear rate (if the shear rate is constant for a certain experimental time, then the time to reach balance may exceed the experimental time). This has two consequences:

1) The flow behavior shows time dependence (i.e., it is thixotropic rather than pseudoplastic)
2) Because the associative interactions lead to a higher effective concentration or a higher phase volume than that of the actual concentration, the onset of shear rate dependence of the viscosity is observed at much lower concentration than 20 % phase volume

Shear rate dependences are shown diagrammatically for dispersions in Figure 12 and for polymer solutions in Figure 18. Well-stabilized dispersions usually show a stronger dependence on shear rate of viscosity than polymer solutions, as indicated.

As the concentration increases, the increasing number of multiple interactions leads to greater increases of viscosity for each equal increment of concentration. Both the upper and the lower Newtonian values increases sharply, but the difference between them also increases (i.e., the shear rate dependence becomes stronger). Finally, a critical volume fraction is reached at which the particles of the dispersion or the solute molecules have so little free volume

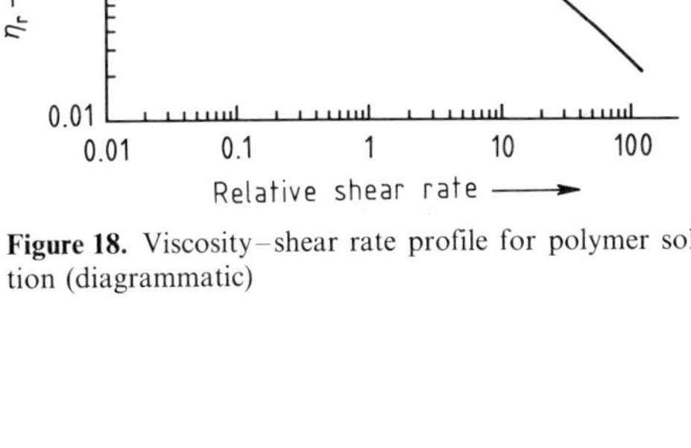

Figure 18. Viscosity–shear rate profile for polymer solution (diagrammatic)

that they are in contact with each other constantly, thus forming a network throughout the volume of the liquid. At this point, the viscosity of the liquid becomes infinite and the liquid exhibits solid-like behavior. Finally, if the volume fraction becomes higher, even these rearrangements under continuous shearing will be impossible, and the material becomes entirely solid-like. These changes normally take place over an effective phase volume fraction of 0.50–0.72.

Differences occur between different systems in this critical volume fraction range. Thus, well-stabilized *hard particle dispersions* exhibit very abrupt changes in viscosity over a narrow concentration range, whereas the changes in *polymer solutions* are far more gradual, due to changes in the polymer molecular volume as the concentration increases. In dilute polymer solutions in a thermodynamically good solvent, the macromolecular coils are fully expanded. As the polymer concentration increases, polymer–polymer interactions becomes stronger than polymer–solvent forces. As a consequence, less solvent is "trapped" within the polymer coil, and the coil contracts in volume until it reaches so-called unperturbed dimensions [95]. Beyond this minimum volume, coils can still overlap or entangle as concentration increases further. Thus the volume shrinkage leads to an initial small change with concentration (dilute solution), then a more rapid change over a fairly wide concentration region where the swollen coils come closer to each other and contract to the unperturbed dimensions (semidilute solution), and then finally a steep viscosity increase with concentration due to overlap (concentrated solution). This last increase is not usually as strong as in concentrated dispersions, due to coil flexibility, and continuous measurements are possible up to 100 % polymer content. Because hard particles are not flexible, continuous shear measurements on concentrated dispersions are not possible beyond a certain critical maximum volume packing fraction, which seems to depend on the nature of the stabilization often around that corresponding to random packing (0.639) for sterically stabilized [96], and higher (0.74) for charge-stabilized dispersion [97], [98].

Some *bimodal particle dispersions* can be fluid up to volume fractions of 0.85 or slightly more, if the relative particle sizes and the concentration of each size are chosen carefully [99]. Apparently, when the large particles reach the critical packing fraction, the smaller particles can fit in the interstices betwen them. Emulsions can also reach similar phase volumes before the overall viscosity becomes very high [100], because the droplets can undergo distortion (due to the flexibility of the interface and the fluidity of their interiors) and pack in hexagonal cross-sectional structures. The continuous phase is then confined to the interfacial surfaces and small pockets where more than two interfacial surfaces meet. The maximum phase volume that can be reached usually depends on the onset of phase inversion. Similar considerations apply to foams (gas-in-liquid dispersions), but in practice, because of the much greater overall fragility of the structure compared to emulsions, they are usually unable to sustain much shear.

Polymer melts exhibit a similar transition in viscosity behavior with (average) molar mass. Thus at low molar mass, the macromolecular coils are not long enough to overlap or entangle and are in their unperturbed dimensions. As molar mass increases, the (low shear rate) viscosity becomes dependent linearly on the molar mass. Beyond a certain critical molar mass, overlap or entanglement commences and the viscosity becomes dependent on molar mass to the (average) power of 3.4. A more or less pronounced transition region may occur, due to the effects of molar mass distribution [101].

Additional effects may occur in *charge-stabilized dispersions,* due to the presence of charged ions of opposite sign to those on the particle that are attached to the particle surface, by either chemical bonds or adsorption of charged stabilizers (surfactants or polymer molecules). Thus a charge double layer is formed at the surface of the particle. The thickness of this layer depends on the electrolyte concentration in the dispersion medium; indeed, if the medium is carefully deionized by dialysis or similar techniques, the double layers can extend so far that even dispersions as dilute as 2% phase volume can show solid-like characteristics and, if the particle size is uniform, quasi-crystalline ordering [102], [103].

Anisotropic particles often exhibit effective phase volumes different from their geometric volume. Such particles may also show specific effects due to interactions when oriented in velocity gradients, for example (rheopectic flow). *Micellar systems* often exhibit effects intermediate between emulsions and dispersions as concentration increases. *Liquid-crystal systems* share some of these properties, but their extreme anisotropy leads to many interesting and complex rheological phenomena [104]. The rheological properties of emulsions, micellar, and liquid-crystal systems are often extremely temperature sensitive.

Finally, net attractive forces between particles in dispersion or between particles and polymers dissolved in the medium can lead to the formation of complex space-filling structures (*flocculates*), whose effective phase volume may vastly exceed the phase volume of the particles alone. If such structures form a network throughout the volume of the liquid, it behaves like a solid at rest or low shear rates (infinite viscosity). If such structures are separated by a reasonable volume of medium, flow will occur but the viscosity will be much higher than that expected on the basis of the dispersion phase volume alone. However, unless the attractive forces involved are very strong or covalent bonding is involved (e.g., in a polymer gel), these structures will be broken down readily by shearing, and continuous shear rate measurements will indicate pseudoplastic or (more likely) thixotropic flow.

3.2. Dynamic (Oscillatory) Measurements

For a Newtonian liquid, the loss component G'' of the complex dynamic modulus of elasticity G^* shows a plot of slope of unity against frequency on double logarithmic axes, whereas the elasticity component G' is much lower [61], often impossible to measure. This is equivalent to a constant real component of the dynamic shear viscosity η' that is independent of frequency (i.e., Newtonian).

Few liquids exhibit viscoelastic responses that are characterized by single relaxation times. Exceptions are some micellar solutions, particularly a number of cationic surfactants, which in the presence of certain organic anions, form long stiff micelles. These solutions possess very definite viscoelasticity, probably due to the formation of transient networks resulting from random interlocking of the very long stiff micellar rods. A typical example is the 0.1 mol/L 1:1 molar ratio cetyltrimethyammonium bromide–salicylic acid solution (see Fig. 13, Section 1.3.3). Apart from these systems, neither simple supercooled liquids (supercooled to bring the transition region within the frequency range of the dynamic mechanical equipment; dielectric measurements do not have this restriction, but even at ambient temperature, nonassociating liquids usually exhibit a range of relaxation times [105], [20]) nor dispersion systems exhibit such behavior, although some dispersions may have very weak viscoelasticity of this kind over a limited concentration range [34]. Most simple liquids, micellar solutions, colloidal solutions, emulsions, and dispersions have a wider transition region from predominantly viscous to elastic properties. Thus, crossover ($G' = G''$) occurs at a lower frequency than the G'' peak, with the crossover frequency corresponding to the longest characteristic time of the system. Such a range of characteristic times reflects a restricted range of free volume and interactions available to each disperse phase particle or to each liquid molecule in the case of supercooled liquids (see Section 1.3.4).

The range of characteristic times is limited by comparison with a *polymer macromolecular chain*, where the chain can be modeled as a string of beads connected by springs, with each bead dissipating energy by frictional resistance with the surrounding medium (whether it be a liquid, as in polymer solutions, or other macromolecules, as in polymer melts). For flexible chains or high molar masses, the number of beads can be large, and the range of characteristic times and the transition region correspondingly wide. The theories developed on this general model by BUECHE [106], [107], ZIMM [108], ROUSE [109], SLONIMSKY [110], and others show a particular distribution of characteristic times, and predictions based on these theories agree well with experimental data. A typical data set for a concentrated dispersion is shown in Figure 15, and for a polymer solution in Figure 19.

If the particles in a dispersion are *flocculated*, the floccules tend to be linear, highly branched, or fractal structures and then the possibility of wide distributions of characteristic times exists, with the maximum times being very large if structures filling the liquid volume are formed. This process becomes analogous to the changes in dynamic mechanical behavior of a gelling polymer system with reaction time (→Polymerization Processes, **A 21**, pp. 314–317). Thus, initially, the dispersion is viscous or viscoelastic with a wider transition region than Maxwell liquids. As flocculation or gelation proceeds, the transition region widens close to the gelation point. At this point, the region is so wide that both components of the complex modulus (G', G'') show power law dependence on frequency with exponents less than unity. Beyond this point, the strength and number of links in the polymer gel network or particle network increase. Because of this, the exponent of the modulus of elasticity G'–frequency power law relationship decreases toward zero, and G' becomes progressively greater than the loss modulus G'' over the entire frequency range (normal frequency ranges, 0.001–40 Hz, supplied by typical commercial rheometers are assumed). The changes and the viscoelastic properties characteristic of each stage of this process are shown in Figure 20.

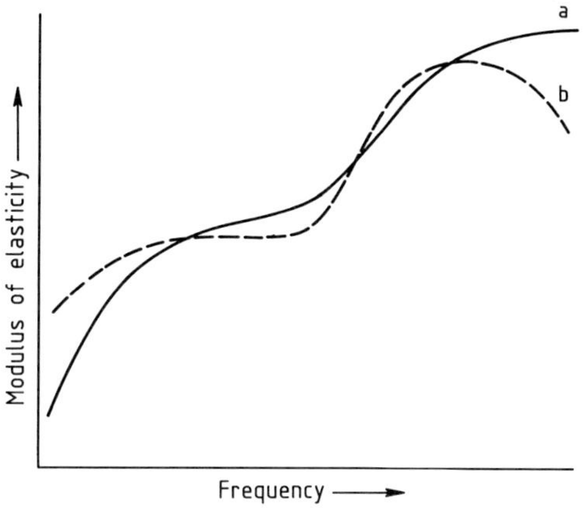

Figure 19. Oscillatory profile for polymer solution (diagrammatic)
a) Dynamic modulus of elasticity G'; b) Loss modulus G''

3.3. Transient Measurements

The two major types of measurements used are constant stress (creep) and constant strain (stress relaxation). Other methods—for example, measurements of stress transients on application of a constant shear rate or of stress relaxation on removal of constant rate of shear—are less frequently employed because of measurement errors and limited applicability. They are used quite often to confirm the predictions of constitutive equations. Of the two main methods, creep often gives more obvious indications of structural change.

Even allowing for instrumental errors, few *solids* exhibit perfect Hookean behavior, except at low stress or strain or short times. Many solids exhibit elastic behavior, as shown in Figure 21 for a creep experiment. This is typical of many

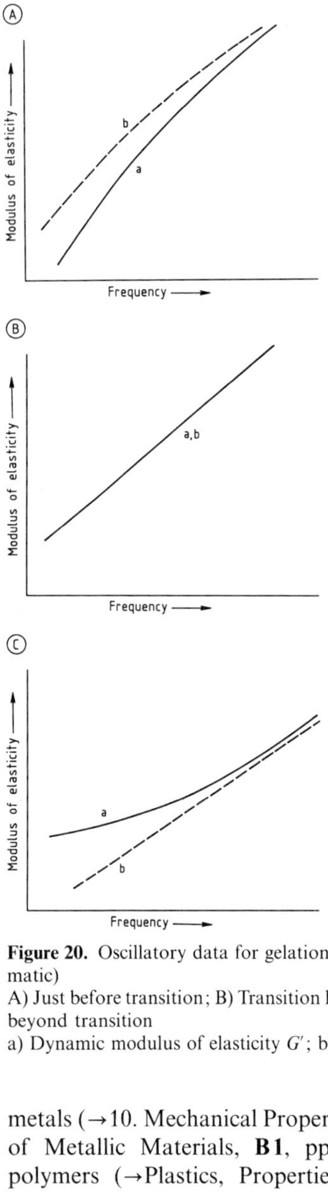

Figure 20. Oscillatory data for gelation process (diagrammatic)
A) Just before transition; B) Transition liquid→gel; C) Just beyond transition
a) Dynamic modulus of elasticity G'; b) Loss modulus G''

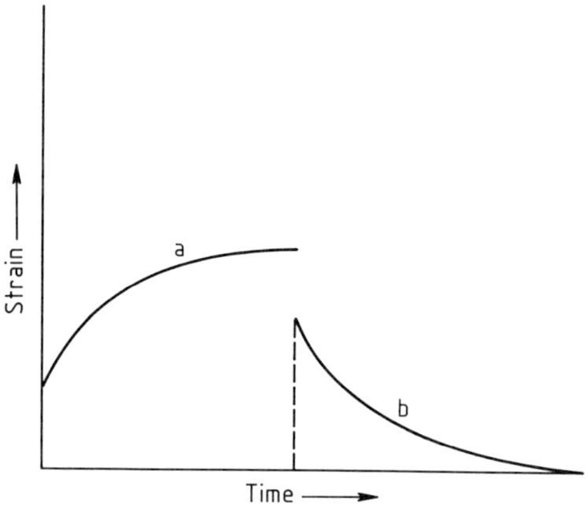

Figure 21. Creep (anelastic behavior)
a) Retardation (stress on); b) Relaxation (stress off)

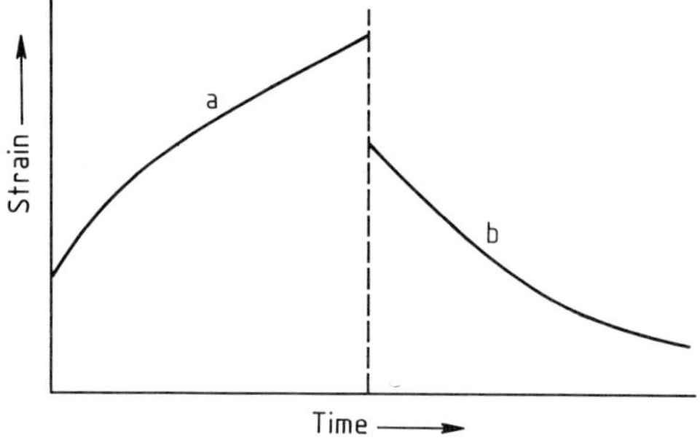

Figure 22. Creep (viscoelastic behavior)
a) Retardation (stress on); b) Relaxation (stress off)

metals (→10. Mechanical Properties and Testing of Metallic Materials, **B1**, pp. **10**-14–**10**-16), polymers (→Plastics, Properties and Testing, **A21**, pp. 14–16), etc. An initial instantaneous deformation is followed by a gradually decreasing deformation, which reaches a constant strain value after sufficient time. When the applied constant stress is removed, an instantaneous reduction in strain value occurs, followed by a slower time-dependent reduction of strain to the initial zero value. No irreversible deformation occurs, and thus no dissipation in the deformation process. Essentially, anelasticity can be defined as Hooke's law behavior extended to include the time dependence of deformation, as first done by ZENER [111]. Two additional constraints are added: (1) at any value of time, the deformation (strain) is proportional to stress; and (2) given sufficient time, the strain will always reach an equilibrium value. Thus, although the modulus or compliance is time dependent, it is not strain or stress dependent.

Structured liquids and *weak solids,* such as paints and swollen gels, exhibit true viscoelastic properties. These are shown in Figure 22, again for creep experiments. Compared to elastic behavior, two essential differences can be seen. First, after a sufficiently long time for stress application, the strain becomes a linear function of time (i.e., some true viscous flow occurs, usually at a very low shear rate, and no equilibrium strain value is reached). Second, when the stress is removed, the strain value tends to an asymptote (i.e., a finite nonzero strain value at long times). So no recovery to zero occurs. However, when these plots are converted to compliance–time plots for different stress values, they may still be identical (i.e., the viscoelasticity is linear in nature). Apparently, in such materials, numerous strong-bonded volumes exist, whose boundaries have weakly bonded interfaces. At low stress, the weak bonds of the interface break, allowing some relative and irreversible movement of the volumes. Many of these interfacial bonds can reform in new positions due to the slow relative motion of the volumes. This ensures some recovery from the deformed state when the stress is removed [111], [112]. However, at higher stress, the response is still viscoelastic, but no longer linear. At still higher stress, total breakdown of structure and rapid flow takes place, as shown by the sharp increase in strain value with time and the slight or zero strain recovery when the stress is removed (Fig. 23).

Similar but less distinct changes can be seen in *stress relaxation plots* as the strain value increases. Thus an initial constant stress value is observed, which then declines either toward zero (flow) or to another constant value (gel). Progressive increase in strain may give modulus–time plots that are identical (linear response), and then the initial stress plateau may decrease or last for a shorter time before disappearing entirely. Finally, a rapid decline to zero stress indicates flow.

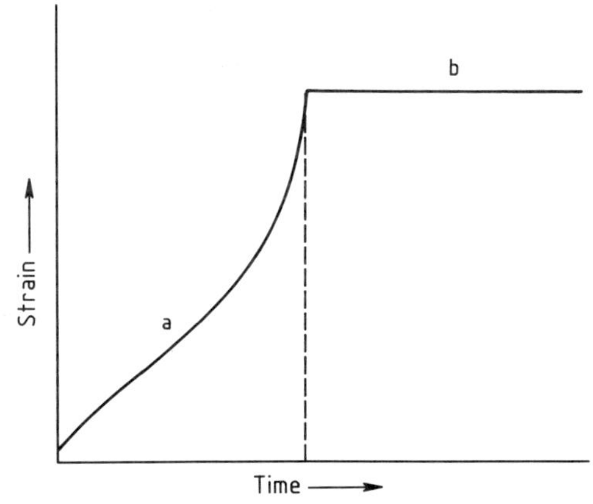

Figure 23. Creep (breakdown to flow)
a) Retardation (stress on); b) Relaxation (stress off)

4. References

General References

[1] H. A. Barnes, J. F. Hutton, K. Walters: *An Introduction to Rheology*, Elsevier, Amsterdam 1989.

[2] K. Walters: *Rheometry*, Chapman & Hall, London 1975.

[3] R. W. Whorlow: *Rheological Techniques*, Ellis Horwood, Chichester 1980.

[4] J. R. van Wazer, J. W. Lyons, K. Y. Kim, R. E. Colwell: *Viscosity and Flow Measurement*, Interscience, New York 1963.

[5] J. D. Ferry: *Viscoelastic Properties of Polymers*, 3rd ed., Wiley-Interscience, New York 1970.

[6] M. Doi, S. F. Edwards: *The Theory of Polymer Dynamics*, Clarendon Press, Oxford 1986.

[7] G. V. Vinogradov, A. Ya. Malkin: *Rheology of Polymers*, Mir Publishers, Springer Verlag, English version 1980.

[8] G. W. Scott-Blair: *A Survey of General & Applied Rheology*, 2nd ed., Pitman, London 1949.

[9] S. Oka in F. R. Eirich (ed.) *Rheology: Theory & Applications*, vol. 3, Academic Press, New York 1960, chap. 2.

[10] J. M. Krieger, W. B. Russel, R. L. Hoffman, Th. F. Tadros in M. S. El-Aasser, R. M. Fitch (eds.): *Future Directions in Polymer Colloids, NATO ASI Ser. E*, no. 138, Martinus Nijhoff, Dordrecht 1987, chap. 6–10.

[11] H. L. Goldsmith, S. G. Mason in F. R. Eirich (ed.): *Rheology: Theory and Applications*, vol. 4, Academic Press, New York 1967, chap. 2.

[12] J. W. Goodwin in: *Specialist Periodical Reports of the Chemical Society (London) Colloid Sci.* **2** (1975) no. 7; *Adv. Colloid Interface Sci.* **42** (1992) 303.

[13] J. Mewis, *J. Non Newtonian Fluid Mech.* **6** (1979) 1.

[14] W. W. Grassley: "The Entanglement Concept in Polymer Rheology," *Adv. Polym. Sci.* **16** (1974); *Comprehensive Polymer Science*, vol. 2, Pergamon Press, Oxford 1989.

Specific References

[15] G. D. Scott-Blair in Ref. [8], pp. 5ff.

[16] D. Halliday, R. Reswick: *Physics*, 3rd ed., Wiley-Interscience, New York 1978, pp. 482–483.

[17] K. Walters in Ref. [2], p. 232.

[18] M. Reiner, *Phys. Today* **17** (1964) 62.

[19] A. B. Metzner, J. L. White, M. M. Denn, *Chem. Eng. Prog.* **62** (1966) 81.

[20] J. Lamb, *Rheol. Acta* **12** (1973) 438.

[21] D. C.-H. Cheng, *Rheol. Acta* **25** (1986) 542.

[22] In Ref. [1], p. 166.

[23] In Ref. [1], p. 159.

[24] D. J. Shaw: *Introduction to Colloid and Surface Chemistry*, 4th ed. Butterworth/Heinemann, Oxford 1992, pp. 210, 220–221, 238–240.

[25] H. L. Goldsmith, S. G. Mason in Ref. [11], sect. V, pp. 161–185.

[26] N. Casson in C. C. Mill (ed.): *Rheology of Disperse Systems*, Pergamon Press, Oxford 1959, pp. 84–104.

[27] British Standard, Glossary of Rheological Terms, BS 5168:1975.

[28] J. M. Dealy, *J. Rheology (N.Y.)* **28** (1984) 181.

[29] In Ref. [1], pp. 159–169.

[30] K. Walters in Ref. [2], p. 233.

[31] T. A. Strivens, *Colloid Polym. Sci.* **267** (1989) 269.

[32] T. Shikata, H. Hirata, T. Kotaka, *Langmuir* **3** (1987) 1081.
[33] K. E. J. Barrett (ed.): *Dispersion Polymerisation in Organic Media,* Wiley-Interscience, London 1975, p. 235.
[34] T. A. Strivens, in press.
[35] F. Mainardi, E. Bonetti, *Prog. Trends in Rheology* (*Suppl. Rheol. Acta*) II (1988) pp. 64–67.
[36] A. P. Metzner, A. B. Metzner, *Rheol. Acta* **9** (1970) 174.
[37] A. B. Metzner, *Rheol. Acta* **10** (1971) 434.
[38] D. R. Oliver, R. Bragg, *Rheol. Acta* **13** (1974) 830.
[39] K. Walters in P. Moldenaers, R. Kennings (eds.): "Theoretical and Applied Rheology," *Proc. Int. Congr. Rheol.*, **11th 1992**, vol. 1, p. 16.
[40] K. Walters in Ref. [2], pp. 210–212.
[41] C. D. Denson, D. L. Crady, *J. Appl. Polym. Sci.* **18** (1974) 1611.
[42] T. A. Strivens in R. Lambourne (ed.): *Paint Technology,* Ellis Horwood, Chichester, 1987, chap. 15.
[43] K. Walters in Ref. [2], pp. 133–160.
[44] R. W. Whorlow in Ref. [3], chap. 3, pp. 113 ff.
[45] T. A. Strivens, *J. Colloid Interface Sci.* **57** (1976) 476.
[46] K. Walters in Ref. [2], Fig. 6.9, 6.10, p. 138.
[47] K. Walters in Ref. [2], p. 47.
[48] T. A. Strivens in Ref. [45].
[49] M. C. Nally, *Br. J. Appl. Phys.* **16** (1965) 1023.
[50] T. A. Strivens, unpublished results.
[51] R. W. Whorlow in Ref. [3], pp. 138–142.
[52] R. W. Whorlow in Ref. [3], p. 255.
[53] H. Markowitz, *J. Appl. Phys.* **23** (1952) 1070.
[54] K. Walters in Ref. [2], pp. 134–136.
[55] K. Walters in Ref. [2], pp. 197–198.
[56] K. Walters in Ref. [2], p. 95.
[57] *Kirk-Othmer,* 3rd ed., **20**, 284.
[58] Methods for the Determination of the Viscosity of Liquids, BS 188:1977 (1988).
[59] ASTM D445.
[60] F. Nazem, C. T. Hill, *Trans. Soc. Rheol.* **18** (1974) 87.
[61] F. N. Cogswell, *Plast. Polym.* **36** (1968) 109.
[62] R. L. Ballman, *Rheol. Acta* **4** (1965) 137.
[63] G. V. Vinogradov, A. I. Leonov, A. N. Prokunin, *Rheol. Acta* **8** (1969) 482.
[64] T. J. Peng, *Jet Propulsions Laboratories Quart, Rev.* **2** (1972) 40.
[65] J. Meissner, *Rheol. Acta* **10** (1971) 230; **8** (1969) 78.
[66] A. Demarmels, J. Meissner, *Rheol. Acta* **24** (1985) 253.
[67] K. Walters in [2], p. 222.
[68] J. Mewis, A. B. Metzer, *J. Fluid. Mech.* **62** (1974) 593.
[69] J. A. Spearof, A. B. Metzner, *Trans. Soc. Rheol.* **16** (1972) 495.
[70] C. A. Moore, J. R. A. Pearson, *Rheol. Acta* **14** (1975) 436.
[71] C. D. Han, L. Segal, *J. Appl. Polym. Sci.* **14** (1970) 2973.
[72] D. Acierno, J. N. Dalton, J. M. Rodriquez, J. L. Little, *J. Appl. Polym. Sci.* **15** (1971) 2395.
[73] N. E. Hudson, J. Ferguson, P. Mackie, *Trans. Soc. Rheol.* **18** (1974) 541.
[74] F. A. Kanel: Ph. D. Thesis, Univ. of Delware, 1972.
[75] G. Astarita, L. Nicodemo, *Chem. Eng. J.* (*Lausanne*) **1** (1970) 57.
[76] G. Fano, *Arch. Fisiol.* **5** (1908) 365.
[77] J. R. A. Pearson, J. T. F. Pickup, *Polymer* **14** (1973) 209.
[78] T. Kizior, F. A. Sayer, *Trans. Soc. Rheol.* **18** (1974) 271.
[79] D. R. Oliver, R. Bragg, *Can. J. Chem. Eng.* **51** (1973) 287.
[80] D. R. Oliver, *Chem. Eng. J.* (*Lausanne*) **6** (1973) 265.
[81] H. Gisekus, *Rheol. Acta* **7** (1968) 127; **8** (1969) 411.
[82] F. N. Cogswell, *Polym. Eng. Sci.* **12** (1972) 64.
[83] A. B. Metzner, E. A. Uebler, C. F. Chan Man Fong, *AIChE J* **15** (1969) 750.
[84] G. Marucci, R. E. Murch, *Ind. Eng. Chem. Fundam.* **9** (1970) 498.
[85] K. Walters in Ref. [2], p. 56; T. A. Strivens, unpublished results.
[86] S. Onogi, T. Matsumoto, *Polym. Eng. Rev.* **1** (1981) 45.
[87] K. Walters in Ref. [2], chap. 6.
[88] K. Walters in [2], pp. 155–160.
[89] T. E. R. Jones, K. Walters, *Rheol. Acta* **10** (1971) 365.
[90] J. B. Enns, J. K. Gilham, *J. Appl. Polym. Sci.* **28** (1983) 2831.
[91] H. A. Barnes, J. F. Hutton, K. Walters in Ref. [1], pp. 60–64.
[92] G. V. Vinogradov, A. Ya. Malkin in Ref. [7], pp. 303–321.
[93] R. W. Whorlow in Ref. [3], pp. 177–187.
[94] G. V. Vinogradov, A. Ya. Malkin in Ref. [7], p. 317.
[95] E. F. Casassa, G. C. Berry in: *Comprehensive Polymer Science,* vol. 2, Pergamon Press, Oxford 1989, chap. 3.
[96] T. A. Strivens, *Colloids Surf.* **18** (1986) 395; *Colloid Polym. Sci.* **261** (1983) 74.
[97] R. Ottewill, *Prog. Colloid Polym. Sci.* **67** (1980) 71.
[98] J. D. Ferry in [5], pp. 519 ff.
[99] C. Parkinson, S. Matsumato, P. Sherman, *J. Colloid Interfac Sci.* **33** (1970) 150.
[100] B. W. Barry, *Adv. Colloid Interfac Sci.* **5** (1975) 37.
[101] T. G. Fox, S. Gratch, S. Loshack in F. R. Eirich (ed.): *Rheology: Theory and Applications,* vol. 1, Academic Press, New York 1956.
[102] N. Ise, H. Matsuoka, K. Ito, H. Yoshida, J. Yamanaka, *Langmuir* **6** (1990) 296.
[103] M. Tomita, K. Takano, T. G. M. van de Ven, *J. Colloid Interfac Sci.* **82** (1983) 367.
[104] V. G. Kulichikhin, V. S. Volkov, N. A. Plate in *Comprehensive Polymer Science,* 1st suppl, Pergamon Press, Oxford 1991, chap. 15.
[105] G. Glarum, *J. Chem. Phys.* **33** (1960) 679.
[106] F. Bueche: *Physical Properties of Polymers,* Interscience, New York 1962.
[107] F. Bueche, *J. Chem. Phys.* **24** (1956) 269.
[108] B. H. Zimm, *J. Chem. Phys.* **24** (1956), 269.
[109] P. Rouse, *J. Chem. Phys.* **21** (1953) 1272.
[110] G. L. Slonimsky, *J. Polym. Sci. Part C* **16** (1967) 1667.
[111] C. Zener: *Elasticity and Anelasticity of Metals* Chicago University Press, Chicago 1948.
[112] J. S. Kochler: *Imperfections in Almost Perfect Crystals,* Wiley-Interscience, New York 1952, p. 197.

Laboratory Information and Management Systems (LIMS)

KLAUS SCHUCHARDT, Veba OEL AG, Gelsenkirchen, Federal Republic of Germany

REINHARD SCHYMIK, Unternehmensberatung Triestram & Partner GmbH, Bochum, Federal Republic of Germany

Laboratory information and management systems (LIMS) support both the administrative tasks of order processing and also the acquisition and assessment of results in the laboratory. LIMS "informs" about results and "manages" order processing.

The laboratory as a production center for discoveries—and so subject to the rules of production planning—is a new and unfamiliar idea. The laboratory is still being recognized as a creative (innovative) work center with a problem-oriented working procedure for nonrecurrent problems. Work of this kind is generally considered to be unpredictable and therefore only plannable to a limited degree. If this was true, however laboratory work could not be combined with the systematic arrangement that supports an information system, and data processing support in the laboratory would only be of minor importance.

In an analytical laboratory, orders and work flow can effectively be dealt with according to a plan and guidelines. For every laboratory there is a limited, yet established array of services depending on the instrumental equipment, the qualifications of personnel, and the objects to be analyzed.

Orders, samples, and analytical methods can be systematically described by a suitable data model.

Laboratory orders are plannable and can be carried out according to the rules of operations scheduling. The uncertainties of time for laboratory orders are greater for material and methodical reasons compared with production planning. In the laboratory operations scheduling is

0-89573-541-5/94/$ 5.00 + .50

usually controlled with less attention to particulars. Nevertheless the preconditions exist for a time, capacity, and personnel plan together with those for an order processing procedure supported by data processing. Laboratory information and management systems undertake these tasks.

1. Definition

1.1. Historical Development

The handling of large data streams such that arise in spectroscopy and chromatography, e.g., autosamplers in chromatography, led to the development of laboratory data systems. Initially these systems took over control functions for instruments, but were later developed into laboratory-wide management systems with the functional areas of data acquisition, evaluation, and order management. LIMS has developed from initially central solutions (host) with a comprehensive problem definition to administrative systems at departmental computer and networking level. Contributing to this development were the decreasing prices for hardware and the promising development of instrumental analysis (on-site information). The availability and use of relational data bank management systems represented a further innovative step in the development of LIMS. It also introduced a transition to the more extensive use of standard software, and solution by an individual program was increasingly supplanted. Most of LIMS development is definitely in the technical field, with little innovation in functionality, i.e., in the rules by which the data are processed. With respect to functionality, in addition to the sample-oriented character that is still preferred today, an extension to order- or project-related applications becomes apparent. At the same time a development fostered by quality assurance can be recognized: the LIMS application is no longer regarded as an isolated solution for the laboratory, but rather is understood to be a component or an element of quality management and assurance for the entire company.

1.2. Outline Solutions

In outline solutions, the goal is to start out from commercial standard software. This is supported by the rapid development of relational data bank management systems. One of the great advantages of such systems lies in a flexible adaptability of the functionality to (future) requirements. However, the use of relational data bank management systems simultaneously makes new and greater demands on the efficiency (speed, security of access) of the hardware.

The data processing implementation of LIMS was pushed further and further into the background in favor of the specific application.

1.3. User Expectations

The user expectations of data processing systems have developed with the available possibilities. The demands for user friendliness and operating convenience have increased. This development is taken into account by graphical user interfaces (GUI) and interactive auxiliary functions, but the user also expects an improved and expanded system functionality. Systems can be adapted and configured to different tasks by the user without programming operations. One of the areas of LIMS development is the integration of software extraneous to LIMS. The specialized software packages are so integrated into the functionality of LIMS that a continuous form of processing is obtained. Globally, the task of LIMS as the exchange of data between LIMS and other data processing applications becomes an absolute necessity.

1.4. Task Areas (see Fig. 1)

LIMS mainly covers the administrative part of support in the laboratory. The main objectives

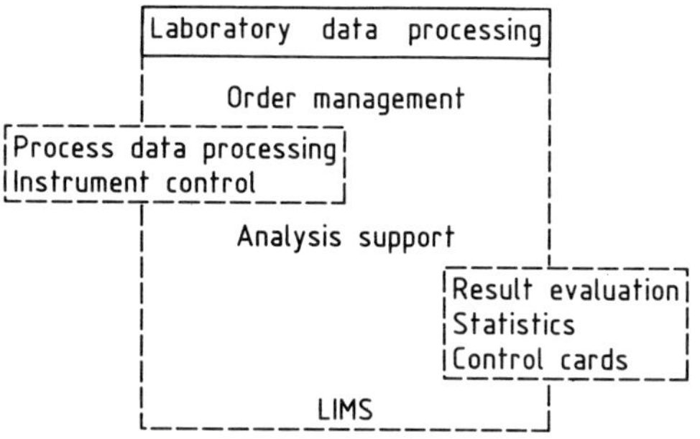

Figure 1. Laboratory data processing using LIMS

are order management and the collection of results together with the resulting archiving functions. Traditionally, LIMS has been a sample- or object-oriented system for an analytical laboratory. Currently, LIMS is being expanded to form a support for an order-oriented or (more comprehensively) a project-oriented operating procedure of research and development laboratories. The *sample-oriented* form can be understood as a subgroup of the project-oriented form. By this definition, the *order-oriented form* is the connecting link between these two types. The *project-oriented form* is based on the processing and examination of samples, but in addition it contains features of a system of project management with respect to time and content and a system of project documentation.

1.5. Functionality

The LIMS functionality is directed to the task of providing a service. Currently, no recognized LIMS standard exists, but there is substantial agreement on the functionality of LIMS applications. The user expects and assumes large functional differences in the individual LIMS applications. However, since the laboratory problems are in fact not that different, the real functional differences between systems result essentially from the limitation of central LIMS functions to individual tasks specific to the company. These include, for example, terminologies, reporting, and forms of data evaluation. The central functionality, i.e., the actual laboratory procedure, shows no fundamental differences. In terms of the LIMS data model, this means that the assumed differences lie mostly in the information that describes tasks and samples and not in that necessary for the operational sequence.

2. Objectives

Compared with other investments in the laboratory, the introduction of a LIMS is exceptional by virtue of the costs and the project duration. In addition to the external costs for hardware, software, and services, further requirements are the preparation of the master data and introduction of the system. This investment requires an economic justification. Measurable and quantifiable benefits are *rationalization effects*, performance increases, and task extensions, which occur both in the laboratory and in other areas of the company. Further positive but not immediately quantifiable consequences result from the higher standards of work expected.

In addition to rationalization, the second aspect, an *improvement in quality* of the laboratory work, is equally important together with the creation of *centralized information* on quality data. Recently, aspects such as *quality assurance* and *data exchange* have also become more significant. Without a LIMS, developments as initiated by computer-integrated manufacturing (CIM) or total quality management (TQM) are scarcely conceivable and not practical.

The systematic approach leads to working rules for the laboratory that are indispensable for a quality assurance system. In addition, the laboratory work becomes both internally and externally comprehensible.

Following the systematic approach required by LIMS, a data base also is created that allows meaningful archive evaluation. Of the secondary reasons for introduction of LIMS, the continual availability of information is especially important.

As a result of appropriate data processing support as provided by a LIMS, conditions are created for evaluating data at a complexity appropriate to a specific problem. Manual data processing, which also is possible in principle, is generally not practiced due to the cost of reliable data gathering, the time required for processing, or the lack of technical competence.

3. LIMS in an Information Network

A laboratory data system must not be considered solely from the perspective of the laboratory; the global company environment also must be taken into account. This results in additional demands for extending communication beyond the laboratory.

Data that are valid throughout the company are thus incorporated in the master data base of a LIMS. In principle, therefore, such general data could be accessed by the LIMS (i.e., not filed in the master data base) to avoid a redundant holding of data (and consequently problems with the data files developing in separate directions). However, the utilization of general data files by LIMS instead of using its own master data base is not yet possible, since the continual availability of data for the LIMS cannot yet be guaranteed. In the same way, organizational difficulties such as systematic arrangement, competence, or responsibility are obstacles to the consistent application of distributed data banks.

Alongside the LIMS computer system in the laboratory, there are usually other computing systems (GC, HPLC, spectroscopic, and workstations). These subsystems are to be subsidiary to the LIMS. At the same time the distributed intelligence at work stations and analytical instruments is used so as to prevent too much interference with analytical processing as a result of malfunctions. Outside the laboratory area, computer applications exist in which laboratory or laboratory-relevant data are processed or managed (e.g., commercial applications, process control systems).

The demarcation of LIMS to systems of the latter kind is relatively easy, because these processing and managing systems do not directly influence current laboratory procedures. In these external systems, data are filled once the processing of an order is complete. Data from these systems are retrieved by LIMS for updating its own data files. This data transfer is merely a technical problem (and a problem of data format conversion). The inclusion of analytical subsystems in LIMS, on the other hand, which lead to different forms of data exchange requires individual, complex solutions. The data flow resulting from this must be fully integrated into the LIMS functionality.

4. Nature of a LIMS

A LIMS is described by its data model and the functionality. Within this framework a LIMS is designed by the user. The description of the analysis and the samples is particularly important since the samples constitute the object of processing, the analysis the form of processing in the laboratory work.

4.1. Data Areas

In principle, three forms of logical data area exist in a LIMS:

1) Master data
2) Dynamic data or order data
3) Archive data

The *master data base* contains the "knowledge" of the laboratory, and thus all information necessary for carrying out the work there. It consists of general, analytical, and task-describing tables. The information contained in the master data base includes among other items the description of the organization, the authorities of personnel, and the descriptions of equipment, analytical methods, and samples. In addition, processing of different orders is described in the master data base. This includes sampling plans, analysis plans, monitoring of testing agents, and processing strategies according to experience. When an order is received, this master data base is always re-accessed and with the information filed there a task is set up in the *dynamic data area*. The raw data are assembled in this temporary data area and, with validation, transferred to the *archive area*. The archive thus originates in the master data.

4.2. Functional Principle

A LIMS is an information system built on experience. Flexibility and adaptability of the system therefore depend on the collection of master data which reflect the user's experience and his "view of the laboratory world". The master data are updated in a dynamic process by new experiences. Due to this continually expanding basic knowledge, LIMS are influenced by experience and is adaptive. If a master data base is of poor quality, therefore, only limited benefit can be expected from the use of even functionally very powerful systems.

4.3. Analysis – Sample

The analytical method is a measuring process that is applied to a sample (object) and whose outcome can comprise one or more results. The sample itself is not characterized by the analytical method, but is described by its own characteristics. The processing of samples (analyses) generates the processing order. The project is made up of various orders (individual steps of a project). The existing logical distinctions between project, order, sample, and analytical method must be reproduced in the data base of the LIMS.

Structure of an Analytical Method. An analytical method is modeled in LIMS in the form of its algorithm (procedure). This procedure must not process nonanalytical information, otherwise the necessary validity is lost.

Test Values–Statistical Characteristics. In addition to the procedural models, method-related statistical evaluations of analytical results must be defined in LIMS. The individual results are characterized by range of application, significance, and precision (e.g., standard deviation, repeatability, reproducibility), which are all filed together with the analytical method.

Calibration. For the calculation of analytical values, prior calibration or access to calibration values for the instruments or methods used is necessary in many cases. These calibration values and the calibration planning are assigned to the method, with consideration taken for the instrumental equipment.

5. Forms of Order

In principle, three forms of order are possible:

1) Frequency-controlled orders
2) Event-controlled orders
3) Random orders

For frequency- and event-controlled (routine) orders, the order data are known within the LIMS, but the *random order* must first be set up on the basis of the master data. It is therefore advisable to provide different functions for recording routine and additional orders.

Production monitoring is usually carried out in a frequency- or event-controlled manner. LIMS uses the corresponding work schedules for several purposes:

1) Sampling schedule
2) Automatic prerecording of orders
3) Order recording
4) Capacity planning
5) Cost accounting

In addition to the routine and medium-term processing of individual samples, all-encompassing mechanisms that enable long-term and project-oriented processing are desirable. These forms of order require individual activities to be planned with regard to time and content. The project orders filed as a network are monitored and controlled by LIMS by means of statements of internal costs and status reports.

From the recorded project schedules, the project activities are described in detail by means of a sample-oriented order-processing schedule. With this, a new sample-oriented order which describes the processing of actual samples, is created for each individual step of the whole project.

In the description of an order the LIMS treats separation operations (phase separations, distillations, preliminary separations, etc.) as conventional sample-preparation techniques. In addition, the individual subsamples are described and the corresponding analysis is allocated to them. Only in this way is an unequivocal allocation of results possible in sample calculations and evaluation methods (and project).

6. Basic Functionality

The functionality of a LIMS is marked by five key areas:

1) Maintenance of master data
2) Order creation
3) Analytical processing
4) Management functions
5) Archive management

6.1. Registration (see Fig. 2)

The work flow in a laboratory begins with the *recording* of an order to examine an object.

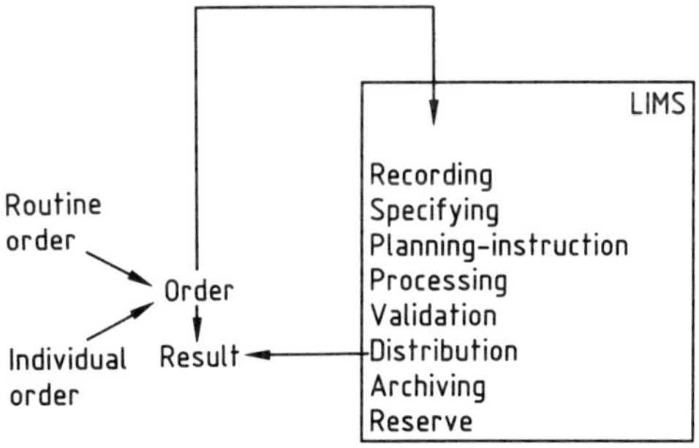

Figure 2. Basic functionality of LIMS

This means a description of the object to be examined, by which it can be identified in the laboratory and used for the subsequent report to the orderer or for subsequent evaluation. The order (i.e., the required analyses, tests, etc.) is allocated to the object identified. Thus with the recording of an order a processing objective must be known. By means of the work order recorded, LIMS draws up the individual orders for the laboratories and work groups involved. The actual implementation of an order is adapted to the nature of the order so that the necessary operating sequence can be efficiently arranged:

Routine Record. With the aid of a routine sample plan, a task list is drawn up automatically from the master data. The registration can, therefore, be carried out by LIMS independently or by the user in confirmation of receipt of the sample (order). For sampling that occurs before the recording of the order, LIMS can draw up task lists for the samplers in advance of the routine sample plans.

Individual Record. Individual orders are recorded by the user, and at the same time master data are re-accessed and the order adapted to the requirement on the basis of the system's proposals.

All data of an order must be modifiable until processing is complete. The modifications must be performed by an authorized person and documented so that they can be understood later on.

6.2. Sample Distribution – Dissemination of Information

Labels, sample waybills, raw data forms, and assorted task lists for work planning can be printed with the aid of LIMS within the laboratory. However, the objective is to ensure the availability of information even without printed lists.

Information about the order, processing situation, and whereabouts of the samples of an order must be available through the LIMS to all those entitled to it. The samples are distributed to the working groups and LIMS upholds the checking of the samples' whereabouts.

6.3. Work Planning

The user decides on the processing sequence with the aid of the task lists drawn up by the LIMS. Since in most currently available LIMS the master data base does not contain information on the capacity of the individual working groups and orders cannot be prepared by LIMS according to their actual processing (network plan), LIMS are not able to carry out independent operations scheduling. If analytical subsystems are attached, generally the person in charge must coordinate the loading of an autosampler between LIMS and subsystem. In principle though, methods exist for automatic sample identification by reading devices.

6.4. Recording of Results

The measurements are carried out and the raw data are recorded in the LIMS. In the next processing step, the LIMS calculates the results from the raw data. If several series of measurements exist, LIMS combines the results to give mean values and calculates the actual precision. The results are examined for their compliance to the range of application and precision as a function of the analytical method. In a further testing stage, the results are compared with expected sample-dependent values (e.g., specifications). Finally, after the completion of all testing stages, the results are released by the person in charge of the analytical method.

6.5. Validation

The results are subjected to a stepwise validation in the LIMS. Already during the first recording, the range of application, precision, and the desired object-related values were controlled. After the release of the individual results, a further checking step follows.

For this purpose, LIMS compiles the results produced at various processing units and tests these results for evaluations. The LIMS constructs the analytical report of the order from the combination of individual validated results.

At the general report stage a final validation occurs with inclusion of possible plausibility tests. In this stage, the LIMS notes the authority and responsibility of the user so as to eliminate unauthorized release.

6.6. Publication

The printout of completed (validated) analytical reports is automatically checked by LIMS. If a printout has to be distributed, then a centralized distribution is preferable to improve the handling efficiency. As distribution is manual the LIMS assists in sorting and copying reports according to requirements. The objective is also, in this case, to make results available to other receiving computer systems (process computers, administration computers, etc.) in order to avoid re-inputs of data.

Although LIMS allows the creation of the individual report as an exception, generally LIMS should produce reports in systematized style which can be reduced to four basic forms:

1) Summary
2) Standard report (e.g., according to EN 45001)
3) Chronological survey
4) Survey by object (plant, product, etc.)

6.7. Archiving

After validation and release, the activity data from the order are converted into archive data and, therefore, withdrawn from any further editing. Herewith the analytical reports acquire document status. Direct amendment of these archived data (reports) is disallowed for documentation reasons. However, well-founded invalidity declarations for individual results and the nomination of new findings with appropriate authorization and justification are allowed. Recipients on the original distribution list of the report are informed automatically by LIMS of any correction (even authorized) of archive data.

6.8. Reserve – Waste Disposal of Samples

As specified in the order record, samples from processing are disposed of as waste or transferred to a reserve, which is also managed by the LIMS.

6.9. Addition to Basic Functions

The basic functionality of a LIMS has to be expanded for actual laboratory requirements. The processing paths described here are certainly adequate for most operations, but further requirements of the functionality are intrinsic to the management of samples during and after processing. An *expanded basic functionality* of order processing includes the correcting and annotating orders as well as assembling order and task lists according to various criteria. Further functions make it possible for the LIMS to produce compilations selected from the archive data and process them graphically. Through the use of interfaces, the selected data can be made available to other data-processing applications. The LIMS provides the user with means of creating reports and defining the reports' contents.

7. Expanded Functionality

7.1. Integration of Extraneous Programs

The compilation of results is followed by the evaluation phase. This means that the LIMS must be able to compile data which can then be processed in other programs. Such an interface is not restricted to the programs mentioned here (e.g., word processing, statistics, and graphical presentation), but is also necessary for integrating the LIMS into a data-processing environment extending throughout the company. Although the same basic approach is followed when integrating LIMS into a global data processing environment, complicated individual solutions also have to be expected.

The integration of a *word processing system* presents a special problem, since the analytical results are not subject to the same protection rules within the word processing system as within LIMS; i.e., they can be modified rather easily. This difference can be desirable, but it raises the question of the reliability of data taken over from LIMS in the outgoing reports. The recipient must therefore be able to distinguish between the different forms of report.

7.2. Cost Accounting

In the cost documentation on laboratory expenses, a LIMS can distinguish between personnel and material costs. The costs per "customer" are assigned through collection at cost centers which makes invoicing possible. Capacity utilization statistics for internal laboratory purposes can be generated from this accounting information. These statistics concerning costs, work load, etc., can provide a basis for medium-term investment decisions.

7.3. Inventory Management

Because all the necessary items of information are available from the order processing, LIMS can accept them into an inventory management system. LIMS continually offers an exact survey of the stock of reserve samples, of stock location, and the storage time.

7.4. Internal Quality System

The internal quality system controls the recording of reference materials and the proper use of analytical instrumentation, maintenance, calibration measurements, reference measurements, repairs, and purchases, etc. In this way LIMS ensures that only sucessfully checked analytical systems and methods are used for the processing of orders.

In addition, the stocks of reference materials and standards are maintained and managed within LIMS.

8. Adaptation of LIMS to Laboratory Tasks

If an electronic data processing system is to be introduced into an existing (laboratory) environment then it must be adapted to it. But since electronic data processing offers the possibility to perform new tasks and modify job processing, changes must also be made in the laboratory field for best utilization of LIMS.

Since the laboratory sequences are similar in all areas of industry, LIMS have developed as broard-based solutions. With similar central functionality, the differences when applied to different laboratory tasks lie particularly in the screen display layout, the logical data structures, and the functional complexity.

Therefore most of the laboratory requirements can be met by a ready-made system if its level of technical development is sufficiently high. A small proportion of the requirements can only be met with limitations; and an even smaller fraction of tasks requires actual changes or extensions. Slight adjustments within the screen display layout and functionality will now and then become necessary after operating experience.

Outline Solutions. After drawing up the performance specifications and examining the profitability, there are two alternatives for the introduction of LIMS:

1) Purchase of a ready-made system (and possibly then adapt)
2) In-house implementation

The purchase of a system must be preferred since, although in-house implementation certainly enables better adaptation to the requirements, it also entails a considerably greater risk. Besides, an individual solution is not subject to general development as a commercially available system. The standard solution demands compromises of the user. These are also necessary in the case of an individual solution, for reasons of cost and efficiency.

With the introduction of LIMS, the desired final state of organization has to be defined and compared with the present state in order to achieve the highest possible degree of systematization. Existing complexities in a laboratory environment are generally determined by past experience and resources and can, in all probability, be simplified or abandoned.

On average, two years are required to implement a LIMS for a medium-size application. Because of the complexity of the system, it is advisable to introduce LIMS stepwise in order to have a consolidation phase after each stage. Implementation can be done in three steps:

1) General LIMS introduction
2) Integration of analytical subsystems
3) Integration into the company's communication network

After the first phase, the LIMS functionality is available to the laboratory. In the second phase, operations are optimized by on-line recording of results. This concludes the adaptation of LIMS within the laboratory. In the third phase, areas outside the laboratory are integrated into the scope of the LIMS.

Adaptations. In principle, LIMS can be extensively adapted to the user's requirements. However, the adaption possibilities available to the user must be clearly distinguished from those available to the manufacturer of the system. In general the *user* can only adapt the system through arrangement of the master data. For the *system manufacturer* it is possible to make both adaptations conforming to the system (arrangement of the display surface) and changes of data structures and functionalities deviating from the system. A change of data structure and functionality can be equated to an individual approach. A LIMS with modified data structure and functionality is not subjected to general further development and maintenance by the manufacturer.

Systems originating from the use of powerful development tools are to be regarded similarly. The advantage here is the more probable minimization of the associated high risk compared with a new individual solution. With any individualistic approach, the future user and the system manufacturer need to be highly competent.

9. LIMS as a Component of Quality Assurance

Extensive responsibilities for documentation are imposed on the operator of a laboratory through the requirements of Good Laboratory Practice, an accreditation according to EN 45001, or in the framework of certification according to ISO 9001-3. A LIMS is basically not a quality assurance system, but rather an object to be itself safeguarded within such a system. That is, the LIMS must be arranged so as to guarantee the requirements of a testable mode of operation. However, within a laboratory a LIMS satisfies many quality requirements through the retention of master data necessary for the system and through the work flows specified by the system. Consequently, the LIMS considerably simplifies the observance of the quality requirements mentioned (and even makes them practically realizable from the expenditure perspective.

A version management of the entries can thus be carried out within the maintenance of the master data. With that, a methods and product archive, among other things, builds up. The handling of testing agents, standards, and reference materials is controlled and documented by LIMS. With proper application of the LIMS a task can be reproduced at any time from the recorded and stored data by reviewing the acquisition of results for an order and the associated decisions.

At the same time the LIMS ensures that in carrying out their duties all users always have the up-to-date information available.

10. Development Prospects

The main emphasis in the development of LIMS will be its further functional integration, into the laboratory itself and within the company. Project-related operating principles and control of the work flows are the main emphases of development in the laboratory. In production control, further possibilities for the laboratory are appearing in the automation of measuring procedures. The penetration of information and function throughout the company will involve the potentialities of a LIMS particularly in quality assurance, production control, and information appraisal. Further technical developments (hardware, standard software) are of secondary importance for the LIMS, since their result is not to develop the application per se of the system but rather primarily to improve its potential uses.

In addition to these lines of development, which mean an updating of the known system functionality, an innovative development step is conceivable. The LIMS, hitherto only passively recording and archiving data, can actively interpret data relationships on a structured knowledge base and consequently influence analytical procedures. Economic considerations will determine which of the directions of development will be favored. Of these, project LIMS, integration within the company, and aspects of quality assurance are emerging as the main points of concern.

11. References

S. Ebel: "Anforderungen an Labordatensysteme," *Nachr. Chem. Tech. Lab.* **36** (1988) M 1 – M 14.

K. Schuchardt, G. Fischer, A. Golloch: "Labor-Informations- und Management-Systeme," *Nachr. Chem. Tech. Lab.* **40** (1992) M 1 – M 20.

R. R. Mahaffey: *LIMS Applied Information Technology for Laboratory*, Van Nostrand Reinhold, New York 1990.

R. D. Mc Dowall: *Laboratory Information Management Systems*, Sigma Press, Wilmslow, UK, 1987.

V. Neitzel: *Labordaten-Verarbeitung*, VCH, Weinheim 1992.

Process Control Engineering

Process Control Engineering

MARTIN POLKE, Lehrstuhl für Prozeßleittechnik, RWTH Aachen, Federal Republic of Germany (Chaps. 1, 2, Sections 5.1–5.3, 6,1, 10.7, 11.1–11.3, Chap. 14)

WOLFGANG AHRENS, Bayer AG, Leverkusen, Federal Republic of Germany (Chap. 2, Section 13.1)

ULRICH PALLASKE, Bayer AG, Leverkusen, Federal Republic of Germany (Sections 3.1, 3.3, 3.4)

RUDOLF METZ, Bayer AG, Leverkusen, Federal Republic of Germany (Section 3.2)

NORBERT KUSCHNERUS, Bayer Ltd., Tokyo, Japan (Section 3.5)

ULRICH EPPLE, GfP, Gesellschaft für Prozeßtechnik, Aßlar, Federal Republic of Germany (Sections 4.1, 4.5, Chaps. 7, 14)

ERNST-DIETER GILLES, Institut für Systemdynamik und Regelungstechnik, Universität Stuttgart, Federal Republic of Germany (Section 4.2)

JÖRG RAISCH, University of Toronto, Department of Electrical Engineering, Ontario, Canada (Section 4.2)

HERBERT WEHLAN, Institut für Systemdynamik und Regelungstechnik, Universität Stuttgart, Federal Republic of Germany (Sections 4.3, 4.4)

MICHAEL HEIM, Lehrstuhl für Prozeßleittechnik, RWTH Aachen, Federal Republic of Germany (Sections 5.1–5.4, 6.1, 11.2, 11.3)

REINER HOTOP, EBERHARD NICKLAUS, WOLFGANG NOERPEL, Bayer AG, Leverkusen, Federal Republic of Germany (Section 5.1)

MICHAEL FREYTAG, Bayer Antwerpen N.V., Antwerpen, The Netherlands (Sections 5.1, 5.2)

KLAUS HARTMANN, Miles Inc., Pittsburgh, PA 15205-9741, United States (Sections 5.1, 5.2)

GÜNTER PINKOWSKI, Krohne Meßtechnik, Duisburg, Federal Republic of Germany (Section 5.5)

HANS-JOSEF SCHNEIDER, Bayer AG, Leverkusen, Federal Republic of Germany (Section 6.2)

MICHAEL BJÖRKMANN, ABB Ströberg Drives, Vaasa, Finland (Section 6.3)

WOLFGANG MUTZ, Bayer AG, Leverkusen, Federal Republic of Germany (Sections 6.4, 10.7)

HARTWIG STEUSLOFF, Fraunhofer Institut für Informations- und Datenverarbeitung, Karlsruhe, Federal Republic of Germany (Chap. 8)

GERD-ULRICH SPOHR, Siemens AG, Köln, Federal Republic of Germany (Chap. 9)

ROLAND VOGELSGESANG, Bayer AG, Leverkusen, Federal Republic of Germany (Sections 10.1–10.6)

NORBERT INGENDAHL, Lehrstuhl für Prozeßleittechnik, RWTH Aachen, Federal Republic of Germany (Section 11.2)

KARL-HEINZ SCHMITT, Bayer AG, Krefeld, Federal Republic of Germany (Section 11.4)

HASSO DRATHEN, Bayer AG, Leverkusen, Federal Republic of Germany (Chap. 12).

DIETRICH BALZER, Elpro AG, Berlin, Federal Republic of Germany (Sections 13.2, 13.3)

0-89573-541-5/94/$ 5.00 + .50

1. Introduction

Since the invention of the steam engine in the 1700s, the world has gone through four long economic cycles, which were described by the Russian economist Nikolai D. Kondratieff (1892–1930) and are known as Kondratieff waves or cycles [1.1].

Each upward phase (identified as "prosperity" in Fig. 1.1) was borne by inventions such as the steam engine (WATT, 1769), the mechanized weaving loom (CARTWRIGHT, 1784–1786), the steam locomotive (STEPHENSON, 1814), the electrodynamic principle (SIEMENS, 1866), the diesel and Otto motors (1893–1897, 1876), and fundamental innovations in chemical process engineering such as the manufacture of basic chemicals.

The world economy entered another long upward phase at the beginning of the 1980s.

The development of the new basic innovation called "information technology", characterized by the invention of the transistor [1.3] and the subsequent rapid development of miniaturization and laser technology, but also by results from the field of informatics such as object orientation [1.4] and Petri networks [1.5], is described in depth in [1.6]. The economic boom identified with the new resource "information" [1.7] is a global phenomenon with heretofore unknown dynamic qualities. International business, national economies, and individual companies find in it growth and innovation potential that could scarcely have been imagined. PORTER describes the importance of information as an essential resource for a successful corporate strategy [1.8]: "*The question today is not whether information technology has significant effects on a company's competitive situation, but merely when and how these effects will become apparent.... Anyone who does not react today will be compelled to follow a course that others have set.*"

The fifth Kondratieff cycle, however, is not only an imposing technical and economic upswing but carries along with it profound social, ecological, cultural, and intellectual changes and challenges. These connections, often revolutionary in scope, were pointed out by SCHUMPETER [1.9].

The fifth Kondratieff cycle will render permanent changes in the way the individual thinks and acts and also into the strategies that corporations and national and international institutions will pursue. A technological advance will become a structural change [1.10].

Along with the classical production factors—energy, labor, raw materials, and proven production technology—information has become not just an additional factor but probably the most important factor.

CIM (computer-integrated manufacturing) and CIP (computer-integrated processing) are not merely slogans today [1.11]. Demands made by society, such as increased environmental awareness, better and more consistent product quality, improved delivery, and more economical management, necessitate production facilities and processes with greater flexibility and ease of operation.

Strategy, that is, the management of the company as a whole, means identifying goals and achieving them. Because a company's definition of goals must always be dynamic, it is important to inquire into what impact changes in the company's environment will have. As trends in socio-

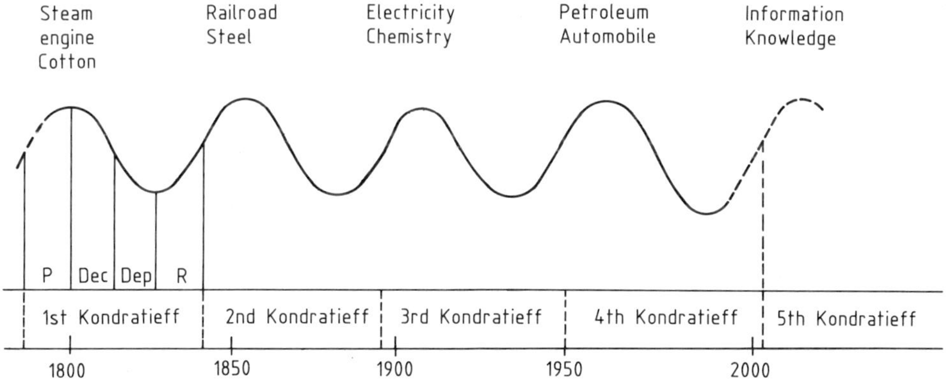

Figure 1.1. The Kondratieff waves [1.2]
P = prosperity; Dec = decline; Dep = depression; R = recovery

cultural, ecological, technological, domestic economic, international economic, and political-legislative settings become stronger and change more rapidly, optimal access to information becomes a central requirement for a successful strategy (e.g., in marketing, sales, and production).

But strategy also means making available the means for business success. An essential success factor is thus the exploitation of new information technologies based on engineering, physics, and information science.

Information is an unlimited resource that can be managed only if it can be structured. In the economic, administrative, and technical spheres, those responsible for information management must optimize the utilization of information by coordinating all parties involved and by creating a technical infrastructure.

The German term *Prozessleittechnik* (process control engineering) was coined at Bayer AG in 1980 as a working title covering the measurement, control, and electrical engineering groups [1.12]. The expression was chosen to denote the integral view of all those concerned with the management or control of production programs based on engineering processes.

Since that time, a number of events have made it clear that classical or signal-oriented measurement and control technology is being supplanted by information-oriented process control engineering.

Long discussions among experts regarding the terms "control" and "automation" came to an end when LAUBER stated a valuable dichotomy, placing humans at the center of the analysis [1.13]: The role of humans can be (1) to specify the framework and the sequence of the process, which then runs automatically, or (2) to gain the maximum information about the process and intervene in it directly, becoming the agent that ultimately controls it. This question is discussed in more depth in Section 4.3. The authors of this article have a clear preference for the more universal term "control" and make use of various levels or degrees of automation (see also [1.14]).

"Essential impetus for development today is coming from information technology, which has become a significant tool in measurement and automation technology. Only information-oriented technology enables the industrial user to operate processes and plants in compliance with the criteria of flexibility, productivity, safety, and environmental protection. The integration of humans into the production process is of great importance if the operator must, for example, perform control functions at the interfaces with the automation system and with the process" [1.15].

The national and international standardization organizations of greatest relevance to process control engineering, such as the International Electrotechnical Commission (IEC) with its Technical Committee TC 65 and the German Electrotechnical Commission (DKE) have also replaced the term "measurement and control technology" with the new complex of "control engineering" (see Chap. 9). The Standardization Working Group on Measurement and Control in the Chemical Industry (NAMUR) had already announced such a new orientation in September 1987 [1.11].

The strong influence of information technology on automation technology and process control engineering is visible not just in the area of devices and systems (especially process control systems; see Section 5.2), but in all aspects of design, construction, operation, maintenance, and so forth (procedures) [1.16].

BASF has made their training programs in operational practice available to the public [1.17]. Detailed procedural instructions can be found in STROHRMANN's [1.18] course on process control engineering (serialized in *atp* since 1984). GRUHN et al. [1.19] and LAUBER [1.13] have published accounts of structuring practices for automation with an emphasis on process engineering and process computing automation technology, respectively. TÖPFER and BESCH have also described principles [1.20].

The time now seems ripe for attempting to present an integrated, information-oriented exposition of this subject.

The information structures employed in control engineering and the architectural principles that generate them are therefore discussed and explained with practical examples (Chap. 2).

Next, the computational and mathematical methods used to acquire, reduce, model, and document process information are presented (Chap. 3).

Information about the process is used for process control (Chap. 4) by applying isolated and integrated control concepts and object-oriented process-control strategies (e.g., recipe management).

The control components of production plants, that is, the electrotechnical devices used for process control (sensor and actuator systems, process control systems including communication technology, process power supply, design of control and power distribution rooms and cable runs), are next analyzed from the information-technology point of view (Chaps. 5–8).

Computer-aided techniques (C-methods) have proved an especially useful information-technological support for the engineering of process control systems. General aspects of such systems and detailed structures of data-model-oriented design aids now in use or under development are described in Chapter 9.

The procedures used for the design, construction, and start up of process control facilities are set forth from a systems viewpoint and on a structural and sequential basis. Decision phases with the most important design document, (requirements and specifications) are clearly distinguished from execution phases. Quality assurance for the design, construction, and start up of process control systems is described in Chapter 10.

In the operation of production plants with process control systems, a change is taking place from signal-oriented man–machine communication to information-oriented, state-based man–process communication. This enables the plant operator to utilize knowledge about the process, formulated in an ergonomically correct way, to control the process. Continuous process analysis and process optimization take on new importance in this context. Strategies for maintaining or restoring the availability of process control functions are described in Chapter 11.

The current status of standardization of process control techniques and systems in national and international bodies, societies, and institutions is presented in Chapter 12.

Finally, the integration of knowledge-based systems and modern concepts of fuzzy logic and neural networks are discussed in Chapter 13.

2. Information Structures in Process Control Engineering

2.1. Principles

Information as a Corporate Resource. In principle, information is an unlimited resource that can be managed only if it is structured. It is the task of information management to optimize the utilization of information by insuring the efficient interplay of all parties involved and providing the technical infrastructure.

The objective of information structuring is to reduce the degree of complexity by appropriate modeling (formation of data models and functional models) and thus create the prerequisites for information processing (input/output, storage, and analysis) by means of information systems. What has to be overcome, above all, is not the material but rather the inadequacy of the human intellect to deal with complexity.

The importance of adopting suitable structuring aids is demonstrated by the way in which technical, administrative, management, and logistical data "circles"—none, as a rule, consistent with the others—frequently come together.

Because they continue to be developed, application systems arise whose functionalities often overlap (Fig. 2.1) [2.1].

What is more, the demand for information depends very heavily on the level of automation present in individual production facilities. Production engineering systems often feature noncommunicating automation islands. In fact, all such information circles are necessary, from the technical and economic standpoint, in order for the corporation to accomplish its task unhindered by departmentalization.

If information as a corporate resource is to be optimally deployed, it is essential to integrate information technology, and this can be done only when information concepts have been harmonized.

Integration of Information/Information Models and of Data/Data Models. Information integration is achieved by consolidating information across projects, leading to a mandatory corporate information model [2.2]–[2.5].

Such information models [2.6] are said to be "semantic" or "conceptual" because, at the user level, they reflect the definitions and interrelationships of concepts but not the realization of

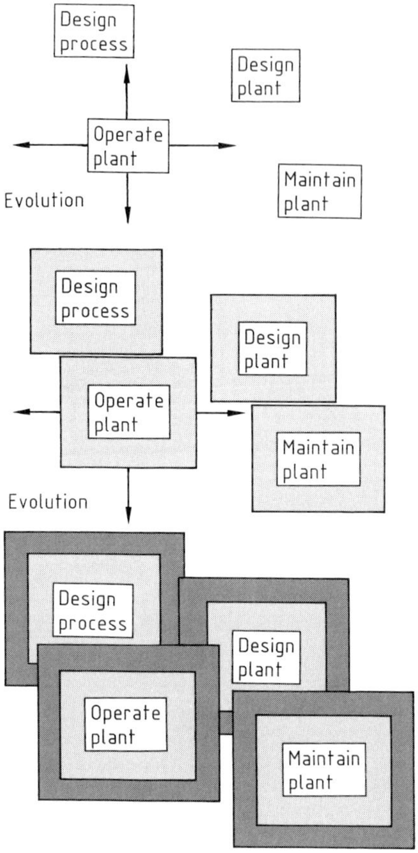

Figure 2.1. Convergence of isolated data "circles" and automation islands [2.1]

information in data-processing (DP) terms or the selection of information for specific applications. In contrast to earlier management information system (MIS) implementations, the result is an extremely stable information model. Appropriate representation techniques yield data from information and data models from information models.

Since 1970, theoretical publications on the form of such data models have been concerned chiefly with how the organizational problem of large data-base systems can be solved. For example, the mathematically grounded relational data model with its normal forms [2.7] was created first; the entity–relationship (ER) model, which stresses the semantic aspect of the data, came later [2.8].

In 1975, the ANSI/X3/SPARC DBMS Study Group introduced the "conceptual scheme", a comprehensive, implementation-neutral structure description of corporate information. This conceptual scheme is of central importance for the understanding and management of information systems [2.2], [2.9]; see also the SPARTEN information and control systems (ISS-$\alpha\alpha$) developed at Bayer AG [2.10].

Further detailed exploration and discussion of the conceptual scheme has taken place in the ISO Working Group with the same title. A detailed report is presented in [2.11].

Work on these aspects has continued; some results have been summarized and various methods, procedures, and tools to aid in the creation of data models have been set forth [2.12]–[2.15].

Today, the state of the art is as follows: In the development of information systems, data properties relating to content, application logic, and organizational aspects are first examined in a "data structure analysis." The results must be described in a semantic or conceptual information model before the data model (which depends on the form of presentation) and thence the data-base design can be derived. The data-base design must take account of requirements and restrictions having to do with data-processing hardware and software. Restrictions based on processing considerations are brought in at another level.

The data-base design can thus be represented in a three-level model:

- Conceptual level: The result is the semantic or conceptual information model
- Logical level: The result is the logical data model, which contains restrictions having to do with data-base hardware and software
- Physical level: The result is the physical data model, which makes allowance for processing restrictions

A conceptual information model is constructed with an eye not to isolated information-processing functions, but to the objects of corporate information processing that are common to all such functions. This is the strategic dimension of data modeling.

Such conceptual corporate data models are now being created in many companies. As a corporation-wide, application-neutral, implementation-neutral structural description of corporate information, such a scheme also constitutes the common linguistic basis for communications between persons involved in DP management. Because it is independent of the allocation of data among the company's computers, independent of physical data storage, and independent of the data-base systems employed, such a scheme can unify information systems based on a wide variety of hardware and software.

Integration of Methods. In the first period of software engineering, the software life-cycle model was in the forefront of discussion. Software development environments were designed in accordance with this philosophy. Alternative philosophies are in use today, such as rapid prototyping and participative system development [2.16]. Both aim at the early involvement of end users, whose acceptance is the single decisive criterion for the success or failure of an application system.

The appearance of data bases gave new impetus to the development of data modeling techniques. Because these are based on a variety of implementations, they are largely incompatible. For example, an elevator control system does not require a data-base design; an information system has no time-critical concurrent processes; and an expert system uses different knowledge representations from a procedural program.

If the data model has a hierarchical, network, or relational structure, different design methods are used.

An integrated methodology must generate both the data design and the functional design (Fig. 2.2).

It has often been overlooked that every design level of this methodological concept deals with a different aspect of the problem.

Thus, data and functional models are built up at the uppermost, conceptual level and then, in the further specification, broken down into a representation that can run on certain hardware.

Integration of Representation Means. The basic prerequisite for the common storage of design data and the transfer of information between methods is an appropriate representation form.

Today, graphical and textual representation means can be thought of as being diametrically opposed.

Each, however, has its justification. Graphical means of representation (flowchart, function diagram, entity-relationship diagram, polling

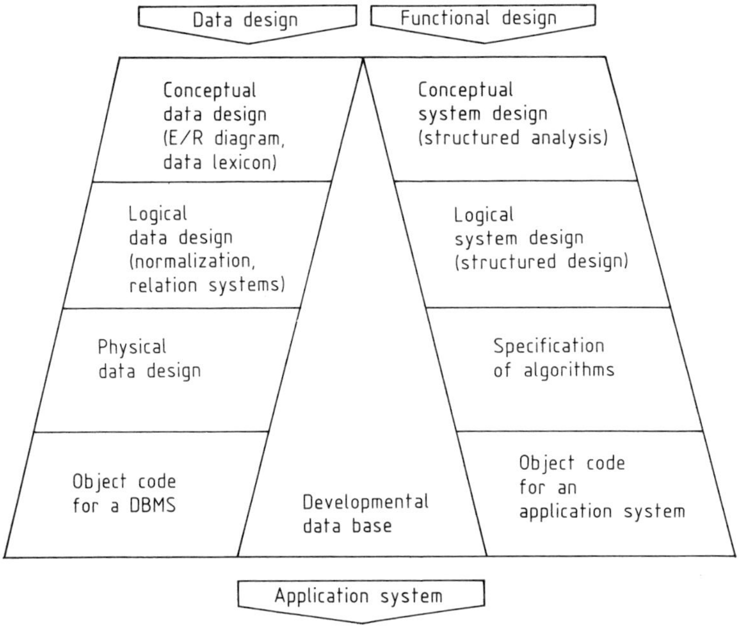

Figure 2.2. Methodological concept [2.17]

hierarchies, etc.) are a suitable idiom for use in the human–computer interface (Table 2.1).

Textual means of representation (pseudocode, structured text, computer languages of various generations) yield expressions that are understandable chiefly to the computer.

There must therefore be a way to translate between graphical and textual means of representation.

The task remains to harmonize these means of representation with one another, obtaining a canonical set of description means that can be used to describe data structures and functional structures with regard to both static and dynamic properties.

Integration of Tools. The most important tools for data integration are data-base systems and data dictionaries [2.4], [2.19].

In order that the data in a data base will represent a consistent system of facts, the permissible and necessary relationships between data are specified by means of integrity conditions and monitored and controlled in real time by a data-base management system. Data-base systems offer a range of concepts that support "information management."

In particular, these include concepts for modeling and manipulating data in an application-oriented fashion (data models), for integrating various applications over the same data base without redundancy (data integration), for making an application neutral to changes in the logical and physical organization of the data (data independence), for insuring data consistency, multiuser synchronization, and data security, and for safeguarding the stored data against unauthorized access.

A data-base system is said to be of hierarchical, network, or relational type, depending on the underlying data model.

Hierarchical and network data-base systems have become established in commercial, administrative applications. Relational data-base systems are gaining ground in engineering applications, such as the support of design processes (in the design of mechanical parts, LSI circuits, process monitoring and control equipment, and software development) and process control. Distributed and object-oriented data-base systems have not yet become widely available from commercial sources.

In contrast to data-base systems, a data dictionary is a system for storing and processing data definitions (data descriptions) for one or more software systems. In its simplest form, a data dictionary can be regarded as a well-organized notebook containing basic information

Table 2.1. Methods and their points of emphasis [2.18]

Method	Points of emphasis
Structured Analysis (SA)	Representation of functions, data regions, interfaces, and data flows connecting them, in the form of hierarchically organizable nets
Structured Design (SD)	Representation of modules and their interfaces in "Structured Charts" and module level model
SADT	Representation of functions, as well as data and control flows connecting them, in the form of hierarchically organizable nets
Entity-Relationship Model	Representation of objects and their quantifiable relations (1:1, 1:n, etc.) in the form of nets
Decision Tables	Representation showing under what combination of conditions certain actions can be performed, in the form of tables
Finite-State Automata	Representation of states and state transitions, and also of inputs initiating changes of state as well as outputs initiated upon changes of state, in the form of nets
Petri Nets	Representation of states and events in the form of nets whose behavior can be dynamically investigated with the aid of the Petri net rules
ISAC	Representation of information sets, as well as information flows and functional sequences, in the form of seven types of nets
Jackson Structured Programming (JSP)	Representation of the data structure of a problem, derivation of the program structure from this data structure
Jackson System Development (JSD)	Representation of objects and their actions in the form of trees, and of the connections between objects in the form of nets
MASCOT	Representation of processes (and devices) and their communication regions in nets
Functional nets	Representation of processes and devices as well as their communication, in the form of nets
HIPO	Tree-type breakdown of a system and definition of every part in terms of input and output data and required operations (steps)

about all the data elements stored in a computer for an application area [2.20]–[2.22].

These support not only a one-time, uniform, consistent documentation of information objects, but also the management of other objects used in the DP environment (such as jobs, projects, users, etc.) as well as their relationships. Data dictionary systems are themselves data-base systems that use a central data base to manage all information needed to describe, process, and use the objects of an application area, regardless of their internal organizational structure. Data dictionary systems also support data management for application systems and thus contribute to the efficient development, implementation, and description of software systems.

A number of other tools have been devised in the course of the software engineering discussion. Each of these tools is suited to a certain class of problems; each either extends over multiple phases or is specific to one phase [2.6], [2.23], [2.24].

For example, editors are for the creation of graphics and texts for use in a variety of methods; compilers, linkers, debuggers, and so forth, for the implementation phase; test tools, for the testing phase.

Tools for retrospective documentation are becoming more and more important at present.

Decades of software development have produced software systems that can scarcely be maintained any more (the "software crisis"). The documentation of these systems is often in a wretched state.

Because of changes in methodology, it is no longer sufficient to document such systems in the context of their own (usually obsolete) approach. What should be aimed for is to create new documentation for old systems, with an eye to future trends in software engineering.

Retrospective documentation is a more and more pressing need in a management context, where decades' worth of investment in software must be safeguarded. This does not necessarily mean that obsolete software systems must continue to be used in their existing form. The minimum requirement is that the information and technical knowledge contained in the system must be revealed and made accessible for new developments.

The integration of tools means that documents can be passed along from phase to phase and from tool to tool.

Integration of tools can succeed only if information, data, and methods have been integrated. This applies not just to all engineering activities during the lifetime of a technical device (see Chap. 10), but also to the design of monitoring and control equipment itself, especially of process control engineering systems (see Chap. 7).

2.2. Architectural Principles for Information Structuring

Systems and Architectural Principles. Real-world systems and elements have internal structures. Along with Aristotle's doctrine of categories [2.25], the Linnaean system—still a valid descriptive model employed in botany—provides an example [2.26]. Darwin's ideas on the origin of species [2.27] also illustrate this structuring process.

In the 1700s, "system" at first meant a collection of linked truths. The links had to be methodologically correct and completely deducible from an underlying principle.

KANT's concept of system stresses the theory of systematics. In his "Critique of Pure Reason," he emphasizes the significance of architecture as the art of systems.

Thus a system, traditionally, is a closed whole whose parts are interrelated, interact, and satisfy certain constraints.

The parts, regarded as elementary, are called elements of the system. A system consists of elements with properties; the elements are structured and linked together by the relationships in accordance with a system principle. Systematization involves specifying system principles and devising system structures. Systems are characterized by the fact that they have properties that cannot always be accounted for by the properties of their elements. Thus simply collecting elements does not form a system. Instead, each system has a structure that is generated by the relationships among its elements.

Systems can be subdivided into parts on as many levels as desired, whereby all parts that are not elements of the system are referred to as subsystems. System elements are the parts of a system whose further subdivision is impossible or undesired.

The sections that follow state principles for the structuring of information. They are called architectural principles because they can be identified at various points in the design process and in various representations of the information system, thus marking the architecture of the information system. The first principles discussed are general ones in the phases of design and implementation.

Principles in the "definition of requirements" phase, quality assurance in parallel with other activities, and operation of a facility might be added here. These are also referred to as construction principles, since they are applied mainly in the construction of DP systems.

This is demonstrated by the example of modern operator interfaces, where structuring of information is used to support human–process communication (see Section 11.2).

Higher concepts of functional and data structuring also make use of the design principles discussed above but represent a more application-oriented aspect of structuring.

The focus is on the object-oriented formulation, which can be applied to both functions and substances [2.28] in an integrated methodological concept. It is the tool for the decomposition principle, with which complicated objects are broken down into their constituent parts.

The next most important structuring principle is abstraction, which is a fundamental approach to concept formation. Adopted recently from expert-systems technology, abstraction is introduced as a construction principle of application systems through class-forming, concept-forming, complex-forming, and functional abstraction. Closely related to the principle of abstraction is that of inheritance [2.29]–[2.31].

The last section deals with transformation principles. Here the various models used for problem-solving are described, starting with procedurally oriented formulations (with their counterpart in functional design) and passing through logic-oriented formulations (as now employed by many expert systems) to state-oriented or event-oriented concepts (which come much closer to process-related DP problems).

State-oriented and event-oriented formulations in turn are closely related to object-oriented ones. Objects are in certain states and communicate with one another via established protocols (see Chap. 8).

The complexity of present-day systems increasingly calls for a holistic, network style of thinking [2.32], [2.33] if the basic laws governing systems, their dependences, and their interactions are to be understood. It will take a new kind of thinking to comprehend the dynamics of a continuous process, plan a monitoring and control system for it, erect and operate the system, employ higher-level information systems, and in the process not overlook human beings [2.34].

The Decomposition Principle. *Design Principles.* Structuring principles occur in all phases of system development. In the design phase, for example, the top-down and bottom-up principles are often encountered. In the top-down approach, a whole is broken down into parts step by step, whereby the complexity of the parts under consideration is systematically decreased (decomposition principle).

In bottom-up design, single parts are assembled step by step and made into a whole.

In practice, mixed forms are often encountered (Fig. 2.3).

Bottom-up design comes into play when prefabricated elements are available from which the overall systems can be assembled. This principle is often used in mathematical and statistical applications, where the elementary mathematical and statistical functions are known and can be made available in program libraries. New applications can be analyzed only by the top-down approach, though there is a trend toward reusable elements here too (see, e.g., Section 4.5).

Implementation Principles. The implementation phase also exhibits principles that ultimately determine the structure and thus the understandability and maintainability of software systems.

The more ordered the structures are, the better they lend themselves to verification, modification, and maintenance. The simplest ordering is linear. Iterations, recursions, and branches generate more complex structures (Fig. 2.4).

Recursive data and functional structures are preferred today because of their mathematical compactness. It can be shown, however, that iterative and recursive structures are equally powerful and can be transformed to one another.

Another type of structure is the directed tree with strict hierarchy or polyhierarchy, which ultimately becomes the general network if the rigorous requirement that every node has only a single input is dropped (Fig. 2.5). The last is es-

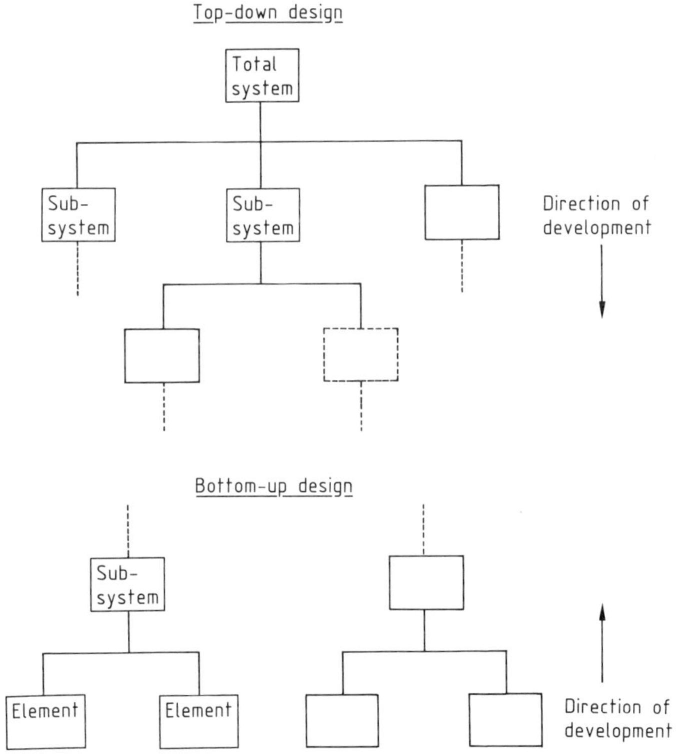

Figure 2.3. Design principles

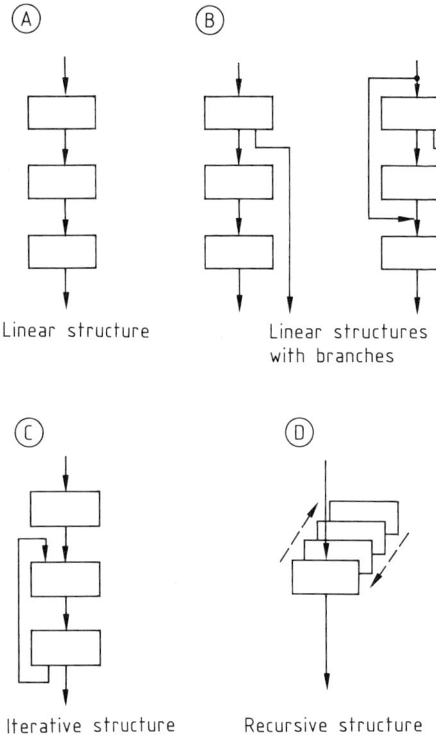

Figure 2.4. Types of structure

pecially important in so-called master–slave structures. The slave should have only one master, since otherwise there can be conflicts in execution. Ring and star structures form a separate class.

Principles for User Interface Design. The user gains access to the system through the user interface. A number of structuring principles apply to this interface, not all of which can be discussed here.

Design principles such as the law of like form, the law of color, or the law of proximity (Fig. 2.6) should be employed in the layout of masks and printouts [2.35], [2.36].

Principles such as conformity to expectation in dialog with the system (also called "principle of minimum surprise"), consistency of dialog elements, and visualization of the application context also play an important part.

Modern software ergonomics focuses on job design for people, a process that is heavily influenced by software [2.37], [2.38].

An important goal in software writing must be the design of complete activities. A complete activity has been defined as an activity that, together with simple execution functions, includes the following:

- Preparatory functions: identification of goals, development of procedures
- Organizational functions: agreement with other workers as to tasks
- Monitoring functions: feedback to the worker on attainment of goals

Incomplete activities are activities where there is essentially no possibility of independent goal identification and decisionmaking, no scope for individual ways of performing tasks, or no adequate feedback. Shortcomings in job design impair the motivational and learning potentials of the work process and affect the worker's welfare and job satisfaction. Intimately linked with the completeness of activities is the degree to which the worker has opportunities to make decisions as to time and substance, that is, freedom of action.

This freedom of action must permit creative users of an information system to devise an individualized procedure in accordance with their experience and working style. At the very least, defining freedom of action should not be part of the program author's responsibility; it is a function of corporate management and personnel practices. Observed acceptance problems often stem from misjudgments in this area.

A new type of human–computer interaction has come into being with the new class of workstations. This user interface has the new feature that it presents the user with graphical images of customary objects belonging to the office world. Such objects can be selected and then manipulated with the aid of a pointing device (mouse, light pen).

SHNEIDERMAN [2.39] introduces "direct manipulation" as a collective term for several new principles in the user interface. The most important principles are the following [2.40], [2.41]:

- Permanent visibility of the object as an icon or pictogram
- Quick, reversible, one-step user actions with immediate feedback
- Replacement of complicated commands with physical actions (see also Section 11.2)

Higher Concepts of Functional and Data Structuring. Historically, the functional structuring of application systems was emphasized in the early period of data processing. A complex function was successively broken down into partial functions until elementary functions realizable in

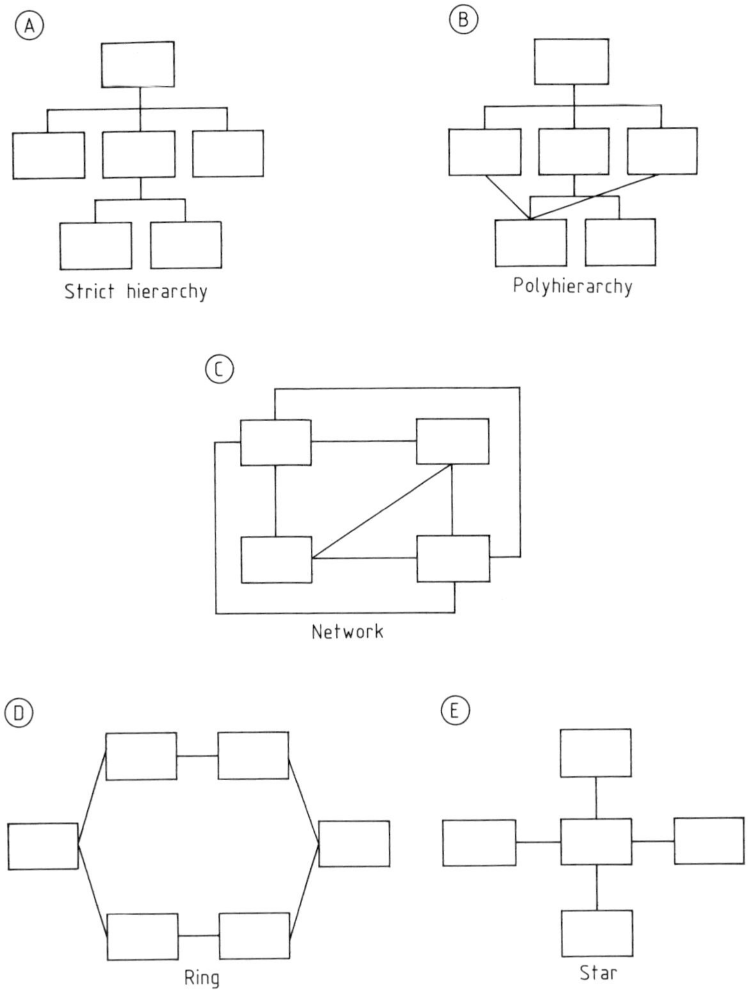

Figure 2.5. Types of tree and network structure

DP hardware and software were arrived at (flowcharts).

Data were fitted into the functional structure but remained meager, as they had to be given in the programming languages known at the time.

The structuring principle was oriented to the procedure employed in problem-solving. As a result, functional structuring led to process-oriented or program-oriented structures (Fig. 2.7).

Even today, functional structuring is still the classical approach, because it leads directly to realizable, procedurally-oriented solutions. At Bayer, for example, all engineering tasks were modeled by the SADT method, somewhat as shown in Figure 9.1. [2.43], [2.44].

Thus the modeling concept (functional structuring) and the implementation concept (procedural programming language) correspond.

Data structuring is of secondary importance: What is interesting is the processing of the data, and data-base systems were not yet in use at this time.

However, functional structuring has several weaknesses:

- There are scarcely any objective criteria for breaking the function down into partial functions. Decomposition is often a matter of the developer's discretion.
- The resulting data structures unavoidably grow out of the functional structure and are

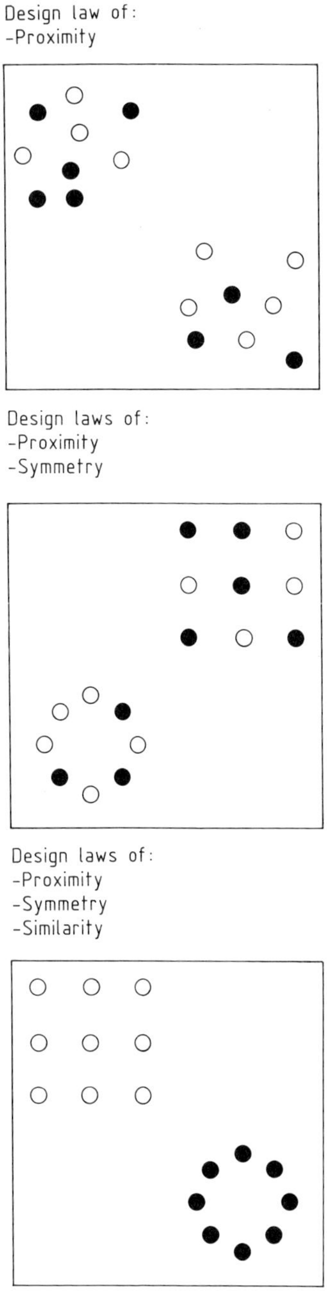

Figure 2.6. Simple design laws [2.35]

monolithically bound to it. Any change in the data structure impacts on the functional structure, and vice versa.

- Frequently, insufficient stress is laid on reusability, since it cannot be known how and in what connection an arbitrarily defined function or an arbitrarily defined module may be used again—in contrast to mathematical and statistical program libraries, where reuse is straightforward. The complexity of such a DP system increases more than proportionally to the functionality.

In the early 1970s, greater value was accorded to data structuring. This was the period of developments such as the PASCAL programming language and the first hierarchical data-base systems. Data structuring became more important.

Subsequent development is marked by attempts to shift the meaning of the data (the information) out of the programs and into the stored data, and to generate program systems in a manner directly opposed to the functional approach. Now the first step in system development is to design the data model on which the later program systems will work.

One speaks of data-oriented or object-oriented structuring, since the analysis is focused on information objects (data structures) as representations of real or abstract objects.

Objects are described and related to one another through their properties, also called attributes. Typical relationships are abstraction and mapping (correspondence) relationships [2.3].

The result of this approach to structuring is the semantic data model, which is designed independently of the functions defined on it [2.12], [2.15].

In his epistemological discussions, Wendt [2.13] classifies objects as concrete and abstract. Concrete objects include things and processes; abstract objects include concepts (see [2.28] and Fig. 2.8).

After the existence of objects and the properties of objects, the relationships of objects are taken up third. A first type of object relationship is described by ordering, which is arrived at through a comparison of some property.

Other important object relationships are part-whole relationships and instrumentality relationships. An instrumentality relationship is said to exist between a thing (e.g., a process facility, a process monitoring and control system) and a process (e.g., a continuous process, a control process) if the thing plays a part in the process.

Inseparably connected to processes are states. States include the initial and final times of processes or subprocesses.

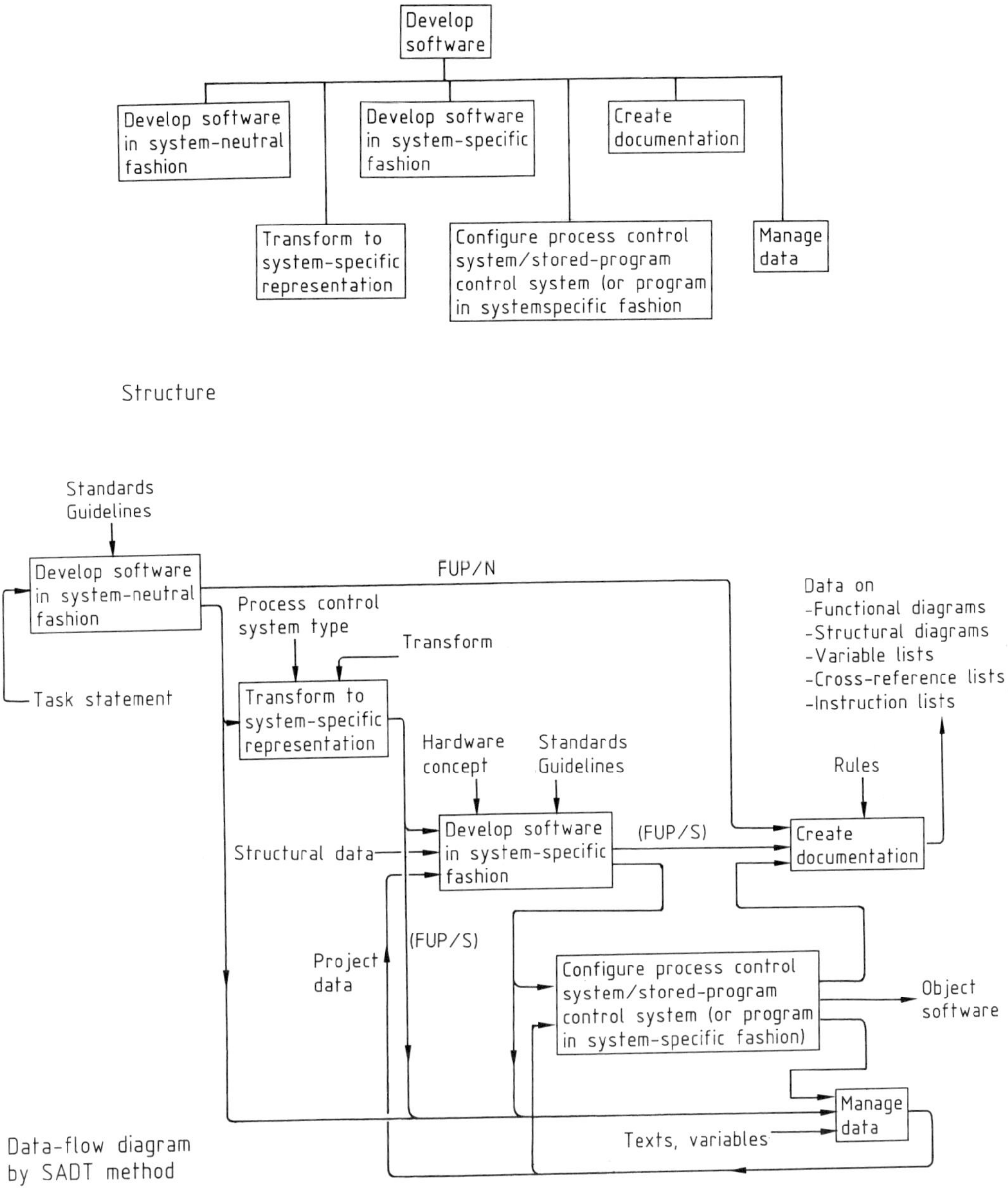

Figure 2.7. Functional structuring for the example "develop software" [2.42]

If countability is associated with the concept of object, and measurability with the concept of property, then the concept associated with the concept of relationship is decidability [2.13]. Objects are counted, properties measured or described (see "Scales" in Chapter 3), and relationships are decided.

Thus the three fundamental concepts—object, properties, and relationship—have been introduced. These make it possible to structure experience, and thus they govern the world of human thought [2.13], [2.45], [2.46].

Concrete objects are the plant, the apparatus, the process monitoring and control system, the production process, the reaction process, the sensor–actuator process, the data-processing process, and so forth. This group of objects is also called "devices"; devices must be designed,

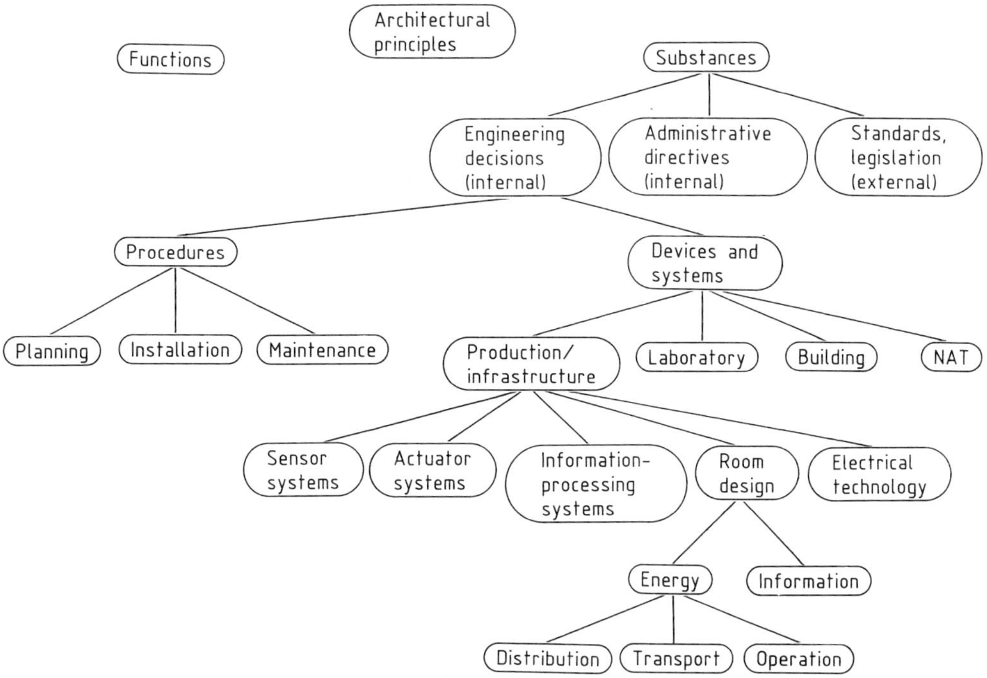

Figure 2.8. Structure of PCE propositions

built operated, maintained, and optimized, these activities being known as "procedures." The outcomes of these procedures are represented as flowcharts, construction drawings, process monitoring and control station diagrams, and so forth. In the Cassirer scheme, both groups of objects are called substances. Functions that describe a task or activity (e.g., plant/apparatus coding; see Section 2.3) also have a representation in the object-oriented formulation (see Figs. 2.22, 2.24, and 3.26).

Functions (i.e., tasks) are implemented through procedures (abstract objects) from devices (concrete objects). Figure 2.8 gives a large-scale survey of "object worlds." For clarity in terminology, relationships are described not with the function concept usual in informatics, but with the epistemologically based relation concept (cf. entity relationship) [2.28].

The abstract representations are the objects that can be manipulated in the computer. They always relate to an object, which they describe from a certain abstract point of view.

The object-oriented approach thus represents a "natural" modularization concept, since the objects can be extracted directly from the application. If the functions are likewise structured in an object-oriented manner, this constitutes a principle of functional structuring that is far less sensitive to modifications.

The Principle of Abstraction. The (provisionally) last step in development combines relational and object-oriented structuring. In the "abstract" data type, data structures and relational structures are merged into a whole. The relations enclose the data as if in a shell; the data can be manipulated only through the relations "guarding" them. The high modularity of these constructs suggests that there may be significant gains in the reusability and maintenance of software. This principle is referred to as data encapsulation or "information hiding" [2.31], [2.47]–[2.49].

The structuring principle is thus concerned with the structuring of data and relations, their (static) structural and (dynamic) sequential structures, and their interactions.

Usually, the sensory or empirical view of things is said to be "concrete," while any description in terms of concepts is said to be "abstract" (see also [2.13]).

Abstraction means reducing complexity by neglecting all inessential properties or features. One also speaks of models.

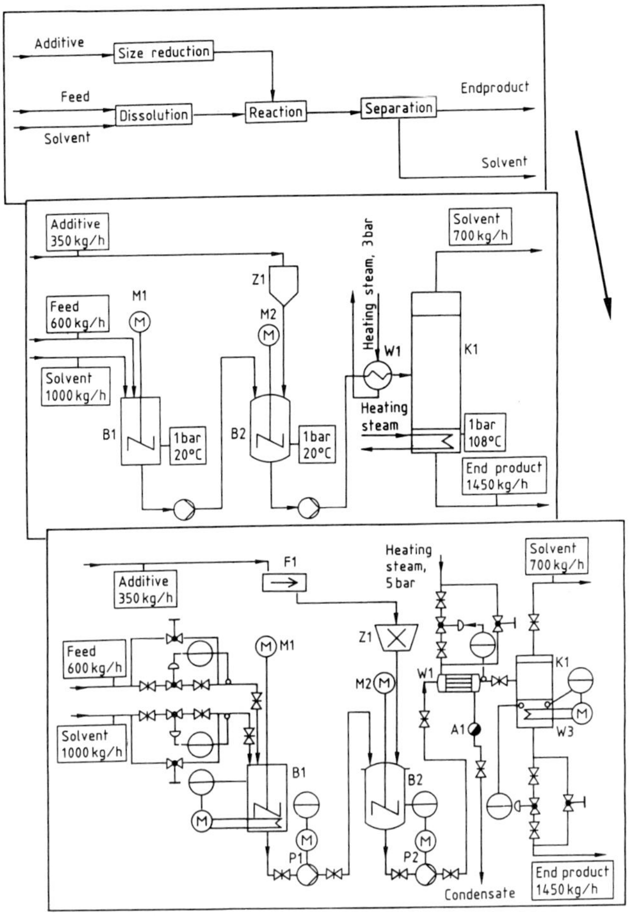

Figure 2.9. Design steps in process engineering [2.50]

Models are always simplified pictures of the real objects (abstractions) as viewed from a certain aspect (see also Sections 3.1 and 3.3).

In process engineering, flowsheets are models of a real facility in graphical form. They are standardized and represent the medium of communication for persons involved in the design, construction, and operation of plants (Fig. 2.9).

Mathematical models in the form of systems of algebraic and differential equations depict partial aspects of the physicochemical and apparatus-specific parts of processes. Such models are used, for example, in model-aided measurement techniques for the control of processes, the detection of faults, and simulation (see Chaps. 3 and 4).

A plant model is a representational picture of the plant that is much favored as an object of study by plant designers and operators.

The three-dimensional representation of plants in CAD systems is a way of getting the advantages of such models while using only DP resources.

Along with mathematics, the field of informatics now offers methods with which complex engineering systems can be modeled, simulated, and optimized.

It must not be overlooked that data models are always reductions from real systems. Models, no matter what kind or in what medium, represent a given system from various points of view. They emphasize essential properties and suppress inessential ones. Within these limitations, however, models are an indispensable part of any modern information technology.

Class-Forming Abstraction. Object orientation means that once the objects have been identified, they are then described in terms of their

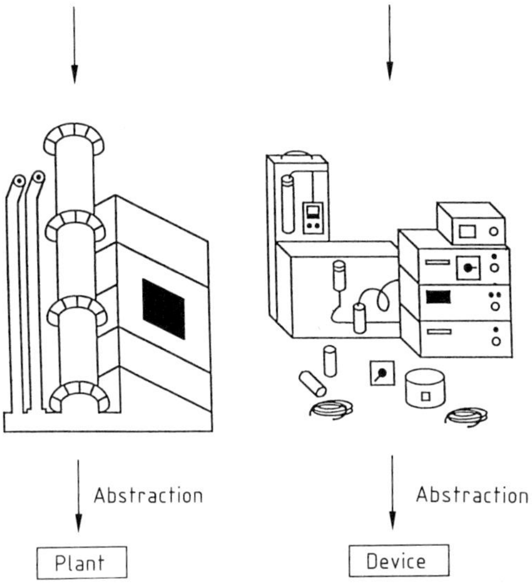

Figure 2.10. From things in the real world to abstract objects

properties or attributes. Each such property consists of a category and a value; these can be scaled either metrically or nonmetrically (see Section 3.1 and Fig. 3.1). The number of properties describing an object is not intrinsically restricted. The treatment is generally limited to the attributes relevant to a mode of examination or an application, or to characteristic attributes; this is equivalent to an abstraction (Fig. 2.10). Such attributes can be listed in a table (Fig. 2.11).

One column in the table describes one characteristic property or attribute of the object. The table represents a "class of objects"; that is, a collection of objects having at least one attribute category in common.

The attributes describing a class simultaneously identify the type of an object. If the same attribute categories are employed for the description of more than one object, these categories are said to be "generic."

A given object in the real world is called an instance of the class (Fig. 2.12). The concept of table, also called relation, leads to the relational data base, which utilizes the relation (= correspondence, dependence, etc.) as its organizing principle. The table as a data structure is purposely not given any further structure, so that it is possible to create a "relational algebra" that is mathematically well-defined. Any manipulation of the data base is thus broken down into set-theory operations.

Apart from class-forming abstraction, the identification of objects is important. Each instance of a class must be identified by a unique name and/or a unique identification number.

Without exception, every one of the classification schemes devised earlier for identification

Table

Name of plant	Identification	Plant complex	Battery or works	Capacity	Utilization

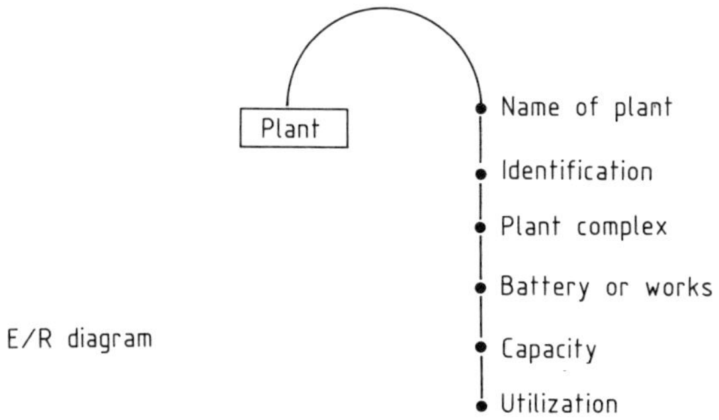

Figure 2.11. The table as a form for describing object properties

Name of plant	Building	Plant complex	Battery or works	Capacity	Utilization
Sulfuric acid unit A	4711	Sulfuric acid	Brazil	100000 t/a	85%

Instances of object class "plant"

Figure 2.12. Instances of a class of objects

purposes has failed, because objects cannot be unambiguously classified (see Concept-Forming Abstraction) and an excessively narrow decimal classification system often proved too restrictive. Classification systems were weak forms of data modeling at a time when data bases and semantic data models were not yet known.

What is recommended today is an unsuggestive, system-free, numeric or alphanumeric code and the description of properties—including the membership of the objects in one or more classes (polyhierarchy)—in a semantic data model.

Another form of identification has been discussed in connection with direct manipulation. An object is identified by "pointing" at it. This principle of identification is extremely important for the ergonomics of user interfaces (cf. the icons first introduced by Apple in the Macintosh). A third option is to describe objects so that they can be found again. The relational data base draws its strength from this approach. Structured query language (SQL) represents an industry standard and makes the results independent of the data-base system.

Concept-Forming Abstraction. Concept formation is an act of abstraction that leads to concepts and hierarchies of concepts. It is a generalizing process, in which particulars are disregarded. Unimportant properties of the object are left out; those remaining govern the class of the objects. Each abstraction rests on a certain objective of thought or speech, for example the objective of being able to think or speak about [2.45]:

- Things, regardless of engineering details
- Procedures, regardless of what principles or concepts apply to them
- Factual statements, regardless of meaning-preserving transformations in the form of circumstances

According to FREGE [2.51], things that are already linguistically available are grouped, from the viewpoint of something they have in common, into a "concept," which offers an invariant way of speaking or thinking about things already linguistically available.

The formation of "abstract" concepts cannot take place in isolation from the intended purpose or objective. This may be a disadvantage if one wishes to do modeling in the most application-independent way possible at an early stage of system design. For example, a biologist might group dogs and cats together as domestic animals, place these in turn under mammals, and so forth, while a jurist might distinguish movable and immovable objects simply because movable objects can be stolen and such taking is a statutory offense. DARWIN, in "The Origin of Species," had already discussed the difficulties involved in devising consistent taxonomies and hierarchies of concepts [2.27].

The concept arising from concept formation is not specified in advance but lies in the discretion of the data modeler. It is therefore quite possible for multiple taxonomies to exist in one and the same realm of objects. This can be illustrated by a simple example: If one abstracts from the manufacturer of certain bimetallic sensors, the concept or class used is that of manufacturer, while the concept introduced is that of bimetallic sensor (Fig. 2.13 A; see also [2.52]). If one wishes to introduce the concept of expansion sensor along with the concept of bimetallic sensor, there is a further abstraction process using the principle of measurement as class (Fig. 2.13 B) and leading to the more abstract concept of temperature sensor.

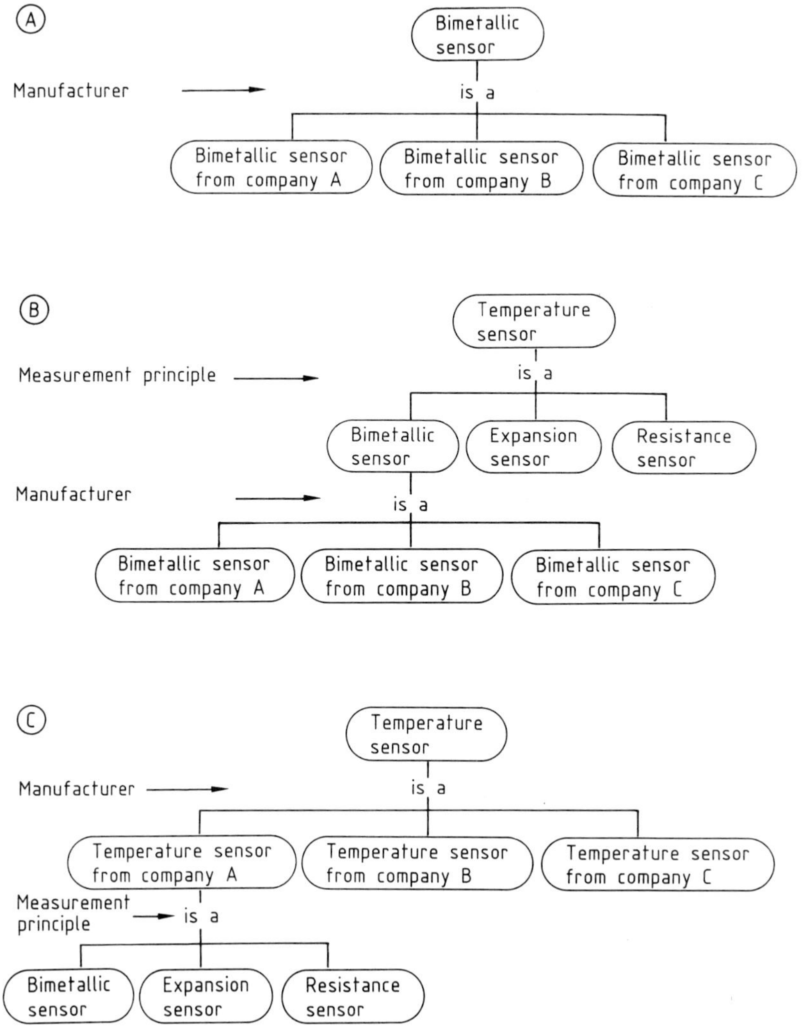

Figure 2.13. Taxonomies of sensor systems
A) Abstraction according to manufacturer; B) Abstraction according to measurement principle; C) Alternative taxonomy

Now it might occur to one to set forth the sensor program in closed form by manufacturer; now the taxonomy looks different again (Fig. 2.13 C; see also Chap. 5).

Complex-Forming Abstraction, Aggregation. In complex-forming abstraction, unlike objects are combined into a new object-concept. The process is also termed aggregation, because the new concept denotes either the static structure of one object or the relationship between objects. The object newly created by aggregation can also have new attributes that are made possible only by the new whole. A distinction can be made between this operation and agglomeration, a simple collection of objects in which the attributes present merely change in value.

Aggregation can also be represented in graphical form. Figure 2.14 A shows the static structure of a sensor system. The concept "customer order" (Fig. 2.14 B), on the other hand, is constructed from the concepts of customer, article, and time. The new customer order concept contains part of the information from each of the original concepts [2.14].

Quite early DP programs were able to handle aggregation in the form of parts lists.

Functional Abstraction. A system can be modeled in greater or lesser detail and in a more

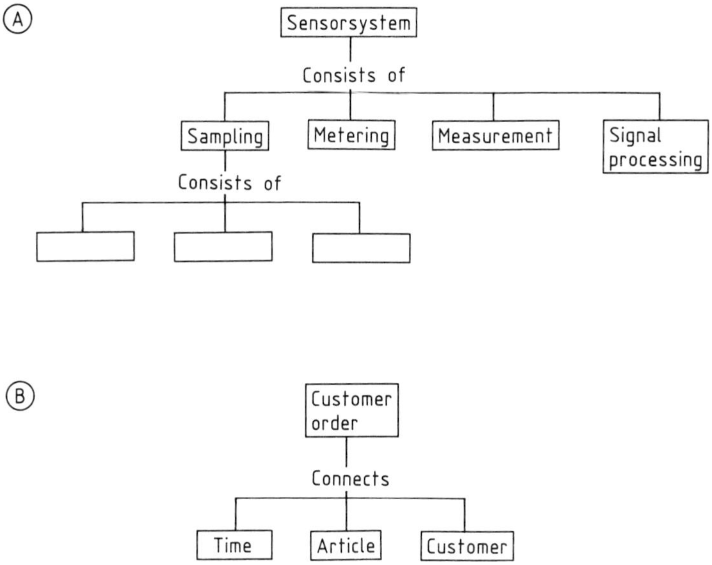

Figure 2.14. Examples of aggregation
A) Structure of a sensor system [2.98]; B) Complex object: customer order

or less abstract fashion, depending on the objective. Different persons, with different tasks and areas of responsibility, take different views of the system. Systems analysts and system modelers always describe a system as an excerpt of reality viewed from one or several of these standpoints. According to POLKE [2.34] (see Fig. 2.28, p. 346), RASMUSSEN introduced systematic order into the more-or-less intuitively grounded models [2.53], describing a system (Fig. 2.15) as functional abstractions over multiple levels. The number of levels can differ from one system to another, depending on the type of system and the objective. Nor do all the levels have to be implemented.

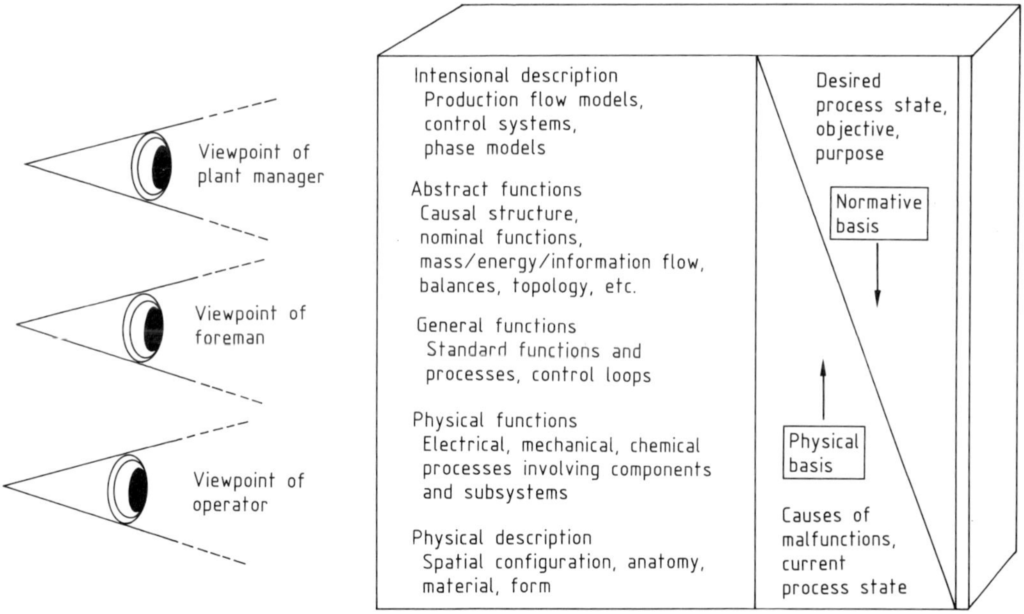

Figure 2.15. RASMUSSEN's scheme of viewpoints [2.53]

The design levels of this model extend from intensional models at the uppermost level to extensional ones at the lowermost.

Extensional models relate to the specific physical world, while higher-level models represent the purpose and objectives that are to be achieved with the system.

Scales for the desired behavior must to be developed from the purpose of the system; in other words, criteria for errors and malfunctions can be formulated only against the background of the process objective.

In certain operating states, for example, alarms may be meaningless; measurement signals may have different nominal and limiting values, depending on the operating mode; and so forth. The causes of errors or malfunctions arise from the physical world (bottom-up), while the criteria for defining errors can only be derived from the process goals (top-down).

This approach to modeling brings order to the great variety of representations of a system. Above all, it demonstrates that no single representation is adequate; a different view of the system may make a different representation useful.

Many technical and economic systems have been created under this architectural principle. Typical examples are computer architectures; architectures of communications equipment, the latter known as the ISO-OSI reference model (e.g., [2.54]); and the level model of a corporation discussed in Section 2.3 [2.34], [2.55]. The structures of monitoring and control systems mentioned in DIN 19222 also fall under this principle [2.56].

The principle of functional abstraction is based on the following rules:

- The levels must offer generally usable functions and relations; that is, they must represent a good abstraction.
- The levels must have clear interfaces and minimal upward and downward connection; they must obey the secrecy principle.
- They must completely implement the required functions; that is, they must be self-contained, with the properties of locality and internal strength.
- A level "uses" the services of the level immediately below it. Jumping over levels is not permitted.

It follows that all functions requiring intensive data exchange with one another should be grouped on the same level (Fig. 2.16).

The design process for engineering systems is also oriented to such a level model. For example, mechanical design begins with the functional structure, proceeds through the operative principle of design synthesis, and finishes with dimensional synthesis.

The Transformation Principle. In chip design, the following levels are recognized: system design, logic design, circuit design, and layout design. The process makes use of detailing algorithms.

Configurable detailing algorithms are not known for the corresponding steps in plant design based on flowsheets (Fig. 2.9).

If the design process within a process monitoring, control, and automation system is to be

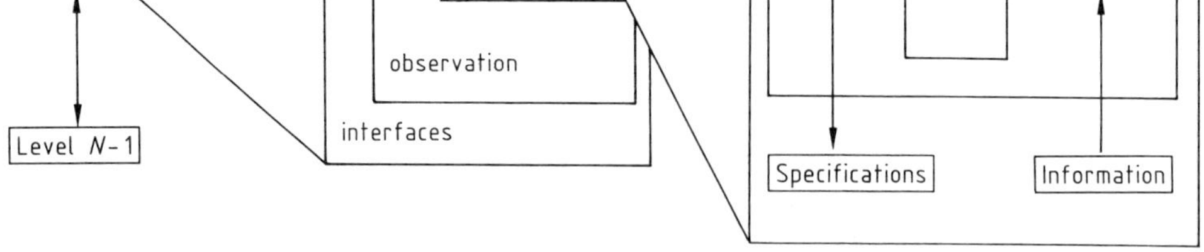

Figure 2.16. Hierarchy of functions (level model)

completely or partly automated, it must be clear above all how the representation form of one design level is to be transformed into the representation form of the next one. For the developer and designer of an automation system, such a concept supports creative design activities, design verification, project planning, and documentation.

This transformation process in informatics is fully automated by problem-oriented languages, assemblers, and on to machine code. Programs that do these transformations are called compilers, assemblers, and so forth.

Sequence-oriented description concepts and representations are required in order to describe the functions of information-processing operations. One speaks of a transformation principle because input data are transformed to output data and initial states to final states.

The following description types are distinguished on the basis of how the operator is implemented:

- Sequence-oriented descriptions
- Logic-oriented descriptions
- State-oriented descriptions

Sequence-Oriented Descriptions. Classically, information-manipulating processes are described in terms of individual, application-neutral elements. It can be shown that any routine or any procedure can be described in terms of the linguistic elements sequence, alternative, and iteration. In a procedural programming language, it would therefore be sufficient to make these three constructs available.

Typical graphical descriptive media for procedural concepts include

- Flowcharts
- Nassi–Shneiderman diagrams
- SADT diagrams
- Michael Jackson diagrams

Textual media include, among others,

- Pseudocode
- Structured text
- Programming languages such as FORTRAN and PASCAL

The analogy between operators in mathematical physics and their importance in the description of state changes (including time-dependent changes) can only be mentioned here [2.57].

Logic-Oriented Descriptions. Logic-oriented concepts relate not so much to the "how" of the process as to the "what." Graphical media used for this purpose include

- Combinational circuits
- Logic diagrams
- Decision tables

Textual media include

- Fourth-generation languages
- Algebraic specifications
- Logic-oriented languages

State-Oriented Descriptions. The automation of technical systems often requires the control of processes that take place in a predominantly step-by-step (i.e., discrete) fashion. The complexity of a discretely controlled system becomes very high when there are processes parallel in the time domain (i.e., concurrent processes). If these processes use shared resources, there is always the risk of "deadlock." The processes block one another in such a way that they cannot continue to operate without outside intervention. Such deadlocks cannot easily be foreseen. The need for an analysis and, possibly, a simulation of these systems is thus manifest (compare the typical situation of management in a multi-project environment [2.58]).

Such problems can be described well in terms of mechanisms or automata.

A finite-state automaton is a system having a specified number n of possible internal states $z_1, \ldots, z_n$. For each state z_i there is a well-defined set of inputs for which the automaton changes to another state. An output can occur when such a transition takes place. Automata can be used to describe the time order of state transitions in technical systems.

State graphs (Fig. 2.17), with circles for states of the system and arcs for transitions between states, are used to visualize these systems.

In practical cases, however, the number of global states increases rapidly, so that the representation of the time order by state graphs becomes unwieldy.

The Petri net, an extension of the state graph concept, offers a more elegant way to describe such systems.

A Petri net is a directed graph whose nodes are of two types, places and transitions. Arcs, which can connect two nodes only if they are of different types, represent logical or causal relationships between nodes [2.38], [2.59]–[2.67].

Using a mode of analysis different from that of state graphs, the Petri net technique allows the

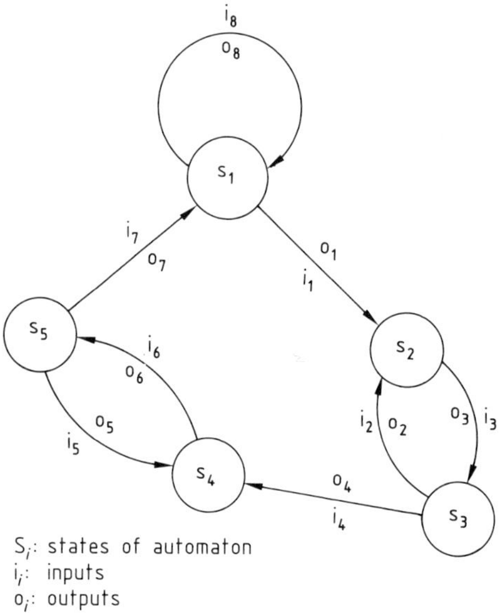

Figure 2.17. State graph illustrating an automaton

description of systems that represent more extensive automata. Thus, the circles (places) represent partial states in the system and, therefore, conditions for the occurrence of certain events that can occur in the system and lead to state transitions.

The partial state represented by a place is in effect if the place holds one (or more) marks or "tokens."

A transition is enabled, that is, it can occur or "fire," when all its input positions (arcs leading to the transition) have tokens while all its output positions (arcs leading away from the transition) have none (Fig. 2.18 A).

When an enabled transition fires, tokens are removed from its input places and placed on its output places, which now may enable transitions in their turn. In larger nets, this rule means that several transitions may be enabled at once; this is known as concurrent execution (Fig. 2.18 B and C).

A Petri net lends itself not just to graphical representation but also to mathematical description. On the basis of such a description, the dynamic behavior of the net can be studied by using computer-aided analytical tools.

Interesting properties of the net include, for example, reachability and the presence of deadlocks: If each possible occupation of the Petri net with tokens (each "marking") is interpreted as a possible global state of an automaton, the associated state graph is also called a reachability graph. It immediately shows whether there are markings of the Petri net that can be reached in just one way or cannot be reached, whether there are deadlocks, or whether the initial marking can be reached again (reversibility; see Fig. 2.18 D).

Petri nets can be broken down in hierarchical fashion and so support top-down and bottom-up design.

In the area of process control, Petri nets are usually represented by functional charts. The technique of functional charts makes it possible to portray directly the structure of a Petri net and offers linguistic elements for the specification of control commands to be issued at the individual steps (nodes of the net).

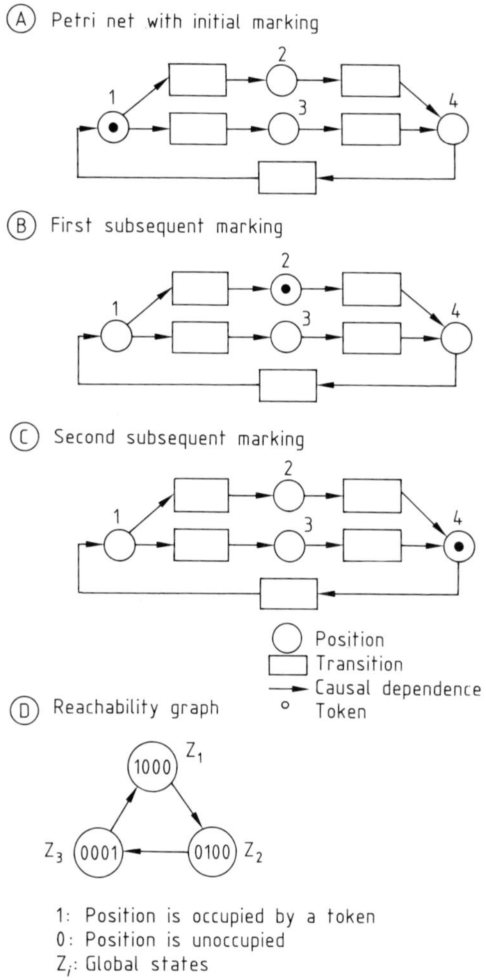

Figure 2.18. Petri net with initial and subsequent markings and the associated reachability graph

One graph that is especially easy to comprehend is the "phase model" of production, which is discussed in more detail in Section 2.3.

2.3. Applications in Process Control Engineering

Object Worlds of Continuous Production Processes. If the system to be designed or operated (whether it is an apparatus, a plant, a control system, a software system, or some other system) with all its subsystems and components is placed at the focus of computer support, the monitoring and control of continuous processes can be broken down into plant engineering, process engineering, and process control engineering worlds.

In order to describe the actual operation, these are supplemented by the world of process communications, logistics, and maintenance (Fig. 2.19) [2.68].

In an object-oriented approach, substances in these worlds can be defined as functions. They represent the uppermost level of a corporate data model within process control engineering [2.64].

Plant Engineering. Typical descriptions in the field of plant engineering take the form of data sheets (Fig. 2.20).

The abstraction principles discussed in Section 2.2 can be seen in this data sheet: The glass-lined stirred tank, as represented in Figure 2.21, consists of drive, clutch, stirrer, tank, stirrer shaft, shaft seal, motor, and gearbox (complex-forming abstraction, aggregation).

Some of the objects are further specified; for example, the stirrer is identified as an impeller, dual-impeller, or anchor type, and the stirrer shaft seal is specified as a stuffing box or a rotating mechanical seal (concept-forming abstraction).

All objects listed on the data sheet are described by a series of attributes and thus stand for a class of objects (class-forming abstraction).

The "plant and apparatus coding" (PAC) scheme can be viewed as a first formulation and crystallization point for a corporate data model [2.69]. This coding system describes the static structure of a plant in terms of the required functionality (Fig. 2.22). It has analogies with the "power plant coding system" [2.70].

While most description systems do not separate the functions (tasks) of a plant (abstract) from the real technical plant (device), the PAC gives a rigorous description only of the function, up to the technical device and its functional components, and is therefore invariant to actual technical implementations (Figs. 2.22 and 3.26).

Earlier DIN representations [2.71] could not employ the object orientation logically utilized in the implementation of the PAC [2.69].

For this reason, stress must be laid on the worldwide harmonization of concepts; this comment applies equally to all three of the "object worlds" under discussion here.

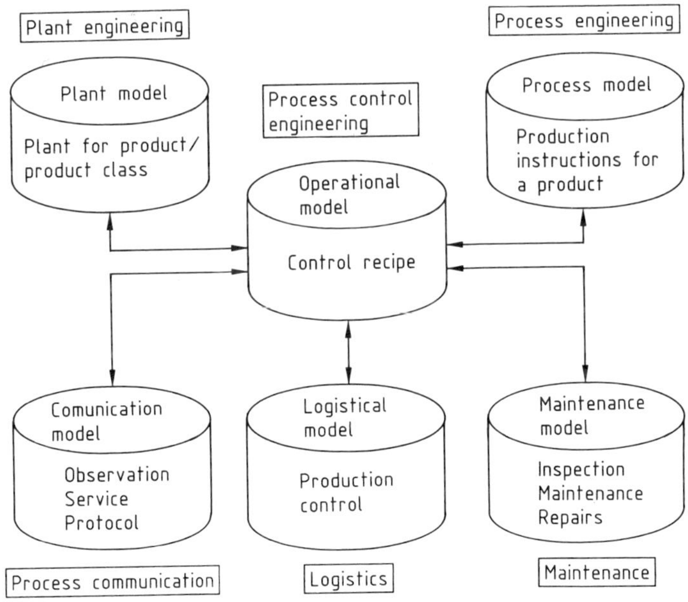

Figure 2.19. Object worlds of continuous production processes [2.68]

Dept.	Process/Plant ____________ Works/battery ______ GB/ZB ______ Building ______ Project No. ______	Section Sheet No. of

No.					
1	**Enameled Stirred Tank, Stirrer with Drive for Stirred Tank,**				Code
2	____ m^3				
3	Customer	File No.		Date	Class/Rev. No.
4	Department	Comm. No.	Inquiry No.		
5	Drawing No.		Order No.		Inventory No.
6	Manufacturer	Stock No.	Bid No.		
7			Contract No.		Code No.
8	Previous Order/Contract Nos.				

No.						
9	Drive	Top drive O	Bottom drive O	Inf. variable from to		Drawing No.
10	Requirements	Speed	rpm	In steps		
11	DIN 28161-A O	Sliding-ring seal removable at side			O	
12	Gearing	Not variable	O	Pole-reversing	O	Drawing No.
13		Speed variable	O	Gear-shift	O	
14		Continuously	Hydraulic O	Change gears	O	
15			Mechanical O	V-belt pulleys	O	
16			Electrical O	Torque transmitted	Nm	
17		Manufacturer		Model, size		
18	Skirt	Companion dimensions DIN 28162 Part 1			O	
19	Clutch	Motor-gearbox		Gearbox-stirrer shaft		
20						
21	Motor	Power	kW	Enclosure		Remark:
22		Speed	rpm	Explosion protection		
23		Voltage	V	Pole-changeable	O	
24		Frequency	Hz	OK for variable frequency	O	
25		Design		Size		
26	Stirrer	Impeller O	DIN 28157 O	One-piece	O	Drawing No.
27		Dual impeller	O	Split	O	
28		Anchor O	DIN 28158 O	Total stirrer length	mm	
29				Stirrer diam.	mm	Enamel grade
30						
31	Stirrer shaft	Diam. in seal	mm	Shaft enameled in seal	O	Drawing No.
32		Shaft split	O	Shaft bushing metallic	O	
33		Shaft integral	O	Material		
34		Shaft end DIN 28159	O			Enamel grade
35						
36	Stirrer shaft seal: Inst. dim.	DIN 28138 Part 2	O			
37	Stuffing box	Oil buffer chamber ring	O	External cooling	O	Drawing No.
38		Leak catching cup	O			
39	Packing rings	Number required		Material		
40		Manufacturer		Model, size		
41	Rot.mech. seal	Manufacturer		Model, size		Remark:
42	WN 1560 Pt 3 O	GL No.				
43	Face ring matl	Product side		Atmosphere side		
44		Flushing medium				
45	Flush pressure	Manufacturer		Model		
46	Recirculation	Thermosiphon	O	Forced circulation	O	
47		Cooler O	No cooler O	Flush pressure	bar	
48				Circulation rate	L/min	
49	Inspections	Factory Standard 6345				
50	**Remarks:** Spare part(s) to fit Stock No.:					Whse. No.
51						
52						
53						
54						Recd. AMV:
55						
56						
57						
58						

No.							
59	Name		Change	1	2	3	4
60	Date						

Figure 2.20. Data sheet for an enameled stirred tank

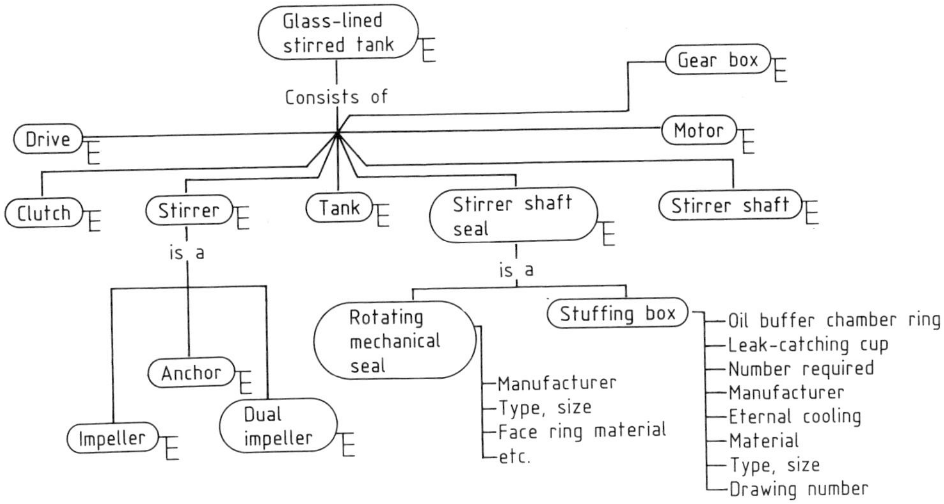

Figure 2.21. Data model of a stirred tank in the entity–relationship (ER) representation

Process Engineering. The following example of process engineering illustrates a taxonomy of unit operations (Fig. 2.23) [2.72].

The classification scheme first looks at the most important operations, such as separation, combination, and so forth. The next criterion applied is the state of aggregation of the starting material (see also [2.73]).

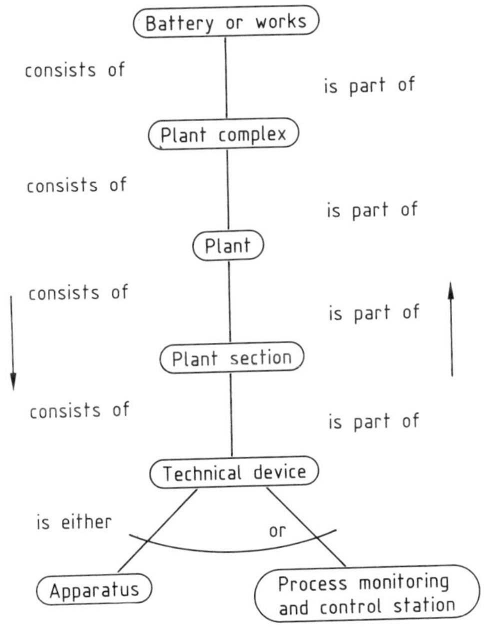

Figure 2.22. Plant model [2.70]

The general production instructions for the making of a given product (Fig. 2.24) also arise from process engineering.

While the first attempts to formalize production instructions were based on the classical structures of thinking in process engineering [2.74], more recent approaches start with the phase model of production derived from the transformation principle [2.75]. The configuration principle of modern process control systems builds on knowledge from object-oriented programming and, more recently, experience in modularization [2.76]–[2.78]. Figure 2.25 illustrates the correspondence of plant model and process model (see also Fig. 3.26).

Process Control Engineering. To illustrate the relationship between process engineering and process control engineering, consider what happens in the transition from a generally valid recipe for making product A to a recipe for making product A in plant X. In a taxonomy of recipes, this is a concretization of the "plant-independent" basic recipe into a plant-dependent control recipe. New research by the ISA, particularly on the topic of batch control systems [2.79], and by NAMUR-AK 2.3 on requirements for recipe-mode operation [2.80] must be taken seriously as a basis for discussion. Figure 2.26 illustrates the principal recipe types [2.80].

If one assumes a self-contained recipe for the individual operating modes (see Fig. 4.48), the taxonomy of recipes shown in Figure 2.27 results.

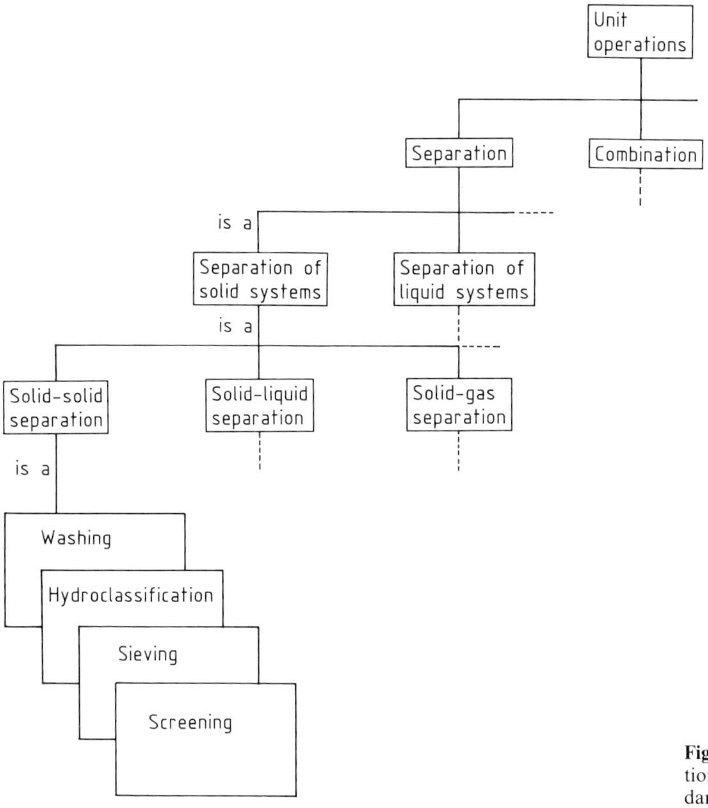

Figure 2.23. Taxonomy of unit operations (process engineering) (GDR standard [2.72])

Level Model of a Corporation. Structuring by hierarchical levels (functional abstraction) is encountered in a variety of forms (e.g., for the description of computer networks or data bases). A corporation can also be represented in this way.

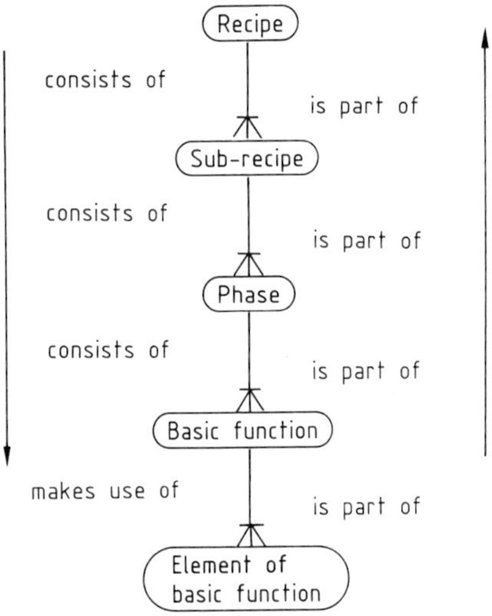

Figure 2.24. Structure of production instructions (process model, recipe) [2.38], [2.64]

According to a suggestion made earlier [2.55], [2.81], if the abstraction principles stated above are applied to a corporation, four levels can be identified. By way of example, Figure 2.28 shows schematically a model of a company involved in the process industry. The four levels are the corporate management level, the production management level, the process control level, and the field level. The dispositive quality of the functions decreases from top to bottom, while the operative quality increases.

In what follows, the corporate, production, and process control levels are discussed for the production division of a chemical industry corporation [2.34], [2.82].

Corporate Management Level. In a world where increasing competition demands rapid and often massive reactions in many markets, the corporate management level, where goal-setting and long-term planning are done, is becoming more and more important.

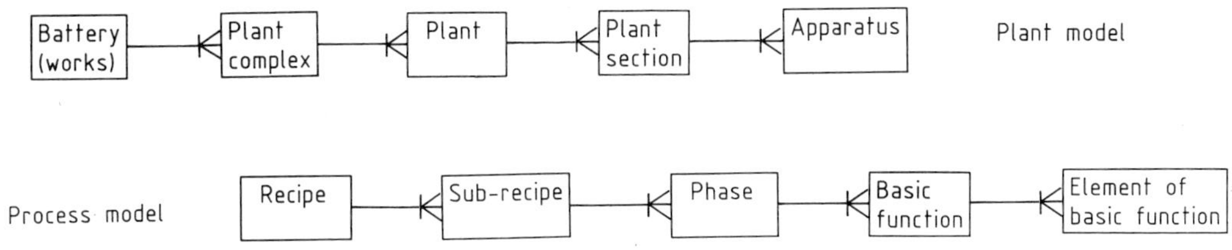

Figure 2.25. Correspondence of process model and plant model

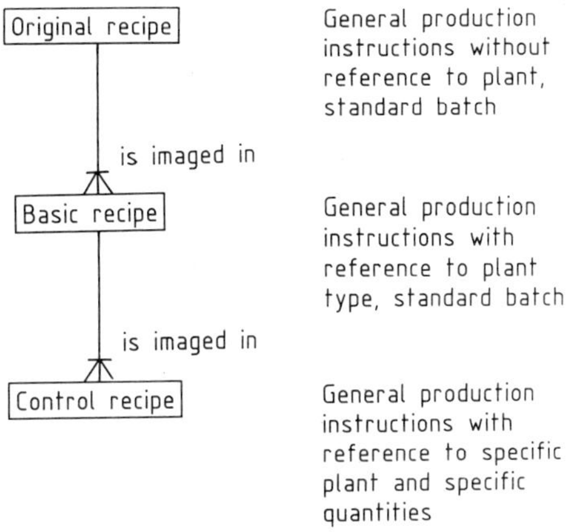

Figure 2.26. Types of recipe [2.80]

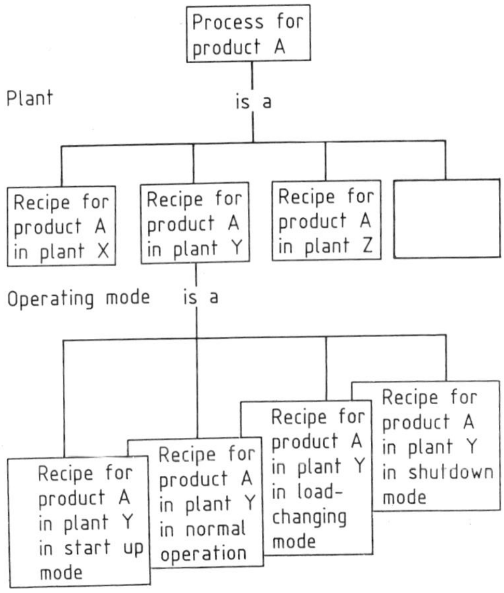

Figure 2.27. Taxonomy of production instructions

The key condition for the support of the corporate management level, once systems-engineering problems have been solved, is effective control of the lower levels, in particular the establishment of conceptual information structures.

For this purpose, the information available on the subordinate levels must be compressed in a way appropriate to its use. A strongly department-oriented view is supplanted by a supra-departmental one; a country-specific view by an increasingly international and global one; and a focus on the individual industry or sector is supplanted by a view that takes in many industries.

The increasing importance of effective product development management, for example, makes clear the need for a supra-departmental view. In order to avoid costs for dead-end development and delays in development, concrete requirement profiles should be derived from consumer needs at an early stage, the existence of competing products should be investigated, and the requirements of downstream processing industries should be taken into consideration (see also Section 11.3 and Figs. 11.38 ff.). This information should be derived with internal development capabilities. Corporate management should be able to examine the development process at any time. Product quality and environmental protection are other examples where new information aggregates, cutting across classical departmental lines and extending all the way up to the corporate management level, are gaining in importance and must therefore be made available (see also Fig. 6 in [2.83]).

Any company that wants to make intelligent decisions on a worldwide basis regarding, for example, research and production sites and markets to be entered, while optimizing its decisions in a supra-departmental fashion, must have consistent, up-to-date information covering all countries of interest. The result will be increased demands on internal accounting in the sense of comprehensive control, that is, a control instrument for corporate management. Highly diverse, historically evolved structures cannot meet these requirements.

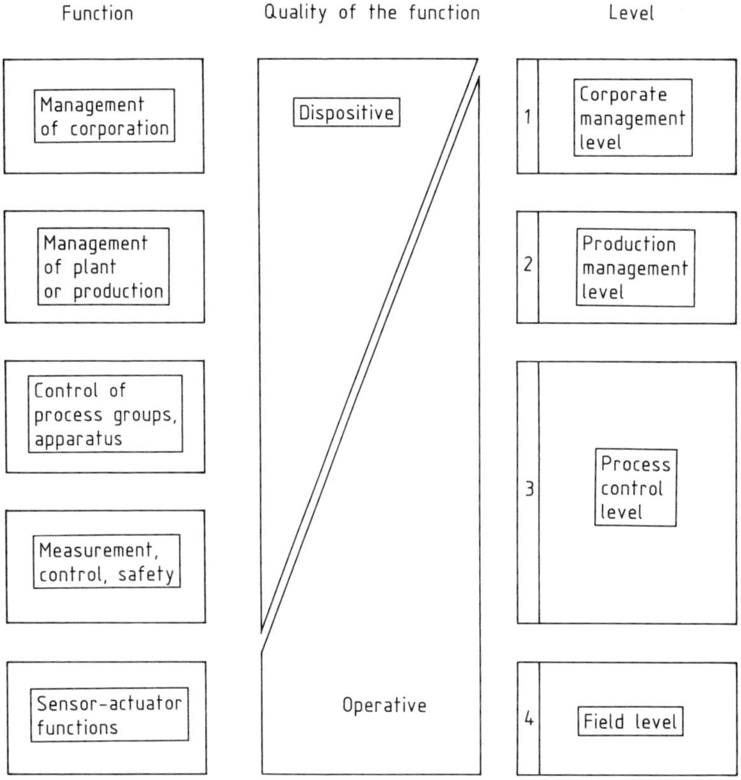

Figure 2.28. Level model of a corporation [2.34]

Particularly in the chemical industry, the product orientation that comes about because of manufacturing conditions must be harmonized with the interests of the consumer market, which often have to do with more than one product. An integrated way of looking at problems can result in synergistic effects in purchasing management as well. A high degree of information transparency at a level higher than the industry or sector is therefore required.

The contribution of information technology at the corporate management level was very limited in the past, because it was restricted to supporting activities directly related to performance or order filling. The technology was marked by an orientation toward single tasks. Data structures were devised in a task-specific manner, in the narrowest sense; data bases were hierarchically structured (see also [2.10]). Data communications between programs also featured a mainly hierarchical structure. The task-oriented application systems based on this technology naturally could make little contribution to supra-departmental views. The core requirements of integration and flexibility could not be met. Personal computers were often made available to staff personnel on the management floors in an attempt to close this gap; most such computers were manually supplied with data from the functional systems and—not surprisingly—produced a wealth of incompatible data.

More recently, however, information technology has reached a stage in which the logical requirements of integration and flexibility can increasingly be realized. This applies both to information storage and to information retrieval and interpretation. The first comprehensive concept was developed for all engineering divisions of Bayer in 1989–1992 [2.44], [2.84].

Requirements imposed on the lower levels are illustrated for the case of production in what follows. A supra-departmental view is selected; that is, the interfaces to other departments are examined (see also Chapter 8).

Production Management Level. The discussion concerning the "factory of the future" has provided an outline of how the industrial production of the future will look. Although each expert's picture of this factory is different in detail, there is a consensus as to the need for increasingly extensive automation (see Fig. 4.49) of production processes for reasons of quality assurance and optimal utilization of raw materials [2.82], [2.85].

Mass production is no longer a prerequisite for automation. The contemplated decentralized automation world, with the self-contained manufacturing islands already in existence, can be realized only if it is connected by unified information and communications systems. The objectives are to improve material flow, shorten storage times, increase the speed of transportation, and cut setup times when the product mix changes. These objectives are essentially the same no matter whether production is a manufacturing or a chemical process. Even so, there is a great difference between the implementations of these two types of systems.

Typical for the manufacturing industry are situations in which one or a few end products are assembled from many individual parts. The assembly process is complicated but lends itself to a virtually complete computer analysis [2.86].

In process industries, such as the chemical industry, the situation is reversed: There is one starting material from which one product or several related products are made (see diagram F8 on p. 379 in [2.82]). This simple, branching flowsheet allows stage-to-stage planning; any change will affect the next stage downstream, but can still be made at any time. The reason is that the process and product properties in each process element are interrelated (see Fig. 2.33). The interrelations, in general, can be described only in a phenomenological way, so that theoretical relations cannot be applied directly.

However, there are also many multiply branched flowsheets in which many end products are produced via several intermediates. These multi-stage batch or continuous production processes have the feature that the central event, the chemical reaction itself, is essentially a probabilistic occurrence and thus calculable only to a limited extent [2.86]. Such processes cannot be handled by production planning methods that are oriented toward parts lists.

Thus the systems integrated in CIM cannot be immediately and directly extended to continuous-process production.

The general structure shown in Figure 2.16 is now implemented in a manner specific to the production management level in Figure 2.29 [2.87], [2.88]. Production specifications supplied by the corporate management level are the following:

- Type of product
- Quantity of product
- Delivery schedule
- Special contract information (e.g., special packaging or formulation)

Detailing procedures are carried out with the aid of level-specific algorithms. Such procedures include adoption of production specifications, selection of basic recipes, comparison of material and apparatus requirements with inventories and free capacities, allotment to production units, assignment of apparatus-specific control recipes, and preparation of schedules. This detailing is often an iterative process that becomes necessary when the nominal/actual comparison in the evaluation algorithm signals disturbances in the production process or when preconditions for detailing are changed during the detailing process itself.

This block of functions has a strongly dispositive character. Once the detailing and disposition process has been completed, the production control function communicates it to the lower process control level in the form of a control recipe (when, what, where, how). The process control level in turn reports the relevant process and product data upward for further compression, either while the process is running or after it has been completed.

These downwardly and upwardly oriented algorithms include specific level data, which take the form of static inventories of apparatus and recipes. The updated execution plans, inventories, and actual product and process data are kept accessible in dynamic data bases.

This type of data management naturally includes a suitable documentation function, which can generate both routine reports and notices of special requirements. Another algorithm, not shown in Figure 2.29, deals with the optimization function. Points such as batch size, tooling time, startup and shutdown procedures, and stockpiles should be optimized. It is also important to modify the basic recipe and control recipe after the interpretation of process and product data (see Section 3.2) if product consequences reveal incompatibilities

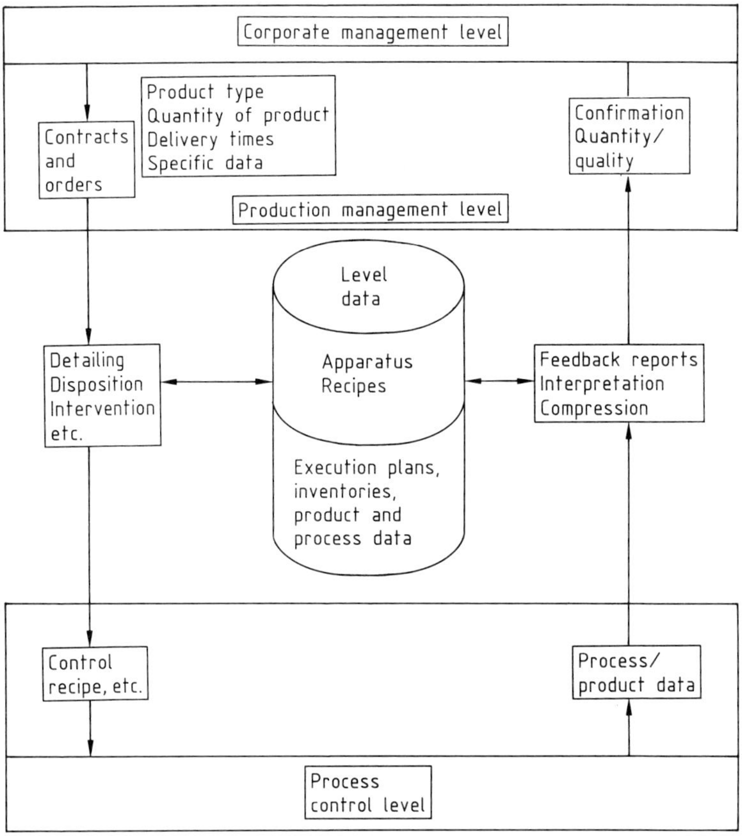

Figure 2.29. Production management level: structure and relationships

or if certain operating points turn out to be sensitive.

In the near future, such a production management system will become a permanent component of all production operations in the process industries. These functions will be implemented in a decentralized manner in individual operations.

Production management systems must be available in a configurable, off-the-shelf form, if only for economic reasons. The individual functions must therefore be implemented in the form of self-contained, comprehensible subsystems (Fig. 2.30).

According to RIEMER, users can do the following on the basis of information that is kept accessible and up-to-date [2.89] (see also [2.87], [2.88]):

- On the basis of sales and operating quantity plans and using inventories and orders, they can determine in advance the net demand for products and feedstocks over all production stages (production planning).
- For this planned demand, on the basis of the instantaneous inventory, purchasing, and orderbook situation, they can ensure the timely procurement of all materials needed for fulfillment of the plan and monitor execution (demand disposition).
- They can keep their own production operations on target as to schedule, quantity, quality, and cost; allot materials and capacities; prepare working documents; continuously monitor production progress; and take prompt action when actual values deviate from nominal values (production control).
- They can accept specifications for production process control; support and monitor the process; and, by direct intervention as well as aggregation and structuring, prepare the continuously acquired production data for production control (process control; see Section 4.5).

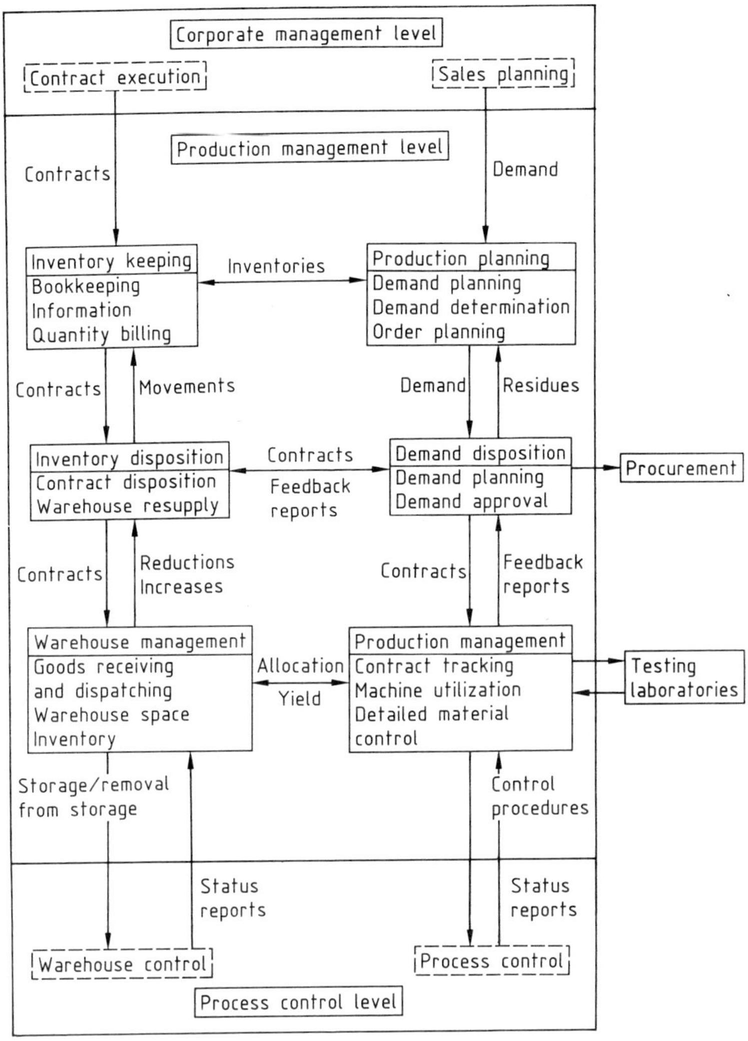

Figure 2.30. Functions of a production management system [2.89]

- They can better support the management of object (products, packaging, etc.) inventories (inventory keeping).
- They can automatically or manually check and plan inventories on the basis of customer, internal, and production orders (inventory disposition).
- They can have the materials at their disposal stored, removed from storage, commissioned, packaged, sealed, and prepared for shipping (warehouse management).
- They can store materials and remove them automatically and with optimal routing of warehouse vehicles (warehouse control).
- They can monitor what is happening in the process, using the minimum number of clear signals and standard information, whether by ad hoc queries on the screen or as printout or graphical output (information and report management).
- They can acquire current data with the required level of detail, store the data, and make them available at various locations for the various functional divisions (corporate planning, controlling, accounting, procurement, duties and taxes, etc.) whether for information or for processing (interfaces).

The example of the production management level shows how the analog of CIM in manufacturing operations is computer integrated processing (CIP) [2.90] in continuous-process operations. Clearly, information has now become a factor of production.

Process Control Level. In the process control level, production orders (i.e., requirements on product, quantity, and delivery time) are transformed to realization processes. This change requires that a processing facility be available, the control recipe (i.e., the recipe adapted to the actual facility being used) has been determined, and all necessary resources are available (see Section 4.5).

The production process is generally subdivided into a sequence of process elements each self-contained in time. The structure of the process monitoring and control system must correspond to this horizontal breakdown of the process. Process control functions for plant sections must be performed in decentralized process control stations and must be suited to handling the information flow in parallel with the process [2.34], [2.91]. The performance of the production order then involves the allotment of partial control recipes to plant sections and the phase-by-phase execution of the recipes through the basic functions of process control engineering (see Figs. 2.24 and 2.37).

Corresponding to the hierarchy of processes and subprocesses, the following functional complexes can be distinguished on the process control level (Fig. 2.31; see also Chaps. 7 and 8) [2.92]:

- Central functions for controlling the process itself (process operating computers, process control computers, process AI computers) and for controlling the process monitoring and control system (process engineering computers for design and diagnostic service). The complex operations involved in diagnosing possible error states of a process control system are merely indicated here (see Chap. 7 and Sections 11.2 and 11.3). The analogy with medical diagnostics should also be mentioned [2.93].
- Individual functions and coordinaton functions: the grouping of all process-level functions in a plant section (process section or process element) in a decentralized process control station (process-level components, PLC).

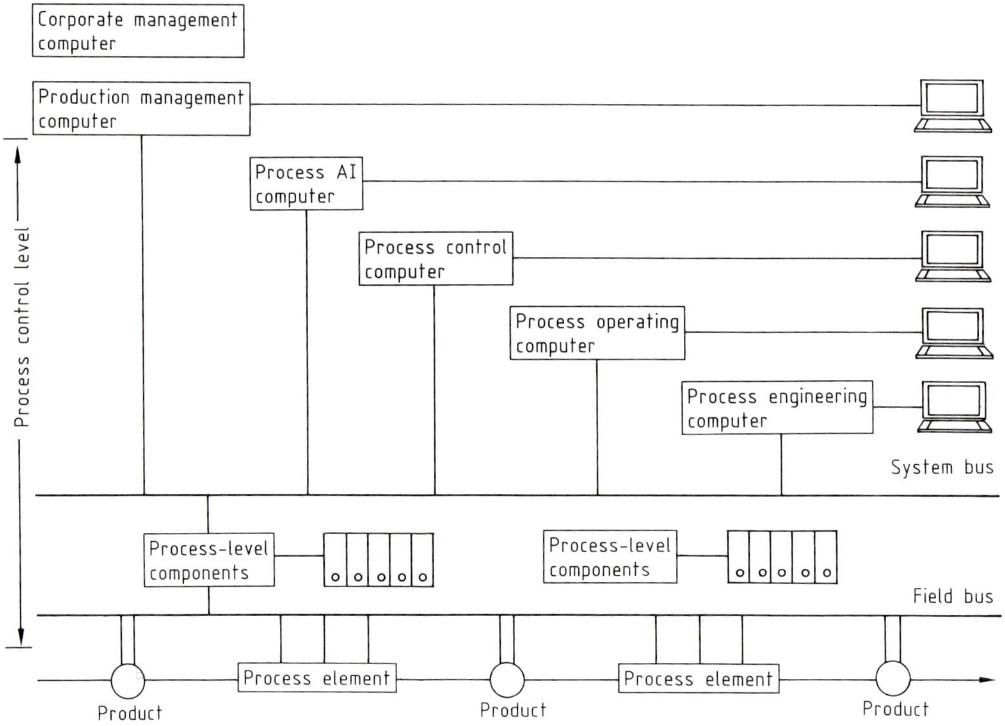

Figure 2.31. Ideal structure of a process/production control system

- System communications functions thought of as interface functions (system bus, field bus).

Phase Model of Production. An important aid for systems analysis of production processes is the phase model of production, which was adopted from software engineering [2.94] and belongs to the category of product nets. Other models lack any representation of product properties in between process elements, which is a prerequisite for constructable quality assurance and process reliability; the phase model supplies this representation. As suggested by BÜSING [2.95], the phase model is readily derived from the basic flowsheet by attaching a label for the products between the individual process elements (Fig. 2.32). Each rectangular box stands for a process element, and each circle for a product. The individual process elements can be regarded as operators that transform the product properties (see "Transformation Principle" in Section 2.2). This method facilitates, among other things, the representation of information flows, which are needed to control the entire process in such a way that it is always held in its instantaneous nominal state. The method also requires a clear picture of what information is (or should be) available in and downstream of each individual process element for process monitoring and quality assurance (see Section 11.3 and [2.96], [2.97]).

Monitoring and control functions relating to certain product and/or process properties must be carried out in each process element. Product and process properties are the object classes of information about the process [2.98]. In Figure 2.33 product and process properties are listed for the example of one portion of a phase model. Every product or process property consists of a category and a value.

Special attention should be paid to scaling (see Section 3.1, Fig. 3.1, and [2.99]).

Process properties are the following:

- State variables such as pressure, temperature, and concentration. The process can be described in terms of these or with derived state functions.
- Process parameters such as heat-transfer coefficient and catalyst activity. These identify the constraints under which the process runs. Process parameters are stationary or at least quasi-stationary quantities.
- Control variables. These characterize intervention in the process.

In theory, the information budget of an entire process can be completely described with an appropriate set of these quantities. In practice, however, it often happens that not all the desired state variables and control variables can be measured or specified. Other information is often employed as a substitute:

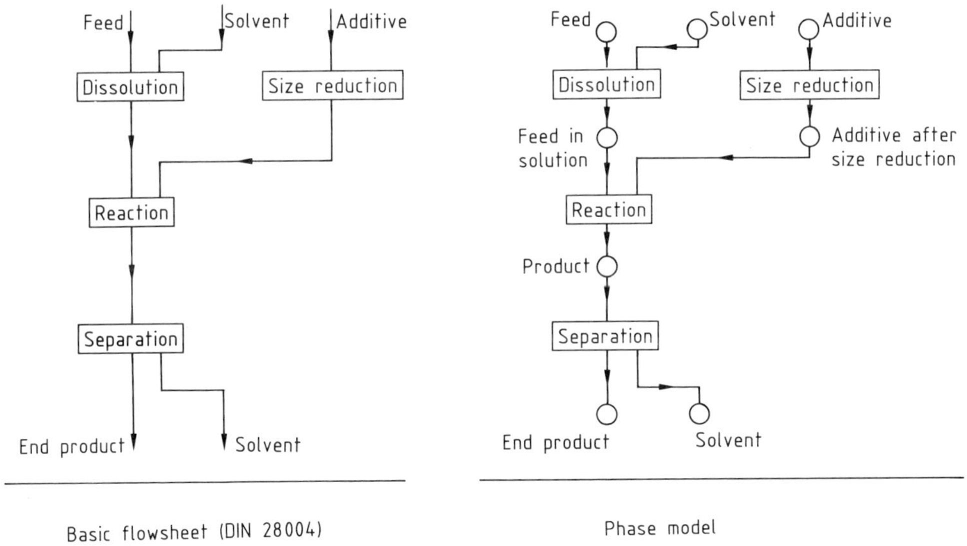

Figure 2.32. Development of the phase model from the basic flowsheet

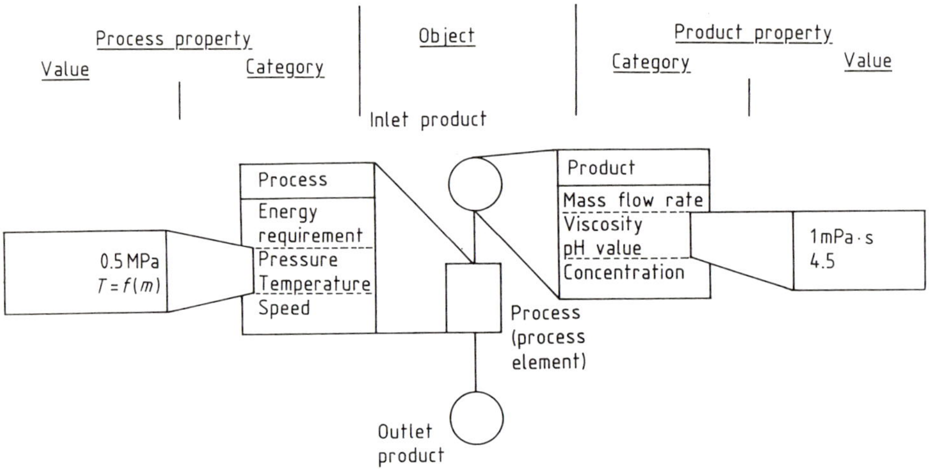

Figure 2.33. Excerpt of the phase model with product and process properties

- Setpoints are correlated with the desired control variables.
- Process indicators are empirically correlated with one or more state variables (these may also be, for example, product properties).

Product properties are the following:

- Physical quantities
- Chemical quantities
- Technological properties
- Product indicators

The last two items, again, are stand-in values. Technological properties are determined in special tests when relevant product properties depend on physical quantities that are difficult to measure or when the relationship to physical quantities is not known. Product indicators are correlated with physical or chemical quantities. Process properties can be employed as product indicators. The importance of this interplay between product properties and process properties was pointed out as early as 1967 when RUMPF introduced the concept of "property functions" [2.100].

Process indicators [2.101] and product indicators are the forerunners of the results that are now obtained by model-aided measurement techniques [2.102].

Product properties and process properties can be represented as property profiles information stream accompanying the material stream in the phase-model [2.34] (see Fig. 2.34). As nominal values, they are an essential part of the production instructions (Section 4.5). As actual values, they are used in process monitoring (Section 11.2), process control (Section 4.5), process analysis (Sections 3.2 and 11.3), and quality assurance (see also [2.82], [2.103], [2.104]). Note that it is not just the values of the product properties that change from stage to stage; categories may also disappear and other new ones may appear. An understanding of this point is necessary in order to follow, for example, the suggestion of GILLES that quality be introduced as a state variable [2.96].

Figure 2.35 shows the property profile "quality" for an intermediate product, with its ranges of validity [2.105]. Figure 2.36 uses the same method to illustrate reliability.

Process engineering is customarily concerned with the material flow (and the energy flow, although this is not illustrated here) in a production operation, whereas the task of process control engineering is to deal with the information flow. The crosslink between the material flow and the information flow is provided by sensors and actuators. Sensor technology (Chap. 5) acquires the information required for process monitoring and control. The function of actuator technology is to derive actions on the process from information extracted from the information stream (see Chap. 6).

A phase model can be hierarchically refined in accordance with structuring principles. This has been done for the example of automating a wastewater treatment plant [2.106]. Figure 2.37 shows the detailing process for chlor-alkali electrolysis [2.75].

Once the process analysis has been refined to the point that products cannot be broken down any further, a transition should be made from a product net to a causal model, either time-based or condition-based.

This transition is accomplished by transforming the interpretation "product with property profile" to the logical proposition, "product present in sufficient quantity and adequate quality at time t_0."

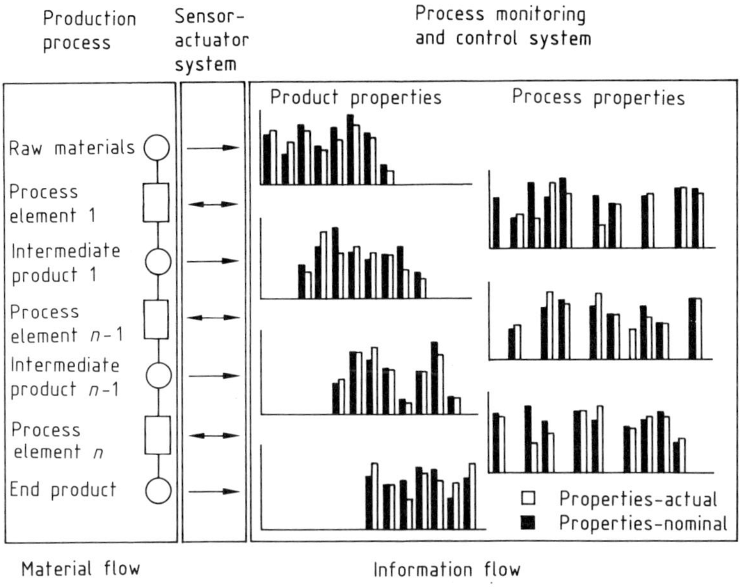

Figure 2.34. Profiles for process and product properties [2.1]

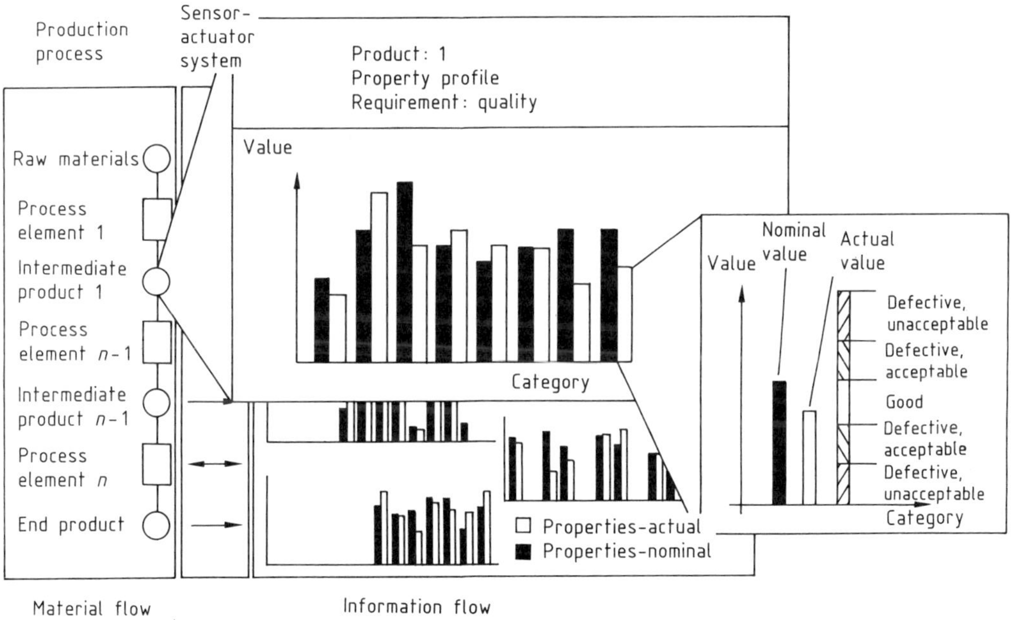

Figure 2.35. Material and information flows in the phase model of quality [2.104]

This logical proposition can be supplemented by further ones, for example:

- All resources needed for carrying out the process (plants, production specifications, monitoring and control system) are available and serviceable at time t_0
- All required energy sources are available in the proper quantity and quality at time t_0

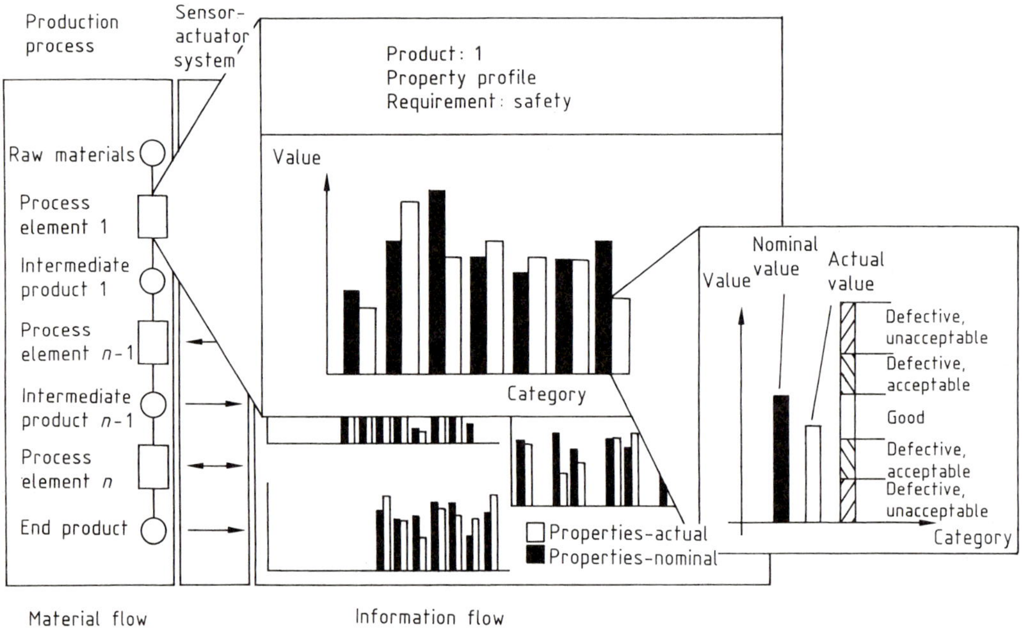

Figure 2.36. Material and information flows in the phase model of safety [2.104]

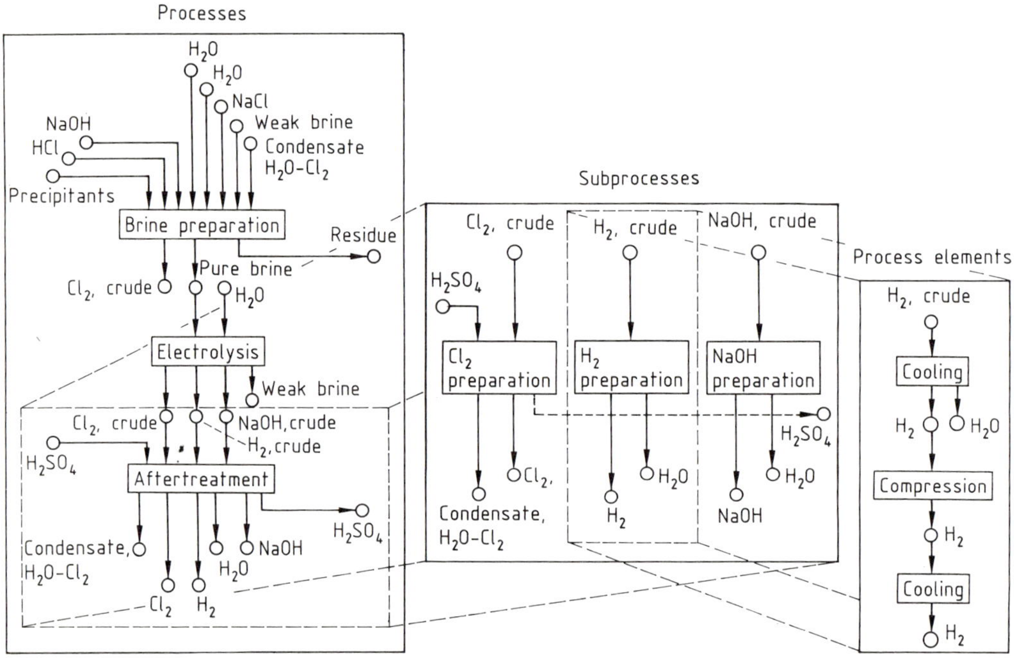

Figure 2.37. Detailing of the phase model for the example of chlor-alkali electrolysis

- All safety devices are operable at time t_0
- Approval for startup has been issued at time t_0

With such a system of conditions applying to each minimal production unit, the way from product net to causal net is evident. In Figure 2.38, the transition is illustrated, starting with the phase model and ending with Petri nets, which in turn make it possible to present a control process in function charts complying with DIN 40 719 [2.107].

The left-hand part of Figure 2.38 shows the phase model of a simple production process. Each box represents a process element and corresponds to a unit operation. On this level of the hierarchy, changes in product properties are in the foreground; the way in which the unit operations are implemented is not of interest yet. If the top-down design process is continued and the technical implementations of the unit operations are specified (e.g., as shown in the diagram of Fig. 2.24), the middle part of Figure 2.38 shows the resulting detailed description. Each unit operation is now represented by its own Petri net relating to the setting of process properties. Basic functions are employed to establish the process properties. The production specification (or recipe) sets forth under what conditions which process property must be set to what value. Basic function elements are utilized in order to execute the basic functions. The resulting function chart describes the detailed solution in process control engineering terms (see also Section 4.5).

The phase model of production is thus the appropriate tool for describing (Chapter 10) and analyzing (Section 11.3) production processes with regard to quality assurance (Figs. 2.33–2.35), process reliability requirements (Fig. 2.36), static structuring (Fig. 2.37), and dynamic structuring (Fig. 2.38).

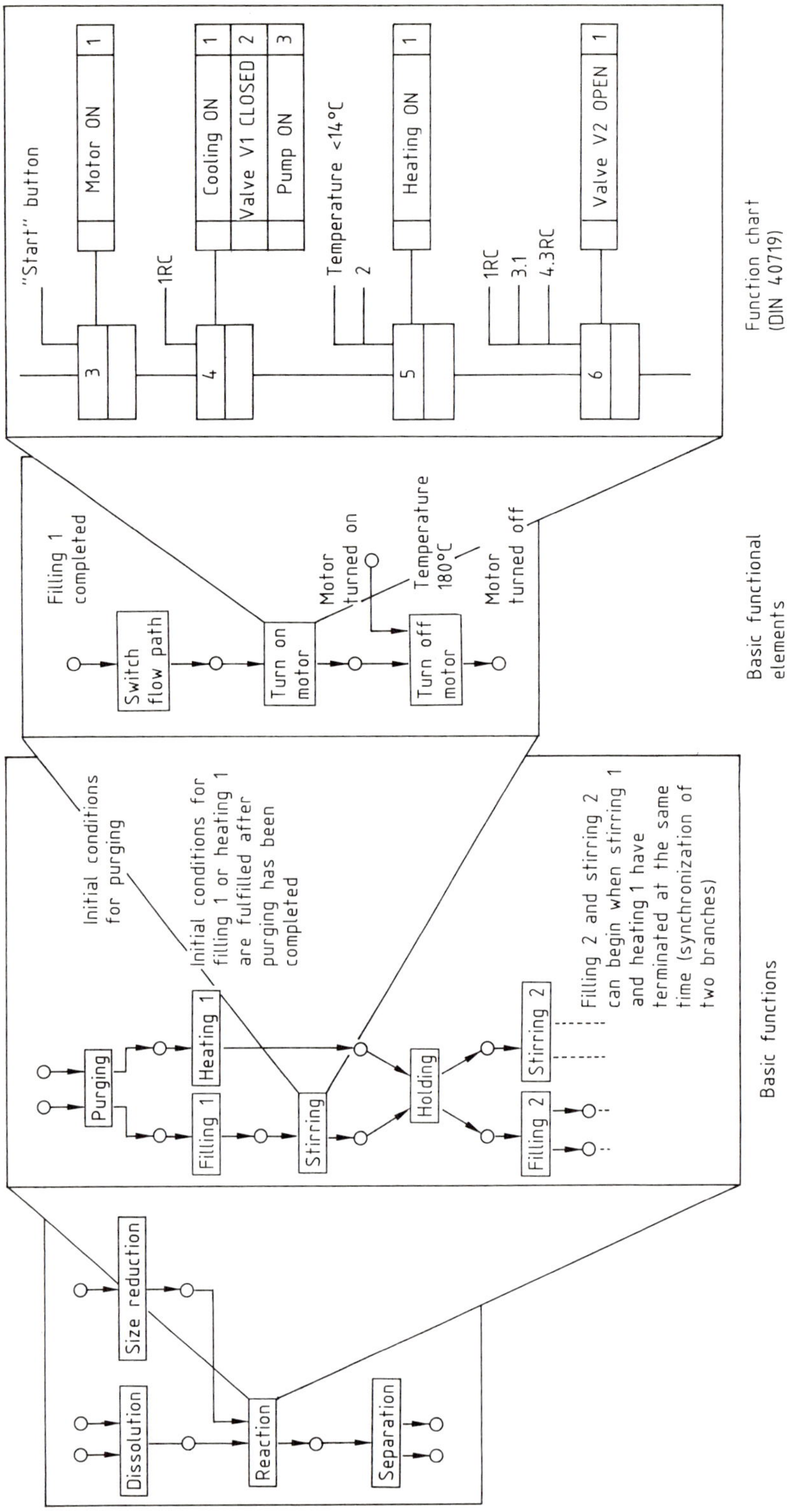

Figure 2.38. Transformation from the phase model via Petri nets to the function chart

3. Knowledge about the Process

3.1. Principles

An essential feature of modern process control engineering is that knowledge about the production process is utilized to the fullest. "Knowledge about the process" does not refer just to knowledge of the static and dynamic attributes or properties of the object classes of the production process (e.g., process or product properties [3.1]), as now represented in entity–relationship models (Chap. 2) and often taking the form of observations and measurements. Instead, this phrase denotes knowledge about how these attributes depend on one another. Because they pertain to the system "production process," they will also be called "system quantities" in what follows.

This knowledge about the interdependence of system quantities can exist in various forms (e.g., characteristic curves, correspondence tables, mathematical process models). The power of such records depends heavily on the type of system quantity.

The type of quantity is established by the kind of scale used to measure it [3.2]. There are four types of scales: nominal, ordinal, interval, and proportional (Fig. 3.1). They differ in information content and interpretability. For example, the nominal scale permits only the relationship like/unlike (or equal/unequal) between two fixed values of a quantity; this applies to quantities such as tank numbers, production line numbers, and so forth. The ordinal scale contains the further information that an order relation exists, as in sensory evaluations or quality levels. With these scales, items can be identified not only by a wide variety of possible symbols but also by numbers, but the intervals between such numbers have no empirical meaning: The scales represent only topological information. Metric scales, where the differences between numerical values are interpretable, are the domain of physical quantities. An example is temperature measured in degrees Celsius, which is an interval scale. If the ratios between the numerical values of a quantity have a physical meaning, as is the case for masses, absolute temperatures, and so forth, the scale is said to be a proportional or ratio scale.

It is evident that the ways of representing knowledge depend on the type of scale applicable to the system quantities. Mathematical process models expressed in systems of differential equations can be employed only when the quantities lend themselves to a metric type of scale.

Before dealing with ordering methods for system quantities and the structure of mathematical process models, first some key concepts of systems theory will be elucidated that are significant in a process control engineering (PCE) context. The notion of "system" presented here contains a deliberate balance between needed generality (to afford sufficient degrees of freedom for separating the overall system into subsystems) and certain properties that are absolutely necessary for arriving at a general system structure. This is a point of great importance in the writing and application of software based on modular princi-

Scale type	Interpretable relations				Examples
	Like/unlike	Ordering	Difference	Ratio	
Nominal	X				Tank no. Production no.
Ordinal	X	X			Quality classes Sensory evaluations
Interval	X	X	X		Temperature (°C) Time (h)
Proportional	X	X	X	X	Temperature (K) mass, energy

Figure 3.1. Types of scales

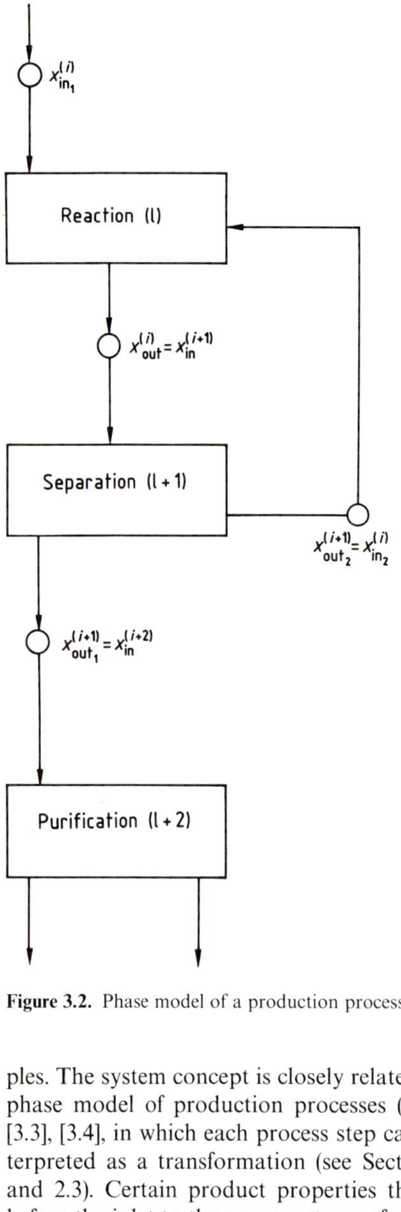

Figure 3.2. Phase model of a production process (excerpt)

ples. The system concept is closely related to the phase model of production processes (Fig. 3.2) [3.3], [3.4], in which each process step can be interpreted as a transformation (see Sections 2.2 and 2.3). Certain product properties that exist before the inlet to the process stage, often represented as an unordered set $\{E_i\}_n$ or as the endpoint of a property vector in product space [3.5], are mapped onto the property vector existing after the process operation n with coordinates $\{E_i\}_{n+1}$.

In simplified form, property vectors are written as x_{in} (inlet to the process stage) and x_{out} (outlet from the stage), as shown in Figure 3.2. The transformation can then be written as:

$$x_{out} = f(x_{in}) \tag{3.1.0}$$

In recent decades, the specialist literature has included many contributions attempting to justify a general theory of systems [3.6]–[3.13]. The axiom-based treatments, like all axiomatizations, represent a formal abstraction of concrete experience. The basic systems-theoretical concepts will be discussed here only in so far as they are essential in a PCE context.

In what follows, all system quantities are assumed to be "measurable," that is, representable on metrical scales.

The extension of this analysis to topological scaling, which has its own descriptive forms, leads to indefinite, conditional statements closely related to "fuzzy logic" and the methods of artificial intelligence (AI). They are not within the scope of this article (see also [3.14]). The scaling must, however, always be considered in the synthesis and analysis of production processes.

If the behavior of a reactor, such as a cascade of stirred tanks, a tubular reactor, or the like, is examined the system quantities can be classified in several categories.

For example, the following classes can be distinguished:

Inputs u_i:
Mass flow rate in, concentrations and temperature in the inlet stream, cooling/heating temperature, etc.

Outputs y_i:
Mass flow rate out, concentrations and temperature in the outlet stream, density of the outlet stream, etc.

System parameters p_i:
Heat-transfer areas, heat-transfer coefficients, specific heats, enthalpies of reaction, kinetic constants, geometric quantities (pipe diameters and lengths), etc.

In this classification, the inputs play the role of independent (environmental) variables acting on the system. They can be thought of as perturbations or, when purposely manipulated, as control variables.

The outputs, on the other hand, are the means by which (at least in principle) the system acts on its environment (see Fig. 3.3).

It may seem surprising that state variables have not been mentioned up to this point. In fact,

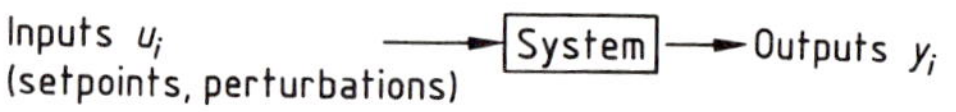

Figure 3.3. Schematic of a system

there is an important class of systems in which only steady-state processes take place, and in principle these can be described without the use of state variables. For example, in a cascade of stirred tanks operating in the steady state, the outputs are uniquely determined by the inputs. The introduction of state variables x_i is necessary only when investigating the time variations of systems. If, for example, the startup behavior of a reactor is to be studied it is not sufficient to know the time dependence $u(t)$ of the inputs; the "internal state" of the system must be known at least at a certain "initial time" t_0 before the outputs $y(t)$ can be predicted. This situation is reflected in the precise mathematical description when the system is formulated as a system of differential equations

$$\dot{x} = f(x, u, t) \tag{3.1.1}$$

which can be solved uniquely only by specifying an initial state x_0 at an initial time t_0.

Formulation (3.1.1) suggests that we generally attempt to describe a system by finding a vector x that has a minimum number of components. In the description of a chemical reactor, for example, one often arrives at a state vector x having the "reaction coordinates" ξ_i and the temperature T as components. All other quantities, such as concentrations, density, etc., are grouped in a vector z and can be stated as functions of x, u, and time t:

$$z = g(x, u, t) \tag{3.1.2}$$

The output vector y can be defined in a similar way:

$$y = C(x, u, t) \tag{3.1.3}$$

If a state vector x with the minimum number n of components is selected, n is said to be the number of degrees of freedom, or the order of the system.

Systems with finite order are also referred to as having lumped (concentrated) parameters. Systems with infinite order are then said to have distributed state variables. The state variables in such systems are generally continuous functions of the position coordinates.

The terms "system" and "process" are frequently used as synonyms. Here, however, "process" is defined as a one-parameter family of system states with the time t as the parameter. Accordingly, processes are generally dynamic; that is, their states are truly time-dependent unless special precautions are taken. Even when the inputs of a system are held constant over time, the system state will be a function of time at least during a startup phase. A special case of extraordinary importance for many problems in practical process control is that of stationary processes, where the system states $x(t)$ persist in a fixed state x_s and thus become independent of time. The state x_s is also said to be steady or stationary. The maintenance of specified steady states is one of the key tasks of process control. Steady states x_s should be considered only in time-invariant systems, where the time does not appear explicitly on the right-hand side of system equations (3.1.1) to (3.1.3). Clearly, the steady states in systems of this kind depend on the (constant) inlet variables u selected, and since $\dot{x}_s = 0$, the stationary system relating to (3.1.1)–(3.1.3) is immediately obtained if (3.1.1) is replaced by

$$0 = f(x_s, u) \tag{3.1.4}$$

For practical purposes, it is important to have a knowledge of the system stability near a stationary state x_s: If a process trajectory begins near state x_s, does the trajectory move farther away from x_s or return to x_s? This raises the question of how the system reacts to small disturbances away from its stationary point x_s. In order to handle this as well as some other questions, it is useful to have a concept of distance in relation to vectors. As is customary, the vector norm $\|x\|$ is used here; the distance between two vectors x_1 and x_2 is taken to be the norm of the difference: $\|x_1 - x_2\|$. In what follows, the commonly encountered Euclidean norm

$$\|x\| = \|x\|_2 = \left(\sum_{i=1}^{n} |\xi_i|^2 \right)^{1/2} \tag{3.1.5}$$

is used, where the ξ_i $(i = 1, 2, \ldots, n)$ are the components of the vector x.

Reconsidering the stability in the vicinity of a stationary state x_s. We first recognize that only the state equations $\dot{x} = f(x, u)$ have any role to play in this respect. LYAPUNOV defines the stability of a stationary solution x_s of $\dot{x} = f(x, u)$ in the following way:

1) x_s is stable if, for any sphere K_r with arbitrary radius r about x_s, there exists a number $\varrho > 0$ such that the following statement holds for a

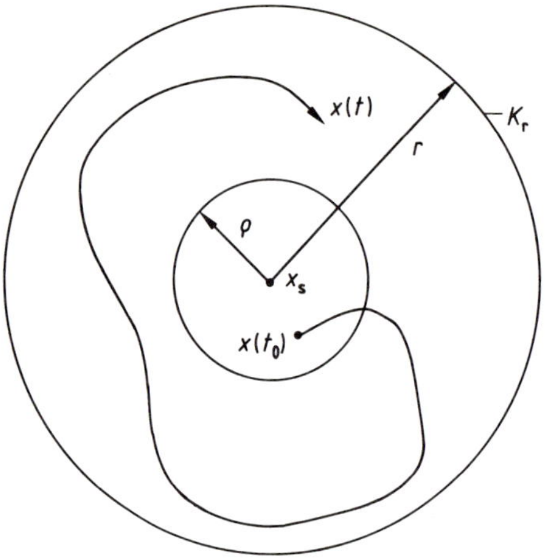

Figure 3.4. Lyapunov stability

process trajectory with initial value $x(t_0)$: $x(t)$ lies in K_r for all $t \geq t_0$ provided that $\|x(t_0) - x_s\| < \varrho$ (Fig. 3.4).

2) x_s is asymptotically stable if x_s is stable and there exists a number ϱ_0 such that $\|x(t_0) - x_s\| < \varrho_0$ implies the convergence of $x(t)$ to x_s; that is, $\|x(t) - x_s\| \to 0$ as $t \to \infty$.
3) x_s is said to be unstable if x_s is not stable.

More on stability and related problems can be found in [3.15]–[3.17].

The property of asymptotic stability says nothing about the rate at which process trajectories beginning near x_s converge to x_s. The notion of the relaxation time T_R serves this purpose: If the estimate

$$\|x(t) - x_s\| \leq \gamma e^{-\frac{t-t_0}{T_R}} \tag{3.1.6}$$

holds for all trajectories beginning sufficiently close to x_s, then x_s is exponentially asymptotically stable with relaxation time T_R.

Systems with a globally valid relaxation time T_R (i.e., T_R independent of specially chosen inputs u) have a remarkable property:

For sufficiently slow variations in the inputs u, they behave in a quasi-stationary manner. More precisely: If the rate of change with time $\dot{u}(t)$ of the input vector satisfies $T_R \cdot \|\dot{u}(t)\| \ll \|u(t)\|$, the time-variant state $x(t)$ can generally be calculated from the formally stationary equation $0 = f(x(t), u(t))$. The quasi-stationary system is purely algebraic (it no longer contains differential equations), and its dynamics depend solely on the time dependence of the inputs $u(t)$. The system thus has no intrinsic dynamics but responds to the time variations of u without inertia (time lag).

The transition to a quasi-stationary system is a radical reduction of the system: The initial system (3.1.1) has n degrees of freedom, but the new system has no internal degrees of freedom. Consider, for example, an ideally mixed stirred tank containing an indicator substance whose time-dependent concentration $u(t)$ in the inlet fulfils $\tau \cdot |\dot{u}(t)| \ll |u(t)|$ (where τ is the mean residence time in the tank). The outlet concentration $y(t)$ then satisfies $y(t) = u(t)$. The intrinsic dynamics of the tank, represented, for example, by its residence-time spectrum, is no longer identifiable.

The quasi-stationary property of subsystems within a specified overall system is not just important for numerical treatment: It is one of the key concepts for explaining "synergetic" effects in nonlinear systems [3.18].

Finally, two terms whose understanding is basic to the mathematical solution of most control engineering problems: controllability and observability [3.13], [3.19]–[3.23].

If at time t_0 a system is in a certain state x_0 and the inputs exhibit a time dependence $u(t)$ such that the system is in a specified state x_1 at a later time t_1, the system is said to be controllable from x_0 to x_1. If any x_0 is controllable to any x_1, the system is said to be completely controllable. Consider, for example, a stirred reactor in which the reaction A → B takes place, whereby the reaction is dependent on the temperature Θ. Let the state vector be $x = (\Theta, [A], [B])^T$, and the inputs u_1 = heating/cooling temperature Θ_K and u_2 = inlet concentration $[A]_{in}$ of species A. Then any state x_0 is controllable to any state x_1 provided the time interval from t_0 to t_1 is sufficiently long. However, if u_2 is not permitted as an input, then clearly x_0 is not controllable to x_1 if $[A]_0$ and $[A]_{in}$ are smaller than $[A]_1$. The controllability concept is of crucial importance for the problem of feedback control (returning the state to the setpoints) and the related problem of pole specification.

In the same way that the controllability links inputs u with state variables x, the observability relates the outputs y to the state variables. The state x is said to be observable at time t_1 if $x(t_1)$ can be determined from a knowledge of $y(t)$ in the finite time interval $t_0 < t < t_1$. In the reaction system discussed above, for example, suppose the state variable Θ is simultaneously taken as a measured quantity y. Then the remaining state variables [A] and [B] can be calculated from the time variation of the temperature Θ, and are hence observable, provided the enthalpy of reaction is nonzero. The observability concept is basic to the concept of the Luenberger observer in control engineering [3.13], [3.19], [3.24], [3.25].

3.2. Analysis Methods for Process Quantities

Introduction. Suppose *k* process quantities (i.e., state variables, inputs, or outputs) are measured at a certain time. These observations can be expressed as a *k*-dimensional vector $V(1, \ldots, k)$. If there are many observations, they form a cluster of points in a *k*-dimensional vector space, which will be called observation space or property space in what follows. Mathematical methods of linear algebra and statistics now become applicable [3.5]. Important uses of these methods include data reduction and transformation. A data reduction can be performed if the point cluster lies, at last approximately, in an *m*-dimensional subspace of the *k*-space. A transformation changes the coordinate system; that is, it generates a more easily handled representation by replacing the measured process quantities with "parameters" calculated from them.

Statistical models are valid only in the region studied; extrapolation beyond this region is not permitted. Furthermore, such models hold only under the assumption that all independent variables influencing the dependent variables have been identified.

Preparation of Process Properties. A statistical analysis can be broken down into the functions of acquisition, archiving, preparation, and interpretation (Fig. 3.5). Statistics proper is part of the interpretation function and will be discussed in depth in the next section. Acquisition and preparation, however, are vital for the practical use of statistical methods and should therefore be explained in further detail.

Acquisition. Statistical research often entails a laborious data collection process. In plants equipped with process monitoring and control systems, the system acquires a large amount of process data for observation and for use in plant operation. Thus it is obviously desirable to use process monitoring and control systems to acquire data for use in statistical studies. When this is done, archiving must be added to the functions of a typical system so that data collected at a given time can be accessed at a later time. What

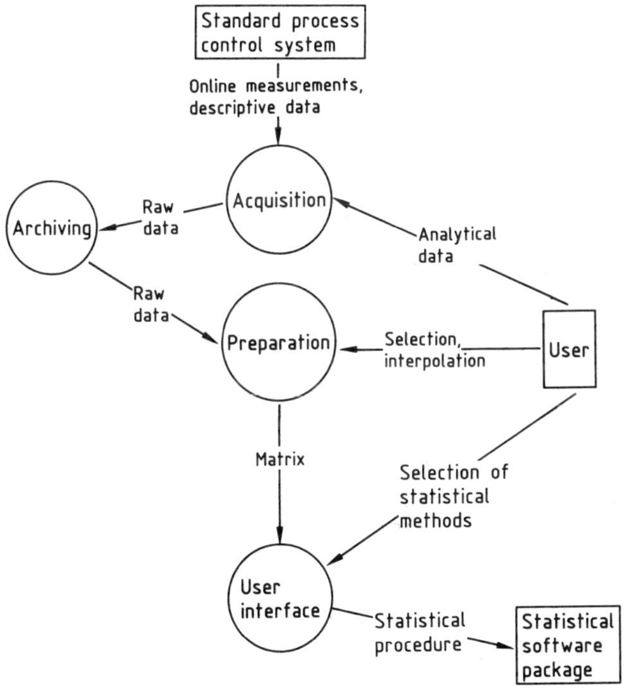

Figure 3.5. Acquisition, archiving, preparation, and interpretation of product and process data

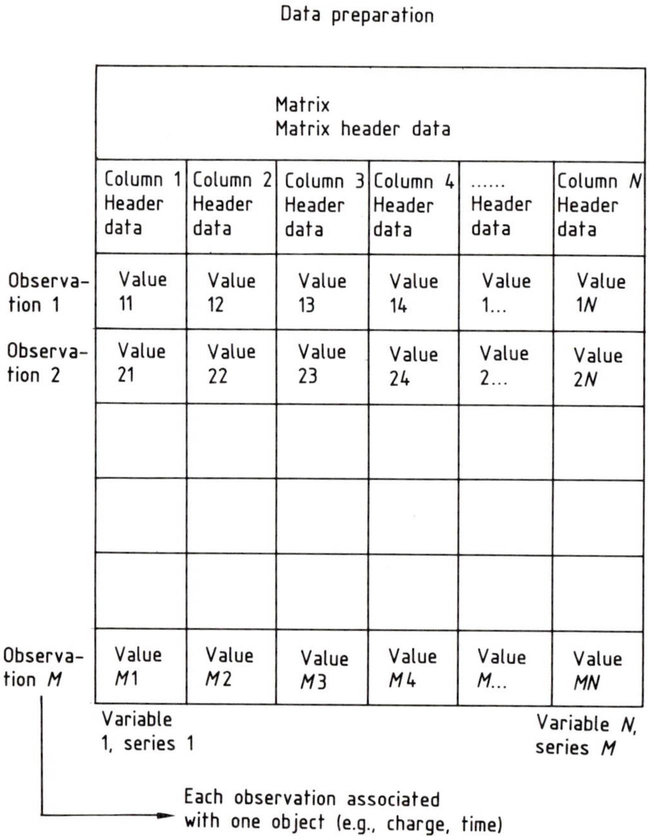

Figure 3.6. Data matrix for observations

is more, certain data on the configuration of the process control system must also be archived so that the archived process data can be properly interpreted even after a reconfiguration. Statistical software packages (e.g., SAS) can be used for the actual interpretation step; such packages for use on PCs offer good functionality at reasonable prices.

Preparation (Formatting). The starting point for many statistical procedures is a matrix W in which the rows represent observations and the columns represent the observed features (Fig. 3.6).

A typical requirement for a process monitoring and control system is that k process quantities be selected in a specified time interval from t_0 to t_m at a specified sampling rate. This would yield an $m \times k$ matrix (i.e., a matrix having m rows and k columns), which can be input directly to statistical analysis.

However, commercial software for process control systems does not to have this functionality. Nor does the simple selection of an $m \times k$ matrix satisfy all practical requirements, since professional presentation requires that descriptive data (name of measurement point, unit of measurement, etc.) be included. The extracted data must be output in a format that can be interpreted by commercial PC software. Because the result of data preparation is essentially a matrix, it makes sense to adopt a file format used in spreadsheet-type analysis.

Software. The preparation (formatting) and statistical analysis of process properties is not a routine activity for the person who does it. This makes it all the more important for the software to have a simple user interface. The format of the measurement station list must offer an easy way of marking the stations chosen for data acquisition (Fig. 3.7) [3.26]. Since all the selected points are sampled at the same time intervals, which as a rule do not coincide with the rhythm of data acquisition, an interpolation type must be specified for each individual station. The interpolation

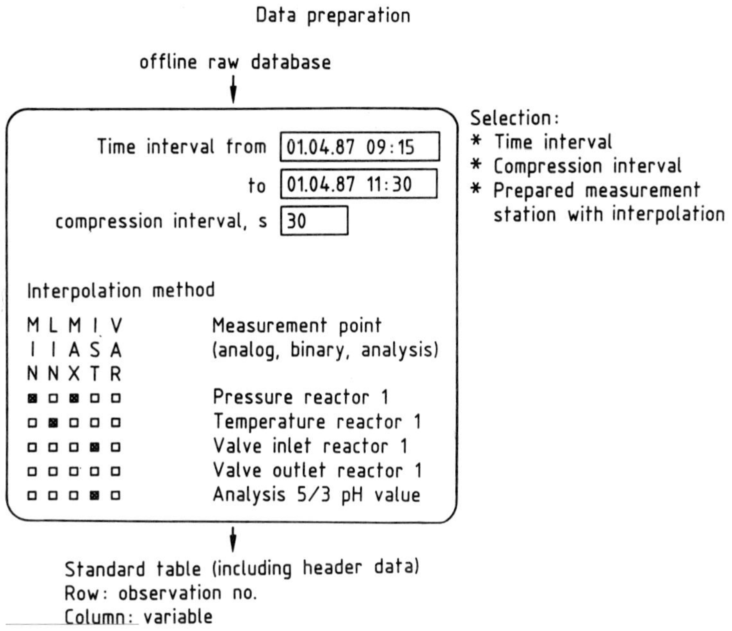

Figure 3.7. Interpolation of raw data to produce a standard table for subsequent analysis

types include linear interpolation; constant incrementing of the measured value up to the next measurement; and mean, minimum, or maximum value (when several acquisition events fall into one data-preparation interval).

Once the process information has been prepared, the data can be transferred to a statistical software package via a suitable interface. As already mentioned, the user expects that not only the measurements, but also the descriptive data, will be extracted from the process control system. The statistical package must then use these descriptive data as captions, axis labels, etc., without any intervention by the user.

Many applications allow the user to define a catalog of statistical methods that can be employed. In such cases, a user interface with the following properties can then be implemented, either as part of the statistical software package or as an add-in:

- Spontaneous selection of a variable as a dependent variable
- Selection of statistical methods by marking a list
- The possibility of depositing rules for conditioned or excluded statistical methods
- Dynamic specification of numbers of variables and observations and identification of descriptive data for statistical procedure

If the statistical package employed does not have this functionality, the user interface can be implemented with additional software. A statistical procedure will then be written in which the numbers of variables and observations, names of dependent variables, and descriptive data are represented by placeholders. The steps in the procedure are defined as statistical methods. The add-on software uses selection rules to interpret user inputs, descriptive data, and files, replaces the placeholders, and ignores methods not selected. The procedure that results can now be handled directly by the statistical package.

Statistical Methods. This survey briefly describes individual statistical methods and discusses their applicability to the interpretation of process information.

Elementary Statistics. The examination of position parameters, such as mean, median, and mode, is meaningful especially when the process is expected to be stationary or pseudostationary (e.g., a constant flow rate in a continuous process). If the usual variability of the process has already been determined on the basis of large data sets, statistical quality control with control charts is possible. The purpose of control charts is to discriminate as accurately as possible between normal fluctuations of a value and unusual variations.

An important point when using a statistical test is to follow a definite methodology. This includes exact formulation of the hypothesis, so that the probabilities of Type I errors (rejection of a correct hypothesis) and Type II errors (acceptance of a false hypothesis) can be precisely studied and evaluated.

In many cases it is useful to determine whether a classical distribution function can be assumed. The important tests used in practice—the F test (comparison of two variances) and the t test (comparison of two means)—assume a normal distribution of the population.

Experimental Design. The concern of experimental design is to determine which of k potential factors $x_1, \ldots, x_k$ actually influence a dependent variable y, and to do so in the most efficient way by means of planned experiments. Two-level factorial experiments are important. First, a realistic minimum value ($-$ level) and a realistic maximum value ($+$ level) are selected for each factor. Not only quantitative factors but also qualitative ones are possible; for example, $+$ denotes product from plant A and $-$ denotes product not from plant A.

The complete design then consists of 2^k experiments. It is obtained by taking all combinations of the $-$ and $+$ levels for the w_i. The effects are the actions of the factors themselves, as well as the interactions of two or more factors. The level of an interaction is obtained by multiplying the levels of the individual factors. For example, if the factors w_1, w_2, w_3 in an experiment have levels $+$, $-$, $-$, then in this test the effect $w_{1,2,3}$ has level $+$.

After performing the experiments, the effects are calculated by taking the mean of all results from trials in which this effect had level $+$ and subtracting the mean of all results from trials in which the effect was $-$. The literature gives formulas for deciding, on the basis of these results, whether an effect is significant.

If the assumption is justified that interactions of more than two factors are not to be expected, then the complete factorial experimental design can be replaced by a fractional factorial design calling for fewer tests.

Methods for generating partial factorial designs can be found in, for example, [3.27]. Software for creating and evaluating designs is available.

The interpretation of two-level factorial experimental designs is a special case of analysis of variance, a technique that can be used with any number of levels. The trick is to use two-level factors first to investigate which potential factors truly affect the experimental result y. The nature of this influence (e.g., whether it is linear) is investigated only after the merely notional factors have been identified and omitted.

Principal Components. On economic grounds, it is often not possible to carry out a designed experiment, and one must be content with random process measurements from the plant. It may be difficult to gain any knowledge from such a collection of data. Principal-component analysis is a method of orienting and, in some cases, reducing high-dimensional observations.

If the observations exist in k-dimensional space, principal-component analysis answers the following question: For any i, $0 < i < k$, how well can the k-dimensional observations be described in a subspace U of dimension i (Fig. 3.8)? While a regression model minimizes only the squared distances in the direction of the dependent variable y, principal-component analysis minimizes the sum of the squared Euclidean distances between U and the individual observations. In this way, all the k process variables are

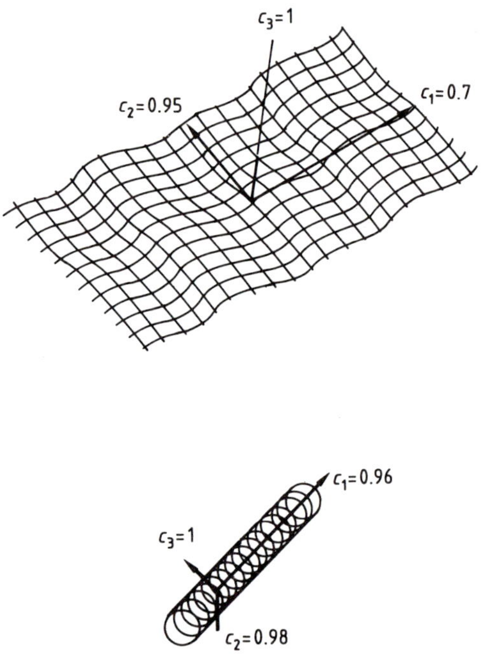

Figure 3.8. Effect of the shape of a point cluster on the principal components

equivalent; none is identified as dependent. Scaling now becomes a problem, especially when a variety of physical units are present. A change of scaling means a numerical change in the Euclidean distance in the k-dimensional observation space, and thus changes the result of the principal-component analysis. In practice, all the process variables are usually scaled so as to have mean 0 and variance 1. When this is done, however, the scaling becomes experiment-dependent. It is also possible to determine a technical minimum and maximum for each process variable, then define the scaling to give these the values -1 and $+1$.

The principal-component analysis determines, for every i, $0 < i < k$, the i-dimensional subspace U_i for which the sum of the squared Euclidean distances to the observations is a minimum. The procedure has the convenient mathematical property that these subspaces are all nested, that is, $U_1 \subseteq U_2 \subseteq \ldots \subseteq U_{k-1}$. They can thus be stated as a sequence of k basis vectors; these are the principal components. The mathematical procedure yields an eigenvalue for each principal component; the sum of all the eigenvalues is k. If, for some i, the sum of the first i eigenvalues $l_1 + l_2 + \ldots l_i$ is close to k, the point cluster of the observations can be regarded as i-dimensional to satisfactory accuracy. Often, $c_i = (l_1 + l_2 + \ldots + l_i)/k$ is used as a parameter, known as the variance contribution of the first i principal components; this has a maximum value of 1.

Time-Series Analysis. A typical problem leading to a statistical study of process variables is the estimation of dead times. The central concept is cross-correlation: The correlation coefficients between the time series $x(t)$ and all time series $y(t + nT)$ are calculated, where T is the sampling rate. If N denotes the value of n that has the largest correlation coefficient, then NT is the estimated dead time.

This method has been extensively developed. Nonlinear relations between $x(t)$ and $y(t)$ can now be included, and processes for which the dead time is not a constant but depends on the nature of $x(t)$ can be investigated.

Systems Engineering. A number of software packages for the statistical analysis and graphical presentation of data have become commercially established in recent years. The methods employed are complex and exist in a wide range of variations, and no practically useful functionality is left out. The publishers of most of this software offer their products for a variety of system platforms (MS-DOS, UNIX, etc.), and data can be shared with other widely used software products (database management programs, spreadsheets, etc.). Examples of established statistical packages are SAS and RS/1. These products are continually being upgraded as standards develop (e.g., X/Windows). While market conditions are generally good for statistical applications software, there are some problems having to do with inadequate management of process data (Fig. 3.9). In practice, statistical analysis includes the functions of data acquisition, archiving, and preparation; these have already been described. But no software standards have yet been adopted in this area, and there are no dominant products in the market. One step in the right direction is The Fix, a process data acquisition system published by Intellution for MS-DOS. The system supports the preparation of historical data in Data Interchange Format (DIF), which many spreadsheet programs, as well as the SAS and RS/1 statistics packages, can handle.

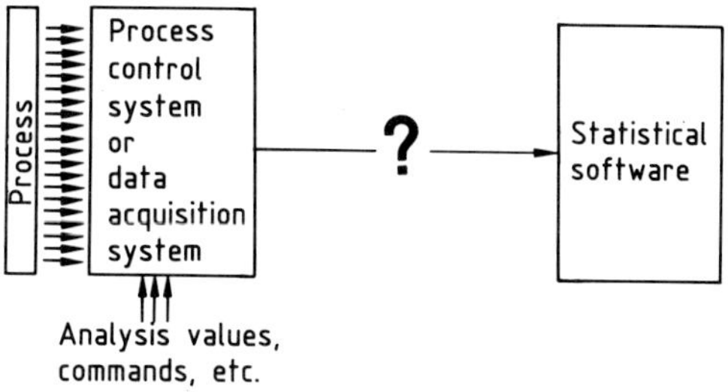

Figure 3.9. Problems in transferring inadequately organized data to statistical software

Process monitoring and control systems published by the established firms, however, do not support the generation of a matrix (Fig. 3.10) in-

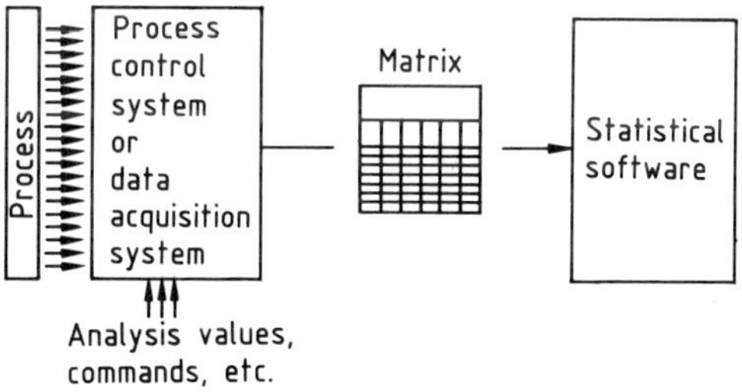

Figure 3.10. The data matrix as interface between process control or data acquisition system and statistical software

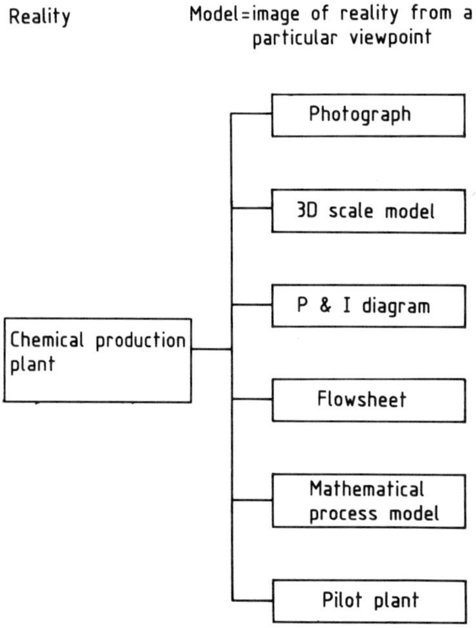

Figure 3.11. Model and reality

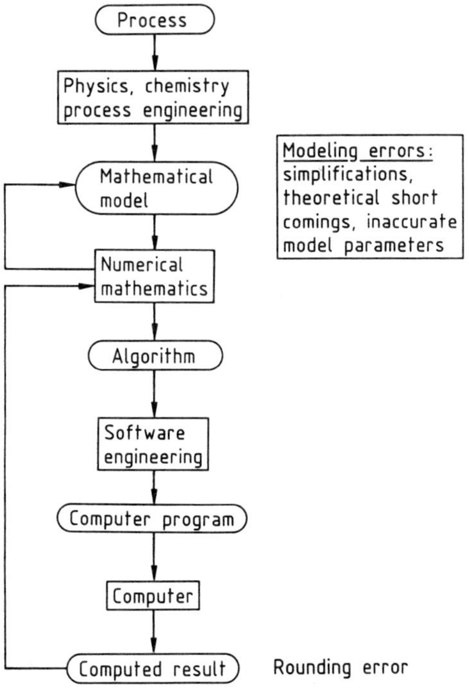

Figure 3.12. From process to computed result

cluding descriptive headers in place of a series of measured values.

3.3. Process Models

Introduction. The general concept "model" is very broad. Any means of analysis and synthesis that is used in learning about reality or in actively designing and fabricating human artifacts can be called a model (see also Section 2.2). Typically, models are simplified pictures from a particular viewpoint of the properties and functions of the real object on which they are based. As for the chemical plant models in Figure 3.11, it can be stated generally that a model reflects only a certain portion of the whole, from a certain aspect and for certain purposes.

The example clearly shows the three main features of all models [3.28]:

1) Mapping.
2) Abbreviation: The model generally lacks many of the attributes of the original. In fact, this information loss is often an advantage because incidentals can be ignored in creating the model. One discards intellectual ballast so as to be able to focus better on the interesting aspect of the whole. These comments are closely related to the third model characteristic.
3) Pragmatism: The model represents the original, aiding in the acquisition of knowledge and permitting a limited range of mental or actual operations.

In contrast to models that represent physical images or simulations of particular originals, mathematical models are mental constructs that generally apply to an entire class of objects and situations. They thus fulfill a key prerequisite for being "scientific" ("Scientific theories are general laws" [3.29]). Mathematical models are formulated (as differential equations, etc.) in terms of variables that cannot enter into a concrete calculation until boundary or initial values and other model parameters have been established. It is in setting these values that the user applies the mathematical model to the "special case."

The mathematical model, as the image of an entire class of processes, constructed with the aid of the basic disciplines (chemistry, physics, process engineering, etc.), starts out with only a theoretical or prospective relationship to actual calculations (Fig. 3.12). Its use in computation depends on the existence of a solution algorithm; in general, a computer program is still needed. The usefulness of a mathematical model is thus determined not only by its consistency with the underlying real object, but also by the bounds set by numerical mathematics and computing capacity set. These limits are still relevant in practice today, because many of the processes found in the chemical industry are highly complicated from the standpoint of mathematical modeling (e.g., emulsion polymerization with its many associated phenomena). Advances in the calculability of mathematical models are essentially governed by advances in computer hardware. An example is the method of finite elements, used for the solution of partial differential equations. Although the fundamental idea was formulated by COURANT in the 1940s [3.30], this publication remained largely unknown, and finite elements were "rediscovered" by engineers in the 1950s. Only then did rapid progress occur in its numerical principles, and these advances accompanied the development of computer technology.

Another example is adaptive control. The first publications on this method appeared in the 1950s [3.31], [3.32], but the impetus for further study came only when the development of fast process control computers made it practically feasible.

The creation of a mathematical model thus means keeping track of numerical problems, the capacity of the computers to be used, and (especially when writing large programs) software engineering considerations. In particular, the three main sources of error (modeling errors, procedural errors, and rounding errors) must be treated as a compromise to minimize the total error. This topic has also been dealt with in [3.33], [3.34].

Mathematical Models: Survey and Structural Properties. This section treats mathematical models that describe the space–time behavior of processes whose model variables and parameters are measured on at least an interval scale (see Section 3.1). The general mathematical description of such processes consists of coupled systems of partial differential equations. The models under consideration are those in which the system is described in terms of model variables classified as inputs u_i, state variables x_i, outputs y_i, and model parameters p_i (see Section 3.1).

In creating and using mathematical models of process apparatus, it is particularly important that the model nearly always consists of two parts, the physicochemical part and the apparatus-specific part (Fig. 3.13). The physicochemical component of the model includes relationships whose validity is totally independent of the particular apparatus in which the modeled operations take place. This self-consistency of the physical component insures that the model can be applied to a variety of apparatus; this feature is naturally a great advantage in the architecture of large applications software packages (material property databases, etc.). The fact that apparatus and facilities can be simulated with off-the-shelf engineering software, without regard to the wide range of variation in the details, is a consequence of the strong structural similarity of the essential apparatus-specific process equations. All space–time processes taking place in continuous-flow equipment, for example, obey balance equations written for conserved quantities (e.g., energy and chemical species), in which one side always contains terms known from the continuity equation while the other side contains diffusive terms, sources, and sinks, which represent only the pure, non-apparatus-specific physics. As special cases, the continuity terms contain both purely spatial variability (steady continuous process) and purely temporal variability (batch process).

1. Physicochemical part (state relations)

Enthalpy	$J=J(x, T)$
Density	$\varrho=\varrho(x, T)$
Ideal gas law	$pV=vRT$

2. Apparatus-specific part (process equations)

Balance equations $\dfrac{\delta x}{\delta T}+\operatorname{div}(\vec{v}\,x)$

$=$ diffusion term + sources − sinks

"Engineering formulas"

Initial and boundary conditions

Figure 3.13. Structure of mathematical process models

"Stage models" are often used in simulation. Such models describe apparatus or parts of apparatus in terms of fictitious or real stages that have no spatial variability of the state quantities in a finite volume. The differential operator $\operatorname{div}(\vec{v}\cdot x)$ is then replaced by appropriate difference operators.

In addition to the balance equations, the typical information includes initial and boundary conditions for the apparatus-specific equations. A number of software products now available make extensive use of these structural properties, both in their design and in the user interface. The conserved quantities are momentum (Navier–

1. Physicochemical part (state relations)

Equilibrium relations	$y^* = y(x, T)$
Boiling curve	$p = p(x, T)$
Ideal gas law	$pV = v R T$
Liquid density	$\varrho = \varrho(x, T)$
Enthalpy of liquid	$J = J(x, T)$
Heats of vaporization	$r = r(T)$
Reaction rates	$R = R(c, T)$

2. Apparatus-specific part (process equations)

Material balances	$(n_i \cdot x_i) = \Sigma$ inputs $- \Sigma$ outputs
Heat balances	$W_i = \Sigma$ inputs $- \Sigma$ outputs
Separation effect of trays	$\frac{y_i - y_{i+1}}{y_i^* - y_{i+1}} = E$
Pressure balance	$\Delta p_i = g_1$ (flow rates, densities, x, T, geometry)
Weir relation	Liquid flow rate $=$ g_2 (liquid volume, geometry)

Figure 3.14. Mathematical model of distillation columns

Stokes equations), mass (continuity equation), and enthalpy (heat balance). Other conserved quantities, as well as sources and sinks, can generally be programmed by the user.

The divergence term $\mathrm{div}(\vec{v} \cdot x)$ has uses that go far beyond describing convection in space. The balance equation can be applied to other pseudospatial propagation phenomena without change.

Two examples:

1) Particle-size distribution
   "Space coordinate" = particle diameter
   Velocity v = rate of growth of particles
   Conserved quantity x = number of particles
2) Particle age distribution
   "Space coordinate" = particle age τ
   Velocity $v = \mathrm{d}\tau/\mathrm{d}t = 1$
   Conserved quantity x = number of particles

As a concrete example to show how the model equations are split, Figure 3.14 shows the mathematical model of a distillation column.

For the physicochemical part, equilibrium relations are needed between the mole fractions in gas and liquid, the relation for the boiling point line (which interelates the pressure with the molar fraction of liquid and the temperature), the ideal gas law, and relations for the liquid density, liquid enthalpy, heat of vaporization, and (if appropriate) reaction rates. All these are purely physical state equations. Given this list of relationships, only an expert might suppose that someone is creating a mathematical model for a distillation column. The apparatus component of the model is required to reveal which apparatus is to be modeled. In the case of a distillation column, this would comprise the material and heat balances for condenser, trays, bottom, and vaporizer; the separating effect of the trays (giving the departure from ideal equilibrium, described by, for example, the Murphree efficiency E); the pressure balances for determining the pressure drop across individual trays (these use basic geometrical data such as perforation diameter, tray thickness, and active tray surface area); and finally a weir relation to give the downflowing liquid flow rate (again based on geometrical data such as weir height).

Spatial discretization with stage models is a special case of a widespread practice (e.g., difference methods, method of finite domains, Galerkin method, with the important special case of finite elements), which meets practical requirements because partial differential equations can generally be solved only by discretizing to finite-dimensional substitute models. These are represented as systems of differential algebraic equations (DAE systems) of the form

$$A(x, z, u)\dot{x} = f(x, z, u) \tag{3.3.1}$$

$$0 = g(x, z, u) \tag{3.3.2}$$

The output equation

$$y = C(x, u) \tag{3.3.3}$$

is also needed.

The special structure of the left-hand side of (3.3.1), where the matrix A depends on x, z, and u, expresses the fact that nearly all process models of practical relevance are linear in the time derivatives $\dot{x}$ of the state variables x. For the numerical solution of DAE systems, see [3.35], [3.36].

In Section 3.1, the state vector x with minimum number n of components was introduced to describe the system, where n is equal to the number of degres of freedom in the system. Unfortunately, the mathematical modeling of a system based directly on the physical phenomena often leads to DAE systems (3.3.1), (3.3.2) in which the number of degrees of freedom is smaller than the number of components of vector x (DAE systems with index greater than 1).

For example, in mathematical models of distillation columns, the boiling-line equation couples the pressure with the differential equation variables temperature and mole fractions. If the pressure is assumed to be constant, the number of degrees of freedom is thus reduced.

The numercial handling of such systems with index greater than 1 is problematical because the solution contains time derivatives of the inputs $u(t)$ up to order $q-1$, where q is the index of the DAE system [3.37]–[3.39]. Thus DAE systems with index greater than 1 are mixed problems involving both integration and differentiation.

Example: The equation $\dot{x}_1 = x_0 - x_1$ (stirred tank with mean residence time $\tau = 1$) with $x_1 = u(t)$ [with $u(t)$ a specified function] has solution $x_0 = u(t) + \dot{u}(t)$. Here the number of degrees of freedom is actually zero; the index is 2.

Design problems such as this example, where certain state or initial values are specified and certain inputs are to be calculated, generally have a high index.

To transform a DAE system with index $q \geq 2$ to an equivalent $q = 1$ system, index reduction methods have been devised [3.37], [3.39]. For large models, however, these are often time consuming since the transformations must be performed with closed formulas, and an expert system that will let the computer do the transformations is not available.

If the system is discretized not only in space but also in time, by considering the system quantities only at specified (generally equally spaced) times $t_i = t_0 + i\Delta t$ $(i = 0, 1, 2, \ldots)$, the time-continuous models (3.3.1)–(3.3.3) become time-discrete models of the form

$$A(x_i, z_i, u_i)x_{i+1} = F(x_i, z_i, u_i) \quad (3.3.4)$$

$$0 = g(x_i, z_i, u_i) \quad (3.3.5)$$

$$y_i = C(x_i, u_i) \quad (3.3.6)$$

Time-discrete systems offer a number of advantages, including simpler numerical solution. Because of their inaccuracy (especially when the system behaves in a strongly nonlinear manner), their use is generally recommended only for on-line adaptation of the system parameters. The structural properties in most cases are invariant to the change from time-continuous to time-discrete models, but there are changes in, for example, the concepts of observability and controllability [3.19].

Up to now we have discussed dynamic models that include both spatial and temporal variations of the system. These models are naturally of interest as aids where the time dependence is an essential feature of the problem. These situations include studies of system behavior by means of:

1) Stability analysis
2) Step response of the system
3) Controller design and tuning of controller parameters; simulation of genuinely dynamic processes such as batch operations
4) Startup and shutdown strategies in continuous equipment
5) Failure analyses

Finally, dynamic models are increasingly used in identifying optimal process control (e.g., in time-optimized control of batch processes when safety and product quality are competing considerations). Besides these advanced simulation techniques, dynamic process models are used on-line in the form of observers, such as the Luenberger observer [3.24], [3.25] or the Kalman–Bucy filter [3.40], [3.41].

As discussed in Section 3.1, the notions of steady-state and quasi-steady-state process models come directly from Equations (3.3.1)–(3.3.3) if $\dot{x}$ is set equal to zero. Naturally, steady-state process models find use chiefly with steady continuous processes and in connection with problems such as

1) The design and optimization of plants and apparatus
2) The analysis of plant characteristics

The state variables can in principle be eliminated from these models, so that a direct relationship appears between the outputs y and the inputs u:

$$y = h(u) \quad (3.3.7)$$

Equation (3.3.7) is the simplest case of a transfer model, in which the value of the output vector y is obtained directly from that of the input vector u by means of a calculation specification. Such a simple link between y and u cannot in principle hold in dynamic process models.

For example, the transfer function between the concentrations u (input) and y (output) for an ideally mixed stirred tank with no reaction is given by

$$y(t) = \frac{1}{\theta} \int_{-\infty}^{t} e^{-\frac{t-\tau}{\theta}} \cdot u(\tau) \mathrm{d}\tau \quad (3.3.8)$$

where θ is the mean residence time.

This formula represents the "steady-state" solution of the dynamic model

$$\dot{x} = \frac{1}{\theta}(u - x) \quad (3.3.9)$$

$$y = x \quad (3.3.10)$$

It is important, however, that in Equation (3.3.8) the value of y at time t does not, as in (3.3.7), depend solely on the value of the input u at time t, but on the time variation of u throughout the past.

Transfer models for dynamic systems are functionals of the form

$$y(t) = h_1(t, u(\tau)_{\tau \leq t}) + h_2(t, u(t)) \quad (3.3.11)$$

The second term $h_2(t,u(t))$ in Equation (3.3.11) takes into account the possibility that the effect on y of time variations in u may be present not just in smoothed form, as in h_1 (discontinuous functions in u become continuous functions y; continuous u become continuously differentiable y), but in unsmoothed form. Then h_2 describes the "feedthrough" of u to y.

The importance of transfer-function models (Eq. 3.3.7 and 3.3.11) is that they make it possible to examine how the inputs u affect the outputs y without knowing the internal structure of the system (represented by the state variables x_i and their dependences on u and on one another). This may be the only approach in, for example, complex systems (e.g., dynamics of drugs in human beings and animals) or in design calculations where only the input/output behavior of the apparatus is known at first and the detailed engineering has not yet been done (e.g., pre-simulations based on linear quantitative balances).

Transfer-function models thus belong to a class of models whose structure is only slightly or not at all influenced by the "nature" of the system being modeled. These purely mathematical (statistical or regression) models therefore offer great freedom in the choice of model structure. The model parameters must always be obtained by fitting to measurements made in the underlying real system. Because such regression models are based on measurements (see Section 3.2) and no attempt is made to derive a predictive theory from them, the model cannot be applied or extrapolated beyond the range of the measured values and the specific apparatus on which the measurements were performed, in contrast to theoretical models. While pure regression models cannot be employed in the design phase, they have an important application in on-line simulation (see Section 11.3). Measurements on the running process make it possible to treat and adjust the model parameters as time-variant quantities, so this accurate knowledge of the model parameters permits the use of regression models with simple structure. When these "adaptive" process models are used, the control parameters can also be modified on-line to fit the current process situation. This technique is called "adaptive control."

Particularly important here are linear process models. In the state–space representation, they have the form

$$\dot{x} = A(t) \cdot x + \mathrm{B}(t) \cdot u(t) \quad (3.3.12)$$

$$y = C(t) \cdot x + \mathrm{D}(t) \cdot u(t) \quad (3.3.13)$$

The system parameters are thus the totality of components in the matrices A (system matrix), B (input or control matrix), C (observation matrix), and D (pass-through matrix), which completely characterize the model of Equations (3.3.12) and (3.3.13). For this model, the block-diagram visualization customary in publications on control engineering takes the form shown in Figure 3.15.

The input x_0 to the integrator block symbolizes the effect of an initial condition $x(t_0) = x_0$ on the solution behavior. The transfer-function model associated with Equations (3.3.12) and (3.3.13) has the form

$$y(t) = \int_{-\infty}^{t} K(t,\tau) \cdot u(\tau)\,\mathrm{d}\tau + D(t) \cdot u(t) \quad (3.3.14)$$

where the matrix $K(t,\tau)$ depends on the time variables t and τ. Equation (3.3.14) gives a mapping $y = F(u)$ from the set of input functions u into the set of outputs y; this mapping satisfies the superposition principle

$$F(u_1 + u_2) = F(u_1) + F(u_2) \quad (3.3.15)$$

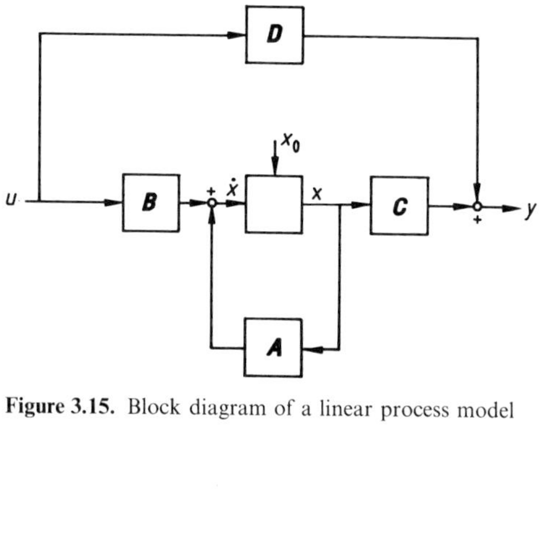

Figure 3.15. Block diagram of a linear process model

(image of sum = sum of images) and

$$F(\alpha \cdot u) = \alpha \cdot F(u) \qquad (3.3.16)$$

(image of α times $u = \alpha$ times image of u).

Linear Process Models. The linear process models introduced above require some more discussion, since they are still very important in both systems analysis and the design and on-line modeling of control systems (despite advances in the calculation of nonlinear systems). Their importance undoubtedly stems from the fact that their simple structure can be completely understood from fundamentals and that many structural properties of nonlinear process models are conserved in linearization (approximation of nonlinear models by linear ones).

What is more, the calculation of linear process models involves far fewer problems than that of nonlinear models. The use of linear models should always be considered when it is certain that the process trajectories in state space do not depart from the region in which linearization is valid. An example is control to maintain a stationary operating point by using an on-line linearized process model.

Special importance attaches to the time-invariant linear models

$$\dot{x} = A \cdot x + B \cdot u(t) \qquad (3.3.17)$$

$$y = C \cdot x + D \cdot u(t) \qquad (3.3.18)$$

which are set by the constant (time-invariant) matrices A, B, C, and D.

As an example of linearization about a stationary operating point, let us analyze the reaction A → B taking place in a stirred tank, using the heat balance. The nonlinear process model is described by the equations

$$\dot{a} = \frac{1}{\theta} \cdot a_E - \frac{1}{\theta} \cdot a - S \cdot e^{-T_A/T} \cdot a \qquad (3.3.19)$$

$$\dot{T} = \alpha(T_c - T) + \beta \cdot S \cdot e^{-T_n/T} \cdot a + \gamma \cdot (T_E - T) \qquad (3.3.20)$$

where a and a_E are the concentrations of A in the tank and in the inlet stream; T is the reaction temperature, T_E the inlet temperature, and T_C the coolant temperature; θ is the mean residence time and S the collision factor; T_A is the activation temperature of the reaction; and α, β, and γ are further model parameters.

Consider given values of the inputs $\bar{a}_E$, $\bar{T}_E$, and $\bar{T}_K$, and the associated stationary values $\bar{a}$ and $\bar{T}$ of the state variable. If the deviations are defined as $a = \bar{a} + x_1$, $T = \bar{T} + x_2$, $a_E = \bar{a}_E + u_1$, $\bar{T}_E = T_E + u_2$, and $T_K = \bar{T}_K + u_3$. Linearization then yields the relations

$$\dot{x}_1 = \frac{1}{\theta} u_1 - \frac{1}{\theta} x_1 - S \cdot e^{-T_A/T} \cdot x_1 - \frac{S \cdot T_A}{\bar{T}^2} e^{-T_A/\bar{T}} \cdot \bar{a} \cdot x_2 \qquad (3.3.21)$$

$$\dot{x}_2 = \alpha \cdot u_3 - \alpha \cdot x_2 + \beta \cdot S \cdot e^{-T_A/\bar{T}} \cdot x_1 + \frac{\beta \cdot S \cdot T_A}{\bar{T}_2^2} e^{-T_A/\bar{T}} \cdot \bar{a} \cdot x_2 + \gamma \cdot u_2 - \gamma \cdot x_2 \qquad (3.3.22)$$

If the temperature is known as an output (measurement) and we accordingly set $y = x_2$, the linearized system has the system matrix

$$A = \begin{bmatrix} -\frac{1}{\theta} - S \cdot e^{-T_A/\bar{T}} & -\frac{S \cdot T_A}{\bar{T}^2} e^{-T_n/\bar{T}} \cdot \bar{a} \\ \beta S e^{-T_A/\bar{T}} & -\alpha - \gamma + \frac{\beta \cdot S \cdot T_A}{\bar{T}_2} e^{-T_A/\bar{T}} \cdot \bar{a} \end{bmatrix} \qquad (3.3.23)$$

the input matrix

$$B = \begin{bmatrix} \frac{1}{\theta} & 0 & 0 \\ 0 & \gamma & \alpha \end{bmatrix} \qquad (3.3.24)$$

observation matrix $C = (0\,1)$, and feed-through matrix $D = (0\,0\,0)$.

The stability (also of nonlinear systems) at a stationary state x_s can be investigated with the system matrix A obtained by linearization about this state. Thus, for example, the system is asymptotically stable at x_s if and only if all the eigenvalues λ_i of A have negative real parts.

In the example described, the eigenvalues of the system matrix A have negative real parts if the conditions

$$\psi = \frac{1}{\theta} + \alpha + \gamma + S\left(1 - \beta \frac{T_A}{\bar{T}^2} \bar{a}\right) e^{-T_A/\bar{T}} > 0 \qquad (3.3.25)$$

and

$$\varphi = \alpha + \gamma + S\left([\theta(\alpha + \gamma)] - p \frac{T_A}{\bar{T}^2} \bar{a}\right) e^{-T/\bar{T}} > 0 \qquad (3.3.26)$$

are fulfilled. In this case the stationariness $\bar{a}$, $\bar{T}$ is asymptotic. Criteria of this type for the stability of linear systems can be generalized to arbitrary n-row system matrices A (e.g., the Hurwitz criterion) [3.16].

The transfer-function model associated with a time-invariant linear model is

$$y(t) = \int_{-\infty}^{t} K(t-\tau) \cdot u(\tau)\mathrm{d}\tau + D \cdot u(t) \quad (3.3.27)$$

The matrix $K(t)$ in Equation (3.3.27), known as the impulse response matrix, is given by

$$K(t) = C \cdot \mathrm{e}^{At} \cdot B \quad (3.3.28)$$

(For the definition of the matrix exponential function appearing in this formula, see [3.19].) If the system has no feed-through (i.e., $D = 0$), and the idealized impulse function $\delta(t)$ (delta function) is taken for $u(t)$, the result is

$$y(t) = \int_{-\infty}^{t} K(t-\tau) \cdot \delta(\tau)\mathrm{d}\tau = K(t) \quad (3.3.29)$$

$K(t)$ itself is thus the response when a unit impulse $\delta(t)$ is input to the system.

Because transfer-function models are often obtained from measurements alone, Equation (3.3.28) cannot be used to determine $K(t)$, since the matrices A, B, and C are unknown in these cases.

Example: The residence-time distribution of a continuous-flow apparatus is $w(t)$, and $w(t)\Delta t$ is the probability that a particle entering the apparatus at time $t_0 = 0$ leaves it in the interval $(t, t + \Delta t)$. By the superposition principle, the following relation is then obtained between the inlet flow rate u and the outlet flow rate y of an indicator substance:

$$y(t) \int_{-\infty}^{t} w(t-\tau) \cdot u(\tau)\mathrm{d}\tau \quad (3.3.30)$$

which is of the type of Equation (3.3.27) with $K(t) = w(t)$.

Accordingly, $w(t)$ can be found as the response to a unit impulse; this is a widely used method for residence-time distributions.

When time-invariant linear process models are investigated, the use of integral transformations is often very productive. The Laplace transformation, which is particularly important, is discussed only briefly here. This transformation, defined by

$$F(s) = \int_{0}^{\infty} \mathrm{e}^{-s \cdot t} \cdot f(t)\mathrm{d}t \quad (3.3.31)$$

assigns to the function $f(t)$, $t \geq 0$, a function $F(s)$ of the independent variable s. Here s can be taken as a complex variable; $F(s)$ is then differentiable (holomorphic) at least in the convergence region of the improper integral in Equation (3.3.31). The convergence region is a half-plane $Re(s) > \sigma$, with boundary $Re(s) = \sigma$ dependent on the function $f(t)$. To make the mapping property of Equation (3.3.31) especially clear, the notation

$$F(s) = L(f(t)) \quad (3.3.32)$$

is also used; $f(t)$ is then called the "original function" and the Laplace transform $F(s)$ is also called the "image function." Of the characteristics of the Laplace transformation, here only three that are particularly important for application to linear process models are mentioned:

1) The transformation is linear. From Equation (3.3.31) it follows that

$$L(a_1 \cdot f_1(t) + a_2 \cdot f_2(t)) = a_1 \cdot L(f_1(t)) + a_2 \cdot L(f_2(t)) \quad (3.3.33)$$

The application of L to linear models thus preserves linearity.

2)

$$L \cdot \frac{\mathrm{d}f(t)}{\mathrm{d}t} = s \cdot L(f(t)) - f(0) \quad (3.3.34)$$

3) If the convolution integral

$$\int_{0}^{t} f_1(t-\tau) \cdot f_2(\tau)\mathrm{d}\tau \quad (3.3.35)$$

is abbreviated as $f_1(t) * f_2(t)$, then

$$L(f_1(t) * f_2(t)) = L(f_1) \cdot L(f_2) \quad (3.3.36)$$

Equations (3.3.34) and (3.3.36), in particular, illustrate the simplifications that result from the use of the Laplace transform. Relatively complicated operations such as differentiation and convolution in the original domain turn into simple algebraic operations.

Computations with the Laplace transform can be schematized as shown in Figure 3.16. Correspondence tables, showing the original and transformed functions corresponding to a num-

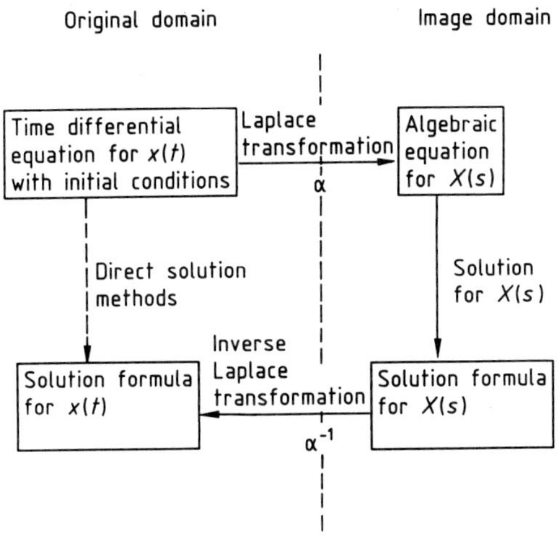

Figure 3.16. Application of the Laplace transformation

ber of elementary functions, are employed. Consider, for example, the linear model

$$\dot{x} = a \cdot x + b \cdot u \quad (3.3.37)$$

with initial condition $x(0) = x_0$. By using Equation (3.3.34):

$$s \cdot X(s) - x_0 = a \cdot X(s) + b \cdot U(s) \quad (3.3.38)$$

From this the following purely algebraic relation between the transforms $X(s)$ *and* $U(s)$ is obtained

$$X(s) = \frac{x_0}{s-a} + \frac{b \cdot U(s)}{s-a} \quad (3.3.39)$$

Together with the correspondence $L(e^{at}) = 1/(s-a)$ and the property (Eq. 3.3.36), transforming (3.3.39) to the original domain directly gives the solution formula

$$x(t) = x_0 \cdot e^{at} + \int_0^t e^{a(t-\tau)} \cdot b \cdot u(\tau) d\tau \quad (3.3.40)$$

For further details on the Laplace transformation, the reader should consult the specialist literature [3.42], [3.43].

The Laplace transformation can also be applied to vector-valued functions such as those appearing in the general model of Equations (3.3.17) and (3.3.18). Consider the case with no feed-through ($D = 0$). Application of the Laplace transformation gives

$$Y(s) = G(s) \cdot U(s) \quad (3.3.41)$$

which offers a simple link between the transforms $U(s)$ of the input and $Y(s)$ of the output in terms of the transfer matrix

$$G(s) = C \cdot (s \cdot I - A)^{-1} B \text{ (I = identity matrix)} \quad (3.3.42)$$

Note that, as usual, we are considering the state that depends only on the temporal behavior of n; that is, $x(0) = 0$ and Equation (3.3.41) holds only for inputs $u(t) = 0$ for $t < 0$. Equation (3.3.41) in the transform domain corresponds to the relation

$$y(t) = \int_0^t K(t-\tau) \cdot u(\tau) d\tau \quad (3.3.43)$$

in the original domain, which follows from Equation (3.3.27) under the stated conditions. Thus the role of the impulse function response matrix $K(t)$ in the original domain is played by the transfer matrix in the transform domain.

A function of great importance in practice is the frequency response $G(j\omega)$, which is obtained by letting s take on imaginary values $j\omega$ (ω real) in Equation (3.3.42). The frequency response satisfies the fundamental relation

$$y(t) = G(j\omega) \cdot e^{j\omega t} \cdot u_0 \quad (3.3.44)$$

which in the original domain gives a simple description of the response to periodic input functions $u(t) = e^{j\omega t} u_0$. Since the frequency response $G(j\omega)$ contains all the information about the process, an obvious way to extract this information is to measure the response functions $y(t)$ over an entire spectrum of input frequencies ω and use Equation (3.3.44) to determine $G(j\omega)$ from these. This "frequency response method" is also a highly sensitive procedure for stability analysis that does not require a knowledge of the system matrix A and its eigenvalues. The determination of the frequency response from measurements with the aid of periodic input functions also makes it possible to obtain the impulse response $K(t)$ more accurately than with the impulse method, even though the amount of measurement required is greater, since $K(t)$ is calculated from the frequency response $G(j\omega)$ through the inverse Fourier transformation:

$$K(t) = \frac{1}{2\pi} \int_{-\infty}^{\infty} G(j\omega) e^{j\omega t} d\omega \quad (3.3.45)$$

This expression can be inverted in accordance with the Fourier integral formula:

$$G(j\omega) = \int_{-\infty}^{\infty} K(t)\,e^{j\omega t}\,dt \qquad (3.3.46)$$

If the time dependence of the transfer function is replaced by a position dependence and, therefore, the temporal frequencies by spatial frequencies, the analogous operations lead to the contrast transfer function (optical transfer function) known from image transmission and photography. This function verifies the transformation principle described in Section 2.2 as well as the phase model derived from that principle in Section 2.3 [3.44].

3.4. Modeling

The derivation of a mathematical model for a given purpose involves several problems. The objective governs

1) The type of model (dynamic, time-continuous, time-discrete, stationary, time-variant, etc.)
2) The structure of the model (theoretical, statistical, transfer-function, linear, etc.)

The restrictions mentioned in Section 3.3, regarding the computability of the model and computing time problems, must also be taken into account. It is often not possible to give an a priori answer to the question of which physical processes should be incorporated into the model and which can be neglected. The addition of a further physical phenomenon to the model may render the former numerical method of solving the model equations useless and may lead to a large increase in total computing time. In such cases, attempts are made to use a simplified version of the model. Ultimately, however, the power of the model depends on the comparison of computed results with experimental data. The modeling process often becomes an iterative procedure in which calculation and experimentation alternate. It is not always the case that successive, improved versions of the model, when compared with the experimental data, will converge to the final model. Occasionally, new measurements are needed for deciding whether to accept or reject certain model assumptions.

These problems suggest a generalization of the problem statement: Given some competing models M_1, M_2, ..., describing a certain system, to find quantitative decision criteria for identifying the "best" model at minimal testing cost. A number of publications on this model discrimination problem have appeared [3.45]–[3.47]; in the well-known Box–Hill method [3.48], an entropy value is assigned to each model based on its agreement with previous experimental results, and the next test point is chosen so as to maximize the entropy difference between models.

Section 3.3 identified discretization as an essential step in obtaining finite-dimensional equivalent models of the type defined in Equations (3.3.1)–(3.3.3). This implies, in general, the problem of choosing the correct state variables in the vector x. Usually there is a wide range of options, and this transformational freedom should be utilized in mathematical models in order to exhaust certain optimization possibilities. The following main objectives can be stated:

- Models as small as possible
- Short computing times
- Model structures as simple as possible

These goals are interrelated and they also compete with the requirement of a minimum model accuracy. Known examples in which suitable coordinate transformations are used to give simpler model structures include the diagonalization of linear systems, with a diagonal matrix A as system matrix, and the change to dimensionless notation, which indeed often works wonders.

The requirement of small but efficient computational models must often be invoked, especially for on-line use. The model reduction or order reduction technique described here involves the replacement of a mathematical model by a smaller model. There are essentially two approaches

- The use of special physicotechnical properties of the system under consideration yields a reduced model that is often highly efficient
- The second approach consists of purely mathematically formal methods for obtaining reduced process models

In both procedures, a key role is played by the coordinate transformation concept. An attempt is made to position the new coordinate system so that the fewest possible coordinates (state variables) are needed and so that these reflect the truly relevant degrees of freedom in the system. The technique involves projecting onto the state–space subspace that is essential for the processes taking place in the system. The idea is closely related to the principal-component analysis method used in Section 3.2 to spot relationships in existing data.

A well-known example of such a projection is the expansion in specified functions in the Ritz–Galerkin method. In this technique, eigenfunctions of the operator or so-called functions with compact support are employed in a finite-elements procedure. Other examples are Fourier series expansion; the method of moments used to solve systems of equations arising in, for example polymerization kinetics; and the use of symmetry properties generally. Mention should also be made of similarity theory [3.49]–[3.51], in which quantities relevant to the problem are determined, a dimensional analysis is performed, and the resulting set of dimensionless quantities is smaller in number than the dimensioned starting quantities.

Another example of order reduction is statistical mechanics, in which the ca. 10^{26} state variables that describe the ensemble of molecules in the subject system are projected onto a few new quantities, such as temperature, pressure, and enthalpy.

In the examples examined up to this point, either the transformations were physically motivated or plausible trial functions, series expansions, or the like were applied, and the coefficients in the expansions were taken as the new state variables. A number of methods have been devised [3.52]–[3.60] which attempt to eliminate this element of arbitrariness. In a certain class of transformations, these techniques identify the one for which a certain figure-of-merit functional is minimized.

Figure 3.17 illustrates the method schematically. Any reducible system has the following property: Every solution trajectory, after quickly passing through an initial branch, merges into a one-dimensional curve. This system therefore has just one relevant degree of freedom, even though it is two-dimensional. The discrepancy between the number of degrees of freedom n and the number of relevant degrees of freedom r is large for most large systems and represents the key to their reducibility. The appropriate coordinate transformation must be found so that z_1 will be the relevant state variable in the new coordinate system (as shown in Fig. 3.17) while z_2 plays a secondary role and can be neglected. If a certain set M of representative solutions $x(t)$ and associated solutions $x^*(t)$ of the reduced system is chosen and the merit function

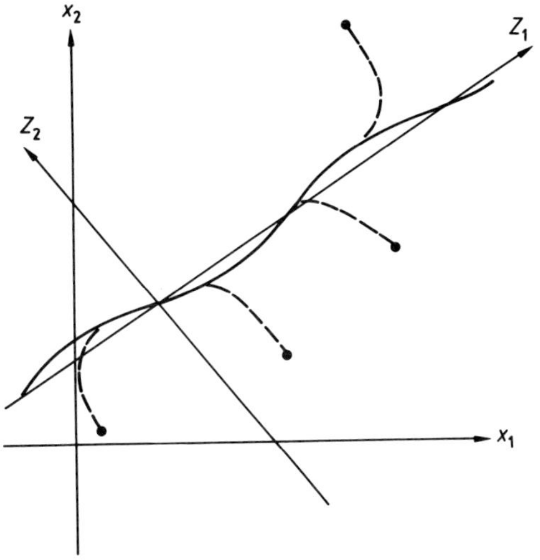

Figure 3.17. Schematic diagram of model reduction

$$J = \sum_{M} \int_{0}^{\infty} \| x(t) - x^*(t) \|^2 \, dt \qquad (3.4.1)$$

[3.58] is examined, then for specified r the reduced system is defined such that J takes on a minimum value J_{min}, whereby

$$\varepsilon_{min} = 100 \, J_{min} / \sum \int_{0}^{\infty} \| x(t) \|^2 \, dt \qquad (3.4.2)$$

is a measure of the mean relative error of the solutions $x^*(t)$ of the reduced system. Figure 3.18 shows ε_{min} as a function of the dimension of the reduced system r, where the original mathematical model was that of a distillation column with 33 trays and three components. The number n of

r	ε_{min}, %
4	0.532
5	0.337
6	0.195
7	0.124
8	0.0730
9	0.0527
.	.
.	.
.	.
68	0

r = number of degrees of freedom of reduced system

ε_{min} = mean relative error of solutions of reduced system

Figure 3.18. Model reduction for a distillation tower

differential equations is 68. The set of all response functions to step inputs was selected as the representative ensemble of solutions. The error ε_{min} decreases with increasing r, becoming zero for $r = 68$, where the unreduced original model is reached. Even for r as small as 4 the error is only ca. 0.5%. Thus the problem requires a reduced model with only four differential equations instead of 68.

3.5. Management and Utilization of Information

Introduction. As early as 1986, the Association of Large Power Plant Operators (Verband der Grosskraftwerksbetreiber) carried out a thorough discussion of changes in process control engineering and the demands imposed on the management of engineering documentation [3.61].

With increasing complexity of plants and processes, the number of monitoring and control devices has increased out of proportion. The main reason has been the growing level of process automation. Rationalization has had little to do with this development; a far more important point is that stringent requirements on product quality not only call for increased information about the process but also demand more powerful process modeling aimed at better reproducibility and direct control of product quality. In addition, a keener awareness of plant safety has led to extensive safety analyses, which in turn have given rise to practices that frequently involve monitoring and control equipment.

These growing demands can easily bring about a documentation conflict between designers and builders of process plants, on the one hand, and operators and maintenance personnel, on the other. The management of data on control devices must enable the use of the documentation right through from concept, through design, execution planning, construction, and operation of the plant. The benefits of such an information system are not limited to rationalization; consistent documentation also simplifies the exchange

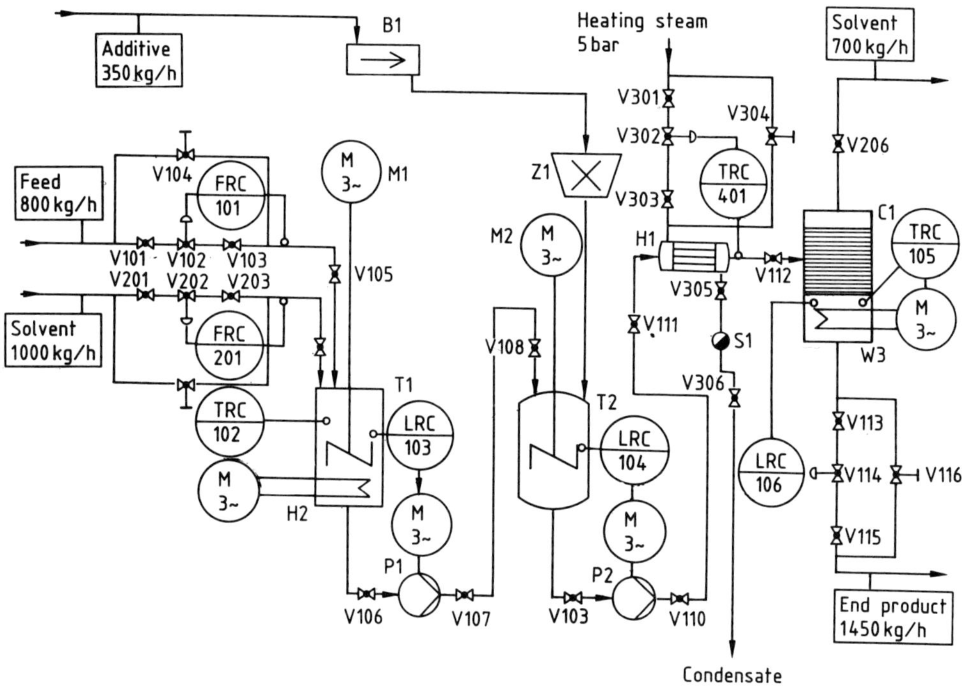

Figure 3.19. Example of a piping and instrumentation diagram (DIN 28004, Part 1) [3.62]
B = belt conveyor; C = column; H = heat exchanger; M = motor; P = pump; S = separator; T = vessel, tank; V = valves; Z = size reduction; FRC = flow registration control; LRC = level registration control; TRC = temperature registration control

of experience between operators and builders of plants.

Many discussions have contributed to the new edition of DIN 40719, Part 2, ("Circuit Documentation: Labeling of Apparatus, Signals, and Documents"). The basic ideas can be extended in large part to the documentation of plant monitoring and control systems.

Forms of Organization. The classical organization of control engineering documents is functional (see Chap. 2). Documents are created in order to accomplish a certain task. The instrumentation of "something" is set forth in, for example, the piping and instrumentation diagram (Fig. 3.19). "Any" control system can be described by functional diagrams (Fig. 3.20) and schematic diagrams (Fig. 3.21) or program flowcharts. Documents for the installation of "any" process control device include schematics as well as terminal diagrams.

This type of documentation does fulfill its intended purpose, but it leaves the user virtually unable to form a comprehensive picture of the properties of the control system. Indeed, the user can hardly get an idea of the system from any other viewpoint than that of documents prepared in the past (see Section 2.2). As a consequence, whenever a new question arises, it is often necessary to create new supplementary

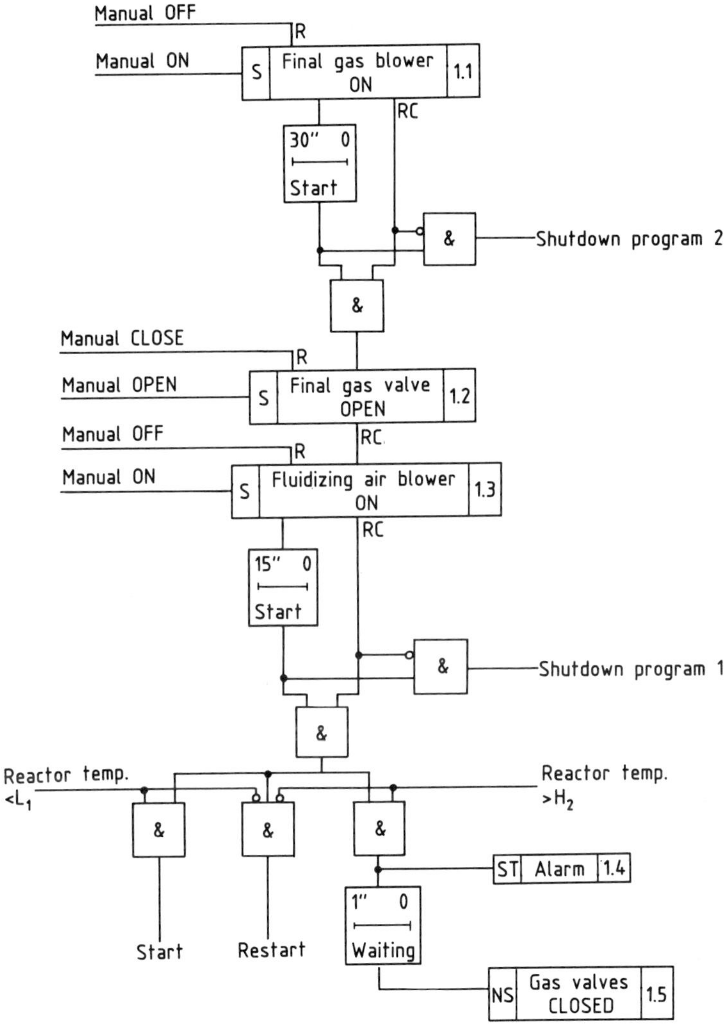

Figure 3.20. Example of a control system function diagram (DIN 40719, Part 6)

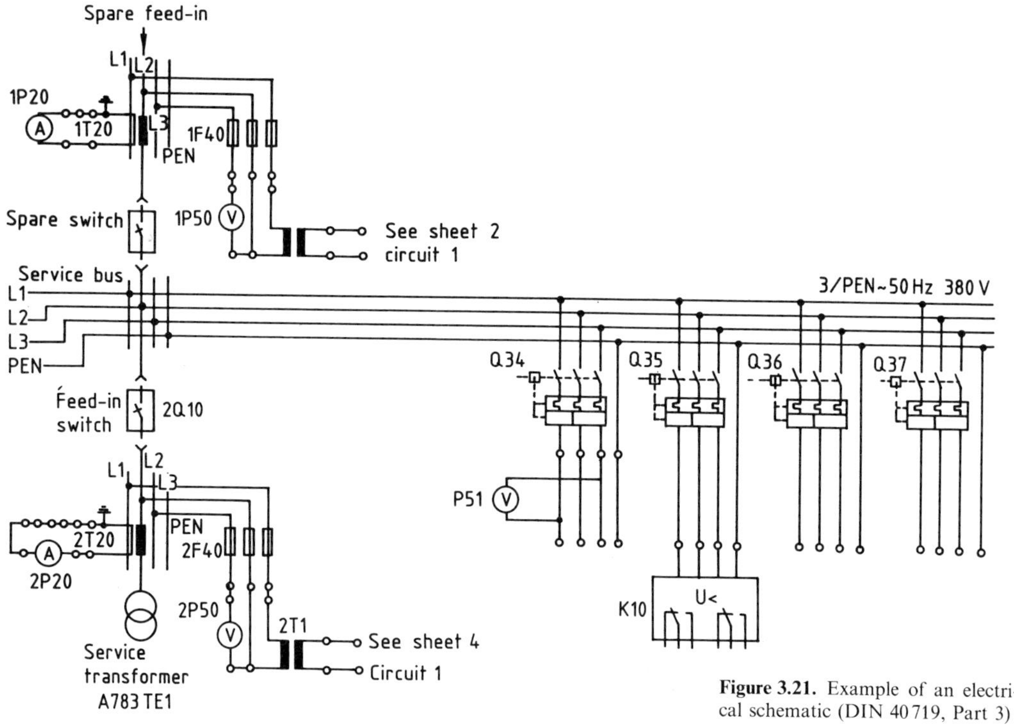

Figure 3.21. Example of an electrical schematic (DIN 40719, Part 3)

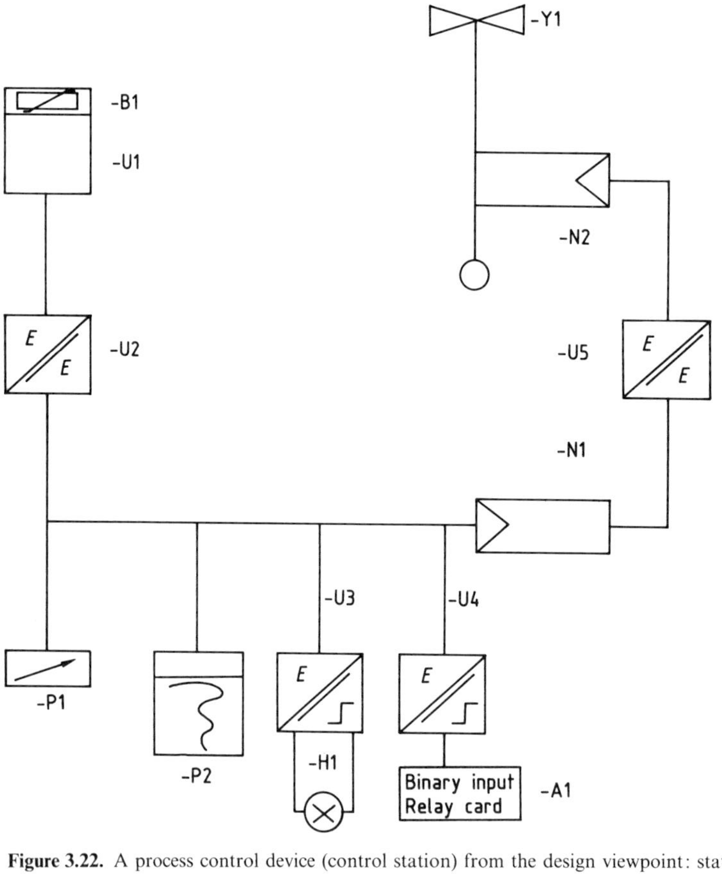

Figure 3.22. A process control device (control station) from the design viewpoint: station (loop) diagram

No.	Control station KACP	Functions checked	Reason for check	Doc.reqd.?	Reqd. condition
1	V105 BA20 P2	I S+ A+	Plant safety	Y	Shut down

Object-related instructions:
Written permission, station setup data sheet, station diagram required

Inspection: Check pressure measurement with I S + A + — Report

1. Perform visual inspection

   Are the elements of the process control station fully labeled, in good mechanical condition, and clean? — OK / OK after repair / Not OK

2. Connect mA test generator in place of transducer.

3. Check measuring range of readout at three points; note actual values.

Nominal values:	Test generator	Readout	Tolerance	Actual values
First value	4.0 mA	0.2 bar	± 0.02 bar	________ bar
Second value	12.0 mA	0.6 bar	± 0.02 bar	________ bar
Third value	20.0 mA	1.0 bar	± 0.02 bar	________ bar

   Are the actual readout values within tolerance? — OK / OK after repair / Not OK

4. Check maximum limit indication.

   Limit: 0.90 bar   Tolerance: ± 0.02 bar

   When limit is exceeded, following indications must occur:

   Does visual indication "P02 max" occur in A 639 W 1 panel 7? — OK / OK after repair / Not OK

   Does acoustic indication occur in A 639 W 1? — OK / OK after repair / Not OK

5. Check maximum limit response.

   Limit: 0.95 bar   Tolerance: ± 0.02 bar

   When limit is exceeded, following responses must occur:

   Does valve P2 – Y1 open? — OK / OK after repair / Not OK

   Does pump P27 shut off? — OK / OK after repair / Not OK

6. Restore to initial condition.

Remarks: ________________________________

Date: ____________ Inspector's name: ____________ ☐ Not inspected

Figure 3.23. A process control device (control station) from the maintenance viewpoint: partial job plan

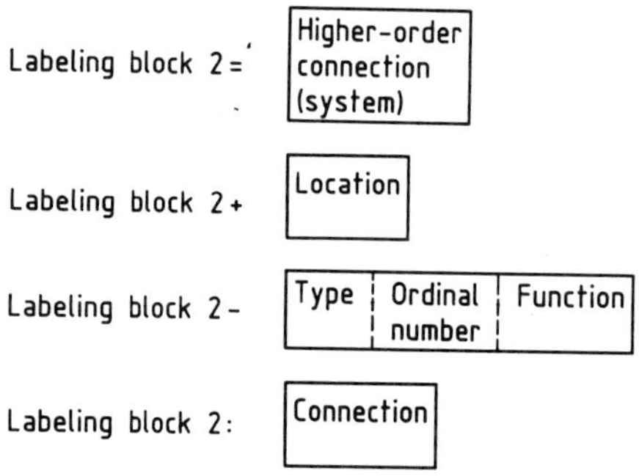

Figure 3.24. Devices (DIN 40 719, Part 2)

documents even though these may contain scarcely any new information.

One remedy for this documentation bottleneck is object-oriented management of data on monitoring and control devices. Documents for a given task are then representations of object-related information, that is, projections from certain viewpoints (Figs. 3.22 and 3.23). Implementation of this concept requires, above all, the devising of a consistent data model for hardware objects (see Section 2.2). It has proved desirable not to model the process control system in isola-

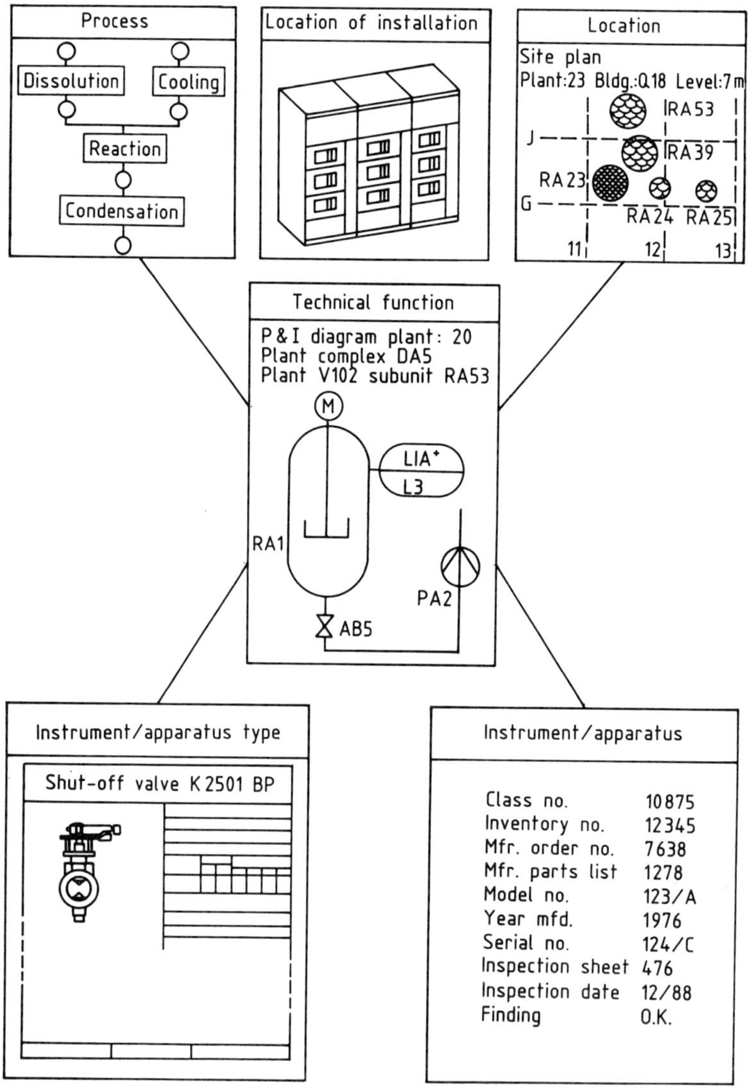

Figure 3.25. Information blocks associated with process plants

tion but to see it as a component of the process plant.

The initial step must be to understand which data blocks are the fundamental ones. DIN 40719, Part 2 defines the most important data blocks for electrical apparatus (Fig. 3.24).

This is not adequate for process plants, since it does not include the purpose of the facility: the process. Incorporating the process into the information structure becomes more important for process control engineering, the more use is made of higher process control functions such as recipe procedures and process models for process control (see Section 4.5).

It has proved desirable to separate detailed description of the engineering properties of a single instrument (substance) from information about its technical function (see also Section 2.3, p. 341). This makes it possible to manage hardware improvements by replacing devices in the plant with others that perform the same function. All these aspects lead to an information structure for process plants and their process control systems as shown in Figure 3.25.

When an object-oriented model is created, these information blocks must be described in terms of interlinked individual objects. Figure 3.26 presents a simplified example in which the "entity-relationship model" described in Chapter 2 was used.

When such a model has been set up, it is possible to file all information by object. For example, data on a process control station is filed only under that station. Data on a higher-order plant subunit are filed only under that subunit. They are automatically relevant to the lower-order control stations. Data on an apparatus or instrument type are filed only under that apparatus or instrument type. They are automatically relevant to the associated process control stations.

Application. After such a data model has been created (provided it is complete), documents can be generated at any time and tailored to any objective. It is at least equally important that given information is managed by object, not in various documents. A computerized revision service can then be set up easily, so that information retrieved at any time will be internally consistent.

Flowsheets are now prepared for a plant (plant unit) or subunit in accordance with this data model (Fig. 3.27).

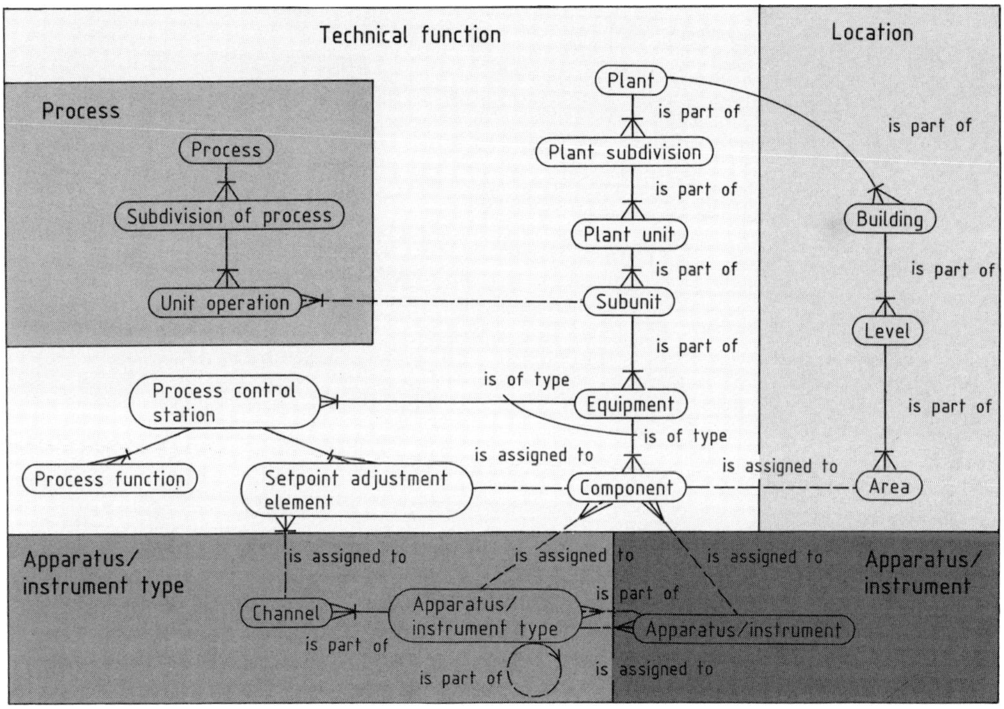

Figure 3.26. Simplified data model for information blocks relating to process plants

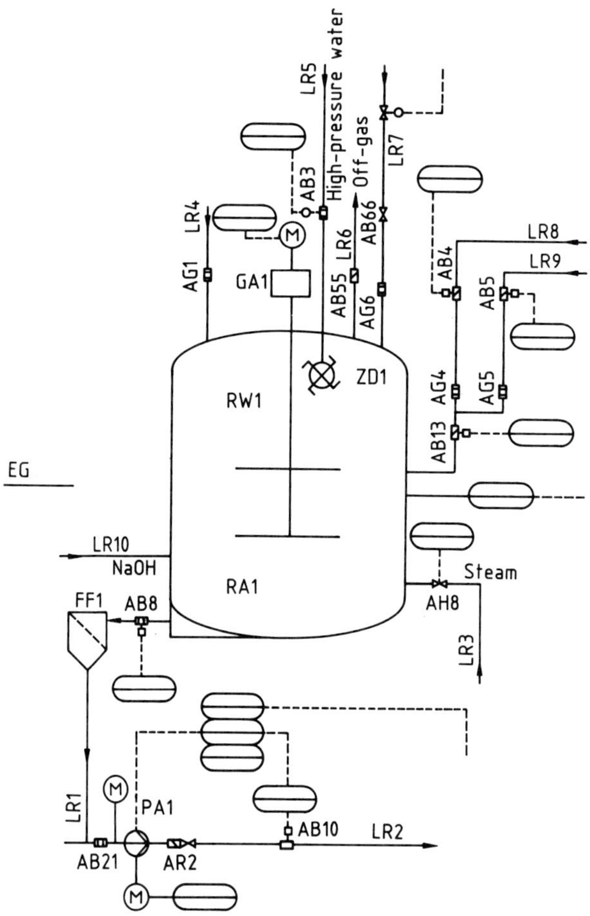

Figure 3.27. Piping and instrumentation diagram of a plant subunit

A functional diagram may refer to a certain subunit (Fig. 3.28) or to a certain control station (it is then known as a station or loop diagram; Fig. 3.29, p. 384).

This object-oriented management of engineering data makes it possible to work with the same documentation at every stage of plant activity and to update the documentation step by step. A glance at the phases of plant activity (Fig. 3.30, p. 385) shows the feasible and necessary level of detail in the documentation at each stage (see also Chap. 7).

Structuring of the processes in terms of unit operations (DIN 28004) is possible and necessary even at the basic research stage. Plants can already be subdivided into units as documented in the basic flowsheet (Fig. 3.31, p. 385). The assignment of unit operations to plant subunits, which is so important for higher-level automation tasks, can take place at this stage.

The phase model derived from experience in software engineering (discussed in detail in Section 2.3) is suitable for this purpose.

Quality assurance, in particular, calls for a systematic representation of product and process requirement profiles, since the needed data acquisition measures as well as sensor and actor systems must be established from this standpoint (Fig. 2.32, p. 386). In this connection, the reader is referred to Japanese Industrial Standard JIS Z 8206-1982, "Graphical Symbols for Process Charts." This goes beyond DIN 28004 by specifying which quantity and quality checks must be performed on product streams. Recent practice is

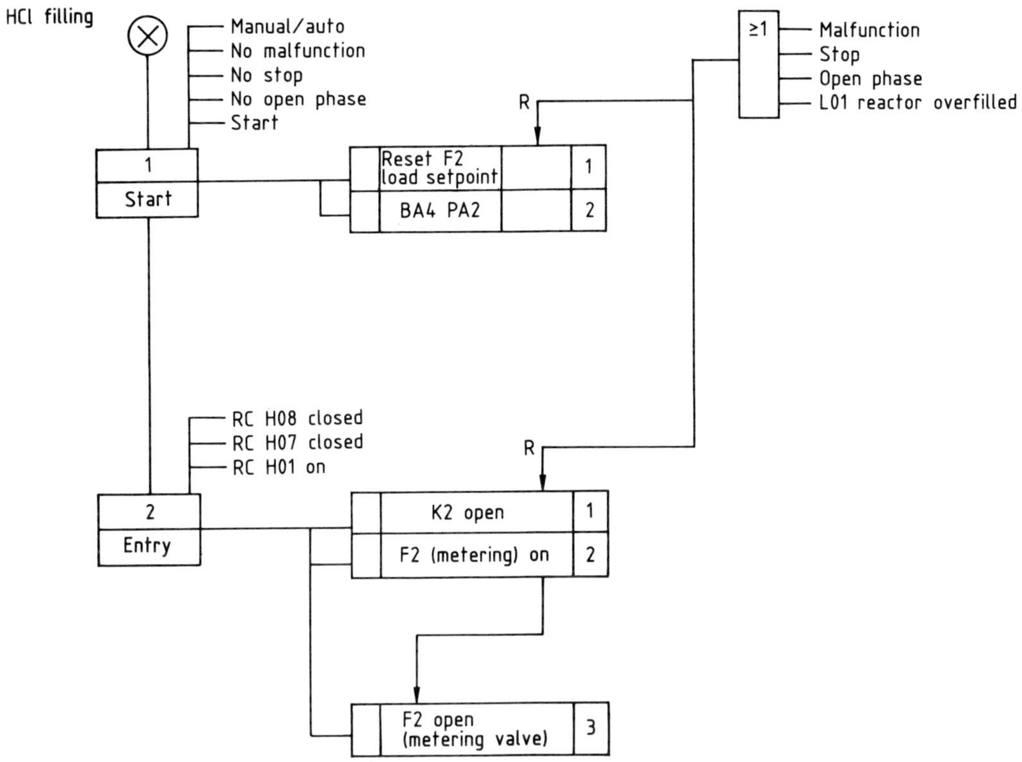

Figure 3.28. Function chart of a plant subunit

to generate process control flowsheets and checklists as tools for quality assurance; these can also serve as evidence that the quality assurance system itself (Fig. 3.32, p. 386) conforms to the standard.

The most important process control equipment is selected in the engineering design stage. Documentation relating to each control device, such as a control station, can be created here.

Of special importance in the approval planning (regulatory engineering) phase is the safety analysis, which establishes particular safety objectives and practices. It is important that all information used in this analysis refer to definite, documented engineering hardware (Fig. 3.33, p. 386), since only then can the engineering of protective measures be assessed before commissioning and during plant operation.

At the stage of execution planning, all engineering details are fixed. The number of implementation options is unlimited in principle, but in any real enterprise the factors of economics and practicability call for the selection of procedures as well as instruments and systems with which, or from which, the requisite functions can be developed. These choices then lead to design and installation standards, but also to equipment and system standards. As discussed in Chapter 2 and at the beginning of this section, these decisions are also documented in the object-oriented framework (Fig. 3.34, p. 387).

Examples of activities during execution planning are the preparation of control station diagrams for process control devices and the specification of requirements on engineering properties. This information is filed and managed in the process control station. Realizable hardware units, for example, are handled as instrument types derived from the instrument standard, while the functional elements are assigned to control devices (Fig. 3.35 p. 387). This has the great advantage that a distinction can be made between requirement and implementation and that, when instruments differing in type are interchanged, only the assignment must be changed. Manual updating of engineering details becomes unnec-

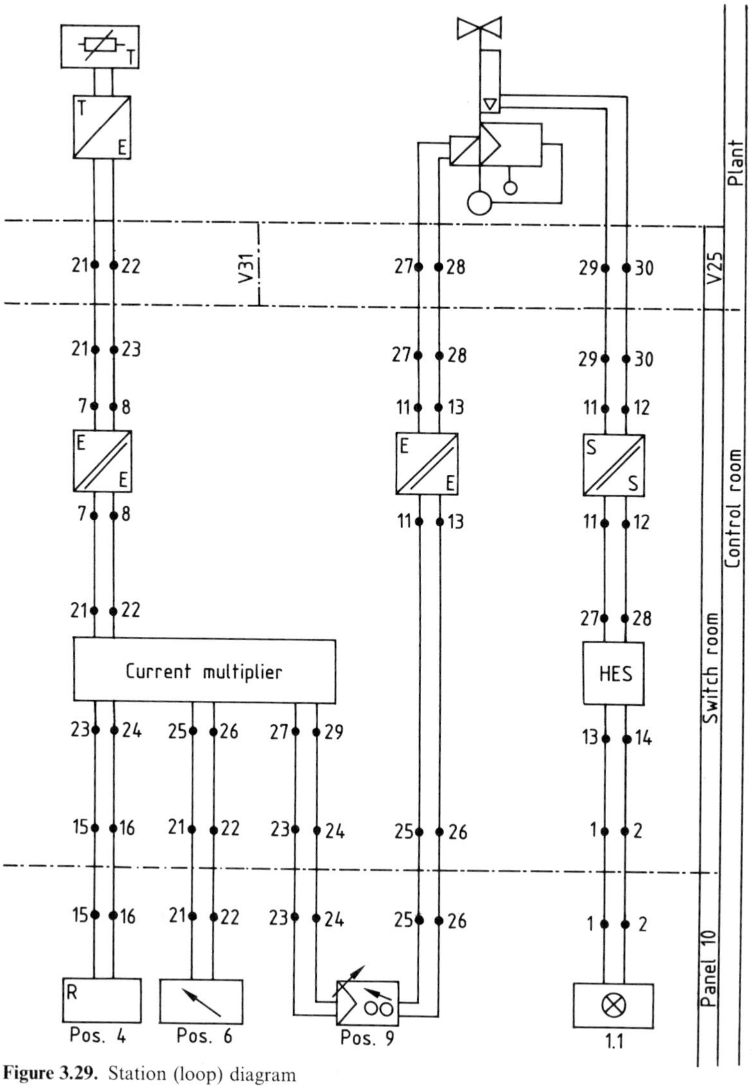

Figure 3.29. Station (loop) diagram

essary. Similarly, the basic functions of the process monitoring and control systems are assigned to the control devices. In addition to a more rational configuration, this offers the advantage of greater reliability with regard to the consistency of engineering data (see Section 4.3).

Similarly, individual devices can be assigned by technical function if necessary. This would be the case, for example, if custom instruments were used instead of off-the-shelf devices, if the instruments were subject to individual inspection, or if maintenance cost records were to be kept for an individual unit.

Such assignments are important items of information for installation and maintenance. When necessary, they must be indicated locally. Maintenance and inspection work is classified by technical function (Fig. 3.36, p. 388). The performance of such work and the results of inspections are documented by technical function. The analysis and repair of defects are done chiefly in reference to the technical function. In the course of

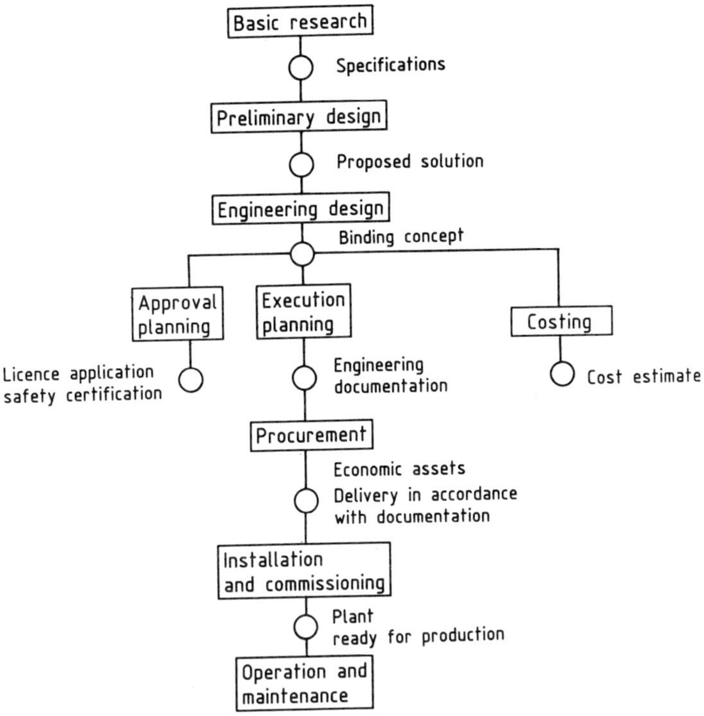

Figure 3.30. Phases of plant activity

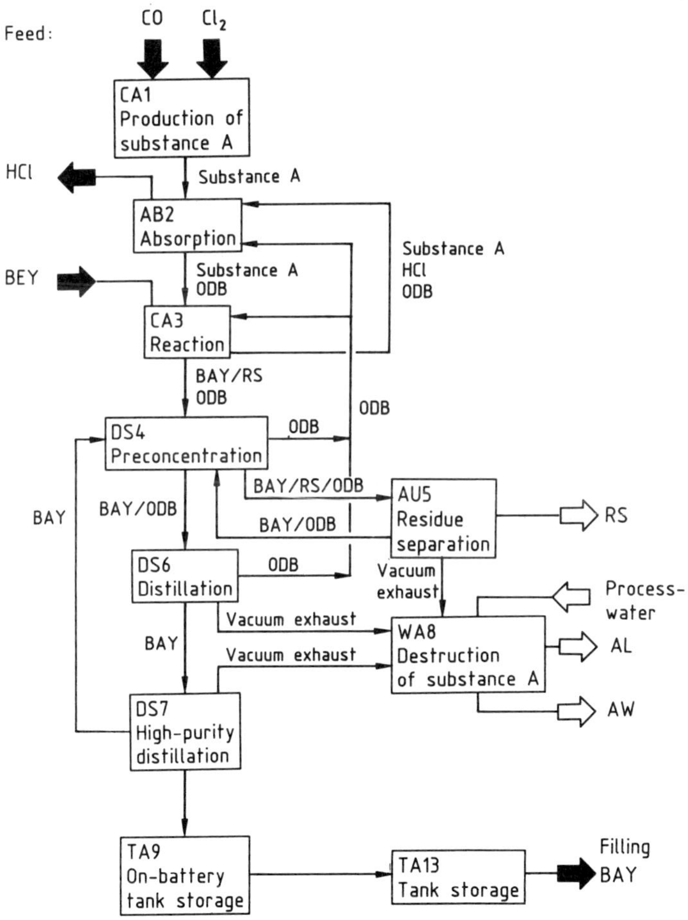

Figure 3.31. Basic flowsheet of a plant

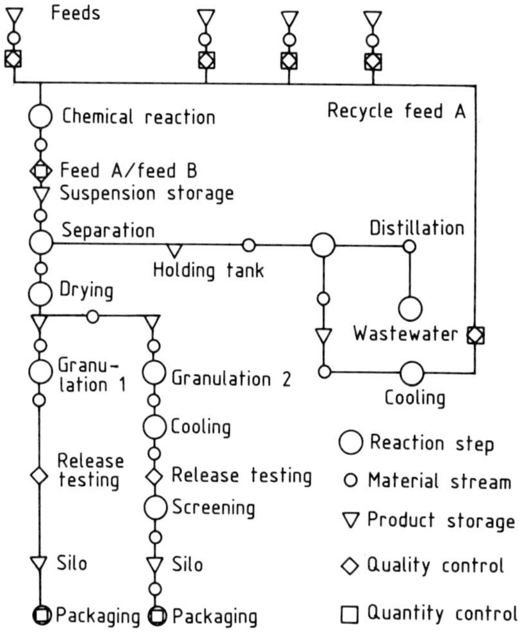

Figure 3.32. Process monitoring flowsheet of a vulcanization accelerator (cf. JIS Z8206)

Safety analysis

Hazard discussion Plant: ____________ Date: 10/27/88 V ____ -A ____ -G Version: __Q__ Page: ______

Facility identification (Internal Standard 9060)

Plant	Plant subdivision	Plant unit	Subunit	Bldg.	Piping and instrumentation diagram	Date
2 2	P P G 6	V 6 9	R A 9 1			

VA No. : __________ Process section: ______________________

AT No. : __________ Plant component: ______________________

Deviation from nominal condition:

DEV.NO.	Description
1	Flow rate (feed) too low

Possible causes

Cause No.	Description
1	Pump malfunction (PA2 or PA3)
2	Fittings closed (AB7 or AB12 or AH17)
3	Flow ratio control defective (F2 and F3)

Figure 3.33. Example of a form used for plant safety analysis
The information from this form must be correlated with the corresponding device (shaded blocks)

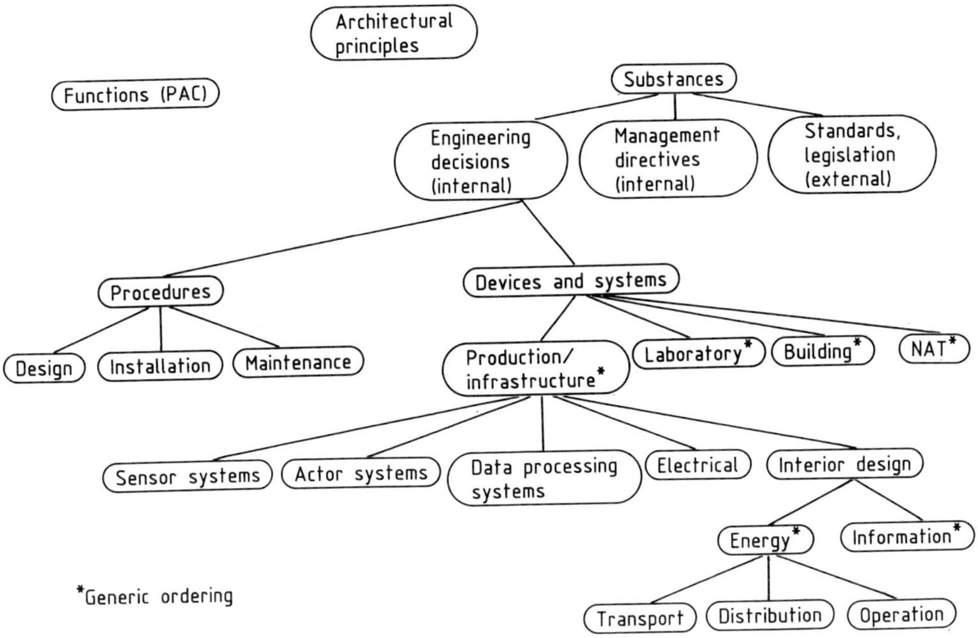

Figure 3.34. Structure of decisions relating to process control stations

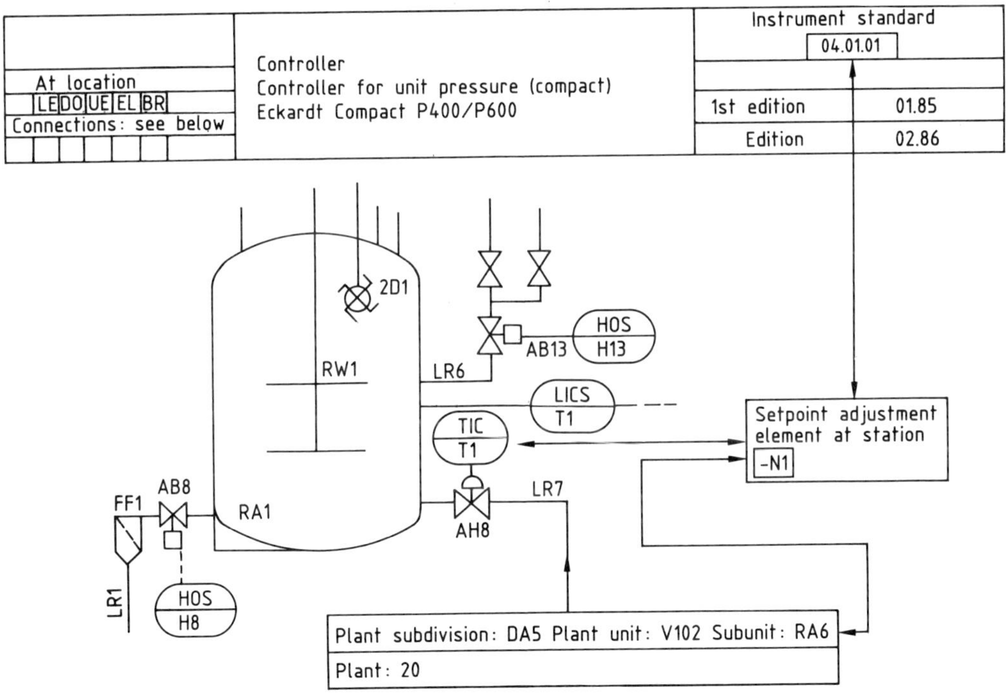

Figure 3.35. Assignment of instrument types to technical function for the example of a process control station

FS-IT Maintenance			
Plant	Plant subdiv.	Plant unit	Subunit
22	FSDL	V40	WS33

Inspection plan
Job specification
Mechanical

Job number 15	Group number 432	As of: 1.89	AKZ 622001	Area: Dralon	Operation: Spinning
Building: E4	Eng.-No.: 075350	Order no.:	Test no.: 3	Registration no. 24	Location no.: 108
Planning date: 15.04.89	Function group: 432	Inconvenience bonus:	Carried out by:		
Type of work:					

Component	Operations:
PA 1	Replace GLRD, adjust coupling
PA 4	Check tightness of motor can, adjust coupling
RM 1	Remove NaOH feed and clean nozzles
VE 1	Inspect/replace V-belts
F8 1	Clean mist collector
WS 31	Use pressure cleaner to remove biological and other deposits; perform visual leak check

Figure 3.36. Example of a maintenance and inspection plan for a component
Each instruction relates to the functioning of one device

examination, the engineering details on the assigned information can be considered; technical defects can be localized through the assignment of installation locations.

Such an information model makes possible an integral analysis of process facilities and, ultimately, of an entire company. The details of the model must be tailored to the technical circumstances. The storage and management of information in relation to engineering objects is crucial for its consistent use throughout the economic life of the plant. Only in this way is it possible to generate documents, at any time and at comparatively low cost, permitting instant analysis from any desired viewpoint.

4. From Process Knowledge to Process Control

4.1. Principles

Industrial processes, in contrast to natural processes, are artificially designed, consciously initiated, and purposely controlled so as to serve a certain end. It is not enough to understand an industrial process and be able to describe its internal dynamics in terms of a model; if the process is to take place at all, some form of external control is needed. This external control includes all functions required for coordinating and controlling the process. The structure, form, and implementation of these process control functions are the subject of the chapter that follows.

The sections on feedback control and optimal control set forth the most important design methods from control theory. Section 4.4 explains the concept of binary control in a variety of applications. In Section 4.5 it is shown how the operational control hierarchy can be structured for a plant as a whole.

Process control functions play a central role in operational process control. For better understanding, these can be explained briefly before going on to the rest of the chapter.

Figure 4.1 shows schematically how process control functions work. The same diagram applies to all control functions. Each control function, on the basis of its own logic, seeks to bring the actual properties of a system into accord with the specified nominal properties. Each control function thus embodies certain assumptions about the behavior of the system. Using this knowledge, a control function can issue control commands in such a way as to influence the system in the desired fashion.

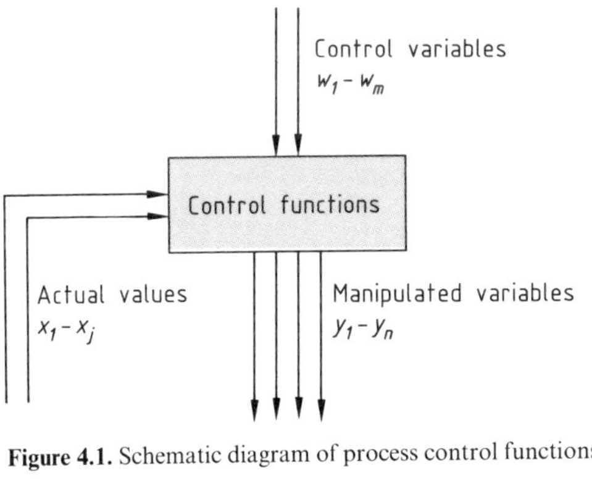

Figure 4.1. Schematic diagram of process control functions

In process control, the control variables and manipulated variables correspond to variable process or product properties. Manipulated variables either impinge directly on actuators or play the role of control variables, acting on lower-level control functions. In this way, process control can be built up in a hierarchical manner.

If we consider, for example, the adjustment of the pH in a reactor by the addition of acid, we obtain the hierarchy illustrated in Figure 4.2.

The task of the production action is to make a product having certain properties at the discharge from the process section. For this purpose, one of the manipulated variables available to it for controlling the process is the pH in the vessel.

The metering device seeks to establish the specified time dependence of pH in the vessel; it therefore passes on the acid flow rate to the lower-level flow controller as a setpoint.

The flow controller, in turn, sets the required flow rate by specifying a certain valve position. The valve position controller exerts a corresponding control pressure to insure that the valve comes into the required position.

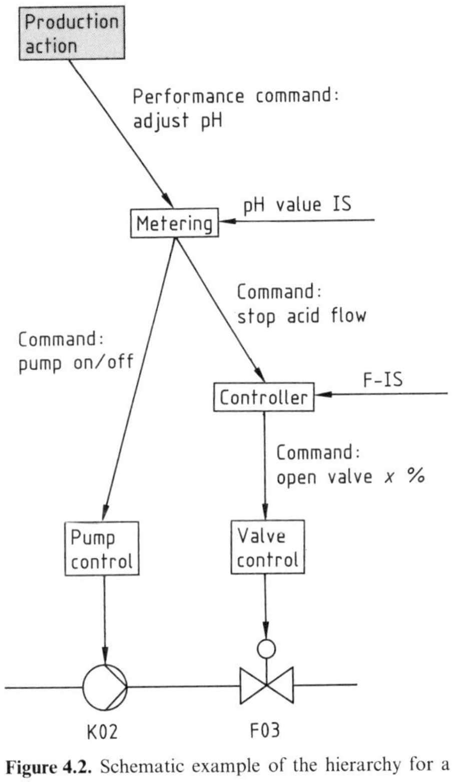

Figure 4.2. Schematic example of the hierarchy for a pH adjustment

In most cases, a process control function has only one control variable; in general, however, *m* control variables and *n* manipulated variables must be allowed for.

The control concept can be customized for any process control function. The methods of control engineering are therefore applicable to process control functions at all levels in the hierarchy.

Process control functions can be classified in four groups by the concept employed:

1) Timed process control functions
2) Event-controlled feedback functions
3) Continuously controlled functions
4) Process control functions (optimized with respect to some quality criterion)

The selection of a concept depends less on the type of process control function and its position in the hierarchy than on the quality requirements, the anticipated severity of disturbances, the possibility of directly or indirectly measuring certain process properties, and (for the fourth group) the availability of suitable process models.

All the concepts are equally capable of reproducing a specified process if it is assumed that there are no disturbances.

Only when malfunctions or variations affect the course of the process will the four groups differ in their capabilities.

In practice, the concepts are combined as needed.

In what follows, the various concepts are illustrated for the case of a pH control. The task for the process control function is to regulate the addition of acid so as to keep the pH in a reactor at the proper value for the process.

1) Timed process control function:
Acid is admitted to the reactor at a fixed, specified flow rate $q(t)$. If there are no disturbances, the desired pH value is established in the reactor. The actual pH value is not fed back.

2) Event-controlled feedback function:
Acid is admitted to the reactor at a fixed, specified flow rate $q(t)$ until a certain specified pH has been reached; the flow is then cut off. If the process takes place without disturbances, the pH in the reactor follows the desired trajectory. If there are disturbances, the actual pH agrees with the desired pH, at least at the end point. The pH must be measured at least once.

3) Continuously controlled function:
The admission of acid is controlled so that the pH in the reactor follows a fixed, specified control trajectory. Control actions compensate for upsets. Continuous pH measurement is required.

4) Optimal process control function:
The control trajectory pH(t) is not specified in fixed form but is determined on the basis of the instantaneous process state and a quality criterion. If the fixed trajectory of concept 3 is assumed optimal in the sense of the quality criterion, functions 3 and 4 initially result in the same control trajectory. No difference is seen until a disturbance causes the actual value to depart from the nominal trajectory. Under concept 3, the control system attempts to force the process back onto the old trajectory; under concept 4, the process is guided along a new trajectory that is optimal when the actual situation is taken into account.

4.2. Feedback Control

Introduction. This section explains the concepts of feedback control and feedforward control. It discusses the distinction between the two as well as what process properties and what process requirements make a feedback structure necessary. The drawbacks to a feedback structure are also elucidated.

Any process is characterized by the action of the inputs on the outputs (Fig. 4.3). Inputs are classified as manipulated variables and disturbances. The former make it possible to intervene in the process, while the latter are undesired environmental effects that act on the process (see Chap. 3). Process outputs are quantities that are to be influenced in an appropriate way. Desired values for the outputs are called setpoints. Time-varying setpoints are called command or reference variables.

The objective of any control system is to interact with the process via the manipulated vari-

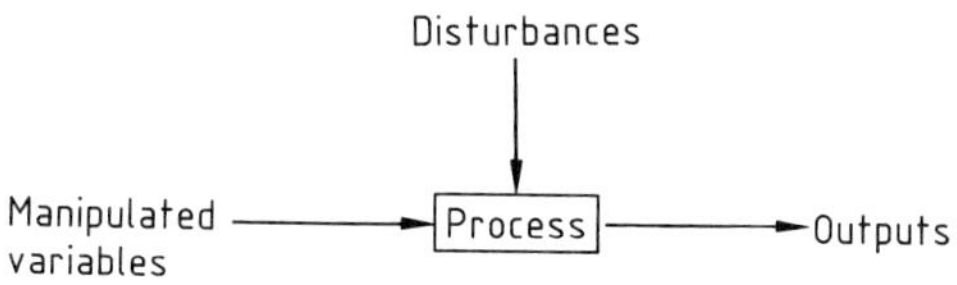

Figure 4.3. Schematic diagram of a process

ables so as to insure stability and good performance.

The second point means that the outputs should react in a desired fashion to a change in the setpoints (reference variables) or the disturbances. In particular, they should react sufficiently fast, and the steady-state difference between setpoints and outputs should be sufficiently small.

Feedforward control can be diagrammed as an open loop (Fig. 4.4). On the basis of externally specified setpoints and measured disturbances, the controller generates manipulated variables as process inputs. No check is made to determine whether the selected manipulated variables actually have the desired effect on the outputs. With this structure the objectives cited can be achieved only if

- All disturbances to the process can be measured
- The effets of control actions on the outputs are known with sufficient precision, so that the mathematical process model underlying the controller design reflects the main characteristics of the real process
- The uncontrolled process is stable (an unstable process cannot be stabilized by feedforward control)

In contrast, good performance can often be achieved with feedback control (Fig. 4.5), even if the above requirements are not met. Some of the outputs to be influenced are measured and compared with their setpoint values. The differences between these values are transmitted to the controller, which uses them in calculating the required control actions. The drawbacks of a feedback control structure are that feeding back the outputs

- Introduces additional measurement errors
- May (if the controller is inappropriately designed) make a stable process unstable
- Fails to react to disturbances until after they have already influenced the outputs

In view of the last point, feedback and feedforward control concepts are often combined. An example is a feedback structure with an additional disturbance input to the controller (Fig. 4.6), in which measurable disturbances are partly compensated by feedforward control but in general they cannot be completely suppressed because of shortcomings in the mathematical process model. Feedback control is used in an attempt to diminish the "residual effects" of these disturbances, as well as the effects of unmeasurable disturbances on the outputs. The difficult aspect of such "mixed" concepts is generally the design of the feedback component. Therefore, the description in what follows will be restricted to this component. Detailed discussions of feedforward and feedback control concepts, along with analyses of their advantages and disadvantages, can be found in [4.1], [4.2].

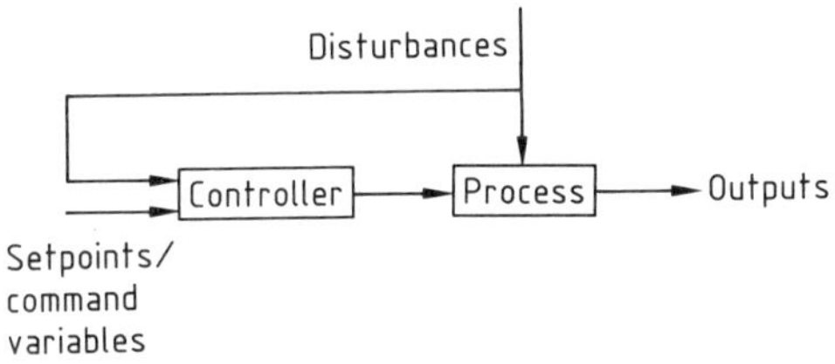

Figure 4.4. Feedforward control

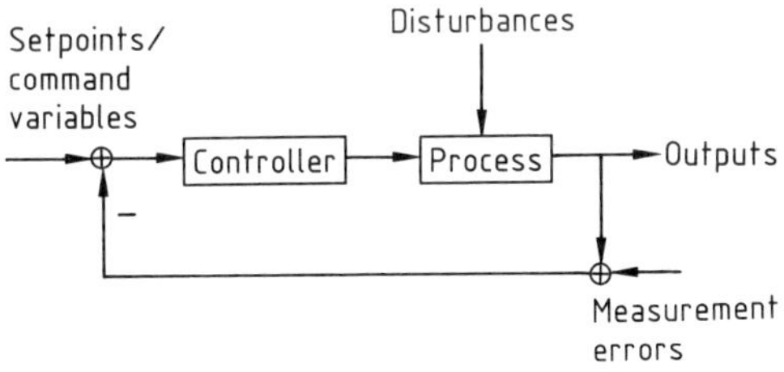

Figure 4.5. Feedback control

Characterization of Desired Control Loop Properties. Before discussing methods of control system design, it is necessary to explain some concepts required for characterizing properties of a control loop.

This section is limited to linear process models (see Section 3.3) and controllers. This restriction is justified in most cases because engineering processes often behave in an approximately lin-

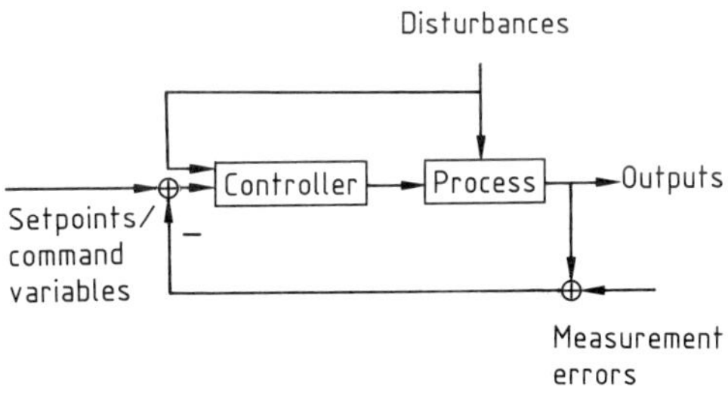

Figure 4.6. Feedback configuration with additional feedforward control of disturbances

ear fashion near the operating point and therefore can be handled by a linear controller. The deviation of the process behavior from the linearity of the model equation can then be treated as part of the model uncertainty, which must be taken into account in controller design anyway. It is also assumed that process model and controller are time invariant. The frequency response is then an ideal tool for investigating control loop properties.

Frequency Response. A linear, time-invariant system can be described by its frequency response $G(j\omega)$ (see Chap. 3). For a single-variable system (i.e., a system having one input and one output), $G(j\omega)$ is a complex-valued scalar, which can be graphically represented in either of two ways. First, it can be plotted in the complex plane for all (positive real) frequencies ω; this gives the Nyquist locus or polar plot of the frequency response. Second, the complex variable $G(j\omega)$ can be split into modulus (magnitude) $|G(j\omega)|$ and phase $\arg[G(j\omega)]$ by polar decomposition,

$$G(j\omega) = |G(j\omega)|\, e^{j\arg[G(j\omega)]} \qquad (4.2.1)$$

and each of these real quantities plotted separately as a function of ω. If logarithmic scales are used for frequency and modulus and a linear scale for phase, the result is the Bode plot of the frequency response. The modulus scale is usually multiplied by 20 and the unit of modulus is then the decibel: $|G(j\omega)|_{\mathrm{dB}} = 20 \log |G(j\omega)|$. Figures 4.7 and 4.8 show the polar plot and the Bode plot of the frequency response for a second-order lag element.

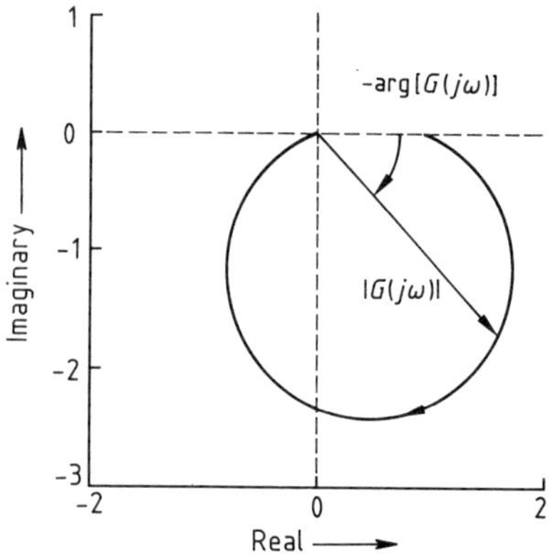

Figure 4.7. Nyquist locus or polar plot of frequency response

The frequency response has a straightforward physical explanation: If an asymptotically stable system characterized by $G(j\omega)$ is excited by a harmonic oscillation with frequency ω and amplitude $\hat{u}$ and the transients are allowed to decay, the output of the system oscillates harmonically at the same frequency, but with a phase shift φ and a different amplitude $\hat{y}$ (Fig. 4.9). The ratio of the output and input signal amplitudes (gain) and the phase shift between the two signals are described by the modulus and phase of $G(j\omega)$, respectively, and can thus be read directly from the Bode plot:

$$|G(j\omega)| = \hat{y}/\hat{u} \qquad (4.2.2)$$

$$\arg\,[G(j\omega)] = \varphi \qquad (4.2.3)$$

The frequency response is analyzed in detail in control engineering textbooks, for example, [4.1]–[4.3].

The requirements imposed on the closed control loop can be quantified with the aid of the frequency responses $G(j\omega)$ of the process and $K(j\omega)$ of the controller.

Stability. The Nyquist criterion represents a necessary and sufficient condition for the stability of a closed control loop [4.1], [4.4]. It requires that the Nyquist locus of the open-loop frequency response

$$Q(j\omega) := G(j\omega)\, K(j\omega) \qquad (4.2.4)$$

encircles the "critical point" $(-1,0)$ $m/2$ times counterclockwise as the frequency ω goes from 0 to ∞, where m is the number of unstable poles of the open loop. Because the Nyquist locus of the example in Figure 4.10 encircles the critical point 1/2 times, the associated closed loop is stable if and only if either $G(j\omega)$ or $K(j\omega)$ (but not both) has one unstable pole.

Response to Reference Inputs and Disturbances; Noise Suppression. To determine whether a control loop exhibits good response to reference inputs and disturbances and has the ability to suppress measurements noise, it is necessary to examine how all quantities acting on the control loop from the outside (Fig. 4.11) influence the control error.

The error $e(j\omega)$ is defined by

$$e(j\omega) := r(j\omega) - y(j\omega) \qquad (4.2.5)$$

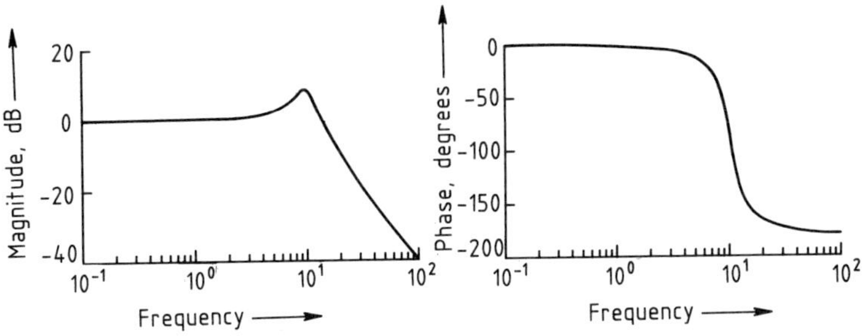

Figure 4.8. Bode plot

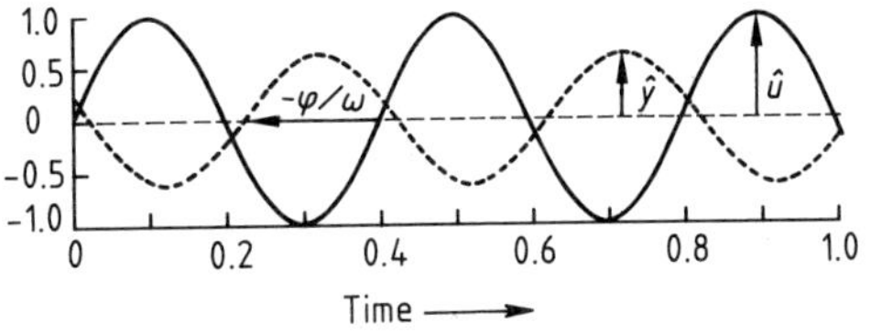

Figure 4.9. Harmonic oscillations at the input and output of a linear time-invariant system

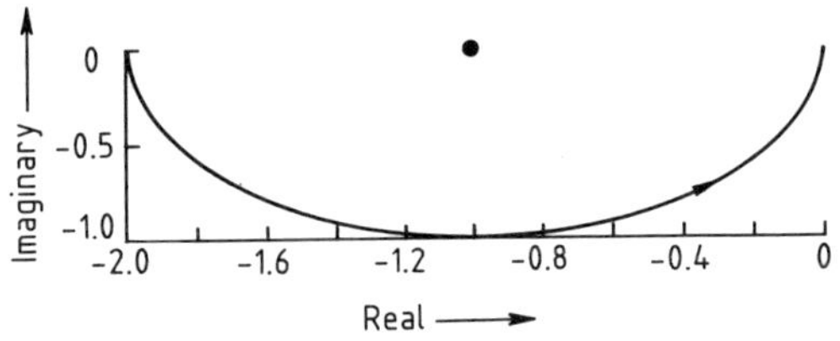

Figure 4.10. Example of Nyquist stability criterion

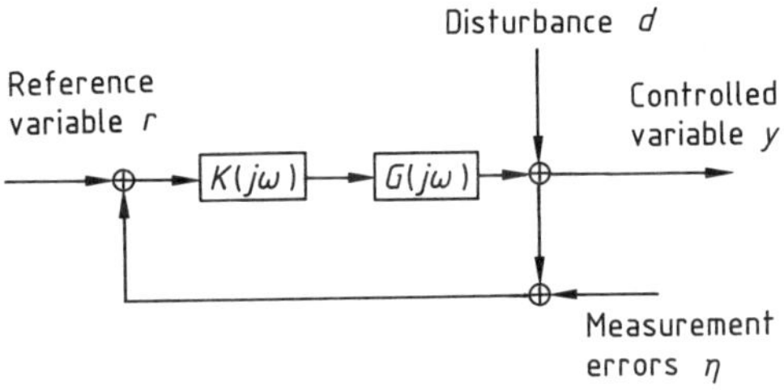

Figure 4.11. Linear feedback loop

Hence

$$e(j\omega) = \underbrace{\frac{1}{1 + G(j\omega)\,K(j\omega)}}_{S(j\omega)}[r(j\omega) - d(j\omega)] + \underbrace{\frac{G(j\omega)\,K(j\omega)}{1 + G(j\omega)\,K(j\omega)}}_{T(j\omega)}\eta(j\omega)$$

The quantity $S(j\omega)$ is called the sensitivity function, because it gives the sensitivity of the control error to a disturbance acting on the output and to the reference variable. $T(j\omega)$ is known as the complementary sensitivity function. The objective is to keep $|S(j\omega)|$ and $|T(j\omega)|$ small, and thus to achieve a small gain between quantities acting on the control loop from outside and the control error. Unfortunately, this is generally impossible, since for any ω

$$S(j\omega) + T(j\omega) = 1 \tag{4.2.7}$$

This "fundamental dilemma" of control engineering, embodied in Equation (4.2.7), is often alleviated because in most casses reference variables and disturbances contain mainly low-frequency components and hence vary comparatively slowly with time, while measurement noise commonly occurs at high frequencies. It is then sufficient to make $|S(j\omega)|$ small at low frequencies and $|T(j\omega)|$ small at high frequencies. Figure 4.12 shows typical plots of the sensitivity and

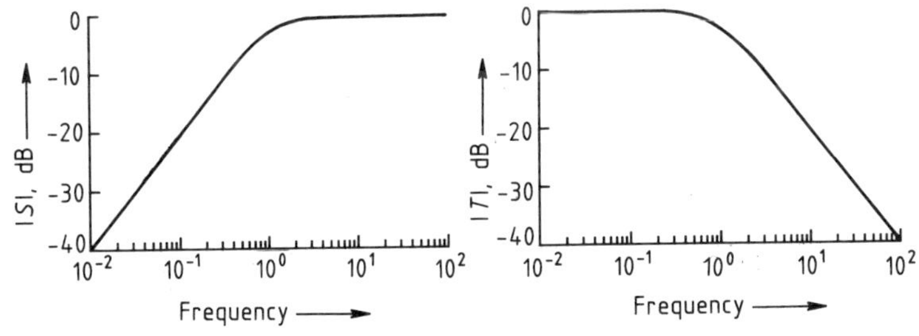

Figure 4.12. Typical shapes of $|S(j\omega)|$ and $|T(j\omega)|$

complementary sensitivity functions. The bandwidth ω_b of the closed loop is the frequency above which

$$|T(j\omega)| < 0.708 \text{ or } |T(j\omega)|_{dB} < -3 \qquad (4.2.8)$$

It "separates" the frequency range in which the closed loop has good response to reference inputs and disturbances from the frequency range in which measurement noise is suppressed. Establishing the desired bandwidth is thus one of the most important decisions in any controller design problem.

Robustness. The discussion up to this point has had to do with how to insure desired properties for the "nominal" control loop (i.e., the system consisting of process model and controller). The objective is to transfer these properties to the real control loop (the system consisting of the real process and the controller). In this context, robustness means that the properties of the nominal control loop should be robust with respect to errors in the model.

Robustness can be investigated only if an upper boundary can be given for the size of the model error. Often an upper bond for the relative model error is estimated by assuming that the real process behaves in a linear and time-invariant fashion near the operating point under consideration. The (unknown) frequency response of the process is represented by $G_r(j\omega)$. If the known frequency response of the model is $G(j\omega)$, the relative model error is then defined by

$$\Delta(j\omega) = \frac{G_r(j\omega) - G(j\omega)}{G(j\omega)} \qquad (4.2.9)$$

Suppose the modulus of the relative model error is bounded by a known real-valued and positive function $p(j\omega)$:

$$|\Delta(j\omega)| < p(j\omega) \qquad (4.2.10)$$

Figure 4.13 shows a typical curve of $p(j\omega)$, which grows rapidly with increasing frequency.

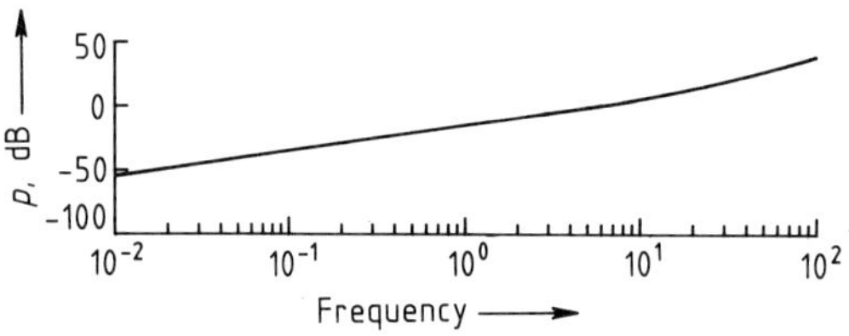

Figure 4.13. Typical error bound

If G_r and G have the same number of unstable poles, it is easy to show [4.5] that the stability of the control loop is robust to all model errors permitted by Equations (4.2.9) and (4.2.10) if and only if the following holds for the nominal complementary sensitivity function

$$|T(j\omega)| < \frac{1}{p(j\omega)} \qquad (4.2.11)$$

To achieve good performance of the real loop the nominal sensitivity function must usually be made sufficiently small at low frequencies while the nominal complementary sensitivity function must be sufficiently small in the frequency range where the model uncertainty is large (see Fig. 4.13) [4.6].

Open-Loop Shaping. The Nyquist criterion makes it possible to guarantee closed-loop stability by appropriate manipulation of the open-loop frequency response $Q(j\omega)$. The sensitivity function and the complementary sensitivity function are used to guarantee the other control-loop properties stated. Since both quantities depend on Q in a simple way (see Eq. 4.1.6), all the desired properties of the control loop can be achieved by suitable "shaping" of the open-loop frequency response $Q(j\omega)$. The term "open-loop shaping" is often used in this context. Good response to reference inputs and disturbances thus requires that

$$|Q(j\omega)| \gg 1 \qquad (4.2.12)$$

at low frequencies, whereas suppression of measurement noise requires that

$$|Q(j\omega)| \ll 1 \qquad (4.2.13)$$

at high frequencies. In the region near the gain crossover frequency ω_d, however, $|Q(j\omega)|$ must not roll off too rapidly. (The gain crossover frequency is the frequency at which $|Q(j\omega)|$ intersects the 0 dB line.) Otherwise, disturbances, reference variables, and measurement noise can produce large control errors in this frequency range. Moreover, rapid gain roll-off near the crossover frequency threatens closed-loop stability. This is a direct consequence of the Bode gain–phase relation [4.1], [4.7]. The ideal would be a 20 dB per decade roll-off for $|Q(j\omega)|$ in this frequency range. If this is the case, the gain crossover frequency ω_d is a good approximation to the bandwidth ω_b of the closed loop.

Multivariable Systems. The discussion to this point has related only to systems with one input and one output. Most industrial processes, however, have multiple manipulated variables and multiple outputs to be suitably influenced. Each of the manipulated variables will, in general, affect several or even all of the controlled variables. The process is then said to be a multivariable (multi-input, multi-output) system, and its dynamics (assuming linearity and time-invariance) are described by the frequency response matrix

$$G(j\omega) = \begin{bmatrix} G_{11}(j\omega) \cdots G_{1q}(j\omega) \\ \vdots \qquad \vdots \\ G_{p1}(j\omega) \cdots G_{pq}(j\omega) \end{bmatrix} \qquad (4.2.14)$$

Most of what has been discussed up to now can be extended to multivariable systems with no difficulty.

The multivariable Nyquist criterion is a necessary and sufficient condition for the stability of the closed loop. It requires that the frequency-dependent eigenvalues of the matrix $Q(j\omega) = G(j\omega)K(j\omega)$ (the "characteristic loci"), taken together, encircle the "critical point" $m/2$ times [4.8], [4.9].

To assess the other control-loop properties, the concept of gain must be generalized [4.10]. In the single-variable case, the gain of an asymptotically stable system corresponds to the modulus of the frequency response. The situation is somewhat more complicated in multivariable systems. Such a system has a gain that depends on the direction of the input vector. The gains of most interest are the minimum and maximum (frequency-dependent) gains. These are characterized by the minimal and maximal singular values of the frequency response matrix:

$$\min_{u \neq 0} \frac{\|y(j\omega)\|}{\|u(j\omega)\|} = \underline{\sigma}[G(j\omega)] = \sqrt{\lambda_{\min}[G^T(-j\omega)\,G(j\omega)]} \qquad (4.2.15)$$

$$\max_{u \neq 0} \frac{\|y(j\omega)\|}{\|u(j\omega)\|} = \bar{\sigma}[G(j\omega)] = \sqrt{\lambda_{\max}[G^T(-j\omega)\,G(j\omega)]} \qquad (4.2.16)$$

Here $\|\ldots\|$ denotes the Euclidean vector norm (see Chap. 3), while $\lambda_{\min}$ and $\lambda_{\max}$ are the smallest and largest eigenvalues.

With these gains, the requirements derived for single-variable systems can readily be extended to the multivariable case [4.5]:

$$\bar{\sigma}[S(j\omega)] \ll 1 \text{ or } \underline{\sigma}[Q(j\omega)] \gg 1 \text{ for } \omega \ll \omega_b \qquad (4.2.17)$$

$$\bar{\sigma}[T(j\omega)] \ll 1 \text{ or } \bar{\sigma}[Q(j\omega)] \ll 1 \text{ for } \omega \gg \omega_b \qquad (4.2.18)$$

The sensitivity matrix S and the complementary sensitivity matrix T are defined similarly to the scalar variables in the single-variable case: $S := (I + GK)^{-1}$ and $T := (I + GK)^{-1}GK$.

If the requirements on the various controlled variables differ sharply, frequency-dependent weighting matrices for S, T, and Q can be used to handle the situation.

Controller Design Methods. This section surveys the most important methods of controller design.

Classical Control Engineering. The methods of classical control engineering continue to be dominant in many areas. In general, these methods assume a linear, time-invariant, single-variable process model. If the requirements on the controlled system are not too stringent, however, these methods can still be used in many cases even where such models give a highly erroneous description of the real process.

The most important design method in classical control engineering is the frequency response method [4.1], [4.3], [4.11]. It has the objective of shaping the open-loop frequency response $Q(j\omega)$ in accordance with what was said in the preceding section. For this purpose compensators

$$K(j\omega) = k\,\frac{1 + j\omega T_1}{1 + j\omega T_2} \qquad (4.2.19)$$

are connected in series with the process. The compensator increases the phase if $T_1 > T_2$ (lead compensator) and decreases it if $T_2 > T_1$ (lag compensator); see Figures 4.14 and 4.15. A special case of the lag compensator for $k = \tilde{k}T_2$ and $T_2 \to \infty$ is the proportional-plus-integral (PI) element. Because of the particular axis scaling, the Bode diagram of a series system can be obtained by simply adding the amplitude and phase curves of the individual elements.

It is thus easy to see what kind of compensating network is required to modify the curves in the desired way. For example, Figure 4.8 shows the frequency response of a process model; the bandwidth of the closed loop is required to be $\omega_b = 1$. In this case a PI element with frequency response $K(j\omega) = (1 + 0.1j\omega)/j\omega$ can be used as

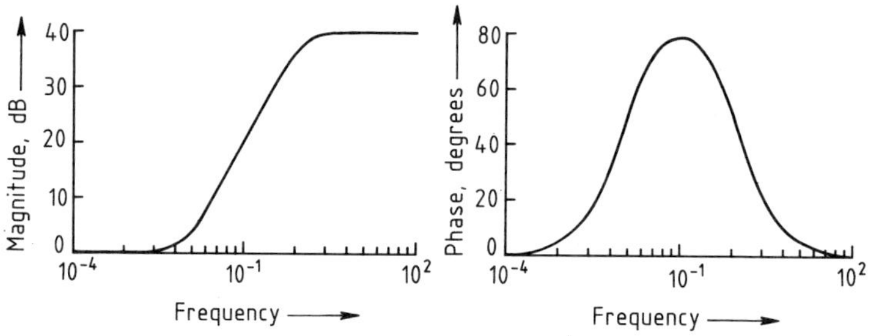

Figure 4.14. Bode plot of a lead compensator

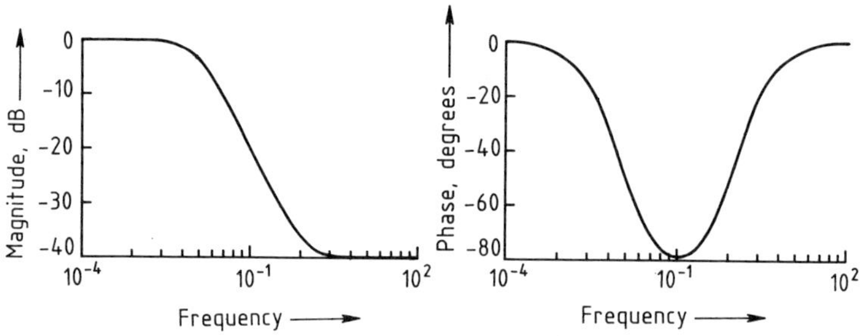

Figure 4.15. Bode plot of a lag compensator

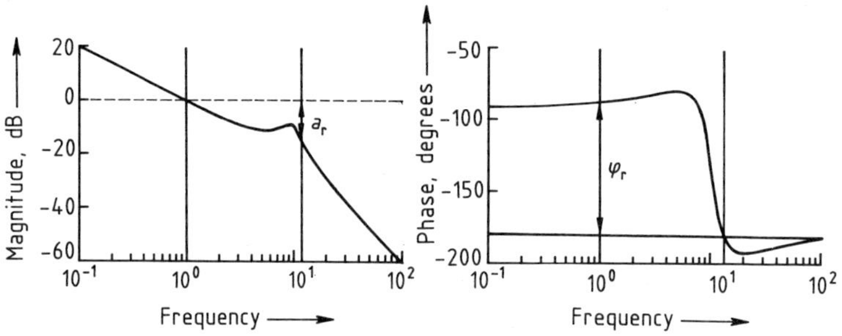

Figure 4.16. Bode plot for $Q(j\omega)$ in the example

the compensator. The resulting amplitude and phase plots for the open loop are shown in Figure 4.16. Clearly, $Q(j\omega)$ satisfies requirements of Equation (4.2.12) at low frequencies and of Equation (4.2.13) at high frequencies. The polar plot of $Q(j\omega)$ can be used to check whether the Nyquist criterion is also satisfied. Figure 4.17 shows the curve near the origin. The quantities ϕ_r ("phase margin") and a_r ("gain margin") give the "distance" at which the polar plot of the frequency response passes the critical point (i.e., how "close" the loop is to instability). The gain and phase margins can also be read directly from the Bode plot in Figure 4.16. The following can serve as a rule of thumb for the use of the frequency-response method: $a_r > 2.5$ or $|a_r|_{dB} > 8$ and $\phi_r > 30°$. Because the gain and phase margins in this example are sufficiently large, a further compensation of $Q(j\omega)$ can be dispensed with.

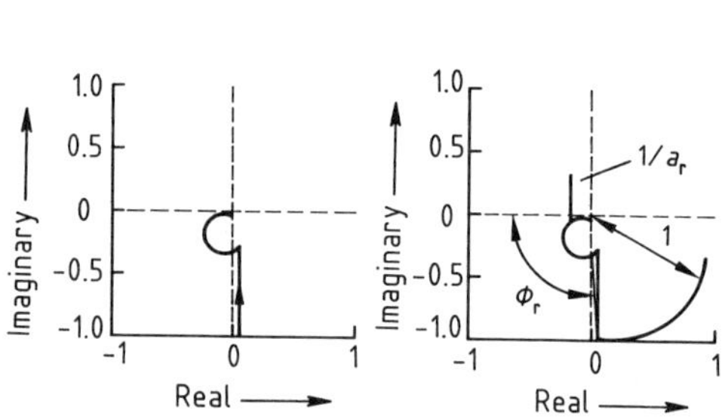

Figure 4.17. Polar plot for $Q(j\omega)$ in the example

Another widely used technique is the root-locus method [4.1], [4.4], [4.11]. This design procedure starts not with the frequency response, but with the transfer function of the plant (process) model. The location of poles and zeroes of this transfer function indicates how the poles, and thus the dynamics of the closed loop, vary as a function of the controller parameters and structure. This method is accordingly suitable for insuring the stability of the control loop. On the other hand, it generally proves quite difficult to insure other desired properties of the control loop. Another disadvantage of this design technique it that it cannot be applied to process modes with time delays.

In process engineering, a controller structure (PID) is often specified and the parameters are obtained by experimenting on the process and evaluating the Ziegler–Nichols rules [4.4], [4.11]. When such a procedure is used in practice, there is no need for an explicit process model; the theoretical justification, however, presumes the existence of linear, time-invariant single-variable models.

Discarding one or more of the three model assumptions in classical control engineering leads to the other three (overlapping) "classes" of control engineering (Fig. 4.18).

Multivariable Control. Of the three model assumptions in classical control theory, the one furthest from reality is usually that the process can be described in terms of several independent single-variable models. Ignoring the links between various manipulated and controlled variables can lead to major problems: input controllers that work against one another (wasting energy) and instability of the controlled process may result. Design methods that take account of the links and thus the multivariable structure of the control loop exist in both the frequency and the time domains. Frequency domain methods begin with a transfer or frequency response matrix as the plant model; time-domain methods are based on state-space models (see Chap. 3). The frequency-domain methods can be classified in two groups: generalizations of the classical frequency-response method, in which an attempt is made to manipulate the open-loop frequency response matrix $Q(j\omega)$ in an appropriate way (open-loop shaping), and approaches that seek to influence directly, or optimize, the closed-loop frequency response matrices (closed-loop shaping). Table 4.1 lists the most commonly used linear, time-invariant multivariable design met ods. Some of these techniques are fully described in textbooks of multivariable control engineering [4.9], [4.12]–[4.15].

The LQG (linear quadratic Gaussian) method is based on a linear, time-invariant state-space model (see Chap. 3) of the control loop:

$$\dot{x}(t) = Ax(t) + Bu(t) + w(t) \qquad (4.2.20)$$

$$y(t) = Cx(t) + v(t) \qquad (4.2.21)$$

where $w(t)$ and $v(t)$ are error terms representing, respectively, disturbances and measurement noise acting on the plant. They are modeled as white (Gaussian) random processes with a mean of zero [4.16], [4.17]. Their spectral densities P and R provide a measure of the "strength" of these random processes. They tell how dependable the deterministic components of the models are. If the goal is to keep the plant as close as possible to the operating point $x = 0$, subject to a restriction on the use of control energy, then it

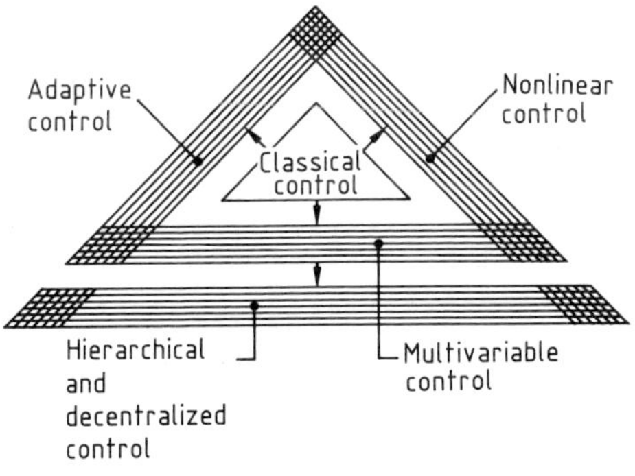

Figure 4.18. Classification of methods in control theory

Table 4.1. Multivariable design method

Time-domain	Frequency domain	
	Direct Nyquist array method Inverse Nyquist array method Characteristic locus method RFN method LQG/LTR method	Open-loop shaping
LQG method Pole assignment	H_2 Minimization H_∞ Minimization IMC method	Closed-loop shaping

is desirable to seek a control strategy that will minimize the cost function

$$J = E\left[\int_0^\infty (x^T L x + u^T M u)\,dt\right] \qquad (4.2.22)$$

The (positive semidefinite) matrix L penalizes deviations of the state variables from zero, and the (positive definite) matrix M penalizes the control energy. The symbol E represents the expected value. This problem can be solved straightforwardly. The separation principle states that the problem can be approached in two independent steps:

- First, ignore the fact that the state is not measurable and suppose it is exactly known. The optimal control strategy is then a constant state feedback:

$$u(t) = -K_c x(t) \qquad (4.2.23)$$

Here K_c can be uniquely calculated from the model matrices A and B and the weight matrices L and M.

- In the second step, determine an optimal (in the sense of minimum error variance) estimate $\hat{x}$ for the unmeasurable state vector x. This is done by the use of a Kalman–Bucy filter [4.16], [4.18] (Fig. 4.19). The Kalman–Bucy filter includes the deterministic part of the plant model and thus attempts to simulate the behavior of the state variables. What is more, however, the model also receives measurement information y, which is used for correcting the estimate (unless it agrees with the measurement $\hat{y}$ predicted by the model internal to the filter). The filter matrix K_f that specifies this correction is uniquely determined by the matrices A and C of the deterministic model component and the spectral densities P and R of the stochastic component. The optimal control strategy is now to replace the true state x in Equation (4.2.23) by the estimate $\hat{x}$.

The derivation and details can be found in, for example, [4.19], [4.20]. Despite its theoretical elegance, the use of the LQG method in practice commonly presents problems. In chemical engineering, state–space models rarely have a simple physical interpretation. Generally, one has to be content with simple frequency-domain models (transfer function matrix, frequency response matrix) derived from the input/output behavior of the real process. While these state–space models can be determined from such input–output models their state variables then often fail to correspond to physically relevant quantities. This fact makes it difficult to determine a suitable weighting matrix L. Finding the spectral densities (especially for the "system noise" w) may well prove to be a matter of luck as well. In such cases, the elements of P and L offer a large number of design parameters whose effects on the desired control-loop behavior are difficult to assess.

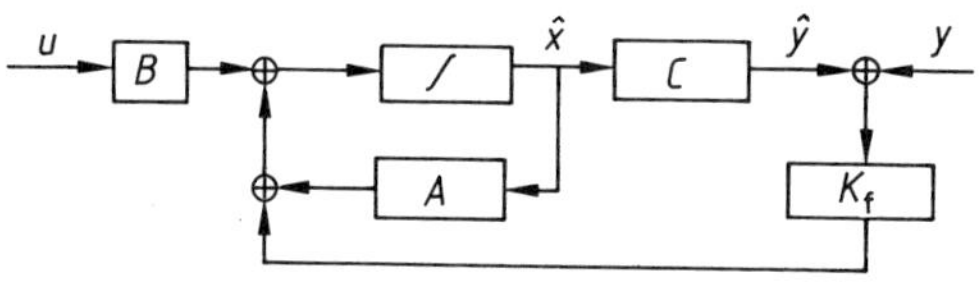

Figure 4.19. Kalman–Bucy filter

One way of avoiding this problem is the LQG/LTR (loop transfer recovery) design method [4.15], [4.21], in which the design parameters P, R, L and M of the LQG design are divorced from any physical meaning and selected such that the open-loop frequency response matrix $Q(j\omega)$ satisfies the requirements stated in the preceding section. For this reason, the LQG/LTR design technique is often classified as a frequency-domain method.

A further time-domain method is pole assignment [4.4], [4.12]. It has the objective of specifying the poles and thus the dynamics of the closed loop by means of appropriate state feedback. If the state is not measurable, it is estimated by a Luenberger observer. The Luenberger observer resembles the Kalman–Bucy filter in that it is an algorithm for reconstructing the system state and in that it contains a model of the plant. It differs from the filter only in the way the correction matrix K_f is computed. In the case of the observer, this is done by specifying the dynamics of the equation for the estimation error. There is again a kind of separation principle in pole assignment: the observer matrix K_f and the state-estimate matrix K_c can be designed independently. The use of this technique may also create problems if the state variables have no direct physical meaning. Furthermore, establishment of the control-loop dynamics does not guarantee satisfactory loop behavior.

The characteristic locus curve method [4.22] seeks to meet the requirements stated in the preceding section by manipulating the eigenvalues of the open-loop frequency response matrix $Q(j\omega)$. These eigenvalues $\lambda_{Qi}(j\omega)$ are called "characteristic frequency responses." If the corresponding eigenvectors are collected in the matrix $V(j\omega)$, the following holds

$$Q(j\omega) = V(j\omega)\ \mathrm{diag}\{\lambda_{Qi}(j\omega)\}\ V^{-1}(j\omega) \qquad (4.2.24)$$

Spectral decompositions for the sensitivity matrix and the complementary sensitivity matrix are then given by

$$S(j\omega) = V(j\omega)\ \mathrm{diag}\left\{\frac{1}{1+\lambda_{Qi}(j\omega)}\right\} V^{-1}(j\omega) \qquad (4.2.25)$$

$$T(j\omega) = V(j\omega)\ \mathrm{diag}\left\{\frac{\lambda_{Qi}(j\omega)}{1+\lambda_{Qi}(j\omega)}\right\} V^{-1}(j\omega) \qquad (4.2.26)$$

The eigenvalues of S and T thus depend in a very simple way on the $\lambda_{Qi}(j\omega)$. Although the gains of a matrix cannot be deduced from the absolute values of its eigenvalues, these do serve as a good indication of the gains in many cases. Therefore, small $|\lambda_{Qi}(j\omega)|$ at high frequencies and large $|\lambda_{Qi}(j\omega)|$ at low frequencies are required. The crossover frequencies for the characteristic frequency responses are selected to correspond to the desired closed-loop bandwidth. What is more, the multivariable Nyquist criterion makes it possible to say how often the locus curves of $\lambda_{Qi}(j\omega)$ (the "characteristic locus curves") must enclose the critical point $(-1,0)$ for the (nominal) control loop to be stable. Once the characteristic frequency responses have been chosen, the behavior of $\bar{\sigma}[S(j\omega)]$ and $\bar{\sigma}[T(j\omega)]$ must be examined to ensure that Equations (4.2.17) and (4.2.18) are satisfied. Now a way must be found to modify the characteristic loci in the desired way. The ideal method is to use a controller frequency response matrix having the same eigenvectors as the plant frequency response matrix. The following would then hold:

$$Q(j\omega) = \underbrace{V\ \mathrm{diag}\{\lambda_{Gi}(j\omega)\}\ V^{-1}}_{G(j\omega)}\ \underbrace{V\ \mathrm{diag}\{\lambda_{Ki}(j\omega)\}\ V^{-1}}_{K(j\omega)} \qquad (4.2.27)$$

$$= V(j\omega)\ \mathrm{diag}\ \underbrace{\{\lambda_{Gi}(j\omega)\,\lambda_{Ki}(j\omega)\}}_{\approx\ \lambda_{Qi}(j\omega)}\ V^{-1}(j\omega)$$

In this case, the $\lambda_{Gi}(j\omega)$ could be treated as single-variable systems, and suitable single-variable controllers $\lambda_{Ki}(j\omega)$ could be designed by the classical frequency-response method. Unfortunately, this is rarely possible, because the elements of V are generally nonrational in $j\omega$, and the resulting controller matrix thus cannot be realized. One therefore has to be content with approximating the ideal controller by approximating the complex matrices $V(j\omega_i)$ with real matrices for selected frequencies ω_i.

A shortcoming of the characteristic locus method is that the $|\lambda_{Qi}(j\omega)|$ provide only an indication of the actual gains of $Q(j\omega)$. In the RFN (reversed-frame normalizing) method [4.23], attempts are made to bring the absolute values of the eigenvalues and the gains (singular values) of $Q(j\omega)$ into coincidence. If this is successful (approximately), all the conditions stated in the preceding section can be met through manipulation of these quantities. A drawback of the characteristic locus and RFN methods is that the characteristic frequency responses cannot be associated with individual plant input and output variables. There is accordingly no systematic way of making different controlled variables display different transient responses or different steady state accuracy. Such difficulties are less pronounced (or nonexistent) in the methods discussed below.

The direct Nyquist array method [4.8] breaks the design process into two steps:

1) First, an attempt is made to approximately diagonalize the plant model by means of a simple—and, if possible, constant—compensator K_c. This means that off-diagonal elements of the frequency-response matrix should be small in magnitude compared to the diagonal elements. In particular, if the elements of the compensated $p \times p$ matrix $Q_c = GK_c$ satisfy

$$\sum_{k=1,k\neq i}^{p} |Q_{C_{ki}}(j\omega)| < |Q_{C_{ii}}(j\omega)| \qquad (4.2.28)$$

for $i = 1, \ldots, p$ and $0 \leq \omega < \infty$, then $Q_c(j\omega)$ is said to be (column-)diagonal-dominant. The diagonal-dominance of a frequency-response matrix can easily be checked graphically. The polar plots of the diagonal elements $Q_{c_{ii}}(j\omega)$ are overlaid with circles whose radii are given by the left-hand side of Equation (4.2.28). The bands formed by these circles are called Gershgorin bands. If Equation (4.2.28) is satisfied, these bands do not cover the origin of the complex plane. A number of methods have been advanced for obtaining narrow Gershgorin bands [4.8], [4.12], [4.15].

2) In the second step, the small off-diagonal elements of Q_c are ignored. Single-variable controllers $K_i(j\omega)$ are designed for the principal diagonal elements of the compensated plant model Q_c; thus the open-loop frequency-response matrix is obtained:

$$Q(j\omega) = \underbrace{G(j\omega)\,K_c(j\omega)}_{Q_c(j\omega)}\ \mathrm{diag}\,\{K_i(j\omega)\} \qquad (4.2.29)$$

This operation alters the form of the Gershgorin bands but leaves their width unchanged, since both the center and the radius of each Gershgorin circle are multiplied by the same factor. The Gershgorin bands of $Q(j\omega)$ make it possible to assess the stability of the closed loop, since they contain the characteristic loci curves. If the bands do not cover the critical point $(-1,0)$, they determine how many times the characteristic locus curves encircle this point. Accordingly, the $K_i(j\omega)$ must be chosen such that the Gershgorin bands of Q do not cover the point $(-1,0)$ and, collectively, they encircle this point counterclockwise $m/2$ times. Figure 4.20 shows an example for $m = 0$. To perform a classical single-variable design for the i-th controller component, the corresponding frequency-response element of the compensated plant model also must be known for the case where the remaining control loops are already closed. Because small off-diagonal elements are still present, this frequency response will not coincide with $Q_{c_{ii}}$. Its polar plot, however, lies in the interior of the Gershgorin band associated with $Q_{c_{ii}}$. This is a further reason for the importance of narrow Gershgorin bands.

Experienced users generally prefer the inverse Nyquist array method [4.8], which has the advantage that the relationship between $y_i(j\omega)$ and $u_i(j\omega)$ becomes sharper when the other control loops are closed. This relationship is localized by Ostrowski bands, which lie in the interior of the Gershgorin bands. The method has the disadvantage that it entails the use of inverse frequency responses. Furthermore, the technique can be applied only to systems in which the number of manipulated variables equals the number of controlled variables.

A design method that has been gaining popularity in the process engineering field is the IMC (internal model control) technique [4.24]. In its simplest form, this method assumes a stable plant. The controller then contains a plant model in parallel with the actual plant (Fig. 4.21). The requirements on the closed control loop are easily fulfilled with this structure, because

1) The closed loop is stable if and only if the controller component $\Gamma(j\omega)$ is stable
2) The sensitivity matrix and the complementary sensitivity matrix depend on Γ in a very simple way:

$$S = I - G\Gamma$$
$$T = G\Gamma \qquad (4.2.30)$$

A promising approach to the design of multivariable controllers is H_∞ minimization [4.25], [4.26]. The discussion in the preceding section showed what kind of behavior should be sought for the maximum singular values of the sensitivity matrix and the complementary sensitivity matrix. In addition, the matrix $K(j\omega)\ S(j\omega)$ can be

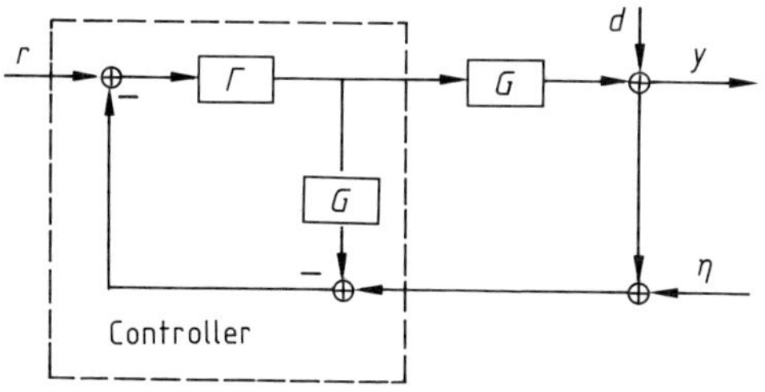

Figure 4.21. IMC structure

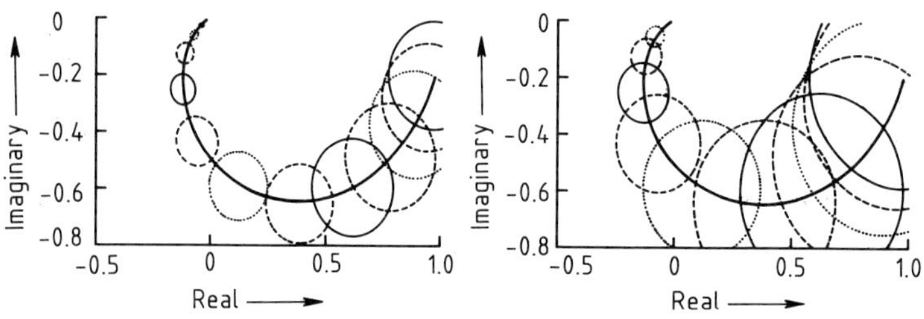

Figure 4.20. Gershgorin bands for a 2 × 2 system

included in the analysis. This matrix indicates how disturbances in the control loop affect the manipulated variables (see Fig. 4.11). Typically, $\bar{\sigma}[K(j\omega)\ S(j\omega)]$ is required to be bounded for all frequencies. This is intended to prevent the manipulated variables being saturated when disturbances occur. To avoid high-frequency oscillation of actuators, a further requirement is that $\bar{\sigma}[K(j\omega)\ S(j\omega)]$ roll off at high ω. For S, T, and KS to have small gains, each in its own frequency range, distinct frequency-dependent weighting matrices W_S, W_T, and W_{KS} are assigned to them. A controller is then sought that stabilizes the control loop and minimizes

$$\sup_{\omega} \bar{\sigma} \begin{bmatrix} W_S(j\omega)\,S(j\omega) \\ W_T(j\omega)\,T(j\omega) \\ W_{KS}(j\omega)\,KS(j\omega) \end{bmatrix} := \left\| \begin{matrix} W_S S \\ W_T T \\ W_{KS} KS \end{matrix} \right\|_{\infty} \quad (4.2.31)$$

$\|\ldots\|_{\infty}$ is known as the H_{∞} norm. Algorithms for solving this problem are complex and computationally intensive. For the user, however, this point is usually of as little importance as the nonuniqueness of the solution. The design engineer need only select the weighting matrices. As a rule, they are chosen to be diagonal matrices with stable, minimum-phase elements. Low-pass elements are commonly used for weighting S, so that small $\bar{\sigma}[S]$ is obtained at low frequencies. Similarly, high-pass elements are used for weighting T and KS. This method has the considerable advantage that the elements of the weighting matrices can be directly assigned to distinct manipulated and controlled variables. Their effect on closed-loop behavior is thus easy to assess. The selection of weights is illustrated on p. 402.

Decentralized and Hierarchical Control. When the known multivariable methods are used, difficulties arise if the process to be controlled has many coupled inputs and outputs. Such large-scale systems should be broken down into subproblems. This can be done heuristically (and the subproblems defined, for example, as physical parts of the plant) or by means of formal techniques (e.g., the block relative-gain array [4.27]). A (multivariable) controller is assigned to each subsystem; in the design of this controller, however, links between the subsystems cannot be completely neglected. If the controllers operate independently of one another, this is known as decentralized control; if they are coordinated at a higher level, the control scheme is said to be hierarchical (Fig. 4.22) [4.28]–[4.32]. Most known hierarchical and decentralized methods belong to the time domain. With regard to their use in practice, essentially the same problems arise as in the LQG or pole assignment methods. The development of frequency-domain decentralized and hierarchical methods promises to be helpful.

Adaptive and Nonlinear Control. Linear control methods reach their limits in processes having no operating point or a continuously variable operating point (batch processes).

Conceptually, the simplest way of dealing with problems involving an operating point that varies slowly over time is to use adaptive control [4.33], [4.34]. Adaptive controllers continuously modify themselves to match the process being controlled. In general, only the simplest class of adaptive methods are employed in practice: self-tuning controllers. An indirect variant of such a controller is made up of a three-level algorithm: on-line estimation of parameters of a process model with specified (usually linear) structure; calculation of parameters for a controller structure, also specified; and action of the resulting controller on the process (Fig. 4.23). In this way,

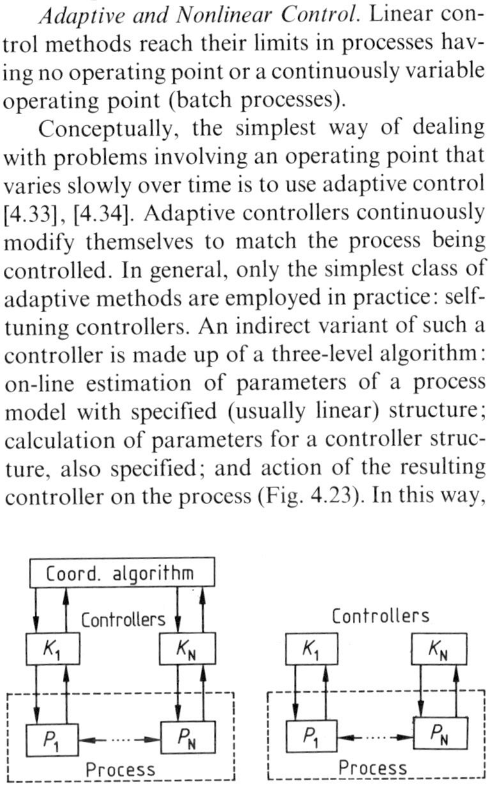

Figure 4.22. Hierarchical and decentralized control structures

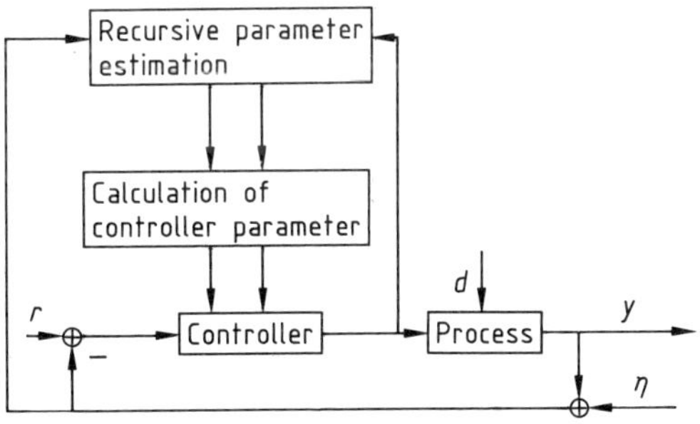

Figure 4.23. Self-tuning controller (indirect)

the entire (nonlinear) control problem is broken down into an estimation problem and a straightforward, generally linear, control problem. Adaptive control structures can therefore be treated as special nonlinear control structures.

If the operating point of the process changes very rapidly or a linear description is altogether impossible, explicitly nonlinear methods must be considered. However, these methods, such as exact linearizaton [4.35], are restricted to a very small class of nonlinear systems. Exceptions are semi-heuristic methods such as nonlinear versions of MPC (model predictive control) [4.36].

Design Aids. Most modern design techniques are so complex that they could not be employed without the help of suitable CACSD (computer-aided control systems design) packages. A continuously updated survey of available software is the ELCS (Extended List of Control Software) [4.37]. A further summary can be found in [4.38].

Design Example. This section shows how the classical frequency-response method and H_∞ minimization can be used in the design of controllers for a pilot-plant distillation column with 40 trays (Fig. 4.24). The device separates a mixture of methanol and propanol. The control objective is to keep the concentrations constant in the bottoms and distillate as the feed rate or feed concentrations change. As is common in binary separations, the concentrations and temperature on an individual tray are related in a unique way. Instead of the concentrations, which are difficult to measure, the temperatures can therefore be controlled. Simulation studies based on a detailed nonlinear model (using the DIVA simulation system [4.39]) have shown that the temperatures on the 14th and 28th trays are most sensitive to disturbances in the feed. These two temperatures are therefore chosen as controlled variables. The manipulated variables are the heat input Q_E and the reflux ratio $\varepsilon = L/V_K$. The controller design is based on a simple linear model obtained by evaluating step responses:

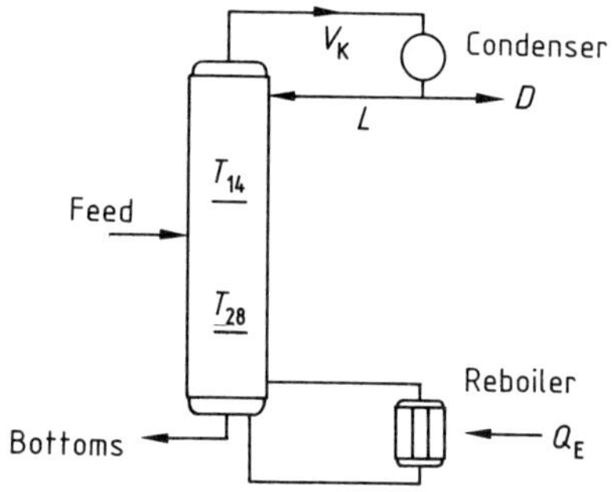

Figure 4.24. Distillation column
V_K = vapor stream; L = reflux stream; D = distillate; Q_E = heat input rate

$$\begin{bmatrix}\Delta T_{14}\\ \Delta T_{28}\end{bmatrix} = \underbrace{\begin{bmatrix} -\dfrac{1220}{6.1j\omega+1} & \dfrac{100}{(1.2j\omega+1)(1.5j\omega+1)} \\ -\dfrac{1500}{2.7j\omega+1} & \dfrac{324}{1.6j\omega+1}\end{bmatrix}}_{G(j\omega)} \begin{bmatrix}\Delta\varepsilon\\ \Delta Q_E\end{bmatrix} \quad (4.2.32)$$

Here Δ denotes deviations of the indicated quantities from the desired stationary operating condition. The time constants of the model are given in hours. The most important specification for linear controller design is the closed-loop bandwidth, which should be close to 40 h^{-1}.

In the classical frequency-response method, the off-diagonal elements are neglected and only the diagonal elements of G are considered. Figure 4.25 shows the magnitudes for these elements. To suppress steady state errors PI controllers are used. Their time constants are selected such that the magnitude of both elements rolls off at a rate of ca. 20 dB per decade over the whole frequency range. The controller gains are set so that the two magnitude curves coincide over a broad frequency range and the bandwidth of the closed multivariable control loop takes on the desired value. The last point is verified by computing the singular values of the complementary sensitivity matrix (Fig. 4.26).

In the H_∞ minimization method, weight matrices are selected for S, T, and KS. Diagonal matrices with minimum-phase elements are employed. Figure 4.27 shows the amplitude characteristics of the elements of W_S (low-pass) and W_T (high-pass). All elements are chosen such that their gains intersect the 0 dB line at 40 h^{-1}, because closed-loop bandwidth should lie close to this frequency. Small constants (0.01) are chosen as diagonal elements of W_{KS}, since in this example no problems with actuator saturation or high-frequency actuator oscillation are expected. The H_∞ cost function is completely defined in this way. It is minimized by a controller of order eight, but this can be reduced to order 5 by model reduction techniques. Figure 4.28 gives plots of the singular values of S and T.

Both controllers were tested in a pilot plant. It was found that the controller obtained by minimization of H_∞ met all practical requirements, while the classical PI controller required on-line tuning of the gains.

Details and comparisons with other multivariable design methods can be found in [4.40].

4.3. Optimal Control

Introduction. Before something can be optimized, the term "optimal" must be defined. In quantitative form, this is done by establishing a performance index (performance criterion, optimization criterion).

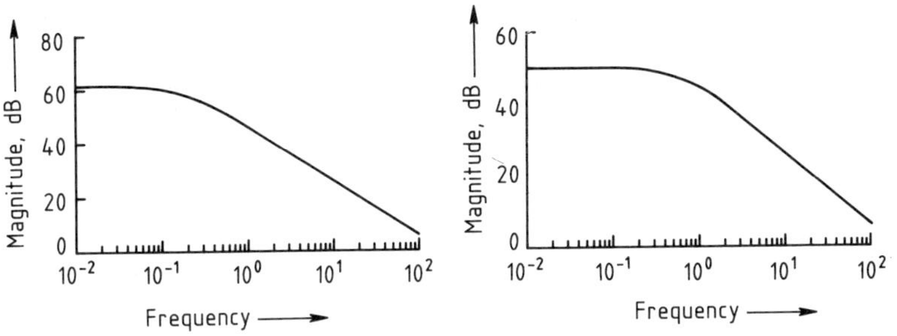

Figure 4.25. Magnitudes of the diagonal elements

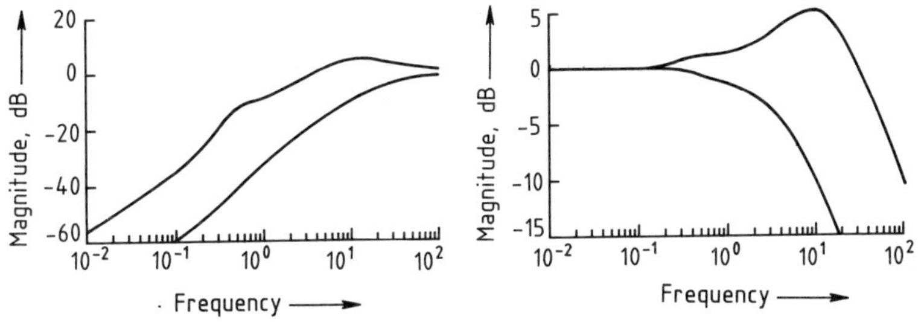

Figure 4.26. Singular values of the sensitivity matrix and complementary sensitivity matrix

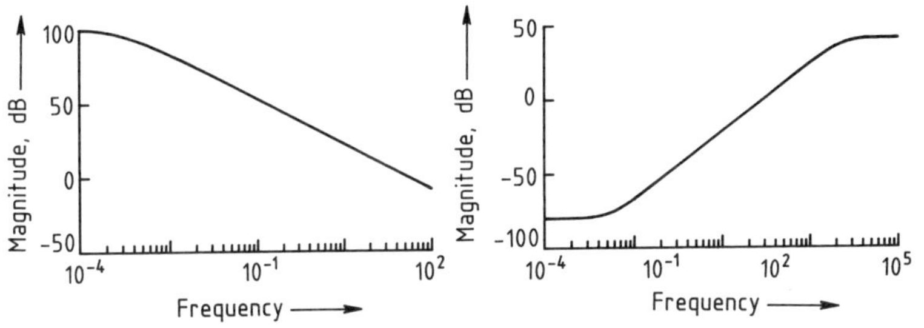

Figure 4.27. Magnitudes of the weight elements

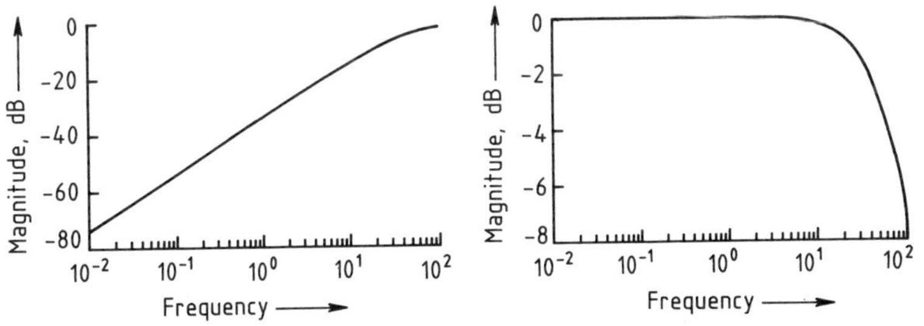

Figure 4.28. Singular values of the sensitivity matrix and complementary sensitivity matrix

The optimization of a dynamic system means formulating the dynamic system behavior in an optimal way under the selected criterion. The system is characterized by the time-dependent processes by which it goes from a given initial state $x(t_0)$ to a final state $x(t_e)$.

The optimization of a controlled dynamic system generally begins with the system already specified; only the control scheme is varied to achieve the optimal system behavior. If the control structure is fixed, so that only its parameters can be varied, we speak of parametric optimization. For example, the Ziegler–Nichols rules (see Section 4.1) yield optimal controller parameters for PI or PID controllers and a class of systems. If, on the other hand, the control scheme—including its structure—can be freely chosen, this is referred to as structural optimization.

In the case of parametric optimization, the numerical values of the variable parameters are assigned a numerical rating based on the performance index; this rating is to be maximized or minimized through the variation and is thus a function of the parameters. In structural optimization, the entire time dependence of the control function $u(t)$ in the time interval $t_0 \le t \le t_e$ is varied. The performance index assigns a numerical rating to this function and is therefore a functional (quality functional, cost functional) [4.41].

Usually, the term "optimal control" is understood as referring to the results of a structural optimization, and this meaning will apply in the discussion that follows.

Optimization Problem. The starting point is a mathematical description of the system in state space (see Section 3.3) in the form of the system equation:

$$\dot{x}(t) = f(x(t), u(t), t) \qquad (4.3.1)$$

What is now sought is the control function $u(t)$ that yields the minimum value of the performance index J for the process by which the system goes from the initial state $x(t_0) = x_0$ to a final state $x(t_e) = x_e$. This is the optimal control function $u^*(t)$. Solving the system equation with given x_0 and $u^*(t)$ gives the optimal state–space trajectory $x^*(t)$ associated with $u^*(t)$.

The final time t_e may be fixed or subject to choice, and similarly for the final state x_e or its components. Quantities that can be freely selected are available for variation; that is, they can be incorporated in the performance index.

Along with specifications of this kind, there are often constraints: State or control variables cannot, or must not, go outside certain value ranges at any time during the transition. For example, a value can be at most completely open or completely closed, the concentration of a species cannot take on values outside the range from zero to unity, and a reactor temperature must not go above a certain upper boundary or below a certain lower boundary.

The problem thus comprises

1) The system equation
2) The initial state x_0
3) The performance index J
4) Possible specifications of t_e and x_e
5) Possible constraints on x and u

Open- and Closed-Loop Optimal Control. For a given initial state x_0, the solution of the optimization problem yields an optimal control function $u^*(t)$. It does not, however, say anything about how the control function can be technically implemented. There are two approaches to doing so:

In open-loop optimal control, the system state x is measured once at time t_0 and $u^*(t)$ is determined from the result. For the remainder of the time interval $t_0 < t \le t_e$, $u^*(t)$ is presented at the control inputs of the system; no further account is taken of the system state. The control function is thus a function of time only from time t_0 onwards; it can be generated, for example, by a computer, which merely plays the role of a function generator.

Such an open-loop structure has, however, the disadvantage that the system follows the optimal trajectory $x^*(t)$ only if no disturbances occur (see Section 4.1). For this reason, open-loop optimal control is often replaced by closed-loop optimal control. The optimal controller continuously generates the control function based on the current system state. If there are no disturbances, the controller holds the system on the optimal trajectory, thus behaving in the same way as the open-loop optimal controller. If disturbances occur, resulting in departures of the system from the optimal trajectory, the controller will initiate corrections so as to offset these departures while the transition process is still under way.

Performance Indices. A general form of performance index, which includes many special cases of importance in engineering [4.4], is the functional

$$J = h(x(t_e), t_e) + \int_{t_0}^{t_e} f_0(x(t), u(t), t)\,\mathrm{d}t \qquad (4.3.2)$$

The first term in this Bolza performance index evaluates the system behavior at the end of the transition process, while the integral term evaluates the entire process. This is known as a Mayer performance index if $f_0 = 0$, and as a Lagrange index for $h = 0$.

The following cases occur under the Lagrange performance index:

- $f_0 = u^T(t)Mu(t)$, the energy-optimal performance index
- $f_0 = x^T(t)Lx(t)$, the ISE performance index (performance index of the integral squared error type)
- $f_0 = 1$, the time-optimal performance index

The diagonal elements of the matrices M and L evaluate the squares of the control and state variables, respectively. All the other matrix elements are often simply set equal to zero.

Methods. This section lists the most important methods and concepts, with references to the literature.

The classical method of minimizing functionals is calculus of variations [4.4], [4.42], [4.43]. This technique is suitable for solving optimization problems without constraints on the control and state variables. An important special case, the closed-loop optimal control of linear systems [4.19], [4.20], [4.44] with quadratic performance index

$$J = x^T(t_e)\,Sx(t_e) + \int_{t_0}^{t_e}\left(x^T(t)\,Lx(t) + u^T(t)\,Mu(t)\right)\mathrm{d}t$$

leads to the so-called Riccati controller.

Systems with constraints on the control variables can be optimized with the Pontryagin maximum principle [4.45]–[4.47]. Such constraints arise, in particular, in time-optimal control problems where the solution is expected to yield control functions that alternate between the upper and lower extreme values of the control variables. Feldbaum's theorem states, for a given class of systems, how many such alternations are required at a maximum.

Dynamic programming [4.48]–[4.50] is a general optimization method that can be used for, among other purposes, the optimization of dynamic systems. It offers the possibility of utilizing the constraints not only on the control variables but also on the state variables. An essential foundation of dynamic programming is Bellman's principle of optimality, which says that any final portion of an optimal trajectory is itself an optimal trajectory. This principle implies the basic procedure: The optimal trajectory and the associated open-loop optimal control scheme are determined in successive backward steps, beginning with the final state x_e.

A thematically organized literature survey of the optimization topics discussed here, as well as others, can be found in [4.4]; see also [4.51]–[4.56].

4.4. Binary Control

Principles. Classical control theory is concerned chiefly with "analog" signals, which are discrete as to time and value. The mathematical description of the process ("plant") and of the automation device ("controller") employs differential equations as well as transformations (Fourier, Laplace).

Industrial control practice, on the other hand, deals with all devices used to influence plants and machinery in a purposeful way. Traditionally, however, the emphasis is on working with binary signals (e.g., in relay and contactor controls) in open-loop structures [4.57]. Analog control loops occur only as low-level elements. Boolean algebra and its adaptation to hardware (switching algebra) are employed for the mathematical description of binary control [4.58].

These contrasting approaches have historical roots: Analog thinking relates to early work on the automation of continuous operations such as those practiced in the process industries, while binary thinking grows out of attempts to automate piecewise operations such as those used in the production industries.

Process automation as now practiced in industry involves the convergence of both approaches. This applies to processes being automated and to the automation equipment itself, but not to the mathematical apparatus.

Operations. Operations that have long been automated by open-loop control methods are now being operated with closed-loop methods as well; an example is the motion programs of industrial robots.

Process engineering, a classical domain of closed-loop control, now makes increasing use of open-loop control, both for the automated startup and shutdown of continuous processes and for automated batch production.

At the same time, a trend toward higher-level optimization of the entire production process, with the goals of improved efficiency and flexibility, is apparent in all areas of industry. The result is generally coupling of subprocesses in the sense of material streams, energy, and information. Process automation thus calls for a holistic examination of continuous and discrete subprocesses, analog and binary process variables, and continuous and sequential operations. (This section does not deal with a third class, object-related operations [4.59], in which individual, identifiable objects pass through the process. As far as process engineering is concerned, this concept figures chiefly in high-level production automation; examples of "objects" are charges and production campaigns. See also Chap. 2 and Section 4.5.)

Automation Equipment. Just as continuous and discrete subprocesses have been linked into "hybrid" processes, automation equipment has also undergone a merger with classical measurement and control devices (Fig. 4.29 A) becoming components of integrated automation systems (Fig. 4.29 B). This development has, however, been driven less by user demand for integrated process automation than by technological advances in instrumentation and control. Mechanical, pneumatic, hydraulic, and electromechanical devices have gradually had to give way to microprocessor-based electronics. Hence the step from consistent instrumentation and control devices to integrated process automation systems was obvious.

Ⓐ

Structure: / Signals:	Without feedback	With feedback
Binary	sequence control	interlock
Analog	measurement	regulatory control

Ⓑ

Structure: / Signals:	Without feedback	With feedback
Binary	programmable logic controller (PLC) ↓	
	(integrated automation system)	
Analog	↑ process control system (PCS)	

Figure 4.29. Application areas of automation systems
A) Classical automation devices; B) Modern automation systems

The adoption of these systems by users has been aided by the appearance of two variants that use software to mimic the operator interface of classical instruments and controls (Fig. 4.29 B):

1) The programmable logic controller (PLC) for users oriented toward open-loop control in the production industries. Programming may be performed graphically with ladder diagrams, which are similar to the schematic diagrams of relay-based control systems.
2) The process control system (PCS) for users oriented toward closed-loop control in the process industries. The PCS interface can, for example, imitate conventional control stations with live flow diagrams and controller front panels (see Section 5.2).

The combination of first-generation PCS and PLC into integrated "hybrid" process automation systems has led to a second generation of process control systems [4.60].

Mathematical Methods. A hybrid system consisting of a hybrid process and a hybrid automation system can be simplified by splitting it into a continuous, "analog" part and a discrete, "binary" part (Fig. 4.30). These subsystems have common structural properties: They contain internal feedback paths and must be treated as dynamic systems with memory elements.

The mathematical methods applied to the two classes of systems have not yet converged [4.61], mainly because the theoretical techniques are at far different stages of development.

For continuous systems (and also for the closely related sampling systems), control theory

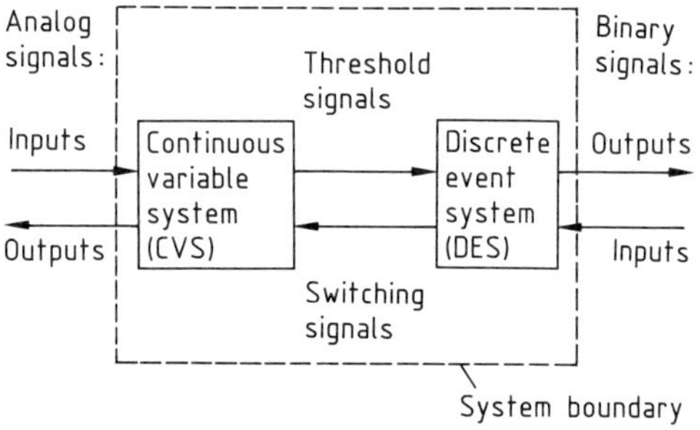

Figure 4.30. Hybrid system
Input and output signals connect the system to surrounding systems (e.g., systems on higher and lower levels; operator interfaces)

offers a polished, complete, and proven mathematical toolkit. For the modeling, analysis, and synthesis of dynamic discrete event systems, in contrast, there is only a rudimentary selection of diverse methods, including switching algebra [4.58], Petri nets [4.62], Boolean differential calculus [4.63], and theory of automata [4.64].

Terminology. The foundation of binary control is switching algebra [4.58], which works on two-valued or logical variables. Its axioms establish the elementary logical operations conjunction (AND), disjunction (OR), and negation. Figure 4.31 shows schematic symbols for the hardware implementations of these operations. Other basic operations are derived from the elementary ones: NAND, NOR, equivalence, antivalence (exclusive OR or EXOR operation), and implication. Combining the basic operations generates switching functions, the implementations of which are called combinatorial circuits or logic controllers. The rules of switching algebra permit the mathematical analysis of networks. The Quine–McCluskey method can be used for the minimization of given logic functions [4.65].

If a combinatorial circuit contains internal feedback paths, memory elements (flipflops) can arise (Fig. 4.32). In this way, a sequential circuit or sequential controller is obtained. There are two classes of flipflops: level-operated and edge-

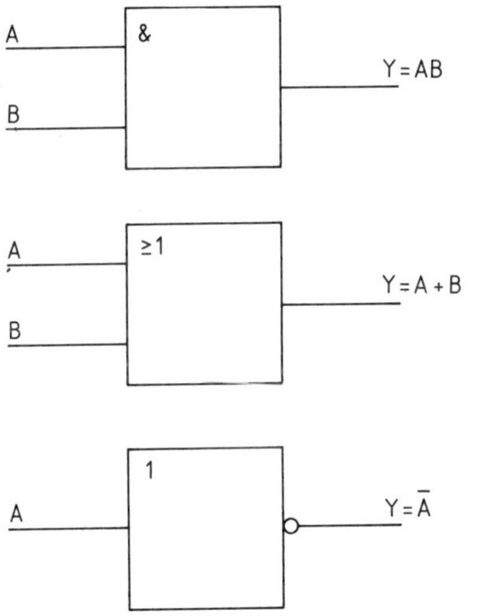

Figure 4.31. Schematic symbols for the elementary logical operations AND, OR, and negation, according to IEC 117-15
A, B inputs; Y output

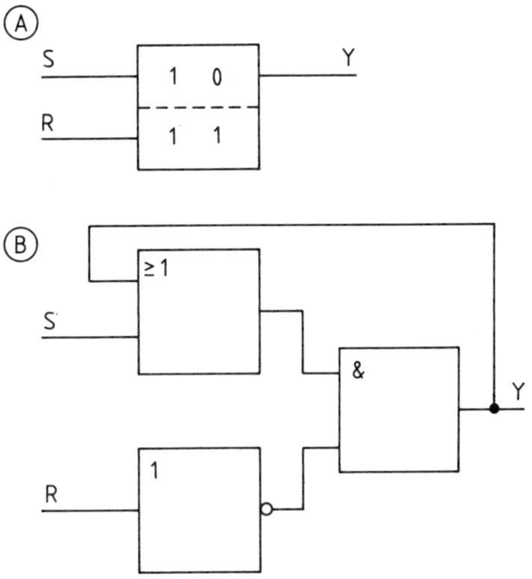

Figure 4.32. RS flipflop
A) Schematic symbol; B) Construction from elementary logical operations
S set input; R reset input (with priority over S)

triggered. In level-operated devices, the output can change its state as soon as one input state changes; in edge-triggered circuits, the output can change its state only when a certain (e.g., positive) edge is presented at a certain input, called the clock input. If, along with the logic functions, a sequential circuit contains only edge-triggered flipflops whose clock inputs are all driven by a central system clock, it is said to be a synchronous sequential circuit (since all the memory states can change only in synchrony with the edge of the clock pulse). Otherwise, the sequential circuit is said to be asynchronous.

A systematic design procedure exists for synchronous sequential circuits. The starting point is a graphical description of the problem, called the state diagram [4.66]. The technique does not, however, insure a minimal realization of the sequential circuit. State diagrams are also unsuitable for the analysis of complex circuits, since a sequential circuit with n memory elements can have up to 2^n states.

Besides the combinatorial and memory elements, there is a third type of basic functional element: timers, which shortens, lengthens, or delays binary impulse signals. They are of very minor importance in the theoretical analysis of binary control systems but are frequently used in practice.

The program of a control system means the establishment of all signal-processing functions.

This can take either of two forms. In a wired-program controller, the functions are established by hardware connection of the individual functional elements. If the connections can be altered in a simple way, the controller is said to be reprogrammable; otherwise it is called fixed-programmed.

In a stored-program controller (or programmable logic controller, PLC), both the functional elements and their connections are simulated in software on a digital computer. The control program is thus part of the computer program. The program for a RAM-programmed controller is stored in read–write memory (random-access memory, RAM); that for a programmable controller with exchangeable memory is stored in read-only memory (ROM).

A PLC includes a digital computer and is thus essentially a synchronous processor. It can, however, be used to simulate combinatorial circuits, asynchronous and synchronous sequential circuits, timers, and also analog signal-processing elements such as controllers, or any combination of these, to a very good approximation. It accordingly represents the most flexible and advanced way of solving all control problems, and it is the most cost-effective technique for all complex control problems.

A number of text-oriented and graphics-oriented languages are available for the specification of control tasks and the programming of PLCs [4.58]:

- Structured text (ST; IEC 1131/3) is particularly suitable for writing implementation-neutral specifications.
- The instruction list (IL; DIN 19239, VDI 2880) is a low-level assembler-like PLC programming language (i.e., similar to machine language).
- With the ladder diagram (LD; DIN 19239, VDI 2880), simulations of relay-based circuit diagrams are employed for PLC programming. This is a graphical language.
- The function block diagram (FBD; DIN 19239, DIN 40719, IEC 117-15) technique, another graphical language, uses elements similar to the circuit symbols for logic gates and memory elements in digital schematics (see Fig. 4.33). If individual logic and memory elements in the logic diagram are combined into macros ("actions" and "steps"; see Fig. 4.33), the result is a sequential function chart (SFC).

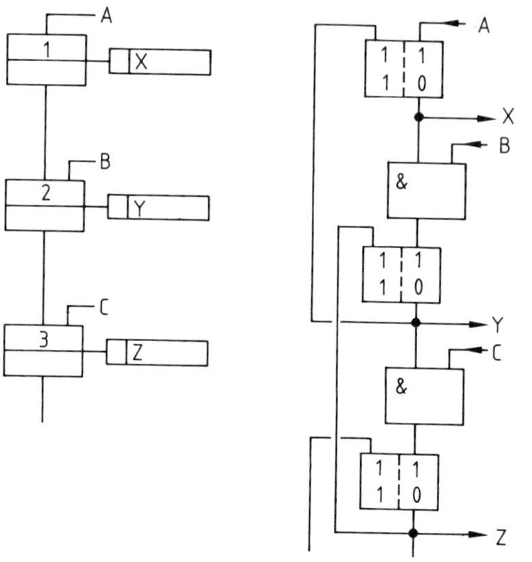

Figure 4.33. Linear step sequence
Macro representation: sequential function chart (left); Explicit representation (right)
1, 2, 3 steps; A, B, C transitions; X, Y, Z actions

- The sequential function chart (DIN 40719/6, VDI 2880) is especially suitable for sequential controls where the individual control goes through states or steps not in an arbitrary sequence but (at least section-wise) in a fixed, linear sequence. This condition does usually hold in practice.

Besides these standardized languages, there are many manufacturer-specific languages, partially constructed from elements of the standardized languages (e.g., Siemens STEP-5).

Sequential function charts include not only sections of linear step sequences, but also branches. A distinction is made between OR (alternative) branches and AND (parallel) branches. In the latter, concurrent control sequences take place in the controller; that is, sequences that proceed in parallel and independently of one another until they are again synchronized to one another at the end of the AND branch. Such controls cannot be usefully diagrammed and analyzed with state diagrams; a much better tool for this purpose is offered by Petri nets [4.62]. The draft of DIN 40719/60 for sequential function charts, in contrast to the old version of DIN 40719/6, includes some important restrictions on branching options to permit Petri net modeling of concurrent controls.

Both Petri nets and sequential function charts are suitable for the specification, simulation, analysis, programming, operator interface, and documentation of controllers. They are particularly useful as support for hierarchical control concepts, which are often important in practice.

Classification of Control Systems. The common language of automation technology includes many control terms, most of which are defined in DIN 19237 and many of which were mentioned in the preceding section (see p. 407). In what follows, these terms are listed under different classification criteria.

- Signal-oriented classification:
  analog, digital, binary control
- Function-oriented classification:
  combinatorial, sequential control
- Structure-oriented classification:
  asynchronous, synchronous control
- Technology-oriented classification:
  wired-program control, stored-program (programmable) control;
  (wired-program control:) fixed-programmed, reprogrammable;
  (stored-program control:) RAM-programmable, exchangeable memory;
  relay, contactor, electronic control;
  mechanical, hydraulic, pneumatic, fluidic control, etc.
- Hierarchy-oriented classification:
  individual, group, coordinating control
- Application-oriented classification:
  interlock control, sequence control;
  burner, scale, transport, recipe control, etc.

Another concept is mode control, which refers to a control level, higher than the signal-processing part of a sequential control, in which an operating mode is selected. VDI/VDE 3683 and DIN 19237 define the following modes (among others): automatic, semiautomatic, single step, step setting, manual. In hierarchical control concepts, the operating modes available may differ from one level of the hierarchy to another and must be coordinated in a suitable way.

4.5. Operational Control of Process Plants

Requirements on the operational control of process plants have grown much more stringent in recent years. The reasons include increasing flexibility in production, shortages in personnel and plant resources, shorter product throughput times, production structures with crosslinks due to energy and product coupling, quality assurance standards, and the greater and greater number of maintenance operations that have to be performed at regular intervals. Optimized production programs and maintenance practices must be integrated into a plant control concept against the background of environmental and occupational safety and health requirements and the reliability standards that they imply.

A number of approaches have been devised and used with success for specific problems.

Brombacher [4.67] and Niese [4.68], [4.69] used the term "process logistics" to describe proposals they developed for managing product streams given the coupling of process elements in normal operation, but also in transient and exceptional situations.

Under the leadership of NAMUR, "recipe" control concepts were developed for batch processes; these have been implemented as standard products by control system manufacturers [4.70].

Riemer has described concepts for production control, which have been turned into tools [4.71].

The same holds for special functions at the plant complex control level [4.72].

A closer analysis reveals that these concepts do accomplish their specific goals but are not consistent with one another in the sense of operational process control.

In future, accordingly, a broader approach must be taken to the operational control of process plants.

The action and resource model discussed here is one such expanded approach to describing the operational control of a process plant. The model is constructed in an object-oriented way. A wide variety of requirements, relating to recipe-based process control, logistical product management, maintenance, and other areas, can be reflected in the model.

The model includes a general formalism for resource allocation and control. Automated tools support the application of the formalism. In the context of the level model, the action model represents a logical standard interface between the production or plant complex level and the process control level.

Introduction. The fundamental task of operational process control is to coordinate and, with the aid of plant resources, carry out various tasks stemming from production requirements, in-plant product logistics, or technical support and plant maintenance.

The action and resource model provides a formalism for systematically finding and implementing solutions.

The action model is based on the idea that every task arising in operational plant control can be completely accomplished by performing individual actions. These actions are performed through orders on the available resources.

The process control problem is thus broken down into three parts:

1) Mapping of tasks into individual actions
2) Planning and scheduling of these actions
3) Control and monitoring of the performance of the actions

Figure 4.34 illustrates the basic structure of the model.

Some important concepts figuring in the model are explained first. These should be understood as working concepts and are used here only in a very restricted sense.

Functional Units. A functional unit is, quite generally, a unit aggregated from various functional elements and, as a whole, accomplishing a self-contained task. Functional elements are selected such that they can be directly implemented with the help of existing plant equipment. Thus they are not fictitious functional objects but concretely represent the capability of a group of process devices. A plant complex includes a variety of functional units, such as plant sections, maintenance foreman groups, auto transport services, energy sources, and so forth.

In general, functional units are organized in a hierarchical manner (see Chap. 2).

Maintenance groups, for example, are organized as follows:

Maintenance technician
Maintenance gang
Maintenance foreman group
Maintenance support group

Process-oriented functional units might be organized as follows:

Individual functional unit
Group functional unit
Plant section unit
Plant unit
Plant complex

The internal structure of these units is discussed in depth below.

Resources. In the concept discussed here "resources" means "that which can perform services in the plant." Resources correspond to functional units that do not functionally overlap and which together represent the capability of the plant. Resources so defined are "intelligent" units to which the plant can pass on orders either directly or through plant actions. Resources are part of the permanent equipment of a plant complex; they are present whether they are processing an order or not. The sum of the resources

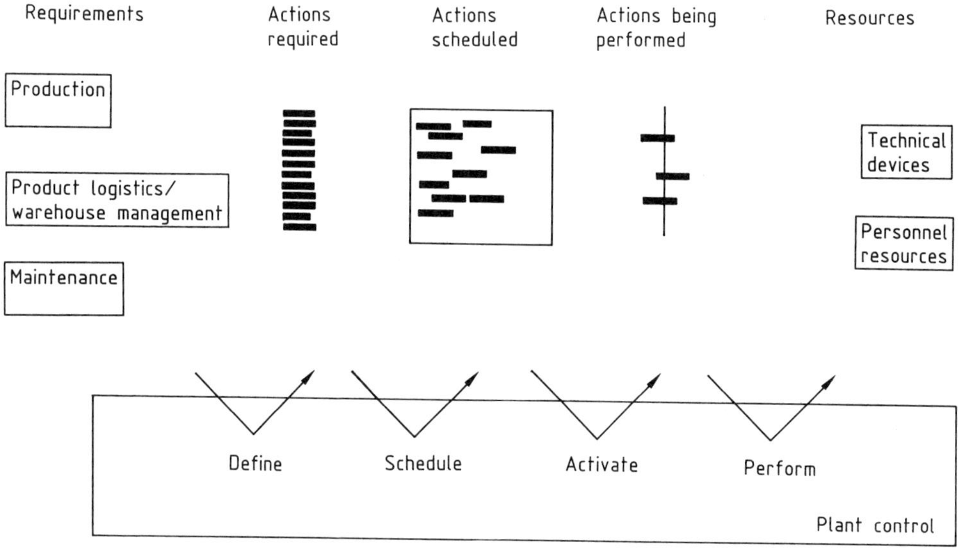

Figure 4.34. Actions as a means of plant control

reflects the functional capability of a plant complex.

Several special rules apply to resources:

- Self-sufficiency:
  Each resource has the capability of managing itself and independently carrying out orders. For the real performance of its task, each resource has its own apparatus and resources to which it has unrestricted access. Each resource can thus be planned, scheduled, and worked independently of all other resources.

- Control interface:
  Each resource has a control interface and an allocation mechanism associated with the interface. A resource may be needed for a variety of plant actions. At any time, however, only one action is allocated to a resource by means of orders. The planning and scheduling of resource utilization is accomplished by higher-level functions. Independently of such planning and scheduling, however, each resource has an allocation mechanism, which locally protects it against improperly being assigned more than one order.

- No parallel processing of orders:
  A specific feature of this model is that each resource can process only one order at any time. Accordingly, there is no concurrent working of orders within a resource. This provision is crucial in determining the overall structure of operational process control.

This "flatness" and the ban on parallel processing of orders are an important advantage in making operational control easy to understand and simple to implement, but they represent a severe restriction on structuring. Functional units, as substantial objects, can be aggregated in accordance with the structural principle. The rules for resource formation now require that one level of aggregation be selected and permanently identified as the resource level used for operational proccss control. Only the functional units on this level are competent to perform operational process control services. If, for example, the maintenance foreman group is identified as the resource for the maintenance hierarchy, the operational process control system can then issue orders only to the foreman group. The issuance of an order to a gang or to a support group would not be permitted. In this example, the rules would demand that a foreman group be processing only one order at any time. This is, however, generally not so in practice. Accordingly, a foreman group is not the proper resource for operational process control. The resources identified should be those on a lower level of aggregation. Gangs, in general, process only one order at any time; hence gangs meet the criteria for a resource and would be a suitable choice as schedulable units for operational control of the maintenance area.

It is important to stress that this restriction to one level of aggregation relates solely to operational process control. All the levels of aggregation are still of interest from the standpoint of other organizational concerns.

Actions. The plant complex can accomplish its task by issuing orders directly to its resources. Such a manual procedure is, however, expensive for the plant complex and also hard to track. The action model offers a more convenient solution. An action is a self-contained control unit that permits the automatic performance of an entire functionally coherent block of tasks. Typical actions in this sense might be

"Pump X liters from tank A to tank B,"
"Do the scheduled maintenance on emergency trip system X,"
or
"Produce X tons of A in accordance with recipe B."

For each action there is a performance specification, which sets forth how the action is to be carried out. Any action may require more than one resource for its performance.

Actions can be combined and grouped at the plant complex control level. At the interface to the lower-order process control system, however, each action is essentially a self-sufficient, independent, self-contained object. The model says nothing about the complexity of an individual action. Both extremely complex actions and elementary actions can be defined and carried out simultaneously in a system. Actions need not be free of redundancy as to their functionality. For example, an action such as "transfer X liters from rail transport to tank A" may well have the same elements as the action "Pump X liters from tank B to tank A." Regardless of the nature of the task or the type of resources employed to accomplish it, a common scheme is used for performing and managing actions. In this way, they form an object class with well-defined properties. In the hierarchy, actions form a standard interface between requirements on plant management and operational performance. This formal decoupling is an essential goal of the action concept, but it should not be taken to imply that task and execution can be analyzed and organized independently of one another. If the structures of resources and those of tasks are not consistent, the result will be a poorly organized package of actions that cannot be efficiently carried out.

Thus it makes no sense, for example, to define special plant engineering resources simply for purposes of process control if such resources cannot be utilized to perform maintenance actions or contribute to in-plant product logistics.

Central importance attaches to the development of a resource structure that is matched (in substance and form) to the internal concerns of various task areas.

Requirements on Operational Plant Control. Operational plant control must meet requirements arising from many areas. These are broken down by the concerns of the areas where they originate. The next section presents examples of "Requirements from Production Planning," "Requirements from In-Plant Product Logistics," and "Requirements from Plant Support" in order to show how actions can be derived from requirements.

Requirements from Production Planning. From production planning, the plant complex receives orders to produce certain products, along with product-related specifications such as product name, nominal quantity, nominal quality, and nominal deadline. In order to accomplish this production, a number of actions must be initiated and carried to completion. These actions can be analyzed in two ways:

- An operational breakdown of the production process into steps corresponding to the subdivisions of the process
- A purely quantitative breakdown of the production process into lots making up the desired (nominal) quantity

- Operational breakdown of the production process

For structuring purposes, a production process can be broken down by using the phase model (see Chap. 2, Figs. 2.32–2.37). Figure 4.35 shows an example for a process A. At the left, the entire process is shown as a closed block with inlet products IP1 to IP3 and outlet products OP1 and OP2. The guiding principle in breaking down the process is that the operations belonging to a process can be separated by simply dividing up the material system by product

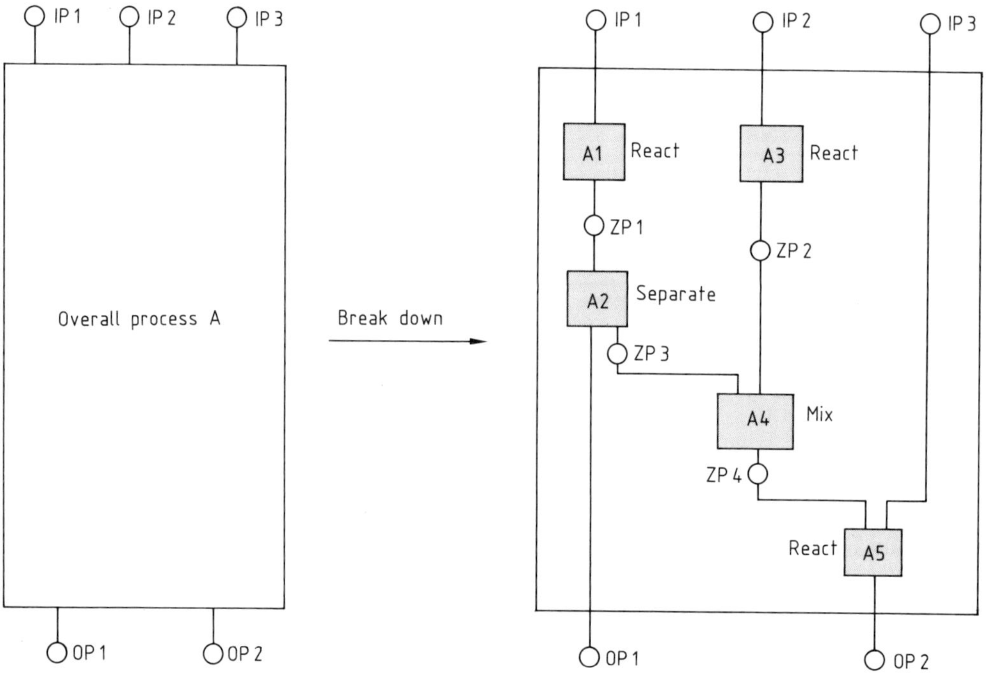

Figure 4.35. Breakdown of a process in the phase model

and dividing up each operation into phases by time.

If this is possible, individual, self-contained subprocesses corresponding to the phases of the phase model are obtained. Each subprocess is characterized by the performance of certain operations. In the example shown, these operations are reaction A1, separation A2, reaction A3, mixing A4, and reaction A5. The products at the beginning and end of a subprocess have well-defined properties. In the example, these products include the inlet and outlet products of the overall process plus intermediate products ZP1 to ZP4. The subprocesses are coupled to one another via these products.

Depending on the complexity of the overall process, the same scheme can be re-applied to break down the subprocesses.

This method reaches its limit when operations cannot be further separated either by a product-related subdivision of the material system or by a time subdivision into phases.

Production specifications are instructions for the execution of a production in a plant. In other words, they are performance specifications for production actions. They extend beyond general process descriptions, stating precise conditions to be maintained and states to be established in order to carry out a certain process in a plant.

These conditions include, for example, properties of products that enter the process. The starting material present in the plant at the beginning of a process should also be regarded as a product occurring in the process.

An important initial condition is the state of the plant at the beginning of the process (see also Section 2.3). Initial condition refers not only to state variables in the narrow sense, such as temperature or control state, but also to general properties such as "degree of contamination," "functional reserve," "modification effected," and so on. These conditions govern the need for tooling or cleaning orders.

The fundamental part of the production specification, however, describes how certain process properties must be controlled in the course of the process so that the desired overall process can be carried out. This also includes the establishment of certain intermediate and end product properties [4.73]. Each production specification is based on a model of a reproducible process.

By analogy to the breakdown of a process, a production specification can also be divided into parts (see Fig. 2.24). It should be noted that every portion of the specification, by virtue of the analysis, contains supplementary conditions on the inlet product (feed) properties and on the states of plant sections at the beginning of the subprocess.

When such a breakdown is carried out, a production specification becomes a number of subspecifications, each of which describes the execution of a subprocess in a plant section (see Fig. 2.25).

It is evident that the analysis of processes and production specifications has a direct effect on the structure of action objects and the way they are managed. This can be illustrated by two extreme situations as shown in Figure 4.36.

In Example 1, the process has not been broken into parts. In other words, the action takes in all partial production steps including intermediate product transport. The overall production process, including the time sequence, is described by a single production specification and, in its entirety, corresponds to a single action.

In Example 2, the overall process is divided into subprocesses, for each of which there is a partial production specification independent of the other subprocesses. An independently schedulable action is assigned to each subprocess. Temporal coordination is achieved through the availability of intermediate products and the initial conditions for the individual subprocesses. Coordination comes under the planning and scheduling functions of the production or plant complex control level.

What level of "granularity" is appropriate for a given production process will depend on the requirements and constraints imposed. The action concept permits arbitrarily broad or granular actions and does not place any restriction on the depth of the breakdown.

- Quantitative breakdown of the production process

If the nominal product quantity is too large to be produced in one step in a closed material system, the overall production can be subdivided into individual batches. A number of batches can be combined into a lot, and a number of lots into a campaign. The NAMUR definition states that a lot is "a delimitable and identifiable quantity of product having a uniform quality. In the case of a batch production process, a lot may include one or more batches" [4.70].

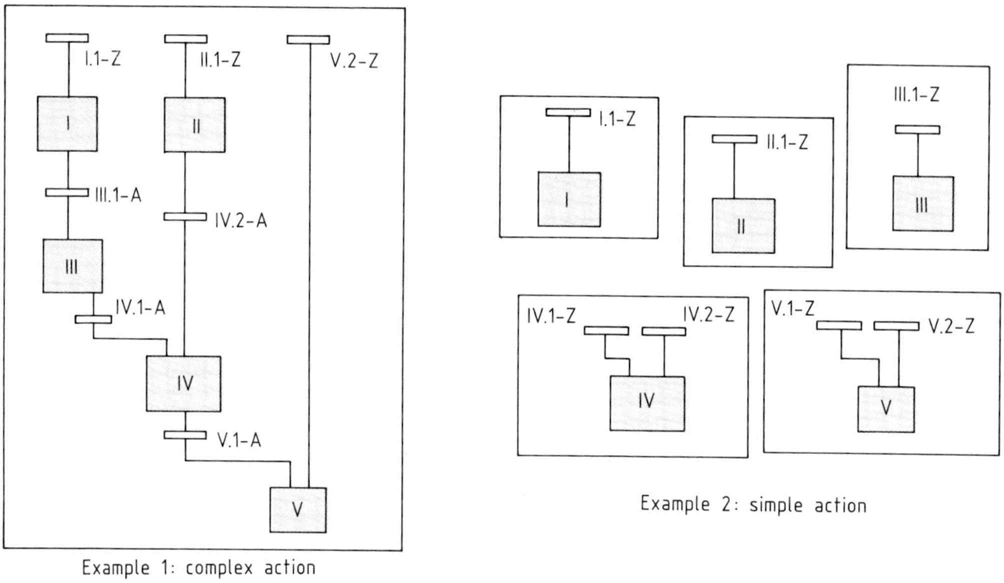

Figure 4.36. Complex and simple actions
Z = prerequisites strictly state-oriented; A = prerequisites also sequence-oriented

Again, the action concept allows any degree of aggregation. In general, actions will relate to the production of one batch; the combination into lots is a planning and scheduling task. In the extreme case, however, a complete lot can be defined as a single action.

Requirements from In-Plant Product Logistics. In-plant product logistics refers to the control of product (material) flow within the plant complex. This activity supplies inlet products (feeds) to the "producing" plant sections and removes the end products and byproducts. The tasks of in-plant product logistics also include the control of internal buffers, tank storage, bunkers, high-bay warehouses, and devices used for storage and reclaiming. This area of responsibility may also extend to the control of tank cars, rail cars, and other mobile storage located on-battery.

Depending on the application, product-logistics activities may take place in the context of production actions or may be independently planned and scheduled tasks executed by separate actions. Such actions might then be:

"Transfer by pumping,"
"Convey from tank X to tank Y,"
"Transfer from tank car unloading station A to tank X,"
or the like.

These actions are required to be highly flexible as to scheduling. Often the orders are not issued until shortly before performance, or rescheduling may be necessary at the last minute (e.g., tank cars are in the wrong sequence on the track, operators at the tank car terminal wish to carry out order A before order B, etc.). This required flexibility in the planning of logistic actions and the requirement that noncentral (local) entities must have access to scheduling (e.g., from the receiving station) must be taken into account in the design of the scheduling system.

For higher-level product management, the granular logistic elements are combined into more complex units. For example, starting products may be delivered to the plant complex, whereby each delivery may consist of a number of tank cars. Within the plant complex, not the receipt of the delivery as a whole but rather the discharge of an individual tank car is generally regarded as a single action. Again, however, the action concept is open and permits modification to suit the organizational setup of the plant. (For example, a complete coal train is unloaded in a different way from a delivery made up of separately marshalled cars.)

Requirements from Plant Support. Plant support, in order to accomplish its tasks, must periodically carry out plant inspection, maintenance, and repair operations. These can also be defined and performed as actions, for example:

"Calibrate probe X,"
"Drain separator sump,"
"Do scheduled maintenance on pump A,"
"Inspect overcharging protection B,"
and so forth.

If a maintenance action renders plant equipment unavailable for production while the action is being performed, the maintenance action must be included in the planning and scheduling system like any production action. The action concept permits a systematic treatment of maintenance actions in resource allocation. What is more, it allows plant-wide monitoring of the progress of planned maintenance. Every maintenance operation defined as an action in the system monitors its own correct performance. If unplanned events occur, the progress status of the action should reflect them. If necessary, an explicit notification should be provided to plant management or the planning and scheduling system (see Fig. 2.19).

Maintenance operations can be broken down in two ways: (1) by the functions or functional units maintained and (2) by expected frequency, for example, on a time or wear basis. Ultimately, these analyses lead to individual activity elements. Job planning combines these elements into packages meaningful from the engineering, functional, and management standpoints, and the packages are treated as maintenance actions for planning, scheduling, and administrative purposes. How this grouping of elements takes place in a particular case will depend in the maintenance concept practiced at the plant complex (see Section 11.4).

Analysis of Plant Equipment and Definition of Resource Structure. The plant complex has technical equipment and personnel resources at its disposal to perform its tasks. This section describes how the apparatus and personnel resources can be broken down and how they can be combined into technologically meaningful functional units. On the basis of the system of functional units created in this way, a suitable resource structure can then be established.

This description is for the example of plant engineering functional units but can be extended to other areas as well.

Breakdown of Plant Equipment. The purpose of analyzing equipment and apparatus available in the plant complex is to identify clearly delimited, nonoverlapping elements. Elements belonging to various devices can then be assembled into meaningful functional units. The process is illustrated for the combination of plant equipment elements and control system elements into plant engineering functional units.

- Breakdown of Process Equipment

Production takes place in process equipment and facilities where the product is conveyed, stored, and transformed. A process plant or facility ultimately consists of valves, pumps, pipes, and so on. Between the plant as a whole and these single elements, however, a number of useful analytical levels can also be found. An important level is the plant section. From some points of view, the plant section is actually more important than the plant. Plant sections are typically delimited in such a way that further subdivision of the material system present in them is impossible or needless. Thus exactly one subprocess can run in a self-sufficient way within every plant section. Here "exactly" means that one, but only one, subprocess can take place in any plant section at a given time.

A plant section has inlets where products enter it and outlets where products leave it. These inlets and outlets serve to connect plant sections to one another. The topological structure implies certain rules for product streams between plant sections. Figure 4.37 shows the most important basic ways of connecting plant sections. In the train, a product moves through one plant section after another (stage by stage). If there are several trains, a product moves through only one of the trains.

In present-day chemical plants, especially plants where several different products are made, these structures are no longer found in pure form. Instead, the plant sections are connected by a complicated network of piping. Figure 4.38 gives an example of such a complex structure. A plant section consists of the components exclusively assigned to it, such as valves, pumps, sensors, piping connections, and so on.

Within a plant section these components can be assembled into groups. In general, a group has the property that it effects a certain flow of matter or energy in the plant section. Figure 4.39 shows a reactor with its groups.

Each group consists of the components exclusively assigned to it.

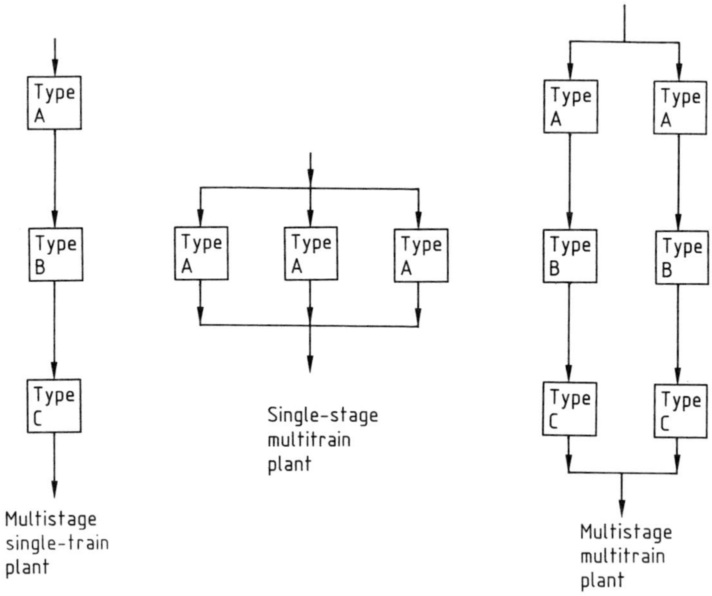

Figure 4.37. Basic connection patterns of plant sections

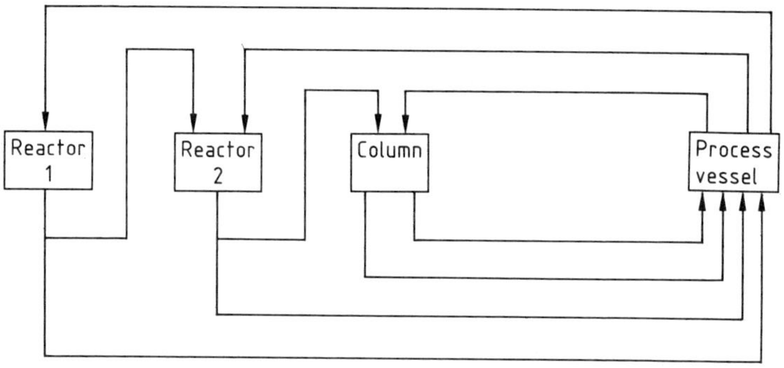

Figure 4.38. Real network of connections between plant sections

Groups are coupled to one another by transfer points for matter or energy, called connections.

- Breakdown of Process Control

The goal of process control is to bring about certain process properties. A certain control logic is required for this purpose. The basic structure of this control logic is described by a control model.

The basis for the control model employed here is a hierarchical control model. As Figure 4.40 shows, a control unit receives orders, commands, and setpoints from a higher-level unit. It executes the order by, in turn, specifying orders, control values, control parameters, and commands to lower-order control units. How it does this is an aspect of its internal logic. The methods of feedforward and feedback control can be used to derive a suitable internal logic.

Grouping of Functional Units. The purpose of forming functional groups is to combine plant equipment in various areas in such a way that it together accomplishes a partial task of the plant. Workable modules for accomplishing the overall task of the plant are obtained in this manner.

This is illustrated by an example of the grouping of plant units and control units.

- Plant (equipment) units and control units can be combined into functional units. The model provides that one control unit is assigned to each plant unit.

At the lowest level, for example, individual control elements such as valves, pumps, electrical heating devices, and so forth, are grouped with individual control units to form "individual functional units." Such an individual functional unit for the blocking and release of a product stream is illustrated in Figure 4.41.

Each individual functional unit permits a certain process property to be set (see also Section 6.2).

The second level comprises group functional units. Figure 4.42 shows schematically how a group functional unit is built up. A group functional unit consists of a group of individual functional units exclusively assigned to it, together with a group control unit, which controls and

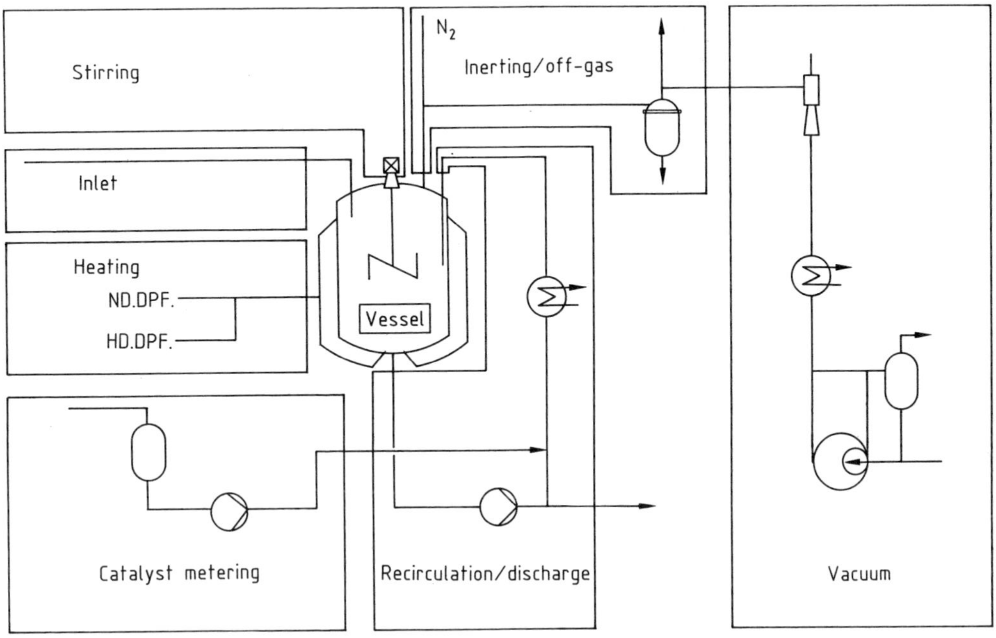

Figure 4.39. Groups of a reactor

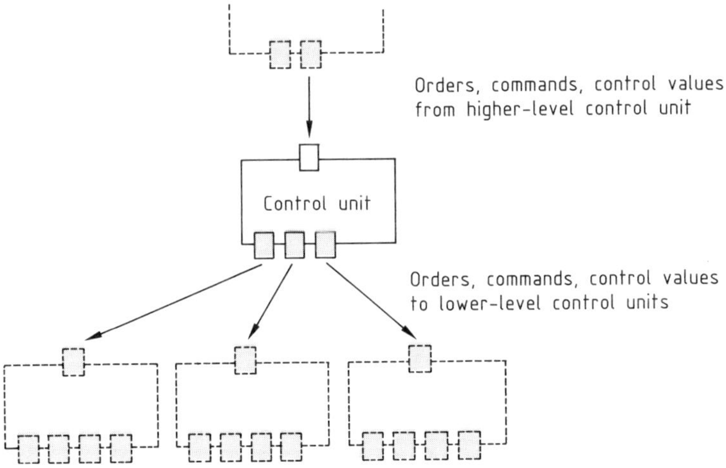

OP 1

Figure 4.40. Hierarchical control model

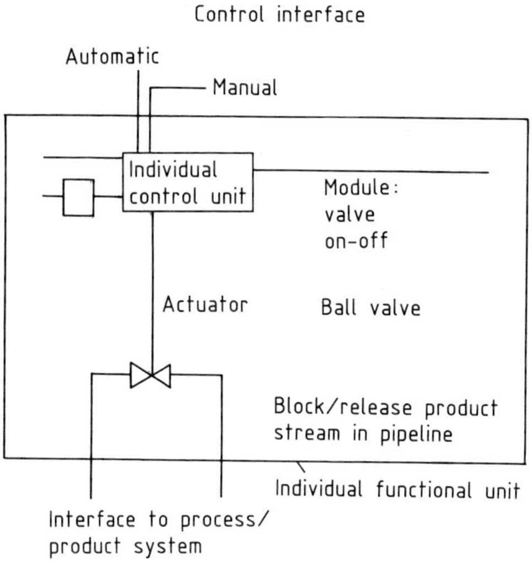

Figure 4.41. Structure of an individual functional unit

coordinates the activities of the individual functional units.

The overall goal of a group functional unit is to combine a group of individual functional units from the control standpoint in such a way that together they act on the process in a manner that is meaningful from the process engineering viewpoint. Individual effects that do not add up to an effect meaningful in terms of process engineering are excluded by the group functional unit.

If several effects useful from the process engineering standpoint can be brought about with a group of individual functional units, the group control unit supports each of these effects with its own control structure. The group functional unit is said to have alternative or selectable "modes."

The functionality of the group functional unit is implemented when the group control unit issues specifications to its assigned individual control units in accordance with the control model.

Figure 4.43 illustrates the structure of a group control unit.

Note that the group functional units discussed here have also been described under the name "basic functions" [4.74], [4.75]. Because this term leads to repeated misunderstandings, however, and is still under discussion in NAMUR [4.70], it is not employed here.

The third level is that of plant section functional units. A plant section functional unit consists of the group functional units of the plant section and a plant section control unit. The plant section control unit reflects all the process control capabilities of the plant section. Typically, group functional units permit an arbitrary number of meaningful program options. Which

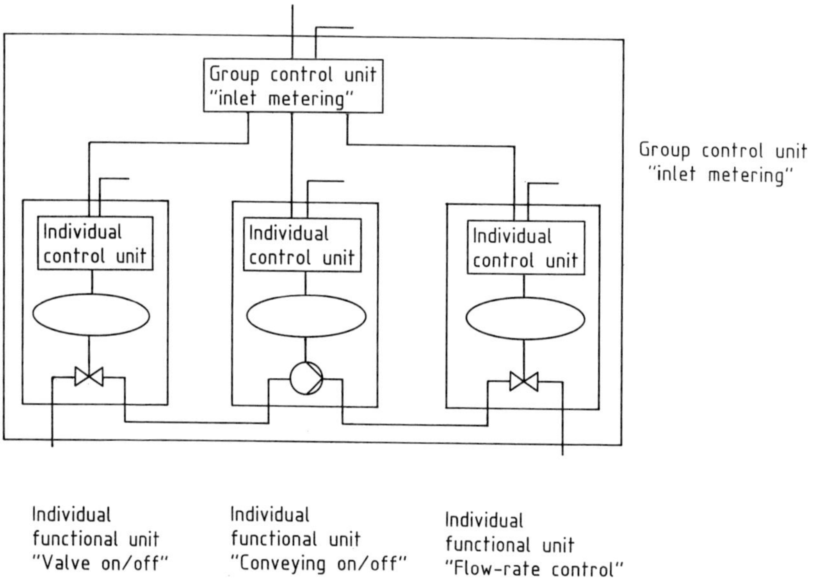

Figure 4.42. Structure of a group functional unit

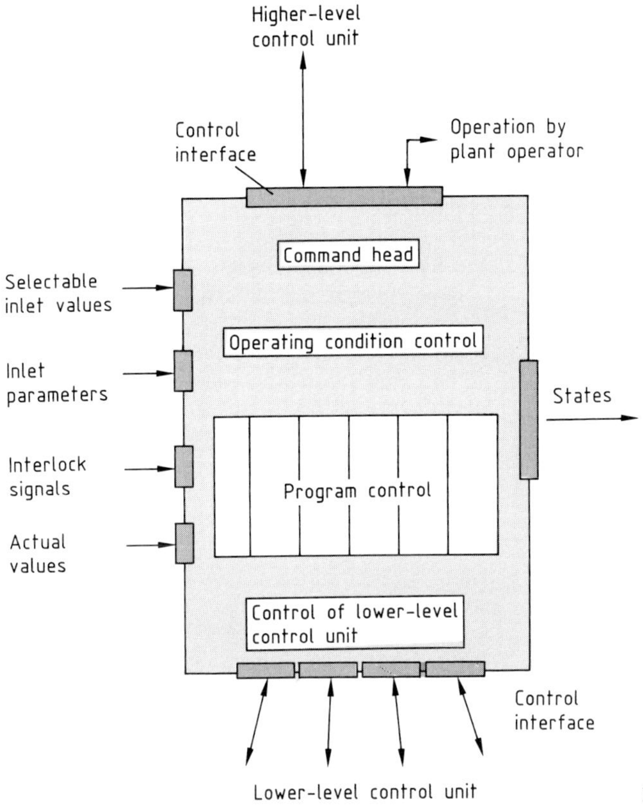

Figure 4.43. Structure of a group control unit

option is needed in a particular case is set forth in the performance order for the individual process. In this way, when issuing an order to the plant section control unit, it is necessary to provide not only control values, control parameters, and commands, but also portions of the performance specification.

To sum up:

Each level offers certain possibilities for functional intervention in the process; that is, each level can carry out certain process control orders. Furthermore, each level completely reflects the process control capability of the plant.

The higher the level, the more complex the task. In the simplest case, the order is to open or close a valve or to set a certain flow rate.

Every functional unit contains control units that carry out specific tasks. A control unit receives orders and/or control values from a control unit on the next higher level and, in turn, issues orders and/or control values as setpoints to control units on the next lower level. The number of levels in the hierarchy can be selected on technological grounds. Criteria might be, for example, the size and complexity of the process plant or facility. For the reference model under consideration here, a two-level structure below the plant section is chosen.

Determination of a Suitable Resource Structure. Establishing a suitable resource structure means selecting an appropriate set from the many functional units for operational plant control. The following criteria must be taken into account:

- The set chosen must fully cover the plant properties relevant to operational plant control.
- The functional units in the chosen set must not overlap. No resource may contain devices belonging to another resource.
- The structure of the resources must have a modularity suitable to the requirements.

In what follows, possible resource structures are discussed for the example of plant functional units.

The criteria of completeness and freedom from overlaps are satisfied at the levels of individual functional units, group functional units, and plant section functional units.

Accordingly, the level most suitable for analysis can be selected.

Maximum flexibility is gained by selecting individual functional units as resources, though at the cost of involved performance specifications. The fact that there are group control functions, which can coordinate the individual functional units independently of the action being carried out, is not included in the treatment. The coordination of individual functional units must be explicitly described in each concrete case by means of instructions contained in the performance specifications for the action.

If group functional units are chosen as resources, virtually all the order-neutral conditions and program steps are contained in the internal logic of the resources. Simple and powerful performance specifications result. Group functional units are an appropriate choice. In the model under consideration, group functional units are identified as plant engineering resources. In this way, every order involved in operational process control references one group functional unit. Orders to individual functional groups are no longer possible. Plant sections are not known to the operational plant control system, and there are no plant section control units.

Figure 4.44 shows the control hierarchy that results from identifying group functional units as resources.

In some process control concepts in use at present, plant section functional units instead of group functional units are the resources for allocation, planning, and scheduling. This notion is suggested by the rule: "At any time, only one subprocess can take place in a plant section." In practice, however, this approach results in some difficulties and disadvantages. Such problems arise, for example, in managing the interconnection of plant sections. Other difficulties come about when actions in the areas of in-plant product logistics or plant support are considered; here it is a hindrance to assume that one, and only one, action can be performed in a plant section at any time. For example, given certain constraints it is quite possible that production, calibration, and maintenance actions can be carried out simultaneously in a plant section.

A component-oriented control structure leads to further difficulties as well. In multifunction facilities, plant sections often carry out many different subprocesses. The needed control specifications can no longer all be established as "modes" embodying all the capabilities of the plant. They must be loaded into the plant section control systems before the subprocesses are performed. This procedure of loading program sections into existing control units entails serious conceptual problems.

On the whole, there is no advantage in the functional coupling of group functional units to form plant section functional units that can be allocated. The reference model accordingly looks directly to group functional units as the allocatable plant engineering resources.

Regardless of the resource structure, the plant section naturally plays a major, if not the

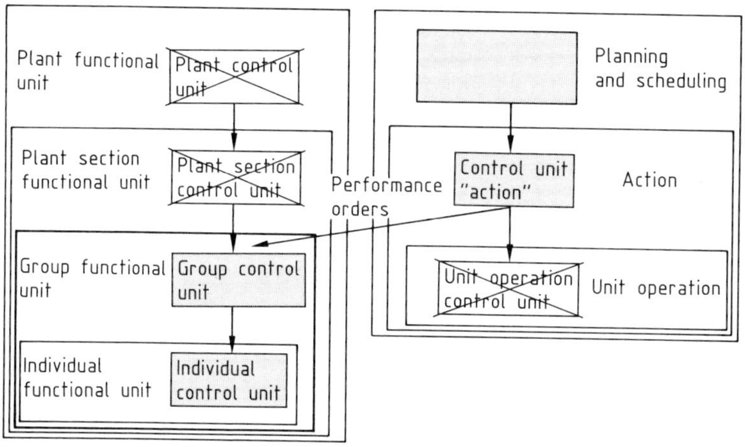

Figure 4.44. Control hierarchy

central, role as the smallest unit of the plant engineering world.

Consequences of the Choice of Resources. The selection of a functional unit as resource not only implies additional formal requirements on the functional unit itself, but also has a strong effect on the semantics of the performance specifications for actions and on the overall structure of the plant control system. These consequences must be considered when making the choice.

Additional Formal Requirements. Resources in this model are self-sufficient, largely self-managing, and self-protecting objects. All resources must exhibit certain standard properties. In this way, the issuance of an order to a resource is standardized and is formally independent of the type of resource. An order to a maintenance gang does not differ formally from the order "heat" given to a group functional unit.

The formal provisions include the following standard properties of a resource:

- Standard order interface

  Each resource must have an interface through which it can receive orders, setpoint values, control parameters, and so forth. The interface is standardized and is, formally, the same for all resources. It is dedicated to communications with the allocating customer (e.g., action X, action Y, manual intervention).

- Standard status interface

  Each resource must have an interface at which it makes certain (generally standardized) significant states available as information for other units. This interface provides information to all interest units and is held available by the resource as a read-only interface. It is formally standardized and is the same for all resources. The value ranges of the state variables "availability state," "type of allocation," and "operating state" are consistent for all types of resources.

- Allocation logic

  Each resource protects itself against multiple allocations. The allocation logic uses a simple principle: The first to allocate an available resource is the customer. A customer must release the resource in order to make it available to other potential customers. Prioritization and the identification of optimal allocation sequences are in the planning and scheduling domain; these operations are not supported by the resource itself. This simple structure appears to be especially appropriate for process engineering, where briefly interrupting one production process to make the resource used for it available for another task is generally not desirable.

- Self-management and self-organized activity

  Each resource is a functional unit that manages itself and operates in a manner internally independent of an allocating customer. Even unallocated resources carry out their control tasks and are continuously active and operational. Resources are responsible for the correct functioning of the equipment subordinate to them and the management of their resources. They are intelligent alarm and report generators.

- Self-sufficient protective logic

  Each resource has protective logic to prevent its creating a forbidden condition. This protective logic operates under its own rules, independently of the order system, and insures priority local protection.

Semantics of Performance Specifications. Once the resources have been chosen, the reference elements are identified for the formulation of actions. The selection of resources thus influences the way of thinking and communicating during operation. The importance of this fact is commonly underestimated when specialists analyze their individual areas.

The linguistic consequences can be illustrated for plant engineering resources.

Three examples of plant engineering resources are considered.

If the input/output channel is chosen as the resource, the performance specification will contain elements such as "set output BA43," "read input AE65," and so forth.

But then both the EMR/process control system engineer who writes an automation function and the chemist who formulates a recipe must ultimately do so on this linguistic level.

It is obvious that this "channel level" is unsuitable as a medium of description.

If individual functional units are selected as resources, the requirements can be formulated in engineering language. The performance specifications now contain elements such as "pump 34 on," "valve 56 closed," "setpoint for controller 36 = 50 °C," and the like. Accordingly, this level permits an engineering formulation and includes device-specific details. But the formulation depends on the specific instrumentation present in the plant.

Finally, if the group functional units are chosen as resources, the instrumentation-related details become internal to the resources. Group functional units represent the functionality of plant (equipment) units. Now the performance specifications will contain elements such as "heat to 40 °C" (vessel heating), "meter in 100 L at 1 L/s" (inlet line), "mix vessel contents" (stirrer), and "transfer up to a maximum of 300 L" (inlet line). In order to avoid terminological confusion, Figure 4.45 shows the situation schematically. The focus is on the "performance order," which corresponds to a concrete instruction to a group functional unit to perform a certain procedure. Every group functional unit has the capability of doing certain procedures. On the other side are the possible performance orders, that is, the linguistic elements from which the performance specifications are assembled.

Group functional units are a convenient, meaningful, and workable level. Choosing this level permits both the formulation of performance specifications for automatic actions and the manual control of actions by issuing single performance orders at a technologically understandable, not excessively complicated level.

Actions. Each action serves to accomplish a self-contained partial task and, within the plant complex, is a separate, identifiable object. How an action is to be carried out is set forth in its performance specification. The system responsible for the performance of an action is identified in planning and scheduling. This performing system must know the performance specification or must be given it. For example, the processing of a production action can be transferred to the plant operator or the process control system. If the action is assigned to the process control system, this system independently performs the action in accordance with the performance specification stored in the system. If the action is assigned to the plant operator, the operator handles the action by manually issuing performance orders in accordance with the performance specification contained in the operating manual. Processing includes issuing performance orders, controlling the allocated resource, collecting archival data, generating the proper reports, giving the ready signal, and monitoring correct operation.

Each action is generated, planned, and scheduled by the plant planning entity and is transferred to a performance system where it is actively processed. It is then deleted from the performance system and the planning and scheduling system, remaining only in the archive system. Figure 4.46 shows the states of an action and the possible transitions.

Lot planning, tank storage planning, maintenance planning, and other activities generate requirements for the performance of certain actions in certain spans of time. In order to schedule the performance of these actions, all actions planned up to a certain time "horizon" are reported to the in-plant planning and scheduling entity. An action thus becomes an object in the planning and scheduling system, which maintains a list of all actions generated. It makes available an operator interface, with

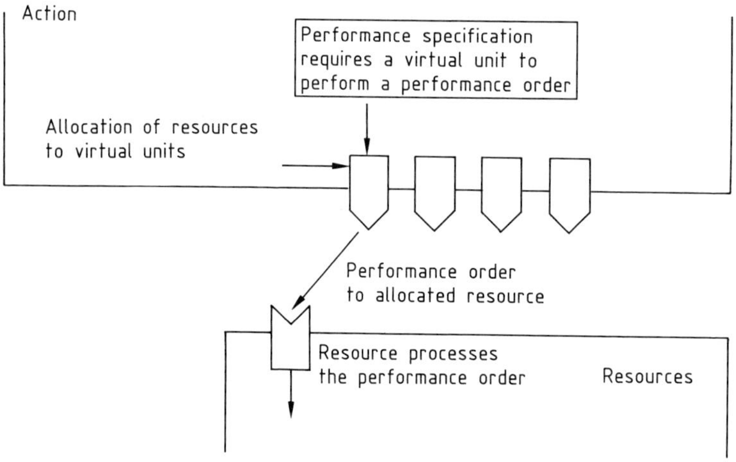

Figure 4.45. From performance specification to implementation

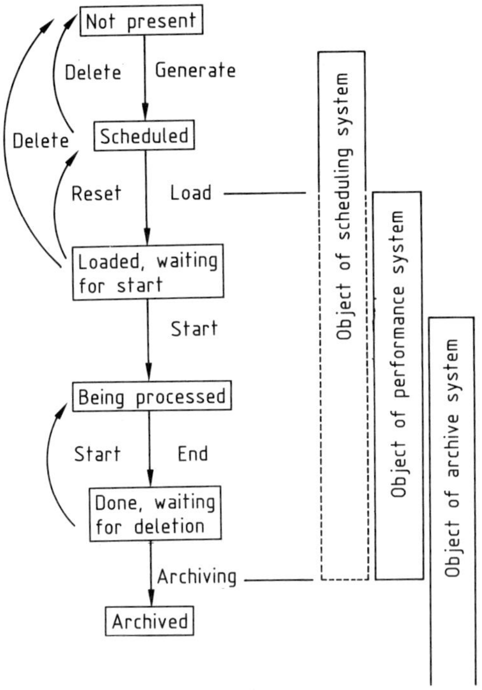

Figure 4.46. Operating states of an action

whose aid further actions can be generated and existing actions can be rescheduled, modified, and deleted if necessary. The planning and scheduling system supports planning by means of prospective calculations, then uses the list of actions to be performed together with the current status to arrive at resource allocations, determine the contents of product storage units, and predict the future states of the actions.

For this purpose, there is a simulation specification for each type of action. With the help of this specification, which treats each action as a planning and scheduling unit, it is possible to forecase when which plant resource will be needed, and to say when and where substances are to be received and discharged (and in what quantities).

On the basis of such a forecast, the action plan can easily be checked for critical resource allocations or critical states of product storage units.

During production, this forecast must take account of the actions now running and those loaded. In other words, the planning and scheduling system must always know the present status of all actions being processed.

Scheduled actions are initially pure scheduling objects in the planning and scheduling system. At this point, they have no access to resources. To be performed, actions must first be loaded. When an action is loaded, a performance system is ordered to perform it; this system receives all the required functions. If there is no instruction to the contrary, the performance system then automatically carries out the action.

An important property of the loading operation is that pre-loading is possible, meaning that an action can be loaded into the performance system long before its scheduled performance. In this way, the requirements on the availability of the planning and scheduling system are eased. An adequate reserve of actions can always be kept loaded in the performance systems. For example, all production actions to be performed automatically in the next 24 h can be loaded into the process control system; the maintenance group is given a list of all maintenance actions to be carried out for the week; the plant operator receives a list of all production actions to be performed manually during the weekend.

A performance system of special interest for process control engineering is the process control system. The loading and automatic execution of actions might be thought of in this way: When an action is loaded, a control unit is created in the process control system and initialized.

The control unit is a self-contained object with the ability to verify conditions and to perform its task automatically. It is no longer dependent on the intervention of higher levels but still can, if necessary, be influenced by higher levels. The loading of an action can be initiated either manually by the plant operator or automatically by the planning and scheduling system.

Once loaded, the action considered as a control unit goes into an initialized condition: "loaded, waiting for start." It has no effect on process control and no access to resources, and it can be deleted or reset at any time without having exerted any influence on the system. The processing of the action begins when the control unit is started; during this operation, the control unit is given access to the needed resources. It gains such access through allocation of group functional units and the construction of the control links required for this purpose. Two concepts should be distinguished here. In the simpler one, all links that the action will need at any time during processing are made at the outset. In a somewhat more complicated concept, only those links called for during the initial phase of processing are present at any time; links that are no longer needed are dynamically broken while the action is still being processed.

Once an action is complete, all links created for it are abolished.

While an action is being processed, the action control unit acts like a normal, permanent control unit (Fig. 4.47).

Once an action has left its initialized state and has been started, it cannot return to the initialized state. The action has been launched, cannot be called back, and must be processed all the way to a correct or incorrect termination. Every action, once started, is stored in an archival system; it cannot be cancelled without entry in the archive. All stored information about the progress of the action and all stored information about group and individual functional units working under its control can be accessed by virtue of this entry in the archival system.

The form of storage depends on the implementation.

It is evident that the action model described here is especially suitable for batch processes, where a one-to-one mapping of actions to production actions is possible. The steps in an action (see the state model in Fig. 4.46) largely correspond to the steps in a segment of a batch production process.

This is not the case in continuous processes, which are characterized by the continuous operation of process steps rather than by steps with definite terminations. The stress is on managing and controlling the modes and transitions between modes as shown in Figure 4.48.

Nonetheless, continuous processes can be described in terms of the action model. One simple

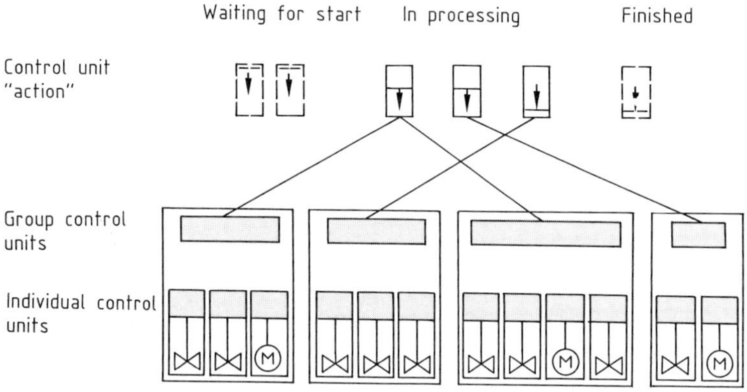

Figure 4.47. The action as control unit

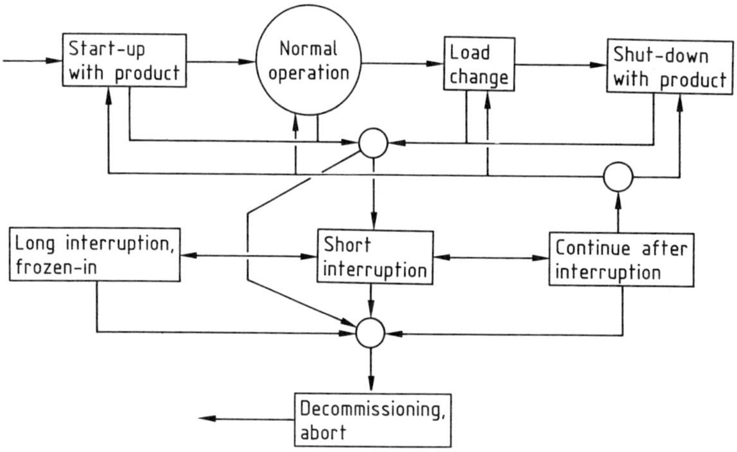

Figure 4.48. Operating modes of a segment of a continuous process [4.67]

option is to "discretize" the continuous process into small increments delimited in time and having definite production objectives.

Automatic Performance of Actions. The automatic performance of actions means implementing the functions required to complete the action by using automatically functioning software or hardware modules. This can be achieved, for example, by implementation in a process monitoring and control system, a maintenance system, or another system for carrying out actions.

The reference model has a structure appropriate for this purpose: Resources and the actions themselves are built up in object-oriented fashion, in two senses: first, that each resource or action contains its own control unit and thus acts as a self-contained unit in the system; second, that the resource or action has properties that can be derived from a type concept. Control units occupy places in a control hierarchy. Each control unit is connected to the next higher-level control unit and the next lower-level control units through well-defined control interfaces. In addition, each control unit has interfaces through which it can exchange information with all other units.

In process control systems, control units can be implemented as modules. This is a common and a standardized practice in process control engineering; for example, the IEC standardized languages (WG65A) are based on an object-oriented modular approach.

Example of a Modular Solution. The following section shows how the action and resource model can be implemented in a modular form in a process control system. The modular model has the following properties:

- Support of the type concept
  The type concept is a special manifestation of a one-level abstraction hierarchy. Each module is of a certain type. The type defines all methods, the description of connectors (input/output interface module), and the description of internal variables. Further, the type includes certain type-specific variable values. The entity, as a management unit, contains not only its name but also its specific data set and the reference to its type. Accordingly, the entity contains no methods or variable descriptions, only parameters.
  In contrast to the individual device approach, in which each compact controller contains the type information (e.g., control algorithm), in process control systems the type information can be contained only once, in one component. Entities then contain only the module state information and module-specific parameters. Under this concept, the type information must already be present in the target system before the creation of an entity. Loading of type information is a separate step from instantiation of the individual module.

- Data exchange through connectors and links
  According to this concept [4.76], data exchange between modules must obey the following principle: Every module offers certain connectors to serve as interfaces to the external world. Two matching connectors of two modules are connected to each other via a link. Once the link is in existence, data can be exchanged. The construction of the link itself is a separate process from data exchange. Links are independent management objects in the system and can be created and again destroyed while production is under way.

- Closure of modules
  Each module as a whole is a unit that can be manipulated. Modules are created by simple instantiation commands. The prerequisite is that the proper type must be present. Once modules have been created, they can be incorporated in the processing by an activation command. They bring along all the prerequisites for their function. Modules can be created at any time, and also deleted again, while production is under way. Once modules have been deleted, no fragments remain in the system.

A particularly simple implementation of the action concept is obtained if all functions required for the performance of the action are contained in the performance specifications (and thus in the type description). The individual action contains not specific functions but only parameters for the control of the performance specification. In this way, there is a strict division between the creation and loading of performance specifications, on the one hand, and the creation and processing of individual actions, on the other (Fig. 4.49). This strict separation is not a practical drawback, since in general the creation of performance specifications takes place separately from the scheduling and operational performance of the actions. The various tasks are often accomplished on different levels: Performance specifications are created and managed in a "recipe" system, while action management and initiation take place in a scheduling system.

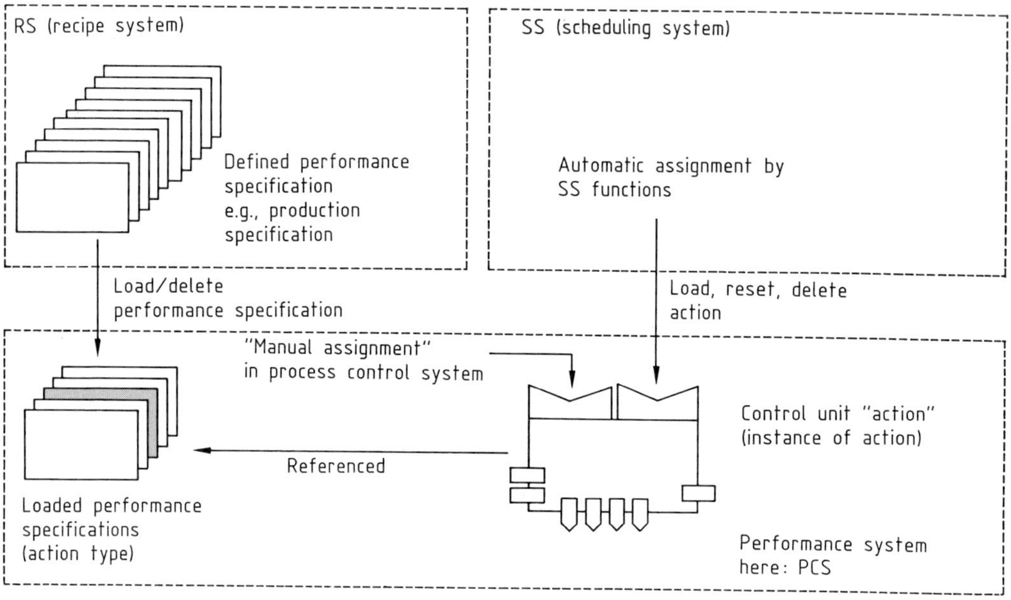

Figure 4.49. Separate loading of performance specification and action (performance system here = process control system)

An important performance feature of a process control system is how convenient it is to create and load a new performance specification, modify an existing performance specification, or generate the associated simulation specifications, original data sets, and parameter masks for actions being created.

Another important point, however, is that it is possible to attribute all required manipulations in the performance system to operations formulated in the modular concept. Thus no loading, writing, or coupling steps on the modular system are necessary. In this way, the action concept can be logically implemented in a modular system.

Automation. A very important property of the model is that automatically performed actions and actions manually performed by the plant operator can be handled concurrently in a process control system. It is thus possible to build up to a higher level of automation in a step-by-step fashion.

When actions can be automatically completed, a system acquires the degree of automation that Polke has called "logistical mode" (see Section 11.1) [4.73].

Summary and Outlook. The action and resource model attempts to describe a general structure of operational plant control. Important features of the model are the resource structure, the standardization of the order interface, and the introduction of "actions" as self-contained control and management objects. This contribution has taken the example of plant engineering resources in order to show how such a structure can be created and implemented in a suitable way.

The concept suggested is, however, a technology-free general mechanism. It is strictly limited to describing the operational performance of actions; thus it contains no specifications on how the planning and scheduling task is to be accomplished or on how recipes are to be generated, and it provides no rules as to how resources are to be selected and organized.

Special models are required for these purposes. By way of example, a resource model for plant engineering resources is described and its consequences explored.

A plant complex must carry out its tasks while fulfilling a variety of requirements and satisfying a number of constraints. "Plant-wide" decisions are needed on a daily basis. The action and resource model enables the plant to formulate its operational decisions in a consistent way and implement them in a systematic manner. The results are not just a simplification of scheduling, but also an improvement in the quality of planning and performance of in-plant actions. The documentation of plant activities is part of the concept and can be synthesized in a systematic way.

The action and resource model is logically object-oriented. It can be extended directly to a corresponding modular system.

With regard to usability, this model has the great advantage that technical devices and personnel can both be used in constructing resources and performance systems. Plants can thus be controlled in a consistent way, regardless of their instantaneous degree of automation.

A special task for the future is to integrate existing process control concepts for production control, plant management, recipe-based process control, and production logistics into the general methodology of the action and resource model, thus arriving at a comprehensive solution for the operational control of process plants.

5. The Process Control System and Its Elements: Process Sensor Systems

5.1. Principles

Sensor, Sensor Technology, and Sensor System: Classification and Standards. Modern process control engineering, with a distributed structure of information processing based on microelectronic devices, has greatly expanded the potential of the information economy for the control, monitoring, optimization, and reliable conduct of processes [5.1].

If this increasing potential of information processing is to be utilized fully, the amount of information acquired directly or indirectly as to the status of the process must also be augmented.

A relative bottleneck exists here, because the development of sensor technology lags behind that of information processing [5.1].

There are a number of reasons for this. Technical functional units in the immediate vicinity of the process are always much more expensive and complicated than those in a protected environment such as the control room. Furthermore, the continued use of conventional, internationally standardized signals together with proven transducer technology stands is hindering further development. Finally, information processing is often regarded as being much more important than information acquisition.

Decentralized process automation is beginning to extend distributed "intelligence" based on the use of microelectronics to the point where measurements are acquired. Therefore, the meanings of "sensor element," "sensor," "sensor technology," "sensor system," and "direct signal processing" have become blurred today.

For example, consider the "critical" link in the action chain associated with measurement. In the current language of process and production engineering, the functional element that embodies the interaction between the material system and the instrumentation is referred to as a "sensor" [5.2].

Here, "sensor" no longer means the same thing as it commonly does in the electronics industry, where it is restricted to the proportional transformation of nonelectrical input quantities to electrical output signals [5.3], [5.4]. The applicable standard (DIN 19226) is therefore in need of revision. A useful starting point for a meaningful definition of the term "sensor" is offered by Sheet 3, "Apparatus Terms," of VDI/VDE Guideline 2600, "Metrology" [5.5]. This document employs "pickup" or "primary element" for the (entire) device that accepts the quantity for measurement at its input and emits a corresponding measurement signal at its output. The "sensor" proper, on the other hand, is the part of the pickup that directly acquires the quantity for measurement and reacts in some sensitive fashion to the quantity. Even this definition, however, is appropriate only for a relatively small fraction of the sensors currently utilized for measuring material quantities. The definition does apply, for example, to semiconductors and piezoelectric sensors; it is difficult to apply to IR photometers and paramagnetic oxygen sensors, because here the interaction between the material system and the instrument does not take place in the sensor, as VDI/VDE 2600 presumes, but via an energy field serving as an auxiliary quantity [5.6].

The present terminology and standards, such as the definition in VDI/VDE 2600, are thus insufficient and require further discussion. An unambiguous resolution of terminology in this area is essential in the long term.

Suggestions for refinement of the terminology have been made in [5.2], [5.4], [5.6], [5.7].

Here, it is proposed to separate information acquisition in and on the process from information acquisition in the laboratory and, as applicable, in the pilot plant. Both classical "process measurement" [5.8], [5.9] and "process analytical techniques" [5.10] must fulfill fundamentally different requirements from the comparable instruments in the laboratory. The shift from information acquisition in the plant laboratory and "process analysis rooms" to acquisition in the production process itself justifies the information-oriented term "process sensor system technology" as opposed to "laboratory measurement technology."

Process Models and Sensor Systems. The concept of "process model" has been known theoretically for more than 20 years; it encompasses "black box," mathematical, physical, and other types of models. Especially important in connection with measurement technology are adaptive process models, also known as observer models.

Hundreds of books and tens of thousands of other publications have appeared on "process models" (see also Chapter 3), "process identification," and "process simulation." There are academic departments concerned with these topics, and many congresses and symposia on them are held every year. Although the theory is advanced, there is a wide gap between theory and practice [5.1], [5.11].

According to PERNE and POLKE [5.12], the process model is ultimately a mapping between product properties and process properties.

Problems of product variation or product development could thus be attributed to a process variation if a suitable process model were available. Identifying such a variation would provide a starting point for "optimization" under a wide range of criteria (see Section 11.3).

This problem can be handled systematically and—most importantly—optimized only if a representative mapping or transformation (i.e., the process model) can be found between the process space and the product space (see Fig. 11.49).

But why use a model instead of the real production process? Simply because the real process is generally much too complex and opaque to permit direct operations based on physical and chemical relationships and algorithmic descriptions of these (see Chapter 3).

However, this chapter is less concerned with the representability of a model in relation to the real process and the desired product qualitites, and more interested in the reaction on sensor elements and sensor system technology.

It is important, above all, that questions of sensor technology not be separated from the topics of process models, process properties, and product properties.

Material Flow, Energy Flow, Information Flow, and Sensor System. As Figure 5.1 shows, there is a prallelism and correlation between material–energy flow and information flow. The link between an event or state in the process and the information system is effected by the sensor system.

Accordingly, the sensor system has to carry out two basic tasks. First, the input quantity resulting from an event or state must be sensed and transformed into a signal quantity suitable for processing. Second, the signal quantity must be set in relation to its cause so that interpretation can extract usable information (see also Section 5.2).

The sensor itself must perform the first task, while the second is in the realm of information processing. Some thought should also be given to the possibility of augmenting the information by serial, parallel, or associative linking of single items, whereby the result may be of far greater relevance than the sum of the individual pieces of information. Typical examples are the temperature profile, the stress tensor, the density field, and the concentration gradient.

This multi-sensor approach is widely used in the field of biological sensors. A typical example for the capabilities of an array of identical sen-

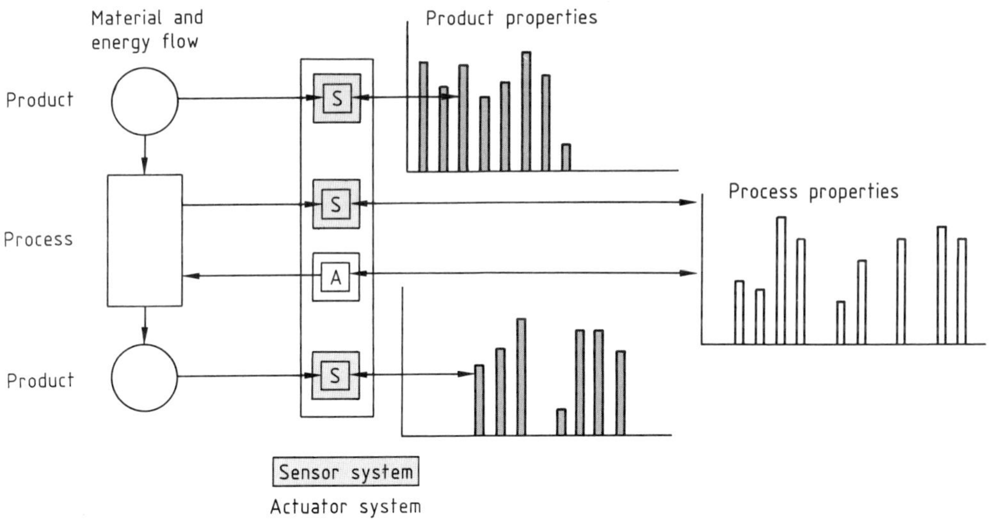

Figure 5.1. Position of sensor systems relative to material, energy, and information streams

sors is image recognition by the eye. A further advantage of arrays becomes apparent in biological sensors but has been little utilized in industrial sensors: While the individual biological sensor is markedly inferior in performance to many industrial sensors, a system with intelligent preprocessing of signals achieves an amazing degree of flexibility in dealing with unforeseen events. Naturally, this flexibility is more vital to biological systems than to industrial ones, where the service loads can be established more precisely. Nonetheless, the gain in flexibility through multi-sensor technology is of interest for industrial systems as well [5.13].

What holds for any economic system also applies to the information economy [5.14]: One can use only what one has. The first step, accordingly, must be information acquisition. It is reasonable to limit the information acquired to that which is important and actually necessary. There are two reasons for such a restriction: to avoid an information flood and to hold the cost of instrumentation within acceptable bounds. Both experience and mathematical and physical methods can be used as guides for limiting the information flow.

Because it is often difficult to identify the relevant properties, that is, the properties that adequately describe the process or the product, more easily measurable properties (indicators) are often resorted to. Process optimization requires better penetration into the process, and this in turn means finding ways to make relevant properties—i.e., those that can serve as significant process indicators—more measurable.

HESSE, in his editorial in the "Sensors 1992" special issue [5.15], used the following terms to describe the shift in emphasis in the development of sensor systems:

"At the beginning of the 1980s, the problems were to systemize technical terminology, to discover effects and technologies that could be exploited in practice, and to find out how medium-sized companies could establish themselves in this field. Later, in the mid-1980s, publications on practical sensors utilizing a wide range of technical principles became dominant. These were still in the foreground in the early 1990s, but systems aspects are now being examined along with applications and on an equal footing.

In fact, systems concepts should have been considered right from the start. These aspects aid in the development of strategic insights and objects, since they describe the overall high-level solution. The introduction of process control engineering in the chemical industry can serve as a model: Sensors are dictated by the system itself and can thus be developed under explicit constraints."

For these reasons, some concepts having to do with the use of sensor (system) technology in process control engineering are discussed first.

From Phase Model of Production to Process Information Model (see Chapter 2). A key aid in describing and analyzing processes is the phase model of production. This object-oriented semantic tool originated in software engineering. The outcome, a directed bipartite graph, belongs first to the category of product nets (see Chapter 2).

The phase model of production consists of two object classes, "product" and "process element." Individual products and process elements are described in terms of attributes. The attributes are the object classes of information about a process; they are defined by category and value (Fig. 5.2 [5.16]).

The value is made up of a numerical value and a unit. Whenever metric scaling is possible (see Chapter 3), the units employed should comply with U.I.P. 20 (1978) [5.17] and ISO 31/0 ("General Principles Concerning Qualities, Units and Symbols") and DIN 1313 [5.18].

Higher-level requirements such as "product quality," "process safety," and so forth are imposed on the production process.

Objective: Product Quality. Product quality refers to a subset of the actual properties of a product, which in terms of category and value form an intersection with the nominal properties demanded by the customer (Fig. 5.3). The development objective is to maximize this intersection within reasonable tolerances.

Ojective: Process Safety and Environmental Safety. Safety tasks such as process safety and environmental safety can be accomplished with the resources of process engineering and process control engineering. This can be achieved by means of technical or organizational actions [5.20] or a combination thereof (see DIN V 19250 [5.21]). This procedure makes it possible to find alternative solutions that are equivalent from the safety standpoint, so that the most economical solution can be applied in implementation [5.22]. In the new NAMUR Recommendation, the required process control safeguards are matched to the risk arising from a process

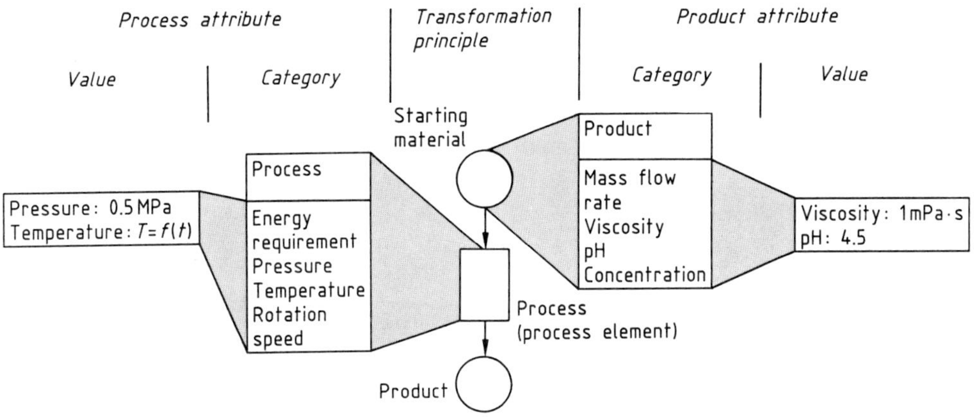

Figure 5.2. Assignment of attributes in the phase model of production [5.16]

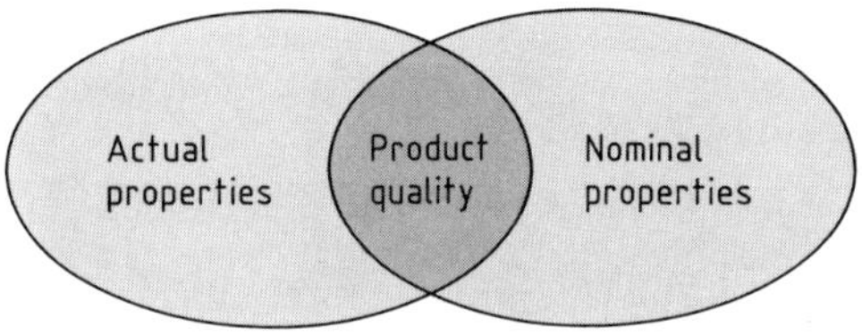

Figure 5.3. Product quality [5.19]

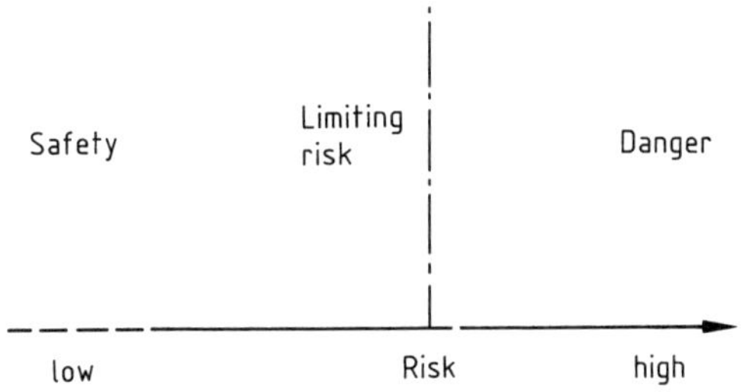

Figure 5.4. Safety, risk, and danger [5.24]

facility or the absence of such process control safeguards. This risk assessment is necessarily qualitative and must be carried out case by case. Other procedures that lead to a comparable level of safety should, however, not be ruled out. The protective objective in the context of this recommendation is to prevent injury to persons as well as major environmental harm and damage to property. In the case of major damage to property, damage assessed as serious danger (e.g., according to [5.23]) should be distinguished from damage evaluated as affecting the company's interests.

Some important concepts should be defined here (see Fig. 5.4), without an in-depth examination of the extensive safety literature [5.22], [5.24]–[5.26] or a workshop on Safety in Chemical Production:

- Danger: A danger is a situation in which the risk exceeds the limiting risk.
- Limiting risk: The limiting risk is the highest still-acceptable risk of a certain technical process or state. In general, the limiting risk cannot be quantified; it is described indirectly or in terms of safety considerations.
- Risk: The risk associated with a particular technical process or state is described in terms of probabilities. Two factors are included: the expected frequency of occurrence of an event leading to damage and the extent of damage expected when the event does occur.
- Safeguards are actions taken to reduce risk by lowering the frequency of occurrence, the extent of damage, or both.
- Safety is a situation in which the risk is lower than the limiting risk.
- Availability: The availability of a technical device is the probability that the proper functionality is present.

It follows that, with regard to the safety-related availability of process control devices, solutions for holding the risk less than the limiting risk should be evaluated from the economic standpoint only if they are equivalent from the safety standpoint.

In contrast, it is useful and necessary to analyze other availability characteristics to determine the economic consequences of malfunctions. This statement applies to the logistical availability of equipment that would impair the ability to deliver (just-in-time, just-in-case) and also to the process availability of production equipment that would impact directly on product quality or plant condition (e.g., production stoppages, market losses, repair costs).

The unambiguous description of proper functioning, that is, the complete set of all nominal attributes needed for such a description, leads to the information model of quality (Fig. 2.35, p. 353) or the information model of safety (Fig. 2.36, p. 354). The histograms illustrated show, for each product object or process object, all attributes necessary for its description, along with the nominal values and their tolerances ("good," "defective, acceptable," and "defective, unacceptable" ranges), which may be scaled either metrically or topologically (see Chapter 3).

This information must be statically and dynamically defined at the maximum level of detail [5.16]. What is more, the spatial and temporal representativeness of the attribute values must be taken into account, especially for ex-line sensor systems (Figs. 5.20 and 5.21, p. 444).

Product and Process Properties. The attributes used to describe the objects of the production process are process properties and product properties.

The term process properties includes the following:

- State variables, such as pressure, temperature, and concentration, with which the process can be described either directly or via derived state functions.
- Process parameters, such as heat-transfer coefficient and catalyst activity, which characterize the constraints under which the process takes place. Process parameters are steady or at least quasi-steady quantities.
- Controlled variables, such as motor speed and setpoints, which characterize actions on the process.

In theory, the process can be described completely if a suitable set of these quantities is available. In practice, however, it often happens that not all the desired state variables and controlled variables can be measured or specified. Substitute information often must be resorted to:

Setpoints, such as stirrer speed, are correlated with the controlled variables or represent residual categories.

Process indicators [5.27] are empirically correlated with one or more state variables (and may also be product properties; see Section 11.3). These represent substitute information.

Controlled variables and setpoints characterize the part of the process state that is imposed from outside. It is the task of the actuator system to convert the specification, provided by the information flow, to an action on the process. The remaining process properties characterize information that is to be acquired from the process with the aid of sensor technology.

Product properties are the following:

- Physical quantities
- Chemical quantities
- Technological properties [5.27]
- Product indicators (see above), which are substitute information

Technological properties are determined by special inspection procedures when relevant product properties depend on physical quantities that can be determined only with great difficulty, or if the relationship with the physical quantities is currently unknown. Product indicators are empirically correlated with physical and chemical quantities, but process properties can also be employed as product indicators. All product properties characterize information that can be acquired by using sensors.

"Technological" and "Physical" Test Procedures. Some fundamental problems in establishing material properties are illustrated for the case of rubbers [5.27].

Technological Test Methods. The testing and characerization of chemical substances is generally limited to assessment based on technological tests or application-oriented model tests, which are single-point measurements which have limited predictive power and do not take account of the functional relationships actually present. Technological tests do not directly lead to conclusions about the physical and chemical properties of chemical substances. This means that such data alone are insufficient as a basis for the optimization of chemical production processes.

The essence of technological testing can be illustrated by a simple diagram (Fig. 5.5). A certain material M_j under a certain load X_i exhibits a technological behavior Z_{ij}.

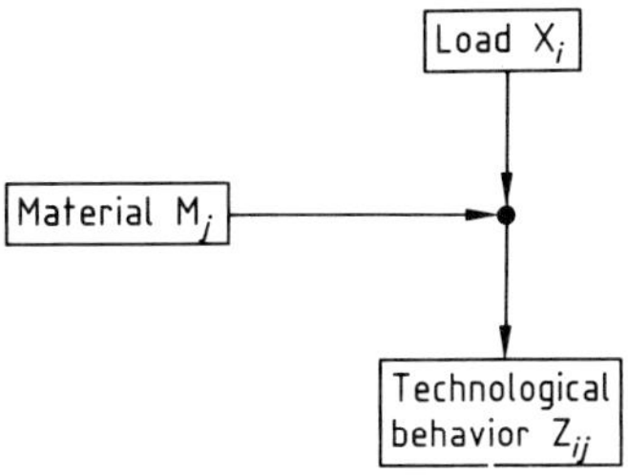

Figure 5.5. Technological inspection [5.27]

When the test load (generally complex) is specified in terms of the arbitrary design of a testing machine, the actual loading often cannot be expressed physically. For example, the Mooney tester [5.28]–[5.30] has an inconsistent shear rate and inhomogeneous stress and temperature distributions in the steady state. Furthermore, the variation of temperature with time during the measurement varies in an uncontrolled way as a function of the specimen used. The single-point measurement performed, namely the torque after 4 min, describes only an integral response of the material relative to the test-chamber geometry as the specimen is subjected to a load that is neither homogeneous nor steady.

The predictive power of this measurement is thus limited to comparison of materials under these conditions. In general, no certain conclusion can be drawn as to the processing qualities of the rubber tested in this way, since the shear loads that occur in industrial processing machinery are several orders of magnitude higher.

Physical Test Methods. A material can be better characterized in terms of its physical properties. These quantities are obtained by applying simple loads and measuring the relationships between load and response. Such simple loads include linear, uniaxial stress, laminar flow at constant shear rate, and a steady homogeneous temperature distribution in the specimen. Complicated loading modes such as occur in practice can generally be synthesized by superimposing simple loading states.

The material properties M_{ij} are material-dependent quantities described a functional relation:

$$Z_{ij} = f(M_{ij}, X_{ij}) \quad (5.1.1)$$

For example, Hooke's law

$$\tau = G\gamma \quad (5.1.2)$$

describes the behavior of many elastic solids. Here, the stress τ is the load ($X_j = \tau$), while the deformation or strain γ represents the material behavior or response ($Z_{ij} = \gamma$). A material constant, the shear modulus $G = \tau/\gamma$, interrelates these two quantities and characterizes the mechanical properties of the material ($M_{ij} = G$).

Figure 5.5 shows that the technological behavior is governed by the material and the loading. The task of a phenomenological theory is to define the formalism that describes the observed relations (Eq. 5.1.1) and the properties M_{ij} of the material (Fig. 5.6).

Idealized models of substances and bodies play a major role here, for example the ideally elastic (Hookean) solid, the Newtonian fluid, and the Maxwell and Voigt bodies (Fig. 5.7). Deviations from the behavior of the idealized body are handled by restricting the range of validity (Hooke range) or by introducing correcting relations. For example, the viscosity of a general non-Newtonian fluid is described by

$$\tau = \eta \mathrm{d}\gamma/\mathrm{d}t \quad (5.1.3)$$

where τ is the shear stress, $\mathrm{d}\gamma/\mathrm{d}t$ is the shear rate, and the viscosity is treated as a function of the shear rate $\mathrm{d}\gamma/\mathrm{d}t$,

$$\eta = \eta(\mathrm{d}\gamma/\mathrm{d}t) \quad (5.1.4)$$

The phenomenological theory describes the physical properties of a material by interrelating the load state and the behavior of the material, whereas polymer physics seeks to devise a theory that correlates the physical properties of a material M_{ij} with the chemical composition and structure of the material or a structure model (e.g.,

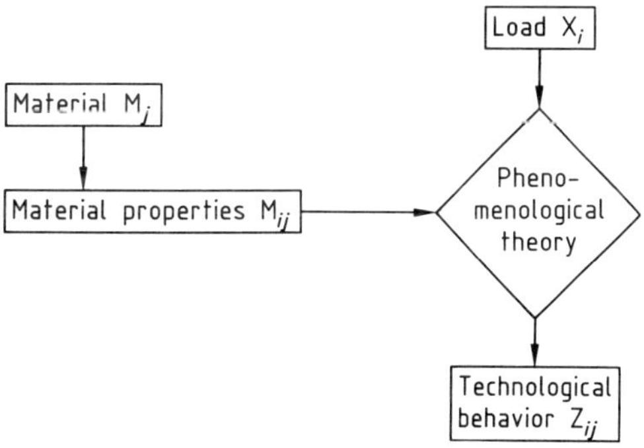

Figure 5.6. Physical inspection [5.27]

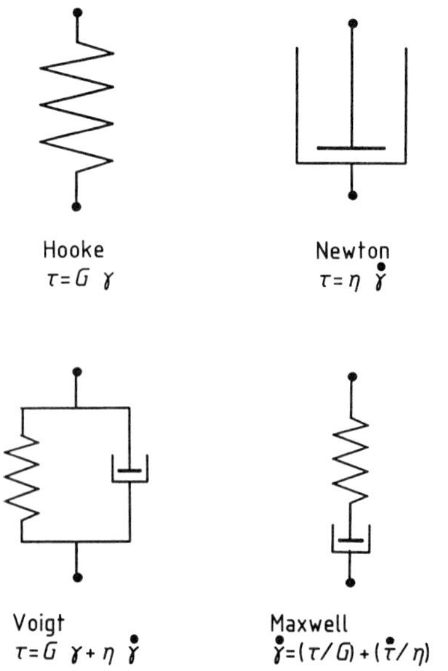

Figure 5.7. Rheological model [5.27]

Kuhn coil model [5.31], Pechold kink theory [5.32]); see Figure 5.8. If this were achieved, there would be a basis for the planned development of particular polymers.

Among the most important properties of elastomers is the viscoelastic behavior. There is still no self-contained phenomenological theory to describe the entire mechanical deformation behavior, including fracture. For small deformations, relaxation spectroscopy [5.32] can be used for complete determination of the viscoelastic behavior of polymeric materials. Under harmonic loads, the complex shear modulus $G^*(\omega, T)$ is measured as a function of frequency ω and temperature T. The results can be used to calculate the relaxation time spectrum $H(\tau_R, T)$ (τ_R = relaxation time), which is closely related to the structure. Using the Boltzmann superposition principle allows the material behavior under complicated, nonharmonic loads to be calculated.

These methods fail at large deformations, where the relation between stress and strain becomes nonlinear. The Boltzmann principle, which states that the effect of a composite load can be treated as the sum of the effects of its composite simple loads, does not apply in such cases. Fundamental findings and results on the material properties of multiphase chemical substances, including these application-relevant ranges, can be found in [5.33]–[5.35].

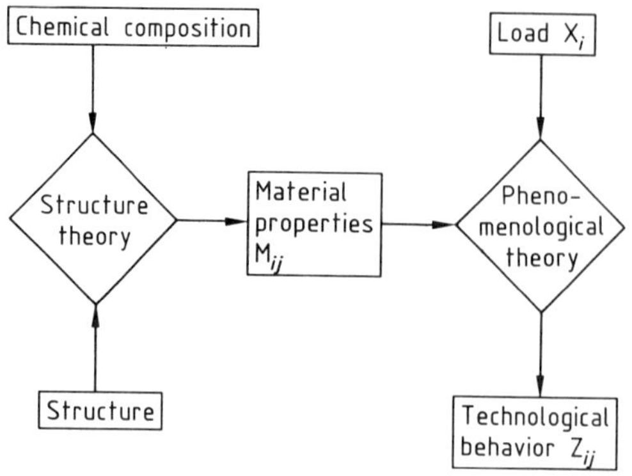

Figure 5.8. Physical inspection and structural theory [5.27]

The question arises to what extent the mechanical behavior of elastomers can be completely characterized despite the lack of a unified nonlinear theory of viscoelasticity. NINOMIYA and FURUTA [5.36] proposed that in general four functions are sufficient to characterize the deformation behavior:

- The relaxation time spectrum $H(\tau_R)$, which characterizes linear viscoelastic behavior at small strains
- The stress–strain curve $\tau(\gamma)$, which characterizes the nonlinearity
- The fracture strain $\varepsilon_B(T)$, which characterizes fracture behavior
- The shift factor $a_T(T)$, which characterizes the temperature dependence [5.37]

Such a description of a polymer, however, requires substantial efforts. To determine the relaxation time spectrum and shift factor, the complex modulus must be measured over a wide range of temperatures and frequencies. Moreover, extensive calculations must be carried out.

The difference between technological and physical test methods can be summed up as follows: Technological tests investigate the behavior of the material in an application-oriented model test, generally involving a single-point measurement under complex loading conditions. In contrast, physical tests are concerned with the fundamental behavior of the material. Simple and easily quantifiable loading modes are selected and then varied such that the fields of properties are determined in multidimensional product spaces (see Fig. 11.45).

Despite the obvious advantages of the physical analysis, technological tests will continue to be important because of their lower cost.

A crucial point in selecting a method is the purpose of testing. Three aims must be considered:

1) Product characterization in general, in the sense of quality assurance (e.g., DIN ISO 9001 [5.38])
2) Acceptance testing and production control
3) Prediction of the practical behavior of a material

The objective of product characterization is to achieve at the most complete possible description of product properties in terms of instrument-neutral physical quantities, to determine the key parameters that affect the product properties, and to quantify functional relationships.

Accordingly, product characterization encompasses the area of physical testing. The demands on instrumentation and the amount of work required for interpretation are so great that these methods are generally only justified for new or as yet unknown products. This is the only way to carry out a reliable, exact comparison between materials.

As a technique for routine inspection, for example in acceptance testing and production control, sampling inspection of material properties is generally adequate. Sampling techniques (representativeness over space and time) are applied to both the quantity of material and the selection of loading mode combinations. The principles of statistical quality control should be observed [5.39].

The creation of correct, optimal experimental designs, however, presupposes a comprehensive knowledge of material properties.

The requirement on acceptance testing procedures is that they can be performed by the vendor at one location and the buyer at another, using different instruments. The parties must agree on the test conditions (establishing the controlling parameters); this is one of the key tasks of international standardization (e.g., ISO). In contrast to product characterization, single-point measurements may be relevant here as well.

In production control, various batches of a product are subjected to spot checks to verify that they have identical physical properties. The important prerequisite is good reproducibility of measurements on a given instrument. It may be desirable to choose complex test conditions in the interest of quick, effective monitoring [5.40].

There are two possible approaches to forecasting the mechanical and technological behavior of a material in practice. First, tests can be performed under the process conditions and loading modes encountered in practice; this step should be preceded by a thorough analysis of the process (see Section 11.3 with Figs. 11.36 and 11.37). Second, in principle it should be possible, given suitable results from the determination of the physical material properties, to calculate in advance how the material will behave under the required loading conditions. The second method, however, often runs into the difficulty already mentioned: the lack of applicable theories. For this reason, the first approach (simulation of practical conditions) is becoming the more important.

By analogy with this example of mechanical product properties, the electromagnetic product properties defined by Maxwell's equations

$$\boldsymbol{D} = \varepsilon \boldsymbol{E} \qquad (5.1.5)$$

$$\boldsymbol{L} = \mu \boldsymbol{H} \qquad (5.1.6)$$

$$\boldsymbol{I} = \sigma \boldsymbol{E} \qquad (5.1.7)$$

can be examined. The first product properties determined here (generic class "application features") are due to molecular motions and are measured as collective phenomena. For example, suitable spectroscopic or chromatographic methods make it possible to resolve the required product reactions over a variable frequency and temperature field:

$$\varepsilon = \varepsilon(T, \omega, \ldots) \qquad (5.1.8)$$

Again, the limit of the linear relations is found by considering the variation of the field quantity:

$$\varepsilon = \varepsilon(T, \omega, E) \qquad (5.1.9)$$

The product properties are in general complex:

$$\varepsilon^* = \varepsilon_r + i\varepsilon_i \qquad (5.1.10)$$

Model-Aided Measurement Procedures for Determining Product and Process Properties. From the connection between process model and information demand, it is possible to identify ways and means of overcoming the limitations of sensor technology by utilizing the present capabilities of information processing (see also [5.1]).

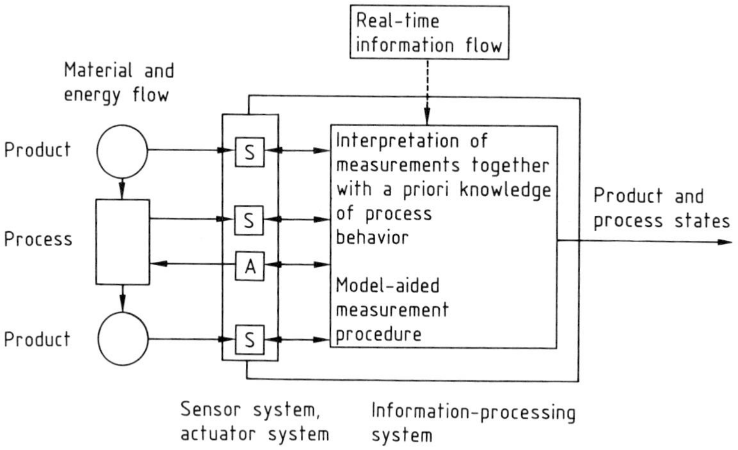

Figure 5.9. Model-aided measurement procedure

The broad potential of model-aided methods, which has been under-exploited in process engineering, should be mentioned first (see Fig. 5.9).

In the context of intelligent sensor systems, this procedure makes it possible to reconstruct unmeasurable process and product properties from measurable quantities by interpreting a priori knowledge about the process in the form of a mathematical model. Ideally, the entire process state can be derived from the available measurements in this way. By integrating a priori knowledge into the measurement process, the predictive power of measured quantities is markedly enhanced in both scope and reliability.

The prerequisite, however, is a sufficiently accurate mathematical model of the process under consideration [5.41]–[5.45]. Especially in chemical process engineering, however, mathematical modeling presents some difficulties. Many processes, because of their complex nature, do not lend themselves to accurate mathematical description. Process facilities often include subprocesses that are not subject to the classical mathematical model-building procedure.

In such cases, for example where the product properties and their relationships can be described only in qualitative terms, new ways of incorporating a priori knowledge about the process into the measurement must be sought. The methods of applied information processing can be expected to reveal new solution approaches. Under some circumstances, verbally formulated algorithms and heuristic methods can lead on to a qualitative description of the process. The use of knowledge-based methods is more and more attractive, for example the inference methods employed in expert systems [5.46]–[5.48] and fuzzy-logic methods [5.49]–[5.51].

Interesting tasks arise here for future research. An initial approach was the project "Technical Expert Systems for Data Interpretation, Diagnostics, and Process Control (Tex-I)," in which several participants (Siemens, Interatom, Bayer, and others) attempted to adapt expert systems techniques to process control. Other projects in this context are SENROB and NERES [5.52].

It is to be hoped, however, that the practical benefits of model theories will be studied and utilized more intensively than in the past. However, casual use of modern or simply fashionable modeling methods to make up for shortage of process knowledge must be avoided [5.1].

Figure 5.10 outlines a model-aided measurement technique for the case where a mathematical process model is available. The mathematical model is set up in parallel with the process and run (simulated) in real time (for example on a microcomputer). The known controlled variables acting on the process are supplied to the model as inputs. The class of model-aided techniques includes the Kalman filter and the Luenberger observer. These are similar in their fundamental structure, differing in that the Kalman filter is designed on a stochastic basis while Luenberger observers are designed by deterministic methods. The structure cited here applies to both continuous and batch processes.

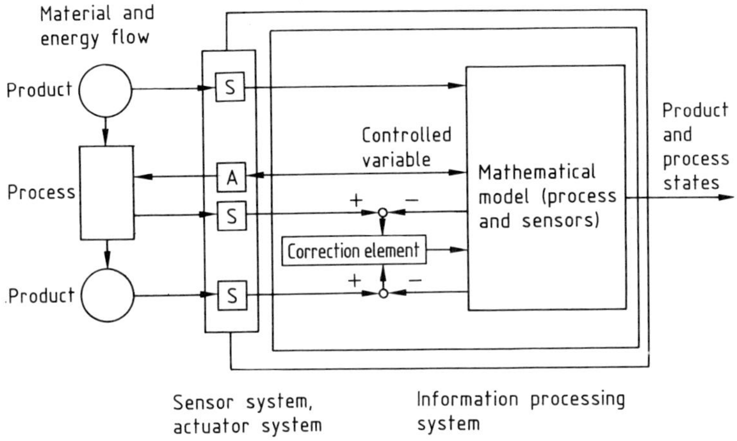

Figure 5.10. Schematic diagram of model-aided measurement procedure

In general, the properties of starting materials are interpreted as inputs, since they fix the starting conditions. They are therefore fed to the model as inputs, together with the controlled variables acting on the process actuators.

It is tempting to use the model to reconstruct the unmeasurable part of the process state in view of all knowledge about the inputs. A variety of problems arise, however. Even if an exact model exists, the simulated process state may coincide with reality only when the initial conditions selected for the model are identical with those of the process. But the initial conditions of the unmeasurable state components are generally also unknown. Further, it must be assumed that unavoidable modeling errors and perturbations acting on the process will lead to increasing divergence between the actual and simulated state variables. Forcing convergence between the model and the process, regardless of the selected initial conditions, perturbations, and error effects, means taking corrective action in the model. This can be done by comparing the actual measured process and end-product properties with the same properties calculated from the simulated process state with sensor models. If the resultant difference signals are then amplified in a suitably designed correction element, the correction can be made in the process model itself. The chief problem is to design the correction element in such a way that the behavior of the model will converge to that of the real process as rapidly as possible.

Figure 5.11 shows the structure of a model-aided measurement method in a simpler form. Here, the product properties and the controlled variables of the process have been combined into a single multidimensional input vector. Similarly, the output vector includes all the measured properties of the process and of the end products. The block diagram clearly shows that the process model, via the measured output variables, is made to track the real process in the manner of a servo follower. The correction element, acting as a controller (closed-loop controller), compensates for differences in behavior between the real process and the model resulting from perturbations, modeling errors, and incorrect initial conditions. In the case of the Kalman filter, it is generally assumed that both process noise and measurement noise act on the plant [5.41], [5.53]. Under the assumption of certain properties for this stochastic perturbing signal, the correction element is designed such as to minimize the expected value of a quadratic function of the difference between the actual and estimated process states. In the design of a Kalman filter, the a priori knowledge thus includes not only the process and sensor models but also models of the two stochastic processes. In the case of a Luenberger observer, in contrast, the design process is purely deterministic [5.42], [5.44].

For this concept of estimating or reconstructing the process state to work in a satisfactory manner, it must be insured that the process can be observed by means of the measurable quantities. In other words, all components of the process state manifest themselves in various ways in the time variation of the measured process and product properties. Criteria are available for in-

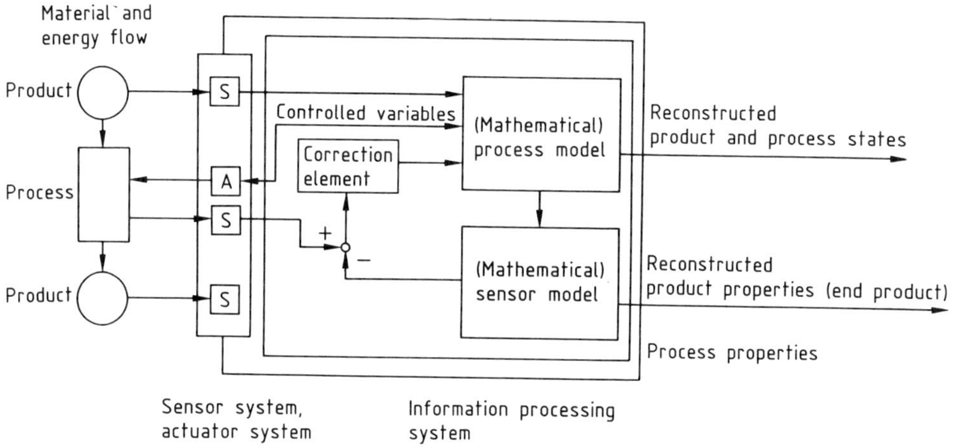

Figure 5.11. Structure of a model-aided measurement procedure

vestigating the observability of a process [5.41], [5.42]. Examples of applications in process engineering can be found in, for example, [5.54].

The discussion to this point has related to the determination of process and product properties in general; measurement and test methods employed in the laboratory, pilot plant, and industrial plant must meet some additional requirements. This chapter is exclusively concerned with measurement in the plant itself, or in an environment close to the process or similar to it. In the past, this subject was referred to as process metrology, or, when analytical questions were being emphasized, process analysis.

The dividing line between these two fields is fluid, and often they are inseparable. It is therefore proposed that only the distinction between process-level (local) and remote sensor system technology be preserved, so that process sensor system technology becomes an appendage of the well-known field of laboratory metrology [5.55] in which special requirements have to be met.

The computer processing of information from the sensor system changes the profile of requirements. In manual operation, measured values are commonly interpreted and processed by an expert plant operator. In this form of information processing the fault tolerance and error rate depend on the plant operator's level of experience and knowledge.

Computer processing imposes the following new requirements on the sensor system:

- Reliability at an acceptable maintenance cost
- Functionality even in extraordinary process states (startup, shutdown, disturbances)
- Accuracy [5.56] and dependability

These requirements result from advances in microelectronics that make computerized information processing a possibility. Microelectronics is also an important tool for the realization of these requirements. After optimization of the instrumentation and installation, the design of intelligent sensors with the ability to report faults and initiate maintenance procedures automatically is in many cases the most economical way to satisfy the requirement of reliability.

The requirement of functionality even in extraordinary process states implies not only robustness but also a thorough understanding of what happens on in the process. Especially in sensor systems used to measure quantities that depend on the composition of the measured stream, many independent variables must be handled, and up to now it was only rarely the case that these were known for all possible process states.

Measurement (from the physicist's standpoint) becomes metrology (from the eingineer's). It is not enough that a method works in principle; it is essential that it always functions effectively [5.57].

The use of computers for information processing, however, makes it possible to acquire information more economically than before. In signal-oriented process control, the motto was: "Acquire information item by item, in parallel, and always," whereas information-oriented process control engineering follows the principle: "Acquire information in the smallest possible amount, but as much as necessary, and only then when it is to be used" [5.58].

The advantage, especially in batch processes, is obvious. Large savings are possible in the wear-out reserve (see Section 11.4 and Figs. 11.61, 11.62) and thus in maintenance costs if, for example, a pH probe in a reaction vessel is brought into contact with the product only when the pH must be measured and otherwise is held in a protected standby condition (see Section 5.3 and [5.59]).

The new requirement profile and the new principle of information acquisition have influenced not only instrumentation but also the interface between the sensor and information-processing systems (Fig. 5.1). Information coupling, for example to a process control system, is no longer limited to the transmission of a measured value. Sensor status messages can also be included (e.g., reports that a unit has failed or malfunctioned or that a maintenance procedure is under way), and it should also be possible for the sensor system to receive information from the process control system (e.g., an instruction to hold itself available; see Figs. 5.14–5.23). The same holds for actuator systems (see Section 6.1).

The definition of the extended interface between the sensor and information-processing systems should also include the signal levels and types and the signal contents.

Besides the change in emphasis as a result of computerized information processing, new measurement functions are being assigned to sensor technology (as they have been throughout the development of this technology). When existing processes, such as polymerizations, are automated and optimized, the need for information is reconsidered and new constraints applied, with the consequence that new measurement tasks are created or an old task is accomplished. The introduction of novel processes (e.g., biotechnology) involves new problems for sensor technology [5.60], [5.61] (see Section 5.3).

The requirements that must be met in the further development of sensor technology for production processes fall into two independent classes (Fig. 5.12):

The changed profile of requirements and the need for higher availability in computerized information processing mean that existing sensor systems must be enhanced. An example is the development of the magnetic induction flowmeter. In 20 years, the power consumption—a key factor in internal heat generation, service life, overall dimensions, etc.—has been reduced by a factor of 20 (Fig. 5.13) [5.62].

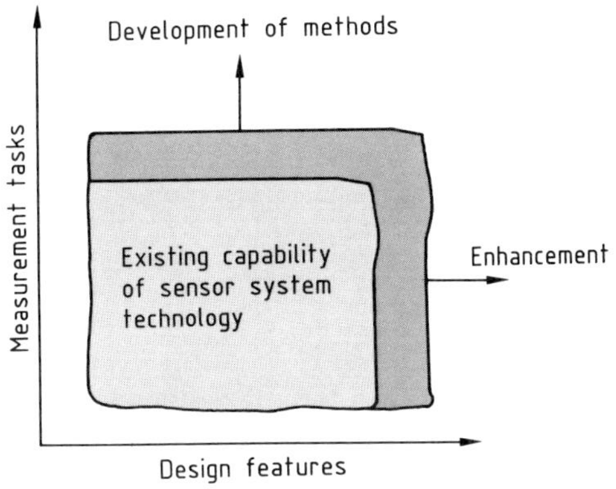

Figure 5.12. Continuing development of sensor system technology

New problems require the development of new methods. Again, the realization of technical systems, often based on well-known methods of physics, is facilitated by the new capabilities of microelectronics and modern materials. Examples are systems using the Coriolis force to determine mass flow rate and others using the radar principle for level measurement.

5.2. Process Sensor System Technology

Information Flow in Sensor Systems. Section 3.5 showed that it is desirable to separate the precise description of the technical properties of an individual instrument from information about its technical function (Fig. 3.26). This makes it possible to maintain required functionality even when instruments are replaced.

Thus it is advantageous to apply the same formalism to information transformation (from the measured property of the material stream to the output signal) that was employed in structuring the material flow in production processes [5.7]. This leads to the idea of information flow in the sensor system, whereby a sharp distinction is made between the form taken by the desired information after each transformation step (corresponding to the product elements in the production process) and the metrological process steps required for information transformation (corresponding to the process elements of production).

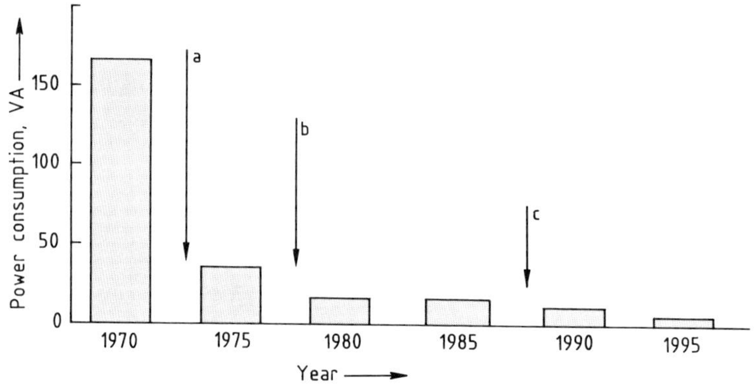

Figure 5.13. Development of the magnetic induction flowmeter (MID) illustrated by power consumption over time [5.62] a) Introduction of switched d.c. field; b) Introduction of compact devices; c) Introduction of microprocessor control

As an example of a simple sensor system, Figure 5.14 illustrates the measurement of temperature with a jacketed thermocouple.

The measured property E of the stream is coupled to an intermediate medium and thus transferred to the latter.

The intermediate medium with property Z now interacts with the sensor element, which transforms Z into an electrical output signal S.

The information flow between the process and a simple sensor system is illustrated, in abstract form, by Figure 5.15 [5.10], [5.63].

It is usually necessary to amplify and condition the electrical output signal of the sensor element (Fig. 5.16):

The electrical output signal of the sensor element S″ is amplified to give a signal S′.

This amplified signal S′ is linearized, mapped onto a unit interval, and made available as a signal S at the sensor system output.

The term *integrated sensor* refers to a system in which the functions of coupling, interaction with sensor element, and signal conditioning are integrated in a single component. If signal processing is included, the term *intelligent sensor* is used, especially when signal processing involves complex calculations and range matching steps [5.64].

In a simple sensor system with compensation (Fig. 5.17), an additional sensor is needed to determine the parameter whose conditioned signal is incorporated into the signal processing. Auxiliary energy is generally required for this purpose; in abstract form, Figure 5.15 is elaborated to Figure 5.18. The energy needed for signal conditioning and signal processing is supplied to the sensor system by a power supply device that

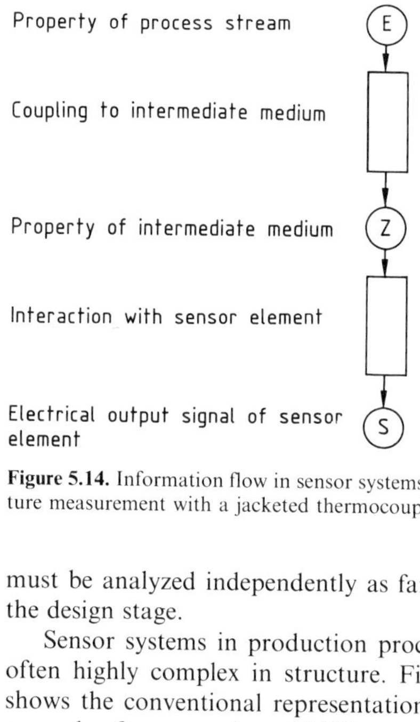

Figure 5.14. Information flow in sensor systems: Temperature measurement with a jacketed thermocouple

must be analyzed independently as far back as the design stage.

Sensor systems in production processes are often highly complex in structure. Figure 5.19 shows the conventional representation, for the example of a gas analyzer [5.10].

Here, the measurement proper is preceded by the functions "sampling," "sample transport," and "sample preparation." Sampling has the task of acquiring a representative sample of the stream to be measured. The sample must be transported without alteration or falsification to the analyzer, which performs the actual measurement. This stage of sample transport often involves conveying the sample over a long distance. The "sample preparation" function involves adapting the properties of the process stream at the sampling point to conditions com-

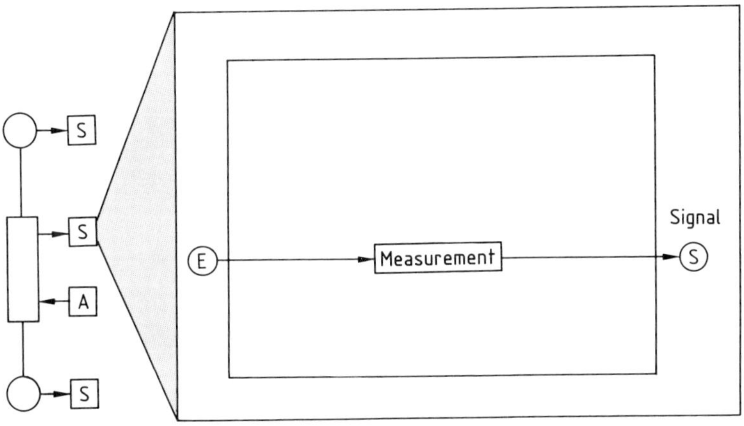

Figure 5.15. Information flow in a simple sensor system [5.10]

patible with the measuring instrument itself. The measured property of the stream—for example, the concentrations of certain components in the stream—must naturally be preserved in a well-defined way. The planning and execution of sampling, sample transport, and sample preparation are strongly dependent on the properties of the stream material, the analytical objective, and the measuring method employed. This strong application dependence is the reason that sensor systems for determining composition-dependent quantities are complex in structure and must be serviced by specialists.

Along with the sensor system modules just described, Figure 5.20 also shows the functions "delivery" and "disposal." The disposal step matches the properties of the sample stream at the analyzer output to the permissible conditions at the point of return to the stream. The delivery function modifies the auxiliary energies and substances such that they can be utilized in the sensor system. Both these functions must be included in the diagram of the entire system because the correctness of the measurement also depends on them and because they are potential sources of error for the sensor system as a whole.

On grounds of economy, it may be desirable to withdraw samples at various points in the process and measure them with a single analyzer. The functions of sampling, preparation, and transport are then preceded by a measuring point selection step (Fig. 5.21). In combination with signal processing for each sampling point, this function correlates the analyzer output signals (S_1 to S_n) with the properties to be determined (E_1 to E_n). Conversely, several analyzers can determine a variety of values on a single sample by accessing the same sample preparation.

As in the case of the production process, monitoring and control systems must also be available for complex sensor systems. All complex sensor systems include for this purpose auxiliary devices such as displays, pressure gauges, flowmeters, or valves for the admission of calibration substances. In more abstract terms: the background functions grouped together as "functional check, control of maintenance procedures, and setting of parameters" must be supported as a minimum [5.63].

These functions are performed manually in conventional sensor systems. At intervals specified by the maintenance strategy, the system is inspected and serviced (e.g., calibrated) by maintenance personnel. However, analogous to the production process, this routine activity is increasingly being automated (Fig. 5.21) and turned over to a dedicated computer system that monitors all the basic functions of the system, controls procedures, and executes corrective actions. For this, the dedicated computer requires system-internal sensors and actuators. The computer furnishes status information about the sensor system and can execute incoming instructions, for example, to come into action in an intermittent mode (on-request or on-event). The outcome is a general structure for complex sensor systems. The information blocks K and S are the basis of the protocols for the future field bus.

"Intelligence" and Sensor Systems. In an intelligent sensor system, the functions discussed above are integrated or are supported in such a

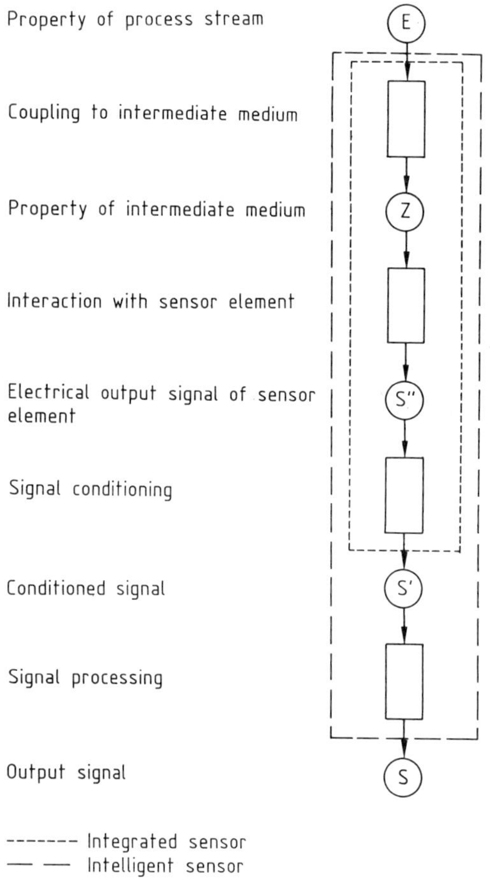

Figure 5.16. Information flow in sensor systems: temperature measurement with jacketed thermocouple and signal processing

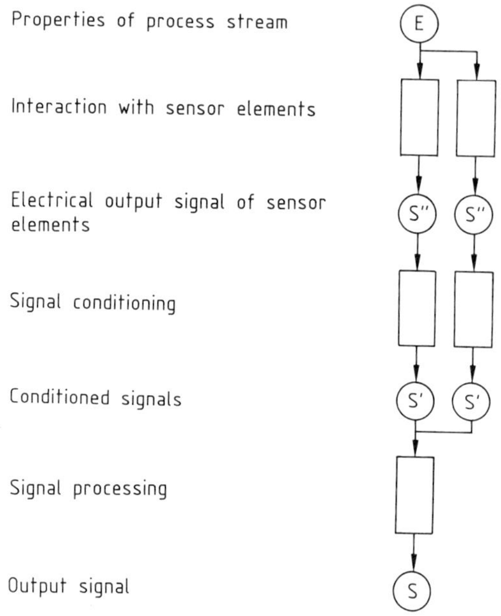

Figure 5.17. Information flow in sensor systems: sensor system with compensation (e.g., pH measurement with temperature compensation)

way that it can be regarded as a sensor system with communication capabilities. In Figure 5.22, these links are provided by a field bus, whose function could equally well be carried out by the use of smart technology based on the HART protocol.

In what follows, the functions of sensor systems are described as classical or intelligent.

Classical functions include:

- Sensing functions, i.e., transformation of physical and chemical effects into a signal suitable for downstream processing
- Signal conditioning (e.g., amplification)
- Signal output (part of the signal processing, such as A/D conversion, takes place in the local or process-level components)

Intelligent functions include:

- Signal processing
- Interconnecting measurements to expand information
- Auxiliary functions for on-line metrology
- Automatic execution of maintenance functions
- Failure strategies (see Fig. 5.16)

Figure 5.22 illustrates classical and intelligent sensor systems and their incorporation in higher-order information-processing systems. The use of a field multiplexer for the connection to the process monitoring and control system bus, which is already a practicable transition approach (even in explosive environments); see Section 5.5.

The intelligent functions of a sensor system can be employed in a five-point program:

- Making all the needed information available
- In a form suited to the application
- At the proper time
- Ready for downstream processing
- Even in extraordinary process states

The goal of this development is to devise autonomous field-level devices that will automatically perform functions such as the maintenance

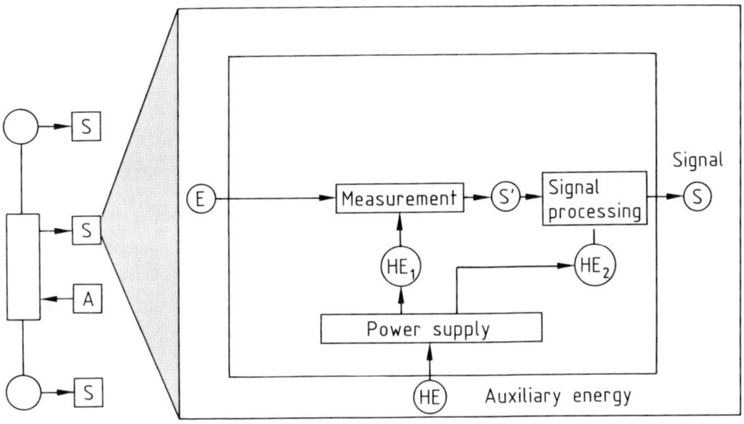

Figure 5.18. Information flow in a sensor system with auxiliary energy and signal processing [5.10]

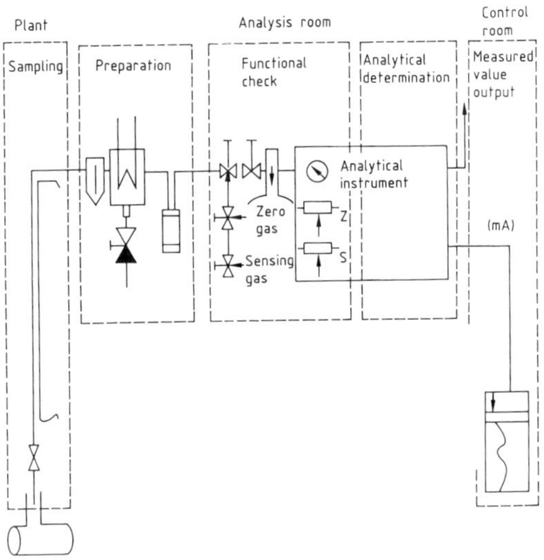

Figure 5.19. Functional groups of a process analyzer (gas analyzer) [5.10]

functions of DIN 31 051 [5.66] discussed above. Intelligent process pH sensor systems can already carry out functions such as electrode cleaning, zeroing, calibration, flushing, and so forth, either "on-event" or "on-request."

Communication with Process Sensor Systems. Consideration of the functional scope of complex sensor systems leads to the question of how the sensor system can appropriately communicate with information-processing systems. The output signal categories M (messages), B (instructions), and S (signals) are the information blocks involved in communications (K) with process control and automation systems. A straight point-to-point connection with communication in one direction does not meet modern requirements. The need for bidirectional, open communications structures, and messaging services is essentially derived from the functions discussed above. More detailed information can be obtained from the literature on field buses [5.67], [5.68].

The connection of a classical sensor system to a local component (LC) is a unidirectional link employing a uniform signal such as 0–20 or 4–20 mA (Fig. 5.2-9). All that is transmitted is the measurement signal or a report of a device failure. Signal processing, that is, conversion to a value with a physical unit, takes place in the LC. Maintenance tasks must be performed locally in the sensor system. According to DIN 31 051 [5.66], these include inspection to evaluate the actual status, maintenance to preserve the nominal status, and repairs to restore the sensor system to nominal status. General discussion of these points can be found in Section 11.4.

If a dedicated computer performs the function monitoring of the intelligent measuring system and carries out simple maintenance procedures, using messages to report on both activities, the process sensor system acquires, functionally, a second interface (in addition to information delivery to the control room via the "process port"), which mediates information exchange with the engineering team via the "engineering port." This functional distinction is illustrated in Figure 5.23.

Given the wide range of methods employed in process sensor system technology and the broad spectrum of application-specific hardware configurations, the only way to achieve system-wide reliability and effective maintenance is by unifying the two interfaces, process port and engineering port. A flexible dedicated computer system, matched to the structure of complex process sensor systems, can organize the required information exchange from and to the sensor system. In this way, expensive custom solutions can be avoided and replaced by acceptable, standard solutions.

Information Exchange with Local Functional Units. According to SCHMITT the following holds for information exchange between the field level and the process monitoring and control system: "As little as possible—as much as necessary" [5.58]. This requirement governs the functional sensor interfaces shown in Figure 5.23 and is discussed further in Figure 5.24 in application to the process port.

In relation to local (process-level) functions, the sensor system has only one task: making the measured value available in proper form. If the sensor system ceases to respond as specified because of an upset of malfunction, this fact should be reported in a group message "Failure" (M_1),

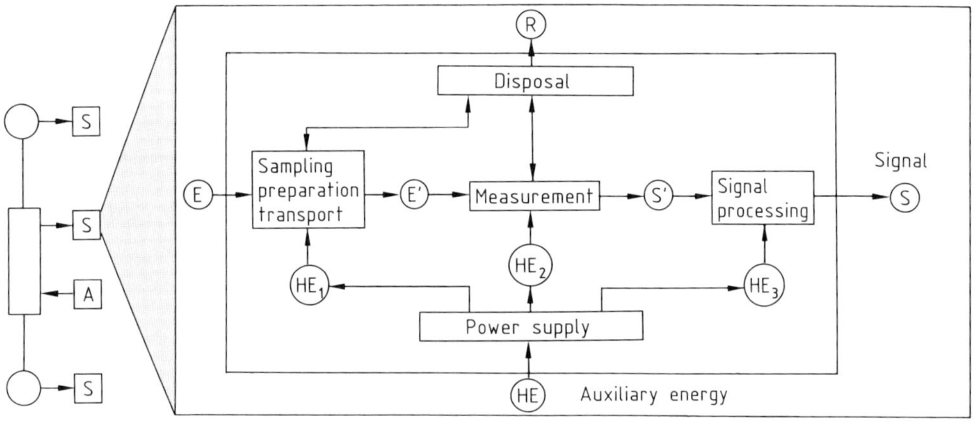

Figure 5.20. Information flow in a sensor system with sampling and other auxiliary functions (disposal, delivery, etc.) [5.65]

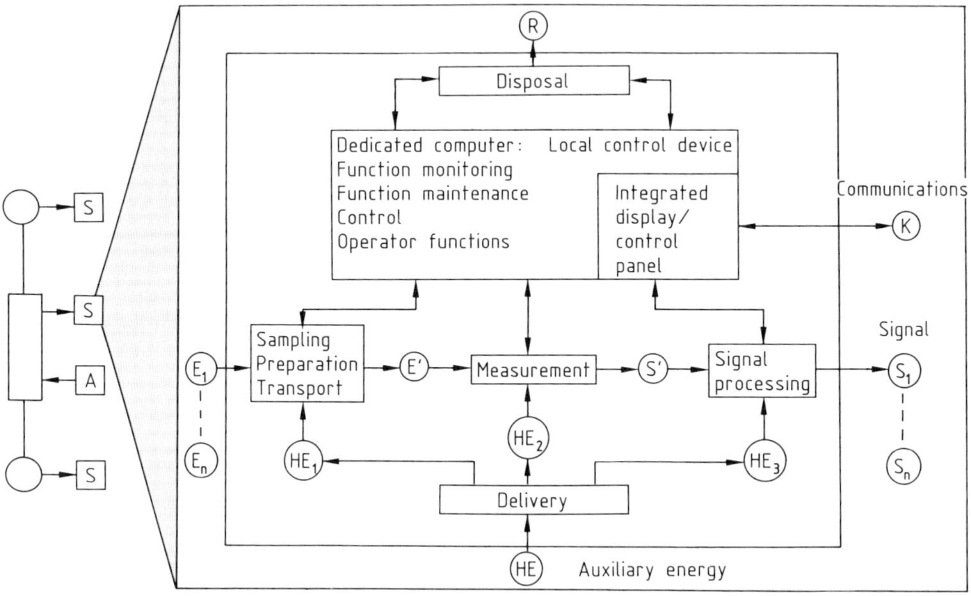

Figure 5.21. Information flow in a sensor system with sampling, auxiliary functions, and integrated dedicated computer [5.65]

which is to be forwarded directly to the display and control component of the control system (see Chapter 7). Any still-pending measurements are then rejected as unusable. This assessment of sensor information presupposes that the sensor system monitors its own internal technical functions.

If internal sensor function monitoring is initiated automatically or manually, measurement signals may appear momentarily that are correct in content but have nothing to do with the process and product properties being measured. This state is indicated by the status message "function monitoring" (M_2). Because of the absolute priority of the information measurement signal, it may be necessary to permit such non-availability due to internal programs in the sensor system only when they can be tolerated by

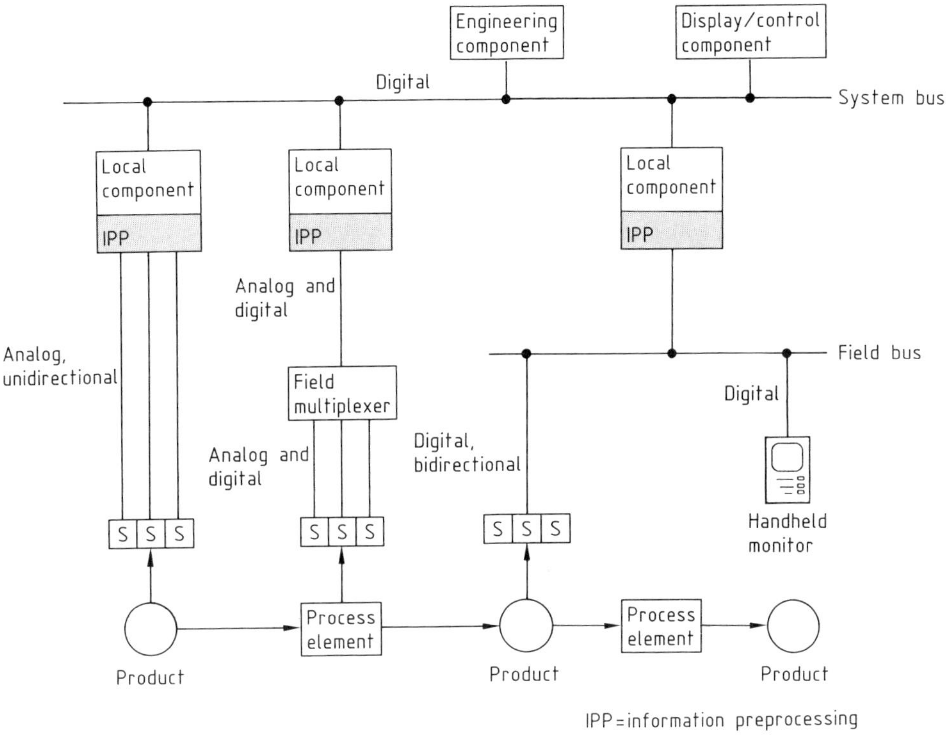

Figure 5.22. Movement of information processing nearer the process level

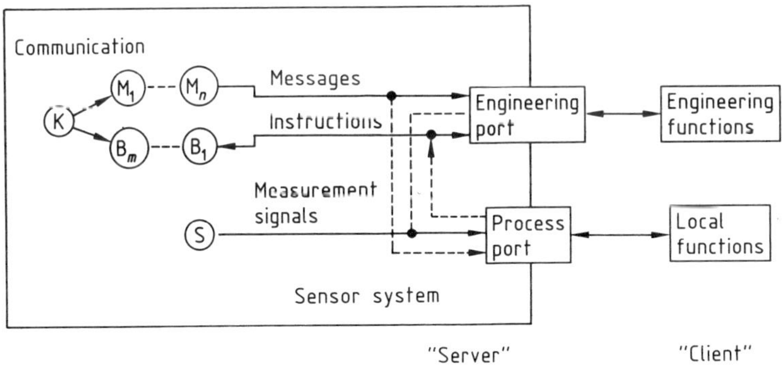

Figure 5.23. Functional breakdown of the sensor interface K (communications; see Fig. 5.21) into "process port" and "engineering port" [5.10]

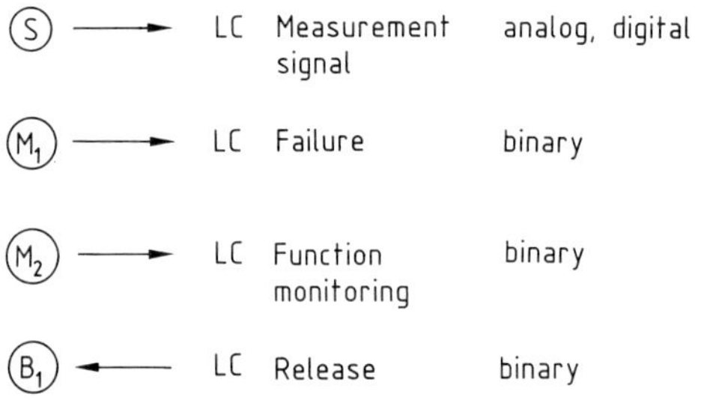

Figure 5.24. Information exchange between sensor system and local process control system component (LC) [5.10]

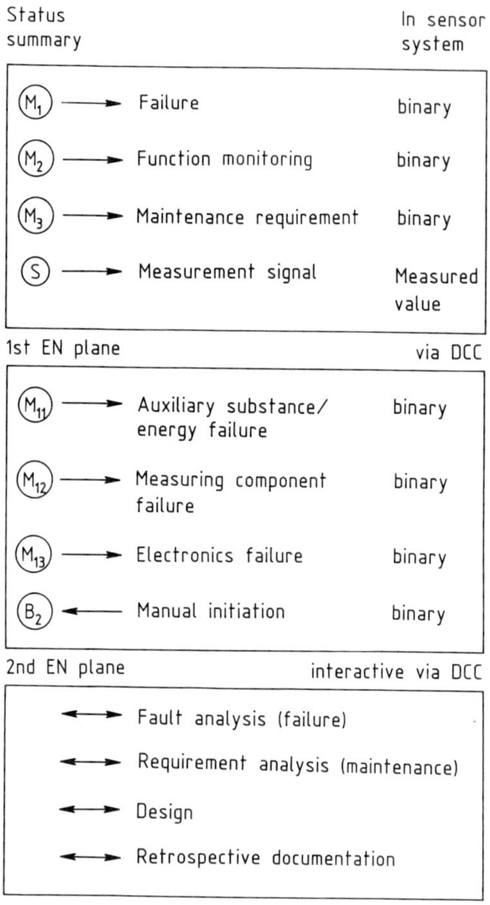

Figure 5.25. Structuring of the interface to engineering [5.10]
DCC = display and control component; EN = Engineering

the control system. An instruction "release" (B_1) from the process control system to the sensor system is a suitable way of handling this situation; for example, the instruction can inhibit or enable the calibration procedure.

The power and low cost of modern microprocessor boards make it possible to use them locally to optimize internal operations in sensor systems without having to shift such functions to central systems. In this way, straightforward system structures can be created. For the sake of manageability, information exchange with process-level functional units of the process monitoring system should be restricted to the measurement signal plus a few status messages and instructions. The handling instructions and procedures associated with each type of status message or instruction should be defined as a strategy in the process control system (for status messages) or in the sensor system (for instructions).

Information Exchange with Engineering Functional Units. The maintenance functions have the task of keeping up reliability and insuring availability. Routine inspection and function monitoring jobs should be automated and turned over to the sensor system itself whenever technically possible at reasonable cost. When service is needed or when a malfunction occurs, the engineering group is requested. From a functional standpoint, access to the sensor system and necessary information exchange take place over a separate interface, labeled "engineering port" in Figure 5.23.

If a process sensor system is fully equipped with self-monitoring functions and devices to maintain the functionality of both the instrument proper and the sample preparation part, the volume of information produced is difficult to manage without structuring. Engineering becomes inefficient and personnel find themselves overloaded if the separate functional concept and distinct operator interface of each instrument type and each manufacturer have to be kept in view.

The capabilities of future field-level communications are discussed here, based on analyses of information structuring for process analyzers [5.10].

For efficiency in engineering, it is proposed that information exchange (see Fig. 5.25) be accomplished in three functionally distinct information "planes" differing in complexity. A status summary is needed for the instrument proper; this includes a group message "failure" when the measuring device is no longer reliably available; a status message "function monitoring" when the sensor system is occupied with itself as a result of a manual intervention or automatic program; and the display "measurement" so that the engineer can make an on-site assessment. If the sensor system is equipped with self-monitoring functions that can detect a failure (for example due to drift in the analyzer or exhaustion of the test gas supply), the status message "maintenance required" should also be displayed.

The inspection, maintenance, and repair of process sensor systems need not always involve specialists. It appears useful to be able to distinguish whether a given malfunction can be remedied by simply replenishing auxiliary substances or replacing an electronic component, or whether the device can be put in service again only after a specialist works on the sample handling system or the physical part of the instrument.

For this purpose, the first engineering (EN) plane includes simple, binary diagnostic indicators. The "failure" message may, for example, be broken down into just three categories: M_{11} if there are problems with the delivery of auxiliary supplies or energy; M_{12} if there are problems in the measuring section, including sample preparation; and M_{13} if a failure occurs in the electronics. For general maintenance, simple instructions such as manual initiation of a function check should be integrated into this EN plane.

This initial diagnostic feature requires communications with the sensor system above and beyond a simple status indication. These communications are carried on through a display and control component near the sensor level. The display and control component can take the form of a keypad with display integrated into the sensor system; an engineering computer some way off; or a portable unit such as the handheld monitor shown in Figure 5.22, which is plugged into the sensor. As to what technical and functional capabilities exist and how they can be exploited, see [5.67], [5.69].

Detailed diagnosis of the causes of malfunctions, which can only be performed by process sensor specialists, should be carried out in a second engineering plane, via one of the functional units. This kind of diagnostics includes single-fault analysis upon a system failure, as well as detailed need analysis upon a maintenance request. The design and commissioning of complex process sensor systems often include configuration and parametrization. These tasks are also performed by specialist personnel and thus form part of the second EN plane. For the manageability of process sensor systems, it is important that the system data be kept updated and that retrospective documentation be maintained.

When such a ternary structure is used in the EN interface, standardization across vendors and instruments appears possible at least for the status summary and the first EN plane [5.70]. For efficient maintenance, it would also be desirable if information exchange through the second plane could be structured in a consistent way and mediated by operator interfaces as similar as possible.

The "sensor-level display and control component" function is increasingly being implemented with a portable computer (laptop PC). If a standardized operating system and graphics environment are employed, vendor- and device-specific parametrization and diagnostic routines can be loaded and run in software. For process analyzer technology at least, such a consistent engineering tool offers substantially greater benefits than a variety of individual solutions with their associated library of manuals (see also [5.71]).

The communication ports of field-level devices should generally be implemented on the basis of the object-oriented model [5.72] defined by the MMS (Manufacturing Message Specification; see Chapter 8 and [5.73], [5.74]). The field device plays the role of server and performs the services requested by the client; see Figure 5.26.

Linked with the cited background functions and their use for process control, maintenance, and design (configuration) tasks is the introduction of general, standardized descriptive languages for field-level devices; these are called Device Description Languages (DDLs). The objective of manufacturer-neutral field-level functions is that the field device to be operated should be addressable without any special knowledge of the physical device, even if devices are modified or replaced by hitherto unknown devices. Standardization efforts toward a universal device description are currently focused on creating a proposal for the ISA SP 50 international field bus standard [5.67], [5.75].

Furthermore, it is not sufficient to solve the problems of field-level communications without taking up questions of what user interfaces are

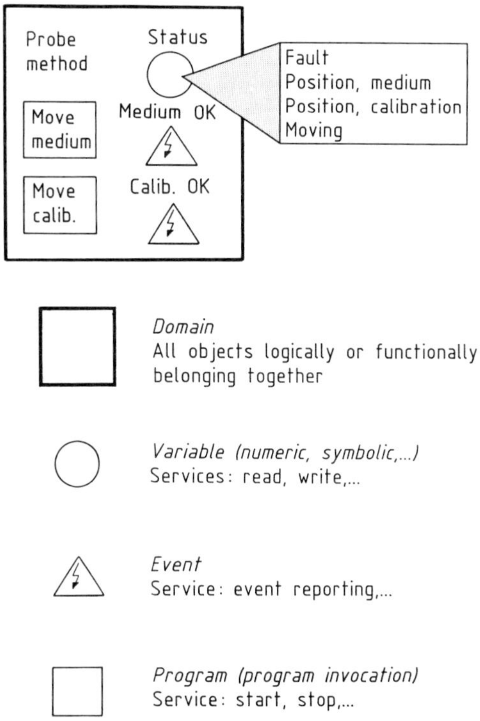

Figure 5.26. MMS communications objects

appropriate for sensor systems and how these interfaces can be standardized [5.71], [5.9]. The application connected above the communications services must be dealt with at the same time. On practical grounds, a variety of devices having the same function must not have unlike operator interfaces [5.67], [5.76].

Moreover, the rapid growth in information demand and the need for auxiliary energy in process sensor systems—which may well exceed the permissible limits in intrinsically safe systems—have made it essential to consider separating the information and energy channels (see Section 5.5, Chapter 8, and [5.68]).

Determination of Instrumentation Cost. Given the information economy of a process modeled with the phase model (see Figs. 5.1 and 2.35), the instrumentation outlay can be determined at any level of detail (see Chapter 2).

Consider, for example, the process step of neutralization in a wastewater treatment process. Among the quantities to be determined are the pH and the inlet flow rate of the wastewater (Fig. 5.27). From the categories and attributes of the process property (here the wastewater pH), the property can be methodically assigned to a suitable sensor system. A CAE tool for this purpose, accessing a source database for sensor–actuator systems, is discussed below.

Figure 5.28 shows an example of how to structure a taxonomy of sensors for process engineering. The user is guided by generic attributes: application features and design features. For the wastewater pH example mentioned above, a sensor system can thus be defined that has the application features on-line measurement, determination of a collective property, measurement in the product stream, and so forth. In general, the requisite instrumentation cost can thus be determined from the information demand obtained from the phase model. Sensor or actuator systems are selected on the basis of the category of process and product properties, with the aid of a taxonomy of such systems (navigation, hypermedia, etc.). Manufacturers still offer monogram-like selection aids [5.77], which will certainly be replaced soon by expert-systems tools.

Taxonomy of Sensor Systems. The classical listing in book form, arranged by apparatus features, is no longer adequate for presenting an up-to-date survey of application options for modern sensor systems.

Nevertheless, such proven classics as Profos [5.78], Strohrmann [5.79], Hengstenberg [5.80], or the Kohlrausch [5.81] are indispensable aids. The journals TM, atp, and MPA have presented thorough surveys of problems and current solutions (see also [5.9], [5.82]–[5.84]). However, in general, encyclopedic treatment must be replaced by structuring.

A priority task for structuring available sensor systems in a way relevant to process control engineering is to emphasize those features that actually bear on potential applications. As discussed previously, these features cannot be identified without a knowledge of process conditions met with in practice; that is, an apparatus-oriented analysis alone is insufficient. Instead, the conceptualizations introduced to structure information characterizing a production process lead to a powerful technique. In future, this approach may serve as a basis for developing powerful CAE tools for the systematic instrumentation of new plants.

Figure 2.8 shows that, if the distinction between function and substance—or function,

function-generating procedures, and devices (Fig. 3.26)—is taken into account, then the world of sensor systems can first be represented in terms of the tasks meaningful in certain industries; see Table 5.1 [5.85] and the lower row of Figure 5.29.

The E–R (entity-relationship) modeling technique described in Chapter 2 is employed, at various levels of hierarchical detail, for this purpose. Such an ordering rule, as explained in Chapter 2, is known as a "taxonomy" (Greek: taxos = tree).

Generic Class: Application Features. The generic class of application features or task features, defined from the functional viewpoint, becomes the uppermost level in this context.

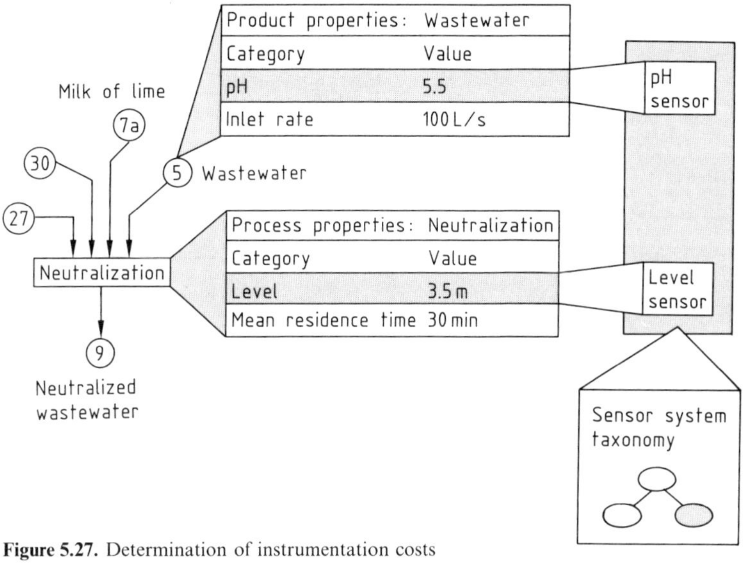

Figure 5.27. Determination of instrumentation costs

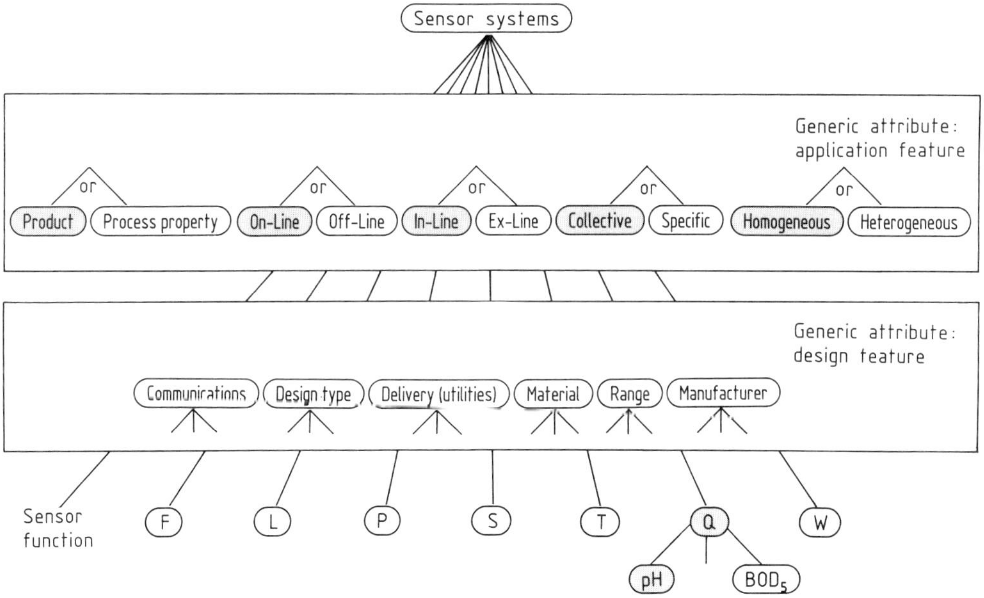

Figure 5.28. Taxonomy of process sensor systems: System properties required for determination of pH and BOD_5

The following are employed as classification criteria:

- The attributes of the object classes of technical processes
- The degree of indirectness with which the measurement signal is determined
- The ratio of relaxation times between the acquisition of the measurement signal and the variations of the property of the stream being measured (see Section 5.1)

Table 5.1. Letter symbols for process control point functions [5.85]

Letter symbol	Group 1: measured values and other input quantities		Group 2: processing (as trailing letter)
	As first letter	As additional letter	
A	alarm		
C			controlling
D	density	difference	
E	all electric variables		sensing element
F	flow rate	ratio	
I			indicating
L	level		low
P	pressure or vacuum		test point connection
Q	quality, e.g., analysis, concentration, conductivity	integrate or totalize	integrating or summating
R	nuclear radiation		recording
T	temperature		transmitting

- Specification of the components of the substance measured

Under these criteria, sensor systems are now classified according to whether they determine product properties or process properties, whether they operate on-line or off-line, whether they are employed in-line or ex-line, whether they can determine collective or specific properties, and whether they are usable in homogeneous or heterogeneous phases.

These generic features are explained briefly below:

- On-line: A continuous correlation is possible between information provided and properties of the process or product. The essential condition for on-line measurement is that the relaxation time of changes in the process or product properties must be markedly longer than the time required to acquire, convert, and deliver the information [$\tau_{R(property)} \gg \tau_{R(sensor)}$] (see also Section 5.1).
- Off-line: Discontinuous measurement; determination of sampling intervals requires knowledge of process characteristics (e.g., kinetics).

This pair of attributes describes the problems the (process) sensor system must handle in terms of communications.

- In-line: The systems used to prepare information about the process or product properties

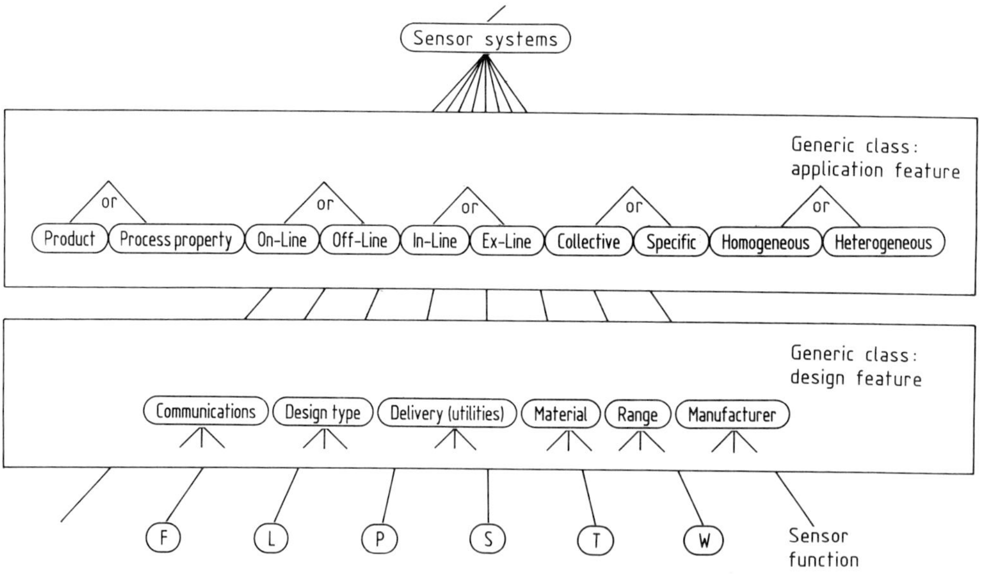

Figure 5.29. Taxonomy of process sensor systems: Breakdown of sensor functions

interact directly with the product stream (or a bypass stream).

- Ex-line: The required sampling operation (extractive) makes it necessary to take into account additional changes in the properties of the stream substance.

Questions of positional and temporal representativeness are extremely important here. This pair describes the problems the system must deal with in terms of metrology.

- Collective: The quantity measured can provide information about the quality of the stream as a whole (e.g., density or flow rate), its reactions (e.g., pH, conductivity), and the relationships between its properties (e.g., relative concentration).
- Specific: The "manifestation of a property" determined makes it possible to gain qualitative and quantitative information about individual components of the stream; such techniques include analytical methods such as spectroscopy and chromatography.

This pair thus states clearly whether the chosen sensor system yields information about atomic/molecular, mechanical, and electromagnetic mechanisms.

- Homogeneous: The sensor system can acquire information only in homogeneous phases (e.g., flow-rate measurement with a mass-flow anemometer).
- Heterogeneous: The sensor system permits the determination of the measured quantity even in heterogeneous states of aggregation, such as dispersions (e.g., flow-rate measurement by the Coriolis principle).

This pair accordingly requires an unambiguous description of the sensor system functionality with regard to its possible use in multi-phase environments.

Generic Class: Design Features. The next level consists of a further class of features which basically summarize information about the design of the sensor systems under consideration.

Sensor systems are classified by communications capability, design type, utilities required, material of construction, range of measurement, and manufacturer. Individual features can be further structured. The dependent lines in Figures 5.29 and 5.30 illustrate this point. In principle, the depth of structuring is unlimited.

To illustrate characteristic features of importance in field-level devices, the safety requirements on process sensor systems are discussed at this point. If "safety" is defined as a risk less than the limiting risk [5.24], as in Section 5.1, safety-relevant properties of a process sensor system can be divided into two classes with regard to possible sources of danger:

- Properties related to insuring the availability and reliability of the sensor system itself
- Properties relating to reducing or preventing danger interactions of the sensor system with the system environment

These contributions to safety are correlated with possible technical safeguards in Figure 5.31. The potential dangers due to interactions with the environment are divided into those that require protection of the sensor system against the environment, and those requiring (possibly indirect) protection of the environment against the effects of the system.

As Figure 5.31 shows, present-day principles, guidelines, and standards on insuring the safety of process control hardware can be classified in the same way:

From a safety standpoint, enhanced availability and reliability gained through fault detection, fault reporting, fault compensation, and automatic calibration figure in the tendency toward increased functionality in field systems through integration of information-processing components, as discussed in this section. An initial and crucial requirement on the safety features of a process sensor system will be the ability of the system to verify and, if appropriate, report its complete functionality.

The protection of live or moving parts of an electrical device against foreign solid bodies or liquids is required by all national and international standards, for example, [5.86]. Because they are represented by the letter code IP (International Protection), these provisions are also referred to as IP protection types.

A topic still under vigorous discussion, in contrast, is how to insure electromagnetic (EM) compatibility. In German and European standards, electromagnetic compatibility is defined as the ability of an electrical device to function satisfactorily in its environment without unduly influencing this environment [5.87], [5.88]. In Figure 5.31, measures to insure electromagnetic compatibility are divided into safeguards against EM interference induced in the device (static dis-

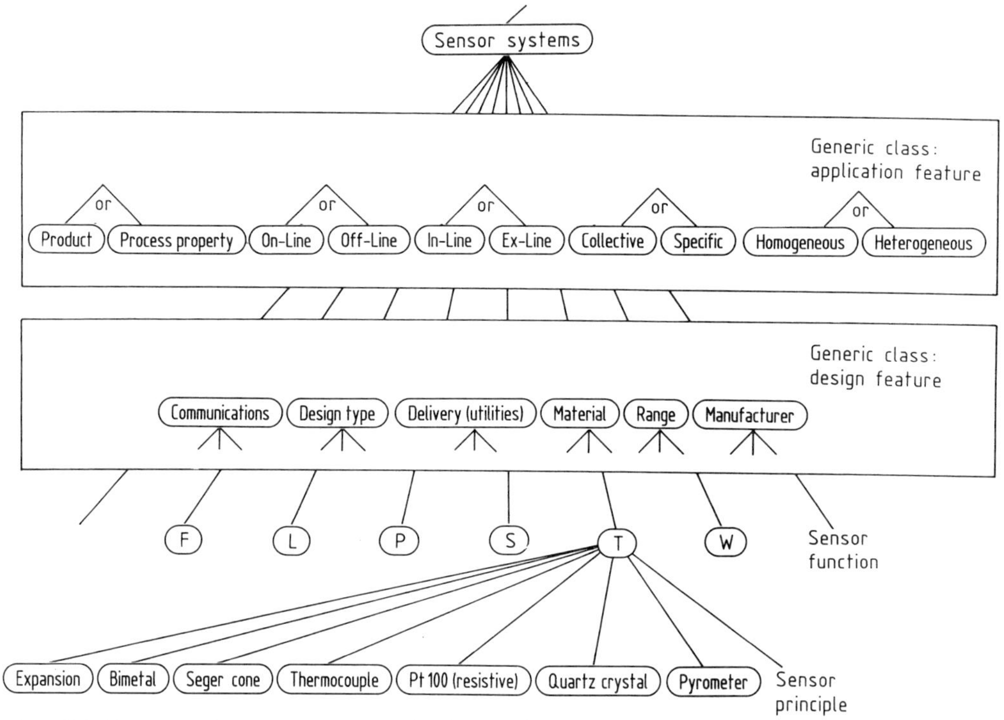

Figure 5.30. Taxonomy of process sensor systems: breakdown of sensor principles for the example of temperature measurement

charges, electromagnetic fields, and fast transients) [5.89] and safeguards against EM emissions into the environment, in particular human exposure, due to fields generated by the device or system [5.90].

Another type of protection incorporated into the field-level device is explosion protection [5.91]. The avoidance of flammable materials and explosive mixtures are usually the primary safeguards against explosions [5.79], [5.92]. With regard to requirements on field-level systems, these practices are of relevance to design and material selection. Primary explosion safeguards must also be considered when analyzing the use of sensor systems, especially the use of process control resources for plant security [5.22]. This area of applications includes all process control hardware employed for monitoring, protection, and damage limitation [5.93]; see also Section 10.4.

Secondary explosion protection in field-level process cotnrol systems includes all measures taken to prevent ignition of explosive substances by preventing energy transfer. The regulation governing electrical devices in explosion hazard areas distinguishes seven types of protection. Among these, intrinsic safety [5.94] sets rigorous criteria on integrated circuits in modern sensor systems (see Section 5.5).

For redundant and fail-safe design, see Section 10.4. Lightning protection practices are generally based on suppressor circuits [5.95].

The lowermost level is concerned with the physical, chemical, and other principles on which sensors operate. The commonly used nomenclature is set forth in DIN 19227. Figures 5.29 and 5.30 show a selection of possible sensor principles; all others are indicated by the dependent lines. As in the case of the design features, further structuring is possible.

The foregoing application-oriented description can serve as a basis for a conceptual, object-oriented classification of sensor systems, differing sharply from the device-oriented and method-oriented structures, as listed below.

Methods of measuring "classical" quantities:

Temperature
- Resistance thermometer
- Thermocouple
- Radiative pyrometer

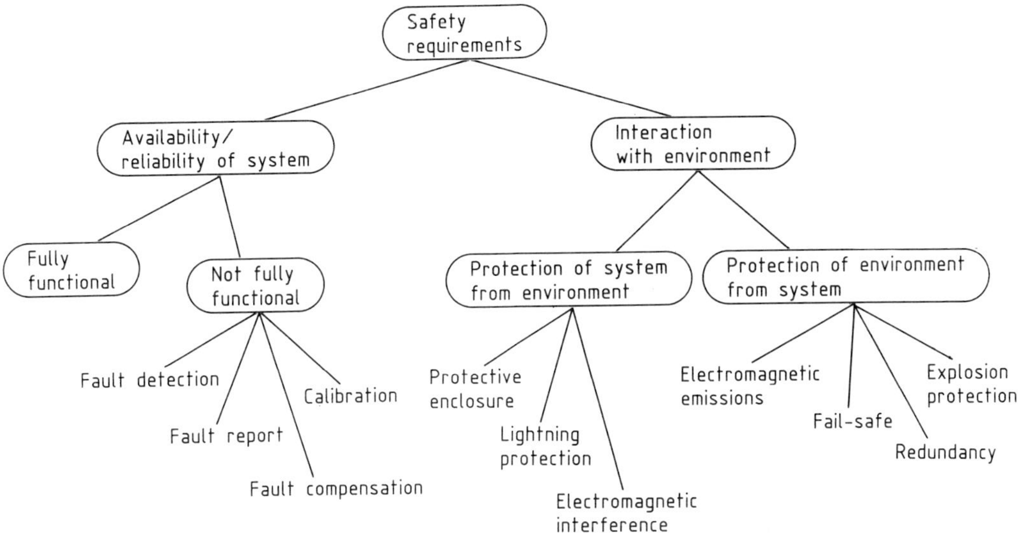

Figure 5.31. Safety requirements on process sensor systems

Pressure
  Elastic elements with capacitive, inductive, piezoresistive or piezoelectric signal conversion
Flow rate
  Magnetic-inductive
  Differential Level
  Hydrostatic
  Electrical and electronic
  Radiometric
pressure

Nonspecific methods:

Density
  Radiometric
  Other
Viscometry
Thermal conductivity
Optical scattering and extinction
Acoustic

Group-specific methods:

Electrical conductivity
Flame ionization
Photoionization
Methods based on group-specific auxiliary reaction
Calorimetry

Specific methods:

Paramagnetism
Photometry
  Absorption (UV, VIS, NIR, IR)
  Diffuse reflectance
Gas chromatography
Liquid chromatography
Mass spectrometry
Electrochemical methods
  Potentiometry
  Amperometry
Titration
Analytical methods based on auxiliary reaction
  Chemoluminescence
  Absorption in aerosols or liquids
Special moisture measuring techniques

The concept presented here makes it possible to create flexible CAE tools, which will gain increasing importance in the design of process monitoring and control systems. For example, interactive and iterative equipment selection by the design engineer is aided by computer-generated querying of the features introduced here [5.96]. There is no need of once-for-all solutions or a priori optimization criteria. Hypermedia concepts can serve as navigation or explanatory aids [5.97].

Determination of Communications Costs. Field-level communications systems (i.e., field bus systems) will be essential components of a networked system. The design of future field bus systems will be based on the information econo my derived from the phase model (Fig. 5.32). A suitable field bus system will be chosen on the basis of a taxonomy, here the taxonomy of field bus systems.

Typical message lengths for sensors and acuators, as well as maximum cycle times derived from the process and product properties, are balanced in order to arrive at the communications costs for the system as a whole. The number of field bus systems and the number of segments in these systems are set on this basis. The result is a

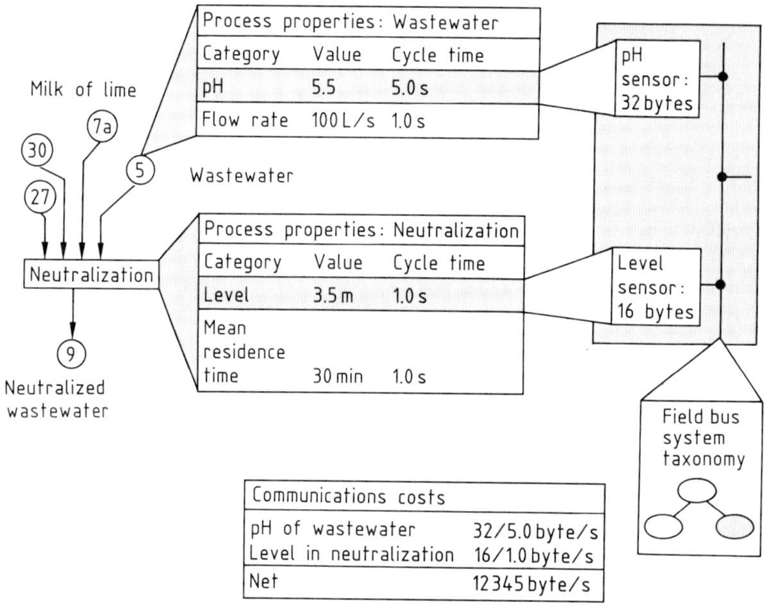

Figure 5.32. Determination of communications costs

picture of a CAE tool for the information-oriented design of communications systems [5.98]. The advantages of information modeling dictated by the process are again manifest here.

A proposal for a taxonomy of field bus systems is shown in Figure 5.33 [5.98]. The generic attributes are architectural features and the communications structure. Monomaster systems have a pure master/slave structure with strictly hierarchical vertical communications between planes. Multimaster or peer-to-peer systems, in contrast, permit horizontal communications within a plane. For the wastewater treatment example of Figure 5.27, both monomaster structures within a plant section and multimaster structures for communications between plant sections are possible. Communications between local components and sensors (e.g., pH and flow sensors) within a plant section might be based on the PROFIBUS or the DIN measuring bus, while communications between plant sections and with higher-level central functions might be implemented on the basis of PROFIBUS.

General Requirements on Process Sensor Systems. A complete introduction to process sensor system technology would also cover selectivities, detection thresholds, time dependences, sensitivities, stabilities, and so forth, all qualities that have to be defined for each sensor system [5.63], [5.99]. These are technical specifications that differ from one measurement task to another because of differences in service conditions, products, perturbations, and so forth.

The profile of requirements on sensor systems is strongly influenced by the downstream computer processing of the information:

- In-line and ex-line sensor systems must not disturb the process.
- It is essential to separate the energy channel from the information channel.
- Communications capabilities and system capabilities must be insured.
- Electromagnetic compatibility is also required; in some cases, external constraints dictate "intrinsic safety" as well [5.7]. See the generic class "Design Features."
- Special conditions in chemical production (corrosion, abrasion, etc.) must be taken into account when choosing the sensor principle and construction material.
- One special requirement for in-line sensor systems is that they must remain functional even under extraordinary process conditions (e.g., upon a change in the state of aggregation).

Economic Aspects. From the standpoint of the process control engineer who designs, commissions, and maintains a sensor system, meeting the stated requirements is closely linked to

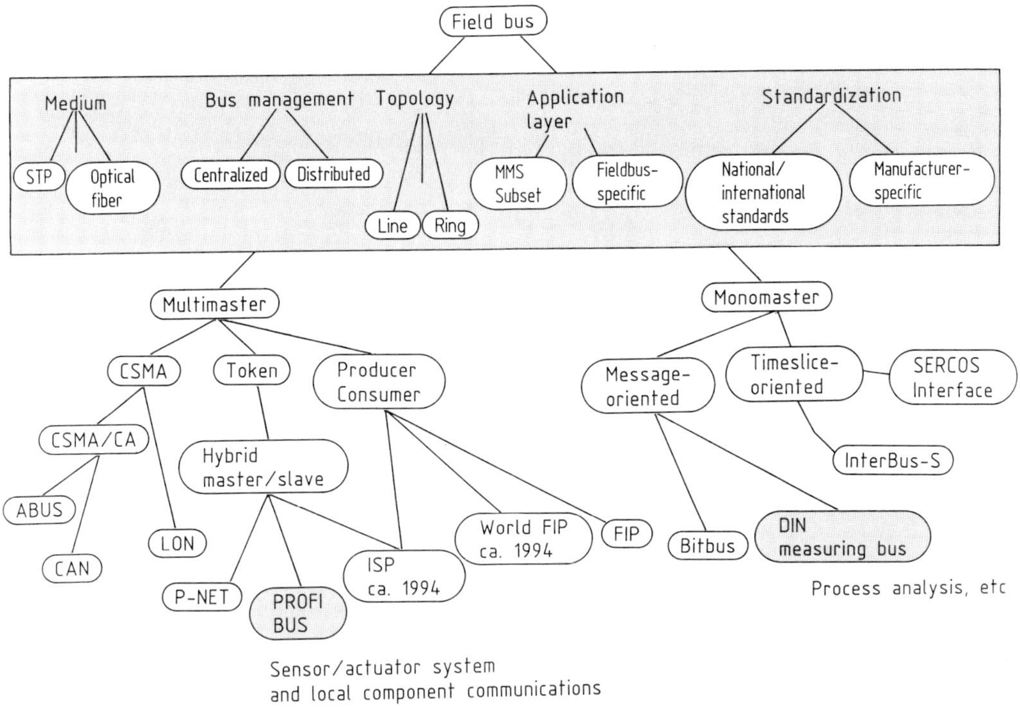

Figure 5.33. Taxonomy of field bus systems [5.98]

the lifetime costs of the sensor system. The term "cost of ownership" expresses the fact that the key criteria include not only the technical specifications and the resulting investment costs, but also longer-term aspects of sensor system use, such as training and maintenance costs. As discussed in Section 2.1, a holistic approach [5.100] must be taken from the outset.

In general, however, the main stress should be on the need for a process control station and a process sensor system. Total costs for a process monitoring and control system are largely determined by this sum. What is decisive is not the initial cost of devices and systems; the field installation costs make up a respectable fraction of total costs (see Section 10.2 and Fig. 10.12). Finally, the costs of control system components depends in part on the number of process control stations.

Often a fear of safety aspects and regulations leads to an exessive number of process control stations being set without a careful and critical analysis of the necessary equipment availability (Section 10.4).

Inadequate knowledge about the process at the time when costs are determined can, on the other hand, result in too few process control stations or incorrect equipment selection. This problem often cannot be corrected until commissioning, by which time costs are very high.

The departures from the original design may give nonspecialists the impression of "overengineering," especially if process control engineering resources must be used to correct solutions that are nonoptimal to poor from the process engineering standpoint. Overengineering due to ignorance is the most common cause of this phenomenon.

It is therefore essential that some tool such as the phase model of Figures 2.35 and 2.36 be employed at the bidding specification stage, in order to take full account of all information needed for the production process.

5.3. Sensor Systems for Special Applications

In accordance with HESSE's suggestion [5.15] to put systems concepts at the beginning, Section 5.1 deals with systems integration of sensors or sensor systems, while Section 5.2 discusses process sensor systems from the standpoint of internal information flows and the taxonomies derived therefrom.

Nevertheless, it is necessary to descend from the lofty Olympus of informatics to the lowlands of the reality that must be dealt with at a pragmatic level.

Sensor systems behave very differently in the laboratory and the plant (see Section 5.2). The tasks they must perform, and therefore the requirements they have to meet, vary from one application to another.

Sensor Systems in Manufacturing [5.101]. For example, a sensor system in a manufacturing industry must provide an industrial robot with information about its surroundings that are relevant to the operation but not predictable. The robot must be able to react to this information in such a way that the desired result is achieved or maintained automatically. This is why a variety of sensors are employed to determine physical quantities and states, transform them into (generally) electrical signals, process them, and interpret them.

As a rule, information transmitted to the robot influences

- The position and orientation of the robot's manipulator or tool
- The speed at which it moves
- The execution of the program

In rough terms, sensors of interest for the field of industrial robots can be classified into

- Tactile sensors
- Noncontact sensors

Another principle of classification is the complexity of the information acquired, allowing sensors to be grouped into

- Sensors that transmit binary signals
- Sensors that transmit one-dimensional measurements (forces in a preferred direction, distances)
- Sensors that transmit multidimensional measurements (forces of arbitrary orientation)
- Sensors capable of pattern recognition (workpiece identification)

Figure 5.34 shows a taxonomy based on these criteria.

Under each of the categories cited, a number of functional principles can be employed, depending on the prevailing boundary conditions (Fig. 5.35).

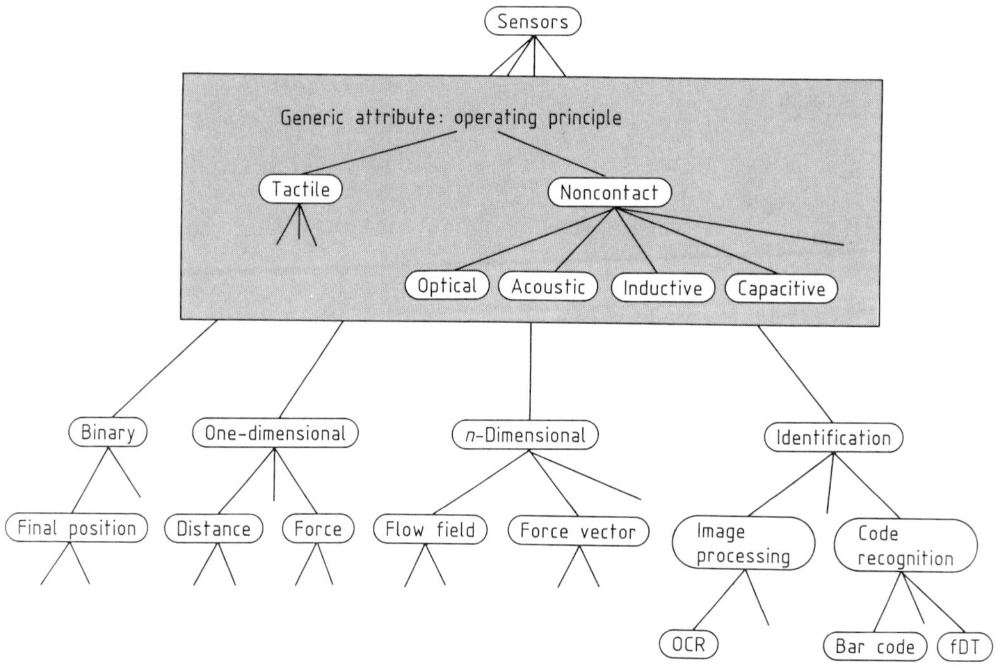

Figure 5.34. Taxonomy of sensors for manufacturing

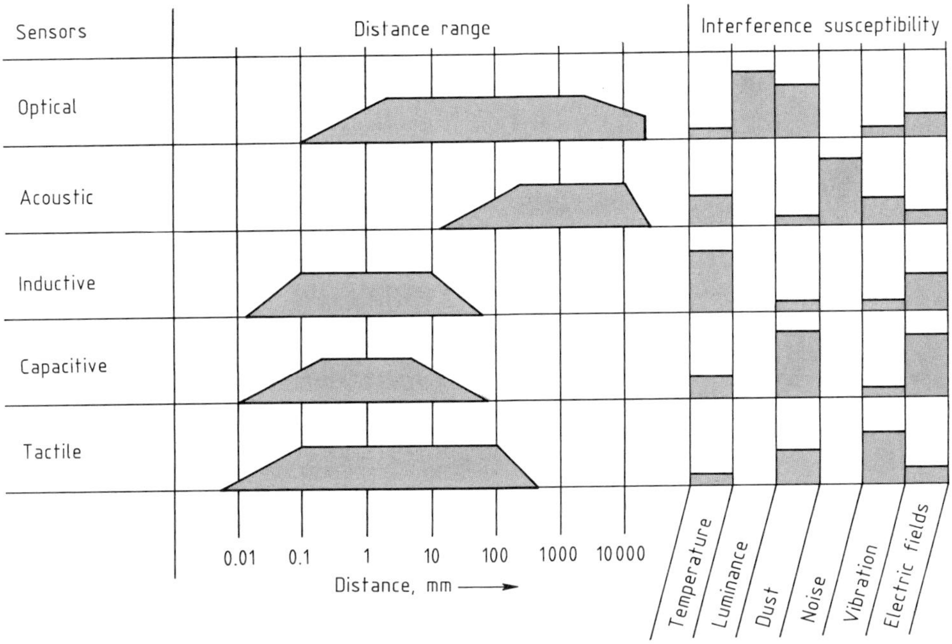

Figure 5.35. Distance measuring systems for industrial robots [5.101]

The linking of robots to image-processing, pattern recognizing sensor systems and to tactile, multidimensional force and moment sensors is an especially urgent problem for future robotics applications.

Important future applications that can be exploited only with the aid of powerful sensors for industrial robots include

- The machining of workpieces
- Assembly
- Emptying of containers
- Continuous joining operations such as welding and seam cementing

The emptying of containers is a typical application for imaging sensors, which have the task of identifying workpieces in the container, measuring their orientations and positions, and providing the industrial robot with the information required to grasp the workpieces correctly.

The areas of machining and assembly are important applications for tactile sensors.

One task that illustrates significant problems and requirements in connection with industrial robots is machining with tolerances specific to the tool, the workpiece, and the robot.

If the tolerances and deviations exceed a certain value, it is no longer possible to achieve satisfactory machining results within a series of workpieces. An optimally designed, and therefore workpiece-variable, control of the machining process is only possible if the most important toleranced quantities are determined by sensors and the process is automatically optimized.

Schweizer and Warnecke distinguish two groups of variables: technological variables oriented to the machining process, such as power consumption and cutting forces, and geometrical variables such as tool wear, geometrical and clamping tolerances, and surface quality. The fundamental problem in such tasks lies in the high data rate and the mathematically demanding simulation process. Adaptation to a variety of tasks also complicates this application, which most commonly involves highly dedicated sensor systems.

Artificial sensors cannot offer the capabilities and flexibility of human senses, but they can reliably perform single functions in manufacturing. A survey of current approaches in the field of quality assurance through production metrology can be found in [5.102].

In assessing future sensor technology, it is useful to compare biological and technical sensor systems [5.13].

In many areas technical sensors can be made substantially more powerful than biological ones. Furthermore, the spectrum of technical sensors is broader; that is, there are technical sensors for a wider variety of physical variables than nature has provided for among biological sensors. For example, technical sensors exist that can detect high- and low-frequency electromagnetic waves, electrical currents, and electric field strengths.

With regard to preprocessing by the sensor system, however, nature is far superior to artifice. Technical sensor systems at present are designed chiefly to detect events in preplanned technical processes. Accordingly, they can be more highly specialized for the expected events. Thus the knowledge that a foreseeable process event has taken place often comes solely from the fact that the sensor has responded. Scarcely any preprocessing is needed; the sensor "speaks for itself." By the same token, however, the system loses flexibility with respect to unexpected events. And this flexibility is particularly important for biological sensor systems, under some circumstances more important than high performance of individual sensors. Biological systems are designed to be more universal, while technical systems are more specialized. This notion is expressed in the term dedicated systems.

Most biological sensor systems are more "intelligently" constructed than artificial systems: Preprocessing takes place in amazingly powerful processing systems, with the result that vastly many individual items of information acquired by many identical sensors are reduced to summary information in a process of "super-sign" formation. It is hardly possible to imagine the demand imposed on the processing system, not just by "concept-forming" logic but also by the coordination of the many sensors in arrays. In the eye, for example, the sensor array of the retina comprises millions of cone and rod cells. Nature again and again offers a very impressive demonstration to technology, showing how flexibly such array systems can react to unforeseen processes of many kinds. The array principle is now being increasingly used in technical systems, specifically where sensors and information-processing systems, given the proper degree of miniaturization and economy, can be integrated to form "intelligent" systems. This incipient symbiosis between electronic sensor technology and the microcomputer is opening up new dimensions for an innovative potential that has scarcely been tapped before [5.13].

Sensor Systems in Medical Technology [5.103]. Sensor systems used in medical diagnosis and monitoring must meet different requirements.

The physician, in examining a patient, used to be dependent on what the patient and his relatives said, the physician's own experience, and above all his senses. Today, medical technology, developing in parallel with general technology, offers valuable help to the physician. The range of his sensory perceptions, in particular, has been hugely expanded by technical resources, sensors (i.e., technical sensory organs) playing an essential role. Milestones in this development include the discovery of X rays by Roentgen in 1896 and the recording of a electrocardiogram by Einthoven in 1902. Procedures based on these techniques remain the cornerstones of clinical and practical diagnosis today.

The principal objective in the use of medical instruments is to support the physician in administering the optimal treatment to the patient. Important narrower objectives in diagnosis and monitoring are to acquire reliable diagnostic information, place the lightest possible burden on the patient, and hold the user's costs and efforts within reasonable limits.

Three areas of medical technology are especially dependent on sensors: imaging techniques [5.104], physiological measurements [5.105], [5.106], and laboratory analytical techniques. In imaging and physiological measurement, the sensor is closely linked to the patient (Fig. 5.36); application aids or transmission media are also employed in many cases.

Only sensors and sensor systems with electrical output signals offer the possibility of instrumental processing, representation, and storage of information. A wide range of methods, drawn from a wide range of sciences, are available for the conditioning and interpretation of these signals; see, for example, the approach to dynamic entropy analysis in [5.107], [5.108].

Nonelectrical sensors that are still of major practical importance include X ray film and mercury thermometers and manometers.

The quality of the signals delivered by the sensor has a substantial influence on the value of the technical system as a whole. Often, no additional useful information can be derived from a poor output signal, no matter how extensive the post-processing. This consideration alone means that the effort of developing and maintaining sensors must be continued along with the work of applying them.

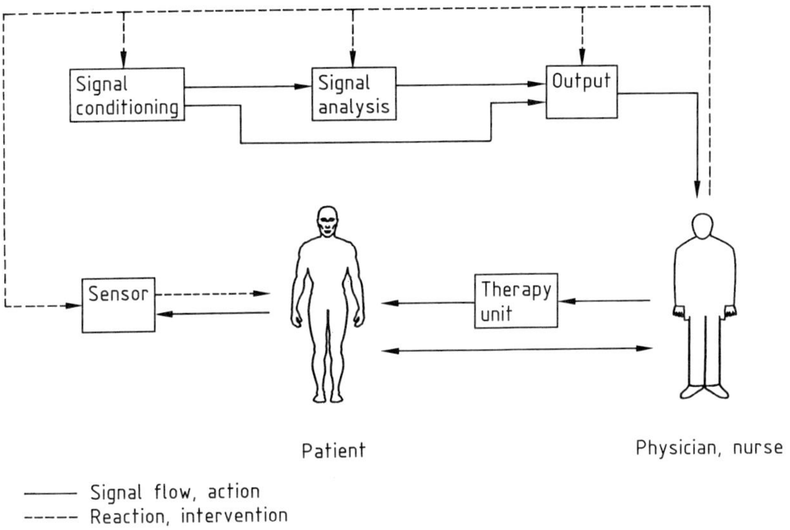

Figure 5.36. Diagnosis and therapy with medical devices [5.103]

Imaging Techniques. Medical imaging techniques can reveal information that is not directly accessible to the human eye, either because of the nature and frequency of the radiation or because of the geometry of the body (Table 5.2). Except for thermography, these are active procedures; that is, the measurement is based on energy introduced into the body in the form of radiation or radiation-emitting substances. Additional contrast-enhancing substances are often applied as well. Crucial advances with regard to exposure time and image quality in some techniques have been achieved by using arrays (with many detectors working in parallel) instead of discrete detectors. Nuclear medicine employs as many as 60 photomultipliers; X-ray computer tomography (CT), over 2000 detectors; and ultrasonic examination methods, up to 200 piezoelements.

More recently, nuclear magnetic resonance (NMR) tomography has become an established technique. It has the advantage of using strong magnetic and high-frequency fields instead of ionizing radiation, so that the patient avoids the risks associated with radiation exposure.

Significant results can be anticipated from positron emission tomography (PET), which can be regarded as a symbiosis of nuclear medicine and computer tomography. This technique permits not only the three-dimensional determination of blood circulation in the brain, but also quantitative studies of metabolism. Together with single-photon emission computer tomogra-

Table 5.2. Sensors for imaging procedures in medical technology [5.103]

Imaging procedure	Sensor	Radiation
Nuclear medicine	scintillation crystal (NaI) + photomultiplier	γ-rays (radioisotope sources)
Classical x-ray (transmission image)	film + intensifier film ($CaWO_3$) image intensifier storage film (Ba halides)	X rays
X-ray computerized tomography, NMR tomography, PET, SPECT	discrete scintillators (CsI) gas detectors (Xe) semiconductors (Ge, CdTe)	X rays
Endoscopy	fiber optics	visible light
Thermography	thermofilm (liquid crystal) bolometer IR vidicon	infrared light (passive)
Nuclear spin tomography	HF antenna	high-frequency pulses, magnetic fields
Ultrasound: B display (real-time, compound) Transmission camera Transmission CT Doppler	piezoceramic piezofilm	acoustic waves (MHz)

phy (SPECT), this procedure raises the hope that diseases in the future may be detected long before anatomical lesions or other symptoms develop [5.109], [5.110].

The images produced by nonelectrical sensor systems, such as X-ray film (with intensifying film) or fiber-optic endoscopy, can be electronically scanned for further processing.

Ultrasonic imaging presents a difficulty not encountered with other imaging techniques: the coupling of the sensor to the body. The contact must be made through a couplant having acoustic properties (impedance) similar to those of water or tissue; serious disturbances result from air inclusions.

Physiological Measurements. The task of physiological metrology is to determine time-dependent states and functions of individual organs and organ systems in the body.

In contrast to imaging techniques, the procedures in physiological metrology are mostly passive; that is, the signals acquired are chiefly mechanical and electrical signals inherent in the body. Quantitative results are needed in most cases. As a consequence, instruments used for temperature and pressure measurement in a medical setting must be calibrated against standards.

Parallels to process sensor system technology, with regard to nearly all application-oriented generic attributes (see Fig. 5.30), abound in this field.

In Table 5.3, important and characteristic applications are listed by groups of organ and the quantities measured. In addition, the relevant sensors and sensor principles are indicated. Further, the table indicates in what setting the techniques are used: D stands for clinical and practical diagnosis, M for monitoring (chiefly intensive monitoring), L for long-term diagnosis or monitoring under everyday conditions [5.111]. If a procedure can also be invasive, that is, if the sensor or parts of the sensor system are introduced into the body for the duration of the measurement or the body must be opened for the measurement, this is indicated by the letter I. Often information can be obtained only by invasive measurement; the decision is then based not only on the balance of benefits and risks for the patient, but also the cost of the sensor, which is influenced by more stringent requirements on quality and miniaturization.

In general, the following points are taken into consideration in the development and selection of sensors:

- Essential precondition: a satisfactory functional principle
- Reliable application with the least possible expenditure of time and materials for routine uses
- Compliance with national and international standards of technical safety
- Compatibility of materials, especially for internal and long-term use
- Resistance to a wide range of mechanical loads and chemical attack (contact agents, body substances, cleaning, disinfection, sterilization)
- Transportability and adequate shelf life under variable ambient conditions
- Economic mass production with reproducible properties

Robust sensors of simple design, which require no involved preparation or post-treatment, are advantageous; the increasing adoption of one-time articles is a direct consequence.

• Electrodes for Surface Potentials: One of the most frequently performed physiological measurements is studying the electrical activity of organs by measuring electric potential differences at the surface of the body. The sensor system used to measure these low voltages (EKG amplitude ca. 1 mV; EEG amplitude ca. 50 μV) includes the electrodes and the electrolyte (each electrode plus electrolyte constitutes a half-cell) together with attachment agents, power cords, and connectors.

The criterion for choosing the electrodes and electrolyte is a well-defined, stable electrode potential with a low noise level. The time required to reach equilibrium after mechanical or electrical disturbances should be in the range of seconds, and the effect of imposed d.c. or a.c. currents should be slight (time-varying overvoltages, corrosion).

Electrolytes and adhesives should be skin-compatible; the electrolyte, as contact agent, must also lower the skin's electrical impedance at the point of the measurement.

For applications in which the required EKG quality is not too high, robust metal electrodes are frequently used. Where the requirements are more stringent or the conditions are more difficult (stress and long-term EKG, intensive care), the Ag–AgCl electrode is the only type employed. It is also the preferred electrode in ready-to-use one-time electrodes precharged with electrolyte.

• Sensors in Intensive Care: The principal intensive care sensor applications are in surgery and neurosurgery, cardiology, internal medicine, obstetrics/gynecology, and pediatrics. Typically the patient is connected to various diagnostic and therapeutic systems (e.g., infusion, respiration, dialysis) and also continuously attended. The sensors or probes attached to the body must not interfere with care and therapeutic measures and, conversely, must be affected as little as possible by these measures; sensors, instruments, and accessories must be as simple and reliable as possible.

A variety of designs are available for EKG, blood pressure, and temperature measurement in all fields:

Table 5.3. Physiological quantities and sensors listed by organ system [5.103]

Quantity	Sensor/ principle	Heart, circulatory	Respiratory	Gastro-intestinal, urogenital	Brain, nervous system, sense organs	Skin, connective
Force, acceleration	piezoresistive, piezoelectric, optical	D M ballistic cardiogram	D M respiratory mechanics, monitoring of apnea		D tremor	D M L motoric, weight
Pressure		D M L I blood pressure	D M respiratory pressure, resistance	D M I bladder pressure, uterine contraction pressure	D M I intracranial pressure, intraocular pressure	
Flow rate	optical, thermal, electromagnetic, ultrasound	D M I blood flow, pulse, stroke volume	D M respiratory flow, volume	D I urine flow, volume		
Sound	microphone	D heart sounds, Korotkoff's sounds	D auscultation	D intestinal sounds, fetal heartbeat	D hearing	
Temperature	thermal (NTC)	D M L integral quantity: body-core, skin temperature				
Electric potential, impedance	electrodes	D M L I EKG, rheogram	D M rheogram	D M fetal EKG, EMG	D M I EEG	D EMG
Magnetic properties	SQUID	D myocardiogram	D silicosis	D iron in the liver	D MEG	
Optical properties	photodetector			D amniotic fluid	D eye	D
Gases	electrochemical (polarography, potentiometry), spectroscopic	D M I O_2, CO_2, pH oximetry	D M O_2, CO_2	D I pH		D M I O_2, CO_2; tissue, transcutaneous
Ions				D dialysis		D I
Organic substances		D I glucose enzymes	D alcohol acetone	D L		D I

* D = diagnostic use; M = monitoring; L = long-term diagnosis or monitoring; I = invasive.

- Ag–AgCl adhesive electrodes for adults and children, as well as for operations
- Blood pressure measurement by intermittent noninvasive means (cuff) or by continuous invasive means (catheter) with external pressure transducer
- Temperature sensors for body core or skin temperature

In addition, specific procedures and transducers are used in certain special areas, for example:

- Photoelectric monitoring of the peripheral pulse
- Monitoring of respiration in adults (thermistor) and in infants (rheographic, noncontact methods)
- Cerebral pressure
- Fetal EKG and contractions in obstetrics
- Monitoring of oxygen supply: CO_2 in respired air (by infrared absorption); transcutaneous measurement of O_2 and CO_2, chiefly in newborns

Sensors and their application must meet higher standards when used with infants, since all the bodily dimensions are much smaller than in adults and both risks and interactions must be more strictly avoided.

• Invasive Measurement of Chemical Quantities: Except for CO_2 measurement in the respiration gas and transcutaneous measurement, the continuous measurement of substances involved in metabolism and other substances in the body has not yet become routine in intensive care medicine. Development of sensors for these purposes, especially in combination with catheters or for use as implants, has been going on for a long time; techniques known from the analytical laboratory, such as spectroscopy, polarography, and potentiometry, are in the foreground.

The use of specific membranes, electrolytes, and enzymes makes it possible to modify electrochemical sensors (oxygen and pH electrodes) for the measurement of other substances, such as glucose, CO_2, or ions.

The miniaturization of sensors and the mechanical and electrical connection of sensors to a catheter are tasks that have already been accomplished or are at least feasible in principle. Applications involving contact with blood have still not been satisfactorily solved, because in its properties and its interaction with materials, blood is probably more complex than any other fluid encountered in technology. The deposition of fibrin and other proteins in the sensor pocket has two consequences: an increased risk of thrombosis for the patient, and disruption of the primary contact surface for the measurement.

The routine adoption of new catheter sensors is also difficult because the station personnel must possess additional knowledge and capabilities, for example, in the operation and calibration of the instruments and in the assessment of the measurements.

Requirements as to the reliability and accuracy of catheter sensors are far surpassed by the requirements imposed on implantable sensors and sensors for controlled therapeutic procedures. Therapeutic systems controlled with chemical sensors are a topic of major interest for the future.

Promising applications for intensive care are respiration, dialysis, and balancing, for example the regulation of the water, electrolyte, and energy economy of the burn victims. Implantable devices for the delivery of medication, such as insulin delivery controlled by a glucose sensor, are a key area for long-term therapy.

The transition from the laboratory analyzer, still routinely used for this purpose, to catheter sensors has not yet been completed, even though the use of integrated circuits has already led to major advances in the direction of miniaturized and multiple sensors. The next step—to implantable sensors with long-term stability—may well be equally large. A contribution toward solving this problem may be anticipated from biosensors, in which integrated electronic structures are built and operated in direct contact with organic materials.

Sensor applications range from imaging systems to the invasive measurement of metabolic variables, with cardiological diagnostics and intensive care as points of emphasis. Metrological methods and application techniques merit special attention, as does the management of the patient–sensor interface.

A comprehensive literature survey can be found in [5.102]. Information on research projects in the methods and applications of microsystems technology can be found in [5.112], which contains a detailed bibliography, and [5.113].

Sensor Systems in Automobiles [5.114]. Automobiles today must satisfy increasingly stringent requirements for pollution control, economy, safety, and comfort. These standards can be met only with the help of electronics. Sensors are the essential link between physical, chemical, and technical processes in the automobile on the one hand and electronics on the other.

Sensors in an automobile must offer high reliability and accuracy in a harsh environment, together with small size and low cost. Costs for transmitters as used in process technology, where quantities are correspondingly small, range from a few hundred to a few thousand Deutschmarks (DM); in contrast, the maximum long-term cost for many types of automobile sensors must be <10 DM.

Classification of Sensors in Automobiles. The sensors needed in automobiles can be classified by use, characteristic curve, output signal type, and interpretation.

- By Use:

- Functional sensors in drive-train control systems and in subsystems including the steering, braking, and body electronics.
- Sensors for monitoring and diagnostics, for example in the cooling, braking, lighting, fuel, and other systems.
- Sensors for safety and security purposes, for example in passenger protection, locking, and anti-theft systems.
- Sensors for information, for example in the measurement of fuel consumption and outside temperature, or in travel route monitoring [5.115].

- By Characteristic Curve: Which of the following characteristics is preferred will depend on the sensor application:

- Continuous, linear: Preferred for systems that perform continuous measurements with a wide control or display range. Uniform sensitivity is desired.
- Continuous, nonlinear: Preferred for systems that perform continuous measurements with a narrow control or display range. High sensitivity over a limited range is desired.
- Discontinuous, limiting value: Used for state-dependent "program variation" in complex systems, for example to enrich the mixture when the throttle valve is full open and to monitor the critical limits in the automobile.
- Discontinuous, stepped: Suitable for monitoring components subject to wear where replacement must be planned ahead of time, such as brake lining replacement before a vacation trip.

- By Output Signal and Interpretation:

- Voltage or current analog
- Amplitude analog
- Frequency analog
- Period analog
- Pulse-length analog
- Digital

Voltage and amplitude analog signals must be submitted to analog/digital (A/D) conversion before they can be processed in digital systems. Recently, however, inexpensive A/D converter circuits have appeared on the market in integrated form or partially integrated on a chip with the microprocessor. Integrated commuting circuits, known as multiplexers, make it possible to use these converters in a quasi-simultaneous manner for several pickups.

Time-varying signals such as frequency, period, and pulse-length analog signals have the advantage that they are not very sensitive to noise on transmission paths and they are easily digitized at the point of reception. The prerequisite, however, is that an appropriate converter circuit is located at the transmitter, because measured variables are not generally in this signal form. What is more, clocks and counters must be provided at the point where the information is to be processed.

Sensors with digital output signals have not been considered because of costs. The downward trend in price for integrated logic circuits, in particular microprocessors, has led to reconsideration of such new sensor configurations, in which microprocessors are included in the sensors. The resulting "smart sensors" offer marked advantages with regard to compensation of undesired variables as well as equalization; in some cases, A/D conversion can be carried out by the sensor itself. Code security procedures can insure reliable signal transmission, even in a noisy environment.

Requirements on Sensors

- Environmental Protection: Sensors used in automobiles operate in a hostile environment, being exposed to heat, cold, moisture, dirt, water spray, salt, electrical interference, mechanical vibrations, and mechanical shock. These factors make the measurement tasks immensely more difficult and represent a serious threat to sensor reliability and service life (Table 5.4).

The most severe conditions are experienced by sensors mounted directly on the engine. A temperature swing from − 40 °C to + 150 °C and even around 800 °C in the exhaust gas stream, as well as 100 *g* accelerations, are common environmental conditions here, but components installed in the engine compartment also experience harsh conditions (Table 5.5).

Whenever possible, sensors are therefore removed from the zones of most severe stress and installed at more favorable locations. For example, pressure sensors used to obtain the motor load signal are usually connected to the intake pipe by means of a hose.

• Reliability and Service Life: Reliability requirements for sensors on board an automobile are in all cases extremely high, but can be ranked as follows: safety functions, mobility functions, diagnostic and monitoring functions, and information. In addition to the needed reliability, there is often a further requirement that a sensor malfunction must also be detected so that the subsystem can switch over to emergency mode and permit at least a limited "limp home" function. Frequently this point has to be considered early in the sensor development process.

As to service life, sensors must also fit in with the trend toward longer-lasting automobiles. Expected values are more than ten years or > 3000 h service.

• Accuracy: Although accuracy requirements in many instances are as high as those in process plants, this accuracy has to be attained at much lower cost. This can only be achieved by mass production in a highly automated manufacturing process. Because the required accuracy is generally the main factor governing selection of sensor principle as well as cost, it is important to take great care in determining the acceptable error. Excessive requirements have a superproportional influence on cost. Accuracies of 2–3 % are needed for many applications.

• Space Requirement and Weight: The trend toward smaller and lighter automobiles, dictated chiefly by fuel economy, means that components have to be made small; accordingly, there is often little space available for the installation of sensors. Lightweight, space-saving sensor designs are therefore preferred. Small sensors, furthermore, have the advantage that they experience smaller accelerations due to vibration and impact and thus are less prone to malfunction.

Table 5.4. Environmental influences on sensors and typical effects [5.114]

Environmental influence	Typical effects
Temperature/temperature change	aging, embrittlement, fatigue, cracking
Atmospheric humidity (with dew and fog)	stray currents, surface and contact corrosion
Precipitation (rain, snow, ice, frost)	electrolysis, swelling
Pressure (air, water)	function
Sand, dust	stray currents
Sunlight	ageing, cracking
Aggressive substances (NaCl, SO_2, H_2S)	surface and contact corrosion
Mildew	stray currents
Salt water and sea mist (road deicing salt, coastal areas)	electrolysis, contact and surface corrosion
Sulfur dioxide (industrial regions)	surface corrosion
Ammonia, perspiration	creep due to stray currents, loss on insulating resistance
Fuel and fuel vapors	surface corrosion, swelling
Brake fluid	surface corrosion, swelling
Battery acid	surface and contact corrosion
Motor oil	swelling

Table 5.5. Service conditions of sensors installed in the engine compartment [5.114]

Mechanical	
Vibrational acceleration	(potholed road) chiefly stochastic chiefly low-frequency (max. 10–30 Hz) effective (time-average) value: 5–25 m/s² max. peak acceleration: 50–80 m/s²
Climatic	
Temperature	normal 120°C; extreme 140°C
Max. humidity	38°C/95% R.H.
Salt fog	24–96 h (DIN 50021), depending on packing density
Water	splash to flooding
Dust and sand	density up to max. 5 g/m³
Chemical	gasoline/oil vapors, alkaline cleansers
Biological	mildews and molds

Quantities Measured in an Automobile. Despite the wide range of electrical and electronic systems used in automobiles, the number of physical quantities measured by sensors is limited. The most frequently used sensors are those measuring distance, angle, rotation speed, pressure, and temperature.

Less common but still important are sensors for flow rate, acceleration, force and torque, partial pressure, and humidity.

Even though the list of quantities is not long, a considerable variety of sensor types are employed because of the diversity of media (streams) measured, measurement ranges, required accuracy, and environmental standards. Two examples are given below:

• Engine Electronics for Gasoline Engines: Fuel injection and ignition must be controlled jointly and matched to each other. In other words, every operating condition must be identified and serve as input to the process of fuel metering and ignition timing. This results in following improvements: increased fuel economy, increased engine power per unit of displacement, and reduction of pollutant levels in the exhaust.

The motor-electronics system needs sensors for engine speed, crankshaft position, air flow rate, air and engine temperatures, and intake pipe pressure (load signal). To insure compliance with stringent exhaust quality regulations, the system is augmented by a λ control with a λ probe. Other possible functions of the electronics include automatic transmission control, idle speed control, knock prevention, and exhaust recycling.

• Electronic Control of Diesel Engines: Diesel engines have traditionally featured mechanical control, but stricter air pollution and noise emission standards as well as demand have led to greater use of electronics. Figure 5.37 shows the process properties involved in optimizing a diesel engine along with the required sensors. Apart from the parameters listed above (e.g., rotation speed, temperature, pressure, and flow rate), quantities such as the beginning of injection at the nozzle and accelerator pedal position are measured.

The example of the sensor detecting the start of injection illustrates how harsh some of the sensor service conditions are. This pickup is directly integrated into the nozzle holder and is thus in the immediate vicinity of the combustion chamber.

Measuring Principles of Automotive Sensors. The technical implementation of sensors for automotive service can be based on a wide range of physical effects (Table 5.6).

Coating techniques are used in temperature sensors (e.g., Ni, Pt), strain gages (e.g., Ta), and rotation speed sensors (e.g., Permalloy).

The constant-resistance heated-wire sensor is used to measure air flow rates. The requisite long-term stability is imparted to the platinum

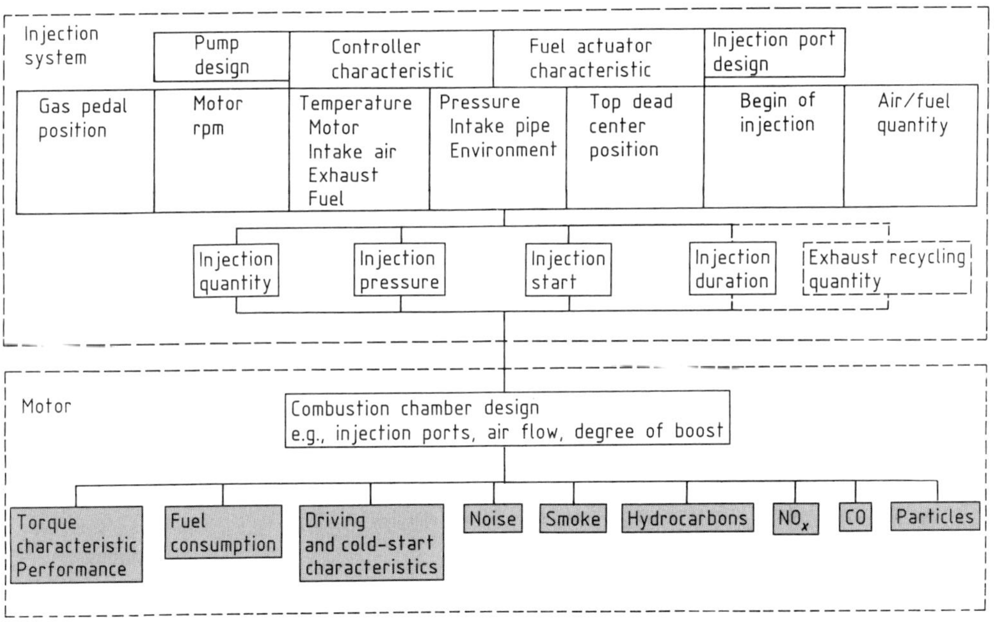

Figure 5.37. Process properties for optimization of a diesel engine [5.114]

Table 5.6. Technologies for automotive sensors [5.114]

Quantity measured	Sensor principle and sensors
Distance/angle	short-circuit ring or disk potentiometer
Rotation speed	galvanomagnetic effects (Hall, magnetoresistive)
Pressure	piezoresistive semiconductor pressure pickups diaphragm unit with Hall, thin-film strain gage, or capacitive pickoff thick-film pressure pickup
Temperature	thin-film metallic sensors semiconductor (Si) sensors
Flow rate	heated wire or film propeller (fuel) circulating-ball sensor
Acceleration	spring/mass system with strain-gage (cast film, thin-film) pickoff piezoelectric sensor piezoresistive semiconductor sensor
Torque	magnetoelastic sensor (torductor) eddy-current sensor optical sensor
Oxygen concentration	O_2 concentration probe with ZrO_2 solid electrolyte

wire by special surface treatment plus regular "refining."

Torque sensors would be useful especially in generating a load signal for electronically controlled automatic transmissions. The torque on the main shaft might be measured, for example. Because the shaft cannot be split, its elastic deformation can be exploited for this purpose. The eddy-current damping principle has proved suitable here. Two metal cylinders with lengthwise slots are slipped one over the other in the axial direction, with a clearance of about 10 cm, and rigidly attached to the shaft. When a torque is applied to the shaft, the cylinders rotate relative to one another. A high-frequency coil placed axially over the cylinder housings experiences unequal degrees of damping as a result.

Continuous acceleration sensors (accelerometers) are used for prompt collision detection so that passive passenger-restraint systems such as an airbag or belt tightener can be triggered. Spring-and-mass systems are a proven type of accelerometer; they are mechanically damped by being operated in silicone oil. A triangular leaf spring, preloaded in one direction, bears a mass at its free end. The inertia of the mass causes flexural loading of the spring in a roughly uniform manner, so that the spring is deflected accordingly. The flexural stress is detected by strain gages, which can also be thin-film devices.

In order to meet exhaust regulations and minimize pollutant emissions, the engine should be supplied with an air/fuel mixture as close as possible to stoichiometric ($\lambda = 1$), so as to insure proper functioning of the downstream catalysts.

Exhaust probes or λ probes permit accurate measurement of the oxygen content and regulation of the λ ratio. These probes have a characteristic curve such that even small departures from $\lambda = 1$ results in a very marked change in the voltage signal. For engines operated over a wider λ range, partial-pressure probes with continuous characteristics are in development.

The situation in the automotive sensor field can be summed up by saying that much has been achieved already but much development is still needed to reach further goals with regard to economy, safety, and exhaust cleanness with the aid of complex system concepts and electronics.

Above all, ways must be found of resolving conflicts between the objectives of performance and cost. One approach involves sensor technologies that offer major economies when put into mass production. The clear favorites at present are semiconductor and thin-film technologies. However, functional packaging, contacts, electronic interpretation, etc. account for a large fraction of sensor costs and largely determine performance and reliability. For these reasons, the industry must not simply further develop but must also invest in the development of these incidental technologies.

Sensor technologies of the same type, strongly oriented to specific requirements, are encountered in rail-mounted vehicles, ships, and aircraft. These application areas cannot be discussed further here.

Sensor Systems in Solid-State Process Engineering [5.116], [5.117]. Solid-state processes offer a further example of the relationships between product properties and process properties (as described in Chapter 3 and Section 5.1), especially with regard to quality assurance standards. The properties of disperse solids are influenced by the chemical composition and the physical state of the product (property function). For development, manufacturing, and use, it is necessary to know how the product state affects the quality attributes. Because quality attributes are determined on the end product, the characterization of the disperse state must be utilized in controlling the process. An important principle of quality assurance is that quality must be definable and thus inspectable. Metrology thus occupies a central place in quality assurance. The practices involved in quality assurance are oriented to the following steps, each of which contains specific problems for disperse products:

- All desired quality attributes are set forth in the profile of requirements
- The process must be structured
- Process control variables must be determined; property functions and modeling are particularly important here

Solid-State Processes. A solid-state process is a sequence of process elements put together in various combinations so that a solid with certain properties is created as the end product. A typical process sequence [5.118], [5.119] is shown in Figure 5.38 in terms of the phase model. As shown in Chapter 2, the phase model contains not only process elements but also the respective product states, and thus takes account of the way the product state changes in the course of the process.

After a chemical reaction, the solid is formed by crystallization or precipitation. It is separated from the liquid phase in a mechanical separation stage, such as filtration, then washed, and finally dried in a thermal operation. A confectioning step may follow, in which the product is put into the form offered for sale. Confectioning may include the unit operations of size reduction, agglomeration, mixing, or fractionation. The end product must ultimately be conveyed, stockpiled, packaged, or metered. The product state changes after each process element: from solution via suspension and filter cake to powder or granular form (depending on the drying or confectioning step). An important feature of the solid-state process is that the product properties do not depend just on the chemical composition but also depend heavily on the disperse properties of the product.

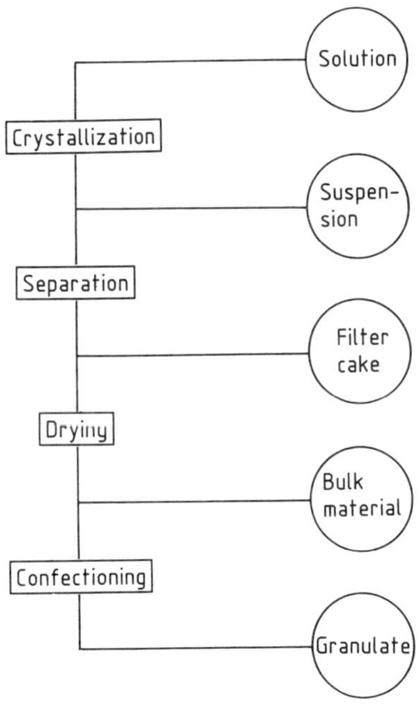

Figure 5.38. A typical solid-state process: process elements and product states [5.116]

RUMPF treats this dependence of the technological properties of a product on the physical properties by introducing the property function [5.120]:

$$\text{Technological property} = f\,(\text{physical property})$$
$$Q_i = f\,(E_{pi})E_{i\,\text{chem}} = \text{const.}$$

Technological properties Q_i	Physical properties E_{pi}
Processing Q_{i1}	Dispersity properties E_{pi}
Filterability	Size
Formability	Shape
Agglomeration behavior	Porosity
Application Q_{i2}	
Activity	
Action	
Instantaneous property	

Property functions apply to the sequence of process steps (processing properties) and also to the end properties, that is, the quality of the product (application properties). In filtration, for example, the fineness of the product is crucial; in the case of an automobile, the color properties depend in part on how finely divided the pigments are.

The product state is not only crucial to the end quality but must also be considered for each process element in solid-state processes. Not only does the product state change in each process element, the product state itself significantly influences the process element. A knowledge of how the quality attributes depend on the product state is essential for the development, manufacture, and use of products. Structuring of the process into units (unit operations) or meaningful areas is necessary if hold points with intermediate goal values are to be defined. These intermediate targets must be checked with appropriate methods of measurement, as a means of early process monitoring. The state of the end product is not simply established in the last process element; every process element throughout the process contributes to it. For this reason, close tolerances must be maintained even in the intermediate stages if the subsequent stages are to remain in the correctable range. The following steps must be considered from a quality assurance standpoint:

- Establishing the quality levels in the profile of requirements
- Structuring the process and identifying the process control variables
- Insuring the reproducibility of the process
- Optimizing the process

The various types of product quality target values are listed below for the example of pigment manufacturing:

Application properties
  Color strength
  Color coordinates
  Purity
  ⋮
Processing properties
  Pourability
  Freeness from dust
  Dispersion hardness
  Gloss
  Interfacial properties
  Rheological properties
  ⋮
Safety
  Low content of side products
  Easy to dispose of
  Nontoxic
  Incapable of dust explosion
  ⋮

Efforts must be made to achieve a number of quality attributes, and these target values relate not just to the final quality (characterized by the application properties) but also to processing properties (see Fig. 11.38), environmental aspects, and safety considerations. All these quality attributes must be monitored (i.e., measured). Two examples indicate how difficult this is.

Suppose that a paint, for example an auto body paint, is to be manufactured from a combination of pigments. The first substantial problems arise in the testing of tinting strength, since this measurement cannot be performed on a dry pigment. In practice, an entire paint factory must be "simulated" before this test can be performed; in other words, a procedure similar to that in the paint factory must be followed on a laboratory scale (Fig. 5.39). The test preparation must be done in the same medium, in the same technical facility, and with the same energy expenditure as would be used by the paint manufacturer; the paint must be applied by the same technique; and the same instrument must be employed for color measurement.

All these requirements for pigment quality inspection cannot, of course, be satisfied when

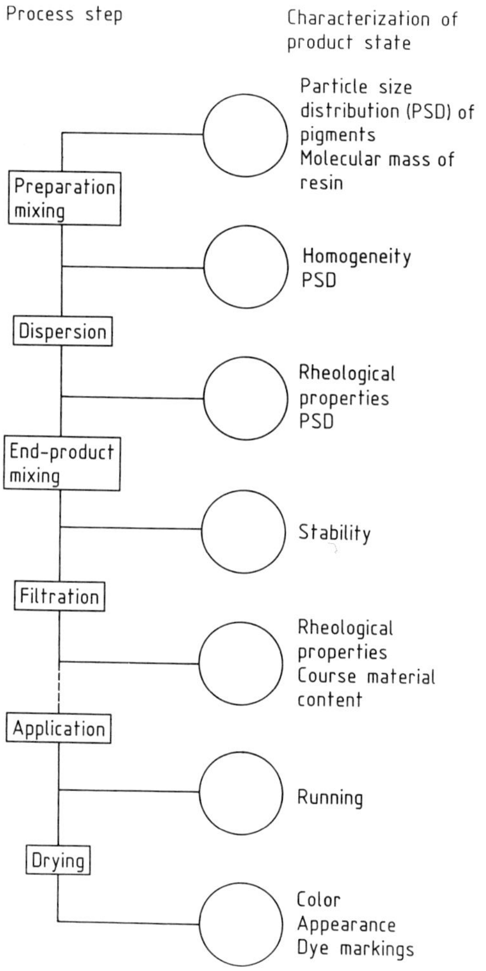

Figure 5.39. Process for manufacturing a paint, from the pigment to the painted surface [5.116]

the pigment is to be used in paint for several users. Instead, from the test in one system, in one apparatus, and so forth (representative testing), deductions must be made about the behavior in other types of apparatus. The development of tinting strength in two unlike dispersion machines illustrate the potential problems of such deductions. In Figure 5.40, the tinting strength is plotted as a function of dispersion time. Unequal energies are applied in different dispersion machines, for example in a laboratory apparatus and the full-scale system. As a rule, tinting strength is determined after one or at most two dispersion times. The result is that the quality attribute "tinting strength" varies widely when the dispersion times and energies cannot be matched, or unambiguously correlated, with those of the user.

This example shows what difficulties arise in quality inspection even when there is just one final target value. A process is designed and optimized for the final target values of a product, and the aim of quality assurance is to maintain these final target values. A standardized testing method leaves open questions of product quality for any individual case. Every user, however, wants a suitable pigment for its own formulation, where "formulation" or "recipe" covers both the paint system (including additives) and the apparatus and operating conditions.

Property Function. An important, necessary, but not sufficient step in quality assurance is quality control. As a rule, quality attributes are measured only on the end product. Thus quality can be controlled but not assured. Assurance, in the sense of "controlling," requires some way of influencing the process, and doing so at the earliest possible process stages. This in turn means that the process must be structured into single steps or meaningful subdivisions and that intermediate goals must be defined or established. These intermediate goals are suitable for early process control. The phase model (Fig. 5.38) and the property functions become especially important for quality assurance. Only when the relationships between quality attributes and physical quantities (property functions) are known is it possible to identify an appropriate process control variable. If the disperse state can be measured and controlled at the point where it is generated or changed, intervention in the process steps can be implemented at the proper time and thus quality can be controlled. In other words, a knowledge of the property functions is a prerequisite for the assurance of the quality of disperse products.

In the example of colored pigments discussed above, some of the property functions are well known. Figure 5.41 shows the target value tinting strength as a function of particle size. For two or more distinct target quantities, a compromise is necessary. The desired particle size should already be established at the crystallization stage.

The property functions must be known if crystallization is to be optimized with respect to

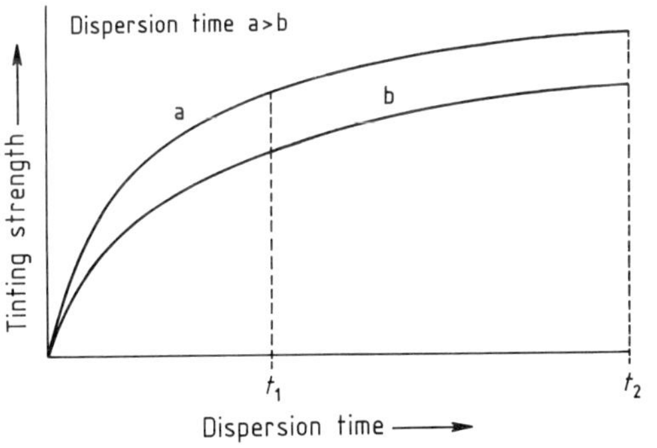

Figure 5.40. Tinting strength development for two dispersion energies [5.116]

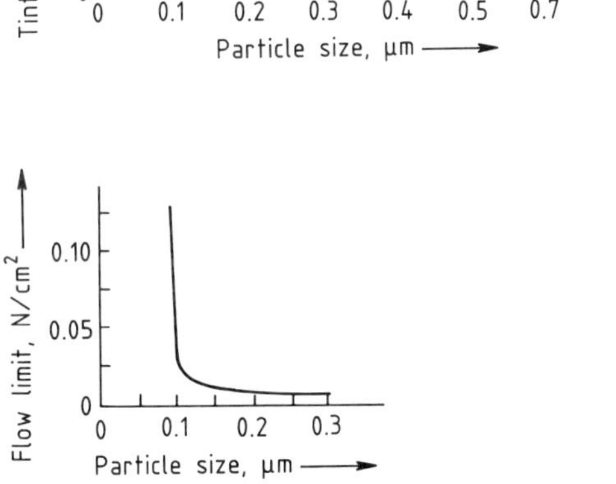

Figure 5.41. Property functions for a colored pigment [5.116]

the application properties. Each subsequent process step changes the particle size and the state of agglomeration [5.121]; that is, the generation of particles by crystallization should be controlled and changes later in the process should be avoided or monitored.

The dispersity values based on the property functions are important control variables and thus make it possible to implement early process control. To determine the property functions, both quantities must be known and measurable, namely the physical quantity (disperseness) and the technological quantity (application or processing quantity). The problems that arise in the measurement of the latter are discussed in Section 5.1. The measurement of dispersity is the focus of particle-size measurement techniques. Some aspects of particle-size measurement will be discussed briefly below.

Examples of particle properties with corresponding test methods are listed below:

Particle size (distribution), agglomerate size (distribution)	diverse methods
Specific surface area	BET method
Flow behavior	shear test
Bulk density	determination of volume and mass
Compressibility	determination of volume and mass, force–distance diagram on compression
Redispersibility	transmission measurements PSD measuremens for various degrees of dispersion
Wetting, porosity	liquid absorption, capillary pressure

Generally, the particle size (primary particles up to agglomerates) and the particle size distribution are of fundamental importance for solid-state process engineering. A number of methods are available for their measurement. These have been described in many publications and analyzed with respect to time, costs, and problems [5.122], [5.123]. The trend is toward measurement techniques and instruments that utilize the optical properties of disperse systems as a tool for characterization.

Instead of going further into individual measurement methods, the problems linked with the use of these methods is discussed here [5.124]; see Figure 5.18. As a rule, a sample must be taken from the process. (The special features of sampling of powders and granular materials have been thoroughly treated [5.125].) The sample must be transported to the measurement location and prepared so that the constraints imposed by the measurement technique can be complied with. Preparation may mean filtering or diluting the sample, cooling or heating it, dispersing it, or adding other substances to it. These measures, which can bring about a change in the product, transform the process state into the measurement state. Moreover, the time between sampling and result is often much too long. A measurement that does not become available until hours or even days after sampling is not useful for process control.

Thus the objective can only be on-line measurement [5.65] to obtain information on the product or process state quickly. Sample preparation must be automated so that the measurement state can be correlated with the process state in the most reproducible manner possible. Ultimately, however, in-line measurement (see "Generic Class: Application Features," p. 449) must be employed to generate the required information about the process state. Two examples of new in-line methods of importance to solid-state processes are presented briefly.

- Laser scanners [5.126]: The particles are scanned with a laser and a chord–length distribution is obtained. The probe can be immersed in any vessel whose contents in are in sufficient motion ($0.01 \leq v \leq 1$ m/s). The particle-size measurement range is 1–1000 µm.
- Photon correlation spectroscopy [5.127]: In this laser technique, particle mobility is measured on the basis of Brownian motion. The Einstein relation links the stochastic motion with the particle size via the diffusion coefficient. This method is in successful use, for example for process characterization in vitamin production.

A comprehensive exposition of particle and particle-system characterization can be found in [5.117], which treats this topic within the framework of sensor system technology. Particle measurement, which extends far beyond the boundaries of particle-size analysis, is completely discussed as to status, trend, and demand in [5.128], which builds on ideas similar to those that underlie quality assurance in solid-state process engineering [5.129].

Sensor Systems for Environmental Protection. The prevention or reduction of environmental pollution caused by industry, commerce, agriculture, households, etc. requires not only technical measures but often also measurements, whereby process sensor systems (see Section 5.2) are used to provide the required information about the properties of the relevant product or process element, about water and air pollution caused, and also about noise pollution.

To enable quick reaction in the sense of environmental responsibility, and also as mandated by environmental legislation, the required information must be available on-line (see Section 5.2).

For this purpose it is especially important that information about the product properties be obtained in undistorted form. This can be achieved by measurements on the relevant product itself, performed in-line, that is, in the actual product stream. This is not a simple matter with many present-day sensor systems.

The special features of sampling (in comparison with solids sampling as discussed elsewhere in this section), with the aim of insuring spatial and temporal representativeness, are discussed further on.

Figure 5.42 shows, in highly simplified form, the tasks of process sensor systems in environmental protection. These systems acquire information about the product and process properties, measure workplace pollutant levels for occupational safety and health reasons, and monitor wastewater, solid and gaseous emissions in stack gases, and noise emissions for environmental protection purposes [5.130].

Flue-Gas Monitoring Sensors. By way of example, some of the requirements for flue-gas monitoring are examined for a coal-fired plant.

- Power-Plant Flue-Gas Monitoring: The German regulations on power station emissions require continuous measurements of the mass concentrations of particulates, CO, halogen compounds, NO_x, and SO_2, and also of the volume concentration of O_2 (Fig. 5.43). All instruments used must be tested for suitability; proper installation of measuring devices must be verified by services designated by the authorities; calibrations must be carried out every five years (plants with thermal power under 300 MW) or every three years (plants over 300 MW). In addition, unannounced functional tests are performed yearly.

The measured values (instantaneous values) are interpreted by

- Taking half-hour averages
- Taking daily averages
- Converting the averages to reference oxygen contents
- Classification of O_2-referred averages and storage in the form of frequency distributions

Pursuant to the regulation, the emission limits are deemed to be complied with if

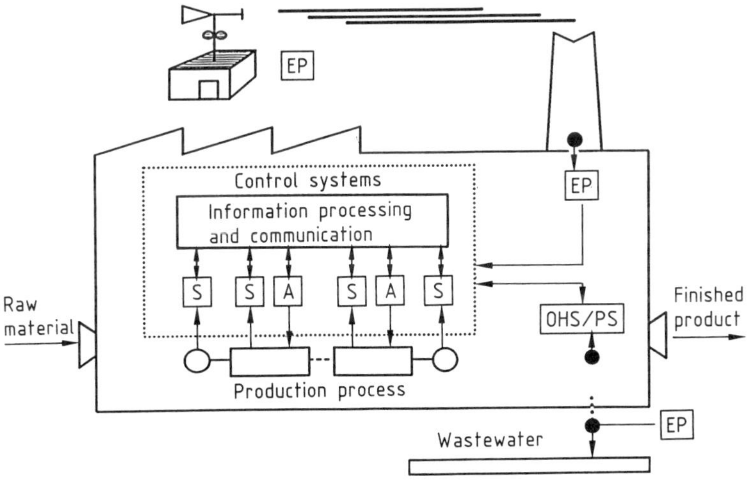

Figure 5.42. Process sensor systems (S) in environmental protection [5.130]
A = actuator system; EP = environmental protection; OHS = occupational health and safety; PS = plant safety

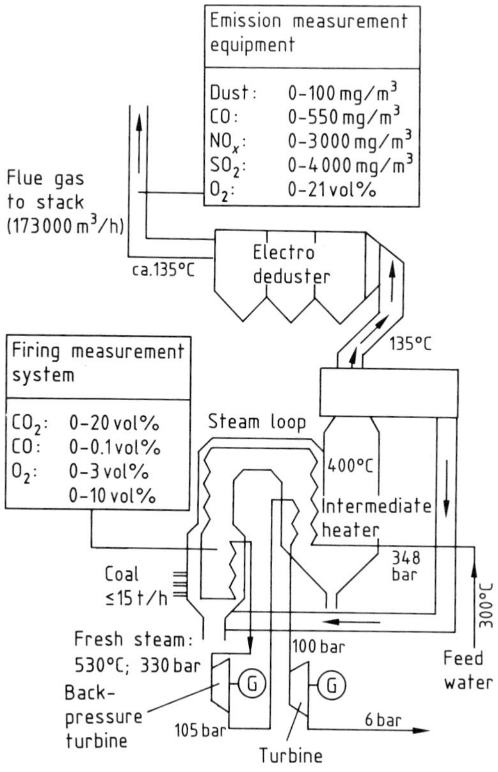

Figure 5.43. Example of a fired facility [5.130]

- All daily averages are ≤ 1.0 EL (EL = emission limit)
- 97 % of all half-hour averages are ≤ 1.2 EL
- 100 % of all half-hour averages are ≤ 2.0 EL

A cross-reference to the concepts of human–process communication (Section 11.2) is relevant here. If compliance with the emission limits is also to be monitored by the operating personnel on behalf of the operator, the levels should be represented by means of suitable presentation and interaction forms. An example of such a form is a cumulative frequency chart updated on-line (see Fig. 5.44) [5.131].

• Sludge Incinerator: Sludge from biological wastewater treatment is dewatered in filter presses and is then forwarded to a furnace for combustion. Such an incineration system includes a hot ("dry") stack gas scrub, which is subject to HCl and SO_2 controls (Fig. 5.45).

The emission measurement hardware is designed to comply with TA Luft. Emissions (concentrations) are referred to wet gas in the measurement stream, so that the content of water vapor must be determined each time. This is done directly on the basis of the measured O_2 levels in the wet gas (O_{2W}) and the dry gas (O_{2D}). The measurements are analyzed by averaging, referring to O_2, and classification. Instruments

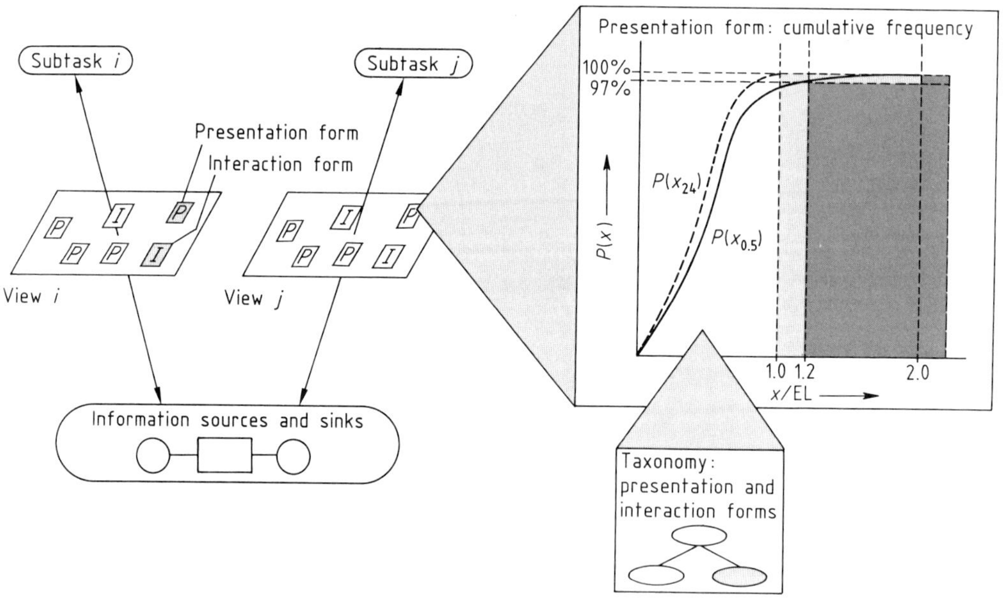

Figure 5.44. Viewpoints, presentation, and interaction [5.131]

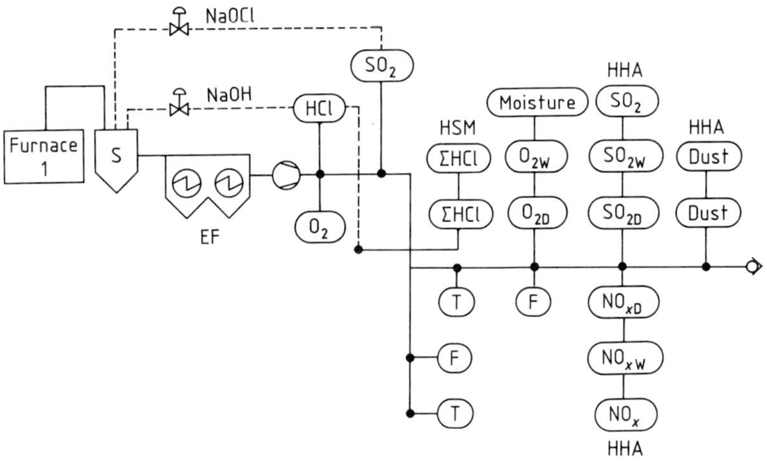

Figure 5.45. Diagram of analytical instrumentation installed in a sludge incinerator plant [5.130]
EF, electrostatic filtration; S, stack-gas scrubbing; HHA, half-hour average

are calibrated by a service holding an official permit.

• General Emission and Impact Monitoring of Gaseous Pollutants: Gaseous pollutants represent a special danger from the environmental and occupational safety and health standpoints.

Formerly, this problem could be handled only with expensive suction-type gas collecting systems, which delivered the mixture of pollutants to product-identifying sensor systems. Examples of such systems were spectroscopes, chromatographs, surface-sensitive metal oxides [5.132] and semiconductors, as well as physicochemical flame ionization detectors [5.133] and ZGSM gas detectors [5.134].

As described in Section 5.2 [5.7], [5.10], the sensor system proper then includes sample preparation, measurement, and application.

As early as 1985, WARNCKE [5.3] reported on optically integrating absorption measurements over long lines in the open atmosphere. The measurement of pollutant components in the infrared by this remote analysis technique is best performed with lasers. Rotating sensor systems place a "detection cone" around the object where emissions are expected to occur (Fig. 5.46).

By way of example, Figure 5.47 shows emission monitoring at a chlorine plant with an "analytical eye," an ultraviolet instrument using a conventional light source. The special innovation here is that the interference due to sunlight is compensated by using a pulsed light source.

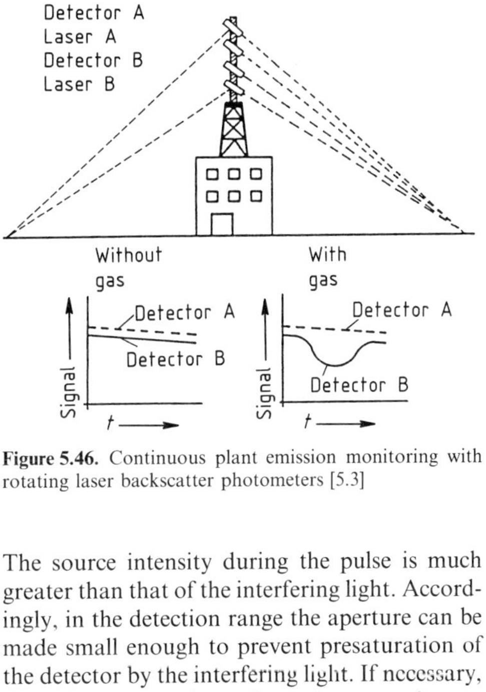

Figure 5.46. Continuous plant emission monitoring with rotating laser backscatter photometers [5.3]

The source intensity during the pulse is much greater than that of the interfering light. Accordingly, in the detection range the aperture can be made small enough to prevent presaturation of the detector by the interfering light. If necessary, measurements can be performed at two reference frequencies to compensate for light scattering by particles.

This procedure amounts to setting up an optical enclosure around the whole production plant. Despite the use of modern lock-in amplifiers, the cross-sensitivity and relatively high cost of the apparatus has kept these developments from gaining wide adoption.

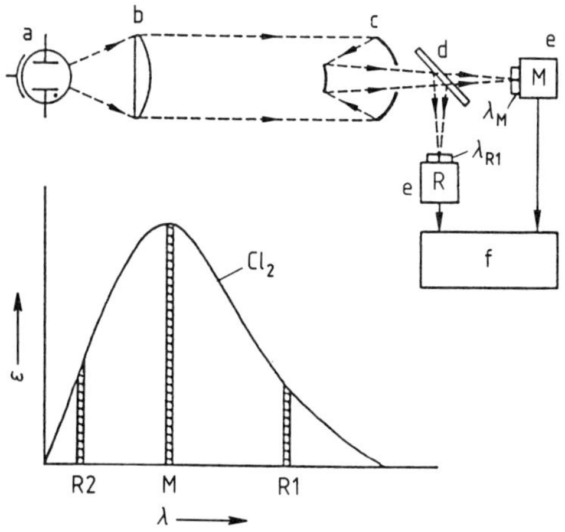

Figure 5.47. Optical monitoring of chlorine emissions [5.3]
a) Flash lamp; b) Lens; c) Telescope; d) Beam splitter; e) Detector; f) Evaluation

The current status of this process sensor system technology is surveyed by [5.135].

Wastewater Monitoring

• Monitoring of Combined Sewer Water: Much of the energy of reaction generated in a process is typically transferred to water (usually withdrawn from a stream) in heat exchangers. Along with storm sewer water, this cooling water is discharged directly into the receiving water through a combined sewer.

Contamination can occur when rain falls on roofs, streets, and other areas containing pollutants; in the ordinary drainage lines; or through leaks in the heat exchangers.

• Process Wastewater Monitoring: All other water used in the plant, including scrub liquor from final cleanup stages, must be conveyed through the now-standard system (including mechanical, chemical, and biological treatment processes) in order to comply with legislative requirements.

Without going into detail on national and international laws and regulations governing wastewater, or on the legal consequences of a violation [5.136] (see Chapter 12), Figure 5.48 illustrates some aspects of the situation in Germany.

The new philosophy of international environmental protection declares the protection of natural waters from pollution to be a high good, regardless of whether the polluting wastewaters originate in households, public facilities, or industry. Accordingly, the availability of treatment functions in treatment plants is declared to be beyond the scope of economic discussion.

Raab and Wachter-Büchner state in particular [5.137]: "In order to comply with the legislative mandate, precautions commensurate with specific potential hazards must be taken. These include, first, technical precautions that reliably prevent contamination of the wastewater or the occurrence of severe pollutant burdens in the first place. An effective complement to these primary technical measures is continuous analytical monitoring of wastewater streams for substances foreign to water. In the chemical industry, many such substances are carbon compounds, which often means organic contaminants. The parameter TOC (total organic carbon), which states the total content of organic carbon in mg C per liter of water, has proved to be a highly suitable quantity to measure for this purpose.

Water monitoring is not, however, limited to the detection of undesired organic contaminants in wastewaters. Increasingly, it is also employed in process monitoring. Along with environmental protection applications, TOC measurement can also be found in production monitoring and control. A company has control of its production only if it also knows the makeup of its wastewater. A topic of recent discussion is the continuous measurement of TOC burdens in wastewater streams (balancing), which is a prerequisite for source apportionment of wastewater streams within a production complex.

Environmental protection regulations and production monitoring needs dictate the measurement mission. If it is taken seriously, this means the continuous determination of TOC. Only if wastewater streams are subject to continuous, on-line monitoring as near as possible to a potential point of entry of water pollutants can inadvertent contaminants be detected promptly and countermeasures put quickly in train. The monitoring of process wastes also requires continuous surveillance, whether to prevent overloading of the treatment plant or to promptly detect improper states in the plant. There is a definite trend toward process TOC measurement, that is, continuous 24-h determination of TOC in the wastewater. From the metrological point of view, this is not a simple task; from the instrumentation standpoint, it requires substantial expenditure.

According to DIN 38409 Part 3 [5.138], TOC is the sum of organically fixed carbon in dissolved and undissolved organic compounds. The DOC (dissolved organic carbon), is the sum of carbon contained in dissolved organic compounds. Other terms in common use are TC and TIC. TC (total carbon) equals the sum of organically and inorganically fixed carbon in dissolved and undissolved compounds. The inorganic component of the TC is called TIC (total inorganic carbon). Each of these is expressed as a concentration in milligrams of carbon per liter (Fig. 5.49 [5.137]).

The basis for all methods used to determine organically fixed carbon in water is the oxidation of carbon to carbon dioxide. This can be accomplished in a thermal process (combustion), by wet chemistry (with suitable oxidizing agents), and/or by irradiation with ultraviolet light. The last method is restricted to solids-free samples.

The carbon dioxide resulting from oxidation is either determined directly or reduced with hydrogen and determined indirectly as methane. The determination can be carried out in the liquid or gas phase. Methods suitable for use in the liquid phase are photometry and conductometry. The most common technique, however, is to determine carbon dioxide in the gas phase by IR spectroscopy. For the TOC measurement methods in ordinary use, the individual process steps are summarized in Table 5.7 [5.137], together with the starting and end products of each phase transformation, up to the acquisition of the measurement.

Depending on the physical locations of the sensor systems and the way they are tied into the sewerage (process discharge, sewer junction, plant discharge, treatment plant), the requirements on the following performance features vary widely:

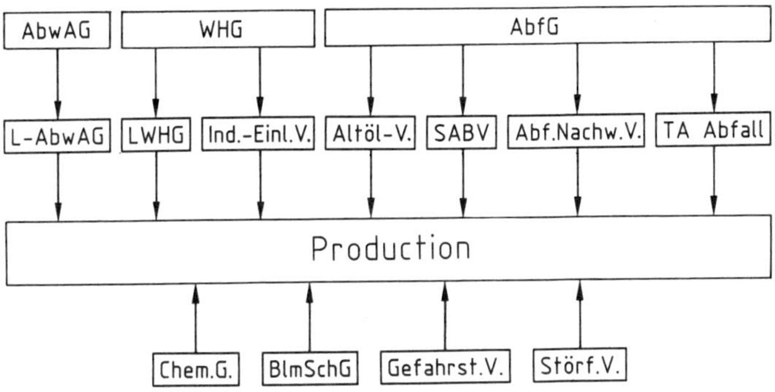

Figure 5.48. German wastewater legislation and regulations [5.136]
AbwAG: Wastewater Levy Act; WHG: Water Resources Management Act; AbfG: Waste Disposal Act; L-AbwAG: State Wastewater Levy Legislation; LWHG: State Water Resources Policy Legislation; Ind.-Einl.V.: Regulation on Indirect Dischargers; Altöl-V.: Waste Oil Regulation; SABV: Regulation on the Definition of Hazardous Waste; Abf.Nachw.V.: Waste and Residual Materials Monitoring Regulation; TA Abfall: Technical Directive on the Disposal of Hazardous Waste; Chem.G.: Chemicals Act; BImSchG: Federal Pollution Control Act; Gefahrst.V.: Hazardous Materials Regulation; Störf.V.: Regulation on Disruptions of Operations

	mg C per liter in			
	organic compounds		inorganic compounds	
	dissolved	undissolved	dissolved	undissolved
TC (total carbon)	O	O	O	O
TIC (total inorganic carbon)			O	O
TOC (total organic carbon)	O	O		
DOC (dissolved organic carbon)	O			

Figure 5.49. Definitions [5.137]

Table 5.7. Steps in conventional TOC measurement

Process step	Starting/intermediate product	Phase transition	Intermediate/end product
I	organically/inorganically fixed carbon (in water)	separation inorganic/organic species	CO_2 (inorganic) + volatile organic carbon compounds (in gas phase) organic carbon (in liquid phase)
II	organic carbon (in liquid + gas phases)	oxidation thermal chemical photochemical (UV)	$CO_2 + H_2O$ + reaction components (in gas phase)
III	$CO_2 + H_2O$ + reaction components (gas phase)	measurement gas preparation (e.g., drying) absorption of CO_2 (in liquid reagent) reaction (hydrogenation of CO_2)	CO_2 (in gas phase) CO_2 (in liquid phase) CH_4 (in gas phase)
IV Analysis	CO_2 (in gas phase) CO_2 (in liquid phase) CH_4 (in gas phase)	selective IR absorption (photometry) color reaction/conductivity change (photometry/conductometry) combustion (flame-ionization detector)	TOC (mg C/L)

- Selectivity
- Detection threshold
- Time behavior
- Availability

Besides the carbon contents TC, TIC, TOC, and DOC, water quality is also determined by other product properties [5.136]:

Physical and chemical product properties
- Temperature
- Conductivity
- pH, redox potential
- Turbidity
- Contaminants from process
- COD
- pO_2
- Salt
- Ammonium
- Nitrate
- Phosphate
- Heavy metals

Biological parameters
- BOD_5
- Saprobic index
- Toxicity

The methodology of wastewater monitoring is shown in Figure 5.50 [5.136].

On-line instrumentation is available for the determination of most product properties. Such fully automated devices take water samples, analyze them, and output measurements in computer-readable form in a continuous manner (or quasi-continuously in the case of cyclical analytical methods); see, for example, [5.139]. The spectrum of on-line measuring instruments that come under consideration ranges from robust devices proven over many years in process monitoring service to complicated and expensive special-purpose apparatus such as TOC analyzers (see above) and toximeters. Similarly broad is the range of performance features and costs for installation and maintenance of the measurement systems.

Representativeness in space and time merits special attention, since the products present in a plant wastewater system generally change from time to time, they are poorly mixed (in laminar flow), and if the pipes do not run full they do not reflect the actual product qualities. Furthermore, it must be ensured that no chemical reactions take place downstream of the sensor system that might lead to other pollutants that then remain undetected. The use of analytical instruments for wastewater analysis must always be examined from the standpoint of explosion hazard.

If there is no usable foreknowledge of the potential pollutants, measurement techniques of universal applicability must be employed; examples are TOC determination (see above) and flame–ionization detectors with phase changers. The apparatus is more expensive and cannot always be installed in an explosion-proof manner; the selection of physical location may also be severely limited.

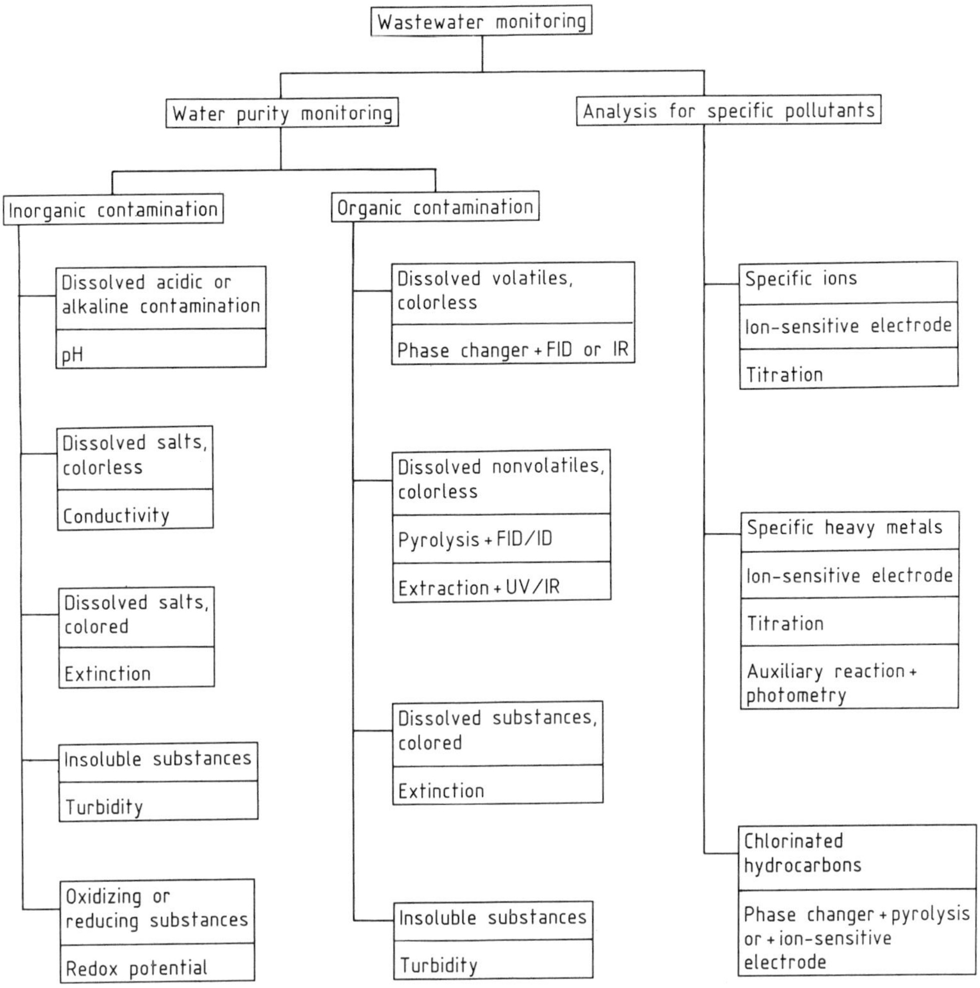

Figure 5.50. Methodology of wastewater quality monitoring [5.136]

If instrumentation is present at the process discharge, expansion is possible by relocation aimed at bringing internal substreams under continuous monitoring. In this way the source of wastewater upsets may be localized [5.130].

Sensor Systems for Noise Pollution Control. The well-known instrumental methods of noise measurement cannot be treated in detail here, but for the sake of completeness a few remarks are offered from an applications standpoint.

The metrological tasks of noise pollution control in the environmental protection context include the following:

Noise level measurements in the vicinity for the determination and monitoring of noise contributed by the plant and by the surroundings, in particular by traffic.

Emission measurements to determine the level of noise radiated from individual pieces of equipment (acoustic power) and prepare an emission register (acoustic power levels of individual sources).

Calculation of impacts from emissions (noise radiation), with allowance for propagation damping between source and test point, to prepare an impact register (acoustic pressure level in the vicinity).

Comparison of measured and calculated impacts (levels) to validate the level mesurement technique or the propagation damping law used in the calculation [5.130].

In the framework of the occupational safety and health effort, these tasks are extended to the workplace proper. The large volume of interpretations require sensor and computer systems that are matched from an information-technology standpoint.

Sensor Systems in Polymer Production. In principle, the same problems arise in establishing the product properties throughout polymer production as are discussed in Section 5.1 [5.27], in Section 11.3 (process analysis and optimization aspects), and specifically in this section (see p. 467) [5.116], [5.128].

Like all production processes, polymerization can be represented in terms of the phase model (see Chapter 2 and Fig. 5.51).

Determination of Relevant Product Properties throughout the Process. The first central problem in determining the product properties is that the properties established or required in the end product (elements of the requirement profile) cannot generally be represented after all the prior steps in the process (see Fig. 2.34); for example, even a change in state of aggregation makes it necessary to examine more than just one category of these product properties.

Determination of Relationships between Process and Product Properties. The second group of problems arises because the relationships between the properties required in the end product or technological application (requirement profile), the actual product properties (qualification profile), and the properties that exist in the prior stages of the production process are not directly known. RUMPF first discussed this question from the standpoint of experience with solid-state processes (see above), coining the term "property function" for these relations [5.120]. Later, ECKER et al. applied terms such as "process indicators" to similar solution approaches [5.27]. Here, as discussed in Section 5.1, the relationship between "technological" [5.7], [5.12], [5.27] and "physical" product properties was formulated. In acrylonitrile polymerization, for example, a viscosity quantity is often used as a control variable (K value) [5.140]–[5.142].

Relaxation Time of Product Properties. The third set of problems is due to the fact that important product properties cannot be determined on-line (see Eq. 5.2-1) because the measurement process has a relaxation time τ_{Rsens}

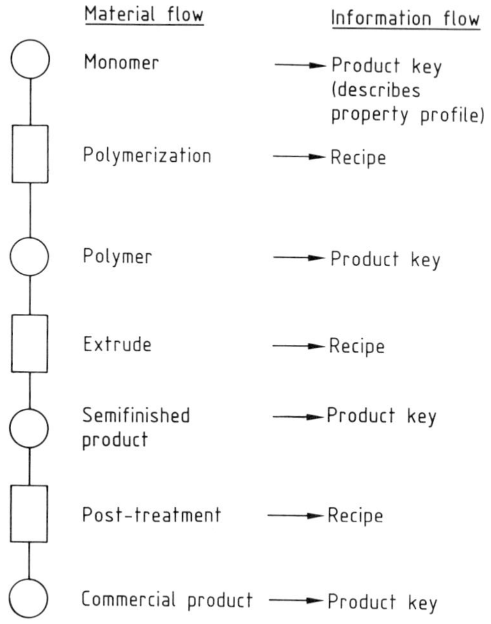

Figure 5.51. Material and information flows for production of a chemical material [5.12]

(required for information acquisition) that is too long or fails to satisfy Equation 5.2-1.

In-Line Measurements. Another important aspect is that information acquisition, in its sensor-linked stage (see Fig. 5.21), often involves not the product itself (in-line, see Section 5.2) but rather an extract, usually prepared after being obtained by appropriate extraction of the product from the production events (ex-line).

HOTOP and KÖNIG have readdressed this problem with application to polymer processes [5.143].

Complex Shear Modulus. The product properties (chiefly the mechanical-technological ones) can be described by the complex shear modulus G^* (see Section 5.1). This "phenomenological" quantity, which statically and dynamically describes the reaction of a product when a mechanical field acts on it, affords a key with which the mechanical behavior can be traced back to elementary molecular motions, provided there is an adequate theoretical framework. KUHN [5.31] and PECHOLD [5.32] carried out useful preliminary work, which MORBIKER et al. [5.33]–[5.35] has developed to the stage of practical application.

Molecular-Mass Distribution. The macroscopic mechanical behavior of polymers is essentially governed by the chain-length distribution or molecular-mass distribution and by the degree of cross-linking. A variety of HPLC methods are used here (Fig. 5.52 [5.144]).

Only when, as discussed above, there is a stable relationship between the molecular properties of the "process product" (in-line) and extracts rendered "measurable" through a series of differential-solubility operations (ex-line) can the conclusion stated above be justified (see also [5.145]).

In any case, a specific relation and a specific model must be sought for each application, so that process control consistent with quality assurance can be implemented. From a modern information-oriented standpoint, all such efforts lead to measurement techniques associated with process indicators [5.27] or supported by models [5.53]. A comprehensive bibliography can be found in [5.11].

Purity Guarantee. Similar aspects must be considered for the second collection of tasks, the guarantee of chemical purity for intermediate and end products.

It is beyond the scope and aim of this article to describe material-identifying spectroscopic and chromatographic methods (see [5.55] for more details).

An illustration of these points is shown in Table 5.8 [5.143].

Sensor Systems in Biotechnical Process Engineering. In the preface to [5.60] SCHÜGERL states that the most important difference between biotechnology and its predecessor, fermentation technology, is that fermentation technology is an assemblage of hard-to-quantify empirical relations, while biotechnology is an assemblage of relations that are still empirical but can be represented in quantitative form. The foundation has been laid for biotechnology to arrive at a point where it can make quantitative predictions with the aid of general laws. The same holds for both microorganisms and cell lines.

Previously, production strains were always selected by tedious screening and then laboriously improved by induced mutation and screening. Today, these strains are increasingly being modified to order by using genetic engineering. Protein design makes possible the purposeful local modification of proteins through appropriate genetic modification of the microorganisms or cells. Metabolite design allows the metabolic pathways detrimental to production to be eliminated and desired pathways to be reinforced.

Analytical techniques play a key role in the industrial exploitation of the full biological potential of the microorganisms and cells obtained in this way. These techniques make possible a precise study of the cell physiology and allow quantitative determination of the regulation of the metabolism. Furthermore, the optimal process conditions for growth and for product formation can be brought about and maintained by means of process analysis.

The same rules apply here as for the other fields discussed in this section, though at a high level of abstraction; see the discussions of solid-state process engineering and polymer production engineering above, as well as Figures 5.1 and 5.2.

Sensor systems commonly used in biotechnical process engineering [5.61] are as follows:

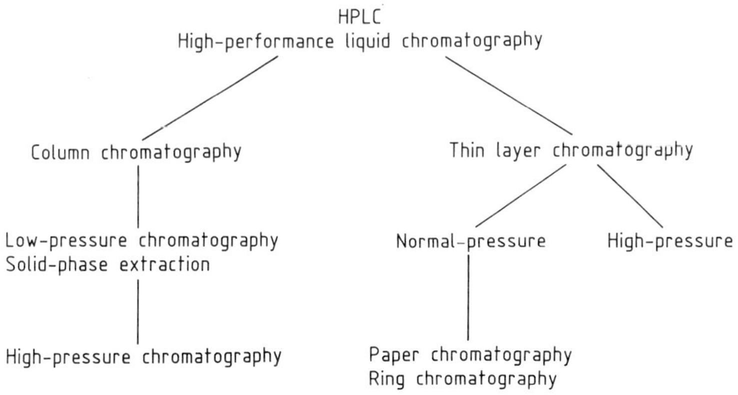

Figure 5.52. Scheme of HPLC methods [5.144]

Table 5.8. On-line measurement tasks in polymer processes [5.143]

Measurement task	On-line analytical methods	
	Conventional	New
Mixture, purity, and trace analysis Additives Side components Residual monomer Polymer End groups	density, refractometry, photometry, titrimetry, GC, photometry after color reaction	spectrometry, liquid chromatography
Molecular mass determination Average molecular masses Molecular-mass distribution Chemical constitution as function of molecular mass	viscosity, light scattering	gel chromatography
Characterization of polymer melt/commercial product Additive content Rheological properties Branching, cross-linking Chemical composition Residual solvent content	rheometry, photometry	spectrometry

On-line and in situ fluorimetry
  NMR spectroscopy
  ESR spectroscopy
  X ray structure analysis
Chromatography
Mass spectrometry
Electrophoresis
Optical and acoustic microscopy
Flow cytometry
Flow injection analysis
Bioelectrodes
Biosensors
  Biocalorimeter
  Bio-FETs
  Piezoelectric biosensors
  Optodes
Antibody/antigen systems

Along with the classical methods, which are known at least in general principle in chemical process engineering, a new class of sensors are now appearing: biosensors, which are used especially to mesure specific biological product properties (see Fig. 5.28). In these sensors, only specific biological reactions yield a sensory effect.

As a rule, however, parallel, serial, or associative linking of information is used [5.3]. A typical example is given by fermenter analysis (see Fig. 5.53): The following specific items of information are available: pressure, flow rate, and oxygen content of inlet air; oxygen, carbon dioxide, and nitrogen content of outlet air; pH and

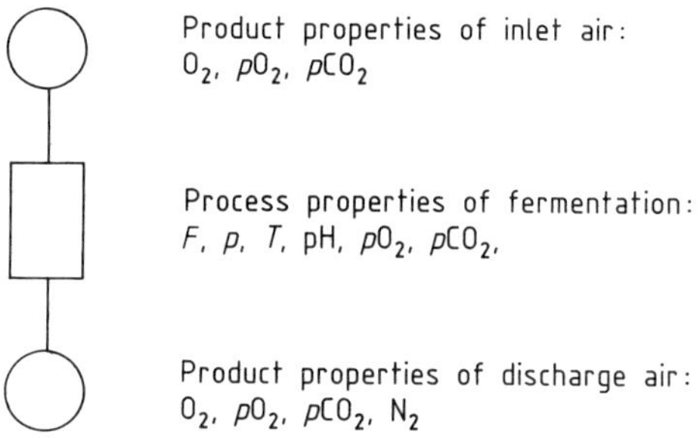

Figure 5.53. Measurements in fermentation [5.3]

turbidity of the solution; and contents of oxygen and carbon dioxide dissolved in the solution. A calculation relating the oxygen and carbon dioxide concentrations to the air throughput (Fig. 5.54) gives the respiratory quotient as the ratio of oxygen uptake rate to carbon dioxide evolution rate. Together with a knowledge of the mass-transfer coefficients from the gas phase via the liquid phase to the cell, this makes possible an optimization of the air throughput at a higher level of assessment [5.3].

As shown by WARNCKE [5.3] and HOTOP and KÖNIG [5.143], the greatest difficulty in biological product property determination lies in sampling and sample preparation, which must be done case by case. Without the aid of robotics in sampling and sample preparation and without information-oriented model support, sensor systems in biotechnical process engineering would be inadequate.

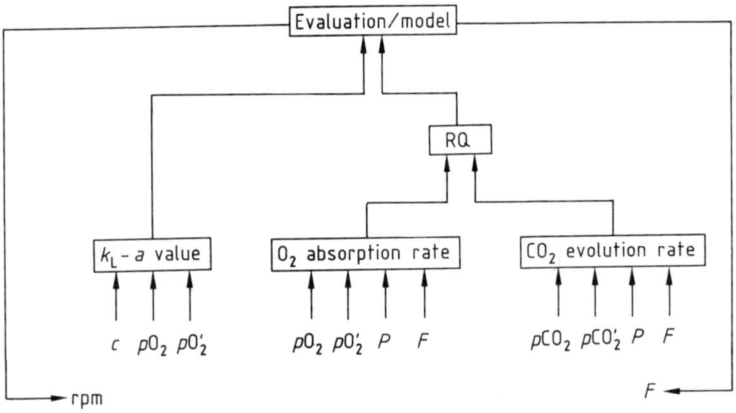

Figure 5.54. Gain in information due to coupling [5.3]

Additional constraints include foreign influences and the disposal of sample material after the measurement (see Fig. 5.20).

Biology itself, however, must increasingly become the guide in the development of biosensors. Multisensors, sensor arrays, and the processing of many simple signals in a manner similar to the neuronal structure of organisms should provide the solution to many problems [5.13].

Interesting results in the identification of process trends have been obtained by ADLER [5.146], who used fuzzy clustering methods, which WINTERSTEIN has tested in practice by applying a multivariate sensor system to wastewater monitoring [5.147].

Outlook. In order to control a production process, it is necessary to look at information on various levels (Fig. 5.55). Sensor and actuator systems link the process with the information processing. Process monitoring and control systems take on the function of information processing and human–process communication.

The information orientation of process control engineering means that proven measurement techniques must be developed toward a sensor system technology.

The necessity of providing interpretable information to the process monitoring and control system at any time means a change in the requirement profile for sensor systems of the future:

- Reliability at reasonable maintenance cost
- Functionality even in extraordinary process states (e.g., startup, shutdown, upsets)
- Adequate accuracy and predictive power
- Easy adaptation to production variation [5.148]

Self-diagnosis and self-maintenance are important qualitites of future sensor systems. The information structure of a universal sensor system is illustrated in Figure 5.21.

Increasingly, manufacturers face the challenge of designing modular, engineered sensor systems. Design, together with easy installation and maintenance, represent untapped potentials for improved efficiency.

Of course, there are not yet any adequately tested sensor systems for polymer and biotechnical production, for the solid and suspension states of aggregation, and for many areas of environmental technology. Essentially, however, it is more important to bring the classical (physical, chemical, and technological) measurement techniques up to the new standard.

HAUPTMANN calls for the following among other points [5.149]:

"In order to develop sensors and bring them to market at low prices, some fundamental requirements must be met:

- Precise knowledge of the market and of international development projects as a basis for innovation
- Concepts, strategies, and visions of future sensor markets and technologies as a basis for long-term success
- Strategic partnership and cooperation with research institutes, universities, users, and manufacturers"

There is a clear trend away from the laboratory towards the process environment [5.3], [5.135].

Already, under the auspices of ISA, IFAC, VDI/VDE-GMA, and other organizations, nu-

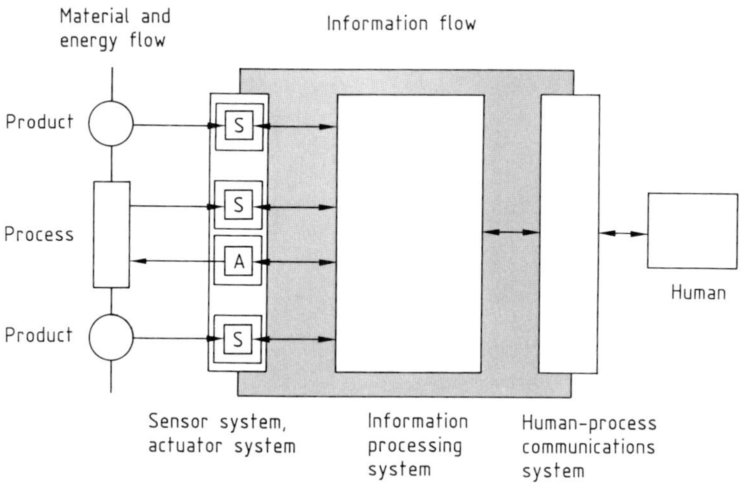

Figure 5.55. Sensor systems in the information chain from material/energy flow to human–process communication

merous forums are available worldwide for the discussion of sensors, sensor systems, technologies, and applications.

These discussions, often academically conducted, are brought down to earth by applications-oriented conferences such as GVC, the Eisenhüttentag (Foundry Conference), and NAMUR, which deal with the harsh conditions of everyday service (see above).

With increasing success, attempts are being made to bring structure into this unbounded variety by using new methods of applied informatics [5.150] and to find routes for effective application through hypermedia (see Fig. 5.30).

Nevertheless, the need to set down in-depth knowledge in encyclopedias [5.151] will still exist, so that this knowledge can be kept up-to-date (with a view to the taxonomies cited above) through supplementation or modification of the assignment of object attributes.

Special sensor technologies are continually being reported on by technical and scientific journals such as TM, IFAC, ISA, and atp. These include:

- Optical sensors (including fiber optics)
- Chemical sensors, such as
  - Liquid electrolyte sensors
  - Solid electrolyte sensors
  - Mass-sensitive sensors
  - Optochemical sensors
  - Calorimetric sensors [5.149]
- Surface-active metal oxide and semiconductor sensors
- Biological sensors

Four general trends can be identified:

Advances in Existing Sensors. The development of existing sensors and the creation of sensor systems continues. Multisensor systems based on commercial sensors, coupled with new signal recognition and signal processing techniques, will open up new application areas. An important aspect is the modeling of the sensor with regard to the effect measured as well as parasitic effects. Pattern recognition in forms ranging from classical to fuzzy logic will promote this development.

Digital Sensors. The resonance sensor is an important step toward digital sensors. This class includes micromechanical resonant sensors as well as sensors based on piezoelectric materials. Analog sensors can be replaced by resonant detection mechanisms with no change in sensor configuration. The coupling of resonant structures with fiber-optic excitation, detection, and signal transmission will increase. The development of entirely novel sensor principles will largely remain an exception.

New or Modified Sensor Materials. Material development is currently a major theme in sensor development. An important role is played by polymers, ceramics, composites, alloys, biomaterials, modified glass-fiber materials, III–V semiconductors, and so forth. Silicon, however, will remain dominant and combinations of silicon with other materials will be increasingly used.

Microsystems. Microsystems technology is having a lasting effect on the field of sensors. A microsystem is a spatial integration of sensor, signal processing, data processing, and actuator. This technology combines methods from microelectronics, micromechanics, microoptics, and information technology. It even makes use of techniques from acoustics, hydraulics, chemistry, and biology. It includes the design, fabrication, and application of technical systems that integrate a variety of functions. The functional units of a microsystem interact closely with one another, and their geometrical dimensions, dictated by functional considerations, are typically in the range from millimeters down to micrometers. An essential criterion for the design of microsystems is optimization of the overall function. This is advantageous from the standpoint of technological effort, because it is not always necessary to optimize the subfunctions per se. But because the interaction between the parameters has the nature of a function itself, the function of a microsystem is more than the sum of the functions of the subcomponents (see also [5.112], [5.152]).

The principle that functionality can be enormously increased by a greater integration density while system reliability and economy are maintained provides the basis and motivation for future microsystems technology. The integration of electrical and nonelectrical subfunctions on a chip or substrate enhances the performance of microsystems and is therefore particularly interesting. The requirements that must be met when integrating sensor and electronics, as well as the actuator whenever possible, are, however, in part contradictory.

The manufacture of future microsystems will require not only suitable principles but also high-performance materials, technologies, and design systems. New assembly and connection techniques will play a crucial role, as will quality assurance. Present-day solutions, technological approaches, and principles of microelectronics, materials research, integrated optics, process engineering, production engineering, chemistry, biology, and medicine must all be examined with regard to their systems capability and integratability [5.153].

The development of sensor systems will, however, be influenced by two other trends:

Transition from Batch to Continuous Processes. As production is changed over from batch to continuous processes, information acquisition cycle times become shorter. Direct coupling or integration of self-contained, autonomous sensor-actuator systems (e.g., for metering pumps, variable-speed and variable-torque motors, and so on) will probably be a necessity (see Chapter 6).

Automation. The problems cited in the preceding sections show how important robot-aided sensor systems will be for sampling, sample preparation, disposal of auxiliary substances, and maintenance actions. For example, Figure 5.56 shows a fully automated pH sensor system with remotely calibrated probe [5.59]. Figure 5.57 shows the sequence of automatic functions in a phase model.

Information utilization leads development in a completely different direction.

"Model-supported measurement methods," which have now become well established, should be mentioned first [5.11], [5.45].

The connection between process model and sensor system implies the ways and means for remedying fundamental shortcomings in measured-value acquisition with the present capabilities of information processing. Given these procedures, unmeasurable process and product properties can be reconstructed from easily measurable quantities through interpretation of a priori knowledge, represented in a mathematical model, about what happens in the process.

Furthermore, experience in petroleum exploration with properly positioned sensors has led to the use of tomographic methods so that, ultimately, spatial representativeness can be achieved on the basis of a one-point (local) measurement.

Modern methods of human–process communication (see Section 11.2) will make it easier to develop temporal single-point measurements by using characteristic curves in product and process configuration spaces (Fig. 5.58).

Sensor and actuator systems, as the link between material flow, energy flow, and information flow, will be designed in future such that they relieve the load on the central processing units of the process control system or actually become subsystems of it.

Developers and users, particularly in the area of sensor–actuator systems, will need to grow beyond the historic borders of their field and

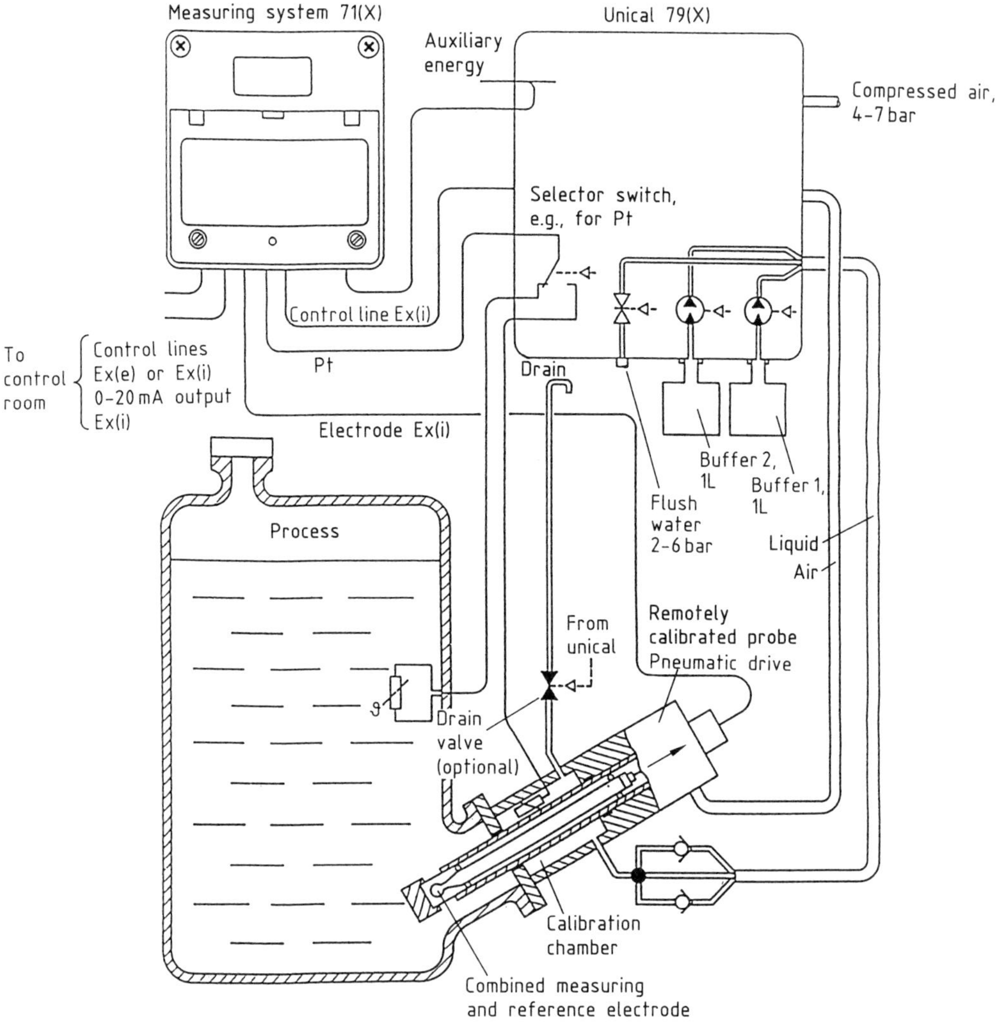

Figure 5.56. Example of an automated pH sensor system with remotely calibrated probe [5.59]

apply a keen and critical sense of analogy to the problem of transferring insights from other experiential worlds into process control engineering.

5.4. The Market for Sensors and Sensor Systems

Extensive market surveys [5.154], [5.155], shows and congresses [5.153], [5.156], [5.157] and seminars held by technical societies and manufacturers all come to one conclusion: The already large sensor market (world volume in 1988: ca. DM 24×10^9) wil expand even further, reaching ca. DM 45×10^9 in 1995 and $>$DM 65×10^9 by the year 2000 [5.158]. Figure 5.59 shows a regional breakdown of the expected boom. The current recession is not expected to alter this trend in a major way.

According to the Prognos figures, process engineering and plant construction are the largest customers of sensor and senosr system manufacturers at present (DM 9×10^9 in 1988). It will be these users that largely motivate and support the market growth; the absolute increase in sales to these customers will be exceeded only by applications in automaking, the relative increase by applications in building and safety engineering as well as in medical technology (Fig. 5.60).

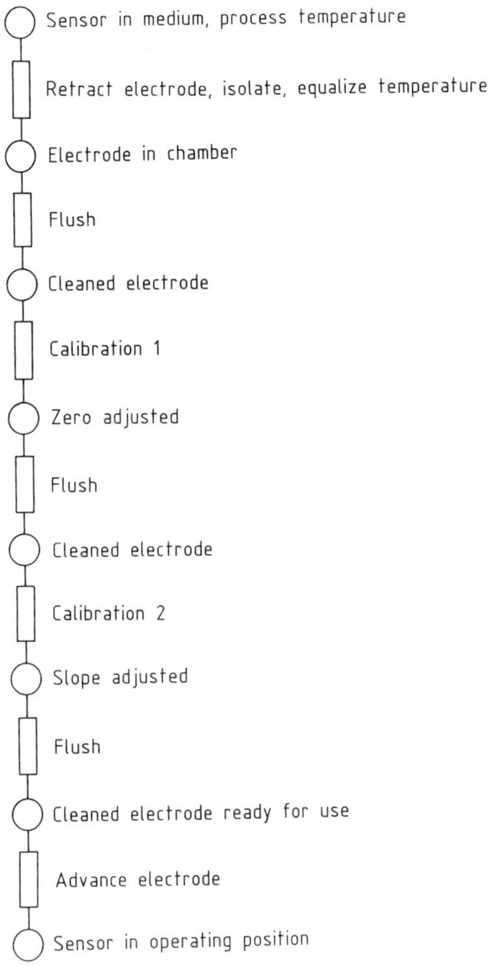

Figure 5.57. Subprocesses in calibration of a pH sensor

If the forecasts are accurate, process engineering and plant construction will account for roughly one-third of sensor sales in 1995 (DM 14.7×10^9). This development will be motivated by requirements for increased product quality and plant availability, and also by more stringent environmental and occupational safety standards.

The Prognos study cited above also shows that plant and process engineering are the dominant users of sensor technology on a region-by-region basis too (Western Europe, United States, Japan) and will probably remain so.

For the sensors used in plant design and construction, Figure 5.61 presents a summmary of application areas, while Figure 5.62 gives a similar breakdown for those used by process plant operators. Notable is the dominance of the petrochemical industry, which is accounted for by the high degree of automation in plants now on-stream.

Figures 5.63 and 5.64 show the sensor market for plant construction and process engineering divided according to the physical quantities measured. Clearly the process properties identified as "classical" in Section 5.2—temperature, pressure, flow rate, and level—account for the highest expenditures, while the properties grouped under "chemical quantities" represent a less important market for the process analysis technology discussed in Section 5.2.

It is interesting to do a cost analysis by the quantities themselves. A comparison of Figures 5.64 and 5.65, the latter showing the distribution of process control stations by quantity measured in medium-sized chemical process plants [5.69], suggests that temperature and level can be measured much more cheaply than pressure and flow rate. This is supported by a market survey [5.159] that compares the numbers and values of standard sensor transmitters sold in the US market (Fig. 5.66).

The reason for this disproportion certainly lies in the use of many relatively simple and robust principles for measuring temperature (for example, with the Pt 100 resistance sensor) and level; with regard to the markedly higher figures for pressure and flow, a detailed comparison with other conventional "classical" technology (e.g., orifice plates) makes it clear that manufacture prices here are motivated by market policies. An intimate knowledge of the market reveals that modern "contactless" methods of temperature and flow measurement, which require costly signal processing arrangements, are being offered at dumping prices in comparison with conventional "mechanical" methods.

A more detailed analysis can be done for sensor systems used in flow rate measurement. According to present figures, these represent cost-intensive process control stations (see also Figs. 5.64, 5.65, and 5.66, in which the third to sixth transmitter categories can be employed in flow measurement).

Newer developments in this field make clear the advantages of intelligent field systems with remote diagnostic and maintenance capabilities. The additional functionalities of these systems, which in conventional technology are rather demanding in maintenance terms, should pay off in the long run by lowering maintenance costs and

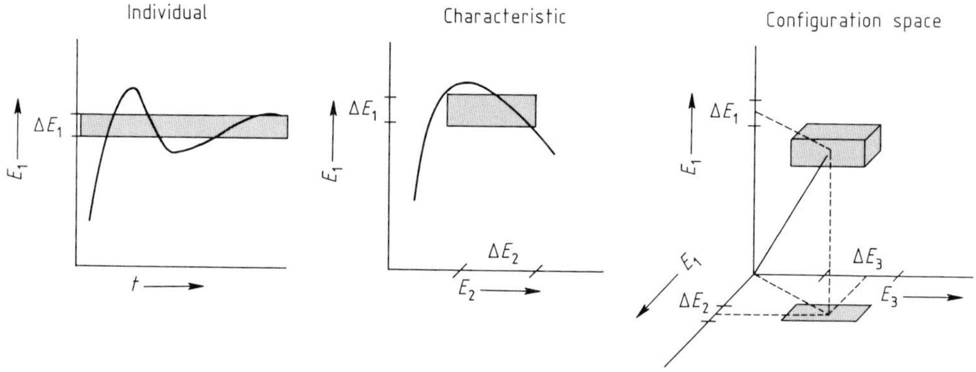

Figure 5.58. From process signal to configuration space (process and product property space)

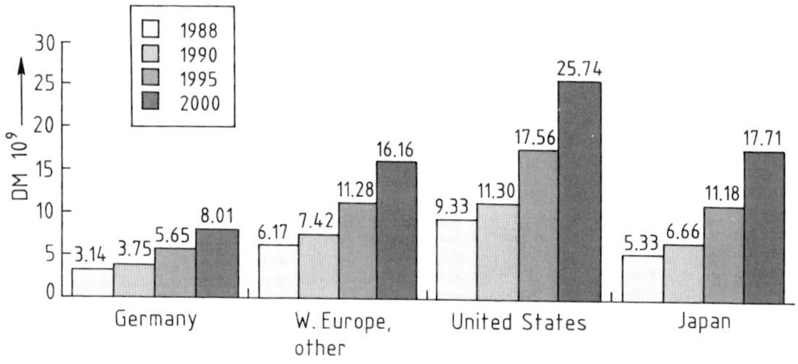

Figure 5.59. Sensor market broken down by region and year up to 2000 [5.158]

reducing energy losses (e.g., diminished pressure losses). There is an increasingly pronounced trend toward methods such as vortex shedding, magnetic induction, and Coriolis methods (direct determination of mass flow rate) and away from the older methods of differential pressure and rotameter-type instruments (see Figs. 5.67 and 5.68).

Modern trends in flow measurement are reported in [5.62], [5.160], [5.161].

All forecasts on applications in plant and process engineering agree on the expected development thrusts and market opportunities: biosensors and microsensor technology [5.152], [5.162].

5.5. Field Installation and Cable Routing

Introduction. Field installation and cable routing denotes the connections and structural work that must be performed to connect the field instrument to the data-processing system in the switch room and thus insure the flow of information between the field level and the process control level.

Modern installation technology is characterized by the use of an individual line from the sensor system to the switch room where the information-processing system (e.g., process control system) is located. Here, "sensor system" means the sensor with sensing element plus the metrology and instrumentation required for data acquisition [5.4]. It is not important whether the sensor system is made up of one unit (compact device) or several parts (e.g., the sensor plus a 19″ rack-mount transducer in the switch room). In the compact unit, sensing element and transducer are conbined in a single instrument. Compact devices have some major advantages.

As to functionality, there is a clear division between sensor system and data-processing unit. The sensor system generally performs the measurement, while all information processing takes place in the control system or stored-program

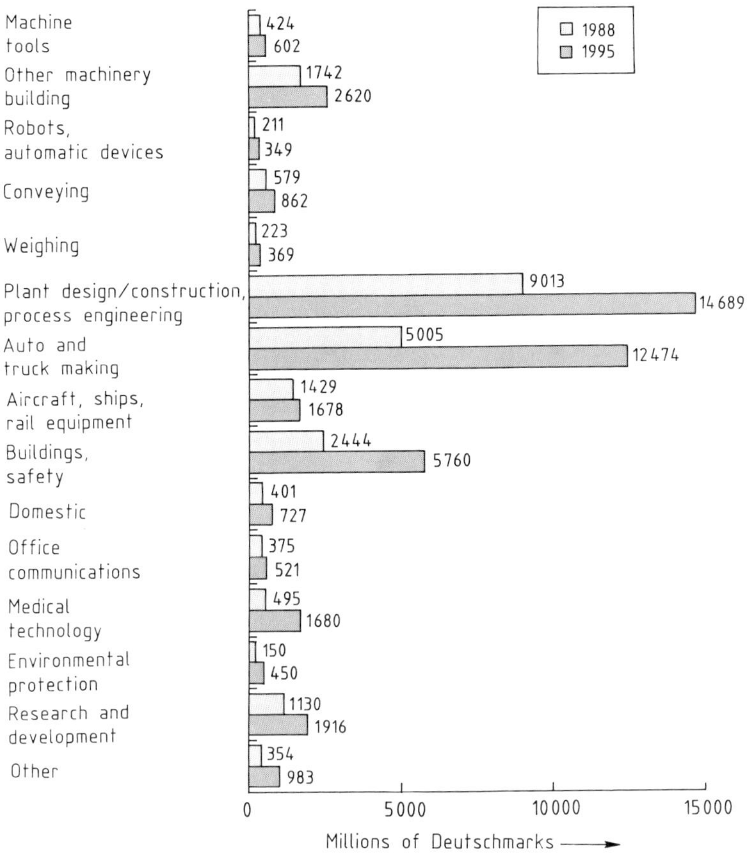

Figure 5.60. World sensor use broken down by application [5.158]

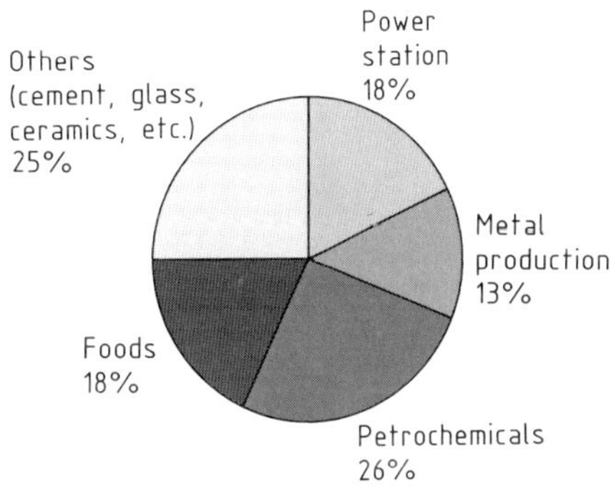

Figure 5.61. Use of sensor systems in plant design and construction, broken down by industry [5.158]

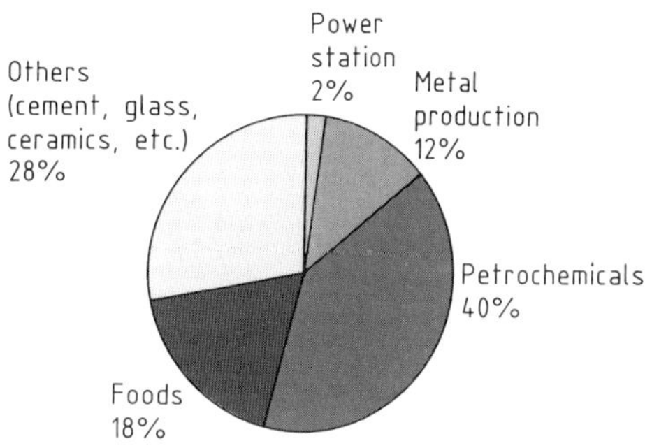

Figure 5.62. Use of sensor systems in process engineering, broken down by industry [5.158]

controllers. The incorporation of microprocessor technology into sensor systems has turned them into "intelligent" systems, even if they show no outward sign of this new capability. The potential of intelligent sensor systems can be realized only through bidirectional communication comprising the transmission of all commands and information. This is also the only way to take advantage of the wide range of possible functional expansions of sensor systems. Field bus technology will not only allow bidirectional communication but also simplify field installation and reduce costs.

The cables needed for field installation are placed along the cable routes provided for them. Cable routing must comply with structural and

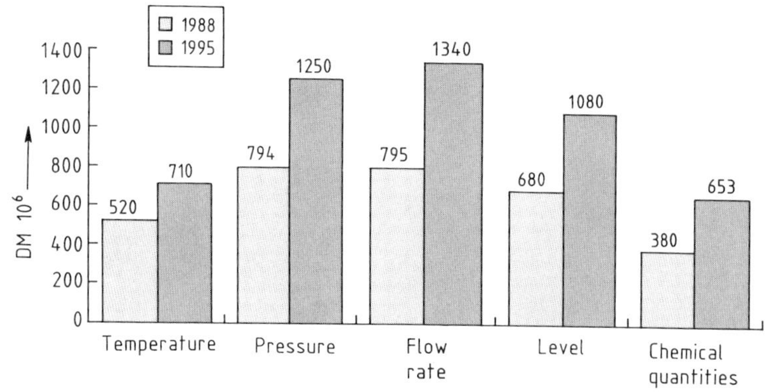

Figure 5.63. Western European sensor market in plant and process engineering, broken down by quantity measured [5.158]

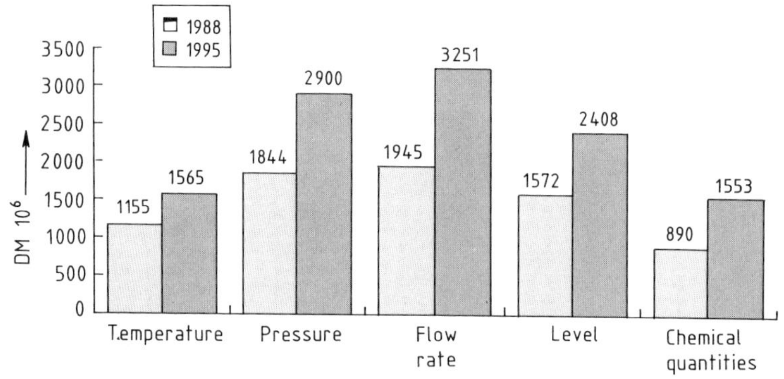

Figure 5.64. World sensor market in plant and process engineering, broken down by quantity measured [5.158]

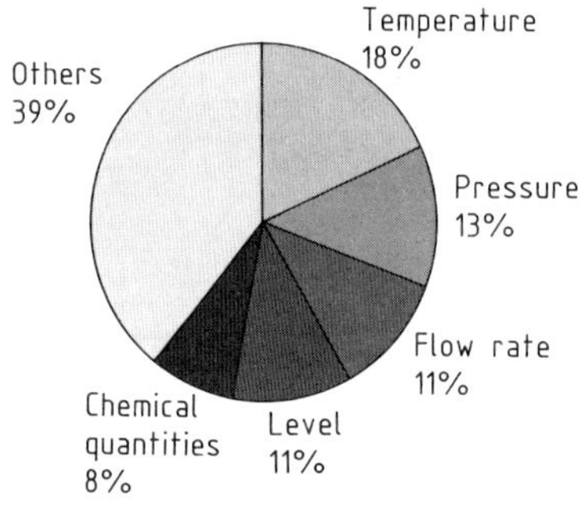

Figure 5.65. Number of process monitoring and control stations in chemical process plants, broken down by quantity measured [5.69]

safety requirements; important points include fire and mechanical protection of the cables. Other practical requirements can be found in the section on cable routing (see p. 497).

Field Installation. *Types of Installation.* Even though the prospects for field installations with field bus and field multiplexer continue to be discussed, new systems and components today are generally installed by proven conventional methods. Figure 5.69 shows five variants of present-day installation practice in Germany [5.67].

In variants (a), (b), and (c) the transducer is located in the switch room, whereas in (d) and (e) it is in the field. On going from (a) to (e), the amount of functionality incorporated in the field-level device increases:

- Variant (a): The sensor element (e.g., a resistance thermometer) contains no electronics. The transducer is located in the switch room and is connected to the sensor by a two- to four-wire cable. If the cable has three or four conductors, the effect of the line resistance can be compensated. Because there are no electronics in the field and no auxiliary energy is supplied to the sensor, the signals depend only on the sensor and are said to be sensor-

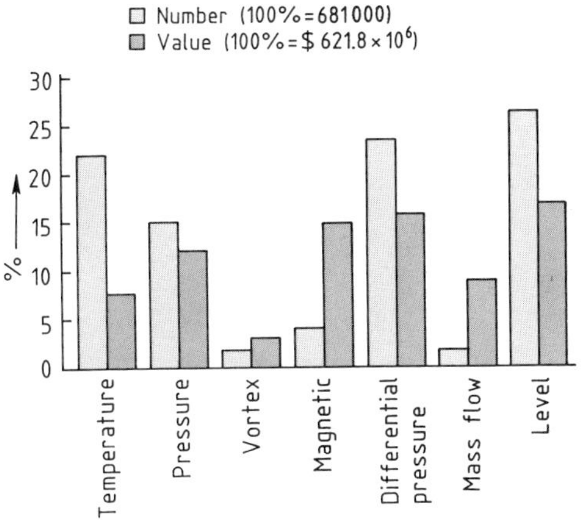

Figure 5.66. Numbers and values of transmitters sold in the U.S. market, 1990 [5.159]

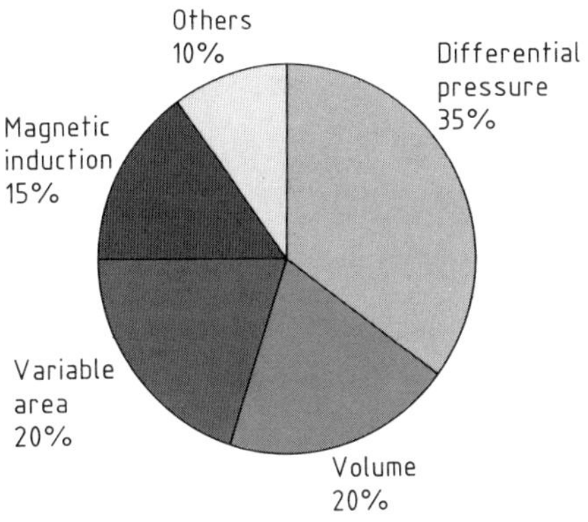

Figure 5.67. Flow rate and mass flow rate sensors broken down by type (estimated 1988 [5.158])

specific. The transducer in this variant is normally installed in a 19″ rack.

- Variant (b): The field-level sensor element (e.g., a pressure sensor) has a two-wire power supply cable and uses a second two-wire cable to transmit a signal (0–20 mV or 4–20 mV) back to the transducer, which is usually located in a 19″ rack. The signals to the transducer are manufacturer-specific, since the sensor signals are altered, but no intelligence resides in the field instrument.
- Variant (c): The sensor element in this variant has electrical preprocessing. In physical terms, the transducer is now present in the field. The manufacturer-specific signal, modulated onto the two-wire cable, is not yet finally pro-

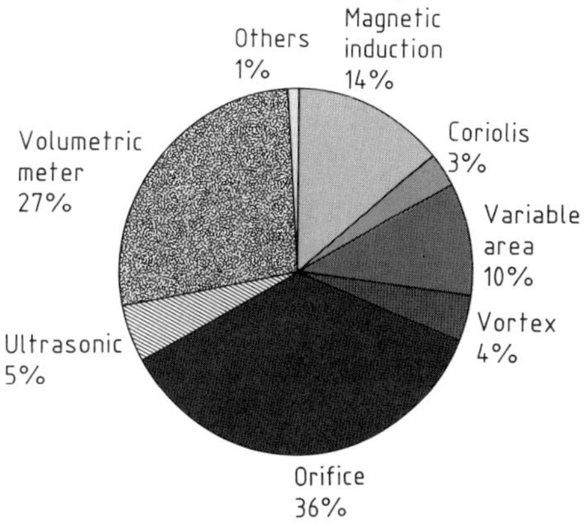

Figure 5.68. Western European sales of flow sensors in 1990, broken down by sensor type [5.62]

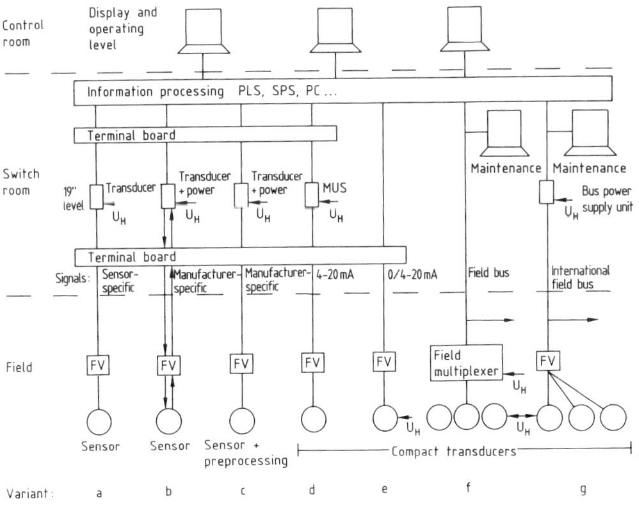

Figure 5.69. Development of field installation
PLS = Process control system; SPS = Stored-program controller; PC = Personal computer; U_H = Auxiliary energy; MUS = Transducer power supply; FV = Field distributor

cessed; for example, linearization and damping have not been performed, and the output has not been converted to a 0–20 mA or 4–20 mA signal. These functions take place in the 19″ rack-mount transducer, whose power supply also supplies the sensor element through the same two-wire cable.

- Variant (d): The transducer is located at the measurement station and is combined with the sensor element to form an overall system (compact instrument). A 19″ rack-mount transducer power supply powers the compact instrument, furnishing 4 mA. The transducer sends the measurement information to the power supply unit as a 0–20 mA or 4–20 mA signal, over the same two-wire cable. The transducer generates this information by varying its internal resistance.
- Variant (e): This variant also employs a compact device. Because of the physical principle used, instruments in this variant are supplied with a relatively large amount of auxiliary energy, so that a separate power supply line is necessary. The output signal corresponding to the measured value lies in the range 0–20 mA or 4–20 mA and can be input directly to postprocessing or used for displays in either the switch room or the control room. Hence no other instrument is needed in the switch room.

All these variants have in common that the information-bearing signals from the field instruments are led to a field distributor or distributor box, to which other field-level instruments are also connected. The cable employed for connecting the field instrument to the field distributor is generally a twisted pair or four wires in two twisted pairs, with 0.5 mm^2 shielded conductors. Data lines and cables are described in DIN VDE 0800-0899.

Consider, for example, the 32-pole field distributor. From it, a 32-wire master cable usually runs to the terminal board in the switch room. Partial allocation is also possible, provided this is allowed by the allocation of master cable signal loops to the interface plane and no future change in the signal loops is permitted. The master-cable specifications could be as follows: 32-wire ($16 \times 2 \times 0.5$ mm^2); i.e., 16 twisted pairs, not shielded pairwise but having a common shield.

In addition to the 19″ interface plane, when required, a connection to a second terminal board is also possible. The second terminal board provides the connection to, for example, the process-level components of the process control system. This connection can be made with a master cable having, for example, the following specifications: 32-wire ($16 \times 2 \times 0.5$ mm^2); i.e., 16 twisted pairs, not shielded pairwise but having a common shield.

Field Installation in Explosion Hazard Areas. Special regulations apply to the installation and operation of equipment in explosion hazard areas. These requirements have not been harmonized on a worldwide basis, so that some discrepancies arise. The discussion that follows applies to Europe.

- Definitions and Terms (VDE 0165): The basis for assessing the scope of the requirements to be imposed is a division of explosion hazard areas into zones by the probability of occurrence of hazardous atmospheres. The gases and vapors present in the plant are classified into explosion groups, while the instruments and other devices to be installed are divided into temperature classes. Each temperature class specifies the maximum surface temperature permitted for the device.
  - Zones: Zone 0 includes regions in which a hazardous explosive atmosphere is present at all times or over prolonged periods.
    Zone 1 includes regions in which hazardous explosive atmosphere is expected to form only rarely and for short periods.
    Zone 2 includes regions in which hazardous explosive atmospheres are expected to occur occasionally.
  - Explosion Groups: Explosive gases and vapors can be classified into explosion groups according to the minimum ignition temperature of their mixtures with air. The groups are designated IIA, IIB, and IIC; gases and vapors in group IIC are most easily ignited. Instruments and devices must be approved for the appropriate explosion group.
  - Temperature Classes: Instruments and devices for installation in explosion hazard areas are classified into temperature classes corresponding to the maximum surface temperature. The classes range from T1 (max. 450°C) to T6 (max. 85°C).
- Cables: Cables and wiring for circuits that are not intrinsically safe must be selected so that they can withstand the anticipated mechanical, chemical, and thermal stresses. The selection must comply with one of the following

standards: DIN VDE 0298 Part 1, 3, or DIN VDE 0891 Part 1, 5, 6.

On mechanical grounds, wire cross sections should not be smaller than certain values (VDE 0165). Conductors should generally be copper or, under certain conditions, aluminum.

- Placement of Cables and Wires: Cable and wiring penetrations to areas free of explosion hazard must be sealed adequately tightly; possible methods include sand seals, approved elastomers, and mortar seals. Cables and wires should have appropriate protection at points where there are especially severe threats due to thermal, mechanical, or chemical attack. One such form of protection is to place the cables and wires inside pipes. This should be done only with approved piping systems; case connections and transitions to cables and wires must also be approved.
  Once the installation for the explosion hazard area has been completed, it must be inspected by electrical specialists for compliance with ElexV § 12. A further inspection for compliance with the accident prevention regulations of UVV 4 § 5 must also be performed by an appropriately licensed person.
- Equipotential Bonding: In explosion hazard areas, equipotential bonding is required to prevent static discharges that might cause ignition. A good foundation earth is incorporated when the plant is first constructed, and the structural steel of the building and, for example, the switch room floor structure are grounded to this in star fashion. The interface cabinets and terminal board cabinets, as well as all other units, are grounded to this floor structure, along the shortest path. The interface cabinets contain the transducer power suppiles and other devices. Between two active devices, such as motors, located not more than 2.5 m apart, the measured resistance should not be greater than 0.2 Ω. The measured resistance between an active device and a passive object such as piping, located a maximum of 2.5 m apart, must not exceed 1 Ω.

An electrical specialist must verify compliance with the equipotential bonding requirements (UVV 4 § 5) visually and by measurement. These measurements must be recorded in an inspection record.

Systems with Intrinsically Safe Circuits. In the "intrinsic safety" type of explosion protection, electrical energy and surface temperature are limited such that sparking and thermal effects that could lead to ignition of certain atmospheres cannot occur. In other words, not even a short circuit can cause sparking. This type of protection exists in all countries that accept the IEC standards, and also in the United States and Canada, which have not yet adopted them.

Only insulated wires (insulation test voltage 500 V a.c.) may be used in intrinsically safe circuits. Several intrinsically safe circuits may be combined in one cable. Cables and wires of intrinsically safe circuits must be labeled or color coded. Sheaths or jackets are color-coded in light blue.

Intrinsically safe circuits and circuits without intrinsic safety must not be laid together in cables, wiring, conduits, or conductor bundles. In cable ducts, insulated spacers must be inserted or wires with certain kinds of sheath or jacket must be selected.

When installing an intrinsically safe circuit with just one apparatus, care should be taken that the maximum permissible inductance and capacitance for the circuit, including cables and wiring, are not exceeded. These permissible values are found on the nameplate of the apparatus. Wires must be treated as lumped capacitances.

Intrinsically safe circuits from the field are generally led to interface planes located in the switch room, which provide the separation from circuits without intrinsic safety. Possible implementations include approved safety barriers and transducer power supplies. The transducer power supply limits the maximum power available on the intrinsically safe side and also supplies a maximum of 4 mA and, say, 15 V to the transducer. The sensor system connected to this device thus has ca. 60 mW available to supply all the electronics.

Comparison of Circuits with and without Intrinsic Safety. The preceding section said that a transducer drawing its power via the 0–20 mA or 4–20 mA signal has ca. 60 mW at its disposal. This implies that only relatively low-power sensor systems can be supplied in this way; however, such sensors make up well over half of all sensor systems. There are, however, other sensor systems that cannot now be operated at such low power levels. The trend is for marginal systems to be supplied with low power at the expense of reliability, accuracy, and/or stability over time. It should be considered whether these drawbacks outweigh the advantages. Such a decision must be made for each application.

The advantages of intrinsically safe sensor systems and their circuits are easy handling and low cable costs, since only two wires per unit are needed. Intrinsically safe instruments and circuits can also be worked on while in service without having to cut off the power.

When nonintrinsically safe apparatus is used, there must be a separate power supply for the sensor system. That is, separate signal and energy channels must be present (see also Sections 5.1 and 5.2), so that at least four wires are required. If the power supply voltage is, for example, 24 V d.c., the supply and signal lines can be put in one cable. This arrangement has the further advantage that the terminal board plane and interface plane can be omitted.

If the two approaches are compared, the intrinsically safe variant has the clear advantage as to handling, while the other offers significant cost benefits. Futhermore, when using intrinsically safe sensor systems that are barely able to function with the power available, some attention must be paid to reliability, accuracy, and time behavior.

Field Installation in the United States. Installation techniques in the United States are quite different from those in Europe. In most cases, the cables are placed in conduits.

- Definitions and Terms: Classes are defined as follows: Class I relates to gases and vapors; Class II, to dusts; and Class III, to fibers or fiber dusts. Each class is further subdivided into Division 1 and Division 2. The discussion that follows concerns only Class I areas.
  - Division 1 is characterized by the permanent, occasional, or periodic occurrence of hazardous concentrations under normal conditions.
  - Division 2 is characterized by the fact that hazardous concentrations can occur only under accident or malfunction conditions.

- Cabling in Class I, Division 1: The only types of cabling permitted in these areas are cables in conduits and mineral-powder-insulated cables of type MI with the end closures in stress-relieving grips. Only materials approved for these areas may be used; for example, conduits and fittings must satisfy U.S. codes, some of which differ from the DIN standards. Conduits are inserted directly into the pressure-tight housing. The conduit system is completely "flameproof," so that a spark originating in the conduit cannot escape to the outside.
- Cabling in Class I, Division 2: Apart from conduits, NEC Article 501-4 also permits sealed track channels or cables or wires of types MI, MC, ALS, CS, TC, or SNM with approved end fittings to be installed in stress-relieving grips.

  Normal wiring is also permitted for circuits that, under normal circumstances, do not liberate enough energy on a wire break, short circuit, or ground fault to ignite an explosive atmosphere.

  Housings, fittings, and connectors generally must not be explosion-protected.
- Installation Practice in the United States: The classification of wiring in the United States is similar to variants (d) + (e) in Figure 5.73, that is, with a field distributor and master cable to the switch room.

Even when cabling without conduit is permitted, common practice is to place all cables in conduits. This has historical reasons: When installation practice was codified in the United States, intrinsic safety did not yet exist, so all cables were run in conduits; in Division 1, the conduits also had to be pressure-tight. Sensor systems, as well as field distributors, were used in pressure-tight designs (as they still are). This type of installation, designed by piping engineers, was reflected in the standards. Changes, even when they appear to make sense, take place very slowly. By now, intrinsic safety does exist in the United States, but its availability has had little influence on installation practice.

The insurance industry actually promotes the retention of this practice. Plants are usually covered by "electrical insurance," written by companies that approve the installation method and generally expect conduits.

Comparison of European and American Field Installation.

- Disadvantages of Conduit Use in the United States: The use of conduits increases the space requirement for the installation, since the routing has to allow for bend radii and offsets.

  The cost of installation is much higher, not only because of the placement of the conduits themselves but also because of the difficulty of pulling the cables through the conduits, their inflexibility, and the clumsy and quite heavy pressure-tight distributor boxes.

Additional gas barriers must be placed in the conduits to prevent the movement of gases and vapors and the propagation of flames from one part of the electrical system to another.

In the long term, problems arise from condensation and the pitting that it causes, as well as from the higher transfer resistances due to corrosion. This is important because the conduit system is also used for equipotential bonding. Condensation is controlled by drains through which the liquid can be removed.

- Advantages of Conduit Use: The conduits afford mechanical protection to the cables.

  As electromagnetic compatibility becomes more and more a factor in plant design, conduits offer much better shielding. Especially for digital data transmission, it is very difficult to separate signals from noise, since they occur in part in the same band of frequencies.

Trends in Field Installation. *Cost Analysis for Present and Future Installations.* An important question in the evaluation of new installation techniques is that of cost. New methods must not only provide technical benefits but should also be, at least, no more expensive. One study has found that future installation practice can make a significant contribution to cost reduction [5.163].

The comparative cost analysis is based on four cabling concepts in a representative plant.

The plant consisted of 400 process control stations. For the first calculation, these were assumed to consist of 240 analog and 160 binary stations. The second calculation was based on 300 binary and 100 analog stations. For the variants with field bus, it was designated that fewer than half the process control stations could be connected to one field bus.

All design, installation, shop, and material costs were included, from the individual line to the field instrument up to and including the input/output cards of the process-level components.

The results are shown in Figure 5.70.

Variant I: conventional, parallel cabling. Cost = 100 % (basis for comparison).

Variant II: field multiplexer hardware used at the field level, so that only a few cables (field buses) had to be run from the field to the switch room. The result shows that there are marked savings. Cost = 82 % (74 % for the plant with the larger fraction of binary stations).

Variant III: field bus (intrinsically safe; transmission rate 31.25 kbaud) for the field instruments that can be directly connected to a field bus. The remainder of the field instruments are connected to a field multiplexer. Cost = 73 % (77 %).

Variant IV: field buses from variant III compressed into one field bus with higher data rate (not intrinsically safe, 1.5 Mbaud). Cost = 69 % (78 %).

When the analysis of "present-day" instrumentation was extended by distinguishing between separate versions and compact devices, it appeared that future trends in instrumentation would see only the compact instruments still in use, with the intelligent transducer generally forming a single unit with the sensor element. Under severe ambient conditions, it often makes sense to install the transducer with its complement of electronics near the sensor element. Because it is not desirable to connect a 19″ transducer on a field bus, this alternative will largely disappear in future.

Field Installations with a Field Multiplexer. The field multiplexer solution is illustrated in Figure 5.69, variant (f), and Figure 5.70. A mixed field installation with field multiplexer and field bus is also possible. The field multiplexer must be approved for service in the explosion hazard area (zone 1). It provides data compression so that the total cost of cabling to the switch room, the required terminal board, the 19″ rack for the transducers and their power supplies, and the necessary plant documentation are all substantially reduced in comparison with conventional installation practice.

Such a field multiplexer can be obtained in a range of sizes. The user must decide the number of channels connected to it above which partial or complete redundancy is to be employed. In the case of complete redundancy, a second field multiplexer is connected in parallel. The line from the field bus directly to the input/output cards of the field-level components in the switch room should not be run together with the existing field bus. The I/O interface module can also be designed with redundancy.

One possibility for the individual lines from field devices to the field multiplexer is an unshielded $2 \times 0.5\ mm^2$ twisted pair. The type of field bus cable depends on the run length and the data transmission rate. A shielded $2 \times 0.5\ mm^2$

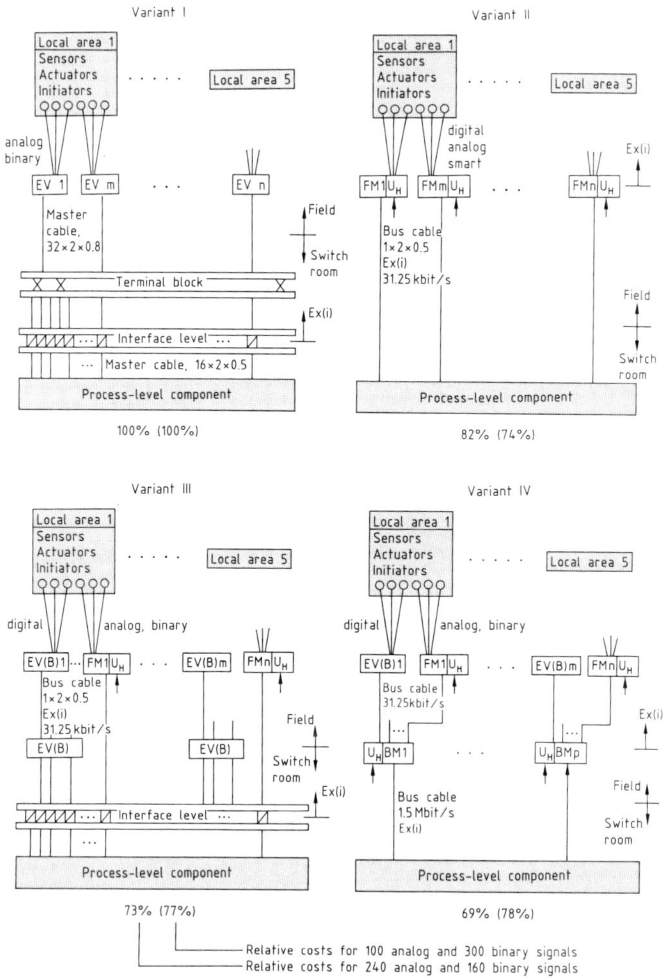

Figure 5.70. Cost comparison for field installation variants [5.163]
EV = Final distributor; FM = Field multiplexer; U_H = Auxiliary energy; BM = Bus multiplexer

twisted pair may be needed for simple applications, a coax cable for more difficult applications. The field multiplexer itself requires external power (e.g., 220 V a.c.).

Field Installation with a Field Bus (see Chapter 8).

- Principles: Figure 5.73, variant (g), shows a possible field installation with field bus. This is a simple example for the direct connection of intelligent field instruments to the field bus. For use in process situations, especially in explosion hazard areas, the field bus must be of the intrinsic-safety type, so that the disconnection and connection of a bus subscriber is possible even while the system is operating. The bus can accommodate a combination of transducers powered from the bus and transducers with external power, as shown in variants (d) + (e). Important is that the bus interface of the sensor system can always be connected to the Ex"i" (explosion protected, intrinsically safe) bus.

The field bus cabling concepts illustrated in Figures 5.69 and 5.70 all relate to a field bus employing copper wire. The field bus can also be realized in the form of an optical fiber cable or a radio link. Radio transmission plays a role in widely dispersed or poorly accessible plants and is not further considered here. Although the "Physical Layer" is now included in the IEC Draft International Standard, optical fiber has not yet been adopted as a physical data transmission medium. Nevertheless, optical fiber technology offers a serious alternative to copper wire.

- Comparison of Copper Wire and Optical Fiber: A field bus installation that seems appropriate for process engineering is that shown in Figure 5.70. The design in question has parallel individual lines from field instrument to field distributor. The several lines are combined into a two-wire line in the field distributor, and the two-wire line is then led to the process-level components. The field distributor is a purely mechanical component. In contrast, the optical fiber system has as its distributor an active element that forwards the signals after appropriate conditioning.

The advantage of optical fiber technology is that the information is unaffected by electromagnetic interference. The separate power supply to the instruments must, however, satisfy the requirements of electromagnetic compatibility. For data transmission via copper wires, it can be expected that proper shielding or the use of special cables will still make it possible to comply with electromagnetic compatibility standards such as those of NAMUR. The price of such a cable is, however, much higher than that of a normal bus line.

Very high data rates can be realized with optical waveguide techniques (5 Mbaud). However, such a rate is not generally needed for the connection of sensor and actuator systems through the field bus.

The number of field instruments that can be connected to the international field bus (now in the definition phase) now appears to be 32. For a copper bus line intended for service in the explosion hazard area and also designed to power the instruments connected to it, the number of instruments is smaller, just nine at present. With optical fiber technology, 32 instruments can be connected to a field bus, even in explosion hazard areas. It depends on the application whether it is desirable to connect this many devices to one I/O card in the process-level component for the sake of redundancy. Present-day I/O cards can accomodate eight channels, roughly the same as the number of instruments that can be connected in explosion hazard areas when copper wire is employed.

One advantage of the copper bus line—that the field instruments can also be powered from the bus line—becomes a drawback in many cases. In accord with the current trend, an attempt is sometimes made to use the bus to supply instruments that could yield more accurate, more reliable, and/or quicker results if more power were available. This dilemma of the instrument designer is often resolved with optical fiber technology, since each field instrument then requires a separate power supply.

Consider, for example, a plant with 200 field devices to be connected to the field bus. A calculation gives the following numbers of lines to be run:

Field bus with copper cable:
  Number of buses: 23 (9 instruments per field bus)
  Number of separate power supply lines: ca. 20 (10% of 200)

Field bus with optical fiber cable:
  Number of buses: 7 (32 instruments per field bus)
  Number of separate power supply lines: 200 (possibly implemented with a field distributor)

- Bus Cable: The definition of a single bus cable for all conceivable applications is unlikely, since it would be too expensive for simple applications. The Draft International Standard for the Physical Layer of the international field bus proposal identifies four cable types for the slow (31.25 kbaud) field bus (ISA/SP50-1992-236P):
  - Type A cable:
    Shielded twisted pair with following specifications:
    - $Z_o = 100\ \Omega \pm 20\%$
    - Maximum damping = 3 dB/km
    - Maximum capacitance = 2 nF/km (wire-shield)
    - Maximum resistance = 22 Ω/km (per wire)
  - Type B cable:
    Twisted pair in a master cable with a common shield and certain specifications
  - Type C cable:
    Twisted pair, possibly in a master cable, but not shielded, with certain specifications
  - Type D cable:
    Simple (not twisted) pair in a master cable, again with established specifications; the worst of the variants
- Field Installation of Safety Circuits: How safety circuits will be handled in future installation practice can be examined on an application-specific basis. Simple binary operations, such as "if limit switch trips, cut off feed," should be distinguished from more complex

conditions for triggering a safety circuit. Such a condition might be a mathematical operation on multiple binary and analog inputs.

The simple binary operation will continue to be wired separately and realized individually by appropriate modules. It may be that an "autonomous" unit in the field multiplexer, powered only by the system, can perform this binary operation, so that the cable need not be run to the switch room. For a more comprehensive safety assessment, it is conceivable that either a hard-wired unit or a stored-program controller could be used.

Field Bus in Explosion Hazard Areas. The situation is analyzed for copper wire as the data transmission medium. A field bus for service in explosion hazard areas must comply with the explosion protection class "intrinsic safety," so that an instrument can be connected or disconnected while the system is in operation. For low-power sensor systems, such as those usually employed for pressure and temperature, it should be possible to use the bus line for both information and power supply to the instruments.

Even externally powered sensor systems must at least carry out the bus connection in the "intrinsic safety" class, so that these can be connected to the Ex"i" field bus together with instruments not requiring external power.

- PTB Tests: A study by PTB has shown that simple descriptive features make it possible to connect instruments made by different manufacturers to an Ex"i" field bus complying with EN 50020 (Design Standard for Intrinsic Safety) with no special system certification. A variety of cable types were tested for a range of cable lengths and bus supply voltages [5.67].

Figure 5.71 illustrates the concept underlying the PTB tests. Only "one active device," the bus power supply, is connected to the field bus. The other devices are "passive" in that they cannot inject power into the bus. In this way, only one device can put power on the line, even in case of a malfunction. The passive devices have low maximum external capacitances and inductances ($C_i < 5$ nF and $L_i < 10$ μH) and minimum input voltages of 9 V. It was further required that each instrument connected must draw a base current of at least 10 mA. The cable length and number of nodes can be optimized by varying the power supply voltage and line length. The concept is applicable to a variety of field buses (Profibus, international field bus, etc.).

The most important finding was that over the wide range of parameters tested, and for various cable types and lengths and various bus power supply voltages, the probability of ignition was not increased by the length of the field bus, the

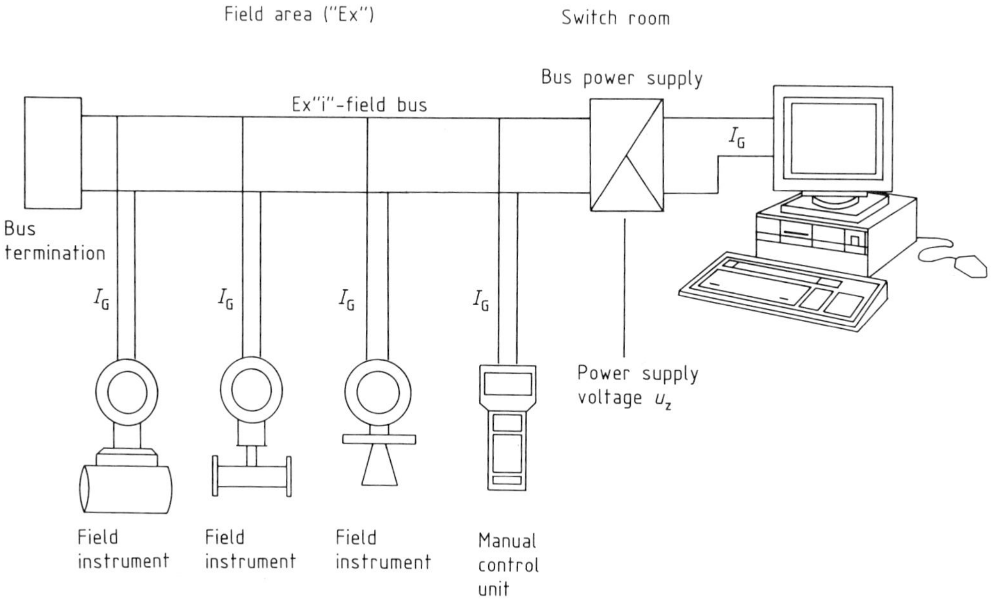

Figure 5.71. Base concept of the PTB study on field bus use in explosion hazard areas
I_G = base current

number of bus connections, or the number of instruments connected to the bus. These results are illustrated in Figures 5.72 (category ib, groupo IIC, $R'_L = 44\ \Omega/km$) and 5.73 (category ib, group IIB).

For a realistic bus power supply voltage of 15 V, the maximum number of nodes is nine for group IIC and 27 for group IIB. The maximum cable length is 787 m for group IIC, and 310 m for group IIB. The bus power supply voltage or conductor cross section can be varied in order to adapt to a wide range of applications.

Important observations from the user's standpoint can be found in [5.68].

Cable Routing. The sensors, actuators, and so forth contained in a plant are usually connected to the process control system by cables. Cable routes are the "streets" along which the cables are laid.

The project manager, in collaboration with fire services and operators, must identify zones of heightened fire hazard and danger zones as defined in Elex V and VbF at the stage of site and building layout [5.164]. The overall concept for the main cable and piping routes must be defined at an early point in time and incorporated into the construction plans. Whenever possible, cables should be located only in areas of slight danger. Accessibility and ease of maintenance must always be kept in mind; this includes later expansions. The spread of fire through lining penetrations into fire hazard areas must be restricted by suitable covers.

When cables and pipes are routed together, the possibility of a harmful interaction through thermal, chemical, and mechanical attack must be borne in mind. Cable and pipe routes should therefore be kept separate when possible. When this is not possible and an adequate standoff to hot plant apparatus cannot be maintained, heat shielding must also be provided. Where areas of severe fire hazard cannot be avoided and the functioning of the process control system must

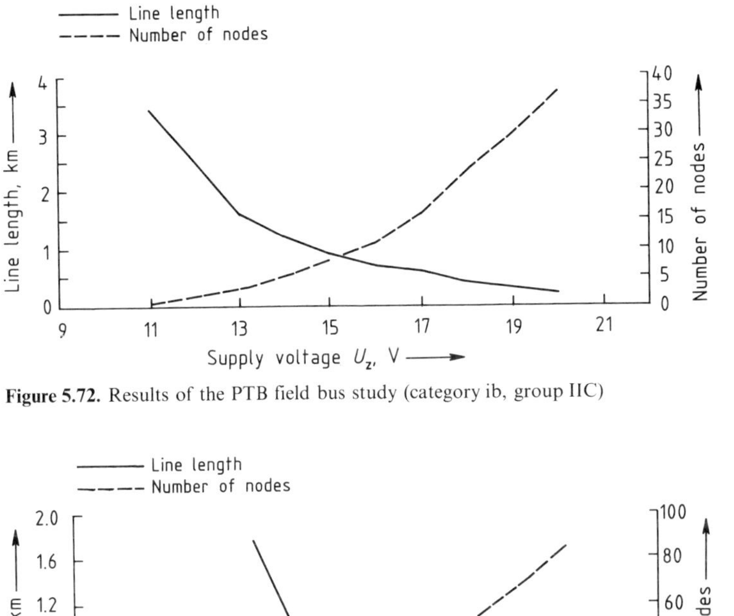

Figure 5.72. Results of the PTB field bus study (category ib, group IIC)

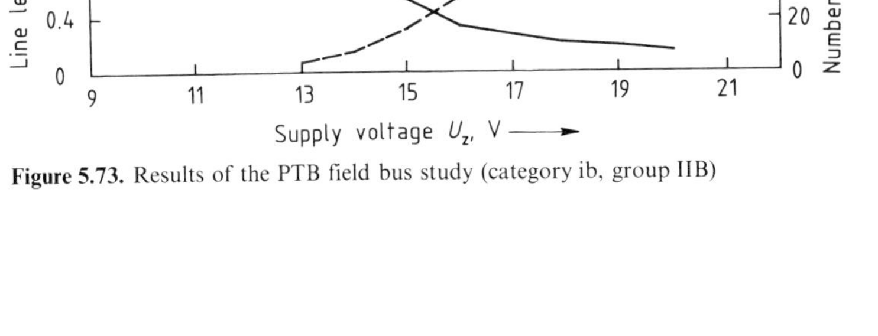

Figure 5.73. Results of the PTB field bus study (category ib, group IIB)

be insured under all conditions, a fire-resistant lining or cladding or the flooding or spraying of cable routes may be necessary.

The hanger design must be suitable to the load. Lining may necessitate cooling of the affected section (forced ventilation). Ventilation systems must not extend over the boundaries of fire compartments, so that the spread of fires into other compartments will be hindered. Reliable extraction of cooling air (fumes and smoke in the case of a cable fire) must be insured. Chemical attack must be handled by appropriate selection of construction materials, coating, or jacketing.

The number and location of process control rooms must be selected to yield shorter and more secure cable routes (see Section 10.7). These rooms should be in areas of little danger. Process control rooms and principal cable routes should always be configured so that fires in the rooms and localized fires in plant areas will affect only a portion of the plant (decentralization). The insertion of cables into process control rooms should take place outside the explosion hazard area whenever possible. Massing of cable routes near such insertions should be avoided (subdivision).

If power, control, and data cables are run together, the thermal load on the cable route is diminished; there is no objection to this arrangement provided electromagnetic compatibility is insured. High-voltage cables (>1 kV), however, should be run by themselves. Dedicated cable routes may also be needed for emergency power supply and safety system cables, whose functioning must be guaranteed for as long as possible during upsets.

Bringing cables from outside to inside to supply a floor or platform is generally safer than bringing them from a central cable tray inside the building. When cables are run on exterior walls or in open recesses indoors, any potential threat from their surroundings should be considered. Working platforms connected by fixed ladders should be built on exterior walls around the height of wall penetrations. In outdoor installations, cables can also be routed on the outside and, if necessary, should be provided with protection against thermal effects.

The cable route must be acccessible to firefighters in case of fire. "Cable walls" must permit access to areas located behind them.

Exposed cables must not be run in staircases or on walkways that serve as escape routes.

Placement of cables in underfloor ducts should always be avoided. It creates the danger that aggressive fluids will penetrate the cables or that explosive gas-air mixtures will form and spread without hindrance.

In cases where the use of underfloor ducts cannot be avoided, possible sources of danger should be eliminated by appropriate precautions, for example by sand filling or ventilation. When sand filling is chosen, cables approved for underground placement and having adequate thermal ratings should be selected.

If fire protection considerations mandate the placement of cables in a cable pit or vault, this must be of fire-resistant design. Suitable access must be provided for both firefighting and the operation of vent and exhaust valves. The chimney effect must be taken into account where cable fires can occur in cable pits. If necessary, the proper design precautions should be taken, such as fire partitions between floors and section-by-section removal of smoke. Compartmentalization should also be considered in horizontal cable routes.

Cable Routing inside Process Control Rooms. The design of cabling and wiring inside process control rooms should allow for some key constraints (see also Section 10.7):

- The penetration to the production facility must be gastight and safe with respect to fire hazards.
- For cables and wires that are to be led into cabinets or cubicles from overhead, or brought from overhead to terminal boards or mounting racks, cable trays should be provided below the ceiling.
- For cables and wires that are to be led into cabinets, cubicles, or racks from beneath, a double floor at least 20 to 30 cm high must be provided (Fig. 10.49). Cables and wires can be laid free on the floor. For larger cable cross sections, in particular, care must be taken that sufficiently large bending radii are allowed.
- On grounds of electromagnetic interference, cables and wires for power and those for data should be spatially separated.
- Massing of cables and lines should be avoided.
- Design must make adequate allowance for retrofit capabilities.

For high power density, continuous air cooling may be needed. The placement of product and energy piping (except process control system piping) in cable pits is always to be avoided.

In areas where there is a danger of dust explosions, design or organizational practices must insure that accumulations of dust are avoided.

Cable Penetration to the Switch Room. Cables running between the field and the switch room must be led into the switch room through a cable bushing in the wall. On grounds of fire protection, a bushing that meets the required fire protection class, such as F90, is selected. The fire department establishes what fire protection class is required. Not only the cable bushing but also the wall and the door to the switch room must comply with this fire protection class. In a low-cost method of penetration, holes of the desired diameter (e.g., 150 mm) are bored at appropriate points in the wall. The holes are half filled with the cables to be led through them. A fireproofing compound is then injected into the spaces between cables. The remainder of the opening is sealed with adequately close-fitting, flexible plugs inserted from both sides (the plugs should foam up in case of fire).

In future, 3-D CAD tools will be important aids in three-dimensional cable routing and piping design.

6. The Process Control System and Its Elements: Process Actuator Systems

6.1. Principles

Chapters 2 and 5 have pointed out the necessity of adopting suitable structuring practices to gain an overview of the wide variety of hardware implementations. The need for currency and completeness of description is complicated by the restriction to essentials (for the sake of clarity) and by the fact that technical progress continues after the manuscript of any publication is submitted. For these reasons, an attempt is made here (as in Chap. 5) to achieve some consistency in description, from the information point of view. This can easily be done, in object-oriented terms, by appropriate assignment of attributes. The viewpoint of the modeler is shaped by the objective, as in any modeling process, and if the objective changes then this viewpoint may (or must) change.

Information-oriented process control engineering uses the term "actuator systems" for all technical systems that enable intervention in the process properties, and ultimately the product properties (see Fig. 6.1). Regardless of the hardware implementation of the system under consideration, this choice of terminology has the advantage that the analysis focuses on the way in which the technical process is influenced or acted on, without necessarily being tied to implementation-dependent nomenclature such as controller, positioner, servo drive, and so forth. Furthermore, the degree of automation, the availability of energy, and the complexity of the actuator system remain initially unaffected by the choice of terms.

If an actuator system is analyzed from this viewpoint, a minimum of three components can always be distinguished (Fig. 6.2). First there is a functional unit that receives and interprets a control command (actuating command), represented by the control signal (actuating signal) from a higher level. The signal may originate in local components (Chap. 7) of a process control system or in programmable controllers. This functional unit of the actuator system is labeled as "control signal processing" in Figure 6.2. An energy-furnishing component, controlled by the output signal from control signal processing, can then be identified. In its simplest form, this component (called "servo drive" in Fig. 6.2) must act as an energy store or else utilize energy supplied via the control signal, passing on this energy in accordance with the control signal. Finally, the third functional unit is the actuator proper, often known as the "final control element." This component is in physical contact with the process properties [6.1]. The components listed here are found in related forms in the classical control devices (control valve, shutoff valve, ball cock, etc.).

The classical terms control element, control drive, and control signal processing indicate the unambiguously defined direction of action; they

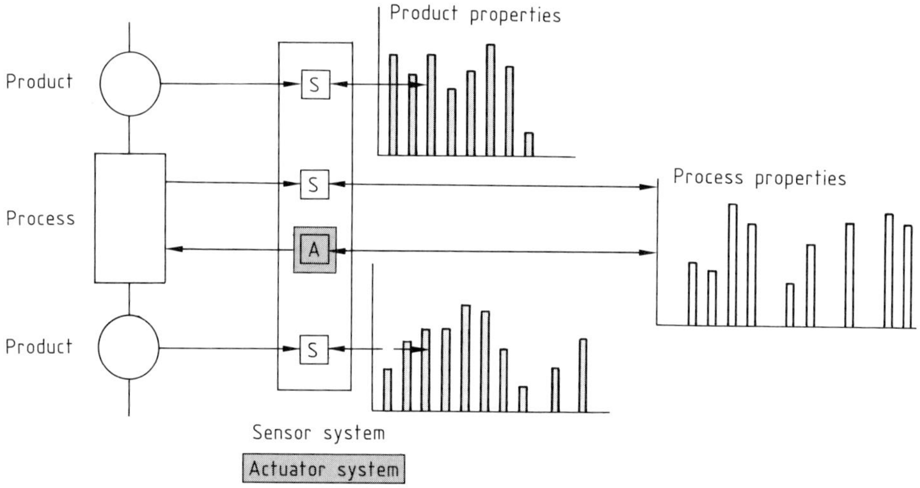

Figure 6.1. Actuator systems: analogy with the phase model of production

do not imply feedback (confirmation of execution or of effect achieved) as to the intervention in the process.

This type of analysis further reveals the problem of separating information flow from energy flow in such simple actuator systems. For example, the energy transmitted by the servo drive to the final control element can simultaneously be treated as a control signal, but one whose action on the process properties is determined by quantities specific to the system and apparatus. It is clear that the resulting dependence of process interventions on unknown effects, such as turbulence, aging, and the like, does not promote goal-oriented process control.

Figure 6.3 shows an actuator system in which an auxiliary energy supply has been added. This configuration makes it possible—at least in principle—to separate information flow from energy flow, to decouple the actuator and the drive from the signal processing, and to achieve flexibility in design.

The recent trend toward "intelligent" actuator systems is illustrated in Figure 6.4. Here an integrated (dedicated) computer takes care of function monitoring, function maintenance as required, control of the process intervention, and other services relevant to design or maintenance purposes [6.2]. Communication with the dedicated computer takes place via a display and control panel, a handheld terminal, a PC, or a display and control component [6.3]. In the long term, communications requirements stemming from additional functionalities will make a field bus solution essential.

The shift of intelligence into the field level permits additional functionalities in the field (local) components. Because the real effect of an actuator operation on the process or product properties is of far greater interest for process

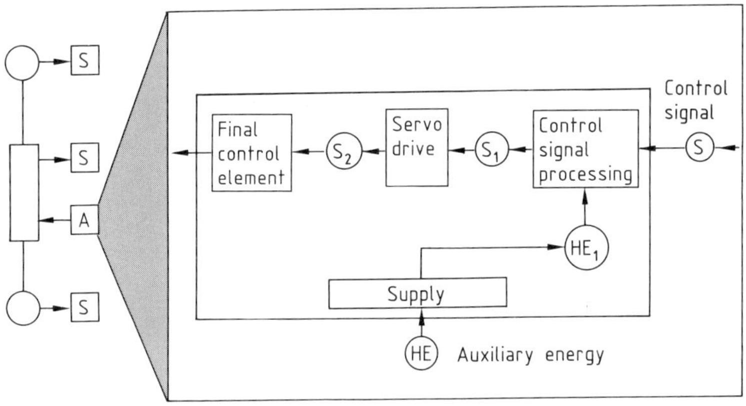

Figure 6.2. Structure of a simple actuator system

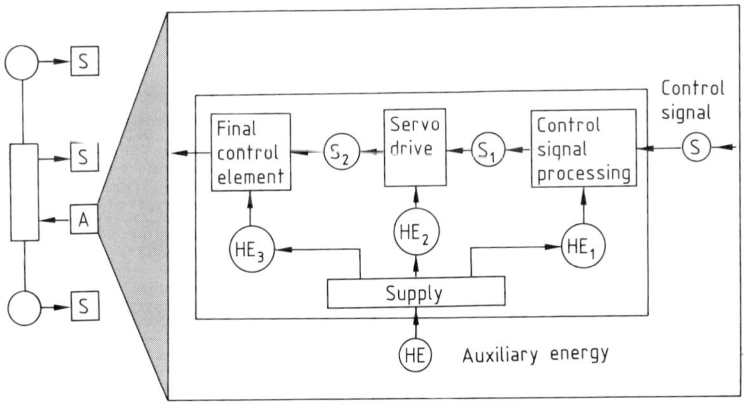

Figure 6.3. Structure of an actuator system with auxiliary energy

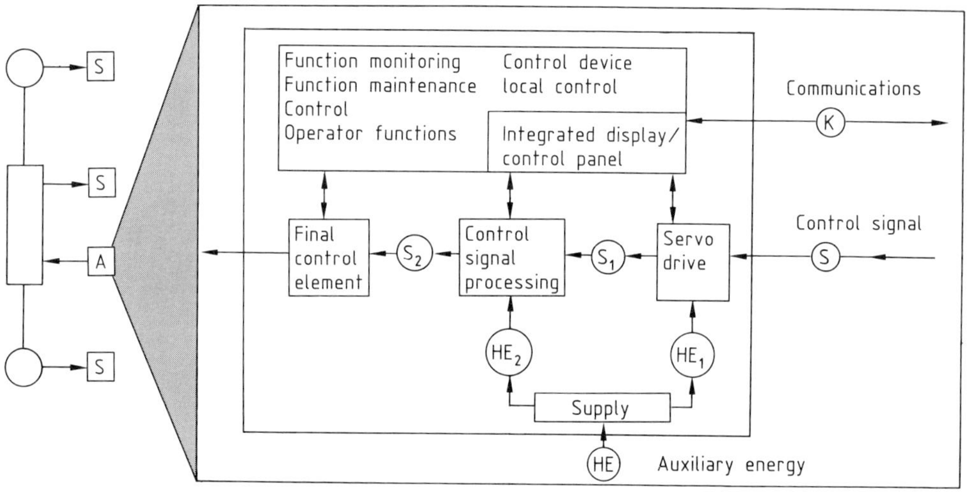

Figure 6.4. Structure of an "intelligent" actuator system

control than information such as the actuator position, there is a trend toward coupling of information directly at the local level. The speed control of a pump as a function of the actual flow rate, as determined with an integrated sensor, is an example of fast control loops on the local level.

In order to survey and classify actuator systems, it is necessary first to define an appropriate viewpoint. In what follows, some criteria are described, leading to the classification presented here.

A breakdown of actuator systems meaningful in the automation context might be based on the degree of automation of process interventions (manual operation, automatic operation), the type of auxiliary energy (electrical, pneumatic, hydraulic, etc.), or even hardware criteria. With regard to information-oriented process control engineering, however, a different type of classification appears more desirable. Actuator systems are analyzed from the standpoint of possible actions on material and energy streams. An attempt to set up such a classification is described here.

The action of an actuator system can be characterized in terms of the possible change in process properties. The manifestation of a property category can be described as changing "continuously" or "discontinuously." The taxonomy that follows therefore uses the criterion of a continuous versus a discontinuous control range.

For example, a mass stream can be controlled in continuous fashion by a control valve or discontinuously by a shutoff valve.

In continuous actuator systems, a further division into proportional and integrating types is possible. Similarly, discontinuous systems can bring about defined absolute changes or incremental changes in the process properties.

An additional feature of actuator systems is important from the standpoint of safety. A system may have an undefined position on failure, where the actuator remains in its most recent position in case of an interruption in the signal or energy flow, or it may have a well-defined safe position (fail-safe). An intermediate group comprises actuator systems in which the drive has a memory characteristic that takes the undefined failure position of the actuator over to a well-defined rest position. For certain processes, this represents an ideal solution from the safety standpoint.

Figure 6.5 shows a taxonomy of actuator systems based on the criteria discussed above as well as the energy flows to be supplied. Because the application and design features here are generic properties, they apply to all the actuator systems discussed here.

Design features—ranging from physical form, material of construction, and control range to communications, safety, and manufacturer—are listed in a second generic class.

Actuator systems are first classified by objective, that is, actuator systems for material streams are distinguished from those for electrical energy streams.

For each group, systems that effect a quantitative change in the streams are next separated from those that essentially bring about a qualitative change (most often defined in energetic terms). The first type, acting quantitatively on the streams, are described as "controlling" systems in Figures 6.5–6.7, while the second type is described as "transforming" systems.

If we consider first actuator systems for material streams (Fig. 6.6), the group of controlling systems include "distribution" systems such as screw conveyors, metering apparatus, etc., and "throttling" systems such as control valves, shutoff valves and gates, cocks, and so forth. In manufacturing and process engineering these two classes are often regarded as independent groups.

The category of transforming systems includes not only the turbine and the burner, but also the pump. This is admissible since the primary use of a pump is to increase the pressure of a material stream; thus the pump, as an actuator system, alters a qualitative property of the material stream.

Other examples of material transforming systems can easily be classified; for example, a stir-

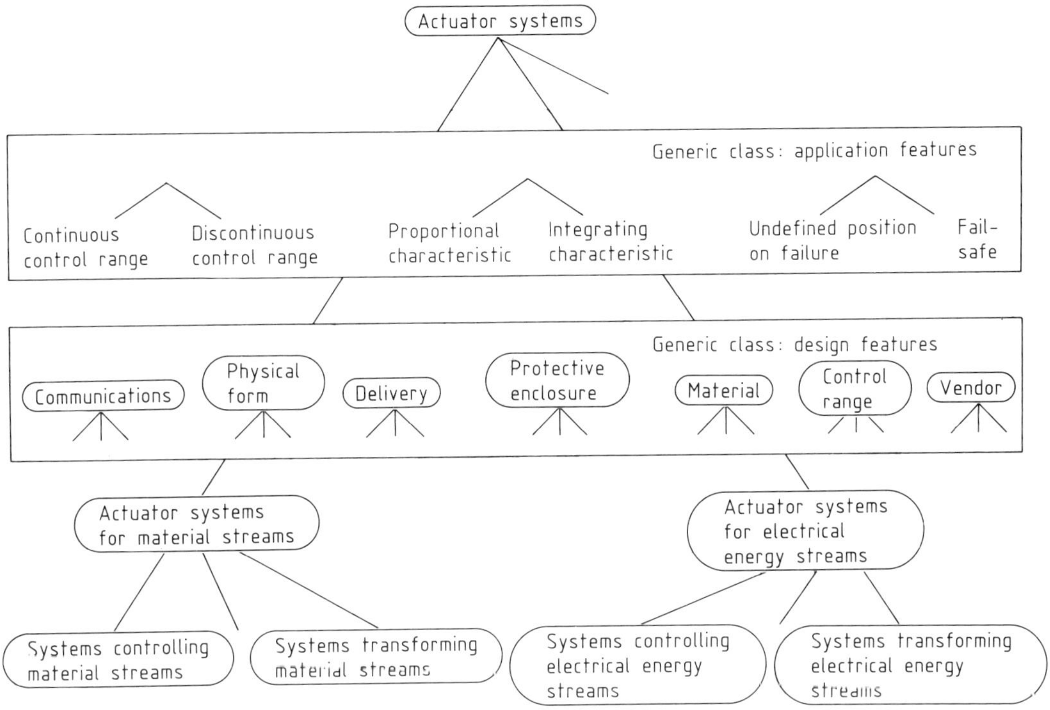

Figure 6.5. Taxonomy of actuator systems for material and energy streams

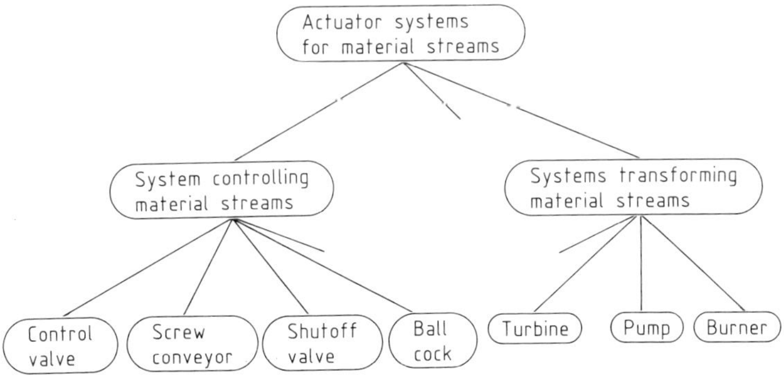

Figure 6.6. Taxonomy of actuator systems for material streams

rer plus drive in a stirred tank reactor, or of a roller, a press, or a tubular reactor.

Figure 6.7, based on the same classification principle, shows a taxonomy for actuator systems working on electrical energy streams. The quantity controlling systems include relays, thyristors, transistors, and gate-turnoff (GTO) devices (see also Section 6.3); the energy transforming systems include generators, lighting, motors, heating, etc.

A more detailed breakdown of energy controlling systems can be found in Section 6.2, which also distinguishes between switch and positioner on the basis of whether the control range is continuous or discontinuous.

Development trends in actuator systems can be briefly described as follows:

- Development from the actuator to the actuator system with integrated intelligence and the incorporation of lower-level control functions
- Expanded communications capability with field multiplexer or field bus [6.4]
- Improved service life, especially for actuators, due to novel materials, designs [6.5], and automatic maintenance techniques [6.6]
- Increased use of electronic frequency converters for motor speed control [6.7]

The trend toward intelligent actuator systems is possible because of advances in microelectronics. For example, a valve position controller with an integrated 16-bit microprocessor is now available [6.8]. Self-parametrization makes it much easier to adapt the device to the drive and the valve; the internal control loops of the actuator are more powerful; and the adaptability to a range of service conditions has led to a reduction in the number of position controller types needed.

The augmented functionality of actuator systems cannot be fully utilized unless there is a suitable facility for communication with higher levels in the process control system. Parametrization, function monitoring, and maintenance require special information, which will be made available by future field bus and communications concepts (see also [6.9] and Chap. 8).

Section 6.3 deals with the trend toward variable-speed drives. It is seen that improved control behavior (linear control characteristic), reduced energy consumption (especially under part-load conditions), motor power and motor losses restricted to the design values, and optimal matching to the process machinery used offer substantial potential for economic and technical optimization.

6.2. Actuator Systems for Material and Energy Streams

Introduction. Together with sensor systems, actuator systems link material, energy, and information streams in processing and manufacturing systems. The control path with sensor system, controller, and actuator system forms a closed control loop. Changes, whether in perturbing values, setpoints, controlled variables, or other quantities, manifest themselves in the system and also in the signals transmitted to and by these devices. With the aid of actuator systems, the material or energy stream is set in the control loop as a function of the system deviation. The

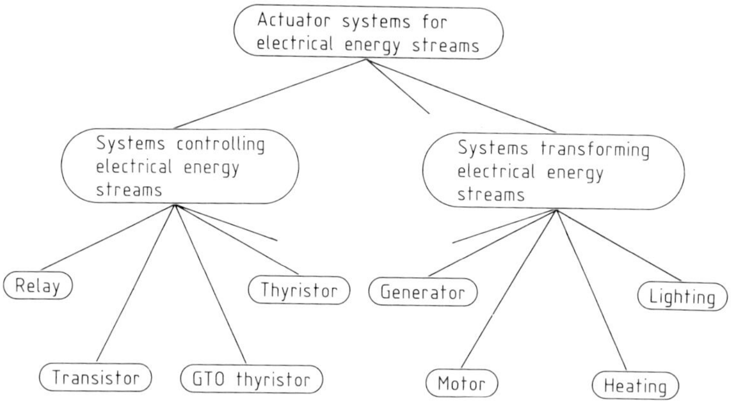

Figure 6.7. Taxonomy of actuator systems for electrical energy streams

action also depends on the characteristics of the control path, the sensor system, and the controller with its control parameters influencing its dynamics. With regard to the dynamics, the control element can be included in the control pathway and the servo drive (actuator drive) can be included in the control device. In the open loop case, the intervention in the process is effected through a control command, which takes the form of a control action or a manual action. The control pathway need not carry feedback.

In actuator systems, the control command (the output of the controller) must be transformed to a control signal (Fig. 6.2). Servo drives in the form of power amplifiers are needed for this purpose. These may be pneumatic, electric, or hydraulic, but hydraulic drives are rarely used in process engineering. Because of their well-known advantages in manufacturing and robotics, however, hydraulic power amplifiers are preferred in these fields. They are robust drives whose high power density results in good control of complex motions, accurate positioning under load, and wide speed ranges. Problems, however, such as poor damping, marked nonlinearity, and load-dependent eigenfrequencies, should be mentioned.

Continuous and switching control elements are distinguished, just as control elements with memory must be distinguished from those with a well-defined rest position. Single action modes or combinations may be required, depending on process or safety requirements (see also Fig. 6.5).

Continuous control elements effect a steady or gradual change in mass or energy flow within the design range, while switching elements act in a step-by-step manner. In process control engineering, there is often only a minor difference between the two types. Control elements of continuous or switching type can be designed by modification of globe and plug valves, cocks, shutoff valves, and the like. The required characteristic determines whether the required accuracy of positioning or control can be achieved under given pressure conditions. The characteristic associated with the system (inherent characteristic) can be matched to the operating characteristic. With regard to accuracy of positioning, a corresponding resolution of the control signal must be ensured. Control elements such as slide valves or solenoid-actuated valves have a design that makes them unsuitable for continuous control and positioning.

Control elements with electric motor drives have memory and remain in the last position adopted if the auxiliary energy goes off or the control signal fails. Ordinarily, control elements take up a well-defined position on a failure of auxiliary energy or control signals. For material streams, this is generally full open or full closed, and for energy streams, full on or full off. In safety devices, this position is called the "safety position." Control elements such as gate valves or stop cocks with double-acting pneumatic drives can be made to exhibit memory by the addition of blocking valves; the use of return springs or pneumatic buffers can result in their exhibiting a well-defined rest position. The size of the pneumatic buffer depends on the number of switching operations and the quality of sealing. The advantage over energy storage in springs is that several switching operations can be performed after the occurrence of a fault.

The term "control" is explained in DIN 19222 [6.10]. Control is the changing of mass, energy, or information flows by means of control elements. Standardized definitions of the terms actuator, actuator system, control element, controller, and control machine have not yet been adopted. They are used at present even though the demarcations between them are often not clearly defined.

Examples of Actuator Systems; Delimitation. Controllers or final control elements are also referred to as actuators or actuator systems, in analogy to sensors or sensor systems. Ordinarily, "sensor" means an element that performs a detection or measurement function. If this element is equipped with signal processing and other modules that permit further options, the result is a "sensor system." The analogy to "actuator" and "actuator system" can be made complete only if the "actuator" is defined exclusively as the part that comes in contact with the product.

Accordingly, the actuator is the final control element or control machine itself, that is, the throttling organ, the pump, the compressor, the blower, the resistance coil (for heating), or the fluorescent tube or incandescent lamp (for lighting). The corresponding actuator drives would be (for controllers of material streams) the diaphragm drive, the piston drive, the rotary drive, the motor, the motor with converter when speed control is required, or the switch or power controller for heating for lighting. The last two items apply to the case of controllers for energy streams. In common usage, however, the terms actuator, actuator system, controller, and final control element are often employed as synonyms, generally for the complete device.

Generally, the actuator system (Fig. 6.2) consists of a final control element (control machine), a servo drive, and a control device. Figure 6.8 shows the setup. The final control element is moved by the servo drive in order to change the flow of material and energy. The servo drive is driven externally with the control signal from a control device. With further devices, parametrization, configuration, status, feedback, and other signals can also be transmitted.

If these terms are applied to two well-known cases—the control valve and the variable-speed pump drive—the consistent representations in Figures 6.9 and 6.10 are obtained. Figure 6.9 shows a controller made up of a control element in the form of a valve (fitting) with a throttling organ, a pneumatic drive, and a control device. The control device is a positioner, which processes the control signal (as setpoint) to the position of the drive (as actual value) in the course of a closed-loop control operation. Along with the control section, the positioner includes a power section, where the control signal is amplified and which is to provide the desired positioning of the control element with the required quality and dynamics. The terms control element, actuator, and fitting are often used synonymously.

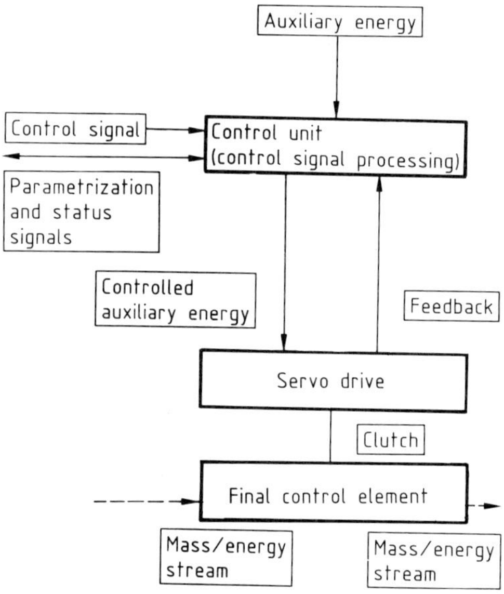

Figure 6.8. Structure of an actuator system

The variable-speed pump drive is a typical interconnection of a control element in the form of a pump, a control drive in the form of a motor with power section integrated into the frequency converter, and a control device in the form of an

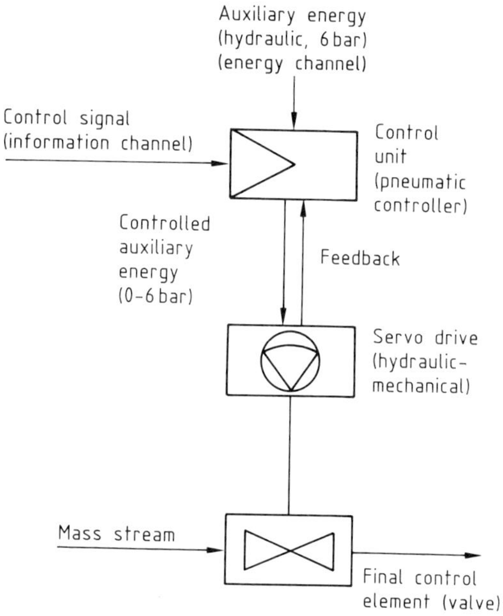

Figure 6.9. Actuator system for mass streams: structure of a control valve

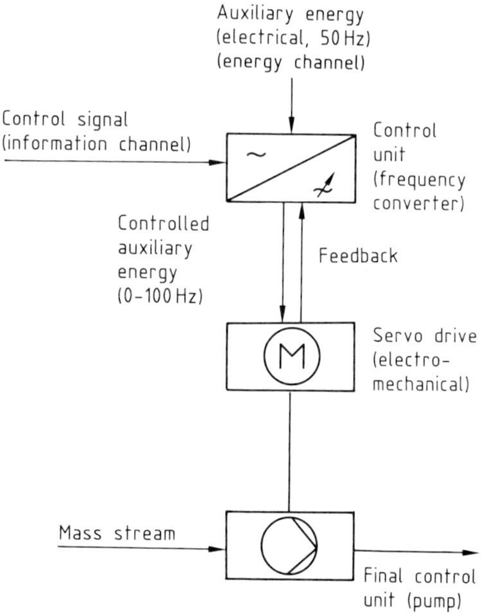

Figure 6.10. Actuator system for mass streams: structure of a variable-speed pump drive

information-processing component. The control signal is fed to the control device, which monitors, reports, limits, diagnoses, tests, generates error messages, archives, establishes a connection as needed, communicates, and so forth; it can also be parametrized and configured (see also Fig. 6.4). Section 6.3 goes into more detail on variable-speed drives with d.c. and a.c. motors.

The simplest case of control of an energy stream is the making and breaking of a circuit to a consumer. Figure 6.11, which is based on Figure 6.8 and analogous to Figures 6.9 and 6.10, shows an actuator system whose control device receives the sensor system signals of interest to the operator, and appropriately switches auxiliary energy to the control drive (coil) in order to activate the control element (switch).

The terms can also be extended to a positioner in a heating system (Fig. 6.12). This is a continuously adjustable heating device, whereby the control signal is generally a unified current signal (0/4–20 mA). The control device, with monitoring, testing, reporting, communications, and display functions among others, drives the control drive or power section. The energy stream to the control element or to the heater coil is adjustable from the control room or locally. It can be set automatically or by hand—depending on what devices are connected in circuit with it—anywhere in a design range specified by the designer.

Points to consider in the design of a controller for a material stream include dynamics, operating point, positioning forces, operating range, characteristic, safety position, operability, service under explosion hazard conditions, and sealing. Design steps include the determination of the nominal diameter (DN), the rated pressure (PN), the construction materials for components in contact with the product, the type of sealing, the positioning forces, the noise emission, the possibility of cavitation, and the matching of the inherent characteristic to the operating and control drive characteristic. In the case of the variable-speed pump drive or a control machine, special consideration should be given to the motor startup characteristics, the bypass circuit, the power rating of the motor and the converter, the motor heating, the torque ratios, the proper matching of the actuator (in the form of a pump) to the motor, the cooling of the converter, the use of tested and certified devices for service in Zone 1, and so forth. In this connection, a number of regulations, specifications, technical guidelines, and so on must be complied with. A process control station data sheet is filled out in which the product, device, process, and design

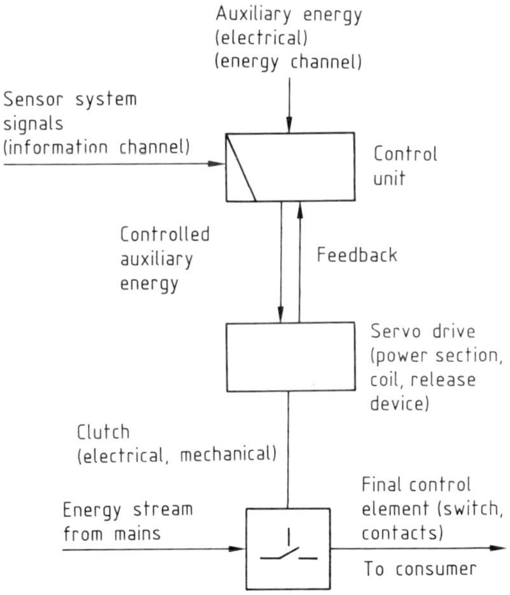

Figure 6.11. Actuator system for energy streams: structure of a switch for control or safety purposes

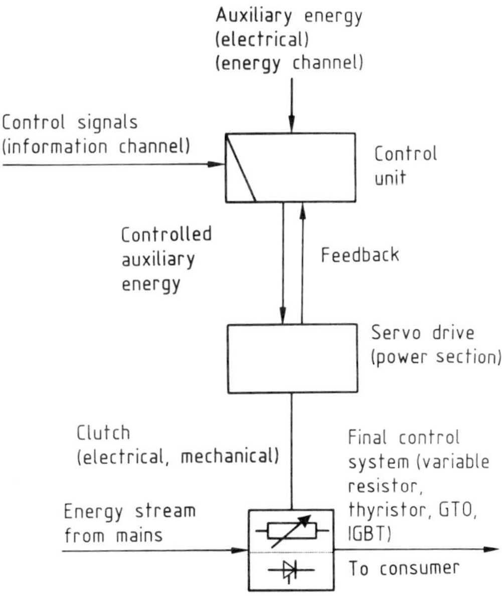

Figure 6.12. Actuator system for mass flow: structure of a positioner for a heating system

data are entered along with data on safety, availability, quality assurance, etc.

With the station data sheet and the requirements from the piping and instrumentation diagrams or the bid specifications, the appropriate design is prepared and the process control system station plan is generated. The procedure is illustrated for a control valve (Fig. 6.13); see also the generic classes in Figure 6.5.

The control valve consists of the valve itself plus a lever or plunger drive, depending on the motion of the throttling organ. The valve can be realized in the form of various devices, such as a globe, peristaltic, or plug valve. The design specification differs from one type of throttling organ to another. The configuration or the way the devices are installed in a process plant can also vary. The best installation position depends on the product, the heating medium, the sealing system, the flange design, and so forth. The design of the lever drive can be based on equipment offered by several manufacturers. As a rule, however, only a single design specification is considered. The same holds for the case of a plunger drive. The devices with their designs, the layouts, the selection, the installation of these devices have been fully described elsewhere. There are also many DIN, IEC, and VDMA standards. In many cases, calculation and design programs are offered with the respective data sheets. Examples of commercially available programs can be found in [6.10].

The path of the control signal via the control signal processing section of the control device, the control drive, and the final control element itself, where it is transformed to a position value, is shown in the middle part of Fig. 6.4. The control element alters a process value in a process element. Generally, one measured quantity is provided to the control device from the process. This quantity is needed as a control or monitoring quantity for the controller or actuator system. Feedback to the process control system or to local components takes the form of messages. In the control device, the actuator computer, together with its communications devices, is shown in Figure 6.4. This device performs function monitoring in the form of self-tests or externally initiated diagnostic tests. The result of these tests is directly displayed locally in error messages, error displays, status messages, and so forth, or (in future technology) will be transmitted over a field bus to a remote location. An integrated display/control panel is always included in complex actuator systems for local parametrization and configuration. Human–machine communication takes place via this interface.

Taxonomies of Actuator Systems for Material and Energy Streams (see Figs. 6.5 to 6.7). In what follows, an attempt is made to devise taxonomies of various actuator systems under various criteria [6.11]. The entity–relationship model is employed, converted directly to file structures (see Chap. 2 and [6.12]).

Controllers (actuator systems) are first classified in terms of their tasks. Controllers for material streams and those for energy streams are grouped separately (Fig. 6.14). At the next level of detail, controllers for energy streams are broken down into switches (e.g., Fig. 6.4) and positioners (Fig. 6.5).

Figure 6.15 shows the classification by technical and functional criteria. Each actuator system consists of a control element, a drive, a control device, and energy storage, if required.

The control element (actuator) may consist, for example, of a valve with a throttling organ, a pump with case and impeller, the inductive coils of an induction furnace, a halogen lamp, the mechanical contacts of a power contactor, or the power thyristors or transistors of a positioner. The drives (power section) are listed in a similar way. In more complex actuator systems, the con-

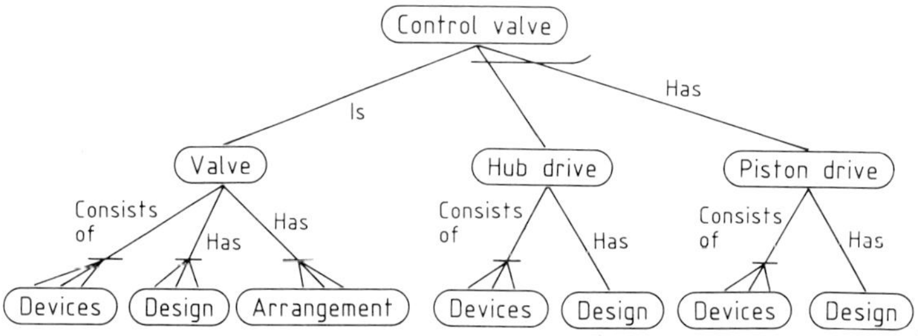

Figure 6.13. Descriptive structure of a control valve for design and installation

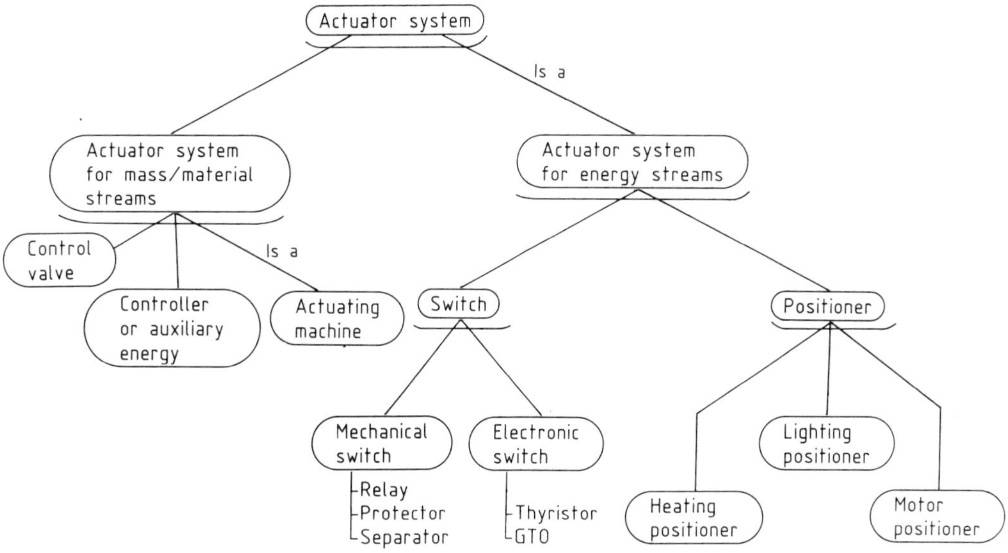

Figure 6.14. Taxonomy of actuator systems for mass and energy streams

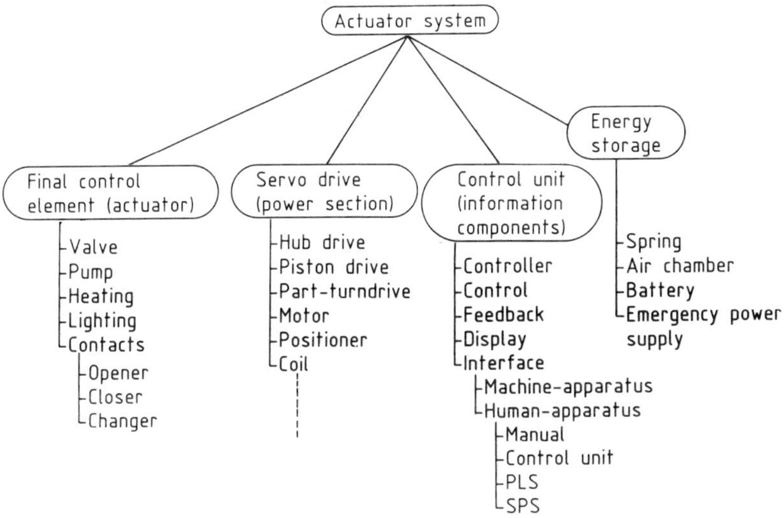

Figure 6.15. Taxonomy of actuator systems for mass streams according to functional criteria

trol element and drive are based on a control device (information-processing component). This component, in the form of an information or signal-processing section, processes, conditions, and transmits the signals to the actuator system and also from the actuator system to and from higher-level devices.

When safety and availability requirements are more stringent, energy storage is required. If the auxiliary energy fails, this stored energy brings the control element to a desired safe position via the drive. Examples are compression and tension springs, which release energy on demand and test the "emergency case" upon every switching operation. A pneumatic buffer can be used as energy store; from which energy stored in compressed air can be withdrawn for a certain time and with which repeated switching operations can be carried out. In other applications, batteries and backup power supplies maintain safe operation. Battery-backed systems can be kept operable without interruption; with emergency generators, on the other hand, interruptions must be reckoned with from the outset. It is often sufficient that the motors continue to run after a short interruption.

Final control elements for material streams are classified as shown in Figure 6.16.

The control element may be a control valve, whereby the function of the throttling organ is carried out by translational or rotational motion. The throttle bodies are said to be straight-line (translational) or rotational.

Another possible classification, in accordance with the terms of DIN IEC 534, Part 1, is illustrated in Figure 6.17.

Mention should be made of the many and varied process connections, sealing systems, fitting types, throttling organ configurations, noise suppression options, drive design possibilities, and control device types.

The control element can also be a controller without auxiliary energy but employing spring or weight loading to insure the admission of a

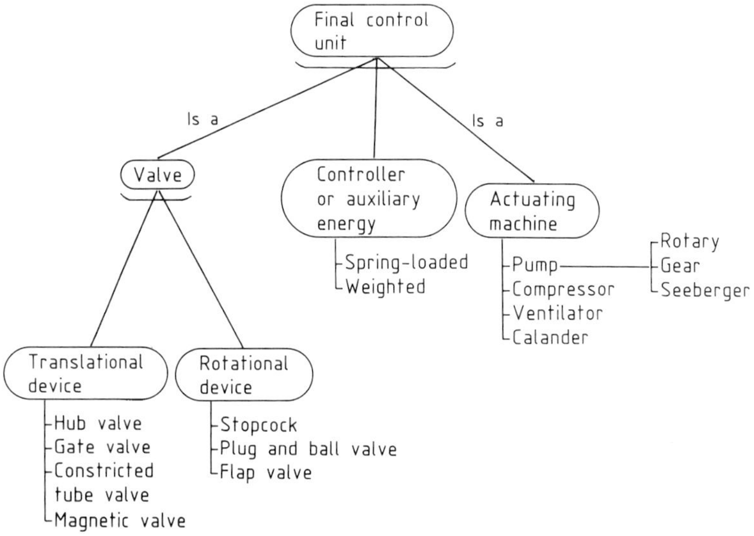

Figure 6.16. Taxonomy of final control elements for mass streams

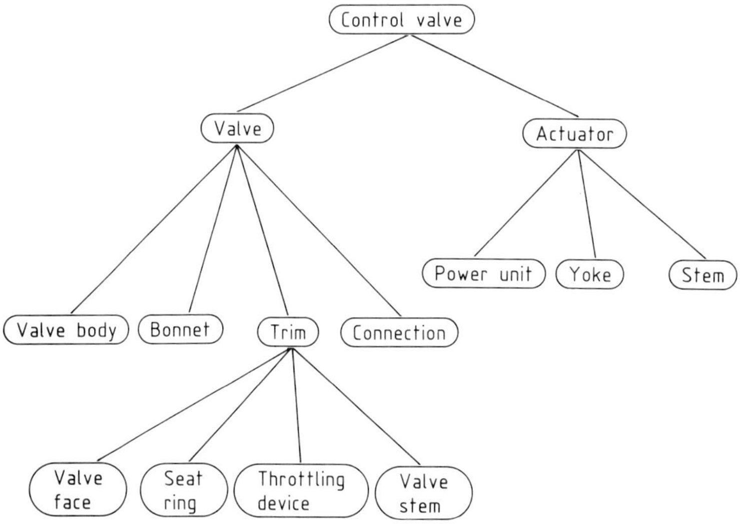

Figure 6.17. Taxonomy of a control valve [DIN IEC, Part 1]

certain quantity (setpoint value) of material per unit time. Large numbers of such devices are used not only in process engineering, but also in heating and air conditioning as well as other branches of industry.

Control machines are becoming increasingly important with the development of modern power electronics, because they can be operated much more efficiently in variable speed mode than controllers with throttling organs. In systems with throttling devices, energy must first be introduced into the material stream, and then part of the energy dissipated by throttling [6.13]. Cavitation can result, with outgassing, phase-change effects, and noise as the consequences. For this reason, variable-speed motors are increasingly used to drive control machines (see Section 6.3 and [6.14]).

Drives. Each actuator system has a power section or drive system (Fig. 6.18), which must be matched to the control element or actuator. In devices with throttling organs, translational and rotational control drives are distinguished. Motors can be classified into d.c., multiphase (Fig. 6.19), compressed-air, and other types.

Depending on the network form, power, required dynamics and quality of control, field and later maintenance conditions, and so forth, the designer selects an asynchronous, a.c., synchronous, or d.c. motor. Attributes of these

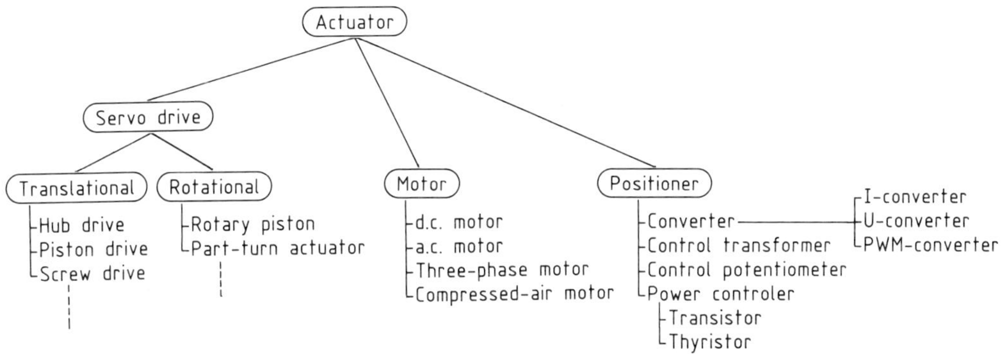

Figure 6.18. Descriptive structure of an actuator or power section

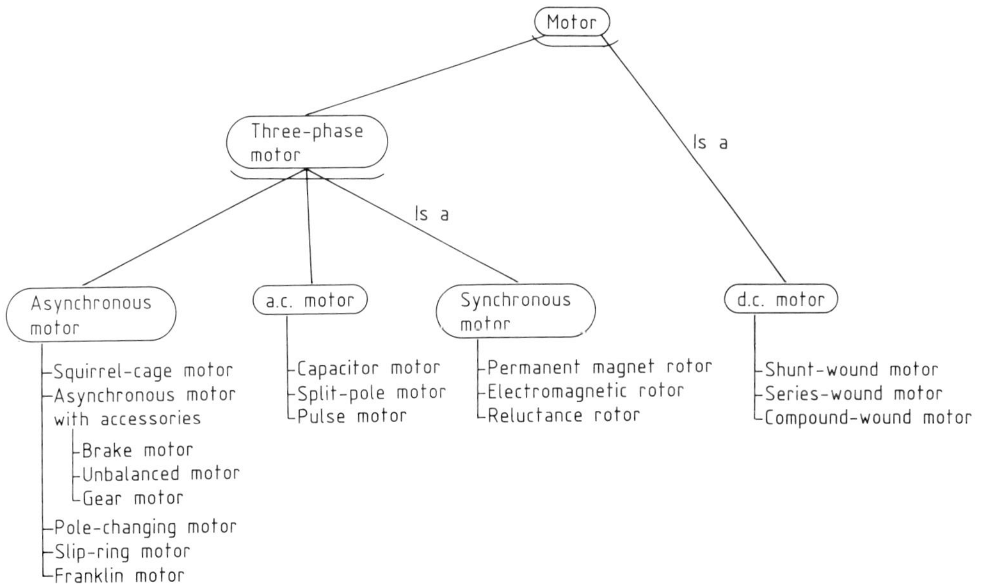

Figure 6.19. Taxonomy of servo drives for the example "motor"

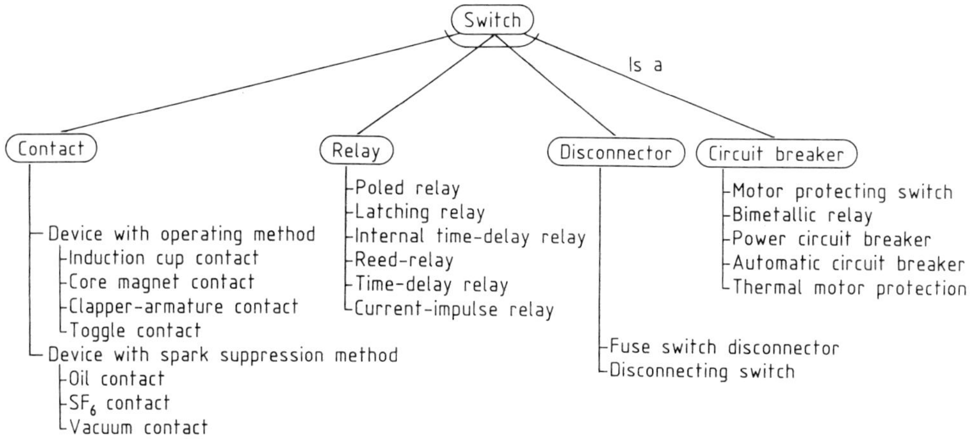

Figure 6.20. Taxonomy of discontinuous actuator systems for energy streams

objects include power, rated voltage, speed, number of pole pairs, motor and explosion protection, case shape and size, type of protective enclosure, weight, ventilation, position of terminal box, manufacturer, supplier, model number, and stock number.

Devices for controlling energy streams can be broken down into switching devices and devices continuously controlling the energy stream (Fig. 6.14). Typical representatives of switching devices or switches are relays, contactors, disconnectors, power circuit breakers, and current-limiting circuit breakers. Each of these consists of a drive (operating mechanism) and a mechanical system, which in the simplest case (e.g., in a low-voltage, high-breaking-capacity breaker) is operated by hand. More complex devices employ a magnet coil or, in the medium- or high-voltage range, a motor drive. Switching functional elements are normally-open, normally-closed, and changeover switches. Figure 6.20 illustrates this taxonomy, but only for the mechanical switch side; electronically operated devices with corresponding functions can be classified similarly.

In the continuously operating devices, the energy stream or power can be set within a certain range, so that matching to a certain operating point within a stated range is possible (see also the schematic diagram in Fig. 6.12). Positioners or position controllers can be used to set the voltage, current, frequency, and other values [6.13], [6.15]. Rectifier technology includes devices to transform or control electrical energy by the use of rectifier valves. The fraction of electrical energy switched, controlled, and trans-

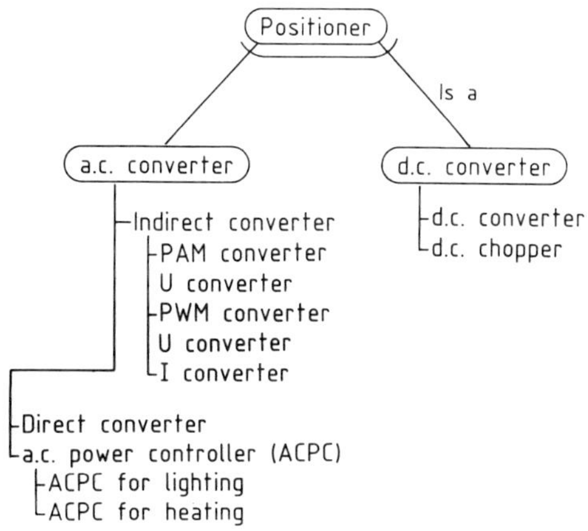

Figure 6.21. Taxonomy of continuous actuator systems for energy streams

formed in the power electronics will increase steadily with growing requirements on the controllability and transformation of electrical energy. Applications for power electronics include industrial drives, electric heating, and energy generation and distribution. The basic functions of positioners can be classified as shown in Figure 6.21 (see also Section 6.3).

The basic functions are rectification, inversion, a.c. conversion, and d.c. conversion. In rectifiers and inverters, which are not regarded as actuator systems, the direction of energy flow is specified. In a.c. converters and d.c. converters, the direction of energy flow can generally be reversed. Variable rectifiers make it possible to control the output voltage and frequency and thus the energy flow. A possible classification of rectifiers by direction of energy flow, into unidirectional and bidireactional rectifiers, would also be conceivable where the energy flow can be reversed. The a.c. and d.c. controls shown in Figure 6.21 should be emphasized; these can be used to control heating and lighting and also in power systems. Further, the various link converters used for motor speed control should be mentioned. There are various pulse-amplitude-modulated frequency converters, pulse-width-modulated frequency converters, and link converters. These devices are commonly referred to as motor controls, heater controls, and lighting controls (see Figs. 6.15 and 6.25). Their attributes as individual objects are power, voltage, current, driving signal, manufacturer, accessories, limiting, torque limiting, type of starting torque enhancement, power and torque transducers, and braking unit.

In general, without going into the individual technical differences between frequency converters, a classification such as that shown in Figure 6.22 can be employed. As shown above, the devices can be broken down into final control element, control device, and power section (control drive). The entire equipment is also called a control machine in accordance with IEC 514. All centrifugal, turbine, gear, and plunger pumps, as well as special-purpose pumps, are included in the category of machines for handling gases, liquefied gases, and liquids. The control drive consists of a power section and the motor. The power section and the control device are commonly integrated, except for the transformer of the rectifier.

Control Device. The control device has two functions: to match the actuator system to various equipment connected before it and to the mode of operation, and to match it to the final control element and the drive itself. Figure 6.23 shows a possible classification.

In more complex actuator systems, control devices are made up of devices such as controllers and monitors with their respective memories, actuator buses, display/communications/feedback devices, accessories, integrated display and control panels, and so on. It can be expected

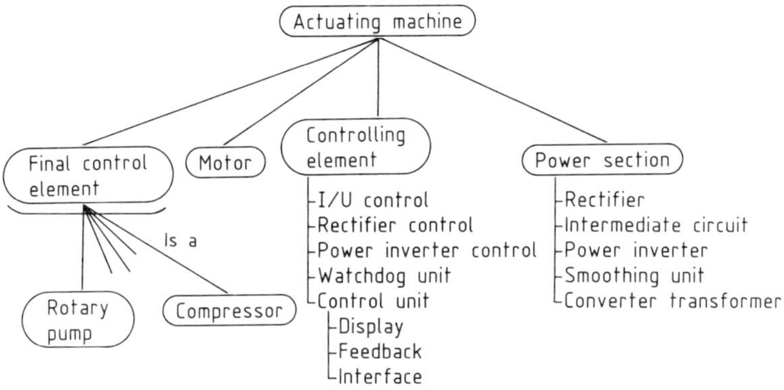

Figure 6.22. Descriptive structure of an actuating machine

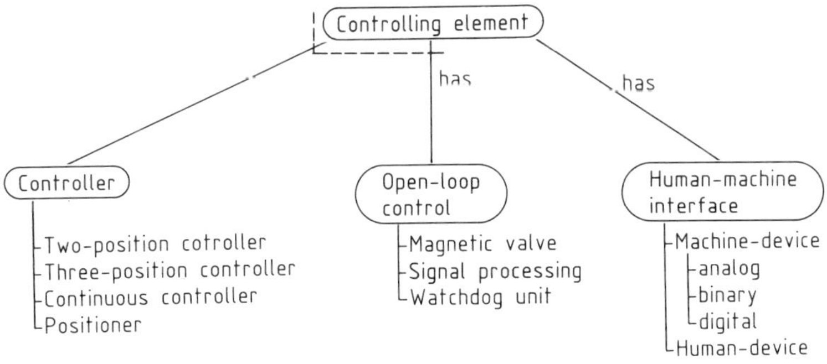

Figure 6.23. Structure of a controlling element

that the control device will go through major changes in the future. The control device should make possible matching to the connector in conventional or smart technology (see Section 5.3 and Chap. 8) or in bidirectional field bus technology. The type of central electronic controls common up to now, with expensive cabling between the sensor and actuator systems and extensive setup operations during commissioning, will be modified for the sake of more easily understood, less expensive, distributed signal processing. For the user, such a device will become a "plug-and-play" unit in spite of its complexity. Not yet fully achieved are networking capabilities and a consistent, manufacturer-neutral, menu-based operator interface as desired by both operator and designer. Futhermore, major efforts are needed in order to bring about a unified field-level communications technology based on internationally accepted standards. Such technology will permit interchangeability of hardware across manufacturers. These interfaces must satisfy increasingly stringent requirements, partly because of the trend to "rationalization" and the desire for a central service. In the near future, this area will become a focus of the activities of manufacturers, users, and developers (see also Section 5.3 and Chap. 8).

The control section, as a module of the control device described above is needed for the human–device interface (Fig. 6.24). In the simplest case, the actuator system can be set with a handwheel, or via a manual control signal from the supervisory device of the controller or from the process control system. In automatic mode, positioning takes place after signal processing with specified setpoint values; this function is performed automatically.

Figure 6.24 illustrates the possibilities of manual and automatic operation. Thus, there are two interfaces: a human–device in manual mode, and machine–device in automatic mode.

In manual operation, interventions can be accomplished by

- A local control station, in the simplest case a handwheel, lever, or the like
- A display and control panel, integrated into the actuator system, on which the system can be set
- A control panel at the supervisory device, for manual and automatic operation
- A screen mask on the display and control component where manual control signals can be set or nominal values and tolerances can be specified

In automatic mode, interventions take place

- Between the final control element and the field controller
- Between the control element and the individual controller in the control room
- Between the control element and the controller component in the process control system, stored-program controller, or PC

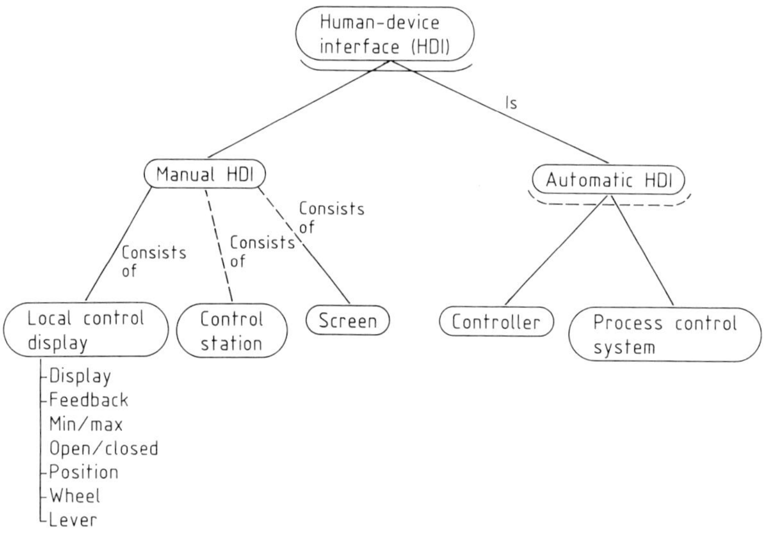

Figure 6.24. Structure of a human–device interface

Outlook. The object-oriented representation in data processing has the following advantages for the user: Objects are presented in an intuitive, structured, transparent form, and searches (e.g., for a certain device) are carried out in prompted mode.

Because devices in this representation are merely sketched, further detailing is necessary at the lower levels if installation, shop, and maintenance workers are to find individual repair or replacement parts with given specifications and given relationships to other devices. Similar requirements arise in the design of a process monitoring and control system, where the specification document dictates the selection of devices; their identification by, say, a CAE plot symbol in a process monitoring and control functional chart; or their representation in a positional world with lower- and higher-level devices.

On the basis of Figure 6.25, this point is illustrated for the example of a device object tree [6.16].

In the object-oriented representation, the devices required for a process control system are described by groups. For example, they can be broken down into structural elements of equal rank: sensor system, actuator system, auxiliary device, information-processing components, electrical, control air system, infrastructure, and installation materials. Sensor systems are classified by physical quantities; actuator systems, by the criteria discussed above; auxiliary equipment, into input and output devices, switches, and protective device; information-processing components, into controllers, supervisory devices, software modules, stored-program controllers, and process control systems with their components, such as central unit, local components, display and control components, and so forth. Note that a "pure" or textbook-type classification is not possible in many cases, and for practical reasons a compromise must be found, especially in the case of actuator systems. Here, continuous control devices are listed under the structural element "actuator system," except for solenoid valves and switching-type control valves, while switching devices for energy streams are classified exclusively as "auxiliary equipment." Naturally, a practical worker seeking a controller without auxiliary energy will look among the other controllers under "information-processing components," and if in need of a reducing station will look under "control air system." Under this object-oriented approach, once a certain device has been found, design specifications, device configuration and installation standards, and even manufacturing standards (if needed) can be found in a further level having the same structure and the same nomenclature of structural elements. After a device has been selected, designed, and configured, the respective activities can be carried out in CAE applications modules, such as process control station data sheet, process control station plan, procurement, and so forth.

Such a system has the further advantages that innovations in hardware and installation can be adopted better and more quickly and that workers in these fields obtain design and manufacturing standards in easily understood form. What is more, changes in standards and directives can be complied with in a better and more economical way (see Chap. 9).

Finally, two more developments should be mentioned:

- Workers in automation, information, and microsystems engineering are energetically pursuing:
  - Piezoelectric actuators [6.17]
  - Shape-memory actuators
  - Magnetoelastic actuators
  - Micromechanical actuators
  - Actuators using electrorheologic and magnetorheologic liquids
  - Precision positioning with electromagnetic actuators

Information on these developments can be found in [6.18].

- The trend in the chemical industry toward smaller-sized batch processes is generating a demand for much faster metering systems, which can only be realized by direct coupling between metering element and flowmeter (see Chap. 5) [6.19].
- In addition to a number of other advantages (energy savings, decrease shear gradient at almost closed position of the control valve—important for biotechnological and polymeric materials—smooth start-up, etc. [6.20]), by means of r.p.m controlled drives, pumps with variable performance can adopt the functions of classical control valves [6.21], [6.22].
- The coupling of sensor and actuator systems in self-sufficient control loops at the field level allows the amount of communication with and the computing load on the process con-

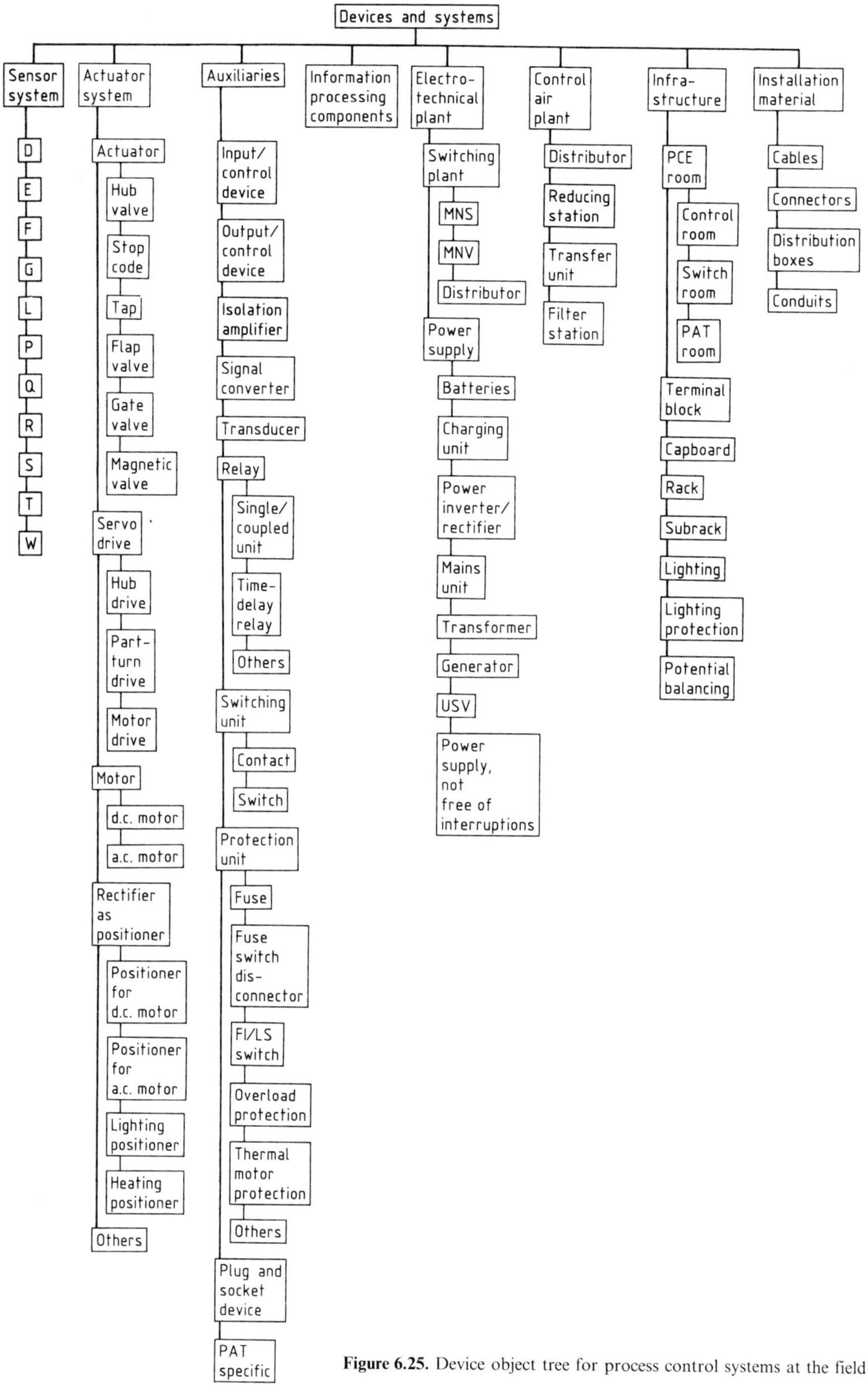

Figure 6.25. Device object tree for process control systems at the field level

trol system to be decreased. Expecially in the area of building automation, interesting application are already available at reasonable cost.

- In the field of sensor/actuator systems in particular it will be necessary for developers and users to move away from the traditionally narrow viewpoint and, by means of analogy, apply insights from other areas to process control engineering [6.21].

6.3. Electrical Drives in the Chemical Industry

Development of Electric Drives. Electric drives have long been used in large quantities in various applications in industry. The majority of the drives are not speed controlled in any way. The first major application for variable speed drives was in traction, applications, especially in locomotives and trams, and later also in metallurgy. Variable speed drives subsequently found their way into a large number of applications in general industry. The general trend today is from fixed-speed to variable-speed drives. Typical applications for modern variable speed drives are, for example, in the field of traction, the main drives for TGV and ICE trains, main propulsion drives for ships, and a large number of process-oriented applications in the pulp and paper, chemical, and metallurgical industries. Other typical application areas cover manufacturing industry, energy production, and water treatment.

Most motors in use today are not speed controlled. About 95% of all motors sold today are asynchronous motors, of which 95% run at fixed speed, connected directly to the mains. In future, the number of asynchronous motors driven by frequency converters is expected to grow rapidly, at ca. 15%/a, while the growth rate for all variable speed drives is ca. 7%/a.

The estimated world market for variable speed drives was ca. \$ 3.3×10^9 in 1991, of which 55% is a.c. and 45% d.c., a trend that will intensify in the future. The a.c. market (Fig. 6.26) is concentrated to Europe, the United States, and Japan. The rest of the world accounts for ca. 10% of the total market. The market split into the various power ranges reflects the differing structures of the industry and the price of energy in the different areas. In terms of numbers, the majority of drives (ca. 90%) are small (< 5 kVA),

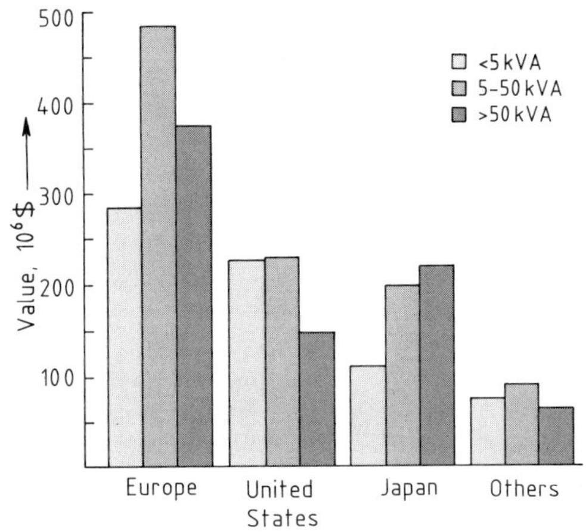

Figure 6.26. World market for variable-speed a.c. drives in 1991

ca. 8% are in the range of 5–50 kVA, and only 2% exceed 50 kVA. The number of motors controlled by variable speed drives in Europe is about 5% for 5 kVA, 9% in the range 5–50 kVA, and 16% for > 50 kVA. The larger the motor is, the more likely it will be driven by a frequency converter. Process demands at higher powers often require variable speed drives, and the realized energy savings are economically more important.

The basis for this development is the a.c. motor, which has a simple, robust, cost-effective, reliable, and rugged design that is relatively easy to modify for use in Ex-areas. Developments in power and control electronics have made drives more technically advanced and cost effective. The price of an a.c. drive has decreased by about 10%/a during the last ten years, while the price for d.c. drives has shown a decrease of ca. 4%/a. The total cost of a drive (drive electronics and motor) is about equal for a.c. and d.c. although a.c. can be slightly more expensive in the middle power ranges. The total life cycle costs for an a.c. drive are in general lower than those of a d.c. drive.

The German chemical industry has a large number of motors in use (Fig. 6.27), one estimate [6.18] is ca. 500000 units. The average power is 7–13 kW, the minimum power is 120 W, and the maximum power is 6800 kW. About 80% of all drives are smaller than 10 kW. Only about 10% of all motors have variable speed; in two-thirds of the cases, this is achieved by pole changing. The large proportion of fixed speed motors indicates that other control methods than variable

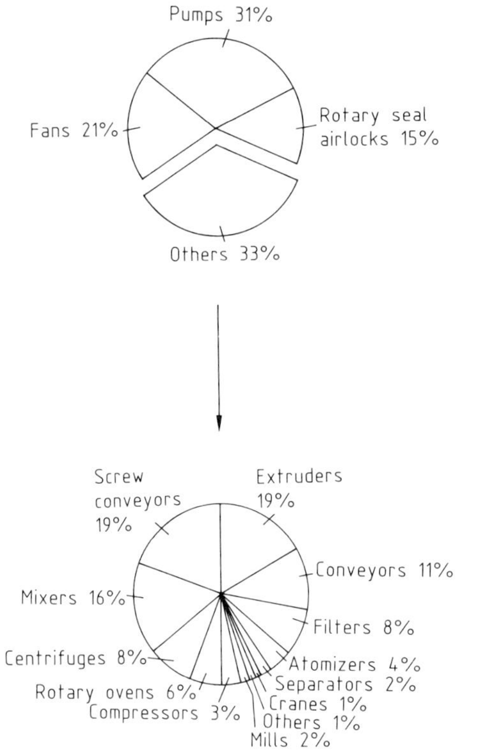

Figure 6.27. Areas of use of variable-speed drives in the German chemical industry

speed are used, usually control valves, bypass control, or mechanical variators.

Most of the drives in the chemical industry are used in the main process, although many drives are used in auxiliary processes in energy production and raw and wastewater treatment. The use of variable speed drives has increased markedly during the last few years. Most of the applications are pumps and fans, but several other applications are also quite common [6.23].

Variable speed drives are used mainly to improve process control [6.23]. By controlling the speed of the prime mover, the material flow can be better controlled; variable-speed pumps can handle logistics and the dosing of materials with low losses and low machine strain, thus improving process control and reducing mechanical wear. Variable speed drives are also used to replace control valves in polymer and biotechnical processes. The soft start feature of variable speed drives eliminates pressure shocks in pipes when motors are started and they also significantly reduce the thermal and mechanical strain on the motors at start, which usually permits more economic dimensioning of the total system. The supply dimensioning is eased by the lack of high starting currents and compensating capacitors. The direct drive energy savings, especially in centrifugal machines, are also of importance when deciding the type of drive required. A 40% saving for variable-speed a.c. drives compared to fixed speed solutions is not uncommon. Variable speed drives also make less noise, as machines are used at optimum speed, rather than at full speed at all times.

Development of A.C. Drive Technology. *Power Semiconductors.* The development of power semiconductors is summarized in Table 6.1. The first commercially successful power semiconductor was the thyristor, developed in the early 1960s. The thyristor is a four-layer device that can be turned on (changed into a conducting state from a nonconducting one) by the application of a gate pulse. Because it works by positive feedback, to turn it off, the current through it must be decreased to zero, after which it regains its blocking, nonconducting state. If the applied voltage across the thyristor does not naturally reverse sign (as happens for components connected to the mains supply), a commu-

Table 6.1. Main electrical data of various power semiconductors

	Thyristor	GTO	Bipolar transistor	IGBT	FET	MCT	FETh
Voltage, V	6000	4500	1600	1600	1000	3500	4500
Current, A	5000	4000	1200	400	10	50	50
Frequency, Hz	400	400	2000	20000	100000	5000	10000
Power, kW	10^4	10^4	10^3	10^3	10^1	10^2	10^2
Chip size, cm^2	80	80	> 10	> 10	5	2	5
Year	1960	1977	1985	1991	1980/90	1993	1995
Processor/ Memory	256 bit	8080/85 4 kbit	8086/186 64 kbit	8096/486 1 Mbit	1 Mbit	80586 16 Mbit	64 Mbit

tation (turn-off) circuit is required. This increases the size and the complexity of a thyristor drive and also limits the maximum switching frequency achievable. Thyristor PWM drives normally switch at ca. 400–600 Hz and the motor is thus fairly loud with a motor loadability of ca. 85 % at 50 Hz. The dU/dt is ca. 500 V/s. Thyristors are now used only in high power drives.

The next major power semiconductor is the GTO (gate turn-off thyristor). As the name implies, it is a thyristor that can be turned off as well as on. This is achieved by careful design of the semiconductor layers and device geometry. As the device can be turned off, no additional commutation circuitry is needed and the size of the drive thus decreases. The GTO is fairly sensitive to a correct handling at turn-off and is today used in high-power drives (200 kW) and at high voltages (> 1 kV). An advantage of GTOs is that they are latching devices, like thyristors, and thus require no control power when they are in a stable state (on or off). In the case of short-circuits in the output, they pose problems, as they do not inherently limit the current and are difficult to turn off in this state. The switching frequency and dU/dt are comparable to thyristors.

In lower power drives today, the main power component used is the bipolar transistor. With this component powers of up to 200 kW can be achieved, and up to 600 kW with several transistors in parallel. The device consists of two Darlington-connected transistors. By controlling its base drive the device can be made to turn on and off. Transistors require a continuous base drive in order to conduct, i.e., they do not latch. In the case of shorts they are more forgiving than GTOs but they require a stiff base drive to turn them off. The simpler turn-off requirements simplify the control circuitry and allows drives to be smaller and more efficient. With Darlingtons, switching frequencies of up to 4 kHz are achieved with a dU/dt of ca. 1–2 kV/s. The higher switching frequency increases the motor loadability to ca. 90 % at 50 Hz and reduces motor noise.

In newer ranges of frequency converters, the power semiconductors are IGBTs (insulated gate base transistors). These devices are a combination of voltage-controlled field effect transistors (FETs) and bipolar transistors, connected so that the control is like the voltage control of a FET and the power side is of the low-drop bipolar type. These components inherently limit the current and they are thus more forgiving and easier to handle in short-circuit situations than transistors. As they are voltage controlled, not current controlled like bipolar Darlingtons, the control circuit is simplified and reduced in volume. There is also a significant reduction in the rating of the auxiliary power supply, as IGBTs require less control power than transistors. The IGBT allows switching frequencies of up to 20 kHz which increases the motor loadability to 100 % at 50 Hz and gives an attainable dU/dt of 5 kV/s.

New components are being developed by power semiconductor manufacturers, for example, MCTs (MOS controlled thyristor) and FETh (field effect controlled thyristor), hybrids between conventional thyristors and FETs, with voltage-controlled switching and a very low state loss, which will reduce the size of future drives.

Frequency Converters. Frequency converters are divided into two main types, depending on their construction. One is the current source inverter (CSI) and the other the voltage source (PWM).

Current source inverters contain a controlled thyristor bridge connected to the supply and a second one connected to the motor. The two bridges are connected by a large inductance. The current to the motor is controlled by the input bridge, and the frequency is controlled by the output bridge. The motor voltage then adjusts itself to the actual current and frequency. In older types, the output bridge consists of thyristors, which means that the motor is a part of the commutating (turn-off) circuit of the thyristors and thus every inverter and motor forms a pair. In newer drives the thyristors are replaced by GTOs or bipolar transistors, which eliminates this coupling. Due to the basic construction with a controlled input bridge, CSIs are suitable for single, stand-alone four-quadrant operation. As they contain a large choke, however, they are usually bulky and the drive may be noisy.

The most common type of drive today is the voltage source inverter. It consists of a diode input rectifier, which generates a constant intermediate circuit voltage. This intermediate circuit consists of a choke and a large capacitor. At lower powers the choke is often omitted. This causes the harmonic content of the supply current in the main to increase, so for installations requiring smooth mains current drives with an intermediate circuit, chokes are recommended.

The rectified and smoothed voltage is supplied to the motor via an output stage, which controls the voltage and the frequency of the waveform applied to the motor. The most commonly used components in this stage today are bipolar transistors at lower powers and GTOs at higher powers. The bipolar transistors are being replaced by IGBTs. The waveform applied to the motor consists of a pulse-train with a sine-weighted duty cycle (pulse-width modulated, PWM). The inductance of the motor filters the current resulting from this pulse-train to near-sinusoidal shape. The effective voltage applied is normally increased in direct proportion to the output frequency to generate a constant flux in the motor. This is possible until the mains voltage is reached (i.e., up to 50 Hz) and with the latest modulation techniques up to about 60 Hz. At higher frequencies the voltage remains constant, and the flux diminishes in inversely proportion to the output frequency. The maximum speeds that can be reached with standard squirrel cage motors are of the order of 40–120 Hz; higher speeds require special motors. The normal PWM inverter described above is used in about 90% of all cases. It is a simple, straightforward approach to variable speed control, requiring no special motors or any sort of feedback. As it is an open loop drive, its dynamics and precision are not of the highest degree. It is, however, sufficient for pumps, fans, extruders, conveyors, and simple machines.

For higher performance, vector-controlled drives are used. In these drives, a feedback signal is required from the rotor position. This information enables the controller to precisely calculate the torque on the motor shaft and also indicates the exact position of the shaft. The information is then fed to a PWM output stage, which controls the current and frequency in the desired manner. With this drive, full load torque at zero speeds and precise position control can be achieved. The drive always requires an encoder on the motor. Typical applications are synchronized lines, paper-machine drives, rolling-mill drives, and all applications where the torque must be precisely controlled.

Power semiconductor development has had a major impact on the size of drives. The average volume per power has decreased to 1/6 in ten years. Today the decisive factor determining the size of a drive is its thermal design. Modern power semiconductors allow a junction temperature of up to 150 °C, corresponding to a case temperature of ca. 100 °C. The case temperature is limited to about 80 °C, sometimes up to 100 °C.

Modern control electronics use 16 bit processors with designs for 32 bit processors or for signal processors in development. This trend of incorporating more computational power in the drive will continue, so that the possibilities for control offered by modern drives will increase. The requisite discrete logic is commonly realized in an ASIC (application specific IC). These are monolithic ICs built using standardized logic cells, so that a defined, application-specific function can be realized. One ASIC can replace several hundred conventional TTL circuits with large savings in space and powering. Lately, technologies have become available that allow a similar approach to analog circuit design.

These developments have also increased the reliability of drives. The MTBF (mean time between failure) for drives in 1983 was ca. 40 000 h, while in 1993 it was ca. 250 000 h. The number of parts necessary to build an 11 kW frequency converter has decreased from 2500 to 500, a trend that will surely continue. The basis for this lies in modern component technology, both for logic components and for power components, and in the better possibilities for simulating and measuring complicated and fast processes.

For critical applications, redundant drives are still recommended, especially in cases where there is a risk of injury to humans, where the process flow is critical and must be controlled at all times.

An aid to achieving high process reliability is the use of switchgear converters, which allow easy changing of drives and of system configuration. Combined with a fully digital commissioning and down-loading of various drive parameters, this allows an extremely rapid change of drives, simply by changing a cassette.

An external sign of quality control and an internal tool is a certification according to ISO 9000, Parts 1–3. This standard defines the requirements regarding standardized working methods and documentation for a company or profit center. Quality is also achieved by standardization of methods in development, production, and testing. Standardization allows a high repetition rate when using various components, either software or hardware. This allows a short development time for new products without endangering product quality.

As an example of standardization, consider solving EMC problems (electromagnetic compatibility), that is the problems occurring when devices have to work in an electrically noisy environment. The main standards in use are the various DIN/VDE and CISPR standards. Generally speaking, two types of interference occur:

- Low frequency: frequencies up to ca. 5 kHz and earth currents
- High frequency: frequencies from ca. 50 kHz to 30 MHz

Low-frequency interference can be divided into periodic interference (caused by nonlinear components such as diodes and thyristors) and aperiodic interference (caused by supply disturbances). High-frequency disturbance is caused by switching phenomena in power switches. The various interferences can influence other devices nearby, either through radiation or conductance. Generally, conducted interference is more problematical.

A standard solution to high- and low-frequency interference is adding filters to the input of the device. At high powers this is usually difficult, bulky, and expensive, especially as the high- and low-frequency ranges require different features. It is also possible to separate the power supply to interfering devices from all other supplies, or to use a d.c. distribution system, in which 12-pulse circuits for supplying the d.c. power distribution minimizes the 5th and 7th harmonic current in the supply. This d.c. power is then supplied to independent output stages in order to drive motors.

Aperiodic disturbances include supply failures, which can last from milliseconds to several seconds. Drives usually use the kinetic energy in the driven load as a supply when the supply fails and then accelerate to the set speed when the supply returns. The possible time the drive can survive depends on the type of load and its kinetic energy. Usually there are also other questions of synchronism: can the drives run when the total process is otherwise down?

To eliminate undesirable influences from the power stage on the control electronics, fiber optics are used. This allows a clean separation of the control circuits, which can be standardized irrespective of rated power, and the power stage, which can be separately optimized and standardized. This technique also allows customized interfaces to be developed for standardized, mass-produced drives.

In future, it will be possible to eliminate a significant number of sources of high- and low-frequency interference by using so-called resonance circuits. For example, with single-phase inputs, a completely sinusoidal supply current can be achieved. A further development allows switching of the power semiconductors in a zero-power state, which reduces disturbances.

Drives as System Elements. A drive is regarded as an actuator from the systems engineering standpoint. It is connected directly to the process and to a field bus. Normally, only commands and status information are exchanged through the bus system, but in special cases the entire commissioning and the parameter setting must be performed associated through it. This can be done either from a remote panel connected to the field bus or from the engineering workstation functioning as bus master. The goal is to move the detailed process control to the sensors and actuators exchanging information through the field bus. To reach this goal intensive work is being done to standardize the field-bus protocols.

A modern a.c. drive (Fig. 6.28) is functionally divided into the actual motor control and the interface with the external world, with control and setting possibilities. This concept uses at least two microprocessors (one for each function). The drive is self-contained, and no external power is necessary. It is in many cases possible to process data locally to control the drive. Such a drive will form a yield unit (Fig. 6.29), decreasing the load on the process controllers and on the control system. This is especially important in fast processes (for example, in dosing applications), so that the central control is not burdened with all details of the process. The drive is then closely connected to the process through local measurement and feedback, making it possible to implement local control of, for example, volume and heat flow. The bus system transmits only commands and status information from the various field units; details are handled locally.

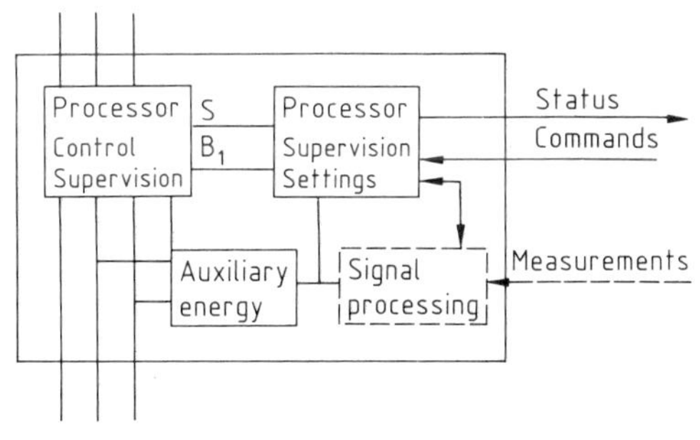

Figure 6.28. Information structure of a modern drive

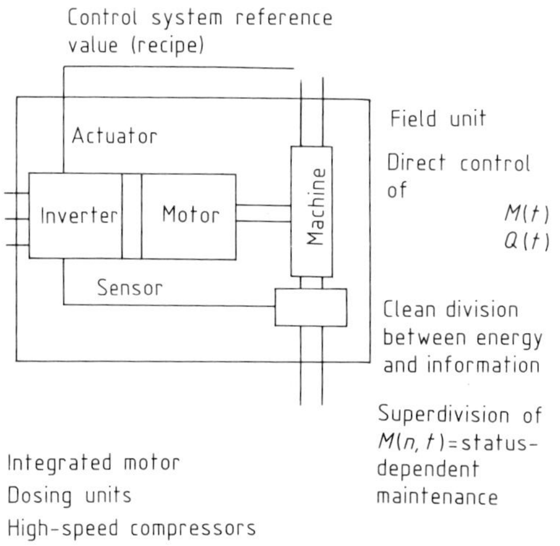

Figure 6.29. Field unit

The planning of a drive system is identical to the general planning of control systems. The basis for the planning is a requirement specification drawn up by the user, describing the function of the drive and its behavior in special cases such as mains failures. This must be carefully thought through and defined, as experience shows that most of the subsequent problems can be avoided here. Based on the requirement, an engineering specification is written, describing with which devices and with which technologies the desired functions will be realized. Here, the economic restrictions of the various projects must be taken into account. This work is usually facilitated by standardization within a plant or company. It is also recommended that a test specification be written for the possible delivery tests, preferably in cooperation with the chosen supplier.

When dimensioning and specifying a drive, the following features must be taken into account: voltage, power, torque requirement as a function of speed, speed range, control signals, braking capability, cabling recommendations, and EMC aspects. The behavior in special cases (short circuits, overloads, underloads, earth faults, etc.) must also be considered. The desired motor features must also be defined (construction, EX class, loadability, etc.).

Drives are considered actuators in modern CAE systems and they are treated as standardized objects with certain, defined features [6.24].

When a drive has been specified and built, an acceptance or delivery test is often required, especially in the case of high-power drives. The goal of a delivery test is to establish that the drive fulfils all the requirements defined in the requirement specification. It is important to specify in detail how the various features must be measured to ensure that the specifications are fulfilled. At lower powers, a type test certificate is usually sufficient. The supplier must be able to test the behavior of the drive at high break-away torques or at mains failures. A test proves that the drive works as specified and allows a first, coarse setting of parameters to be done. It is usually not possible to optimize the drive; this must be done under true working conditions on site.

This optimization is made easier by digital technology, which also allows data transfer and trouble shooting over telephone lines. A multichannel recorder integrated into the drive also facilitates trouble shooting. For trend analysis, the drive itself stores and displays certain key data as a function of time. This can be done by a process control system, but it is simpler if the drive makes it itself and transmits only the results to the control system. The analysis of the trends makes it easier to judge when maintenance is needed, and avoids being bound to a fixed time interval. One example of this is the calculation of when bearings need to be lubricated. In this case, the drive calculates the lubrication interval and indicates when it is necessary to lubricate the bearings. This time depends on the speed at which the motor is running and can be appreciably longer than the time calculated from the nominal speed of the motor.

Trends. Two directions can be distinguished in the development of drives. On the one hand motors, and on the other hand power electronics and information processing electronics are continually being developed. In addition, a development in the mechanical construction of electric drives can be traced.

New motor ranges better suited to the use of converters are being developed. Standard motors are designed for mains use, as 95% of motors are connected directly to the mains. Use with a converter makes different demands on the motor. FEM methods help when new, converter optimized designs are being simulated and built. They also aid in developing new high-speed motor concepts by optimizing the rotor slots for inverter use. The new slot shape reduces the losses for a 100 kW motor from 4 kW to 0.8 kW. This motor is designed for a compressor drive with a speed of 10000 rpm.

The main problems of high-speed drives are mechanical and concern bearings, lubrication, and life time. Conventional lubrication is sufficient up to about 16000 rpm above which it is necessary to use oil fog lubrication, which increases cost.

The high speeds also allow better material utilization. The high speed motor allows a higher power for the same motor size, or a smaller motor at the same power rating. In high-speed applications, gears are completely eliminated leading to lower investments, less maintenance, and better total efficiency.

The development of new magnets makes it possible to manufacture motors with permanent magnet rotos. The materials development has been rapid, making it possible to obtain realistic performance at a reasonable cost. The cost for neodynium magnets is today about 300 \$/kg, allowing the development and sale of special motors. These motors have high efficiency, a power factor equal to unity, and a simple positioning drive without feedback can be realized, as the motor has no slip. Powers up to 100 kW have been reached under laboratory conditions. Research in this sector is concentrated on production methods, as it is difficult to work permanent magnets.

A motor of this type is also well suited to integral motor applications. In these motors, the motor and the drive are mechanically and thermally integrated, making it possible to talk about integrated field units consisting of motor, power electronics, and information processing electronics.

In the inverter, several areas are being developed. The development of microprocessors for the consumption and automotive industries makes powerful and cost efficient microprocessors available for use in drives as well. Semiconductor manufacturing technology also allows the design and manufacture of large ASIC circuits to replace discrete logic. The high processing power available makes it possible to realize complex algorithms, for example, the so-called flux control of a.c. motors. The magnetic vectors in the motor are calculated based on the position of the switches in the output power stage and on measurements of the currents and voltages. This requires a very powerful processor if it is to be performed in real time. This control method gives a standard a.c. motor dynamics superior to those of a d.c. motor.

The powerful processors make it possible to use adaptive control in motor control as well as in a possible controller built into a field unit. It is possible to optimize the motor control as a function of load. These adaptive methods make commissioning easier and are also a step in making drives more user-friendly.

Other trends are control algorithms that take into account the mechanical features of the shaft between motor and the driven machine. With an elastic or possibly an electroviscous clutch between motor and machine it is possible to reduce the impact of torque shocks in the load on the motor, but control of the load is more difficult. An algorithm of this type allows the use of smaller motors. Applications can be found in rolling mills and in long and heavy conveyor drives.

To reduce the size of the inverter and to modularize it, several approaches are being pursued. One is the so-called matrix inverter [6.26], which connects a three-phase motor directly to a three-phase supply and allows the motor speed to be controlled. The motor voltage does not correspond to the full mains voltage and thus a special motor is needed. One difficulty with this concept is that switches able to conduct and block in both directions are needed, and such components are not yet available in monolithic form. Semiconductor manufacturers are developing these components. As the matrix inverter does not contain any bulky components, intermediate circuit chokes, or and capacitors, it is a good candidate for integrated solutions.

Another solution for standardizing drives is d.c. distribution (Fig. 6.30) [6.25]. Instead of having one rectifier bridge, choke, and capacitor bank in each drive, the separate drives are supplied with a d.c. voltage from a common source. This larger unit contains one rectifier, choke, and capacitor bank. It can be smaller than the combined power of the individual units, as it is often possible to use the regenerative power from braking drives to supply driving ones. Typically, considerable savings are achieved, for example, in centrifuge applications, where by braking one centrifuge the energy can be used to accelerate others. The mains rectifier must only supply the energy lost. At higher powers it is also possible to use multiple-pulse connections to reduce the strain on the supply. Galvanic, periodic, and aperiodic disturbances are also easier to control.

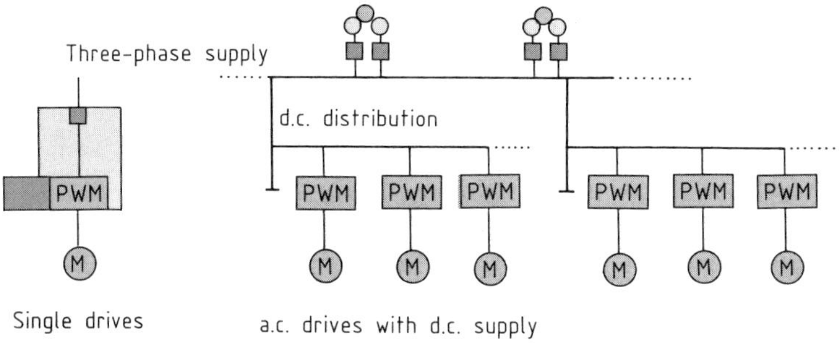

Figure 6.30. d.c. distribution

A further step in modularization is the combination of power switches and their controlling circuits in one package. Modules are today available containing power switches and their drivers and protection circuits [6.26]. Such modules reduce the size of the inverter and its reliability increases as the number of separate components decreases.

This trend towards higher integration has also reached the chip level. Chips will soon be available with several functions integrated. This is called SMART power. Further in the future lies chips integrating power switches and total control of the drive. There are no theoretical difficulties in integrating the matrix inverter onto one chip but many practical difficulties. The limits of what can be done today lie at one power switch and its immediate drivers and protection circuits at a voltage of 600 V. These chips are very expensive, as volumes are small, the production process complex, and yields low.

The developments in the fields of power electronics, information electronics, silicon technology, and mechanics will combine in the form of integrated devices, where most of the necessary functions are integrated. The devices will integrate all necessary sensors, actuators, controls, and protections in a single mechanical unit.

The energy savings made possible by using variable speed drives, not only in the drive itself but through the total energy supply chain can be quite substantial. When a control valve is replaced by a variable speed drive, the efficiency of the total system is significantly increased, especially at low volume flows. The total efficiency can increase from 11.5 % to 23 %, an increase of 100 %. These savings reduce environmental pollution and require smaller investments, as the available energy is used more efficiently.

6.4. Electric Power Supply Systems

Principles. A useful way to visualize the structure of electric power supply systems in industrial plants is provided by the level model (Fig. 6.31). This scheme assigns a central role to the process power supply system, which connects actuators and sensors on the "consumer" level with power plants via medium-voltage distribution units on the "transport" level.

The process power supply system is tailored to the needs of each plant and must satisfy requirements under the following three headings [6.27]:

• Adequate capacity under normal conditions. The specific requirements are as follows:

- Design for operation under rated conditions
- Availability of electric power at any desired location
- Flexible delivery of electric power under various service conditions
- Voltages suitable for various consumer interfaces

• Availability requirements fall into three groups, each with its own set of objectives:

- Availability from the safety standpoint (occupational safety and health, environmental protection)
- Availability from the process engineering standpoint (product quality, safety of equipment)
- Availability from the logistics standpoint (capacity of vendors, production costs)

In engineering terms, this means:

- Limited effects on the process when power disruptions occur
- No interference of one electrical consumer with another

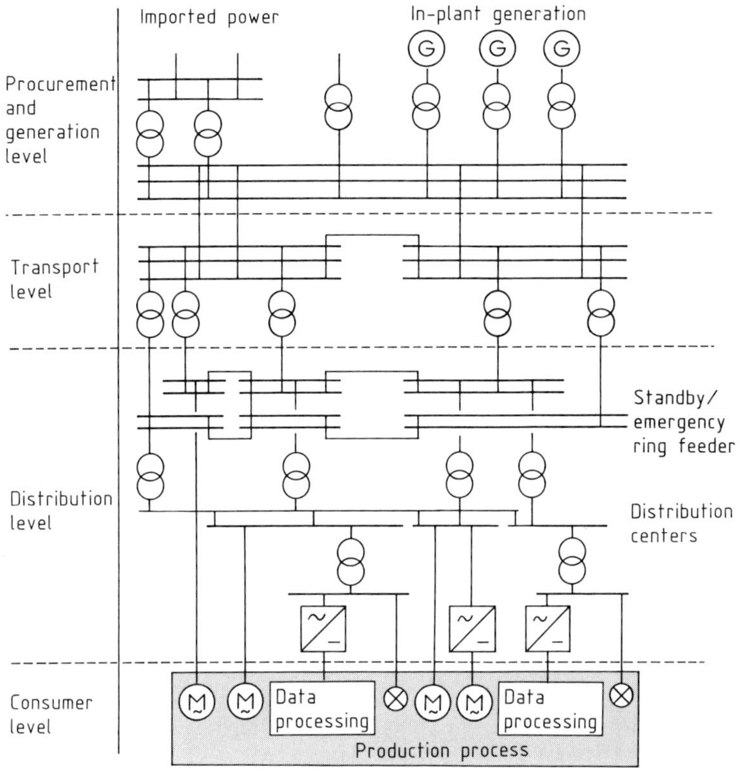

Figure 6.31. Level model of power supply

- Limited effects on the power supply system in case of operation upsets (e.g., fires).

• Safety of electrical equipment. The specific requirements are as follows:

- No danger to personnel
  No environmental hazards

These three sets of requirements completely describe the profile of requirements for the process power supply system.

This section describes the essential concepts employed to meet the stated requirements [6.27], [6.28]. By way of illustration, the process power supply system for a wastewater treatment plant is described in [6.29].

Implementation of Normal Power Supply. The basis for supplying electric power for the nominal operation of a plant is established in the design phase. Here, the three-phase switchgear and control gear is dimensioned in view of the predicted requirements of consumers; utilization and demand coefficients are employed for this purpose. The numbers and sizes of power transformers needed are also determined at this point. Direct-current power supply systems are designed by a similar method, with the closed-circuit current and estimated component peak loads as starting points.

Under normal operating conditions, the entire power supply system of a plant is secured by splitting up the power demand among a number of sources, using in-plant generating capacity and imported power.

The relationship between in-plant and imported power depends on a number of factors, including the size of the plant and seasonal variations in steam demand.

In the chemical industry, the principle of cogeneration (combined heat and power generation) is used in creating the optimal power generation system.

The fact that power to run the plant comes from two sources does not affect the design of electrical equipment inside the plant. The production process is embedded in a complex, two-dimensional network of cables that covers the entire battery area, connects supply to consumption, protects the process systems, and insures any redundancy that may be required.

Cable networks in industrial plants may be radial or meshed. The radial design is more important today. The meshed network, which offers the advantage of two-way supply, is declining in importance, in part because problems can arise in dealing with the short-circuit power when existing facilities are expanded.

For special cases, distributed generation with diesel-generator sets or battery-backed converters serves to deliver reliable power. This approach should be used only when the solution for the normal case is not technically feasible or when the economics are very unfavorable.

Direct current systems are generally designed in distributed form within the plant unit or section. In future, the increasing number of variable-speed drives may also lead to other structures, such as d.c. networks.

Cable and installation costs can be controlled if the location of switch rooms and battery rooms in the central plant buildings is tailored to the process and/or such facilities are distributed in accordance with demand concentration.

Modern switchgear technology can react quickly to service changes by virtue of

- Modular switchgear design
- Busbar distributors, which are especially suitable for laboratories and workshops
- Worksite distribution boards for installation jobs

Process power supply systems in normal service are designed for standard voltages. Commonly used voltages are as follows:

Data acquisition and processing	24 V d.c.
Lighting Wall receptacles Low-voltage, low-power motors	230/400 V a.c.
Low-voltage motors, general	500/690 V a.c.
High-voltage motors	5 kV, 6 kV, or 10 kV a.c.

Special situations may require special voltages, such as 25 kV or 30 kV for electrolysis processes.

Implementation of High Availability (Conceptual Solution). Before the adoption of measures to enhance process power supply systems, two points must be considered:

- The availability of power supply systems in general is normally higher than 99%. Outage time per year, as a rule, is in the range of minutes. This holds for every point in the network and for each consumer.
- The cost of upgrading, with the aim of approaching even closer to 100% availability, increases with rising power level, increasing number of hierarchical levels in the system, and decreasing outage length. It is not, however, possible to achieve 100% availability.

The profile of requirements for the process power supply system, as set forth in the bidding specifications, is met by standard and special practices.

Standard practices include the following:

- Use of network architecture typical of industrial plants, with multiple pathways at the supply and generation levels and complex network structure at the transport and distribution level (Fig. 6.31)
- Use of the IT network
- Use of modular switchgear
- Installation of A–B machinery, with one machine always on stream and the other on standby

Special practices include:

- Inclusion of a standby power supply system
- Upgrading of d.c. power supply system to a protected d.c. power supply system
- Splitting up both switchgear and cable installations into two subsystems
- Compliance with interference immunity requirements

In that follows, practices of a special nature are discussed in more depth; among this group of practices, only those involved in standby power supply and protected d.c. power supply are covered.

Two types of standby power are important for process power supply systems in industrial plants (Fig. 6.32):

- Standby power from a second, independent network
- Standby power from a standby generating system (diversity of power sources)

Standby power supply from a second, independent network is implemented by means of a reserve, backup, standby, or emergency ring feeder (Fig. 6.31), continuously supplied via automatic voltage detectors.

In general, when standby power supply systems are employed, the bidding specifications must list consumers by permissible outage time and assign them to classes of low-voltage switchgear. Figure 6.33 shows a classifiction that has proved acceptable as a first rough scheme.

Figures 6.34 and 6.35 identify types of low-voltage switchgear and the classes to which they are assigned.

Class 1 consumers must be further broken down for in-depth analysis.

Protected d.c. power supply systems (Fig. 6.36) feature 100% reserve capacity because of

- The dual feeder arrangement
- Additional backup of each feeder in case of line outage

This redundancy must be extended to the data transmission and processing system, that is, to the cards in the subracks if possible.

Implementation of Safety. The terms safety, protection, and danger are interlinked by the concept of risk. Safety practices are intended to insure that the limiting risk is not exceeded. Such practices are mandated by legislation, regula-

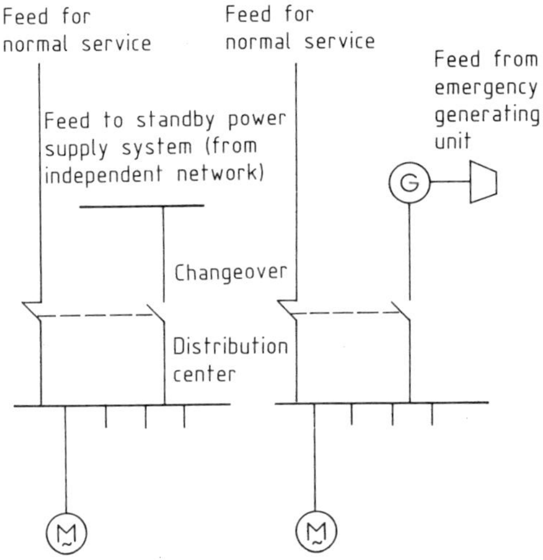

Figure 6.32. Standby power supply system

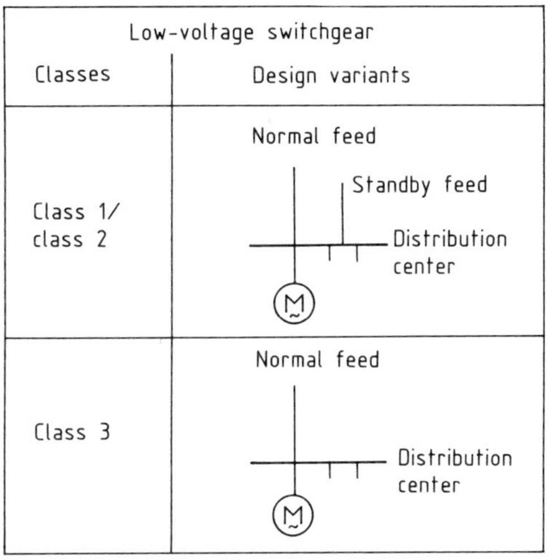

Figure 6.34. Classes and variants of low-voltage switchgear

Classes of low-voltage switchgear	Possible outage times to consumers on failure of normal power supply	Details of low-voltage switchgear	
		Number of feeders	Devices actuated on failure of normal power supply
Class 1	0–15 s	2	Backup system continuously on line and ready
Class 2	>15 s	2	Backup system automatically switched on line
Class 3	At least until restoration of normal power	1	

Figure 6.33. Classes of low-voltage switchgear, possible outage times, and details of implementation

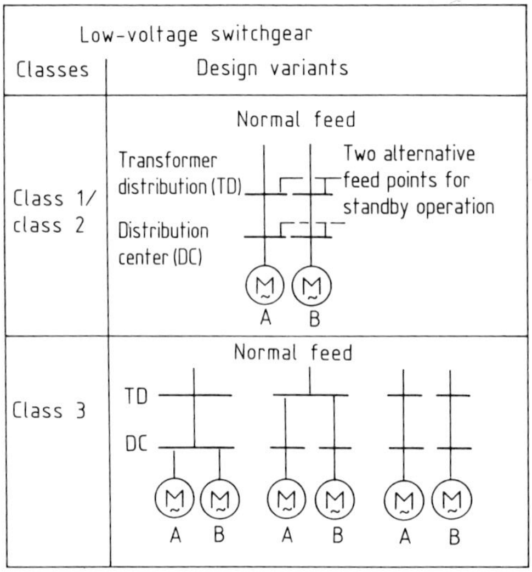

Figure 6.35. Classes and design variants of low-voltage switchgear in plant with A and B machines

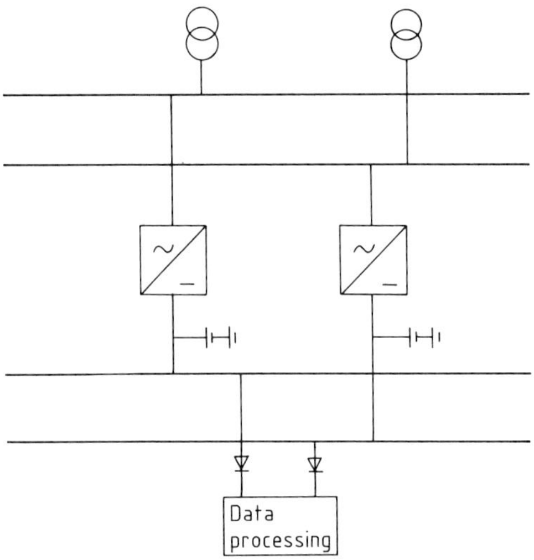

Figure 6.36. Protected d.c. power supply system

tion, engineering codes, and plant-specific directives issued by operators.

In the case of power supply equipment or power supply systems in general, "safety" refers to two independent objectives:

- Protection against electric shock (personnel protection).
- Protection against failure of the electric power supply system, which is ultimately a question of "reliability" (safety of equipment, protection of product and plant). This kind of

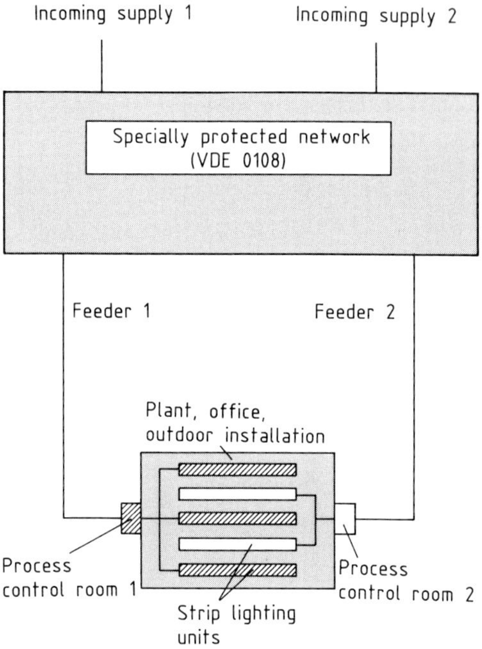

Figure 6.37. Lighting system fed from the specially protected network

protection is achieved by means of the measures employed for high availability.

Apart from functionality, protection against electric shock has been the second dominant theme in power supplies since the start of electrical engineering [6.30]. Notable among the many standards now in existence is the extensive VDE code, which is continually updated.

The practices listed below serve as examples of how personnel protection is achieved in process power supply systems for the chemical industry:

- Increased use of 30 mA ground-fault circuit interrupters. This switch cuts off the flow of fault currents that could cause fibrillation.
- Installation of devices safe against touch by the hand.
- Application of the accident prevention specification VBG 4, "Electrical Installations and Equipment."
- Wider use of safety switches on drives.
- Application of power supply options originally intended for escape route lighting to overall production power supply systems (Fig. 6.37) [6.31].

Environmental considerations apply chiefly to process engineering. In general, these requirements do no impact on the process power supply system, which comes into the picture only when "reliable" power is required for devices affording protection or limiting damage to process control systems [6.32]. It must be verified whether a standby power supply system or a proteted d.c. system is adequate. One type of such a "secure" power supply system is a high-availability power supply system.

High-Availability Process Power Supply System for a Wastewater Treatment Plant (Example). If a plant unit does not have its own waste disposal facilities, a central plant with biological treatment stages is of fundamental importance for maintaining production in a chemical plant. Such a treatment plant requires a high-availability power supply system, which must comprise all the levels of the power supply scheme and extend all the way down to the process control components in the field as well as control rooms [6.29].

The quantity structure of the plant as built is as follows:

- Distributed process control with dual host computer system
- 1800 process control stations
- 530 uncontrolled 3-phase drives with total power consumption of 10 MW
- 70 controlled electrical drives with total power consumption of 3.5 MW

The following voltages are available (Fig. 6.38):

– High voltage:	220 and 110 kV
– Medium voltage:	25, 10, and 5 kV
– Low voltage:	500 V for motors; 380/220 V for lighting, wall receptacles, computers, and hand tools; 24 V for measurement and control apparatus

Comparison with the level model (Fig. 6.31) shows the relationships to be as follows: The high-voltage network corresponds to the import and generation level; the medium-voltage network, to the transport level; the low-voltage network, to the distribution level.

The dual-bus structure illustrated in Figure 6.38, with the consumers separated into two subsystems, insures that the failure of one unit (e.g., a transformer or feeder) affects only some of the consumers at levels lower in the hierarchy. The failure of both system would, however, lead to shutdown of all consumers. However, this must be avoided for particularly important consumers. In such cases, further standby power supply practices come into play:

- Use of diesel generators
- Use of battery-backed uninterruptible power supplies
- Use of protected d.c. power supplies

Figures 6.39 to 6.41 show these backup arrangements for the low-voltage (500, 220, and 24 V) systems in the wastewater treatment plant.

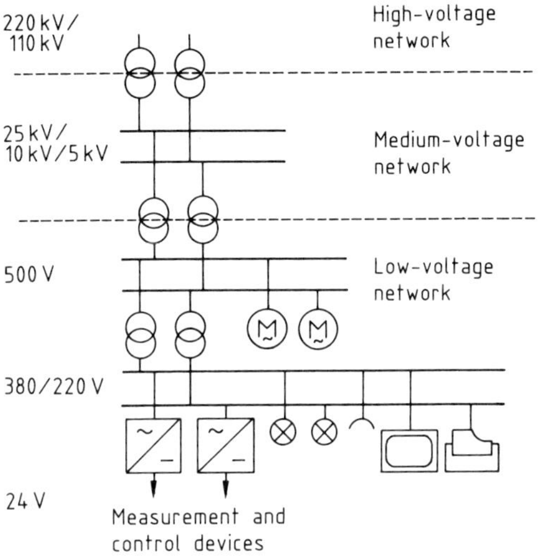

Figure 6.38. Voltage levels in a chemical plant

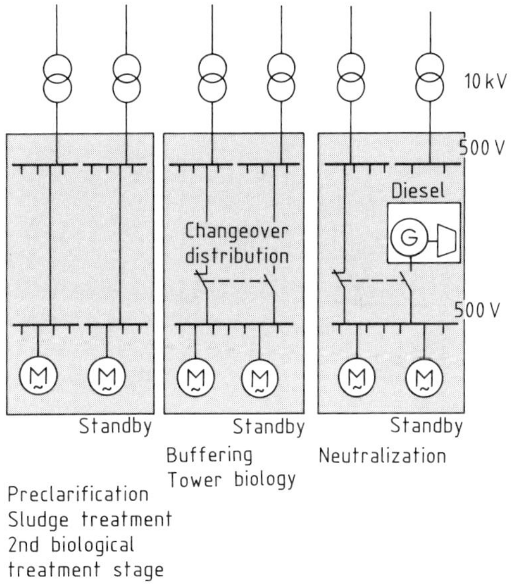

Figure 6.39. 500 V low-voltage supply system in a wastewater treatment plant

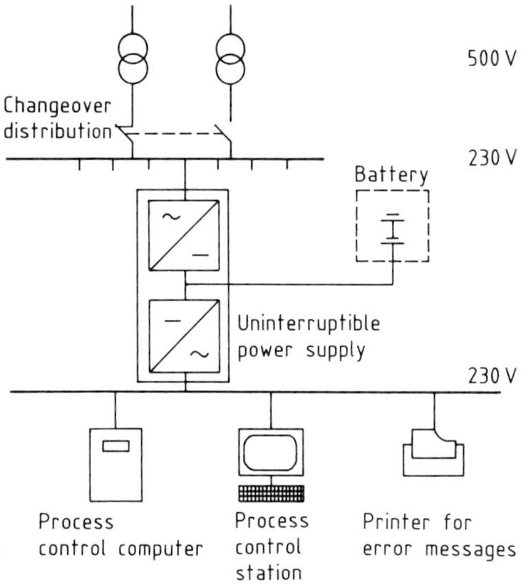

Figure 6.40. 230 V low-voltage supply system in a wastewater treatment plant

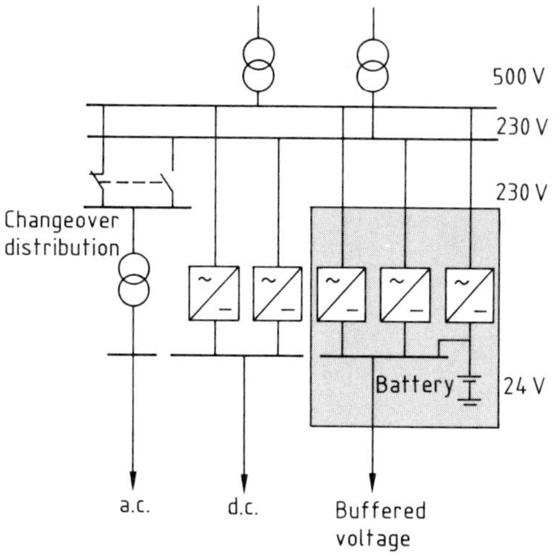

Figure 6.41. 24 V low-voltage supply system in a wastewater treatment plant

7. The Process Control System and Its Elements: Distributed Control Systems

7.1. Principles

The task of the process instrumentation and control hardware is to assist plant management in the control of its plants and in the control and monitoring of its production processes [7.1].

This chapter discusses the structure and function of process instrumentation and control hardware. The structure is discussed on the basis of a reference model corresponding to the fundamental structures of current systems.

In a general sense, process instrumentation and control hardware comprise all devices between the sensors and actuators and the operator controlling the process (see Fig. 7.1 and 11.4).

The process instrumentation and control hardware must perform a number of distinct functions. Figure 7.2, for example, shows a metering device. The required functions can be classified as follows:

- Controlling the process
  Automation functions that permit the self-contained control of plant units and process steps. Example:
  – If action "metering on," then meter from B 1 to B 2 *x* liters at *q* L/s.
- Enhancing the safety and reliability of the process
  Functions prior in time, independent of control function specifications, that intervene in the process in order to enhance its safety and reliability. Because of high required availabilities and special performance provisions, these functions are often implemented in a separate "protective level." Example:
  – If L 2 > L 2max, pump K 1 is disabled.
- Providing displays
  Display functions that report plant and process status to the operator in a concentrated and easily understood form. Example:
  – Display level L 1.
- Generating messages
  Message and alarm functions that explicitly notify the operator of certain events and provide information to the central message and alarm processing system. Example:
  – If L 2 exceeds L 2max, transmit a message. (Class: "Alarm"; text: "L 2max alarm")
- Archiving
  Archiving functions to store records of events and status logs. Examples:

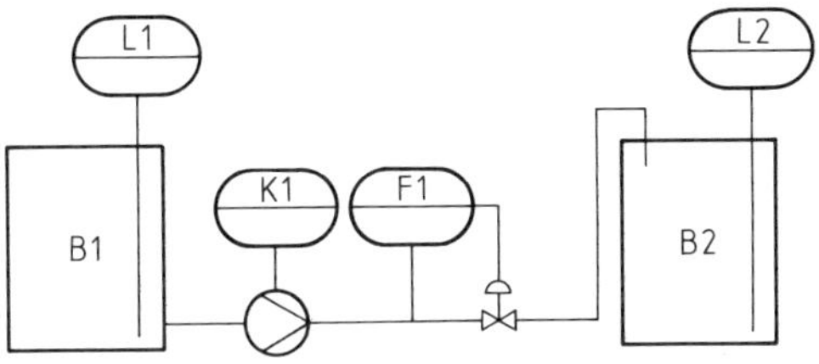

Figure 7.2. Example: metering system

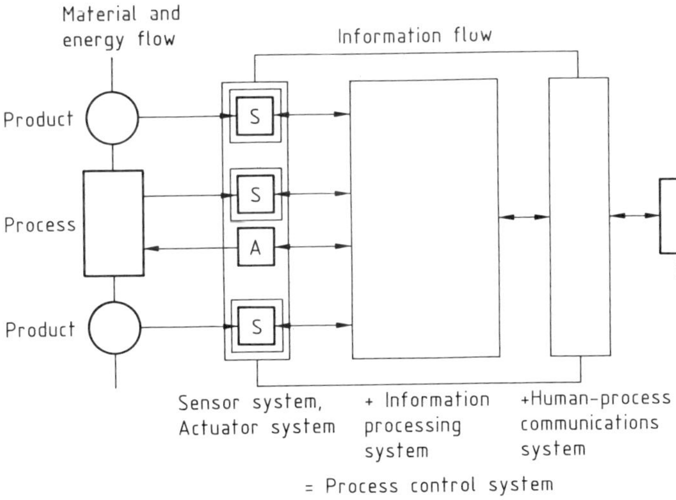

Figure 7.1. Process monitoring and control system

- Archive all reports.
- Archive level L 1 every 10 s.

- Generating and interpreting reports
Recording and evaluating (interpreting) functions that retrieve, evaluate, and print out archival and current data when a certain condition or command occurs. Example:
  - Log the last metering operation performed (starting time, ending time, quantity, events).

If all functions to be performed by the instrumentation and control hardware are arranged in accord with the functionally abstract "level" model of the process control functions (see Section 2.2 and Fig. 2.16), some basic conclusions can be drawn as to the automation of this system.

The lower levels of the hierarchy contain many parallel functions with high intervention frequencies. These functions are chiefly suitable for automatic processing by process control devices. At higher levels, the intervention frequency generally decreases, but certain functions (e.g., load recipes) require the transfer of large amounts of data.

The automation of these functions generally proceeds from bottom to top, in keeping with the hierarchical breakdown of process control functions.

Different plant sections can differ in degree of automation. The degree of automation can be divided into four levels [7.2] (Table 11.1). Each degree includes not only the automation of the respective process control functions but also the automation of the associated safety/reliability functions. A safety- or reliability-enhancing function is assigned to the level at which it intervenes in the instruction branch. A safety function "if pump temperature > permissible maximum → turn pump off" is assigned to the individual function level; a safety function "if coolant temperature > permissible maximum → turn column off" is assigned to the plant section level.

The functions that a specific process instrumentation and control system must perform can be set down in a requirements list.

In principle, a project specification consists of a list of such required individual functions. Some of these functions, as shown in Figure 7.3 [7.3], are carried out by the plant operator, the rest by the process control system. The term "automatic" refers to a function that is performed

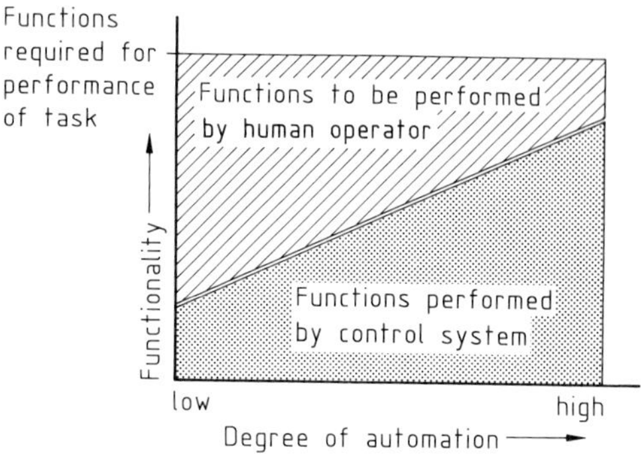

Figure 7.3. Degree of automation

by the process control system without external action. The degree of automation is the fraction of all functions that can be represented by automatic functions. Nevertheless, the overall scope of the control task must be considered regardless of the degree of automation. Each function that does not take place automatically must be performed by the plant operator. By changing the operating mode, the instantaneous degree of automation can be made lower than the maximum degree of plant automation provided for in the design (see Fig. 7.4).

When examining the properties of a process monitoring and control system, a distinction must be made between the external functional viewpoint and the internal structural viewpoint. The external functional view regards the finally configured process instrumentation and control system as a "black box" having certain functional qualities. The internal structural viewpoint regards how the system is structured and how the individual functions are implemented.

From the process control standpoint, the internal structure of the process instrumentation and control hardware is often of secondary interest. The required automation functions, safety and reliability functions, archiving functions, operator and monitoring interfaces, and so forth can be implemented under a wide variety of system concepts. The differences, however, manifest themselves in important supplemental properties. The plant operator, for example, might see differences in system reaction times, the way in which data surges are controlled, the failure characteristics of components, or the central coordination of workstations. Important from the plant management viewpoint, then, are additional properties that cannot be implemented simply as individual functions but result from the struc-

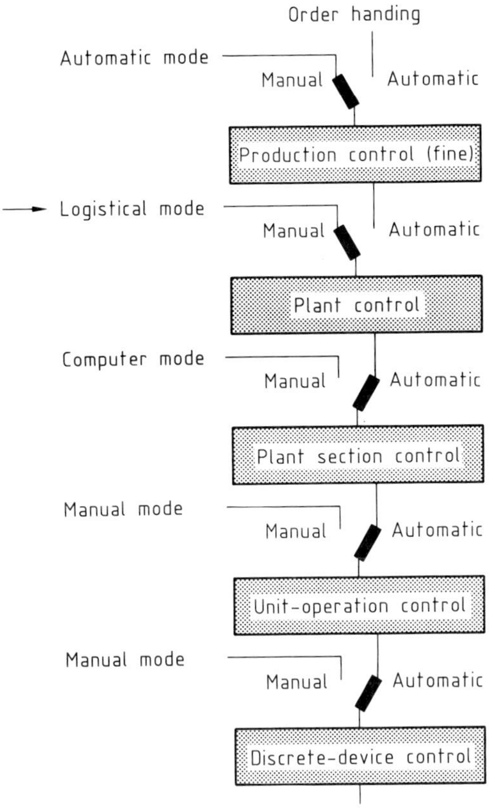

Figure 7.4. Auto/manual selection

ture of the process instrumentation and control system as a whole. The same applies even more so to the planning and operation of a monitoring and control system. Such additional properties of importance include the following:

- Transparency of structure
- Ease of operation
- Structure with few defects
- Capability of modular expansion
- High availability of individual functions

These can bc realized only if they are incorporated in the control system design, i.e., in the internal structure, right from the start.

Historical Development. Today, the phrase "modern" process control systems refers more precisely to distributed or decentralized systems. The world of process control engineering before the development of the type of system discussed here (i.e., before ca. 1975) featured two extreme structures, both of which still exist.

In the *totally parallel structure*, each control task (control action, display, recording, etc.) is performed by a discrete device or combination of discrete devices. An example is the compact controller used in the implementation of a control loop. The parallel system terminates in the control room with a great number of instruments, buttons, and lights. It is left to the designer to ensure that the operator can scan the panels without becoming overstressed in critical situations (see Section 8.2).

The decentralized structure means that a plant so equipped has a fairly high availability, since the failure of a discrete device generally affects only the control task directly assigned to it. The solution of complex control problems, however, requires considerable interlinking of the individual process control stations (see Fig. 3.26). With discrete devices, this networking can be implemented only through explicit interconnection of devices. Subsequent changes of task are therefore expensive because the system must be rewired, and the flexibility of such a plant is low.

For more complex applications, there was the *totally centralized structure* implemented by the process control computer. This structure, which features the maximum degree of centralization on a single device, has all the process monitoring and control functions concentrated in one device, which is intended to alleviate the disadvantages of discrete-device technology.

All control tasks are carried out by a central computer with interfaces to the process. Individual process control stations are linked together by software, so that the scheme can easily be modified, for example, if the recipe is changed.

The resulting flexibility is partly offset by the lower availability of the plant. A failure of the central computer affects the entire process. Therefore, this structure has generally been used only in conjunction with special practices such as hardware redundancy or subordinate backup systems, which have lead to higher costs for such plants. Nevertheless, the centralized structure in this form still plays a role in the landscape of system structures [7.4].

Modern decentralized process control systems now fall between these two extreme structures (Fig. 7.5). The search for an optimal structure has led the manufacturers of these systems to solutions that differ in the degree of decentralization.

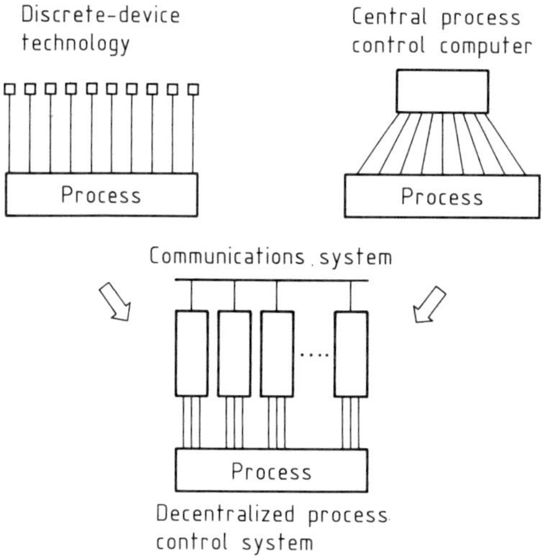

Figure 7.5. Decentralized process monitoring and control system compared with central process control computer and discrete-device structures

In simplified terms, a decentralized process control system arises through the logical subdivision of a centralized system. This approach is useful because only the use of a process control computer provides the conditions for implementing process monitoring and control functions through digital computer technology.

The development of microprocessors has made the desired or necessary computing power economically viable. Thus, another important prerequisite for the creation of decentralized process monitoring and control systems was fulfilled [7.5]. Such a system consists of individual computer components, each of which exhibits the properties of a process control computer to a limited extent. These components are linked together by a communication system. As a whole, they have the same capacity as a large central process control computer. However, the decentralized structure leads to higher availability than a centralized design.

The development of process instrumentation and control systems has followed three distinct lines:

- Decentralized stored-program controller systems
- Integrated decentralized process monitoring and control systems
- Intelligent central systems with dumb peripherals

Most process monitoring and control systems can be placed fairly unambiguously in one of these classes. Differences in historical development of the three groups have led to variations in characteristic features such as size of decentralized components, structure of system communication, type of internal processing, integration of operator control and monitoring into the overall concept, and type of design.

The Decentralized Stored-Program Controller (SPC) System. The starting point for SPC systems was the individual SPC (or programmable controller), which is a self-contained unit performing a particular automation task. The emphasis was on the solution of binary networking and sequencing problems. SPCs were conceived as an alternative to relay controllers, which were becoming increasingly complex, costly, and inflexible (as to wiring). For this purpose, the SPCs were for optimized fast execution of binary logic functions, the integration of control and monitoring functions initially playing a very minor role. Elements such as buttons, switches, displays, and lights were employed to operate the SPC and observe its functions. These were connected to the SPC via conventional field-level inputs and outputs. Outwardly, the SPC thus behaved as a wired discrete control device.

The development of SPC systems has led to three major new capabilities:

- Binary-word processing for handling integer and real variables
- Connection to serial bus systems for data exchange with other SPC components
- Support of display-aided operator control and monitoring systems

Compatibility requirements have dictated consistency of internal structure in the development of SPC-aided systems. Present-day process monitoring and control systems based on SPCs thus show traces of their original functions.

SPCs have gained wide acceptance. Other apparatus frequently contains self-sufficient automation components based on them and is sold in pre-assembled package form. Software creation and design are performed by the apparatus manufacturer, so that special in-house control system know-how can be exploited and protected.

From the plant engineering standpoint, manufacturing automation has been and still is the domain of SPCs. An early development was the linking of SPCs and bus systems to form large networks. The purpose of these communications systems was to permit the sharing of data, especially control data and status messages, between SPCs and between the central control computers and the SPCs. A typical feature is simple variable sharing, in which the communication event and all shared variables must be individually designed. An adequate integrated operating and monitoring structure with appropriate system-supported accesses has not yet become available (see Section 11.2).

SPCs are also being used increasingly for plant automation in the field of process engineering. The starting points are often logistical side areas such as formulating lines, packaging lines, and high-bay warehouses. Starting from these applications, SPCs are spreading into chemistry-related areas. Especially in small and medium-sized plants, SPCs are often preferred to decentralized process monitoring and control systems because of their lower hardware costs.

Although such SPC-based solutions are often more expensive than complete distributed control systems when a total cost analysis including engineering costs for design, maintenance, and technical support is carried out, the low cost of adopting them may make them more attractive to plant management when investment funds are being sought.

CRT-aided operator control and monitoring features are available today for SPC systems (see Section 11.2). Generally intended for use at individual workstations with limited quantity structure, these devices offer good functionality. Problems arise in larger systems with coordinated multi-station operation, increased availability requirements, and central system services (e.g., network clock synchronization).

The separate design of SPC automation functions, communications functions, and operator control and monitoring functions remains an unsolved problem. The lack of certain central functions that would be present in a process control operating system has proved especially serious in SPC-based systems.

Integrated Decentralized Process Control System. The development of integrated decentralized control systems began with the control of large continuous process plants. In contrast to SPCs, the initial motivation was not so much the need to increase the degree of automation as the wish to improve functionality at the individual-function level by making design and installation simpler and cheaper and by making operation and monitoring more straightforward.

The starting point came when discrete-device technology ran into its quantitative limits. In this approach, each individual function is implemented in a separate, individual device; every motor has its own start button, its own run light, and its own contactor control. Figure 7.6 shows a control panel with discrete devices, which are wired with electrical signal lines on the back.

The increasing number of individual functions and the presence of up to 50 control panels in central control rooms meant that an immense collection of controls and indicators had to be put together, creating space problems and also difficulties in comprehensibility. The primary task of the plant operator was to maintain a steady watch over the continuous process. To read the instruments, the operator had to walk along meters of panel bays.

The main goal of the decentralized process control system was to replace this discrete technology with a system combining the advantages of distributed structure and easy operation with the advantages of centralized data management and easily understood, display-supported operation and monitoring.

Figure 7.7 illustrates the component structure employed in decentralized process control systems. The distributed components contain the automation functions for the individual plant areas. They are organized in parallel with the process. The central components contain operator control, monitoring, archiving, evaluation, and central coordination functions [7.6]–[7.8].

From the very first, decentralized process control systems were characterized by ease of design and effectiveness of communications between process-level components and central control and monitoring components. Distributed systems have the following features: high-availability redundant bus systems, components that function reliably, low error rate in component design, and good technical support at the component level.

An important property of process control systems is the functional-module approach, in which software modules are handled in the same way as discrete devices. Functional modules can be inserted, switched on, removed, and connected to other functional modules while the plant is on stream. These properties result in simpler and

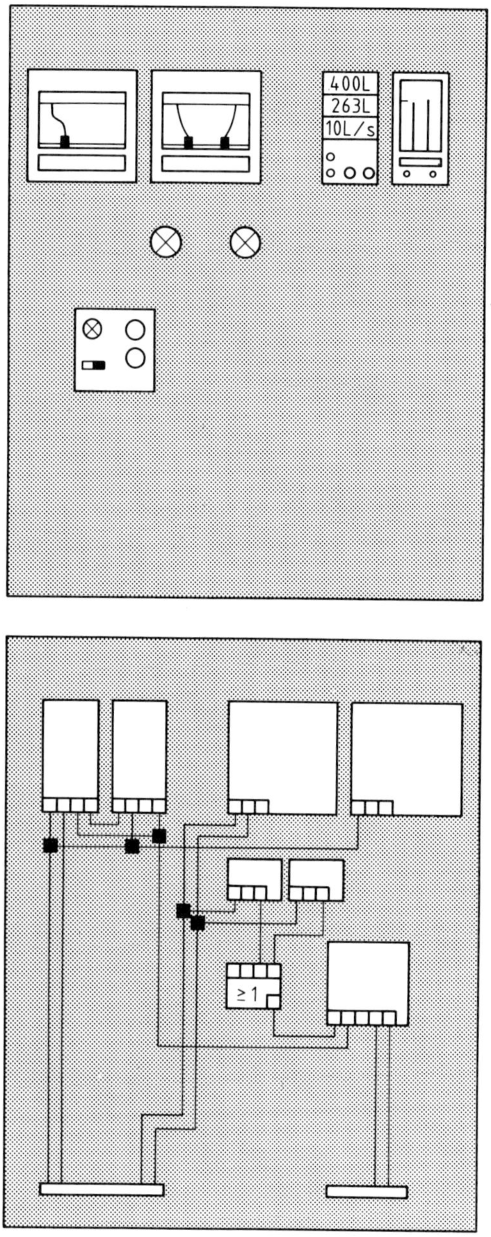

Figure 7.6. Discrete-device technology
Front and back of a control panel

hence cheaper design as well as high system reliability.

The operator interface of functional modules is designed to correspond to the discrete-device operator interface. Because of the similarity in operating properties, operating and maintenance personnel quickly learned to work with the new technology. This was an important reason for the successful introduction of decentralized process control systems over a broad front.

The successful introduction of distributed process control systems led to a demand for a higher degree of process automation. Concepts that have come to the fore in recent years have led to the integration of higher-order control functions, up to the level of recipe control and plant management, into the structures of distributed process control systems (see Chap. 2).

Intelligent Central Systems with Unintelligent Peripherals. The basis for this line of system development is the technology of process control computers, which became available for industrial service in the 1960s, that is, before stored-program systems and decentralized process control systems. Their introduction was greated with a certain technological euphoria, and due to the lack of alternatives this approach was adopted for the full range of plant automation, operation, and monitoring tasks. However, this concept was unsuccessful because of availability problems (due to the centralized structure), the fault susceptibility of the computers then in use, and the fact that both the hardware and the applications software could be supported and serviced only by specialists.

Today, the only situations in which this concept is used are those where high-quality functions are required sporadically and access to installed monitoring and control systems is not possible. An example is the case of miniplants and pilot plants, where availability requirements may not be high but it is necessary to test higher-performance control algorithms, implement model-based measuring techniques, or evaluate particular data. Furthermore, in research-level activities, personnel are often able to operate a process control computer themselves.

On the whole, however, this technology is declining in importance. It is increasingly common to use small, decentralized process monitoring and control systems or SPC-based systems for the control and operation of pilot plants. Special "research" functions can be implemented and tested on networks of flexible process control computers.

Structural Models. A process monitoring and control system is a complex pattern made up of many object classes; it can be structured in a wide variety of ways. As a result, a system de-

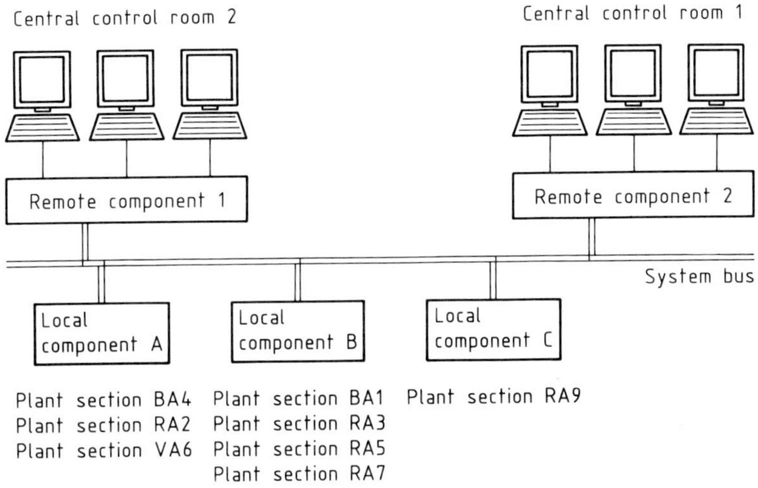

Figure 7.7. Component structure of a distributed process control system

scription is difficult to understand and is always incomplete. Individual features can be explained only from a certain viewpoint, with other aspects being ignored. Nonetheless—and this is precisely the art of structural analysis—all aspects must be equally kept in view.

To provide a framework for the properties briefly discussed in the sections that follow, the most important structural models and the most important ways of looking at a process control system are described briefly.

Building-Block Model. The building-block model represents the structure of a system unit in terms of disjoint elements and connections (Fig. 7.8). This model is generally applicable; it can be employed, for example, to describe the structure of mechanical, electrical, and functional system units. The building-block model plays an important role in the structuring of process control systems.

The building-block model has some properties that set it apart from general system decomposition and aggregation schemes (see Section 2.2, Fig. 2.1):

- Nesting principle: Each element used in building up a system unit is completely and exclusively allocated to this system unit. It is, so to speak, "built into" the system unit and is no longer available outside the system unit.
- External interfaces: Each element and each system unit exhibits its properties at certain well-defined external interfaces (e.g., electrical terminals, functional module inputs/outputs).

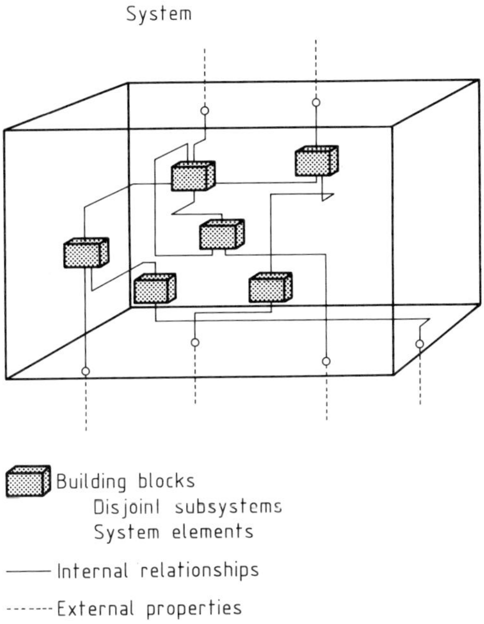

Figure 7.8. The building-block model

External connections are possible only through these interfaces, which can in turn relate it to various worlds. A compact controller, for example, has terminals as its electrical interfaces, displays and buttons as its human–machine interface, and bore holes for mechanical mounting.

- Implementation background: Building-block models describe realizable system structures. Hence it is not meaningful to define fictitious (dummy) elements with no realistic background and thus to built up fictitious systems. Instead, the objective must be to identify elements such that they are realizable but also represent manipulable units in the context of the overall structure.

Building-block models are an effective aid to describing the structure of system units of all types. In process control systems, they are well-suited to describing the structure of the hardware, describing the structure of the logical components, and describing the structure of the system and applications software.

Types, Roles, and Functional Units. When one speaks of a system unit or a building block, one must bear in mind that the unit referred to is a self-contained object that can be examined from various standpoints: role (function), functional unit (device; see Fig. 7.9), type of role, and type of functional unit.

When a control system manufacturer refers to its control system, it means the functional types offered. The design of a control structure means interconnecting functional units. These units correspond to roles that are selected from a list of well-defined types of roles. Ultimately, what is installed are functional units. This distinction is just as important for design and commissioning as it is for the overall understanding of the system. The relationships are shown schematically in Figure 7.9.

Functional units are the concrete devices, components, or software modules that actually perform a function.

The type of functional unit describes the essential qualities of a functional unit. Manufacturers' catalogs list, for example, the types of functional units available. If a catalog lists a P-REG compact controller, then one can order x units of this type. The manufacturer will then deliver x functional units of this type.

The type description of functional units goes beyond the purely functional description, taking in electrical properties, overall dimensions, weigths, prices, etc. The functional properties are a subset of the type properties of the real functional units.

Roles are fictitious functional units. An example of a role is flow controller F 1 in plant section B 12. This role can be defined—and used in building up the role system—independent of which functional unit (i.e., which device or software module) will later fulfill the role.

The model employed here specifies that each role is of a certain type. The type defines all the essential properties of a role. Obviously, any number of roles of the same type can be defined in a plant.

Role systems are systems built up from discrete roles. In structure, role systems correspond to the building-block model. A functional role system is illustrated in Figure 7.10. This role system describes exclusively functional properties. It can be implemented with a wide variety of technologies involving a wide range of functinonal unit types, for example with software modules or in discrete devices. There are also more concrete formulations of role systems. For example, the control panel shown in Figure 7.6 can be thougth of as a concrete instance of the functional role system illustrated in Figure 7.10. In this role system, the roles are established much more fully. The description of each role includes not just functional properties but also electrical and constructional properties. These definitions can be made concrete if only functional units of a certain type are considered for its implementation.

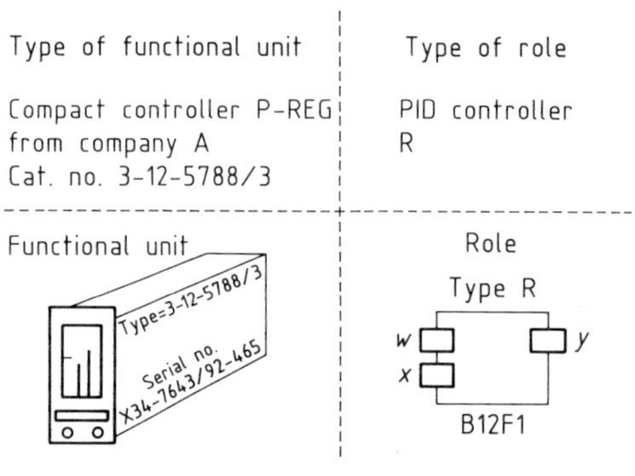

Figure 7.9. Roles and functional units

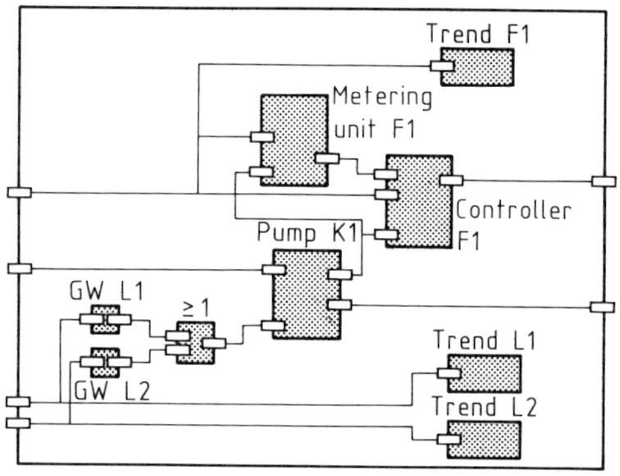

Figure 7.10. The system of roles

Section 7.5. gives a detailed account of how role systems are built up, concretized, and implemented.

Hierarchical Structural Models. The object worlds of a process monitoring and control system can be arranged in hierarchically related functional levels. Each functional level represents a level of abstraction on which certain functional elements are available and certain tasks (or functions) can be carried out (see Section 2.2).

The modeling of a process control system involves not just one hierarchical structure, but hierarchical structures based on various criteria.

● Operational levels: The hierarchical structure of the operational levels includes interfaces for

- The control system manufacturer who develops the system
- The project engineer who configures the process control system for the specific application
- The plant operator who ultimately guides the process

Figure 7.11 illustrates such a structure. The lowest level is the base system level, which contains units and functionalities provided by the control system manufacturer, that is, not specially developed for the control system. These might include PCs; workstations; operating systems such as DOS, UNIX, or VMS; database management systems such as dBase, Informix, or Oracle; and all the hardware not specially designed for the control system.

The second level contains the process control operating system. This level makes up the functionality of a manufacturer's control system. The user receives its units and functionalities when the control system is delivered.

The third level is the process control level. It is custom-designed and is available to plant management for active control of the process. It satisfies all requirements on the control system as set forth in the project specifications.

In the model just described, the various levels are regarded as dense. This means, for example, that the designer does not need to access any level other than the process control operating system level (such as the base system). The operating system can include any required elements of the base system in its functional scope.

Thus the levels of functional abstraction correspond to self-contained working platforms. An

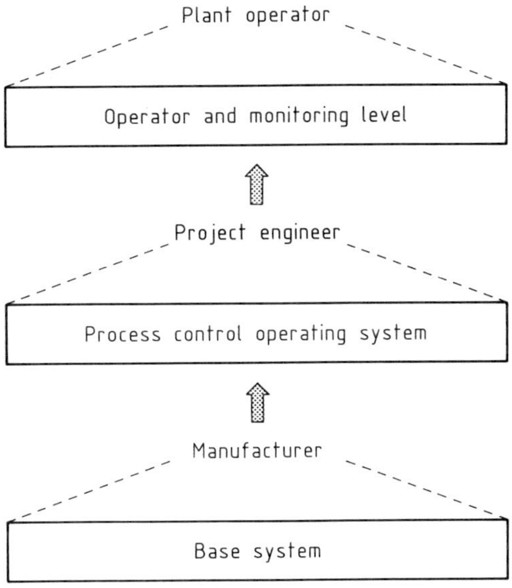

Figure 7.11. Operational levels

important criterion for assessing a process control system is the functionality of the process control operating system level. Functional elements not available on this level must be programmed by the designer as required (see Figs. 2.15 and 2.16).

● Virtual machines: The execution functions necessary for the accomplishment of the tasks imposed can also be arranged in a hierarchical model by analogy to the operational levels. During its execution a function makes use of functions of the lower levels. The result is a system of virtual machines, each of which (as shown in Fig. 7.12) offers the higher-level functions an interface with certain services [7.9].

The hierarchical structure of execution functions is crucial for the internal implementation of program structures as well as the startup/restart behavior of the system.

At this point it becomes clear that the hierarchical structures illustrated are not to be viewed simply as fictitious structures but as concrete conceptual and real structure generating programs. This is true of all the organizational schemes presented here.

According to a bottom-up principle, the functions on one level can be implemented with functional elements made available by the next lower level. In this way, more complex functional elements can be created and, in turn, made available for use by higher-level functions.

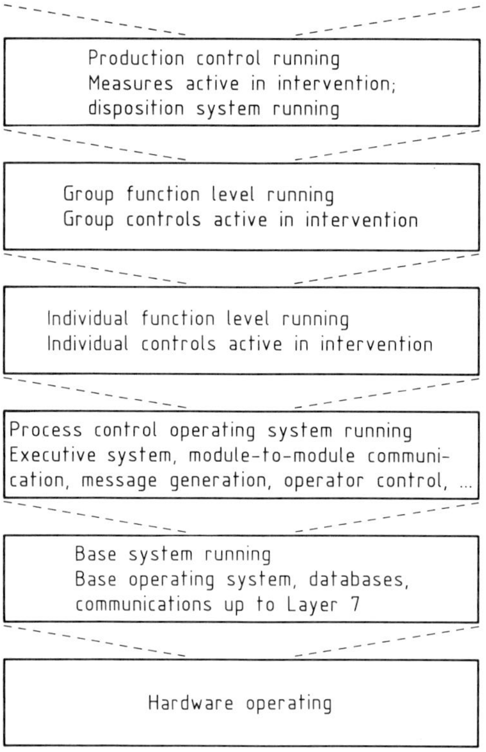

Figure 7.12. Virtual machines in a process control system

3) Communications protocols: Communication between two applications can also be represented by a level model. Each protocol level corresponds to a "virtual medium" having certain properties, which the next higher protocol level can utilize in implementing its functions. A well-known scheme is the OSI reference model [7.10], which consits of seven layers and the overlying user functions and user data (see Fig. 8.3):

- 7 Application layer
- 6 Presentation layer
- 5 Session layer
- 4 Transport layer
- 3 Network layer
- 2 Data link layer
- 1 Physical layer

The structure of the communications layers (levels) is discussed in Section 7.2 for the example of systems communications and then described in general and comprehensive terms in Section 8.4.

4) Process control: Process control tasks can be arranged into a hierarchical control model.

In broad terms, the control model can be broken down into the following hierarchical levels: process control level, production management level, and corporate management level (see Section 2.3., Figs. 2.30 and 2.31).

The process control level can be divided into the following sublevels in accordance with the model of Section 4.5:

- Action level
- Group control level
- Individual control level
- Channel/signal level

Viewpoints. In the structural analysis of a process control system, the diverse viewpoints of the teams involved must be taken into account. For example, the following aspects are among those that must be considered:

- Development
- Sales
- Plant management
- Design
- Plant technical support and maintenance
- Device management

Each group represented by a viewpoint has an interest only in certain objects with certain properties. It does not make sense, however, to develop an isolated model from each individual viewpoint, covering only the objects of interest. Instead, it should be possible to relate all viewpoints to a single building-block model of a process monitoring and control system. What is needed are balanced model structures that will yield a useful mode of organization for all groups involved.

7.2. System and Component Structure

Overview. A decentralized process control system consists of individual components connected to one another through a system bus. The division of labor among the components varies in organization from system to system. One general distinction is between process-level or local components and remote components. Any component that has an I/O interface to the field is regarded as local, while any component that has no I/O interface to the field is considered remote. Figure 7.7 illustrates a typical functional structure.

Process-level components are generally organized in the same way as the plant. Each plant section is permanently assigned to a local com-

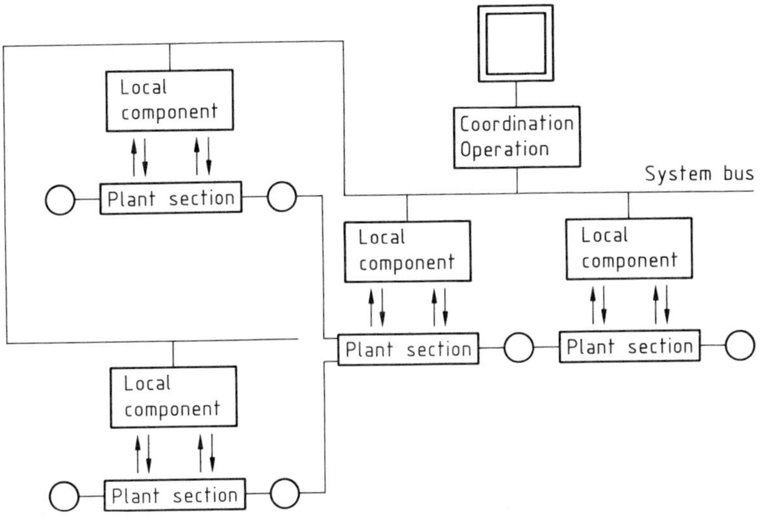

Figure 7.13. Organization of local components in parallel with plant organization

ponent which controls and monitors the plant sections assigned to it. The component is autarkic (self-sufficient) and viable in the sense that it can perform its control and monitoring tasks independently. The result is a typical decentralized structure for process control systems as shown in Figure 7.13.

Remote components embody functions that are not directly a part of process-level control and monitoring. Examples of such functions are central operator functions, archiving functions, evaluation and interpretation functions, logging functions, production control functions, recipe management functions, and also engineering functions and diagnostic functions [7.6].

This division of labor can be modified from case to case. Many systems today still have "entry-level" versions in which design functions, operator and monitoring functions, and archiving functions, for example, are integrated into the local components. Such a local component can be used alone, with no remote components. The introduction of economical remote components based on PC solutions has led to a perceptible decline in the use of such discrete component systems. In future, local components will perform only control and monitoring tasks, and even entry-level versions will feature a separate, small remote component.

Local components. A local component consists of a central unit and a peripheral unit, which in turn consists of peripheral or signal modules (Fig. 7.14)

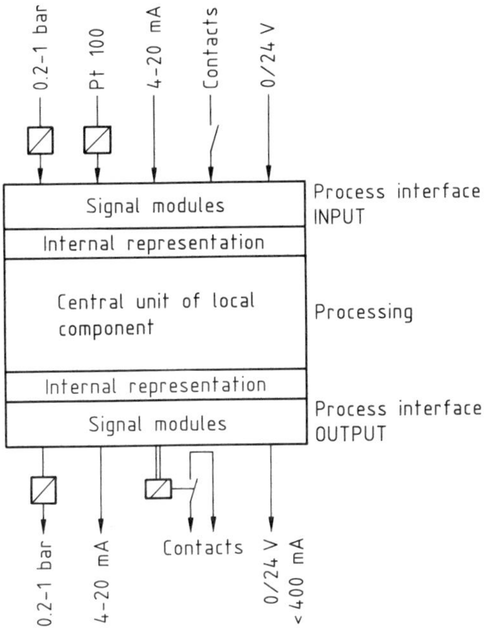

Figure 7.14. Coupling of field signals by signal modules

Peripheral (signal) modules differ in functional form and can accordingly be placed in four categories:

1) Intelligent discrete control modules, which completely match the functionality of an individual control unit
2) Straight interface modules, which serve only as signal converters

3) Subsystem connection modules, with which subsystems can be linked
4) Field bus modules, which connect intelligent actuators and sensors through a field bus

Intelligent peripheral modules include, for example, controller modules, metering modules, and control modules specific to motors and other devices. The functionality of the individual control level as a whole can thus be realized by separate, independent modules. This added decentrality does, however, have tradeoffs both in cost and in space requirements. If intelligent modules are not absolutely necessary for reasons having to do with the scanning rate or special availability requirements, they are not used at present; instead, their functions are implemented in software in the central unit of the local component. The function of the peripheral unit is thus reduced to simply the input and output of signals, which can be achieved in compact fashion by using multichannel interface modules. For example, input modules are needed in order to feed and read two-wire transducers; to process the signals from NAMUR initiators, strain gauges, and mechanical contacts; to acquire and linearize thermocouple and Pt 100 signals; to handle multirange frequency inputs; and to process 0(4)–20 mA, 0–10 V, 0–24 V, BCD, and other input signals from units. The corresponding output modules provide the following functions: output of current/voltage signals, possibly into standard loads (up to 24 V); power outputs for solenoid-actuated valves; floating outputs; pulsed outputs for integrating positioning drives; and so forth.

In process control engineering, it is important that terminals always be floating or isolated. Inputs for intrinsically safe measuring circuits are a desirable option.

The conversion of measurements and control information to electrical signals can be relocated into local hardware by using field multiplexers. These devices still provide conversions to and from standard electrical signals, but in this case the operation takes place in the field rather than in the control room.

Subsystem coupling modules allow the connection of intelligent subsystems such as scale controls, converters, process analyzers, press controls, filter controls, and other machine-specific controls. In most cases, subsystem coupling functions support serial point-to-point connections, but there are also bus-type structures with the process control system as permanent master.

Field buses provide logical connections of actuators and sensors with no conversion to static electrical signals. Numerous proposals have been made for defining and standardizing field buses and marketing them as standard systems. Up to the present, it has not been resolved which solutions will finally prevail in the marketplace. Figure 7.15 illustrates some important field bus systems currently under discussion [7.11]. In addition to national criteria and corporate tactical decisions, differences in functional requirements also contribute to difficulties in the standardization of field buses.

The *central unit* of a local (process-level) component generally consists of a special microprocessor, ca. 1–4 MB RAM, and a connection to the system communications. Additional modules are available in many systems for implementing a local operator control and monitoring function or to provide a background memory for archiving purposes. It is important that the local component should have a process control operating system and be integrated—with its object and communications structure—into the overall system. This means that the system functions of the local components permit, for example, the support of central design and the provision of server services for the operator control, monitoring, archiving, alarm, and message generation functions.

Remote Components. Figure 7.16 illustrates the tasks (functions) to be performed by a process control system independent of how the system is implemented [7.12]. Two basic structures exist at present:

- First-generation control systems have no information model that applies throughout the system. The individual components are specially designed and optimized for the tasks they implement. Thus, operation, computer, and engineering components generally have different system architectures and can neither be configured and maintained in a uniform manner, nor controlled through a consistent user interface.
- Second-generation process control systems have an information model consistent throughout the system. Data processing management (operating system, data handling, communications, operator interface) for all

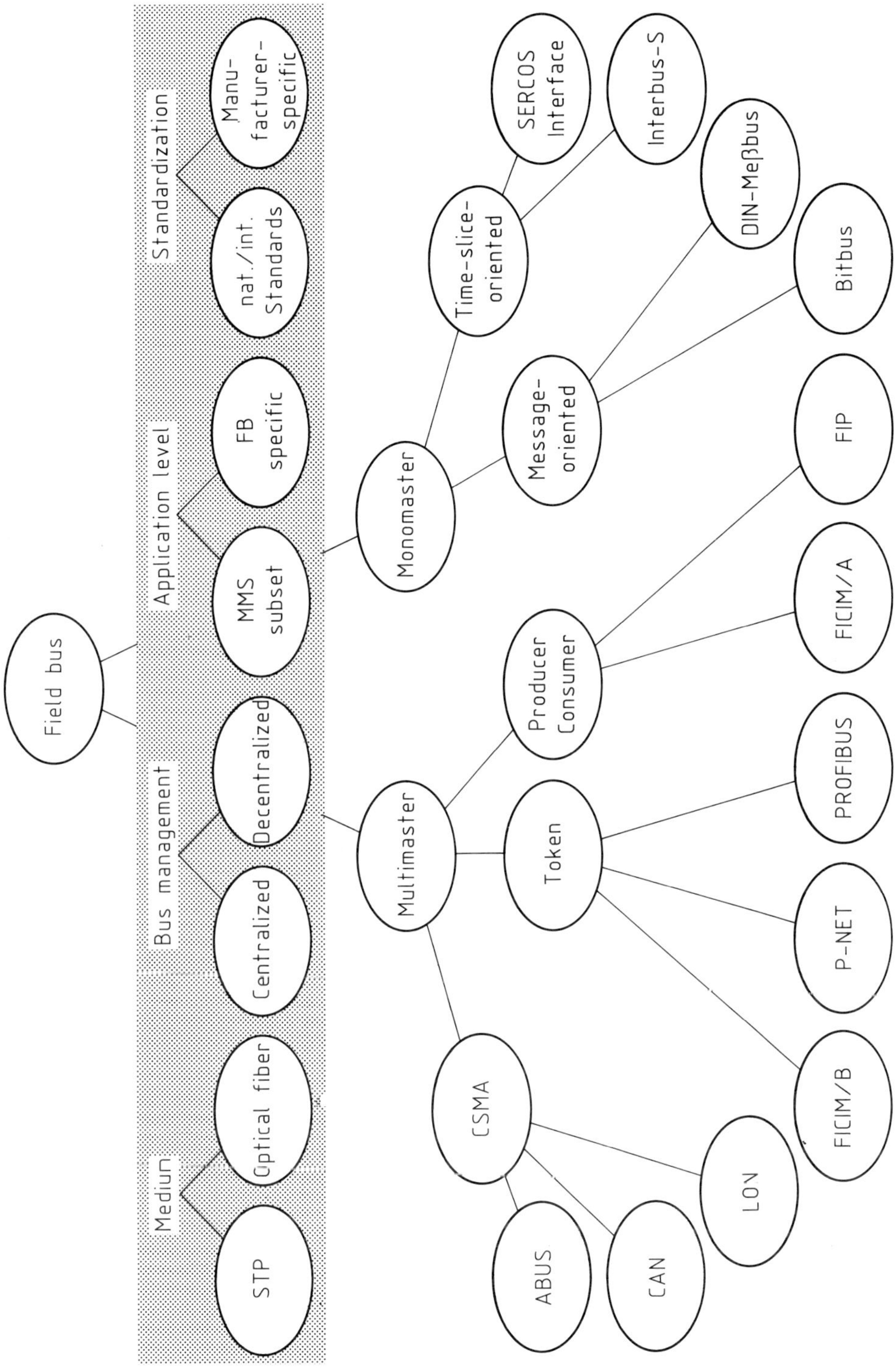

Figure 7.15. Classification of field bus systems

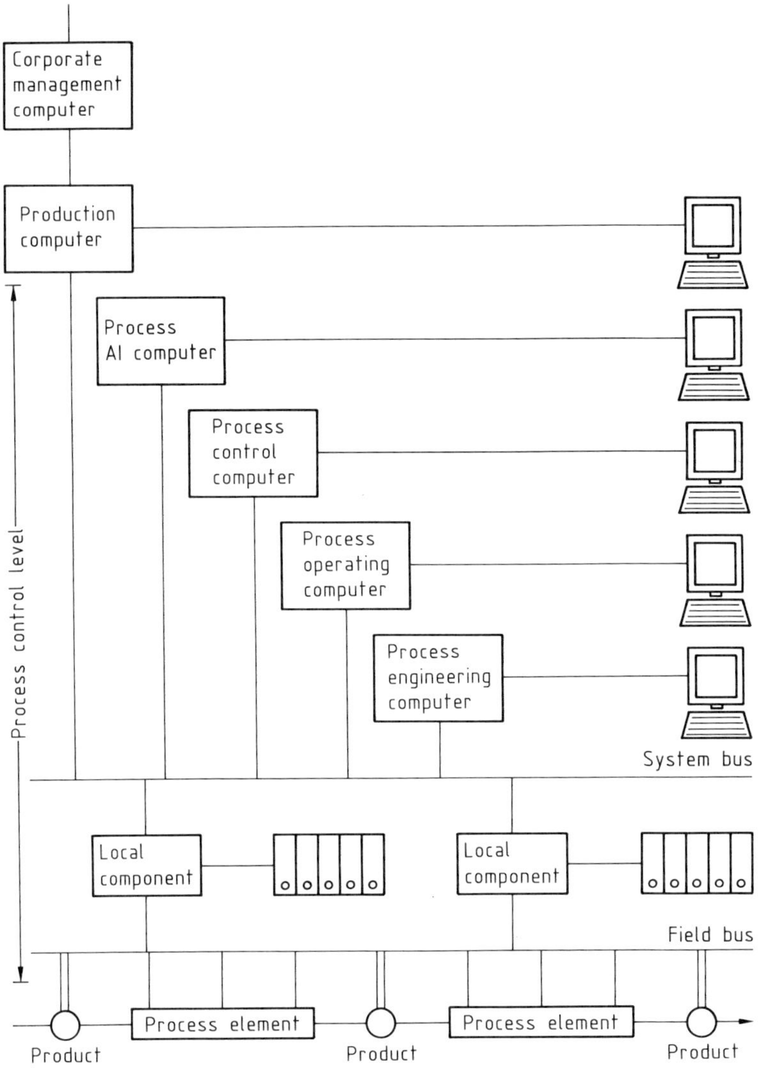

Figure 7.16. Ideal structure of a process/production control system

remote components is implemented in the most uniform way possible (Fig. 7.17).

The advantage of this concept is that each function can, in principle, be performed in any component. The allotment of required functions to components in a specific case is carried out with regard to the following criteria: decentrality, uniform processor loading, minimal data exchange, and realization of certain redundancies. The operator interface is uniform across the system. Any function in any remote component can be accessed from any terminal. Thus, from a given terminal, one can configure a component x in the system, use an operator function to perform a certain action in a component y, view the archive stored in component z, and effect an input to the production control system located in component z. Which interventions are possible in a particular case is governed solely by access authorization. The extent to which the human–machine interface is consistent between packages will depend on the state of development of the individual system packages. The need for a consistent concept, however, requires a consistent systems approach from the outset. In small

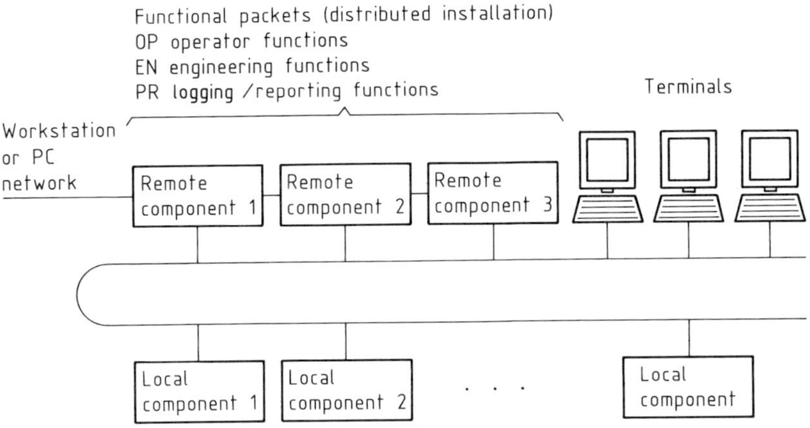

Figure 7.17. Process control system with continuous operating system platform for distributed use

plants, this concept has the important advantage that a modest hardware platform can make available the entire system functionality; thus the hardware governs primarily the realizable quantity structures (quantitative) rather than the functions that can be performed (qualitative). If the requirements are increased, the power of the system can be adapted to the new tasks by the modular addition of hardware. The presence of a consistent information or data model is, however, an essential prerequisite (see Chap. 2).

System Bus (see also Chap. 8). The system bus is the nerve center of the process control system. The following services, among others, are performed via the system bus:

- Continuous delivery of data from local and remote components to the monitoring functions
- Transmission of operator commands, acknowledgments, and instructions from central process and production control functions
- Provision of required data to the central archive
- Delivery of messages and alarms to the message receiving services
- Object manipulation and object information to assist central engineering
- Loading of actions (production specifications, execution specifications)
- Internal update functions when individual components go out and are restarted
- Temporal synchronization

The system bus of a process control must have the following technical qualities:

- High availability
- Insensitivity to electrical and electromagnetic interference
- Deterministic behavior, even when under load
- Open architecture (compliance with international and de facto standards)
- Hardware and software capable of adding or deleting subscribers while bus is in operation

The following features are possible ways to meet these requirements:

- Use of optical fibers for data transmission. Optical-fiber technology permits high transfer rates even in large-scale networks. Electromagnetic interference and potential problems in the bus system are physically excluded.
- Transparent redundancy of medium. The required high availability is possible only if the transport medium is redundant in design. Its redundancy must be transparent; that is, higher communication levels must not be affected by a switch from a pathway to the redundant pathway.
- Use of token-passing protocols. The use of the token method for transfers means that every subscriber can send important messages within a fixed length of time, even when the bus is heavily loaded.

Currently there are no standard system buses having these implementation features in the middle- (ca. 1 Mbit/s) and high-performance (ca. 10 Mbit/s) ranges. It is therefore no surprise that all system buses now available, while they do use standards for certain levels, still represent manufacturer-specific solutions on the whole.

Unfortunately this commercially important range of medium and high transfer rates is not currently experiencing any development that could lead to a standard solution with the properties needed in a process control system. It appears that there is still some time to wait for an "open" system bus for use in process control engineering.

The only hope at the moment is FDDI in the ultrahigh-performance field. FDDI has some prospect of acceptance as a general standard and has all the technical qualities expected in a system bus for process control systems. It is currently handicapped by its very high cost and by the space and power requirements of the interface groups to handle the high transfer rate (100 Mbit/s). Nonetheless, this approach is going to be a focus of attention in process control engineering circles.

The following comment should, however, be made on the evaluation of data-transfer rates.

If the transfer requirements imposed on a system bus in a process control system for controlling the process are considered, it can be seen that a gross transfer rate of 1 Mbit/s is sufficient for such tasks, even in larger networks. The net transfer rate strongly depends on a choice of individual transmission services and the network data distribution model.

Bottlenecks can occur, for example, when archives are updating one another, when static image information is being transmitted over the bus, or if an unsuitable data model has been chosen. A typical example of a transfer-intensive data model is the centralized display of all local component status information at remote component locations.

In the assessment of a bus system, therefore, the gross transfer rate is less important than a well-thought-out model of data management and data sharing on the network.

Component Structure of Commercial Process Control Systems. Figures are intended to explain the component structures of several process control systems. The selection is meant to provide examples and does not reflect either market position or technical ranking.

In Figures 7.18–7.31, the notation used for the components complies with the manufacturers' product designations.

The following symbols are used for functional packages:

AR Archiving functions
CO Control functions
EN Engineering functions
GA Gateway functions

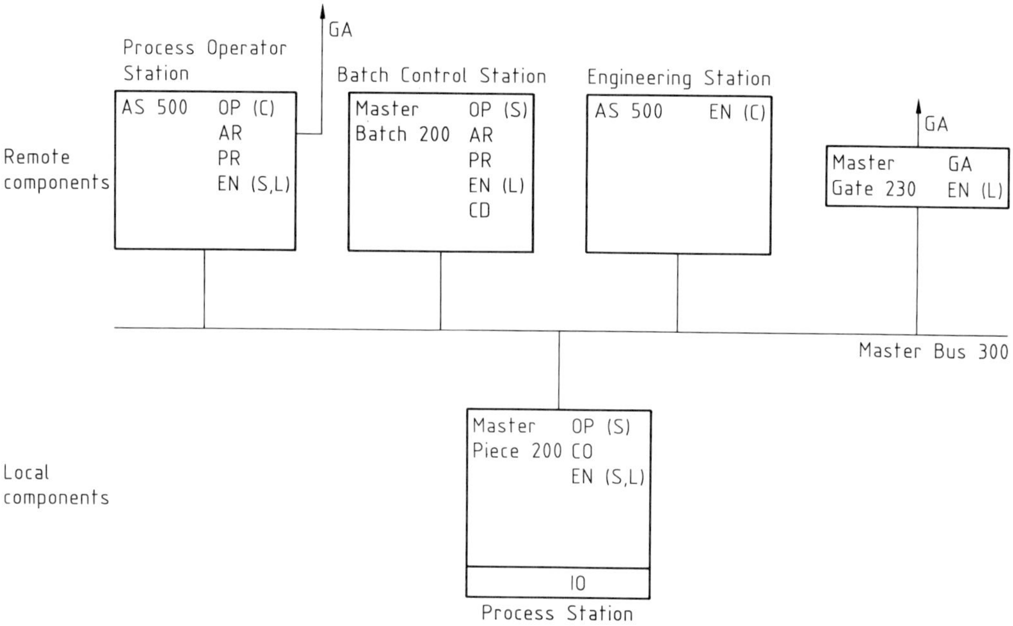

Figure 7.18. Functional structure, ABB Master

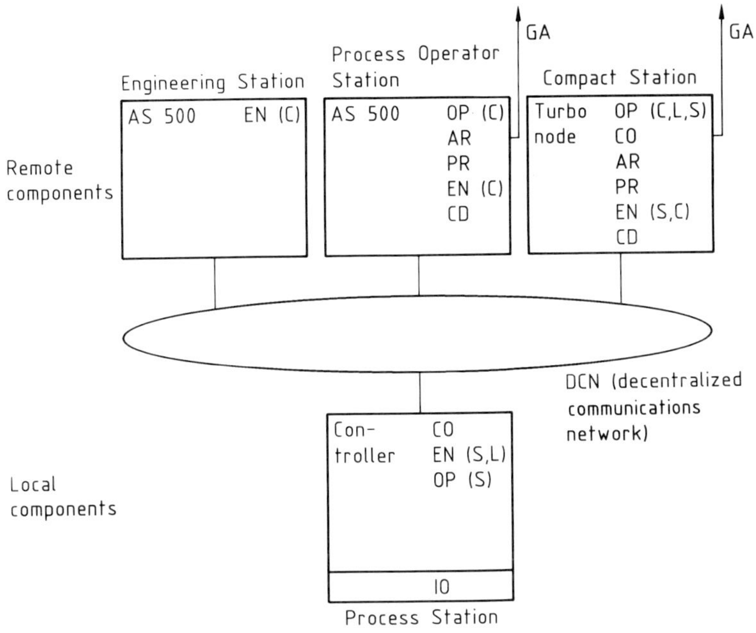

Figure 7.19. Functional structure, ABB MOD 300

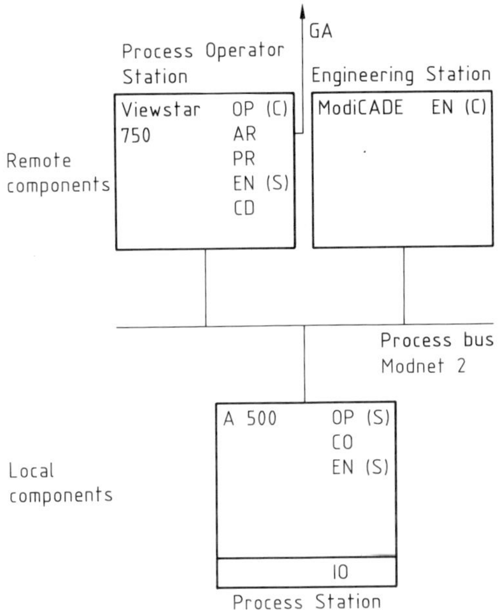

Figure 7.20. Functional structure, AEG Geamatics P

IO Process input/output functions
OP Operator functions
PR Protocol (reporting/logging) functions
CD Coordination functions

Note that some functions relate to not a single component but several. In such cases, the packages are separated into client and server portions and a letter (C, S, or L) is appended in parentheses:

(C) Acts in the system as client of the function
(S) Acts in the system as server of the function
(L) Function is implemented locally with no involvement of system communications

The codes C and S are used only for components that support the function on the system side, that is, when communications need not be separately programmed.

Examples:

Engineering Functions. The engineering component in which design activities are carried out acts as a client in the system. The target system designed by the central engineering function acts as a server in the system. The engineering station would thus have (C) appended to its symbol, while the target systems designed by the central engineering functions via system communications would have (S) appended. Locally configurable components would have an (L).

Operator Functions. Functions available troughout the system provide for operator control, monitoring, and the transmission of messages and alarms. With these functions, information is shared between individual applications in the target system and the operator (via the hu-

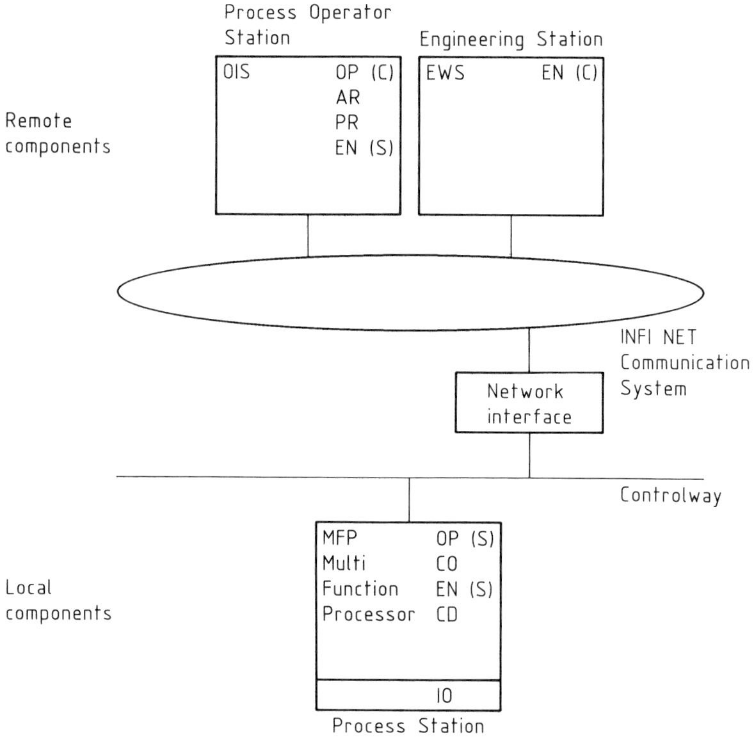

Figure 7.21. Functional structure, Bailey INFI 90

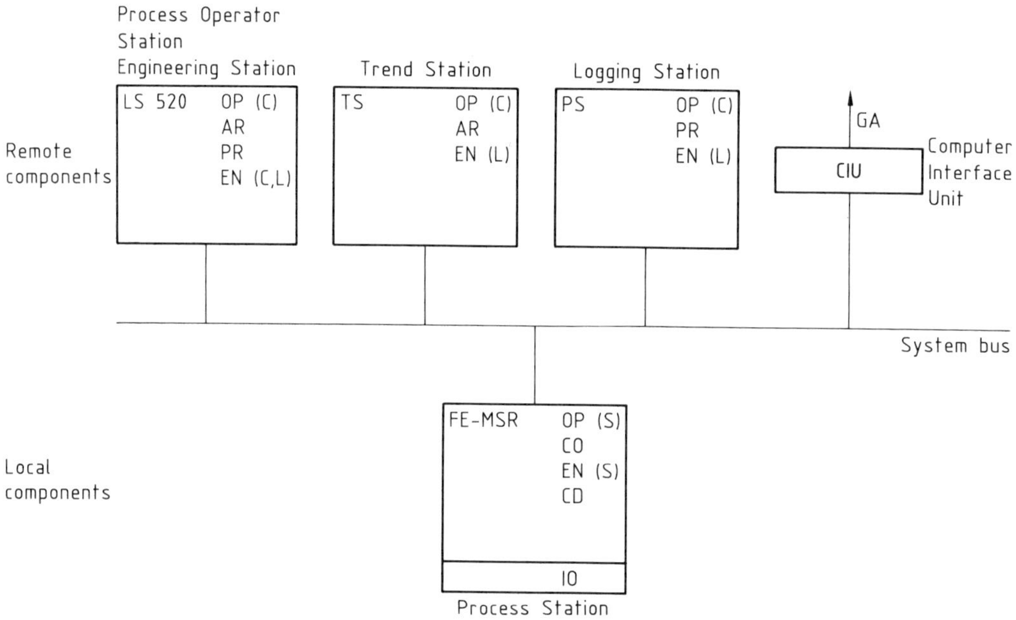

Figure 7.22. Functional structure, Eckardt PLS 80 E

man–machine interface). The display and operator components act as clients, while the target component in which the operation takes place or in which a status is interrogated is regarded as the server.

If one examines the structures in detail (see Figs. 7.18–7.31), one sees that most of the process control systems illustrated offer extensive systems solutions for the handling of local components. The major shortcomings are in communications between remote components. Server properties are largely lacking in these remote components. In other words, remote components often cannot be centrally designed (configured), and the normal operator control and monitoring functions do not make it possible to operate and monitor applications in other remote components.

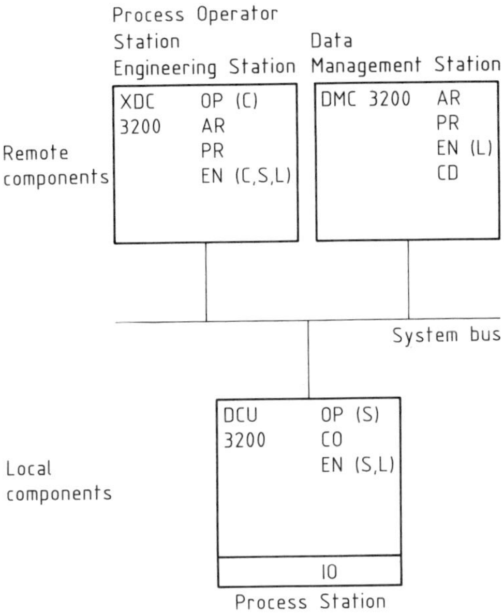

Figure 7.23. Functional structure, Fischer & Porter DCI System SIX

Redundancy. The following concepts can be applied to redundant design:

- Necessity of redundancy (see also Section 10.4)

The availability of the process control system depends on the frequency and duration of failures of its components. In general, when the process control system or a portion of it fails, the production plant or the respective plant section also fails.

If an especially high availability is required of a plant section, or if the failure probability of this plant section must be particularly low, the local components controlling it must be designed with redundancy. However, this is an exceptional case. In many situations, it is acceptable for individual plant sections to fail. Production can be briefly interrupted, or plant sections redundant from the production engineering standpoint can

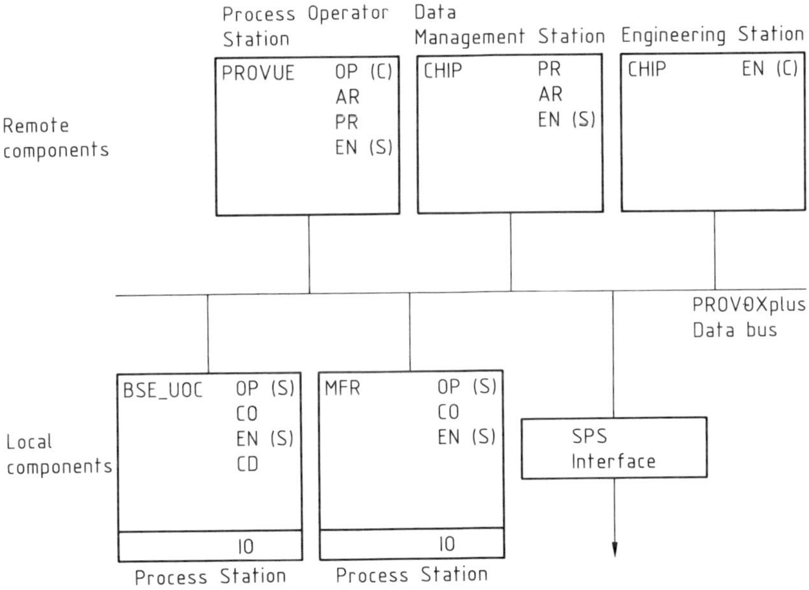

Figure 7.24. Functional structure, Fisher Controls PROVOXplus

be switched over. The decentralized structure of the control system means that the failure of one local component affects only one of the redundant plant sections, provided the allocation has been chosen correctly. Here, the process control system need not exhibit additional redundancy.

In the case of continuous processes that run nonstop for 3–5 years, as is common in some industries, scheduled maintenance of the monitoring and control hardware becomes a problem. The performance of scheduled maintenance generally involves removing modules, while inspec-

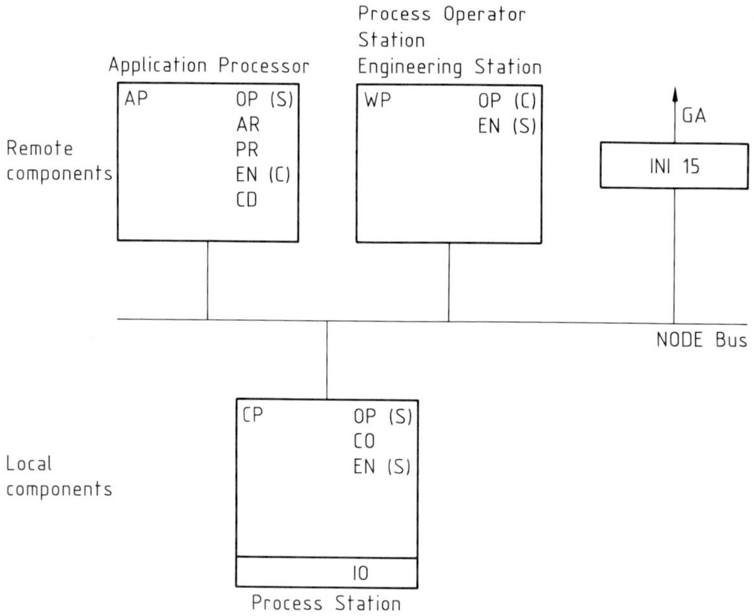

Figure 7.25. Functional structure, Foxboro IAS

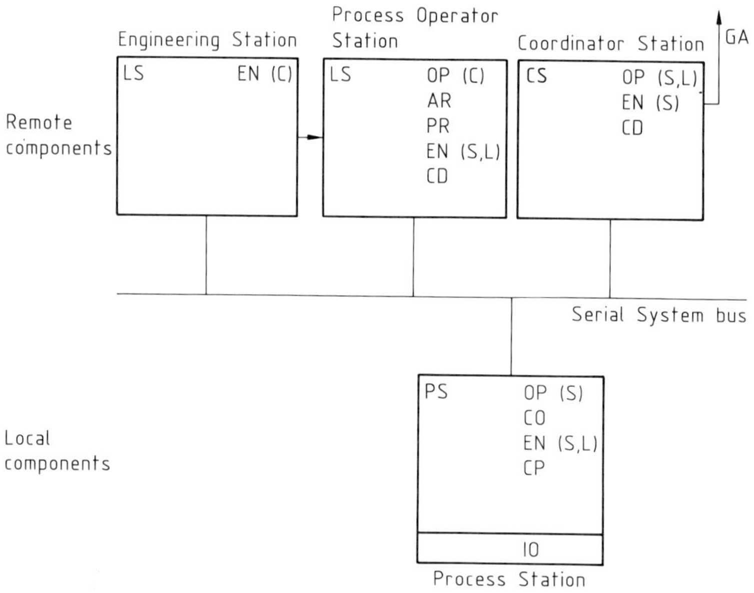

Figure 7.26. Functional structure, H & B Contronic P

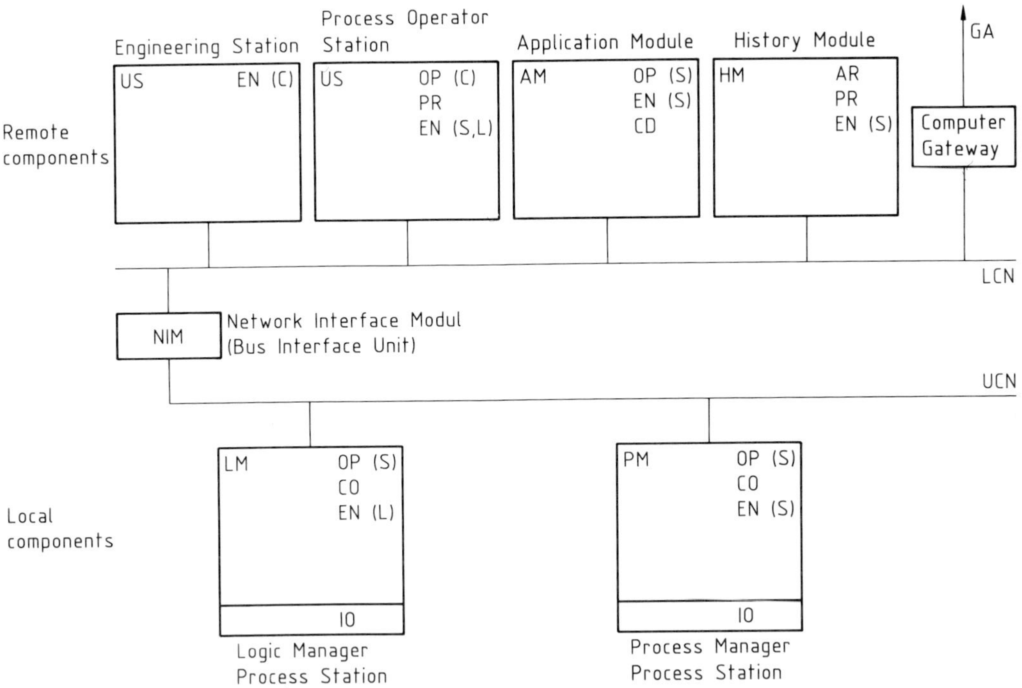

Figure 7.27. Functional structure, Honeywell TDC 3000

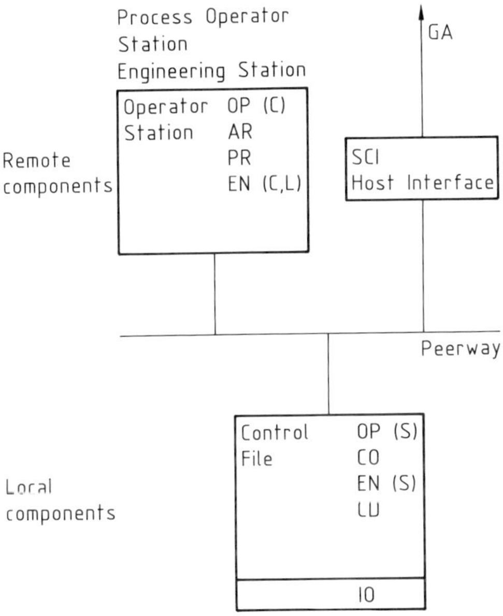

Figure 7.28. Functional structure, Rosemount System 3

tion entails removing entire system portions from active service. Redundancy is essential if this is to be done.

Dispensing with integrated operator and monitoring functions in the decentralized local components makes it necessary to have a redundant bus system with redundant implementation of operator functions in the remote components. Back-up operation with local intervention at the local components is generally no longer possible.

- Modular redundancy concept

In the modular concept, the decision whether to make a unit redundant or not is made separately for each unit in a system. Three points favor modular redundancy:

- Redundancy is only desirable for well-defined, self-contained units
- Modularization creates redundancy nodes, which separate the individual redundant units from one another and increase the availability of the system
- Redundancy is expensive and it is therefore sensible to employ redundant design only where necessary for a specific application

Figure 7.32 illustrates a modular redundancy concept for the system bus and local (process-level) components. This is an instance of transparent redundancy of the medium, which means that the individual data-transfer and processing units implement redundancy in such a way that higher-level system and user functions do not see it logically.

The situation is different in the area of remote components (i.e., on the "client side"), where standard components (workstations, industrial PCs) are employed in a nonredundant system structure. Redundant design is aided by special functions in the process control operating system and general system services and is ultimately realized in the structure of the user functions.

- Quality of changeover to redundant units

The redundant layout of units means that if a unit A fails, it is possible to change over to another unit B which continues to implement the required functions.

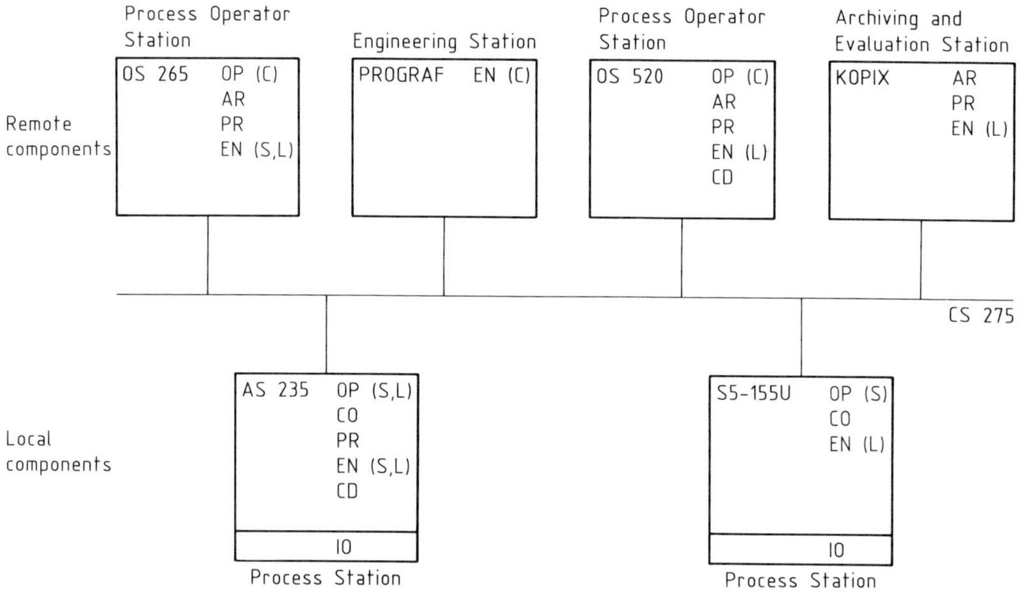

Figure 7.29. Functional structure, Siemens Teleperm M

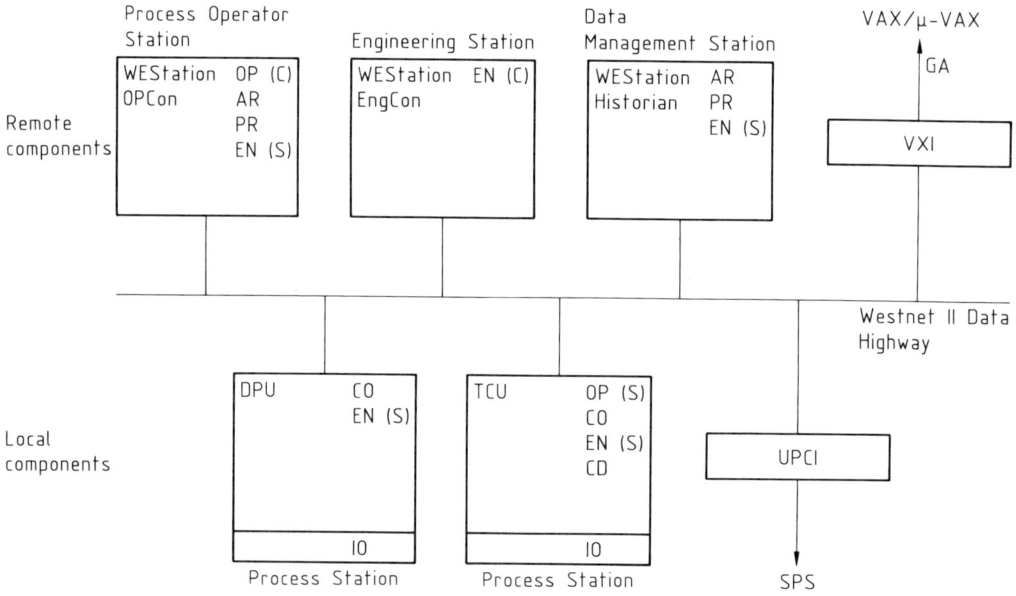

Figure 7.30. Functional structure, Westinghouse Controlmatic WDPF II

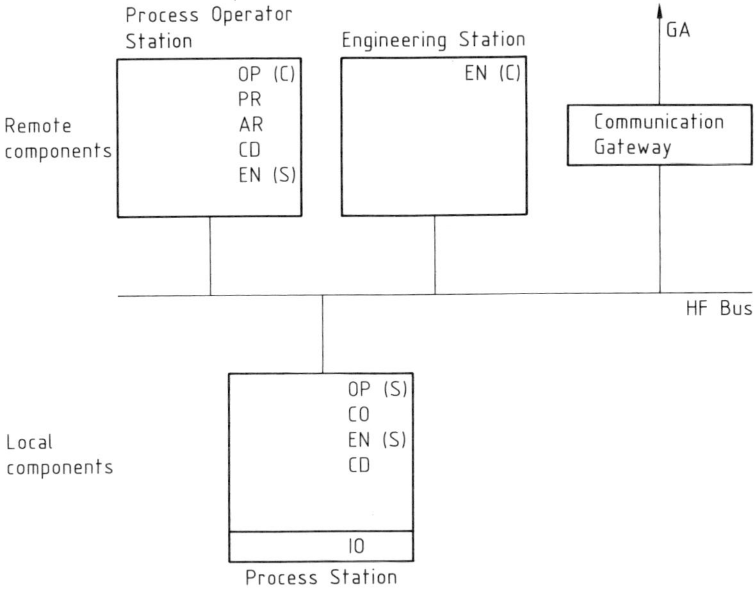

Figure 7.31. Functional structure, Yokogawa Centum XL

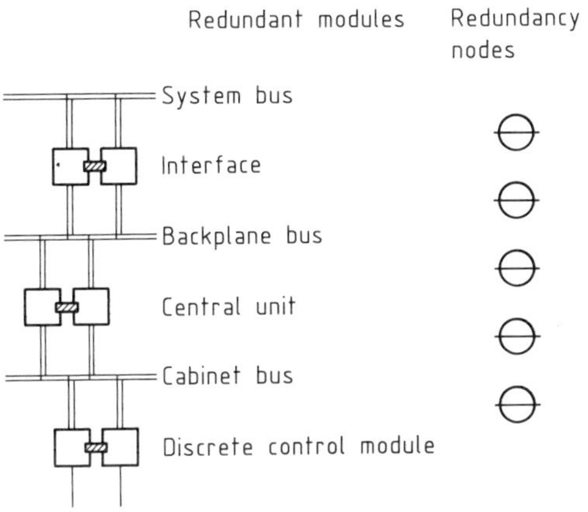

Figure 7.32. Modular redundancy concept

An ideal changeover is as follows:

- A malfunction in A is correctly detected as soon as it occurs, and the changeover takes place before a single datum in A is altered
- Prior to the changeover, B is in exactly the same state as A
- The changeover takes place with no time delay
- The changeover is "bumpless"

This is, of course, an idealization. In real systems, a number of problems arise:

– Detection of the Malfunction. A malfunction can be detected by self-diagnosis in A itself or by external monitoring. External monitoring systems are commonly based on the comparator principle. In a 1-out-of-2 redundant setup, however, localizing the malfunction is a problem. If the two units do not agree, which is wrong, A or B? Another problem is the time between the occurrence of a malfunction and its detection. Has A improperly changed its state in this interval as a result of the malfunction?

– Synchronization of B. B must begin working at the point where A quits. The question is how B, at an acceptable cost, can track A so as to hit the starting point as exactly as possible. The designer must then ask: How is the synchronization to be done? If A and B are clock-synchronized, or if there are synchronization points in the software, what deviations in the instantaneous states are tolerated and what deviations are unacceptable? Are data transfers in progress terminated and restarted, or is control passed "on the fly"? And so forth.

– Changeover. When the changeover takes place, all communications links of unit A are broken and replaced by links of unit B. The following problems arise at this point: How is the logical changeover accomplished? How long does B take to come into sync? How long will it take before all partners of A/B have dropped the old connections and made the new ones? Is the changeover automatic or manual? Does unit B remain on permanent standby, or must it be turned on or even fetched from storage?

– Restart. Once the malfunction has been corrected, how is the automatic restart of A (up to hot standby) supported without impairing the function of B?

The implementation of a modular redundant process control system is essential if the stated requirements are to be met. A good redundancy concept is the mark of distinction in a process control system. The questions stated under "Quality of Changeover to Redundant Unit" are not intended to lead the designer to a particular solution but rather to indicate the complexity of the problem and afford some warning against obvious but half-baked solutions.

Electrical and Mechanical Design. Process monitoring and control systems must function reliably, feature robust construction, and be adaptable and modifiable even while in service. Special importance therefore attaches to the mechanical and electrical design of these systems.

Mechanical Design. Local components are generally rack-mounted inside steel cabinets. Central modules are located in one plug-in unit; peripheral modules, either along with the central modules or in additional plug-in units. A module can be a simple card or an enclosed rack-mount device. The technique used to connect signal lines leading to the field varies from system to system; plugs on the front and backside are common, as is direct wiring to terminal posts or screw terminals in a special terminal bay.

An important mechanical design question is whether the cabinet must be accessible from one side or both sides. Practice has shown that access to the back of the plug-in units is necessary. This can be achieved by leaving a walkway at the back or, when this is not possible, by using a swing-out cage.

Remote components must also be considered in the course of mechanical design. Because these are generally off-the-shelf hardware, they must be mounted in an appropriately stable and secure way, either in control room panels or in properly equipped cabinets in the process control rooms (see also Sections 5.3 and 10.7).

Electrical design includes the overall electrical behavior of the process control system components. Important aspects of electrical design are power supply, isolation and earthing, protective enclosures and safety requirements, electromagnetic compatibility, and response to external conditions such as climatic, seismic, and chemical effects (see Section 6.4).

- Power supply.

The process control system is responsible not only for controlling the plant but also for monitoring it. It is desirable to provide a highly reliable uninterruptible power supply for the process control system; this makes it possible at least to acquire and record information about the process and plant status even when power to the processing facility is interrupted. It is precisely these exceptional situations that make the process control system so vital as an information system (see Section 6.4).

- Isolation; earthing concept.

A process control system communicates with a number of partners, some of them from different manufacturers. This is particularly true of local (process-level) components as well as the subsystem and field areas. It must be possible to adapt the control system to the local grounding concept. The goal is thus a control system "open" from the grounding standpoint. This requirement will be fulfilled if each input and output in the process monitoring and control system is electrically isolated. The earthing concept determines which potentials are connected in a specific case (see Section 5.3).

- Protective enclosures and safety requirements.

Process control systems are electrical systems and must satisfy regulations on shock-hazard protection and protection against the ingress of solid bodies and water. These regulations are based on IEC 529 (protection against foreign bodies and water) and IEC 536 (shock-hazard protection). In process control systems it is particularly important that the following sections be spatially separated from one another: low voltage; functional extra-low voltage with safe separation (PELV as specified in DIN 100, Part 410); and intrinsically safe circuits.

- Electromagnetic compatibility.

Process control systems are operated in an electrically harsh environment. In many situations today, these systems are located next to static converters and other power electronics, in the same room or even cabinet to cabinet. Operating personnel use handheld radios when the cabinets are open, and the rebranching field cables can pick up interference. Therefore, electromagnetic compatibility is a special concern in process control systems. Electromagnetic compatibility is a measure of resistance to noise resulting from

- Static discharges (IEC 801-2)
- Electromagnetic fields (IEC 801-3)
- Bursts [IEC 65(Co)39]
- Hybrid voltage transients [IEC 801-5 (Draft)]

Other guidelines, recommendations, and standards have been issued by NAMUR and DIN; see, for example, [7.13]–[7.15].

For each device, the severity of the effect can be measured and classified. Four classes can be defined:

1) No impairment of operation or function
2) Momentary impairment of operation or function, self-correcting
3) Momentary impairment of operation or function in which restoration of service requires resetting or operator intervention
4) Permanent loss of function resulting from damage to the device (or its components)

In general, it is expected that the maximum effects under moderately severe testing (IEC Severity 3) will fall into Class 2.

• Other conditions.

This heading takes in all other environmental effects that might lead to impairment of the functioning of the process control system. They include, among others:

- Temperature and humidity
- Seismic stress
- Chemical attack (e.g., acidic gases)

Environmental conditions are defined and classified in IEC 721; test methods are set forth in IEC 68.

7.3. Process Control Operating System

A process monitoring and control system is a distributed system in which modular control structures are managed and executed in the most efficient way possible. This applies in particular to the local (process-level) area. The process control system should make it easy for the user to work with a modular system (as illustrated in Fig. 7.10, p. 538) in a similar way to a system based on discrete devices.

For example, it should be possible to install or remove a module without interfering with the operation of other modules, and to make or break a connection or modify a parameter value while the system is in operation. Another important point is reliability. For example, the making of a connection should not inadvertently cause internal interference with a modular algorithm, while the manipulation of a module should not result in the accidental change of a parameter value in another module.

These requirements can be satisfied only if the process control system has an operating system with appropriate object manipulation and object-oriented programming capabilities; the operating system should also support the creation of modular software structures.

Three approaches have been taken to implementing these requirements.

In the first approach, the entire "modular intelligence" is concentrated in the design system. The local components responsible for control system operation do not have a modular structure. The design system makes a translator run, generating a transparent code from the objects managed, and loads this code into the local components for execution. The advantage of this technique is that the requirements on local components are modest; the local components can be structured easily and thus economically. On the whole, however, this concept has not yielded any satisfactory solutions. The main reason is that changing a single object requires retranslation and reloading of the entire program. Such a system is extremely resistant to changes.

When partial translation or partial loading is desirable, the design system must have a detailed knowledge of, and partial authority over, the data organization of the local components. The cost of developing such a design system is high and cannot be controlled, especially when the local components are further developed.

For these reasons, all successful SPC (stored-program controller) systems are decentralized process control systems that include management structures for objects in the local components themselves.

It is necessary to distinguish between the "procedure-oriented" modular approach used in SPC systems and the "object-oriented" modular approach used in distributed process control systems.

Figure 7.33 illustrates the difference. In the procedure-oriented modular technology, data modules and procedure modules are the objects. Distinct procedure modules communicate through data modules. The procedures describe certain data elements of the data modules, which

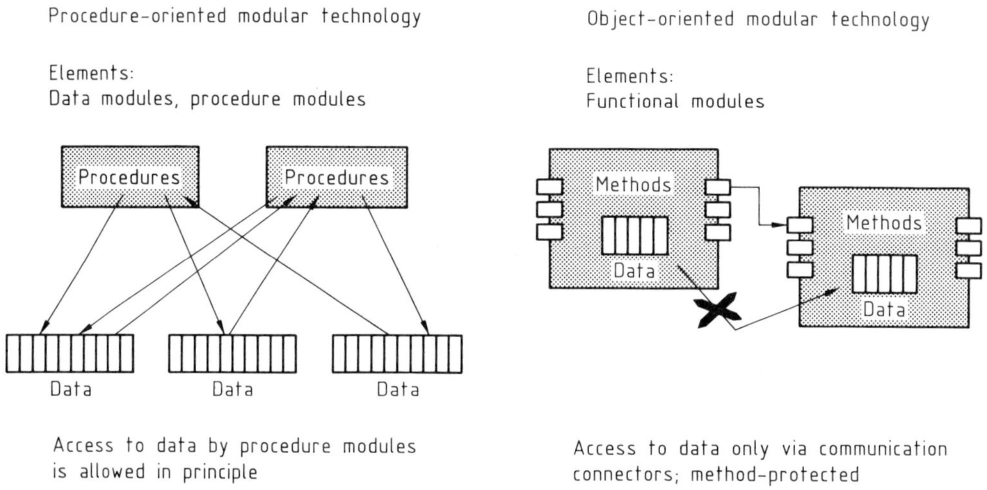

Figure 7.33. Procedure-oriented and object-oriented modular technologies

in turn are read by their procedures. The connection does not exist as an explicit object. In this method, the system does not afford protection of access to the data areas; that is, all data can be read by any procedure, and also written if the procedure is write-enabled.

The right-hand part of Figure 7.33 shows the object-oriented modular technology, where data and procedures (here also called methods) are combined in a single module that can be handled only as a whole. Data sharing takes place via connections between the module connectors. Direct writing and reading of internal module data is not possible at all. The only way to access these internal data is locally, through the methods inherent to each module.

It is important to maintain the distinction between these two technologies. In practice, many misunderstandings arise when the term "modular technology" is used. In the SPC world, procedures are also referred to as "functional modules" and data as "data modules."

The forms of office organization shown in Figure 7.34 illustrate the properties and applications of the procedure-oriented and object-oriented approaches. On the left, Müller and Schulz work together in a project team. They sit across the table from each other, in a room containing a cabinet with a shared set of files. This arrangement corresponds to a procedure orientation. Müller and Schulz correspond to the procedures; the set of files, to the jointly available and processed data. This form of organization has the advantage that every worker has quick, direct access to the common data. The setup does, however, require intensive communication and agreement between the two team members: each must know exactly what the other is doing and what data are being changed. While this organizational form is feasible for small, easily over-

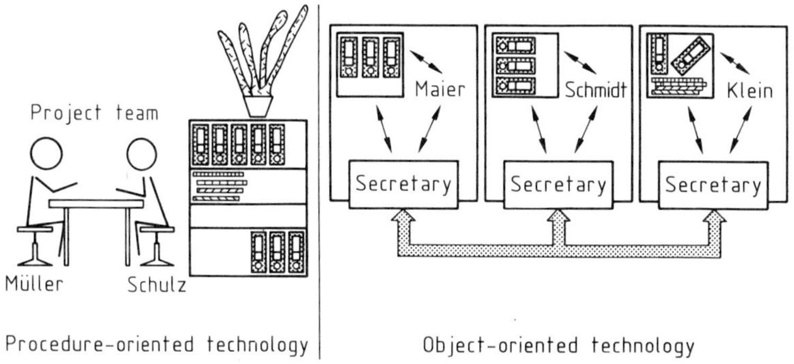

Figure 7.34. Examples of procedure-oriented and object-oriented setups

seen units, in larger and more complex structures it is no longer possible.

More complex structures require local data safeguards. Such a form of organization is shown on the right in Figure 7.34. Maier has his own set of documents, to which only he has access. No other team member can enter his office to view a file or enter anything in a file; anyone wishing to obtain such information or make such changes must ask Maier to provide information or make an entry. Maier himself decides whether and how to respond to this request. Maier and his data form a unit that, as a whole, performs certain tasks. This form of organization corresponds to an object orientation.

If the parallel is drawn to process control systems, it becomes clear that the traditional approach in SPC-based systems—which historically arose from small automation systems with fast access requirements—is procedure orientation. Decentralized process control systems, on the other hand, were designed for large, complex networks from the very beginning, and in these systems the object-oriented modular approach predominates.

The remainder of this section will describe the basic properties of a process control operating system supporting an object-oriented modular system.

Software Modules. One characteristic of a modular system is that a module—with all its parts, data records, connectors, and methods—is treated as a unit. In production service, from the communications standpoint, the module can be identified by a single reference.

Such an object-oriented modular concept can be implemented in either of two ways: by realizing the modules as flat, completely autonomous objects, or by realizing them as entities belonging to a class hierarchy.

Methods using discrete devices correspond to the first approach. Each compact controller has its own data storage, its own interfaces, and its own logic for implementing the PID algorithm. In a software-type concept, which follows the second approach, the modules are realized as instances of module types; this corresponds to a special, one-level abstraction scheme.

When it is realized in the functional system, each module is therefore split into two objects, a type object and an instance object. The type object contains the processing methods, a description of the data structure, and a description of the connector structure. The instance object (entity) contains the module name, the individual status information, and a reference to the type object.

This functional structure implies two properties that are significant for module handling in process control systems:

- Type objects and entities can be created and loaded separately
- More than one entity can be defined for one type object

Figure 7.35 shows realizations of three controller modules. All three use the common type object "PID controller"; modules TA 12 L 1, TA 14 F 1, and TA 3 T 2 are three entities.

The creation and loading of types is an operation that generally calls for profound system knowledge and can be performed only by certain persons. In many process control systems, the user cannot create new module types with their own data structure, connector structure, and methods. The user must make do with the set of types provided by the manufacturer.

In contrast, it is very easy to work with instances under the software-type concept. Because all data structures, connector structures, and methods of a module are defined when the type is created, the instances need only be generated as data records, parametrized, and inserted in the processing sequence. These are simple operations, and in virtually every process control system they are supported by the operating system of the local components, even during production operation.

The software-type concept has proved effective for the handling, management, and sequential organization of the modular system in process control systems. In a variety of components, the concept has been embodied and optimized for the requirements of a real-time system with many objects to be processed concurrently.

Certain constraints relative to a general class hierarchy are unavoidable. They include the following:

- The functional system supports only one level of abstraction (entity–type).
- Type data are present in multiple copies and are thus not free of redundancy. The type data for a module are available locally, as loaded type objects, with respect to every processor.

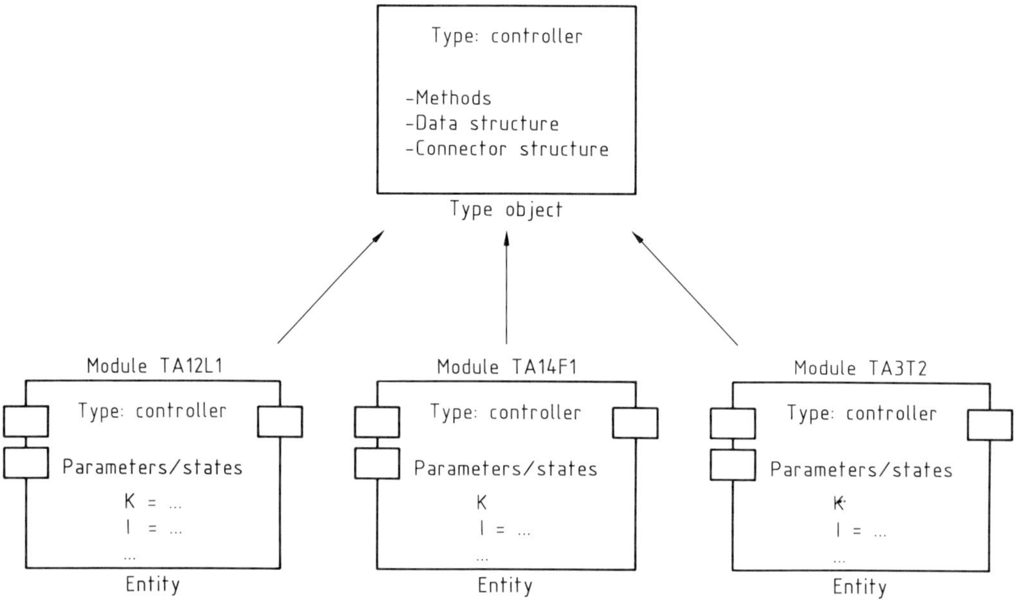

Figure 7.35. Software implementation of three modules of the type "controller"

- Each instance has a direct reference to the type object of its logical environment and knows what information it will find where.

As far as the external function of a module in the modular network is concerned, it makes no difference whether the module is realized as an autonomous object or as an instance of a type. In the software-type concept, the instances provide the external representation of the module with all of its properties.

Communications Network. A major task of the process control operating system is to manage and dynamically perform communications in the network of modules. Each module has connectors serving as external interfaces, through which it exchanges information with its surroundings. All communications in the modular network can be depicted in terms of individual communications links, each between one module input and one module output.

From the applications standpoint, there are two basic forms of information exchange: the "get" relation and the "send" relation.

Figure 7.36 illustrates a *get* relation. B is interested in the present state of A and reads this state from A without feedback. Such a relation corresponds, in discrete technology, to acquiring a signal from an output terminal. A given output A can be read by various inputs.

Figure 7.36. Get relationship

In this kind of relation, only the recipient is interested in exchanging information. The recipient needs a state that another unit is furnishing as an output, and needs it as quickly as possible. Regardless of the concrete form of this transfer—a network may well contain sending and event-driven system services—the relation as seen from the user's viewpoint corresponds to a get relation.

Relations of this type lead to low-error-rate structures that are easy to reconstruct in dynamic fashion. The fact of access without feedback means that get connections can be made without further safeguards.

Within the component, a get connection corresponds to a read access of an output variable and, accordingly, can be implemented efficiently.

Many process control systems permit the treatment of get connections as objects in the local components, or at least get connections internal to a component. Their management and general implementation are taken care of by the operating system.

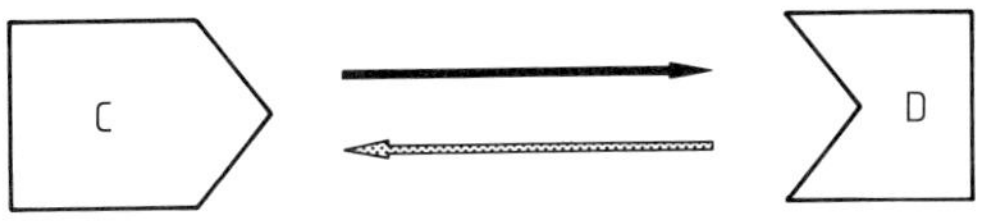

Figure 7.37. Send relationship

Figure 7.37 shows a *send* relation. Here one application has a functional interest in communicating something to another application. This would be the case, for example, when higher-level units transmit orders to lower-level units; where monitoring units have the task of automatically generating messages and alarms; or where operator control units wish to send a new control value to an application.

The attribute send here does not refer to the type of transmission; it represents the fact that the initiative and interest in data exchange is primarily on the part of the sender.

A send relation is characterized by the following features:

- The application behind connector C knows under what conditions C is to send what information to whom and activates connector C accordingly.
- The message from C leads to a write access in D; that is, the state of D changes.
- The associated service in D must therefore have a protective mechanism that receives the message, checks it, and performs the write only after successful checking. The protective mechanism checks both sender and content.
- To avoid unpredictable response times, the server service responsible for checking the message and processing it on receipt must be initiated as soon as the message arrives.

By way of example, Figure 7.38 shows an instruction input connector. When a message containing a new instruction arrives, the availability of the connector is first checked. If the connector is available, it is then occupied. The message is next checked for formal correctness (type of message, variable types, etc.). If the result is positive, the message is correct. This does not mean that the instruction can also be accepted by the module in terms of its content. The receiving service therefore has a link to a part of the module method specially allotted to the connector. This checks whether the content of the instruction is permissible from the application viewpoint. If the result is positive, the instruction is finally accepted and written as a new effective instruction. The sender of the instruction is notified of the acceptance, or the reasons for nonacceptance, of the instruction.

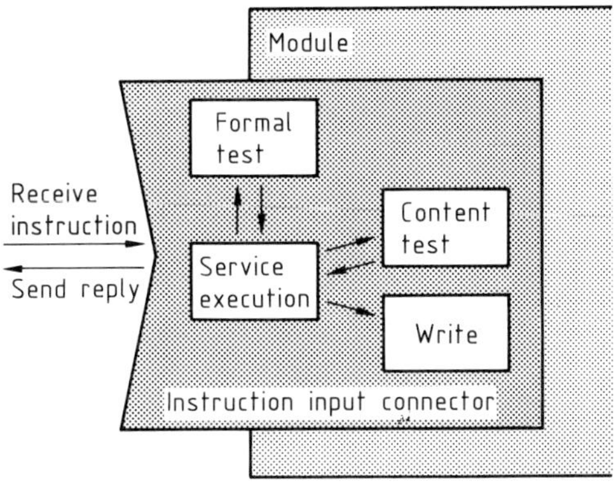

Figure 7.38. Instruction input connector

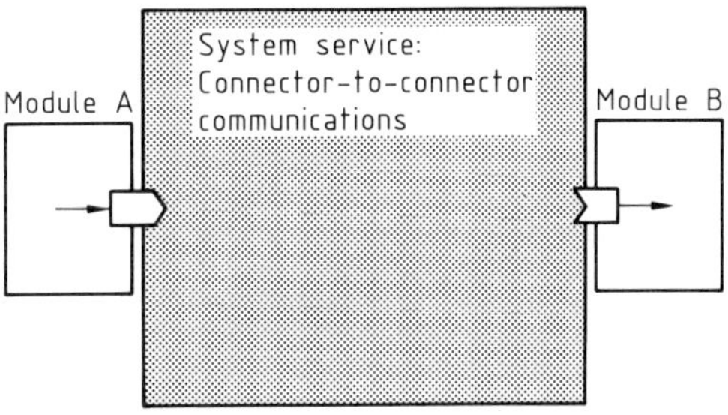

Figure 7.39. Communications as a system service

Most process control operating systems also permit unprotected writing into effective module input areas. Some users make intensive use of this capability, transforming the process control system into a quasi-procedure-oriented system. This is not, however, in keeping with "good programming practice" and should be avoided whenever possible.

A process control operating system must have the goal, as illustrated in Figure 7.39, of managing all communications relations between connectors in the modular network as objects and embodying their implementation, as seen from the user's viewpoint, in a system service. Many process control systems already do this within a local component, at least for relations internal to the component. The critical relations, however, are those between connectors belonging to modules implemented in different components. Here the user must explicitly design (configure) the communication in many cases. For example, it may be necessary to integrate and wire certain LAN communications modules. The message orientation and the strict formalization

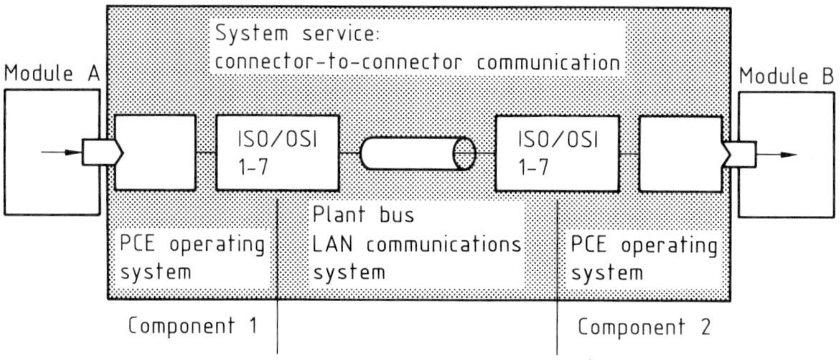

Figure 7.40. Connector-to-connector communications via the LAN

mean, however, that the connector model will later enable the operating system to manage and execute connector-to-connector relations over the LAN.

As a consequence of the special two-level module → connector object structure and the client/server character of the relations defined, it is possible to make efficient use of MMI services (see Section 8.5) for imaging. Open communications (see Fig. 7.40) can thus become a basic feature integrated into a communications network extending over many components.

Connector-to-connector communications can thus be organized as a network-wide message exchange system. A crucial point is that the simple local relations, with their efficient execution methods, are incorporated as an integral part of this overall system. If, as shown in Figure 7.41, two applications are located in the same component, the process control operating system can implement connector-to-connector communications with optimized local access functions.

This technique can greatly increase the processing power of the execution system.

At the process (local) level, "get" relations within a component are the most frequent type of communications and are crucial for the loading of the communications network.

If all these relations were executed through a general message exchange system, the capacities of the functional units would be far too small. This level of performance can only be achieved by effective use of internal pointers and direct reading, both organized by the operating system.

Addressing. The objects and name spaces contained in a process control system depend on the organizational viewpoint and scheme. It is necessary, for example, to distinguish between

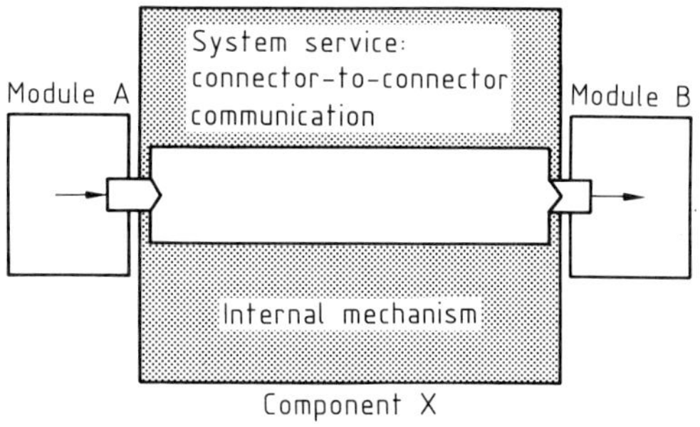

Figure 7.41. Communications internal to a component

the hardware viewpoint (component, module, unit), the access viewpoint (bus segment, subscriber, module, unit, connector, variable), the open communications viewpoint (VMD, domain, variable), the programming viewpoint (unit, connector), and the engineering viewpoints relating to the structure of the process plant, the breakdown of the process to be performed, or the structure of the production sequence.

Each way of analyzing the system has its own hierarchical name space, in terms of which the individual objects in the system so analyzed can be addressed. There are many mappings and assignments among the objects in various system breakdowns, and hence many object management structures and addressing options. These are not presented in detail here.

A question that is of interest, however, is addressing for LAN communications in the modular network. Most process control operating systems have not integrated communication via LAN into their module communications, and, therefore, either special communications functions must be installed (READ, WRITE, SEND, RECEIVE,...) or communications must be structured with direct, unprotected read or write access to the module variables.

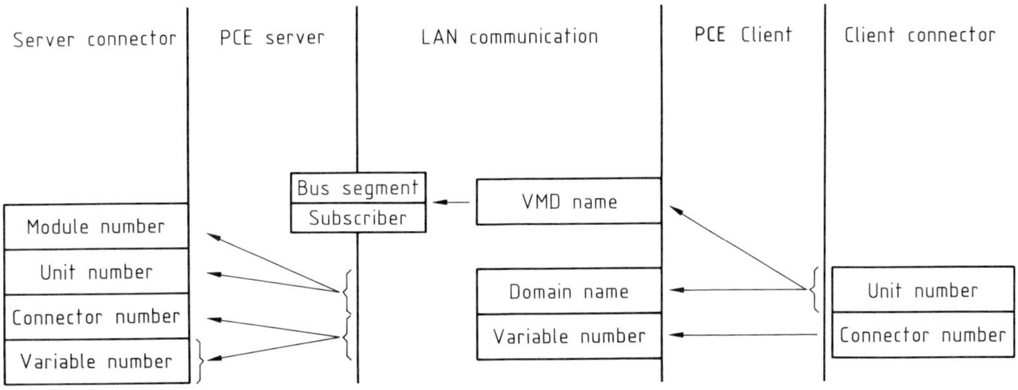

Figure 7.42. Example of connector addressing via LAN with names

Figure 7.42 shows one possible addressing scheme in the modular approach. The scheme is shaped by three aspects, which must be mapped into one another.

Addressing in the client is done from the programming viewpoint. The client connector wishes to make a connection to a certain server connector. The server connector is identified by its name and the name of the module (unit).

The process control operating system (OS) recognizes that this unit is located in another VMD and maps the requirements onto an MMI function. The unit is thus mapped into an MMI domain and the connector into an MMI variable in 1:1 fashion.

In the target VMD, the process control OS performs the MMI server functions and exchanges data with the server connector. The OS uses its internal access addressing scheme, for example the module number, unit number, connector number, and individual variable numbers.

This is just one possibility. There are others, such as access through abbreviated addresses. The important point for effective open communications is that modules (units) and connectors can be mapped 1:1 into addressable communications abjects.

Execution Properties. An important task of the process control operating system is to control the sequence of module processing. In general, a module consists of a main method and various connector-specific methods.

The main method is started by a call to the module by the executive system. In process control systems, calls typically come in cycles at fixed time intervals. Each module can be allotted such a fixed cycle for itself. Cyclic method calling is independent of whether or not changes of state have taken place at the input connectors. Cycles of between 1 and 4 seconds are typical in process engineering.

Connector-specific methods are permanently allocated to the input connectors. They are started in acyclic mode when messages arrive at the respective input connectors.

Asynchronous Operation. The use of fixed cycles generates deterministic sequences between certain modules. Even so, the modular network as a whole must be regarded as a network of modules in asynchronous operation. The processing of a module can be interrupted, for example, by the processing of a higher-priority cycle sequence or by connector-specific methods started in acyclic mode. Furthermore, if module processing is shared among several processors, cycles of equal priority also run asynchronously.

Asynchronous operation results in data consistency problems:

- While one module is running, the outputs of other modules may change
- While one module is running, processing of another module may be interrupted, so that its outputs represent results from different cycles

These problems are not generally relevant. The output states requested must be as current as possible. Inconsistencies are not a difficulty for the processing algorithms. Applications requirements, for example in module type creation, may call for appropriate measures to insure consistency in certain cases.

The efficiency-oriented concept of process control operating systems manifests itself again here: avoidance of unnecessary management overhead.

Synchronization methods are available to insure the consistency of output and input data. However, they impose a substantial burden on the executive system and markedly impair the performance of the local components. Most process control operating systems purposely dispense with system measures to impose consistency. Note that all the devices in a discrete control system also function in asynchronous mode without consistency-insuring measures.

7.4. General System Services

A control system is expected to offer, as an integral feature, certain general system services for production operations. These should be implemented as firmware in the process control operating system (OS).

The pertinent services include the following:

- Communications and organization services for operation, monitoring, the generation of messages and alarms, archiving, and report creation
- Time management and network synchronization services
- Monitoring and diagnostic services for process control hardware
- Services supporting the changeover between redundant units and the matching of their states

Integration of these services into the process control OS by the manufacturer is necessary because only in this way can they be implemented in a meaningful way and performed with the necessary effectiveness.

Unfortunately, there are currently no standards governing the functionalities of these system services; the available, manufacturer-specific solutions differ all the way back to their principles. The user must consider the structure of these manufacturer-specific system services and fit the applications software to it.

A manufacturer-neutral design procedure is therefore possible only within narrow limits. In principle, this design approach is restricted to the configuration of certain processing functions with standard languages such as FUP, module language, and so forth.

This part of the configuration job often makes up only 40 % of the whole. The remaining configuration work must deal with configuring and parametrizing manufacturer-specific system services.

This section will describe in detail the tasks and contents of the system services. Each service is treated as an individual functional module.

Monitoring (see also Section 11.2). Today, processes are generally monitored by using color graphics displays. The interface can be based on standard solutions from the PC or workstation realm. The monitoring services provide the monitor interface with the appropriate video and status information. One monitoring service includes transmission of the state value from its source to the monitoring system, maintaining an internal state map (state or status image) in the monitoring system, and displaying the state value at the interface. Figure 7.43 shows schematically a monitoring service.

The monitoring system always displays the current state values. If a state is dynamically changing, it is important to display its most recent value. Past values and timestamps are not of interest. On arrival, each new value overwrites the old value on the screen.

All monitoring interfaces are implemented by the monitoring system. The same holds for the display of archival data, messages, records, and so forth.

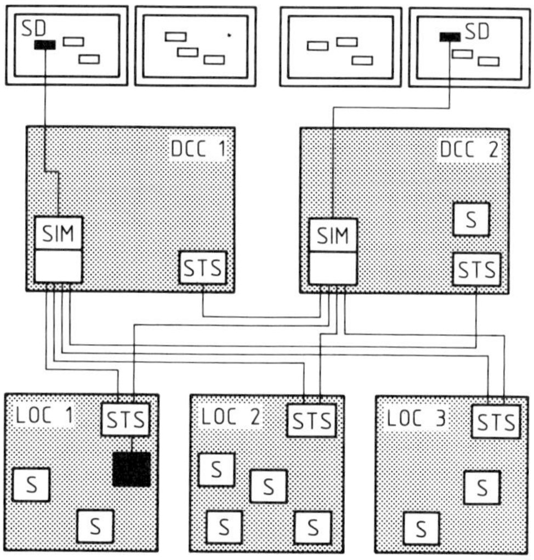

Figure 7.43. Schematic diagram of a monitoring service S = state; SD = state display; SIM = state image; STS = state transmission server; DCC = display and control component; LOC = local component

When the message log is displayed, for example, the message system is the source of the states displayed. The timestamp of a message is a state of this system and is retrieved and displayed by the monitoring system in the same way as any other state.

State Displays. A state can be displayed in a variety of ways, and standard interfaces offer all the possibilities. It is therefore a fundamental question of human–machine or human–process communication design how a state is represented in each particular case.

Figure 7.44 presents examples of ways of depicting a binary state. Often, a binary state is represented by a color change in a text field or by the presence of a mark. In flowsheets, the process state that the binary state represents is indicated symbolically. For example, a curved arrow or a color applied to the valve symbol can be used to indicate whether a valve is open.

Figure 7.45 shows several representations for an integer or floating-point value. Here, standard systems offer a wide range of easily implemented display options (see Figs. 11.13 to 11.21).

Maintenance of State Image. The various ways of maintaining the state image in the monitoring system give rise to basic differences in process control system data models.

Important goals in managing and maintaining the state image (map) in the monitoring system include the following:

- Keeping the displayed states updated: The current value of a source state should be displayed within one second if possible (or at least within two seconds in exceptional cases).
- Rapid preparation of a new image with the proper state values: When a new image is selected, the image together with all updated states should be constructed within 2 (within 4 in exceptional cases).
- Minimizing the data transfer cost: In most situations, the state sources are not located in the same display and control component as the monitoring function itself. The required states must then be transferred via the bus. In large systems in particular, or when the depth of detail is great, measures must be taken to reduce the amount of data transferred.
- Minimizing the cost of preparing the state information in the source system: Some effort must be expended in the source system to prepare the state data and organize the data transfer. This extra work must be considered from the outset when calculating the load on the source system. Problems arise, in particular, when several display and control components are used, each with its monitoring interface and all accessing states of a relatively small local component, or when a change-driven data transfer service is implemented.
- Modular organization: The state image maintenance procedure must be set up in such a way that new images can easily be inserted in a monitoring system and an additional display and control component, with its own monitoring system, can be connected to the system communications at any time.

The simplest way to organize and maintain the state image for monitoring is by using a 100% image in the display and control components (DCCs). All states that might ever be re-

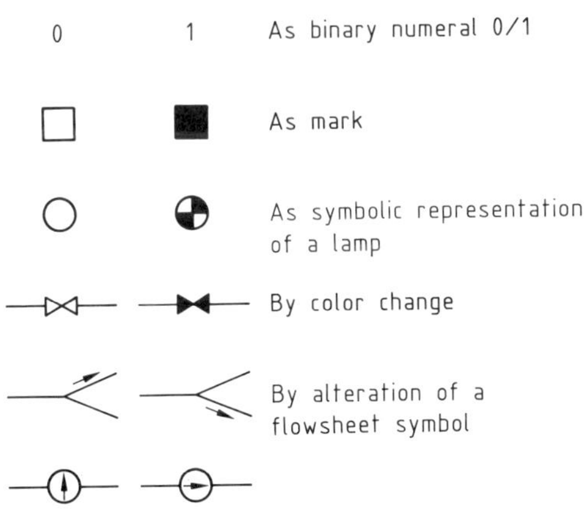

Figure 7.44. Representations of binary states

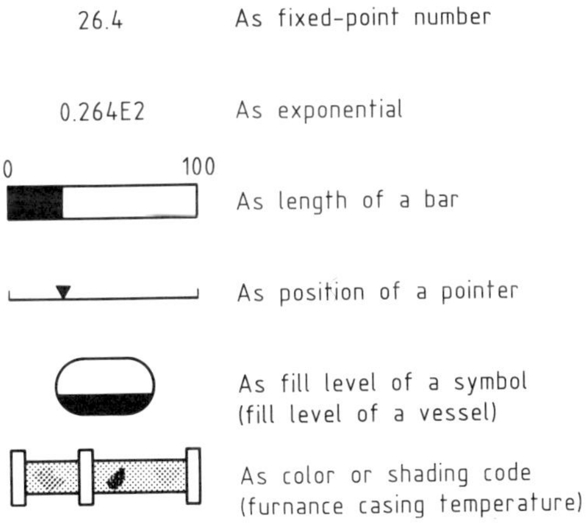

Figure 7.45. Representations of a floating-point value

quired for monitoring are held in every DCC at all times. This solution, which is practicable in small one-station systems, leads to performance problems in large multistation setups. If the possibility of insetting detailed state information is not to be lost, data transfer must be optimized, for example by selective or change-driven transmission.

In selective transmission (Fig. 7.46), only those states that are currently displayed are updated in the state image. The data-transfer load thus depends on the scope of the images currently displayed, not on the scope of all images configured. The image can be carried to any depth, down to the smallest details, without any change in the preparation and transmission load. Selecting a new image means a change in the set of states to be updated. This must also be reported to the source systems and new arrangements made there. The process control operating systems of the components involved must therefore include an effective service for this purpose. Process control systems with the option of selective state images can generally be upgraded to large multi-workstation installations without difficulty.

The second possibility is change-driven transmission. In normal operation, many states undergo changes infrequently or not at all. This leads to the possibility of transmitting only when a state has changed significantly. The number of data to be transmitted under normal conditions is drastically decreased. The detection of state

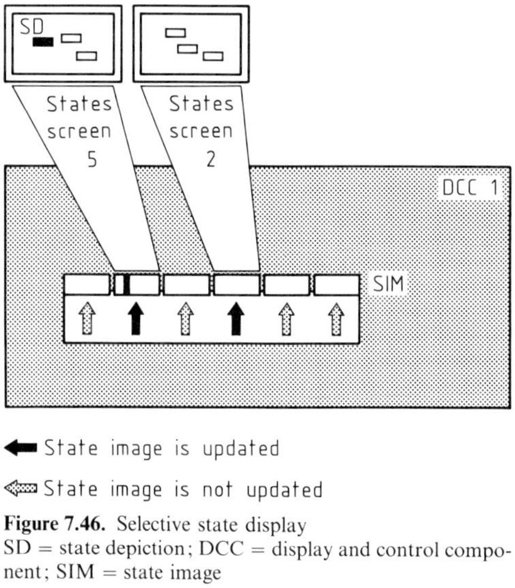

Figure 7.46. Selective state display
SD = state depiction; DCC = display and control component; SIM = state image

changes in the source system presents a problem. Any change of state in a binary value is significant, but in the case of an integer or floating-point value, a change is only significant if it exceeds a certain set distance from the old value. Detecting deviations leads to a perceptible load on the source system. A further drawback to change-driven transmission is that load fluctuations are difficult to calculate. Large numbers of switching operations take place in the plant precisely under critical conditions, so that there are simultaneous changes in nearly all process parameters. A change-driven system must be designed for these extreme peak loads.

Operator control refers to the specification of values by an operator via the system's own human–machine interface. The following units are involved:

- The operator, who communicates with the operator control system through the input device.
- The operator control system, which handles the selection and execution of instructions.
- The instruction acceptor, which checks the value entered. If the result is positive, the value is accepted and furnished as a new effective value to the device addressed.

Each operator console has its own operator control system, which functions independently of the others. The topology of the operator control system is shown in Figure 7.47.

An operator intervention is generally distinguished from a design intervention. The plant operator operates while the design engineer designs. The distinction is that an operator control action always communicates with an instruction acceptor (as an existing entity). Design (configuration) functions define, delete, and manipulate all possible types of objects. Linguistic usage aside, the actions of the designer can also be understood as "operating." Design functions are initiated and, on the basis of the operator instructions, cause appropriate object manipulations to be performed.

The sequence of events in a typical operator action is as follows:

Selection. The operator action is generally initiated from an operator window, which provides the operator with all information needed for operation, such as the current value of the variable to be manipulated, local enabling of operator control, control limits to be maintained,

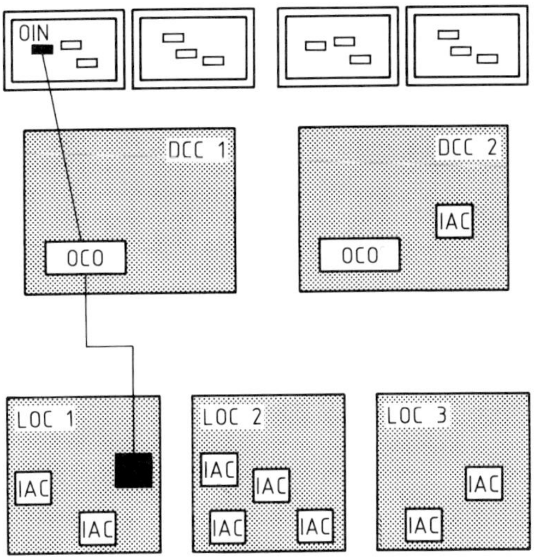

Figure 7.47. Structure of the operator control system
OIN = operator input; OCO = operator control; IAC = instruction acceptor; DCC = display and control component; LOC = local component

or process states relevant for operation. The structure of the window and the information furnished in it are governed by monitoring functions. An operator window should be openable regardless of operator authorization. The variables that can be manipulated by the operator are indicated in the window (e.g., shading as in a pull-down menu system).

The first step in the actual operator action takes place when the operator selects a variable or variable group for manipulation. The operator control system checks whether the selected variable is manipulable and whether the operator is permitted to manipulate it. Several options are available for the nature and syntax of the selection. It can be effected, for example, by entering a command on the keyboard, or by clicking on a symbol marking the variable. Once a selection has been made, the operator control system is in instruction mode. Now the operator can identify and manipulate one of the selected variables or can terminate the control action. In general, concurrent selections of an instruction acceptor are permitted. The instruction acceptor does not know whether it is selected or not in an operator control system. The instruction acceptor is not occupied by being selected. The same variable may be concurrently selected for manipulation in several operator control systems.

Input. The control value is input in a specified input syntax. A wide variety of possibilities exist, depending on the type of input device, the type of value, and the control philosophy employed. Some possibilities are as follows:

- Input the value on the keyboard and press ENTER
- Click on the appropriate symbol to change a binary value
- Simultaneously press the command key and the release key (two-hand operation)
- Press an UP or DOWN arrow key to increase or decrease an analog value
- Move a slider

All these actions ultimately constitute a control instruction in the operator control system. Once the input is complete, the instruction is sent directly to the instruction acceptor concerned.

Control Instruction. The operator control system transmits the control instruction to the instruction acceptor, which may be located in the same component as the operator control system or in another component of the system. If the operator control system does not succeed in properly transmitting the instruction to the instruction acceptor, the manipulation is considered not possible or rejected. If the instruction is properly transmitted to the instruction acceptor, the acceptor begins to process its methods. First it checks whether the manipulation is permitted from the standpoint of the application. During the processing of these acceptor methods, no other control instruction can be transmitted to the acceptor. Although such a collision is extremely unlikely, the process control OS must exclude this possibility.

Processing of Instruction Acceptor Methods. Data security in a process monitoring and control system requires that the instruction acceptor have its own methods, which it uses to check whether it will accept a control instruction or not. System solutions in which the effective value in the instruction acceptor is directly overwritten by a control action should be avoided.

In the context of the connector model, an instruction acceptor is an input connector of a write connection. The events at the instruction acceptor can be described as follows in this model: The connector consists of two parts, a system unit and an application unit. The control instruction is first received by the system unit, which checks whether the connector is enabled and whether the instruction is formally correct. The system unit then starts the application-specific method component of the instruction acceptor.

In this section, customized for the application, a check is made to determine whether the specified value is or is not permitted on the basis of the instantaneous state of the application. If the value is acceptable, it is taken over as the new effective value; if not, the old effective value is retained. The execution of these steps is recorded in a message stating that a manipulation was attempted and whether it succeeded or failed.

Reaction on the Part of the Process. The operator control function described here is strictly limited to the transmission of the control instruction. The consequences of such an instruction have nothing to do with the control function itself. Once the control instruction has been entered in the instruction acceptor, the control action is terminated. The process consequences of the control instruction can be tracked by, for example, the monitoring system.

An important point for the operation of a process control system is that the operator always be notified of the acceptance or nonacceptance of a manipulation. A control instruction must therefore be processed and passed on to the instruction acceptor immediately after its input. Processing in the component that houses the acceptor must be interrupted so that the checking methods of the instruction acceptor can be processed. The total time between the end of input and the report from the acceptor must be no more than 2 s, or 4 s in the worst case.

In order to maintain this timing, many systems simply acknowledge the input directly from the operator control system, only supplying it to the instruction acceptor later. Such solutions should always be rejected. They give the operator a false picture of the facts, and if difficulties arise, they result in a loss of confidence in the correctness of the status display.

Message Handling. "Messages", in general terms, are explicit reports on important events. Not every change of state in the system leads to a message.

Among the purposes served by messages are the following:

- To inform the plant operator about events in the process
- To provide the maintenance personnel with information about malfunctions and changes of operating status in the plant
- To furnish the production manager with key data on the production process
- To document the process in the plant

The following restrictions apply:

- Messages do not serve to replace records of status, switching logs, and detailed event reports needed for event analysis.
- The recipients of messages are always special message system units. Messages do not have the task of notifying other applications that an event has occurred so that automatic control processes can be initiated.
- The message functions do not update the state image in the display and control components during startup or operation.

The functional unit in which a message is generated is referred to as the "message source". The inclusion of a message source establishes that a certain event is so significant that a message is to be transmitted when it occurs. Two distinct concepts are employed: decentralized (distributed) and central message generation.

Decentralized Message Generation. In the decentralized message generation, the message sources are built into the lowermost functional level at which the significance of a signal can be determined. Message sources can be integrated directly into the process functional units. Examples include the following:

- Malfunction of a motor interlock: Interlock malfunctions are generally classified and reported as significant events. A malfunction of the interlock is detected in the motor control and can be reported directly from this unit. In other words, the functional unit "motor" acts automatically and independently to generate an appropriate message.
- Channel failure in a module: Channel failures are always reported as control hardware malfunctions. An intelligent module can independently generate a message if such a malfunction occurs.
- Arrival at a mark in a recipe sequence: When a significant mark is reached in a recipe operation (e.g., "preliminary reaction completed"), the recipe control unit can automatically generate an operational message.

Under a concept of this type, messages are created in all local functional units as well as in various modules and components. The messages must first be conveyed securely to the respective message receivers. In the distributed concept—and this is an important advantage—message transportation can be effected by system services.

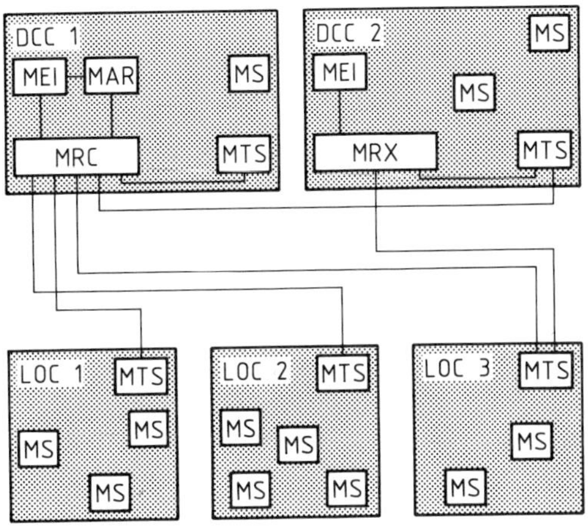

Figure 7.48. Structure of the message system in a system with decentralized message generation
DCC = display and control component; LOC = local component; MAR = message archive; MEI = message interpretation; MRC = message receiver (central unit); MRX = message receiver (auxiliary unit); MS = message source; MTS = message transmission receiver

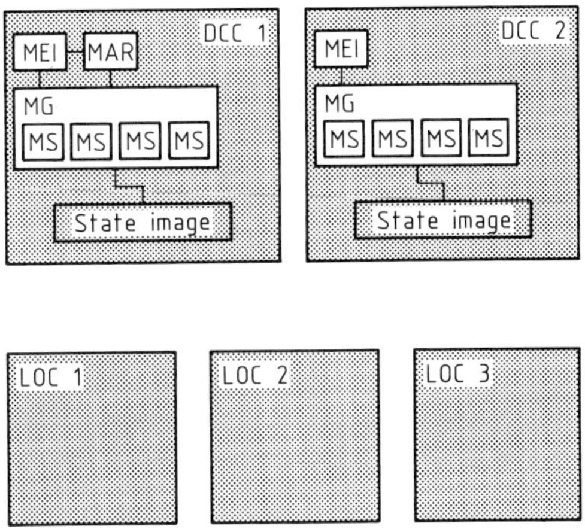

Figure 7.49. Structure of the message system in a system with centralized message generation
DCC = display and control component; LOC = local component; MAR = message archive; MEI = message interpretation; MG = message generation; MS = message source

Figure 7.48 illustrates one configuration. Messages are always generated in message sources and locally delivered to a message transmission server. Each logical unit has its own local message transmission server. The task of this device is to collect the messages from a component, hold them in interim (buffer) storage, and transmit them by secure means to designated message receiving and archiving units.

Centralized Message Generation. A concept based on centralized message generation is shown in Figure 7.49. Messages are generated exclusively in the central component, where message generation, message archiving, and archive interpretation are generally combined.

The centralized concept is based on a different information model from the decentralized one. In the distributed concept, not all events must be reported to the central units via system communications. For example, a change of state of the control valve of a two-position controller is a normal operational event of secondary importance. This switching operation need not be reported and documented as an event. In the decentralized concept, reporting of the state (instantaneous valve position) for representation in an image is handled by services independent of message transmission.

The centralized concept links state and event reporting. The central component usually maintains a 100 % image of all required states for this purpose. When an event occurs in a central component, the state image is updated to reflect the state affected by the event. A check is run in parallel fashion to determine whether the event is one that must be reported. One problem in the centralized message concept is the management of information between multiple central stations. For every message, there must be no more than one message source in the entire system. The acknowledgement status is managed at this message source. Regardless of where it takes place, every acknowledgement must lead to resetting of the acknowledgement request, so that the acknowledgement is central for the entire system. In this respect, many control systems with centralized message generation (in particular SPC-based systems) do not have a clean structure. They do not support the sharing of message functions among central components and do not offer the possibility of central acknowledgement.

Content of a Message. The following information must be contained in a message:

- Who sent the message? Each message contains the identification of the message source and of the process unit in which the event being reported took place.
- When was the message sent? Every message contains a timestamp giving the time when the event reported took place.
- What was reported? Every message contains a precise description of its meaning. This de-

scription can be a plain text or a pointer to a plain text.

- How important is the message? Each message has a certain significance class, for example alarms, warnings, out of tolerance, sequential event report, and so forth.
- What change of state occurred? A message is sent whenever a change of state occurs in the message source. In general, a message contains the state in effect after the change and the nature of the change. Types of changes include, for example, "incoming", "outgoing", "pulse", "acknowledged", and so forth.

Along with the message source ID, the timestamp, the state, and the type of state change must be entered as instantaneous values at the moment when the message is created. Other information can be appended to the message later. The key here is the message source ID.

Message Transmission and Message Distribution. The task of the message transmission server is to receive, buffer, distribute, and transmit messages to the proper receiving and interpretation units (see Fig. 7.48). Message transmission servers must therefore insure that no message gets lost, even during a flurry of messages.

Message distribution can be described by the following general picture: The message receiving unit logs on to the message transmission servers every time a message is received.

There is one "main" message receiving unit. Other "secondary" message receiving units, for certain message groups, can log on to the message transmission servers in parallel. If the main message receiving unit fails, the message transmission server attempts to buffer messages until a replacement unit has logged on as the new main receiving unit.

Message Receiving and Archive Units. The task of a message receiving unit is to receive incoming messages and enter them in the logged-on archive. The most important archive is the sequential event archive.

All incoming messages are chronologically entered into the sequential event archive. Figure 7.50 shows a sequential event display, in this case an excerpt from the last messages entered in the archive. From the sequential event archive, messages can be grouped under a wide variety of criteria, for example sorted by charges, plant sections, significance classes, time intervals, and so forth, or by combinations of such criteria (see Fig. 3.26).

Besides the sequential event archive, there is usually a list containing all message sources that have generated current messages. Figure 7.51 illustrates such a list. A message source appears on this list (it is said to be "signaled") if either the event-generating bit is still set (value still out of range) or acknowledgement has not taken place yet.

May 28, 1991 — Sequential event display, Group V13RA4 — 16:54:05

Time	Unit	Text	Class	Type
9:41:23	V13RA4T3	temperature high	A	incoming
9:41:33	V13RA4T3	temperature high	B	acknowledge
9:43:22	V13RA4L2	level high	W	incoming
9:43:31	V13RA4L2	level high	B	acknowledge
9:44:38	V13RA4T3	temperature high	A	outgoing
9:44:46	V13RA4T3	temperature high	B	acknowledge
9:46:65	V13RA4L2	level high	W	outgoing
9:47:41	V13RA4F6	flow low	A	incoming
9:47:53	V13RA4F6	flow low	B	acknowledge
14:34:23	V13RA4K1	interlock malfunction	A	incoming
16:33:06	V13RA4F7	invalid measurement	A	incoming
16:53:53	V13RA4S1	please confirm recipe data	I	incoming

Figure 7.50. Sequential event display

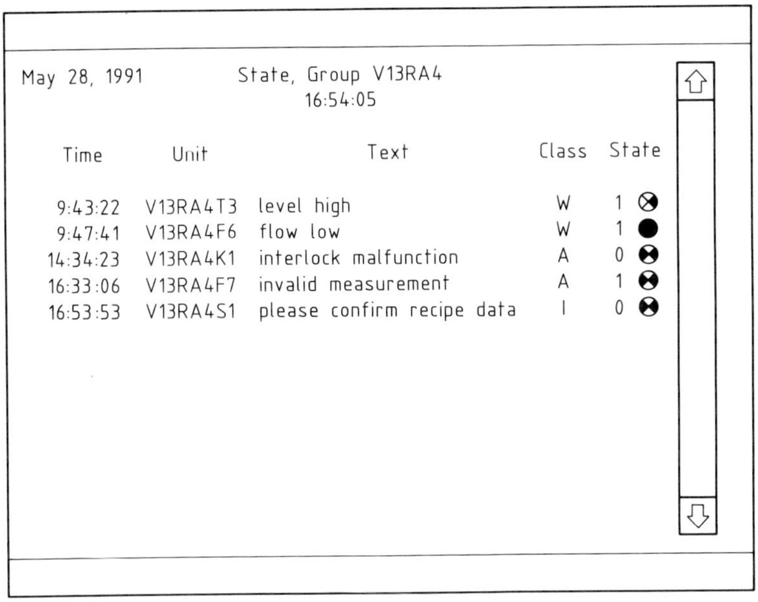

Figure 7.51. List of message-originating sources

Archiving. If processes are to be studied in retrospect, certain events and the time variations of certain states in the system must be archived. The resulting data can then be accessed by interpretive functions. The outcome of an interpretation can be output in the form of, for example, plots or lists. In modern systems, a strict distinction is observed between the data stored in the archive, the interpretive function per se, and the form and representation of the printout. While data storage in the archive is structured in keeping with considerations internal to the system, the user can define what the interpretive functions are to do and in what form the reports are to be output. Archiving presents a number of problems that can be solved only with the aid of special system services. Examples are:

- To append on-line process data quickly and reliably
- To carry out informational functions in parallel with the continuous addition of new data
- To safeguard the data automatically
- To implement the archiving functions in redundant fashion (see Section 3.2)

Time Synchronization. Time (clock) synchronization is an essential system function in a distributed process control system. If events are to be listed in time order, the messages must be timestamped as near as possible to their place of origin. This may be in a local component or may be in a module where, for example, a control hardware malfunction is detected. A sequence of two events in two distinct system units can be correctly detected only if the clocks in the two units "tell the same time". Synchronization insures that the time discrepancy over all clocks in the system will fall within a narrow time interval. A typical value is 10 ms; specially equipped systems achieve discrepancies of <2 ms.

The time is absolutely specified by a central system clock. Additionally, radio clocks are employed as system clocks, with the advantage that time changes (e.g., from standard to daylight-saving time) are performed automatically.

Process Control Hardware Diagnostics. The diagnostic functions monitor the operation of the control hardware and include tools for analyzing the state of the monitoring and control equipment. System diagnostic functions fall into three groups.

The first group includes *control hardware monitoring functions*, which are integral constituents of the individual control system modules and functional units. These functions run automatically and continuously. The result of each check is written into the disturbance status of the unit concerned and can be read from there. If the trouble status of a unit changes, the unit automatically sends a hardware message. Hardware messages get special treatment in many systems, although this is not really necessary. That a message is a hardware message can easily be determined, either directly by examining the message source ID or indirectly by means of a special flag in the message. In this way, control system hardware messages can be treated in the same way as process messages. Typical hardware messages include:

- Failure of an I/O channel (alarm)
- Failure of a redundant bus connection (warning)

The second group takes in *diagnostic functions* of interest to the *designer* or to the on-site *technician*. These have to do with special system analyses that are initiated by the operator only as needed. The functions are supported, in many cases, by external diagnostic equipment. Examples from this group are functions for analyzing the following (and comparable functions):

- The current configuration
- Memory utilization and processor load
- Run-time measurements

The third group consists of *diagnostic functions* used only by specialists at the *manufacturing firm* which permit detailed analysis of system behavior.

7.5. Design and Commissioning

Before a plant is operated, it must be designed and commissioned. The engineering work necessary for this purpose is a key item in the total cost of a process control system. The distribution of the work varies widely. In a decentralized process control system with a well-designed OS and the proper design and commissioning support, pure system procurement costs are relatively high but design and commissioning costs are low. In a "cheap" system (most SPC-based systems fall into this category), the hardware is favorably priced but design and commissioning costs are substantial.

The discussion in this section is based on the structural model of design and commissioning shown in Figure 7.52.

Design. The design process begins with the technological task statement. In practice, however, all that is usually available is a piping and instrumentation diagram. Informative documentation on the functional relationships is either nonexistent or, for the project team at least, difficult to access. Functional data are often found scattered throughout the process development reports or in descriptions provided by the equipment fitter. Accordingly, one of the most important tasks for the project team is first to gather together all such information, develop the functions, and describe them in a formalized way in the project specification (see Chaps. 2 and 10). This formal summary may well take into account the structure of the later implementation. For process control, this means that the specification takes the form of not a list of requirements, but a role system with certain functional properties. This role system can be refined stepwise and ultimately can be employed as the basis for the implementation. Thus the process control specification can be expanded later, thus becoming the documentation of the process control system.

An important advantage of this approach is the systematic procedure. Each function discussed in project conferences can be directly incorporated into the role system as a property. As a result, efficient design work with few errors is possible, and the operational aspect of the procedure is implementation-oriented.

In what follows, the design steps required in such a procedure are presented in detail in terms of the structural model of Figure 7.52.

Definition of Technological Module Types Required. The definition of a suitable collection of module types generally precedes the design work. Industry and corporate standards apply in many cases. The GMA is now working to define a standard set of module types, and the IEC has recently taken up this project.

The following criteria must be considered in establishing the module types:

- Manageable functional scope.

Functional modules should be neither too small nor too large in terms of functional scope.

Small modules permit very flexible structuring of functionality and thus can be used very generally. The disadvantage is the large number of small modules and connections required, resulting in increased design and object management costs. The error rate increases, and the efficiency of processing falls off drastically when the system is later mapped onto the functional system. Another drawback is the low degree of functional standardization that inevitably results from the element types defined.

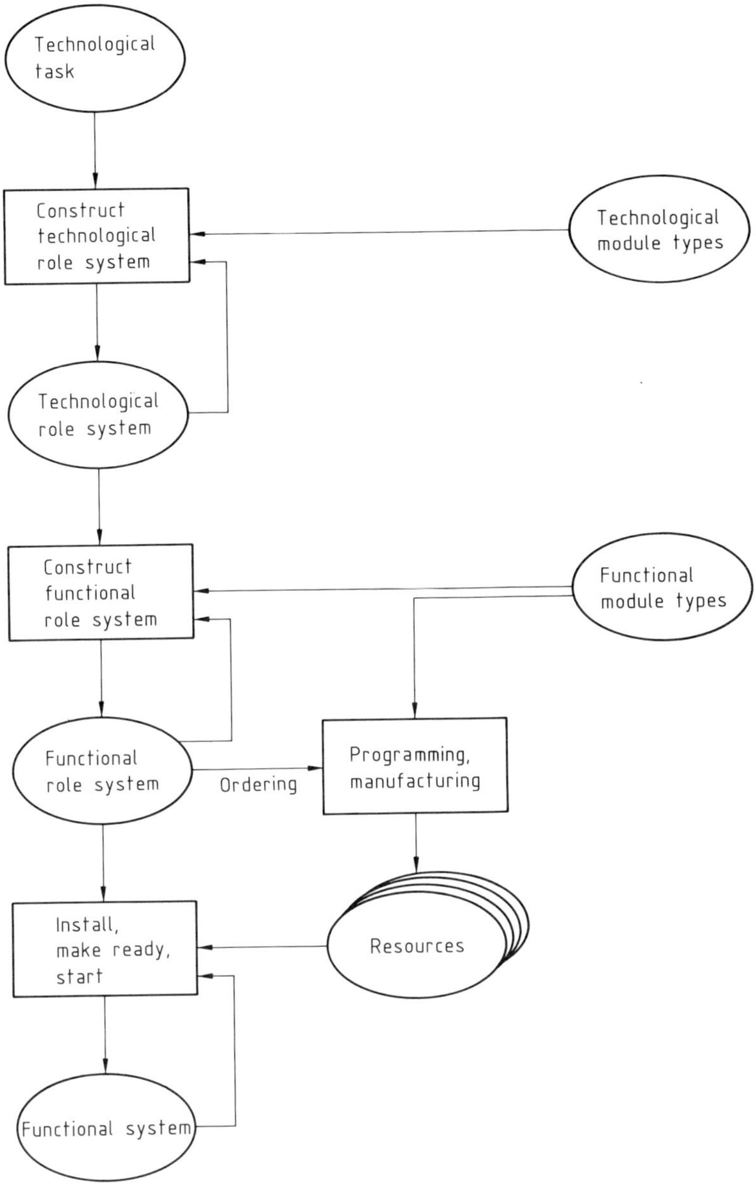

Figure 7.52. Structural model of system design and commissioning

Large modules permit fairly standardized structuring with a few modules. Large modules can be efficiently processed in the later mapping in the functional system. The disadvantages of large modules are the cost of defining them, their functional rigidity, and the possibility of an excessively broad spectrum of models. If large modules are to be used, a consensus must be found as to their functionality. Because these modules embody technological standards, this is a much costlier operation than for generic small modules. To cover all technological cases, either the modules must be made very extensive, with many selectable and parametrizable options, or a great number of types must be defined for fairly similar technological tasks. (What functions must a motor control module contain, and how many distinct motor control modules are desirable?)

- Feasibility of using functional module types.

It makes little sense to define technological module types such that their later mapping onto functional module types is difficult or impossible. It is therefore essential to check, while fixing the technological module types, how they can be realized in various systems.

Configuration of the Technological Role System. From the spectrum of types, the required modules are identified, instantiated, connected, and parametrized. The result is a technological role system in modular form. In accordance with a structure such as the hierarchical structure of Section 4.5, this role system is broken down into individual function, group function, plant coordination, action, and other levels. This technological role system comprehensively and formally reflects all requirements derived from the technological task statement.

The documentation of the technological role system can be regarded as a formalized specification. In addition, it directly serves as the basic documentation for further design work specific to the process control system. The technological role system itself should be formulated in a manner largely independent of the control system (see also Chaps. 2 and 10).

Definition of Functional Module Types. Along with the firmware or special-purpose software, most manufacturers of process control systems supply a collection of functional module types. If the user remains with these prefabricated types, the complicated procedure of creating customer software module types can often be dispensed with. Indeed, many process control systems do not allow user programming of software modules.

Thus it is chiefly the manufacturer that defines the functional module types. The user must manage with these manufacturer-specific collections of types. Only in special cases can a competent user define and program new sets of types; such a user will attempt whenever possible to create software module types that exhibit a 1:1 correspondence with the functionalities of the technological module types used. Otherwise, the manufacturer must attempt to satisfy user requirements.

Configuration of the Functional Role System. The functions embodied in the technological role system are transformed into software structures specific to the control system. In this process, the structures of the technological role system are preserved in unchanged form as far as possible. A frequently used aid is the creation of "typicals."

Typicals are pre-assembled structures made up of functional module types. They are not types but structures having special naming conventions and which can be easily copied in the design interface. Typicals can readily be put together by the user. If typicals are created that functionally correspond to the technological module types, the functional role system can easily be configured with the aid of these typicals.

The configuration of the functional role system falls into two parts:

- Mapping the technological structure onto the functional structure specific to the control system
- Establishing the locations of individual functional units

"Location" here refers to the logical location of a unit. For a module, for example, this is the logical card slot; for a software module, it is the processor board in which the module is run.

Once the network of software modules has been defined, it is therefore necessary to establish which modules will be run on which processor boards. The component and module structure of the process monitoring and control system must be known, at least in outline, before this step can be done.

The result is the finally configured functional system. In a design system, functional units are handled as objects; they can be represented and documented in accordance with certain criteria and be loaded directly into the functional unit for execution.

Commissioning (see also Sections 10.5 and 10.6). Commissioning includes all actions required to bring a system or system unit from an initial state to an active working state. A single functional unit, a group of functional units, a whole system component, or a complete system can be commissioned. In defining what will be regarded as the initial state, one commonly takes an organizational (management) view, defining the initial state to be whatever state the commissioner finds when the operation begins. In a process control system, for example, the initial state is the system fully installed but not yet energized. When a compact device is purchased, commissioning starts with unpacking. Certain simple installation tasks may thus be included in commissioning.

From a systems engineering standpoint, commissioning is the last step involved in making the functional system available for service. The functional system includes all subsystems needed for active implementation of production operation.

Availability of a Functional Unit. The functional system is the system that actively realizes production operation. Its units fulfill roles whose interplay results in the desired system functionality.

The separation of role systems and functional systems is of central importance for an understanding of commissioning.

A functional role system is set up in such a way that for every role there is a functional unit that plays this role. For a functional unit to carry out a role, it must be allocated to this role and must be made available at the defined location of the role. This part of the operation is called "installation".

Figure 7.53 illustrates the availability states of a functional unit along with the transitions between these states.

- Not present: There is no suitable functional unit at the location. For example, the card is not plugged in or the software module is not loaded.
- Install: Installation means making a functional unit available; for example, assembling a cabinet rack, plugging in a card, or loading a software module. There are two key events in installation: when the functional unit is physically present at the location, and when the functional unit has been configured for the task it is to perform. The order of these events is not significant here and can be determined case by case. For example, in installing a card, the jumpers are placed first and the card is then plugged in, whereas for installing a software module entity, the entity is created first and only then are values assigned for the parameters.
- Present: The functional unit is said to be present after installation; i.e., it is finally configured for its role and is available at the location designated for the role.
- Make ready: The state of the object after installation, for example of a newly loaded software module, is generally undefined. A special function called "make ready" must be carried out in order to bring the object to a well-defined "ready" state adapted to the instantaneous system state.
- Ready: The availability state "ready" means that the unit is not only present but also ready to start. In other words, at this instant all conditions exist for connecting the unit actively into the network. The "ready" state plays an important role specifically in distributed process control systems. Here the situation often arises where certain units have already been activated and begun carrying out their tasks. A further unit is now to be connected into this running structure. Naturally, such a connection must not be made except under certain conditions. The "ready" state indicates that these synchronization conditions exist at this moment and the unit can be connected into active operation.
- Start: From the "ready" state, a functional unit can be started. It then goes into the "operating" state.
- Operating: In this state, the unit is effective in the functional system as an active element. It is performing the role intended for it.

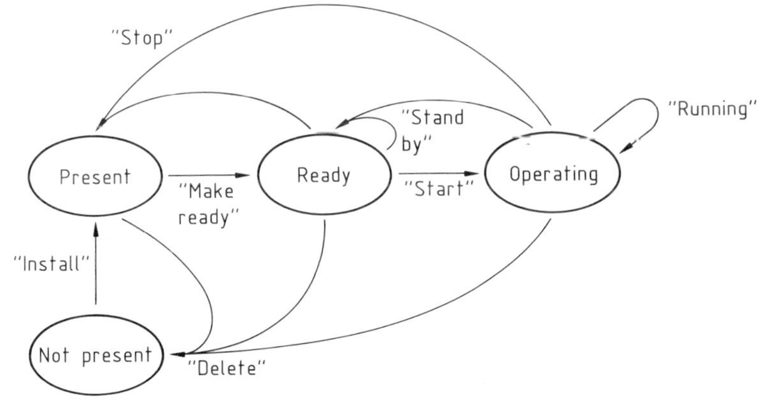

Figure 7.53. Availability states of a functional unit

A typical situation in a distributed process control system is where the functional systems are operating in certain regions and not in others (see Fig. 7.54). The process control system is purposely designed so that functional units can be handled in a "granular" fashion. If a functional unit is not operating, no role that is performed by this unit can be realized. In the example shown, modules C and E are not operating; modules A, B, D, and F are operating. To the extent allowed by disrupted communications relations I, II, III, and IV, the operating modules are fulfilling the role intended for them.

Hierarchy of the Functional System. The functional system of a control system should be regarded as having a multilevel structure. Each level satisfies the role requirements of the next higher level.

The structuring and implementation of a functional system are always carried from the bottom up, starting from the physical hardware units. This strict bottom-up principle underlies all commissioning and startup operations.

Figure 7.55 illustrates a typical functional system structure for control system hardware. Each functional unit can operate only when all functional units are operating in the lower levels on which it is built. If a functional unit fails, no functional unit in any higher level that requires the failed unit for the execution of its role will remain in the "operating" availability state.

- Hardware level: The hardware level furnishes devices, modules, displays, cables, and so forth as functional units. The lowest level of an industrial functional system is always the physical level. Without a physical basis, no industrial functional system can operate.

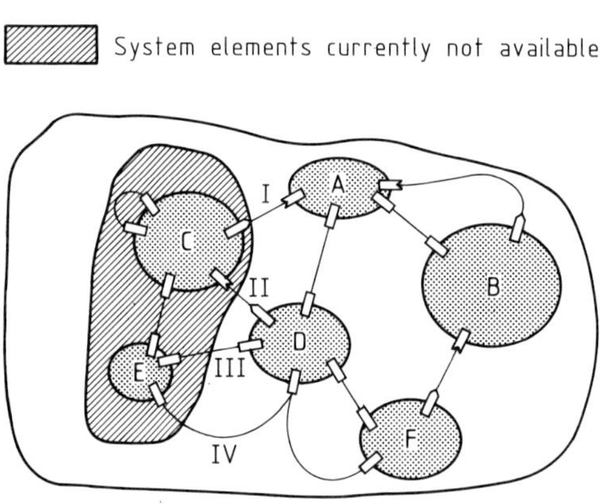

Figure 7.54. Functional system with several unavailable units

- Base system level: The base system level provides the base operating system, the required databases, and other standard tools (e.g., standard communications tools) as functional units. The control system functions are built up as actively available units from these standard tools.
- Process control level: The process control operating system level includes all functional units needed for the management and executive control of technological functional units. Examples from this class are system functions such as module processing, module-to-module communications, message generation, operator control, display, archiving, and so forth.
- Application level: The application level implements the process control functions on the

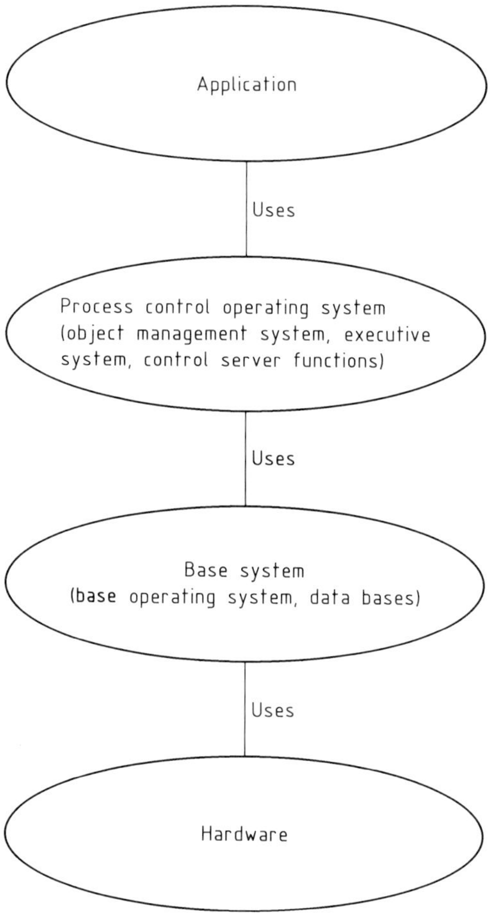

Figure 7.55. Hierarchical structure of the functional system

basis of a functioning process control operating system. If process control is structured in a hierarchical fashion, the application level itself can be regarded as a hierarchical system of process control functions, one built on another. Commissioning of the process control system based on operating software modules is a special aspect that will not be discussed here. From the standpoint of the process monitoring and control hardware, the application level consists of a flat network of intercommunicating modules.

8. The Process Control System and Its Elements: Information Logistics

8.1. Principles

The need for information integration in production companies is generally acknowledged today, but this effort is subject to two kinds of limitations. The first are due to historical circumstances surrounding the introduction of computer-aided information processing ("island" solutions); the second have to do with the heterogeneity of the computer systems employed, which in turn stemmed from functional requirements. The starting point for solving the associated problems is the derivation of the requisite integration functions along with their scope; the process is aided by proper interface definitions and standards. Two worlds are involved here: the world of production engineering in the definition of what functions are needed, and the world of information and data processing in the implementation of these functions. Informatics is called on to create suitable methods and tools for the second range of problems. All experience to date shows that optimal solutions demand close cooperation between the two fields; this necessity represents a special challenge because of, among other factors, difficulties in understanding functional requirements and differences in terminology.

The new field of "information logistics" will contribute to an integrated view of methods and tools for information distribution and processing in production plants. By analogy with the logistics of materials, energy, or personnel, information logistics combines methods that are concerned with the modeling, distribution, storage, and retrieval of information. The new field must identify the interrelationships of these individual disciplines with regard to the tasks of information processing in production and, with the aid of tools, must put them to work in optimizing information-logistic systems. The methodological edifice of information logistics is not yet complete [8.1], [8.2], but separate disciplines do already exist in distinct form. One such is distribution—the "communication" of information within the company—which is treated in depth in this chapter.

8.2. Functional Structures and Information Flow in Production Companies

For a better understanding of what requirements manufacturing information systems must meet [8.3], [8.4], it is desirable to work out models of information flow in the company. If such information flows in production companies are examined more closely, an information hierarchy such as the one shown in Figure 8.1 is revealed [8.5]. The subdivision of production companies into various control and management levels arises from the way corporations are organized, and so the details will differ from one industry to another [8.6]. A consistent feature, however, is that the lowest level of the structure includes many single items of information about

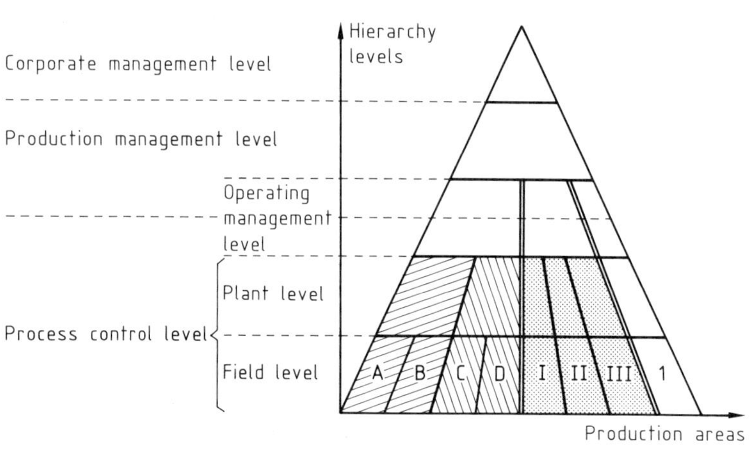

Figure 8.1. Levels of control in production

processes, while this information is more and more strongly condensed in the higher levels (all the way up to corporate management). Conversely, the objectives of the higher levels of management and control give rise to items of information about production steps or stages at the lower levels. Figure 8.1 shows that these processes can extend to various levels and take a variety of forms. Again, structural differences between industries and companies are a factor.

The subdivision into corporate, plant, and process levels (Fig. 8.1) is specific to the process industries, because here it is often the case that independent facilities on the process control level are supplied with production specifications (e.g., recipes) directly from the production management level. In other industries, such as iron and steel, the products move in sequence through a number of plants or units where the raw material is upgraded, step by step, and ultimately converted to a salable product. This production structure is advantageously described by introducing an extra or intermediate level of management or control. This intermediate level breaks the general instructions from the production management level down into detailed instructions for the individual production units.

This structure implies that information flows both vertically (between control levels) and horizontally (within each control level). Both the transport and the storage of information reflect these information streams. Information logistics has to supply methods for the modeling and optimization of communications and information storage systems.

In general, the various levels employ disparate computer systems for information storage and processing. Information logistics, however, demands suitable high-capacity information pathways within and between the levels of control. It should be pointed out that information logistics is not just one of the tasks involved in information integration for production companies; it is a methodological aid to the planning and implementation of a corporate information utilization policy. Data communications and storage facilities do not become information communications and storage facilities until the meaning and utilization of the (digital) data have been established.

At first, information integration across the control levels of Figure 8.1 was thought of chiefly as a problem in manufacturing engineering. One indication of this is the wide adoption of the term CIM (computer-integrated manufacturing). Nevertheless, process engineering requires the same kind of information integration and has a functional structure wholly congruent with Figure 8.1. The term CIP (computer-integrated processing) is seen more and more often in this connection, showing that information integration is increasingly regarded as vital in both branches of production. It is therefore suggested that CIP be taken to mean computer-integrated production; this term expresses the fact that the same methodological aids will be applicable, and must be available, for information integration in all kinds of production operations. An analysis of the steps involved in manufacturing further shows that there are not only the discrete, piece-by-piece operations but also completely continuous "process-like" operations. If information integration becomes a way for methods of manufacturing engineering and process engineering to come together, this will be in accord with the situation in the plant. The integrative techniques of information logistics can then be put to optimal use.

8.3. Computer Communications Between and Within Control Levels

The need for information communication between the levels of Figure 8.1, as well as the requirements of information logistics, demand that appropriate aids be made available in the production company. As Figure 8.2 shows, the implementation of computer-aided information communication calls for a hierarchy of computer networks [8.5]. The information-handling facilities at each level of control must meet the following two requirements:

- There must be dedicated computers at each level. Information processing for more than one level can be combined in one computer, but the unavoidable increase in system complexity will result in a loss of flexibility, efficiency, and maintainability.
- The communications facilities at each level must be tailored to the particular needs of the level. The communications equipment must also be able to connect heterogeneous computer systems together when a level uses computers of different generations, types, and capabilities because of special requirements, procurement strategies, or historical circumstances.

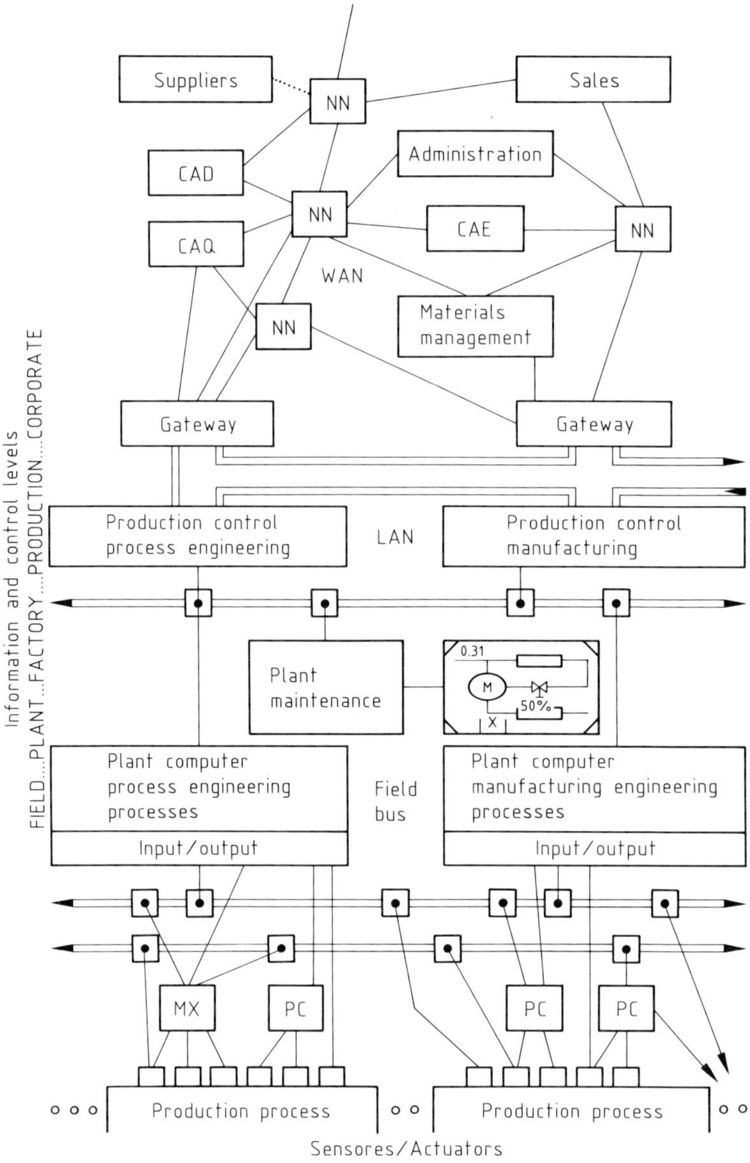

Figure 8.2. Networks for computer integration

The networking concept illustrated in Figure 8.2 is one approach to meeting these demands. The following points may be important in individual cases:

The communications facilities must be designed and implemented with attention to the necessary time and consistency conditions. The effort of determining and modeling such conditions is justified by the optimal economics and flexibility of the communications system.

The communications system must be fault-tolerant as well as safe and secure with respect to intentional and inadvertent falsification of information. The redundancy and coding procedures employed for this purpose should be selected to fit each application [8.7].

Because production facilities are continually being changed, information logistics must be correspondingly flexible. Well-defined and, when possible, standardized interfaces and communications rules (protocols) are needed in order to meet this requirement economically, especially in heterogeneous systems.

Information integration across all levels of management and control is possible only if all information is in digital form.

However, information integration does not mean just data communications between one computer and another. Information logistics, in particular the transport and storage of digital data, must be supplemented by conventions on the meaning (semantics) of the digital data. The producer and consumer, who may be in different areas of activity, must interpret the digital data in the same way; only then do the data become information. Thus, when disparate computer systems are formed into an information network, it is particularly important that protocols be defined and standardized for semantic transition areas between applications systems. Formal methods of syntactical and semantic notation, based on experience and results from informatics, should be fully utilized for this purpose.

8.4. Computer Communications in Industrial Production; Standards

Standards already exist for the implementation of computer networks. These are discussed from a technical standpoint in what follows. The equally important economic aspects of various approaches call for an exact knowledge of each situation and its constraints. Computer-integrated production does not exist as an off-the-shelf solution, but the existence of standard protocols, however, makes it easier to handle specific applications problems.

Communications at the Process Control Level. Plant computers and the technical process often communicate (Fig. 8.2) through intelligent subsystems, for example stored-program control systems (SPS) or field-level multiplexers (MUX). Other direct methods of communication include digital input/digital output (DIDO) and standard serial and parallel buses (SCSI, IEEE 488, RS 232/422/485). Field buses, which permit serial data transfer under field conditions, are especially important and are treated in depth in Section 8.6.

Local Area Networks (LANs) at Higher Control Levels. Local networks have become vital in connecting computers with near real-time functions, such as plant unit, plant, and control-room computers, at the process and plant control levels. These setups allow both protected communication between two partners and unprotected simultaneous communication with more than one partner in the network (broadcasting). Physically, these networks are available based on either copper wire or optical fiber. The standard for levels 1 and 2 is IEEE 802, which has now been adopted by ISO as well. Line access protocols are CSMA/CD and token-passing. The CSMA/CD protocol (ISO IS 8802/3) has found wide acceptance in office systems, and a number of industrial products are on the market. This protocol is easy to implement, but the possibility of collisions means that transmission times are not well-defined. Its applicability under real-time conditions is therefore controversial.

In the token-ring protocol (ISO IS 8802/4), permission to transmit is linked with possession of the "token," a special data structure sent in the form of a telegram. There is no possibility of colliding transmissions. Certain participants can be assigned priorities for permission to transmit through special token distribution mechanisms, so that even worst-case communications times are guaranteed. As a result, the token-ring protocol is particularly suitable for real-time applications. On the other hand, special fault-tolerance methods should be noted, such as the replacement of a lost token, which can markedly increase the administrative effort involved in running the token protocol compared with the CSMA/CD protocol.

Open Networks (WAN = Wide Area Networks). At the production and corporate levels of control, computers must communicate over distances longer than the few kilometers to which local area networks are restricted. Wide area networks (WAN) can link two communications partners over any distance, worldwide. In contrast to the broadcast mode of LANs, connections between nodes in open networks must be explicitly opened and closed. A variety of transmission media are available: all kinds of copper wires, wireless links (radio, satellite, infrared), and optical fibers. Open networks are characterized by the possibility of adding new participants while the network is in operation; therefore, communications standards are especially important in this area. ISO is making its contribution in the well-known OSI model [8.8]. Both LANs and WANs can be employed at all levels shown in Figure 8.1; special applications requirements may make the use of an open network desirable even at the process level of control.

Communication through Private Branch Exchanges (PBXs). The private branch exchange (PBX) has been a common installation in all kinds of companies for decades. For computer communications, the PBX has many of the properties of an open network [8.9]; for example, it sets up a point-to-point connection between two participants. A point to consider is the nonnegligible, and unpredictable, time needed to establish and drop connections, which are thus only partially suitable for real-time applications.

An important development is the transmission of digital data in parallel with telephone traffic, which is still partly analog but will be increasingly digital in future. With equipment now on the market, the Integrated Services Digital Network (ISDN) is limited to 64 kbit/s for the transmission of digital data. The PBX is therefore more suited to the transmission of digital data in parallel with speech at present, and isthus interesting chiefly at the corporate and production levels of control. The introduction of faster call-processing devices (with processing times in the millisecond range) will allow the PBX to penetrate local area network applications in the medium term.

8.5. MAP/TOP: Protocol Standards for Information Integration in Production Companies

Experience with the ISO standard protocols for data communications between computers [8.9] has shown that the standards are not adequate for true information communications. Up to the beginning of the 1980s, level 7 of the OSI model (the applications level) included only fairly general functions (file transfer, virtual terminal, etc.) with no reference to the meaning (semantics) of the transferred data. What was defined was thus tools with which the user could construct an information integration.

While this was a significant advance for computer communications applications, with their initially commercial orientation, the broad functionality and thus the semantic variety of the digital information that must be handled when computers are integrated into production meant that further standardization was needed on level 7 of the OSI model (see Fig. 8.3). This situation has led, since 1980, to proposals for MAP (Manufacturing Automation Protocols [8.10]) and TOP (Technical and Office Protocols). These proposals originated with users of automation and process control systems. MAP was developed by General Motors for purposes of manufacturing automation and, TOP, by Boeing as an approach to automating the engineering office.

MAP and TOP together provide for comprehensive information communications in industrial companies, from the administrative and resource allocation aspect through design and development (CAD) to computer-aided production and quality control (CAQ). Both protocol environments constitute a key tool for CIP (Computer-Integrated Production).

The impetus for MAP and TOP came from users of industrial automation and process control systems. It is in the user's interest to select from the widest possible range of options the optimal solution to each problem while insuring the integrability of the new features. This is the alternative to the "single-source" solution often put together by one computer manufacturer. The interests of the suppliers and users of automation and process control systems are in conflict here: Complete standardization of interfaces for a wide variety of information-processing devices in a production company will breach the protective walls around computer networks, which have commonly been supplied by one vendor. On the other hand, this standardization trend expands the market for each vendor. The prerequisites are precise and strictly consistent standardization of protocols and a user with sufficient know-how and capacity to cope with the interoperability problems.

Objectives and States of MAP/TOP. The full definition and application of the MAP/TOP protocol hierarchy should guarantee direct communication between computers, control units, and their programming systems, even when these are supplied by different manufacturers. The objectives of MAP/TOP take in the following points:

The protocols should allow for the special properties and requirements of local plant and office networks. Further, they should be as compatible as possible with the respective

Application functions and data				
Assembly	CAE	Manufacturing	Quality control	Statistics
CAD	Office	Material management	Process control	Blue prints

Communications standards	CNMA		Testsystems
	MAP	TOP	
7 Application (Standard services)		Message handling system (MHS; CCITT X.400)	available
	ISO File transfer, access and management	(FTAM-subset; DIS 8571)	available
	ISO Manufacturing message service (MMS; DIS 9506)		available
	MAP Directory services (DS)		available
	MAP Network management (NM)		available
	ISO Application control service elements (ACSE; Subset: DIS 8649/1-2 and DIS 8650/1-2		
6 Presentation	ISO 8822, 8823		development
5 Session	ISO Session kernel	(IS 8326, IS 8327)	available
4 Transport	ISO Transport class 4	(IS 8072, IS 8073)	available
3 Network	ISO Connectionless internet	(DIS 8348, DIS 8473)	available
2 Data link	ISO Logical link control (DIS 8802/2); Type 1 | Type 3	Type 1	available
1 Physical	ISO Token passing bus (DIS 8802/4) (IEEE 802.4); Broadband | Carrierband	ISO CSMA/CD (DIS 8802/3); Baseband (IEEE 802.3)	

Figure 8.3. MAP/TOP protocol hierarchy

open network protocols, so that communications between the two types of networks can be simplified (Fig. 8.2).

In contrast to the ISO protocols for open networks, with the few generally applicable functions of applications level 7, the MAP/TOP protocol hierarchy for level 7 includes special applications functions tailored for manufacturing and office automation (e.g., MMS = Manufacturing Message Specification and MHS = Message Handling Specification).

In support of these ambitious objectives, the MAP/TOP definition includes quality-assurance measures such as conformity tests, which insure and certify the ability of protocol products from different manufacturers to communicate with one another.

Thus the objectives of MAP/TOP go beyond those of the ISO open network protocols, taking on the difficult task of identifying and standardizing applications functions in the manufacturing and office area at level 7. This makes it clear why the development of MAP/TOP, especially the applications protocols, is still in flux and will not be completed in the foreseeable future.

Figure 8.3 illustrates the MAP protocol environment as now available. At levels 1 and 2, the LAN protocols discussed in Section 8.3 are standardized. Levels 2 to 6 contain protocol definitions equivalent to WAN protocols (the same for MAP and TOP). The use of these protocol standards at levels 1 to 6 insures data communications between applications functions in production and office.

On applications level 7, three protocol types are distinguished. The first corresponds to the ISO protocols for WANs; these are essentially ISO standard applications protocols such as file transfer.

The second protocol type covers application-specific functions. Examples are the MMS protocol family in MAP and the MHS protocol system in TOP (Fig. 8.3). A third protocol type on level 7 deals with the functions of network administration; examples are Network Management (NM) and Directory Service (DS). These protocols also offer functions for handling errors as well as fault-tolerance practices.

Figure 8.3 shows the intention of the ESPRIT CNMA (Communication Network for Manufacturing Applications) project, a European undertaking that supports MAP and TOP in a similar way [8.11].

While the MAP specification focuses on manufacturing operations, a similar problem is evident in process operations. Proposals for a PMS (Processing Message Service) are therefore coming from a variety of sources. Its functions will correspond to the communications requirements of computers in the processing industries.

Generally, there is the problem of linking various functional environments to the basic MMS standard. Figure 8.4 shows the "Companion Standards" concept that has now been chosen. A standardized MMS core includes generally usable concepts and structural specifications for applications standards that will serve as the basis for a variety of Companion Standards relating to technical specialties. This procedure allows the step-by-step construction and expansion of the applications function standards on the basis of user experience.

8.6. Field Bus Systems

Basis and Objectives of Field Bus Systems. Communication at the field level links field devices with control-level computers. The devices in question include sensor systems that locally acquire complex information about the production process; actuator systems that influence material and energy flows; and programmable (stored-program) controllers that perform local control and optimization tasks. The same holds, with suitable adaptations, for both processing and manufacturing operations.

Distribution, as a basic property of present-day control and instrumentation, extends to information processing at the field level and, to an increasing degree, blurs the distinction between sensors and actuators, on the one hand, and control computers on the other. This means that powerful communications systems meeting a variety of requirements are also needed at the field level. Examples include direct communication between intelligent sensors and actuators in local control loops, as well as communications links used to transmit product data between plant control systems and control computers.

Today, most signal transmission at the field level is by field devices individually wired to the control computers. The standard 0–20 mA and 4–20 mA signals are dominant in process engineering and have become standards in chemical engineering. In manufacturing, sensors and actu-

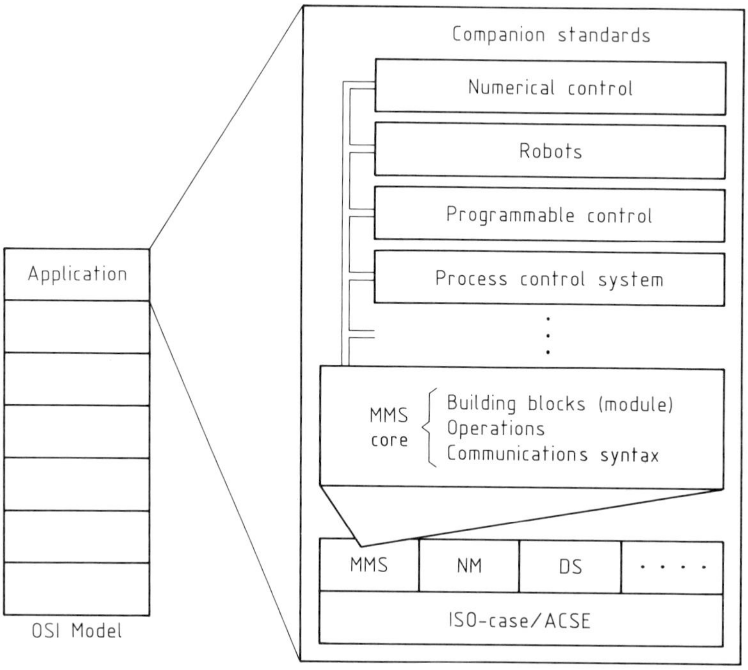

Figure 8.4. Application-oriented Companion Standards

ators on machines are also singly and directly linked to the cell computer or to programmable controllers.

The variety of information involved in modern production processes sets limits to this form of communication with regard to plannability, maintainability, immunity to electromagnetic interference, and economy. Field bus systems should overcome these limitations:

- The star configuration of individual lines is replaced by a digital bus known as the field bus. It is cabled to the various field devices and transmits digitized data via digital interfaces.
- Such a bus connects some tens of field devices among themselves and to a plant computer. If the devices are close together, the cost of cabling is substantially less than for individual cabling (Fig. 8.5). If field devices are assumed to be evenly distributed over a specified area, the savings in cabling costs can be estimated from Figure 8.6.
- The growing intelligence of field devices calls for the transmission not only of process measurements and control information but also of digital status and control data between field devices and control computers. The additional cabling that used to be required for this purpose can be omitted when the field bus is employed.
- The conversion of analog signals to digital locally, that is, at the connection to the field bus, prevents interference to the signal on route to the control computer. The redundancy of digital codes also makes it possible to reconstruct erroneous data.
- The standardization of field bus protocols permits standard information interfaces between field devices and a wide variety of control computers, so that open communication results at the field level. This is not possible at present with the wide range of information transmission media at the field level (Fig. 8.6).
- Diverse media can be used for digital data transmission (e.g., copper twisted pair, coaxial cable, optical fiber, radio waves), so that economic and technical optimization is possible (transmission rate, noise immunity, energy expenditure/intrinsic safety, electrical isolation, etc.).

These objectives and potential features make up the basis for field bus standards now being devised. The many degrees of freedom in the specification and implementation of field buses lead to a variety of approaches and have worked against the desired uniformity in field-level information interfaces.

Field Bus Systems in Development. In recent years, teams in several countries have made a start toward specifying and implementing field bus systems. The properties of an architecture for all these projects is described first. The various approaches to implementing this architecture employed in the several projects are then discussed.

General Field Bus Architecture. A field bus system is a local network having a restricted number of nodes, which communicate exclusively among themselves or with a control computer. There is accordingly no need for the services of ISO levels 3 to 5. Presentation level services are also unnecessary in general form, since they can be performed in special form by the services of level 7 or 2. For open communication between field bus nodes, however, the services of level 7 must be specified. A general field bus architecture thus includes ISO levels 1 (physical layer), 2 (link layer), and 7 (application layer). This reduced architecture must be supplemented by network management services as well as application-specific services on level 7, such as those already introduced as Companion Standards [8.12] in MAP/TOP.

The American standards group ISA SP 50 has defined two performance classes (H1 and H2) for the *physical layer*. The scope of H1 is the replacement of existing analog standard interfaces (0–20 mA and 4–20 mA); the H1 specification must permit adequately fast and error-proof transmission of process, measurement, and control signals. Accordingly, the H1 specification is largely governed by process engineering requirements.

The H2 specification is to extend the H1 concept to applications with higher performance demand. This includes the fast transmission of large data sets, as is needed for the loading of machining programs in manufacturing engineering. H2 is therefore specified for manufacturing.

IEC TC65C has created specifications for a field bus standard for use in industrial control and instrumentation. In addition to the physical layer performance classes already cited, it describes other properties: the topology (linear bus or tree), the transmission medium (twisted pair

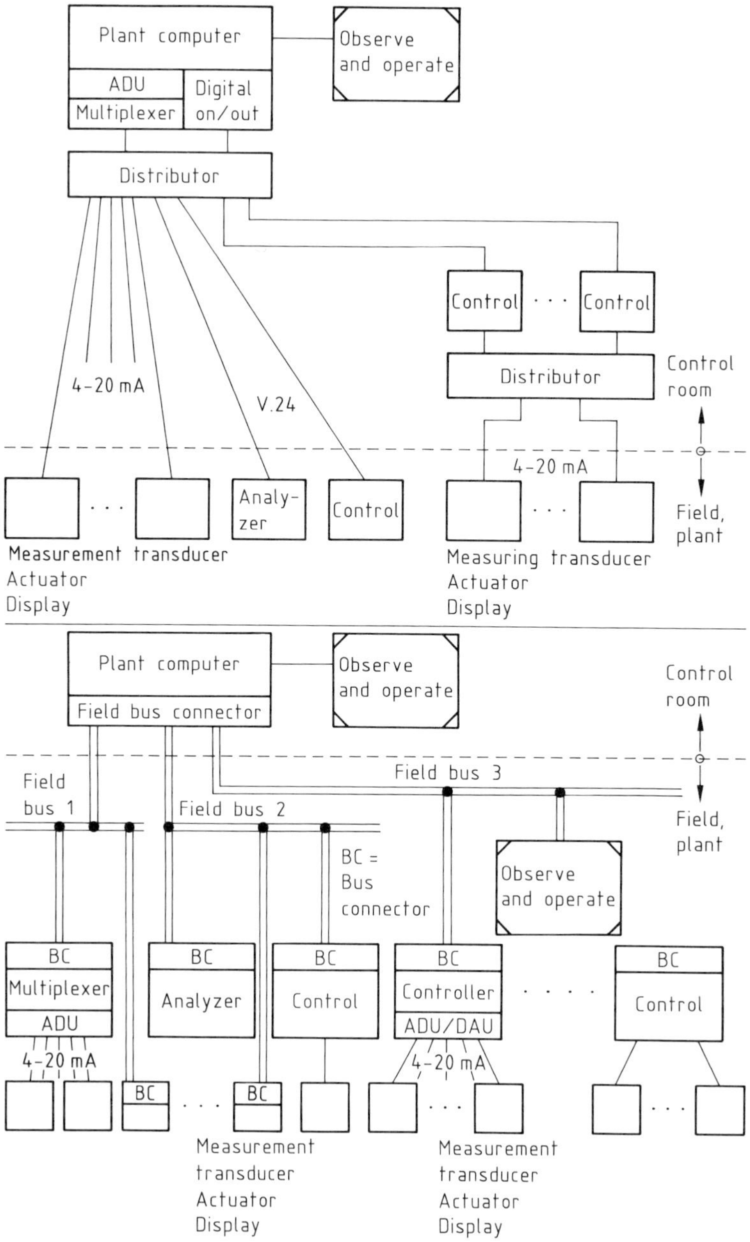

Figure 8.5. Transformation from star-configured individual cabling to field bus structures

or, later, optical fiber or radio link), and finally the number of nodes on each field bus segment (32). For performance class H1, a data rate of 31.25 kbit/s and a maximum cable length of 1900 m are set forth. Performance class H2 has a data rate of 1 Mbit/s and cable lengths of up to 750 m. The intrinsic safety of the physical layer is handled by limiting the electric power that can be carried on the bus. The H1 specification allows enough power to run at least five field devices, at least three of which can operate in a hazardous environment.

The H2 specification provides for at least three field devices to be supplied on the bus and permits AC transmission over the bus lines. Intrinsic safety is not explicitly considered in the

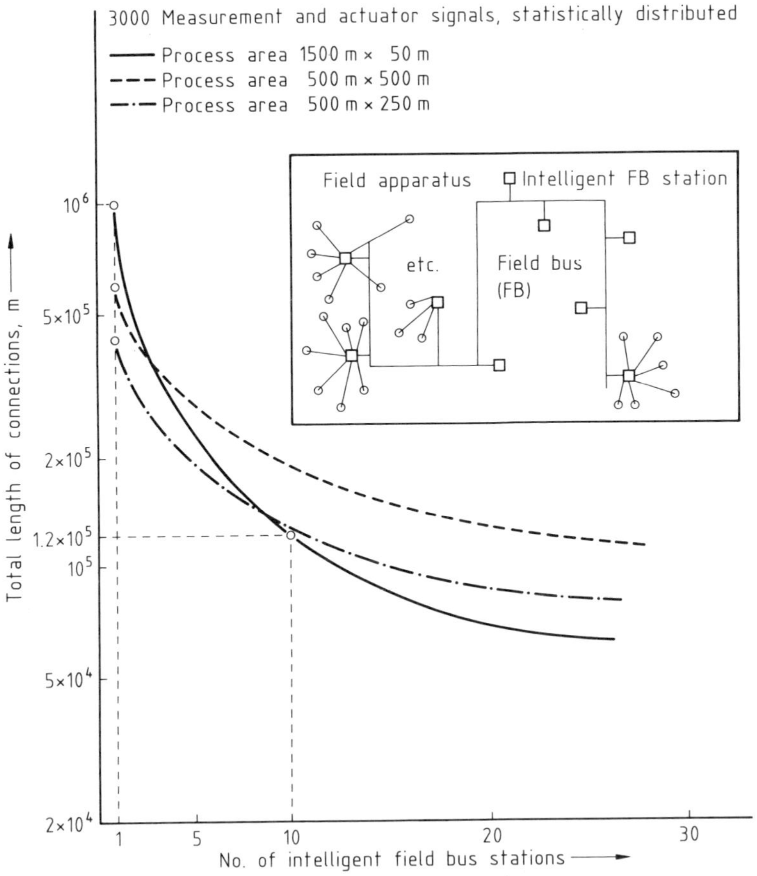

Figure 8.6. Estimation of cabling costs in field bus structures [8.25]

H2 specification, since it is of no importance as a property of manufacturing systems. At present, power supply to field devices, even with auxiliary energy sources, is not a function of field bus systems; over the long term, however, separating the information channel from the (auxiliary) energy channel in process operations is a goal to be sought [8.13].

The specification of the *link layer* in field bus designs corresponds to the ISO/OSI Data Link Layer (DLL). In this way, the two sublayers LLC (Logic Link Control) and MAC (Medium Access Control) are established.

For the application area of field bus systems in general, the *link layer* must provide adequate efficiency (real-time environment) and also support access priorities as well as point-to-point connection or simultaneous connection of one node with several or all other nodes (multicast, broadcast). The essential novelty of the field bus concept as described, however, lies in the procedures by which nodes access the common data transmission medium (MAC). While the French FIP proposal [8.14] features a centrally organized mechanism in the sense of bus allocation, other proposals (PROFIBUS [8.15]) utilize a distributed access mechanism based on token passing as set forth in IEEE 802.4. These differences exist because FIP assumes better real-time properties for the central bus control. PROFIBUS uses standard microelectronics for the efficient performance of the more complex token-passing protocol, which allows more flexible parametrizing of the bus allocation for individual nodes [8.16].

The *application layer* must furnish application functions without regard to the protocols of the lower layers. These functions must insure the exchange of information in a distributed system and with adequate efficiency. Because applications in industry, whether processing or manufacturing, are subject to strong timing conditions, efficiency is particularly important. This holds for cyclic information transmission as well as the detection of spontaneous events. While the functionality of the protocol services on the application layer can be specified independently of the lower protocol layers, their performance under real-time conditions is crucially dependent on the efficiency of all protocol layers.

The application layer for field bus systems must offer cyclic and noncyclic services. Cyclic services are, for example, the reading and writing of process variables or parameters as well as access to information on the validity and consistency of such variables. Noncyclic services include the sending of reports based on an event, the sending of management information, and the initialization or configuration of subsystems during startup of technical processes.

A variety of approaches can be taken to the implementation of such services. The essential difference lies in the relationship between the information-generating and information-consuming processes. For example, client/server models as well as producer/consumer models can be used [8.17]. Requirements on the field bus for applications in process engineering have been thoroughly described by PFLEGER [8.18] and DRATHEN [8.19].

Already in MAP and TOP it was seen that the specification of complete function sets for all conceivable applications in industrial production is not possible in a uniform standard. On the basis of an MMS core standard (Manufacturing Message Specification, ISO IS 9506), special function sets have therefore been standardized for various application areas; these are called *companion standards* [8.12]. A similar procedure is also provided for the standardization of field bus systems. Companion standards define function blocks, such as those for the control of processes (analog and digital input/output, PID controllers, etc.) or for the utilization of production facilities (configuration, object description, maintenance, startup, etc.). The companion standards for field bus systems will resemble those for MAP wherever appropriate in terms of functionality. BÜSING has described the importance of MAP for process control engineering [8.20].

In the ISO/OSI model, the *network management* functions supplement the transmission protocol functions and communicate with them. As in any case where intelligent subsystems are networked, network management is also required for a field bus system.

Of the five management areas of the ISO/OSI model, the following are pertinent to field bus systems:

- Performance management
- Configuration management
- Fault management

The management functions for a field bus system are implemented in a network management station. Its functions and services include system initialization, the loading of programs and parameters, the starting and stopping of the communications system, the detection and localization of faults, and the reconfiguration of the field bus system in reaction to faults identified. Furthermore, the management station monitors the performance and loading of individual communications channels, identifies bottlenecks, and generates planning data for network upgrading. Such a network management component appears in all the field bus system now in development.

Incorporation of Field Bus Systems into the Information Hierarchy. As shown in Figures 8.5 and 8.6, field bus systems connect intelligent field devices with one another and with plant computers. They must support information transport at the field level but are also incorporated, through the plant computers, into the overall information pyramid. Since field bus systems are used in both processing and manufacturing, their real-time performance and message formats are subject to a variety of requirements.

Within the field level, real-time requirements are critical when information transmission over a field bus is within a control loop. Depending on the type of process being controlled, guaranteed transmission times in the range of a few milliseconds are needed; the messages to be transmitted are, however, short. The situation is similar for field bus systems in manufacturing cells. Longer messages and less critical real-time conditions occur when information is exchanged between CAD/CAM systems and manufacturing cells. The guaranteeable response time over a field bus depends on the types of services implemented.

When field bus systems are integrated into the information pyramid, it must be kept in mind that levels 3 to 6 do not exist or are emulated only to the extent necessary for direct field bus operation. This means that a special field bus node must offer the necessary network functions and also must have the requisite information about the communications partners at higher levels of the information pyramid. As in reduced MAP architectures, there must be a gateway that, on the one hand, is fully networkable and, on the other, can execute the field bus protocols (Fig. 8.7).

Constraints Arising from Applications. The ESPRIT CNMA project (Communication Network for Manufacturing Applications) carried out a survey to find out what operational constraints would apply to a field bus standard in the fields of processing and manufacturing engineering:

Processing:

Field devices to be connected:
Programmable controllers, sensor and actuator systems, minicomputers, field multiplexers, and power supplies for simple sensors and actuators

Information to be transmitted:
Cyclic (length ≤ 4 bytes, transmission time between 100 ms and 2 s)
Alarms (length ≤ 2 bytes, transmission in a few ms)
Overall field bus transmission rate > 250 kbit/s

Process dimensions:
Tens of meters up to roughly 2 km; distance to control room between 50 m and 3 km; total bus cable length between 75 m and 10 km

Ambient conditions:
High temperatures (> 80 °C), humidity, dust, vibration, electromagnetic interference; the transmission medium must be resistant to oil, steam, and ultraviolet; insulation must not contain halogens

Safety:
Intrinsic safety is an essential property; safety classes IP 65, NEC Class 1 Division 2

Conformance to protocol specifications
Certification of conformance in order to insure connectability, interoperability, and interchangeability of field devices on the field bus is regarded as important

Manufacturing:

Field devices to be connected:
Programmable controllers, numerical controllers, robot controllers; intelligent sensor and actuator systems are of minor importance

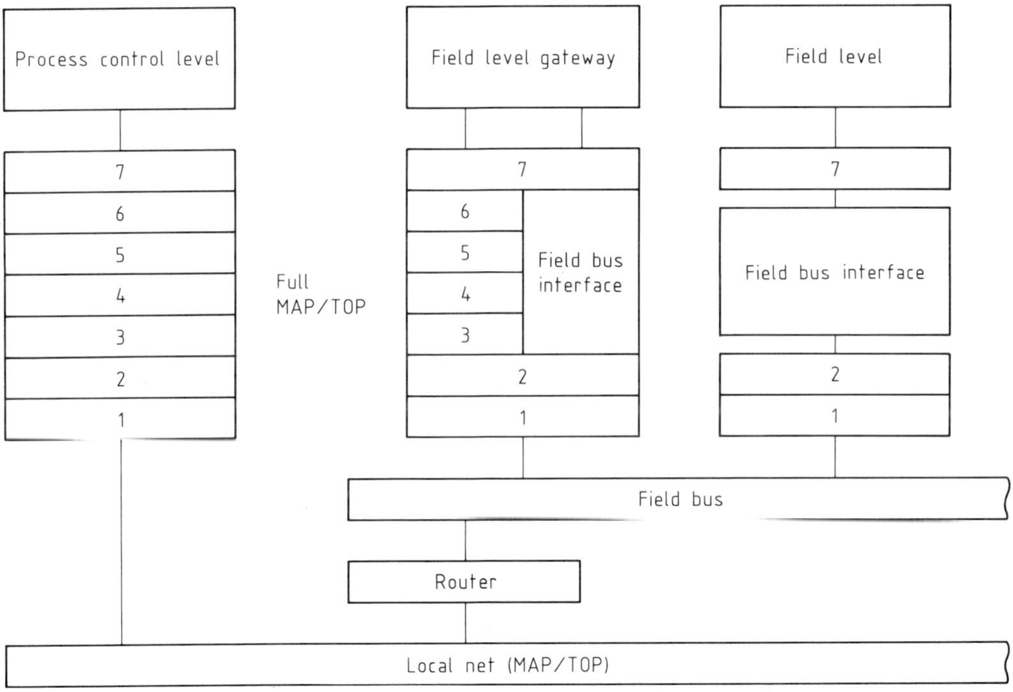

Figure 8.7. Linking of field bus and MAP LAN

Information:
Cyclic data (length 1–64 bytes, transmission time between 10 ms and 1 min)
Alarms (length 2–8 bytes, response time ca. 1 s)
Production data (length 2–256 bytes, time interval between 10 s and 8 h)

Process dimensions:
The diameter of manufacturing applications is ca. 10–80 m; usually the control room is nearby; bus cable length between 5 m and 50 m

Ambient conditions:
Severe electromagnetic interference, wide fluctuations in power supply voltage, static discharges

Safety:
No special requirements on safety or intrinsic safety of bus system

Protocol conformance:
Certification of field bus products for conformance to protocol is regarded as important

These survey results are reflected in different ways in existing field bus standards and those in preparation. The discussion that follows will deal with the most important properties of two field bus definitions; these properties typify the current situation in the establishment of field bus standards. As early as 1989, POLKE pointed out the needed separation into energy and information channels and questioned the requirement of "intrinsic safety" against the background of modern sensor and actuator systems [8.13].

PROFIBUS (Process Field Bus). PROFIBUS is a joint industry/science project in Germany, supported by the Federal Ministry of Research and Technology (BMFT), which prepared a field bus definition that is now a German standard (DIN 19245). The PROFIBUS user organization (PNO) is charged to maintain and further develop the PROFIBUS standard.

PROFIBUS corresponds to the architectural principles discussed above and was conceived for use in both process and manufacturing industries. Its essential properties include the use of the token-passing access procedure on layer 2 and the Field Bus Message Specification (FMS) including network management over all layers. FMS [8.21] is derived from ISO IS 9506 (MMS = Manufacturing Message Specification) so that it can communicate with MAP applications in a simple manner and make use of experience gained in the development of MMS. A Lower Layer Interface (level 7a) replaces the functions of ISO layers 3–6 needed for field bus systems. A conformity test is included in the PROFIBUS definition.

At the physical layer, PROFIBUS defines a twisted pair cable with a maximum length of 1200 m (without repeaters) or 4800 m (with repeaters). If repeaters are used, the bus topology can be straight bus, tree, or star. The transmission rate can be between 9.6 and 500 kbit/s. The number of nodes can be up to 122, of which a maximum of 32 can be master stations. Redundancy, intrinsic safety, and transmission of auxiliary energy comply with the specifications of ISA SP 50.

Layer 2 of PROFIBUS (MAC) includes two protocols, the distributed token-passing protocol (IEEE 802.4 or IEC IS 88204) and the centralized access protocol. In this way, communication among a number of master stations, or between one or more master stations and the substations (slave stations), is possible.

A special characteristic of the PROFIBUS definition is the use of standard processors for the protocol implementation, with a crucial effect on the efficiency of data transmission. On the basis of this efficiency, PROFIBUS at present is only conditionally suitable for service as a signal path between closed control loops with critical real-time conditions.

FIP (Factory Instrumentation Protocol). The FIP protocol definition has been in preparation since 1985, with support from the French ministry of higher education and research ministry. Its purpose is information transmission in a manufacturing setting. The FIP bus consortium includes more than 70 companies located in several European countries. There are at present three French standards for layers 1, 2, and 7. Other French standards for network management, MAP/MMS-compatible functions at layer 7, and optical fibers on the physical layer are in preparation [8.14].

The physical level allows for a maximum of 256 nodes. Three transmission rates are provided on the twisted-pair cable (31.25 kbit/s, 1 Mbit/s, 2.5 Mbit/s). The cable can be up to 3 km long at 1 Mbit/s. For optical fiber systems, the maximum length between two node connections is 1600 m (glass fiber) or 125 m (plastic fiber).

On layer 2, there is a centralized access control procedure (MAC), which supports both cyclic data transfer and the transmission of aperiodic data or longer messages.

The FIP definition, based on centralized bus control, assumes that data can also be transmitted within closed control loops. To achieve the required efficiency, protocols are implemented in specific ASICs differing in functionality. The FULLFIP ASICs include the currently standardized functions of layer 7, layer 2, and (in part) layer 1. This approach supports the connection of simple and inexpensive field devices (simple sensors) and minimizes the energy requirements of the interfaces. The decision to use special ASICs means, however, that they must be continuously modified to keep up with changing technology and standards. Given that international standardization of field buses has not yet stabilized, this point may be a problem.

The Route to Open Communication at the Field Level. Open communication at the field level means a consistent, internationally standardized protocol environment allowing field devices and plant computers made by a wide range of manufacturers to be linked to one another. This is particularly important at the field level, because a division of labor between manufacturers is natural and unavoidable in the area of field devices. Open communication has not yet penetrated into the production control and process control levels, but this fact cannot conceal the fundamentally different situation at the field level. Accordingly, manufacturers of field devices have not failed to come up with suggestions to remedy the lack of unity in the world of field bus definitions.

For example, the EUREKA field bus project has the goal of evaluating present field bus concepts and thus promoting the standardization of device interfaces to field buses. Another approach is embodied in the ISP project (Interoperable Systems Project), an international initiative by leading manufacturers of instrumentation and automation hardware wishing to create and standardize a field bus definition optimized for industrial applications on the basis of experience with existing field bus definitions.

The international standardization of field bus systems is proceeding in IEC TC65/SC65C, which is collaborating with the American standards body ISA. The ISA SP50 working group is creating a field bus specification for use in international standardization. The H1 and H2 performance classes defined by ISA SP50 have already been mentioned. In 1989, ISA SP50 attempted to harmonize the discrepancies between the access control methods (layer 2, MAC) in various field bus definitions. No final result has been achieved, but there is a trend toward "distributed" access control as set forth in IEEE 802.4 or ISO IS 88024 and used in PROFIBUS. The actual linking functions on layer 2 (LLC) have also been specified; the objective is to unify the most important mediation functions in all field bus proposals while simplifying these functions and thus improving their efficiency. On the whole, there is a trend toward the harmonization of field bus proposals, although the effect of existing national standards and international industrial cooperation on the likelihood of an international standard being accepted must not be underestimated.

The European Commission supported (under ESPRIT) the FICIM project (Fieldbus Integration into CIM) in order to promote the harmonization efforts discussed. From November 1990 to November 1992, the properties and interoperability of the two European field bus specifications (PROFIBUS and FIP) were tested in a pilot installation. The project had the participation of industrial companies as well as NAMUR, the Standards Working Group for Measurement and Control Engineering in the German Chemical Industry. The application functions and application standards were to support and insure the integration of information from these two field bus systems into a CIM system, independently of the lower field bus layers. The pilot phase of this project showed that such an integration goal is indeed achievable. The pilot facility is therefore being used for further studies of the interoperability of field bus systems.

The certification of field bus implementations for conformance to the standardized protocols is regarded as important in all field bus projects. Specifications and implementations of conformance test systems for all layers now exist for the FIP bus and the PROFIBUS. The Upper Tester, that is, the part of the test system above the protocol layer being tested, presents a special difficulty here. Because it is expected that field bus protocol implementations will be optimized for medium-sized and small intelligent field systems, the formerly usual (in MAP) configuration of the Upper Tester on the field bus device under test is no longer possible. The standardized conformance test for such devices thus includes an "Upper Tester Agent" (UTA) on the field bus device; through the field bus, this communicates with the actual Upper Tester on the separate test system.

The wide range of efforts to unify standardization work now in progress, and also to insure conformance of field bus components to the standards, suggests that open communication at the field level can be realized despite the variety of approaches now used. Manufacturers and users in the field of industrial technology should provide intensive support to this development work, since it is an important prerequisite for managing the complexity of future industrial control systems.

Again, a meaningful solution will be obtained only when the information model providing a semantic description of field-level problems has been adopted (see Chap. 2 and Section 5.1).

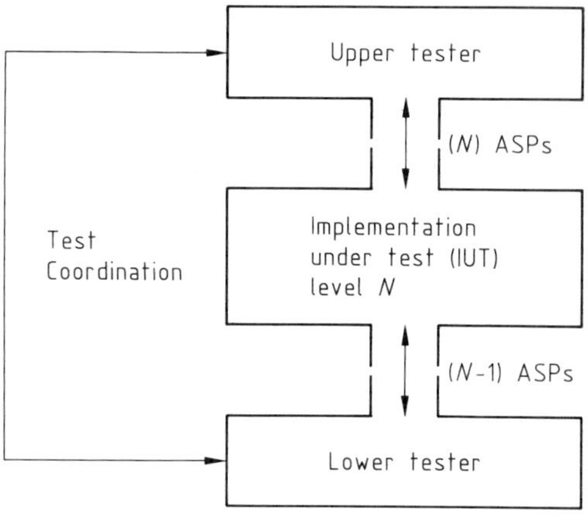

Figure 8.8. Principle of conformance testing [8.22]
ASP = Abstract Service Primitive

8.7. Quality Assurance: Conformance and Interoperability Tests

When computer communications equipment made by different manufacturers is used, the user must be able to rely on correct implementation of the protocols. The current state of the art does not allow a formal specification and thus an automatic verification of complex protocols, and so protocol products have to be tested with a standardized set of test cases. These "scenarios" define status data, parameters, and function calls, with whose aid all relevant protocol functions and protocol states can be tested as completely as possible or necessary. A protocol product that passes these tests is said to "conform" to the protocol specifications. If these specifications are taken to be complete and correct, a conformant protocol product will be able to communicate with any other corresponding protocol product.

In a full MAP/TOP or field bus protocol implementation, all protocols at the various levels must be tested. Figure 8.8 shows the general procedure for conformance tests [8.22]. The protocol being tested ("implementation under test") is enclosed by a test frame consisting of an upper tester (UT) and a lower tester (LT). While the LT responds to the test protocol in a well-defined way and detects the reactions of the protocol, the UT interacts with the protocol as if it were a protocol product on the next higher level. The UT is assumed to function without errors; this supposition is justifiable only because the UT merely provides certain interface conditions (ASP = Abstract Service Primitive) and need not exhibit the full functionality of a protocol.

A test coordinator makes certain that the test scenarios set forth in the conformity test are run, causes the responses of the specimen to be stored, and interprets the test results. Interpretation consists of comparing the reactions of the specimen with a predefined set of expected reactions. The conformance of the protocol product or the presence of implementation errors can be deduced from this comparison.

The fact that protocol products have passed the conformance test does not necessarily guarantee their interoperability. Like any other test, the conformance test is incomplete; what is more, it cannot be more complete and correct than a possibly incomplete and inconsistent protocol specification. This uncertainty can be reduced only by additional interoperability tests.

An interoperability test studies the actual interoperability of protocol products, that is, their real ability to interact in open networks. Interoperability depends on conformance to protocols but also on the matching of parameters that can be adjusted when the protocols are run. In this context, the term profile means an application-specific selection of parameters and functions of a protocol in accordance with a standard. Depending on the ranges of variation of such parameters, different profiles may conform to the standard but not be interoperable. The problems associated with the range of options in profile definition are not discussed further here.

A final key aspect of the serviceability of protocol products is efficiency and performance. Even given correct and conformant protocol implementation, interoperability can be significantly impaired by, for example, wide differences in timing between protocol products. In future, accordingly, standardized interoperability and performance tests will be defined.

An interoperability test system must establish a connection between the protocol components being tested and monitor the resulting flow of data and information. The use of expanded lower testers (cooperating with a communications observer) is conceivable. The interoperability tester compiles the results from the lower tester and the communications observer.

While conformance tests are done against a (possibly partly formal) protocol specification, the interoperability test deals with communications aspects that cannot be so well specified. It is thus much harder to standardize this kind of test. The user of communications protocols, however, sees interoperability as the proper acceptance criterion for protocol products; this is why work is under way to standardize these tests.

8.8. Methods and Tools for Protocol Specification

The preceding sections have made it clear how crucial the degree of formalization of protocol specifications is for the correct implementation of protocol standards. This section describes some procedures and aids now being used in the specification and implementation of protocols and in test systems. The protocols in question are those in the context of the ISO–OSI reference model [8.8].

These protocols [8.8] define services that can be described by the functions invoked and the data exchanged. Services are composed of service elements such as establishing and dropping connections, transferring data, and exchanging acknowledgements. Such service elements are deemed indivisible sequences of events and must be implemented in a proper way with regard to fault tolerance and consistency requirements. They are neutral with respect to the data or function calls mediated by them.

Service elements are based on "service primitives," that is, functions such as "request," "response," "indicate," or "confirm." In present-day protocol specifications, these service primitives and thus the service elements built up from them are commonly described in natural language. In this way, a formal specification of the functional (procedural) part of protocols is not usually stated, notwithstanding the fact that these functional (procedural) parts of protocols represent finite automata. The form of description as a finite automaton is, however, used only in exceptional cases and for subfunctions. One important reason for this situation is the complexity of the automata, whose many and varied options make them hard and inconvenient to describe in the form of states and changes of state.

Protocol Data Units (PDUs) exchanged with the aid of the service primitives, on the other hand, are now specified formally. A "language" of abstract syntactical notation is available for this purpose; ISO has standardized it under IS 8824, and it has become known under the name "ASN.1" (Abstract Syntax Notation One) [8.23]. Advantage has also been taken of experience gained in the informatics field with various higher-level programming aids and their data structure description methods.

At present, ASN.1 is used chiefly to define PDUs in protocols at the applications level of the ISO model [8.24]. As Figure 8.9 shows, applications programs or applications entities exchange information by means of PDUs. The ASN.1 language makes it possible to formulate these PDUs in the abstract manner standardized by IS 8824. It should be ensured that the meaning (semantics) of the PDUs in a given local applications syntax and in the abstract ASN.1 syntax remains unchanged. Thus, semantics-preserving transformations between the syntax of any local language and the abstract ASN.1 syntax must be defined.

The transmission of these PDUs via a communications system requires a concrete syntax, called the transfer syntax, that is understood by both communications partners. The transfer syntax must be defined at level 6 of the ISO model. Accordingly, ISO has standardized a "concrete transfer syntax" under IS 8825 [8.23], just as the abstract syntax is standardized under IS 8824. Thus, for PDUs actually to be transferrable between two applications entities, the abstract PDU definitions notated in ASN.1 must be translated from the real local syntax of each applications program into the IS 8825 transfer syntax. This assignment of a transfer syntax to

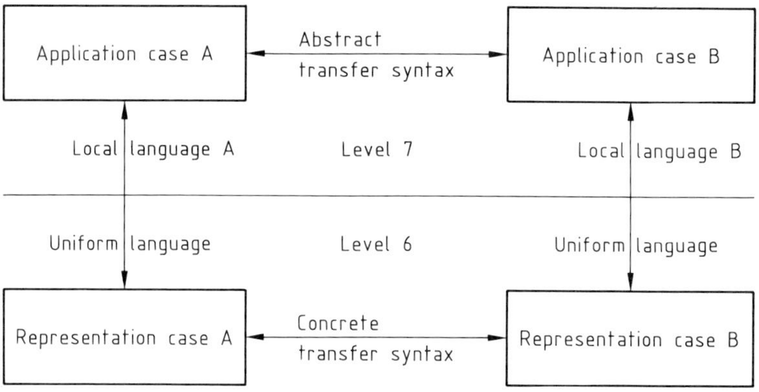

Figure 8.9. Representation context [8.26]

an abstract syntax is called "representation context."

It is significant that the representation context currently standardized by ISO for ASN.1 is not a unique standard valid for all time. Other representation contexts especially suitable for certain applications areas can be defined later on; this is particularly important in connection with the companion standards discussed in Section 8.4. Thus, in order that one applications-specific representation context can be replaced by another for communications between applications program, every communications process begins with the negotiation of the proper representation context. One example might be an exchange of communications of applications program systems between partners in the realm of process control engineering and partners in the realm of office communications. The difference in encoding conventions (e.g., measured data or CAD data) means that the representation context must be adapted.

There is already an alternative to the representation context ASN.1/Basic Encoding Rules, in the form of the "default context," which is used whenever no special representation context has been negotiated. This is the case in particular when a connection is being established. The default context, however, contains neither an abstract syntax notation nor a concrete transfer syntax. The unstructured bit stream between the two applications programs must be directly interpreted by the programs.

Because the PDUs of important MAP applications protocols, especially the MMS (Manufacturing Message Specification) services, have been specified in ASN.1, it is an obvious step to employ the same representation context for specifying the conformance tests for MMS. As Figure 8.10 shows, the conformance tests for the protocols MMS, Network Management (NM), and Directory Service (DS) are based on a set of ASN.1 tools with whose aid the test parameters and dynamic test cases for these three protocol groups have been formulated and processed. The system of tools for conformity tests, as shown in Figure 8.10, were devised by the Institute for Information and Data Processing of the Fraunhofer Society. Though not completed at the time, they were used successfully in the conformity test at IENE '88 (International Enterprise Networking Event) in Baltimore, June 1988. It is expected that the representation context based on ASN.1 is sufficiently powerful to cover other Companion Standards as well. The standardization of PDUs, at least for the protocols named, is thus formalized to the extent that conformance tests can be derived from the actual protocol definitions in a direct manner, without substantial recourse to natural-language definitions.

8.9. Steps toward Computer-Integrated Production

Consistent information integration in production companies is the objective of the computer communication utilities derived and discussed in this chapter in the context of information logistics. While very important, they are by no means the only tools available for reaching this objective. Computer-Integrated Production (CIP) can be achieved only through collaboration between disciplines: production engineering (manufacturing engineering, control

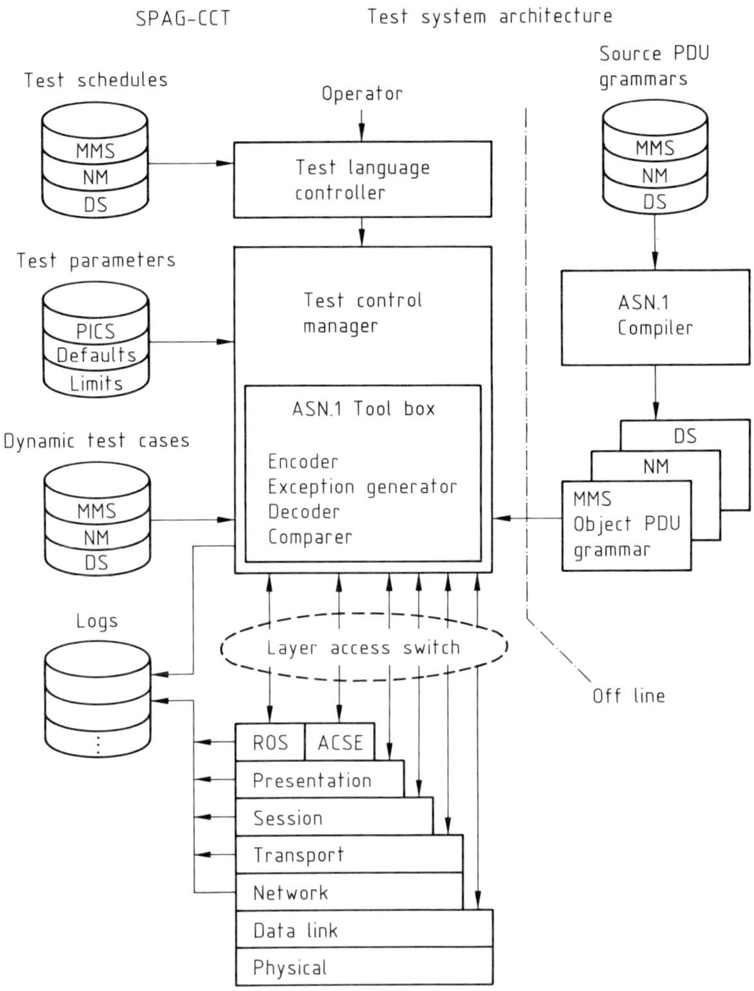

Figure 8.10. Implemented architecture of conformance testing

engineering, logistics, production planning, etc.), communications engineering (digital data transmission), and informatics (systems engineering, hardware and software technology). This integration of methods, in which information logistics is merely one part of the needed methodological edifice, is only just starting to gain support from products on the market and certainly cannot be achieved through standardized communications protocols alone. The custom CIP solution for a given plant often constitutes an analysis and modification of internal structures and modes of working.

The translation of such an internal reorientation to the use of computer communications techniques must take place in a continuously changing environment. Thus, MAP Versions 2.1 and 3.0 are not compatible; an early involvement in this protocol world gives rise to valuable experience, but also to a continuing need for new investment in more-advanced protocol products. Even within General Motors, initial experience with MAP was problematical. But it would be wrong to think that these unavoidable startup difficulties in a very complex field represent a failure of information integration. Outcomes of this experience have included more protocols on the applications level (or on all protocol levels at once), such as Network Management or Directory Service. In this way, gaps in the MAP/TOP definition are closed and the use of computer communications is facilitated.

The same applies to the present status of field bus standardization. Again, international differences in the view of applications requirements and product line definition must be ironed out in such a way that a single set of international standards can be devised in a usable form. The reaction of industrial firms to the very protracted standardization process shows once more that industrial users and manufacturer of communications systems are interested in open systems of information logistics.

All this development of tools will, however, be fruitless if users of CIP do not acquire the knowledge and training they will need to understand the total system. Along with education within the corporation, occupational training has a role to play. At the same time, there is a need for competent, neutral sources of advice. It is precisely small and medium-sized companies that lack the personnel resources to keep up with the rapid development of information integration and turn it to their own use, even though this assimilation is urgently needed to insure the future of the company's production and products. Regional sources of advice (chambers of commerce and industry) as well as research institutions cooperating with industry are increasingly available.

The path to computer-integrated production can be descerned today. It will lead, with much difficulty and many setbacks, to an improvement in the competitiveness and market position of those companies that have the commitment, creativitiy, and perseverance to follow this path.

9. Computer-Aided Methods

9.1. Principles

Coming years will see greater and greater importance attached to the use of computer-aided methods and computerized tools for the design, erection, operation, and maintenance of processing facilities. The steadily increasing number and complexity of industrial plants and the requirement for complete, current documentation will generate a volume of work that could scarcely be handled without computer support. At present, however, the ratio of computer-aided tools to demand is still fairly small.

As the needed computer-aided engineering (CAE) systems are developed and selected, it is crucial that the engineering disciplines involved must be treated not as individual, independent working groups but as the intermeshed system they really make up, in which the results achieved in one discipline influence the work of the others. This does not mean that one "universal" tool has to be devised for all engineers. The model of a well-filled and organized "toolbox" appears to promise better success. Such a box contains a special tool for every job. The tools, however, are designed so that they all share consistent interfaces and thus can be combined with one another at will. Any problem that arises can thus be solved in optimal fashion; if the objective is changed, the "toolbox" can be adapted to the new situation through the modification of an existing tool or the addition of a new one.

This book does not deal with computer-aided design tools for plant engineering (e.g., INTERGRAPH).

9.2. System Analysis

The background for system selection must include the formulation of overall objectives and possible interfaces.

The overall objective is broken down into partial objectives, which in turn are refined further into tasks. This procedure is continued until the individual tasks are manageable in their complexity and the functions needed for accomplishing them can be described along with the resulting data flow patterns.

To make this task analysis easier for the user and permit an unambiguous interpretation of the results, a variety of methods have been devised for displaying the process in graphical form with defined symbols. The best-known methods are structured analysis and design technique (SADT) [9.1] and structured analysis (SA) [9.2]. Either method can be carried out manually with pencil and paper, but increasingly they are embodied in software packages that make it possible to run these analytical tools on PCs and graphics workstations.

In what follows, the SADT method is briefly illustrated in a simplified example.

The essential elements of the SADT diagram are labeled boxes and arrows. Boxes symbolize activities (functions), while arrows stand for data and information and thus characterize the interfaces between boxes.

The following rules also apply:

- Input information always enters a box from the left, is transformed into output information in the box, and leaves the box on the right.
- Controlling information influences this transformation and may also be used per se in producing the output. Such information always approaches a box from above.
- The function of a box may be supported by mechanisms, which are optionally represented as arrows entering from below.

A model generated by SADT is a structured breakdown of a problem, represented as a hierarchically ordered sequence of diagrams. The topmost level of the hierarchy is a diagram with a single "black box" representing the total system, showing its connections to the outside. On the next level, this black box is broken down into a number of boxes; the same procedure is continued down the levels.

Figure 9.1 shows, in simplified form, the top two levels of an SADT model representing the engineering functions involved in the design, erection, and operation of a process plant.

The upper part of the figure shows the black box representing the overall system, with interfaces to the outside world. In the bottom part, the overall problem is broken down into the following subproblems:

- Design the process
- Design the plant
- Erect the plant
- Operate the plant

In the graphical presentation, the functions (boxes) are normally arranged on the diagonal

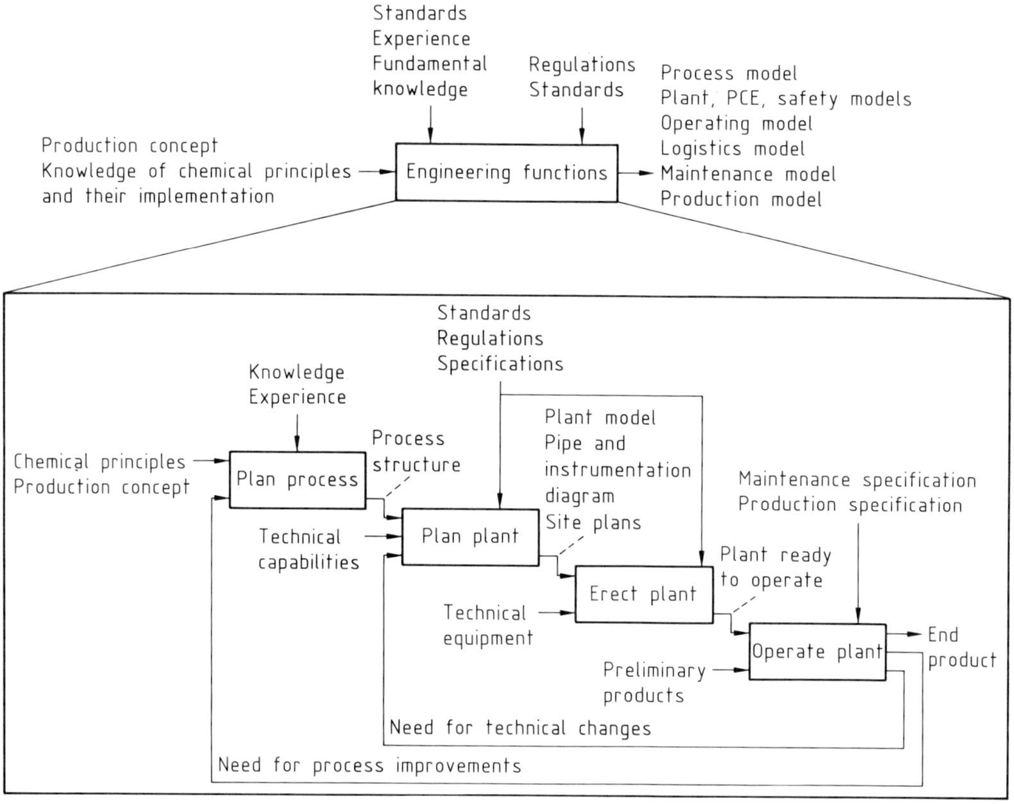

Figure 9.1. A SADT diagram (top two levels)

from top left to bottom right. The order of functions is governed by the required flow of information and data, shown as arrows between the boxes. Feedback of information may also occur, in which case the procedure becomes interative. Interfaces to the outside world are also shown in greater detail.

Now each of the functions is further broken down in diagrams and documented. Obviously, complex problems yield many diagrams, all partially interdependent. Hence, if changes are made, these relationships must be taken into account and the appropriate corrections performed. The use of such a method thus calls for a considerable amount of administrative effort; on large projects, this is difficult to accomplish without suitable computer support. However, the implementation of a software project is virtually impossible without a prior system analysis.

In parallel to the design of the function model just described, it is necessary to define the overall system data model that applies to all functions. Only in this way can proper sharing of information among functions and consistent data management be insured. As a rule, the data model is set up as an entity-relationship diagram with data dictionary. Programs for this design task are also on the market; in many cases they are already combined with SA or SADT tools to make up software development systems (see also Chap. 2).

9.3. CAE System for Process Control Engineering

Within the overall system, the individual CAE tools used in the several engineering disciplines must have compatible interfaces. In addition each tool should have the optimal functional scope for the objectives of the field in which it is used. The special requirements on such a CAE system for use in process control engineering are therefore discussed in what follows.

A CAE system for process control engineering must support both the design and the later operation of the process measurement and control system, including maintenance (Section 11.4). This latter aspect is particularly important here. Over a span of years, the design changes in an onstream plant due to detail improvements or process changes can easily loom as large as the original design. At the same time, the documentation must meet stringent standards of currency and consistency, since these changes generally must be performed while the plant is on stream or during brief shutdowns (see Section 3.5).

The CAE system must therefore be able to handle data having a variable and complex structure and to afford assistance in the processing of such data.

Process measurement and control functions are implemented in hardware and software. In practice, it has been found that a subdivision of the task into hardware design and software design is useful, and in fact these sub-tasks can even be carried out by different persons in the case of large projects. Both areas, of course, have access to a jointly defined interface, but are otherwise not temporally linked and can therefore work independently of one other.

The hardware data to be managed in the CAE system can be broken down into three classes:

- Station data: all information describing the task and the functional design of a process control station. As a rule, these are documented in station data sheets and station diagrams (see Figs. 10.31–10.34).
- Device data contain the exact technical specifications of the devices employed. These serve as the basis for later procurement.
- Position data describe the configuration of the devices in the facility and contain the infrastructure created by the master cable. Typical forms of documentation are setup plans, terminal allocations, and cable plans.

The data describing the software have a different structure. Generally, modern process control systems have modular software. The modules are linked to one another and arranged into various levels of a hierarchy, depending on the task. The lowermost level is made up of modules that implement the functional design of a single process control station (Section 4.4). Higher-level modules coordinate the work of lower-level ones. Each module is composed of functional building blocks, which are the smallest software elements of a process control system. The most important documentation forms for software are function charts, logic diagrams, and element structural plans.

All these data groups are closely linked; together, they represent all the information about the process control equipment. In what follows, they are treated together under the term "project data." As a rule, these project data are managed by a database system (Chap. 10).

9.4. Structure of a CAE System

The ideal structure of a CAE system as shown in Figure 9.2 reflects the constraints discussed above.

The heart of the system is a relational database, which contains the computer model of the planned equipment. All manipulations of the database take place interactively; depending on the task, graphical interfaces may be used.

At the start of the design work, basic information for the project is acquired from the other departments involved. The most important source is the process control tender specifications, which set forth requirements on the pro-

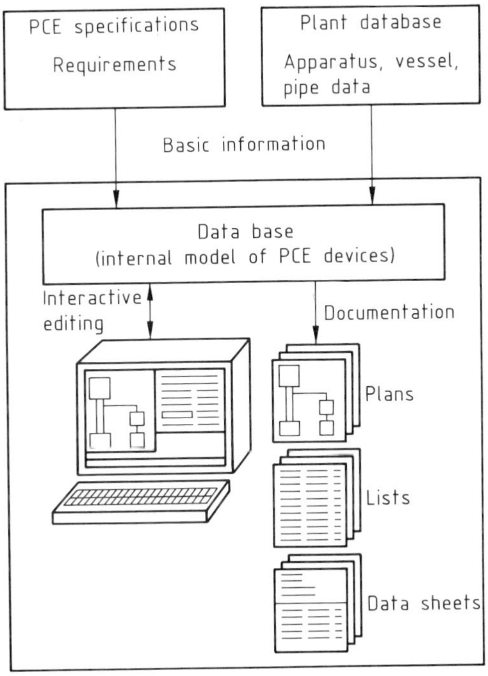

Figure 9.2. Structure of a CAE system for process control engineering

cess control devices to be implemented as well as their interconnections. Information is also required from the plant database, which supplies technical data on the plant sections affected as well as ambient conditions where the process control device is to be installed.

On this basis, the designer begins with the detail design of process control devices. This procedure is interactive and employs a graphical user interface. The system offers a number of tools that can, for example, prepare plans, enter specification data, define cables, or copy entire plans or cabinet terminal allocations. These manipulations gradually build up an internal model of the process control equipment in the database. Every interaction on the screen brings about a corresponding modification in the database.

This internal (computer) model also serves as the basis for all documentation. The various documents are merely the results of various inquiries formulated to the same database under various criteria. In this way it is insured that any subsequent changes will automatically be incorporated into the next round of documentation.

A CAE system with this level of complexity may consist of a number of largely independent modules, all of which communicate with one another through the common database. One possible subdivision into task areas is as follows:

- Hardware design
- Software design
- Procurement
- Installation

These are largely independent as far as processing is concerned and have access to one another only through a few well-defined interfaces. Accordingly, independent solutions are now on the market for certain of these task areas. When such products are employed, however, special attention should be focused on the possibility of later integration into an overall system.

Because process control devices cannot be designed and operated in isolation from the plant as a whole, smooth communication should be insured between this process and other departments, such as plant design, process engineering, purchasing, and maintenance [9.3]. The only way to be certain of this is to define a common data structure within the company: the company data model. Without such a data model, the danger exists that terms and information might be differently interpreted in the several systems, so that consistent data management would be impossible.

A crucial problem for the acceptance and efficiency of a CAE system is whether and to what extent the system user interface is adapted to the user's methods. Typical users impose varying, and in part inconsistent, requirements.

One user is the design engineer, who quickly becomes an "expert" through daily work with the system in a variety of modes, can operate it "in his sleep," and therefore wishes to get results quickly, with the fewest possible manipulations, and without having to read help messages.

In contrast, there is the operating engineer, who is confronted with the system only occasionally and briefly, for example during process upsets and when design changes are being worked out. The system should prompt this user in an intuitive way, with extra help functions, to avoid a prolonged learning process and the study of manuals. Because the drawing is the most comprehensible "language" in conventional design work too, it is an obvious step to design the entire user interface and system tools for largely graphical operation. In this way, the user's previous working methods are largely preserved, so that learning the system can be accelerated.

The use of hierarchically structured menu trees can drastically simplify the operation of choosing from a large number of devices or symbols. If the menus and the underlying database are cleverly designed, progressive narrowing of the field quickly leads to certain individuals without the use of lengthy catalogs and lists. This technique is widely used in modern systems [9.4].

9.5. Aids for Hardware Design

A number of CAE systems for process control hardware design have appeared on the market. They take two distinct approaches to implementing the task. One is to combine independent standard programs for database applications and 2-D graphics, using suitable auxiliary programs to link and supplement the commercial software. The other approach leads to an integrated system, in which the graphical tool is directly linked to the database and each is independent of the other. Both solution approaches will be described in what follows, and their advantages and disadvantages will be discussed.

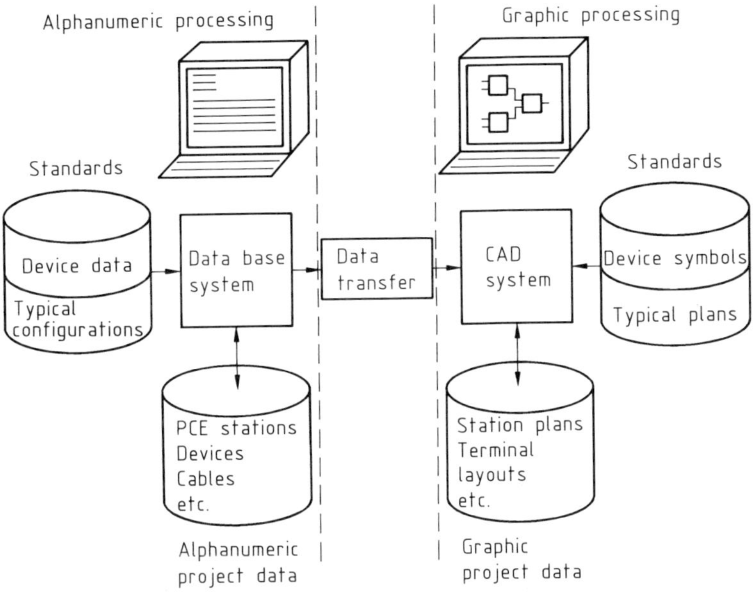

Figure 9.3. Structure of a combined CAE system

Combined CAE System. A combined system includes a free-standing, independent software package for alphanumeric tasks and one for graphical tasks, each package having its own data management features (Fig. 9.3).

Basic information contained in the alphanumeric package includes specifications on standard devices (process control equipment data; Figs. 2.9 and 10.31). Typical configurations, with information on combinations of devices for stated purposes, are also included as standards. In parallel, the graphical program package must have the respective symbols for the individual devices as well as typical plans for device combinations. It has to be explicitly verified that the typical configuration and the typical plan are logically consistent, since this forms the basis for subsequent automated processing (see also Figs. 10.32–10.34 and 3.22).

The first step in design is to input the process control stations, in the form of a station list, into the alphanumeric software and assign a typical configuration to each station on the basis of the task statement. These data are supplemented by structural information, such as the terminal allocations of cabinets and racks and the master cable connections. The devices used may also be specified more closely. Once the station and configuration data have been prepared, they can be exported to the graphical side with the aid on an auxiliary program. The respective plans are now prepared. Infrastructure information, such as terminal codes and the like, is entered into prepared text placeholders. The information managed in the alphanumeric module can thus be output as documentation along with the graphics.

There are a number of advantages to this approach. The use of common standard programs makes it possible to assemble such a system at relatively low cost. Because the programs do not have to run in parallel, the system can be and usually is executed on a PC with the MS-DOS operating system. The system as a whole is highly modular in structure, and the user can generally expand it with additional functions.

These advantages are partially offset by several drawbacks. The use of several software packages from different publishers means that the user must be familiar with the diverse operating concepts of the programs. Because configurations and plans are generated in different programs, there is a danger of input errors, which can lead to inconsistencies.

A combined CAE system thus has its strengths in the design areas, where a high degree of standardization is possible. If subsequent changes not complying with the standard are made, however, corrections must be made in both the alphanumeric and graphical components of the documentation.

Integrated CAE System. The integrated solution features a direct link between the graphics package and the database. Graphical manipulations thus influence the database contents immediately, while alphanumeric inputs influence the graphical representation on the basis of certain rules. This close connectedness both insures the consistency of the data and makes it possible to carry out plausibility checks. Interactive operation builds up an internal model of the process control equipment, which serves as the basis for all interpretations. All documentation, whether in graphical or alphanumeric form, is derived from this model (Fig. 9.4).

The close coupling of graphics and database system usually requires the use of a multitasking operating system. The graphics packages employed are specially developed software with special provision for interaction with the database. UNIX graphics workstations serve as the hardware component.

The advantages of the integrated approach lie in the consistency of the data. In case of subsequent changes, corrections are essentially needed at just one place.

Because of their more complex structure, integrated systems are necessarily more expensive than combined ones. What is more, high-performance hardware is required if an acceptable response time is to be achieved [9.4]–[9.7].

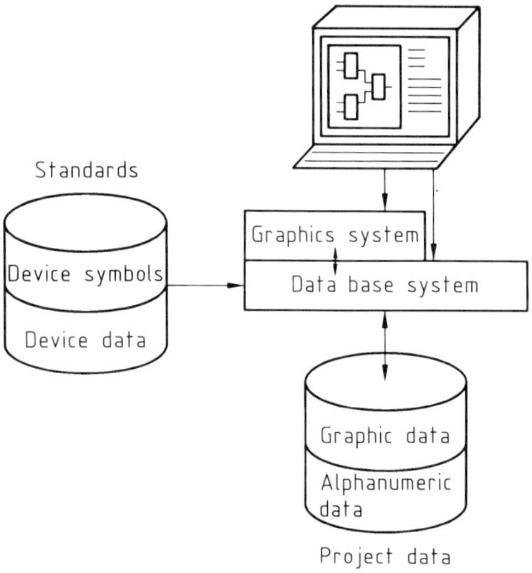

Figure 9.4. Structure of an integrated CAE system

Over the long term, the integrated CAE system is the better solution. Its present cost disadvantages are being offset by technical developments in the workstation market.

9.6. Aids for Software Design

Hierarchical Organization of Software. The principles of functional organization and progressive refinement have been tested in practice for the description of complex relationships (see Chap. 2). A good example of such a procedure are the design steps involved in the development of a complex integrated circuit such as a CPU. First, the required functions are represented in the form of modules at a fairly abstract level, and the requisite interactions between these modules are described. Next, the internal structures of the individual modules are established; the same principle is applied recursively in this step. The objective is thus narrowed down to the level of single transistor functions. In the subsequent fabrication operations, the complex functioning of the circuit is achieved through the combination and interconnection of many very simple subfunctions.

A similar procedure can be applied to the preparation of software for process control systems. The overall task is broken down into subtasks with well-defined interfaces (see Section 5.2). By narrowing down the description, one ultimately reaches the lowermost level of abstraction, which in the case of process control systems is formed by the components and logic functions available. The software for a system is finally written by appropriately combining these basic elements (Fig. 9.5).

If there is so much similarity in the procedures, the question arises whether parts of the highly advanced commercial CAE tools for circuit design might not be usable in modified form for process control system software development. In what follows, the minimum requirements on such a tool are stated briefly.

Graphical Generation of Function Charts and Structural Diagrams. The building blocks or logic functions of the process control system, as furnished by the publishers, form the basis of the applications software. The stated task is accomplished by combining these, in suitable form, to make structures. The most important aid in design and subsequent maintenance is the func-

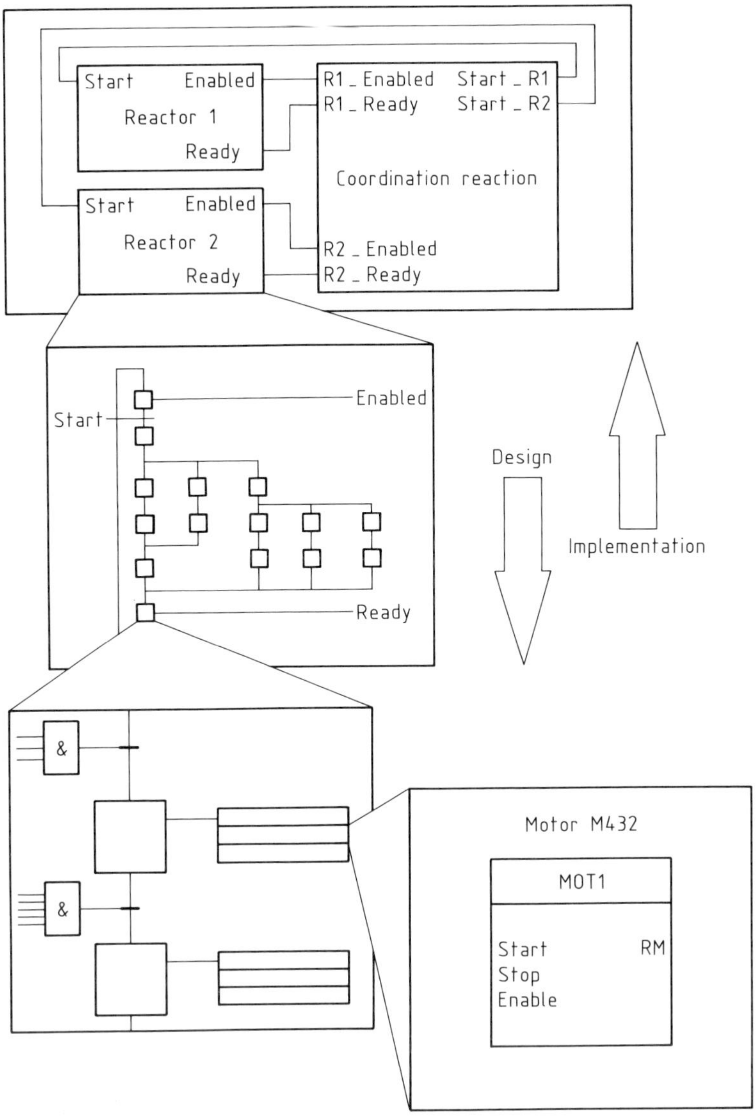

Figure 9.5. Progressive refinement of a software structure

tion chart or structural diagram (Figs. 3.20 and 10.35). It shows graphically all the components used, with their interconnections and parameters. The creation of a graphical tool for generating these diagrams is crucial. The element symbols are selected from a library and placed on the screen interactively. The connections between elements are generated by constructing lines between the element terminals. Parameter values can also be interactively assigned to specific element terminals. The logical content of these diagrams must be determined by the system and automatically translated into a sequence of programming instructions for the process control system. Plausibility checks can be performed to eliminate programming errors directly at the source. This procedure simultaneously insures consistency between the program and the documentation (see Section 10.6).

The CAE tool must support the method of successive refinement described above. The user must be able to generate abstract modules and their connections in a plan without needing to know their internal structures. Internal structure can be represented in a later diagram; the system automatically takes care of cross-referencing. In this way, complex structures can be detailed step by step, down to the component level. During code generation, the system must recognize and take into account information about the entire hierarchy of diagrams.

Another important point is the use of "typicals." These are multicomponent structures that can be used without alteration in many diagrams. Constructing a good library of typicals can drastically reduce the required design effort. At the same time, it leads to standardized software, since a fixed combination of elements can always be used for like functions. Such a capability must therefore be implemented in a CAE tool (see also Sections 4.4 and 5.7).

Apart from building blocks defined specifically for each process control system, there are system-neutral descriptive forms for logic and sequential control systems. The appearance and interpretation of these are specified in ISO and DIN standards. It should therefore be possible to program controls with the aids described above, using the standard symbols. The translation to the language of a particular process control system is another function of the CAE tool.

Construction of Screen Diagrams. Like the preparation of component-oriented software, the programming of the user interface for a process control system is a particularly critical point (see Section 11.2). At present, this represents a major design effort, requiring support by CAE tools. Graphical plant diagrams used in interactive mode have come into widespread use in the user interfaces of process control systems. The graphical design of such images should be performed with tools whose quality and ease of use are consistent with those of commercial CAD systems. The definition of dynamic displays or controls might be done by marking the position and assigning the variable name. Diagrams constructed in this way should be usable for all components of the process control system without case-to-case modifications.

Documentation of Existing Facilities. A special problem arises when existing plants must subsequently be documented (see Section 3.5 and Chap. 10). Often, the current program no longer corresponds with the original documentation because changes have been carried out commissioning or maintenance. All process measurement and control systems can document the current program structure in the form of component and configuration lists. This form of documentation is, however, difficult to understand, and in practice it cannot be employed for troubleshooting. As a rule, therefore, there is no alternative to changing or regenerating function charts and structural diagrams on the basis of the current lists, but this procedure is time-consuming and expensive. Users are therefore demanding automatic graphical documentation of the current state of a process control system.

Synthetic diagram generation on the basis of information present in the control system constitutes a basic problem. Because the system contains no information about the original form of the diagram, the component and terminal positions needed for the representation must be determined by using algorithms such as "information flow top to bottom" and "minimize feedback paths." The consequence is that a synthetic diagram with given content may have a different graphical organization from a hand-drawn diagram. Furthermore, even small changes may alter the positions of some components relative to the old version because of differences in evaluation in the placement operation. This property may interfere with the immediate re-identification of a circuit. Nevertheless, retrospective graphical documentation can insure complete consistency between the documentation and the current program. For this reason, users are willing to tolerate the disadvantages of retrospective documentation until a better alternative becomes available.

System-Neutral Programming. In recent years, various manufacturers of process control systems have developed CAE tools, each tailored to a particular system. Although such a package provides effective support to the user, it also poses a new problem. Large firms and large engineering offices, which usually prepare designs for a variety of control systems, would have to employ a large number of diverse, system-specific CAE systems in order to simplify software writing. Because these systems are mutually incompatible, porting existing application software to a new system ultimately means doing a complete redesign. Users have therefore welcomed the efforts of the IEC to devise a neutral, standardized programming language for process control systems. With such a language, system-independent and thus portable technology applications could be developed. Using such packages, the user could eventually perform tasks in a problem-oriented way regardless of which CAE system was in use.

9.7. Outlook

The development of CAE systems has been strongly influenced by rapid advances in computing hardware and software. If users are not to find themselves isolated in a few years and restricted to possibly obsolete computer systems, modern CAE systems must be made independent of both hardware and software. This objective can be achieved only by means of international standards.

In the operating systems area, UNIX Sys. V or OSF can serve as a standard. Most publishers have now adopted the MIT X-Windows System as the basis for the graphical user interface. A number of competing tools are currently available for the actual design of the user interface, but OSF MOTIF is emerging as the favorite. A CAE system developed with these tools can thus be regarded as having a good future.

One particularly critical point is the data management problem. Because plant design data generally must be kept for 20 to 30 years, stringent requirements apply to the dependability and development potential of the database program publisher. At the same time, exotic data structures should be avoided so that data can, at least in principle, be imported into a different database system. In the context of workstation networking, the potential of decentralized data storage should be investigated.

Graphical workstations, connected to one another and to service computers by local networks, will become the standard hardware. The simultaneous rapid rise of computer performance and decreasing hardware costs will allow use of high-performance solutions. The high requirement for storage space will be partly answered by advances in rewritable optomagnetic discs, which offer high data security and long-term stability.

10. Design and Construction of Process Control Systems

10.1. Principles

The previous chapters have described the functional and hardware elements of process control engineering. This chapter deals with the design and construction of process control facilities.

It is seldom that process control problems arise in isolation. As a rule, they are embedded in an overall problem. This may concern an investment project in the field of chemical plant construction; process control engineering then represents one component of the problem, while other disciplines such as machinery and apparatus, building construction, and piping are also involved.

In the design and construction of plants, a distinction is made between new construction and extension or reconstruction.

Extension and reconstruction activities are generally more difficult to plan than new construction projects, since interfaces to existing plants have to be taken into account and the existing infrastructure and engineering must be modified. The overall extent of the work usually only becomes apparent when planning is already under way. A complicating factor is that economics often dictates the incorporation of changes while the plant is on stream or during shutdowns. Because the scheduling and duration of shutdowns is governed chiefly by operational constraints, however, planning of schedules and resource utilization must be done with special care (Section 10.2).

New construction projects based on processes already in use within the company are generally simpler.

The handling of processes new to the company (i.e., created in process development or purchased from a licensor) is more difficult. Scaleup problems can involve substantial risks.

Figure 10.1 gives a model for the implementation of chemical plant construction projects. The main blocks represent

- Project management
- Project planning and execution
- Quality assurance

The design and construction of plants is carried out in the framework of projects. Project management (Section 10.2) accompanies project activities, coordinates the work, and monitors compliance with directives.

In Germany, the usual approach project planning and execution is based on the regulation governing architects' and engineers' fees [10.1]. It is useful to illustrate the procedure in the form of a phase model, which breaks tasks or functions down into phases and steps and shows their outcomes (Fig. 10.1).

Note that the phase model represents an ideal. In actuality, the several phases merge continuously and may even overlap. Feedback and iteration between phases are also common. Interfaces between disciplines further complicate the process.

The phases can be grouped into two large phases on the basis of the problems to be solved; these are called the decision phase and the execution phase [10.2].

The decision phase (Section 10.3) involves the conceptual work. The task is formulated here; it can be thought of as the mission to be performed in the course of the project. The written description of the problem is called the performance specifications (Section 10.4); it forms an essential component of the decision phase.

The performance specifications give rise to the "binding concept," embodied in the specifications, which describes how and with what resources the stated task is to be performed (Section 10.4).

The concept has long-term consequences, since it influences the economics of the plant throughout its service life. Identifying the concept is therefore accorded special importance. This step requires great care and is supported by analyses of alternative solutions as well as an examination of the overall project.

One task may admit of several solutions, among which the optimal one for the company (from both economic and technical standpoints) is to be selected. The selection criteria may include

- Attainment of targets
- Fulfillment of economic criteria (Section 10.4)
- Compliance with regulations (Section 10.6)
- Availability of plant
- User-friendliness
- Use of proven techniques

The total analysis of all disciplines involved in an investment helps to prevent the overall pro-

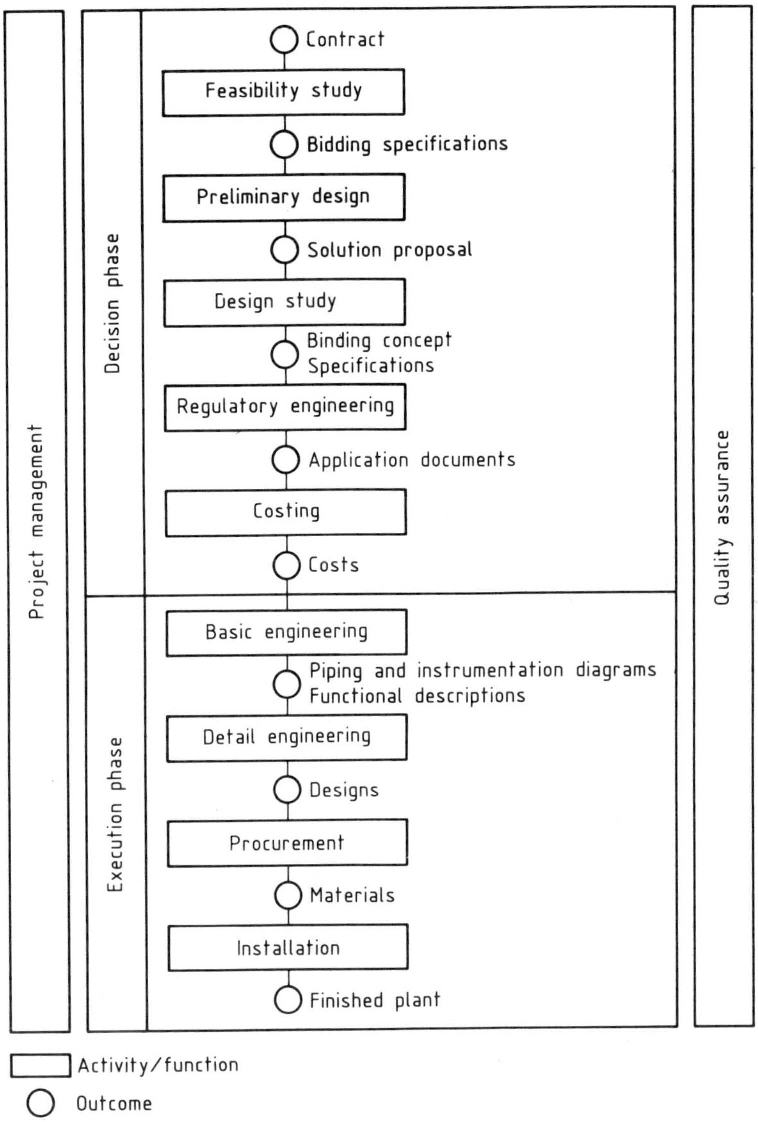

Figure 10.1. Phase model for chemical plant construction projects

ject being harmed by suboptimization of individual disciplines.

The binding concept is the basis for corporate management's decision to proceed further.

The execution phase (Section 10.5) is oriented toward workmanship. Using results derived in the decision phase, it creates a functional plant.

When the chemical plant design and construction process is divided into decision and execution phases, with the stress on holistic analysis and conceptualization, the result is that the problem is handled in a fundamentally different way from what used to be common (see also Chaps. 2 and 8).

In the past, planning for certain functional specialties, including process control engineering, often caused delays in other parts of the process, such as plant design. The tasks of the "trailing" discipline were sometimes formulated from the standpoint of the "leading" one.

A holistic view of the task could scarcely come about in this way. If such a view was formed in the course of the project, any corrections arrived at through such an analysis, no matter how logical and necessary, were not carried out, simply because the planning processes in the several disciplines had reached such very discrepant stages by that time that revision of existing plans by the methods then in use could not be justified in terms of cost and lost time. The systematic use of data-model-oriented planning in plant construction, such as the Intergraph plant design system, makes it possible to implement changes such as those described in Chapter 9.

In the past, compromise solutions replaced logical, holistic solutions; suboptimization of individual disciplines was the rule.

The complexity and diversity of techniques now in use, together with the resulting interdependence, have led to continuous interaction between disciplines, virtually barring such an approach.

Only the analysis of the project in its totality, coupled with the conceptualizing approach and an intensive quality-assurance effort (Section 10.6) going along with the execution of the task, can satisfy the requirements imposed on the economics, safety, and environmental compatibility of engineering solutions today.

10.2. Organizational Requirements

The organizational concept in the execution of projects is called project management.

The term "project" refers to the performance of individual tasks bounded in time and having clear objectives; these tasks may further be characterized by features such as interorganizational components, many involved divisions, high level of complexity, and possibly heightened risk [10.3].

Various types of project exist, including

- Construction projects
- Plant projects
- Organizational (management) projects
- Research and development (R & D) projects
- Software projects

In general a project is the complete temporal framework of conditions of all substances for the realization of tasks represented by functions [10.4]. "Substances" are suitable procedures, systems, and elements (Figs. 2.9 and 3.26) and are also called "resources"; Figure 10.2 illustrates this relationship.

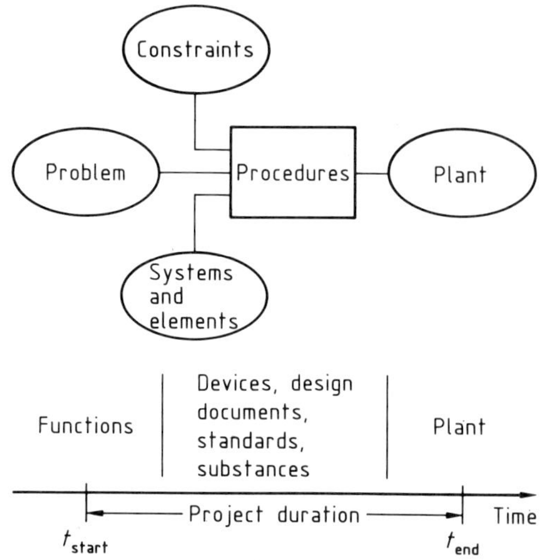

Figure 10.2. From problem to plant

In order to achieve the project objective, these resources must be procured, combined, coordinated, and utilized. Project management is responsible for these functions.

Organization. The organizational setup of project management differs from the "line" organization generally found in companies.

Organizational Forms. A variety of organizational forms are seen in project management. The three most important are briefly described below [10.5].

An across-lines form, also called "matrix" project organization, has proved especially successful (Fig. 10.3). The customary line organization is retained but is enlarged by a further dimension—project management. Certain aspects of the project, such as technical responsibility, remain under line jurisdiction; others, such as project responsibility, are transferred to project management.

This form of organization offers the advantage that personnel resources are optimally utilized, since employees remain in the project only as long as they are needed. As a result, specialist knowledge and special experience can be more easily brought in as well as transferred to subsequent projects. These advantages are partly offset by some drawbracks, in particular jurisdictional conflicts between line and project management and the relatively great effort required to demarcate responsibility.

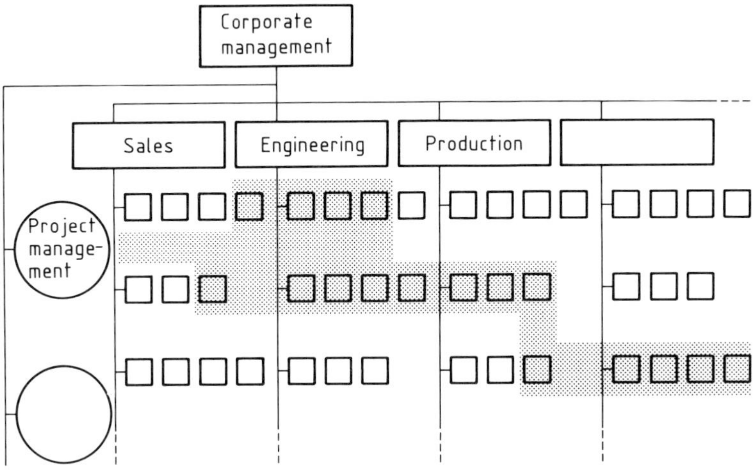

Figure 10.3. "Matrix" project organization [10.5]

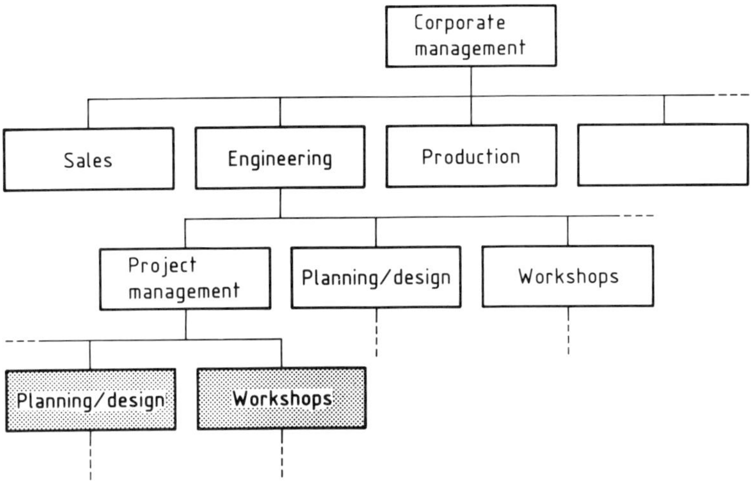

Figure 10.4. "Straight" project organization [10.5]

The advantages, however, outweigh the disadvantages. The matrix form of project organization has therefore found a wide variety of applications. It has been broadly accepted, in particular, by the chemical industry.

In "straight" project organization (Fig. 10.4), a separate, independent project line is created in addition to the main lines; project management has its own organization partly similar to those of the main lines. Project responsibility and technical jurisdiction are combined at a single point. Quick reactions are possible when difficulties arise in the planning and execution of the project. Against these advantages stand difficulties in recruiting personnel at the start of the project and in keeping these people occupied after its completion. Straight project organization is mainly used in large, long-term projects.

In "influence" project organization (Fig. 10.5), project management is simply a coordination instrument, without the authority to issue orders. It tracks progress with respect to schedules and costs and, if necessary, suggests actions to the respective line jurisdictions. Project management has no responsibility for the project; it is responsible only for providing timely information to the line jurisdictions. This arrangement does not favor successful cooperation across lines. When difficulties arise, reactions usually come slowly and delays result.

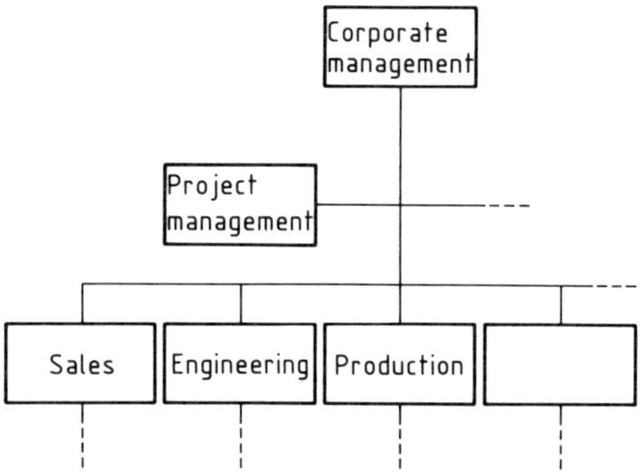

Figure 10.5. "Influence" project organization

This form of project organization is usually practiced on smaller projects and projects involving several companies.

Organizational Structure. For project management with matrix organization, especially in the chemical industry, a threefold division of responsibility has proved logical and useful (Fig. 10.6):

- Project management committee
- Project manager
- Project engineer

The project managment committee is the actual supervisory and decision-making body. It includes the decision makers of the crucial lines involved in the project, i.e., those issuing the order for it and those executing it. The project management committee formulates global mission targets. It monitors the complete and on-schedule attainment of these targets, makes the needed decisions when targets come into conflict, and causes these decisions to be implemented in the lines. The committee insures the availability of material and personnel resources.

The project manager is responsible for structuring the project. His duties include the definition of sub-tasks in accordance with the project structure. Principal activities are the planning and oversight of schedules, costs, materials and other resources used, and progress. In addition, the project manager sets up and implements the reporting system.

The project engineers support the project manager by providing organizational and technical guidance and oversight to the project teams, which are detailed by the lines to perform sub-tasks within the project.

The makeup of project management depends on the size and importance of the project (Fig. 10.7). The "size" of a project can be defined in a variety of ways; for investment projects, the level of investment is the key criterion.

In large projects, the project management committee is made up of delegated decision makers from top management. For small to medium sized projects, on the other hand, decision makers from lower strata of the hierarchy form the committee.

For large projects, a head-office project manager is designated; this person always performs management functions. Smaller projects do not require so much expense for the project management function, and a number of alternatives exist here. One manager can manage several projects at once, or one manager can focus on a single project but add specialist functions to management ones. The manager then becomes a participant in the project. The same holds for the project engineers: They have 100% of their time

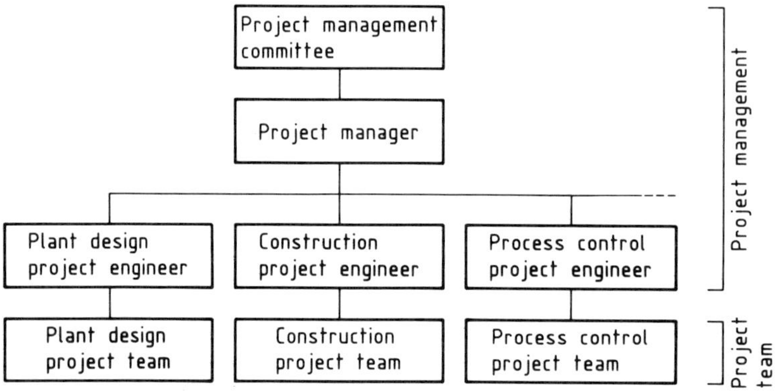

Figure 10.6. Organizational structure of project management

	Project management committee	Project manager	Project engineer
Key projects, projects of strategic importance	Decisionmakers from top levels in the hierarchy (top management)	o for one project: with pure management function	o for one project: with pure control and monitoring functions
Medium-sized projects	Decisionmakers from middle levels of the hierarchy	o for several projects: with pure management function or o for one project: with management and technical specialist function	o for several projects: with pure control and monitoring functions o for one project: with pure control and monitoring functions
Small projects	Decisionmakers from lower levels of the hierarchy		

Figure 10.7. Project management organization as a function of project size and importance

Project management committee	Project manager	Project engineer
o Monitoring and control o Formulation of global targets o Resolution of target conflicts o Implementation of decisions in the lines o Securing of material and personnel resources	o Project structuring o Definition of sub-tasks o Schedule planning and monitoring o Cost planning and monitoring o Capacity planning and monitoring o Progress planning and monitoring o Reporting system	o Technical and organizational control and monitoring of project teams

Figure 10.8. Tasks of project management

assigned to large projects, a lower percentage for smaller projects.

Tasks. Figure 10.8 summarizes the functions of the project management committee, project manager, and project engineer.

Major management activities include project structuring as well as cost, schedule, and resource management. Performing these functions requires close cooperation between all the activities involved in executing the project.

Project Structuring (see also Chap. 2 and Sections 3.5 and 4.3). The nature, scope, and complexity of a project dictate the manner and depth of structuring. This activity results in structural charts and can be based on a variety of approaches, with a corresponding variety of structural charts as outcomes [10.6].

The object-oriented structural chart breaks down the material and immaterial objects to be created by the project into components in hierarchical fashion. Figure 10.9 shows a portion of such a structural chart for the erection of a chemical plant.

The function-oriented structural chart shows the breakdown of project activities.

The structural chart of technical scope of performance shows what intermediate and final outcomes are to be generated by the project. All documents to be produced, such as reports, plans, drawings, lists, descriptions, certificates, contracts, and so forth are included, regardless of whether they will still be needed after the end of the project.

The last two types of structural chart are seldom made up by themselves; a combined form is usual.

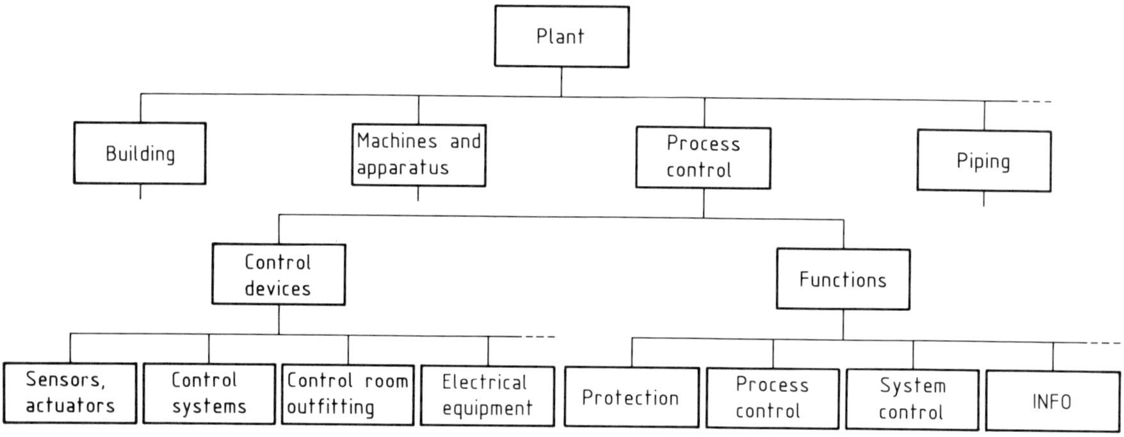

Figure 10.9. Example of an object-oriented structural chart

The main goal of project structuring is to determine the total volume of work necessary for the success of the project. In this way, the preconditions for further project planning are fulfilled, since project structural charts influence, among other activities, the planning of schedules, costs, and resource utilization [10.6].

The transparency achieved through project structuring makes is possible to identify key tasks. It becomes easier to coordinate project teams and delegate their sub-tasks. Economic, technical, and also organizational risks, such as jurisdictional conflicts between project management and lines, can be delimited. Project structural charts are a prerequisite for effective project monitoring.

Cost management refers to all cost-related activities in the course of a project, such as costing, planning, monitoring, and billing.

Cost management is of central importance for a project. It must cover in detail

- Procurement costs for machinery, apparatus, buildings, process control equipment and systems, infrastructure, and other materials
- Erection and installation costs
- Engineering costs due to planning, design, and administrative activities
- Start-up costs

The specialist literature offers a number of methods for determining procurement costs in chemical plant construction [10.7]. The accuracy of the results depends on the method used, the depth of knowledge about the project, and other factors.

It is beyond the scope of this chapter to describe all methods in detail. Just two, widely used in plant construction, merit discussion.

The first can be called the "method of multiplication (or ratio) factors." The simplest application of the technique requires a rough, object-oriented breakdown of the overall costs into, say, the categories:

- Machinery and apparatus
- Buildings
- Piping
- Process control equipment and systems
- Miscellaneous

The machinery and apparatus needed for the realization of a process are the first things to become known during an investment project. They can be calculated from offers or empirical figures and adjusted for inflation and market trends.

The principle of the method is that the costs of all other categories involved in the investment can be determined by using ratio factors based on the costs of apparatus and machinery (Fig. 10.10).

The person practicing the method needs experience to obtain usable results, since the factors include scope-dependent ranges. For example, office structures must be costed differently from industrial structures, and larger factors should be used for new "green-field" plant investments than for extensions based on existing buildings. The assessment of piping costs depends on, among other things, the complexity of the piping structure and the material of construc-

Total investment
- Apparatus, machinery
- Buildings — Factor $\varphi_1 = f$(structure type, plant type, ...)
- Piping — Factor $\varphi_2 = f$(complexity, materials of construction, ...)
- Process measurement and control — Factor $\varphi_3 = f$(functional scope, required availability, safety aspects, ...)
- Miscellaneous — Factor $\varphi_4 = f$(...)

$$\text{Total costs} = (1 + \sum_{i=1}^{n} \varphi_i) \cdot S_0$$

S_0 = total machinery and apparatus cost
S_i = total cost of category i
$\varphi_i = S_i / S_0$

Figure 10.10. Cost estimation by the multiplication factor method

tion. Influences on the ratio factors for process control equipment include functional scope, required availability, and safety aspects.

Experience has shown that the method of ratio factors makes it possible to estimate costs with an accuracy of ca. $\pm 20\,\%$.

The second cost-estimation technique, is "unit-cost estimation." Like the ratio-factor method, it is based on an object structure, but a more detailed one.

The simplest case for cost estimation is the procurement of "substances" (see Fig. 2.9), provided the quantity structure is known accurately enough (see also Fig. 10.24).

To every object at a structural level, the technique assigns a price referred to an object-specific unit (per item, unit area, etc.). The total price is obtained by multiplying this unit price by the quantity. For example, a possible cost structure for process measurement and control is shown in Figure 10.11.

This method is suited both to coarse and fine estimates, since cost estimates can be obtained for virtually any degree of subdivision. The accuracy of estimation increases with the level of detail. At the same time, however, more effort is required because the detailed breakdown calls for thorough knowledge, which can be acquired only gradually and after the project has been started.

The amount of effort can be reduced if the in-depth analysis is restricted to cost-relevant items while only a rough estimate is done for items of less relevance to the cost. This procedure does not result in any marked loss of accuracy. Cost-relevant items are those that make a large contribution to the total cost, for example by virtue of large quantities and/or high unit prices.

Incorrectly determined unit prices have less effect on the estimated cost than does an incorrect quantity structure. Experience indicates that this is where most errors occur.

Multiplication factors and unit prices are obtained by evaluating the invoices for a number of investment projects already carried out, extracting the cost structures, and then taking averages.

Today, installation is rarely carried out by in-house departments, and is mostly performed by contractors.

Installation can be paid for in any of three ways [10.8]: by expenses, by unit price, or by fixed price.

When charges are based on expenses, only the price per hour of work is established; the fee is then calculated from the number of hours worked. This form of payment requires careful, continuous oversight.

When the unit price basis is used, a price is agreed on for each unit of performance. The total charge is the sum of these unit prices. This form of payment is suitable especially when changes and additions to the cost category are accepted. This billing system makes these particularly easy to handle. The amount of bookkeeping effort involved, however, is not inconsiderable.

The fixed price system calls for an exact description of the work to be done, largely completed plans, and installation conditions free of disruptions. The administrative cost is low. Changes in required installation may, however, lead to substantial cost overruns.

Engineering costs include all costs arising from

- Planning design activities
- Contract administration
- Supervision of installation work
- Start-up

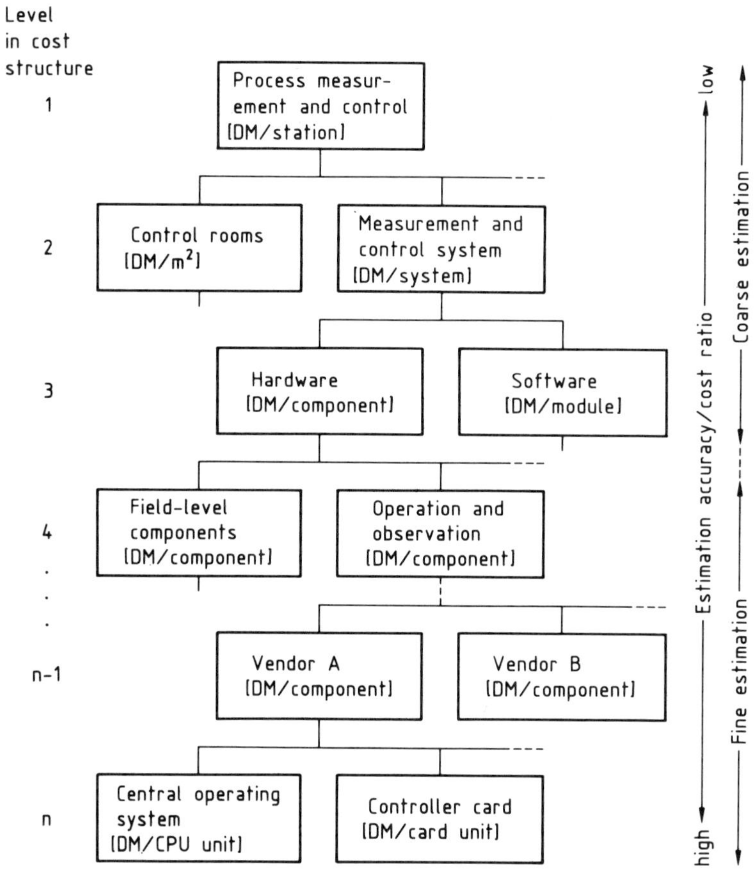

Figure 10.11. Example of a cost structure for unit-cost estimation

In Germany, for example, engineering costs are largely set by the regulation governing architects' and engineers' fees [10.1].

Design costs are generally estimated on the basis of unit prices for stated functions. These activities can, however, vary in difficulty, and this fact is reflected in variable ratio factors. Some planning and design work, for example in the areas of feasibility study and project management, will be roughly equal in volume for any design project. This weighs more heavily on small projects than on large ones, so that small projects are more costly than large ones.

Contract administration and installation supervision are calculated on a percentage basis; commissioning is calculated in accordance with the man-hours required.

By way of example, Figure 10.12 shows the cost structure of an investment for erecting a green-field chemical plant.

Cost estimates are made in parallel with the planning and design phase of a project. They become more accurate as the planning progresses.

The results of cost estimation form the basis for a profitability analysis and become part of the application for approval of the investment. Upon approval, the estimated costs become the binding cost framework for the project. Overruns are permitted only in a few justifiable (and exceptional) cases.

Adherence to the cost schedule requires continuous monitoring of costs over the project; an accounting of costs must be presented at the end of the project.

Schedule Management. In simple cases, scheduling can be carried out by using bar charts (Fig. 10.13, top), which show the beginning and duration of each activity as well as the earliest

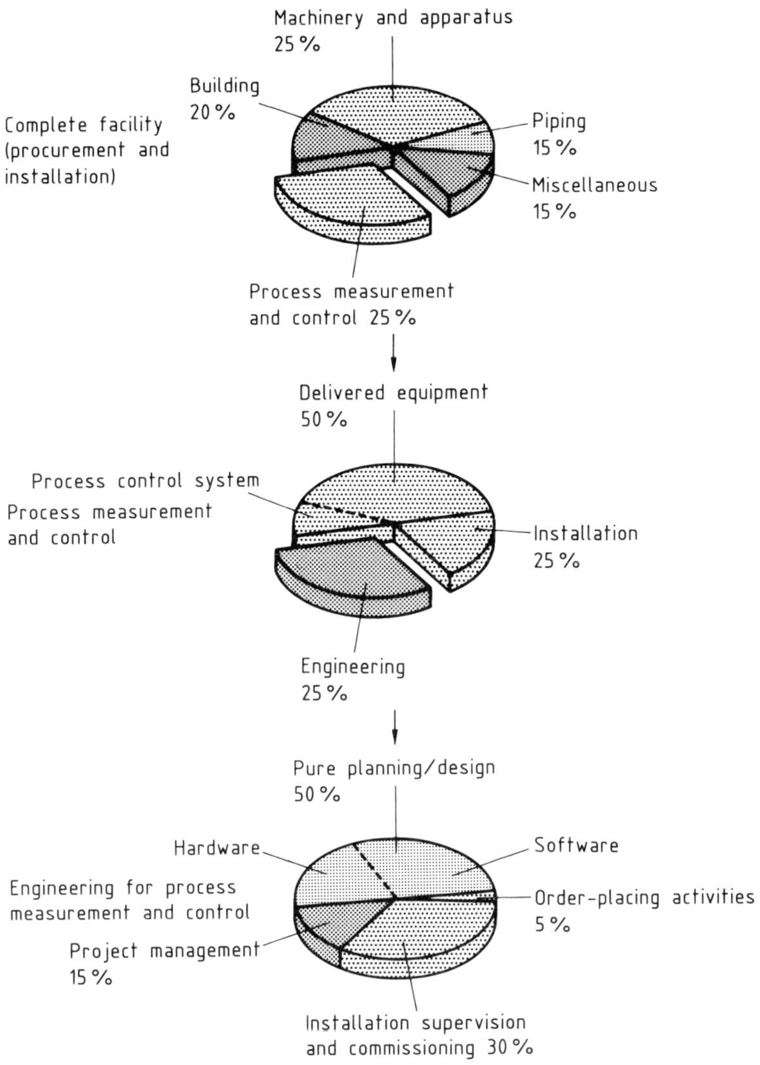

Figure 10.12. Cost structure of a chemical plant investment

and latest occurrence of procedures along with buffer times. Such charts can be prepared at various levels of detail; if they are properly maintained, they provide a good overview of the schedule situation at any time. However, their critical disadvantage is that they do not show the interdependences between activities. For projects having a complex sequential structure, scheduling cannot be planned or overseen on this basis. Network charts can be of use in these cases [10.9].

In network charting, activities and relationships are stored in network files and graphically plotted, for example as shown in Figure 10.13, bottom. This technique makes it possible to determine not only the earliest possible but also the latest acceptable inception points for activities. Furthermore, if buffer times are analyzed, "critical paths" can be identified. A critical path is a path through the network that has no slack time.

Projects may contain more than one critical path, and these require especially close supervision during project execution. Schedule changes may open up new critical paths, but their influence on the overall schedule situation can be cal-

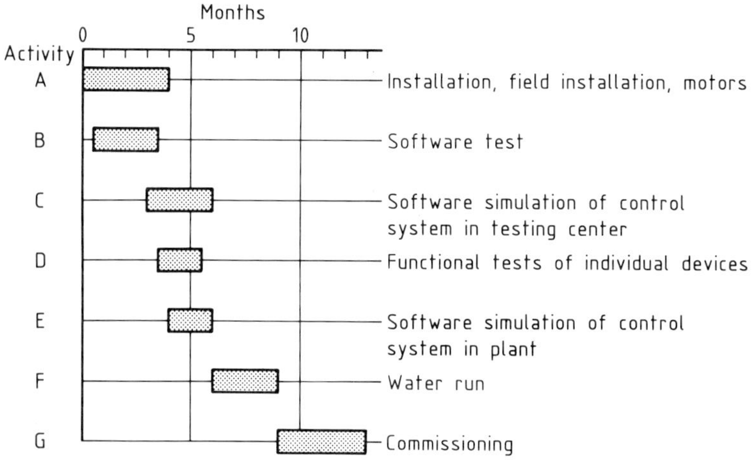

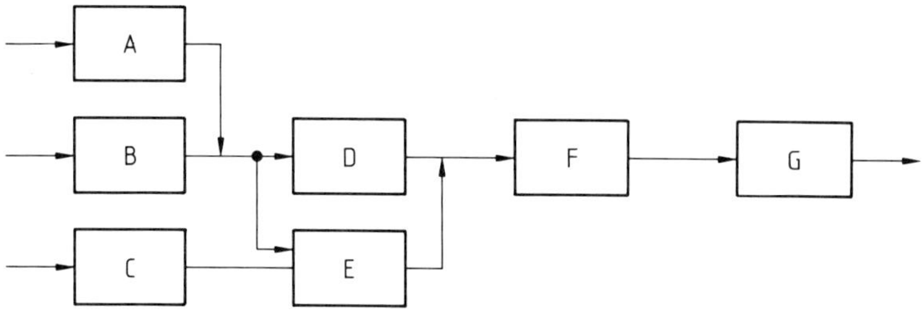

Figure 10.13. Bar chart (Gantt chart) and network chart

culated in advance, allowing prompt reaction to such changes.

Network methods offer the project manager an aid in focusing on essentials in schedule supervision. The manager thus becomes free to perform other management tasks.

The charts used for this purpose must be continually updated. Because of the work involved in updating and the complexity that is sometimes associated with network charts, expecially in large projects, it is recommended that specially trained personnel be assigned to this job. A hierarchical procedure is also applied to network charts, where each decision maker is furnished with the requisite information structure. Petri networks (see Chap. 2) are the basis for all types of network charts.

Resource Management. Resource management (resource allocation) is an extension of schedule management [10.6], whereby "resources" are personnel, machinery, funds, etc.

In general, schedules are drawn up as if the needed resources were available in adequate quantity at any time; however, only rarely is this actually so. Generally, in-house and outside resources are limited. Resource shortfalls or the complete unavailability of resources will result in delays if, for example, certain activities must be carried out later or not at all, or if activities that should be performed in parallel and have been so scheduled, have to be carried out in sequence instead.

Such imponderables in scheduling can be avoided if scheduling and resource planning are combined. There are two approaches: resource-constrained allocation and resource leveling.

Resource-constrained allocation is based on what resources are available. Deviations from the original schedule are allowed, but the project duration must be made as short as possible.

In resource leveling, the specified completion date is the top priority and additional resources are procured to the extent required to achieve it.

In practice, an intermediate course is generally followed.

The resource "personnel" is often a cause of special concern. In boom times, for example, a

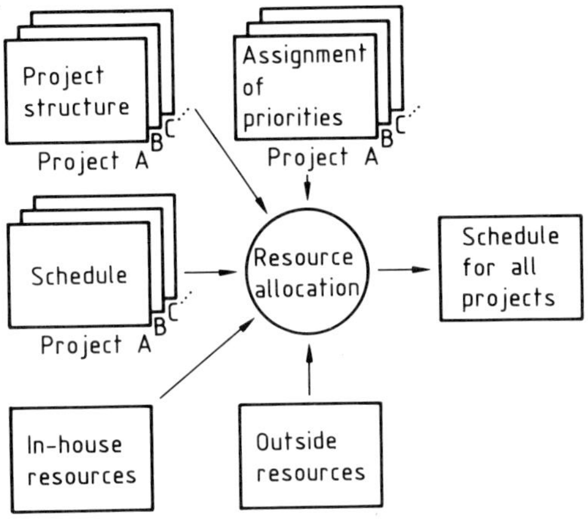

Figure 10.14. Resource management

resource leveling approach is questionable because not enough of the specialists needed for execution of a project may be available. One way out of this conflict, though it is seldom considered, is to reduce the originally stated scope of work. The prerequisites for doing so are that the quality of the work will not suffer and that retrofitting will be possible later on at no significant extra cost.

The problem of resource management becomes more complicated if several projects differing in structure must be planned at the same time, as often occurs in the chemical industry. The problem can often be solved only by assigning priorities to the projects (Fig. 10.14).

Project activities are supported in a variety of ways by computer-aided management tools. These have been reviewed in [10.10] and [10.11].

Recently, the Microsoft project management system MS Project [10.12] appears to be finding acceptance in multiproject management situations.

10.3. Decision Phase

The fine structure of the decision phase can be seen in Figure 10.1. The purpose of the feasibility survey is to elucidate and develop the statement of tasks in close collaboration with the customer. Figure 10.15 shows possible points covered by a feasibility study for a chemical plant construction project.

The economic aspects are just as important as the chemical and process principles. These are supplemented by substance data and constraints.

- o Economic aspects
  - Market interest
  - Contemplated production level
  - Nature of investment project
    - * New plant
    - * Expansion
  - Intended site
  - Intended capital investment
  - Economics or profitability
- o Chemical principles
  - Reaction equations
  - Mole ratios
  - Reaction stages
  - Enthalpies of reaction
- o Process principles
  - Structure
  - Product and heat streams
  - Process conditions
  - Mode of operation
    - * Batch/continuous
    - * Single-component/multicomponent
- o Substance properties for feeds, intermediates, byproducts, and end products
  - Physical data
  - Toxicological data
  - Ecological data
  - Engineering data
  - Safety characteristics
- o Constraints
  - Legislation and internal requirements
    - * Environmental protection
    - * Plant safety
    - * Occupational safety
  - Patents
  - Local conditions
- o Description of plant
  - Structure
  - State of preservation
  - Utilization
  - Availability
  - Operation and monitoring
  - Engineering functions
  - Safety requirements
  - Infrastructure

Figure 10.15. Possible topics covered by feasibility study

A detailed description of the plant completes the picture of the project. An additional analysis of weak points will reveal any flaws and thus contribute to making the task statement more concrete and precise (see Chapters 2–4).

This stage results in the performance specifications, which set forth the solution-invariant description of the problem to be solved (requirements). Its structure and content, in particular as it relates to plant safety, is discussed in more detail in Section 10.4.

In the preliminary design, a number of alternative solutions can be offered to the stated problem; from these, one is to be chosen as optimal for the company from both economic and technical standpoints (see Section 10.1), whereby a cost/benefit analysis may be useful.

Any optimization potential in the selected solution can then be revealed by sensitivity analysis, which investigates the influence of important, highly sensitive variables on the process by means of, for example:

- Mathematical models
- Application of empirical and theoretical knowledge
- Plausibility analysis

The optimization potential thus found can be realized in the design study. The result is a binding concept for both process and plant.

The binding concept is described in the specifications, whose structure and content are discussed further in Section 10.4, with particular regard to plant safety and economics.

The decision phase ends when the documents required for obtaining regulatory approval have been prepared and costing has been done (Section 10.2).

Errors in project planning and execution can arise from a wide variety of causes. The more serious are these problems and the later they are discoverd, they higher is the cost of correcting them.

10.4. Specifications

Bidding Specifications. The specifications describe the task in structured form, mainly from the user's viewpoint, and state what effort must be brought to bear [10.2], [10.13].

The bidding specifications can be regarded as the basis for a contract betweeen the party designing and constructing the plant and the later plant operator. They are the basis of all further planning and design.

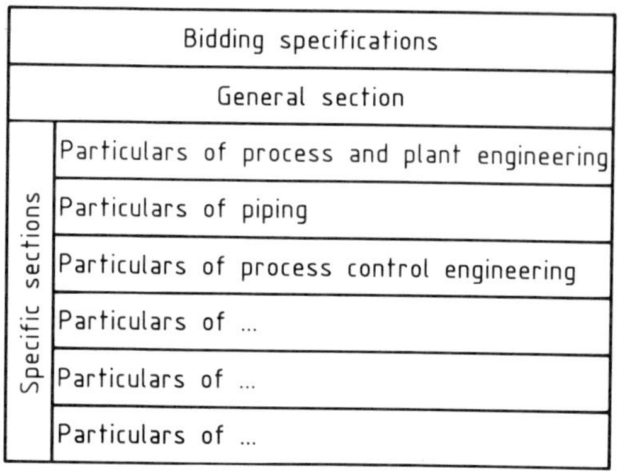

Figure 10.16. Breakdown of bidding specifications

Structure and Content. The bidding specifications are broken down in parallel with the topics in the feasibility study (Fig. 10.15). They are broken down into a general section relating to all components and a specific section for each component (Fig. 10.16).

The general section includes important information applicable to all components. Possible topics in the general section of the bidding specifications are:

1) General information on the project
2) Current position and shortcomings
3) Objective
4) Structure of the process and plant
5) Safety and availability requirements
6) Miscellaneous conditions to be considered
7) Framework concept

A vital part of the general section of the bidding specifications is the "framework concept," which describes the real environment of the plant as well as expected expansions. This section, in particular, describes interfaces to other plants (not just for process measurement and control).

The component sections of the specifications discuss these points only briefly, or may simply refer to the general section. Instead, these sections state the points of importance for the execution of the particular tasks. Possible contents of the section on process measurement and control are:

1) Typical requirements for process control station functions
2) Special control and monitoring functions

Corporate management	Plant management	Process management
■ Strategic planning ■ Logistical systems Contract performance Maintenance Operations planning Procurement ■ Information systems • Sales and marketing Sales Partners Sales planning • Production Production planning (overall) Capacity planning Investment planning ■ Quality assurance ■ Internal cost accounting ■ Personnel management	■ Administration and production logistics Production planning (detailed) Plant section operations planning Personnel planning Production flow monitoring Batch approval Schedule monitoring ■ Economic tasks Cost functions Cost analyses Statistics ■ Engineering tasks Recipe handling Assessment of process and product properties Machine changeovers Process engineering procedures Quality assurance	■ Higher-level functions: Management of processes and plants • Recipe-mode operation • Process operation Monitoring Intervention Documentation • Process logistics Startup and shutdown Load and product changing Buffer strategies • Disturbance handling ■ Field-level functions: Measurement, control, safety • Determination of process properties • Determination of product properties • Setpoint (closed-loop) control • Operational chart (open-loop) control • Reporting of deviations • Shutdown • Operation and monitoring

Figure 10.17. Functions broken down by level of the production model [10.14]

3) Process control functions
4) Human/process communications
5) Off-line interpretation functions
6) Infrastructure; room and building equipment
7) Communications hardware
8) Process analysis
9) Safety requirements
10) Availability requirements
11) Quality assurance
12) Miscellaneous

An important item in the bidding specifications is the description of functions used to control the process. The particulars of process measurement and control are described (Fig. 10.17, column 3) [10.14].

"Field-level" functions for the performance of control and safety tasks are distinguished from "higher-level" functions involved in the management of the process and the plant. The later include, for example, process operation, process logistics, and recipe-mode operation (Chaps. 2 and 4).

These functions in turn are interfaced to other higher-level functions involved in the management of a works/plant complex or corporate management (Fig. 10.17, columns 1 and 2) [10.14].

The breakdown of functions is based on the level model of production (Fig. 10.18) [10.15] (see also Chap. 2).

The totality of functions to be performed and their interfaces to one another defines the type and size of the process control systems. For example, Figure 10.19 shows a system that handles the functions of all levels of the production model (see also Section 5.2).

The holistic view and conceptual procedure require prompt acquisition of information about necessary and desired functions for the later operation of the plant. Particulars of various pro-

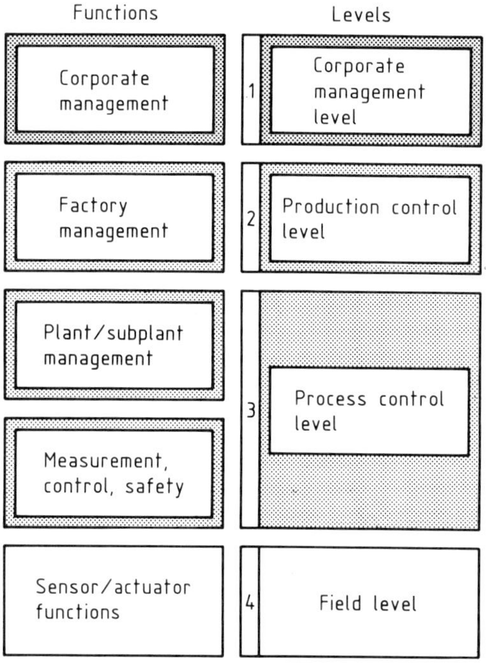

Figure 10.18. Level model of production [10.14]

cess management tasks are treated in depth in Section 4.3.

The rules stated in Chapter 2 apply in a special way to the description of recipe-mode operation [10.13], [10.16] (see Fig. 10.20), [10.17], [10.18].

As proposed earlier [10.13], the bidding specifications are now prepared in checklist form [10.19].

Process engineering functions may also include room and building functions (heating, climate control, ventilation, lighting, etc.) as well as communications (e.g., telephone).

Specifications. The specifications give a detailed description of the nominal condition, that is, how and with the requirements of the bidding specifications are to be realized. The specifications set forth the solution-variant realization. These documents are binding on the contractor (in-house planning and design department or outside firm). The specifications form the basis for the contract, especially when the planning and design work is contracted out [10.2].

Structure and Content. The specifications are organized similarly to the bidding specifications and consist of a general section and component-specific sections.

The process-engineering aspects governed by the specifications are illustrated in Figure 10.21. For the sake of comparison, provisions for the process and for the plant are listed in parallel [10.8], [10.20], [10.21].

Provisions on process and plant are first briefly explained (left column of Fig. 10.21).

The structuring of the process and of the plant is the same as set forth in the bidding specifications (see also "Phase Model of Production" in Chapter 2, as well as Figs. 10.22–10.24). The breakdown is taken over into the specifications unless changes have been made; otherwise, it is modified to accord with the new requirements.

The breakdown of the process is independent of whether it is already known or new within the company.

Even known processes must be restructured since a new investment always entails adaptation to the state of the art, and adaptation can strongly affect the process structure.

According to [10.22], the process can be broken down into process stages and unit operations. Some concepts in [10.22] must be modified, because there is no guarantee of completeness, uniqueness, or freedom from conflicts. "Process stages" are generally self-contained portions of a process.

Unit operations are the smallest steps in a process and correspond to process elements. Any unit operation converts inlet products with particular properties to outlet products with different product properties. As was discussed in Chapter 2, product properties are those qualities that characterize the product, such as physical and chemical values, engineering properties, and product indicators [2.102], [10.23].

A unit operation can be specified as shown in, for example, Figure 2.23. It can be specified, for example, that a separation is to be carried out by distillation. Such a statement of the method of performance also establishes the basic form of the trajectories that describe the transformation of inlet products with their properties into outlet products with theirs. The properties that characterize this transformation process are the process properties.

Figure 10.22 shows a unit operation with the inlet and outlet products, their product properties, and the process properties. The product and process properties must be set forth in the specifications for the process. As Figure 10.23 shows,

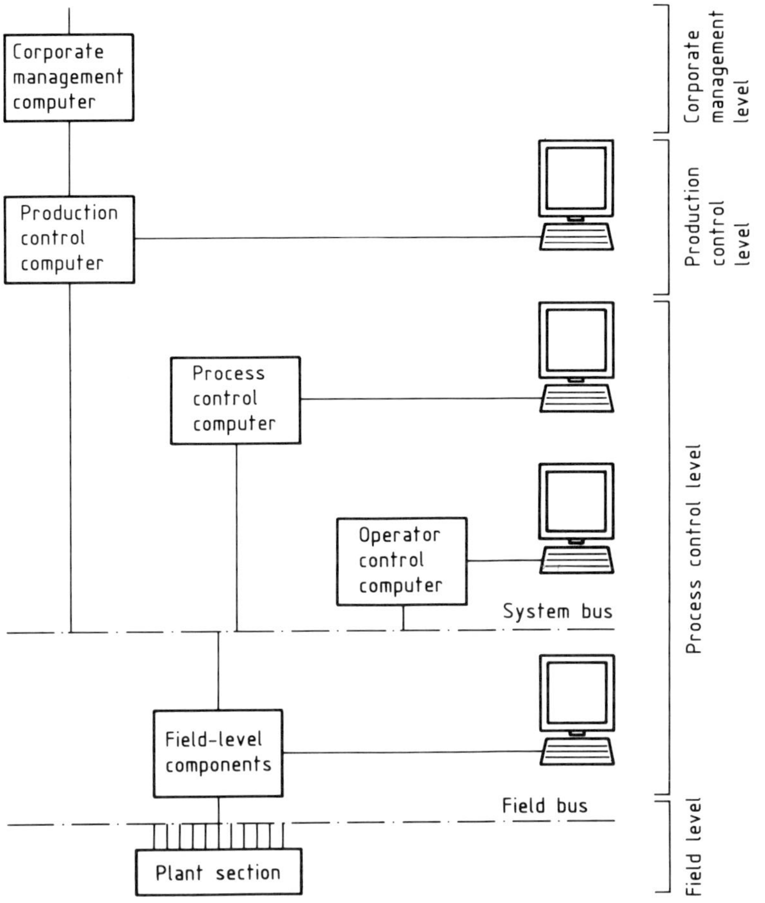

Figure 10.19. Schematic diagram of a control system [10.14]

the acceptable ranges of variation (tolerances) should also be given if possible. In this way, a profile of requirements on thc overall process is obtained; this should cover all operating modes (startup, shutdown, load change, etc.; see Fig. 10.24). This description, coupled with actual values, is the basis for a quality information system (QIS).

The profile of requirements is the essential condition for creating a reproducible operation [10.2]. The overall sensor and actuator system is later defined and quantified on the basis of this profile. Further more, possible ciritcal states of product properties, which might affect the availability of the sensor–actuator system, can be identified in this way (plugging due to solid/liquid phase separation, etc.).

Material and thermal balances indicate ways of insuring the supply of substances and energy to the process and the removal of substances and energy from it. They also give the consumption rates per unit amount of product, which are required for the profitability calculation at the end of the decision phase (Section 10.4).

Once the process structure has been completed, the plant structure is broken down by assigning unit operations to appropriate plant sections. A plant section is a functional unit of a plant that can be operated independently (corresponding to a unit operation). Examples of plant sections in chemical process plants are absorbers, chemical reactors, crystallizers, tank storage facilities, and so forth.

The totality of all plant sections required for the performance of a process is the plant.

Technical equipment in a plant section includes both devices involved in the process, such as vessels, machinery, apparatus, piping, and fittings, and process measurement and control units [10.24]. The detailed description of a plant section is based on these elements (see Fig. 2.22).

Safety and economic considerations result in a variety of requirements being imposed on the objects of a plant, such as plant sections and technical equipment. Such requirements must be taken into account in planning and design, construction, and operation.

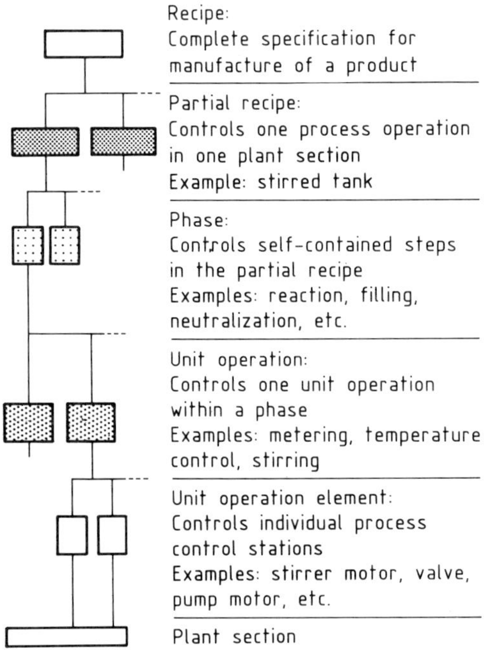

Figure 10.20. Example of recipe control [10.16]

In the lifetime of a plant, a large quantity of data is generated, used, interchanged, and if necessary modified. If such data and other information is to be exchanged without losses, each item of information must be uniquely associated with the plant object to which it relates. This can only be achieved if the objects are uniquely labeled [10.25].

Plants and apparatus can be identified in various ways. For example, each can be assigned an unstructured designation.

In a complex system such as a chemical plant, it is better to structure the system in a hierarchical fashion, and then have the system of identification codes reflect this structure. In this way, an informative and easy-to-use identification of plant objects is obtained [10.26], [10.27].

An engineering object can be "named" in many ways; a few examples are: by purpose, by technical function, by location, or by device type.

If all such features were included at once in the object identification, an unwieldy code would result. For this reason, just a single attribute is employed. It is desirable to choose a feature that is completely and permanently identifies the object (e.g., the technical function

Process/plant specifications	Process control specifications
o Process - Structure - Material and thermal balances - Unit operations - Product and process properties	o Field-level and associated devices - Type of station and interconnection - Types of devices - Safety and availability measures
o Plant - Structure - Plant sections - Safety concept	o Systems hardware and software - Level of automation - System structure - Safety and availability measures - Operation and monitoring - Software specifications
o Infrastructure o Miscellaneous - Construction concept - Provisional site designation	o Infrastructure and building equipment - Process control rooms - Power supply - Safety and availability measures
Rough cost estimation	

Figure 10.21. Structure and content of specifications

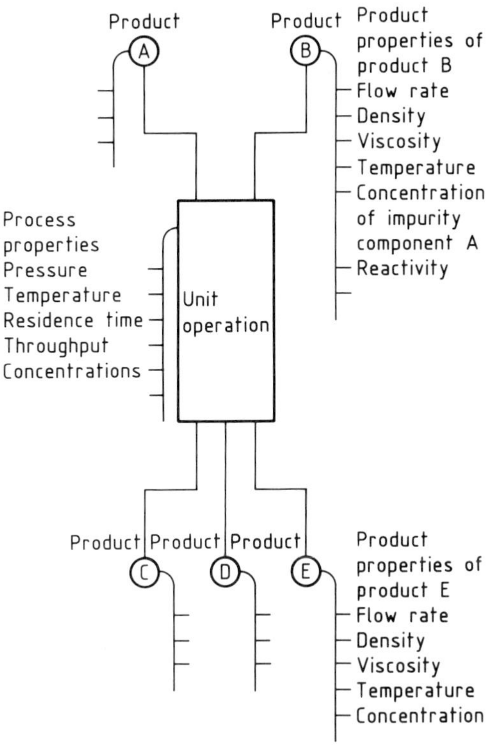

Figure 10.22. Extract from a phase model

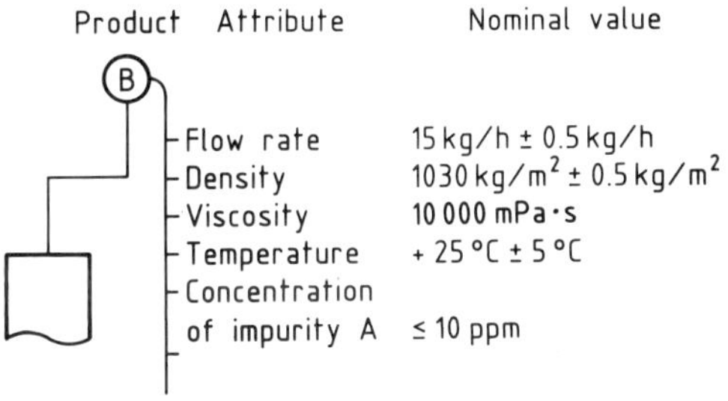

Figure 10.23. Example of a list of product properties

[2.69]). This can be illustrated for a temperature measurement and control station.

Suppose the purpose of the station is to maintain a constant temperature in a certain unit operation, say mixing. The technical function is to measure, record, indicate, and control (TIRC) at a certain point in the plant, a Pt 100 instrument being used as the measuring sensor.

If this sensor is now replaced by a thermocouple or if the plant is changed over to a different process with different unit operations (e.g., reaction instead of mixing), the station still retains its technical function.

The coding of technical objects is a planning/design task. It is carried out as early as possible, but at the latest when the specifications are prepared. Today this is done in entity–relationship (E/R) models, which are generated in a consistent way throughout the company.

On the basis of the safety provisions in the bidding specifications, the safety concept is derived from the operating conditions together with the potential hazard presented to persons and the environment by substances involved in the process.

A construction concept and a provisional identification of the site round off the process and plant concept.

The overall concept is documented in phase models plus the overall and process flowsheets discussed in Chapter 2, which contain fundamental and supplemental information.

The process measurement and control concept follows from the process and plant structure, the profile of requirements on the process, and the task statement contained in the bidding specifications.

The specifications set forth the types of process measurement and control stations and, roughly, the way in which they are interconnected.

The process and plant control functions performed by the control system determine the extent to which the plant is automated (see Section 4.3).

If functions on the plant control level are also to be implemented, a production control computer must be provided. The interfaces between functions at different levels must be described and specified as exactly as possible (see Chap. 2).

The number of field-level components and their assignment to plant sections is based on the process structure, the required functionalities, and the safety and availability requirements.

The operation and monitoring concept (see Section 11.2) essentially comprises the specification of operator displays, reports, and logs. The operator station concepts must also be worked out. These define the scope of tasks to be performed at the various control stations and the association of control stations with plants and plant sections [10.28].

The information structure should be hierarchical, with high-level diagrams containing few

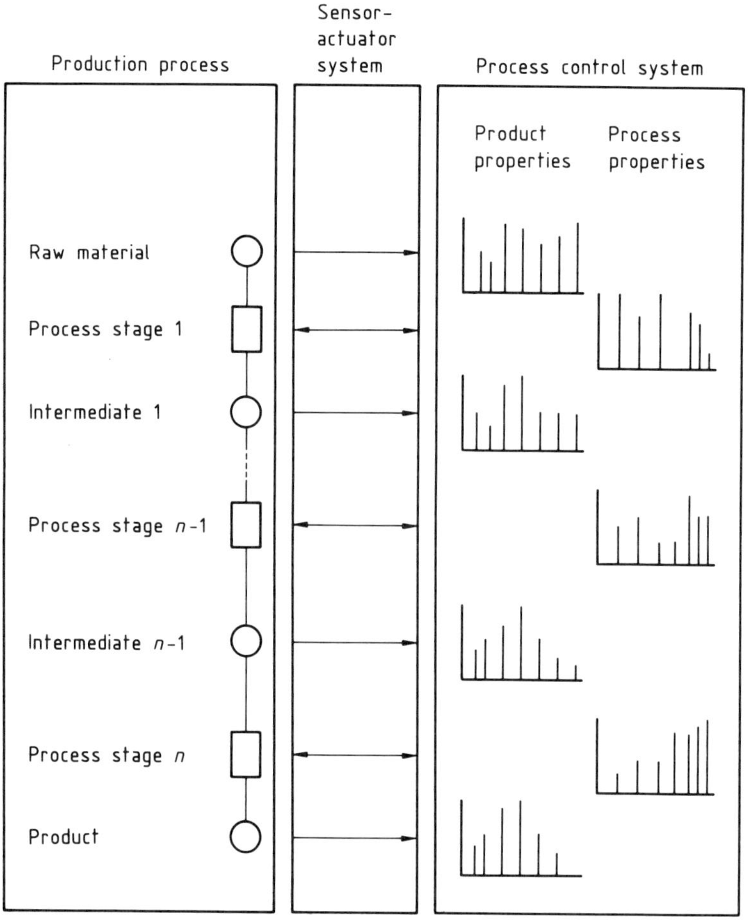

Figure 10.24. Property profile of an overall process [10.23]

items of information, and station-level diagrams, extensive information.

Reports should be specified and classified as follows:

- Alarms
- Malfunctions (in the measurement and control system)
- Warnings (advance notice of alarms)
- Errors (in the peripheral hardware of the measurement and control system)

Safety and availability requirements govern how priorities are to be assigned to the messages and how they are to be presented in the control system (see also "Availability Aspects" below).

Logs and records are subdivided into, for example, batch records, shift logs, daily logs, and alarm logs.

If, for example, the bidding specifications call for recipe-mode operation, the software specifications (Fig. 10.21) should include the appropriate large-scale function charts, specify the selectable parameters of the recipes, and state how the recipes will be maintained and archived.

At this point, the task to be performed has been explained enough that a concrete solution proposal can be submitted with the aid of a process control system. This proposal must, however, take account of the measurement and control provisions described in the framework concept of the operation in which the process will be realized. For example, if a process control system from a certain manufacturer is already in use, then for maintenance reasons it is recommended that the planner stay with this manufacturer unless the adaptation to the task carries unacceptable risks with it.

Availability Aspects. *Basic Classification of Process Measurement and Control Equipment.* Availability requirements must be examined from several viewpoints:

- Safety
- Process engineering
- Logistics

Safety is mandated by legislation and regulations. The safety of persons in the plant (worker protection) and of the surroundings (environmental protection) must be guaranteed.

The safety of engineering processes is a key criterion. A high level of safety results from continual and intensive work on safety aspects. Once a high degree of safety has been achieved, it must be maintained and improved. The special handling of safety aspects in the bidding specifications and the specifications is meant to stress this requirement.

The term "process" is understood to mean the totality of interacting operations in a system by which matter, energy, or information is transformed, transported, or stored. In the case of technical processes, the physical quantities are determined and controlled by technical means [10.29].

The safety of a process is determined by the processes, plants, and process control equipment (Fig. 10.25). An essential contribution to processs safety is due to process control equipment [10.30]. Such equipment can be classified as shown in Figure 10.25.

In assessing process safety, the following three types of operational conditions are distinguished [10.30]:

- Good range
- Acceptable range
- Unacceptable range

The good range is the intended range for the values of a process parameter. It is defined such that the plant does not sustain any damage beyond ordinary wear and tear, while the quality and quantity of goods produced satisfy the requirements imposed.

The acceptable range is adjacent to the good range. In it, the plant still sustains no damage beyond ordinary wear and tear; the quality and quantity of goods produced still satisfy the requirements to within extended tolerances.

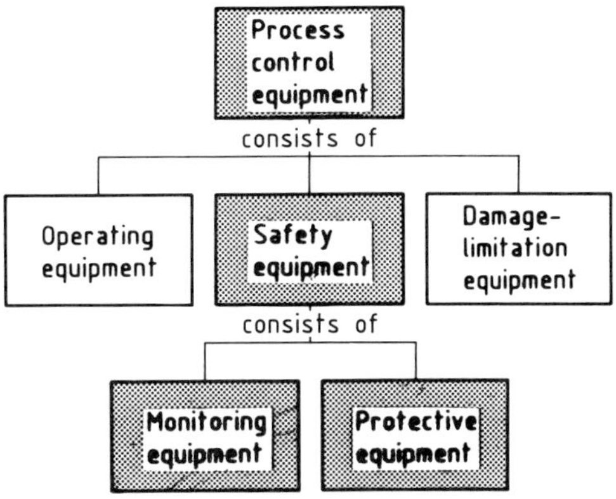

Figure 10.25. Summary of measurement and control equipment performing process safety functions

In the unacceptable range, the value of a process parameter is such that the plant may sustain or cause greater damage, or the quality and quantity of goods produced no longer satisfy the requirements stated.

These parameter ranges can now be correlated with process measurement and control equipment. Devices used for measurement, control, reporting, and so forth serve to keep the plants operating as intended (in the good range). They continually intervene in the process. Only in the acceptable range do the process control safety devices first come into operation. On entry into this range, the monitoring devices respond, prompting the operator to take suitable action to bring the process back into the good range.

If these actions are not taken as expected, protective features in the process measurement and control system, such as safety valves, rupture disks, and the like, prevent entry into the unacceptable range.

Processes can further be provided with damage-limitation features such as spray-blocking walls, warning systems, and special traffic control systems that block off the site of the danger.

Figure 10.26 summarizes the correlation between the ranges of process parameters and the process measurement and control devices, showing the intervention levels and modes of action [10.30].

The safety requirements mandated by legislation for the overall process in a plant to be erected, and in particular for the process control equipment, must be precisely described in the

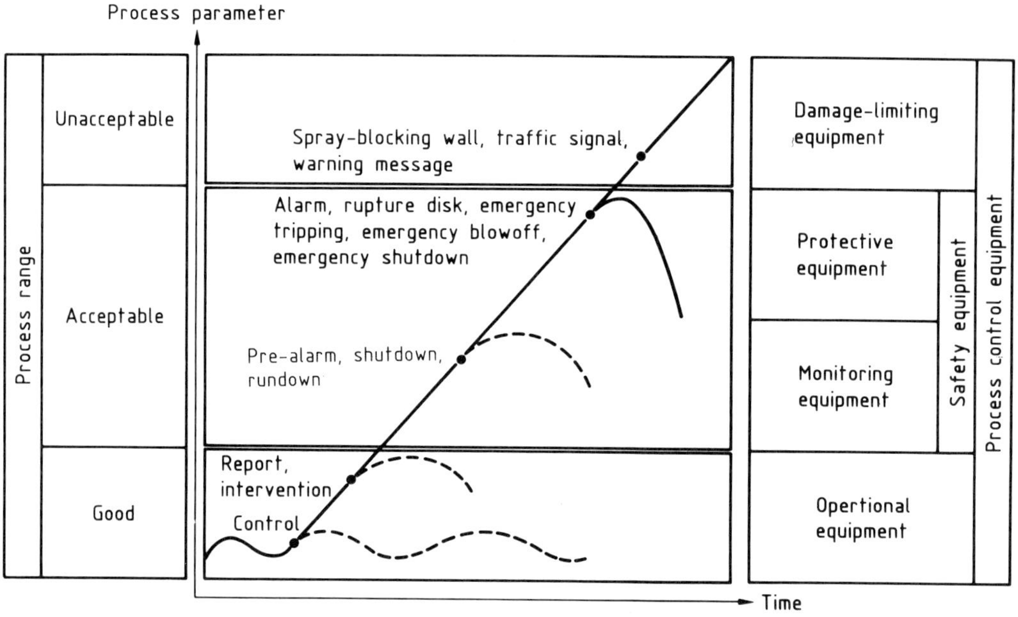

Figure 10.26. Process control devices and process safety [10.30]

bidding specifications, since they influence the process design as well as the design of the process measurement and control system.

The required availability from the process engineering standpoint affects both product quality and the protection of the facilities. The logistics-based availability bears on the ability to deliver products and has direct effects on manufacturing costs. The availability requirements imposed by logistics and process engineering must be included in the economic picture, in contrast to the safety requirements.

The measures taken to comply with these requirements are set forth in the specifications.

Implementation. The safety and availability requirements are key factors affecting the structure and organization of field-level and system-level components. These lead to, among others, the need for the process sections and/or plant sections to be largely autonomous relative to one another and relative to malfunctions in higher-level operating and/or control computers.

To insure the protection of workers and the environment as mandated by law, a number of other technical measures set forth in codes and regulations must also be undertaken. A few of these are listed in Figure 10.37.

Allowance must always be made for defects that limit the proper functioning of technical equipment or even take it out of action. In the case of safety equipment, care must therefore be taken that no impairment or failure of the safety function occurs even when there are defects in the technical equipment performing such safety function.

This goal can be achieved through two distinct approaches to safety equipment design, called "fail-safe" and "redundant." In fail-safe design, all malfunctions deemed possible cause the safety devices to fail in a single well-defined sense, namely so as to trigger the safety or protective function. Redundant design means including more than one component for a certain task. If one component fails, a backup component takes over its function.

Redundancy in process control systems means that system components, such as field-level, operating and monitoring units, and/or bus systems, are present two or more times. The types of redundancy practiced in technical systems are as follows [10.31]:

- Static redundancy
  Instead of a single component, several components are installed in parallel. The system

as a whole performs in the intended way as long as at least one unit is free of defects.

- Dynamic backup redundancy
  In case of malfunction, an additional component is switched in to replace the defective component. The additional component is not in operation when no malfunction is present.
  - Local dynamic backup redundnacy
    Assigned to every component is a redundant component, which is continuously reconfigured (hot standby).
  - Global dynamic backup redundancy
    A redundant component is assigned to several components. In case of a malfunction, the redundant component is reconfigured once (warm standby).
- Dynamic functional redundancy
  When no malfunction is present, the redundant component performs its own tasks; if a malfunction occurs, it takes over the functions of the defective unit by restricting the capacity or shutting down inessential functions.
- Backup
  The general term "backup" refers to dynamic backup redundancy for a single process measurement and control function, such as control.

There are a variety of methods for acquiring redundant information about the process. Figure 10.27 presents some typical examples.

The term homogeneous redundancy is used when the same physical quantities and the same measurement techniques are empolyed for data acquisition. Physical diversity uses distinct physical quantities. Methodological diversity means using the same physical quantity for data acquisition, but measuring it by different methods.

In process measurement and control systems, functions are implemented through hardware and software. Process control systems should be used to realize safety devices only when they can be validated, that is, when it can be demonstrated that the software involved in the safety function always performs properly and reliably under all circumstances. If this cannot be demonstrated, safety functions must be implemented by hardwired control units.

Redundancy implies a high-availability concept. Despite, or indeed precisely because of, the presence of multiple process components, such high availability is not always realized. Problems arising from the number, interconnection, and

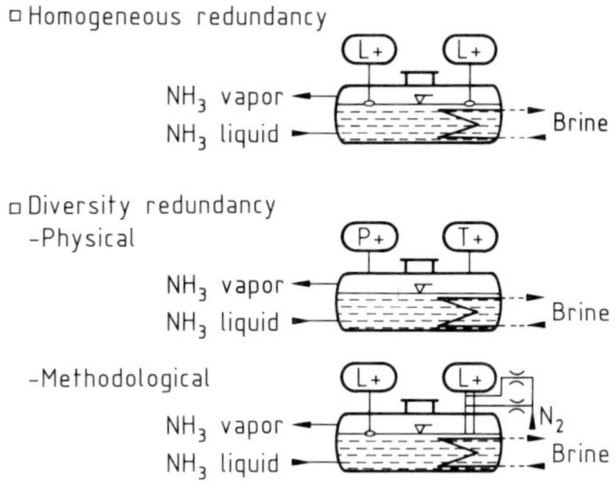

Figure 10.27. Redundancy in measurement [10.30]

synchronization of components in a redundant system may work against it.

Safety concepts and availability requirements must also affect the security of the process power supply system. The number and type of incoming supply lines, multibus configurations, switching devices, and standby power systems depend on it [10.32]. Section 6.4 surveys the most important options.

The specifications contain additional provisions dealing with the infrastructure (Fig. 10.21, p. 620), including process control areas, such as control and switching rooms, as well as overall building equipment.

The specifications end by giving cost figures and the profitability calculations based on them. These are crucial for the decision to continue with the investment project. If the economic criteria established by corporate management are satisfied, the project is continued; otherwise, either its scope is modified or its is terminated.

Economic Aspects. The purpose of any investment project is to generate profits. The specifications must therefore demonstrate that the proposed actions to carry out the tasks described in the bidding specifications will contribute to profits. The question of economics is vital, especially with regard to the desired level of automation (Section 4.3) [10.2]. A separate economic calculation can be done for each automation level.

- Automation of field-level functions of the process control level (Fig. 10.17, p. 617)

This level is mainly concerned with the functions of measurement, control, operation, and

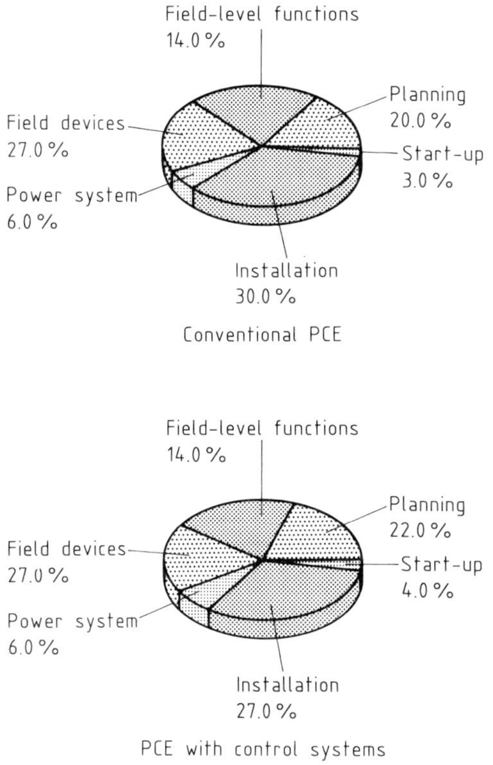

Figure 10.28. Comparison of cost breakdowns for the implementation of field-level functions of the process control levels with conventional process control technology and with process control systems [10.14]

reporting. These can be implemented by conventional method or by process control systems. An analysis of several projects has shown that there is, in principle, no difference in financial outlay between these two options. Both investment costs and the breakdown of costs (Fig. 10.28) are roughly comparable. On the basis of this knowledge, control tasks of the type discussed here should always be implemented with process control systems, since these have potential for expansion that is generally lacking with conventional methods.

- Automation of field-level and higher-level functions of the process control level

As Figure 10.17 shows, process control systems make it possible to carry out recipe-mode operation, tasks having to do with process logistics, and higher-level control functions. The automation of packaging lines or storage facilities can be included under this heading.

Performing these functions generally entails additional investment costs, which are, however, justified only if the annual profit gained through their use exceeds a certain value specified in advance by corporate management. The profit is calculated as the savings achieved through the investment minus fixed costs.

The profit expressed as a percentage of the total investment is called the return on investment (ROI).

$$\text{ROI}\,[\%/\text{a}] = \frac{\text{savings [DM/a]} - \text{[fixed costs [DM/a]}}{\text{total investment [DM]}} \times 100$$

The reciprocal 100/ROI is the payback time, i.e., after how many years the total investment is paid back in profits:

Either figure can be used as a rough measure of the profitability of an action.

Table 10.1 shows a cost analysis done in this way. The benefits can be broken down in a variety of ways, but essentially into five categories, although not all of these can be quantified. If only the quantifiable items are included, satisfactorily high ROI values or satisfactorily short payback times (all < 2.5 years) are obtained.

These results favor the use of process measurement and control systems for the implementation of higher-level functions of the process control level.

- Automation of functions of the production control levels (Figs. 10.17 and 10.18, pp. 617–618)

The economic analysis is problematic because the simple investment return calculation methods described above are no longer adequate. The associated benefits of better information (i.e., comprehensive, up-to-the-minute, error-free, and perhaps condensed) being available as the basis for analyzing both engineering and management weak points and also for the real-time control of the operation cannot be quantified.

Ultimately, corporate objectives will dictate the implementation of a production control level. Failure to do so, however, may result in long-term competitive disadvantages.

10.5. Execution Phase

The fine structure of the execution phase is indicated in Figure 10.1, p. 605.

Table 10.1. Cost analysis for the implementation of higher-level functions at the process control level using process measurement and control systems [10.33]

Task	No. of projects	Savings*, 10^3 DM/a							Total investment, 10^3 DM	Fixed costs, 10^3 DM/a	ROI**, %/a
		Improved product quality	Reduced inspection	Energy savings	Personnel savings	Improved yield	Enhanced plant utilization	Sum			
Recipe-mode operation	2	–	+	–	455	–	385	840	1300	285	40
		70	+	–	650	500	210	1360	1900	375	50
Process logistics	1	+	+	+	400	1200	+	1600	2500	595	40
Higher-level control technology	1	+	–	160	220	–	–	380	540	145	40
Automated packing/storage	2	+	–	–	+	–	+	480	455	115	80
								4080	2400	580	140

* + Beneficial but not quantifiable; – No benefits identifiable. ** Return on investment.

Basic Engineering. In the basic engineering phase, the individual tasks are formulated in their final form, in collaboration with the customer, and the functional solution is described. The work to be done here is listed below:

Planning and design for process and plant engineering:

Design of plant sections (apparatus) in accordance with the process
Decision on materials of contruction
Preparation of site plans and provisional layouts
Preparation of piping and instrumentation diagrams

Planning and design for process control engineering:

Design of process control stations
  Establishment of structure
  Establishment of devices to be used
Design of process measurement and control system
  Dimensioning of process interface
  Design of data processing software
  Dimensioning of the control system
  Dimensioning of control and other rooms
  Spatial disposition in the plant
Preparation of piping and instrumentation diagrams

The process and plant engineering work includes plant design in accordance with the process, sizing of the plant sections (apparatus), and elucidation of materials questions. With regard to documentation, see in particular Section 3.5.

Preparation of Piping and Instrumentation Diagrams. A key task is the preparation of P & I (piping and instrumentation) flowsheets [10.22] (see Fig. 10.29). Provisional layouts and site plans are also drawn up.

The process control engineering organization is involved in the preparation of P & I diagrams, since its planning and design work will be based on these. Process control tasks are described in the station symbols (Fig. 10.30) of the P & I diagrams. This description should be sufficiently exact enough that it, together with the binding concept, makes possible the design of the PCE stations.

The process control station includes all measurement and control elements needed to perform the functions specified, such as sensors and/or actuators, signal transducers, power supplies, and processing functions, for example indicating (I), recording (R), controlling (C), etc. [10.34].

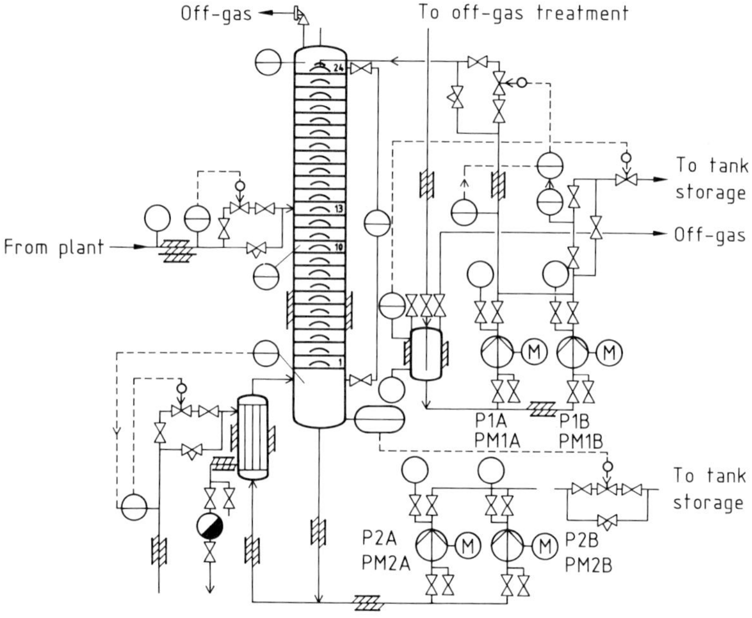

Figure 10.29. P & I flowsheet of a plant

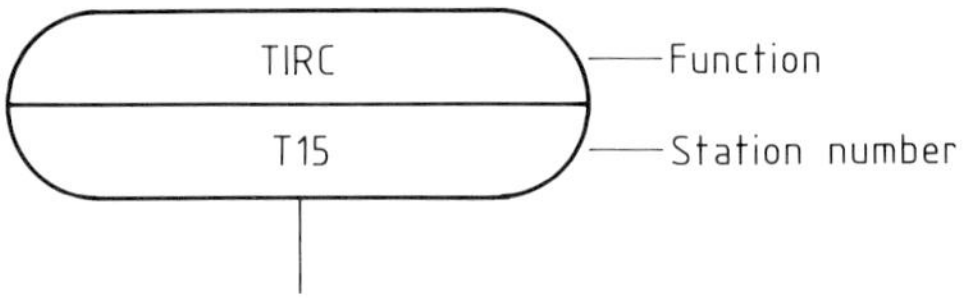

Figure 10.30. Representation of a process control station in the P & I flowsheet [10.24]

Process Control Station Data Sheet. The station design is set forth in a "data sheet," which must list all elements making up the station and also provide information on the substances with which the sensors and actuators come in contact and on the installation location (Fig. 10.31).

Given the measurement and control ranges of the sensors and actuators, suitable equipment is identified for the individual elements on the basis of instrument standards (see Chaps. 7 and 8 and Section 10.6).

Process Control Station Diagram. Station diagrams document how the equipment is wired together (Fig. 10.32). At this point, the planning and design work on stations utilizing conventional technology is finished. If data processing functions are implemented in process measurement and control systems, additional design work is required for the hardware structure of the process interface as well as the software structure incorporated into the control system proper.

Analog and binary signals coming into the process control system must be transformed into information that this system can understand. The conversion takes place in the process interface and is performed by input/output components for the various types of signal. Backup controllers are also implemented with components. These components are shown in the station data sheet as well as the station hardware diagram (Fig. 10.33; see also Sections 7.2 and 10.2).

Station Software Diagram. The station hardware diagrams are supplemented by station software diagrams (Fig. 10.34; see also Section 9.6), which show how standard software modules, provided by the process control system manufacturer, are linked so as to build up the PCE station processing functions.

Function Chart. The data processing functions of a station must be performed in a coordinated manner; this requires a coordination function at a higher level in the hierarchy. The essence of this function is a task-specific sequence control that must be specially prepared for each field-level component of a measurement and control system. Sequence controls are for-

Header

	Station data sheet	Prepared: 20.10.1989

Facility			Process control station	*F1.2*
Plant complex	*AB-Reak.* Plant	Plant section	Type of station	*FIC*
Building	*A123*		Interconnected with	*T1.1*
Project number	*1234567*		P&I diagram No.	*831227*

Station identification	*Steam inlet R01*
Station is labeled	*Steam*
Circuit type	
Installation standard	

Description of substance

Substance measured		R	Substance controlled		R
Designation of substance	*Steam 5 bar*		Name of substance	*Steam 5 bar*	
Composition			Composition		
State of aggregation	*Vapor*		State of aggregation	*Vapor*	
Corrosive constituents	*None*		Corrosive constituents	*None*	
Suspended solids			Suspended solids	kg/m^3	
Density in operating state	kg/m^3		Density in operating state	kg/m^3	
Density at standard conditions	kg/m^3		Density at standard conditions		
Dynamic viscosity	mPa s		Dynamic viscosity	mPa s	
Electrical conductivity	μS/cm				

	min.	norm.	max.	units			min.	norm.	max.	units
Pressure		*5*	*5.5*	*bar*		Pressure		*5*	*5.5*	*bar*
Temperature		*180*	*220*	*°C*		Temperature		*180*	*220*	*°C*
Flow rate		*500*				Flow rate				
Limits										

Substance measured/minus side			
Density/minus side			
Behavior in sampling line			

Location

Measuring location		R	Controll location/device		R
Installation location	*Steam R01*		Installation location	*Steam R01*	
Piping/apparatus Nom. diam. DN *50*	Nom. press. PN *16*		Piping Nom. diam. DN *50*	Nom. press. PN *16*	
Material	*GS-C25*		Material	*GS-C25*	
Heating			Insulation thickness	mm	
Insulation thickness	mm		Safety code		
Measuring span (M)	mm				
Floor clearance (HO)	mm				
Overall length (vessel) (HE)	mm		Control device, type	*241-1*	
Flange, top/bottom DN/	DN		Control device Nom. diam. DN *50*	Nom. press. PN *16*	
Ambient temp. from to	°C		Overall length/height *230* mm	*305* mm	
Overall length (piping)	mm		Mode of operation	*Closed by spring*	
Pressure drop	bar		Seat/critical volume flow rate	*31* mm /16	
Safety code			Pressure drop	bar	

List of station elements

No.	Code	Text	Input	Output	Location	Remarks	R
1	Own	Standard orifice			plant B1		
2	03.42.21	Transducer		4–20 mA	plant B1		
3	08.07.01	Diaphragm valve	0–6 bar		plant V5		
4	08.02.01	Solenoid valve	DC 24 V	0–15 bar	plant Y1		
5	04.21.01	Position controller	4–20 mA	1.4–6 bar	plant N1		
6	14.22.05	Power supply	4–20 mA	4–20 mA	K1 E2 A2		
7	03.08.03	Signal transducer	4–20 mA	4–20 mA	K1 E3 A1	MUS/TVA	
8	3.8.1918	Signal transducer	2×0–20 mA	8×0–20 mA	K1 E4 A1	GUE	
9	Own	Control K1.1			K2 E3	IV 02	
10	45.45.45	Voltage distribution	DC 24 V	DC 24 V	K2 E4	SIV 60, F2	

Figure 10.31. Example of a station data sheet

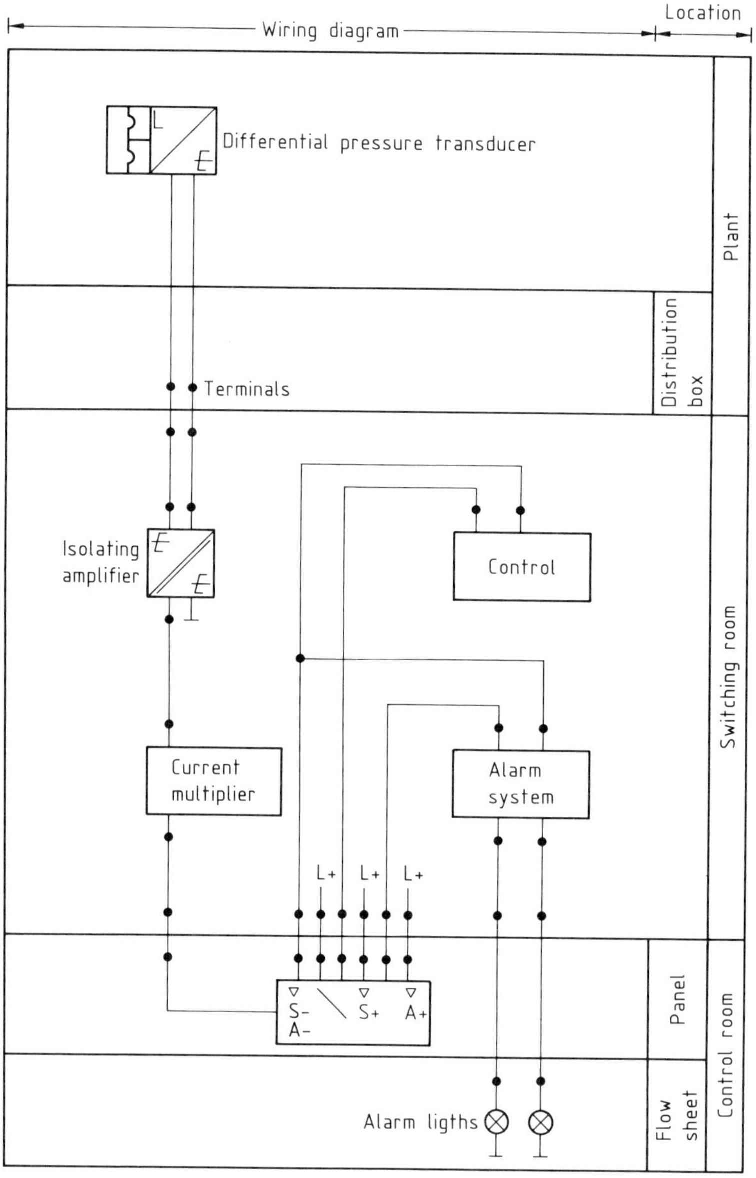

Figure 10.32. Example of a station plan for a conventional process control station

mulated in a system-neutral way by using function charts (Fig. 10.35) [10.35]; the same holds for station-level interlocks.

The function chart is a process-oriented depiction of the control function, independent of the way in which it is implemented. It replaces or supplements the verbal description of the function. The function chart clearly and unambiguously describes the function and thus aids in understanding between manufacturer and user. Furthermore, it facilitates collaboration between specialists [10.36]. In future projects, Petri networks will be employed at an early stage, from which function charts can be automatically generated (see Chap. 2 and Section 4.3).

Dimensioning of Process Measurement and Control Components. Generally, a process measurement and control system consists of a number of field-level components. Like the station-level functions of field-level components, these

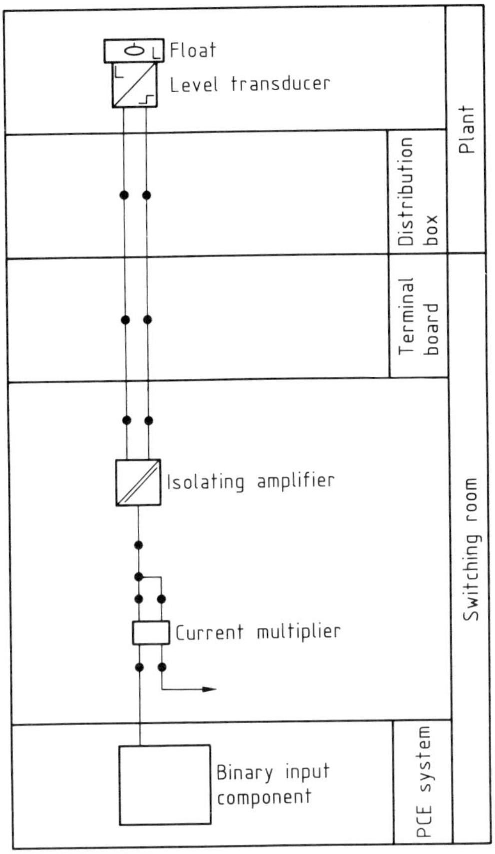

Figure 10.33. Example of a hardware station diagram for a station with signal processing (LS+)

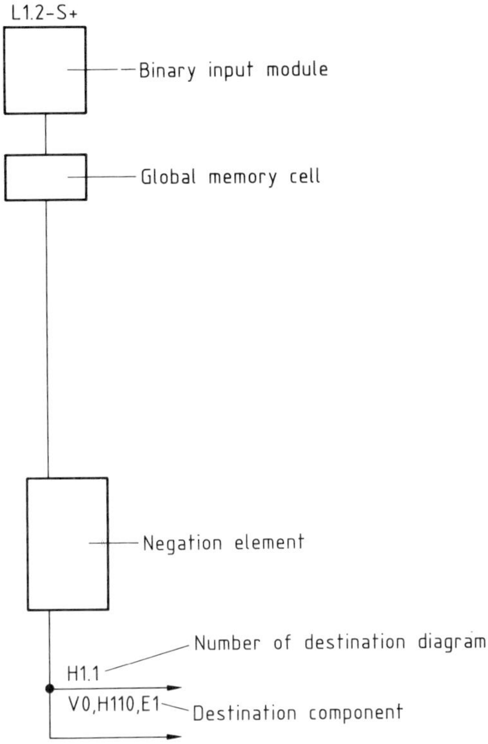

Figure 10.34. Example of a station software diagram for a station with signal processing (LS+) in a process measurement and control system

must also be coordinated, but at a higher level, for example, in the context of recipe-mode operation. A process control computer is used for this coordination function.

The size and type of control computer will depend on the scope of the coordination function. The selection of the operating computer is based on the number of graphical operating displays (Section 11.2).

The number and size of the field-level components assigned to a process section is influenced by the number of stations to be implemented, the functional scope, and the requirements on availability and safety (Section 10.4).

The station-level functions are built up from software modules. These differ in memory requirement and machine time, so that flow-through and pressure-control loops assembled from such modules may require 3–4 times as much memory and 4–6 times as much processing time as pure measurement loops.

If a large number of time-comsuming functions must be implemented in a single section of the process, field-level components whose time properties allow this must be used. Today, memory problems are of little importance.

The tasks of basic engineering also include dimensioning the rooms for the installation of process measurement and control equipment (see Section 10.7).

Detail Engineering. In the detail engineering phase, detailed and comprehensive documents are prepared for use in procurement, construction, operation, and maintenance of a plant. Consistent documentation is a vital consideration at this stage (see Section 3.5). The planning and design work can be broken down between process and plant engineering and process control as follows:

Planning and design for process and plant engineering

Detailed layout of plant sections, apparatus
Construction drawings

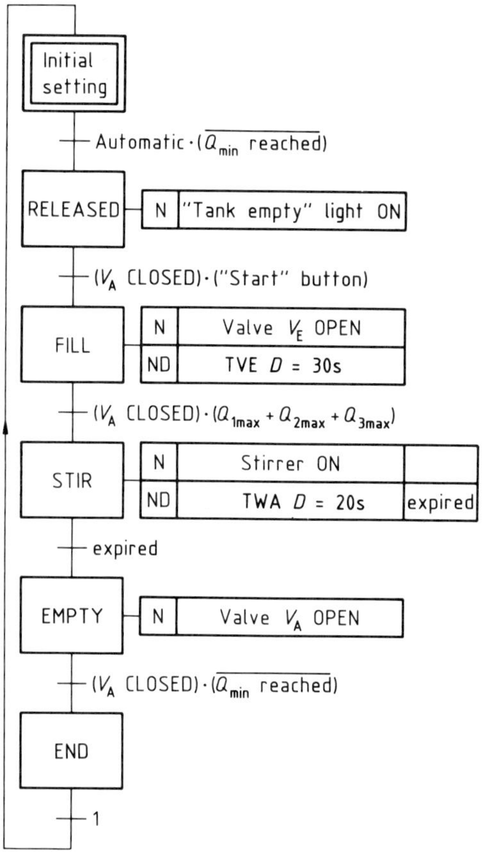

Figure 10.35. Example of a function chart for tank filling

Site plans, layouts
Piping design

Planning and design for process control engineering
Preparation of data sheets
Completion of station diagrams and function charts
Programming
Working drawings of mimic diagrams for operation and monitoring
Preparation of circuit and configuration diagrams
Preparation of commissioning checklists

Preparation of operating handbook
Preparation for letting contract to outside firms
Orders and contracts
Scheduling and resource allocation for installation

The plant design activity prepares design drawings for machinery and apparatus. Construction design supplies building plans as well as calculations of foundations, floor loads, vibrational loads due to machinery, and so forth.

The piping activity supplements its design documents with piping isometrics and parts lists. Care should be taken here to achieve smooth cooperation with local field instrumentation and cable routing activities (e.g., with the aid of suitable 3D CAE methods).

Planning and design of the process measurement and control system completes the documentation generated in the detail engineering phase.

The PCE station hardware diagrams contain additional information making it possible to localize the station elements by giving, for example, the switchbox identification, rack-mount module number, card slot, and also to identify the incoming and outgoing signal lines by referring to the terminal numbers. Figure 10.36 shows a detail of the station diagram of Figure 10.32 with this supplementary information.

The station software diagrams, function charts, designs for recipe-mode control operations, and other higher functionalities are transformed into programs understandable by the process control system.

Because screen displays are associated with variables in the software modules, the preparation of screens for process monitoring and control is an expensive process. It can account for a substantial fraction of software costs. Before this step is performed, the screen must be designed. A useful tool for the process of agreement with the later operator of the plant is working drawings, usually prepared by CAD methods, which are easy to change and supplement. Special ergonomic requirements for state-oriented opertion are discussed in Section 8.2.

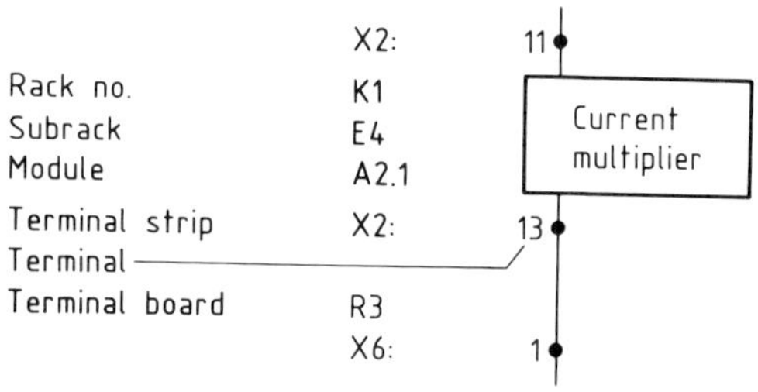

Figure 10.36. Excerpt from a supplemented station diagram

An operating handbook must be written during detail engineering. It should assist in the early detection of design errors and the taking of proper corrective actions. Design errors can be remedied in a relatively cost-neutral way at this point. Corrections upon commissioning are expensive. Unfortunately, little use is made of this capability at present.

The following activities are carried out in collaboration with the purchasing departments:

- Requests for quotations to vendors and/or installation firms
- Orders

When contracting with outside firms, the nature and content of the contracts are agreed on with the legal and purchasing departments.

For installation, schedules should be prepared with allowance for delivery times (see also the network chart in Fig. 10.37).

It is also desirable to generate a commissioning concept, which should use diagrams and lists to document the following points (among others):

1) Personnel requirement
2) Schedule
3) Procedures
4) Type of functional tests, etc.

Recent years have seen more and more computer-aided tools for use as planning and design aids. These support not so much the conceptual phase as the manual or material phase of design, where the plans needed for the erection and operation of the plant are drawn up (see Chap. 9).

Installation. Installation includes the erection of all classes of equipment involved in an investment project and the interconnection of all components into a completed, operational plant. The stages in installation are as follows [10.8]:

Preparation
Site preparation
Site organization

Execution
Steel construction
Installation of machinery, apparatus, tanks, etc.
Installation of piping
Installation of process measurement and control equipment
Installation in process control rooms
Installation of cable trays in the field
Field installation

Inspection
Machinery, apparatus, tanks
Operability
Tightness
⋮
Process measurement and control
Visual inspection
Wiring inspection
Functional testing

The preparations include the contruction of a site with site offices, storage areas, and installation areas, as well as the creation of a site organization to handle scheduling, resource allocation, costing, and the monitoring of performance relative to these plans.

Good installation planning reveals itself in that the installation of individual categories of assets is coordinated so as to insure smooth, reliable progress. Interferences between categories should not occur. What is more, all resources — such as installation personnel, plant sections, components, and other materials required for the installation of individual asset categories — and all working plans and related documents should be available on time and in adequate quantities and should meet stated quality standards.

The on-schedule installation of process measurement and control equipment requires special care, since as a rule it cannot be started until significant parts of the other categories have been installed. Detailed network charts (see above) can be helpful here.

The installation of process measurement and control equipment falls into three parts:

- Installation in control rooms
- Installation of cable trays in the field
- Installation of process measurement and control devices in the field

Installation "in the field" means all installation activities outside the process control rooms.

Installation work in the control rooms can begin at a very early time in the overall installation process, immediately after the buildings have been completed. Some 20 % of all process control installation work falls under this heading. Cable tray installation can proceed roughly in parallel with piping installation. But the installation of instruments cannot start until ca. 80 % of the piping installation has been done.

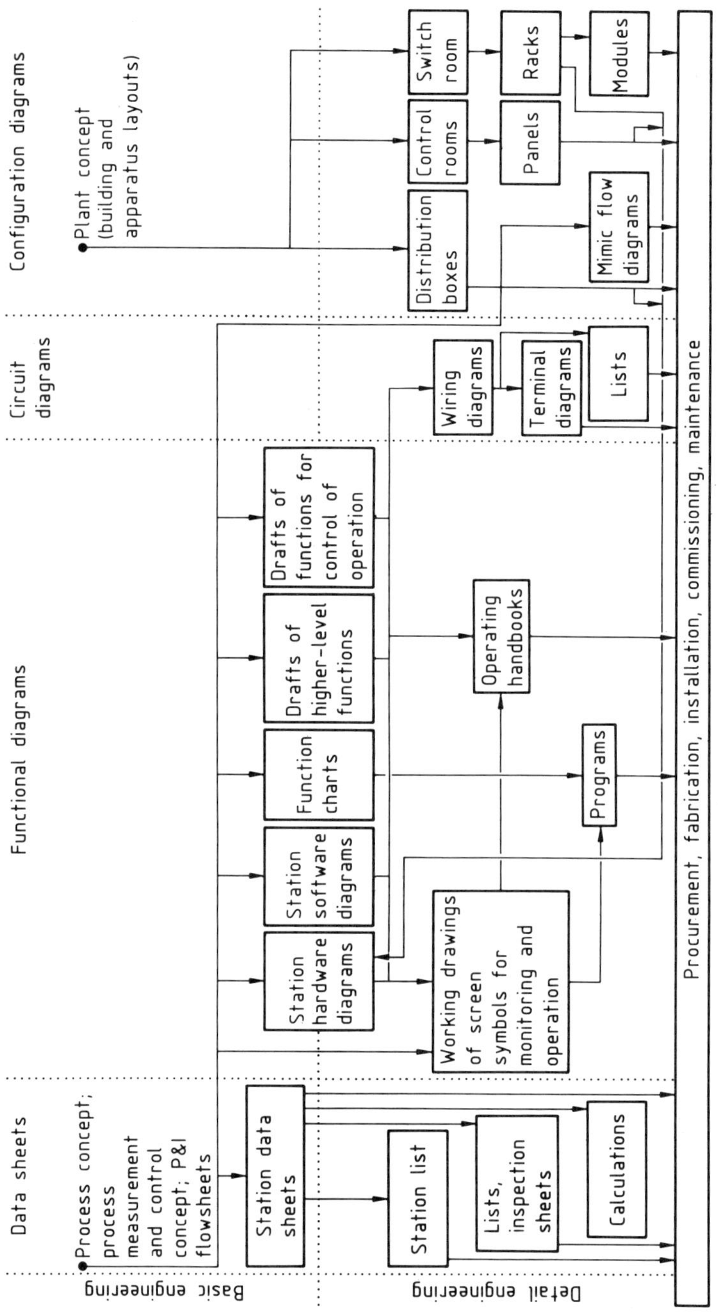

Figure 10.37. Information flow and process control engineering documentation in the basic and detail engineering phases [10.8]

Installation ends with an inspection, which is part of the quality assurance (see Section 10.6).

The installation activity must be thoroughly cognizant of the "concurrence" of events in the sense of Petri network theory (see Chap. 2).

Installation offers a broad range of rationalization possibilities for later maintenance practices (see Section 11.4). The equipment manufacturers still have some way to go here (see Section 11.1).

10.6 Quality Assurance

The quality of the items prepared must be continuously assured during the project. These include plans, software specifications, programs, plant sections, installation, and so forth. Consistent documentation must be provided for all these activities (Section 3.5, [10.37], [10.38]).

Major elements of quality assurance are

- Provisions governing planning, design, and installation
- Controls
- Inspection and testing

Today, inspection and testing includes data-model-oriented consistency checks, as known from software engineering, as well as final installation inspection and commissioning.

Governing provisions are, for example, legislation, standards, and internal corporate directives.

Legislative provisions describe the constraints under which a plant is to be erected and operated. They safeguard the environment, in the broadest sense, against possible threats (see Fig. 10.37).

In addition to legislation, a number of national, regional, and international standards must be considered. These standards essentially serve to ensure consistent planning, design, and execution (see also Chap. 12).

Internal corporate guidelines interpret and give concrete form to the provisions of laws and standards as well as corporate objectives (see Fig. 2.9).

Examples of provisions for process control engineering are given below:

- General planning/design guidelines
  - Documentation
  - Nomenclature, identifications, abbreviations

    ⋮
- Provisions governing project planning and execution
  - Description of scope of work
  - Costs; ordering
  - Schedules

    ⋮
- Substance data
- Basis for calculations and design
- Technical provisions
  - Interconnection of devices
  - Protective, safety, reporting systems
  - Hardware and software structure of control systems
  - Outfitting of process control rooms
  - Equipment for building, labs, shops

    ⋮
- Provisions for fabrication and installation
  - Design of standard modules
  - Design of protective enclosures, distribution boxes, instrument housings

    ⋮

"Design standard" is understood here to mean technical provisions for the function "planning and design." The design standard for process control engineering is supplemented by an equipment standard, which narrows the range of devices available on the market to a few choices, in particular those that have proved reliable in many years' service or through qualification tests (see Fig. 2.9).

The scope of work often goes through marked changes in the planning and design of a plant. The reason is inadequate preparation of the investment project. The nature and quantity of the technology to be implemented are not described exactly. Concepts change later on, and designs already adopted must be reworked. The resulting "patchwork" can hardly improve quality and, as a rule, actually degrades it.

Situations of this kind cannot be avoided, but proper preparation of the bidding specifications and specifications can significantly improve them. They must therefore be regarded as an important contribution to improving the quality of planning and design.

Design quality control is a major undertaking, because of the large volume of documents generated. It becomes more difficult if the design work is done by outside firms. For this reason, quality control is often reduced to just a spot-check of results. An object-oriented data model for the plant as a whole is thus all the more important (see Section 3.5 and Chap. 9).

It is shown below for the example of the hardware documentation of a process control engineering project that this procedure is quite adequate for practical use. A differentiated assessment of process control engineering documentation requires breaking it down into units for examination. In the case of hardware documentation, the logical unit is the documentation of a single process control station, comprising the station data sheet, the station diagram, and the corresponding information from the interconnection lists. These documents can be examined as soon as the execution planning has been completed; if the complete job (planning, design, installation supervision, and commissioning) is carried out by one contractor, they can be inspected after the process facility has been commissioned. In the case of a complete contract, the contractor therefore has to examine the documentation only once.

For the documentation of a process control station to be deemed correct after commissioning, the documents included must meet all the following quality criteria:

- The station identification must be correct
- The documentation must correctly reflect the equipment installed
- Any cross-references required must be present
- Each terminal designation must be correct; that is, the information in the process control station diagram and in the associated interconnection lists must agree with the acutal interconnection of the devices

The results of two sampling inspections of process control engineering documentation for a project comprising 324 PCE stations are shown in Figure 10.38. In the first examination, the sample size was increased in steps of 32 PCE stations (roughly 10% of the total number of stations). Above a relative sample size of ca. 30%, the relative error is nearly constant, and it appeared unnecessary to continue the inspection in this case.

In a second inspection by a different inspector, only those errors overlooked in the first inspection were noted. The result is around 5% and must be regarded as the errors of the first

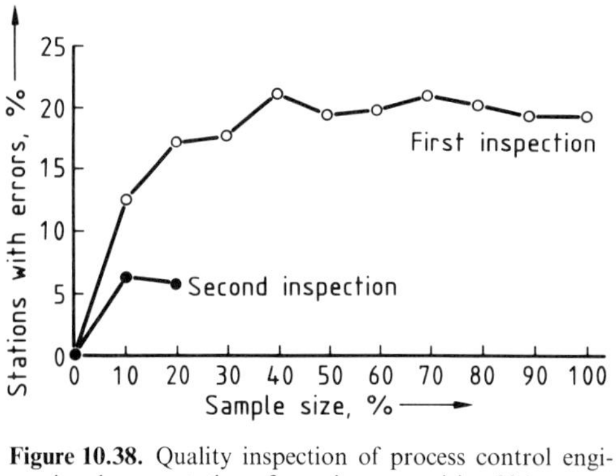

Figure 10.38. Quality inspection of process control engineering documentation of a project comprising 324 process control stations

inspector ("slip-through"). Similar slip-through values were found in studies of the manufacturing industries [10.39].

The results of this examination show that inspection of the process control engineering documentation by the contractor is essential [10.40]. In practice, the following procedure has proven suitable: If the customer's examination reveals an error lying above a predetermined threshold (e.g., 5%), the customer returns the entire documentation to the contractor for complete revision. However, if the relative error is < 5%, the contractor need only remedy the errors found [10.41].

On economic grounds, only spot checks can be carried out on the process control engineering documentation. Therefore, the risk exists that the error found in the spot check is greater (risk for contractor) or less (risk for customer) than the "true" error that would be found in a complete examination. To resolve this conflict, model calculations have been performed in order to determine the sample size for various project sizes (number of process control stations). The following assumptions were made:

- The error limit for acceptance of process control engineering documentation is 5%
- The sample must be large enough that the probability of accepting the documentation (contractor's risk) when the documentation contains a relative error > 10% (twice the acceptable error limit) is < 20%
- The "hypergeometric distribution" is the appropriate mathematical description of the sampling insepction described [10.42]

The results of these model calculations are presented in Table 10.2. The most important result of these model calculations is that in projects with > 100 process control stations, given an acceptable limiting error rate of 5%, sampling inspection of 50 stations is sufficient to decide on the acceptance of the process control engineering

Table 10.2. Sample size for quality inspection of process control engineering documentation

Number of PCE stations	Sample size	
	Absolute	Relative
20	14	70%
50	35	66%
100	50	50%
500	50	10%
1000	50	5%

documentation or assurance measures. The advantages of sampling inspection in large projects are greatly reduced insepction costs.

The monitoring of software design presents major problems. In principle, software design can be checked only during commissioning. A modular architecture is helpful if it permits step-by-step testing, from the single module up to the entire system. This type of examination cannot, however, eliminate software defects that lie outside the scope of the check. Software always hides certain imponderables; hence the recommendation that functions of immediate safety significance never be performed by software alone, but always in combination with a conventional safety level (Section 10.4 and [10.30]).

It can be expected that the quality of planning and design results will improve markedly if the computer methods still in development (see Chap. 9) are adopted.

Quality assurance also requires installation practices such as the following:

- Issuance of installation standards (technical provisions for the function "installation")
- Incoming inspection
- Supervision of installation activities
- Functional tests

An example for the contents of an installation standard for process measurement and control equipment is listed below:

Provisions on the mechanical construction of process measurement and control equipment in control rooms and other rooms

Panel designs
Cabinet and rack-mount designs
Subracks
Cooling air supply and distribution
Cable basements
Cable lead-throughs
Cable trays and their routing on pipe bridges
Distribution boxes
Transducer boxes design and installation
Terminal boxes

Provisions on electrical design

Auxiliary power supply
Grounding/equipotential bonding
Cable routing
Power supply and distribution
Hookup methods

Provisions on the construction of process measurement and control stations

Installation material in configuration and interconnection

⋮

It contains, for example, provisions on

- The mechanical construction of process measurement and control equipment in control rooms and other rooms
- Electrical design
- The construction of individual stations

These standards should be supplemented by installation layout diagrams (Fig. 10.39), which use drawings and lists to set forth the installation materials needed for certain tasks as well as their configuration and interconnection.

The quality of an installation activity is also enhanced if standardized, prefabricated, and pretested hardware components and other equipment are employed.

In process control engineering, the use of such standard elements in the form of panels, racks, cabinets, cable trays, subracks for alarm devices, power-supply equipment, and so forth has long been common.

Incoming inspections during installation relate not only to the work plans but also to the materials and equipment to be used. The plans must be complete and correct; materials and devices must comply with specifications.

It is desirable to have specialist personnel—for example, project engineers with knowledge of the process or installation foremen with competence in the work to be done—to supervise the installation activities. Such personnel should be designated by the customer of the project. Only in this way can high-quality execution be guaranteed.

An essential element of quality assurance is the final installation inspection (see "Installation" in Section 10.5, p. 633, and [10.8]).

The process control engineering activity carries out three levels of inspections:

1) Visual inspection
2) Wiring checks
3) Functional testing

This procedure insures that the individual process measurement and control devices perform correctly and interact properly.

Visual inspection establishes correct mounting. Proper interconnection of the elements is determined by wiring checks.

The aim of functional testing is to ensure that the active operating sequences of the process

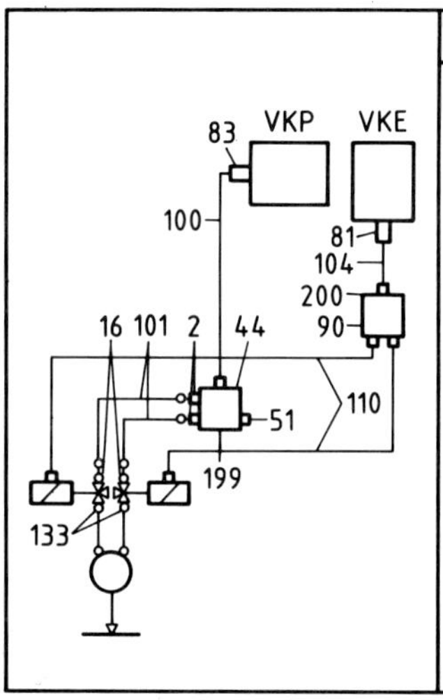

No.	Pos.	Name	Amount	Data sheet	Material
1	44	Hose-pipe coupling, triple	1	7.5 Bl. 1	Polymer
2	51	Cylinder-head screws R 1/4" × 12	1	7.5 Bl. 1	Polymer
3	81	Compression gland PG 13.5	1	6.9 Bl. 1	plastic
4	83	Compression gland PG 21	1	6.9 Bl. 1	plastic
5	100	Pressure transmission cable, 3 tubes, Type D2YBY	20 m	7.1 Bl. 1	
6	104	Wire NYM-I 5 × 1.5	20 m	6.3 Bl. 1	
7	110	Wire NYMHY-I 5 × 1.5	4 m	6.3 Bl. 1	
8	133	Seal R 3/8"	4		
9	2	GE 6-L-1/4" NPT	2	7.7 Bl. 1	steel/cd
10	16	GE 6-L-3/8"	4	7.7 Bl. 1	steel/cd
11	90	Branch box 8102/1 EXE G5 4 × 2.5	1	6.1 Bl. 1	
12	101	Pressure hose Nd4 with 6 mm pipe connection	4 m	7.1 Bl. 1	
13	199	Mount for pneumatic terminal box	1		steel
14	200	Mount for electrical terminal box	1		steel

Figure 10.39. Example of an installation configuration plan [10.8]

measurement and control devices accomplish the functions specified for the process tasks. On the basis of the planning and design documents, the individual process control stations are systematically inspected from sensor to actuator. Process control functions are usually tested by first verifying the lower-level functions and then proceeding to successively higher levels.

A hierarchically organized process simulation would be desirable for functional testing, since only such a tool makes it possible to draw unambiguous conclusions as to whether the plant as a whole will also operate in the intended way. If such a simulation cannot be performed, as is still the case for most processes, this information cannot be acquired until commissioning (see Chaps. 3 and 11).

Commissioning is the initial operation of the entire plant, including process measurement and control equipment, by the operator. It is the last test showing that the process runs in the intended manner. Commissioning can thus be regarded as further quality assurance practice.

Commissioning is performed in three steps:

- Preparation
- Operation with "inert" media ("water run")
- Operation with chemicals

The first two steps are carried out by the departments involved in the planning, design, and erection of the plant. If the project has been contracted out, these steps are done by the outside firms together with the plant operators. The functions to be performed during commissioning are [10.30]:

- Preparation
  - Technical training of operating personnel (operating instructions)
  - Technical training of maintenance personnel
  - Safety training
  - Stocking of spare parts and standby equipment
  - Securing of feeds, auxiliary media, energy supplies
  - Preparation of the plant (purging, drainage, venting, drying)
  - Loading of the process measurement and control system

  ⋮
- Operation with "inert" media ("water run")
  - Commissioning of auxiliary systems (steam, cooling water, etc.);
    loops (heating water loops, etc.)
  - Testing of safety features
  - Simulated operation
  - Testing in operating modes (startup, shutdown, load changing, etc.)
  - Performance of needed amendments

  ⋮
- Operation with chemicals
  - Production test run

- Modification of process measurement and control equipment
- Guarantee run
- ⋮

After preparations are complete, the plant is started up with inert (i.e., unreactive) products, beginning with the auxiliary and service facilities. Next, circuits such as heating loops are commissioned. After the entire plant has been placed in operation, the safety devices are inspected.

The opportunity presents itself for a comprehensive test of all modes that will be encountered in later operation (startup, shutdown, load change). On the basis of the experience gathered, any amendments needed are now carried out. The documents are revised at the same time.

The next step is operation with chemicals, that is, a production test run. Responsibility for the plant ultimately passes over to the operator.

Ordinarily, a plant is started up at a low throughput and gradually brought up to rated output. The final modification of the process measurement and control equipment to the plant is carried out during this phase. Included in this modification are the adjustment of controller parameters, the input of limiting values, and so forth.

In particular, if the plant has been designed and built by an outside firm, evidence must be presented that the plant complies with the design as to capacity, yield, availability, and quality. This evidence takes the form of an extended test run. The plant documentation (see above) must be finalized at this time.

When commissioning has established quality, that is, shown that the process runs in the contemplated way, the project is over. A detailed final report is mandatory.

10.7. Process Control Rooms

Introduction. This section describes the requirements that process control rooms must meet, offers suggestions for their appropriate design, and reports operating experience [10.45].

In the context of optimal human–process communication, it is indicated what requirements, stemming from technical, ergonomic, and other knowledge, govern the architecture, construction, and outfitting of control rooms. The design suggestions are based on scientific publications, standards, guidelines, and operating experience. Details of execution can be found in vendors literature [10.46].

Architecture. *Relationship of Process Control Rooms to the Plant.* It is advantageous to locate process control rooms relative to the production facilities of the plant so as to create short paths for installation and for access to the plant. The control room should be on a level with those parts of the plant most often accessed. Control room location and structure must afford adequate protection [10.47].

The "main wing" layout (Fig. 10.40) offers an optimal way of meeting these requirements. Furthermore, direct connection between laboratory, superintendent's office, and social spaces promotes necessary communication. Protective functions are also taken care of, for example in relation to hygiene or safeguards against undesirable external influences.

In large-area combinations of facilities (Fig. 10.41), it may be useful to erect a freestand-

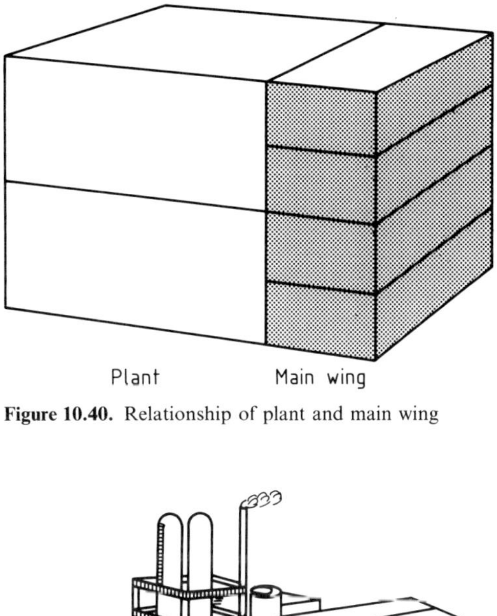

Figure 10.40. Relationship of plant and main wing

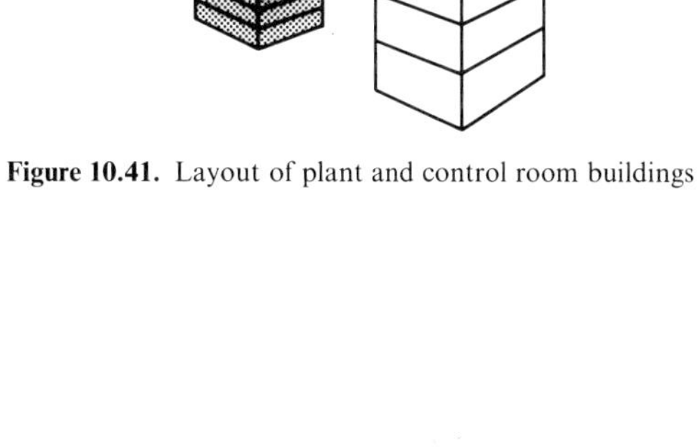

Figure 10.41. Layout of plant and control room buildings

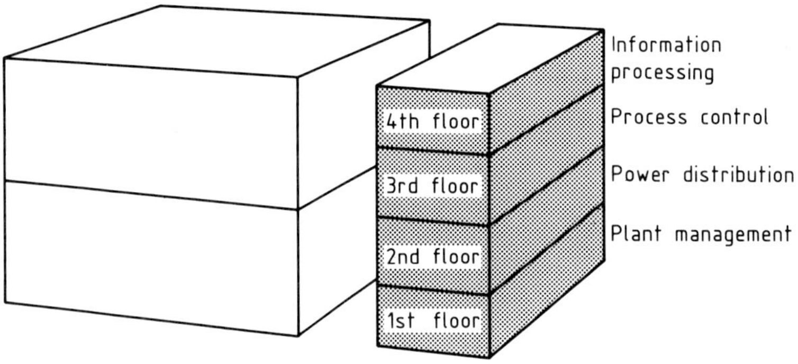

Figure 10.42. Breakdown of the main wing

ing control building to take over a variety of operational functions.

Arrangement of Control Rooms. Process control rooms and other rooms that house, for example, plant management should be laid out in a functional way so as to permit an optimal flow of operations. This can be done whether these spaces are located in a main wing (Fig. 10.42) or in a separate building.

The control room is the central area for the performance of essential process control engineering functions. Such tasks are described in the "level" model of information (see Section 2.3). The switch room for information processing holds the equipment necessary for processing and interpreting information. The switch room for electrical power distribution is the site of central process electrical equipment.

The control room and switch room for information processing should be located close together. The electrical switch room should also be placed in the main wing. If distances are long, however, electrical distribution for the plant can be housed in more than one switch room (to minimize cabling costs).

The central communications equipment for a plant needs its own room. Rooms for process analytical instruments should be placed at crucial sampling points in the plant. Special requirements apply to these rooms [10.48].

The following points, in particular, must be considered when positioning control rooms relative to the plant and one another:

- Safe access and escape routes
- Optimal cable and utilities routing
- Fire safety
- Favorable location for firefighting if necessary

Space Requirement for Process Control Rooms. Space requirements can be determined with graphs or ratios derived from operating experience, which offer guidelines (Fig. 10.43) for initial control room design. The increase in individual functions expected through the end of the project has to be estimated and reserve space must be provided.

Starting with some minimum size (ca. 25 m^2), the space required for a control room is worked out from the number of process control stations to be observed and controlled. The figure of 0.1 m^2 per process control station is a lower limit for conventional instrumentation. The use of process control systems and process control computers will reduce the area needed.

The space required for the information processing switch room is also governed by the number of process control stations. A minimum figure is again 25 m^2. With conventional instrumentation, the minimum additional area required is 0.2 m^2 per process control station. The use of process control systems will cut the space requirement.

The installed power is the essential factor determining the needed area for the power distribution switch room; the minimum size is 15 m^2. About one square meter is required for every 10–20 kW of installed capacity. If there are many variable-speed drives, however, the required area per kilowatt may increase.

Ergonomics. Ergonomics is concerned with the interrelations of human beings and the world in which they work [10.43]. Due attention to ergonomic factors in the design and construction of control rooms is a precondition for good human–process communication (see also Section 11.2).

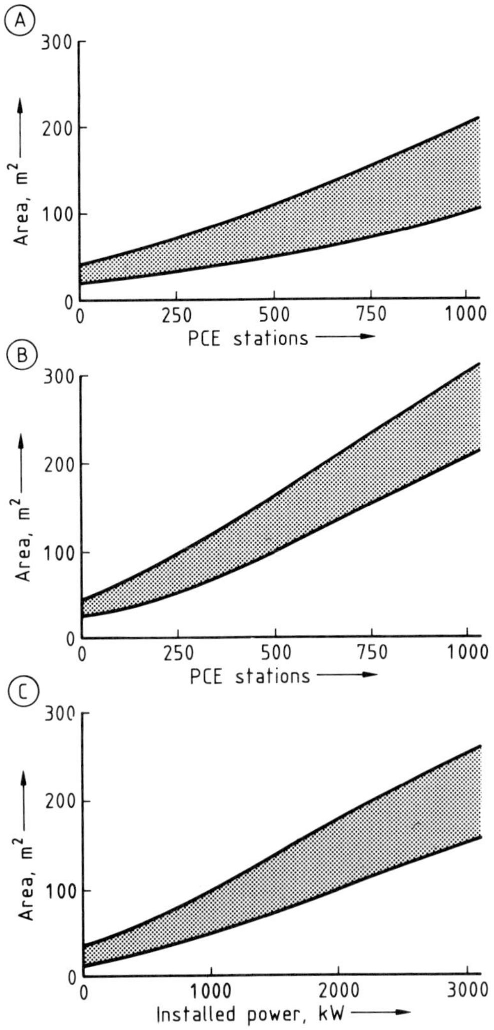

Figure 10.43. Space requirements for process control rooms

Anthropometry. Anthropometric considerations lead to guidelines for spacings, openings, and clearances in the placement of individual devices and equipment in control rooms.

Pushbutton controls, switches, and display elements should be arranged so as to allow optimal access and visibility (Fig. 10.44) [10.49], [10.50]. An important factor in determining the appearance and size of labels is the distance from which they will be viewed. When locating display and control panels, care should be taken that they are in easy view of the operator.

Lighting. A variety of visual tasks must be performed in control rooms (Fig. 10.45). The

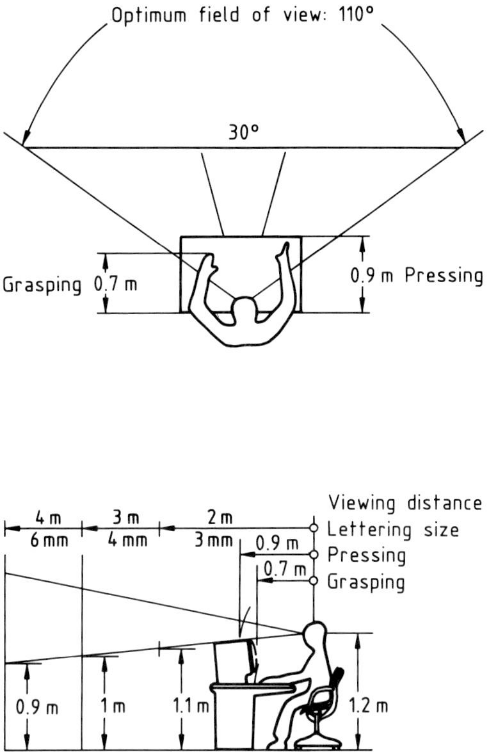

Figure 10.44. Optimal manipulation and viewing areas

Figure 10.45. Control room lighting

type and intensity of illumination must be adapted to each such task [10.52]. Illuminances between 200 and 500 lux are recommended for error-free reading of meters and other readouts, illuminated symbols on panels, and characters on screens and displays. Lighting properly adapted to each visual task features separate, continuously variable brightness adjustments for

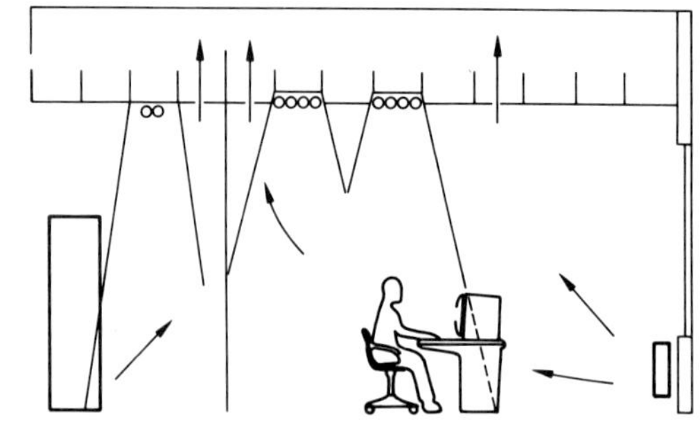

Figure 10.46. Ambient conditions

the two zones. Fill-in lighting serves to enhance the local illuminance on printers, desks, and so forth. An appropriate configuration of lighting fixtures in combination with glare screens should help to prevent reflection from displays. Since control rooms are generally occupied day and night, the lamps should provide daylight-balanced lights in order to insure consistency in color and illumination.

Switchgear rooms should have an illuminance of 300 lux.

Backup lighting must be provided in control rooms and is recommended in switchgear rooms. Safety lighting must be provided for purposes of accident prevention. Details of lighting design can be found in company literature [10.44].

Climate Control. Ambient conditions in the control room must be held constant within a fairly narrow range [10.51]. Furthermore, because personnel are present at all times, there must be a continuous, draft-free supply of fresh air (5 to 10 air changes per hour; Fig. 10.46).

Switchgear rooms are not continuously occupied, and the tolerance range for the ambient conditions is therefore wider. In contrast to the control room, much heat is evolved here and must be removed by a waste-heat removal system. Computer components belonging to process monitoring and control systems may have stringent climatic requirements, which can best be satisfied by separate climate control devices.

Undesirable pollutants must be kept out of all rooms used for process control functions. A slight overpressure is therefore maintained in such areas; filtered fresh air from pollutant-free areas is delivered through special floor registers and extracted through ceiling grilles. The control room must have the capability of shutting down the ventilation of all rooms at any time.

Acoustics. Acoustic conditions in the control room should be such that personnel there experience no difficulty in understanding one another or in communicating with the outside world by telephone or intercom, and also such that acoustic signals can easily be detected by the human ear. Care should be taken that the noise level does not exceed the 55 dB(A) ambient noise, the reverberation time in the middle frequency range (500 to 3000 Hz) is ≤ 1 s, and the level of acoustical danger signals in an octave range lies a minimum of 10 dB(A) and a maximum of 20 dB(A) above the interference.

Control Room Design. *Doors and Windows.* On grounds of safety and health, pathways between the production facility and rooms that house process control functions may be equipped with airlocks. Airlock doors must be fireproof, self-closing, and gasketed, and they must open in the escape direction. Larger monitoring and control rooms must have a second emergency exit.

Because communication with the process takes place exclusively via process monitoring and control devices, windows opening into the plant are not needed for communication. Windows may, however, be required by workplace regulations [10.53]. When control rooms are located near explosion hazard areas, appropriately rated exterior windows must be provided.

Floors. Control rooms should be constructed with floating floors. Cables are led through floor penetrations or cable ducts placed in the floor (Fig. 10.47). A variety of special plastic floor tiles are available. Ceramic tile is used where contamination is severe. Double flooring in computer and switchgear rooms (Fig. 10.48) permits free cable routing.

Ceilings and Walls. False ceilings are particularly suitable for control rooms. They offer a range of options both for concealing the unfinished ceiling and for the placement of functional elements such as fire alarm and fire-suppression systems, air conditioning, cable ducting, and lighting fixtures. Esthetic, acoustical, and climatic considerations require attention to not only the ceiling but also the walls. All other rooms that house process monitoring and control functions have ceilings and walls painted in light colors.

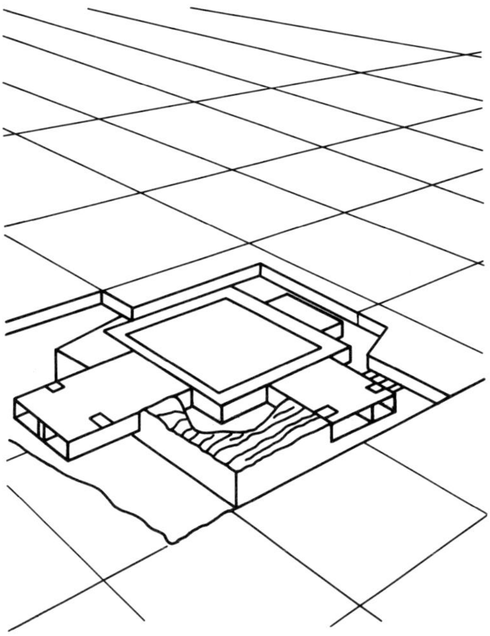

Figure 10.47. Cable duct installation

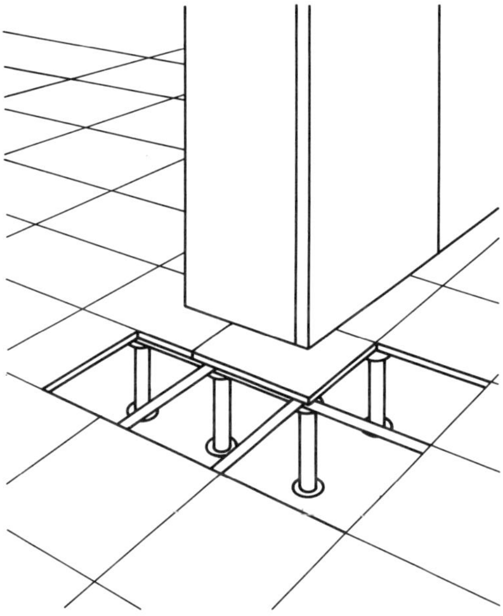

Figure 10.48. Double floor

The control room should have an unfinished height of 4 m (to bottom face of ceiling joists). The height of the false ceiling should be 3 to 3.50 m, depending on the size of the room. A height of 3 m (to bottom face of air ducts) is recommended for switchgear rooms.

Figure 10.49. Operator desk

Figure 10.50. Operator station

Ceilings and walls must comply with pertinent fire-protection codes (F 30 or F 90); this applies particularly to cable penetrations.

Desks (Fig. 10.49) hold visual displays and controls such as keyboards, light pens, joysticks, mice, and touch-screen input devices. Horizontally and vertically pivoting display bases permit adjustment to suit individual preferences [10.50]. Additional free space is needed, for example, to accommodate a radio and telephones and to permit occasional writing.

Operator stations in control rooms are used for monitoring and controlling the process. They enable the operator to intervene in the process as required. An operator station (Fig. 10.50) generally consists of a color graphics monitor or monitors and controls. The controls usually consist of a keyboard (with movable or touch-sensitive keys) for the input of commands and/or a light pen with which the process can be controlled directly on the screen.

The light pen has the advantage that input errors are largely eliminated and performance

Figure 10.51. Process communications wall, furniture, and accessories

speed is roughly doubled. Fatigue-free operation with the light pen is, however, possible only with the elbow supported, so that the monitor should be located only ca. 40 cm away from the operator. Other control arrangements employ tracking devices such as joystick and mouse or permit operation by touching the front of the screen (touch-screen devices).

Hard-copy devices and printers are used to document the process in printed form. They should be installed in noiseproof, dustproof enclosures in a readily accessible location (unless they have to be placed in a separate room).

The *process communications wall* comprises panels holding extra monitors as well as individual displays and controls (Fig. 10.51). The preferred configurations are straight, L-shaped, and U-shaped. Mimic flowsheets can be laid out along the top. The various functional elements are modular devices. To make the spatial organization clear to the observer, they can be arranged in accord with the production flow.

Furniture and Accessories. Besides the monitoring and control devices, the control room should be equipped with the following (Fig. 10.51):

- Bookcases for documentation as well as storage of supplies such as printer paper
- Desk/worktable
- Bulletin board
- Flip chart
- Wall clock showing date, day of the week, and time
- Storage for protective clothing and safety devices (e.g., respiratory protection, fire extinguishers). This should be placed where the operating personnel can access it quickly and without hindrance on the way to the plant.

11. Operation

11.1. Principles

This chapter discusses some points relating to process control engineering that must be borne in mind when operating process plants.

Along with reliability (environmental, occupational safety, and health issues) [11.1]–[11.3], the most important quality of process operation is reproducibility. In other words, the required product properties [11.4] of the end products must lie within the respective tolerances (see also Fig. 2.33). The same applies to all relevant intermediate species, of course. Product properties should always be regarded as "primary nominal values" [11.5].

As shown in detail in Chapter 2, product properties are described in terms of individual process elements. These in turn are described in terms of process properties [11.4]. Process and product properties are linked by transformations expressed in "operators," and the product properties can thus be represented as functions of the process properties [11.6], [11.7]. Gilles [11.8] takes this a step further and, in a first formulation, allows the set of product properties in the production process to be categorized as a state variable.

Because the above properties or attributes are mostly multidimensional aggregations, it is meaningful to refer to "product spaces" and "process spaces."

Thus, to achieve product properties complying with the requirement profile, the respective process properties must be maintained in each process element. These in turn have their own requirement profiles with corresponding tolerances. These requirement profiles of the process properties in each process elements are to be understood as "secondary" nominal values or setpoints. They are constituents of the control loops (generally involving thermodynamic and reaction-kinetic quantities) in the process control system.

Figure 11.1 shows a nominal/actual comparison for a production process. This representation is the most important piece of information required for controlling the process. Detailed discussions can be found in Sections 2.3 and 11.2.

When production processes are to be optimized with respect to certain target quantities, process models must be created. These are mathematical descriptions including the interdependences of process properties as well as the dependences of product properties on process properties. As shown in Chapter 3, models can be

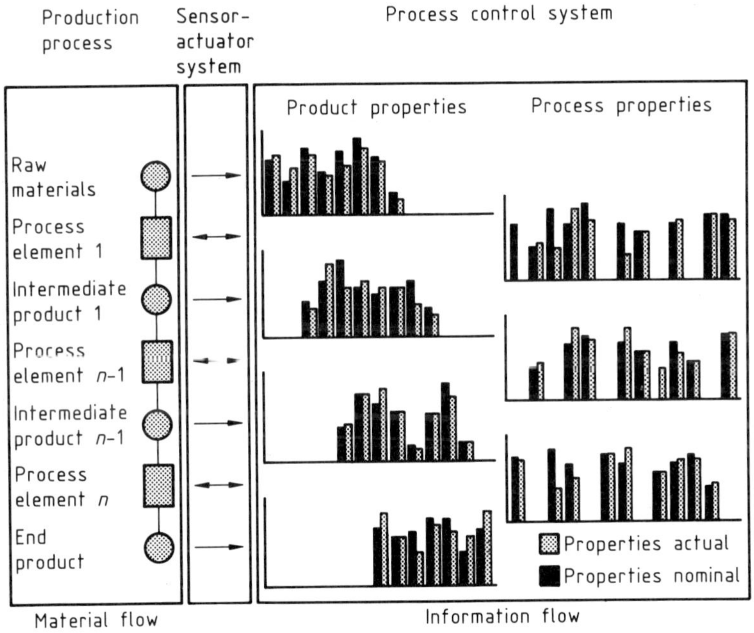

Figure 11.1. Nominal/actual comparison for a production process

subdivided into purely empirical or statistical models and "theoretical" models.

Statistical models are derived from the actual (measured) product and process properties in the real process running in a pilot or full-scale facility (see Fig. 11.1 and Section 3.2). Its equations are purely mathematical in nature. It is difficult to assign physical interpretations to its variables by using techniques to reduce the dimensionality [11.9], [11.10].

Static and dynamic process models are essential for effective simulations.

In general, "simulation" refers to the use of information technology to forecast the behavior of the independent variables in a system when the independent variables or the initial and boundary conditions change [11.11], [11.12].

Provided the basic model (see Chap. 3) can be formulated in explicit mathematical form, the fundamental problem is to find suitable numerical procedures for the economical solution of the mathematical problem. As interactive communication between humans and the process becomes more important, the graphical presentation of the data model or functional model (see Chap. 2 and [11.13]) is gaining in significance [11.6].

Problem-oriented "object spaces," with the "attributes" of the objects as their coordinates, should be designed so as to promote human thought and learning (see Section 11.2). Information reduction is a key tool to doing this.

This holds both for training a user concerned with a technical, economic, or socio-cultural system (see Chap. 2) and for devising solution aids for finding "optimal" operating points. Zeitz [11.11] is therefore justified in according the highest importance to such simulation. The framework for system modeling must be extended beyond the customary metric scaling into the realm of topology. In contrast to current well-known solution approaches, simulation thus takes in predominantly algebraic and differential-equation-oriented structures as unified topologies of knowledge, as suggested by Gilles [11.14], in his discussion of knowledge-based process control engineering. Numerical and knowledge-based solution tools will have to be harmonized with one another in the future. For example, the NAMUR Status Reports include noteworthy articles offering many suggestions and solution approaches [11.15]–[11.21].

The modern model-aided procedure will also lead to new and expanded safety concepts. These include concrete approaches to the early detection of hazardous conditions based on model-aided measurement methods [11.4]. Experience in biotechnology also suggests that applications will appear in prognostic medical diagnostics.

One of the principal tasks of process control engineering is to increase the degree of computability in each process step and to represent it in suitable process models. However, only a purposeful collaboration with process engineering can lead to a fruitful symbiosis here. This holistic mode of analysis will be established in a new engineering discipline, process technology [11.22]; see also Section 11.3.

Representations that are closely related to the phase model of production are the Petri nets [11.23]. While the phase model represents the topology of product flow, Petri nets describe the system of conditions governing product flow. In the context of Petri net theory, any technical system can be represented in terms of two description categories:

- Existence of a state
- Occurrence of an event

Each state is terminated by at least one event and/or brought into being by at least one event [11.24]. States can also be interpreted as conditions that must be fulfilled for a transformation specification to be carried out. In order to describe the dynamic sequence, a further description element — the token — is introduced. The token moves through the static net as a traffic unit. The total population of the net by tokens characterizes a global state of the net.

The central modeling aspect of Petri nets is the concept of concurrency. Two events are said to be concurrent if they are causally independent of one another [11.25].

In processes, buffers are used in an attempt to decouple the flow of material and prevent disturbances propagating through the entire plant. Subsystems decoupled via buffers can operate independently of one another, that is, concurrently, within certain limits. In this way, Petri nets can be applied to production processes. The net theory offers the possibility of modeling such systems and determining their dynamic properties by computation or simulation. Local analysis and modeling has two effects: It reduces the complexity and it creates the possibility of using a process-logistical procedure along with decoupling via buffers [11.26].

In automatic mode (for which recipe-mode operation is a prerequisite), the process-logistical problem (Table 11.1) [11.27], [11.28] is that multiple intercoupled process elements operating in distinct modes must be managed in terms of their mass flow. The definition of operating modes is less well-known than the "recipe" concept discussed above. It describes all modes of operation within a process element, including exceptional and disturbance situations, in discrete form. It must be set in phase-specific fashion for the batch case and section-specific fashion for the continuous case. For a continuous process element and its transitions, the various operating modes are shown in Figure 11.2 [11.29]. Because of the wide range of possible mode specifications for various process elements, the process states desired in disturbance-free operation and the process states sought when disturbances are present are specified as functions of the instantaneous state, in well-defined steps, or are calculated in real time and set down in control tables. The contents of these mode-specific tables are derived from simulation calculations or experimental data.

Figure 11.3 shows schematically how the categories of importance for operation—reproduction, optimization, and automation—are interrelated.

Wüchner [11.30] has discussed which requirements apply to personnel when modern process control systems are used. Krauss [11.31] sees the acceptance of process control systems operating in information-oriented fashion as essentially insured by correct design of the display and control components of the control systems.

What applies to communication between humans and the process control system is equally important with regard to the sensor systems, which serve as "eyes" on the process [11.32].

After this general discussion of operation, the following sections deal with human–process communications (Section 11.2), process analysis and process optimization (Section 11.3), and maintenance strategies (Section 11.4).

Human–process communication comprises the integration of information acquisition (sensor technology), information processing (control system), information and knowledge utilization (human component), and information feedback to the process (actuator systems). The totality of these processes (Fig. 11.4) insures proper operation of the process.

Process analysis and process optimization aimed at achieving permanent gains in reproducibility, safety, and economics is a single engineering task from the viewpoint of process control engineering.

Table 11.1. Levels of automation

Process elements	Mode	Necessary information management	System hardware	Characteristics	Realization
Decoupled	1 manual	all flow paths and all process/product properties manually input as setpoints, managed, and documented	all sensors, all actuators, individual controls and control systems	reproducible production in all operating modes (process models)	≥75%
	2 computer	all flow paths and all process/product properties input to computer as setpoints, managed, and structured in documented form BF*: switch net; set parameters; BF* provided with setpoints	as in 1, plus PCS* with BFE* and BF*, PCC*	reproducible production in all operating modes with recipe operation, process models, simulation, optimization	≤20%
Coupled	3 logistical	as in 2, plus buffer management, buffer strategy, source-point management	as in 2	buffer strategy-oriented recipe mode for all operating modes	≤1%
	4 automatic	as in 2 and 3, plus control of divergence of process elements through variation of process properties	as in 2, plus net orientation (Petri nets)	divergence-oriented recipe mode for all operating modes	1995

* BF = basic function; BFE = basic functional element; PCC = process control computer; PCS = process control system.

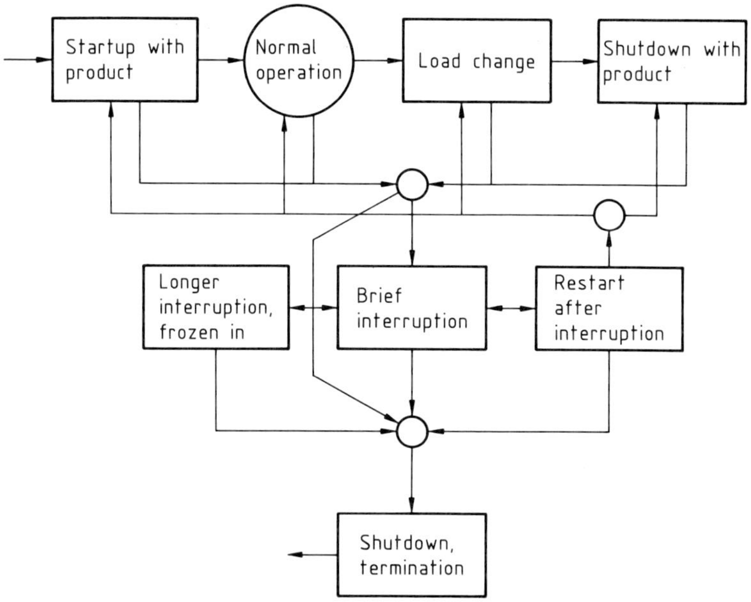

Figure 11.2. Operating modes of a continuous process section and their possible transitions [11.29]

Maintenance strategies aim to preserve and improve the availability of the production facility and its process control components.

11.2. Human–Process Communications

Introduction. Humans today are surrounded by a great number of technical processes. In production, new products are made from many elements by combining operations. Production arrives at new products through the reaction of feed materials. Energy generation and distribution processes are of major importance. Other important technical processes permit the transportation of people, goods, and information on land, by water, and in the air.

The human role in all these processes is to monitor and control. This can be carried out in a "planning and forethought" mode. The human optimizes technical processes as to structure and sequence, detects hazardous situations in the processes at an early stage, and can then intervene promptly to remedy them.

The complete interaction between human and process, which is gathered together under the phrase "human–process communication" in what follows, is governed by the operator, working at an operator station, who runs apparatus and devices, observes measurements, and alternately abstracts and concretizes in an attempt to comprehend what is happening in the process on the basis of individual-signal observations.

In a production process, information relating to an individual process apparatus is accessible at the individual measurement stations still to be seen everywhere (Fig. 11.5). There are also very large control rooms (Fig. 11.6) in which information from large plants is gathered together and displayed on consoles and panels. The color graphics display (Fig. 11.7) is a modern alternative offering an easily and quickly comprehensible representation, even for complicated processes.

Any production process has an a priori dependence on the combined functioning of physical, chemical, and other sensors, which provide the human operator with knowledge about the properties of products and processes by means of appropriate information processing. Automation of production processes, in contrast, leaves process events in a state that can ultimately be comprehended by human senses, except as required by systems-engineering and social-psychological exigencies [11.33].

The preceding chapters have taken up the problem of what information should be determined for process control purposes (see Chaps.

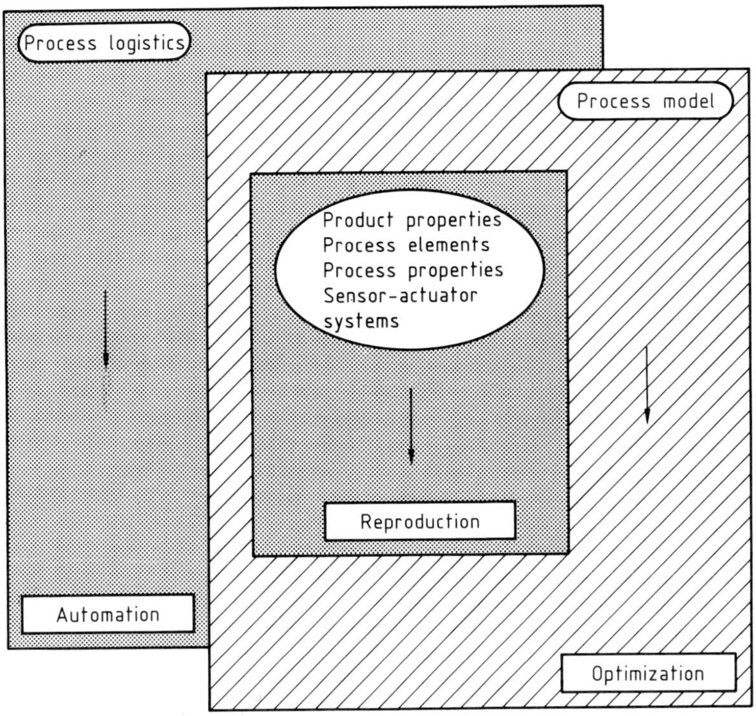

Figure 11.3. Reproduction, optimization, and automation

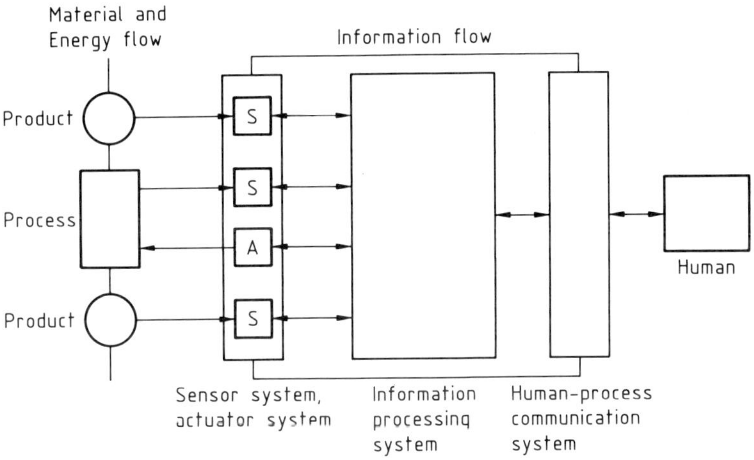

Figure 11.4. Human–process communication

2, 3, and 7 and Section 4.5). Now to these is added the question of which presentation of the available information is best suited to human communication abilities. Modern resources of information technology hold much promise in this area. As yet, however, there has been no exact study of their capabilities in the context of information-oriented process control.

Shortcomings are often seen in the selection and presentation of information as commonly performed at present. For example, an intermittently severe cognitive stress is imposed on operators in "exceptional" situations, which include not only changes in operating values in continuous plants (startup, shutdown, load change), but also unforeseeable changes at the dispositive and operative levels. Ironically, plant automation often leads to a high degree of monotony during normal operation [11.34], which can be blamed on the supply of information being inadequate for the performance of additional duties within the sphere of competence [11.35].

Figure 11.5. Individual measuring stand in process engineering

Figure 11.6. Large control room for monitoring operations

Figure 11.7. Color display with light pen and keyboard (Siemens Teleperm M [11.86])

Besides the proper distribution of tasks between human and information-processing systems (see "Degree of Automation" in Chap. 7, p. 533), the responsibility for the control of a process is a question of some importance. The question of what information is "appropriate" becomes significant, although it must not be forgotten that furnishing such information is a necessary but far from a sufficient condition for any assumption of responsibility [11.36].

This section discusses the general problem of human–process communications for the example of a production process. In discussing the procedures and systems used, reference is also be made to other technical processes.

The special problems of machine-level monitoring and control, grouped under the heading of MMI (Man–Machine Interface), is not the point of focus here, although at times it will be necessary to point out crucial differences between MMI and MPC (Man–Process Communication). A helpful introduction to human–machine communications can be found in [11.37]; a useful reference work is [11.38].

First, it will be beneficial to survey briefly the capabilities of humans in their interaction with large bodies of information today. In the commercial and administrative sphere, the high-speed printer serving as output medium in computing centers generates long columns of numbers and symbols on tall stacks of paper (Fig. 11.8); this device continues to play a central role. The term "business graphics" refers to a mode of presentation that is more suitable for summarizing complex information (Fig. 11.9).

Figure 11.8. High-speed printer with tables

Until recently, engineering devices first took form on the drawing board. Computer-aided design (CAD) marks the beginning of a new world in which design is supported by all the resources of modern computer technology.

Acquisition and Presentation of Information. In earlier centralized process control as well as current decentralized systems, all relevant information about the technical process (and today also about the distributed process control system) must be assembled at a central location so that process control can be optimized under holistic criteria. This central process control location must also have a high availability so that operating personnel can carry out monitoring and intervention at any time. High availability is also required because process condition documentation is often mandated by law.

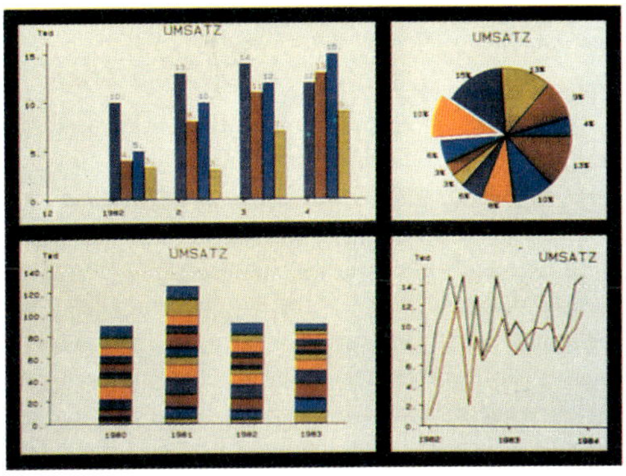

Figure 11.9. Business graphics

Human–process communication began in the early 1900s with direct display and control by field-level devices (Fig. 11.10). The results gained, together with the development of auxiliary-energy field devices, led to a concentration of such devices in a local control stand. Increased plant size and more stringent documentation requirements, aided by unit controllers and modular control systems, led to the central control stand with a flowsheet-like representation of the process around 1950. This flowsheet technique has several advantages for human–process communications:

- The process structure (or "process topology") and the condition of the process are displayed together. An example is the combination track diagram and signal/control box used by the German Federal Railway (Fig. 11.11).

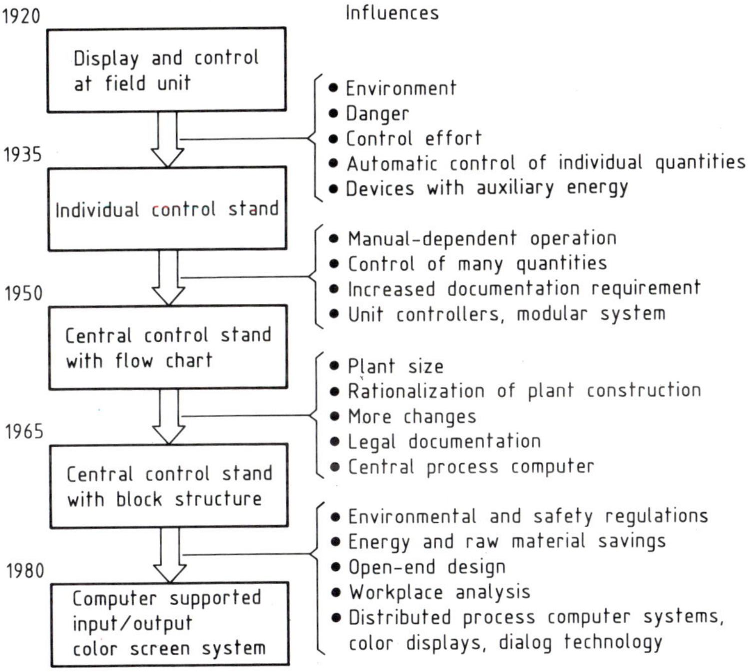

Figure 11.10. Development of process control room design [11.34]

Figure 11.11. Combination track diagram and signal/control box used by the German Federal Railway

Figure 11.12. Central control stand with block structure

- The assignment of control and display elements to the process is implemented through a suitable configuration in an abstract, graphically represented process structure. As a result, control elements can be directly addressed and actuated while their place in the process structure is clearly shown.

With increasing process size and number of process states, the space requirements and costs of hardware process flowsheets have grown. In the case of extensive processes such as those in the petrochemical industry, process control rooms employing flowsheet displays are up to 100 m long and can be managed only by a correspondingly large and mobile staff (Fig. 11.6), hence the names "bicycle" and "rollerskate" contol rooms. High costs of creating such a display are followed by high costs of modifying it, so it is difficult to alter the flowsheet-type control room to keep up with changes in the technical process.

From 1965 on, increasing plant size and the growing number of display and control devices in the central process control room—along with the availability of central process control computers—led to the central control stand with block structure (Fig. 11.12), similar to the quick-scan display technique used in aircraft design. Control rooms employing the block-structure principle have the following features:

- Like instruments are placed in groups
- The relationship between operating and display elements and the technical process is identified by names (nameplates)
- The control and display elements are selected and actuated indirectly; there is no direct topological relationship to a graphical representation of the process.

The operating personnel must therefore remember the process structure. It takes several steps to select and actuate the control and display elements, and this operation calls for good memory and abstraction abilities. The purely technical capabilities of the block-structure control room as to modification are certainly superior to—and more user-friendly than—those of the flowsheet-type room, but it is more difficult for the operating personnel to keep up with frequent changes in the room.

With the availability of powerful and economical computer-supported color displays, the representation methods of the block-structure control room can be retained and adapted to the capabilities of the display screen.

The most important types of video display now in use are discussed on the basis of the control and monitoring devices employed in the process control systems TELEPERM M [11.39], MASTER [11.40], and Geamatics [11.41].

Figure 11.13 shows the general block-structure principle with color graphics displays, with details presented in Figures 11.14–11.16. These are standard or predefined displays and exist as prefabricated data records with the basic representation of images for plant areas, groups, and individual control loops. All the designer must do is adding plant-specific data to this data record (see Chap. 8).

The so-called free display is the exact opposite of the standard displays. Here the designer has a totally free hand. Standard symbol sets offer an aid; elementary symbols can be combined to make new symbols or a toolbox with elementary graphics objects such as circle and rectangle. The designer can thus represent anything from a valve to a steam boiler. If certain symbols are not included, a new symbol set can

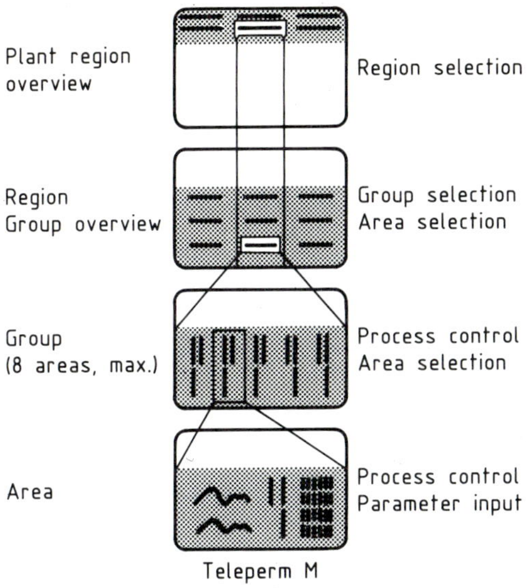

Figure 11.13. Block-structure principle with color displays [11.27]

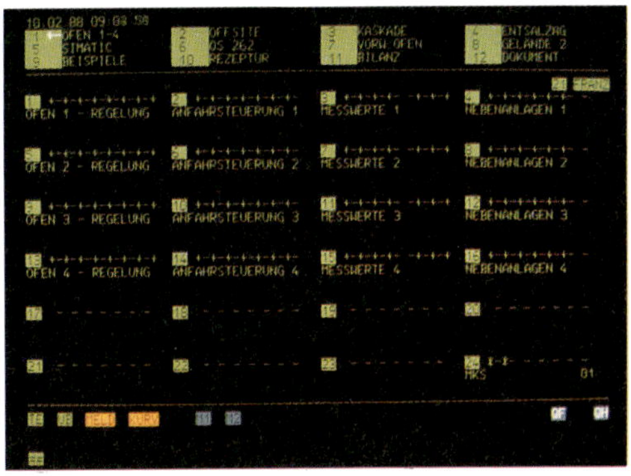

Figure 11.14. Summary diagram [11.39]

be put up on the screen. Separately stored sub-images can be inserted into various overall images, thus saving much time. State displays are also easy to create. In on-line operation, they appear on the screen with the true values, thus facilitating a nominal/actual comparison. Many such free displays can be stored in the operating computer as a data record and can be scrolled through sequentially (Figs. 11.17 and 11.18).

Standardized menus simplify the creation of curves and groups of curves, the assignment of curves to measurement stations, the assembly of curves into full plots, and the specification of storage form, display form, measurement range, and unit of measurement. The system may request all these details in interactive mode.

Curve plots can be presented in a standard or free display. Depending on the physical input device, the selection is made by pointing to a digital or a bar-graph display with a light pen (Fig. 11.19), touching a touch screen, or using a mouse or trackball to position a cursor. Curves are shown in windows with the required deeper level of detail.

Complete standard images can also be incorporated in these windows, for example, the standard representation of a plant group. Window support makes process displays far easier to understand, since only the permanently needed information is always visible; details appear in the window only when needed. The window size and position are determined by the designer; in modern systems, they can be modified on-line by the user (Figs. 11.20 and 11.21).

What is more, user procedures and reports can be freely designed. In doing this, the user determines everything: size and layout as well as dynamic information up to the way in which report creation is initiated. Reports can be generated by specifying a start time and a printing cycle (shift report) or certain events (charge report). All reports can be retrieved at any time, for example, in order to reconstruct a certain plant state (Fig. 11.22).

Thus the windows technique pioneered by Xerox PARC, Apple Macintosh, and Microsoft has found use in color display components of process control systems.

The supplementing or replacement of the block-structure control room by the introduction of color video displays has also changed the mode of operation. In keeping with the concept of furnishing information about the process and product properties, apparatus, devices, and field-level (local) and process control systems to a central location at all times as needed for process control, additional analysis of the identification and selection of the desired information was required. The search for input techniques that place an a priori limitation on the number of possible errors has led, for example, to the selection of well-defined screen regions by touching with a light pen, thus avoiding erroneous digital input. The touch screen is another approach to the same problem. Efforts to optimize "input technology" are related to the topics of acuracy, speed, and physiological and psychological burden on the operator. These points are discussed in [11.42], [11.43].

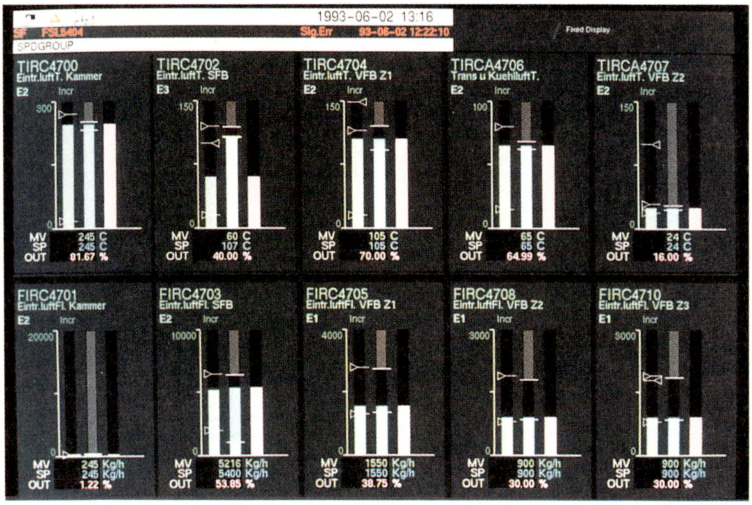

Figure 11.15. Group diagram [11.41]

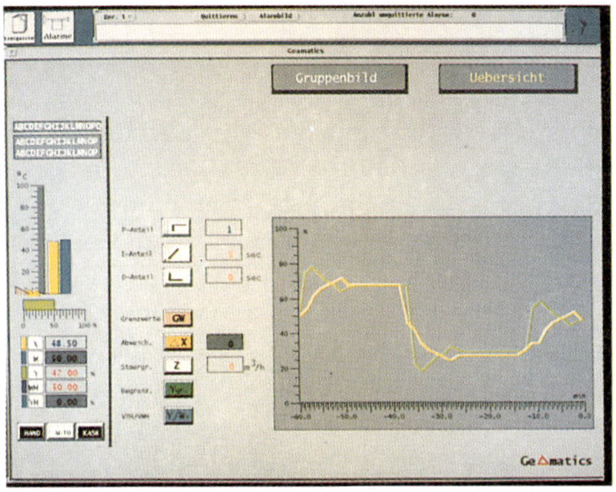

Figure 11.16. Loop diagram [11.40]

At this point, a fundamental difference between human–machine (man–machine) communication (MMC) and human–process (man–process) communication (MPC) should be emphasized. While MMC is based on a specified content and context of the information and goes on to formulate conditions such that the information flow through the human–machine interface will be suited to human capabilities, MPC is concerned with tasks arising in the course of process control and the information that must be furnished so that these tasks can be performed. From, this standpoint, the first questions raised in MPC relate to historically accreted information structures, which are largely oriented toward apparatus, devices, and control loops, and which supports the interpretation of information in precisely this context.

After a thorough analysis of the contributions made by the operator to process control (see Section 7.1, Fig. 7.3, and Section 11.1, Table 11.1) and the information that must be made available for this purpose, the design of human–process communication must take account of the "cognitive faculties" of humans, which also serve as the basis for the delivery of task-relevant information. Findings from psychology have a role to play here.

Interface design in software engineering and cognitive engineering employs an approach borrowed from cognitive psychology in which human beings and their communications capabilities are examined in terms of the "information processor" paradigm. An introduction to the principles and limitations of this approach can be found in [11.44].

According to the conceptual models of cognitive engineering, humans act on three planes (Fig. 11.23) [11.45]. First, from sensory afferents and associated arrays of features, signals are derived through training and directly transformed into actions (intuitive behavior). The second level of behavior is based on rules selected in accordance with a recognition procedure and the association of state and task. Finally, knowledge and experience are turned into goal-dependent decisions, which lead via a planning process to new rules of behavior. Clearly, the last two cognitive levels require a greater information-processing effort.

Any arrangement for human–process communication must therefore use an ingenious ar-

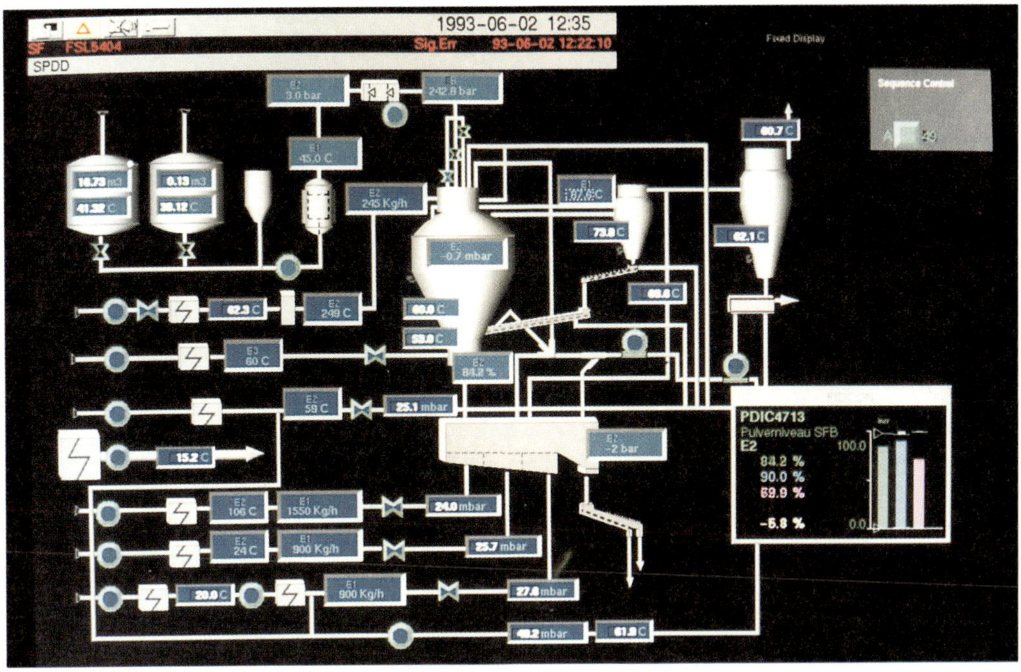

Figure 11.17. Free display [11.41]

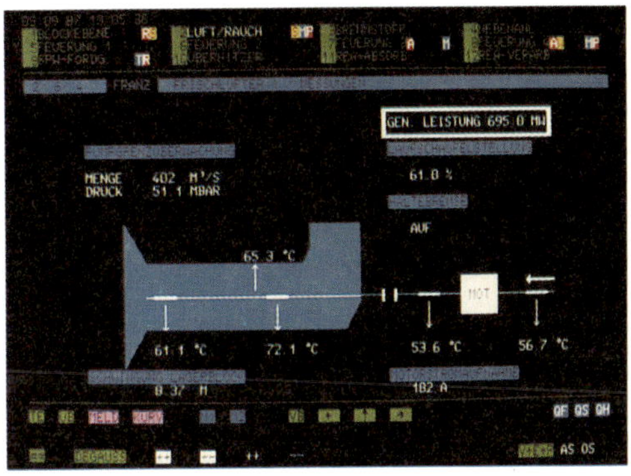

Figure 11.18. Free display [11.39]

ray of features and representation of the process state to provide especially good support to, for example, the "intuitive" level if the human operator is required to react quickly in the course of a monitoring task (such as to issue an alarm). In tasks having to do with quality, where links between process and product properties must be discovered and analyzed, possibly over wide spans of process sections, the cognitive faculties of the two higher levels should be supported in the model. According to ideas of cognitive engineering [11.46], [11.47], [11.48], this can be accomplished by providing information at various levels of detail or by using a rule-based advisory system (expert system).

How individual mental activities involved in operating and monitoring a (partially) automated technical process can be supported by appropriate software has been shown in [11.35]; see Figure 11.24.

In a paper titled "Mozart, or Our Dilemma When Dealing with Genius," EIGEN [11.49] quotes from ECCLES' work "The Human Brain" [11.50]. Here, a similar distinction is made between the dominant hemisphere, which controls conscious behavior in an analytical, logical, comprehending way, and the subdominant hemisphere, which reacts an a musical, graphical, holistic way: This can be understood as implicitly relating to various human information-processing cababilities. Thus, pictorial representations, including graphics, make information comprehensible in a chiefly synthetic (holistic) manner, while texts appear to stimulate analytical, sequential thought.

Deepening the understanding of the perceptive, cognitive, and memory procedures of the human brain is the realm of psychologists. A guide to the abundant literature of psychology is given in [11.51].

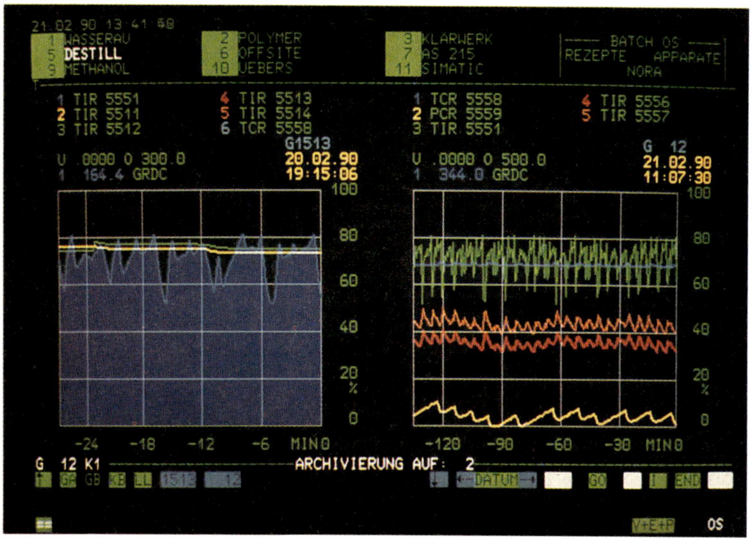

Figure 11.19. Curve plot [11.39]

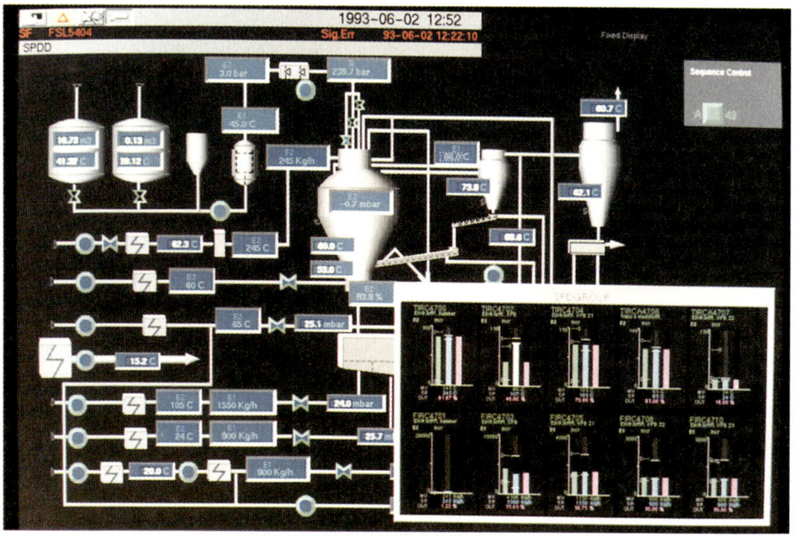

Figure 11.20. Windowing: standard displays in free display [11.36]

Psychophysics deals with the quantification and scaling (see Section 2.3) of psychological meaningful phenomena. It studies the relationships between the "objective" features of the thing perceived, the underlying physiological processes, and the experience of perception. Gestalt psychology has also made contributions in this area [11.52]. Findings from these sciences find use today in software ergonomics among other fields [11.53]. Recommended introductions to physiology are [11.54], [11.55]; an exposition of perceptual psychology is given in [11.56].

Now that magnetic-resonance and positron-emission imaging techniques have made it possible to localize functions of the human brain by functional activation of localizable glucose metabolism [11.57]–[11.60], neuropsychology may provide a bridge between physiology and psychology. Results from this discipline that

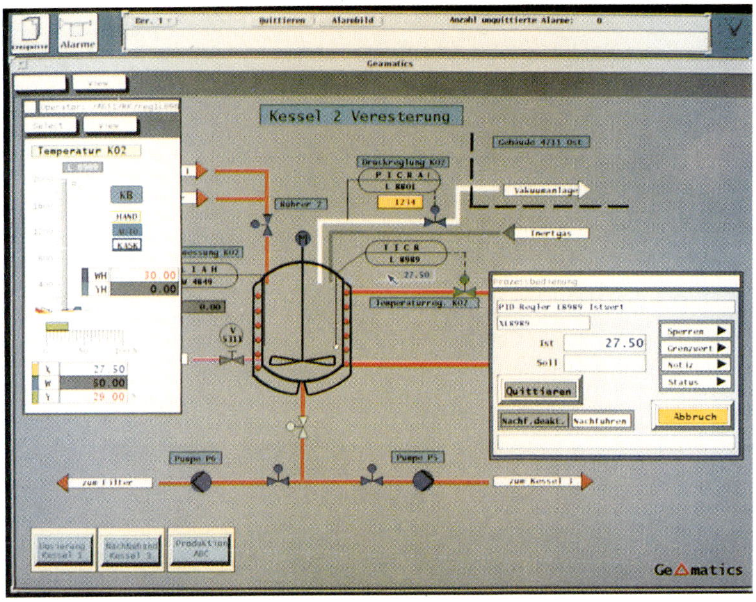

Figure 11.21. Windowing: standard displays and control panel in free display [11.40]

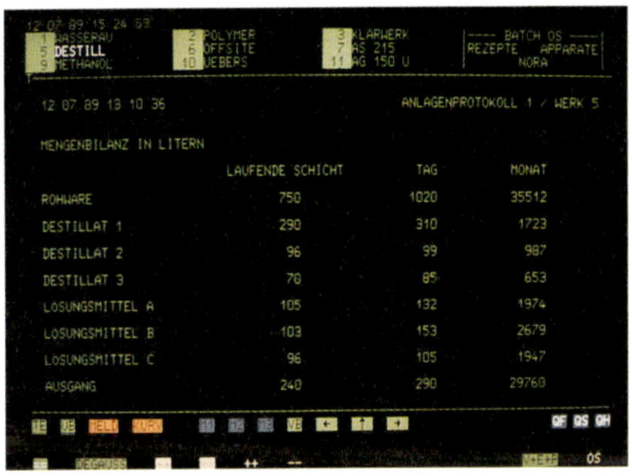

Figure 11.22. User report [11.35]

may be significant for human–process communication are eagerly anticipated in [11.61].

A fundamental dispute between purely scientific-positivist and phenomenological-anthropological modes of analyzing psychological processes, in particular the concerns of perceptual psychology, was shown by STRAUS as early as 1936 [11.62] STRAUS warns urgently against allowing an anthropomorphic interpretation of mechanical activities to be turned into mechanomorphic interpretations of human and animal behavior. Similar statements have been made by other members of the "holistic" school, such as ECCLES, POPPER, LORENZ, and GIBSON [11.63]. They point to fundamental problems in the reductionist approach when regarded from the standpoints of neurology, philosophy, biology, and perceptual psychology (see, e.g., [11.50, pp. 34 ff.]) and show how specific human abilities might be understood in terms of the integration of subfunctions and how the preconditions could thus be created for participation in supra-individual "worlds" such as culture and science (see [11.64, pp. 148 ff.]).

The presentation of messages relating to the technical process must be oriented toward specific target groups of users. This does not simply refer to the experience, for example, that the message coding must correspond to the importance of the message itself—which is reflected in varying reaction times (Fig. 11.25)—or to the limited field of view of human vision (Fig. 11.26); such experiences must be taken into consideration when designing process control rooms and in particular when setting up the configuration of monitors [11.65] (see also Section 10.7).

The presentation of information must also take into account that process engineers, for example, need a different view of the process from day-shift or night-shift operating personnel, whose spectrum of tasks extends from process analysis and optimization to straight operation

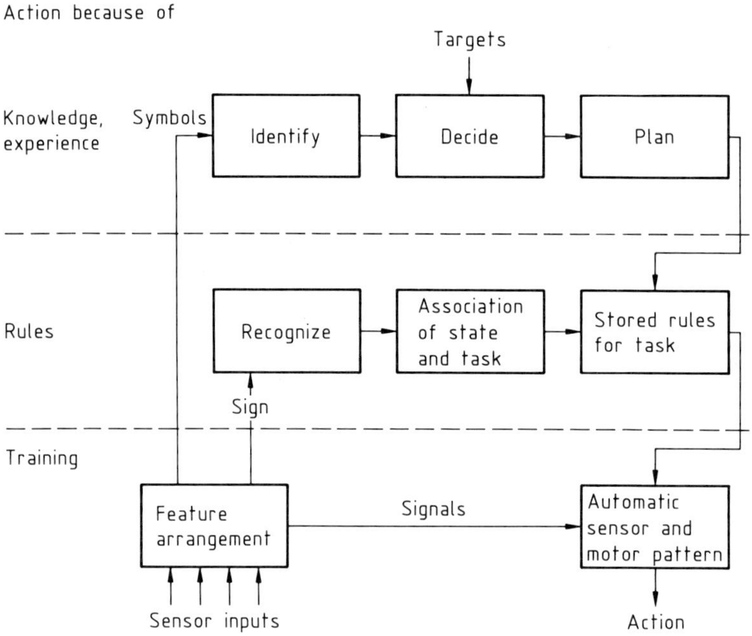

Figure 11.23. Levels of human behavior [11.45]

of the process in accordance with stated rules (Figs. 2.15 and 11.27).

Modern process control systems, with their display and control components, furnish a flexible tool with which, using the many individual process signals available, process states can be represented in an anthropotechnically and economically favorable fashion. As Figures 11.13–11.22 show, operating personnel build their own models of the technical process, based on the individual signals, so that they are then able to interpret these signals correctly and take appropriate action to control the process state.

The availability of high-performance computers suggests using the computer itself to compress individual signals into an overall view of the process—an image or, mathematically speaking, a model of the process—and conveying to the operating personnel this compressed information based on the individual signals. This would complete the transition to "information-oriented" process control engineering. In signal-oriented process control, the information philosophy was: "Acquire and present information item by item, in parallel, and always."

In information-oriented process control engineering, the philosophy can be reformulated as: "Acquire information in the smallest possible amount, but as much as necessary, and not until it is to be used, and in a form that is clearest for the operating personnel".

Thus a holistic point of view will make it possible to enhance the competence of the operating personnel with the objective of "state-oriented" human–process communication.

Information Reduction (see also Section 3.2). Complicated technical processes generate a large number of signals, which the human operator cannot take in and interpret in direct form. For example, if two dozen signals are plotted in bar-graph form, the task of deducing certain process states from certain patterns can generally not be performed by a human. Procedures of signal processing and pattern recognition must therefore be employed to reduce this complexity.

Several methods for reducing the dimensionality are known, for example, from speech processing [11.66]. Orthogonal transformations, the best-known of which is the Fourier transformation, play an important role in this procedure. The idea is to represent an original signal by a linear combination of sine and cosine signals differing in frequency and phase. The weighting factors in the linear combination are the new dimensions, the number of which can be reduced by leaving out, say, high or low frequencies, with no substantial effect on the signal form.

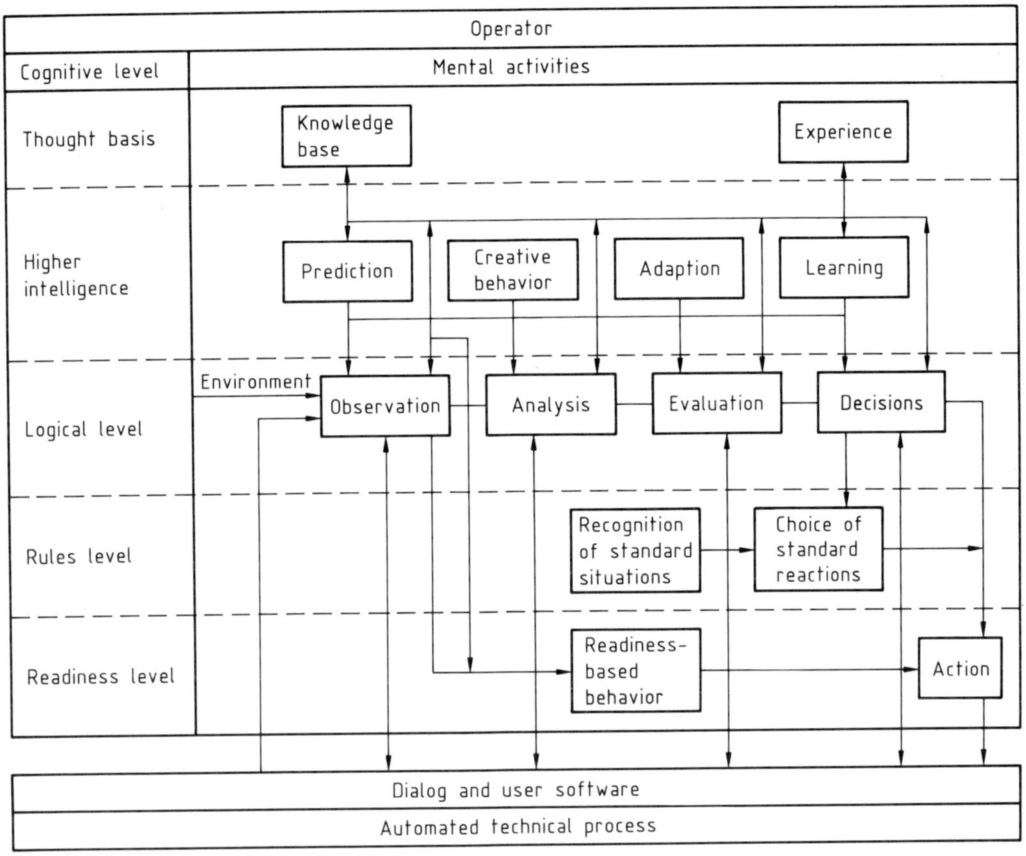

Figure 11.24. Support of operator's mental activities by interactive and custom software in the control and monitoring of an automated technological process [11.35]

Another orthogonal transformation is the Karhunen-Loève transformation, in which sines and cosines are replaced by functions derived from the original signal. These are represented as vectors whose contributions to the total signal decrease with increasing ordinal number. By forming a linear combination of a few such vectors, the original signal can be reconstructed well; the weighting factors in the linear combination are the new transformed signals with smaller dimensionality [11.67], [11.68].

The method of factor analysis takes the large amount of possible information about the product and the process and generates a system of mutually independent new variables that can be ordered, so that it is possible to say how much each of these new parameters contributes to the total information.

If such methods are used uncritically, the problem arises that the new dimensions may not lend themselves to ready interpretation by the user. For example, signal mean values and signal ratios are highly suitable and are also subjectively accepted as indicators.

This discussion has mentioned and briefly described just a few methods for reducing the dimension. Applying such a method also reduces the redundancy, since as a rule a redundant signal can be represented as a combination of other signals, so that recovering them does not supply any new knowledge about the process state.

This situation typically occurs in an overinstrumented system where there are, for example, more sensors than are required for an unambiguous determination of the state. However, overinstrumentation also makes it possible to carry out signal reconstruction for signal sources that may have failed because of some technical defect (redundant instrumentation).

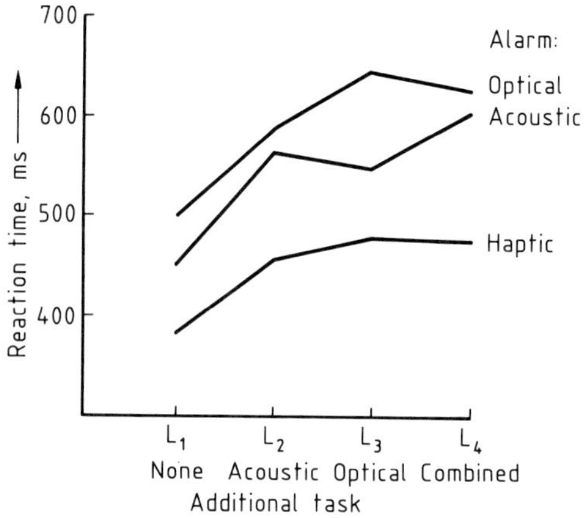

Figure 11.25. Reaction times to messages encoded in various ways [11.59]

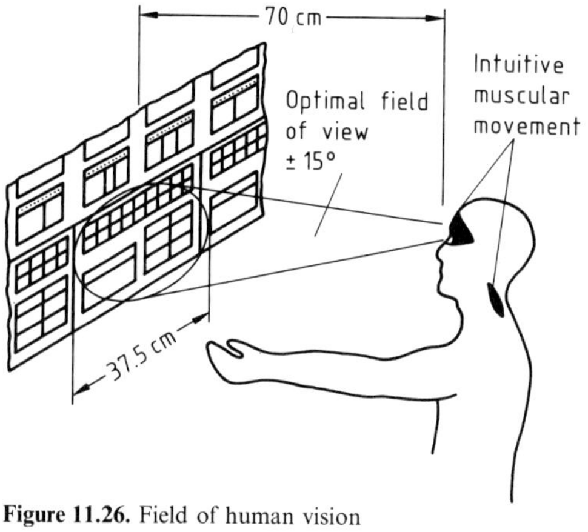

Figure 11.26. Field of human vision

Even though there may be substantial time shifts between signals, especially in extended technical processes, there are deterministic relationships between them (e.g., measurements upstream and downstream of a lag element). In many cases, such shifts can be detected by cross-correlation analysis, but in general the result for a signal source is that not only the measurement at a time T but also some number N of measurements obtained at earlier times must be taken into consideration (see Fig. 11.28).

Pattern-recognition techniques afford still more support to the operator. For example, hazardous states in a process can be detected on the basis of a pattern in the signal. Figure 11.28 illustrates the principle of pattern-recognition systems. The original signals are first subjected to feature extraction. This is analogous to the di-

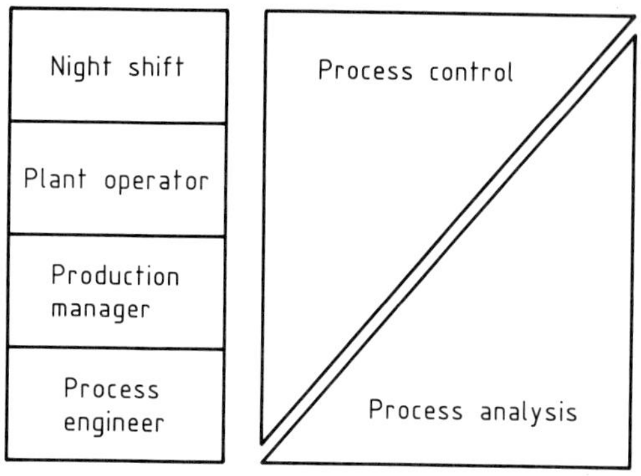

Figure 11.27. Process communication

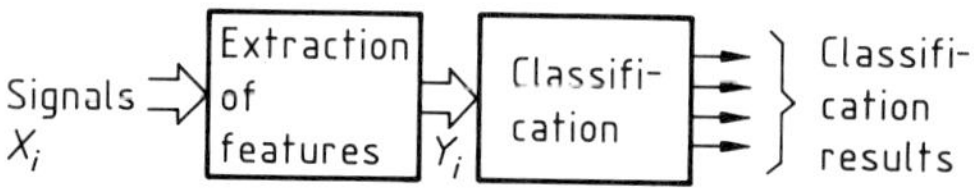

Figure 11.28. Schematic diagram of pattern recognition [11.59]

mension-reduction operation described above; the smallest possible number of features, and hence highly informative features, are derived from the original signals (see also Chap. 2).

This means that the subsequent classification can be done in a space of few dimensions. Classification involves forming regions in the N-dimensional feature space, under the assumption that patterns lying close together in feature space belong to the same class. In a two-dimensional feature space, such regions can be delimited by, for example, a chain of line segments.

If these lines are drawn automatically, possibly with operator interaction, the classifier is said to "learn." Under certain circumstances, depending on operator behavior, such a classifier may be able to adapt to changed process situations.

By using cluster analysis [11.69], [11.70], class formation can be carried out automatically without external supervision. Each new pattern is assigned to the region (cluster) to which it is closest. Cluster analysis can be used, for example, to find automatically and subsequently identify particularly frequent and typical process states. In future, this type of procedure will undoubtedly provide significant aid to the operator (see also Fig. 11.63).

These order-reduction procedures make it possible to represent process states in three-dimensional form, making it possible to visualize the product or process space (Fig. 11.29).

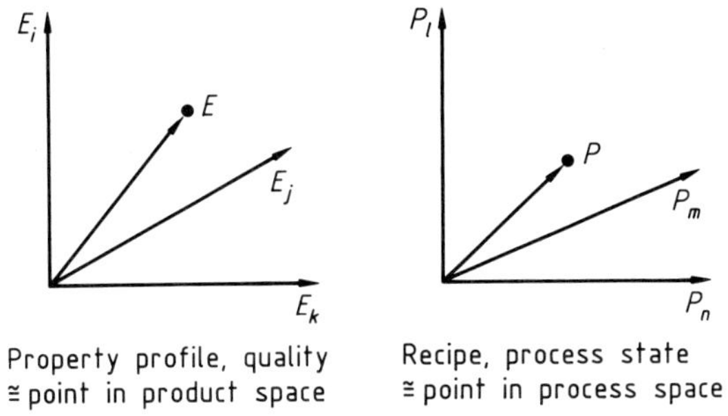

Figure 11.29. Analogy between product and process spaces [11.67], [11.68]

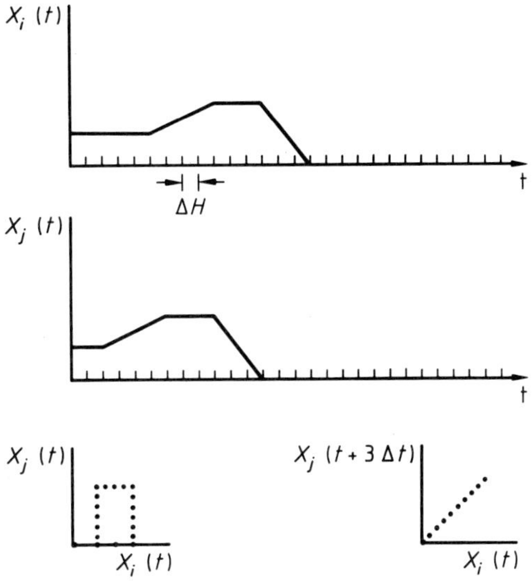

Figure 11.30. Effect of time delays

A similar mode of analysis is arrived at if the methods of similarity theory are applied to a signal pool [11.9], [11.10]. The direct presentation of the dimensioned X-quantities in X-space with X-relationships is replaced by the dimensionless presentation of π-quantities in π-space with π-relationships.

The presentation of state spaces, as a prerequisite for "state-oriented" human–process communication [11.35], [11.71], [11.72], requires procedures by which relationships between the process and product properties can be obtained from the time dependences of these properties (as conventionally supplied by the instruments currently used for process monitoring).

Thus the problem is to derive the function $X_i(X_j)$ from the time dependences $X_i(t)$ and $X_j(t)$ of the measurements.

It is straightforward to plot the last N pairs of values (X_i, X_j) acquired to give a two-dimensional scatter diagram in coordinates X_j versus X_i. This representation may actually make it possible to derive the desired relationship. Under some circumstances, regression methods can also be used; for example, the mean and dispersion of X_i can be obtained for each X_j product. Utmost care is advisable, however, because no a priori prediction can be made as to the form of the curve.

Two further difficult time problems must be solved:

- In general, the function $X_i(X_j)$ is not stationary. Otherwise it could be determined once for all and would have no predictive power for the operator. The time constant that describes the behavior governs the size of the "window" over which the function $X_i(X_j)$ can be determined. The window may be fixed (a new measurement vanishes every time a new scan takes place) or decaying (e.g., exponential); the computing effort required is then less. The operator receives a picture of the function that shows a continuous slight variation.
- X_i and X_j may be very closely related but with a (possibly long) time shift or bias. If the X_i and X_j measured at the same time are plotted, the result may have no predictive power at all (Fig. 11.30).

The time shift can be identified by cross-correlation analysis, after which measured values $X_i(t)$ and $X_j(t + \Delta t)$ can be paired. It is also possible to create a three-dimensional representation with the time offset as the third dimension. The resulting "mountain range" would reveal where the dependence was most distinct. Changes in time offset, in any case, provide an important indicator of the process state (see also Section 3.2).

Information Storage under Real-Time Conditions. In future human–process communication systems, a major role will be played by the computer's internal data model [11.73]–[11.76]. This model must contain all data acquired and stored during the operation of the process control system (see Chaps. 2, 9, and particularly 8). Such a procedure will require data storage under real-time conditions, and this objective has given rise to the conception and development of real-time databases. The following typical features distinguish these from classical databases:

- They contain instantaneous data generated and required in real-time process operation. In addition, there are data to be archived for infrequent later interpretation. These archival data will replace the process logger charts common today; the large bodies of data acquired will typically be stored on video-disk memory devices. A real-time database, in the narrow sense, is concerned with the management of instantaneous data.
- These databases are very small in comparison with commercial databases.
- Reaction-time requirements are stringent, both for the storage of data (which must always be carried out such that a consistent data model is preserved) and for its retrieval (since such requests may be made by both the process and the operator). Reaction times in the millisecond range and very short time intervals for insuring data consistency are needed here.

In the implementation of real-time databases, the result is extensive utilization of main memory.

The use of a real-time database in human–process communication systems will lead to major simplifications in programming. With the software tools belonging to a data base, individual queries can easily be handled; support of a query language with graphical output resources will simplify interaction with the operator.

Graphical Technology. EIGEN [11.9] has shown that pictorial representations (graphics) make it easier to comprehend information than does text.

The use of computer graphics in mathematics and the natural sciences is surveyed in [11.77]. Graphical kernel systems such as GKS/GKS-3D or PHIGS/PHIGS$^+$ are device-independent basic software systems for graphical programming; implemented on workstations, they increasingly form the basis for CAD tools [11.76], [11.77]. Multidimensional state-space representation, in particular, requires images with almost photographic realism; today these can be generated with efficient algorithms and data structures or with ray-optics techniques from physics (ray tracing) [11.78].

Finally, computer graphics can handle demanding visualization tasks only with the aid of knowledge-based methods. Conversely, as discussed in Chapter 13, increasingly complex artificial-intelligence (AI) systems depend on graphical visualization aids to create viewable images of models and states. As a result, the disciplines of graphical data processing and artificial intelligence are becoming more and more interwoven, as is reported in [11.79].

What follows is a discussion of the graphics techniques available in current process monitoring and control systems. All graphical output devices now in use for human–process communication are color displays. Graphics systems in first-generation process monitoring and control systems feature medium resolution (roughly equivalent to the broadcast television standard) and permit the semigraphical representation of process states and process flowsheets. "Semigraphical" means that the screen is subdivided into predefined fields containing, for example, 8×8 pixels, in each of which alphanumeric characters or graphical image elements can be placed. A character generator driven in an 8-bit code contains the bit pattern for the characters and the selected graphical elements.

The refresh memories in modern systems are too large to be displayed on the screen. Thus the process image can be moved back and forth so that the screen appears to represent a segment of a large landscape (rolling image).

On the input side, the keyboard is complemented by a variety of pointing devices. These facilitate both graphical input and selection from menus shown on the screen. Typical devices for this purpose include trackballs; joysticks; touch screens, in which various physical methods are used to detect the location of the finger touching the screen; and light pens.

Each of these systems gives x/y coordinates on the screen, which are accepted as inputs to the operator software.

The trend in development of graphical devices for display and control components is toward pixel-oriented color displays. These offer much better resolution as well as physical parameters (19″ and 21″ screens with resolutions of 1280×1024 dots, displaying 256 colors simultaneously out of 16.7 million). There is a trend toward the standardization of hardware and software. The devices are based on PCs or workstations with RISC processors and run under operating systems such as DOS, UNIX, or their derivatives.

The principle embodied in these user interfaces is splitting into components that are relevant for the user interface and those that are specific to the processing tasks in an application [11.80]. The distinguishing feature from the user's standpoint is the structuring of the entire screen area in movable, variable-size, iconizable subregions or windows, which can accept user input and display representations independently of one another. Screen windowing and network-wide window management tasks are performed by basic windowing systems, which are included in the operating system or added on as utility libraries (e.g., X Windows [11.81]). "Network-wide" means that the user, from any network node, can interact with an application running on any other computer(s) in the network. One possible principle for such interaction is the client-server principle. Above the window management level are graphical user interfaces (OSF-Motif, Open Look in the UNIX environment, MS Windows or Presentation Manager in the DOS environment).

The problems relating to control and monitoring components (see Chap. 7) have now impelled a number of manufacturers to create their own process visualization features for certain information-processing systems that have no display and control components of their own, especially systems based on stored-program controllers. One of the main components of these visualization devices, which can simultaneously carry out control functions, is a database system permanently linked to the process and hence running in real time. Because only a few of the user interfaces based on the well-known operating systems have been standardized (e.g., X-Motiv in the UNIX realm or Windows in the DOS environment), a number of very diverse "artistically" designed tools have appeared in this area. These require an extremely critical examination before they are put to use.

Several systems based on this architecture have been offered especially for process visualization (e.g., InTouch [11.82], Factory Link [11.83], APROL [11.84], Dynavis-X [11.85] COROS-LS-B [11.86]). Some process monitoring and control systems have integrated display and control components (e.g., SATTLINE [11.87], Viewstar 750 [11.88], Advant-Station [11.90]). The discussion that follows introduces special characteristics of these systems without describing individual products.

The functionality of these systems is outstanding because elementary, interactive, and scalable graphics objects are available. System resources, such as fonts, fill patterns, and color tables, can be utilized, and raster or bitmap graphics, either scanned in or generated by other graphics software, can be incorporated. With the aid of these design resources, predefined graphical input and output objects specific to process control engineering can be created and placed in libraries for later use. These include virtual keys, switch panels, slider controls, analog displays, and so forth. Another aspect of the graphical functionality is the "logical zoom" function, in which excerpts of images can be displayed in varying sizes and levels of detail and information is added as the zoom level increases [11.85].

Visualization systems based on graphical user interfaces are further characterized by expanded capabilities for defining the interaction (dialog) between user and process. These features include the interpretation of user inputs, the execution of appropriate commands, the creation of graphical objects when certain events take place, and so on. Along with well-known event-oriented and data-oriented concepts, complex control structures made up of command sequences, branches, and loops can be initiated, windows can be opened and closed, and graphical objects can be created and deleted. Flexible, state-dependent image structures can be built up in this way and need not be oriented to fixed image hierarchies.

Another characteristic is that these systems can be set up to utilize open system services present in any hardware and software platform so as to expand the process visualization functionality. Such mechanisms include system services for data exchange, command transmission, and simple integration of application-specific software upgrades. Examples are Dynamic Data Exchange, the Dynamic Link Library concept and multimedia add-ons under Windows and (in part) OS/2, database query languages such as SQL, and pipe mechanisms under UNIX. In this way, open visualization systems offer better integration capabilities than the display and control components in first-generation process control systems. One example is the connection of the visualization system to a spreadsheet analysis in which process data can be linked with mathematical functions on-line. Unified operator control concepts in a user interface come into play here because different programs then present themselves to the user with the same appearance mode of interaction (dialog). The utilization of system services also includes the possibility of creating help systems and defining macros.

For the designer of a process visualization utility, ease of use dictates that the capabilities of the graphical user interface be employed in design. "Toolkits" or interface modules are not adequate for this purpose. They merely provide development interfaces at the programming-language level and require much specialized knowledge. Abstraction levels suitable for the design of display and control components offer classes of tools, such as CAD-like editors for graphical interfaces in combination with editors for formulating "interactive" control, so that interface objects can be selected, constructed, furnished with attributes, and linked to user actions and process data (interface builder and user interface management systems [11.80]). Resources taken over from the CAD world, such as traps in conjunction with underlaid rasters, further facilitate the work.

From the process control engineering standpoint, the high flexibility of visualization systems also holds some risks. In the design of process visualization, the assignment of appropriate parameters should be used to restrict the user's options in interactive mode. For example, it should not be possible to place other windows on top of alarm windows. Similarly, a basic subdivision of the screen into fixed-position, fixed-size summary and detail windows should be permitted. Application-specific software add-ons and process data manipulations in other software packages may prove far more successful if access rights questions are not handled as functions of user or process state. In "closed" process monitoring and control systems, such mechanisms are made available as system services.

Of particular importance in process control engineering is the seamless integration of visualization design into the engineering process as a whole. This includes, for example, the linking of process visualization and dynamic process data generated in the design of process-level (local) components. Capabilities should be provided that make available the engineering data which indicates which process data are used in which visualization functions (design database).

Anticipated developments in graphics hardware for human–process communication are discussed below. Modern visualization systems intended for use as display and control components in process control systems are based on 2D graphics. Three-dimensional images can also be displayed on a 2D screen. The representation of the Space Shuttle in Figure 11.31 shows how hidden lines and surfaces can be removed from the three-dimensional picture by using computer graphics. However, mapping a three-dimensional model onto a two-dimensional surface (well-defined portion of the image, well-defined angle of view) places a considerable load on the computer that constructs the individual points of the image. Accordingly, 3D graphics on a 2D screen is severely limited in speed. Graphical representation capabilities in process control systems with modern 2D graphics systems are illustrated in Figure 11.17.

The performance of graphics systems is improving continuously. Graphics systems from the CAD realm and scientific data processing already hold an internal 3D representation, which they map onto the selected viewport. This implies a substantially larger graphics memory (a complex volume model, for example, requires 500 000 vectors, each represented by several bytes, so that the total memory demand is several megabytes). The enormously high performance required of graphics systems in these applications stem from close-to-reality graphical representation combined with animation capabilities. An example is the simulation of poor visibility in the training of aircraft pilots [11.91]. Graphics systems, usually implemented on a dedicated board or even as multiprocessor systems, can now plot up to 10^6 3D vectors per second. The workstation CPU and the graphics processors split the work up in a rendering pipeline [11.92].

Figure 11.31. Three-dimensional representation of the Space Shuttle on a two-dimensional screen [11.89]

Data transmission over an internal bus is a bottleneck, as in many other situations. Modern graphics workstations have main memory capacities of 16–128 MB and even more. The capabilities of modern high-performance graphics hold the promise of real-time visualization of complex data collections in n dimensions, a procedure that has now become the state of the art in scientific data processing. Isolines, textures, and other resources are used along with color to handle the n-th dimension or variant in graphical representations. Photographs of glowing metal slabs or bars provide an example. From the color of a glowing piece of metal, it is easy to determine the temperatures of the individual zones. The color scale runs from white (highest temperature) to black (lowest temperature). In Figure 11.31, color is used as the fourth dimension in visualizing the temperature distribution of the Space Shuttle. Color can also be used to visualize other quantities such as height, material density, etc. The use of color as a supplementary element in graphics is advantageous chiefly in raster graphics, since colors can be gradated extremely finely in this kind of representation (high color resolution).

There have also been new developments in pointing devices. The mouse is now commonplace in office automation. In conjunction with the use of workstations as display and control components, it will be used in the process control room environment as well. Futuristic concepts of interactive devices go yet further: An accelerometer on the fingertip makes it possible to determine the direction in which the finger is pointing, and observation of the pupil tells where the operator is gazing. Electronic gloves equipped with sensors (datagloves) are already in use, chiefly for research and in virtual reality systems [11.93]. The first commercial applications have come about in the CAD field. In the n-dimensional data or feature space concepts discussed above, an operator using such an interactive tool can manipulate data and freely select a point for viewing images, thus wandering through the feature space.

Multimedia Techniques. Control and monitoring functions in process control systems have recently been expanded by techniques broadly referred to as multimedia. Steinmetz has given a qualitative, application-neutral definition of this term [11.94]. The types of media supported and the options for media processing on a computer system are key features. Multimedia systems must be able to process discrete media (e.g., text and graphics) and continuous media (e.g., audio and video) independently of one another. The computer is extending its capacities to every medium by which information can be propagated and presented. The central points in multimedia are integration of media, range of media, distribution of multimedia information over a computer network, and user interaction with a multimedia system.

From the standpoint of information technology, crucial tasks for media integration are to create digital representations of the media, to develop suitable storage media, and to augment communications networks. The main problem is the high flow rate of time-critical multimedia data. For example, the digitization of a video sequence at normal resolution requires data rates of over 20 MB/s. Key technologies for implementation are digital signal processors for data compression, optical storage media, and high-data-rate networks.

Ultimately, multimedia systems make sense only if efficient system management insures the on-time processing of, in particular, continuous media. Requirements in this area apply to real-time processing in multimedia systems, specific service reservation, and the synchronization of linked information units or information streams relating to discrete and continuous media. Finally, the developers and users of a multimedia system should be able to operate in the way they have become accustomed to with discrete media. This means that multimedia functionality must be properly abstracted in development systems, user interfaces, and database systems. The object-oriented approach finds use here. Examples of the abstraction of multimedia capabilities can be found in multimedia add-ons to window-oriented graphical user interfaces running on personal computers and workstations. The desktop metaphor has been expanded to include, for example, a telephone, a dictating machine, and a video conference. These points will not be discussed at length here, nor will the associated problems, in particular the management of large bodies of data, diverse data formats, links between distinct media objects, and content-oriented searching. Reference should be made to the specialist literature [11.94]–[11.99]. In what follows, special requirements on the use of multimedia are presented, along with possible application areas in process control engineering.

Real-time requirements in process control mean that the correlation of data in a variety of media—especially continuous media—to objects, events, states, alarms, etc., is a central problem. Furthermore, because events in technical processes occur in stochastic fashion, it is impossible to predict at what time which multimedia input/output devices need to be available. Conflicts, such as when several audio sequences must be output at the same time, must not occur, although it is entirely possible for the operator to be simultaneously occupied with several distinct tasks. These points can be traced back to the problem of multimedia resource management. A concept for the integration of multimedia in an information and diagnostic system for use in process control is proposed in [11.100]. A major role is played by the presentation system, which is responsible for synchronized input and output matched to system and user requirements.

Already common in the process industries is the use of cameras and microphones to monitor machinery, plant sections, and products. Generally, this involves self-sufficient analong monitoring systems with no connection to the process monitoring and control system, or else hybrid systems in which analog recording devices are controlled by computers through interfaces. In both cases, the data stream remains unprocessed and isolated from other information-processing functions. In digital multimedia systems of the future, integrated processing of all media will be possible. First, other information can be appended to audio and video signals (timestamp, links to process data, comments, etc.). The result will be new capabilities for messaging, reporting, and archiving systems. Second, digital image processing will offer new possibilities for human–process communication. Digital image processing is already employed in industrial robots (machine vision), in computer-aided quality inspection of finished parts, and in the monitoring of machinery [11.101], [11.102]. In human–process communication, digital image processing, perhaps in conjunction with CCD cameras, will provide a route to alternative methods by which product and process states can be determined and classified and the results presented in a visually informative way. An example is the assessment of temperature profiles in metal working by determining gray-scale distributions [11.103]. Finally, the operations of digital image processing have the goal of presenting the operator with visual information in an appropriate and easily remembered form. A closely related topic is pattern recognition, a set of techniques which permit the identification and grouping of logically like image contents and thus the identification of single objects [11.104].

In addition to the question of possible applications for multimedia systems, another point of interest for human–process communication is how these systems can be put to use. First of all, the modality—i.e., the way in which the operator's sensory organs take part in the interaction—of the operator–system interaction is expaned, for example by speech, but also by futuristic concepts such as the interpretation of body language [11.105], [11.106]. The objective is to make user interfaces more "natural" by letting them use a broader sensory spectrum. Information acquisition can be facilitated and supported through input and output in the original medium without transformation, as in voice mail, or through multimodal presentation of information in more than one medium. An application is the audiovisual notebook, in which video records with added comments provide information about special events.

A second system-related aspect is "active" user behavior. The pertinent concept here is hypertext or hypermedia [11.107], which is a concept for organizing and actively "exploring" information.

Essentially, a hypertext or hypermedia system has two components: networks consisting of elementary information units (nodes) and flexible, directed references (links), and tools for creating these networks, reading them, and so forth. Hypermedia differs from hypertext in that the nodes contain multimedia information. The user wanders through the network in survey fashion (browsing) or with a definite destination (navigation). The contents of nodes and links need to be static. The concept also includes executable code as well as the on-line determination of the destination of dynamic links as an aid to navigation. More complex structures can be generated by aggregating networks. For a more in-depth treatment and discussion of further aspects, such as programming and tool/system support, see [11.53], [11.108], [11.109].

The first applications of hypertext concepts came in the organization of documents in "information systems" (information kiosks) and in education. Hypermedia systems have found industrial use as technical information systems for commissioning, repair, and service documentation [11.110]. In technical processes, the operator of the future will obtain required knowledge, such as safety aspects, through information systems of this kind. At present, work in this area is being carried on in a EUREKA project [11.111]. Hypertext concepts in human–process communication can also be put to work in on-line operator support or operator training systems, in particular systems that will guide the unpracticed operator to the correct control actions [11.112]. Similar considerations relating to ship steering are presented in [11.113]. Closely related to hypermedia concepts in process control is the linkage to knowledge-based methods; hypermedia systems are used as user interfaces for diagnostic systems, since operator behavior can be better supported with them [11.114] (see also Chap. 13).

Trend Forecasts. Instances of information-oriented process control engineering can be found today in aeronautics, the power-plant industry, and automaking.

Flying involves very complicated changes in attitude resulting from course changes. In modern aircraft, these trimming or dressing actions are performed for the pilot by a computer-supported control system. Thus the pilot can fully concentrate on determining the course, using a joystick to specify course changes to the autopilot, which in turn generates the control instructions needed to alter the attitude. In order to reduce further the cognitive burden on the pilot, aircraft such as the Airbus A310 include a display showing the instantaneous position with the planned course and a proposed course deviation. The justification for the deviation is incorporated into the display; for example, if it is an area of bad weather, a weather radar display is superimposed (Fig. 11.32) [11.115].

Similar developments are under way in automotive engineering. Navigation devices use graphics to display the position and the proposed course to the desired destination, taking account of current traffic conditions [11.116]. Required maintenance is determined by using models and reported to the driver; an example is the service-interval display developed by BMW [11.117].

Another example of information-oriented process communication is a display showing ignition angle plots for a digital motor ignition system [11.118].

In power-plant control engineering, the status of a boiler or turbine is displayed as an instantaneous operating point on a plot of characteristic curves. Any change in operating point can be effected by using a crosshair pointer to specify a new nominal operating point. Individual setpoints are then obtained from a process model [11.119].

New approaches to information-oriented process control engineering are also seen in the chemical industry. In general terms, product properties E can be represented as functions g of process properties P. This procedure is readily illustrated in process space (Fig. 11.33) [11.120]. For a certain product property E_0, there exists a function g^* of the process proper-

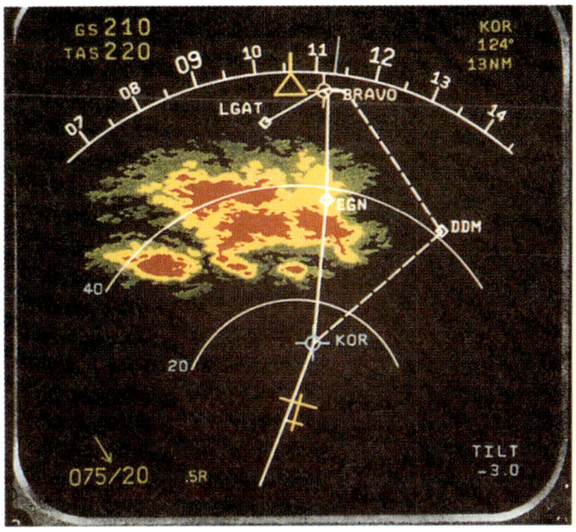

Figure 11.32. Airbus A310, navigational display [11.115]

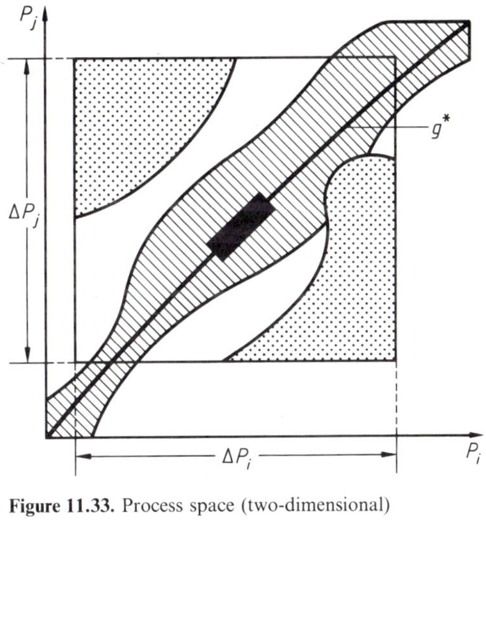

Figure 11.33. Process space (two-dimensional)

ties P_i and P_j, as shown by the curve. The thicker portion of the line characterizes the optimal region for the process, characterized by compliance with additional constraints. If certain tolerances ($E_0 \pm \Delta E_0$) are set when the values of the product properties are established, then in the process-space representation this corresponds to a certain variation range of the process properties (shaded region). In a specified technical apparatus, of course, the process properties can be varied only within certain ranges ($\Delta P_i, \Delta P_j$); the rectangle shown in the figure is obtained as a result. Safety and environmental restrictions narrow the acceptable region of the process space still further, leading to the areas enclosed by dashed lines in the figure.

Figure 11.34 shows another way of thinking about a real process. In process space (normally the coordinates are configuration parameters determined by order reduction), the hatched region shows the normal production condition for a certain product. The white rectangle located on this is the economically optimum operating region. The variation in the process properties allowed by a specified apparatus is indicated by the outline of a parallelepiped. The upper and lower gray areas show the restrictions imposed for safety and environmental reasons.

Figure 11.34 also shows how the operating point on the screen can be moved into the target region, for example with a joystick. The change so specified is then automatically transformed to setpoint (control) signals [11.121].

Tracking the operating points in a process as just described indicates the distance from critical regions and, eventually, may also make it possible to initiate prompt computer-aided action to remedy disturbances. The analytical preconditions for doing this have long been known from fault-tree analysis.

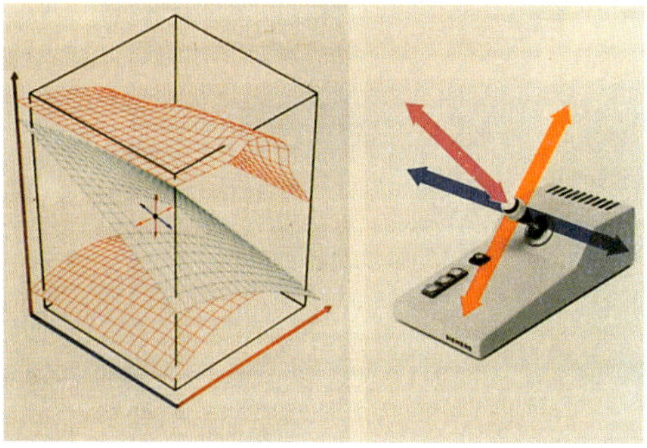

Figure 11.34. Process control with the joystick

Another type of process communication can be implemented with the help of Wolf diagrams [11.122].

These examples from a variety of technical application areas suffice to illustrate the present and foreseeable capabilities of information-oriented communication.

Information-oriented process control engineering will relieve operating personnel of some burden, help them avoid errors in disturbance situations, and promote the optimization of processes through improved modeling. This innovation will also make it easier to produce better and more consistent quality in accordance with specified nominal profiles. The operating personnel will be able to put more effort into studying the process as a basis for modeling. Furthermore, it will be possible to employ operating personnel with a broader spectrum of skills, and at the same time the monotony of continuously watching single signals will be broken up. Finally, the early detection of incipient transitions to undesirable operating states will lead to improved operational safety and reliability.

These are all good reasons for supporting information-oriented process control engineering. "Signal-oriented" process control engineering, which recognized only human–machine and instrument–machine modes of communication, will be replaced by "information-oriented" process control engineering, which includes the concept of human–process communication.

11.3. Process Analysis and Process Optimization

In 1984, Blass [11.123] proposed a method that requires a methodically structured procedure in the development of complex systems so that a clear, comprehensible framework can be obtained for the application of specialist knowledge and creativity and the range of possible approaches to finding a strong solution can be developed and exhausted. The method of choice should always be a systems approach, which furnishes methods, principles, and proven resources in a relatively abstract and easily interpreted fashion, in the form of the system concept, the systems procedure, and project management.

Nevertheless, there are still processes whose development has not had the benefit of the systems approach. Thus, retrospective methods were sought for determining the validity of existing processes so that they can be optimized.

An example is the electrophotographic reproduction of images, for which BIXBY [11.124] and BICKMORE [11.125] found an approach that involved breaking complex processes down into steps and interpreting these in physical, chemical, and process-engineering terms.

Figure 11.35 shows how the transfer process can be analyzed into single, physically comprehensible steps and how information can be gained in this way about the overall process.

This representation, often called a Jones diagram [11.126], is an effective way of taking individual transformation operations into account and converting object information into pictorial form [11.127].

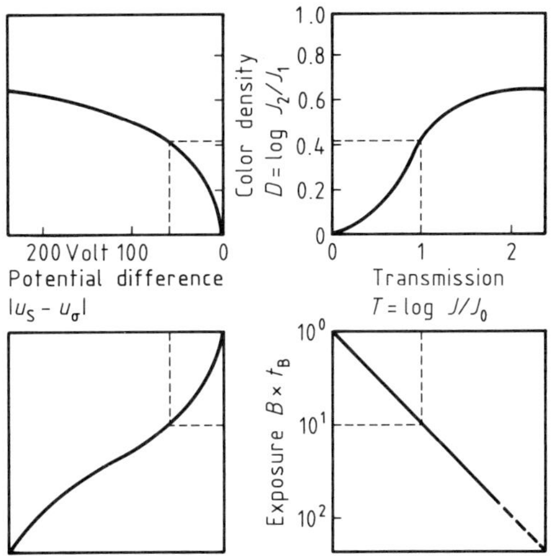

Figure 11.35. Analysis of the electrophotographic copying process into steps [11.124], [11.25]

Transfer Theory and Transformation Principle (see Chap. 2). Any optical copying process has the task of transforming the optical density distribution of the object $T(x)$ to an optical density distribution of the image $D(x)$. This process is described by transfer theory. Let the object $T(x)$ be expressed in terms of spatial frequencies R by the Fourier transform method:

$$T(x) = \int_R h(R)\,e^{2iRx}\,dR \tag{11.3.1}$$

The frequency R denotes lattice periods per unit length. The image $D(x)$ is then represented by

$$D(x) = \int_R K(R)\,h(R)\,e^{2iRx}\,dR \tag{11.3.2}$$

where $K(R)$ is the modulation transfer function (MTF), generally a complex-valued function. Its modulus describes the contrast with which the sublattice of frequency R is transferred from the object to the image.

The advantages of formulating the imaging process in this way is that the transfer fucntion collates all the properties of the transfer system in a single function; by multiplication of further transfer functions, processes can be represented in simple form and complicated processes can be broken down into steps.

At a high level of abstraction, profiles of product properties (Fig. 11.1) can be compared to optical density (blackening) distributions. Each process element then indicates (as discussed in Chap. 2) how each element of the product property profile is transferred to a new element, namely a property element of the end product, by a process step or process segment.

The analogy to the concepts advanced by GILLES [11.8], [11.128] and POLKE [11.129] is then logical.

In this relatively simple procedure, in which the modulation transfer function plays an effective modeling role, it was possible to describe and analyze the individual physical operations in the complex copying process. Later, the term "process analysis" was introduced [11.130] for quite different processes, such as the vulcanization of rubber. Existing processes were examined, from starting products through all process steps to end products, including the operational logistics, technical equipment, and process control engineering.

The entire process, the plant, and also the operational surroundings are thus examined with the objectives of separating dependent from independent variables; identifying possible improvements on the basis of the relations found; and devising and utilizing new—and possibly unconventional—solution approaches.

Figure 11.36 shows the material and data flows in a process, and Figure 11.37 shows the same relationships for the example of a vulcanization process. As early as 1972, a "process indicator," a precursor of the model-aided measurement technique [11.131], was in use.

When there are obvious weak points in the operation, improvements are often carried out

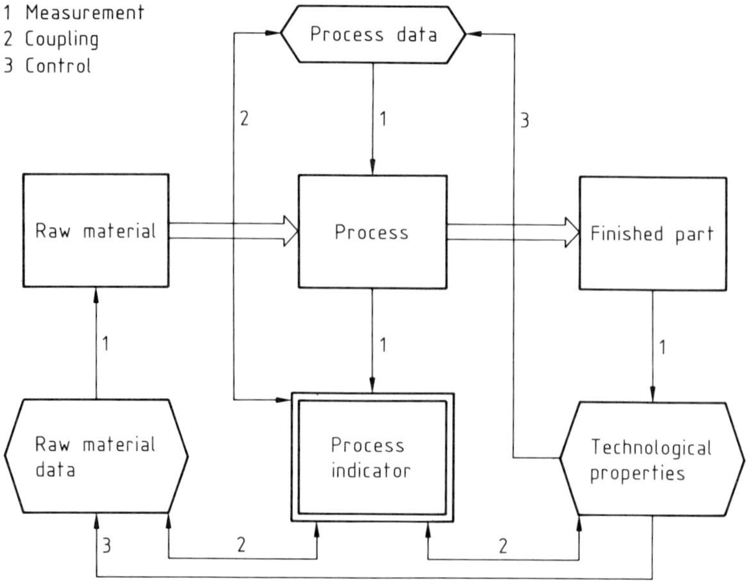

Figure 11.36. Process analysis (material and data flows)

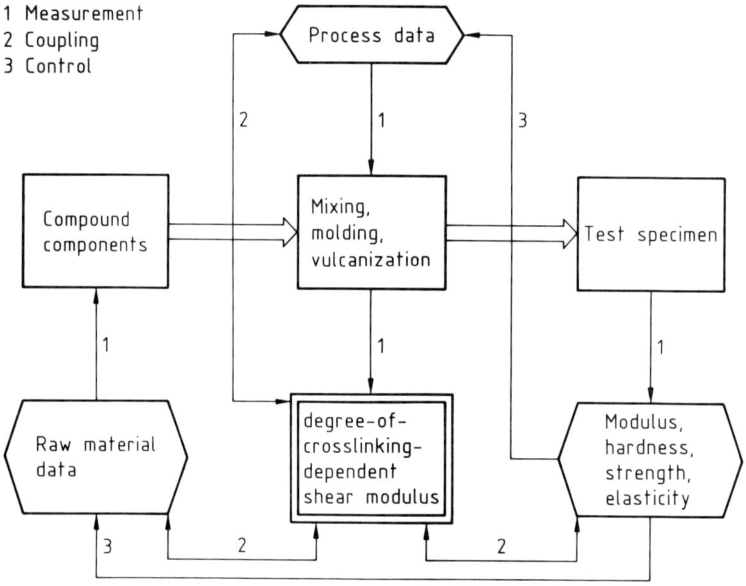

Figure 11.37. Process analysis (rubber vulcanization)

by taking single actions based on day-to-day experience. Such improvements have to do with the optimization of a unit operation, a plant section, or a single apparatus. Along with the advantage of solving the problem relatively quickly, this procedure has the drawback that the real cause of the problem remains undiscovered and substantial potential for improvement remains unexploited. As a rule, therefore, process analysis must always be performed from a holistic point of view so that improvements throughout the process and the plant can be effected. Finally, process analysis is a tributary activity of process development (see Fig. 11.57, p. 679).

The first step in a process analysis is to break down the overall process into individual process elements in such a way that the product property changes brought about by each element are covered by the phase model described in Chapter 2 (see Fig. 2.32).

As stated above, the transformation principle of Chapter 2 is used extensively (see also Fig. 11.58, p. 679).

These relationships were also pointed out by RUMPF, though he used a different vocabulary, speaking of property functions in the course of the process [11.128].

This fundamental insight into the holistic analysis of a complete process is reinforced when process control systems are employed; very much more than previously, these presuppose a thorough knowledge of the process. This point was made in 1979 in the documentation of a correspondence course published by the Chamber of Engineering of the German Democratic Republic [11.132]. The form description of the dependent variables in a production process, of the product properties, as a consequence of the variation of the independent variables, of the process properties, has since come to be referred to as modeling (see also Chap. 3).

Here, the problems of process analysis are treated from the viewpoint of the process control engineer, who must acquire knowledge about the controlled process from the process engineer, the designer, the equipment fabricator, and others; that is, from the people who have the main responsibility for designing the apparatus, the facilities, and the processes to be run there, and whose work therefore must precede as well as supplement that of the process control engineer. This applies, in particular, to the analysis and description of processes in a form that is usable by the process control engineer and accessible to his methods and procedures.

The result of such a process analysis with a process control engineering orientation, however, not only provides the process control engineer with information about the process but also provides the process specialist with pointers on the active design of the process. This specialist is then better able to develop the process from the outset in such a way that it lends itself to the process control engineer's approach. This will greatly facilitate the design of the process control system. The outcome of such a process analysis can also be a considerable aid in the commissioning and maintenance of the plant [11.132].

The restriction to a process-control orientation of the process analysis has the result that the informational context becomes a key factor in the analysis. The work of BLASS [11.123] has set the direction for the overall aspect of process analysis, which must always be kept in mind even when dealing with details. Together with presentations at the annual conferences of the VDI Association for Process Engineering and Chemical Engineering (GVC), BLASS' publications provide an ever-fresh source of new methodological and technical information [11.133]. BLASS also stresses the need for heuristic procedures, especially in engineering and scientific fields.

For this reason, what follows is a presentation of one such method that has been tested in practice, both for process analysis and for process development [11.7], [11.134].

Under certain conditions, product properties and process properties can be plotted as points in appropriately defined mathematical spaces. Some mappings between these spaces can be interpreted as production models that permit the formal mathematical description of product development in terms of process variation. There are tools, many of them tested in the field, for implementing this generally valid concept [11.7].

A company that wishes to succeed in the market must adapt the quality of its products to the requirements of the market. This adaptation can be interpreted as a measurement and decision problem.

The properties of the products produced by the company and required by the market should be quantifiable and their relative separation should be measurable. In principle, this objective can be achieved by reducing all "technological" properties to physical ones [11.130]. The physical properties assigned to a certain product are then interpretable as a point in a metric space. The nearness of the product quality to a required level can be objectivized in this way. The process of deciding whether product or process development should be carried out thus becomes transparent.

The question with what resources a quality improvement can be effected is first of all a question of operational capabilities: With what processes are the current products made? Will changes of recipe in the context of existing processes result in the desired quality changes, or will process modifications or even new process development have to be carried out.

An analytical approach here affords some help in ordering the operational reaction possibilities. If a process is represented in terms of the set of assigned and independently variable process properties, then in analogy to the product space these span the process space, in which a metric is defined (provided the properties are measurable). A recipe is a certain expression of all the properties characterizing the process and can thus be interpreted as a point in this space (process state; see Fig. 11.29, p. 661).

Now product development can be carried out in a controllable manner if it is possible to construct a mapping that assigns points in process space (recipes) to points in product space (product qualities). It is shown how the relationships between these spaces must be utilized in order to achieve product development through process variation.

It is further shown that process optimization, viewed from the standpoint just developed, generally reduces to a variational problem with constraints. The mapping from process to product space describes, in model terms, the interaction between plant and product during the production process. Accordingly, this mapping can also be used as an aid to solving control problems. The importance of such process models for the construction and utilization of technical information and control systems, as well as for the solution of automation problems with the aid of modern computer systems in the chemical industry, should be expressly reemphasized in this connection.

Product and Process Properties (see Section 2.3). The market exerts a variety of influences on product development. For example, consumer habits change over time, often because of changes in the priority of their needs [11.135]. New technologies in the converting industries impose other (usually more stringent) requirements on products. The market demand for products with novel or improved properties can also be triggered by the appearance of competing products (Fig. 11.38).

The company, for its part, offers to this market products whose properties are governed by the raw materials and energies used, the number and skills of the operating personnel, and by the technological condition of the apparatus available. Figure 11.39 illustrates the decision prob-

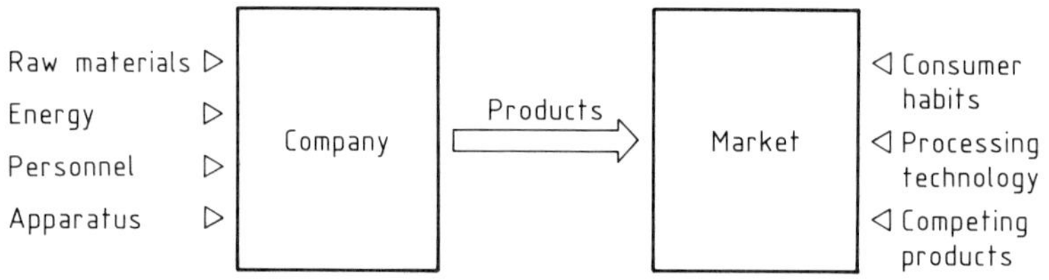

Figure 11.38. Factors affecting product quality (left) and market requirements (right)

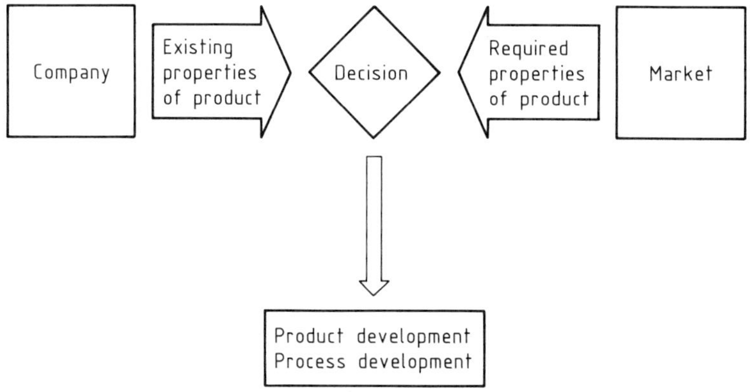

Figure 11.39. Decision problem for/against product development and/or process development

lem arising from the situation just described. If the existing properties of the products made by the company are no longer adequate to satisfy the needs of the market, the company must decide whether to risk the loss of competitiveness or to react to the new situation by undertaking product and/or process development. To render this decision process objective and intuitive, it is useful to define the "properties" of the products more exactly, on the one hand, and on the other to quantify the company's ability to react ("process") [11.134].

Product Properties. Beginning with the idealization that product properties can be measured directly and immediately, every product property can be plotted in a histogram, along with its tolerance range, and the full histogram can be interpreted as the property profile (Fig. 11.40). In a number of studies [11.136]–[11.143], this method has proved extremely helpful in the analysis of requirement and qualification profiles (see also Fig. 11.41).

Applying this idealization to the concept of product properties often reveals the practical difficulty of relating distinct characterizations of one and the same property as determined by subjective assessments. In principle, of course, even the "technological" properties can be reduced to physical ones, measured, and mapped on metrical scales. For this to happen, however, the usually complex relationships must be elucidated, and only in a few cases can this be expected. The

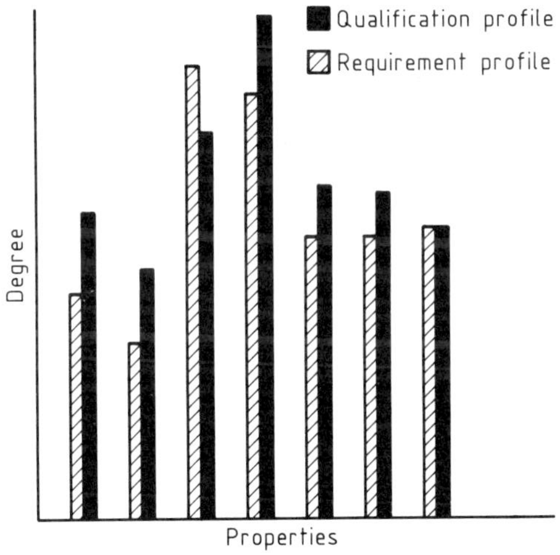

Figure 11.40. Plot showing qualification profile and requirement profile

assessment problems arising here are not dealt with further; reference is made to the literature [11.144] and to Figure 3.1.

If the treatment is initially limited to the case where all properties are measurable, at least one distance function is found that satisfies the conditions for a metric. Let E denote the set of product properties and d the distance function; then the ordered pair $\{E,d\}$ is a metric space. In accordance with standard mathematical practice, the distance measure is immediately dropped from the symbol and E is referred to as the product property space. The concept of "space" can be retained even when there are nonmetrical types of attributes, provided the distance concept is suitably generalized (e.g., topology and separation axiom → Hausdorff space).

As a rule, the property profile of a product includes redundancy. An analysis of the deterministic and/or statistical relationships of the profile leads to a reduction in the number of attributes needed to describe the object "product." This process is briefly described in Figure 11.42 and illustrated in Figures 11.43 and 11.44.

The result is that any product property profile can, with the aid of the concepts just developed, be treated as a point in a suitably selected product space. These concepts also make possible an exact definition of the hard-to-pin-down term "quality": Quality is a point in product space. If the components of the profile are also linearly independent and their realizations can be plotted on the real axis, then the property space is a vector space over the real numbers, and any profile can be interpreted as a vector in this space. Tolerances of the individual product properties allow the point to grow into an elementary volume (Fig. 11.45; see also Fig. 11.30).

Process Properties. A "process" is the operation of creating products having certain properties. Its operationalization leads to the elements of the process profile, the process properties P_j. A given realization of all the quantities characterizing the process (e.g., state variables, parameters, control variables, etc.) is called a process state. By construction, so to speak, all relevant process properties involved in process development and plant design will be measurable (monitorable). In canonical fashion, the independently variable process properties span a vector space. By analogy with the product property

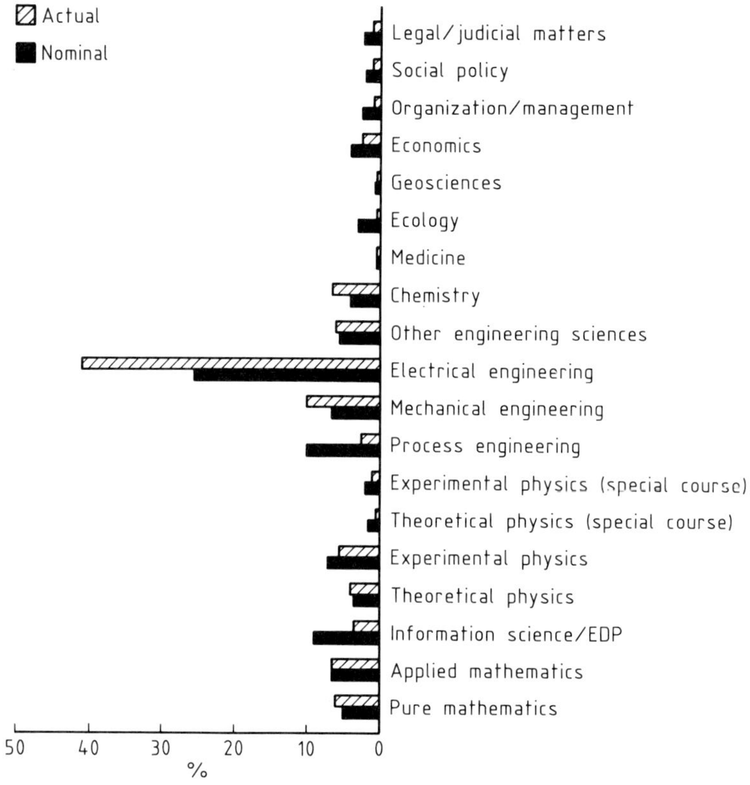

Figure 11.41. Qualification profiles of electrical engineers in process control engineering [11.141]

Reduction in number of independent components of profiles by identifying

a) deterministic relationships
b) statistical relationships

Examples	Relationship	Explanatory value
Electrodynamics	Physical theory	High
Diffusion equation	Semi-phenomenological model	↓
$J = \sigma E$	Empirical model	↓
$y = \alpha + \beta x$	Regression model	Low

Figure 11.42. Analysis of product property profile (see also Section 3.4)

profile, a recipe can now be interpreted as a point in the process space (Fig. 11.30).

Since the region of definition (range of variation) of each process property in real systems is an interval of finite length, in analyzing a particular plant the treatment can be restricted to a finite region in process space. All that is necessary is to replace E_j by P_j in Figure 11.45.

Product and Process Development. The entire production process is handled as a successive transformation of product property profiles (see Section 2.3). A process, specified by the process properties and represented by a point in process space, maps an initial property profile to a final property profile and thus links two points in property space. Now the procedure of "product

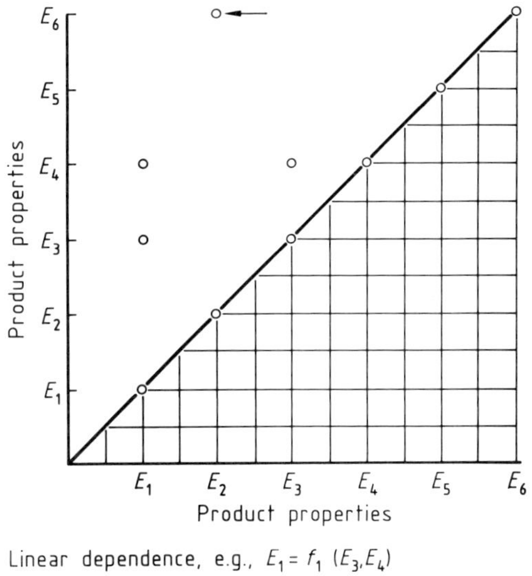

Figure 11.43. Correlation between components of the product property profile

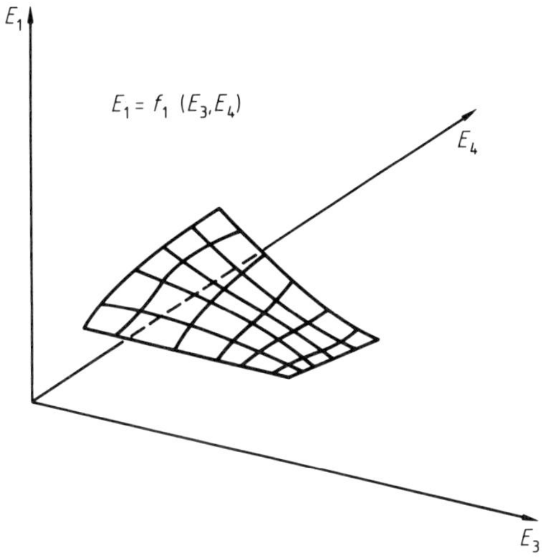

Figure 11.44. Functional dependences between components of the product property profile

development by process variation" can be formally represented as a function

$$F[P_1, P_2, \ldots, P_j;\ E_1, E_2, \ldots, E_i;\ E_A] = 0 \qquad (11.3.3)$$

where P_j are the process properties in process space, E_i are the product properties in product space, and E_A denotes the initial product properties upstream of the process (see also Figs. 11.1 and 11.58).

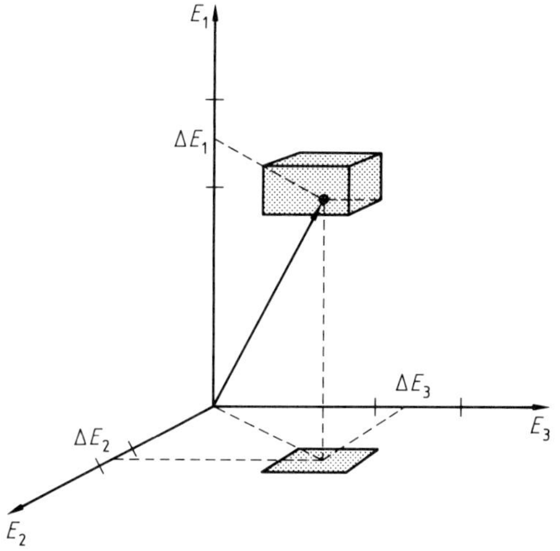

Figure 11.45. The product property profile as a multidimensional vector in product space

After these general preliminary comments, the particular question of concern in this discussion—the form of the product property profile—comes to the fore. The initial profile is taken as given and fixed. Equation (11.3.3) then depends on the variables P and E. If the market now demands from a product a certain nominal property profile E_S, this has its formal expression in the condition:

$$g(P_j) = 0 \qquad (11.3.4)$$

where g is specified by the initial property profile E_A and the nominal property profile E_S and by the interaction of plant, apparatus, and product taking place in the process step under examination.

Specifying one fixed property profile (a point in property space) reduces the number of degrees of freedom in process space by one. The set of permitted product states then lies on a hyperplane in process space (Figs. 11.46 and 11.47). After this analysis of the relationships between product and process spaces, an attempt is now made to answer the question of how a product property profile imposed by the market can be realized with the company's resources, taking both safety and economics into account.

Interpolation. In the first step, this objective is to be achieved and the possible experiential horizon of an existing plant is to be exhausted.

A "recipe change of the first kind," which does not lead out of the original hyperplane in process space (Fig. 11.47), does not change the product properties (Fig. 11.46); such a change is therefore ruled out if the required profile E_S and the initial profile E_A are different. If, however, one moves off the hyperplane ("recipe change of the second kind"), products are now obtained with new property profiles (Figs. 11.48 and 11.49).

If one of these points lies sufficiently close to E_S, product development can be regarded as successful and no plant change is required.

If the requirements imposed cannot be satisfied in the context of the existing capabilities, the initial experiential horizon in process space must be enlarged.

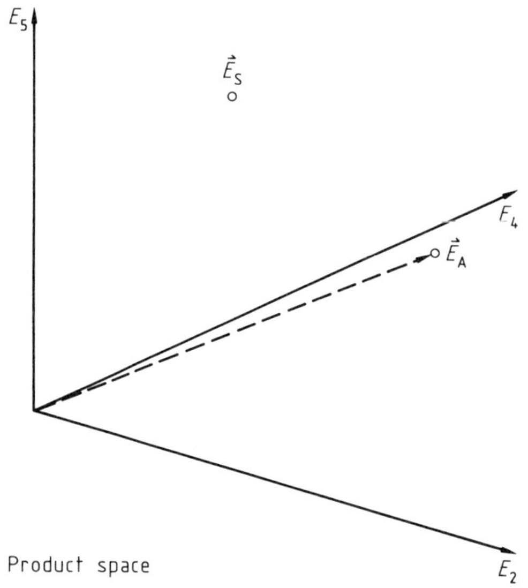

Figure 11.46. Initial product property profile E_A and nominal product property profile E_S as points in product space

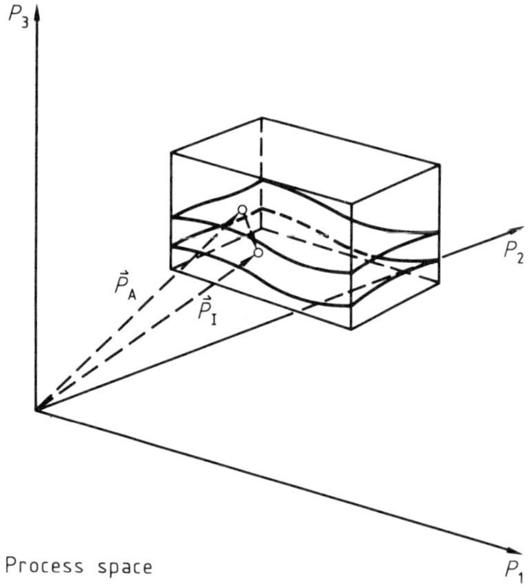

Figure 11.48. Property-changing process-state variations with no modification of plant

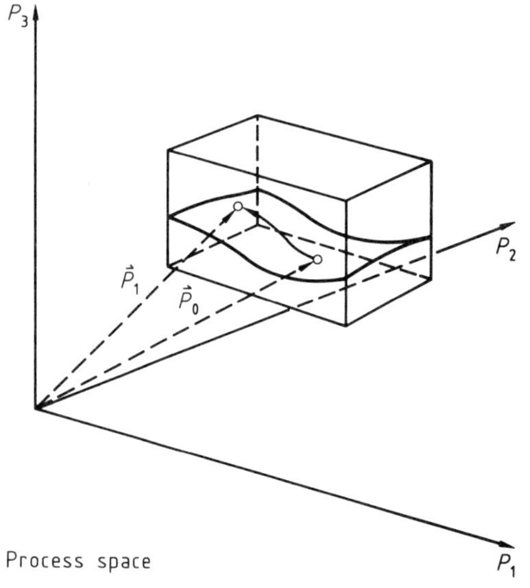

Figure 11.47. Hypersurface of property-equivalent process states in process space

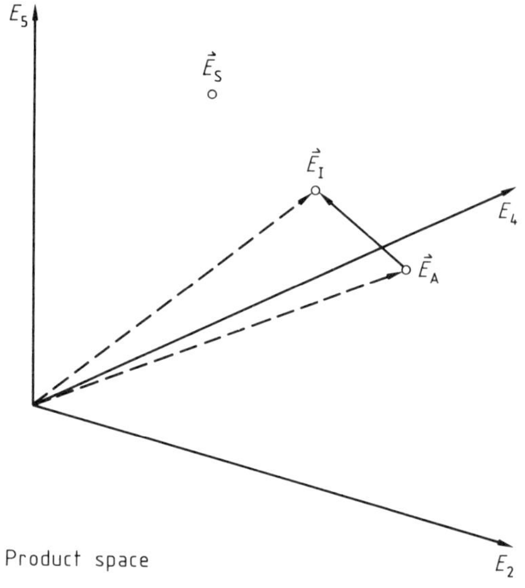

Figure 11.49. Possible property changes after the action described in Fig. 11.48

Extrapolation. In a second step, this is to happen while the dependences in process space are preserved. The same set of process properties are selected as above. If, in the development of products with new properties, the ability to produce the original assortment is to be preserved, then the range of variation of the individual process properties must be expanded by constructive actions. If the strategy requires the substitution of individual products in case of (partial) abandonment of the original assortment, subsequent process development comes down to generating a new range of variation of the process properties (upgrading, "reboring," debottlenecking).

Figures 11.50 and 11.51 illustrate this extrapolation step. If such an extrapolation does not lead to the target, the experiential horizon must be expanded in a third step by development of a new process or a new plant.

Under some circumstances, this leads to new relationships in process space (Figs. 11.52 and 11.53). It remains valid that a hypersurface (generally of a different nature) in process space corresponds to the (desired) property profile.

Optimization. Subject to the condition of generating equivalent property profiles, all process states on the respective process-space surface are equivalent. This fact can now be used in selecting a process state optimal with respect to the extremalization of a target quantity (cost, energy consumption, yield, etc.). From the mathematical standpoint, this commonly reduces to a variational problem with constraints (Fig. 11.54).

Noted that the concept developed above makes the actions to be taken in product development by process variation, as well as the way

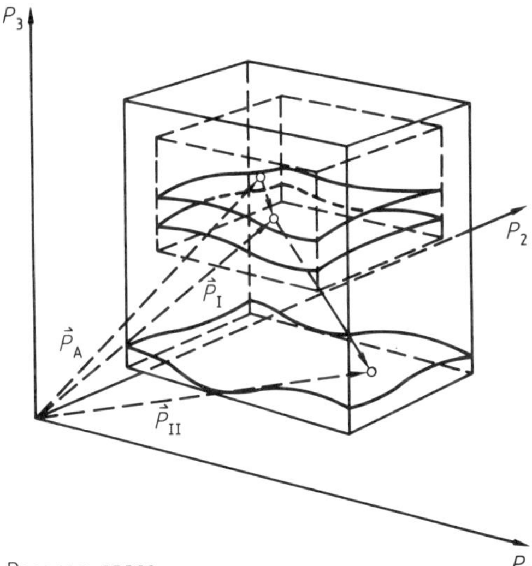

Figure 11.50. Property-changing process-state variations with modification of plant

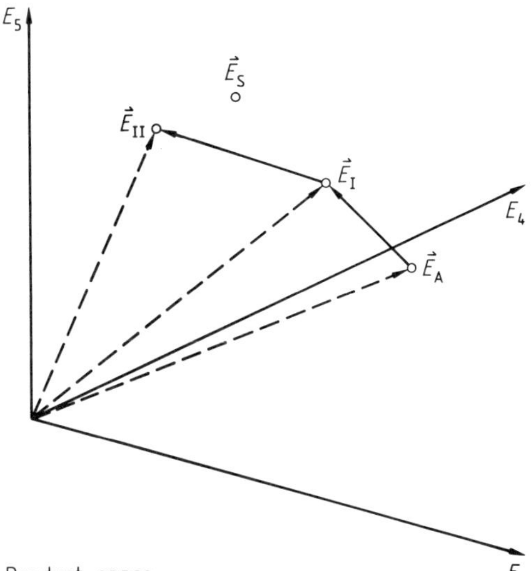

Figure 11.51. Possible property change after the action described in Figure 11.50

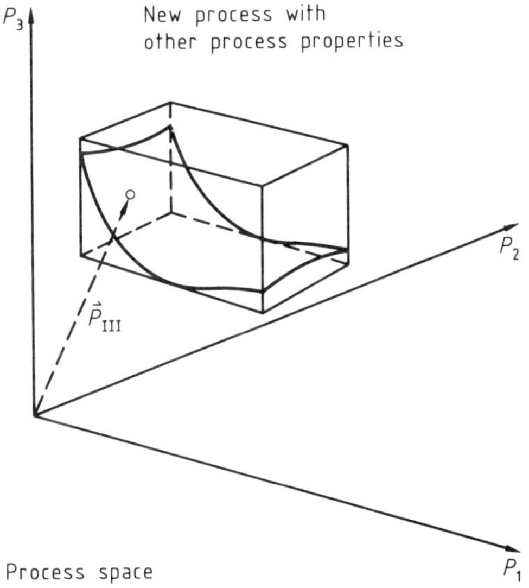

Figure 11.52. Hypersurface of property-equivalent process states in the new process space generated by process development

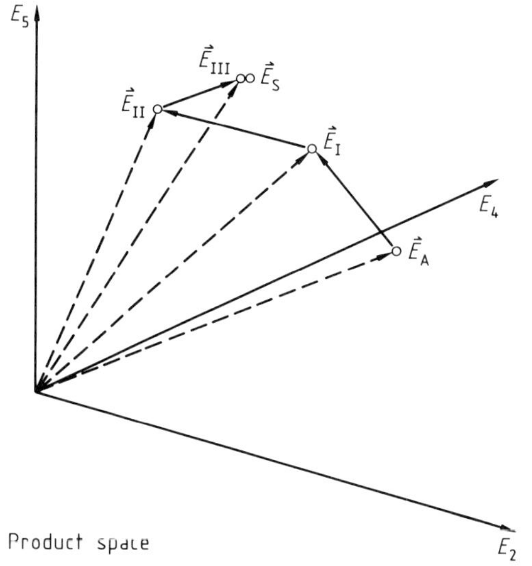

Figure 11.53. Possible property change after the new process development described in Fig. 11.52

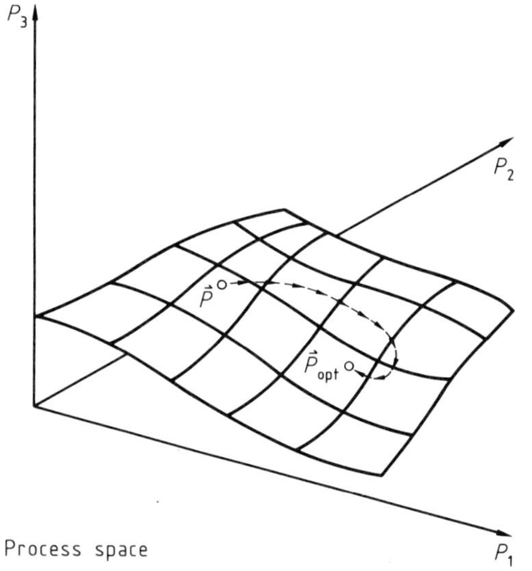

Figure 11.54. Optimization of the process with respect to a goal quantity

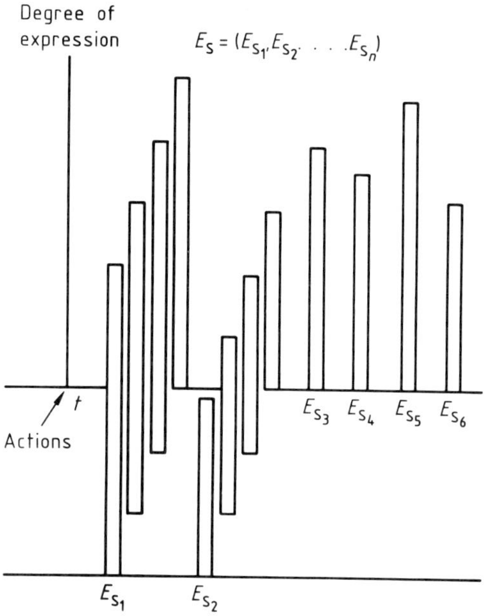

Figure 11.55. The product property profile as a function of time

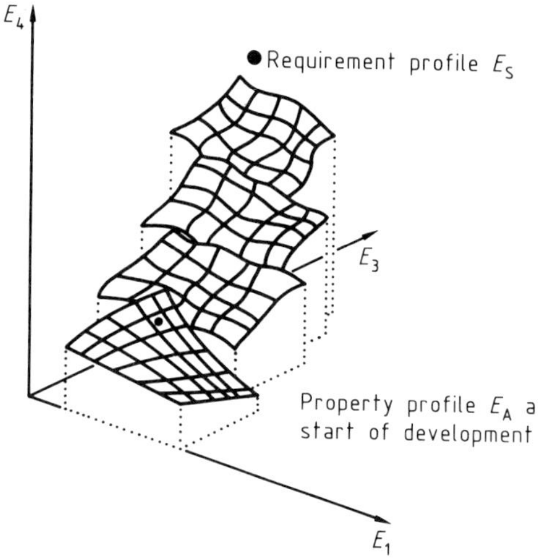

Figure 11.56. Dynamics of process development can be traced in product space: the portions of surfaces represent sub-goals of development

of evaluating their success, more intuitive, and also makes it possible to check them step by step (see also Figs. 11.55 and 11.56). However, a certain effort must be taken in order to prepare the required tools.

The question whether product development is necessary is thus answered by comparing the qualification profile E_A and the requirement profile E_S of the marketed product, that is, by deciding whether the existing properties of the product correspond to the properties required by the market.

If the qualification profile and the requirement profile are in agreement to a great extent (that is, if the point in product space does not change), then a merely desirable improvement in the result does not demand any product change;

if the process structure remains the same, however, it will be useful to optimize the possible changes in process properties (top row in Fig. 11.57) without moving off the equiproduct surface (see also Fig. 11.54).

A product modification required by the market can be effected in either of two different ways:

- If the requirement profile is close to the qualification profile, then it is generally possible to continue working with the same process structure. For a product modification, the process properties are changed within the technically possible range of variation, but this necessitates moving off the equiproduct surface (middle row of Fig. 11.57).
- If the requirement profile and the qualification profile of the desired product are far apart, then as a rule it is also necessary to change the process structure, that is, carry out a new development (bottom row in Fig. 11.57).

Measures for Realization. *Production Model.* The material flow of a typical multistage production process in the chemical industry (phase model) is shown in Fig. 11.1, p. 645). By abstraction with the aid of the concepts developed above, this material flow is mapped onto an information flow. The elements of the product space are described by the product properties and the points in the process space correspond to the process properties. Further details can be found in [11.131] and Figure 11.58 (see also Sections 2.3 and 3.5). Figure 11.59 shows how the information acquired can be utilized to discover the explicit relationships between plant, machine, apparatus, and product (i.e., between the process properties and the process) for each process step.

In particular, methods of "system identification" make it possible to derive statistical process models from production data. Such models are needed for tasks in process control engineering (e.g., process indicators [11.130], observer model). Multivariate methods (such as factor analysis, cluster analysis, and analysis of variance) make it possible to establish functional relationships between process data and quality data (see Section 3.2).

In summary, it can be stated that the central objective of any process analysis and process optimization is to achieve reproducible nominal product properties. Naturally, the first step is to describe with care the product properties, from the beginning up to the end of the entire produc-

	Point in product space	Point in process space	Process structure
Process optimization	does not change	changes, stays on equiproduct surface	remains unchanged
Product modification	changes slightly	moves off equiproduct surface*	remains unchanged
New product development	changes greatly	moves off equiproduct surface	changes

*The equiproduct surface is the locus of all process property profiles that lead to the same product property profile.

Figure 11.57. Process/product variations

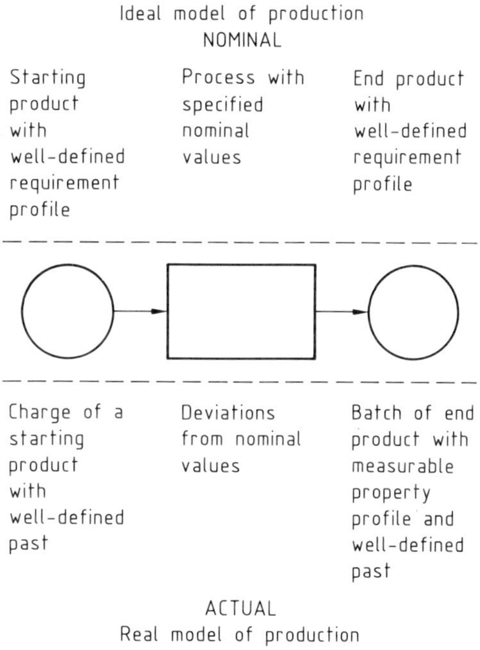

Figure 11.58. Ideal and real models of production (phase model)

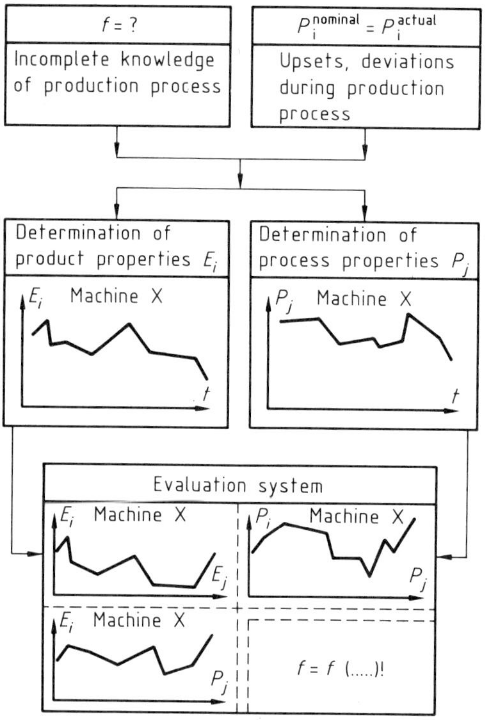

Figure 11.59. Evaluation system for quantifying the relationship between process and product properties (see Section 3.2)

tion process, and of course also for each intermediate step. The creation of the phase model as shown in Figure 11.57 is therefore the central task of the process technologist. A holistic viewpoint is of fundamental importance (see also Chap. 10). This information-technical step is the most important task.

Execution planning, execution, and quality assurance are described in depth in Chapter 10. They are to be applied, with appropriate modifications, to the problem and task discussed here.

Process analysis and thus process knowledge are also necessary conditions (as shown in Section 11.2) for state-oriented human–process communication so that the production process can be managed and controlled at all times and under all conditions.

11.4. Maintenance Strategies

Introduction. Before maintenance strategies are discussed, the key objectives of maintenance should be defined [11.145].

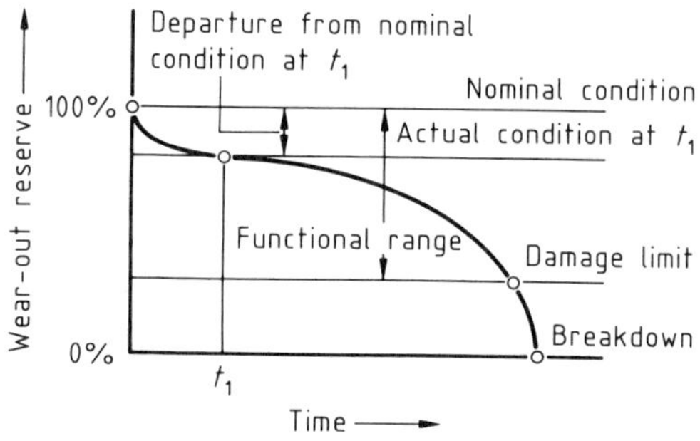

Figure 11.60. Wear-out reserve versus time

Production facilities are used for making consumer goods. During this production process, the facilities wear out in various ways. In order to remain functional despite wear, production facilities have a reserve of possible function fulfillments, called "wear-out reserve," which is gradually diminished during service (Fig. 11.60) and ultimately leads to breakdown.

In the nominal condition, the production facility has a 100% degree of function fulfillment. After a certain time, it loses its wear-out reserve. The production facility remains operable, however, over a broad range.

The objective of maintenance is to keep the wear-out reserve always above the "damage limit" in order to prevent breakdowns. Maintenance includes three activities [11.145]:

- Routine maintenance to preserve the nominal condition
- Inspection to measure and assess the actual condition
- Corrective maintenance to restore the nominal condition

It is recommended that corrective maintenance be subdivided into scheduled maintenance and forced repairs (unscheduled maintenance) (Fig. 11.61).

This distinction relates solely to the management of maintenance work, not to the quality of the work. In the case of unscheduled maintenance, including forced repairs, the schedule is dictated by the event, and the maintenance action is generally predetermined as well. The maintenance worker can thus do nothing but react. In the case of scheduled maintenance, the sequence of actions is determined in advance by the maintenance worker himself, who also influences the actions taken. In other words, the worker can take the initiative. Naturally, it is an

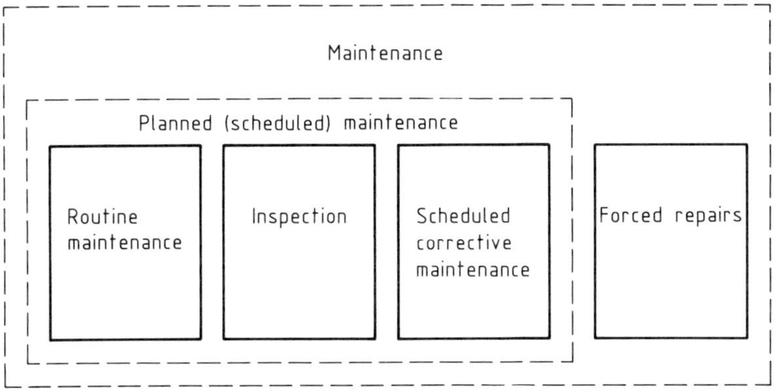

Figure 11.61. Scheduled and unscheduled maintenance

important goal to put as much maintenance as possible into the "scheduled" category.

Maintenance will become even more important. The reasons are the increasing complexity of plants, the greater depth and breadth of automation, the growing size of production facilities, and the increasingly stringent requirements imposed on product quality, safety, and environmental protection.

Modern maintenance practices for process control and automation systems are discussed in [11.3].

As the tasks of maintenance change, its objectives also change. The goals of a maintenance organization have been stated as follows [11.146]: To guarantee the retention of value and functionality of production facilities while insuring compliance with requirements relating to

- Increasing safety and quality levels
- Reliability of on-time performance
- Economically optimal operation

Only when objectives have been clearly formulated can the "correct" strategies be identified and embodied in company policies so that economy measures in the maintenance area do not cause certain maintenance tasks to be neglected.

It is the increasingly high required safety and quality levels that mandate a change of priorities in cost-reduction strategies. Two common cost-reduction practices are

- Economical performance of maintenance activities
- Optimization of material and supply use in maintenance

But the potential of these has been largely exhausted. Now a new requirement has taken on top priority: reduction or even prevention of maintenance.

This practice, which involves the elimination of weak points, still holds a substantial potential for improvement. Production facilities needing fewer repairs have higher availability and can be operated both more economically and more safely. Finally, "maintenance" in a holistic sense has become a topic in the design process (Chap. 10) and must be considered at that stage.

Strategies for Performance of Maintenance Activities. Maintenance strategies serve to make maintenance a more systematic operation. They are oriented toward maintenance activities (Fig. 11.61). One speaks of a certain maintenance strategy, say an inspection-oriented strategy, if this activity (inspection) is purposely emphasized in the context of changing objectives and is not sacrificed when economy measures are introduced. Naturally, this does not mean that other maintenance activities are wholly de-emphasized.

Breakdown Strategy. The dominant role in the past was played by the breakdown strategy, which is based on forced repairs and, by definition, unscheduled maintenance. Despite the high cost of devices used in process control engineering, this strategy is still fairly important because many of the sensors and actuators come into contact with products and hence experience contamination, deposits, plugging, and corrosion. It can be shown statistically that such breakdowns are not greatly reduced by more intensive scheduled maintenance. Some improvement is gained with measuring techniques, such as radiometry, that do not involve contact with the product.

Indeed, the breakdown strategy does have advantages: It entails low costs, the work can be done on site by foreman/craft teams, and downtime is generally short.

In recent years, however, some significant drawbacks have also emerged:

- Costs increase with increasing degree of plant utilization and investment level
- Because of the "workshop" orientation, it may take much effort and expense to maintain operation at weak points such as those resulting from interaction with products; this strategy thus achieves the exact opposite of maintenance prevention
- More stringent requirements on safety, occupational health, and environmental protection, coupled with heightened public sensitivity, mean that a breakdown maintenance strategy cannot be practiced exclusively

Scheduled maintenance strategies have thus become more important.

Preventive Maintenance Strategy. Much was expected of preventive maintenance when scheduled maintenance strategies were first introduced. It is not just a matter of cost that has made this approach obsolete. The reliability of many devices (service lives of valves and motors) today is so great that preventive maintenance is justified only where an element wears out after a known time and can then be replaced as a unit, for example in the case of fluorescent tubes. Preventive maintenance is still necessary when there are special factors, such as plant safety or quality assurance considerations.

Inspection-Oriented Strategy. As safety and environmental requirements grow more stringent, the inspection-oriented strategy has come to play an important role. This applies to many devices significant for damage prevention or limitation, such as rupture disks, leakage retaining vessels, and protective and damage-limiting devices in process control systems. Legislation, such as the Störfallverordnung in Germany [Accident Regulation], mandates the continuous surveillance (inspection) of such devices in the form of periodic function checks. The performance and results of these actions must be documented.

These requirements have had dramatic effects. As a one-time engineering cost, they have given rise to extensive safety analyses with plant upgrades and danger aversion plans (e.g., functional inspection charts). More importantly, they have led to continuing additional labor costs for plant monitoring and safeguarding and for the documentation of actions carried out (see Section 3.5).

The specific cost of monitoring and documentation may decline over a span of years as new and improved inspection methods appear on the market. These might include instruments with automatic self-testing and auxiliary diagnostic devices in electronic systems (see Chap. 5), as well as model-aided early fault detection methods (e.g., [11.147]).

Another type of inspection-oriented strategy, visual inspection, is the basis for determining and assessing the condition of production facilities and process control equipment. The "preservative register" is a suitable technique for documenting the results [11.148].

Depending on the economic and political situation, another strategy may gain in importance: Maximization of plant availability, and thus safety, by application of the redundancy principle, as is already required in nuclear power plants [11.149]. The redundancy principle requires installation of spare devices, which automatically take up the function of broken-down units and report the malfunctions. This is not, however, a typical maintenance strategy but rather a practice that must be adopted at the design stage (see Chap. 10).

High availability, such as is achieved under the redundancy principle, improves the operating economics. The application of this principle leads, however, to higher costs, at least initially. Additional investment is required, which in turn gives rise to additional maintenance costs, since the redundant units must also be kept operable even when on standby status.

Scheduled Corrective Maintenance Strategy. The analysis of strategies would not be complete without the strategy of scheduled corrective maintenance. Scheduling corrective maintenance (repairs) does not have the aim of restoring the nominal condition; instead, alternative solutions must be sought, with an eye to economic, ecological, and process-engineering considerations. The following question must be examined: maintenance, investment, or both? This applies to plants, plant sections, apparatus, and process control equipment. The procedure, also referred to as "intelligent" maintenance, regards maintenance in the context of the plant economy, that is, holistically. The "plant economy" includes

- Design
- Fabrication, installation, setup
- Use (operation, maintenance, upgrading)
- Replacement

The centerpiece of the method is a decision model that provides a sort of value analysis to support decisionmaking about maintenance practices. On the basis of a situation analysis covering certain independent variables and constraints, the technique compares alternative actions and rates them through cost–benefit analysis. Decisionmaking is not, however, automatic; an educated interpretation of events is needed.

Maintenance Control Loop. Maintenance is concerned with the actual and nominal conditions of the facilities and equipment being maintained. Maintenance can thus be regarded as a control loop, as illustrated in Figure 11.62, which shows a breakdown of maintenance activities. Findings, damage, and malfunctions must be acquired and evaluated; weak points must be identified by a process of closing in. Finally, the loop must generate actions, which function as control actions. They react on the facilities being maintained (except for unschedulable breakdown maintenance).

In the maintenance control loop, planning (design) work is an essential component of the plant economy. It is at this stage that maintenance costs are set by the choice of process and plant design. Planning yields the plant and device documentation, which serve as controlling data, but the working plans for scheduled maintenance can also be generated in this stage. Maintenance-oriented installation is crucially

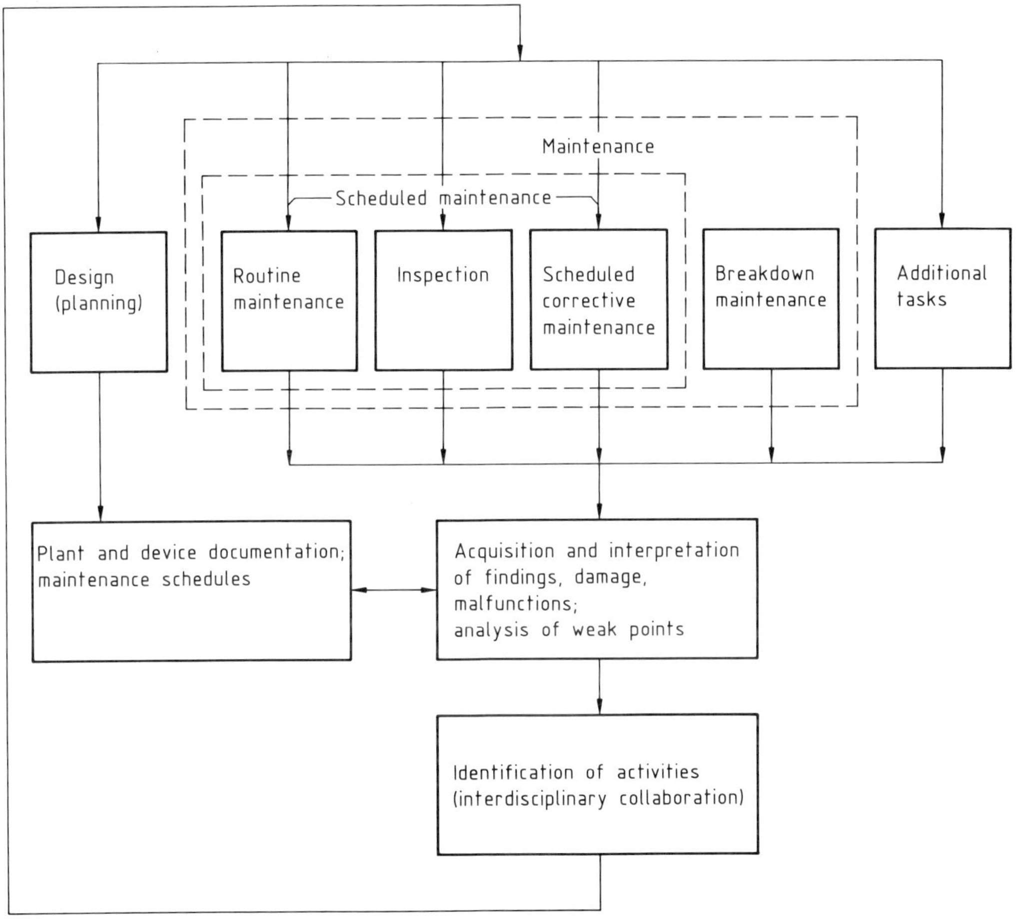

Figure 11.62. The maintenance control loop [11.146]

important, along with the maintainable design of apparatus and equipment (see Section 3.5 and Chap. 10).

A holistic maintenance program requires the support of a powerful information system. Maintenance-relevant information stems from all maintenance activities. The documentation, however, must focus more on movement data generated on site by the operating engineers in the course of their work. This information goes into object-related history files, which can then provide the informational basis for documenting maintenance activities or analyzing weak points.

Figure 11.63 illustrates the identification of weak points in process control systems. The ordinate is the total time spent in remedying problems, and the abscissa is the number of problems arising with individual objects in the system. Each entry characterizes one object. The weak points are located at top right. They stand out because of the large maintenance time requirement and the large number of problems. On the basis of the history files, the numbered groups can be formed by an computational technique called cluster analysis [11.150].

Outlook. Breakdown maintenance continues to play by far the major role in the chemical industry. What is lacking is the logical application of scheduled maintenance strategies. This has to do with a shortage of methods and tools with which a maintenance program oriented chiefly toward preserving and restoring productive capacity can be conceptually and informationally integrated into the plant economy. There are, however, difficulties in achieving this kind of integration; the staggered and often uncoordinated adoption of information technology (hardware and software systems) has meant that bodies of engineering, administration, economic, and logistical data are mutually inconsistent. This underlines the importance of using the methods described above for structuring the

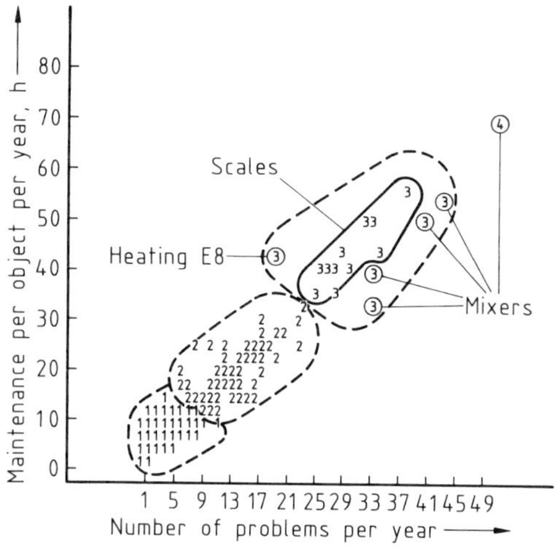

Figure 11.63. Use of cluster analysis to identify weak points in a process control system

plant economy, and only then implementing it technically (see Chap. 2).

In an integrated data processing system, maintenance should be a component of the plant economy with at least the following elements [11.147]:

1) A documentation system for the acquisition, structuring, editing, and storage of maintenance-relevant plant data
2) Planning and control systems for scheduled maintenance
3) A control system for generating action- and object-based cost and performance reports and calculating maintenance performance figures

A maintenance information system comprising these blocks should have compatible interfaces to other in-company data processing systems, especially to the production planning and production control, bookkeeping and cost accounting, materials records, and engineering subsystems (see also Fig. 2.19).

12. Standards, Committees, and Associations

12.1. Principles

The committees and associations that are either directly or indirectly concerned with process measurement and control are so many and diverse that it is impossible to present a complete list. This chapter merely indicates the points of concentration, functions, and objectives of the most important such standardization bodies.

What is Standardization? A clear answer to this question appears in DIN 820 Part 1; 01.86; Paragraph 2: "Standardization is the unification of material and immaterial objects for the benefit of the general public, carried out jointly and in a planned manner by the interested circles. It must not lead to a special economic advantage for individual parties. It promotes rationalization and quality assurance in business, technology, science, and administration. It contributes to the safety of persons and property as well as the improvement of quality in all areas of life. What is more, it contributes to good order and the dissemination of information in the specific area standardized. Standardization is practiced at the national, regional, and international levels."

Who May Create Standards? In Germany, only DIN and DKE are permitted to issue mandatory standards. Under the standardization agreement between DIN and the Federal Republic of Germany, May 5, 1975, DIN is to represent the standardization interests of the Federal Republic in regional and international standardization bodies. CEN and CENELEC European standards also become effective only after they have been published by DIN or DKE. IEC and ISO international standards can be adopted as national or European standards, but there is no obligation to do so.

German-speakers distinguish between the terms "norm" and "standard." Norms are always somewhat official, while standards are private in nature and represent agreement among certain groups. Both terms relate to technical specifications. In the case of a norm, such specifications are adopted by a governmental or internationally recognized standardization organization. Common usage, unfortunately, does not always maintain this precise distinction, and one often speaks of ISO standards. In the German sense, these are norms, since they are issued by ISO.

VDE and VDI standards (called "VDE-Bestimmungen" and "VDI-Richtlinien") do not have the status of norms in law. They are designated as the state of the art. The same applies to the recommendations and work sheets published from time to time by other bodies such as NAMUR, ZVEI, or VIK.

The purpose of these publications is to make results in science and technology known to a broad public as promptly as possible, since experience has shown that the issuance of norms is a very slow process. These papers generally become proposed norms for consideration by standardization bodies, are discussed among a broad public, and after some years appear in modified form as norms. The essential function of these associations is thus to create draft norms and evaluate draft norms.

Both norms and standards are described by the single word "standard" in American English. A "norm" in Germany (and many other countries) might be equated, quite roughly, to a mandatory standard or a standard embodied in a governmental regulation, while German "standard" might correspond to a "voluntary" standard. Standards issued by ANSI, ASTM, etc., are considered voluntary in the United States, but elsewhere they may be regarded as closer in legal status to "norms." The remainder of this chapter employs "standard" for both classes but provides clarification when necessary.

So-called industry standards are in essence neither standards nor norms, but successful products or specifications that have become established in the market.

Situation in the United States. The ISA occupies a dual position in the United States: As a technical-scientific organization (comparable with VDE in Germany), it not only takes part in the preparation of standards but also has the power to issue its own standards, having been granted this right, for the field of process control engineering, by the American National Standards Institute (ANSI; see Section 12.3).

Also deserving of mention is the Institute of Electrical and Electronics Engineers (IEEE), which develops and generates international standards. Its publications, however, are not mandatory standards but recommendations.

Changes in the Standardization of Process Control Engineering. The 1980s saw a marked change in technology, from signal-oriented measurement and control to information-oriented process control engineering. Not only companies but also national and international standardization bodies have had to deal with this change.

While standardization bodies in the 1970s were concerned with embodying the state of the art in standards, in the early 1980s the emphasis changed to standardization in parallel with research. To lower research costs, a context for development was established in national and international standardization bodies.

The changes in technology and in standards emphasis also necessitated a new structure. This was acknowledged at the end of the 1980s when three new bodies were set up: IEC TC65 (Technical Committee 65—Industrial Process Measurement and Control), DKE FB9 ("Special Area 9—Measurement and Control"), and NAMUR ("Standards Working Group for Measurement and Control Engineering in the Chemical Industry").

12.2. Standardization Bodies and Other Organizations Involved in Standardization

The most important standards bodies for process control engineering, with their functions and objectives, are listed below:

CCITT. The Comité Consultatif Internationale Télégraphique et Téléphonique (International Consultative Committee on Telegraphy and Telephony) unites national postal and telecommunications organizations and has jurisdiction over telecommunications standards.

CEN. The Comité Européen de Normalisation (European Committee for Standardization) has the European national standardization bodies as its members. Together with CENELEC, it creates European standards and provisional standards.

CENELEC. The Comité Européen de Normalisation Electrique (European Committee for Electrical Standardization) includes European national standardization bodies concerned with electrical technology.

CEPT. The Comité Européen des Administrations des Postes et des Télécommunications (European Committee of Postal and Telecommunications Administrations) is the conference of European postal and telecommunications agencies.

DIN. The Deutsches Institut für Normung (German Standardization Institute) is the national institution that develops and adopts standards (mandatory standards).

DKE. The Deutsche elektrotechnische Kommission im DIN und VDE (German Electrotechnical Commission of DIN and VDE, VDE = German Electrical Engineers' Society) is the national organization that develops standards and safety provisions in the electrotechnical field.

ETSI. The European Telecommunications Standards Institute was established by European telecommunications agencies together with the EC Commission after substantial deficiencies had been revealed in the assimilation of standards in the telecommunications field. ETSI's task is European harmonization of standards for public-sector telecommunications services.

IEC. The International Electrotechnical Commission develops international standards in the electrotechnical field.

ISO. The International Organization for Standardization functions as an umbrella organization for nearly 90 national standardization bodies. In the interest of flexibility, ISO standards are formulated with many degrees of freedom and must be made more specific by lower-level institutions.

The *International Electrotechnical Commission (IEC)* was founded in 1906. Its membership includes more than 40 national committees, and the official languages are English, French, and Russian. The purpose of the IEC is to support international collaboration in all matters of standardization and certification in the electrical and electronics fields. It aims to achieve this by means of publications, recommendations, and international standards serving as models for incorporation into national legal codes. IEC standards serve manufacturers, consumers, research institutes, and governmental agencies concerned with international trade and electrical safety issues. Any country, without restriction, can become an IEC member. Figure 12.1 summarizes the structure of the IEC.

The IEC Suborganization "Conformity Testing to Standards for Safety of Electrical Equipment" (IECEE) has the purpose of certifying products for conformity to IEC standards. The certification system applies to electrical devices used in the home, the office, and the workplace. For further information, see [12.1].

The IEC Technical Committee TC65, "Industrial Process Measurement and Control" (Fig. 12.2), has as its scope the creation of international standards for systems and elements of industrial process control, including batch and continuous processes, as well as the coordination of standards for measurement and control devices and systems. Subcommittee SC65A, "System Aspects," is concerned with standardizing generic aspects of systems, such as operating conditions, assessment of systems, functional reliability, and so forth. SC65B, "Devices," deals with standardization of devices and systems in the field, the switch room, and the control room. In SC65C, "Digital Communication," standards are developed for communications between subsystems.

The *European Committee for Standardization* (CEN) and the *European Committee for Electrical Standardization* (CENELEC) came into being in the early 1960s as regional standards bodies (Fig. 12.3). The membership of CEN/CENELEC includes the national standards institutes of all EC and EFTA countries. These institutes are also members of ISO (International Organization for Standardization) IEC (International Electrotechnical Commission).

CENELEC has the objective of harmonizing technical standards in Europe. In order to achieve these goals, the harmonization efforts of the Community are (under Article 100 of the EEC Treaty) to concentrate on solutions that establish binding criteria and make standards the basis for determining the technical details of products. Moreover, with regard to problems specific to new technologies, it is important that standardization at the trans-European level become effective at an early stage.

Publications created by the specialist activity of CEN/CENELEC and issued by CEN/CENELEC are called European Standards (EN), Harmonization Documents (HD), Provisional European Standards (ENV), or CEN/CENELEC Reports. European Standards and Harmonization Documents are referred to as "regional standards" in the sense of the ISO/IEC definition. The various publications differ in the following ways:

- The basic rule is to create European Standards, since it is important that the national standards of the member countries be identical wherever possible.
- Harmonization Documents are issued when the change to identical national standards is unnecessary or impractical, and especially

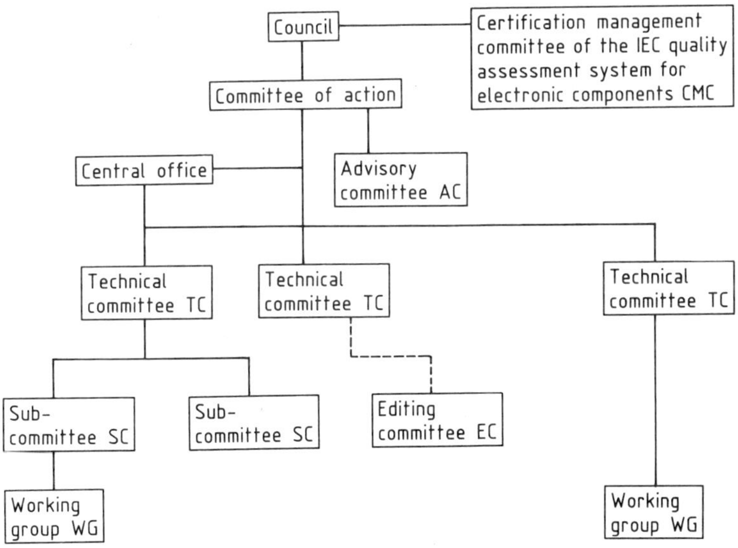

Figure 12.1. Structure of the IEC (simplified)

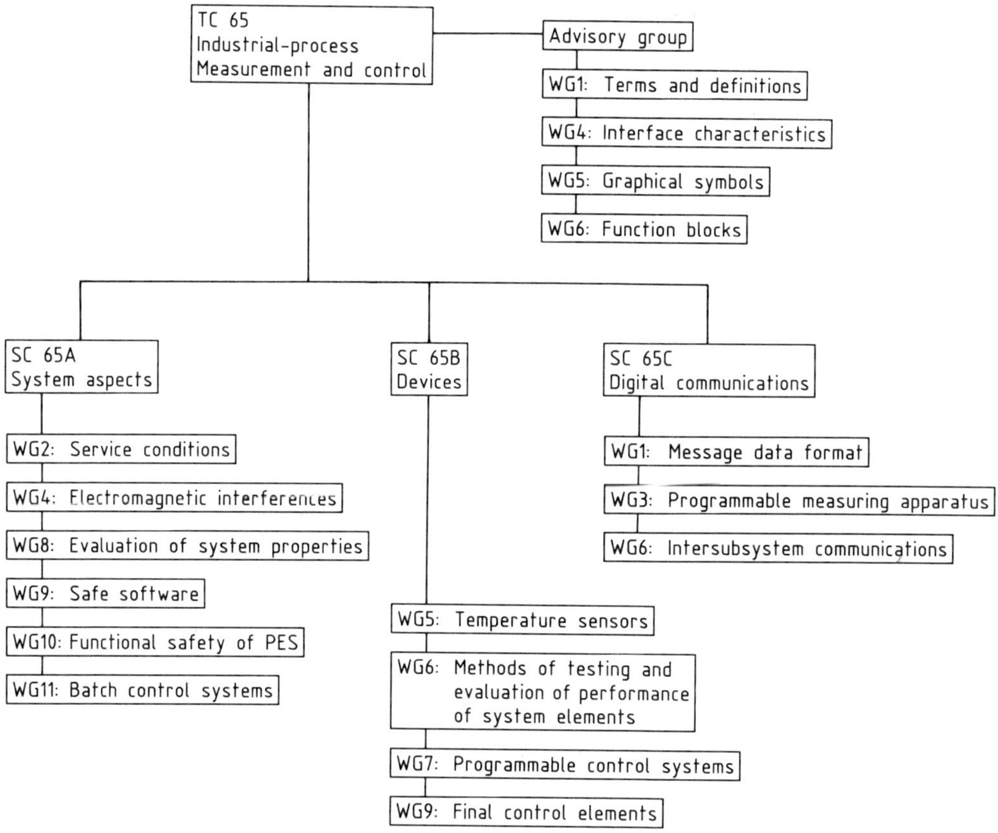

Figure 12.2. Structure of IEC TC 65

when a unification can be achieved only by granting national variances.

- Provisional European Standards are planned standards for use on a preliminary basis.
- CEN/CENELEC Reports are intended to provide information.

The Vilamoura procedure has the goal of promoting collaboration between national committees on a regional level by decentralizing the earliest stages of work. All new intentions to standardize on the national level should be disclosed in detailed form, so that all members are acquainted with the content of the proposed project and can indicate their interest in active participation if they wish to do so.

"New intentions to standardize on the national level" means all intentions to standardize at the national level except for:

- Revision of existing national standards (see explanation below)
- Adoption of the results of CENELEC work
- Adoption of the results of IEC work

The results of national notifications lead to an EN, an HD, or an ENV. Notifications must be sent to all parties concerned (national committees, the CENELEC Secretariat, and the Technical Committee) for response. If no positive responses are received, the notifying national committee is free to develop a national standard.

The notification process for the revision of national standards is based on the assumption that the intention to revise is justified by the necessity of updating the standards. The revision of a national standard for the purpose of taking over an IEC standard, IEC provisional standard, or IEC publication, or for the purpose of bringing the national standard into conformity with an HD or an EN need not to be disclosed under this procedure.

The Technical Committees of CENELEC are numbered to correspond with those of the IEC.

The member countries of CENELEC are listed below:

Austria: Österreichisches Elektrotechnisches Komitee (ÖEK) beim Österreichischen Verband für Elektrotechnik (ÖVE) (Austrian Electrotechnical Committee of the Austrian Electrotechnical Society)

Belgium: Comité Electrotechnique Belge (CEB), Belgisch Electrotechnisch Comité (BEC) (Belgian Electrotechnical Committee)

Switzerland: Comité Electrotechnique Suisse (CES) (Swiss Electrotechnical Committee)

Germany: Deutsche Elektrotechnische Kommission im DIN und VDE (DKE) (German Electrotechnical Commission of DIN and VDE)

Denmark: Dansk Elektroteknisk Komite (DEK) (Danish Electrotechnical Committee)

Spain: Asociación Española de Normalización y Certificación (AENOR) (Spanish Standardization and Certification Association)

Finland: Finnish Electrotechnical Standard Association (SESKO)

France: Union Technique de l'Electricité (UTE) (Technical Union of Electricity)

United Kingdom: British Electrotechnical Committee (BEC), British Standards Institution (BSI)

Greece: Hellenic Organization for Standardization (ELOT)

Ireland: Electro-Technical Council of Ireland (ETCI)

Iceland: The Icelandic Council for Standardization (STRI), Technological Institute of Iceland

Italy: Comitato Elettrotecnico Italiano (CEI) (Italian Electrotechnical Committee)

Luxembourg: Service de l'Energie de l'Etat (SEE) (State Energy Service)

Netherlands: Nederlands Elektrotechnisch Comité (NEC) (Netherlands Electrotechnical Committee)

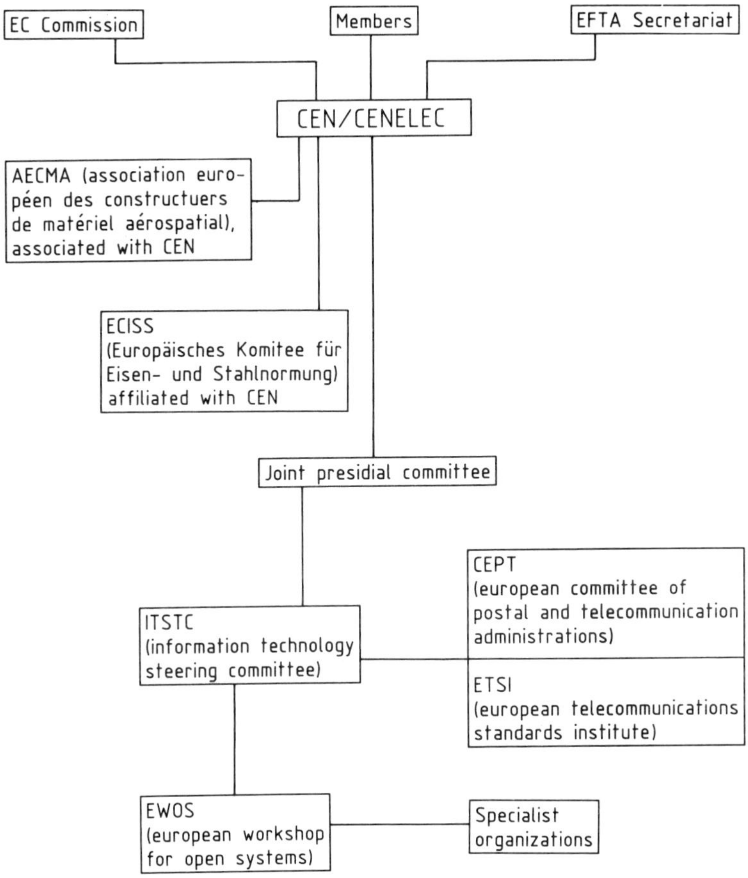

Figure 12.3. Organizational chart of CENELEC

Norway: Norsk Elektroteknisk Komite (NEK) (Norwegian Electrotechnical Committee)

Portugal: Instituto Português da Qualidade (IPQ) (Portuguese Quality Institute)

Sweden: Svenska Elektriska Kommissionen (SEK) (Swedish Electrical Commission)

The EC Commission has therefore devoted special attention to the development of standardized test methods for verifying compliance with the standards. The ultimate objective is a European certification framework that would award a publicly recognized certificate to products that conform to the standards. The ECITI (European Committee for Information Technology Testing and Certification) is a body created by the EC Commission for the purpose of promoting the testing of standard-conforming products in Europe.

CENELEC always follows the principle of using the results of IEC work whenever possible and also of carrying out new standardization projects within the IEC when there is a prospect of accomplishing this within a given time span. In this way, multiple treatments of the same topic are avoided. The extensive collection of IEC publications and the continuation of standardization in the IEC are important components of this effort. With the information agreement of November 1989 and the cooperation agreement that came into force on January 1, 1991, the essential tools for further collaboration between IEC and CENELEC and their members were created. The cooperation agreement aims, above all, to accelerate the standardization process by causing a draft prepared on one organization level to be submitted for agreement simultaneously and in parallel on the international and European levels.

DIN. The German standards system deals with all technical disciplines. DIN standards reflect actuality and document the level of technical experience. Each standard is investigated with regard to its economic impact, and only those that are absolutely necessary are issued. DIN standards contribute to rationalization, quality assurance, safety, and understanding in business, engineering, science, and administration, and also among the public. DIN and its agencies serve as the national representation in international and European standards institutions (ISO/IEC, CEN/CENELEC).

DKE. DKE is the German organization that develops standards and safety rules in the field of electrical technology. DKE is dedicated to the safety of persons using electricity on the job, at home, and during leisure time. DKE is further involved in solving problems of environmental protection, answering consumer questions, and creating nuclear safety regulations. DKE has jurisdiction over the harmonization of electrotechnical standards within the European Community.

In Germany, mandatory standards can be created only by DIN or DKE. Each European Standard must be incorporated by all CEN/CENELEC members into their own standards system. EFTA countries that have decided against the European Standards are free of this obligation. Adoption requirements such as those that apply to EN and HD documents do not exist for ISO and IEC international standards. On the European level, ISO/IEC standards are subjected to a questionnaire procedure; if they meet with agreement, they are uniformly adopted in all EC and EFTA countries.

Results of the work of DKE are issued primarily under the titles

- VDE-Bestimmung (VDE voluntary standard)
- DIN-Norm (DIN mandatory standard)
- DIN-EN-Norm (DIN European Standard), with DKE number if applicable
- DIN-IEC-Norm (DIN–IEC world standard), with DKE number if applicable

DKE's "Special Area" 9 (FB9) bore the name "Measurement and Control" up through 1990 and was then renamed "Process Control Engineering." FB9 (Fig. 12.4) creates basic standards for industrial process control as well as specialized standards for the following application areas: process engineering, power plants, networks, buildings, pipelines, ironmaking and iron founding, mining, shipbuilding, instruction and training, and laboratory practice.

It is not the job of standardization bodies to carry out preliminary technical and scientific work. This must be done by the societies that cover the various areas (e.g., NAMUR, GMA, ZVEI). This function includes the development of technical concepts, technical and scientific groundworks for standardization projects, and engineering studies and assessments. The procedure of standardization in parallel with development can only become an effective aid to both

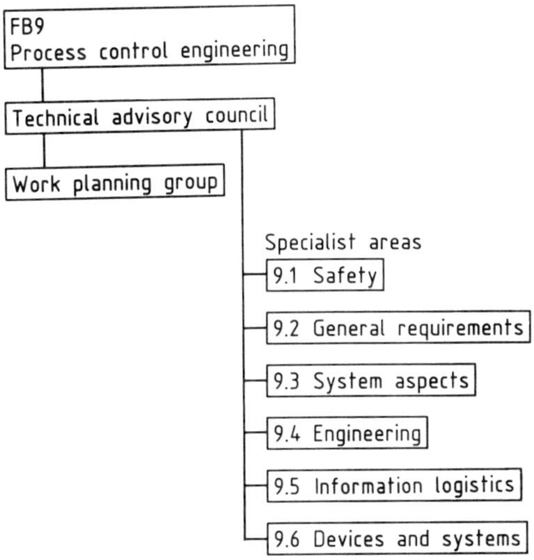

Figure 12.4. Areas of interest of DKE FB9

parties when close cooperation is achieved between these technical societies and the standards bodies.

VDE. The transfer of technical and scientific knowledge, as a service to all professional electrical and electronics engineers, is the first task of VDE. The spectrum ranges from historical aspects of electrical engineering, through occupational and educational problems, to accident research, and the drafting of standards.

VDI. The transfer of technical and scientific knowledge is the primary objective of technical and scientific work performed by the Association of German Engineers (VDI). This is accomplished chiefly in the VDI specialist subdivisions.

An important emphasis in these subdivisions is the creation of standards documents known as *Richtlinien* (guidelines), which reflect the state of the art but do not have the legal status of mandatory standards.

VCI. Among the diverse committees and working teams of the Verband der Chemischen Industrie (VCI, Association of the Chemical Industry), the Engineering and Law Committee merits special mention. Through the "Measurement and Control" working team, VCI makes use of NAMUR specialist knowledge. This working team is identical with the governing body of *NAMUR* (see below).

NAMUR. The Normenarbeitsgemeinschaft für Mess- und Regelungstechnik in der Chemischen Industrie (NAMUR, Standards Working Group for Measurement and Control Engineering in the Chemical Industry; Fig. 12.5) takes in users of process control engineering (electrical,

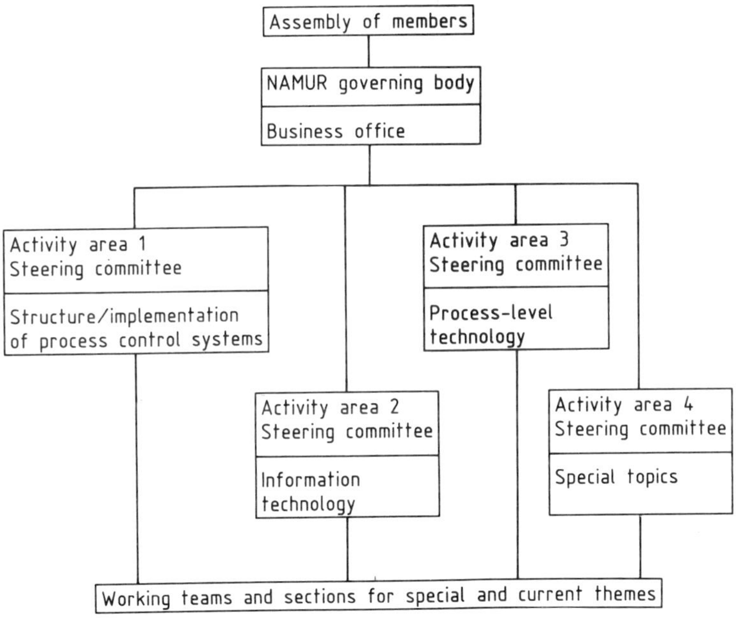

Figure 12.5. Organizational chart of NAMUR

measurement, and control engineering) in the chemical and allied industries. It was founded in 1949 by BASF, Chemische Werke Hüls, DAG Troisdorf, Glanzstoff-Courtaulds, Henkel, Ruhrchemie, UK Wesseling, and Farbenfabriken Bayer. At present, NAMUR has 54 member companies, a third of them located outside Germany (in Switzerland, Austria, Belgium, and the Netherlands).

Activities of NAMUR relate chiefly to:

- Participation in the standards committees of DIN and DKE (in particular, DKE FB9, "Process Control Engineering")
- Participation and input in the creation of standards and specifications at the voluntary standards level (e.g., VDI/VDE-Richtlinien documents)
- Exchange of experience with process control devices, especially with newly developed instrumentation and other hardware
- Publication of NAMUR recommendations in all fields of process control engineering when no other standards exist
- Scientific support of technical journals
- Publication of NAMUR Status reports to document the state of the art in the field of process control engineering

Further details can be found in [2.90] and in [12.2].

VIK. The Vereinigung Industrielle Kraftwirtschaft (VIK, Industrial Energy Management Alliance) was founded in 1947.

The principal concerns of the Electrotechnical Committee are as follows:

- Exchange of experience
- Influence on the preparation of devices suitable for consumer use
- Influence on the modification of VDE standards to conform with consumer requirements
- Influence on decreasing the number of designs
- Efforts to establish optimal conditions for accident prevention
- Attainment of economic benefit in the use of electrotechnical devices

This applies to electrical installations and systems in industrial power supply and production service.

ZVEI. The Zentralverband Elektrotechnik- und Elektronikindustrie e.V. (ZVEI, Central Federation of the Electrical and Electronics Industry) represents the interests of the German electrical and electronics industry in matters of economic and techology policy. ZVEI has the following tasks:

- To safeguard the common professional and business interests of its members before all agencies, in particular by cooperating with regulatory, economic, and other agencies and institutions and seeking to provide input to legislative measures
- To promote the general economic and technical development of the electrical and electronics industries

ISA. The Instrument Society of America (ISA) was established in 1945 in Pittsburgh, Pennsylvania.

It is a nonprofit organization active in the technology, science, and educational sector. Its concerns are with the theory, design, manufacture, and application of process control devices, systems, and computers. It issues the monthly International Journal of Instrumentation and Control (INTECH).

In the United States, the ISA made its name with an extensive training program and companion publications such as books, journals, software packages, videos, and CBT (computer-based training) packages. Moreover, ANSI (American National Standards Institute) has delegated to ISA the right to issue mandatory standards in the field of process control engineering. Valuable comments can be found in numerous ANSI/ISA standards sections.

12.3. Technical and Scientific Bodies

It is difficult and often impossible to distinguish between societies that are active in the standardization field and bodies whose work is primarily technical and scientific. The emphasis for technical and scientific committees is on the sharing of experience in research and development, production and technology, training and continuing education, and safety. Such bodies can frequently be identified by the fact that they regularly organize high-level congresses. In this way they contribute to forming public opinion, and this in turn influences standardization.

In Germany, for example, the DVT unites a number of technical and scientific associations, such as DPG, DIN, GI, VDE, and VDI. Also important for process control engineering are GMA, Gesellschaft für Informatik (GI, Society for Information Processing), and Gesellschaft für Mathematik und Datenverarbeitung (GMD, Society for Mathematics and Data Processing), a public-sector research institution.

At the international level, there are a variety of national societies with special objectives. One such society oriented toward automation technology is IFAC; toward metrology, IMEKO; and toward the type testing of measurement and control equipment, the International Instrument Users Association.

DVT. The Deutsche Verband technisch-wissenschaftlicher Vereine (DVT, German Federation of Technical and Scientific Societies) currently includes 100 technical and scientific organizations. It has the objective of analyzing high-level problems in science and technology and representing the interests of engineers and natural scientists in scientific, economic, social, political, governmental, and administrative circles. Its tasks include the promotion of applied sciences, the unification of common technical principles, the improvement of technical education, and cooperation in shaping legislation in the field of technology and technical management. The DVT regards itself as a link between members and also between the membership and business societies whenever technical and scientific problems are addressed.

AMA. The Arbeitsgemeinschaft Messwertaufnehmer e.V. (AMA, Data Sensors Working Group) is an association of manufacturing firms in the areas of sensor technology, sensor-based measurement and automation, and microsystems technology.

DPG. The Deutsche Physikalische Gesellschaft e.V. (DPG, German Physical Society) has the following objectives: to advance pure and applied physics, to bring its members and all physicists living in Germany closer to one another, to represent physicists as a body to other members of society, and to promote the exchange of experience among physicists in Germany and between them and their colleagues in other countries.

Notable special activities in the field of process control engineering are the conferences held by the Beratende Ausschuss der Industriephysiker (BAI, Advisory Committee of Physicists in Industry): the sensor technology conference in Münster and the automation conference in Bonn.

GMA. The Gesellschaft Mess- und Automatisierungstechnik im VDI und VDE (GMA, Metrology and Automation Society of the VDI and VDE) is active in measurement and control technology, automation, and process control engineering. The spectrum of related applications extends from the basic chemicals industry to the process industries, industrial and process engineering, energy generation and distribution, the automotive, transportation, and aerospace industries, construction engineering, medical technology, public utilities and waste disposal operations, and special applications in machinery, apparatus, and instruments.

Measurement and control, as a basic process-level technology, forms the basis for the functionally oriented, hierarchically structured field of automation and process control engineering. The principal points of emphasis in the action chain between process, automation system, and human being, as well as the higher-order, associative disciplines and engineering technologies, are communication, design, and simulation, together with information processing and, on this basis, the techniques of process operation and monitoring, automation equipment and systems, the technology of actuators and sensors, and also the dynamics of machinery and processes.

Areas of activity of the GMA:

- Exchange and evaluation of experience and information
- Holding conferences and lectures
- Cooperation with national and international bodies working in this field
- Stimulation and support of research
- Publication and support of scientific journals
- Development of guidelines and other recommendations
- Sponsoring young engineers
- Involvement in education

National collaborations:

DGK	Deutsche Gesellschaft für Kybernetik
DGQ	Deutsche Gesellschaft für Qualität e.V.
DIN	Deutsches Institut für Normung
DKE	Deutsche Elektrotechnische Kommission im DIN und VDE

GI Gesellschaft für Informatik e. V.
NAMUR Normenarbeitsgemeinschaft für Meß- und Regelungstechnik in der Chemischen Industrie

International collaborations:

IMEKO Internationale Meßtechnische Konföderation
IFAC International Federation of Automatic Control

The organization of the GMA is shown in Figure 12.7.

GI. The Gesellschaft für Informatik (GI, Society for Information Processing) has as its objective the interchange of technical and scientific information in the field of "informatics" or information processing.

GMD. The Gesellschaft für Mathematik und Datenverarbeitung mbH (GMD, Society for Mathematics and Data Processing) is a public-sector research institution that uses basic research in the field of informatics to further the development of information technology and its applications. On the basis of wide-ranging basic research, conceptual work is performed and methods are devised that will lead to tools.

GVC. The VDI-Gesellschaft Verfahrenstechnik und Chemieingenieurwesen (GVC, the VDE Society for Process Engineering and Chemical Engineering) is the specialist organization of engineers active in process and chemical engineering as well as allied occupational groups. One of its central activities is the annual congress of process engineers. Its objectives are as follows:

- Sharing of technical and scientific experience
- Employment of specialist competence of the GVC technical committees
- Training and continuing education
- Collaboration with other institutions
- Social policy

IFAC. The International Federation of Automatic Control was founded in 1957 as a multinational organization. Its members are technical and scientific societies from 45 countries. Its objective is to support control science and technology in a holistic sense, that is, technically, physically, biologically, socially, and economically, both in theory and in practice. The task of IFAC is to support information exchange between national and international organizations active in the fields of automation and process control engineering. IFAC holds a world congress every three years; between congresses, it organizes symposia and workshops on special themes.

Figure 12.6 shows the organizational structure of IFAC.

IFAC is a member of the "Five International Associations Coordinating Committee" (FIACC), which also includes the following bodies:

IFIP International Federation for Information Processing

IFORS International Federation of Operational Research Societies

IMACS International Association for Mathematics and Computers in Simulation

IMEKO International Measurement Confederation

Member societies of IFAC by country are as follows:

Argentina: Asociacion Argentina de Control Automatico (AADECA)

Australia: The Institution of Engineers

Austria: Arbeitsgemeinschaft für Automatisierung

Belgium: Federation IBRA/BIRA

Brazil: Sociedade Brasileira de Automatica (SBA)

Bulgaria: National Council of Automation

Canada: Canadian National Committee for IFAC

Chile: Asociacion Chilena de Control Automatico (ACCA)

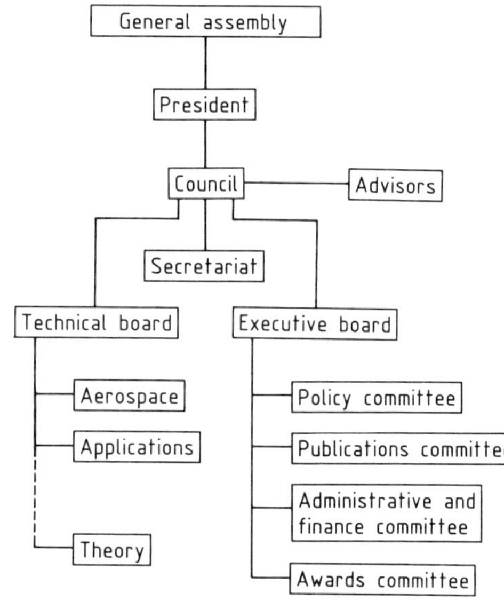

Figure 12.6. Organizational chart of IFAC

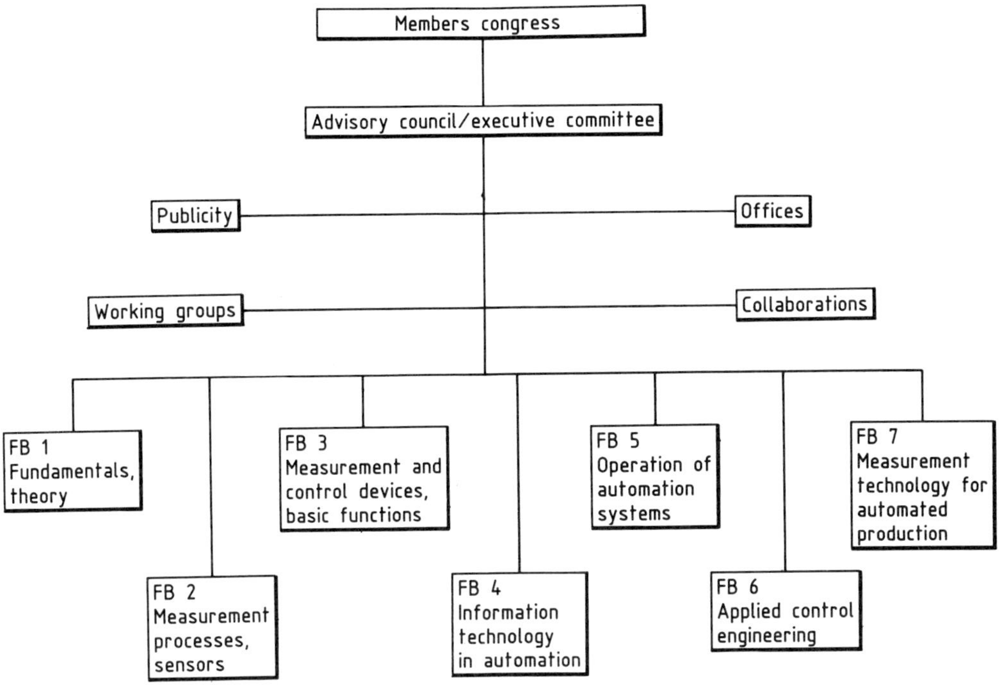

Figure 12.7. Organizational chart of GMA

China: Chinese Association of Automation, Institute of Automation

Czechoslovakia: CSFR National Committee of Automatic Control

Denmark: El. Power Engineering Department, Denmark Technical University, Lyngby

Egypt: Egyptian High Committee of Automatic Control

Finland: Finnish Society of Automatic Control

France: Association Française pour la Cybernetique, Economique et Technique—AFCET

Germany: VDI/VDE Gesellschaft Mess- und Automatisierungstechnik (GMA)

Greece: Technical Chamber of Greece

Hong Kong: The Hong Kong Institution of Engineers

Hungary: IFAC National Member Organization—Computer and Automation Institute

India: India Institution of Engineers

Israel: Israel Association of Automatic Control

Italy: CNR Commissione IFAC

Japan: Science Council of Japan

Kuwait: Kuwait Society of Engineering

Mexico: Asociacion de Mexico de Control Automatica—AMCA

Morocco: Association Marocaine pour le Développement de l'Electronique, de l'Informatique et de L'Automatique—A.M.A.D.E.I.A.

Netherlands: Royal Institution of Engineers

New Zealand: Institution of Professional Engineers

North Korea: Korean General Federation for Science and Technology, Korean Association of Electronics & Automation

Norway: Norsk Forening for Automatisiering (NFA)

Pakistan: The Institution of Engineers

Poland: Polski Komitet Pomiarow i Automatyki i Robotyki

Portugal: Associacao Portuguesa para o Desenvolvimento da Investigacao Operacional—APDIO, CESUR-IST

Romania: Comisia de Automatizari, Inst. de Cercetari si Proiectari Automatizari—IPA

Singapore: Instrumentation and Control Society

South Africa: South African Council for Automation

South Korea: Korean Association of Automatic Control

Spain: Comite Español de la IFAC

Sweden: Kommiten Svenska IFAC

Switzerland: SGA, Schweizerische Gesellschaft für Automatik

Turkey: Turkish National Member Organization—Otomatik Kontrol Türk Milli Comitesi (TOK)

United Kingdom: United Kingdom Automatic Control Council—UKACC

United States: American Automatic Control Council

CIS: National Committee of Automatic Control

Yugoslavia: Electrotechnic Faculty, Automatics and Systems Engineering Division

IMEKO. The Internationale Messtechnische Konföderation (IMEKO, International Measurement Confederation) was founded in 1958. It includes technical and scientific organizations in 31 countries. Its objectives are as follows:

- To support international sharing of experience in science and technology for the development of metrology, manufacturing and design of devices, and application of devices in research and industry
- To promote cooperation of scientists and engineers
- To hold a world congress every three years
- To establish Technical Committees
- To organize symposia, workshops, etc., on special themes and to publish the proceedings
- To cooperate with other international organizations having similar objectives (see IFAC)

The current list of Technical Committees is as follows:

TC 1	Higher Education
TC 2	Photonic Measurement
TC 3	Measurement of Force and Mass
TC 4	Measurement of Electrical Quantities
TC 5	Hardness Measurement
TC 7	Measurement Theory
TC 8	Metrology
TC 9	Flow Measurement
TC 10	Technical Diagnostics
TC 11	Metrological Requirement for Developing Countries
TC 12	Temperature and Thermal Measurement
TC 13	Measurement in Biology and Medicine
TC 14	Measurement of Geometrical Quantities
TC 15	Experimental Mechanics
TC 16	Pressure Measurement
TC 17	Measurement in Robotics

Member societies of IMEKO by country are as follows:

Australia: Institute of Instrumentation and Control

Austria: ÖVE-ÖIAV Fachgruppe Messtechnik

Belgium: Belgian Measurement Confederation

Brazil: Brazilian Association of Mechanical Sciences

Bulgaria: Committee for Measurement of the Association of Science

China: Chinese Society for Measurement

Czechoslovakia: Czechoslovak Scientific and Technical Society—CSVTS

Denmark: The Danish Society for Engineering Metrology

Egypt: Egyptian Organization for Standardization

Finland: Finnish Society of Automatic Control

France: Association Française pour la Cybernetique Economique et Technique—AFCET

Germany: VDI/VDE Gesellschaft Mess- und Automatisierungstechnik (GMA)

Hungary: Scientific Society of Measurement and Automation

India: Institution of Instrumentation Scientists and Technologists

Italy: National Research Council

Japan: The Society of Instrument and Control Engineers

Korea: Korea Standards Research Institute

Morocco: Society for Automation and Control

Netherlands: Royal Institution of Engineers Division for Automatic Control

New Zealand: The Institute of Measurement and Control (Inc.)

Norway: Norwegian Society of Automatic Control

Philippines: Philippine Instrumentation and Control Society

Poland: Committee for Measurement, Robotics and Automation—NOT

Romania: Romanian Measurement Society

Spain: Committee for Metrology, Spanish Association for Quality Control

Sweden: Instrument Society of Sweden

Switzerland: Schweizerische Gesellschaft für Automatik

United Kingdom: The Institute of Measurement and Control

United States: Instrument Society of America—ISA

CIS: All-Union Scientific-Technical Society of the Instrument Industry

Yugoslavia: JUREMA

International Instrument Users Association. The International Instrument Users Association unites three societies, "SIREP-WIB-EXERA" (EXERA = Association des Exploitants d'Equipment de Mesure, de Régulation et d'Automatisme, Association of Users of Equipment for Measurement, Control, and Automation), with the objective of performing suitability tests on measurement and control devices and systems. The concerns in this type testing are operability, safety, availability, accuracy, freedom from interference, ease of maintenance, etc. The results are communicated only to the member firms and the manufacturing firm. Membership is limited to user firms.

Addresses of Organizations Mentioned in Sections 12.2 and 12.3:

Arbeitsgemeinschaft Messwertaufnehmer e.V. (AMA)
Linprunstrasse 49
80335 München
Germany

CENELEC Central Secretariat
Rue de Strassart 35
B-1050 Bruxelles
Belgium
Telephone: + 32-2/519-6871
Fax: + 32-2/519-6919
Telex: 26257
Teletex: 206/2210097 = CENCEL

DIN Deutsches Institut für Normung e.V.
Burggrafenstrasse 6
10787 Berlin
Germany

DKE Deutsche Elektrotechnische Kommission
im DIN und VDE
Stresemannallee 15
60596 Frankfurt/Main
Germany

Deutsche Physikalische Gesellschaft e.V. (DPG)
Hauptstrasse 5
53604 Bad Honnef
Germany

Deutscher Verband technisch-wissenschaftlicher Vereine (DVT)
Graf-Recke-Strasse 84
Postfach 10 11 39
40239 Düsseldorf
Germany

Association des Exploitants d'Equipment de Mesure, de Régulation et d'Automatisme (EXERA)
BP Cerchar No 2
60550 Verneuil en Halatte
France

Gesellschaft für Informatik e.V. (GI)
Godesberger Allee 99
53175 Bonn
Germany

VDI/VDE-Gesellschaft Mess- und Automatisierungstechnik (GMA)
Graf-Recke-Strasse 84
Postfach 10 11 39
40239 Düsseldorf
Germany

GMD-Birlinghoven
Postfach 12 40
Schloss Birlinghoven
53757 Sankt Augustin
Germany

VDI Gesellschaft Verfahrenstechnik und Chemieingenieurwesen (GVC)
Graf-Recke-Strasse 84
Postfach 10 11 39
40239 Düsseldorf
Germany

Internationale Elektrotechnische Kommission (IEC)
3, rue de Varembé
1211 Geneva 20
Switzerland

IFAC Secretariat
Schlossplatz 12
A-2361 Laxenburg
Austria

IMEKO Secretariat
H-1371 Budapest
P.O. Box 457
Hungary

Instrument Society of America (ISA)
67 Alexander Drive
P.O. Box 12277
Research Triangle Park NC 27709
U.S.A.

Contact address in Europe:
ISA International
Avenue Marcel Thiry 204
1200 Brussels
Belgium

NAMUR Normenarbeitsgemeinschaft für Mess- und Regelungstechnik in der chemischen Industrie
51368 Leverkusen, Bayerwerk
IN-PLT, Geb. H 1
Germany

SIREP International Instrument Users' Association
South Hill
Chislehurst, Kent BR 7 5 EH
England

Verband der Chemischen Industrie e.V. (VCI)
Karlstrasse 21
60329 Frankfurt
Germany

Verband Deutscher Elektrotechniker (VDE) e.V.
VDE-Haus, Stresemannallee 15
60596 Frankfurt/Main
Germany
Telephone: 069/6308-0

Verein Deutscher Ingenieure (VDI)
Graf-Recke-Strasse 84
Postfach 10 11 39
40239 Düsseldorf
Germany

Vereinigung Industrielle Kraftwirtschaft (VIK)
Ausschuss "Elektrotechnik"
Richard-Wagner-Strasse 41
45128 Essen
Germany

International Instrument Users' Association WIB
Barronielaan 15
4818 PA Breda
Netherlands

Zentralverband Elektrotechnik- und
Elektronikindustrie e.V. (ZVEI)
Stresemannallee 19
60596 Frankfurt/Main
Germany

12.4. Shows and Fairs

Shows and fairs in the field of process control engineering can be categorized into three major groups:

- Shows devoted to one specialty (e.g., process engineering, manufacturing engineering) and including process control engineering as a sideline. Significant here are ACHEMA for process engineering and the Hannover Messe Industrie for manufacturing engineering.
- Shows oriented toward process control engineering as a whole. In the forefront of this group are INTERKAMA in Düsseldorf, BIAS in Milan, MESUCORA in Paris, HET INSTRUMENT in Utrecht, ILMAC/INELTEC in Basel, INSTRURAMA in Brussels, C & I in Birmingham, and the ISA Show in the United States.
- Shows that concentrate on a single area of process control engineering, such as Sensor in Nuremberg and Systec, Systems, Laser, and Analytica, all in Munich.

The alphabetical listing that follows is certainly not complete but does give more information on the shows mentioned above, citing the emphasis and a contact address for each.

ACHEMA—International Meeting for Chemical Technology and Biotechnology; includes a congress in the field of chemistry, with special emphasis on process control engineering

Areas of concentration: research and innovation; measurement and control; process control engineering; materials science and materials testing; laboratory and analytical technology; mechanical and thermal processes; packaging and storage technology; safety engineering; occupational safety and health; pumps, compressors, and fittings; biotechnology; environmental protection

Contact address:

DECHEMA
Deutsche Gesellschaft für Chemisches
Apparatewesen, Chemische Technik und
Biotechnologie
Postfach 15 01 04
60061 Frankfurt/Main
Germany

Analytica—International Technical Show for Biochemical and Instrumental Analytical Technique, with International Meeting

Technical show focusing on analytical laboratory techniques

Areas of concentration: analytical instruments and methods, laboratory technique and apparatus, laboratory data acquisition and processing, industrial chemistry, industrial analytic technique, materials testing, quality control, environmental analysis, biotechnology, medical diagnostics

Contact address:

Münchener Messe- und Ausstellungsgesellschaft mbH
Postfach 12 10 09
80034 München
Germany

BIAS—International Show and Congress in the Field of Automation, Instrumentation, and Microelectronics

Areas of concentration: process computers and process control systems, measuring instruments, controllers, valves, sensors, transducers, transmitters, components and accessories, control devices for processing and fabrication operations, logistics, communications, bus systems, remote control, robotic systems

Contact address:

E.I.O.M.
BIAS Secretariat
Viale Premuda, 2
I-20129 Milan
Italy

CeBIT—World Center for Office, Information, Telecommunications

International technical show for computer techniques

Areas of concentration: screen communications, office communications, text processing, computer technology, data processing, information technology, software, telecommunications, communications technology, communications engineering

Contact address:

Deutsche Messe AG
Messegelände
30521 Hannover
Germany

Control & Instrumentation—National technical show and congress for process control engineering, with emphasis on the chemical industry; held at Birmingham, England

Areas of concentration: process control engineering, computer systems, control technology, actuators, sensors, warning systems, analytical technology, weighing, fiber optics, testing and inspection, CAE/CAD/CAM software, displays and operator interfaces

Contact address:
MGB Exhibitions Ltd.
C & I Secretariat
Marlowe House
109 Station Road
Sidcup, Kent DA 15 7 ET
England

Hannover Messe Industrie—The world's largest industrial show, focusing chiefly on control and process drive engineering but with process control engineering as a secondary theme

Areas of concentration: electrical and electronics engineering; microelectronics; energy, air pollution, and environmental engineering; industrial optics and lasers; materials of construction; surface technology; process drive engineering; control engineering; installation engineering; industrial robots; material-flow engineering

Contact address:
Deutsche Messe AG
Messegelände
30521 Hannover
Germany

Het Instrument—Technical Show for Industrial Instrumentation

Dutch national show in the field of process control engineering; technical show for health care, science, and industry; held in Utrecht

Areas of concentration: devices and systems (sensors, measurement, testing and inspection, weighing, and control engineering); medical technology; laboratory technology (analysis); electronics; research and innovation

Contact address:
Sekretariat Het Instrument
Postfach 1 52
Birkenstraat 108
3760 AD Soest
Netherlands

ILMAC—International chemical show for laboratory technology, chemical engineering, and measurement and automation

Areas of concentration: analytical laboratory techniques; isotope and radiation chemistry; process engineering; measurement, control, and automation engineering; biotechnology; laboratory animal management; environmental protection; health care

Contact address:
Schweizer Mustermesse
ILMAC Secretariat
Messeplatz
CH-4021 Basel
Switzerland

INELTEC—International electrical and electronics engineering show

Areas of concentration: power supply and distribution; energy supply systems; communications; control systems; components; measurement, testing, and inspection; power electronics

Contact address:
Schweizer Mustermesse
INELTEC Secretariat
Messeplatz
CH-4021 Basel
Switzerland

INSTRURAMA—Technical Show for Metrology, Instrumentation, Control, and Automation National show for the field of process control engineering and laboratory technology

Areas of concentration: apparatus and laboratory equipment; devices, systems, and components for process control and management of industrial automation and informatics in the fields of energy, production, refining, communications, environmental protection

Contact address:
UDIAS
INSTRURAMA Secretariat
Avenue Slegers 203
B-1200 Brussels
Belgium

INTERKAMA—Innovation Market for Measurement and Automation

Largest international show in the field of process control engineering, taking in all aspects of the field but focusing on research institutions, professional qualifications, exhibitor seminars, and congress

Areas of concentration: process control and monitoring systems; measurement and control engineering; sensor technology; analysis and weighing; actuators; communications; data processing; displays and output devices; components and accessories; tools; testing and inspection; services

Contact address:
Düsseldorfer Messegesellschaft bmH—NOWEA-INTERKAMA Referat
Postfach 32 02 03
40417 Düsseldorf
Germany

ISA Show—American technical show for process control engineering, with emphasis on a congress; held at various sites

Areas of concentration: measurement and automation technology (see, e.g., INTECH/ISA 91)

Contact address:
Instrument Society of America (ISA)
67 Alexander Drive
P.O. Box 12277
Research Triangle Park NC 27709
United States

LASER—Innovative and Applied Optoelectronics International Technical Show and International Congress

Specialty show with stress on laser technolgy

Areas of concentration: laser and laser systems technology for manufacturing, communications, environment, medicine, science, optronics, components, optical measurement and inspection, holography, pattern recognition, image processing, fiber optics, sensor technology, microwave engineering

Contact address:
Münchener Messe- und Ausstellungsgesellschaft mbH
Postfach 12 10 09
80034 München
Germany

MESUCORA—International technical show and congress for measurement and control engineering, automation, and industrial data processing, as well as engineering

Areas of concentration: dimensional and mechanical measurements, electrical measurements, process control and automation devices, inspection and test equipment, industrial and scientific data processing, telecommunications systems, engineering, laboratory apparatus, analytical instruments, biological and biomedical metrology, optics and electrooptics, nuclear measurements, vacuum and cryotechnology, instructional and learning aids

Contact address:

Association MESUCORA
11, rue Hamelin
F-75783 Paris Cedex 16
France

SENSOR, Nürnberg—Specialty show focusing on sensor technology
Areas of concentration: sensors, measurement and testing, analytical and weighing technology

Contact address:

Nürnberg Messe GmbH
Messezentrum
90471 Nürnberg
Germany

SYSTEC—International Technical Show and International Congress for Computer Integration in Industry
Technical show focusing on CIM
Areas of concentration: research and experimentation, development, and design of CAD, CAE, and CAM software; consulting; services; system components; networks; workstations

Contact address:

Münchener Messe- undAusstellungsgesellschaft mbH
Postfach 12 10 09
80043 München
Germany

SYSTEMS—Computers and Communications—International Technical Show and International Congress
Specialty show for computer technology (from PC to the networked data processing scene)
Areas of concentration: hardware, software, communications technology, systems components, user-oriented DP problem solving

Contact address:

Münchener Messe- und Ausstellungsgesellschaft mbH
Postfach 121009
80043 München
Germany

13. Integration of Knowledge-Based Systems in Process Control Engineering

13.1. Principles

Information technology now permeates the field of process control engineering, not just in the form of computers, storage media, and communications devices, but also via concepts, methods, and techniques for structuring a wide variety of information. These concepts, methods, and techniques now mark every facet of the field, from the structuring of knowledge about processes to be automated, to the use of increasingly powerful microprocessors in control facilities (sensor/actuator systems, process monitoring and control systems), the structuring of the process control engineer's work during the design, construction, and operation of control facilities, and the availability of computer-aided design tools. Automation is no longer conceivable without them. They are discussed in depth in preceding chapters.

The more advanced technologies of knowledge-based and adaptive systems are already becoming important.

Expert systems, or knowlegde-based systems, arose in the field of AI (artificial intelligence) research, chiefly at American universities and research institutions. The first expert systems appeared in the early 1970s, but their high demand for computer resources prevented their becoming established at that time.

Knowledge-based systems in the sense of a software technology are distinguished by the architectural principle that application-oriented knowledge and processing logic are separate. This can be referred to as a software technology because the field incorporates novel programming methods and techniques. It can also be said that a "paradigm shift" has taken place in informatics, since procedural programming (supported by programming languages such as FORTRAN, PASCAL, and PL/1) has now been joined, on an equal footing, by other paradigms: logic programming (PROLOG = programming in logic), rule programming (OPS 5), functional programming (LISP = list processing), and object-oriented programming (FLAVOR, SMALLTALK, C++, etc.).

Integrated in a hybrid programming system, all these paradigms can be utilized in parallel, depending on what representation or type of knowledge appears most suitable.

The aim of knowledge-based expert systems is to explicitly make available, in a knowledge base, the knowledge of experts as embodied in standards, regulations, and guidelines, but also the knowledge they have gained through experience.

Only later was it recognized that the simple, application-neutral processing logic of "inference machines" (or inference engines) is insufficient for handling certain types of problem.

In addition to the representation of knowledge, an ontological analysis of the problem type is needed. Thus another problem-solving strategy comes into play as used for example for seeking errors in a technical system [13.5] (diagnostic strategies) or for designing or planning of a technical system. The basis for the latter activity is "planning strategies," which are generally top-down, beginning with functional requirements and proceeding through design alternatives to engineering specifications.

Diagnostic strategies are relatively well understood today, and special expert-systems tools employing them are available. Planning strategies are not so well developed. A point of importance in the problem of planning is "nonmonotonic inference": During planning or design, the planner may find that an approach already adopted is not leading to success, so that design decisions must be withdrawn. Expert systems must be able to keep up with this nonmonotonic decision-making process.

The crucial question in expert systems is still how to acquire experience. In contrast to encoded technical knowledge, which is formalized, standardized, documented, and based on a consensus in the technical world, experience is not formalized, not standardized, not documented, and often not a matter of consensus. Nor can knowledge acquisition be approached by methods of classical systems analysis, since it is concerned with the understanding of mental processes.

For a certain class of problems, adaptive systems or neural networks offer a solution.

Neural networks are now being used successfully for pattern recognition and classification. Pattern recognition refers not just to the typical recognition of images, handwriting, and sounds, but also to the recognition of relationships between physical, chemical, and apparatus-specific quantities.

A neural network is a large, highly connected system of simple processors. Whereas the human brain consists of ca. 10^{12} neurons, artificial neural networks today have few neurons, generally no more than 100.

In such a network, a neuron receives inputs from several others. Only when the sum of the weighted input signals exceeds a threshold value does the neuron transmit a signal to the neurons connected to it.

The "programming" of such a network, also called learning, consists of modifying the weights of the connections inside the network such that a pattern presented at the input causes the desired response at the output. One of many programming methods is to modify the internal weights on the basis of the difference between the actual output and the nominal output. When the learning phase is complete, the neural network can interpret new patterns correctly, albeit with a residual error.

New network designer's art consists of finding a design, a learning rule, and suitable representative examples so that an optimal neural network will result at the end of the learning phase.

Another approach to dealing with uncertainty and indistinctness is fuzzy logic. This technique, which has been brought to industrial maturity in Japan, is based on the mathematical theory of fuzzy sets.

In classical logic, a proposition can take only the values "true" or "false," 0 or 1. In fuzzy logic, a proposition or "linguistic variable" is said to have such-and-such percent of value *a* and such-and-such percent of value *b*.

For example, consider pressure as the variable. In an engineering system, a process control engineer might say that the pressure can take on qualitative values such as very low, low, moderte, high, and very high. Now a given pressure—whereby "pressure" is now understood as a linguistic variable in fuzzy logic—can be characterized by the degree to which it belongs to the set of "high" pressures. This is expressed by a numerical value between 0 and 1. Zero means that the pressure does not belong to the set of high pressures at all, while 1 means that it belongs to that set entirely.

A variable might be located, for example, in an intermediate range from high to very high. It no longer takes on a unique value but can be described by, for example, the expression "pressure = (very low = 0.0, low = 0.0, moderate = 0.0, high = 0.3, very high = 0.8)."

Astonishingly, fuzzy logic has found broad applications in control engineering. This should not, however, be surprising, because many quantities cannot be measured exactly and noise of various kinds is often present. A study performed by the Frauenhofer Gesellschaft gives a comprehensive view of the potential applications of fuzzy control, in which an extensive literature collection can be found [13.18].

Fuzzy-logic control systems in practice often show more stable behavior than conventional methods. Interesting developments have arisen through the combination of classical deterministic control algorithms with fuzzy logic. Combinations with neural networks make is possible to gain the advantages of both approaches.

Another technique for finding and retrieving information involves "hypertext/hypermedia" systems. The basis for hypertext is that electronically stored knowledge bases are accessed not with formal query languages or commands but by associative searching in a network consisting of information nodes and arbitrary links between these nodes. Employing the principle of direct manipulation, the user follows a train of associations by clicking the mouse on interesting objects or interesting passages that appear in a document. Nonlinearity is the guiding principle of hypertext. The freedom to navigate within such a network can also lead to disorientation. Tools must therefore aid the user in moving around information space.

The technique of hypertext systems, often combined with object-oriented data bases, is used particularly for technical documentation, standards, and regulations.

Hypermedia systems result when the information nodes are not exclusively text nodes but can also offer graphics, audio, animation, and video sequences.

To summarize, information-oriented process control engineering is already fully operational with no end to its development in sight. This quick review of novel software technologies suggests that the immediate future will see remarkable new advances in information processing. This trend has led to many seminars and tutorials, with special mention being made here of the tutorials at the GMA Congress 1993, in which fuzzy control, neural nets modelling dynamic systems, ER/SADT, and information-oriented modelling concepts were reported in detail.

13.2. Knowledge-Based Approach

Introduction. The ideal structure of a process control system is presented in Chapter 7. In Figure 7.16, the "remote" components include an AI (artificial intelligence) computer, which forms the hardware basis for knowledge-based systems. In process control engineering, the latter term is understood as referring to a combination of conventional process automation approaches and AI methods.

The need for knowledge-based systems using AI computers in process control engineering arises from the development of the object being automated, that is, the processing system.

For example, the continuing trend toward more intensive process control — a consequence of technological progress in process, manufacturing, and automation engineering — has led to a number of qualitatively new and distinct features in modern production facilities. These include large product throughputs, extreme process parameters, operation up to the stability limits, high parametric sensitivity, direct coupling of some process units without intermediate storage, and a high degree of material, energetic, and informational interconnection between production areas and operational divisions. The objectives of process control have come to include process securing (enhancement of process reliability), plant availability, process efficiency, product quality, production flexibility, and environmental safety. For these reasons, complex control tasks at higher levels are steadily gaining in importance for process automation. Most of these tasks involve the solution of the following problems [13.19]:

- The complexity and high dimensionality of control tasks (more than five control variables that can be freely selected)
- The incomplete observability and the nonlinearity of the subsystems
- Compliance with complicated side conditions (including a number of logical conditions)

The classical methods of modeling and control often fail on grounds of expense or because of parameter identification problems or real-time requirements. Generally, higher-level control tasks can be accomplished only when their complexity and dimensionality are reduced by situation recognition techniques and when experiential knowledge (to delimit the solution domain) is combined with situational models (to define the solution point).

In the context of knowledge-based systems for process control engineering, the principal AI elements are expert systems, artificial neural networks, and fuzzy systems. Expert systems and fuzzy systems (both employing methods of mathematical logic) are similar in their information structure. They are used as rule-based systems in process automation, in both open loops (consultative systems) and in closed control loops. The sole difference is that expert systems employ "distinct" knowledge (precisely measured process signals, unambiguous and either/or rules) while fuzzy systems employ "indistinct" knowledge (membership functions, fuzzy rules). From the applications standpoint, artificial neural networks are very similar to statistical methods of process modeling.

This section deals mainly with the use of expert systems and neural networks. AI methods are just beginning to find wide use in process control engineering [13.19].

Rule-Based Systems. The main types of rule-based systems used in process control engineering are expert systems and fuzzy systems. These seek to imitate the process of human thought on a phenomenological level.

If these systems are to be used as control elements in process control engineering, it is necessary to analyze the process employed by the operator of a production plant in solving problems. Figure 13.1 shows the subtasks and the solution process for the example of a reaction to disturbances in the process plant.

An analysis of this problem-solving process shows the following:

- The solution process involves the steps of diagnosis, optimization, and planning. Each step is separated from the next by a simulation phase. In this way, solutions are validated on simulation models, which may be both heuristic models and, often, mathematical models "in the operator's head" (thought models). These simulation techniques, whether practiced consciously or not, form part of a continuous iterative mental process, which ultimately leads to optimal conduct of the process. In other words, the suitability or optimality of the target situation is investigated by simulation and then corrected, leading to a corrected target situation, which in turn is simulated, and so on until the best result has been attained according to the operator's judgment.

- A nominal/actual comparison, the construction of control algorithms, and the determination of setpoint values, all as components of the diagnosis, simulation, and planning steps, can be performed with conventional control resources. The result is a higher degree of automation and hence the accomplishment of process stabilization tasks.

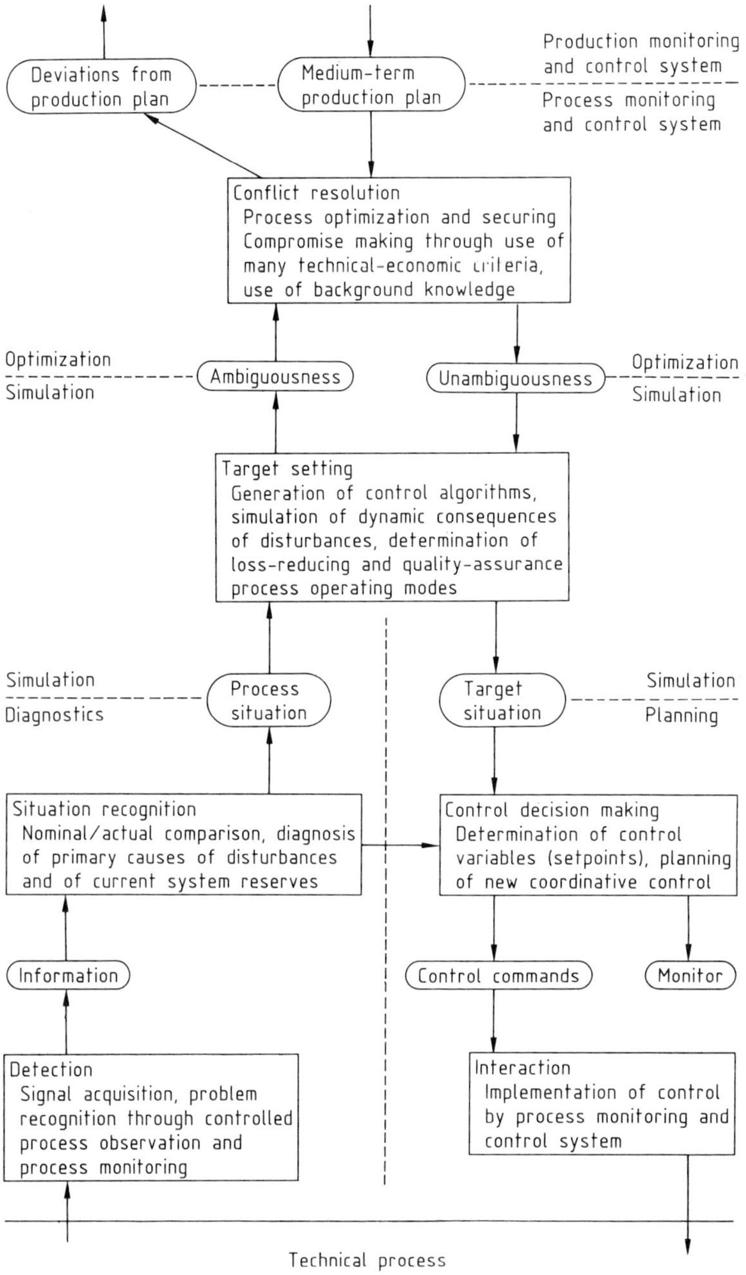

Figure 13.1. Tasks in knowledge-supported process control

- Conflict-resolution strategies in the context of optimization tasks can generally be implemented only by using artificial intelligence methods, namely the accomplishment of tasks relating to process optimization and securing. As a rule, knowledge-based systems here operate in consultative mode, with no increase in the degree of automation. If these tasks in all of the solution space are algebraically describable, however, it is possible to dispense with AI components.
- Knowledge-based systems cannot handle background knowledge (e.g., historical knowledge about the design process, knowledge of maintenance strategies or of the market situation). This part of the problem-solving process thus cannot be supported by knowledge-based process control engineering.

There are no fixed procedures for modeling the problem-solving process as a mental process. This is the fundamental distinction to the mathematical modeling of engineering processes, in which the application of balance equations (mass, enthalpy, and momentum balance) leads to an unambiguous and exact mathematical model.

Analyzing the problem-solving process from the standpoint of process control theory allows two classes of tasks to be identified: disturbance analysis and control synthesis.

Disturbance analysis encompasses solving the following tasks:

- Determination of primary casues of disturbances: Disturbances arise through slow parameter drift due to wear, contamination, and corrosion, or through sudden structural changes due to the failure of subsystems.
- Determination of secondary consequences of disturbances: Frequently occurring secondary effects are bottlenecks in feedstocks, energy-transfer media, intermediate products, or recirculating streams in adjacent process sections or in downstream process steps, which (depending on the process dynamics) most often manifest themselves immediately but, ultimately, further increase the resulting production losses.
  On the basis of the primary causes of disturbances, a simulation of the dynamic consequences of disturbances thus provides important supplemental information about the objectives of measures taken against disturbances.

The following tasks must be carried out in control synthesis:

- Identification of primary repair measures means deriving the repair measures by which the primary causes of disturbances can be directly remedied.
- Determination of current system reserves: Along with the medium-term removal of primary causes of disturbances, in the short term it is necessary to make a control decision that will bridge over the failure of the malfunctioning subsystems for the duration of the repair measures and thus minimize the resulting production losses.
- Generation of a provisional operating regime: On the basis of the current system reserves, a new process regime is generated that will bridge over the system failure for the time required to remedy it and thus minimize the resulting production losses.
- Restoration of operation: Once the primary causes of disturbances have been remedied, the system must be restored to its normal state. This also requires for coordination of all subsystems.
- Process optimization: Current system reserves, as well as feedstock and energy resources, are deployed such that a technical-economic criterion takes on an optimum value while compliance with technical and economic constraints is maintained.

In what follows, some theoretical principles of rule-based systems are presented. These are essentially based on mathematical logic.

Neural Networks. In contrast to expert systems, artificial neural networks imitate the process of human thought at a biological level [13.19], [13.20]. The aim is a functional simulation of the human brain or of the human nervous system, which comprises some 25 billion nerve cells (neurons). Figure 13.2 shows a single neuron and the way in which neurons are interconnected.

A neuron receives its input information via the dendrites, which are connected by synapses to the information outputs of other neurons. Each neuron has only a single information output, the axon, which may be connected to as many as 10^4 "downstream" neurons through dendrites. From the sum of the inputs, an output dependent on the state of the neuron is generated and passed on to the downstream neurons by

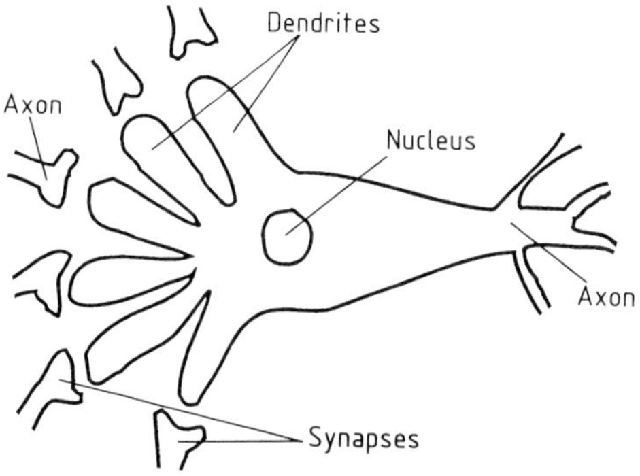

Figure 13.2. A biological neuron

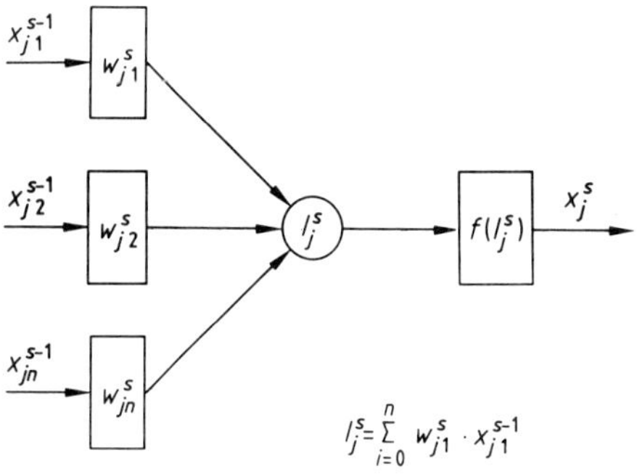

Figure 13.3. Functional diagram of an artificial neuron

way of the axon. Biological information is thus transported throughout the network (nervous system), whereby it may also be modified. A significant feature of the process is the massive parallelism of information transport and processing. The above properties of neurons allow an abstract model of an artificial neuron (processing element) to be devised, which is described by the following rules (see Fig. 13.3):

- A neuron can adopt either of two states, the resting state and the action state
- A neuron j in layer s has multiple inputs x_{ji}^{s-1}, ..., x_{jn}^{s-1} (synaptic connections) and one output (axon) x_j^s
- The output of the neuron is connected to the inputs of other neurons, and in this way a neural network is formed
- Some inputs and outputs of the neurons are connected to the outside world
- A neuron enters the action (excited) state when a sufficient number of its inputs have been excited

In this discussion,

x_{js}^s denotes the instantaneous output states of the j-th neuron of layer s,
w_{ji}^s the weightings of the connections of the i-th neuron in layer $s-1$ to the j-th neuron in layer s, and
I_j^s the weighted sum of the inputs of the j-th neuron in layer s.

The last rule, in particular, implies that the state of an artificial neural network depends only on the state of its elements. The individual neurons thus operate independently of one another.

For process control engineering the features of interest are the basic functions of neural networks (pattern association, pattern reconstruction). Here "patterns" are vectors (e.g., sets of measurements used to characterize a plant state or to assess product quality).

Three types of processing elements (units) are distinguished in an artificial neural network:

- Input units, which acquire their input information from the surroundings of the network (e.g., sensors) and are "visible" from outside the system
- Hidden units, which process incoming information and are "invisible" from outside the system
- Output units, which deliver output information to the surroundings of the network (e.g., actuators)

The individual units are grouped by function into layers, which are interconnected in a manner that depends on the problem. Figure 13.4 shows a typical multilayer network. Such networks are used principally for classification tasks.

Note that the behavior of the artificial neural network is strongly influenced by the network structure along with other factors.

A distinction is often made between the normal working phase (recall phase) and the learning or training phase of a neural network. Before the network becomes able to carry out given tasks, it must be trained for those tasks. Learning may be supervised or unsupervised. The first mode involves learning on the basis of specified examples. The back-propagation method is well known. In unsupervised learning, the network independently trains itself on the basis of rules in the input data (e.g., patterns).

In back-propagation, the output of the j-th element of layers s is determined by transforma-

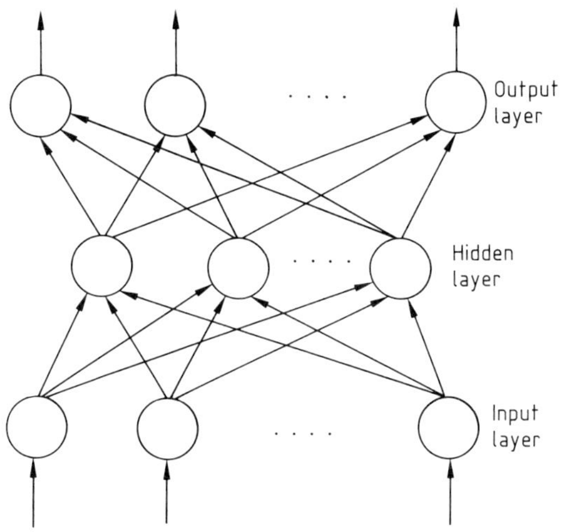

Figure 13.4. Multilayer network with one hidden layer

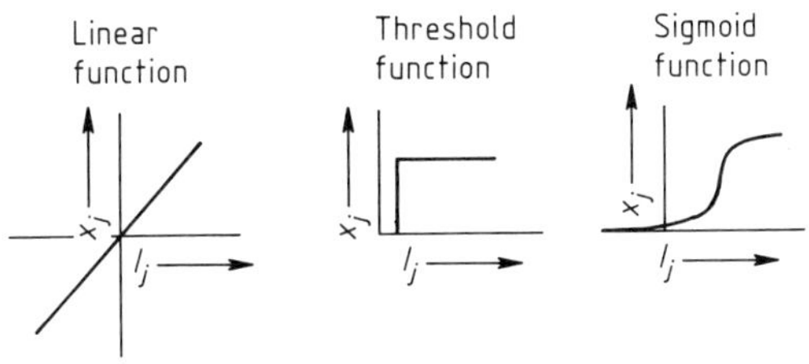

Figure 13.5. Possible transfer functions of a processing unit in a neural network

tion with transfer function $f(I_j^s)$ (see Fig. 13.5 for an example).

$$x_j^s = f\left[\sum_{i=0}^{n} (w_{ji}^s \cdot x_i^{s-1})\right] = f(I_j^s) \tag{13.2.1}$$

In back-propagation networks, the quality of the neural network is usually evaluated by a global error function E, which is needed in order to define local errors in the output layer and thus initiate the back-propagation of errors.

If it is assumed that vectors $\boldsymbol{i}$ and $\boldsymbol{d}$ are given to the network as the input and output patterns during the learning phase, while $\boldsymbol{o}$ represents the output vector calculated with the instantaneous weights, a global error function for the artificial neural network referred to the expected output $\boldsymbol{d}$ can be defined as

$$E = \frac{1}{2} \sum_{i=1}^{m} (\boldsymbol{d}_i - \boldsymbol{o}_i)^2 \tag{13.2.2}$$

where m is the number of components of vectors $\boldsymbol{d}$ and $\boldsymbol{o}$.

The aim of the learning process is to minimize the global error E of the network by selecting weights w_{ji}^s. Given the current set of weights w_{ji}^s, the problem is to determine whether and by how much these weights must be changed. The formula

$$\Delta w_{ji}^s = -\alpha \cdot \frac{\delta E}{\delta w_{ji}^s} \tag{13.2.3}$$

is used for this purpose, where α is the learning coefficient.

Each weight is thus corrected in accordance with the magnitude and direction of the negative gradient on the surface of E. The partial derivative in Equation (13.2.3) is calculated as

$$\frac{\delta E}{\delta w_{ij}^s} = \frac{\delta E}{\delta I_j^s} \cdot \frac{\delta I^s}{\delta w_{ij}^s} \tag{13.2.4}$$

Setting

$$e_j^s = -\frac{\delta E}{\delta I_j^s} \tag{13.2.5}$$

and applying the chain rule twice gives the following relation between e^s and e^{s+1}, needed for the iterative computation:

$$e_j^s = f(I_j^s) \cdot \sum_{k=1}^{m} (e_k^{s+1} \cdot w_{kj}^{s+1}) \tag{13.2.6}$$

where m is the number of processing elements in layer $s+1$.

For the output layer

$$e_k^o = -\frac{\delta E}{\delta I_k^o} = -\frac{\delta E}{\delta o_k} \cdot \frac{\delta o^k}{\delta I_k^o}$$
$$= (d_k - o_k) \cdot f(I_k^o) \tag{13.2.7}$$

Combining equations (13.2.3) and (13.2.4) and using Figure 13.3 gives

$$\Delta w_{ji}^s = x \cdot e_j^s \cdot x_i^{s-1} \tag{13.2.8}$$

Equation (13.2.8) represents the delta learning algorithm of the back-propagation method. Now the mechanism of back-propagation in the artificial neural network can be made explicit:

- Forward propagation of the inputs up to the output layer with the aid of Equation (13.2.1) and determination of the error at the output layer
- Successive correction of the weightings w_{ji} (with Eq. 13.2.8) during the process of back-propagation toward the input layer
- Recalculation of the output layer error with Equation (13.2.1) and recorrection of the w_{ji} with Equation (13.2.8), and so forth

Algorithmic Integration of Knowledge-Based Systems. Expert systems and neural networks have both advantages and disadvantages for process control (see Table 13.1).

A useful way of compensating for the drawbacks is to employ both kinds of systems together. Figure 13.6 illustrates this point; the formulas presented there are explained in the text. This also enhances the software reliability. The artificial intelligence components determine the solution domain, while the conventional algebraic methods of process control give the solution point. From the standpoint of automation theory, the result is thus a localization of the process control task.

Table 13.1. Comparison of expert systems and neural networks

System	Advantages	Disadvantages
Expert systems, fuzzy systems	use of surface knowledge	difficulties in knowledge acquisition (e.g., rules, membership functions)
Neural networks	ability to learn	difficulties in interpretation of solution (e.g., no possibility of explanation)

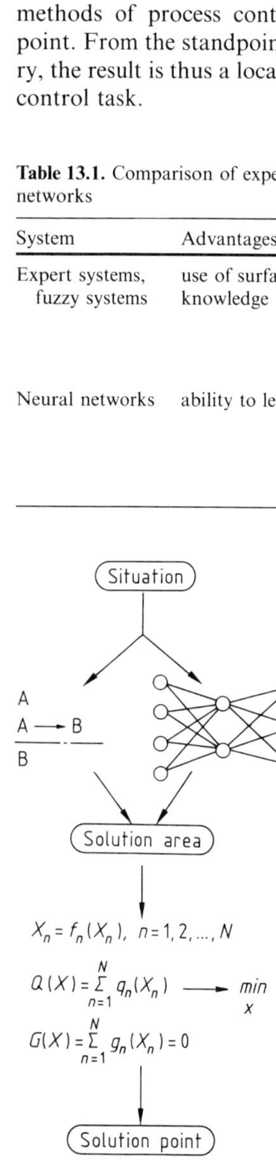

Figure 13.6. Algorithmic integration of knowledge-based systems

The mathematical models are significantly simplified in this way, because a mathematical description is needed only in the vicinity of the solution point. Linear models may well be adequate. Thus the control algorithms relating to process optimization, stabilization, and securing must be run under real-time-capable software, either on the process monitoring and control computer or on a special AI computer.

The definition of the solution domain by expert systems and neural networks (essentially in the context of classification tasks) is treated above. Now let us examine in more detail the application of algebraic methods to the determination of the solution point. It is assumed that a processing system can be broken down into subsystems (SS_n), interconnected by material and energy streams (see Figs. 13.7 and 13.8). This means simply that the phase model of production is employed in developing the software structure (see Chap. 2).

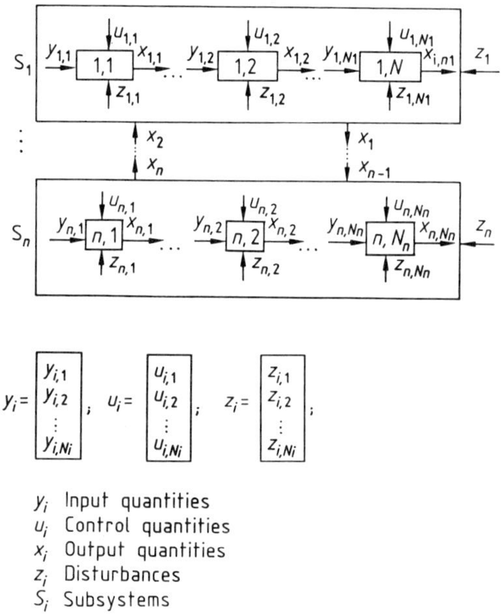

Figure 13.7. Decomposition of a technological system

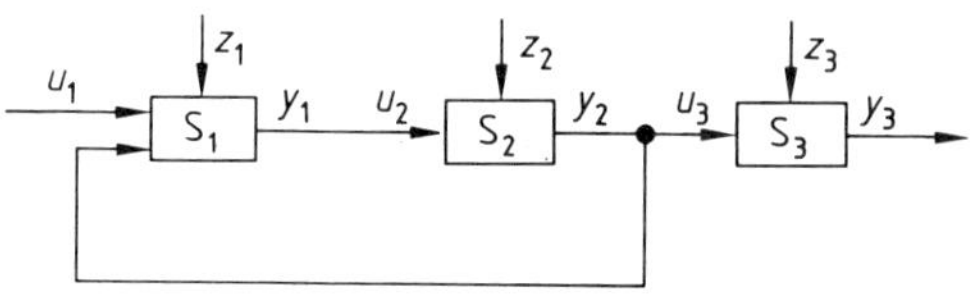

Figure 13.8. Example of a coupled total system

The following aspects are crucial in analyzing the system into subsystems:

- Topological (breakdown into technological elements, apparatus groups, etc.)
- Dynamic (breakdown into subsystems by dynamic properties, e.g., duration of dynamic transient processes, time lag)
- Combination of topological and dynamic aspects

The following notation is used:

$X_n = (U_n, Z_n, Y_n)$

represents the vector of the process parameters of the n-th subsystem;

$\dot{x} = (X_1, X_2, \ldots, X_N)$

the vector of the process parameters of the entire system;

$X_n = f_n(X_n)$

the simplified mathematical model of the n-th subsystem;

$G(X) = 0$

the coupling conditions of the subsystems; and

$Q(X)$

the goal function in process optimization, the Lyapunov function in process stabilization, and the negative first derivative of the Lyapunov function with respect to time in process reliability enhancement and quality assurance.

Thus the algorithmic structure of all process-control tasks is unified. As a consequence, marked simplifications have been achieved in the applications software (fewer software errors, better software maintenance).

The use of the optimization condition $Q(X) \underset{X}{\rightarrow} \min$ naturally requires some explanation.

With regard to the algorithmic integration of knowledge-based systems, conventional mathematical models are used as simulation systems for generating examples (in the case of neural networks) and of unmeasurable process information (in the case of expert systems). This point will be treated in more depth below.

Technical Integration of Knowledge-Based Systems. *General Principles of Integration.* Figure 13.9 shows the technical integration of expert systems and neural networks into the process monitoring and control plane. Knowledge-based process control can be represented in three levels:

- Elementary situation recognition with conventional methods of limit monitoring, deterministic or statistical class ordering, and trend analysis.
- Expert systems and neural networks combined with simulation systems based on mathematical models of the technological process. The objective is to identify the domain of the desired solution to the control problem.
- Conventional process control with the use of local or simplified mathematical models.

Integration of Expert Systems. The recommended procedure for technical integration of expert systems is to implement a real-time-capable expert system shell (expert system with empty knowledge base, i.e., without rule knowledge) on the control or AI computer, then fill it with concrete process-control knowledge in the course of knowledge acquisition (see Section 13.3). Such expert system shells must imitate the problem-solving process described here. A successful example is the PROCON expert system shell, which has been implemented in several cases [13.21]–[13.24]. This shell will be described in more detail in what follows.

- System components: PROCON is subdivided into a development system and a runtime system with the following components (Fig. 13.10):

- Knowledge representation: object-oriented representation concept for explicit structuring of problem spaces
- Knowledge manipulation: cooperative problem-solving processes for diagnosis and application of expert knowledge toward efficient delimitation of a solution
- Knowledge acquisition: problem-specific development environment for input, inspection, and transformation of knowledge into a runtime-efficient representation
- Knowledge consultation: user environment for verification and consultation of implemented knowledge bases

PROCON incorporates process, program, and dialog interfaces with the following features:

- Preparation of all needed process variables that can be acquired on-line, through coupling with the process control system
- Preparation of all needed process variables that can be acquired off-line, through interactive dialog with operating personnel

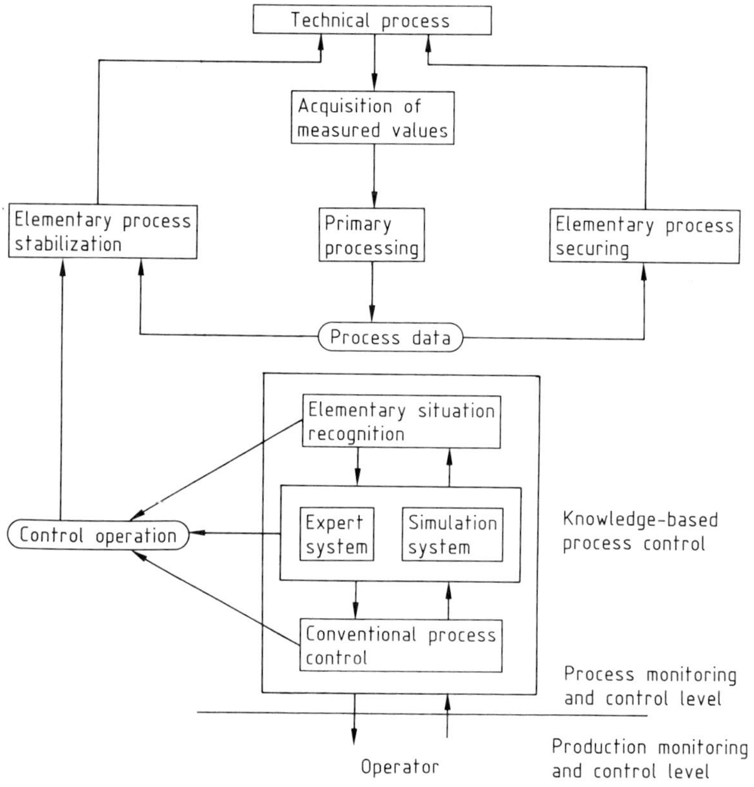

Figure 13.9. Technical integration of knowledge-based systems

- Automatic interrogation of redundant information sources by the expert system whenever failures of measurement stations are reported
- Manual configuration of charge-dependent process parameters with parameter sets or in interactive dialog with operating personnel
- Rule-governed activation of external simulation and optimization programs via the program interface
- Manual, timed cyclical, or external consultation start

Requirements on the Knowledge Base. Applying expert systems to process control tasks means keeping pace with the process dynamics so that the integration of the knowledge base can be insured and critical reaction times met. Heavy modularization of expert knowledge into small knowledge units is dictated by both the need for adequate runtime efficiency of knowledge processing and the need for easy maintainability of the knowledge base as technological conditions change and operational experience is gathered.

One useful modularization principle results from a division into disturbance and control classes. The set of all disturbances can be imaged on a set of disturbance classes that are qualitatively distinguished by primary causes and secondary consequences of disturbances. Similarly, all control actions can be placed in control classes on the basis of the disturbance class to be compensated and the available system reserves. If the two classifications are successively refined, a tree-like structure of disturbance and control classes is created. In the terminology of expert systems technology, this is referred to as a class hierarchy (taxonomy) of prototypical problem classes (objects). Descending through the problem hierarchy thus corresponds to localizing the current system disturbance and the resulting therapeutic control action. On a path from the root to a leaf of the taxonomy, each object encountered makes a contribution to the total solution of a problem. Each object contains a partial knowledge base made up of declarative (facts), functional (functions), and procedural (rules) knowledge specifications (Fig. 13.11).

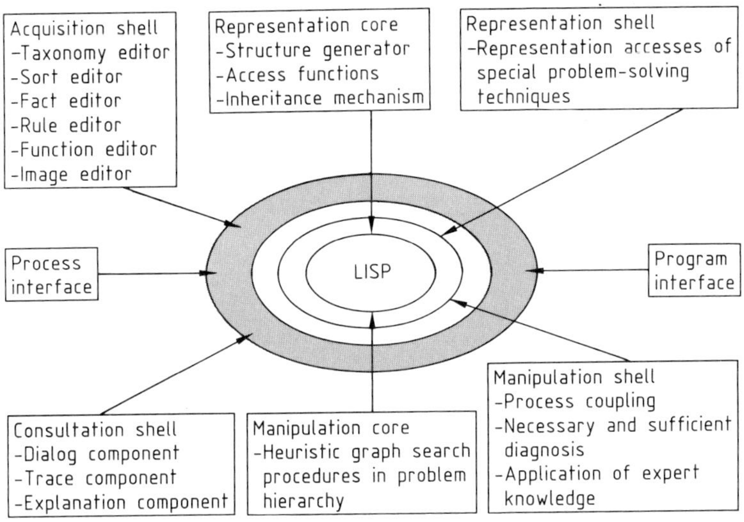

Figure 13.10. PROCON system components

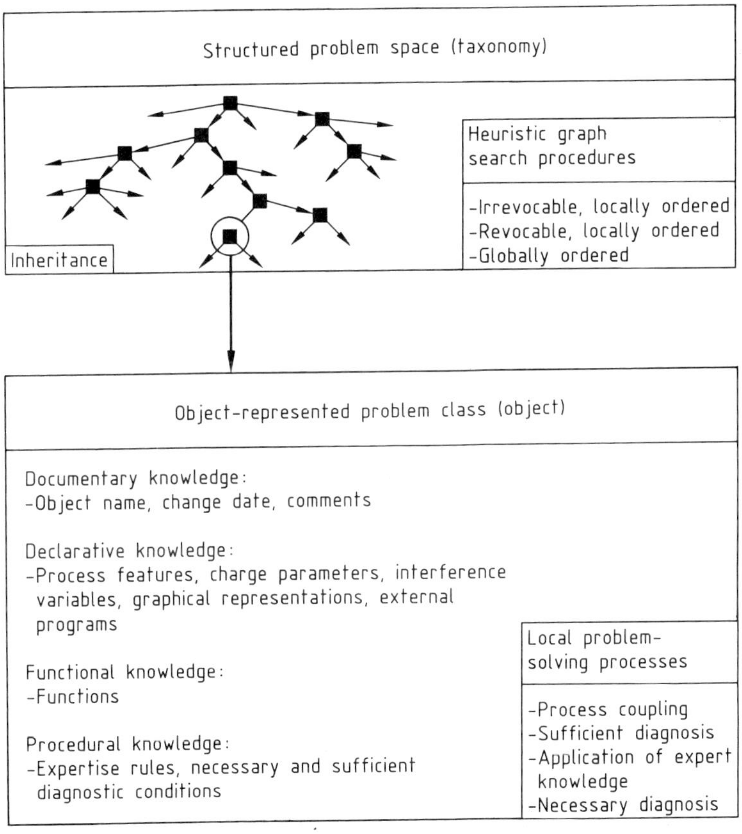

Figure 13.11. Concept of knowledge representation in PROCON

Each object has access to the knowledge of all its predecessors in the problem hierarchy (inheritance):

- Declarative knowledge includes process information that can be automatically updated (values that can be acquired by on-line measurement, values that can be acquired by off-line laboratory work, etc.); user-configurable (technological) constants; interference variables derivable in the course of problem-solving; and representations parametrizable and external programs invokable by the rules of expert knowledge application (simulation and optimization models).
- Procedural knowledge. The following rule bases can be defined in each object:

– Rules for application of expert knowledge ("expertise" rules) are forward-chained production rules for the derivation of interferences from a given problem class (Fig. 13.11). They can derive interference variables, form solution texts to be displayed, parametrize representations for output, and invoke external programs for simulation and optimization.
– Diagnostic rules are classification rules for the derivation of so-called sympathy ratings for the objects in a problem hierarchy (Fig. 13.12). The expert system uses these ratings to decide what route to follow through the problem hierarchy, that is, what problem classes actually need to be examined in a concrete disturbance situation.

```
Rule D1

If
 the paste pump is not in operation &
 the level in the paste mixer is <5% &
 the level in VE1 is under 5% &
 the level in VE2 is under 5%

Then
 METER_DIST_1 90
 METER_DIST_2 50

Rule D2

If
 the paste valve is set to recycle &
 the level in the paste mixer is at Max-Alarm &
 the level in VE1 is over 5% &
 the level in VE2 is over 5%

Then
 METER_DIST_1 40
 METER_DIST_2 80
```

Figure 13.12. Examples of diagnostic rules (METER_DIST_1 and METER_DIST_2 are two lower-level objects of the current problem class)

Problem Solving. The object-oriented structuring of the problem space permits efficient problem reduction through two problem-solving levels:

- Higher problem-solving level: Starting from the root of the problem hierarchy, a heuristic graph searching procedure determines a route through the structure of the problem space. In so doing, it establishes which problem classes actually need to be examined in a concrete disturbance situation, and in what order. The examination of a problem class is initiated by an activation message to its local problem-solving process.
- Lower problem-solving level: The local problem-solving of an activated object begins with a process coupling, then continues with an expertise and diagnostic process. Through the process coupling, all process features defined in the given object (with respective coupling parameters) are first updated. The expertise process (application of expert knowledge) that follows next evaluates the expertise rule base of the object in order to derive interferences from the problem class. Finally, the diagnostic process uses the diagnostic rule base of the object to rate its successors in the problem hierarchy.
  The remainder of the problem-solving process follows the rules of mathematical logic as described above.

Development Environment. The work of developing practically relevant knowledge bases is commonly divided among specialist experts (process engineers, automation engineers, expert operators) and, in the long term, makes possible the maintenance of the knowledge base as technological conditions change and operational experience is gathered. For this reason, special value attaches to interactive tools for the support of incremental, problem-specific knowledge acquisition. In PROCON, the acquisition component integrates the following aids in a self-contained development environment:

- Knowledge input: Graphically supported structure editor for taxonomy input. Mask editors for fact input, screen editors for rule and function definition.

- Knowledge inspection: Automatic search for inheritable knowledge specifications anywhere in the knowledge base; compressed explanation of defined knowledge contents of the active inheritance hierarchy during knowledge input.
- Knowledge transformation: Transformation of a problem-specific external knowledge representation into a runtime-efficient internal representation; formal-language or natural-language rule notation can be selected.
- Knowledge verification: Testing for syntactic closure and semantic consistency in the active inheritance hierarchy during rule translation; testing for adequacy and completeness in a comprehensive trace environment with case data management.

This development environment is the basis for knowledge acquisition as described in Section 13.3.

Integration of Neural Networks [13.25]. In comparison with the conventional methods of process control engineering, neural information processing requires high-performance storage media as well as fast processing units (and possibly the use of special processors). Furthermore, the practical deployment of neural networks calls for ingenuity in network configuration as well as a training technique optimally suited to the application.

Hardware and Software Implementations. The recommended hardware support for neural network algorithms differs from the classical Von Neumann structure. The following solution approaches are of interest:

- The use of special neural processors to speed up serial operation. These units feature a separate floating-point processor, a supplemental cache memory, and complete primary storage capacity and employ the pipelining principle. Examples are the PC boards made by SAIC Corp.
- Multiple-instruction, multiple-data structures, which can be implemented with available "transputer" components; a number of software packages are already on the market for transputer boards.
- The use of vector processors. If these devices have a sufficient number of processor components, they make it possible to determine network activity layer by layer (i.e., to determine the activation vectors section by section with a high degree of concurrency).

At present, software simulators are used to convert the various network types to the computing and monitoring/control hardware currently used in process automation. Their implementation on general-purpose processors (e.g., Intel 8086) meets the requirements of flexibility, portability, and high integrability.

Only in exceptional cases can a dedicated implementation be recommended, especially since there are a number of powerful neural shells on the market. These systems generally have communications interfaces to the user and file levels and offer a wide range of distinct network types. Particularly important for integration into process control systems are runtime routines that enable customized, memory-efficient source code runtime modules to be integrated into the applications software. One example of such a software simulator is a development system called Neural Works Professional II, published by Neural Ware Inc. and intended for PC and workstation environments.

Note, however, that straight software implementations have the drawback of limited capacity, both in the training and in the implementation phase. A combination of dedicated hardware and dedicated software should therefore be preferred. The integration of parallel processing structures (e.g., in the form of transputers) into monitoring and control systems is anticipated.

13.3. Knowledge Engineering

Procedures and Methods. The creation of the knowledge base is the key step in the integration of knowledge-based systems into process control engineering. This procedure employs methods of knowledge engineering.

The term knowledge engineering denotes the totality of the processes involved in the design, implementation, and maintenance of expert systems (when expert system shells are used, this essentially means filling a knowledge base). Knowledge engineering thus continues throughout the life cycle of an expert system. The analogy to software engineering is clear. Knowledge engineering is the bottleneck in expert systems development (the same applies to automation engineering). Thus there is a parallel to the "software crisis" in the realm of conventional software.

Knowledge engineering, which can also be termed systems analysis, involves close collaboration among process control engineers, process engineers, and information specialists (Fig. 13.13).

In empirical social research, many years' experience and a broad methodological potential are available for the acquisition of knowledge and the traditional, nonformalized use of expert knowledge. Because these are also available to knowledge engineering, it is not surprising that the techniques currently in use for the acquisition of expert knowledge are largely derived from the methodological inventory of empirical social research. The techniques must, however, be modified to suit the needs of this special application area of process control engineering. Desirable, above all, is an appropriate combination of these techniques with the methodological potential built up over decades in theoretical and experimental process analysis. The knowledge to be incorporated in expert systems is generally found in the most diverse knowledge sources: technical manuals, operating instructions, case studies, and reports, but also data files and databases that may contain operational data for concrete process situations. In addition, personal experience and the knowledge of technical specialists (experts) is also sought. It is the last of these knowledge sources that is crucial in process automation. Moreover, a theoretical process analysis may also involve the creation of mathematical models that are then available as secondary knowledge sources.

Accordingly, a complex utilization of all knowledge sources—theoretical mathematical models, process data, and operator experience—is to be recommended as a way of reducing effort (see Fig. 13.14).

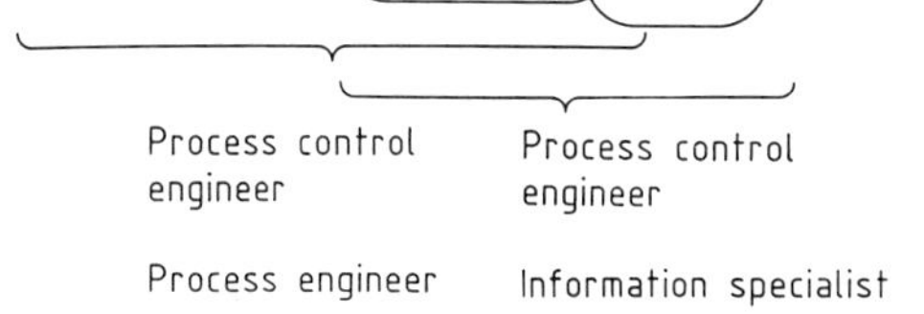

Figure 13.13. Interdisciplinary cooperation in knowledge engineering

Formulated in general terms, the knowledge acquisition component of an expert system must support the transformation of this very heterogeneous knowledge to a homogeneous form, so that it corresponds to the representation scheme defined by the structure of the knowledge base and by the interference component. Thus the knowledge acquisition component aids in the description of information or knowledge from the area of the user's tasks, building up a knowledge base from this information. The wide variety of knowledge sources makes this an exceedingly complicated problem, which entails a considerable effort in the case of an expert system for process automation. In present-day expert systems, various "manual" techniques are often employed for knowledge extraction; editors support only the final input to the knowledge base. When knowledge is prepared manually, however, relevant information is frequently lost or is not acquired at all. It is thus all the more important to develop knowledge acquisition components that will interactively support and assist the knowledge engineer (who is, as a rule, the process control engineer) in performing this task.

On the basis of these general aspects, a four-phase concept of knowledge engineering is recommended. This concept is oriented primarily toward process automation [13.26], [13.27]. The fundamental idea is to subdivide the iterative process of developing an expert system into the following phases (see Fig. 13.15):

- Definition
- Acquisition
- Operationalization
- Maintenance

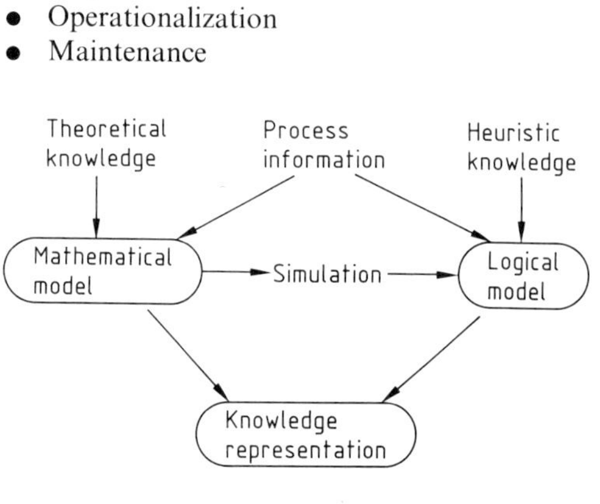

Figure 13.14. Reduction of effort by complex utilization of all knowledge sources

The definition phase provides a first global survey of the application domain. A specification of requirements must be created; that is, the problems for solution must be stated, their transformation into objectives and tasks of the expert system must be defined, and the feasibility of the expert system project must be investigated. In other words, the control task is established in this phase. The definition of the control task may be supported, for example, by comparing the disturbance amplitude D_z and frequency f_z for each control object. An analysis of this kind is done for the dominant disturbance in each control problem. Figure 13.16 shows that closed regions can be defined in the D_z–f_z plane, in which process securing, optimization, and stabilization are to be performed.

These process control components are defined as follows (see also Section 4.2):

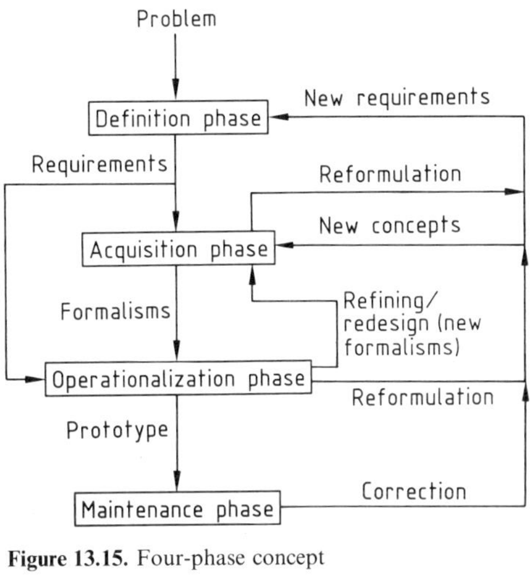

Figure 13.15. Four-phase concept

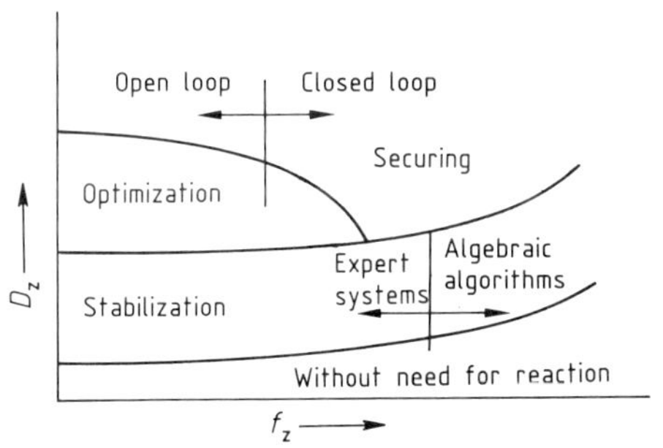

Figure 13.16. Relationship between automation tasks and properties of the disturbances

- Process stabilization: Compensation of correctable disturbances with the aim of holding the process parameters constant at specified values
- Process optimization: Determination and setting of control variables so as to optimize a specified criterion while complying with specified restrictions
- Process securing: Compensation of uncorrectable disturbances with the aim of preventing unacceptable process states and product qualities or of minimizing their consequences

Figure 13.16 also shows, as a function of f_z, the trend in the use of expert systems and mathematical algorithms (conventional process control) as well as the use of consultative systems (open loop) and automatic systems (closed loop). In this way, it is possible to identify a certain number of standard situations to which a particular control task applies. Expert systems are especially well-suited to the solution of process securing tasks, since as a rule there are no mathematical models for large disturbance amplitudes (emergencies), although a certain amount of empirical knowledge exists. The requirements worked out in the definition phase must be set forth in an appropriate form (specifications).

The acquisition phase includes analysis of disturbances, causal relationships, and possible control responses, as well as the aggregation of results into situation and control classes. The process analysis yields a formal model of the problem space in which the general questions to be answered in the definition phase should be detailed. The real difficulties in this phase lie in the design of knowledge bases. The knowledge should be structured to match the technology. It must directly mirror the operator's experience or mode of action and be determined chiefly by automation theory. In the operation phase, the knowledge is structured to match the system, as specified by the representation mechanisms that the expert system shell provides. This transformation of knowledge requires a knowledge engineer with experience and knowledge in many scientific disciplines.

Now let us take a closer look at "technology-like" structuring. The structure of the knowledge in this case is derived from the structure of the control object; a different form of structuring is generally meaningful in each disturbance class. Experience has shown that the technological system can be broken down into subsystems, each of which is controlled in autonomous fashion while all the subsystems are simultaneously coordinated. The decomposition of the entire technological system is based on the phase model of production (see Chap. 2). This principle can be illustrated for the example of process securing. The system is decomposed into a certain number of subsystems or levels, depending on the disturbance class; these subsystems show relevant differences in the dynamic properties relative to the dominant disturbances (Fig. 13.7). The upper (i-th) plane reacts with virtually no time lag to signals from the lower (i + 1-th) plane. Each level contains a number of subsystems, which are connected to one another by energy, material, and information streams and exhibit similar dynamic properties.

On the basis of the structure of the control object, it is recommended that a hierarchical structure with multiple coordination levels be used for the process securing system. The subsystems should have the greatest possible autonomy (see Fig. 13.17). The subtasks of knowledge-based process control that are solved are the ones described above under "knowledge-based systems": determination of causes and consequences of disturbances, identification of control actions for the subsystems and for the system as a whole. One of the fundamental ideas in the technology-oriented structuring of knowledge is this: When an uncorrectable disturbance z_i takes place, the application of algorithm A_{1i} detects and compensates this disturbance (determination of primary causes of disturbance, derivation of primary repair measures). The "compensation" of a disturbance z_i means controlling a subsystem such that the given system is

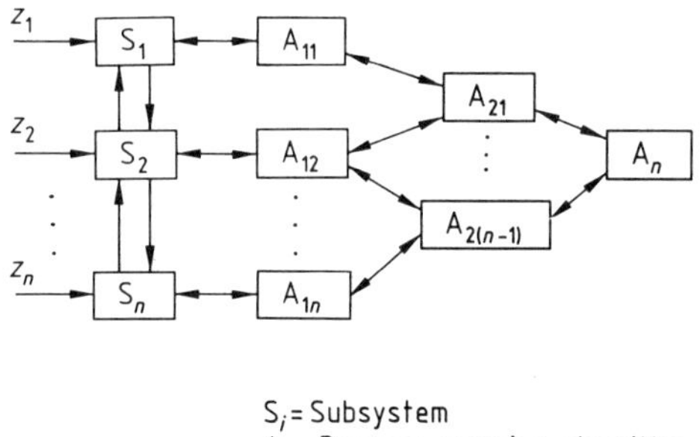

Figure 13.17. Structuring of a process securing system

in a failure state for a time Δt shorter than some acceptable value Δt_{acc} ($\Delta t < \Delta t_{acc}$). If this fails, coordination of two adjacent subsystems S_i and S_{i-1} is effected by using $A_{2(i-1)}$. First, the effect of the disturbed or failed subsystem S_i on S_{i-1} is identified (determination of secondary consequences of disturbance); second, the instantaneous dynamic reserves of S_{i-1} for compensating the failure of S_i are determined; third, a provisional regime (e.g., part-load or turndown regime) is generated. If even these steps do not compensate the disturbance, three subsystems are coordinated (for example, S_i, S_{i+1}, and S_{i+2} by using A_{3i}), then four subsystems, and so forth. This process is continued until the disturbance z_i has been compensated. Whenever a new subsystem is included, the above steps are repeated: identification of secondary consequences of disturbance, determination of instantaneous system reserves, and generation of a provisional regime. If successful compensation of the dangerous disturbance has not been achieved despite coordination of all subsystems, an emergency shutdown of the technological system is performed. In the context of knowledge-based process control, these coordinated shutdown modes can also be treated as provisional regimes. The normal regime is restored when the disturbance z_i is no longer present.

The operationalization phase includes the "system-like" structuring of the knowledge as well as steps to make the expert system functional.

Newly acquired knowledge is added during the maintenance phase (e.g., in the form of new "rules"). In addition, portions of the knowledge basis may have to be replaced altogether if changes take place in the technology. Expert systems are considerably more maintainable than conventional software systems, which is largely due to the fact that software errors do not necessitate complete reprogramming (software technology aspect). This leads to enhanced reliability of the process control system and lower maintenance costs.

Application Example. A chemical plant for the production of poly(ethylene terephthalate) fibers (trade name Grisuten) is equipped with a TDC 3000 process control system. The main tasks of the plant operator fall into the following categories:

- Process monitoring, the principal tasks being the detection and diagnosis of disturbances

- Process securing, i.e., the prevention of situations that would endanger the plant

In case of a disturbance, the operator must identify and implement a suitable reaction on the basis of available process information. Frequent operator interventions are needed in the following situations:

- Startup and shutdown
- Modifications of the operating regime
- Failure of plant sections
- Occurrence of critical process states that impair more than one plant section

Localized disturbances can generally be compensated by the existing automation system. In the above situations, the quality of operator decisions varies widely. Along with qualitatively good decisions by some operators, wrong decisions—some of them serious—are made by other, less experienced operators, especially when they are under psychological stress. Furthermore, the required specialist knowledge is commonly split up among several operators. The basis for an operator's decisions is essentially his own experience; theoretical background is less important. The main feedstocks are ethylene glycol and terephthalic acid (TPA). The process is continuous and can be broken down into the following phases (see also Fig. 13.18).

Metering/homogenizing. The feed materials are metered in fixed proportions into the paste mixer, where they are homogenized. Metering errors or inadequate homogenization of the feeds can lead to serious disturbances in the downstream system and to a major degradation of fiber quality.

Esterification. The reaction mixture is converted to diglycol terephthalate in a two-stage esterification reaction. The reaction is strongly endothermic, so heat has to be supplied. This process is carried out under vacuum, and the extracted vapors are distilled to recover ethylene glycol, which is recycled to the metering/homogenization stage. The crucial point here is to insure a minimum residence time of the reaction mixture in the reactors, to attain an appropriate degree of esterification.

Condensation. The diglycol terephthalate from the esterification stage is polycondensed to poly(ethylene terephthalate) in four condensation stages. The strongly endothermic reaction is carried out under vacuum with heating. The important quality criterion is that an appropriate degree of polycondensation must be attained after final condensation.

Spinning/finishing/extrusion. The polymer melt is spun and/or extruded in a melt-spinning stage. A number of treatments are performed on the spun filaments, after which they are ready for shipment. Every process step in the condensation and spinning/finishing/extrusion stages results in a significant increase in the viscosity of the reaction mixture.

The task of the expert system is to serve as a decision aid to the operator in carrying out pro-

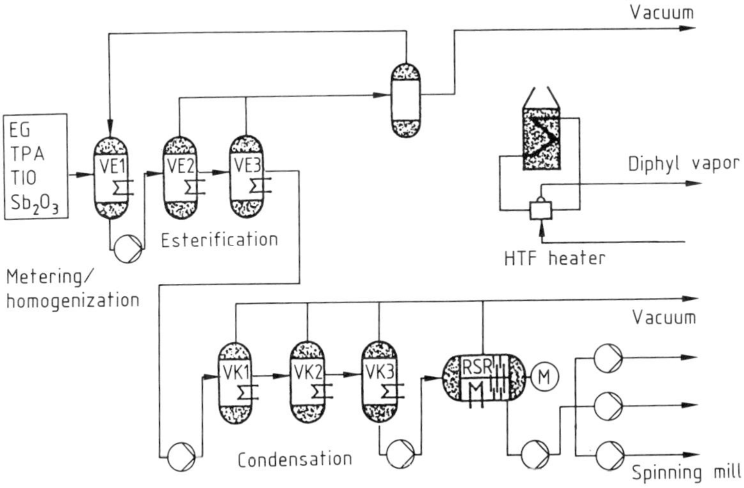

Figure 13.18. Process flowsheet
VE = esterification; VK = precondensation; RSR = ring-and-disk reactor; EG = ethylene glycol; TPA = terephthalic acid; HTF = heat-transfer fluid

cess monitoring and securing tasks at the higher level. In particular, these tasks include:

- Startup:
  - Cold startup of the chemical section of the plant (mean product temperature < T_{min})
  - Warm startup of the chemical section of the plant (mean product temperature > T_{min})
- Failure of plant sections:
  - Failure of heating system for heat-transfer fluid (HTF)
- Occurrence of critical process states:
  - Vacuum disturbance
  - Viscosity disturbance
  - Disturbance of mole ratio/COOH groups
  - Level and flow rate disturbances

The aim in reacting to disturbances is to insure high process reliability and product quality with the minimum reduction in plant capacity. Whenever possible, process data should be acquired from the process on-line, without placing extra demands on the operator. Interviews with experts have shown that all disturbance classes are interrelated in a complex manner (see Fig. 13.19). For this reason, in particular, they have not yet been completely brought under control.

Now let us examine the practical creation of a knowledge base more closely. The four-phase concept described above is applied to the disturbance class "HTF heater failure." The heat-transfer-fluid heating unit supplies heat to the strongly endothermic reactions in the VE and VK/RSR reactor groups (Fig. 13.18). A failure or loss of capacity in the HTF heater always results in severe disturbances (Fig. 13.19):

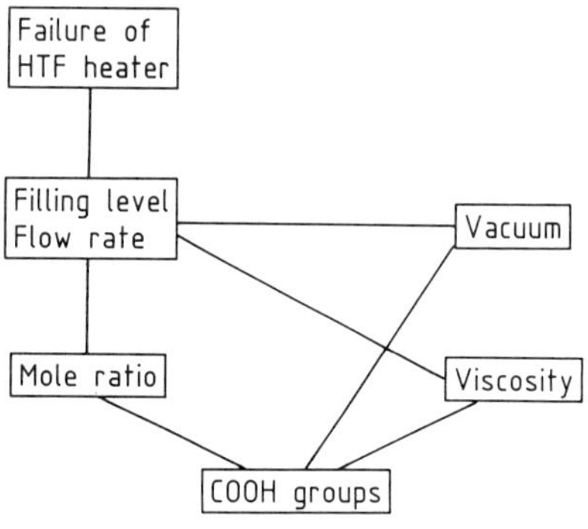

Figure 13.19. Disturbance classes and their interrelationships

- Level and flow rate disturbances: Because of the rapidly cooling product lines, continuous flow of product between the individual reactors is not insured after just a short time. The consequences are unstable levels and residence-time disturbances.
- Vacuum disturbances: Unstable filling levels result in foaming in the reactors, which can result in product being carried into the vacuum system.
- Mole ratio disturbances: Residence time disturbances lead to an excessively high or low conversion of the reaction medium. The consequence is a severe decrease in fiber quality and possible disturbances to the viscosity or COOH groups of the product.
- Viscosity disturbances: If the viscosity of the reaction process departs from the acceptable tolerance range. Both the processability and the quality of the product fall off drastically.
- COOH group disturbances: Because of excessive residence times, much of the feed ethylene glycol flashes off, leaving an excess of TPA and thus of COOH groups. As a result, major decreases in fiber quality occur.

For each of these disturbance classes, the proper responses to offset the disturbance (process securing) were ascertained by interviews with experts. The typical procedure for process securing (see above) was used as the basis for these interviews. Starting from the dynamic properties relevant in the case of an HTF heater failure, the technological system was broken down into subsystems S_1 (HTF heater) and S_2 (homogenizing, esterification, condensation, RSR reactor). The process securing system, in analogy to Figure 13.17, has the structure shown in Figure 13.20. This structure also describes the procedure used by the plant operator to compensate for the disturbance. The operator moves down a tree structure, level by level, until a solution to the process securing problem has been found.

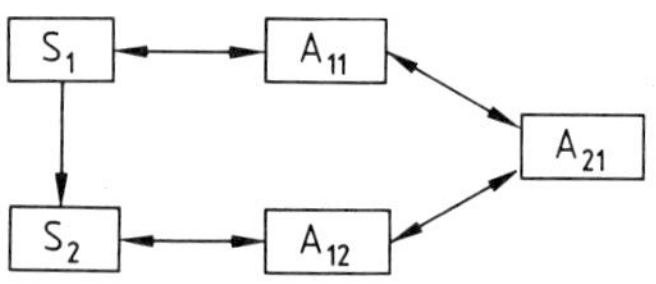

Figure 13.20. Structure of the process securing system upon a failure of the HTF heater

Figure 13.21 shows this procedure schematically. The respective decisions and actions are listed in accordance with the algorithms A_{ij} in Figure 13.9.

Figure 13.21 also demonstrates the structure of the technology-oriented knowledge base; reactions to possible downstream disturbances (level, vacuum, etc.) are merely indicated. Direct conversion to the expert system would have the consequence that many rules or knowledge items would have to be input more than once (e.g., 12 times for the cold and warm startup knowledge

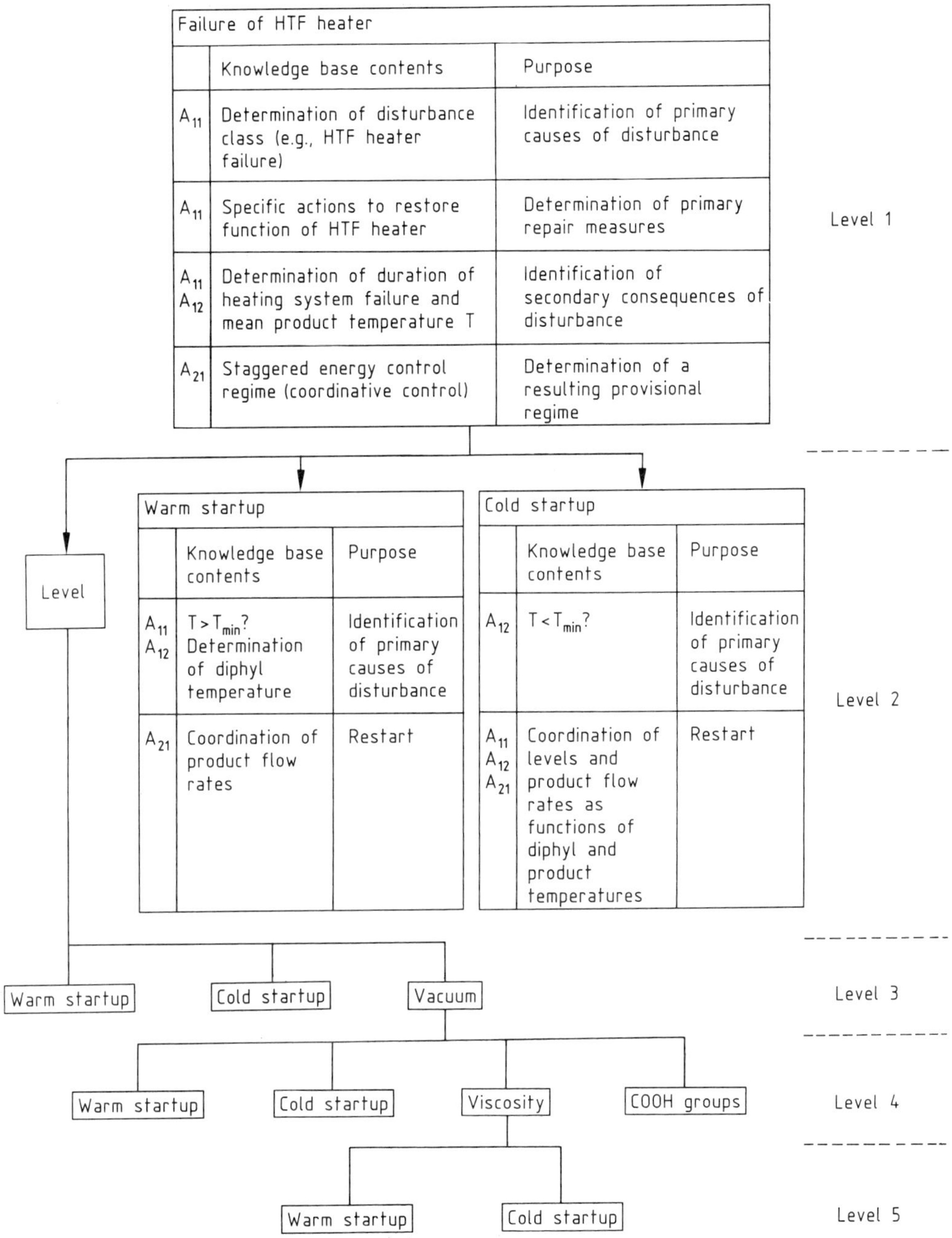

Figure 13.21. Technology-oriented knowledge base

sub-base). This expresses the fact that several disturbance classes can be correlated with a control class. Large memory problems would result. Furthermore, searching problem space with such a tree would narrow down the problem very slowly, so that the solution process would not be efficient enough (real-time problem). For this reason, the knowledge base was structured in a situation-oriented fashion, as shown in Figure 13.22. The drawback to this structuring is that conclusions about possible consequent disturbances must be drawn at the very first level. Thus warm startup, cold startup, and other situations are shown not as consequent situations of an HTF heater failure but as general and equivalent disturbance classes. Technology-oriented structuring makes it possible to include the complete knowledge. Situation-oriented structuring, in contrast, improves the efficiency of knowledge processing for process securing problems in which the focus is on coordination of more than one processing unit.

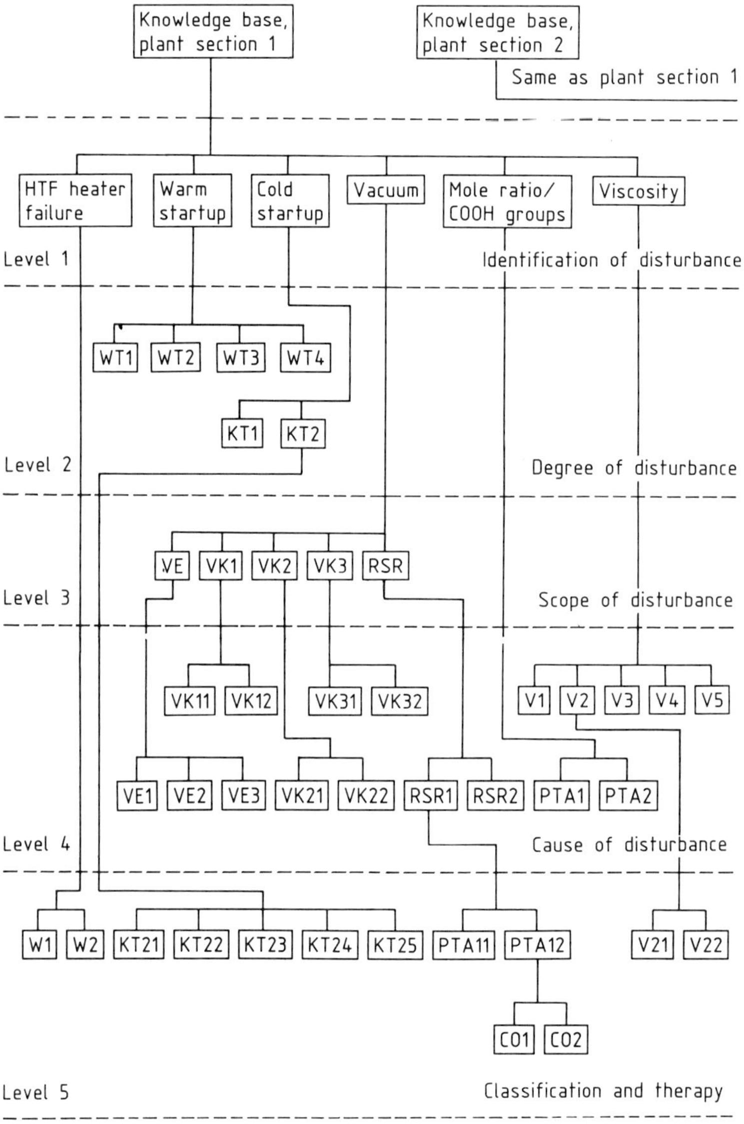

Figure 13.22. Structure of the knowledge base in the expert system

In general, the final structure of the knowledge base (see Fig. 13.23) is governed by the procedure learned from the plant operator, by automation theory, and also by the structuring capabilities of the expert system. The reasons for this structure are as follows:

- The decision tree is simplified (fewer levels)
- Knowledge items are represented only once
- Memory requirements are reduced
- Reliability is improved
- The system gains in maintainability

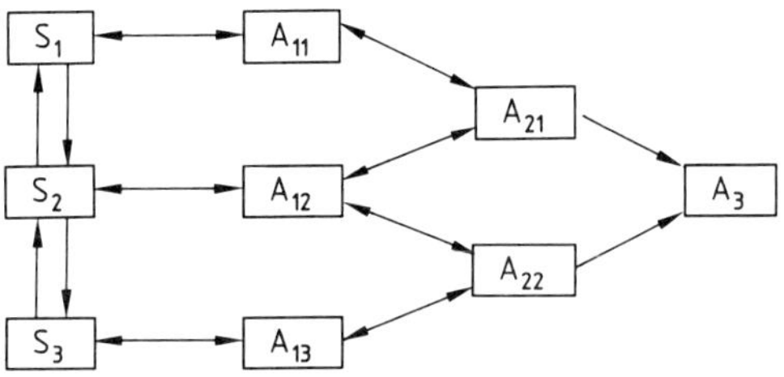

Figure 13.23. Structure of the process securing system upon a vacuum disturbance
S_1 = esterification; S_2 = condensation; S_3 = ring-and-disk reactor

As has been mentioned, each primary cause of disturbances leads to a different decomposition of the technological plant, which logically yields a different structure for the process securing system. For example, consider a vacuum disturbance, which can also be a secondary consequence of a disturbance. The coordinative control action to be generated (see Fig. 13.23) will depend on how widespread the disturbance is within the chemical section of the plant, which, dynamically and topologically, can be broken down into three subsystems: esterification (S_1), condensation (S_2), and ring-and-disk reactor (S_3). The method of knowledge acquisition is the same as described above for the case of an HTF heater failure.

The knowledge base for process securing (improvement of process reliability) in the chemical fiber plant includes, for one plant section, some 100 objects (problem classes) in six levels with about 220 process attributes, 40 interference variables as well as process constants, and 500 rules, of which 160 are diagnostic and 340 govern the application of expert knowledge. The expert system developed is coupled to the existing TDC 3000 process control system and is in permanent service. Typical response times of the

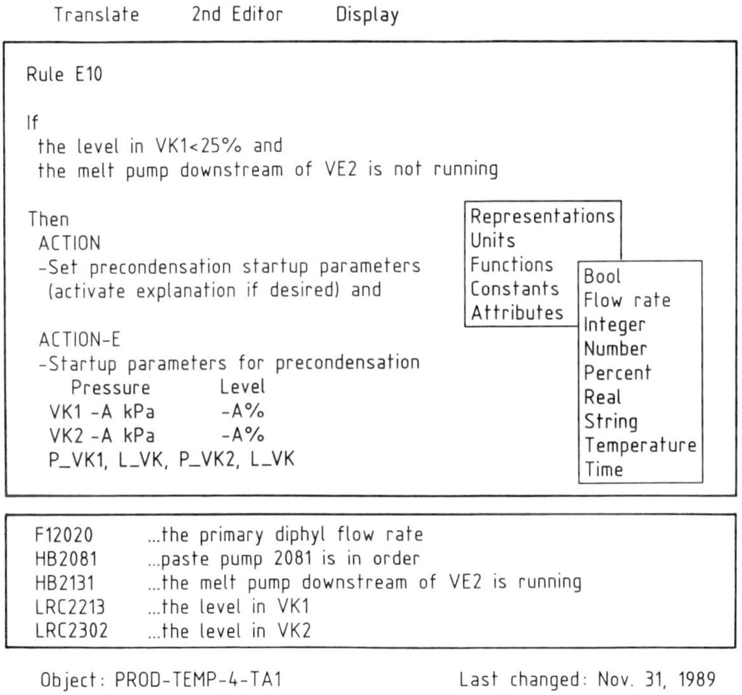

Figure 13.24. PROCON expert system shell, acquisition component, definition of a rule base for application of expert knowledge (expertise rule base)

system are in the range of 4–10 s; roughly 70–80 % of the time spent is for on-line acquisition of process data needed for expertise. The development extended over a period of about one year, during which time, 11 experts were interviewed. Figures 13.24–13.26 illustrate the consultation of the expert system by the plant operator.

```
Arguments   Rule text   State   Results   Secure   Quit

Argument values   Evaluation of expertise rules in PROD_TEMP_4_TA1
Argument texts

Condition 1:
 LRC2213          =20%

Condition 2:
 HB2131           =FALSE

RULE E10
 Because
  the level in VK1<25% and
  the melt pump downstream of VE2 is not running
 RECOMMENDED ACTION=Set precondensation startup parameters
                    (activate explanation if desired)

 RECOMMENDED ACTION-E=Startup parameters for precondensation:

                        Pressure   Level
                  VK1   100 kPa    50.0%
                  VK2   120 kPa    50.0%

Knowledge base: GRISUTEN                         Test consultation
```

Figure 13.25. PROCON expert system shell, consultation component, expertise rule trace with inspection

```
Solutions   Explanations   Representations   Securing   Quit
Solution 1 with rating 100                              23:59:03

   The following results were derived:

   RECOMMENDED ACTION:

   -As precaution, check grades being spun (line assignment)

   -Set TRC 2461, TRC 2501 and TRC 2811 to MAN

   -Wait for furnace startup test to end
   -Turn off position 113, lines 1, 2, and 3
   -Turn off viscometer (electrical)
   -The paste valve must be set to recirculate
   -Positions 107.1 and 107.2 must be turned off

   -Set precondensation startup parameters
    (activate explanation if desired)

Knowledge base: GRISUTEN                         Test consultation
```

Figure 13.26. PROCON expert system shell, consultation component, presentation of results

14. Appendix

14.1. Glossary

Prefatory Note. During the period of transition to a holistic mode of analysis in process control engineering, the terminologies of individual specialties are not adapted to one another, and contradictions may even occur.

The development of an internally consistent, generally accepted terminological structure for process control engineering is a critical and ambitious task. It is the concern of the standardization bodies with jurisdiction over this field.

The following list of terms cannot and is not intended to anticipate such undertakings. The purpose here is to create a unified platform of concepts to serve as a basis for the treatment here. The list defines only a sampling of terms, the ones most important for understanding, and makes no claim of completeness.

The following abbreviations are used:

DF Definition
EP Explanation
LI Literature (references)
EX Example
SN Synonym
CL Classification

Abstraction

DF Abstraction denotes reducing complexity by neglecting all inessential properties or attributes. Which properties and attributes are essential and which can be regarded as inessential will depend primarily on the viewpoint adopted and on the question or problem stated. Both viewpoint and formulation of problem will determine the result of any abstraction.

There are several forms of abstraction:

DF 1 Class-forming abstraction

Class-forming abstraction takes at set of unrelated objects and orders the set in accordance with common properties and attributes. All objects that possess a common set of attributes can be assigned to one class. All objects that belong to this class are referred to as "instances" of the class.

EP 1 Abstraction can take place in several stages.

Associated with any abstraction hierarchy is a taxonomy of concepts or terms denoting the objects in the individual classes.

DF 2 Complex-forming abstraction

Related *objects* can be combined into a *system* with its own characteristic properties. When unlike objects are joined into a system, the diversity of properties of the individual objects is mapped onto a few external properties of the system. This procedure is called complex-forming abstraction.

SN 2 Special case: Aggregation

DF 3 Functional abstraction

When aggregation is perform, typically not just the complexity of the object but also the level (plane) of examination changes. With every aggregation step, the examination becomes more "abstract" [14.1]. The levels may be characterized, for example, as physical, operative, tactical, strategic, or intensional. Such a hierarchical structure is also referred to as functional abstraction.

Aggregation

DF Combination of related *objects* into a *system* having characteristic properties. Aggregation is a case of complex-forming abstraction. The objects examined are *classes* or *instances*.

EX By means of assembly, a collection of initially noncoherent objects such as wheels, body, engine, transmission, etc. becomes a system, in this case an automobile. The system exhibits its own characteristic properties, which can only be derived indirectly from the interaction of the element properties. For example, an engine is characterized by the following properties: number of cylinders, way in which valves are operated, number of valves, weight, displacement, etc. The transmission is characterized, among others, by the gear ratios. If the total system "automobile" is examined, the interaction of the properties of the individual elements generates a huge variety of overall system properties. From a given standpoint, however, only a certain subset of these will be of interest. For example, if an automobile is regarded as an element of a transportation system, its behavior can be characterized by a few properties, such as maximum number of passengers, maximum speed, fuel economy, and range. The properties of the individual objects from which the automobile was assembled are no longer of interest.

EP Aggregation forms complex objects by stepwise combination of simple objects.

SN Complex-forming abstraction

Attribute

DF A characteristic or feature of an object (see *Property*).

SN Property

Automatic

DF Performance of a task by a self-contained technical system without outside aid.

LI DIN 19 233

Basic Function

DF 1 Function of process control. Establishment of an independently set process property in a plant section.

DF 2 Functional unit with the capabilities of establishing certain process properties in a plant section. The functionality that a plant section possesses for intervention in the process is disjointly distributed over the basic functions. At a higher level of abstraction, the functionality of a plant section for process control is obtained by superposition of the functionalities of its basic functions.

DF 3 Software/hardware module with the capability of realizing the functionality of a basic function (DF 2) (see also *Functional Unit*).
The term "basic function" is also used with another meaning in process control engineering:

DF 4 In the sense of DIN 19 243, Part 1, basic functions are the smallest building blocks for the acquisition, processing, and output of analog and binary quantities. Here, basic functions include, for example, acquisition via analog input, BCD encoding, and the formation of a reference value.

DF 5 Synonymous with *group control.*

DF 6 Synonymous with *monitoring and control function.*

EP Here, "basic function" is exclusively used in the definitions given in DF 1 and DF 2.

Basic Functional Element

DF 1 Function, element of a *basic function* (DF 1).

DF 2 Functional unit or *functional element* (DF 2) from which a *basic function* (DF 2) is constructed.
In this sense, a basic functional element can be a process intervention, a discrete control device, a group control, or any *functional element* or *functional unit* that is used to construct a *basic function*.

DF 3 Software/hardware module with the capability of implementing a single basic functional element (DF 2) (see note under *Functional Unit*).

Breakdown (Analysis)

DF Elucidation of certain properties of an *object* through a system of simpler objects interacting in an ordered fashion.
The breakdown process is subject to the nesting principle.

EP The system is to be viewed as a model such that, for the question or problem stated, it can explain the external behavior on the basis of the interaction of system elements.
Quite different breakdowns may be obtained, depending on the question formulated and the model selected.

EX An automobile might be broken down
- from the assembly standpoint, into wheels, engine, body, etc.
- from the disposal standpoint, into plastic parts, metal parts, textiles, etc.

Call

DF One-time activation of an algorithm (better: the message that initiates the activation).

EX An analog controller operates continuously and need not be called; a discrete controller, on the other hand, operates under a cyclically (periodically) called algorithm.

Class

DF Group of objects characterized by a set of common attributes.

Command Mode

DF The command mode is a characteristic property of technical units of a command hierarchy. It indicates whether the unit in question receives its control values and commands from the plant operator ("manual" command mode) or from a higher-level technical unit ("automatic" command mode).

Connector

DF Connectors are subobjects of functional modules. Through them, the functional module mediates its message traffic with the surroundings. Each connector thus incorporates certain client and/or server properties of the module in a task-oriented fashion. A connector has an interface to the communications system, its own method for handling the message traffic, and an internal interface with access to the information held and used in the module.

EP Module inputs and outputs, as defined, for example, in the languages of IEC 1131, are special connector types. They serve primarily to handle "get" connections, in which the recipient, through an input, reads information from the output of another module. The IEC languages provide no possibility of handling more complex communications functions (e.g., semaphore-protected write) at the module surface. Separate "communications modules" must be defined for this purpose. With the introduction of connectors, these functions can be incorporated in the semantics of the module (component) language.

LI [14.2]

Control

The term "control" is used with five distinct meanings:

DF 1 A sequence of action: "Control" or "feedforward control" denotes an open sequence of action over a single transmission element. This meaning is contrasted with "feedback control," which denotes a closed sequence of action.

LI 1 DIN 19226

DF 2 A binary operation: "Binary control" denotes binary processing of input signals yielding binary output signals.

LI 2 DIN 19237

EP Here analog processing is contrasted with binary control. There are two forms of binary control: *logic control* and *sequence control.*

DF 3 A functional unit with the capability of performing all functions occurring on a control level (plane, layer). In the control system, it receives tasks from a functional unit located higher in the hierarchy, executes these tasks on the basis of a specified method or recipe, and concludes them by specifying appropriate instructions to the next lower level.

CL *Group control, plant section control, plant control, production control*, etc.

DF 4 Hardware/software modules that realize the functional unit of DF 3 (see note under *Functional Unit*).

DF 5 In a hardware context, a programmable logic controller (PLC).

Control Value

DF A control value is a time-varying setpoint of a controlled *process property.* The expression "controlled process property" signifies that there is a concrete *process control function* that (in normal operation) guides the actual value of the process property so as to track the nominal value in a specified fashion.

EP The control task can be carried out in various ways, for example
- by closed-loop control
- with event-controlled feedback
- by open-loop control

Decentralized Control System

DF 1 Totality of all technical apparatus that makes possible the function "monitoring and control".

DF 2 In a hardware sense, the monitoring and control system is part of the process monitoring and control hardware.

Degree of Automation

DF The degree of automation is the fraction of the process control tasks that is performed automatically by the monitoring and control system.

LI [14.3]

Device Control Function

DF *Function* for the control and monitoring of devices included in the monitoring and control system but not acting directly on the process. In the case of a sensor, for example, the following functions are carried out: select measuring range, initiate calibration, select place of measurement, etc. In the case of an actuator, the device control functions might include the selection of certain actuator characteristics (e.g., converter), the initiation of a calibration (e.g., metering scales), or the selection of a redundant resource (e.g., dual pump).

EP The functional unit *discrete control device* possesses the functionality to take over the device control function.

Discrete (Control) Device; Discrete Device Technology

DF *Functional unit* with the capability of controlling and monitoring an actuator or sensor. This functional unit generally contains both functional elements for controlling and monitoring the device itself and functional elements for the processing of useful information, for example, to output control commands or to accept and prepare measured values.
In the actuator branch, the discrete device acts as a *group control.* Here, the task is to bring about certain process properties in accordance with the specified *control values.*

Element

DF *Object* that is in a well-defined relationship to other objects in the context of a *system* and that cannot or is not to be further decomposed in the context of the present analysis.

EP In contrast to an object in general, associated with an element per se are the relations (relationships, dependences) to the other elements in the system.
Certain functions are transferred to an element in the context of the system.

EX A screw in a 100-pack is an object. In a certain system, the screw that holds two parts A and B together is an element of the system. In the system, the *function* "connect A and B" is transferred to this element.

SN System element

Entity

DF 1 A certain state of being of an existing thing; also the thing itself.

DF 2 In connection with object orientation, another name for an object that exists in reality or conceptually and is of essential importance to the problem under consideration. Entities thus form an excerpt of the real world that is necessary for the handling of the problem under consideration. Abstract concepts such as "function" can also be entities.

LI [14.4]

Entity–Relationship Model

DF The entity–relationship (E/R) model describes the *entities* of an analysis space together with the relations existing between these entities.

EP E/R models are suitable for describing static relations between entities (consists of..., is a..., is derived from...).

EP E/R models in general offer a graphical syntax for the intuitive representation of entities and their relations.

LI [14.5]

Execution Unit (see also Functional Unit)

DF Execution units are functional units that are capable of realizing a certain role. They are based on resources that possess the required functionality; spatially and logically, they are placed in a suitable layer; and they are configured so that they exhibit the special functionality needed for the role.

EP See *Role*.

SN 1 Device
Here the term "device" should, however, be understood as including units realized logically and in software.

SN 2 Here, execution units are referred to as functional units.

Framework Concept

DF Document describing the structure within which long-term development is to be accomplished.

Function

DF 1 Clearly defined task within a larger totality.
Functions are a separate category of objects.

EP 1 Note: A function is the task proper, and must be strictly distinguished from the capability (functionality) of a unit to perform a task.
In the context of the structure of a system, functions are mapped onto units (*functional units*) that possess the functionalities required for the functions. This transfer makes the units into elements of the technical system, which now implement the functions.
(See also *Functionality, Functional Unit, Element, Substance*).

EP 2 The term "function" is also used informally in relation to concepts that do not have to do with functions (see, e.g., *Functional Unit, Functional Element, Basic Function*; also the commonly used term "functional module" frequently refers to a program module).
Note: Although these expressions are inappropriate with regard to clear categorization of objects, they are not changed here but are used in accordance with common practice.

DF 2 In the process control engineering context, the only other meaning of "function" that is of any interest is the mathematical function. A function is a variable quantity whose value depends in an unambiguous way on the value of certain other quantities. (It is a mapping between algebraically structured sets.)

Functional Element

An ambivalent term denoting both the elements of functions and the elements of functional units.

DF 1 Element of a *function*.

EX 1 Display analog value, transform signal, link values, form limiting value, verify status signal, control valve, block motor, control,...

DF 2 Functional unit with elementary functionality.

EX 2 Functional elements, defined as the smallest units of functionality under consideration, include display elements, signal-transforming elements, linking elements, and limit pickups; or, specified from a technological standpoint, process monitoring and control station element, basic functional element, ..., motor control, valve control, PID controller,...

EP Functional elements can be regarded as the bridge between the top-down analysis of functions and the bottom-up synthesis of functionality. A function must be broken down in such a way that it ultimately consists of functional elements (*Functions*, DF 1), each of which can be fully implemented by functional elements (*Functional Units*, DF 2).

Functionality

DF The capability of an object to perform certain tasks (*functions*).

EP Functionality denotes a fundamental capability of an object and in this sense is a resource. This resource can be utilized by employing the object as an element of a system. In the system context, certain tasks are assigned to the object (as an element of the system). In this way, a part of the functionality of the object is used.

EX A drive, as an object (*functional unit*), has the following capabilities: "can be at a stop," "can turn to the left," and "can turn to the right." The drive M 2 associated with roller R 4 of a roller conveyor has the tasks of turning to the right or of being at a stop. The capability of the object "drive M 2" expressed as "can turn to the left" is not utilized in this system environment.

CL Functionality denotes not an object but a set of properties of an object.

Functional Unit (see also *Execution Unit*)

DF An object with a certain required functionality; that is, an object having certain properties that are necessary for the implementation of the tasks to be performed.

CL Functional units are not functions but play the role of substances. Functional units have the character of resources and can be called on once and only once at any given time.

EX In the planning of a roller conveyor system (see the example under *Functionality*), the system is built up, say, from functional units such as drives, rollers, and so on. These are neither functions nor devices but postulated objects with well-defined functionality. In the planning of the roller conveyor system, these objects, viewed as available resources, are used as system elements; they are combined and linked in such a way that the roller conveyor system as a whole can fulfill the task allotted to it.

EP The definition of functional units is arbitrary. It would be possible to define any functional units with any functionality. It is, however, expedient to choose functional units in such a way that the set of properties (functionality) coincides with, or is a subset of, the property set of available substances. In the software realm, for example, this means that, in constructing a system, functional units may be used only if there are program modules corresponding to them.
Because of the correspondence between functional unit and object, synonymous expressions are used for the two even though this is not wholly correct (see *Function*).
Examples include the expressions "plant control," "discrete control device," "group control," and so forth.
An even more unfortunate situation arises when a synonymous term is used for the function, the functional unit, and the object. This happens with, for example, *basic functions* and *basic functional elements*. Metering is a function; there is a functional unit "metering device" for it; ultimately, there is a software module that implements the functionality of the metering device. In common usage, however, the expression "basic function" is applied to all three objects (see Section 4.5).

EP 1 Examples of the naming of functional units:
Classes of functional units include *production control, plant control, plant section control, group control, discrete device control, basic function.*
Types of functional units include metering device, motor, valve, level control, temperature measurement, reactor control, tower control, and so on.
Instances of functional units include (standard plant and apparatus coding) motor V 13 R 4 K 12 and level control V 13 R 4 L 6.

EP 2 Here, the expression "functional unit" is also used for a subordinate concept, defined under *Execution Unit*.

Group Control

DF 1 *Functional unit* with the capability of establishing certain process properties in accordance with specified control values. The group control uses internal methods to calculate setpoint values, which it passes on to group controls or discrete control devices at the next lower level.

DF 2 Software/hardware module with the capability of implementing the *functional unit* "group control" (DF 1) (see note under *Functional Unit*).

DF 3 Functional unit with permanently defined, generally applicable control functionality.

LI [14.6]

Heuristics

DF Doctrine of the methodical acquisition of new knowledge that leads to acceptable results even though not guaranteeing a definite solution of a problem (since there is no proof of convergence). Heuristic procedures are frequently used in the engineering sciences and in operations research.

Inspection

DF Determination and evaluation of the actual state of a plant or apparatus.

LI DIN 31 051 (1985)

Instance

DF 1 All objects that belong to a class are referred to as instances of the class.

EP 1 This definition is general. In a hierarchy of classes, each class can be regarded as an instance of the next higher class. The definition makes no restriction as to the form of class formation.

EX 1 The electric motor is an instance of the class of motors. The asynchronous motor is an instance of the class of electric motors.
A certrain model of motor (manufacturer X, model Z) is an instance of the class of asynchronous motors. The motor with serial number XXX is an instance of a model of motor.

DF 2 Objects representing the lowermost level of a hierarchy of classes. An instance in this meaning cannot be further instantiated.

EX 2 The motor with serial number XXX is an instance in this sense. Every feature, at every time, has a certain expression (serial number, power, construction, voltage, explosion protection, instantaneous rotation angle and rotation speed, ...).

EP 2 The instance is itself an *abstraction* of the concrete object and takes account only of the properties needed for classification and identification. Other properties of the real object do not become properties of the instance.

EP 2 Note: The instance of a *functional unit*, even though it has many of the same *properties* (*functionality*), is not identical with the instance of the corresponding physical object. They are correlated only temporarily, for the time of installation. If the physical object is removed (for example, taken into the workshop), the two objects may under some circumstances have different relationships.

Instruction Acceptor

DF The instruction acceptor is the connector through which a functional module receives a control instruction.

EP Each functional module can have more than one instruction acceptor.

The control instruction is a command specification with a write character. The instruction acceptor must accept the control instruction immediately after its arrival, independently of cyclic module processing. It checks the content of the control instruction from a formal and a semantic standpoint, sends a reply to the operating system; on acceptance of the command, it writes the new values to the module as effective values.

SN Operator control, control instruction.

Logic Control

DF Control using Boolean logical operations to assign certain output signal states to the input signal states.

LI DIN 19 237

Maintenance

DF 1 Measures taken to keep a plant in functioning condition.

DF 2 Measures to insure the nominal state of a plant.

LI DIN 31 051 (1985)

Model

DF Representation of the important relations between the properties of an object.

SN Model system.

Module

DF In general, a module is an object that forms a closed unit and can be manipulated as such.

EP The concept of "module" should not be further delimited. Subtle distinctions to other concepts, such as block, building block, unit, and so forth, based on the degree of complexity or other attributes, are not meaningful.

SN Apparatus module: device unit
Software module: functional module
Functional module: functional unit

Nesting Principle

DF Specification relating to the *aggregation* of *functional units. Instances* of *functional units* are resources that are available once. In the framework of an overall structure, they can be used only once and thus also aggregated only once.

EX 1 If gear Z 1 is installed in gearbox G 1, the gear is no longer available; the gearbox as a whole, however, is still available and can be installed in vehicle F 1. Once this has been done, again gearbox G 1 is no longer available; that is, it cannot simultaneously be installed in vehicle F 2.

EX 2 If a green box is packed inside a yellow box, the green box is no longer available and cannot also be packed inside a blue box. It is, however, possible to pack the yellow box (with the hidden green box inside it) in the blue box.

EX 3 In the case of physical objects, as in examples EX 1 and EX 2, the nesting principle is trivial. The important point is that it also applies to instances of functional units considered as "resources of functionality." If a functional unit A is an element of another functional unit B, functional unit A is no longer available for other functional units (in B, a certain task has been allotted to it that makes claims on its resources).

EP The nesting principle applies only to *instances* of *functional units*. It does not apply to the *aggregation* of *classes* or to the *aggregation* of *instances* of *functions*. A *functional element* can be simultaneously an *element* of more than one *function*.

Object

DF Subject of knowledge, perception, thinking, and action. In the context of object orientation: identifiable patterns that are describable in terms of properties. The internal structure of objects is not directly perceivable but may be deduced from the properties describing the object.

EP Both functions and substances can be treated as objects in this sense [14.7]. Object is not an antonym of function.

Operating Mode

This term is used with several meanings.

DF 1 Degree of automation of a plant.
Here, the operating mode is the plane (level) of operator control currently selected.

EX 1 A coarse classification includes the following modes: manual, computer, logistical, and automatic [14.3].
- Manual mode: the monitoring and control system takes over only certain process control tasks for the automatic execution of individual loops.
- Computer mode: the monitoring and control system takes over all process control tasks for the automatic performance of a process section in a plant section.
- Logistical mode: the monitoring and control system takes over all process

control tasks for the automatic fulfillment of scheduled orders within a plant.
- Automatic mode: the monitoring and control system takes over all process control tasks including fine production control for the automatic scheduling and fulfillment of orders within a plant.

DF 2 One of the fundamentally different behavior patterns of functional units.

Alternatively, certain technological function units can carry out distinct (selectable) tasks.

EX 2 The functional unit "inert-gas blanketing" has the modes "pressurizing" and "purging." In the first, the task is to establish the pressure in accordance with the specified control value; in the second, to establish the flow of nitrogen through the system.

DF 3 Way in which a plant is run in dependence on the control of the operating value or throughput-to-capacity ratio (startup, normal running, load changing, emergency shutdown, etc.).

LI 3 [14.6]

Operating Value
(Throughput-to-Capacity Ratio)

DF The ratio of product throughput through a plant to design throughput.

LI [14.6]

Parameter

DF 1 Parameter (of an equation). In functions and equations, an auxiliary quantity appearing alongside the variables proper and either left undefined or held constant.

DF 2 Parameter (of a path or trajectory). Special case of a parameter of an equation.

DF 3 Parameter (of a process). Variable that influences a technical, economic, or other process.

EP In the process control context, this term is often used in an imprecise and poorly understood way, and it should be largely avoided.

Partial Recipe

DF Complete description for making a certain end product or intermediate product in a particular type of plant section.
A partial recipe consists of:
- A specification as to the properties of the inlet products
- A specification as to the state of the plant section at the beginning of the process
- A manufacturing specification for the control of the process.

Phase

DF Phases subdivide *process elements* into a strict temporal order.

EP Each process element, at each point in time, is in precisely one phase. There is no parallelism of phases in a plant section.

EP Remark: In many cases, the phases are named after the unit operations that predominate in them. This often leads to misunderstanding. In a phase referred to as "meter starting substances," the unit operations "meter A," "meter B," "temper," and "mix" (agitate) can take place in parallel. The sequence of phases, however, is strictly serial.

EP The subdivisions of the *phase model* at the lowermost level of refinement correspond to the phases.

Phase Model

DF The phase model is a special network model with which a procedure can be systematically described. The phase model consists of two elements:
- Process elements in which n inlet products are transformed to m end products by means of operations.
- Product nodes with well-defined product properties, which connect the process elements to one another.

EP The phase model offers two substantial advantages:
- Logical network structure
The network structure permits systematic refinement (top-down method) from a coarse structure to a detailed description.

The specification of nodes in terms of products makes it possible to describe and refine the individual process elements independently of one another.
The phase model is not concerned with the form of the temporal and engineering realization of the coupling.

- The possibility of establishing conditions that must be observed, by specifying profiles of process and product properties in parallel with the design process.

The specification of profiles for the process properties to be brought about in the process elements and the product properties required in the nodes can be done in advance and refined with the network structure.
The phase model is the basis for the construction of quality assurance systems.

LI [14.3]

Plant

DF Combination of plant sections into a delimited technological unit.

EP The context in which the unit is delimited is not fixed. The important point is that each plant section can be assigned to only one plant.

EP An appropriate delimitation can be found if, for example, the plant is viewed as an analytical space for a fine production control system.

LI DIN 28004

Plant Complex

DF Organizational combination of plants and other units into an overall heterogeneous unit.

Plant Control

DF Functional unit with the capability of initiating and coordinating the processing elements in the plant sections of a plant on the basis of a specified operating plan.

Plant Section

DF Self-contained unit for the realization of a technical task (see Fig. 2.25).

EP Among the points characterizing a plant section is that the product treated in it is to be regarded as substantially located in a balance space. This balance space may well be distributed (tower, fixed-bed reactor, etc.) but still forms one unit from the standpoint of process engineering.
Two stirred-tank reactors should be viewed as two separate plant sections because, in general, each is characterized by its own balance space.

LI DIN 28004

Plant Section Control

DF 1 *Functional unit* with the capability of controlling the *group controls* or *basic functions* of a *plant section* and of automatically controlling a process section in accordance with a specified recipe.

DF 2 Hardware/software module for the realization of the "functional unit plant section control" (see note under *Function*).

Process (General)

DF A totality of interacting trains of events in a system by which matter, energy, or information is transformed, transported, or stored. The smallest unit for analysis is the process element.

EP Note: The term is used to refer to the actual train of events taking place in the real world, not to the system characterized by this process or the specification underlying the process.

EP In a processing system, "process" denotes the dynamic variation of product properties as a result of the action of unit operations or the variation of process properties due to the action of basic functions.

LI DIN 66201

Process Control

DF Process control takes in all measures that effect the desired course of a process in the sense of established objectives. These measures either can be executed automatically by the process monitoring and control hardware itself, or may have to be performed by human operators.

LI DIN 19222

Process Control Function

DF *Process monitoring and control function* with the task of influencing the course of a process in such a way that certain *process properties* will track specified *control values*. A process control function is realized by hierarchically structured process control units in the feedforward control branch and a network of process information processing units in the feedback branches (see also *Control Value*).

Process Control Station (Process Monitoring and Control Station; formerly Measurement and Control Station)

The concept of the process control station is a key one in process control engineering. Even so, it is difficult to define and delimit. In common usage, the term refers to a variety of concepts:

DF 1 A point of connection of the process. In this sense, a process control station is the interface between the process monitoring and control functional unit and an engineering functional unit (used so, at least in part, in the station data sheet or in the station detail of the piping and instrumentation diagram).

DF 2 A functional unit of process control. This functional unit is characterized by the fact that it possesses at least one process connection point. The task of the functional unit "process control station" is primarily to execute the assigned processing functions (O, I, C, S, ...). Two remarks should be made, however:

- Only in a few cases is a processing function realized, as a whole, by a functional unit. In many cases (e.g., the S function, central A, I, O evaluating functional units), multiple functional units (and multiple process control stations, in the case of the S function) are involved in realizing a processing function.
- The differentiation between the functional unit "process control station" and more complex interconnections is indistinct (at what complexity does a functional unit with process connection points cease to be a process control station?).

DF 3 A combination of functions. In this sense, the process control station is given by the sum of the defined processing functions assigned to the process connection points (used so, at least in part, in the station detail of the piping and instrumentation diagram).

LI DIN 19226

Process Control Station Element

DF Process control station elements are the smallest functional units into which a process control station (DF 2) is broken down. Process control station elements have functionality established such that they can be realized by available substances (devices, software modules, etc.). Each process control station element is needed for the performance of one or more processing functions.

Process Element

DF Smallest unit used in analysis of a process.

EP A process element can be realized by a process step or a process section, depending on whether the analysis is time-based or space-based.

Processing Function

DF Term denoting process monitoring and control functions allotted to a process control station. The processing functions are represented in the station detail.

EP Note that a processing function is generally realized through more than one functional unit and, in certain cases, through more than one process control station, despite the unambiguous allocation (see note under *Process Control Station*).

Process Intervention

DF Monitoring and control functional unit that combines all functional elements (process control station elements) of control hardware between discrete control device and process.

Process Monitoring and Control Function

DF *Function* realized with the aid of the monitoring and control hardware. In general, monitoring and control functions are closed lines of action between process elements, between operator and process, between operator input and operator output, or in general the response of the outputs of the process control system to certain input signals.

CL Monitoring and control functions in process control engineering can be specified by control action in the following ways (among others):
Process control functions
- After operation DOS_3_ON, meter in x liters in t minutes
- After operation T5_ON, turn on plant section 5

Process securing (stabilization) functions
- If temperature T 15 > 80 °C, turn off pump 3

Process information functions
- Display temperature T 15 in on-screen flowsheet

Device control functions
- Calibrate pH probe Q 14

System control functions
- Select screen page plant section ST 3

Monitoring and control functions also occur in other settings such as manufacturing engineering or building management engineering.

Process Monitoring and Control Hardware

DF All devices employed for the task of process monitoring and control.

LI DIN 19 222

SN Process control system

Process Property

DF Process properties are the categories of information about a process. They include state variables, process parameters, control variables, setpoint parameters, and process indicators.

LI [14.8]

Process State

DF Expression of the process property vector at a given time (see Section 11.3).

EP See also *Quality*.

Production Control

DF 1 Functional unit with the task of scheduling production requirements, preparing a production schedule, allocating the needed resources, and monitoring the performance of the schedule.

DF 2 Software/hardware module with the capabilities of realizing precisely the function "production control" (DF 1) (see note under *Function*; see also Fig. 2.30).

Product Properties

DF Product properties are the categories of information about a product. They include physical quantities, chemical quantities, technological properties, and product indicators.

LI [14.8]

Project

DF Expressed generally in the terminology used here, a project is the complete temporal framework of conditions governing all *substances* for the realization of tasks represented by *functions*. The substances consist of procedures, systems, and elements (see Figs. 2.9 and 3.26) and are also referred to as resources.

Property

DF A characteristic or feature of an object. A property consists of the type of feature (category, size) and the expression of the feature (intensity, value). The type of expression (metrical, topological, etc.) depends on the type of feature.

EX A house might have the properties

Type of feature:	Expression:
Address	Hauptstraße 3 D-11111 Kleinhausen
Owner	Mrs. Meier
Height	8.25 m
Building style	Jugendstil (Art Nouveau)

EP Classes are marked by common feature types. The expression is an instance quantity.

SN *Attribute*

Quality

DF Expression of the product property vector at a given time (see Section 11.3).

EP See also *Process State*.

Recipe

DF A recipe consists of the sum of the partial recipes needed to make an end product and the explicitly stated relationships between the partial recipes.

EP The term "recipe" has been introduced but is still subject to various interpretations. "Production instructions" or "production specification" would be a more precise term and should be given preference in the future.

Redundancy (General)

DF 1 Concept in information theory, denoting that a message contains elements that can be omitted as not providing any additional information but merely supporting the desired basic information.
- Beneficial redundancy: constituents that can be omitted would back up the information if other constituents were omitted.
- Empty redundancy: constituents that can be omitted would not back up the information content if other constituents were omitted.

DF 2 Severalfold presence of a functionality in order to enhance the availability of a function.

Redundant Measurement

DF The measurement of a certain process property is said to be redundant when the actual value of this process property can be acquired from at least two disjoint groups of available measurements.

EP If a measurement becomes unavailable, the disjointness means that only one group is affected, and the actual value continues to be available to the system. This is an example of beneficial redundancy.

- Homogeneous redundancy: the actual value of the desired process property is determined by using the same measuring principle and by multiple measurements of the same auxiliary process properties.
- Diversity: the term "diversity" (diverse redundancy) is used when:
  - The actual value of the desired process property is determined from measurements of different auxiliary process properties (example: determination of flow rate via pressure drop and with magnetic flowmeter).
  - The actual (instantaneous) value of the desired process property is determined on the basis of the same auxiliary process properties, but the auxiliary process properties are measured by different methods (example: level measurement with float and ultrasonic instruments).

Repair(s)

DF The restoration of the nominal condition of a plant.

LI DIN 31051 (1985)

Requirements (see also *Specifications*)

DF Document describing the tasks (*functions*) that are to be implemented in the framework of a project. The initial description is solution-invariant.

SN Bidding specifications

Role

DF Roles are unambiguously described tasks within a larger context. The realization and performance of a role requires a functional unit (execution unit), which realizes the role in a certain layer at a certain time.

EX The relationships between roles, resources, functional (execution) units, and layer can be illustrated with three examples:
- Theatrical performance
  Consider the role of "Gessler" in the play "William Tell" (virtual system).
  The realization of this role demands an actor. Any actor who knows the text and is capable of playing Gessler is a

potential candidate and thus belongs to the pool of resources. Now suppose the play is to be performed on October 16, 1995, at venue X. The role system thus requires that an actor suitable for the representation of Gessler be available at X on October 16, 1995. For the theater manager, this means that an actor Y must be sought and signed for the performance; that is, a resource must be selected and allocated. This can take place ahead of time when the theater season is being planned. For the play to be actually performed, it is important that Mr. Y be available and prepared for the role at X on October 16, 1995. He is configured for his role through makeup and appropriate costuming. Mr. Y is then an available function (execution) unit for the role of "Gessler" at X on October 16, 1995.

- Hardware component
  Consider the following role: discrete control component for reactor X. It belongs to the process monitoring and control system being planned (virtual system). Each component of type Y made by manufacturer A possesses the required functionality and is thus a potential resource. As a functional (execution) unit, it is present (available) if, as a hardware part, it has been delivered, configured (e.g., jumpers placed), and installed in the proper slot. (This job is performed by installation.) In this state, the component can take over the role envisioned for it in the process monitoring and control system.
- Motor control module
  Consider the role of the control of motor V 12 TA 45 K 2. The process control being designed is the virtual system. For the realization of the role, a software module must be placed in the process monitoring and control system and made available as a functional (execution) unit. Any instance of the motor control type MOT becomes a candidate. Thus, an instance MOT_V 12 TA 45 K 2 is defined, configured (here this may mean parameter assignment: number of feedback messages, startup time, etc.), and installed in the component provided (which here corresponds to the layer). Now this software module is capable of realizing the role of the motor control V 12 TA 45 K 2.

SN Function

Securing (Stabilization) Functions

DF *Functions* for preventing unacceptable deviations of the *process*, of the *process control*, or of the *process monitoring and control hardware* from the nominal state. Securing functions do not intervene in the sequence of events under normal operating conditions and permit the *process control functions* to perform their task. Before a deviation enters the unacceptable range, the securing functions are activated, interrupting the access of the process control functions.

EX Typical securing functions are interlocking of functional units and STOP conditions in sequence controls.

EP Other expressions of this concept refer to improving or enhancing the reliability or safety of the process.

Sequence Control

DF *Control* in which the program is constrained to progress step by step and the advance from one step to the next step in the program depends on advancing conditions.

LI DIN 19237
Depending on the type of conditions for advance, the sequence control is said to be time-dependent or process-dependent. One also speaks simply of timed control or event control.

Specifications (see also *Requirements*)

DF 1 Document that contains all information required for the realization of a certain system.

EP 1 The specifications describe how the functionality of a system as set forth in the requirements can be realized with the help of given resources. "Given resources" mean functional units to be used, objects (hardware/software) to be used, rules and procedures to be applied, and constraints to be observed (e.g., limitations on time, money, material, space, and personnel resources). The specifications are linked to the solution adopted.

DF 2 Separate document describing the constraints, rules, and data that must be followed in the procedure.

Substances

In philosophy, the concept "substance" is employed with a variety of meanings. Here it is used in the sense of CASSIRER [14.7]:

DF Substances are objects in the real world that can be described and identified in terms of attributes.
Substances either are real objects themselves or correspond virtually to real objects.
Functions always require appropriate substances for their realization.

EP Substances include systems and their elements (devices), as well as skills and methods (procedures). With the aid of procedures, functions are realized from systems and their elements through the use of a part of their properties.
Of the great number of possible substances (see Fig. 2.8), "standards" (e.g., design, installation, and hardware standards) are derived by means of appropriate definitions. Substances, as real objects, can be either hardware objects or software objects.

Subsystem

DF Part of a *system*, having the following properties:
- Elements of a subsystem are always elements of the surrounding system as well.
- The relationship between elements in a subsystem always corresponds to the same relationship between the same elements in the surrounding system.
- Distinct subsystems of a system can overlap.

EP A special case is the complete subdivision of a system into nonoverlapping subsystems. This case corresponds to a *break down* in accordance with the *nesting principle*.

LI [14.9]

System

DF A delimited, ordered totality of interacting subsystems or system elements. This configuration is separated, or regarded as being separated, from its surroundings by an envelope.

EP One characteristic of a system is that it simultaneously admits of an external and an internal view.
In the external view, the system is examined from outside its envelope, which is thought of as opaque. The system behaves as an aggregated object (black box).
In the internal view, the system is examined from inside its envelope, which is thought of as opaque. From this standpoint, the structure and mode of operation of the system can be seen.
The properties of a system cannot generally be elucidated by superposing or aggregating the properties of its elements.

EX A gearbox is a system of gears, locking mechanisms, bearings, and shafts. From the outside, in the technological view, all that is interesting about a gearbox is the transmission ratio between the driven shaft and the driving shaft. The internal geometry and the number of gears are not important. From the inside, on the other hand, it is precisely the geometrical structure of shafts and gears that is of interest. The only way to determine the transmission ratio effective at the envelope is through a complicated computation based on the individual gears of the active branch.

EP It is important that a system admits of both views at the same time. It can be said to represent an object at the moment of aggregation or disaggregation.

LI DIN 66201, DIN 19226, [14.3]

Taxonomy

DF A methodology used in science, consisting in the assignment of concepts or individual phenomena to a general ordering scheme.

Type

DF Class whose objects exhibit a high degree of interrelatedness.

Unit Operations

DF The smallest physical or chemical operation performed, which acts on a material system in a plant section.

EP By means of physical or chemical operations, the material system in a plant section is transformed from a state A to a state B. These operations can be broken down into unit operations. As a rule, unit operations are carried out one after another, but there are unit operations that take place in parallel. Examples: conveying and drying in the pneumatic conveying dryer, reaction and mixing in the stirred-tank reactor, etc.

CL Unit operations can be classified under conveying, mechanical processes, force-field processes, processes with molecular driving force, and chemical processes.

LI [14.6]

14.2. Abbreviations

AI	Analog Input, Artificial Intelligence
AMA	Arbeitsgemeinschaft Meßwertaufnehmer
ANSI	American National Standards Institute
AO	Analog Output
AR	Archival Functions
ASN.1	Abstract Syntax Notation One
ASP	Abstract Service Primitive
C^{++}	Programming language (object oriented)
C Methods	Computer-supported methods
CAD	Computer Aided Drafting, Drawing, Design
CAE	Computer Aided Engineering
CASE	Computer Aided Software Engineering
CCIT	Comité Consultatif Internationale Télégraphique et Téléphonique
CEN	Comité Européen de Normalisation
CENELEC	Comité Européen de Normalisation Electrique
CEPT	Comité Européen des Administrations des Postes et des Télécommunications
CIM	Computer Integrated Manufacturing
CO	Control Functions
CIP	Computer Integrated Processing; Cleaning in Process; Cleaning in Place
CPU	Central Processing Unit
CSMA/CA	Carrier Sense, Multiple Access, Collision Avoidance
CSMA/CD	Carrier Sense, Multiple Access, Collision Detection
DBMS	Data Base Management System
DCS	Distributed Control System
DEC	Digital Equipment Corporation
DES	Discrete Event System
DI	Digital Input
DIF	Data Interchange Format
DGK	Deutsche Gesellschaft für Kybernetik
DGQ	Deutsche Gesellschaft für Qualität
DKE	Deutsche Elektrotechnische Kommission
DLL	Data Link Layer
DO	Digital Output
DOS	Disk Operating System
DPG	Deutsche Physikalische Gesellschaft
DS	Directory Service
DVT	Deutscher Verbund Technisch-wissenschaftlicher Vereine
EAPROM	Electrically Alterable Programmable Read Only Memory
ECITC	European Committee for Information Technology Testing and Certification
EEPROM	Electrically Erasable Programmable Read Only Memory
EIA	Electronics Industries Association
ELCS	Extended List of Control Software
ElexV	Verordnung über elektrische Anlagen in explosionsgefährdeten Räumen
EPROM	Erasable Programmable Read Only Memory
EPS	European Physical Society
E/R	Entity–Relationship

ESPRIT	European Strategic Program for Research in Information Technology
ETSI	European Telecommunications Standards Institute
EUREKA	European Research Coordination Agency
EXERA	Association des Exploitants d'Equipement de Mesure, de Régulation et d'Automatisme
FIP	Factory Instrumentation Protocol
FLAVOR	Programming language
GA	Gateway Functions
GI	Gesellschaft für Informatik
GMA	VDI/VDE-Gesellschaft für Meß- und Automatisierungstechnik
GMD	Gesellschaft für Mathematik und Datenverarbeitung
GMR	VDI/VDE-Gesellschaft für Meß- und Regelungstechnik, (now: GMA)
GVC	VDI-Gesellschaft Verfahrenstechnik und Chemieingenieurwesen
HART	Highway Adressable Remote Transducer
HD	Harmonization Documents
HDLC	High Level Data Link Control
HHT	Hand Held Terminal
HPC	Human–Process Communication
IEC	International Electrotechnical Commission
IECEE	IEC Sub-Committee: Conformity Testing to Standard for Safety of Electrical Equipment
IEEE	Institute of Electrical and Electronics Engineers
IFAC	International Federation of Automatic Control
IFC	International Fieldbus Consortium
IFIP	International Federation for Information Processing
IFORS	International Federation of Operational Research Societies
IMACS	International Association for Mathematics and Computers in Simulation
IMEKO	Internationale Meßtechnische Konföderation
IMC	Internal Model Control
IMS	Information Management System
IP	Input Product
ISA	Instrument Society of America
ISDN	Integrated Services Digital Network
ISO	International Organization for Standardization
ISP	International Systems Project
LAN	Local Area Network
LISP	List Processing Language
LLC	Logical Link Control
LTR	Loop Transfer Recovery
MAC	Medium Access Control
MAP	Manufacturing Automation Protocol
MIS	Management Information System
MMC	Man–Machine Communication
MMS	Manufacturing Message Specification
MODEM	Modulator and Demodulator
MPC	Man–Process Communication
MVS	Multi-Virtual System
NAMUR	Normenarbeitsgesellschaft für Meß- und Regeltechnik in der chemischen Industrie
NIR	Near Infrared
OOA	Object Oriented Analysis
OOD	Object Oriented Design
OP	Output Product
OS	Operating System or Open System
OSF	Open Software Foundation
OSI	Open Systems Interconnection
P	Proportional (controller)
PBX	Private Branch Exchange
PC	Personal Computer
PDS	Plant Design System
PDU	Protocol Data Unit
PI	Proportional Integral (controller)
PID	Proportional Integral Differential (controller)
PL/1	Programming Language Number 1
PLC	Programmable Logical Controller
RFN	Reversed Frame Normalizing Process
PROFIBUS	Process Field Bus
PROLOG	Programming in Logic (programming language)
PROM	Programmable Read Only Memory

PTB	Physikalisch Technische Bundesanstalt
RAM	Random Access Memory
ROM	Read Only Memory
ROI	Return on Investment
SA	Structured Analysis
SADT	Structured Analysis and Design Technique
SAP	Service Access Point
SAS	Statistical Analysis System, Statistikprogramm der Firma SAS Institute
SIREP	International Instrument User's Association
SNA	Systems Network Architecture (IBM)
SPARC	Standards Planning and Requirements Committee (RISC Architecture)
ST	Structured Text
STEP 5	Programming language
SQL	Standard Query Language
TC	Technical Committee (IEC)
TIRC	Temperature Indicating Recording Controlling (DIN 19227)
UNIX	Multitasking and Multiuser operating system
UTE	Union Technique de l'Electricité
UVV	Unfallverhütungsvorschrift
VbF	Verordnung über brennbare Flüssigkeiten
VCI	Verband der Chemischen Industrie
VDI	Verein Deutscher Ingenieure
VDE	Verband Deutscher Elektrotechniker
VIK	Vereinigung Industrielle Kraftwirtschaft
VLSI	Very Large Scale Integration
VMD	Virtual Manufactoring Device
VMS	Operating system from DEC
VOB	Verdingungsordnung für Bauleistungen
WAN	Wide Area Network
XP	Expert System
ZVEI	Zentralverband Elektrotechnik- und Elektronikindustrie

15. References

General References

Automation and Process Control Engineering

D. Balzer: *Wissensbasierte Systeme in der Automatisierungstechnik*, Carl Hanser Verlag, München 1992.

K. Bender (ed.): *Profibus – Der Feldbus für die Automation*, Carl Hanser Verlag, München 1990.

G. Brack: *Automatisierungstechnik für Anwender*, VEB Deutscher Verlag für Grundstoffindustrie, Leipzig–Stuttgart 1993.

G. Brack, A. Helms: *Automatisierungstechnik*, VEB Deutscher Verlag für Grundstoffindustrie, Leipzig 1985.

G. Färber: *Prozeßrechentechnik: Allgemeines, Hard- und Software, Planungshinweise*, 2nd ed., Springer Verlag, Berlin 1992.

E. L. Grant, W. G. Ireson, R. S. Leavenworth: *Principles of Engineering Economy*, John Wiley and Sons, New York 1982.

J. Hengstenberg, B. Sturm, O. Winkler, K. H. Schmitt (eds.): *Messen, Steuern und Regeln in der Chemischen Technik*, vols. 1–5, Springer Verlag, Berlin 1980.

K. Hiroi: *Basis and Application of Digital Control Systems*, Japan.

ISA: Human Engineering for Control Centers, ISA Recommended Practice RP 60.3, ISA 1985.

R. Lauber: *Prozeßautomatisierung*, 2nd ed., vol. 1, Springer Verlag, Berlin 1989.

R. Magin, W. Wüchner: *Digitale Prozeßleittechnik*, Verlag Vogel, Würzburg 1987.

P. Metzing, D. Balzer, H. Reinhardt: *Prozeßsteuerung*, VEB Deutscher Verlag für Grundstoffindustrie, Leipzig 1988.

I. Morishita: *Digital Control Systems*, Society of Instrumentation and Process Control, Japan.

W. Motsch: *Prozeßrechner – Strukturen*, Verlag Vieweg, Braunschweig 1993.

NAMUR (Hrsg.): NAMUR-Statusberichte 1985, 1987, 1990, 1993, R. Oldenbourg Verlag, München.

G. Platt: *Process Control, a Primer for the Nonspecialist and the Newcomer*, ISA 1988.

M. Polke: Prozeßleittechnik, 2nd ed., Oldenbourg, München 1984.

R. Rück, A. Stockert, F. O. Vogel: *CIM und LOGISTIK im Unternehmen*, Carl Hanser Verlag, München 1992.

E. Schrüfer (ed.): *VDI-Lexikon Meß- und Automatisierungstechnik*, VDI-Verlag, Düsseldorf 1991.

G. G. Seip (ed.): *Elektrische Installationstechnik*, vols. 1–3, Siemens AG, Berlin 1971.

A. Shoten: *PID Control*, Systems Control and Information Society, Japan.

G. Strohrmann: *Automatisierungstechnik*, vols. I and II, R. Oldenbourg Verlag, München 1990.

M. Syrbe, M. Thoma (eds.): *Fachberichte Messen-Steuern-Regeln*, vols. 1–10, Springer Verlag, Berlin.

H. Töpfer, P. Besch: *Grundlagen der Automatisierungstechnik*, VEB Deutscher Verlag Technik, Berlin 1990.

H. Töpfer, W. Kriesel: *Funktionseinheiten der Automatisierungstechnik*, VEB Deutscher Verlag Technik, Berlin 1988.

Kompaktseminar Prozeßleittechnik: Themen in Einzeldarstellungen, R. Oldenbourg Verlag, München 1986ff.

Systems Engineering

M. Bruns: *Systemtechnik*, Springer Verlag, Heidelberg 1991.

W. F. Daenzer (ed.): *Systems Engineering*, 6nd ed., Verlag Industrielle Organisation, Zürich 1988.

J. Raasch: *Systementwicklung mit strukturierten Methoden*, Carl Hanser Verlag, München 1992.

R. Unbehauen: *Systemtheorie: Grundlagen für Ingenieure*, R. Oldenbourg Verlag, München 1993.

Computer Science

D. Abel: *Petri-Netze für Ingenieure*, Springer Verlag, Berlin 1990.

L. Christov: *Objektorientierte Programmiersprache C+++: Vollständige Einführung anhand von Beispielen*, VDI-Verlag, Düsseldorf 1992.

J. Fiedler, K. F. Rix, H. Zöller: *Objekt-orientierte Programmierung in der Automatisierung*, VDI-Verlag, Düsseldorf 1991.

G. Glöe (ed.): *Software-Zuverlässigkeit*, VDI-Verlag, Düsseldorf 1992.

M. Maser: *Grundlagen der allgemeinen Kommunikationstheorie*, Verlag Berliner Union, Stuttgart and Kohlhammer, Stuttgart 1971.

Expertensysteme: Einführung in Technik und Anwendung, Siemens AG, Berlin 1987.

U. Rembold: *Einführung in die Informatik*, Carl Hanser Verlag, München 1991.

E. Schnieder (ed.): *Prozeßinformation: Einführung mit Petrinetzen*, 2nd ed., Verlag Vieweg, Braunschweig 1993.

S. Wendt: *Nichtphysikalische Grundlagen der Informationstechnik*, Springer Verlag, Berlin 1989.

Various editors: Reihe Informatik Fachberichte, Springer Verlag, Berlin.

Process Engineering

E. Blaß: *Entwicklung verfahrenstechnischer Prozesse*, Salle und Sauerländer Verlag, Frankfurt 1989.

W. Büchner, R. Schliebs, G. Winter, K. H. Büchel: *Industrielle Anorganische Chemie*, 2nd ed., Verlag Chemie, Weinheim 1986.

M. Engshuber, R. Müller: *Grundlagen der Verfahrenstechnik für Automatisierungsingenieure*, VEB Deutscher Verlag für Grundstoffindustrie, Leipzig–Stuttgart 1993.

E. Ignatowitz (ed.): *Chemietechnik*, Europa-Lehrmittel Verlag, Wuppertal 1980.

H. Robel et al.: *Lehrbuch der chemischen Verfahrenstechnik*, VEB Deutscher Verlag für Grundstoffindustrie, Leipzig 1983.

Ullmann, 4th ed., **5**, 891.

K. Weissermel, H.-J. Arpe: *Industrielle Organische Chemie*, 3rd ed., VCH Verlagsgesellschaft, Weinheim 1990.

Control Engineering and Simulation

A. Auer: *Speicherprogrammierte Regelung*, Hüthig, Heidelberg 1990.

A. Böttiger: *Regelungstechnik: Eine Einführung für Ingenieure und Naturwissenschaftler*, 2nd ed., R. Oldenbourg Verlag, München 1991.

F. Dörrscheidt, W. Latzel: *Grundlagen der Regelungstechnik*, 2nd ed., B. G. Teubner Verlag, Stuttgart 1992.

T. Ebel: *Regelungstechnik*, 6th ed., B. G. Teubner Verlag (Studienskripten), Stuttgart 1991.

O. Föllinger: *Nichtlineare Regelungen*, vols. I and II, R. Oldenbourg Verlag, München, vol. I (5th ed.) 1989, vol. II (6th ed.) 1991.

O. Föllinger: *Regelungstechnik: Einführung in die Methoden und ihre Anwendung*, 7th ed., Hüthig, Heidelberg 1992.

J. C. Gille, M. Pelegrin, P. Decaulne: *Lehrgang der Regelungstechnik*, R. Oldenbourg Verlag, München 1963.

E.-D. Gilles: *Systeme mit verteilten Parametern*, R. Oldenbourg Verlag, München 1973.

H.-M. Hanisch: *Petri-Netze in der Verfahrenstechnik: Modellierung und Steuerung verfahrenstechnischer Systeme*, R. Oldenbourg Verlag, München 1992.

R. Isermann: *Identifikation dynamischer Systeme 1: Grundlegende Methoden*, 2nd ed., Springer Verlag, Berlin 1992.

R. Isermann: *Identifikation dynamischer Systeme 2: Besondere Methoden, Anwendungen*, 2nd ed., Springer Verlag, Berlin 1992.

D. P. F. Möller: *Modellbildung, Simulation und Identifikation dynamischer Systeme*, Springer Verlag, Berlin 1992.

W. Oppelt: *Kleines Handbuch technischer Regelvorgänge*, Verlag Chemie, Weinheim 1972.

M. Papageorgiou: *Optimierung: Statische, dynamische, stochastische Verfahren für die Anwendung*, R. Oldenbourg Verlag, München 1991.

E. Samal: *Grundriß der praktischen Regelungstechnik*, 17th ed., R. Oldenbourg Verlag, München 1991.

G. Schmidt: *Grundlagen der Regelungstechnik*, Springer Verlag, Berlin 1989.

G. Schmidt: *Simulationstechnik*, R. Oldenbourg Verlag, München 1980.

J. R. Schwarz, B. Friedland (eds.) *Linear Systems*, McGraw-Hill, New York 1965.

W. W. Solodownikow: *Grundlagen der selbständigen Regelung*, vols. 1–2, R. Oldenbourg Verlag, München 1959.

M. Thoma: *Theorie linearer Regelsysteme*, Verlag Vieweg, Braunschweig 1973.

J. G. Truxal: *Entwurf automatischer Regelsysteme*, R. Oldenbourg Verlag, München 1960.

G. Wunsch, H. Schreiber: *Stochastische Systeme*, 3rd ed., Springer Verlag, Berlin 1992.

Control Technology

A. Auer: *SPS – Progammierung: Beispiele und Aufgaben*, 3rd ed., Hüthig, Heidelberg 1991.

A. Auer: *SPS – Aufbau und Programmierung*, 4th ed., Hüthig, Heidelberg 1991.

P. Conrads: *Vom Umgang mit SPS*, Hüthig, Heidelberg 1991.

R. Haug: *Pneumatische Steuerungstechnik*, 2nd ed., B. G. Teubner (Studienskripten), Stuttgart 1991.

J. Petry: *SPS – Projektierung und Programmierung*, 2nd ed., Hüthig, Heidelberg 1990.

G. Wellenreuther, D. Zastrow: *Steuerungstechnik mit SPS*, Verlag Vieweg, Braunschweig 1991.

G. Wellenreuther, D. Zastrow: *Lösungsbuch Steuerungstechnik mit SPS*, Verlag Vieweg, Braunschweig 1992.

Metrology

R. Best: *Digitale Meßwertverarbeitung*, R. Oldenbourg Verlag, München 1991.

W. H. Cubberly (ed.): *Comprehensive Dictionary of Measurement and Control*, 2nd ed., ISA 1992.

O. Fiedler: *Strömungs- und Durchflußmeßtechnik*, R. Oldenbourg Verlag, München 1992.

H. Gatzmanga et al.: *Prozeßmeßtechnik*, VEB Deutscher Verlag für Grundstoffindustrie, Leipzig 1989.

W. Göpel, J. Hesse, J. N. Zemel (eds.): *Sensors – A Comprehensive Survey*, vols. 1–8, VCH Verlagsgesellschaft, Weinheim 1992.

K.-H. Gonschorek, H. Singer (eds.): *Elektromagnetische Verträglichkeit: Grundlagen, Analysen, Maßnahmen*, B. G. Teubner Verlag, Stuttgart 1992.

E. Hering, K. Bressler, J. Gutekunst: *Elektronik für Ingenieure*, VDI-Verlag, Düsseldorf 1992.

J. W. Klein, P. Dullenkopf, A. Glasmachers: *Elektronische Meßtechnik: Meßsysteme und Schaltungen* (Teubner Studienbücher), B. G. Teubner Verlag, Stuttgart 1992.

H. Kronmüller: *Prinzipien der Prozeßmeßtechnik*, vols. 1 and 2, Schnäker Verlag, Karlsruhe 1986.

S. Miyazaki et al.: *Instumentation and Control for Process*, Nikkan Kohgyo, Shuppan-sha, Japan.

P. Profos, T. Pfeifer (eds.): *Grundlagen der Meßtechnik*, 3rd ed., R. Oldenbourg Verlag, München 1992.

P. Profos, T. Pfeifer (eds.): *Handbuch der industriellen Meßtechnik*, 5th ed., R. Oldenbourg Verlag, München 1992.

W. Richter: *Grundlagen der elektrischen Meßtechnik*, Verlag Technik, Berlin 1989.

G. W. Schanz: *Sensoren – Fühler der Meßtechnik*, 3rd ed., Hüthig, Heidelberg 1993.

H. Schaumburg: *Sensoren*, B. G. Teubner Verlag, Stuttgart 1992.

G. Schnell (ed.): *Sensoren in der Automatisierungstechnik*, 2nd ed., Verlag Vieweg, Braunschweig 1993.

G. Strohrmann: *Meßtechnik im Chemiebereich*, R. Oldenbourg Verlag, München 1991.

H.-R. Tränkler: *Taschenbuch der Meßtechnik mit Schwerpunkt Sensortechnik*, R. Oldenbourg Verlag, München 1992.

Ullmann, 4th ed., **5**, 797–890.

Instrumentation Handbook, Instrumentation and Process Control Engineer's Association, Japan.

Conference Proceedings (see also Chap. 12)

Beratender Ausschuß der Industriephysiker der Deutschen Physikalischen Gesellschaft: Physikertagung, e.g., Proceedings of: Der Physiker in der Automatisierungstechnik (1989); Sensorphysik und Sensortechnik (1984).

IFAC: IFAC-World Congress: Automatic-Control, Selected Papers of the IFAC World Congress, Pergamon Press, Oxford.

INTERKAMA: INTERKAMA-proceedings as series: "Fachberichte Messen, Steuern, Regeln", P. Thoma (ed.), Springer Verlag, Berlin, since 1989: G. Schmidt, H. Steusloff (eds.), R. Oldenbourg Verlag, München.

GVC: Jahrestreffen der Verfahrensingenieure, Kurzfassung der Vorträge und Poster.

VDI: Automatisierungstechnik Tagungsband, VDI-Verlag, Düsseldorf.

Journals

automatica
Automatisierungstechnik (at)
Automatisierungstechnische Praxis (atp)
Chemie Ingenieur Technik (CIT)
Computers & Chemical Engineering
Control Engineering Practice
Informatik Spektrum
INTECH (ISA)
Journal of Process Control
messen & prüfen
Process Control and Quality
Technisches Messen (tm)
VDI-Z
ZWFCIM

References for Chapter 1

[1.1] N. D. Kondratieff: "Die langen Wellen der Konjunktur," in: *Archiv für Sozialwissenschaft und Sozialpolitik*, vol. 56, Mohr Verlag, Tübingen 1926.

[1.2] Leo A. Nefiodow: "Der fünfte Kondratieff," in: *Strategien zum Strukturwandel in Wirtschaft und Gesellschaft*, Gabler-Verlag, Wiesbaden 1990.

[1.3] J. Bardeen, W. H. Brattain: "Transistor, A Semiconductor Triode," *Phys. Rev.* **74** (1948) 230–231.

[1.4] D. P. Chen: "The Entity Relationship Model," *ACM Trans. Database Syst.* **1** (1976) 9–36.

[1.5] C. A. Petri: "Kommunikation mit Automaten," Dissertation, TH Darmstadt 1961.

[1.6] H. Nottebohm: "Entwicklungen der Informationstechnik," *Automatisierungstechnische Praxis (atp)* (1985) nos. 5, 6, 224–230, 267–273.

[1.7] M. Polke: "Information als kritische Ressource," in H. A. Henzler (ed.): *Handbuch der strategischen Führung*, Gabler-Verlag, Wiesbaden 1988, pp. 353–378.

[1.8] M. E. Porter: *Wettbewerbsvorteile*, Campus Verlag, New York–Frankfurt 1986.

[1.9] J. Schumpeter: *Theorie der wirtschaftlichen Entwicklung*, 6th ed., Duncker und Humbolt Verlag, Berlin 1964.

[1.10] M. Polke: *Technology and Structural Change*, presented at Scottish Council, International Forum, 1986.

[1.11] B. Sturm, B. Will, M. Polke: "Die NAMUR – gestern, heute und morgen," *Automatisierungstechnische Praxis (atp)* **30** (1988) 11–29.

[1.12] A. Lippert et al.: Bayerinterne Untersuchungen und Organisationsanweisungen, Leverkusen 1979–80.

[1.13] R. Lauber: "Prozeßautomatisierung, Prozeßleittechnik oder Prozeßdatenverarbeitung," *Automatisierungstechnische Praxis (atp)* **28** (1986) 244–245.

[1.14] R. Lauber: *Prozeßautomatisierung als Hochschultext*, 2nd ed., Springer Verlag, Berlin–Heidelberg–New York 1989.

[1.15] VDI/VDE Gesellschaft Meß- und Automatisierungstechnik (GMA): *Aufgaben – Struktur – Aktivitäten*, Düsseldorf 1991.

[1.16] M. Polke: *Der Physiker in der Automatisierungstechnik*, DPG, Kongreß-, Ausstellungs- und Verwaltungs-GmbH, Bad Honnef 1989.

[1.17] R. Magin, W. Wüchner: *Digitale Prozeßleittechnik*, Vogel-Buch Verlag, Würzburg 1987.

[1.18] G. Strohrmann: "Prozeßleittechnik," *rtp-Seminar, Regulungstechnische Praxis (rtp)* **26** (1989) no. 10, 543–550.

[1.19] G. Gruhn et al.: *Verfahrenstechnik – Prozeßmeßtechnik, Prozeßsteuerung, Automatisierungstechnik*, VEB Deutscher Verlag für Grundstoffindustrie, Leipzig 1985.

[1.20] H. Töpfer, P. Besch: *Grundlagen der Automatisierungstechnik*, 2nd ed., Verlag Technik, Berlin 1989.

References for Chapter 2

[2.1] M. Polke: "Erforderliche Informationsstrukturen für die Qualitätssicherung," *GVC-Fachtagung "Qualität"*, Heidelberg, June 1992.

[2.2] A.-W. Scheer: *Wirtschaftsinformatik*, Springer Verlag, Berlin 1988.

[2.3] M. Vetter: *Aufbau betrieblicher Informationssysteme mittels konzeptioneller Datenmodellierung*, Teubner, Stuttgart 1989.

[2.4] G. Schlageter, W. Stucky: *Datenbanksysteme: Konzepte und Modelle*, Teubner, Stuttgart 1983.

[2.5] H. Wedeking, E. Ortner: *Systematisches Konstruieren von Datenbankanwendungen: Zur Methodologie der angewandten Informatik*, Hanser Verlag, München 1980.

[2.6] P. Kruck: "Methodische Software-Entwicklung," Dissertation, Universität-Gesamthochschule Wuppertal, 1987.

[2.7] E. F. Codd: "A Relational Model of Data for Large Shared Data Banks," *Commun. ACM* **13** (1970) 377–387.

[2.8] P. P.-S. Chen: The Entity-Relationship-Model – Toward a Unified View of Data, *ACM Transactions on Database Systems*, vol. 1, Cambridge, Mass., March 1976, pp. 9–36.

[2.9] A.-W. Scheer: *EDV-orientierte Betriebswirtschaftslehre*, Springer Verlag, Berlin 1990.

[2.10] H. Vossberg et al.: "Die Informations- und Steuerungssysteme der Sparten," Bayer, Leverkusen, pp. 1970 ff.

[2.11] R. Durchholz: "Konzeptionelles Schema," *Informatik Spektrum* **7** (1984) 245–247.

[2.12] E. Ortner: "Semantische Modellierung – Datenbankentwurf auf der Ebene der Benutzer," *Informatik-Spektrum* **8** (1985) 20–28.

[2.13] W. Wendt: *Nichtphysikalische Grundlage der Informationsstechnik*, Springer Verlag, Berlin 1989.

[2.14] E. Ortner, B. Söllner: "Semantische Datenmodellierung nach der Objekttypenmethode," *Informatik-Spektrum* **12** (1989) 31–42.

[2.15] F. Gebhart: "Semantisches Wissen in Datenbanken – ein Literaturbericht," *Informatik-Spektrum* **10** (1987) 79–88.

[2.16] H. Peschke: "Partizipative Entwicklung und Einführung von Informationssystemen," in H. Balzert (ed.): *Einführung in die Software-Ergonomie*, De Gruyter, Berlin 1988, pp. 299–322.

[2.17] D. B. Preßmar: "Strategien für den Einsatz von Methoden und Werkzeugen in der Softwaretechnologie," in T. Gutzwiller, H. Österle (eds.): *Anleitung zu einer praxisorientierten Software-Entwicklungsumgebung*, vol. 2, Angewandte Informationstechnik, Verlag, Hallbergmoos 1988, pp. 9–24.

[2.18] S. Löffler, A. Warner: "Integration von Software-Entwicklungswerkzeugen," in H. Österle (ed.): *Anleitung zu einer praxisorientierten Software-Entwicklungsumgebung*, vol. 1, Angewandte Informationstechnik, Verlag, Hallbergmoos 1988, pp. 29–37.

[2.19] U. Rembold (ed.): *Einführung in die Informatik*, Hanser Verlag, München 1987.

[2.20] K. Heuer: "Die Rolle von Dictionary-Systemen in Software-Produktionsumgebungen," in H. Österle (ed.): *Anleitung zu einer praxisorientierten Software-Entwicklungsumgebung*, vol. 1, Angewandte Informatik, Verlag, Hallbergmoos 1988, pp. 117–133.

[2.21] S. Holloway: "Data Dictionary as an Administrative Tool – How to Succeed in the Nineties," in: Pergamon Infotech (ed.): *Information Update* **1** (1987) no. 1, 17–25.

[2.22] E. Ortner, J. Rössner, B. Söllner: "Entwicklung und Verwaltung standardisierter Datenelemente," *Informatik-Spektrum* **13** (1990) 17–30.

[2.23] H. Balzert: *Die Entwicklung von Software-Systemen; Prinzipien, Methoden, Sprachen, Werkzeuge*, Bibliographisches Institut, Mannheim 1990.

[2.24] H. Balzert (ed.): *CASE-Systeme und Werkzeuge*, Wissenschaftsverlag, Mannheim 1990.

[2.25] Aristoteles: *Kategorien*, Akademie Verlag, Berlin 1984.

[2.26] C. v. Linné: *Systema naturae*, British Museum of Natural History Publishers, London 1958.

[2.27] C. Darwin: *Die Entstehung der Arten* (In der Übersetzung von C. W. Neumann), Reclam, Stuttgart 1963.

[2.28] E. Cassirer: *Substanzbegriff und Funktionsbegriff*, 6th ed., Wissenschaftliche Buchgesellschaft, Darmstadt 1990.

[2.29] W. Ahrens: "Einsatz von Expertensystemen in der Prozeßleittechnik," *Automatisierungstechnische Praxis atp* **10** (1987) 475–485.

[2.30] W. Gotthard: "Datenbanksysteme für Software-Produktionsumgebungen," *Informatik, Fachberichte 1993*, Springer Verlag, Berlin 1988.

[2.31] H. Stoyan, H. Wedekind: *Objektorientierte Software- und Hardwarearchitekturen*, Teubner, Stuttgart 1983.

[2.32] F. Vester: *Unsere Welt – ein vernetztes System*, Deutscher Taschenbuch Verlag, München 1983.

[2.33] F. Vester: *Leitmotiv vernetztes Denken*, Heine, Verlag, München 1988.

[2.34] M. Polke: "Informationshaushalt technischer Prozesse, in *Prozeßrechner 1984*, Springer Verlag, Berlin 1984.

[2.35] H Zwerina, P. Haubner: "Gestaltung von Informationen auf Bildschirmen," in K. P. Fähnrich (ed.): *Software-Ergonomie*, R. Oldenbourg Verlag, München 1987.

[2.36] H. Zwerina: "Masken und Formulare," in H. Balzert (ed.): *Einführung in die Software-Ergonomie*, De Gruyter, Berlin 1988, pp. 163–174.

[2.37] W. Hacker: "Software-Ergonomie: Gestalten rechnergestützter geistiger Arbeit, in W. Schänpflug (ed.): *Software-Ergonomie 87*, Teubner, Stuttgart 1987, pp. 31–54.

[2.38] W. Ahrens, M. Polke: "Netzmodelle als systemtechnische Informationsbasis für die Prozeßleittechnik," *Automatisierungstechnische Praxis atp* **37** (1989) no. 3, 94–103; **37** (1989) no. 4, 138–144.

[2.39] B. Shneiderman: "Direct Manipulation: A Step beyond Programming Languages," *IEEE Computer* **16** (1983) Nov. 8, 57–69.

[2.40] R. Ilg, J. Ziegler: "Direkte Manipulation, in H. Balzert (ed.): *Einführung in die Software-Ergonomie*, De Gruyter, Berlin 1988, pp. 135–194.

[2.41] K.-P. Fähnrich, J. Ziegler: "Direkte Manipulation als Interaktionsform an Arbeitsplatzrechnern," in H. J. Bullinger (ed.): *Software-Ergonomie 85*, Teubner, Stuttgart 1985.

[2.42] K. Klose: "Software-Entwicklung für Prozeßleitsysteme," *NAMUR-Statusbericht 87, Prozeßleittechnik für die Chemische Industrie*, R. Oldenbourg Verlag, München 1987, pp. 224–229.

[2.43] CAE 2000: Bayerinterner Code für die Modellierung der Ingenieurfunktion, Leverkusen 1989–1992.

[2.44] M. Chaineux: "Ausrichtung und Konception von DV-Anwendungssystemen auf der Basis von Unternehmungsmodellen, CAE 2000," *GMA Kongreß*, Dresden, Sept. 1993.

[2.45] A. L. Luft: *Informatik als Technik-Wissenschaft. Eine Orientierungshilfe für das Informatik-Studium. BI*, Wissenschaftsverlag, Mannheim 1988.

[2.46] R. Hagemüller, L. Max, J. Wolf: "Das Entwicklungssystem DOMINO von Siemens," in T. Gutzwiller, H. Österle (eds.): *Anleitung zu einer praxisorientierten Software-Entwicklungsumgebung*, vol. 2, Angewandte Informationstechnik Verlag, Hallbergmoos 1988, pp. 41–61.

[2.47] W. Ahrens: *Objekt-orientierte Wissensrepräsentation in der Prozeßleittechnik. Prozeßrechensysteme '88*, Springer Verlag, Berlin 1989, pp. 498–513.

[2.48] E. F. Cox: *Object Oriented Programming*, Addison Wesley, Reading, Mass., 1986.

[2.49] H. Stoyan: *Programmiermethoden der Künstlichen Intelligenz*, vol. 1 (1988), vol. 2 (1991), Springer Verlag, Berlin.

[2.50] E. Ignatowitz: *Chemietechnik*, Verlag Europa-Lehrmittel, Wuppertal 1982.

[2.51] G. Frege in G. Patzing (ed.): *Funktion, Begriff, Bedeutung. Fünf logische Studien*, G. Patzing Verlag, Göttingen 1966.

[2.52] DIN 32705, Klassifikationssysteme, Beuth-Verlag, Berlin 1987.

[2.53] I. Rasmussen: The Role of Hierarchical Knowledge Representation in Decision Making on System Management. IEEE Transactions on Systems, *Man and Cybernetics*, **SMC-15** (1985) no. 2, 234–243.

[2.54] F.-J. Kaufels: *Rechnernetzwerk-Systemarchitekturen und Datenkommunikation*, Wissenschaftsverlag, Mannheim 1989.

[2.55] R. Perne, M. Polke: "Produktentwicklung durch Verfahrensvariation," *Regelungstechnik* **30** (1982) no. 5, 147–156.

[2.56] DIN 19222, Leittechnik, Begriffe, Beuth-Verlag, Berlin 1985.

[2.57] B. Polke, G. Rauprich: *Operatoren zur Beschreibung zeitabhängiger Zustandsänderungen*, in press.

[2.58] T. Suckut: "Informationsmodell für Mehrprojekt-Management," Dissertation RWTH, Aachen 1993.

[2.59] E. Schnieder: *Prozeßinformatik, Einführung mit Petri-Netzen*, Vieweg & Sons, Braunschweig 1986.

[2.60] C. A. Petri: "Kommunikation mit Automaten," Dissertation, TH Darmstadt 1961.

[2.61] W. Reisig: *Petri-Netze*, Springer Verlag, Berlin 1982.

[2.62] E. Schnieder, H. Gückel: "Petri-Netze in der Automatisierungstechnik," *Automatisierungstechnik at* **37** (1989) no. 5, 173–181; no. 6, 234–241.

[2.63] D. Abel: "Modellbildung und Analyse diskret gesteuerter Systeme mit Petri-Netzen," *Automatisierungstechnik at* **36** (1988) no. 12, 455–462.

[2.64] W. Ahrens, M. Polke: "Informationsstrukturen in der Automatisierungstechnik," in G. Schmidt, H. Steusloff (eds.): "Mit vernetzten, intelligenten Komponenten zu leistungsfähigen Meß- und Automatisierungssystemen", *INTERKAMA-Kongreß 89*, R. Oldenbourg Verlag, München 1989, pp. 47–67.

[2.65] E. C. Yamalidou, J. C. Kantor: "Modeling and Analysis of Discrete-Event Processes Using Petri Net Theory" *ATCHE Annual Meeting*, San Francisco 1989.

[2.66] P. Winkler: "Anforderungsbeschreibung mit Netzmodellen," *Automatisierungstechnische Praxis (atp)* **28** (1986) no. 1, 32–39.

[2.67] J. C. Kantor, E. P. Patsidou: "On the Sceduling and Control of Multipurpose Batch Plants," *AICHE Annual Meeting 1989*, Paper no. 1436, San Francisco 1989.

[2.68] A. Hartramph: personal communication 1989.

[2.69] Anlagen- und Apparatekennzeichen der Bayer-AG, Werknorm 9060, Leverkusen 1989.

[2.70] *KKS-Kraftwerk-Kennzeichensystem*, 4th ed., VGB Kraftwerkstechnik GmbH, Essen 1991.

[2.71] DIN 28 004, Fließbilder Verfahrenstechnischer Anlagen, Beuth Verlag, Berlin 1977.

[2.72] TGL 25 000, März 1974 Verfahrenstechnik, Grundoperationen, Klassifikation.

[2.73] *Ullmann*, 4th ed., **4**, 1.

[2.74] R. J. Uhlig: "Erstellen von Ablaufsteuerungen für Chargenprozesse mit wechselnden Rezepturen, *Automatisierungstechnische Praxis (atp)* **29**, (1987) no. 1, 17–23.

[2.75] J. Lauber, M. Polke: "Das Phasenmodell der Produktion: Ein semantisches CAE-Werkzeug zur Prozeßbeschreibung," *Regelungstechnischen Kolloquium*, Boppard, March 1993.

[2.76] O. Graf: "Standardisierte Grundfunktionen verringern Softwareaufwand bei Teleperm M," *Eng. & Autom.* **3** (1991) 2–5.

[2.77] U. Epple: "Funktionsorientierte Strukturierung prozeßleittechnischer Aufgaben." *Automatisierungstechnische Praxis atp* **31** (1989) no. 10, 468–474.

[2.78] U. Epple, H. Kopec, R. Schmidt; "Strukturierung von Prozeßführungsaufgaben und Leitsystemsoftware," *Automatisierungstechnische Praxis (atp)* **31** (1989) no. 10, 468–474.

[2.79] T. G. Fisher: "Batch Control Systems – Design, Application and Implementation, ISA, Paris 1990.

[2.80] F. J. Kersting et al.: "Anforderungen an Systeme zur Rezeptfahrweise," NAMUR Empfehlung NE 25, Version vom 19.5.1992.

[2.81] M. Polke, F. Portele: "Stand und Entwicklung der Prozeßleittechnik für die Textilindustrie," *Melliand Textilber.* **65** (1984) 291–297.

[2.82] M. Polke: "Produktionsplanung in der Chemischen Industrie, in D. Hahn, G. Laßmann (eds.): *Produktionswirtschaft*, vol. 2, Physica-Verlag, Heidelberg 1989, pp. 369–413.

[2.83] M. Polke: "Prozeßleittechnik: Trend zur Automatisierung in der Chemischen Industrie hält ungebrochen an Achema 91," *Automatisierungstechnische Praxis (atp)* **33** (1991) no. 5, 219–222.

[2.84] Bayer AG: CAE 2000 Bayerinterne Dokumentation, Leverkusen 1992.

[2.85] M. Polke: "Information als kritische Ressource," in A. Henzler (ed.): *Handbuch Strategische Führung*, Betriebswirtschaftlicher Verlag, Wiesbaden 1988, pp. 353–378.

[2.86] G. Schmidt: "Was sind und wie entstehen komplexe Systeme und welche spezifischen Aufgaben stellen sie für die Regelungstechnik?", *Regelungstechnik* **30** (1982) no. 10, 331–339.

[2.87] W. Hofmann: "Aufgabe der Produktionsleitebene in der Chemischen Industrie," *NAMUR-Statusbericht 85, Prozeßleittechnik für die Chemische Industrie*, R. Oldenbourg Verlag, München 1985, pp. 44–50.

[2.88] R. W. Peters: "Informationshaushalte in Labor und Produktion," *NAMUR-Statusbericht 85, Prozeßleittechnik für die Chemische Industrie*, R. Oldenbourg Verlag, München 1985, 26–35.

[2.89] S. Riemer: "Funktionen der Produktionslogistik in der Verfahrenstechnik," in O. Schmidt, H. Steusloff (eds.): "Mit vernetzten, intelligenten Komponenten zu leistungsfähigen Meß- und Automatisierungssystemen, *INTERKAMA-Kongreß 89*, R. Oldenbourg Verlag, München 1989, pp. 381–390.

[2.90] B. Sturm, B. Will, M. Polke: "Die NAMUR – gestern, heute und morgen," *Automatisierungstechnische Praxis (atp)* **30** (1988) no. 1, 11–29.

[2.91] K. F. Früh: "Optimalstruktur dezentraler Prozeßleitsysteme," *RTP Regelungstech. Praxis* **21** (1979) 273 ff.

[2.92] W. Körner, P. Moll, M. Polke: "Zur funktionalen Gliederung von Leitsystemen, *Automatisierungstechnische Praxis (atp)* **30** (1987) no. 9, 393–400.

[2.93] W. Kaufmann (ed.): *Diagnostische Entscheidungsprozesse in der inneren Medizin*, Schattaur Verlag, Stuttgart 1986.

[2.94] M. Wienand: "Mündliche und unveröffentlichte Hinweise auf das DV-Phasenmodell bei der Konzeption von Unternehmens-Informationsmodellen," Bayer AG, Leverkusen 1974 ff.

[2.95] W. Büsing: *Ergebnis vieler Diskussionen zur Klärung des von Polke/Wienand intuitiv vorgeschlagenen Phasenmodells*, personal communication, 1983.

[2.96] E. D. Gilles, M. Wurst: "Prozeßführung zur Sicherung reproduzierbarer Qualität in der verfahrenstechnischen Produktion," lecture held on *GVC-Jahrestreffen*, Tagungsband, Stuttgart 1990, pp. 138, 139.

[2.97] E. D. Gilles: "Prozeßführung zur Sicherung reproduzierbarer Qualität in der verfahrenstechnischen Produktion," *Chem. Ing. Tech.* **63** (1991) no. 9, 910–918.

[2.98] E. D. Gilles, E. Nicklaus, M. Polke: "Sensortechnik in der Chemie – Status und Trend," *Chem.-Ing.-Tech.* **58** (1986) 557–564.

[2.99] K. H. Padberg, P.-Th. Wilrich: "Die Auswertung von Daten und ihre Abhängigkeit von der Merkmalsart, part 1, Skalentypen," *QZ* **26** (1981) no. 6, 179 ff.

[2.100] H. Rumpf: "Über die Eigenschaften von Nutzstäuben," *Staub Reinhalt. Luft* **27** (1967) 3–13.

[2.101] R. Ecker, H. Kramer, H. K. Müller, M. Polke: "Neue Aspekte bei der Prüfung von Kautschuk und Gummi," *Kautsch. Gummi Kunstst.* **25** (1972) no. 1, 5–10.

[2.102] E. D. Gilles, E. Nicklaus, M. Polke: "Sensortechnik in der Chemie – Status und Trend, *Automatisierungstechnische Praxis atp* **28** (1986) no. 9, 423–437; no. 10, 479–484.

[2.103] M. Polke: "CAE zwischen Betreibern und Anlagenbauern," *Partnerschaft im Anlagenbau*, Haus der Technik, Essen, Nov. 1992.

[2.104] M. Polke: "Prozeßleittechnik; Sicherheit und Qualität in der Prozeßindustrie," *Vortrag auf dem VDE-Kongreß 1993*, Berlin, Januar 1993.

[2.105] J. König: "Sicherheitstechnische Aspekte der Prozeßleittechnik," *Chem. Ing. Tech.* **59** (1987) no. 3, 196–204.

[2.106] B. Bieker, H. Rückert: *Entwicklung eines Expertensystems zur Prozeßführung von Bioreaktoren einer Kläranlage der Bayer AG*, Leverkusen, Aug. 1989.

[2.107] DIN 40 719, Part 6, Schaltungsunterlagen, Regeln und graphische Symbole für Funktionspläne, Beuth-Verlag, Berlin, Mai 1992.

References for Chapter 3

[3.1] E. D. Gilles, E. Nicklaus, M. Polke: "Sensortechnik in der Chemie – Status und Trend," *atp* **28** (1986) no. 6, 423–431.

[3.2] K. H. Padberg, P.-Th. Wilrich: "Die Auswertung von Daten und ihre Abhängigkeit von der Merkmalsart," *Z. Qualität Sicherheit* **26** (1981) no. 179—183.
[3.3] M. Polke: "Prozeßleittechnik für die Chemie – Status und Trend," *Automatisierungstechnische Praxis atp* **28** (1985) no. 5, 214–223.
[3.4] E. D. Gilles: "Auf dem Wege zu einer modellgestützten Prozeßleittechnik," *Automatisierungstechnische Praxis atp* **31** (1888) 265–270, 441–449.
[3.5] R. Perne, M. Polke: "Produktentwicklung durch Verfahrensvariation," *Regelungstechnik* **30** (1982) no. 5, 147–156.
[3.6] I. L. Casti: *Dynamical Systems and their Applications*, Academic Press, New York 1977.
[3.7] C. T. Chen: *Introduction to Linear System Theory*, Holt Rinehart and Winston, New York 1970.
[3.8] L. A. Zadeh, C. A. Desoer: *Linear System Theory*, Mc Graw-Hill, New York 1963.
[3.9] R. Unbehauen: *Systemtheorie*, R. Oldenbourg Verlag, München 1969.
[3.10] R. E. Kalman, P. L. Falb, M. Arbib: *Topics in Mathematical Systems Theory*, Mc Graw-Hill, New York 1969.
[3.11] L. Padulo, M. Arbib: *System Theory*, Hemisphere, Washington 1974.
[3.12] F. Pichler: *Mathematische Systemtheorie: Dynamische Konstruktionen*, De Gruyter, Berlin 1975.
[3.13] H. W. Knobloch, H. Kwakernaak: *Lineare Kontrolltheorie*, Springer Verlag, Berlin 1985.
[3.14] S. Wendt: *Nichtphysikalische Grundlagen der Informationstechnik*, Springer Verlag, Berlin-Heidelberg-New York 1989.
[3.15] W. Hahn: *Theorie und Anwendung der direkten Methode von Ljapunow*, Springer Verlag, Berlin 1959.
[3.16] W. Hahn, *Stability of Motion*, Springer Verlag, Berlin 1967.
[3.17] J. La Salle, S. Lefschetz: *Die Stabilitätstheorie von Ljapunow*, Bibliographisches Institut, Mannheim 1967.
[3.18] H. Haken: *Synergetik*, Springer Verlag, Berlin 1983.
[3.19] T. Kailath: *Linear Systems*, Prentice Hall, Engelwood, N. J., 1980.
[3.20] E. Gilbert: "Controllability and Observability in Multivariable Control Systems," *J. Soc. Ind. Appl. Math. Ser. A* **1** (1963) 128–151.
[3.21] R. E. Kalman: "On the General Theory of Control Systems," *Proc. 1st. Cong. Autom. Control*, **1960**, 481–493.
[3.22] R. E. Kalman: "Mathematical Description of Linear Systems," *J. Soc. Ind. Appl. Math. Ser. A* **1** (1963) 152–192.
[3.23] R. E. Kalman: *Lectures on Controllability and Observability*, C.I.M.E., Bologna 1968.
[3.24] D. G. Luenberger: "Observers for Multivariate Systems," *IEEE Trans. Autom. Control* **AC-11** (1966) 190–197.
[3.25] D. G. Luenberger: "An Introduction to Observers," *IEEE Trans. Autom. Control* **AC-16** (1971) 596–603.
[3.26] R. Metz: *Auswertesysteme. Der Physiker in der Automatisierungstechnik*, Deutsche Physikalische Gesellschaft e.V., Bad Honnef, 1989, pp. 125–141.
[3.27] W. Masing: *Handbuch der Qualitätssicherung*, Hanser Verlag, München – Wien 1988.
[3.28] H. Stachowiak: *Allgemeine Modelltheorie*, Springer Verlag, Berlin 1973.
[3.29] K. R. Popper: *Logik der Forschung*, J. C. B. Mohr (Paul Siebeck), Tübingen 1982.
[3.30] R. Courant: "Variational Methods for the Solution of Problems of Equilibrium and Vibration," *Bull. Am. Math. Soc.* **49** (1943) 1–23.
[3.31] F. A. Aseltine et al.: "A Survey of Adaptive Control Systems," *IRE Trans. Autom. Control* **6** (1958) 102–108.
[3.32] R. E. Kalman: "Design of a Self-Optimizing Control System," *Trans. ASME* **80** (1958) 468–478.
[3.33] A. Rosenbluth, N. Wiener: "The Role of Model in Science," *Philos. Science* **10** (1945), 316–321.
[3.34] P. Eykhoff: *System Identification*, J. Wiley, London 1974.
[3.35] K. E. Brennan, S. L. Campbell, L. R. Petzold: *Numerical Solution of Initial-Value Problems in Differential-Algebraic Equations*, North-Holland, Amsterdam 1989.
[3.36] C. W. Gear: "Simultaneous Numerical Solution of Differential-Algebraic Equations," *IEEE Trans. Circuit Theory* **18** (1971) 89–95.
[3.37] R. Bachmann, L. Brüll, U. Pallaske: "A Contribution of the Numerical Treatment of Differential Algebraic Equations Arising in Chemical Engineering," *DECHEMA-Monogr.* **116** (1989) 343–349.
[3.38] L. R. Petzold: "Differential-Algebraic Equations are not ODE's," *SIAM J. Sci. Stat. Comput.* **3** (1982) 367–384.
[3.39] C. W. Gear, L. R. Petzold: "ODE Methods for the Solution of Differential-Algebraic Systems," *SIAM J. Numer. Anal.* **21** (1984) 716–728.
[3.40] R. E. Kalman, R. S. Bucy: "New Results in Linear Filtering and Prediction Theory," *J. Basic. Eng.* **83** (1961) 95–108.
[3.41] G. O. Kallianpu: *Stochastic Filtering Theory*, Springer Verlag, Berlin 1980.
[3.42] G. Doetsch: *Einführung in Theorie und Anwendung der Laplace-Transformation*, Birkhäuser-Verlag, Berlin 1976.
[3.43] O. Föllinger: *Laplace- und Fourier-Transformation*, Elitera-Verlag, Berlin 1977.
[3.44] M. Polke: "Zur Physik der Elektrophotographie," in: *Festkörperprobleme VIII*, Vieweg/Pergamon Press, Oxford 1968, pp. 131–174.
[3.45] W. G. Hunter, A. M. Reiner: *Technometrics* **7** (1965) 307–323.
[3.46] A. C. Atkinson, V. V. Fedorov: *Biometrika* **62** (1975) 57–70.
[3.47] F. J. Dumez, G. F. Froment: *Ind. Eng. Chem. Process Des. Dev.* **15** (1976) 291–301.
[3.48] G. E. P. Box, W. J. Hill: "Discrimination Among Mechanistic Models," *Technometrics* **9** (1967) 57–71.
[3.49] J. Pawlowski: *Die Ähnlichkeitstheorie in der physikalisch-technischen Forschung*, Springer Verlag, Berlin 1971.
[3.50] H. Görtler: *Dimensionsanalyse*, Springer Verlag, Berlin 1975.
[3.51] A. Betz: "Ähnlichkeitstheorie und Modelltechnik," in: *Physikhüttte*, vol. I: "Mechanik," Ernst u. Sohn, Berlin-München-Düsseldorf 1971.
[3.52] E. Eitelberg: "Modellreduktion durch Minimieren des Gleichungsfehlers," *Regelungstechnik* **26** (1978) 320–322.
[3.53] F. D. Galiana: "On the Approximation of Multiple Input – Multiple Output Constant Linear Systems," *Int. J. Control* **17** (1973) no. 6, 1313–1324.

M. Koch, T. Kuhn, J. Wernstedt: "Ein neues Entwurfskonzept für Fuzzy-Regelungen," *Automatisierungstech. (at)* **41** (1993) no. 5, 152–158.

H. Kiendl, K. D. Meyer-Gramann, B. Cuno, M. Fritsch, Chr. Frenck: "Fuzzy Control," Beiträge zur "Theorie für Anwender," *Automatisierungstechnik (at)* **41** (1993) no. 2. A 5.

W. Brauer, D. Hernàndes: "Verteilte künstliche Intelligenz und kooperatives Arbeiten," *4. Internationaler GI-Kongreß Wissensbasierte Systeme*, München, Oct. 1991. *Informatik Fachberichte*, Springer Verlag, Berlin 1991.

E. D. Gilles: "Modellbildung, ein Hilfsmittel der Meßtechnik," *Tech. Mess.* **46** (1979) 225–232, 271–274.

E. D. Gilles: "Application of Model Based Measuring Techniques in Chemical Engineering," *Proceedings of International Conference on Industrial Process Modelling and Control*, Hangzhou, China, 6–9 June, 1985.

llmann, 4th ed., 5.

Pfeifer, P. Profos (eds.): *Handbuch der Industriel-n Meßtechnik*, R. Oldenbourg Verlag, München 992.

H. Feltgen: personal communication, 1972.

H. Schmitt: "Die Automatisierung von Chemie-lagen mit Mitteln der Systemtechnik – Analysen d Trends," *Chem. Ing. Tech.* **53** (1981) no. 8, 620–5.

ick Elektronische Meßgeräte GmbH & Co.: Meß-System 71 pH," company brochure no. 7959 150 00, Berlin 1993.

Schügerl (ed.): *Analytische Methoden in der Bio-hnologie*, Vieweg Verlag, Braunschweig 1991.

Bailey, D. F. Ollis: *Biochemical Engineering ndamentals*, McGraw-Hill Book Company, New rk 1986.

Martens: "Sensor-Systeme für Durchfluß und lstand," *55. NAMUR* Lahnstein Nov. 1992.

Nicklaus: "Auswirkungen moderner Prozeß-echnik auf Sensorsysteme," *Chem. Ing. Tech.* **59** 7) no. 8, 615–621.

Hesse, H. Küttner: "Entwicklungstendenzen der ortechnik," *Ind.-Elektronik + Elektronik* **28** 3) no. 5, 36–44.

Polke: "On-line Betriebsmeßtechnik," Einführungsvortrag in den Themenkreis der gleichnamigen ersession auf dem Jahrestreffen der GVC, Wien, 1, 1992.

31 051, Instandhalten: Begriffe und Maßnah- Beuth Verlag, Berlin 1985.

nkowski: "Instrumentierungstechnik im Wan- *Automatisierungstechnische Praxis (atp)* **34**) no. 5, 247–254, 325–327.

nkowski: "Fortführung der Ex-Feldbus-Un-hung," *Automatisierungstechnische Praxis* **34** (1992) no. 11, 617–622.

nkowski: "Einheitliche, herstellerunabhängige ge-Bedienoberfläche für kommunikations- Meßumformer," *Automatisierungstechnische (atp)* **34** (1992) no. 2, 530–537.

hten: "Der Feldbus aus Sicht des Anwen- *Automatisierungstechnische Praxis (atp)* **33** no. 2, 55-64.

b: "Die Rolle der Sensoren in der Verfah-hnik," *Chem. Ing. Tech.* **61** (1989) 199–

[5.70] W. Stieler: "Sensorprinzipien zur Messung physikalischer Produkteigenschaften," company report, Bayer AG, Leverkusen, Nov. 1982.

[5.71] H. Raab: "Zur Struktur der Anzeige-Bedienoberfläche digitaler Sensorsysteme," *Automatisierungstechnische Praxis (atp)* **33** (1991) no. 2, 64–71.

[5.72] ISO: MMS 9506, part 1, Service Definitions; MMS, part 2, Protocol Specification, 1988.

[5.73] C. Nøkleby: *The MMS Implementation on the P-NET Fieldbus System, 2nd International Conference on P-Net Fieldbus Systems*, Deggendorf 1992.

[5.74] A. Valenzano, C. Demartini, L. Climiniera: *MAP and TOP Communications, Standards and Applications*, Addison-Wesley Publishing Company, Workingham 1992.

[5.75] W. Stadter: "Device-Description-Language. Ein wichtiger Beitrag zur herstellerunabhängigen Austauschbarkeit von Feld- und Bediengeräten," *Automatisierungstechnische Praxis (atp)* **34** (1992) no. 1, 21 ff.

[5.76] VDI/VDE Richtlinie: Einheitliche Anzeige- und Bedienoberfläche auf Personalcomputern für digitale Feldgeräte, Gründruck 1992.

[5.77] Krohne America Inc.: "Flowmeter Selection Guide," company brochure, Peabody, Mass. 1992.

[5.78] P. Profos, T. Pfeifer: *Handbuch der Industriellen Meßtechnik*, R. Oldenbourg Verlag, München 1992.

[5.79] G. Strohrmann: *Einführung in die Meßtechnik im Chemiebetrieb*, R. Oldenbourg Verlag, München 1990.

[5.80] I. Hengstenberg, B. Sturm, O. Winkler: *Messen, Steuern und Regeln in der chemischen Technik*, Springer Verlag, Berlin 1980.

[5.81] F. Kohlrausch: *Praktische Physik*, Teubner Verlagsgesellschaft, Leipzig 1950.

[5.82] A. Mangold: "Stellgeräte für verfahrenstechnische Anlagen," *Automatisierungstechnische Praxis (atp)* **33** (1991) no. 11, 585 ff.

[5.83] M. Huk: "Stellgeräte für verfahrenstechnische Anlagen," *Automatisierungstechnische Praxis (atp)* **3** (1993) no 3, 131 ff.

[5.84] H.-J. Schneider: "Sensorsysteme und die Kommunikation im Feld," *Automatisierungstechnische Praxis (atp)* **35** (1993) no. 2, 71 ff.

[5.85] DIN 19227, Graphische Symbole und Kennbuchstaben für die Prozeßleittechnik. T1: Darstellung von Aufgaben, Beuth Verlag, Berlin 1990 (draft).

[5.86] DIN 40050, IP-Schutzarten. Berührungs-, Fremdkörper- und Wasserschutz für elektrische Betriebsmittel, Beuth Verlag, Berlin 1980.

[5.87] DIN VDE 0838, Rückwirkungen in Stromversorgungsnetzen, die durch Haushaltsgeräte und ähnliche elektrische Einrichtungen versursacht werden, T1–T3, Beuth Verlag, Berlin 1987.

[5.88] NAMUR-Empfehlung 21, Störfestigkeitsanforderungen, Part 1: EMV von Betriebsmitteln der Prozeß- und Laborleittechnik. NAMUR, Darmstadt 1988.

[5.89] DIN VDE 0843, Elektromagnetische Verträglichkeit von Meß-, Steuer- und Regeleinrichtungen in der industriellen Prozeßtechnik, T1–T4, Beuth Verlag, Berlin 1984–1988.

[5.90] DIN VDE 0848, T1–T3: Gefährdung durch elektromagnetische Felder and T4–T4 A1: Sicherheit bei elektromagnetischen Feldern, Beuth Verlag, Berlin 1982–1990.

[3.54] K. Glover: "All Optimal Hankel-Norm Approximations of Linear Multivariable Systems and Their L_∞-Frror Bounds," *Int. J. Control* **39** (1985) no. 6, 1115–1193.

[3.55] D. C. Hyland, D. S. Bernstein: "The Optimal Projection Equations for Model Reduction and the Relationship among the Methods of Wilson, Skelton and Moore," *IEEE Trans. Autom. Control* **AC-30** (1985) no. 12, 1201–1211.

[3.56] L. Litz: *Reduktion der Ordnung linearer Zustandsraummodelle mittels modaler Verfahren*, Hochschul-Verlag, Stuttgart 1979.

[3.57] B. C. Moore: "Principal Component Analysis in Linear Systems: Controllability, Observability, and Model Reduction," *IEEE Trans. Autom. Control* **AC-26** (1981) 17–31.

[3.58] U. Pallaske: "Ein Verfahren zur Ordnungsreduktion mathematischer Prozeßmodelle," *Chem. Ing. Tech.* **59** (1987) no. 7, 604–605.

[3.59] R. E. Skelton: "Cost Decomposition of Linear Systems With Application to Model Reduction," *Int. J. Control* **32** (1980) 1031–1055.

[3.60] D. A. Wilson: "Optimum Solution of Model-Reduction Problem," *Proc. IEEE* **58** (1970) 1161–1165.

[3.61] E. Welfonder: "Leittechnikdokumentation aus Betreibersicht," *ETG/VGB-Fachtagung Betriebsgerechte Dokumentation in Kraftwerken, Einfluß der modernen Leittechnik und Planungsmittel*, Baden-Baden 1986.

[3.62] E. Ignatowitz: *Chemietechnik*, Verlag Europa-Lehrmittel, Wuppertal 1980.

References for Chapter 4

[4.1] I. M. Horowitz: *Synthesis of Feedback Systems*, Academic Press, New York 1963.

[4.2] W. Oppelt: *Kleines Handbuch technischer Regelvorgänge*, Verlag Chemie, Weinheim 1972.

[4.3] O. Föllinger: *Regelungstechnik*, Hüthig, Heidelberg 1985.

[4.4] O. Föllinger: *Optimierung dynamischer Systeme. Eine Einführung für Ingenieure*, Oldenbourg Verlag, München 1985.

[4.5] J. C. Doyle, G. Stein: "Multivariable Feedback Design: Concepts for a Classical/Modern Synthesis," *IEEE Trans. Autom. Control* **26** (1981) 4–16.

[4.6] M. Morari, J. C. Doyle: "A Unifying Framework for Control System Design under Uncertainty and its Implications for Chemical Process Control," *Proc. Chem. Proc. Control* **III** (1986) 5–51.

[4.7] J. S. Freudenberg, D. P. Looze: *Frequency Domain Properties of Scalar and Multivariable Feedback Systems*, Springer Verlag, Berlin 1988.

[4.8] H. H. Rosenbrock: *Computer-Aided Control System Design*, Academic Press, London 1974.

[4.9] D. H. Owens: *Feedback and Multivariable Systems*, Peter Peregrinus, Stevenage 1978.

[4.10] A. G. J. MacFarlane, D. F. A. Scott-Jones: "Vector Gain," *Int. J. Control* **29** (1979) 65–91.

[4.11] Y. Takahashi, M. J. Rabins, D. M. Auslander: *Control and Dynamic Systems*, Addison-Wesley, Reading, Mass. 1970.

[4.12] R. V. Patel, N. Munro: *Multivariable Systems Theory and Design*, Pergamon Press, Oxford 1982.

[4.13] U. Korn, H. Wilfert: *Mehrgrößenregelungen*, VEB Verlag Technik, Berlin 1982.

[4.14] H. Tolle: *Mehrgrößen-Regelkreissynthese*, Oldenbourg Verlag, München 1983.

[4.15] J. M. Maciejowski: *Multivariable Feedback Design*, Addison-Wesley, Reading, Mass., 1989.

[4.16] A. Gelb (ed.): *Applied Optimal Estimation*, M.I.T. Press, Cambridge, Mass., 1974.

[4.17] K. Brammer, G. Siffling: *Stochastische Grundlagen des Kalman-Bucy-Filters: Wahrscheinlichkeitsrechnung und Zufallprozesse*, Oldenbourg Verlag, München 1975.

[4.18] K. Brammer, G. Siffling: *Kalman-Bucy-Filters Deterministische Beobachtung und stochastische Filterung*, Oldenbourg Verlag, München 1975.

[4.19] H. Kwakernaak, R. Siwan: *Linear Optimal Control Systems*, J. Wiley, New York 1972.

[4.20] B. D. O. Anderson, J. B. Moore: *Linear Optimal Control*, Prentice-Hall, Englewood Cliffs, NJ, 1972.

[4.21] G. Stein, M. Athans: "The LQG/LTR Procedure for Multivariable Feedback Control Design," *IEEE Trans. Autom. Control* **32** (1987) 105–114.

[4.22] A. G. J. MacFarlane, B. Kouvaritakis: "A Design Technique for Linear Multivariable Feedback Systems," *Int. J. Control* **25** (1977) 837–874.

[4.23] Y. S. Hung, A. G. J. MacFarlane: *Multivariable Fedback: A Quasi-Classical Approach*, Springer Verlag, Berlin 1982.

[4.24] C. E. Garcia, M. Morari: "Inernal Model Control: 1. A Unifying Review and Some New Results," *Ind. Eng. Chem. Process Des. Dev.* **21** (1982) 308–323.

[4.25] B. A. Francis: *A Course in H_∞ Control Theory*, Springer Verlag, Berlin 1987.

[4.26] J. Doyle, K. Glover, P. Khargonekar, B. Francis: "State-Space Solutions to Standard H_2 and H_∞ Control Problems," *IEEE Trans. Autom. Control* **34** (1989) 831–847.

[4.27] V. Manousiouthakis, R. Savage, Y. Arkun: "Synthesis of Decentralized Control Structures Using the Concept of Block Relative Gain," *Am. Inst. Chem. Eng.* **32** (1986) 991–1003.

[4.28] D. D. Siljak: *Large-Scale Dynamic Systems*, North-Holland, New York 1978.

[4.29] M. G. Singh, T. Titli: *Systems: Decomposition, Optimisation and Control*, Pergamon Press, Oxford 1978.

[4.30] M. G. Singh: *Dynamical Hierarchical Control*, North-Holland, Amsterdam 1980.

[4.31] M. G. Singh: *Decentralised Control*, North-Holland, Amsterdam 1981.

[4.32] L. Litz: *Dezentrale Regelung*, Oldenbourg Verlag, München 1983.

[4.33] C. J. Harris, S. A. Billings (eds.): *Self-tuning and Adaptive Control: Theory and Applications*, Peter Peregrinus, Stevenage 1981.

[4.34] M. M. Gupta, C. T. Chen (eds.): *Adaptive Methods for Control System Design*, IEEE Press, New York 1985.

[4.35] A. Isidori: *Nonlinear Control Systems: An Introduction*, Springer Verlag, Berlin 1985.

[4.36] C. E. Garcia, D. M. Prett, M. Morari: "Model Predictive Control: Theory and Practice – a Survey," *Automatica* **25** (1989) 335–347.

[4.37] D. K. Frederick, C. J. Herget, R. Kool, M. Rimvall: "ELCS – The Extended List of Control Software," unpublished, WGS, Eindhoven University of Technology, Eindhoven 1987.

[4.38] M. Jamshidi, C. Herget: *Computer-Aided Control Systems Engineering*, North-Holland, Amsterdam 1985.
[4.39] P. Holl, W. Marquardt, E. D. Gilles: "Diva – a Powerful Tool for Dynamic Process Simulation," *Comput. Chem. Eng.* **12** (1988) 421–425.
[4.40] F. Allgöwer, J. Raisch: "Multivariable Controller Design for an Industrial Distillation Column," *Proceedings of 5th IMA International Conference on Control Theory*, 1988.
[4.41] A. Duschek: *Vorlesungen über höhere Mathematik*, vol. 3, Springer Verlag, Wien 1960.
[4.42] P. Funk: *Variationsrechnung und ihre Anwendung in Physik und Technik*, Springer Verlag, Berlin 1962.
[4.43] E. Klingbeil: *Variationsrechnung*, Bibliographisches Institut, Mannheim 1977.
[4.44] H. Schwarz: *Optimale Regelung linearer Systeme*, Bibliographisches Institut, Mannheim 1976.
[4.45] W. Boltjanski: *Mathematische Methoden der optimalen Steuerung*, Carl Hanser Verlag, München 1972.
[4.46] J. Hsu, A. Meyer: *Modern Control Principles and Applications*, part III, McGraw-Hill, New York 1968.
[4.47] L. Pontrjagin et al.: *Mathematische Theorie optimaler Prozesse*, Oldenbourg Verlag, München 1967.
[4.48] R. Bellmann: *Dynamic Programming*, Princeton University Press, Princeton 1965.
[4.49] R. Bellmann, S. Dreyfus: *Applied Dynamic Programming*, Princeton University Press, Princeton 1964.
[4.50] G. Schneider, H. Mikolcic: *Einführung in die Methode der dynamischen Programmierung*, Oldenbourg Verlag, München 1972.
[4.51] M. Athans, P. Falb: *Optimal Control*, McGraw-Hill, New York 1966.
[4.52] G. Bieß et al.: *Optimale Prozesse und Systeme*, Verlag Harry Deutsch, Thun–Frankfurt 1980.
[4.53] A. Bryson, Y. Ho: *Applied Optimal Control*, J. Wiley, New York 1975.
[4.54] E. Hofer, R. Lunderstädt: *Numerische Methoden der Optimierung*, Oldenbourg Verlag, München 1975.
[4.55] U. Hoffmann, H. Hofmann: *Einführung in die Optimierung*, Verlag Chemie, Weinheim 1971.
[4.56] H. Tolle: *Optimierungsverfahren*, Springer Verlag, Berlin 1971.
[4.57] K. Stahl: *Industrielle Steuerungstechnik in schaltalgebraischer Behandlung*, Oldenbourg Verlag, München 1965.
[4.58] K. H. Fasol: *Binäre Steuerungstechnik*, Springer Verlag, Berlin 1988.
[4.59] R. Lauber: *Prozeßautomatisierung*, vol. 1, Springer Verlag, Berlin 1989.
[4.60] W. Körner, P. Moll, M. Polke: "Zur funktionalen Gliederung von Leitsystemen," *Automatisierungstechnische Praxis (atp)* **29** (1987) no. 9, 393–400.
[4.61] A. H. Levis: "Challenges to Control: a Collective View," *IEEE Trans. Autom. Control* **AC-32** (1987) no. 4, 275–285.
[4.62] B. Rosenstengel, U. Wienand: *Petri-Netze – Eine anwendungsorientierte Einführung*, Vieweg Verlag, Wiesbaden 1982.
[4.63] D. Bochmann, C. Posthoff: *Binäre dynamische Systeme*, Oldenbourg Verlag, München 1981.
[4.64] R. Herschel: *Einführung in die Theorie der Automaten, Sprachen und Algorithmen*, Oldenbourg Verlag, München 1974.
[4.65] E. Leonhardt: *Grundlagen der Digitaltechnik. Eine systematische Einführung*, Carl Hanser Verlag, München 1982.
[4.66] S. Wendt: *Entwurf komplexer Schaltwerke*, Springer Verlag, Berlin 1974.
[4.67] M. Brombacher, M. Polke: "Perspektiven der Prozeßleittechnik," *Automatisierungstechnische Praxis (atp)* **29** (1987) 501–509, 558–567, 1987, pp. 22–35.
[4.68] M. Niese, EP 0104411, 1986.
[4.69] M. Niese: "Struktureller Aufbau und dezentrale Führung in Produktionsanlagen," *INTERKAMA Kongreß*, München 1989, pp. 339–349.
[4.70] M. Bruns, M. Eibl, F. J. Kersting, R. J. Uhlig: "Anforderungen an Systeme zur Rezeptfahrweise," *Automatisierungstechnische Praxis (atp)* **35** (1993) no. 1, 40–44; no. 2, 104–108.
[4.71] S. Riemer: "Funktionen der Produktionslogistik in der Verfahrenstechnik," *INTERKAMA Kongreß*, München 1989, pp. 381–390.
[4.72] K.-F. Geibig: "Funktionen der Betriebsleitebene," *Automatisierungstechnische Praxis (atp)* **34** (1992) no. 2, 68–72.
[4.73] M. Polke: *Prozeßleittechnik*, Universität Stuttgart, 1984 ff.
[4.74] L. Prins, J. Stäheli: "Zur Automatisierung von diskontinuierlichen Vielzweckanlagen in Chemiebetrieben," *Regelungstech. Prax.* **20** (1978) no. 9, 256–261.
[4.75] U. Epple, H. Kopec, R. Schmidt: "Strukturierung von Prozeßführungsaufgaben und Leitsystemsoftware," *Automatisierungstechnische Praxis (atp)* **34** (1992) no. 2, 59–67.
[4.76] U. Epple: "Die Bausteintechnik – Grundlage einer objektorientierten Netzstruktur für die Prozeßleittechnik, GMA-Fachbericht," *VDE-Kongreß*, Berlin 20.–23.1.1993.

References for Chapter 5

[5.1] E. Pavlik: "Sensoren als Quelle für den Informationshaushalt technischer Prozesse," *46. NAMUR Hauptsitzung*, Darmstadt, 7–9 Dec. 1983, 48. Physikertagung and Frühjahrstagung, 15 March 1984, Münster, pp. 1322 ff, Physik-Verlag GmbH, Weinheim 1984.
[5.2] R. Weisbeck: "Sensoren für die Chemische Industrie – eine kritische Bewertung des Standes der Technik und derzeitiger Sensorentwicklung," *46. NAMUR Hauptsitzung*, Darmstadt, 7–9 Dec. 1983.
[5.3] H. Warncke: "Sensortechnik in der Chemischen Industrie: Aufgaben und Möglichkeiten der Stoffgrößenmeßtechnik," *Tech. Mess.* 52 (1985) no. 4.
[5.4] H. Raab: "Digitale Sensoren und Sensorsysteme – Welche Informationen sollen sie liefern?," *Automatisierungstechnische Praxis (atp)* **30** (1988) no. 11, 534–544.
[5.5] VDI/VDE 2600, Blatt 3: Metrologie, Gerätetechnische Begriffe, Beuth Verlag, Berlin 1973.
[5.6] H. Warncke: Vorteile, Probleme und Möglichkeiten der in-situ-Messung von Stoffgrößen, Bayer-interner Bericht, Leverkusen, Nov. 1982.
[5.7] E. D. Gilles, E. Nicklaus, M. Polke: "Sensortechnik in der Chemie – Status und Trends," *Automatisierungstechnische Praxis (atp)* **28** (1986) no. 9/10, 423–432, 479–484.
[5.8] *Ullmann*, 4th ed., **5**, pp. 797–890.

[5.9] H.-J. Schneider: "Sensorsysteme für die Betriebsmeßtechnik und Kommunikation im Feld," *Automatisierungstechnische Praxis (atp)* **33** (1991) no. 10, 511–528.
[5.10] K. Hartmann: "On-line Analytik aus funktionaler Sicht, *Automatisierungstechnische Praxis (atp)* **31** (1989) no. 11, 510–515.
[5.11] H. Schuler, Ch.-U. Schmidt: "Modellgestütztes Messen in chemischen Reaktoren," *Chem. Ing. Tech.* **63** (1991) no. 3, 189–203.
[5.12] R. Perne, M. Polke: "Produktentwicklung durch Verfahrensvariation," *Regelungstech. (rt)* **30** (1982) no. 5, 147–156.
[5.13] K. Garbrecht, A. Prommer: *Mikroelektronik. Eine Technik macht Epoche*, Siemens, Berlin 1981.
[5.14] M. Polke: "Informationshaushalt technischer Prozesse," *Automatisierungstechnische Praxis (atp)* **27** (1985) no. 4, 214–217.
[5.15] J. Hesse: *Sensoren auf dem Weg zur Systemintegration*, special edition, Oldenbourg Verlag, München 1992.
[5.16] J. Lauber: "Das Phasenmodell der Produktion: Ein semantisches CAE-Werkzeug zur Prozeßbeschreibung," *Regelungstechnischen Kolloquium*, Boppard, Germany, Feb. 1993.
[5.17] International Union for Pure and Applied Physics (IUPAP): Symbols, Units and Nomenclature in Physics, Document U.I.P.20 (1978), Physik Verlag GmbH, Weinheim 1980.
[5.18] DIN 1313, Physikalische Größen und Gleichungen. Begriffe und Schreibweisen, Beuth Verlag, Berlin 1978.
[5.19] M. Polke: "Erforderliche Informationsstrukturen für die Qualitätssicherung," *Chem. Ing. Tech.* **65** (1993) no. 7, 791–796.
[5.20] W. König: Sicherheitstechnische Aspekte zur Prozeßleittechnik, *12. Arbeitsschutz-Symposium der Bayer AG*, Leverkusen, 22 Nov. 1985.
[5.21] DIN V 19250, Messen – Steuern – Regeln. Grundlegende Sicherheitsbetrachtungen für MSR-Schutzeinrichtungen, Beuth Verlag, Berlin 1989.
[5.22] NAMUR-Empfehlung 31: Anlagensicherung mit Mitteln der Prozeßleittechnik, NAMUR, Darmstadt 1992.
[5.23] Verordnung zur Durchführung des Bundesimmissionsschutzgesetzes (Störfall-Verordnung) of 19 May 1988 and 5 July 1991.
[5.24] DIN VDE 31000: Allgemeine Leitsätze für das sicherheitsgerechte Gestalten technischer Erzeugnisse. Begriffe der Sicherheitstechnik. Grundbegriffe, part 2, Beuth Verlag, Berlin 1987.
[5.25] 2nd IUPAC-Workshop on Safety in Chemical Production, Pacifico Yokohama, Japan, May 31–June 4, 1993.
[5.26] M. Polke: "Process Automation in Chemical Industry," *2nd IUPAC-Workshop on Safety in Chemical Production*, Yokohama, Japan, June 1993.
[5.27] R. Ecker, H. Kramer, H.K. Müller, M. Polke: "Neue Aspekte bei der Prüfung von Kautschuk und Gummi," *Kautsch. Gummi Kunstst.* **25** (1972) no. 1, 5–10.
[5.28] M. Mooney, H. Rolla, H. Taylor, J. H. Fielding: "Development and Standardization of Test for Evaluating Processibility of Rubber," *Symposium on Rubber Testing*, American Society for Testing Special Technical Materials, 1974.
[5.29] DIN 53523, Bestimm Mooney im Warmsc Berlin 1965.
[5.30] ASTM D927/49 T, M Rubber and Rubberl Disc-viscometer, 1962
[5.31] W. Kuhn: "Beziehun statistischer Molekül schaften hochpolym (1936) 3.
[5.32] W. Pechold, S. Blase *Polym.* **189** (1965) 1
[5.33] L. Morbitzer, H. J. "Struktur und Ei kunststoffen," *Ange* 19–42.
[5.34] L. Morbitzer, D. "Structure and Pr fluence of Particle Modulus Tempera tion Behavior," *J.* 2704.
[5.35] L. Morbitzer, U. physikalischen A *Makromol. Chem*
[5.36] K. Ninomiya, J. Data from Diffe *ing of IISPR*, M
[5.37] J. D. Ferry: *Vi* John Wiley & S
[5.38] DIN ISO 9001, ell zur Darlegu Entwicklung, dienst, Beuth
[5.39] K. H. Padberg von Daten Merkmalsart, 214.
[5.40] DIN ISO 900 tätssicherung Anwendung,
[5.41] R. E. Kalma ing and Pre (1960) 35–4
[5.42] D. G. Luen Systems," 190–197.
[5.43] A. Gelb (e Press, Can
[5.44] W. Bramm *terministis terung*, R.
[5.45] E. D. Gil tomatisie *Verhandl schaft*, s
[5.46] N. R. Iu *tech. Pr*
[5.47] D. Balz *tisierun*
[5.48] H. Fitt der P *sche P*
[5.49] H.-P. lung r *techni* 239–
[5.50]
[5.51]
[5.52]
[5.53]
[5.54]
[5.55]
[5.56] *l* 1
[5.57] K
[5.58] K a u 6
[5.59] K " 71
[5.60] K. *te*
[5.61] J. *Fu* Yo
[5.62] U. Fü
[5.63] E. leit (19
[5.64] D. Sen (19
[5.65] M. run Post Oct.
[5.66] DIN men,
[5.67] G. P del," (199 G. Pi tersu *(atp)* G. Pi Anzei fähige *Praxi*
[5.68] H. Dr ders," (1991)
[5.69] H. Ra renste 205.

[5.91] NAMUR-Empfehlung 12, Explosionsschutz in Analysengeräteräumen, NAMUR, Darmstadt 1983.

[5.92] G. Strohrmann: *Anlagensicherung mit Mitteln der MSR-Technik*, Oldenbourg Verlag, München 1983.

[5.93] P. Netter: "Anlagensicherung mit Mitteln der Prozeßleittechnik," *Automatisierungstechnische Praxis (atp)* **35** (1993) no. 1, 45–51.

[5.94] NAMUR Empfehlung 24, Anforderungen an Meßeinsätze für Temperatursensoren (Widerstandsthermometer, Thermoelemente) zum Einsatz in eigensicheren Stromkreisen, NAMUR, Darmstadt 1991.

[5.95] F. Pigler: "Blitzschutz für Leittechnische Anlagen," *Automatisierungstechnische Praxis (atp)* **28** (1987) no. 5, 200–205.

[5.96] R.-W. Peters, D. Plieninger: "Bayer-interne Arbeiten zu CAE-gestützten Arbeitsmethoden," Leverkusen 1993.

[5.97] G. Heeg: "Objektorientierte Systeme," company brochure, Georg Heeg Verlag, Dortmund 1993.

[5.98] M. Arnold: "Analyse der Kommunikation zwischen Simulationsmodellen technischer Prozesse und den Prozeßnahen Komponenten von Prozeßleitsystemen," Diplomarbeit, RWTH, Aachen 1993.

[5.99] E. Nicklaus: "Sensortechnik in der Chemischen Produktion," *Mess. Prüf. Auto* (1987) 518–522.

[5.100] M. Polke: "Prozeßtechnik," Symposium der RWTH Aachen and Bayer Leverkusen 1988.

[5.101] M. Schweizer, H. J. Warnecke: "Sensoren für die Fertigungstechnik," *48. Physikertagung Münster 1984*, Physik-Verlag, Weinheim 1984, pp. 1331 ff.

[5.102] T. Pfeifer: "Fertigungsmesstechnik und Qualitätssicherung," Tagungsband zum Fachkongreß Microtecnic '92, *VDI Ber.* **1006**, Oct. 1992.

[5.103] H. Reichenberger: "Sensoren für Diagnose und Überwachung in der Medizintechnik," *48. Physikertagung Münster 1984*, Physik-Verlag GmbH, Weinheim 1984, pp. 1357 ff.

[5.104] E. Krestel (ed): "Bildgebende Systeme für die medizinische Diagnostik," Siemens AG, Berlin 1980.

[5.105] W. König (ed.): *Klinische-physiologische Untersuchungsmethoden*, Thieme Verlag, Stuttgart 1972.

[5.106] H. Kresse (ed.): *Kompendium Elektromedizin – Grundlagen, Anwendungen, Geräte*, Siemens AG, Berlin 1982.

[5.107] S. Kutscha: "Statistische Bewertungskriterien für die Entropieanalyse dynamischer Systeme," *Fortschrittsber. VDI Z. Reihe 8*, **178** (1989).

[5.108] S. Kutscha, H. Haubrich, K. Henning: "Lokalisierung von Epilepsieherden durch Einsatz der dynamischen Entropieanalyse," *Automatisierungstech. (at)* **40** (1992) no. 2, 70–76.

[5.109] H.-W. Delank: *Neurologie*, Enke Verlag, Stuttgart 1991.

[5.110] K. Poeck: *Neurologie*, Springer Verlag, Berlin 1987.

[5.111] F. D. Stott et al. (eds.): "ISAM-Gent-1981," *Proceedings of the 4th International Symposium on Ambulatory Monitoring and the 2nd Gent Workshop on Blood Pressure Variability*, Academic Press, London 1982.

[5.112] Bundesministerium für Forschung und Technologie: Mikrosystemtechnik 1994–1999 – Programm im Rahmen des Zukunftkonzeptes Informationstechnik, Referat 414, Bonn 1994.

[5.113] AMA Fachverband für Sensorik e.V.: Sensor '93 Kongreß, Technologie-Forum Mikrosystemtechnik, Nürnberg, Oct. 1993.

[5.114] P. Egelhaaf, F. Heintz, E. Zabler: "Sensoren im Kraftfahrzeug," *48. Physikertagung Münster 1984*, Physik-Verlag GmbH, Weinheim 1984, pp. 1367 ff.

[5.115] U. Kiencke: "Sensorik im Kraftfahrzeug – vom Sensor zum System," *VDI Ber.* **939** (1992).

[5.116] R. Polke: "Qualitätssicherung in der Feststoffverfahrenstechnik," *Chem. Ing. Tech.* **62** (1990) no. 10, 813–821.

[5.117] R. Polke: "The Relevance of Particle Measurement Technology to Process Engineering," *Particle and Particle Systems Characterization* **10** (1993) 86–91.

[5.118] R. Polke, N. Scholz, *Part. Charact.* **2** (1985) 39–40.

[5.119] Ullmann's Encyclopedia of Industrial Chemistry, Vol. B 2, VCH, Weinheim-New York 1988, pp. **1**-1–**1**-7.

[5.120] H. Rumpf: *Staub Reinhalt. Luft* **27** (1967) 3–13.

[5.121] R. Polke: *6. Fachtagung Granulometrie*, Dresden 1989, vol. 1, 39–48.

[5.122] K. Leschonski, W. Alex, B. Koglin: *Chem. Ing. Tech.* **46** (1974) 23–26; **47** (1975) 97–100.

[5.123] G. König, R. Polke, N. Scholz: *Verfahrenstechnik (Mainz)* **3** (1986) 10–17.

[5.124] R. Polke, M. Schäfer, N. Scholz: *4. Europ. Symp. Partikelmeßtechnik PARTEC*, Nürnberg 1989, vol. I, pp. 47–61.

[5.125] K. Sommer: *Probenahme von Pulvern und körnigen Massengütern: Grundlagen, Verfahren, Geräte*, Springer Verlag, Berlin 1979.

[5.126] Laser Sensor Techn. Inc.: "Lasentec," company brochure, Bellevue, WA 98009.

[5.127] D. Horn, *Angew. Makromol. Chem.* **166/167** (1989) 139–153.

[5.128] R. Polke, M. Rädle, N. Scholz: "Partikelmeßtechnik Status, Trend – Bedarf," *Chem. Ing. Tech.* **65** (1993) no. 10, 1191–1199.

[5.129] J. Krekel, R. Polke: "Qualitätssicherung bei der Verfahrensentwicklung," *Chem. Ing. Tech.* **64** (1992) no. 6, 528–535.

[5.130] H. H. Meyer, W. Noerpel, P. J. Schröder, W. Stieler: "Meßtechnik im Umweltschutz," company report, Bayer AG, Leverkusen, July 1988.

[5.131] M. Arnold, N. Ingendahl, M. Polke: "Prozeßautomatisierung in der Chemischen Industrie," *Innovationsdynamik prozeßanalytischer, automatisierungstechnischer und verfahrenstechnischer Methoden in der chemischen Industrie*, VDI-Hallescher Bezirksverein, Halle 14–15 Oct. 1993.

[5.132] G. Heiland: "Sensoren auf Metalloxidbasis," *48. Physikertagung Münster 1984*, Physik-Verlag GmbH, Weinheim 1984, pp. 1393 ff.

[5.133] H. Warncke, M. Kahl, H. H. Meyer, P. Schürmeyer, DE 2932436, 1984.

[5.134] D. Kitzelmann: "Elektrochemische Sensoren zur Erfassung von Schadstoffen in Industrie und Umwelt," *48. Physikertagung Münster 1984*, Physik-Verlag GmbH, Weinheim 1984, 1463 ff.

[5.135] R. Hotop: "Neue Methoden der Prozeßanalysetechnik für die verfahrenstechnische Produktion, *Chem. Ing. Tech.* **65** (1993) no. 8, 921–928.

[5.136] W. Noerpel: "Abwasserüberwachung mittels online Analytik," company report, Bayer AG, Leverkusen 1990.

[5.137] H. Raab, A. Wachter-Büchner: "Kontinuierliche TOC-Meßtechnik in der Abwasserüberwachung," *Automatisierungstechnische Praxis (atp)* **34** (1992) no. 3, 128–136.

[5.138] DIN 38409, Summarische Wartungs- und Stoffkenngrößen (Gruppe): Bestimmung des gesamten organisch gebundenen Kohlenstoffs (TOC), Beuth Verlag, Berlin 1983.

[5.139] Endress + Hauser, International Holding AG (eds.): *Abwasserhandbuch*, Reinach 1992.

[5.140] G. Fikentscher: *Cellulosechemie* **13** (1932) 58.

[5.141] M. Hoffmann, H. Krömer, R. Kuhn: *Polymeranalytik I*, Thieme Verlag, Stuttgart 1977.

[5.142] H.-G. Elias: *Makromoleküle*, Verlag Hüthig & Wepf, Heidelberg 1975.

[5.143] R. Hotop, J. König: "On-line Analytik in Polymerprozessen: Derzeitige Situation und Zukunftsaufgaben," *Automatisierungstechnische Praxis (atp)* **33** (1991) no. 12, 610–616.

[5.144] W. Günther: "Die HPLC im praktischen Vergleich," *Messen, Regeln, Automatisieren 1993*, Verlag Hoppenstedt & Co., Berlin 1993, pp. 17–24.

[5.145] B. von Falkai (ed.): *Synthesefasern*, Verlag Chemie, Weinheim 1981.

[5.146] B. Adler: "Erkennen von Prozeßtrends durch Methoden der Fuzzy Clusterung und Fuzzy Logic," lecture at VDI Tagung *Innovationsdynamik prozeßanalytischer, automatisierungstechnischer und verfahrenstechnischer Methoden in der chemischen Industrie*, Preprints GVC-VDI, Düsseldorf, Oct. 1993.

[5.147] M. Winterstein, G. Brückner, B. Adler: "Einsatz eines multivarianten Sensorsystems zur Abwasserüberwachung," lecture held at the GVC-VDI Tagung *Innovationsdynamik prozeßanalytischer, automatisierungstechnischer und verfahrenstechnischer Methoden in der chemischen Industrie*, Preprints GVC-VDI, Düsseldorf, Oct. 1993.

[5.148] M. Polke: "Trends der Prozeßleittechnik," *Automatisierungstechnische Praxis (atp)* **31** (1989) no. 9, 408–415.

[5.149] P. Hauptmann, R. Lucklum: "Sensoren für die Prozeßtechnik," lecture held at VDI Tagung *Innovationsdynamik prozeßanalytischer, automatisierungstechnischer und verfahrenstechnischer Methoden in der chemischen Industrie*, Preprints GVC-VDI, Düsseldorf, Oct. 1993.

[5.150] R.-W. Peters, Dissertation, RWTH, Aachen 1993.

[5.151] W. Göpel, J. Hesse, J. N. Zemel: *Sensors*, vols. 1–8, VCH Verlagsgesellschaft, Weinheim 1989.

[5.152] G. Tschulena: "Von Mikrosensoren zu Mikrosystemen – Stand und Perspektiven," *TM Tech. Mess.* **60** (1993) no. 9, 313–318.

[5.153] AMA: Sensor 91 – Kongreßband Marketing, Fachverband für Sensorik e.V., Nürnberg 1991.

[5.154] ISA Directory 1993: Comprehensive Guide for Measurement and Control Market, Instrument Society of America, North Carolina, 1993.

[5.155] *Messen, Regeln, Automatisieren; Einkaufsführer der Meß-, Regelungs- und Automatisierungstechnik*, Verlag Hoppenstedt & Co., Darmstadt 1993.

[5.156] *6. Internationale Fachmesse mit Kongreß für Sensorik & Systemtechnik*, Nürnberg, Oct. 1993.

[5.157] *MessComp '93-7*, Kongreßmesse für industrielle Meßtechnik, Wiesbaden, Sept. 1993.

[5.158] Prognos AG: *Sensortechnik 2000*, Prognos Weltreport, vol. 1, Basel, Sept. 1988.

[5.159] DMCW Associates: "U.S. Process Control Marketplace, Opportunities and Trends in the 1990's," final report, ref. no. MC-10, Menlo Park, CA 1991.

[5.160] K. Günzel: "Trends in der Prozeßmeßtechnik," *Tech. Mess.* **59** (1992) no. 9, 333–339.

[5.161] A. Brucker, H. Umbach: "Tendenzen bei der Messung von Durchfluß," *Automatisierungstechnische Praxis (atp)* **34** (1992) no. 3, 123–127.

[5.162] Batelle-Institut: *Sensors, Miniturisation and Integration*, Batelle-Institut e.V., Frankfurt 1988.

[5.163] G. Mark, R. Heidel, K. P. Frehse: "Wirtschaftlichkeit des Feldbuseinsatzes: Analyse des Installationsaufwandes unterschiedlicher Feldkommunikationssysteme," *Automatisierungstechnische Praxis (atp)* **34** (1992) no. 9, 507–512.

[5.164] Bayer AG: *Kabel- und Rohrwege*, company report, Bayer AG, Leverkusen, Oct. 1987.

References for Chapter 6

[6.1] H. Janocha (ed.): *Aktoren*, Springer Verlag, Berlin 1992.

[6.2] H. Drahten: "Der Feldbus aus Sicht des Anwenders," *Automatisierungstechnische Praxis (atp)* **33** (1991) no. 2, 55–64.

[6.3] G. Pinkowski: "Instrumentierungstechnik im Wandel. Von der Smart-Technik zum herstellerunabhängigen Bedienkonzept," *Automatisierungstechnische Praxis (atp)* **6** (1992) no. 5, 247–254.

[6.4] H. Junginger, H. Wehlan: "Der Feldmultiplexer aus Anwendersicht," *Automatisierungstechnische Praxis (atp)* **31** (1989) 12.

[6.5] A. Mangold: ACHEMA '91: "Stellgeräte für verfahrenstechnische Anlagen," *Automatisierungstechnische Praxis (atp)* **33** (1991) no. 11, 585–593.

[6.6] DIN 31051: Instandhaltung, Begriffe und Maßnahmen, Beuth Verlag GmbH, Berlin 1985.

[6.7] M. Björkmann: Elektrische Antriebstechnik in der Chemischen Industrie, NAMUR Hauptversammlung 1991, Lahnstein.

[6.8] B. Koenig, O. Ohligschläger: "Ein intelligenter elektropneumatischer Stellungsregler," *Automatisierungstechnische Praxis (atp)* **31** (1989) 8.

[6.9] M. Pandit, J. König, H. Hoffman: "Ein intelligenter elektropneumatischer Stellungsregler," *Automatisierungstechnische Praxis (atp)* **31** (1989) 367–373.

[6.10] DIN 19222: Leittechnik, Begriffe, March 1985.

[6.11] M. Polke: "On-line Betriebsmeßtechnik," lecture at the *GVC Jahrestreffen*, Wien 1992.

[6.12] P. Kruck: "Methodische Software-Entwicklung (dargestellt am Beispiel eines CAD-Systems zur Symbolischen Konstruktion), Dissertation, Bergische Universität GH Wuppertal 1987.

[6.13] H.-J. Dorn: "Drehzahlveränderliche Pumpenantriebe: Eine Alternative zu Stellventilen?" *Automatisierungstechnische Praxis (atp)* **29** (1987) no. 6.

[6.14] H. Linnenbrink: "Drehzahlveränderbare Antriebe, ein neuer Arbeitskreis im Arbeitsfeld," *3. NAMUR Hauptbericht, Hauptsitzung 53*, Lahnstein 1991.

[6.15] Antriebstechnik von AEG, Fachbereich Antriebssysteme und Leistungselektronik A 95, Gleichstrom-, Drehstrom- und Servoantriebe, Kurzdokumentation, Bestell-Nr. 029.129951, 1991.

[6.16] Bayer-interner Arbeitskreis PLT-Festlegungen: Einführung von PLT-Festlegungen nach Strukturierung auf Basis von E/R-Diagrammen, Leverkusen 1990.
[6.17] Canon, various company brochures, EOS Cameras, 1989.
[6.18] Veranstaltungsprogramm zu Internationale Konferenz zu neuen Aktoren – Actuator 1994, Bremen.
[6.19] V. Wagner et al., Bayer Internal Report, Leverkusen 1990.
[6.20] K. Hartmann, W. Schepers: Drehzahlgeregelte Pumpenanlagen, Bayer internal lecture, Leverkusen, Oct. 12, 1988.
[6.21] M. Polke: "Trends in der Prozeßleittechnik," *Automatisierungstechnische Praxis (atp)* **31** (1989) no. 9, 408–415.
[6.22] N. Haby: "Drehzahlveränderbare Pumpen als Stellglieder im Regelkreisen," *Regelungstechnische Praxis (atp)* **22** (1986) 6.
[6.23] H. Linnenbrink: "Drehzahlveränderbare Antriebe. Ein neuer Arbeitskreis im Arbeitsfeld," *3. NAMUR-Hauptbericht, Hauptsitzung 53*, 1990.
[6.24] M. Polke: "Trend zur Automatisierung in der Chemischen Industrie hält ungebrochen an: Gedanken zur ACHEMA 1991," *Automatisierungstechnische Praxis (atp)* **33** (1991) no. 5, 219.
[6.25] M. Harmoinen: "Säädetyt AC – käytöt tulevaisuudessa," (Variable speed AC drives in the future), ABB Strömberg Oy, Power Electronics, Vaasa, Finland 1991.
[6.26] Sanyo Electric, Semiconductor Division, Sanyo MST Hybrid System, Japan, Oct. 1989.
[6.27] W. Mutz: "Prozeßnahe Stromversorgung," *Automatisierungstechnische Praxis (atp)* Sonderheft NAMUR Statusbericht 87, 1987.
[6.28] W. Mutz: "Niederspannungstechnik für die Betriebe," EN-PLT-Kolloquium "Prozeßelektrotechnik," Bayer AG, Leverkusen 1987.
[6.29] M. Freytag: "Sichere Stromversorgung für PLT-Einrichtungen," PLT-Kolloquium "Erfahrungen mit dem Einsatz moderner Prozeßleittechnik," Bayer AG, Leverkusen 1984.
[6.30] K. Kling: Stand und Entwicklung der Errichtungsbestimmungen VDE 0100.etz b.H.6, pp. 128–131, 1974.
[6.31] Bayer AG: "Beleuchtungsanlagen," PLT-Festlegungen 06.53.01, Leverkusen 1984.
[6.32] Bayer AG: "Die Bedeutung des Sicherheitsschalters bei Arbeiten in Gefahrenbereichen technischer Arbeitsmittel," Arbeitsschutzfestlegung LE2, Leverkusen 1984.

References for Chapter 7

[7.1] G. U. Spohr: "Prozeßleitsysteme," in: *Prozeßleittechnik*, 1st ed., Oldenbourg Verlag, München 1992, Chap. 5.2.
[7.2] M. Polke: "Prozeßleittechnik," Skriptum zur Vorlesung WS84/85 Universität Stuttgart, Ausgabe 1984 und folgende.
[7.3] U. Epple: "Prozeßführung," in: *Prozeßleittechnik*, 1st ed., Oldenbourg Verlag, München 1992, Chap. 4.3.
[7.4] M. Brombacher, M. Polke: "Perspektiven der Prozeßleittechnik," *Chem. Ing. Tech.* **60** (1988) no. 7, 539–546.
[7.5] K. F. Früh: "Optimalstruktur dezentraler Prozeßleitsysteme," *RTP Regelungstech. Prax.* **21** (1979) 273 ff.
[7.6] W. Körner, P. Moll, M. Polke: "Zur funktionalen Gliederung von Leitsystemen," *Automatisierungstechnische Praxis (atp)* **29** (1987) no. 9, 393–400.
[7.7] U. Epple: "Funktionsorientierte Strukturierung prozeßleittechnischer Aufgaben," *Automatisierungstechnische Praxis (atp)* **31** (1989) no. 10, 468–474.
[7.8] L. Litz: "Anforderungen an die Prozeßleitsysteme der nächsten Generation," *53. NAMUR Hauptsitzung*, Lahnstein 1990.
[7.9] ISO: MMS, Part 1: Service Definitions, MMS Part 2: Protocol Specification, 1988.
[7.10] ISO: ISO/DIS 74498 Open Systems Interconnection Basic Reference Model, May 1983.
[7.11] M. Arnold: "Analyse der Kommunikation zwischen Simulationsmodellen technischer Prozesse und den prozeßnahen Komponenten von Prozeßleitsystemen," Diplomarbeit, Lehrstuhl für Prozeßleittechnik, RWTH Aachen, 1993.
[7.12] M. Polke: "Information als kritische Ressource," in H. A. Henzler (ed.): *Handbuch strategische Führung*, Gabler, Wiesbaden 1988.
[7.13] NAMUR Empfehlung 21: Elektromagnetische Verträglichkeit von Betriebsmitteln der Prozeß- und Laborleittechnik, NAMUR Geschäftsstelle, Leverkusen Dec. 1990.
[7.14] DIN VDE 848 T1–T3: Gefährdung durch elektromagnetische Felder, T4: Sicherheit bei elektromagnetischen Feldern, Beuth Verlag, Berlin 1982.
[7.15] DIN VDE 843 T1–T4: Elektromagnetische Verträglichkeit von Meß-, Steuer- und Regeleinrichtungen in der industriellen Prozeßtechnik, Beuth Verlag, Berlin 1987.

References for Chapter 8

[8.1] J. König: "Datenkommunikation in der Leittechnik," *Automatisierungstechnische Praxis (atp)* **28** (1986) no. 5, 213–214.
[8.2] M. Freytag: "Datenkommunikation in der Leittechnik – Status und Trend," *Automatisierungstechnische Praxis (atp)* **28** (1986) no. 5, 215–222.
[8.3] M. Polke: "Informationshaushalt technischer Systeme," in Prozeßrechner 1984, Prozeßdatenverarbeitung im Wandel 4, *Prozeßrechnertagung 1984, Karlsruhe*, Springer Verlag, Berlin-Heidelberg 1984, pp. 62–70.
[8.4] M. Syrbe: "Informationstechnik, Chance und Verantwortung," *Vortrag Produktionstechnisches Kolloquium*, Berlin 1986, Fraunhofer-Gesellschaft, München.
[8.5] H. Steusloff: "Kommunikation in technischen Systemen," *GI/NTG Tagung Kommunikation in verteilten Systemen*, Karlsruhe, 11.–15.3.1985, *Informatik-Fachberichte* **111**, Springer Verlag, Berlin 1985, pp. 33–55.
[8.6] R. W. Peters: "Funktionale Gliederung der Informationssysteme für Produktion, Labor und Infrastruktur; Funktionen der Produktionsleitebene," *48. NAMUR Hauptsitzung*, Lahnstein 1985; Berichtsband pp. 210–232.
[8.7] D. Powell et al.: "The Delta-4 Approach to Dependability in Open Distributed Computing Systems," *FTCS 18*, Tokio 1988.
[8.8] ISO: Open Systems Interconnection Basic Reference Model, ISO/DIS 7498, May 1983.

[8.9] O. Gihr, E.-H. Göldner, P. J. Kühn, K. Sauer: "Lokale Netze und ISDN-Nebenstellenanlagen, Stand und Entwicklungstendenzen," *PIK (Praxis der Informationsverarbeitung und Kommunikation)*, **3**, Hanser Verlag, München 1986, pp. 26–34.

[8.10] General Motors (GM): "Manufacturing Automation Protocol (MAP) Document Version 2," GM EMCC-APMES A/MD-39, General Motors Technical Center, Warren 1984, MI 48090-9040.

[8.11] ESPRIT Project 5104 (CNMA): CNMA Implementation Guide, Revision 5.1, Volume 1–3. University of Porto (ed.), Ref. 5104.92.35/F1.PD, 1992.

[8.12] H. Steusloff: "Technische Kommunikation in der industriellen Automation. INFINA '89, Informatik für die industrielle Automation," *VDI-Berichte* **723**, VDI-Verlag, 1989, pp. 319–338.

[8.13] M. Polke: "Trends der Prozeßleittechnik," *Automatisierungstechnische Praxis (atp)* **31** (1989) 408–415.

[8.14] Factory Instrumentation Protocol C46-602 Application Layer (MPS), C 46-603 Data Link Layer, C46-604, Twisted Pair Physical Layer. Französische Normungsorganisation AFNOR, Paris 1988.

[8.15] DIN 19245: Messen, Steuern, Regeln, PROFIBUS, Process Field Bus. Beuth Verlag, Berlin.

[8.16] D. Janetzky, K. Watson: Performance Evaluation of the MAP Token Bus in Real Time Applications. Advances in Local Area Networks (eds.: K. Kümmerle, F. Tobagi, J. Limb). IEEE, New York 1987, pp. 411–425.

[8.17] Fieldbus; ESPRIT-CIM-Report: CNMA, Phase 4, Addendum 2, Chapter 4, in Vorbereitung; Kommission der Europäischen Gemeinschaften, DG XIII, 200 Rue de la Loi, Brüssel, Belgien.

[8.18] J. A. H. Pfleger: "Kommunikationssystem Feldbus," *Automatisierungstechnische Praxis (atp)* **28** (1986) 223–227.

[8.19] H. Drathen: "Neuorientierung der leittechnischen Normung auf nationaler und internationaler Ebene," *Automatisierungstechnische Praxis (atp)* **33** (1991) no. 11, 580–585.

[8.20] W. Büsing: "MAP für die Prozeßleittechnik: Die Herausforderung bleibt," *NAMUR Statusbericht '87, Prozeßleittechnik in der Chemischen Industrie*, R. Oldenbourg Verlag, München 1987, pp. 101–106.

[8.21] G. Scheidenberger: "Fieldbus Message Specification (FMS)". *Automatisierungstechnische Praxis (atp)* **31** (1989) 152.

[8.22] D. Heger: "Konformitätsprüfung offener Kommunikationsprotokolle für die Automatisierungstechnik. Fachbericht Messen, Steuern, Regeln. Fortschritte in der Meß- und Automatisierungstechnik durch Informationstechnik," vol. 14, *INTERKAMA-Kongreß*, Oldenbourg Verlag, München 1986.

[8.23] ISO: "Specification of Abstract Syntax Notation One (ASN.1)," IS 8824, May 1987; "Specification of Basic Encoding Rules for Abstract Syntax Notation One (ASN.1)," IS 8825, May 1987.

[8.24] W. Gora, R. Speyerer: *ASN.1, Abstract Syntax Notation One*, DATACOM Buchverlag, Pulheim 1987.

[8.25] M. Syrbe: "Basic Principles of Advanced Process Control System Structures and a Realization with Distributed Microcomputers," *IFAC World Congress*, June 1978, Helsinki, Pergamon Press, New York 1979, pp. 393–407.

[8.26] M. Siefert: "Entwurf und Implementierung eines ASN. 1-Compliers," Diplomarbeit, Fakultät für Informatik, Universität Karlsruhe/Frauenhofer-Institut für Informations- und Datenverarbeitung, Karlsruhe 1988.

References for Chapter 9

[9.1] D. T. Ross, K. E. Schomann, Jr.: "Structured Analysis for Requirements Definition," *IEEE Transactions on Software Engineering* **SE-3** (1977) no. 1, 6–15.

[9.2] T. De Marco: *Structured Analysis and System Specification*, Yourdon Press, New York 1978.

[9.3] M. Polke: "Prozeßleittechnik; Trend zur Automatisierung in der Chemischen Industrie hält ungebrochen an," *Automatisierungstechnische Praxis (atp)* **33** (1991) no. 5, 219–222.

[9.4] mbp: Produktbeschreibung des CAE-Systems HW-PLAN, Dortmund 1990.

[9.5] G.-U. Spohr: "CAE für die Prozeßleittechnik," *51. NAMUR-Hauptsitzung*, Lahnstein, Dec. 1988.

[9.6] G.-U. Spohr, R. Schrieber: "Anforderungen an CAE-Systeme," *INTERKAMA-Kongreß*, Düsseldorf, Oct. 1989, in: G. Schmidt, H. Steusloff (eds.): *Mit vernetzten intelligenten Komponenten zu leistungsfähigeren Meß- und Automatisierungssystemen*, Oldenbourg Verlag, München 1989, pp. 575–584.

[9.7] G.-U. Spohr: "CAE in der Prozeßleittechnik," Vorlesungsmanuskript, RWTH Aachen, Lehrstuhl für Prozeßleittechnik, 1991.

References for Chapter 10

[10.1] HOAI: Rechtsverordnung zum Gesetz zur Regelung von Ingenieur- und Architektenleistungen vom 04.11.1971.

[10.2] M. Polke: "Einführung in die Prozeßleittechnik," Manuskript zur Vorlesung, Universität Stuttgart 1984.

[10.3] DIN 69901, Projektwirtschaft; Projektmanagement; Begriffe, Beuth Verlag, Berlin, August 1987.

[10.4] E. Cassirer: *Substanzbegriff und Funktionsbegriff*, 6th ed., Wissenschaftliche Buchgesellschaft, Darmstadt 1990.

[10.5] W. F. Daenzer: *Systems Engineering*, Verlag Industrielle Organisation, Zürich 1988.

[10.6] D. Reschke, A. Schelle, K. Schnopp: *Handbuch "Projektmanagement,"* Verlag TÜV Rheinland, Köln 1989.

[10.7] H. H. Schulze, A. Schembra: "Schätzung der Investitionskosten und Produktionskosten bei der Verfahrensentwicklung und Anlagenplanung," lecture held at the *GVC-Tagung* über Entwicklung und Auslegung verfahrenstechnischer Prozesse, Baden-Baden, 12/13 June 1989.

[10.8] E. Willems, H. Hennecke: "Planung, Montage, Inbetriebnahme und Ingenieurbetreuung von Einrichtungen zum Messen, Steuern, Regeln," in J. Hengstenberg, K. H. Schmitt, B. Sturm, D. Winkler (eds.): *Messen, Steuern, Regeln in der chemischen Technik*, 3rd ed., vol. 5, Springer Verlag, München 1985.

[10.9] H. Groh, R. Gutsch: *Netzplantechnik*, VDI-Verlag, Düsseldorf 1982.

[10.10] S. Dworatschek et al.: "Project Management Software," *The International GPM/INTERNET-Symposium*, Garmisch, 15–18 June 1986.

[10.11] A. Tampier: "Dezentrale Einzelprojektplanung und zentrale Multiprojektplanung im automatischen Verbund VAX-PC-VAX mit IST-Kosten-Übernahme," in D. Reschke, A. Schelle (eds.): *Beiträge zum Projektmanagement-Forum 1989*, Tagung Garmisch, 4–6 Oct. 1989.

[10.12] MS-Project, software product von Micro-Soft.

[10.13] M. Brombacher: "Das Lastenheft als Grundlage der Automatisierung chemischer Verfahren und seine Darstellung als Expertensystem," Dissertation, Technische Universität, München 1985.

[10.14] M. Polke: "Informationshaushalt technischer Prozesse," *Automatisierungstechnische Praxis (atp)* **27** (1985) 161–171.

[10.15] M. Polke: "Information als kritische Ressource," in H. A. Henzler (ed.): *Handbuch "Strategische Führung,"* Gabler-Verlag, Wiesbaden 1988.

[10.16] T H. Müller-Heinzerling: "Flexible Standardsoftware zur Automatisierung von Chargenprozessen mit Prozeßleitsystem TELEPERM M," *Automatisierungstechnische Praxis*, **30** (1988) no. 6.

[10.17] R. Schmidt, Diverse Bayer-interne Mitteilungen, Leverkusen 1989–1990.

[10.18] M. Niese, EP 0104411, 1986.

[10.19] N. Becker: Diverse Bayer-interne Mitteilungen, Leverkusen 1989–1990.

[10.20] *Ullmann*, 4th ed., **4**, 70.

[10.21] G. Bernecker: *Planung und Bau verfahrenstechnischer Anlagen*, VDI-Verlag Düsseldorf 1984.

[10.22] DIN 28004, Fließbilder verfahrenstechnischer Anlagen; Begriffe, Fließbildarten, Informationsinhalt, part 1, Beuth Verlag, Berlin, May 1988.

[10.23] M. Polke: "Prozeßleittechnik in der Produktion; Voraussetzung für Diversifikation von Chemiefasern," *Lenzinger Ber.* **67** (1989) 12–23.

[10.24] DIN 19227, Bildzeichen und Kennbuchstaben für Messen, Steuern, Regeln in der Verfahrenstechnik; Zeichen für die funktionelle Darstellung, part 1, Beuth Verlag, Berlin, Sept. 1973.

[10.25] W. Ahrens, M. Polke: "Informationsstrukturen in der Automatisierungstechnik," lecture held at Interkama-Kongreß 1989, in G. Schmidt, H. Steusloff (eds.): *Mit vernetzten, intelligenten Komponenten zu leistungsfähigen Meß- und Automatisierungssystemen*, R. Oldenbourg Verlag, München 1989, pp. 47–68.

[10.26] E. Welfonder: "Leittechnik – Dokumentation aus Betreibersicht," *ETG/VGB-Fachtagung Betriebsgerechte Dokumentation in Kraftwerken, Einfluß der modernen Leittechnik und Planungsmittel*, Baden-Baden 1986.

[10.27] DIN 32705, Klassifikationssysteme; Erstellung und Weiterentwicklung von Klassifikationssystemen, Beuth Verlag, Berlin 1987.

[10.28] R. Kessler: "Prozeßorientierte Strukturierung von verfahrenstechnischen Anlagen," *Automatisierungstechnische Praxis (atp)* **31** (1989) no. 10, 461–467.

[10.29] DIN 66201, Prozeßrechensysteme; Begriffe, part 1, Beuth Verlag, Berlin, May 1981.

[10.30] J. König: "Sicherheitstechnische Aspekte der Prozeßleittechnik," *Chem.-Ing. Tech.* **49** (1987) no. 3, 196–204.

[10.31] R. Magin, W. Wüchner: *Digitale Prozeßleittechnik*, Vogel Buchverlag, Würzburg 1987.

[10.32] M. Freytag: "USV und Notstromversorgung," *VDI-Tagung*, Fellbach/Stuttgart, 25.–26.10.1989.

[10.33] R. Peters: Diverse Bayer-interne Mitteilungen, Leverkusen 1985.

[10.34] E. Ignatowitz: *Chemietechnik*, Verlag Europa-Lehrmittel, Wuppertal 1982, p. 112.

[10.35] DIN/IEC 3B (Sec)40, Ausführung von Funktionsplänen für Steuerungen, Identisch mit IEC 3B (Sec)40, Oct. 1985.

[10.36] DIN 40719, Kennzeichnungsregeln zur funktions- und ortsbezogenen Kennzeichnung von Systemen und Anlagen, Einrichtungen und Betriebsmitteln, Anschlüssen, Signalen und technischer Dokumentation, Teil 6, Entwurf, Beuth Verlag, March 1989.

[10.37] D. Plieninger: "Qualitätssicherung in der leittechnischen Planung," NAMUR-Hauptsitzung, Lahnstein 1991.

[10.38] K. H. Schmitt: Qualitätssicherung in der Leittechnik, NAMUR-Hauptsitzung, Lahnstein 1991.

[10.39] W. Masing: *Qualitätslehre, DGQ*, 5th ed., Beuth Verlag, Berlin 1977.

[10.40] A. Müller: Qualitätsprüfung von PLT-Dokumentationen als Vertragsbestandteil bei der Vergabe von Ingenieurleistungen an Fremdfirmen, Bayer-interne Festlegung, Leverkusen 1992.

[10.41] D. Plieninger, A. Müller: "Ansätze zur Qualitätssicherung in der leittechnischen Planung," *Automatisierungstechnische Praxis (atp)* **34** (1992) no. 10, 564ff.

[10.42] H. Rinne, H.-J. Mittag: *Statistische Methoden der Qualitätssicherung*, 2nd ed., Hanser Verlag, München 1991.

[10.43] G. Färber, M. Polke, H. Steusloff: "Mensch-Prozeß-Kommunikation," *Chem. Ing. Tech.* **57** (1985) no. 4, 307–317.

[10.44] Bayer AG, Beleuchtungsanlagen, PLT-Festlegungen, 06.53.01, Leverkusen 1984.

[10.45] Bayer AG: Prozeßleitwarten, Leitstände, Nebenräume. Richtlinien für die konstruktive Gestaltung, PLT-Festlegungen, Leverkusen, 1989.

[10.46] R. Beuchelt, U. Fierus, H. Hennecke, E. Willems: PLT-Raumdesign, interne Richtlinie der Bayer AG Leverkusen 1985.

[10.47] VDI/VDE 3546: Konstruktive Gestaltung von Prozeßleitwarten. Blatt 1 (1987): Allgemeiner Teil; Blatt 2 (1981): Bautechnische Maßnahmen; Blatt 3 (1988): Ausführung des Leitstandes; Blatt 4 (1986): Nebenräume; Blatt 5 (1991): Anordnung von Monitoren, Beuth Verlag, Berlin.

[10.48] NAMUR Empfehlung NE 12: "Explosionsschutz von Analysegeräteräumen". NAMUR Geschäftsstelle Leverkusen, Bayerwerk, Leverkusen 1983.

[10.49] DIN 33402: Körpermaße des Menschen. Beuth Verlag, Berlin 1986.

[10.50] DIN 33414: Ergonomische Gestaltung von Warten. Beuth Verlag, Berlin 1988.

[10.51] DIN 33403: Klima am Arbeitsplatz und in der Arbeitsumgebung, Beuth Verlag, Berlin 1988.

[10.52] DIN 5035: Beleuchtung mit künstlichem Licht, Teil 2: Richtwerte für Arbeitsstätten in Innenräumen und in Freien, Beuth Verlag. Berlin, 1990.

[10.53] Verordnung über Arbeitsstätten. Bundesministerium für Arbeit und Sozialordnung, Referat Öffentlichkeitsarbeit, Bonn 1975.

References for Chapter 11

[11.1] J. König: "Sicherheitstechnische Aspekte der Prozeßleittechnik," *Chem. Ing. Tech.* **59** (1987) no. 3 216–218.

[11.2] R. Bachmann, J. König: "Towards Increased Safety – Some Future Aspects of Modern Process Control for Chemical Plants," in *Automatic Control World Congress 1987*, Pergamon Press, Oxford.

[11.3] R. Isermann (ed.): *IFAC-Symposium: Safeprocess*, Baden-Baden 1991.

[11.4] E. D. Gilles: "Von der MSR-Technik zur wissensbasierten Prozeßleittechnik," *Chem. Anlagen + Verfahren* **4** (1985) 62–64.

[11.5] M. Polke: "Erforderliche Informationsstrukturen für die Qualitätssicherung," *Chem. Ing. Tech.* **65** (1993) no. 7, 791–796.

[11.6] G. Färber, M. Polke, H. Steusloff: "Mensch-Prozeß-Kommunikation," *Chem. Ing. Tech.* **57** (1985) no. 4, 307–317.

[11.7] R. Perne, M. Polke: "Produktentwicklung durch Verfahrensvariation," *Regelungstechnik* **30** (1982) no. 5, 147–156.

[11.8] E. D. Gilles, M. Wurst: "Prozeßführung zur Sicherung reproduzierbarer Qualität in der verfahrenstechnischen Produktion," lecture at *GVC-Jahrestreffen*, Stuttgart 1990, pp. 138–139.

[11.9] J. Pawlowski: *Die Ähnlichkeitstheorie in der physikalisch-technischen Forschung*, Springer Verlag, Berlin 1971.

[11.10] J. Pawlowski: "Ähnlichkeitstheorie – ein wirksames Instrument der Verfahrenstechnik," *GVC-Tagung Ähnlichkeitstheoretische Methoden und ihr verstärkter Einsatz in der Praxis*, Mainz, June 15, 1984.

[11.11] M. Zeitz: "Simulationstechnik," *Jahrestreffen der Verfahrensingenieure*, Straßburg 1986, Oldenbourg Verlag, München 1987, pp. 164–169.

[11.12] H. Schuler: Methoden zur Prozeßführung mit Simulationsmodellen, *Automatisierungstechnische Praxis (atp)* **31** (1989) no. 10, 475–481; no. 11, 519–523.

[11.13] W. Ahrens, M. Polke: "Netzmodelle als systemtechnische Informationsbasis für die Prozeßleittechnik," *Automatisierungstechnik (at)* **37** (1989) no. 3, 94–103.

[11.14] E. D. Gilles: "Auf dem Wege zu einer modellgestützten Prozeßleittechnik," *50. Jahrestagung NAMUR*, Lahnstein 1987, Oldenbourg Verlag, München 1987, pp. 35–45.

[11.15] E. D. Gilles, W. Marquardt: "Dynamische Anlagensimulation in der chemischen Verfahrenstechnik," *Automatisierungstechnische Praxis (atp)* **34** (1986) no. 2, 49–58.

[11.16] W. Ahrens: "Einsatz von Expertensystemen in der Prozeßleittechnik," *Automatisierungstechnische Praxis (atp)* **29** (1987) no. 10, 475–485.

[11.17] M. Brombacher: "Expertensysteme zur Unterstützung der Planung (Lastenheft)," *Automatisierungstechnische Praxis (atp)* **30** (1988) no. 6, 280–291.

[11.18] W. Marquardt: "Rechnergestützte Erstellung verfahrenstechnischer Prozeßmodelle," *Chem. Ing. Tech.* **64** (1992) no. 1, 25–40.

[11.19] H. Fittler: "Anwendung von Expertensysteme in der Prozeßautomation," *Automatisierungstechnische Praxis (atp)* **33** (1991) no. 4, 199–205.

[11.20] E. D. Gilles et al.: "Ein Trainingssimulator zur Ausbildung von Betriebspersonal in der chemischen Industrie," *Automatisierungstechnische Praxis (atp)* **32** (1990) no. 7, 343–350.

[11.21] J. Laube, M. Polke: "Das Phasenmodell der Produktion: Ein semantisches CAE-Werkzeug zur Prozessbeschreibung," Lecture at *Regelungstechnische Kolloquim*, Boppard, March 1993.

[11.22] A. Lippert et al.: "Gedanken zur Neugestaltung der Verfahrenstechnik," company report, Bayer, Leverkusen, Nov. 1988.

[11.23] C. A. Petri: "Kommunikation mit Automaten," Dissertation, Technische Hochschule Darmstadt 1961.

[11.24] W. Reisig: *Systementwurf mit Netzen*, Springer Verlag, Berlin 1985.

[11.25] G. Scheschong: "Eine auf Petrinetzen basierende Konstruktionsanalyse und (Teile-)verifikationsmethode zur Modellierungsunterstützung bei Entwicklung von Informationssystemen," Dissertation, TU Berlin 1984.

[11.26] M. Brombacher, M. Polke: "Perspektiven der Prozeßleittechnik," *Automatisierungstechnische Praxis (atp)* **29** (1987) no. 11, 503–509; no. 12, 558–561.

[11.27] U. Epple: "Funktionsorientierte Strukturierung Prozeßleittechnischer Aufgaben," *Automatisierungstechnische Praxis (atp)* **31** (1989) no. 10, 468–474.

[11.28] O. Graf: "Standartisierte Grundfunktionen verringern Softwareaufwand bei Teleperm M.," *eng & auto* **3** (1991) 2–5.

[11.29] Bayer AG, EPO 104 411, 1986 (M. Niese).

[11.30] W. Wüchner: "Anforderungen an das Personal beim Einsatz moderner Prozeßleitsysteme, *Automatisierungstechnische Praxis (atp)* **28** (1986) no. 1 13–18.

[11.31] G. Krauss: "Akzeptanz der Prozeßleitsysteme durch die Betriebe," *Automatisierungstechnische Praxis (atp)* **10** (1985) no. 10 470–476.

[11.32] H. Raab: "Die Rolle der Sensoren in der Verfahrenstechnik, und: Digitale Sensoren und Sensorsysteme: Welche Informationen sollen sie liefern," *Chem. Ing. Tech.* **61** (1989) no. 3, 41–47.

[11.33] Verein Deutscher Ingenieure: "Der Ingenieur in Beruf und Gesellschaft," Handlungsempfehlung: Sozialvertragliche Gestaltung von Automatisierungsvorhaben, Düsseldorf, Dec. 1989.

[11.34] L. Bainbridge: "Ironies of Automation," *Automatica* **19** (1983) no. 6, 775–559.

[11.35] H. Grauel, H. Bauer, P. Metzing: "Erhöhung der Kompetenz des Betreiberpersonals – Ziel der zustandsorientierten Mensch-Prozeß-Kommunikation in Chemieanlagen," *MSR (Berlin)* **30** (1987) no. 10, 448–455.

[11.36] H. Lenk: "Können Informationssysteme moralisch verantwortlich sein?" *Informatik Spektrum* **12** (1989) 248–255.

[11.37] H. J. Charwart: *Lexikon der Mensch–Maschine–Kommunikation*, Oldenbourg Verlag, München 1992.

[11.38] B. Geiser: "Was Window-Manager in der Leittechnik leisten sollen," *Elektronik* **16** (1991) 52–59.

[11.39] Siemens AG: "Teleperm M, company brochure, no. A 19100 E816-B677 100045 PA 10895, 1991.

[11.40] AEG: Schirmbilder für Viewstar 750, Frankfurt.

[11.41] ABB Prozeßautomation GmbH: Schirmbilder für Advant Station.

[11.42] K. R. Boff, J. E. Lincoln: *Engineering Data Compendium – Human Perception and Performance*, Harry G. Armstrong Aerospace Medical Research Lab., Ohio 1988.

[11.43] H. Schmidtke: *Handbuch der Ergonomie*, 2nd ed., Carl Hanser Verlag, München 1989.

[11.44] N. Heineken: "Aufnahme und Verarbeitung von Information," to be published in H. Schmidtke: *Handbuch der Ergonomie*, 2nd ed., Carl Hanser Verlag, München 1989.

[11.45] H. Steusloff: "Mensch-Prozeß-Kommunikation," *DPG-Arbeitstagung*, Bad Honnef, Dec. 1983, pp. 1–23.

[11.46] R. W. Bailey: *Human Performance Engineering, A Guide for Systems Designers*, Bell Telephone Laboratories, Inc., Prentice-Hall, Englewood Cliffs, N.J. 1982.

[11.47] I. Rasmussen: "The Role of Hierarchical Knowledge Representation in Decision Making in System Management," IEEE Transations on Systems, Man and Cybernetics, vol., SMC -15(1985) 2, pp. 234–243.

[11.48] G. Johannsen: "Towards a New Quality of Automation in Complex Man-Machine Systems," *Automatica* **28** (1992) no. 2, 355–373.

[11.49] M. Eigen: "Mozart – oder unsere Verlegenheit im Umgang mit dem Genie," in *Perspektiven der Wissenschaft*, DVA, Stuttgart 1988, pp. 76–104.

[11.50] J. C. Eccles: *Das Gehirn des Menschen*, 6th ed., R. Piper, München 1975.

[11.51] H. and K. Benesch: *dtv-Atlas zur Psychologie*, dtv, München 1987, pp. 90–123.

[11.52] W. Metzger (ed.): "1. Wahrnehmung und Bewußtsein," *Handbuch der Psychologie*, vol. I, Hogrete, Göttingen 1966.

[11.53] N. A. Streitz, P. A. Gloor (eds.): "Hypertext und Hypermedia: Von theoretischen Konzepten zur praktischen Anwendung," *Informatik-Fachberichte* **249**, Springer Verlag, Heidelberg 1990.

[11.54] A. Despopoulos, S. Silbernagel: *Taschenatlas der Physiologie*, Georg Thieme Verlag, Stuttgart 1988.

[11.55] R. F. Schmidt, G. Thews: *Physiologie des Menschen*, Springer Verlag, Berlin 1983.

[11.56] A. Hajos: *Einführung in die Wahrnehmungspsychologie*, Wissenschaftliche Buchgesellschaft, Darmstadt 1980.

[11.57] W. D. Heiss: "Kernspintomographie und Positronen-Emissions-Tomographie zur Lokalisation menschlicher Hirnfunktion," *Ver. Ges. Dtsch. Naturforsch. Ärzte* **114** (1986) 241–269.

[11.58] G. Deutsch, S. P. Springer: *Linkes und rechtes Gehirn: Funktionale Asymmetrien*, 2nd ed., Spektrum der Wissenschaft – Verlagsgesellschaft, Heidelberg 1988.

[11.59] H. Ostertag: "Positronen-Emissions-Tomographie (PET)," *Phys. Bl.* **48** (1992) no. 2, 77.

[11.60] K. Poeck (ed.): *Klinische Neuropsychologie*, Georg Thieme Verlag, Stuttgart 1989.

[11.61] R. J. Trotter: "Das Jahrzehnt des Gehirns," *Psychologie Heute*, Jan. 1992, 57–59.

[11.62] E. Straus: *Vom Sinn der Sinne*, Springer Verlag, Berlin 1978.

[11.63] J. J. Gibson: *Wahrnehmung und Umwelt*, Urban & Schwarzenberg, München 1982.

[11.64] K. Lorenz: *Die Rückseite des Spiegels. Versuch einer Naturgeschichte menschlichen Erkennens*, dtv, München 1977.

[11.65] VDI/VDE Gesellschaft Meß- und Automatisierungstechnik (GMA), VDI/VDE Richtlinie 3536, Düsseldorf, Jan. 1990.

[11.66] D. O'Shaughnessy: *Speech Communication*, Addison-Wesley, Reading, Mass. 1987.

[11.67] D. Fukunada, W. L. G. Koontz, *Inf. Control* **16** (1970) 85–101.

[11.68] E. Paulus, D. Längle: "Die Anwendung der Karhunen-Loève-Entwicklung für die digitale Sprachanalyse und -synthese," *NTG-Fachtagung Signalverarbeitung*, Erlangen 1973, pp. 362–369.

[11.69] R. M. Anderberg: *Cluster Analysis for Applications*, Academic Press, New York 1973.

[11.70] H. Späth: *Cluster-Analyse-Algorithmen*, 2nd ed., Oldenbourg Verlag, München 1977.

[11.71] K. Stephan, P. Metzing: "Automatisierung komplexer Chargenprozesse," *MSR (Berlin)* **29** (1986) 511–515.

[11.72] B. Böhme: "Anwendung der Situationssteuerung in komplexen verfahrenstechnischen Systemen," in Wissenschaftliche Berichte, vol. 15, TH Leipzig, 1985, pp. 48–81.

[11.73] R. Müller, F. Müller, A. Parusel: "Planning Man-Machine-Communication in Process Control Systems," *Proceedings of the BIAS 1990*, Milan, pp. 407–417.

[11.74] G. S. Axelby, A. P. Sage (eds.): Zusammenfassung der Tagungsbeiträge der IFAC/IFIP/IFORS/IEA Conference on Analysis, Design and Evaluation on Man-Machine-Systems, *Automatica* **19** (1982) no. 6, 593–779.

[11.75] J. Richter: "Experimentelle Untersuchung zur aufgabenbezogenen Gestaltung von Informationssystemen in zentralen Überwachungseinheiten," Dissertation, Humbold-Universität, Berlin 1982.

[11.76] A. De Meyer: "Software verfahrensgerecht strukturieren," company report, Bayer Antwerpen NV, Antwerpen 1991.

[11.77] H. Jürgens, D. Saupe (eds.): *Visualisierung in Mathematik und Naturwissenschaften*, Springer Verlag, Berlin 1989.

[11.78] H. Müller: *Realistische Computergraphik*, Springer Verlag, Berlin 1988.

[11.79] K. Kansy, P. Wisskirchen (eds.): *Graphik und KI*, Springer Verlag, Berlin 1990.

[11.80] K. P. Fähnrich, C. Janssen, G. Groh: "Entwicklungswerkzeuge für graphische Benutzerschnittstellen," *Computer Magazin* **2** (1992) 6–13.

[11.81] D. A. Young: *X Window Systems*, Prentice Hall, Englewood Cliffs, N.J. 1989.

[11.82] Eckelmann: Industrieautomation, InTouch, company brochure, Wiesbaden-Erbenheim 1991.

[11.83] PROLINK AG, Factory Link, company brochure, Seengen 1992.

[11.84] Prozeß-Leittechnik GmbH, APROL, company brochure, Hilden 1991.

[11.85] G. Geiser: *Mensch–Maschine–Kommunikation*, Oldenbourg Verlag, München 1990.

[11.86] Siemens AG: Teleperm M: Das Prozeßleitsystem, company brochure no. A 19100-E815-A2-V1, 1983.

[11.87] SattControl GmbH, SATTLINE, company brochure, Wuppertal 1990.

[11.88] AEG, GeAmatics, Company Brochure Frankfurt 1992.

[11.89] Eckardt AG: LS 130 Leistation des PLS 80 E, Systemübersicht PLS 180 t, Druckschrift B 133/0692, Frankfurt 1992.

[11.90] ABB Prozeßautomation GmbH: ABB Master, company brochure, Mannheim 1992.

[11.91] J. Encarnacao: "Die perfekte Täuschung als hochgestecktes Ziel," forschung – Mitteilungen der DFG **2** (1992) 4–7.

[11.92] M. Utermöhle, H. G. Pagendarm: "Sichtweisen – Visualisierung," *Grundlagen und Entscheidungskriterien* **IX** (1992) 92–98.

[11.93] H. W. Franke: "Welten aus Bits and Bytes," *VDI Nachr.* **9** (1992) 13–17.

[11.94] R. Steinmetz, R. G. Herrtwich: "Integrierte verteile Multimedia-Systeme," *Informatik-Spektrum* **14** (1991) 249–260.

[11.95] R. G. Herrtwich: "Betriebsmittelvergabe unter Echtzeitgesichtspunkten," *Informatik-Spektrum* **14** (1991).

[11.96] D. P. Anderson: "Resource Management for Digital Audio and Video," IEEE Workshop on Real-Time-Operating Systems and Software, Charlottesville, May 1990.

[11.97] P. Hoepner: "Synchronizing the Presentation of Multimedia Objects – ODA Extensions," *Multimedia – Systems, Interaction and Applications, 1st Eurographics Workshop*, Stockholm, April 1991, pp. 87–100.

[11.98] N. S. Borenstein: *Multimedia Application Development with the Andrew Toolkit*, Prentice Hall, Englewood Cliffs 1990.

[11.99] K. Meyer-Wegener: *Multimedia-Datenbanken*, B. G. Teubner, Stuttgart 1991.

[11.100] J. L. Alty, C. D. C. McCartney: "Managing Multimedia Resources in Process Control," *Multimedia – Systems, Interaction and Applications, 1st Eurographics Workshop*, Stockholm, April 1991, pp. 293–306.

[11.101] E. Golla: "Multimedia-Anwendung für die Industrie," *CIM-Praxis* **1** (1993) 46–47.

[11.102] W. Michaeli, K. Borgschulte, M. Philip: "Neue Konzepte bei der Automation in der kunststoffverarbeitenden Industrie durch Anwendung von Bildverarbeitungssystemen," *Proceedings of the 9th International Conference on Automated Inspection and Product Control*, Stuttgart, May 1989, pp. 115–123.

[11.103] F. J. Kleiner, M. Liewld: "Einsatz von CCD-Array-Kameras für optoelektronische Messungen in der Umformtechnik," in: *Umformtechnik – Ideen, Konzepte und Entwicklungen*, Teubner Verlag, Stuttgart 1992, pp. 243–260.

[11.104] P. Haberäcker: *Digitale Bildverarbeitung*, Carl Hanser Verlag, München 1989.

[11.105] D. Weimar: "A Synthetical Sicual Environment with Hand Gesturing and Voice Input," *Proceedings of CHI' 89*, May 1989, pp. 235–240.

[11.106] D. M. Frohlich: "The Design Space of Interface," in *Multimedia – Systems, Interaction and Applications*, 1st *Eurographics Workshop*, Stockholm, April 1991, pp. 53–69.

[11.107] H. Marmolin: "Multimedia from the Perspectives of Psychology," in: *Multimedia – Systems, Interaction and Applications, 1st Eurographics Workshop*, Stockholm, April 1991, pp. 39–52.

[11.108] R. Bogaschewsky, "Hypertext-/Hypermedia-Systeme – Ein Überblick," *Informatik-Spektrum* **15** (1992) 127–143.

[11.109] M. Mühlhäuser: "Hypermedia-Konzepte zur Verarbeitung multimedialer Information," *Informatik-Spektrum* **14** (1991) 281–290.

[11.110] D. Roller: "Technical Information System for Assembly, Test and Service Documentation," *Proc. of the 3rd Int. Conf. on Human Aspects of Advanced Manufacturing and Hybrid Automation*, Gelsenkirchen, Aug. 1992, pp. 273–278.

[11.111] A. Lieberam: "Expertensysteme in der Chemischen Technik," *Chem. Ing. Tech.* **65** (1993) no. 4, 362–363.

[11.112] H. Kumano: "Learning with Interactive Multimedia by Plant Operators," *IFAC Workshop on Production in Process Industry*, Düsseldorf, March 1993, pp. 127–134.

[11.113] M. Z. Conklin: "Operator Performance Support System (OPPS)," *Proceedings of Technology* 2002, Feb. 1992.

[11.114] J. A. Mullens, L. C. Williams, W. L. Zabriskie: "Integrated Diagnostic System for a Process Controller," *International Mashinery Monitoring and Diagnostic Conference*, Las Vegas, September 1989.

[11.115] J. P. Hack, P. H. Heldt: "Das Cockpit des Airbus A 310," *Spektrum der Wissenschaft*, March 1984, 38–52.

[11.116] Blaupunkt GmbH, ALI, company brochure, Frankfurt 1984.

[11.117] BMW AG, company brochure no. VS-202 IX/83 311 500410, München 1983.

[11.118] BMW AG, company brochure no. 411 070111-1/84 VM, München 1984.

[11.119] Siemens AG, Teleperm M – Das Prozeßleitsystem für Kraftwerke, company brochure no. A 19100 E661-A2-V1, 1984.

[11.120] M. Polke, M. Wienand: "Anforderungs- und Qualifikationsprofile," Festschrift zum 60. Geburtstag von Prof. Dr. Stöckmann, Karlsruhe, Oct. 1978.

[11.121] G. Farber, M. Polke, H. Steusloff: "Mensch–Prozeß–Kommunikation," *Chem. Ing. Tech.* **57** (1985) no. 4 307–309.

[11.122] P. I. Burton, *Chem. Eng. (Rugby, Engl.)*, March 1981, 103.

[11.123] E. Blass: "Methodische Entwicklung verfahrenstechnischer Prozesse," *Chem. Ing. Tech.* **57** (1985) no. 3, 201–210.

[11.124] W. E. Bixby, P. G. Andrus, L. E. Walkup, *Photogr. Eng.* **5** (1954) 195–201.

[11.125] I. T. Bickmore, R. E. Hayford, H. E. Clark, *Photogr. Sci. Eng.* **3** (1959) 120–214.

[11.126] L. A. Jones, *J. Franklin-Inst.* **187** (1920) 29.

[11.127] M. Polke: "Zur Physik der Elektrophotographie," in: *Festkörperprobleme*, vol. VIII, Vieweg/Pergamon Press, Oxford 1968, pp. 131–174.

[11.128] H. Rumpf: "Über die Eigenschaften von Nutzstäuben," *Staub Reinh. Luft* **27** (1967) 3–13.

[11.129] R. Polke, N. Scholz: "Quality Control in Solids Process Engineering," *Part. Charact.* **2** (1985) 39–48.

[11.130] R. Ecker, H. Kramer, H. K. Müller, M. Polke: "Neue Aspekte bei der Prüfung von Kautschuk und Gummi," *Kautsch. Gummi Kunstst.* **1** (1972) 5–10.

[11.131] E. D. Gilles, E. Nicklaus, M. Polke: "Sensortechnik in der Chemie, Status und Trend," *Automatisierungstechnische Praxis (atp)* **28** (1986) no. 9, 423–431; no. 10, 479–484.

[11.132] H. Töpfer, H. Bischoff: "Grundlagen der Prozeßanalyse," KTD. Wissenschaftlich-Technische Gesellschaft für Meß- und Automatisierungstechnik, Lehrbrief 1 bis 4, Dresden 1979.

[11.133] J. Lauber, M. Polke: "Das Phasermodell der Produktion als methodisches Hilfsmittel für die Prozeßentwicklung," *GVC-Sitzung Prozeß- und Anlagentechnik*, Oct. 1992, Wiesbaden.

[11.134] M. Breitenfelder, M. Polke, M. Wienand: "Der Weg eines Produktes vom Konzept zur Kontrolle," lecture held by M. Polke at the Arbeitstagung *Forschungsmanagement in der Physik* 9.–12. June 1976, Universität Bielefeld.

[11.135] M. Polke, P. Siara: "Chemiefasern – Chancen und Probleme in der Zukunft," Vortrag anläßlich des Zukunftssymposiums – Textil "Märkte morgen", 1975, Verband der Textilindustrie Westfalen, Schriftenreihe Heft 22.

[11.136] M. Polke, M. Wienand: "Der Physiker in der Industrie – Ausbildung und Beruf," *Phys. Bl.* **6** (1975) and **1** (1976).

[11.137] R. W. Peters, M. Polke: "Verfahren zur Objektivierung der Studienreform," VDI-Hauptgruppe Ingenieurausbildung, Düsseldorf 1972.

[11.138] R. W. Peters, M. Polke: "Beruf und Ausbildung der Ingenieure – Eine empirische Untersuchung," VDI-Hauptgruppe Der Ingenieur in Beruf und Gesellschaft, Düsseldorf 1972.

[11.139] H. Hillmer, R. W. Peters, M. Polke: *Studium, Beruf und Qualifikation der Ingenieure*, VDI-Verlag, Düsseldorf 1976.

[11.140] M. Polke: "Die Bedeutung des Qualifikationsprofils des Ingenieurs für die Gestaltung der Studiengänge in der Ingenieur-Aus- und Weiterbildung," *FEANI-OECP-Seminar*, München 1973.

[11.141] H. Drahten, M. Polke: "Anforderungsprofil und Qualifikationsprofil als Basis zur Ermittlung des Weiterbildungsbedarfs – Ergebnis einer Umfrage," in G. Schmidt, H. Steusloff (eds.): "Mit vernetzten, intelligenten Komponenten zu leistungsfähigen Meß- und Automatisierungssystemen," *Tagungsband Interkama-Kongreß 1989*, Oldenbourg Verlag, München, pp. 643–651.

[11.142] M. Polke: "Aus- und Weiterbildung in und an moderner Prozeßleittechnik," in W. Brückers, N. Meyer (eds.): *Arbeit, Technik und berufliche Bildung im Metallbereich*, vol. 3, Verlagsgesellschaft, Köln 1988, pp. 27–53.

[11.143] M. Polke: "Prozeßleittechnik in der Produktion – Voraussetzung für Diversifikation von Chemiefasern," *Lenzinger Ber.* **67** (1989) no. 8, 12–24.

[11.144] K. H. Padberg, P.-Th. Wilrich: Die Auswertung von Daten und ihre Abhängigkeit von der Merkmalsart, part 1, *Qualität und Zuverlässigkeit* **26** (1981) 179–183.

[11.145] DIN 31 051, Begriffe und Maßnahmen, Beuth Verlag, Berlin 1985.

[11.146] F. Zutt, H.-P. Hubich: "Sinnvolles Agieren war früher kaum möglich; Praxisbeispiel. Instandhaltung aus der Sicht der chemischen Industrie. Instandhaltungsmanagement der 90er Jahre," in W. Schulte, G. Küffner (eds.): *Blick durch die Wirtschaft*, Frankfurter Allgemeine Zeitung GmbH, Frankfurt 1988, pp. 287–298.

[11.147] R. Isermann: Model Based Fault Diagnosis of Machines via Parameter Estimation Methods and Knowledge Processing," *IFAC-Symposium*, Baden-Baden 1991.

[11.148] H. Strickling et al.: "Erstellung von Erhaltungskatastern," company report, Bayer, Leverkusen FS 1975, PLT 1987.

[11.149] W. Aleite, D. Wach: "Supervision and fault diagnosis," *IFAC-Symposium*, Baden-Baden 1991.

[11.150] N. Kuschnerus: "Die EDV-unterstützte Planung der Instandhaltung: Instandhaltung als Teil der Anlagenwirtschaft," company report, Bayer AG, Leverkusen 1988.

[11.151] W. Männel: "Modernes Instandhaltungsmanagement: Anforderungsgerechte IH-Konzepte mehren den Unternehmenserfolg," *IH-Markt*, 1989, pp. 4–14.

References for Chapter 12

[12.1] IEC: Status and Rules of Procedure of the IEC, 01 (Central Office), 809, 809 A, July 1990.

[12.2] M. Polke, B. Will: "Tradition und Fortschritt – 40 Jahre NAMUR," *Automatisierungstechnische Praxis (atp)* **32** (1990) no. 3, 105–109.

References for Chapter 13

General References

Expert Systems

[13.1] W. Ahrens: "Einsatz von Expertensystemen in der Prozeßleittechnik," *Automatisierungstechnische Praxis (atp)* **29** (1987) 475–485.

[13.2] W. Ahrens: "Expertensysteme für die Prozeßführung, Erfahrungen aus dem Verbundprojekt TEX-I," *Chem. Ing. Tech.* **62** (1990) no. 8, 635–644.

[13.3] R. Bernhardt, C. Koppermann, U. Müller-Nebler, K.-H. Wietschorke: "Expertensystem für betriebsnahe Problemstellungen," *Automatisierungstechnische Praxis (atp)* **31** (1989) 580–587.

[13.4] H. Fittler: "Anwendung von Expertensystemen in der Prozeßautomation," *Automatisierungstechnische Praxis (atp)* **31** (1989) 74–80.

[13.5] W. Kaufmann (ed.): *Diagnostische Entscheidungsprozesse in der inneren Medizin*, Schattauer, Stuttgart 1986.

[13.6] R. Soltysiak: "Praktische Anwendung von Expertensystemen in der Prozeßleittechnik," *Automatisierungstechnische Praxis (atp)* **32** (1990) 247–257.

[13.7] R. Soltysiak: "HEPROX, ein Expertensystemshell für Prozeßführungsaufgaben," *Automatisierungstechnische Praxis (atp)* **33** (1991) 74–80.

Fuzzy Logic

[13.8] T. Mc Cusker: "Neural Networks and Fuzzy Logic, Tools of Promise for Controls," *Control Eng.* **37** (1990) May, 84–85.

[13.9] B. Kosko: *Neural Networks and Fuzzy Systems*, Prentice Hall, Englewood Cliffs 1990.

[13.10] L. A. Zadeh: *Fuzzy Sets, Information and Control*, **8** (1965) 338–353.

[13.11] H.-I. Zimmermann: *Fuzzy Set Theory – and Its Applications*, 2nd ed., Klumer-Nijhoff Publ., Boston 1990.

Hypertext/Hypermedia

[13.12] I. Conklin: Hypertext: A Introduction and Survey, *IEEE Comp. Mag.* **20** (1987) no. 9, 17–41.
[13.13] P. A. Gloor, N. A. Streitz (eds.): "Hypertext und Hypermedia," *Informatik Fachberichte 249*, Springer Verlag, Berlin 1990.
[13.14] P. A. Gloor: *Hypermedia-Anwendungsentwicklung*, Teubner, Stuttgart 1990.
[13.15] T. Nelsen: "A File Structure of the Complex, the Changing and the Indeterminate," *Proc. 20th Nat. ACM-Conf.* **1965**, 84–100.

Neural Networks

[13.16] R. Eckmiller, G. Hartmann, G. Hanske (eds.): *Parallel Processing in Neural Systems and Computers*, Elsevier, Amsterdam 1990.
[13.17] H. Ritter: *Neuronale Netze*, Addison Wesley, Reading, Mass., 1989.
[13.18] G. Bretthauer, G. Straube: "Einsatzmöglichkeiten für Fuzzy-Regler," Frauenhofer-Institut für Informations- und Datenverarbeitung, Dresden 1993.

Specific References

[13.19] D. Balzer, V. Kirbach, V. May: "Knowledge based Process Control," in: H. Steusloff, M. Polke (eds.): *Integration of Design, Implementation and Application in Measurement, Automation and Control*, INTERKAMA Tagungsband, Oldenbourg Verlag, München 1992.
[13.20] D. Balzer, V. May, U. Starke: "Wissensgestützte Prozeßführung," part 1: *Automatisierungstechnische Praxis (atp)* **34** (1992) no. 1, 36–43; part 2: *Automatisierungstechnische Praxis (atp)* **34** (1992) no. 2, 85–92.
[13.21] D. Balzer, V. Kirchbach, V. May: "A Real-Time Expert System for Process Control," *Proceedings of the 2nd IFAC Workshop on Artificial Intelligenze in Real-Time Control, AIRTC '89*, Shenyang 1989.
[13.22] B. Böhme, U. Starke: "Expert Systems in Process Control," *Proceedings IFAC Symposium on Large Scale Systems*, Berlin 1989.
[13.23] V. May, R. Wieland, B. Böhme, D. Balzer: "PROCON-I ein Expertensystemshell zur Lösung höherer Automatisierungsaufgaben," *Proceedings of the 33rd IWK*, Ilmenau 1988.
[13.24] V. May: "Grundlagen und Entwicklung von Rahmen-Expertensystemen für Aufgaben der Prozeßführung -das Expertensystemshell PROCON," Dissertation, TH Leipzig 1989.
[13.25] D. Balzer, V. Kirbach: "Neuronale Netze in der Prozeßautomatisierung," *atp* **34** (1992) no. 7, 395–401.
[13.26] N. Kunde: "Methoden und Werkzeuge zur Wissensakquisition in der Prozeßautomatisierung," Dissertation, TH Leipzig, 1991.
[13.27] U. Starke: "Eine Methodologie für das Knowledge Engineering in der Prozeßsteuerung," Dissertation, TH Leipzig 1990.

References for Chapter 14

[14.1] I. Rasmussen: "The Role of Hierarchical Knowledge Representation in Decision Making on System Management," IEEE Transaction on Systems Man and Cybernetics, vol. SMC-15 (1985) no. 2, 234–243.
[14.2] U. Epple: "Modular Technology, Basis of an Object-Oriented Network Structure for Process Control Engineering," *GMA-Fachbericht 4*, VDE-Verlag, Berlin 1993.
[14.3] M. Niese, EP 0104411, 1986.
[14.4] A. Strickling: "CAE in der Prozeßleittechnik," *NAMUR-Statusbericht 1987*, pp. 211–217.
[14.5] P. Chen: "The Entity-Relationship Model – Towards a Unified View of Data," *ACM Transactions on Database Systems* (1976) 1.
[14.6] H. Robel et al.: *Lehrbuch der chemischen Verfahrenstechnik*, VEB Deutscher Verlag für Grundstoffindustrie, Leipzig 1983.
[14.7] E. Cassirer: *Substanzbegriff und Funktionsbegriff*, Springer Verlag, Berlin 1910.
[14.8] E. D. Gilles, E. Nicklaus, M. Polke: "Sensortechnik in der Chemie – Status und Trend," *Automatisierungstechnische Praxis (atp)* **9** (1986) 423–431; **10** (1986) 479–484.
[14.9] M. Polke: Prozeßleittechnik, Manuskript der Vorlesung, Universität Stuttgart.

[3.54] K. Glover: "All Optimal Hankel-Norm Approximations of Linear Multivariable Systems and Their L_∞-Error Bounds," *Int. J. Control* **39** (1985) no. 6, 1115–1193.

[3.55] D. C. Hyland, D. S. Bernstein: "The Optimal Projection Equations for Model Reduction and the Relationship among the Methods of Wilson, Skelton and Moore," *IEEE Trans. Autom. Control* **AC-30** (1985) no. 12, 1201–1211.

[3.56] L. Litz: *Reduktion der Ordnung linearer Zustandsraummodelle mittels modaler Verfahren*, Hochschul-Verlag, Stuttgart 1979.

[3.57] B. C. Moore: "Principal Component Analysis in Linear Systems: Controllability, Observability, and Model Reduction," *IEEE Trans. Autom. Control* **AC-26** (1981) 17–31.

[3.58] U. Pallaske: "Ein Verfahren zur Ordnungsreduktion mathematischer Prozeßmodelle," *Chem. Ing. Tech.* **59** (1987) no. 7, 604–605.

[3.59] R. E. Skelton: "Cost Decomposition of Linear Systems With Application to Model Reduction," *Int. J. Control* **32** (1980) 1031–1055.

[3.60] D. A. Wilson: "Optimum Solution of Model-Reduction Problem," *Proc. IEEE* **58** (1970) 1161–1165.

[3.61] E. Welfonder: "Leittechnikdokumentation aus Betreibersicht," *ETG/VGB-Fachtagung Betriebsgerechte Dokumentation in Kraftwerken, Einfluß der modernen Leittechnik und Planungsmittel*, Baden-Baden 1986.

[3.62] E. Ignatowitz: *Chemietechnik*, Verlag Europa-Lehrmittel, Wuppertal 1980.

References for Chapter 4

[4.1] I. M. Horowitz: *Synthesis of Feedback Systems*, Academic Press, New York 1963.

[4.2] W. Oppelt: *Kleines Handbuch technischer Regelvorgänge*, Verlag Chemie, Weinheim 1972.

[4.3] O. Föllinger: *Regelungstechnik*, Hüthig, Heidelberg 1985.

[4.4] O. Föllinger: *Optimierung dynamischer Systeme. Eine Einführung für Ingenieure*, Oldenbourg Verlag, München 1985.

[4.5] J. C. Doyle, G. Stein: "Multivariable Feedback Design: Concepts for a Classical/Modern Synthesis," *IEEE Trans. Autom. Control* **26** (1981) 4–16.

[4.6] M. Morari, J. C. Doyle: "A Unifying Framework for Control System Design under Uncertainty and its Implications for Chemical Process Control," *Proc. Chem. Proc. Control* **III** (1986) 5–51.

[4.7] J. S. Freudenberg, D. P. Looze: *Frequency Domain Properties of Scalar and Multivariable Feedback Systems*, Springer Verlag, Berlin 1988.

[4.8] H. H. Rosenbrock: *Computer-Aided Control System Design*, Academic Press, London 1974.

[4.9] D. H. Owens: *Feedback and Multivariable Systems*, Peter Peregrinus, Stevenage 1978.

[4.10] A. G. J. MacFarlane, D. F. A. Scott-Jones: "Vector Gain," *Int. J. Control* **29** (1979) 65–91.

[4.11] Y. Takahashi, M. J. Rabins, D. M. Auslander: *Control and Dynamic Systems*, Addison-Wesley, Reading, Mass. 1970.

[4.12] R. V. Patel, N. Munro: *Multivariable Systems Theory and Design*, Pergamon Press, Oxford 1982.

[4.13] U. Korn, H. Wilfert: *Mehrgrößenregelungen*, VEB Verlag Technik, Berlin 1982.

[4.14] H. Tolle: *Mehrgrößen-Regelkreissynthese*, Oldenbourg Verlag, München 1983.

[4.15] J. M. Maciejowski: *Multivariable Feedback Design*, Addison-Wesley, Reading, Mass., 1989.

[4.16] A. Gelb (ed.): *Applied Optimal Estimation*, M.I.T. Press, Cambridge, Mass., 1974.

[4.17] K. Brammer, G. Siffling: *Stochastische Grundlagen des Kalman-Bucy-Filters: Wahrscheinlichkeitsrechnung und Zufallprozesse*, Oldenbourg Verlag, München 1975.

[4.18] K. Brammer, G. Siffling: *Kalman-Bucy-Filters Deterministische Beobachtung und stochastische Filterung*, Oldenbourg Verlag, München 1975.

[4.19] H. Kwakernaak, R. Siwan: *Linear Optimal Control Systems*, J. Wiley, New York 1972.

[4.20] B. D. O. Anderson, J. B. Moore: *Linear Optimal Control*, Prentice-Hall, Englewood Cliffs, NJ, 1972.

[4.21] G. Stein, M. Athans: "The LQG/LTR Procedure for Multivariable Feedback Control Design," *IEEE Trans. Autom. Control* **32** (1987) 105–114.

[4.22] A. G. J. MacFarlane, B. Kouvaritakis: "A Design Technique for Linear Multivariable Feedback Systems," *Int. J. Control* **25** (1977) 837–874.

[4.23] Y. S. Hung, A. G. J. MacFarlane: *Multivariable Fedback: A Quasi-Classical Approach*, Springer Verlag, Berlin 1982.

[4.24] C. E. Garcia, M. Morari: "Inernal Model Control: 1. A Unifying Review and Some New Results," *Ind. Eng. Chem. Process Des. Dev.* **21** (1982) 308–323.

[4.25] B. A. Francis: *A Course in H_∞ Control Theory*, Springer Verlag, Berlin 1987.

[4.26] J. Doyle, K. Glover, P. Khargonekar, B. Francis: "State-Space Solutions to Standard H_2 and H_∞ Control Problems," *IEEE Trans. Autom. Control* **34** (1989) 831–847.

[4.27] V. Manousiouthakis, R. Savage, Y. Arkun: "Synthesis of Decentralized Control Structures Using the Concept of Block Relative Gain," *Am. Inst. Chem. Eng.* **32** (1986) 991–1003.

[4.28] D. D. Siljak: *Large-Scale Dynamic Systems*, North-Holland, New York 1978.

[4.29] M. G. Singh, T. Titli: *Systems: Decomposition, Optimisation and Control*, Pergamon Press, Oxford 1978.

[4.30] M. G. Singh: *Dynamical Hierarchical Control*, North-Holland, Amsterdam 1980.

[4.31] M. G. Singh: *Decentralised Control*, North-Holland, Amsterdam 1981.

[4.32] L. Litz: *Dezentrale Regelung*, Oldenbourg Verlag, München 1983.

[4.33] C. J. Harris, S. A. Billings (eds.): *Self-tuning and Adaptive Control: Theory and Applications*, Peter Peregrinus, Stevenage 1981.

[4.34] M. M. Gupta, C. T. Chen (eds.): *Adaptive Methods for Control System Design*, IEEE Press, New York 1985.

[4.35] A. Isidori: *Nonlinear Control Systems: An Introduction*, Springer Verlag, Berlin 1985.

[4.36] C. E. Garcia, D. M. Prett, M. Morari: "Model Predictive Control: Theory and Practice – a Survey," *Automatica* **25** (1989) 335–347.

[4.37] D. K. Frederick, C. J. Herget, R. Kool, M. Rimvall: "ELCS – The Extended List of Control Software," unpublished, WGS, Eindhoven University of Technology, Eindhoven 1987.

[4.38] M. Jamshidi, C. Herget: *Computer-Aided Control Systems Engineering*, North-Holland, Amsterdam 1985.

[4.39] P. Holl, W. Marquardt, E. D. Gilles: "Diva – a Powerful Tool for Dynamic Process Simulation," *Comput. Chem. Eng.* **12** (1988) 421–425.

[4.40] F. Allgöwer, J. Raisch: "Multivariable Controller Design for an Industrial Distillation Column," *Proceedings of 5th IMA International Conference on Control Theory*, 1988.

[4.41] A. Duschek: *Vorlesungen über höhere Mathematik*, vol. 3, Springer Verlag, Wien 1960.

[4.42] P. Funk: *Variationsrechnung und ihre Anwendung in Physik und Technik*, Springer Verlag, Berlin 1962.

[4.43] E. Klingbeil: *Variationsrechnung*, Bibliographisches Institut, Mannheim 1977.

[4.44] H. Schwarz: *Optimale Regelung linearer Systeme*, Bibliographisches Institut, Mannheim 1976.

[4.45] W. Boltjanski: *Mathematische Methoden der optimalen Steuerung*, Carl Hanser Verlag, München 1972.

[4.46] J. Hsu, A. Meyer: *Modern Control Principles and Applications*, part III, McGraw-Hill, New York 1968.

[4.47] L. Pontrjagin et al.: *Mathematische Theorie optimaler Prozesse*, Oldenbourg Verlag, München 1967.

[4.48] R. Bellmann: *Dynamic Programming*, Princeton University Press, Princeton 1965.

[4.49] R. Bellmann, S. Dreyfus: *Applied Dynamic Programming*, Princeton University Press, Princeton 1964.

[4.50] G. Schneider, H. Mikolcic: *Einführung in die Methode der dynamischen Programmierung*, Oldenbourg Verlag, München 1972.

[4.51] M. Athans, P. Falb: *Optimal Control*, McGraw-Hill, New York 1966.

[4.52] G. Bieß et al.: *Optimale Prozesse und Systeme*, Verlag Harry Deutsch, Thun–Frankfurt 1980.

[4.53] A. Bryson, Y. Ho: *Applied Optimal Control*, J. Wiley, New York 1975.

[4.54] E. Hofer, R. Lunderstädt: *Numerische Methoden der Optimierung*, Oldenbourg Verlag, München 1975.

[4.55] U. Hoffmann, H. Hofmann: *Einführung in die Optimierung*, Verlag Chemie, Weinheim 1971.

[4.56] H. Tolle: *Optimierungsverfahren*, Springer Verlag, Berlin 1971.

[4.57] K. Stahl: *Industrielle Steuerungstechnik in schaltalgebraischer Behandlung*, Oldenbourg Verlag, München 1965.

[4.58] K. H. Fasol: *Binäre Steuerungstechnik*, Springer Verlag, Berlin 1988.

[4.59] R. Lauber: *Prozeßautomatisierung*, vol. 1, Springer Verlag, Berlin 1989.

[4.60] W. Körner, P. Moll, M. Polke: "Zur funktionalen Gliederung von Leitsystemen," *Automatisierungstechnische Praxis (atp)* **29** (1987) no. 9, 393–400.

[4.61] A. H. Levis: "Challenges to Control: a Collective View," *IEEE Trans. Autom. Control* **AC-32** (1987) no. 4, 275–285.

[4.62] B. Rosenstengel, U. Wienand: *Petri-Netze – Eine anwendungsorientierte Einführung*, Vieweg Verlag, Wiesbaden 1982.

[4.63] D. Bochmann, C. Posthoff: *Binäre dynamische Systeme*, Oldenbourg Verlag, München 1981.

[4.64] R. Herschel: *Einführung in die Theorie der Automaten, Sprachen und Algorithmen*, Oldenbourg Verlag, München 1974.

[4.65] E. Leonhardt: *Grundlagen der Digitaltechnik. Eine systematische Einführung*, Carl Hanser Verlag, München 1982.

[4.66] S. Wendt: *Entwurf komplexer Schaltwerke*, Springer Verlag, Berlin 1974.

[4.67] M. Brombacher, M. Polke: "Perspektiven der Prozeßleittechnik," *Automatisierungstechnische Praxis (atp)* **29** (1987) 501–509, 558–567, 1987, pp. 22–35.

[4.68] M. Niese, EP 0104411, 1986.

[4.69] M. Niese: "Struktureller Aufbau und dezentrale Führung in Produktionsanlagen," *INTERKAMA Kongreß*, München 1989, pp. 339–349.

[4.70] M. Bruns, M. Eibl, F. J. Kersting, R. J. Uhlig: "Anforderungen an Systeme zur Rezeptfahrweise," *Automatisierungstechnische Praxis (atp)* **35** (1993) no. 1, 40–44; no. 2, 104–108.

[4.71] S. Riemer: "Funktionen der Produktionslogistik in der Verfahrenstechnik," *INTERKAMA Kongreß*, München 1989, pp. 381–390.

[4.72] K.-F. Geibig: "Funktionen der Betriebsleitebene," *Automatisierungstechnische Praxis (atp)* **34** (1992) no. 2, 68–72.

[4.73] M. Polke: *Prozeßleittechnik*, Universität Stuttgart, 1984 ff.

[4.74] L. Prins, J. Stäheli: "Zur Automatisierung von diskontinuierlichen Vielzweckanlagen in Chemiebetrieben," *Regelungstech. Prax.* **20** (1978) no. 9, 256–261.

[4.75] U. Epple, H. Kopec, R. Schmidt: "Strukturierung von Prozeßführungsaufgaben und Leitsystemsoftware," *Automatisierungstechnische Praxis (atp)* **34** (1992) no. 2, 59–67.

[4.76] U. Epple: "Die Bausteintechnik – Grundlage einer objektorientierten Netzstruktur für die Prozeßleittechnik, GMA-Fachbericht," *VDE-Kongreß*, Berlin 20.–23.1.1993.

References for Chapter 5

[5.1] E. Pavlik: "Sensoren als Quelle für den Informationshaushalt technischer Prozesse," *46. NAMUR Hauptsitzung*, Darmstadt, 7–9 Dec. 1983, 48. Physikertagung and Frühjahrstagung, 15 March 1984, Münster, pp. 1322 ff, Physik-Verlag GmbH, Weinheim 1984.

[5.2] R. Weisbeck: "Sensoren für die Chemische Industrie – eine kritische Bewertung des Standes der Technik und derzeitiger Sensorentwicklung," *46. NAMUR Hauptsitzung*, Darmstadt, 7–9 Dec. 1983.

[5.3] H. Warncke: "Sensortechnik in der Chemischen Industrie: Aufgaben und Möglichkeiten der Stoffgrößenmeßtechnik," *Tech. Mess.* 52 (1985) no. 4.

[5.4] H. Raab: "Digitale Sensoren und Sensorsysteme – Welche Informationen sollen sie liefern?," *Automatisierungstechnische Praxis (atp)* **30** (1988) no. 11, 534–544.

[5.5] VDI/VDE 2600, Blatt 3: Metrologie, Gerätetechnische Begriffe, Beuth Verlag, Berlin 1973.

[5.6] H. Warncke: Vorteile, Probleme und Möglichkeiten der in-situ-Messung von Stoffgrößen, Bayer-interner Bericht, Leverkusen, Nov. 1982.

[5.7] E. D. Gilles, E. Nicklaus, M. Polke: "Sensortechnik in der Chemie – Status und Trends," *Automatisierungstechnische Praxis (atp)* **28** (1986) no. 9/10, 423–432, 479–484.

[5.8] *Ullmann*, 4th ed., **5**, pp. 797–890.

[5.9] H.-J. Schneider: "Sensorsysteme für die Betriebsmeßtechnik und Kommunikation im Feld," *Automatisierungstechnische Praxis (atp)* **33** (1991) no. 10, 511–528.

[5.10] K. Hartmann: "On-line Analytik aus funktionaler Sicht, *Automatisierungstechnische Praxis (atp)* **31** (1989) no. 11, 510–515.

[5.11] H. Schuler, Ch.-U. Schmidt: "Modellgestütztes Messen in chemischen Reaktoren," *Chem. Ing. Tech.* **63** (1991) no. 3, 189–203.

[5.12] R. Perne, M. Polke: "Produktentwicklung durch Verfahrensvariation," *Regelungstech. (rt)* **30** (1982) no. 5, 147–156.

[5.13] K. Garbrecht, A. Prommer: *Mikroelektronik. Eine Technik macht Epoche*, Siemens, Berlin 1981.

[5.14] M. Polke: "Informationshaushalt technischer Prozesse," *Automatisierungstechnische Praxis (atp)* **27** (1985) no. 4, 214–217.

[5.15] J. Hesse: *Sensoren auf dem Weg zur Systemintegration*, special edition, Oldenbourg Verlag, München 1992.

[5.16] J. Lauber: "Das Phasenmodell der Produktion: Ein semantisches CAE-Werkzeug zur Prozeßbeschreibung," *Regelungstechnischen Kolloquium*, Boppard, Germany, Feb. 1993.

[5.17] International Union for Pure and Applied Physics (IUPAP): Symbols, Units and Nomenclature in Physics, Document U.I.P.20 (1978), Physik Verlag GmbH, Weinheim 1980.

[5.18] DIN 1313, Physikalische Größen und Gleichungen. Begriffe und Schreibweisen, Beuth Verlag, Berlin 1978.

[5.19] M. Polke: "Erforderliche Informationsstrukturen für die Qualitätssicherung," *Chem. Ing. Tech.* **65** (1993) no. 7, 791–796.

[5.20] W. König: Sicherheitstechnische Aspekte zur Prozeßleittechnik, *12. Arbeitsschutz-Symposium der Bayer AG*, Leverkusen, 22 Nov. 1985.

[5.21] DIN V 19250, Messen – Steuern – Regeln. Grundlegende Sicherheitsbetrachtungen für MSR-Schutzeinrichtungen, Beuth Verlag, Berlin 1989.

[5.22] NAMUR-Empfehlung 31: Anlagensicherung mit Mitteln der Prozeßleittechnik, NAMUR, Darmstadt 1992.

[5.23] Verordnung zur Durchführung des Bundesimmissionsschutzgesetzes (Störfall-Verordnung) of 19 May 1988 and 5 July 1991.

[5.24] DIN VDE 31000: Allgemeine Leitsätze für das sicherheitsgerechte Gestalten technischer Erzeugnisse. Begriffe der Sicherheitstechnik. Grundbegriffe, part 2, Beuth Verlag, Berlin 1987.

[5.25] 2nd IUPAC-Workshop on Safety in Chemical Production, Pacifico Yokohama, Japan, May 31–June 4, 1993.

[5.26] M. Polke: "Process Automation in Chemical Industry," *2nd IUPAC-Workshop on Safety in Chemical Production*, Yokohama, Japan, June 1993.

[5.27] R. Ecker, H. Kramer, H.K. Müller, M. Polke: "Neue Aspekte bei der Prüfung von Kautschuk und Gummi," *Kautsch. Gummi Kunstst.* **25** (1972) no. 1, 5–10.

[5.28] M. Mooney, H. Rolla, H. Taylor, J. H. Fielding: "Development and Standardization of Test for Evaluating Processibility of Rubber," *Symposium on Rubber Testing*, American Society for Testing Special Technical Materials, 1974.

[5.29] DIN 53523, Bestimmung der Plastizität nach Mooney im Warmscherversuch, Beuth Verlag, Berlin 1965.

[5.30] ASTM D927/49 T, Method of Test for Viscosity of Rubber and Rubberlike Material by the Sharing Disc-viscometer, 1962.

[5.31] W. Kuhn: "Beziehungen zwischen Molekülgröße, statistischer Molekülgestalt und elastischen Eigenschaften hochpolymerer Stoffe," *Kolloid-Z.* **76** (1936) 3.

[5.32] W. Pechold, S. Blasenbrey, S. Woerner, *Kolloid-Z. Polym.* **189** (1965) 14.

[5.33] L. Morbitzer, H. J. Kreß, C. Lindner, K. H. Ott: "Struktur und Eigenschaften von Mehrphasenkunststoffen," *Angew. Makromol. Chem.* **132** (1985) 19–42.

[5.34] L. Morbitzer, D. Kranz, G. Humme, K. H. Ott: "Structure and Properties of ABS Polymers. X. Influence of Particle Size and Graft Structure on Loss Modulus Temperature Dependence and Deformation Behavior," *J. Appl. Polym. Sci.* **20** (1986) 2691–2704.

[5.35] L. Morbitzer, U. Grigo: "Die wichtigsten polymerphysikalischen Aspekte des Polycarbonats," *Angew. Makromol. Chem.* **162** (1988) 87–107.

[5.36] K. Ninomiya, J. Furuta: "Interrelation Between Data from Different Physical Tests," *Annual Meeting of IISPR*, Mexico City, May 1970.

[5.37] J. D. Ferry: *Viscoelastic Properties of Polymers*, John Wiley & Sons, New York 1961.

[5.38] DIN ISO 9001, Qualitätssicherungssysteme – Modell zur Darlegung der Qualitätssicherung in Design/Entwicklung, Produktion, Montage und Kundendienst, Beuth Verlag, Berlin 1990.

[5.39] K. H. Padberg, P.-Th. Wilrich: "Die Auswertung von Daten und ihre Abhängigkeit von der Merkmalsart," *QZ* **26** (1981) no. 6, 179–184, 210–214.

[5.40] DIN ISO 9000, Qualitätsmanagement- und Qualitätssicherungsnormen, Leitfaden zur Auswahl und Anwendung, Beuth Verlag, Berlin 1990.

[5.41] R. E. Kalman: "A New Approach to Linear Filtering and Prediction Problems," *J. Basic. Eng.* **82D** (1960) 35–45.

[5.42] D. G. Luenberger: "Observers for Multivariable Systems," *IEEE Trans. Autom. Control* **11** (1966) 190–197.

[5.43] A. Gelb (ed.): *Applied Optimal Estimation*, The MIT Press, Cambridge, Mass., 1974.

[5.44] W. Brammer, C. Siffling: *Kalman-Bucy-Filter. Deterministische Beobachtung und stochastische Filterung*, R. Oldenbourg Verlag, München 1975.

[5.45] E. D. Gilles: "Modellgestützte Verfahren in der Automatisierungstechnik," *Spring Meeting of DPG, Verhandlungen der Deutschen Physikalischen Gesellschaft*, series VI, vol. 24 (1989).

[5.46] N. R. Iudica: "Expertensysteme," *RTP Regelungstech. Prax.* **26** (1984) no. 9, 402–407.

[5.47] D. Balzer: *Wissensbasierte Systeme in der Automatisierungstechnik*, Hanser Verlag, München 1991.

[5.48] H. Fittler: "Anwendung von Expertensystemen in der Prozeßautomation," *Automatisierungstechnische Praxis (atp)* **33** (1991) no. 4, 199–205.

[5.49] H.-P. Preuß: "Fuzzy Control – heuristische Regelung mittels unscharfer Logik," *Automatisierungstechnische Praxis (atp)* **34** (1992) no. 4, 176–184, 239–246.

[5.50] M. Koch, T. Kuhn, J. Wernstedt: "Ein neues Entwurfskonzept für Fuzzy-Regelungen," *Automatisierungstech. (at)* **41** (1993) no. 5, 152–158.

[5.51] H. Kiendl, K. D. Meyer-Gramann, B. Cuno, M. Fritsch, Chr. Frenck: "Fuzzy Control," Beiträge zur "Theorie für Anwender," *Automatisierungstechnik (at)* **41** (1993) no. 2. A 5.

[5.52] W. Brauer, D. Hernàndes: "Verteilte künstliche Intelligenz und kooperatives Arbeiten," *4. Internationaler GI-Kongreß Wissensbasierte Systeme*, München, Oct. 1991. *Informatik Fachberichte*, Springer Verlag, Berlin 1991.

[5.53] E. D. Gilles: "Modellbildung, ein Hilfsmittel der Meßtechnik," *Tech. Mess.* **46** (1979) 225–232, 271–274.

[5.54] E. D. Gilles: "Application of Model Based Measuring Techniques in Chemical Engineering," *Proceedings of International Conference on Industrial Process Modelling and Control*, Hangzhou, China, 6–9 June, 1985.

[5.55] *Ullmann*, 4th ed., **5**.

[5.56] T. Pfeifer, P. Profos (eds.): *Handbuch der Industriellen Meßtechnik*, R. Oldenbourg Verlag, München 1992.

[5.57] K. H. Feltgen: personal communication, 1972.

[5.58] K. H. Schmitt: "Die Automatisierung von Chemieanlagen mit Mitteln der Systemtechnik – Analysen und Trends," *Chem. Ing. Tech.* **53** (1981) no. 8, 620–625.

[5.59] Knick Elektronische Meßgeräte GmbH & Co.: "Meß-System 71 pH," company brochure no. 717959150 00, Berlin 1993.

[5.60] K. Schügerl (ed.): *Analytische Methoden in der Biotechnologie*, Vieweg Verlag, Braunschweig 1991.

[5.61] J. E. Bailey, D. F. Ollis: *Biochemical Engineering Fundamentals*, McGraw-Hill Book Company, New York 1986.

[5.62] U. Martens: "Sensor-Systeme für Durchfluß und Füllstand," *55. NAMUR* Lahnstein Nov. 1992.

[5.63] E. Nicklaus: "Auswirkungen moderner Prozeßleittechnik auf Sensorsysteme," *Chem. Ing. Tech.* **59** (1987) no. 8, 615–621.

[5.64] D. Hesse, H. Küttner: "Entwicklungstendenzen der Sensortechnik," *Ind.-Elektronik + Elektronik* **28** (1983) no. 5, 36–44.

[5.65] M. Polke: "On-line Betriebsmeßtechnik," Einführungsvortrag in den Themenkreis der gleichnamigen Postersession auf dem Jahrestreffen der GVC, Wien, Oct. 1, 1992.

[5.66] DIN 31 051, Instandhalten: Begriffe und Maßnahmen, Beuth Verlag, Berlin 1985.

[5.67] G. Pinkowski: "Instrumentierungstechnik im Wandel," *Automatisierungstechnische Praxis (atp)* **34** (1992) no. 5, 247–254, 325–327.
G. Pinkowski: "Fortführung der Ex-Feldbus-Untersuchung," *Automatisierungstechnische Praxis (atp)* **34** (1992) no. 11, 617–622.
G. Pinkowski: "Einheitliche, herstellerunabhängige Anzeige-Bedienoberfläche für kommunikationsfähige Meßumformer," *Automatisierungstechnische Praxis (atp)* **34** (1992) no. 2, 530–537.

[5.68] H. Drahten: "Der Feldbus aus Sicht des Anwenders," *Automatisierungstechnische Praxis (atp)* **33** (1991) no. 2, 55-64.

[5.69] H. Raab: "Die Rolle der Sensoren in der Verfahrenstechnik," *Chem. Ing. Tech.* **61** (1989) 199–205.

[5.70] W. Stieler: "Sensorprinzipien zur Messung physikalischer Produkteigenschaften," company report, Bayer AG, Leverkusen, Nov. 1982.

[5.71] H. Raab: "Zur Struktur der Anzeige-Bedienoberfläche digitaler Sensorsysteme," *Automatisierungstechnische Praxis (atp)* **33** (1991) no. 2, 64–71.

[5.72] ISO: MMS 9506, part 1, Service Definitions; MMS, part 2, Protocol Specification, 1988.

[5.73] C. Nøkleby: *The MMS Implementation on the P-NET Fieldbus System, 2nd International Conference on P-Net Fieldbus Systems*, Deggendorf 1992.

[5.74] A. Valenzano, C. Demartini, L. Climiniera: *MAP and TOP Communications, Standards and Applications*, Addison-Wesley Publishing Company, Workingham 1992.

[5.75] W. Stadter: "Device-Description-Language. Ein wichtiger Beitrag zur herstellerunabhängigen Austauschbarkeit von Feld- und Bediengeräten," *Automatisierungstechnische Praxis (atp)* **34** (1992) no. 1, 21 ff.

[5.76] VDI/VDE Richtlinie: Einheitliche Anzeige- und Bedienoberfläche auf Personalcomputern für digitale Feldgeräte, Gründruck 1992.

[5.77] Krohne America Inc.: "Flowmeter Selection Guide," company brochure, Peabody, Mass. 1992.

[5.78] P. Profos, T. Pfeifer: *Handbuch der Industriellen Meßtechnik*, R. Oldenbourg Verlag, München 1992.

[5.79] G. Strohrmann: *Einführung in die Meßtechnik im Chemiebetrieb*, R. Oldenbourg Verlag, München 1990.

[5.80] I. Hengstenberg, B. Sturm, O. Winkler: *Messen, Steuern und Regeln in der chemischen Technik*, Springer Verlag, Berlin 1980.

[5.81] F. Kohlrausch: *Praktische Physik*, Teubner Verlagsgesellschaft, Leipzig 1950.

[5.82] A. Mangold: "Stellgeräte für verfahrenstechnische Anlagen," *Automatisierungstechnische Praxis (atp)* **33** (1991) no. 11, 585 ff.

[5.83] M. Huk: "Stellgeräte für verfahrenstechnische Anlagen," *Automatisierungstechnische Praxis (atp)* **3** (1993) no 3, 131 ff.

[5.84] H.-J. Schneider: "Sensorsysteme und die Kommunikation im Feld," *Automatisierungstechnische Praxis (atp)* **35** (1993) no. 2, 71 ff.

[5.85] DIN 19 227, Graphische Symbole und Kennbuchstaben für die Prozeßleittechnik. T1: Darstellung von Aufgaben, Beuth Verlag, Berlin 1990 (draft).

[5.86] DIN 40 050, IP-Schutzarten. Berührungs-, Fremdkörper- und Wasserschutz für elektrische Betriebsmittel, Beuth Verlag, Berlin 1980.

[5.87] DIN VDE 0838, Rückwirkungen in Stromversorgungsnetzen, die durch Haushaltsgeräte und ähnliche elektrische Einrichtungen versursacht werden, T1–T3, Beuth Verlag, Berlin 1987.

[5.88] NAMUR-Empfehlung 21, Störfestigkeitsanforderungen, Part 1: EMV von Betriebsmitteln der Prozeß- und Laborleittechnik. NAMUR, Darmstadt 1988.

[5.89] DIN VDE 0843, Elektromagnetische Verträglichkeit von Meß-, Steuer- und Regeleinrichtungen in der industriellen Prozeßtechnik, T1–T4, Beuth Verlag, Berlin 1984–1988.

[5.90] DIN VDE 0848, T1–T3: Gefährdung durch elektromagnetische Felder and T4–T4 A1: Sicherheit bei elektromagnetischen Feldern, Beuth Verlag, Berlin 1982–1990.